Advanced Textbooks in Co
and Signal Processing

Series editors
Michael J. Grimble, Glasgow, UK
Michael A. Johnson, Kidlington, UK

To Wada
with regards

More information about this series at http://www.springer.com/series/4045

Stephen J. Dodds

Feedback Control

Linear, Nonlinear and Robust Techniques
and Design with Industrial Applications

Springer

Stephen J. Dodds
School of Architecture, Computing and Engineering
University of East London
London, UK

Additional material to this book can be downloaded from http://extras.springer.com.

ISSN 1439-2232
Advanced Textbooks in Control and Signal Processing
ISBN 978-1-4471-6674-0   ISBN 978-1-4471-6675-7   (eBook)
DOI 10.1007/978-1-4471-6675-7

Library of Congress Control Number: 2015944090

Springer London Heidelberg New York Dordrecht
© Springer-Verlag London 2015
This work is subject to copyright. All rights are reserved by the Publisher, whether the whole or part of the material is concerned, specifically the rights of translation, reprinting, reuse of illustrations, recitation, broadcasting, reproduction on microfilms or in any other physical way, and transmission or information storage and retrieval, electronic adaptation, computer software, or by similar or dissimilar methodology now known or hereafter developed.
The use of general descriptive names, registered names, trademarks, service marks, etc. in this publication does not imply, even in the absence of a specific statement, that such names are exempt from the relevant protective laws and regulations and therefore free for general use.
The publisher, the authors and the editors are safe to assume that the advice and information in this book are believed to be true and accurate at the date of publication. Neither the publisher nor the authors or the editors give a warranty, express or implied, with respect to the material contained herein or for any errors or omissions that may have been made.

Printed on acid-free paper

Springer-Verlag London Ltd. is part of Springer Science+Business Media (www.springer.com)

To Margaret

Preface

Subject Matter

This book is on feedback control systems covering many applications such as spacecraft orientation, positioning of mechanisms in industry, liquid level and temperature control, and engine management. The general approach taken is to commence with the simplest tentative solution to a control problem and in some cases to demonstrate and explain why this is insufficient and introduce appropriate measures to ensure that satisfactory performance is achieved, thereby giving the reader a full understanding of the features of different controllers. This approach enables the reader to apply traditional control techniques where these are sufficient but also encourages the creation of more sophisticated control systems that work better for other applications through taking advantage of modern digital implementation. For ease of understanding, the book is self-contained with very little reliance on references, much attention being paid to explanations of the underlying concepts and detailed mathematical derivations being given where necessary, showing every step. Ample use is made of diagrams to aid the conceptual explanations. The reader's interest in the subject matter is maintained by giving reasons for the inclusion of every topic and bringing the material to life by frequent reference to real applications. The numerous control system examples are backed up by simulations. The reader's understanding is reinforced by a set of problems and solutions on each chapter provided in the book website, designed to develop the reader's ability to tackle original and sometimes unusual control problems. The reader's understanding is developed further by the opportunity to experiment with any of the fully commented MATLAB®–SIMULINK® control system simulations that may be downloaded from the book website. These are also intended to help readers to develop simulations of their own control applications.

Purpose and Readership

The main benefit of reading the book is to develop the understanding and skills needed to pursue a rewarding career as a creative control engineer. The readers will include (a) undergraduates in the final year of an engineering degree, (b) master's students studying feedback control, (c) PhD students carrying out research in the control techniques covered or developing feedback control systems to support their projects and (d) research and development engineers in industry wishing to create control systems fully benefitting from the modern digital implementation media.

The numerous control system examples together with the simulations would be suitable for seeding and/or supporting final-year undergraduate or master's projects.

Content of the Chapters

Chapter 1. Introduction

After defining the notation and nomenclature used throughout the book, a review of the traditional industrial controllers is given, commencing with the simplest. This ensures some continuity between the elementary studies that will already have been undertaken by the reader and the more advanced material. Then a comprehensive treatment of the correlation between the relative pole and zero locations of the Laplace transfer function of a linear time-invariant system and its dynamic characteristics is given, quantified by the pole-to-pole and pole-to-zero dominance ratios.

Chapter 2. Plant Modelling

This chapter develops the background theory and provides the knowledge needed to generate plant models. After an introduction to the basic character of plants and their components, a subsection on physical modelling is presented. This is based on the underlying science of the various applications. Within the space limitations, the main emphasis is on mechanical systems and electric motors as actuators to cater for a large proportion of the applications. Some introductory material on thermal and fluid systems is also given. This is followed by a substantial section of identification of plant models from input and output signals in the frequency and time domains.

The appendix contains a comprehensive treatment of the kinematics of vehicle attitude control, relevant to applications such as spacecraft, aircraft and underwater vehicles. This is followed by a presentation of plant model determination from frequency response data including procedures for identifying plants with relatively close poles and/or zeros. Finally, a case study of plant modelling in the automotive industry is presented that embodies some of the techniques covered in the chapter.

Chapter 3. Plant Model Manipulation and Analysis

Further to the physical modelling and identified transfer functions developed in Chap. 2, this chapter presents means of converting these to other forms of plant model needed for the application of specific control techniques, including transfer function block diagrams and state space models, both continuous and discrete.

The theory of the z-transform is reviewed, and means of directly generating z-transfer function models including the sample/hold are derived that are directly useful for control system design. These are included in Table 3 of the Tables section preceding the index at the end of the book alongside the z-transforms usually presented elsewhere.

State representation and the associated transformations are covered. The observer canonical form, the controller canonical form and the modal form for single-input, single-output and multivariable plants are studied. Controllability and observability analyses for both continuous and discrete state space models are included.

Chapter 4. Traditional Controllers: Model Based Design

This chapter commences with the simplest feedback control systems to ensure continuity and provide some revision for readers who have only undertaken one year of undergraduate study of linear control systems. As the chapter progresses, various performance demands are introduced together with increases in the plant order. Controllers are selected through the needs of application examples. At each stage, features, either in the control structure or design methodology, are introduced that meet the specification. With this approach, the reader will fully understand the features and be able to select the simplest suitable traditional controller for a plant of first or second order and calculate its gains, based on pole assignment, to meet a given performance specification in terms of settling time, steady state error and sensitivity/robustness.

The behaviour of linear systems of the third and higher order is studied in preparation for designing control systems for second-order plants using traditional controllers containing integral terms and the more general control systems of Chap. 5. The author's settling time formulae are derived for use in conjunction with the pole assignment design of systems of arbitrary order.

Finally, connections between performance specifications in the time domain and the frequency domain are established.

The appendix contains an unusual but useful adaptation of Mason's formula for direct application to linear system block diagrams to derive transfer functions and characteristic polynomials. This is followed by a pole placement procedure for cancellation of zeros introduced by traditional controllers, together with statements of its limitations. Finally, partial pole assignment is presented for linear control systems whose order exceeds the number of independently adjustable controller gains.

Chapter 5. Linear Controllers for LTI SISO Plants of Arbitrary Order: Model-Based Design

The model-based control system design approach based on pole assignment introduced in Chap. 4 is extended beyond systems of second order first by means of linear state feedback control and subsequently by means of polynomial control.

The state is assumed available for use with the linear state feedback control system designs derived in this chapter, these being rendered practicable when used with the state estimation techniques presented in Chap. 8.

The effects of closed-loop transfer function zeros are studied, and means of taking them into account or eliminating them in the control system design using dynamic pre-compensators to achieve satisfactory responses are developed.

The generic technique of polynomial control is introduced in a straightforward manner, simple means of determining suitable polynomial degrees for a given plant being devised. The solution of the Diophantine equations to calculate the polynomial coefficients for the pole assignment is expressed as a linear matrix equation suitable for computer-aided design.

The appendix contains two aids to computer-aided design. The first is Ackermann's gain formulae for the pole assignment design using any state representation for linear state feedback control, also for observers to be read in conjunction with Chap. 8. The second aid is linear characteristic polynomial interpolation for computing the adjustable parameters for the pole assignment design of any linear system whose characteristic polynomial coefficients are linear with respect to the parameters.

Chapter 6. Discrete Control of LTI SISO Plants

The general structure, timing, algorithms and flow charts of discrete controllers are first discussed. Then the correlation of the behaviour of discrete dynamical systems with the z-plane pole locations is studied, including stability analysis. The effects of transfer function zeros are also considered.

A simple procedure is presented for converting continuous controller designs for digital implementation provided the sampling period is sufficiently small. A criterion of applicability of continuous linear time-invariant system theory is developed that determines whether or not this procedure is valid.

The remainder of the chapter is devoted to a design method applicable, in theory, with unlimited sampling periods and which caters for plants of arbitrary order. This entails pole placement in the z-plane in which a specified settling time is nearly realised if it is considerably larger than the sampling period. As the demanded settling time approaches zero for a fixed sampling time, a dead beat response is approached, which has the shortest possible settling time and is therefore the best compromise.

Discrete polynomial control is presented, following the continuous version of Chap. 5. This has the same structure as the RST controller but a different design procedure. It is shown how this methodology enables computational delay allowance.

Finally, discrete polynomial control of plants containing pure time delays, aided by a Smith predictor, is addressed.

Chapter 7. Model Based Control of Nonlinear and Linear Plants

First, the focus is on the control of nonlinear plants. This commences with traditional linearisation about the operating point, which enables linear control system design provided the plant states are restricted to lie in the region of the operating point. This is followed by feedback linearising control, which removes the operating point restriction and is applicable to multivariable as well as single-input, single-output plants.

The underlying principle of feedback linearising control, which forces the closed-loop system to obey a prescribed differential equation, is extended in two directions. First, feedback linearising control is applied to linear plants, which is found to be straightforward for multivariable plants. This is further extended to the discrete domain. Second, the prescribed closed-loop differential equation is allowed to be nonlinear, catering for control strategies such as near time-optimal control. In both these cases, the title, feedback linearising control, is replaced by the more appropriate title, forced dynamic control.

Chapter 8. State Estimation

The basic full state observer for linear, time-invariant, single-input, single-output plants is first developed. The separation principle and transparency property are covered and the design procedure given. The full state observer is then extended for the estimation of external disturbances together with the plant state. The discrete version is then developed together with the design procedure. The continuous full state observer for linear time-invariant multivariable plants and its design procedure are then presented.

The remainder of the chapter is devoted to the effects of measurement noise and plant noise on the state estimate and how this may be taken into account in observer design using power spectral density and variance information. The discrete Kalman filter algorithm is then introduced and comparisons made with the discrete observer algorithm for linear time-invariant multivariable plants. A derivation of the discrete Kalman gain algorithm is given. Comparisons are made with the continuous version.

The appendix contains two approaches to nonlinear observer design restricted to plants of full relative degree. The first comprises a set of filtered output derivative estimators constituting a state estimate, practicable provided the measurement noise levels are not too high. The second affords more measurement noise filtering by using the output derivative estimates of the first approach as raw measurements for a special observer in which the nonlinear elements of the plant model are excluded from the correction loop.

Chapter 9. Switched and Saturating Control Techniques

The first technique presented is pulse modulation that enables controllers designed for continuous control variables to be utilised, highlighting applications such as power electronic drives.

Switched state feedback control based on the switching function and the associated switching boundary is then introduced. The behaviour of second-order systems is studied using phase portraits. This is extended to saturating control with continuous control variables and the boundary layer.

Optimal open-loop control is introduced via Pontryagin's maximum principle. The special case of time-optimal control of a linear time-invariant plant is studied, and information from this is used to synthesis switched feedback time-optimal control laws for first-order plants and second-order plants with switching boundaries derived using the back tracing method. Limit cycling control is studied for first-order plants.

Switched feedback control of higher-order plants is exemplified by first deriving the time-optimal switching boundary for a triple integrator plant and applying it for spacecraft attitude control using variable geometry panels with solar radiation pressure. This is followed by posicast control of plants containing lightly damped oscillatory modes.

The appendix contains limit cycling control for switched state feedback control of second-order plants. An example is given on attitude control of a rigid body spacecraft actuated by on-off thrusters using piecewise parabolic switching boundaries with acceleration parameters adapting to a disturbance torque estimate to maintain a limit cycle of constant amplitude.

Preface   xiii

Chapter 10. Sliding Mode Control and Its Relatives

First, standard sliding mode control of single-input, single-output plants is developed from the material of Chap. 9 on switched state feedback control, first with second-order plants by study of the closed-loop phase portraits. The purpose of sliding mode control in achieving robustness is emphasised. The equivalent control is defined. Control chatter is identified. The conditions for the existence of sliding motion are derived, and the reaching of the sliding region of the switching boundary from arbitrary initial states is studied. The system behaviour with external disturbances is considered.

The introduction of nonlinear switching boundaries to prevent overshooting in second-order systems with arbitrary initial states is presented. This is followed by sub-time-optimal sliding mode control of second-order plants using a nonlinear boundary layer based on a double parabolic switching boundary tangential to a linear region passing through the origin of the phase plane.

Sliding mode control of single-input, single-output plants of arbitrary order is introduced, for which the design procedure for the linear switching boundary yielding specified closed-loop dynamics is presented.

Next, three methods of control chatter avoidance are presented. These are (1) the pseudo sliding mode controller using the boundary layer method, following from the material in Chap. 9; (2) the control smoothing integrator method entailing the augmentation of the plant with one or more pure integrators to remove the switching to a primary control variable in the controller software; and (3) higher-order sliding mode control, introducing a control structure in which not only the switching function is driven to zero in the sliding mode but also its derivatives up to a specified order.

The relatives of the sliding mode controller are presented, which stem from the pseudo sliding mode controller with the boundary layer. They comprise all the linear controllers that may be designed by robust pole placement.

The remainder of the chapter introduces multivariable sliding mode control by pairing controlled outputs with control inputs via study of the equations derived to determine the relative degrees. This is followed by discrete sliding mode control formulated for multivariable linear plants, single-input, single-output plants being included as special cases. This also constitutes a fourth method for eliminating control chatter as its control variables are piecewise constant approximations to the equivalent control in the sliding mode.

The appendix presents observer-based robust control, a technique developed by the author, which is a distant relative of sliding mode control in that it forces desired behaviour by means of judiciously applied high gains. Although requiring adjustment of controller parameters, it is capable of similar robustness to sliding mode control and, importantly, is capable of accommodating model order uncertainty while yielding a specified closed-loop dynamics with a fixed order.

Chapter 11. Motion Control

The general-purpose jointed-arm robot is first introduced together with a model that applies also to other mechanisms whose motion is to be controlled. A generalised feedback linearising control law is then given. Modelling simplifications applicable to geared mechanisms are then developed.

Dynamic lag pre-compensation is presented, including a polynomial controller with inbuilt derivative feedforward to assist in this pre-compensation.

Next, the important topic of frictional energy minimisation is introduced, which if implemented on a large scale can drastically reduce the carbon footprint. The optimal control strategy is first formulated with the aid of Pontryagin's method. This is used to derive an optimal reference input function that can be followed using a controller with a dynamic lag pre-compensator to implement optimal feedback control. The performance improvement over traditional control methods is assessed.

The appendix presents reference input function planning using cubic and quintic splines, a method enabling exact derivatives to be computed for dynamic lag pre-compensator implementation.

Recommended Reading to Support Courses

Undergraduates in their final year would benefit from reading Chaps. 1 and 2; the sections of Chap. 3 on single-input, single-output plants; Chaps. 4, 5 and 6; the sections of Chap. 7 on linearisation about the operating point and feedback linearising control of single-input, single-output plants; and the sections of Chap. 8 on observers for single-input, single-output plants. The remaining material of Chaps. 3, 7 and 8 together with Chaps. 9, 10 and 11 would be suitable for graduate students studying to master's level. All will benefit from studying the examples and working with the simulations that may be downloaded from the book website, this also providing material for establishing final-year undergraduate and master's projects.

Acknowledgements

First, I would like to thank my numerous students and academic contacts for encouraging me to produce this book. Their comments and suggestions have been a positive contribution. I would like to extend special thanks to my industrial contacts, A Fallahi, P Stadler and J L Pedersen, who were also my research students, for reading the material during its development and providing application examples and feedback that has resulted in a more readable and useful text than otherwise would have been possible. People of the past who I thank especially for their help in various ways including inspiration to strive for a career in engineering, full of creativity and innovation, are J A Cross, R A Rawlins, N P Small, E Royser, S Upson, G Elmes, T Collier, W F Lovering, N Ream, E Watson, T. Konwerski, S E Williamson, D Atherton, P D Roberts, J F Coales, A T Fuller, J Billingsley, J M Maciejowski, K Glover, B J Oke, P E G Cope, S Armstrong, A Sarnecki, M Noton, W M Hosny, R A Savill, A P Bedding, G Harvey, J Vittek, T Orlowska Kowalska, K Szabat, B Grzezik, W Koczara, P C Hughes, J L Junkins, A G Loukianov, V I Utkin, V A Utkin, V Rutkovsky, S D Zemlyakov, V M Sukhanov, V Glumov and Y Pyatnitsky.

Special thanks go to my grandfather, H Cook, for interesting me in engineering at an early stage in life and my own son Alasdair, for his continuing encouragement to complete the book through his unwavering belief in my ability to inspire and teach others aspiring to work in my field.

Last, but not least, I deeply thank my loving wife, Margaret, to whom I dedicate the book, for her patience and understanding during the many months of preparation.

Contents

1	**Introduction**		1
	1.1 Overview		1
	1.2 Notation and Nomenclature		1
		1.2.1 Scalars, Vectors and Matrices	1
		1.2.2 Subscripts and Superscripts	2
		1.2.3 Constants and Variables	2
		1.2.4 Nomenclature and Standard Symbols	3
		1.2.5 Variables and Their Laplace Transforms	4
	1.3 Review of Traditional PID Controllers and Their Variants		4
		1.3.1 Traditional Error-Actuated Controllers	4
		1.3.2 Zero-Less Versions of the Traditional Controllers	19
		1.3.3 Traditional Controller Selection Guidelines	22
		1.3.4 Measurement Noise Filtering for Derivative Term	25
		1.3.5 Anti-windup Loop for Integral Term	32
	1.4 Dominance in the Pole–Zero Distribution		43
		1.4.1 Background	43
		1.4.2 Modes of Linear Systems	43
		1.4.3 Dominance in Pole Distributions	49
		1.4.4 Dominance in Systems with Zeros	53
	1.5 The Steps of Control System Design		69
	1.6 The Flexibility of Digital Implementation		71
	Reference		71
2	**Plant Modelling**		73
	2.1 Introduction		73
		2.1.1 Overview	73
		2.1.2 Dynamical and Non-Dynamical Systems	73
		2.1.3 Linearity and Nonlinearity	76
		2.1.4 Modelling Categories and Basic Forms of Model	80

2.2	Physical Modelling		81
	2.2.1	Introduction	81
	2.2.2	Mechanical Modeling Principles	81
	2.2.3	Two Basic Mechanical Components	92
	2.2.4	Modelling for Vehicle Attitude and Position Control	97
	2.2.5	Electric Motors	104
	2.2.6	Vector-Controlled AC Motors as Control Actuators	109
	2.2.7	Fluid and Thermal Subsystems	117
2.3	Identification of LTI Plants from Measurements		121
	2.3.1	Overview	121
	2.3.2	Plant Model Determination from Step Response	122
	2.3.3	Plant Model Determination from Frequency Response	127
	2.3.4	Recursive Parameter Estimation: An Introduction	151
References			167

3 Plant Model Manipulation and Analysis 169

3.1	Introduction		169
3.2	The State Space Model		170
	3.2.1	Introduction	170
	3.2.2	Forming a State-Space Model	171
	3.2.3	The General State-Space Model	172
	3.2.4	The General LTI State-Space Model	173
	3.2.5	Some Preliminary Control Theory	175
	3.2.6	Controllability Analysis of Continuous LTI Plant Models	179
	3.2.7	Observability Analysis of Continuous LTI Plant Models	183
	3.2.8	The State-Variable Block Diagram	187
	3.2.9	Transfer Function from Continuous LTI State Space Model	190
	3.2.10	Relative Degree	192
3.3	State Representation		194
	3.3.1	Introduction	194
	3.3.2	LTI SISO State-Space Models from Transfer Functions	200
	3.3.3	Transformation Matrices Connecting Linear Models	211
	3.3.4	Modal Forms for Multivariable LTI Plants: Transformations	217
	3.3.5	SISO Controller and Observer Canonical Forms	237
	3.3.6	Multivariable Controller and Observer Canonical Forms	245

	3.4	Discrete LTI Plant Models	255
		3.4.1 Formation of the Discrete State Space Model	255
		3.4.2 State Space Model Derivation from Modal Basis Functions...	259
		3.4.3 Plant z-Transfer Function Model	274
		3.4.4 Change of Sampling Period for z-Transfer Function Models	282
		3.4.5 Controllability Analysis of Discrete LTI Plant Models..	287
		3.4.6 Analysis of Discrete LTI Plant Models.................	290
	References...		293
4	**Traditional Controllers: Model Based Design**		**295**
	4.1	Approach	295
	4.2	Pole Assignment	297
	4.3	Definition of Settling Time	299
	4.4	PID Controllers and Their Variants.............................	300
		4.4.1 First Order Systems	300
		4.4.2 Second Order Systems	307
		4.4.3 Cascade Control Structure	319
	4.5	Systems of Third and Higher Order	325
		4.5.1 Attainable Closed Loop Dynamics	325
		4.5.2 The Laplace to Time Domain Inverse Scaling Law ...	327
		4.5.3 Step Responses with Coincident Closed Loop Poles..	329
		4.5.4 Derivation of the Settling Time Formulae	331
		4.5.5 Settling Time Formula Error Determination and Correction...	333
		4.5.6 Closed Loop Poles for Given Overshoot and Settling Time.......................................	335
	4.6	Performance Specifications in the Frequency Domain	337
		4.6.1 Background..	337
		4.6.2 Closed Loop System Bandwidth	337
		4.6.3 Sensitivity and Robustness..............................	339
		4.6.4 Stability Analysis in the Frequency Domain...........	347
	References...		354
5	**Linear Controllers for LTI SISO Plants of Arbitrary Order: Model-Based Design**...		**355**
	5.1	Overview...	355
	5.2	Linear Continuous State Feedback Control	356
		5.2.1 Introduction...	356
		5.2.2 Linear State Feedback Control Law	357
		5.2.3 Matrix–Vector Formulation	358
		5.2.4 Closed-Loop Transfer Function	360

		5.2.5	Pole Assignment Using the Matrix–Vector Formulation	361
		5.2.6	Pole Assignment Using Mason's Formula	364
		5.2.7	Pole Assignment for Plants with Significant Zeros	368
		5.2.8	State Feedback Controllers with Additional Integral Terms	387
	5.3	Polynomial Control		398
		5.3.1	Introduction	398
		5.3.2	Formulation of Polynomial Controller Structure	399
		5.3.3	Constraints on Controller Polynomial Degrees	401
		5.3.4	Determination of the Controller Parameters	403
		5.3.5	The Polynomial Integral Controller	407
	References			414
6	**Discrete Control of LTI SISO Plants**			**415**
	6.1	Introduction		415
	6.2	Real-Time Operation of Digital Controllers		416
	6.3	Dynamics of Discrete Linear Systems		417
		6.3.1	Stability Analysis in the z-Plane	417
		6.3.2	Connection Between Dynamic Behaviour and the z-Plane Pole Locations	425
		6.3.3	The Effects of Zeros in the z-Plane	432
	6.4	Criterion for Applicability of Continuous LTI System Theory		435
	6.5	Discrete Control for Small Iteration Intervals		439
		6.5.1	Introduction	439
		6.5.2	Discrete Equations of the Basic Elements	439
		6.5.3	Discrete Controller Block Diagrams for Simulation	443
		6.5.4	Control Algorithms and Flow Charts	447
	6.6	Discrete Control with Unlimited Iteration Intervals		452
		6.6.1	Pole Placement Design with the Settling Time Formulae	452
		6.6.2	Pole Placement for Negligible Digital Processing Time	456
		6.6.3	Computational Delay Allowance	464
		6.6.4	Discrete Integral Polynomial Control	472
		6.6.5	Control of Plants Containing Pure Time Delays	476
	References			480
7	**Model Based Control of Nonlinear and Linear Plants**			**481**
	7.1	Introduction		481
	7.2	Linearisation About an Operating Point		482
		7.2.1	Basic Principle	482
		7.2.2	Linear State-Space Model	484
		7.2.3	Limitation	491

	7.3	Feedback Linearising and Forced Dynamic Control	491
		7.3.1 Preliminaries	491
		7.3.2 Feedback Linearising Control of Plants with Full Relative Degree	495
		7.3.3 Feedback Linearising Control of Plants Less Than Full Relative Degree	509
		7.3.4 Forced Dynamic Control of Continuous LTI Plants ...	522
		7.3.5 FDC and FLC Using Discrete Plant Models	543
		7.3.6 Near-Time-Optimal Position Control Through FDC	550
	References		560
8	**State Estimation**		**561**
	8.1	Introduction	561
	8.2	The Full State Continuous Observer for LTI SISO Plants	562
		8.2.1 Introduction	562
		8.2.2 The Separation Principle and the Transparency Property	565
		8.2.3 Design of the Real-Time Model Correction Loop	567
		8.2.4 Estimation of Disturbances	575
	8.3	The Full State Discrete Observer for LTI SISO Plants	587
		8.3.1 Introduction	587
		8.3.2 Observer Algorithm and Design Procedure	587
	8.4	The Full State Observer for Multivariable Plants	592
		8.4.1 Introduction	592
		8.4.2 Matrix–Vector Design Method for SISO LTI Plants	592
		8.4.3 Matrix–Vector Design Method for Multivariable LTI Plants	594
	8.5	The Noise Filtering Property of the Observer	596
		8.5.1 Background	596
		8.5.2 Lumped Plant Noise and Measurement Noise Sources	597
		8.5.3 State Estimation Error Variation with Observer Gains	598
		8.5.4 State Estimation Error Transfer Function Relationship	599
		8.5.5 Considering Noise Levels in Observer Design	602
	8.6	The Kalman Filter	613
		8.6.1 Introduction	613
		8.6.2 The Discrete Observer	614
		8.6.3 The Kalman Filter: State Difference and Error Equations	615
		8.6.4 Derivation of the Discrete Kalman Gain Algorithm ...	617

		8.6.5	The Steady-State Kalman Filter	621
		8.6.6	The Kalman–Bucy Filter	622
	References			624
9	**Switched and Saturating Control Techniques**			**625**
	9.1	Introduction		625
		9.1.1	Switched Control	625
		9.1.2	Saturating Control	627
	9.2	Pulse Modulation for Use with Continuous Controllers		628
		9.2.1	Basic Concept	628
		9.2.2	Implementation	629
	9.3	Switched State Feedback Control: Basic Concepts		638
	9.4	Switching Function Sign Convention		641
	9.5	Boundary Layer for Saturating Control Systems		642
	9.6	Supporting Theory		644
		9.6.1	Background	644
		9.6.2	Optimal Control Through Pontryagin's Maximum Principle	644
	9.7	Feedback Control of First-Order Plants		651
		9.7.1	Time-Optimal Feedback Control: Analytical Method	651
		9.7.2	Time-Optimal Feedback Control: Graphical Approach	654
		9.7.3	Limit Cycling and Its Control	655
		9.7.4	Control with Time-Varying Reference Inputs	658
		9.7.5	Continuous Control with Saturation	662
	9.8	Feedback Control of Second-Order Plants		666
		9.8.1	Introduction	666
		9.8.2	State Trajectories and State Portraits	666
		9.8.3	Time-Optimal Feedback Control of the Double Integrator Plant	669
		9.8.4	Time-Optimal Control Law Synthesis Using State Portraits	672
		9.8.5	Continuous Control with Saturation	676
		9.8.6	Limit Cycling Control	687
	9.9	Feedback Control of Third and Higher-Order Plants		687
		9.9.1	Overview	687
		9.9.2	Time-Optimal Control of the Triple Integrator Plant	687
		9.9.3	Posicast Control of Fourth-Order Plants with Oscillatory Modes	696
	References			704

Contents   xxiii

10 Sliding Mode Control and Its Relatives 705
    10.1   Introduction 705
            10.1.1   Purpose and Origin 705
            10.1.2   Basic Principle 705
            10.1.3   Implementation for Robustness 707
    10.2   Control of SISO Second-Order Plants of Full Relative
           Degree 713
            10.2.1   The Plant Model 713
            10.2.2   Phase Portraits 714
            10.2.3   Sliding Motion 716
            10.2.4   The Equivalent Control 718
            10.2.5   Control Chatter 718
            10.2.6   Conditions for the Existence of Sliding Motion 719
            10.2.7   Reaching the Sliding Condition 720
            10.2.8   Closed-Loop Dynamics in the Sliding Mode 723
            10.2.9   Control with Time-Varying Disturbances
                      and Reference Inputs 723
            10.2.10 Rate-Limiting Switching Boundary for Zero
                      Overshoot 725
            10.2.11 Sub-Time-Optimal Control 726
    10.3   Control of SISO Plants of Arbitrary Order 730
            10.3.1   Control of Plants Having Full Relative Degree 730
            10.3.2   Control of Plants Less Than Full Relative Degree 738
    10.4   Methods for Elimination of Control Chatter 742
            10.4.1   The Boundary Layer Method 742
            10.4.2   The Control Smoothing Integrator Method 746
            10.4.3   Higher-Order Sliding Mode Control 756
    10.5   Controllers with Robust Pole Placement 766
            10.5.1   Introduction 766
            10.5.2   Output Derivative State Feedback Controller 767
            10.5.3   Dynamic Controllers with Robust Pole Placement 770
    10.6   Multivariable Sliding Mode Control: An Introduction 773
            10.6.1   Overview 773
            10.6.2   Simple Approach with Minimum Plant
                      Information 774
            10.6.3   Discrete Sliding Mode Control 778
    References 792

11 Motion Control 793
    11.1   Introduction 793
    11.2   Controlled Mechanisms 793
            11.2.1   The General-Purpose Jointed-Arm Robot 793
            11.2.2   General Model 794

		11.2.3	Feedback Linearising Control Law	798
		11.2.4	Simplified Model for Mechanisms with Geared Actuators	799
	11.3	Dynamic Lag Pre-compensation		805
		11.3.1	Definition of Dynamic Lag	805
		11.3.2	Derivative Feedforward Pre-compensation	806
		11.3.3	Implementation with Feedback Linearising Control	822
	11.4	Optimal Control for Minimising Frictional Energy Loss		826
		11.4.1	Motivation	826
		11.4.2	Formulation of Optimal Control	827
		11.4.3	Minimum Frictional Energy State Feedback Control Law	830
		11.4.4	Higher-Order Mechanisms	832
		11.4.5	Near-Optimal Control Using a Reference Input Generator	832
	References			846

Tables 847

Laplace Transforms and z-Transfer Functions 847
Characteristic Polynomial Coefficients of the Settling
Time Formulae 850

Appendices 853

A2	Appendix to Chap. 2			853
	A2.1	Kinematics of Vehicle Attitude Control		853
	A2.2	Plant Model Determination from Frequency Response		867
	A2.3	A Case Study of Plant Modelling Undertaken in Industry: Modelling for a Throttle Valve Servomechanism		881
	References			899
A4	Appendix to Chap. 4			900
	A4.1	Application of Mason's Formula Using Block Diagrams		900
	A4.2	Traditional Controller Zero Cancellation by Pole Assignment		908
	A4.3	Partial Pole Assignment for Traditional Controllers		915
	Reference			919
A5	Appendix to Chap. 5			920
	A5.1	Computer Aided Pole Assignment		920
	A5.2	Linear Characteristic Polynomial Interpolation		927
	A5.3	Routh Stability Criterion		936

A8	Appendix to Chap. 8		940
	A8.1	An Approach for State Estimation for Nonlinear Plants..	940
	A8.1.2	Observer Based on Linearised Plant Model............	941
	A8.1.3	Output Derivative Based State Estimator	941
A9	Appendix to Chap. 9		954
	A9.1	Limit Cycling Control for Second Order Plants	954
	References		970
A10	Appendix to Chap. 10		971
	A10.1	Observer Based Robust Control	971
	References		984
A11	Appendix to Chap. 11		985
	A11.1	Path Planning and Reference Input Trajectory Generation	985
	Reference		1002

Index 1003

Chapter 1
Introduction

1.1 Overview

After defining the notation and nomenclature used throughout the book, this chapter gives a review of the traditional controllers that have been used in industry for many years, commencing with the simplest. This ensures some continuity between the elementary studies in control engineering that will already have been undertaken by the reader and the more advanced material. There will inevitably be an overlap with these elementary studies but this should serve as useful revision. The chapter ends with a comprehensive treatment of the correlation between the relative pole and zero locations of the transfer function of a linear time-invariant (LTI) system and its dynamic characteristics, including a quantitative approach that will be useful in the design of control systems.

1.2 Notation and Nomenclature

1.2.1 Scalars, Vectors and Matrices

Scalar quantities are shown in italics, such as v for velocity or C for a constant. Vector quantities are shown bold, lower case and non-italic, an example being the state vector, **x**. Matrices are shown bold and upper case, such as the plant matrix, **A**, or the state transition matrix, **Φ**. If a matrix is time varying then the functional notation is always used to indicate this, an example being $\mathbf{M}(t)$.

1.2.2 Subscripts and Superscripts

Subscripts used for classification are non-italic, non-bold, Roman characters. Examples are the maximum value of the control variable, u_{max}, and the proportional gain, K_{P}, of a controller.

Subscripts used to numerically order a quantity are either lower-case numerals or non-bold, italic, Roman characters that are usually lower case. For example, the third component of a state vector is shown as x_3 and the general i^{th} component of this vector is shown as x_i.

Superscripts usually represent mathematical exponentiation. Scalar exponents are shown as numerals or italic non-bold Roman or lower-case Greek characters. Examples are ω^2, T^n, s^N and $e^{j\omega t}$. Matrix exponents comply with Sect. 1.2.1. An example is the state transition matrix expressed as $e^{\mathbf{A}t}$. An exception is the compact notation for derivatives of variables with respect to time. Thus, $\frac{d^n x}{dt^n}$ is written as $x^{(n)}$, the parentheses being used to avoid confusion with exponentiation.

If a quantity requires two classifications, then one can be a subscript and the other a superscript using any symbol that could not be used for exponentiation, to avoid confusion. For example, the model state trajectory obtained by applying the positive saturated control level could be represented by $\mathbf{x}_{\text{m}}^{+}(t)$.

1.2.3 Constants and Variables

As in most other publications on feedback control, only Roman or Greek characters represent constants and variables. Also, certain subsets of the alphabets are used for each category as shown in Tables 1.1 and 1.2. The '–' entries indicate that the character is not generally used for this purpose.

An exception to Table 1.1 is the use of upper-case Roman characters for Laplace transforms of time-varying quantities, which is explained in Sect. 1.2.5.

Table 1.1 Roman character usage for constants (c), continuous variables (v) and integers (i)

a	(c)	A	(c)	j	(i)	J	(c)	s	(v)	S	(c)
b	(c)	B	(c)	k	(i)	K	(c)	t	(v)	T	(c)
c	(c)	C	(c)	l	(i)	L	(c)	u	(v)	U	(c)
d	(v)	D	(c)	m	(i)	M	(c)	v	(v)	V	(c)
e	(v)	E	(c)	n	(i)	N	(c)	w	(v)	W	(c)
f	(v)	F	(c)	o	–	O	–	x	(v)	X	(c)
g	(c)	G	(c)	p	(c or i)	P	(c)	y	(v)	Y	(c)
h	(v)	H	(c)	q	(c or i)	Q	(c)	z	(v)	Z	(c)
i	(i)	I	(c)	r	(c or i)	R	(c)				

1.2 Notation and Nomenclature

Table 1.2 Greek character usage for constants (c), continuous variables (v) and integers (i)

α	(v)	A	–	φ	(v)	ϑ	(v)	σ	(v)	Σ	–
β	(v)	B	–	κ	–	K	–	τ	(v)	T	–
χ	(v)	X	–	λ	(v)	Λ	(v)	υ	(v)	Y	–
δ	–	Δ	–	μ	(v)	M	–	ϖ	(v)	ς	–
ε	(v)	E	–	ν	(v)	N	–	ω	(v)	Ω	(v)
ϕ	(v)	Φ	(v)	o	–	O	–	ξ	(v)	Ξ	(v)
γ	(v)	Γ	(v)	π	(c)	Π	–	ψ	(v)	Ψ	(v)
η	(c)	H	–	θ	(v)	Θ	(v)	ζ	(c)	Z	–
ι	–	I	–	ρ	(v)	P	–				

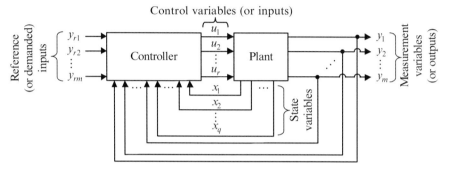

Fig. 1.1 General control system block diagram for introduction of nomenclature

1.2.4 Nomenclature and Standard Symbols

The nomenclature to be used throughout the book will now be introduced. First, the object to be controlled is referred to as the *plant*, a term originating in industry describing various processes. It is widely used, however, for any controlled object.

The basic terms are shown in the control system block diagram of Fig. 1.1. This shows a plant and a controller together with the four categories of variables associated with them. The measurement variables are usually the controlled variables, each of which has to respond to the corresponding reference inputs. There are, however, a few exceptions in which some or all of the controlled variables are not measurement variables, as in the so-called sensorless electric drive applications. Systems with more than one controlled variable are usually referred to as *multivariable* or *multiple input, multiple output* (MIMO) systems. In many cases, however, such as in the following section, there is only one controlled variable. Such systems are usually referred to as *single input, single output* [SISO] systems.

The corresponding descriptions also apply to the plant in isolation. So the terms multivariable (or MIMO) and SISO plant are often used.

The state variables are very important because a complete set of them, equal in number to the plant order, represent the instantaneous behaviour of the plant (Chap. 2). Some state variables are shown in Fig. 1.1 as physical signals fed back to the controller in addition to the measurement variables, but these are not always present. Some of the more sophisticated controllers make use of all these state variables to obtain the best possible control for the application in hand, within the hardware limitations and the accuracy of the available plant model. Estimates of any state variables needed whose direct measurement is impracticable can be obtained using state estimation algorithms (Chap. 4).

1.2.5 Variables and Their Laplace Transforms

The notation, $\mathcal{L}\{x(t)\} = x(s)$, is sometimes used. While the intended meaning should be obvious to the reader, according to the standard functional notation, it could be taken that the Laplace transform is obtained by replacing t by s in $\alpha(t)$ which is clearly incorrect. To avoid this issue, in keeping with much of the literature, the Laplace transforms of time-varying quantities represented by lower-case Roman italic characters are represented by their upper-case counterparts. For example, $\mathcal{L}\{y(t)\} = Y(s)$, or for a vector, $\mathcal{L}\{\mathbf{x}(t)\} = \mathbf{X}(s)$. Sometimes, however, lower-case Greek symbols are used for time-varying quantities whose upper-case counterparts are the same as Roman ones that represent constants. For example, $\mathcal{L}\{\alpha(t)\} = A(s)$. To avoid this without altering the notation substantially, the same character is used for the transformed variable but enlarged. Thus, $\mathcal{L}\{\alpha(t)\} = \alpha(s)$. The alternative notation, $\mathcal{L}\{x(t)\} = \bar{x}(s)$, also used elsewhere, is not adopted as a similar notation is used in Chap. 7 to represent operating point values. For example, if a variable, $x(t)$, is expressed as $x(t) = \bar{x} + \tilde{x}(t)$, then $\bar{x}$ is a constant operating point value and $\tilde{x}(t)$ is a variation about this operating point value.

A variable, $x(t)$, may be just shown as x but the functional notation is always shown in the Laplace transform, $X(s)$. This avoids any confusion with constants.

1.3 Review of Traditional PID Controllers and Their Variants

1.3.1 Traditional Error-Actuated Controllers

1.3.1.1 The PID Controller

The Proportional Integral Derivative (PID) controller is the generic controller from which all the traditional controllers can be derived. This has three adjustable parameters and two forms, as shown in Fig. 1.2.

The single disturbance input, $D(s)$, acting at the same point as $U(s)$, is equivalent to the set of physical disturbances acting on the real plant, in the sense that it has the same effect on $Y(s)$. This is significant in many applications.

1.3 Review of Traditional PID Controllers and Their Variants

Fig. 1.2 The PID controller in a closed-loop system. (**a**) in terms of individual gains and (**b**) in terms of proportional gain and action times

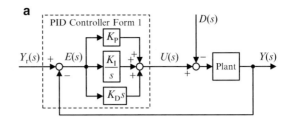

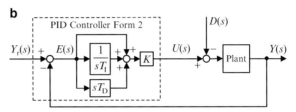

The convention of $D(s)$ acting via a negative summing junction input, which loses no generality, originates in the field of electric drives in which an external load torque applied using a passive brake to a motor reduces its speed.

Form 1 is mainly referred to in this book, the adjustable parameters being the proportional gain, K_P, the integral gain, K_I, and the differential gain, K_D. In Form 2, the adjustable parameters are the forward path gain, K, the integral action time, T_I, and the derivative action time, T_D. It may easily be shown that the two controllers of Fig. 1.2 are equivalent by writing down the controller equations applying to the block diagrams. Thus,

$$E(s) = Y_r(s) - Y(s), \tag{1.1}$$

$$U(s) = \left(K_P + \frac{K_I}{s} + K_D s \right) E(s) \tag{1.2}$$

and

$$U(s) = K \left(1 + \frac{1}{sT_I} + sT_D \right) E(s). \tag{1.3}$$

Both controllers can be made to produce the same values of $U(s)$ for the same error input, $E(s)$, and the same initial value of the integrator output (usually zero) by equating the right-hand sides of (1.2) and (1.3) to yield

$$K_P = K, \quad K_I = \frac{K}{T_I} \left(\text{or } T_I = \frac{K_P}{K_I} \right) \text{ and } K_D = K T_D \left(\text{or } T_D = \frac{K_D}{K_P} \right). \tag{1.4}$$

The PID controller is sometimes referred to as the three-term controller since it comprises the proportional, integral and derivative terms, identifiable in Fig. 1.2. The three terms are often described as implementing different *control actions*. In the time domain, the PID controller equations corresponding to (1.2) and (1.3) are

$$u(t) = \underbrace{K_P e(t)}_{\text{Proportional control action}} + \underbrace{K_I \int_0^t e(\tau)\,d\tau}_{\text{Integral control action}} + \underbrace{K_D \frac{d}{dt} e(t)}_{\text{Derivative control action}}$$

or

$$u(t) = \overbrace{K e(t)}^{} + \overbrace{K \cdot \frac{1}{T_I} \int_0^t e(\tau)\,d\tau}^{} + \overbrace{K \cdot T_D \frac{d}{dt} e(t)}^{} \qquad (1.5)$$

Next, the purposes of these control actions will be discussed. The overall objective is to drive $e(t)$ to zero but the manner in which the PID controller either succeeds (or fails) to do this depends on the plant characteristics and the weightings of the control actions present in $u(t)$, determined by the settings of the three gains, K_P, K_I and K_D (or the three parameters, K, T_I and T_D). Sometimes not all the control actions are needed and the appropriate gains can be set to zero, resulting in the well-known proportional, integral, PD or PI controllers obtained, respectively, by setting $K_I = K_D = 0$, $K_P = K_D = 0$, $K_I = 0$ or $K_D = 0$.

1.3.1.2 The Proportional Control Action

The proportional control action may be regarded as the basic one needed to control a plant in nearly every case, sometimes being sufficient alone.

1.3.1.3 The Integral Control Action

The integral control action is used to eliminate the *steady-state error*. The term *steady state* applies to stable dynamical systems and may be regarded as the *state variables* (Chap. 2) reaching constant values as $t \to \infty$. If, however, *any* variable, $q(t)$, tends to a constant value, it is referred to as the *steady-state value*, whether or not it is a state variable, and is denoted q_{ss}. If

$$\lim_{t \to \infty} [e(t) = y_r(t) - y(t)] = e_{ss} = \text{const.}, \qquad (1.6)$$

then e_{ss} is the *steady-state error*, noting that $y(t)$ is a state variable but $y_r(t)$ is not. Some control systems will be met in Chap. 5 that have finite or zero steady-state errors with $y_r(t)$ a polynomial in t, but for this explanation, it will be assumed that

1.3 Review of Traditional PID Controllers and Their Variants

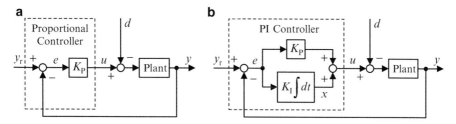

Fig. 1.3 Control of a plant subject to external disturbances. (**a**) with a proportional controller and (**b**) with a PI controller

$y_r = \text{const.} = Y_r$ and also $d = \text{const.} = D$. The action of the integral term in yielding $e_{ss} = 0$ may be explained by considering the systems of Fig. 1.3.

The plant transfer function relationship,

$$Y(s) = \frac{G(s)}{s^q}[U(s) - D(s)], \qquad (1.7)$$

is assumed to apply, where $G(0)$ is finite and q is an integer defined as the *plant type*. This is slightly different from the *open-loop system type* that includes the controller. Assuming closed-loop stability, if y_r and d are finite and constant, all the variables of both systems in Fig. 1.3 will settle to constant, finite values, some of which may be zero in the steady state.

First consider the system of Fig. 1.3a. If $q = 0$, the plant has no integral action of its own. Then even if $d = 0$, $u_{ss} \neq 0$ would be needed to maintain $y_{ss} \neq 0 = \text{const.}$ and since the system is linear,

$$y_{ss} = K_{DC} u_{ss}, \qquad (1.8)$$

where $K_{DC} = G(0)$ is the DC (direct current) gain of the plant that relates the constant steady-state output to the constant steady state-input. The term originates in the field of electrical engineering where direct current is defined as a *constant current*. This requires $e_{ss} \neq 0$, since

$$u_{ss} = K_P e_{ss} \Rightarrow e_{ss} = \frac{u_{ss}}{K_P}. \qquad (1.9)$$

where

$$e_{ss} = y_r - y_{ss}. \qquad (1.10)$$

Substituting for (a) y_{ss} in (1.10) using (1.8) and then (b) u_{ss} using (1.9) yields

$$e_{ss} = y_r - K_{DC} K_P e_{ss} \Rightarrow e_{ss} = \frac{y_r}{1 + K_{DC} K_P}, \qquad (1.11)$$

indicating that it would be possible to reduce but not eliminate the steady-state error by increasing the proportional gain, K_p, to a finite practicable value.

Now suppose $q > 0$ in (1.7). Referring again to Fig. 1.3a, if $d = 0$, then $e_{ss} = 0$ due to the integral action of the plant. If $e_{ss} \neq 0$, y would be changing, which is not the case. If, however, $d \neq 0$,

$$u_{ss} = d \qquad (1.12)$$

is needed to counteract d. Otherwise, y would be changing due to the net plant input being $u_{ss} - d \neq 0$. Substituting for u_{ss} in (1.9) using (1.12) then yields

$$e_{ss} = \frac{d}{K_P} \qquad (1.13)$$

which, as previously, can be *reduced but not eliminated* by increasing K_P. So integral action in the plant cannot remove a steady-state error due to a constant disturbance. Only integral action in the *controller* can eliminate the steady-state error.

If $q = 0$ and $d = $ const., then (1.8) would be replaced by

$$y_{ss} = K_{DC}(u_{ss} - d). \qquad (1.14)$$

Substituting for y_{ss} in (1.10) using (1.14) and then u_{ss} using (1.9) yields

$$e_{ss} = y_r - K_{DC}(K_P e_{ss} - d) \Rightarrow e_{ss} = \frac{y_r + K_{DC} d}{1 + K_{DC} K_P} \qquad (1.15)$$

As previously, this can be reduced, but not eliminated, by increasing K_P.

Turning attention to Fig. 1.3b, let $y_r = $ const. and $d = $ const.. First suppose $e_{ss} = $ const. $\neq 0$ and the closed-loop system is stable. Then the output, $x(t)$, of the integrator would be increasing at a constant rate of $K_I e_{ss}$, contradicting the fact that x_{ss} must be constant if the system is stable. Only $e_{ss} = 0$ yields $x_{ss} = $ const.. The integral term therefore eliminates the steady-state error.

1.3.1.4 The Derivative Control Action

Next, qualitative reasoning for needing derivative action for the control of some plants will be given. Many plants, excluding those of first order, exhibit a property that could be described as *inertia*, by analogy with the behaviour of mechanical systems. Suppose such a plant is disconnected from the controller and the initial conditions are zero. Now the behaviour of the plant with inputs provided by a manual operator will be considered. First, if a step control input is applied, the output, $y(t)$, and its first derivative, $\dot{y}(t)$, will grow from zero at a finite rate and initially have the same sign. Once this has occurred, let the control input be reversed in sign. Then for many plants, $\dot{y}(t)$ will not immediately change sign. To be precise,

1.3 Review of Traditional PID Controllers and Their Variants

this will happen if the *relative degree*, sometimes called the *rank*, of the plant is $r > 2$. This property is fully discussed in Chap. 2. For a linear plant, $r = n - m$, where n is the number of poles and m is the number of zeros in the transfer function. A mechanical analogy is the application of an applied force to accelerate a rigid body to a certain velocity and the continuation of the motion in the same direction, due to the momentum of the body, after removal of the force. Now suppose that the operator undertakes the task of applying $u(t)$ to take the plant output from zero to a constant reference value, Y_r, by monitoring the error, $e(t) = Y_r - y(t)$. An inexperienced operator might simply apply $u(t)$ with such a sign as to reduce $e(t)$ until it is zero and then set $u(t) = 0$. This would result in the error, $e(t)$, changing sign and increasing in magnitude due to $\dot{y}$ maintaining its original sign. The operator would then apply $u(t)$ again but with the opposite sign in a second attempt to bring $e(t)$ to zero. So $y(t)$ would overshoot the reference value. If the operator continued with this simple strategy, then $y(t)$ would oscillate about Y_r, hopefully with a decaying amplitude but possibly with an increasing amplitude, indicating instability! A simple proportional feedback controller would act in a similar manner. After this experience, the operator might attempt the better strategy of predicting the overshoot by observing $\dot{e}(t)$ as well as $e(t)$ and use this extra information to 'apply the brake' in time by reversing the sign of $u(t)$ before $e(t)$ reaches zero. The derivative term in a PID controller performs this function. Whether or not it succeeds in entirely preventing an overshoot depends on the specific plant and the setting of the derivative gain, K_D, and the other controller gains, K_P and K_I, it generally has the effect of reducing overshooting or oscillations of the controlled output.

1.3.1.5 Mechanical Analogy of PD Control Loop

Since the derivative term of a controller and higher derivative terms in other controllers to be introduced later play an important part in achieving the required stability and specified performance of feedback control systems, I will be informative to study a mechanical analogy with a more analytical approach. This is an automobile suspension system. Figure 1.4 shows a quarter-vehicle model consisting of a mass, representing the vehicle body, suspended on a spring connected to the wheel hub.

Figure 1.4a, b respectively, show the system with and without a damper. A damper produces a force acting on the vehicle body proportional to the relative vertical velocity, $\dot{y}$, between the body and the wheel hub and opposing the motion. In the force balance equations shown, g is the acceleration due to gravity [m/s/s], K_s is the spring constant [N/m], K_d is the viscous damping coefficient [Ns/m] and y_0 is the height of the upper end of the suspension spring above the ground if the vehicle body were to be removed. Also, ω_n [rad/s] is the undamped natural frequency, i.e. the frequency of oscillation if the damper were to be removed and ζ is the damping ratio. The plant in this analogy is the vehicle mass and the controlled variable is its vertical position, y, the reference input being $y_r = 0$ and therefore not shown. Figure 1.4a is analogous to the application of the proportional control

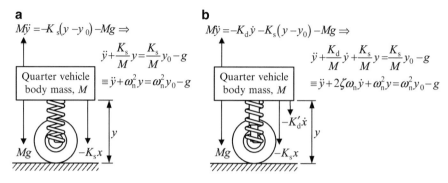

Fig. 1.4 Vehicle suspension analogy (quarter-vehicle model). (**a**) without damper and (**b**) with damper

action alone, the spring providing this by producing a force acting on the vehicle body that opposes any displacement from the equilibrium position, y_{eq}, satisfying $\dot{y}_{eq} = \ddot{y}_{eq} = 0 \Rightarrow y_{eq} = y_0 - g/\omega_n^2$.

$$y_{eq} = y_0 - g/\omega_n^2. \tag{1.16}$$

The gravitational force is analogous to a constant external disturbance acting on the plant. With the undamped suspension, the differential equation of motion is

$$\ddot{y} + \omega_n^2 y = \omega_n^2 y_0 - g, \tag{1.17}$$

which, in view of (1.16), may be written

$$\ddot{y} + \omega_n^2 y = \omega_n^2 y_{eq}. \tag{1.18}$$

In the unforced case, i.e. with $\omega_n^2 y_{eq} = 0$, (1.18) may be recognised as the equation of simple harmonic motion, $\ddot{y} = -\omega_n^2 y$. If the vehicle were to have an arbitrary initial displacement of $y(0)$ with $\dot{y}(0) = 0$, deriving the solution of (1.18) using Laplace transforms (Table 1 in Tables) yields

$$\mathcal{L}\{\ddot{y} + \omega_n^2 y\} = \mathcal{L}\{\omega_n^2 y_{eq}\} \Rightarrow s^2 Y(s) - sy(0) - \dot{y}(0) + \omega_n^2 Y(s) = \omega_n^2 y_{eq}/s.$$

Since $\dot{y}(0) = 0$, this yields

$$Y(s) = \frac{1}{s^2 + \omega_n^2} \cdot \left[sy(0) + \frac{\omega_n^2 y_{eq}}{s} \right] \Rightarrow$$

$$y(t) = \mathcal{L}^{-1}\{Y(s)\} = y(0) \cos(\omega_n t) + y_{eq}[1 - \cos(\omega_n t)] \tag{1.19}$$

which, as expected, indicates a continuous oscillation with a mean value of y_{eq}. In fact, attempting the position control of a mass moving without any viscous friction using only the proportional control action would produce such an oscillation.

1.3 Review of Traditional PID Controllers and Their Variants

In Fig. 1.4b, the viscous damping coefficient, K_d, is analogous to the derivative feedback gain, K_D, of the PID controller. The solution to the differential equation of motion, with the same initial conditions as for the undamped case, is given by

$$\mathcal{L}\{\ddot{y} + 2\zeta\omega_n\dot{y} + \omega_n^2 y\} = \mathcal{L}\{\omega_n^2 y_{eq}\} \Rightarrow$$
$$s^2 Y(s) - sy(0) + 2\zeta\omega_n [sY(s) - y(0)] + \omega_n^2 Y(s) = \omega_n^2 y_{eq}/s \Rightarrow$$
$$Y(s) = \frac{1}{s^2 + 2\zeta\omega_n s + \omega_n^2} \cdot \left[sy(0) + 2\zeta\omega_n y(0) + \frac{\omega_n^2 y_{eq}}{s} \right]$$

Then $y(t) = \mathcal{L}^{-1}\{Y(s)\}$. Hence,

$$y(t) = \frac{2\zeta\omega_n y(0)}{\omega_d} e^{-\zeta\omega_n t} \sin(\omega_d t) + y(0) e^{-\zeta\omega_n t} \left[\cos(\omega_d t) - \frac{\zeta\omega_n}{\omega_d} \sin(\omega_d t) \right]$$
$$+ y_{eq} \left[1 - \frac{\omega_n}{\omega_d} e^{-\zeta\omega_n t} \sin(\omega_d t + \phi) \right], \qquad (1.20)$$

where $\omega_d = \omega_n \sqrt{1 - \zeta^2}$ and $\phi = \cos^{-1}(\zeta)$. This is a damped oscillation at a lower frequency, ω_d, than in the undamped case, i.e. the damped natural frequency. The vehicle displacement has a steady-state value of

$$y_{ss} = \lim_{t \to \infty} y(t) = y_{eq} = y_0 - g/\omega_n^2,$$

which is also the mean value of the oscillation without the damping indicated by (1.19), as would be expected. It should be noted that (1.20) is valid for $0 < \zeta < 1$, which is an underdamped system. For $\zeta > 1$, the basic dynamic character of the system is determined by two exponential terms, $\exp\left[-\left(\zeta \pm \sqrt{\zeta^2 - 1}\right)\omega_n t\right]$ instead of the exponentially decaying sinusoidal terms and no oscillations occur.

The following example is a control system directly demonstrating the validity of the vehicle suspension analogy.

Example 1.1 Single-axis PD attitude control of a rigid-body spacecraft

Consider a PD controller applied to control the attitude of a rigid-body spacecraft about a single axis (Chap. 2) as shown in Fig. 1.5.

Applying elementary block diagram algebra yields:

$$\frac{Y(s)}{Y_r(s)} = \frac{(K_P + K_D s)(b_0/s^2)}{1 + (K_P + K_D s)(b_0/s^2)} = \frac{(K_D s + K_P) b_0}{s^2 + (K_D s + K_P) b_0} \equiv \frac{2\zeta\omega_n s + \omega_n^2}{s^2 + 2\zeta\omega_n s + \omega_n^2}. \qquad (1.21)$$

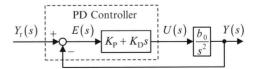

Fig. 1.5 PD single-axis attitude control of a spacecraft

Despite the PD controller being commonplace, the standard form of this second-order closed-loop transfer function in terms of the undamped natural frequency, ω_n, and the damping ratio, ζ, differs from the one often quoted, i.e.

$$\frac{\omega_n^2}{s^2 + 2\zeta\omega_n s + \omega_n^2} \qquad (1.22)$$

due to the zero at $s = -\omega_n/(2\zeta)$ introduced by the controller. It is important to note, however, that while such zeros do affect the dynamic response to $y_r(t)$ (Sect. 1.4.4), they do not affect the basic control system dynamic character that depends only on the closed-loop poles, i.e. the roots of the denominator polynomial. If the reference input is $y_r(t) = Y_r h(t)$, where Y_r is the constant step value and $h(t)$ is the unit step function defined by $h(t) \triangleq \{0 \text{ for } t < 0, 1 \text{ for } t \geq 0\}$, then using a table of Laplace transforms (Table 1 in Tables), the response of the closed-loop system (1.21) with zero initial conditions is

$$y(t) = \mathcal{L}^{-1}\left\{\frac{Y_r}{s} \cdot \frac{2\zeta\omega_n s + \omega_n^2}{s^2 + 2\zeta\omega_n s + \omega_n^2}\right\} = \mathcal{L}^{-1}\left\{\frac{2\zeta\omega_n Y_r}{s^2 + 2\zeta\omega_n s + \omega_n^2}\right\}$$

$$+ \mathcal{L}^{-1}\left\{\frac{1}{s} \cdot \frac{\omega_n^2 Y_r}{s^2 + 2\zeta\omega_n s + \omega_n^2}\right\}$$

$$= \underbrace{\frac{2\zeta\omega_n}{\omega_d} Y_r e^{-\zeta\omega_n t} \sin(\omega_d t)}_{\text{term due to zero}} + \underbrace{Y_r \left[1 - \frac{\omega_n}{\omega_d} e^{-\zeta\omega_n t} \sin(\omega_d t + \phi)\right]}_{\text{response without zero}}, \qquad (1.23)$$

where $\omega_d = \omega_n\sqrt{1-\zeta^2}$ and $\phi = \cos^{-1}(\zeta)$ for $0 < \zeta < 1$. For $\zeta > 1$, (1.23) becomes

$$y(t) = \mathcal{L}^{-1}\left\{\frac{Y_r}{s} \cdot \frac{2\zeta\omega_n s + \omega_n^2}{(s+a)(s+b)}\right\} \text{ where } a = \left(\zeta - \sqrt{\zeta^2 - 1}\right)\omega_n \text{ and }$$

$$b = \left(\zeta + \sqrt{\zeta^2 - 1}\right)\omega_n$$

$$= 2\zeta\omega_n Y_r \mathcal{L}^{-1}\left\{\frac{s+c}{s(s+a)(s+b)}\right\}$$

$$= \frac{2\zeta\omega_n Y_r}{ab}\left[c - \frac{b(c-a)}{b-a}e^{-at} + \frac{a(c-b)}{b-a}e^{-bt}\right] \qquad (1.24)$$

and therefore no oscillation of the step response occurs. The well-known critically damped case for which $\zeta = 1$ corresponds to coincidence of the two negative real closed loop poles. This is covered in Chap. 4 where the general case of multiple closed loop poles is considered.

The similarity between the step response of this PD control loop and the transient behaviour of the motor vehicle suspension system with non-zero initial conditions

1.3 Review of Traditional PID Controllers and Their Variants

is apparent by comparing (1.20) with (1.23). By inspection of the force balance equations of Fig. 1.4b,

$$K_d = 2\zeta\omega_n M, \quad K_s = \omega_n^2 M \tag{1.25}$$

and from (1.21),

$$K_D = 2\zeta\omega_n/b_0, \quad K_P = \omega_n^2/b_0 \tag{1.26}$$

The accuracy of the suspension system analogy is then apparent, the spring constant, K_s, being equivalent to the proportional gain, K_P, and the viscous damping coefficient, K_d, being equivalent to the derivative gain, K_D.

The basic dynamic character of the control loop is determined by the oscillatory terms and the exponential decay terms of (1.23) if the controller gains are adjusted to yield an underdamped system or by the two exponential terms of (1.24) if the system is overdamped. It is evident from (1.23) and (1.24) that the zero introduced by the controller just alters the relative weights of these terms in $y(t)$.

Regarding the effects of changing the controller gains, from (1.26),

$$\omega_n = \sqrt{b_0 K_P} \text{ and } \zeta = b_0 K_D/(2\omega_n) = K_D\sqrt{b_0}/\left(2\sqrt{K_P}\right). \tag{1.27}$$

Hence, increasing K_D increases the damping ratio, ζ, and therefore increases the decay rate of the exponential term, $\exp(-\zeta\omega_n t)$ if the system is underdamped. An important observation in the step response (1.23) is that t is multiplied by ω_n wherever it occurs. It follows that increasing ω_n by a factor of λ reduces the timescale of the step response by a factor of λ without altering its shape. Bearing this in mind, referring again to (1.27), increasing K_P increases ω_n and therefore speeds up the response but at the same time reduces ζ, necessitating a further increase of K_D to restore the original degree of damping. This illustrates that tuning the PD controller by trial and error, which is commonly practised in industry, is an iterative process that could be time consuming, even for this simple system.

The following example is a first-order plant that, in contrast with Example 1.1, requires no active damping. Hence, no derivative action is required.

Example 1.2 PI control of a heating process

Consider a PI controller applied to a first-order linear plant such as the heating process with one dominant time constant (Chap. 2) as shown in Fig. 1.6.

Fig. 1.6 PI controller applied to a first-order heating process (Chap. 2)

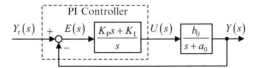

In this case, the closed-loop system is of second order since the plant and controller are both of first order, the closed-loop transfer function being K_I

$$\frac{Y(s)}{Y_r(s)} = \frac{(K_P s + K_I) b_0}{s^2 + a_0 s} \bigg/ \left[1 + \frac{(K_P s + K_I) b_0}{s^2 + a_0 s} \right]$$

$$= \frac{(K_P s + K_I) b_0}{s^2 + (a_0 + K_P b_0) s + K_I b_0} = \frac{(2\zeta\omega_n - a_0) s + \omega_n^2}{s^2 + 2\zeta\omega_n s + \omega_n^2}. \quad (1.28)$$

It follows that

$$\omega_n = \sqrt{K_I b_0} \text{ and } \zeta = (a_0 + K_P b_0) / (2\omega_n) = (a_0 + K_P b_0) / \left(2\sqrt{K_I b_0} \right). \quad (1.29)$$

Comparison of Example 1.2 with Example 1.1 reveals that K_I now determines the speed of response instead of K_P, while K_P has changed its role from determining the speed of response to that of determining the degree of damping. This demonstrates the dependence of the effects of the controller gains on the plant under control. The following section, however, gives an indication of the expected effects of PID controller gain adjustments in applications to second-order plants.

1.3.1.6 Effects of PID Controller Gains

Many plants to which PID control is applied are second order without finite zeros and for this reason a general indication of the effects of the individual gain adjustments will be provided for this category of plant, by means of simulations. The general control loop is shown in Fig. 1.7.

In this case the controller is of first order so the closed-loop system is of third order, with transfer function

$$\frac{Y(s)}{Y_r(s)} = \frac{\left(K_D s^2 + K_P s + K_I\right) b_0}{s^3 + a_1 s^2 + a_0 s} \bigg/ \left[1 + \frac{\left(K_D s^2 + K_P s + K_I\right) b_0}{s^3 + a_1 s^2 + a_0 s} \right].$$

$$= \frac{\left(K_D s^2 + K_P s + K_I\right) b_0}{s^3 + (a_1 + K_D b_0) s^2 + (a_0 + K_P b_0) s + K_I b_0} \quad (1.30)$$

Fig. 1.7 PID control of a second-order plant

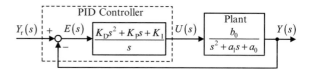

1.3 Review of Traditional PID Controllers and Their Variants

Depending on the controller gain settings, either all the three closed-loop poles are real or one is real and the other two are complex conjugates. In the latter case, the system can be decomposed into second- and first-order subsystems for the purpose of analysing its dynamic characteristics, by partial fraction expansion, yielding

$$\frac{Y(s)}{Y_r(s)} = \frac{\omega_n^2 p}{\left(s^2 + 2\zeta\omega_n s + \omega_n^2\right)(s+p)} = \frac{A_1 s + A_0}{s^2 + 2\zeta\omega_n s + \omega_n^2} + \frac{B_0}{s+p} \quad (1.31)$$

where $0 < \zeta < 1$ and

$$A_1 = -\omega_n^2 p/D, \ A_0 = (p - 2\zeta\omega_n)\omega_n^2 p/D, \ B_0 = \omega_n^2 p/D, \ D = p^2 - 2\zeta\omega_n p + \omega_n^2. \quad (1.32)$$

Hence, the step response of this third-order system is the sum of the step responses of first- and second-order subsystems, whose dynamic characteristics are well known. The second-order step response is similar to (1.23) and the first-order step response is $B_0 \left[1 - \exp(-pt)\right]$. If, on the other hand, the three closed-loop poles are real, the system can be decomposed into three first-order subsystems yielding

$$\frac{Y(s)}{Y_r(s)} = \frac{p_1 p_2 p_3}{(s+p_1)(s+p_2)(s+p_3)} = \frac{C_0}{s+p_1} + \frac{D_0}{s+p_2} + \frac{E_0}{s+p_3} \quad (1.33)$$

where

$$C_0 = \frac{p_1 p_2 p_3}{(p_2 - p_1)(p_3 - p_1)}, \ D_0 = \frac{p_1 p_2 p_3}{(p_1 - p_2)(p_3 - p_2)}, \ E_0 = \frac{p_1 p_2 p_3}{(p_1 - p_3)(p_2 - p_3)} \quad (1.34)$$

but (1.31) is still valid if $\zeta > 1$. To consider the effects of the controller gains, equating the denominators of (1.30) and (1.31) yields

$$s^3 + (a_1 + K_D b_0) s^2 + (a_0 + K_P b_0) s + K_I b_0$$
$$= s^3 + (2\zeta\omega_n + p) s^2 + \left(2\zeta\omega_n p + \omega_n^2\right) s + \omega_n^2 p \Rightarrow$$
$$\omega_n^2 p = K_I b_0, \ (2\zeta p + \omega_n) \omega_n = a_0 + K_P b_0, \ 2\zeta\omega_n + p = a_1 + K_d b_0. \quad (1.35)$$

It is evident that all the three parameters, p, ζ and ω_n, depend on all three controller gains and do so in a nonlinear and complex way. It is no surprise, therefore, that PID controllers are not always straightforward to tune by trial and error, particularly for higher-order plants. For this reason, the preferred approach of the author, assuming a reasonably accurate plant model is available, is to first choose the closed-loop system parameters, p, ζ and ω_n, that yield an acceptable step response. This is quite straightforward. Suppose ζ and ω_n are first chosen as if the system is of second order to achieve a specified maximum percentage overshoot and settling time. This is a standard procedure. Then the presence of the pole at $-p + j0$ will cause the overshoot to be less than specified, which is harmless. The settling time, however, will be longer than specified, but this can easily be compensated by increasing ω_n. This, in turn, will increase the overshoot but if this exceeds the specified maximum value, ζ can be increased to reduce the overshoot to an acceptable value. Once suitable values of p, ω_n and ζ have been determined, the controller gains that realise this are calculated using the following gain formulae derived by solving (1.35) to yield

$$K_I = \omega_n^2 p/b_0, \quad K_P = [(2\zeta p + \omega_n)\omega_n - a_0]/b_0 \text{ and } K_D = (2\zeta\omega_n + p - a_1)/b_0. \tag{1.36}$$

This does not require the individual effects of the controller gains to be known. It should be mentioned, however, that the gains of traditional PID controllers, particularly those implemented with analogue electronics, could only be adjusted to positive values, the expectation being that they should always be positive. Industrial practitioners would tune the gains by combining experience with trial and error to obtain an acceptable step response within this constraint. The model-based approach based on (1.36), however, allows negative values of K_P and K_D, which are easily accommodated in the software of modern digital implementation. Although this might appear incorrect in the light of the traditional approach, selecting p, ζ and ω_n with positive values will guarantee closed-loop poles in the left half of the s-plane and therefore result in the intended performance. In most cases, the gains will turn out to be positive but occasionally one or more of the gains will be negative, depending on the values of a_0 and a_1. This is a simple introduction to model-based control system design. Negative gains can occur in any model-based control technique.

Since an accurate plant model may not always be available, the reader may occasionally need to undertake traditional tuning by trial and error. For this reason, an attempt is made to demonstrate the general effects of individual changes in the controller parameters, but it must be born in mind that these effects depend on the plant, particularly its order, as already demonstrated by comparing the example of Fig. 1.6 with that of Fig. 1.5.

Returning to the system of Fig. 1.7, the plant to be simulated is the single attitude control axis of the spacecraft (Chap. 2) as in Fig. 1.5 where the PID controller is used instead of the PD controller in order to eliminate steady-state errors due to constant external disturbance torque components. So in this case, $a_0 = a_1 = 0$ and the gains yielded by (1.36) can only be positive. The spacecraft parameters are typical of a moderately sized satellite, a moment of inertia of $J = 200 \left[\text{Kg m}^2\right]$ and a reaction wheel torque constant of $K_w = 0.05$ [Nm/V] being taken, yielding $b_0 = K_w/J = 2.5 \times 10^{-04}$ [rad/s/s/V]. The output of the attitude sensor will be converted to radians in the software of the computer implementing the control algorithm (Chap. 6) and therefore the measurement, y, will be taken in units of radians. ω_n is chosen as 0.2 [rad/s] to yield a step response settling time (Chap. 3) of the order of 40 s, which is practicable due to the limited control torques from the reaction wheels, and the damping ratio of the complex conjugate pole pair is set

1.3 Review of Traditional PID Controllers and Their Variants

to $\zeta = 0.5$, which is purposely on the low side to enable any improvements in the damping due to the gain adjustments to be visible. The gains, K_P, K_I and K_D will first be set to specific values, $\overline{K}_P$, $\overline{K}_I$ and $\overline{K}_D$, using (1.36) to yield a baseline step response. Then, simulations are run in which the gains are varied, one at a time, above and below the calculated values. The influence of the first-order subsystem will be made similar to that of the second-order oscillatory subsystem by setting its pole value, $-p$, equal to the negative real part, $-\zeta\omega_n$, of the complex conjugate pole pair (Sect. 1.4), so (1.36) becomes

$$\begin{cases} K_I = \lambda_I \overline{K}_I, \ \overline{K}_I = \zeta\omega_n^3/b_0 \\ K_P = \lambda_P \overline{K}_P, \ \overline{K}_P = \left[(2\zeta^2 + 1)\omega_n^2 - a_0\right]/b_0 \\ K_D = \lambda_D \overline{K}_D, \ \overline{K}_D = (3\zeta\omega_n - a_1)/b_0. \end{cases} \quad (1.37)$$

The plant parameters above yield $\overline{K}_P = 80$, $\overline{K}_I = 5.\dot{3}$ and $\overline{K}_D = 400$. The gain adjustment parameters are chosen within the constraints $\lambda_q > 1$ to increase gain and $0 < \lambda_q < 1$, q = P, I, D, to reduce gain in order to keep it positive. Each of the controller gains is varied to two levels, firstly above the nominal values with $\lambda_P = \lambda_I = \lambda_D = 2$ and secondly below the nominal values, with $\lambda_P = \lambda_I = \lambda_D = 1/2$. The results are shown in Fig. 1.8.

Observing these step responses, increasing K_P reduces the settling time and also increases the damping of the oscillations. Increasing K_I reduces the damping of the oscillations, which, of course, does not improve the transient performance, but this term is necessary to guarantee zero steady-state error with a constant component of the external disturbance torque. Increasing K_D increases the damping of the oscillations, which is true in many applications.

If it is required to achieve a specified settling time with minimal or zero overshoot, then the adjustment of K_P, K_I and K_D by trial and error could be very time consuming, but the method of control system design by pole assignment presented in Chap. 3 could be used to achieve this quickly for applications in which accurate plant models are available, the spacecraft example being one (Chap. 2).

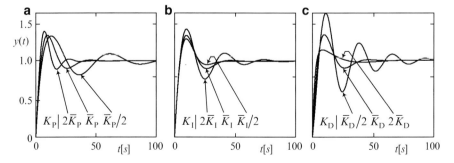

Fig. 1.8 Effects of gain adjustments on step responses of PID controller for spacecraft. (**a**) $K_P \updownarrow$; $(K_I, K_D) = (\overline{K}_I, \overline{K}_D)$; (**b**) $K_I \updownarrow$; $(K_P, K_D) = (\overline{K}_P, \overline{K}_D)$; (**c**) $K_D \updownarrow$; $(K_P, K_I) = (\overline{K}_P, \overline{K}_I)$

It may be observed that an initial overshoot occurs in all the step responses of Fig. 1.8 regardless of the degree of damping of the oscillations. There are, in fact, two causes of overshooting in linear control systems. The well-known one is complex conjugate poles. These, however, will only sometimes cause this, depending on the presence and values of other real closed-loop poles. In the third-order system under study, the real pole has to have a value smaller than a certain threshold, which is of the same order of magnitude as the complex conjugate poles, in order for an overshoot to occur in the step response. As will be explained in Sect. 1.4.4, the second cause of overshooting is the zeros introduced by the controller. In order for a zero to cause an overshoot, its magnitude must be below a certain threshold, which is of the same order of magnitude as the closed-loop poles nearest the origin of the s-plane. In the example of Fig. 1.8, the overshooting is attributable to both causes.

Overshooting is sometimes considered desirable by practitioners of the traditional controllers, since it ensures that $y(t)$ actually reaches a new value of a constant reference if a step change occurs in $y_r(t)$, bearing in mind that linear systems without overshoot, in theory, never reach the reference input with certain initial conditions, including zero initial conditions. When the overshooting is caused by zeros, it is sometimes called *derivative kick* since it depends upon the derivative action of the controller on the reference input when it steps from one constant value to another. This becomes clear after manipulating the block diagram of Fig. 1.7 to yield Fig. 1.9. Note that the integrator in the controller has been combined with the plant by moving the factor, $1/s$, forward.

Then, the input to the third-order block so formed is the control input derivative, $\dot{U}(s) = sU(s)$. This has been decomposed into a feedback component, $\dot{U}_{fb}(s)$, and a feedforward component, $\dot{U}_{ff}(s)$. So in the time domain,

$$\dot{u}_{ff}(t) = K_D \ddot{y}_r(t) + K_P \dot{y}_r(t) + K_I y_r(t). \tag{1.38}$$

If $y_r(t) = Y_r h(t)$, then $\dot{y}_r(t)$ is a positive infinite impulse for $Y_r > 0$ and $\ddot{y}_r(t)$ is an infinite impulse doublet, i.e. a positive infinite impulse followed by a negative infinite impulse after an infinitesimal delay. $\dot{u}_{ff}(t)$ does not exist as a signal in the physical system as clearly it would not be realisable in the hardware, but its effect is manifest in the plant output, $y(t)$, as the initial overshoot. Although $y(t)$ reaches the constant reference level, Y_r, earlier with the derivative kick than without, the ensuing overshoot delays the final settling of $y(t)$ towards Y_r. In fact, as will be seen in Sect. 1.4.4 and Chap. 4, if the step responses of two systems with identical real

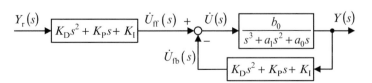

Fig. 1.9 Block diagram manipulation to demonstrate derivative action on reference input

1.3 Review of Traditional PID Controllers and Their Variants

poles are compared, one with finite zeros and the other without, then if the zeros are considerably smaller in magnitude than the poles, the settling time of the system with the zeros can be longer. It is therefore debatable whether the derivative kick associated with the PID, PI and PD controllers is really useful. Some practitioners of position control servomechanisms, however, recommend a small overshoot in the linear model of the system to minimise steady-state errors due to *stick–slip friction* in practice. This is the nonlinear friction due to the imperfect machining of relatively moving surfaces in a mechanism, which means that the control torque or force has to exceed a certain minimum magnitude to cause any motion (Chap. 2). It possible, of course, for the small overshoot to be achieved in a system without finite zeros by designing it so that there is a complex conjugate pair of closed-loop poles with relatively small imaginary parts. If desired, the derivative kick effect can be eliminated by employing the zero-less versions of the traditional controllers presented in the following section.

1.3.2 Zero-Less Versions of the Traditional Controllers

Provided the plant transfer function has no finite zeros, the overshooting and/or undershooting caused by zeros can be eliminated by using controllers that do not introduce zeros. These have an equally simple structure to the controllers of Sect. 1.3.1. These could have been easily implemented by analogue electronics in the previous era and are also straightforward to implement digitally (Chap. 6). The one shown in Fig. 1.10a is equivalent to the PID controller shown again in Fig. 1.10b for comparison. The difference is that only the integral term acts on the error, $E(s)$, while the proportional and derivative terms act only on the controlled output. This controller is distinguished from the PID using the acronym, IPD.

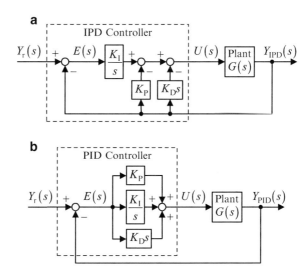

Fig. 1.10 Comparison of PID control loop with IDP control loop. (**a**) IPD control loop without controller zeros. (**b**) PID control loop with controller zeros

Before comparing the closed-loop transfer functions, the plant transfer function will be expressed (Chap. 2) as

$$G(s) = K_{DC}\frac{N(s)}{D(s)} = \frac{K_{DC}\sum_{j=0}^{m}b_j s^j}{s^n + \sum_{j=0}^{m}a_i s^i}, \quad n > m \qquad (1.39)$$

where K_{DC} is the DC gain; a_i, $i = 1, 2, \ldots, n$ and b_j, $j = 0, 1, \ldots, m$, are constant coefficients, some of which may be zero; and $b_m = 1$.

The IPD control loop is not of the classical canonical structure to which the well-known formula,

$$\frac{Y(s)}{Y_r(s)} = \frac{G(s)}{1 + G(s)H(s)},$$

applies, where $G(s)$ is the forward path transfer function and $H(s)$ is the feedback transfer function. This formula could be applied, however, but *three times* after applying the rules of block diagram reduction. Instead, the application of Mason's formula (Appendix A4) can achieve the same result much more quickly. Hence,

$$\frac{Y_{IPD}(s)}{Y_r(s)} = \frac{\frac{K_I}{s}.K_{DC}\frac{N(s)}{D(s)}}{1 - \left\{-K_{DC}\frac{N(s)}{D(s)}\left[K_D s + K_P + \frac{K_I}{s}\right]\right\}}$$

$$= \frac{K_I K_{DC} N(s)}{sD(s) + K_{DC}\left(K_D s^2 + K_P s + K_I\right) N(s)} \qquad (1.40)$$

Familiarity with the application of Mason's formula is advisable since many different block diagram structures will be met in which this is useful.

For comparison, the closed-loop transfer function of the PID control loop is

$$\frac{Y_{PID}(s)}{Y_r(s)} = \frac{\left(K_D s + K_P + \frac{K_I}{s}\right).K_{DC}\frac{N(s)}{D(s)}}{1 + \left(K_D s + K_P + \frac{K_I}{s}\right).K_{DC}\frac{N(s)}{D(s)}}$$

$$= \frac{\left(K_D s^2 + K_P s + K_I\right) K_{DC} N(s)}{sD(s) + K_{DC}\left(K_D s^2 + K_P s + K_I\right) N(s)}. \qquad (1.41)$$

Thus, the PID controller introduces zeros that are the roots of $K_D s^2 + K_P s + K_I = 0$, while the IPD controller does not, and the characteristic polynomial, $sD(s) + K_{DC}\left(K_D s^2 + K_P s + K_I\right) N(s)$, is the same for the IPD and PID control loops, meaning that for the same settings of K_P, K_I and K_D, both control loops have the same closed-loop poles and therefore have the same basic dynamic character, the only difference being the relative weightings of the system modes (Chap. 2 and Sect. 1.4).

1.3 Review of Traditional PID Controllers and Their Variants

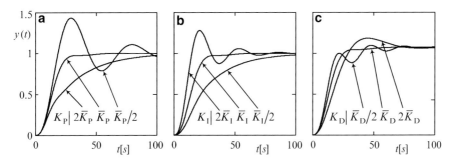

Fig. 1.11 Effects of gain adjustments on step responses of IPD controller for spacecraft. (**a**) $K_P \updownarrow$; $(K_I, K_D) = (\overline{K}_I, \overline{K}_D)$; (**b**) $K_I \updownarrow$; $(K_P, K_D) = (\overline{K}_P, \overline{K}_D)$; (**c**) $K_D \updownarrow$; $(K_P, K_I) = (\overline{K}_P, \overline{K}_I)$

It should also be observed that any finite plant zeros, i.e. the roots of $N(s) = 0$, are present in both control loops and their effects may be apparent in the closed-loop system (Sect. 1.4.4) unless they are cancelled by pole placement or pre-compensation (Chap. 4).

To demonstrate the differences between the step responses of the IPD and PID control loops for the same gain settings, the simulations of Fig. 1.8 are repeated for the IPD controller and the results are shown in Fig. 1.11.

Since the characteristic polynomials are the same for both the IPD and the PID control loops, (1.35), (1.36) and (1.37) still hold, and the damping ratio and undamped natural frequency of the complex conjugate pole pair of the nominal system are set, respectively, to $\zeta = 0.5$ and $\omega_n = 0.2$ [rad/s], as previously. Comparison with Fig. 1.8 reveals differences due to the absence of derivative kick. Since the numerator of the closed-loop transfer function (1.41) using the PID controller has a quadratic numerator polynomial, the step response comprises that of the zero-less version with closed-loop transfer function (1.40) using the IPD controller plus a weighted sum of the first and second derivatives. Specifically

$$Y_{PID}(s) = \left(K_D s^2 + K_P s + K_I\right) \cdot \frac{1}{K_I} \cdot Y_{IPD}(s) \Rightarrow$$
$$y_{PID}(t) = y_{IPD}(t) + (K_P/K_I)\,\dot{y}_{IPD}(t) + (K_D/K_I)\,\ddot{y}_{IPD}(t). \quad (1.42)$$

Even for the cases of Fig. 1.11 where the oscillation of $y_{IPD}(t)$ due to the closed-loop poles is nearly absent, the derivative effect of (1.42) causes at least one overshoot followed by one undershoot of $y_{PID}(t)$, as evident in Fig. 1.8. This is readily understandable considering that $\dot{y}_{IPD}(t)$ is the instantaneous slope of the graph of $y_{IPD}(t)$ and $\ddot{y}_{IPD}(t)$ is the instantaneous slope of the graph of $\dot{y}_{IPD}(t)$.

Regarding the effects of the IPD gain adjustments, they are similar to but not identical to those for the PID controller, due to the derivative kick effect. With reference to Fig. 1.8, reducing K_P increases the settling time but increasing it reduces the damping of the oscillations and increases the settling time. Increasing K_I reduces the damping of the oscillations, but as previously, it is necessary to keep $K_I > 0$, to guarantee zero steady-state error with a constant component of the external disturbance torque. Increasing K_D increases the damping of the oscillations, as before.

1.3.3 Traditional Controller Selection Guidelines

This section summarises the traditional controller variants, their closed-loop transfer function relationships and guidelines for selection, in tabular form. The reader may wish to prove the statements made in this section as an exercise.

1.3.3.1 The Proportional Controller

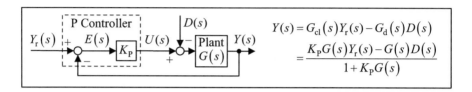

If $G(s)$ is of first order, then a prescribed first-order $G_{cl}(s)$ can be realised. If $G(s)$ is of order $n \geq 2$, it must have at least $n-1$ poles in the left half of the s-plane for closed-loop stability to be attainable. Attainment, in addition, of a specified transient performance is only possible in some cases. If zero steady-state error is required with $y_r = $ const., then $d = 0$ and $G(s)$ must be of type '1'.

1.3.3.2 The PI and IP Controllers

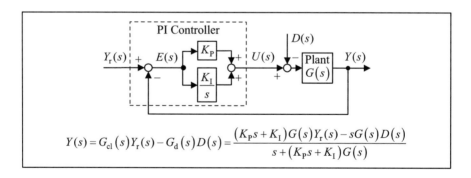

Specific to the PI controller: If $G(s)$ is of first order, a prescribed first-order $G_{cl}(s)$ can be realised by placing one closed-loop pole to cancel the controller zero but only if this takes place in the right half of the s-plane. Otherwise, an overshoot may occur in the step response due to the zero.

1.3 Review of Traditional PID Controllers and Their Variants

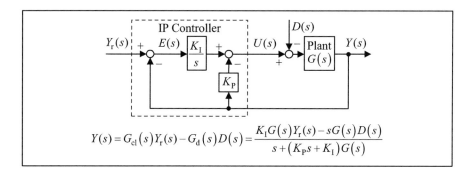

Specific to the IP controller: No controller zero is introduced. If $G(s)$ is of first order, a prescribed second-order $G_{cl}(s)$ can be realised.

In common with the PI and IP controllers: Both controllers may be applied to plants of order $n \geq 2$, but there must be at least $n - 1$ plant poles in the left half of the s-plane for closed-loop stability to be attainable. Achieving a specified transient performance can entail time-consuming tuning by trial and error and is only possible in some cases. If $y_r = $ const. and $d = $ const., then $e_{ss} = 0$.

1.3.3.3 The PD and DP Controllers

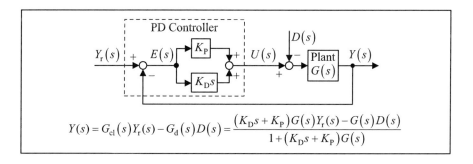

Specific to the PD controller: $G(s)$ has to be of order $n \geq 2$. If $n = 2$, a prescribed first-order $G_{cl}(s)$ can be realised by placing one closed-loop pole to cancel the controller zero but provided this takes place in the left half of the s-plane. Otherwise, an overshoot in the step response may occur due to the zero.

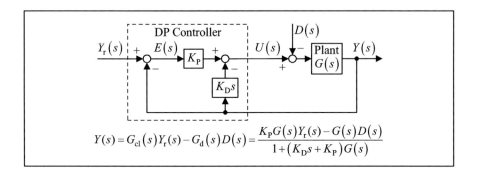

Specific to the DP controller: No controller zero is introduced. If $G(s)$ is of second order without a finite zero, a prescribed second-order $G_{cl}(s)$ can be realised.

In common with the PD and DP controllers: Both controllers may be applied if $G(s)$ has a rank (i.e. relative degree) of $r \geq 2$. If $G(s)$ is of order $n \geq 3$, it must have at least $n - 2$ poles in the left half of the s-plane for closed-loop stability to be attainable. Achieving a specified transient performance can entail time-consuming tuning by trial and error and is only possible in some cases. If $y_r = $ const. and $d = 0$, then $e_{ss} = 0$ if the plant is of type '1' or greater.

1.3.3.4 The PID and IPD Controllers

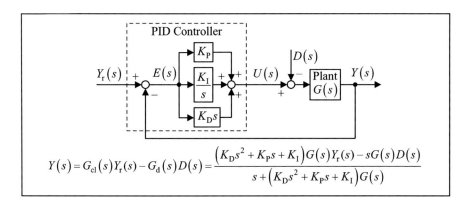

Specific to the PID Controller: $G(s)$ has to be of order $n \geq 2$. If $n = 2$, a prescribed first-order $G_{cl}(s)$ can be realised by placing two of the closed-loop poles to cancel the two controller zeros but provided this takes place in the left half of the s-plane. Otherwise, an overshoot and possibly an undershoot may occur in the step response due to the zeros.

1.3 Review of Traditional PID Controllers and Their Variants

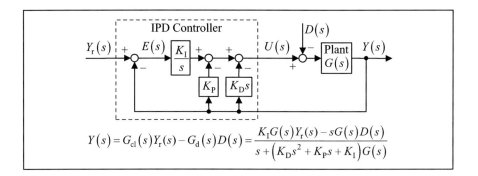

Specific to the IPD Controller: No controller zeros are introduced. If $G(s)$ is of second order without a finite zero, a prescribed third-order $G_{cl}(s)$ can be realised.

In Common with the PID and IDP Controllers: Both controllers may be applied if $G(s)$ has a rank (i.e., relative degree) of $r \geq 2$. If $G(s)$ is of order $n \geq 3$, it must have at least $n - 2$ poles in the left half of the s-plane for closed-loop stability to be attainable. Achieving a specified transient performance can entail time-consuming tuning by trial and error and is only possible in some cases. If $y_r = $ const. and $d = $ const., then $e_{ss} = 0$.

The following two subsections give two refinements of the traditional controllers sometimes needed in the presence of certain hardware imperfections in practice.

1.3.4 Measurement Noise Filtering for Derivative Term

The random noise generated in measurement hardware that contaminates the measurement, $y(t)$, is sometimes significant. It can propagate through a control loop and cause excessive control actuator activity giving rise to wear, overheating and/or audible noise, depending on the application. It is well known that electronic amplifiers, which are often used in the measurement hardware of control systems, generate unwanted noise. If the input to an amplifier is grounded, then if the output voltage is examined on a smaller and smaller scale, then eventually, it appears as a random variable (with zero mean value if any DC offset has been compensated for). This random voltage is due to noise voltages generated by individual components of the amplifier, such as resistors and semiconductor devices. Measurement noise can be troublesome in high-precision applications. For example, space telescopes usually use charge coupled device (CCD) star sensors for absolute attitude measurements, and the level of random noise contaminating the measurements for the dimmer stars can be of the order of arcseconds (1 arcsec $\equiv 1/3,600$ deg.), and the attitude control accuracy required is of the order of 0.1 arcsec. Under these circumstances, it is essential to carry out filtering to obtain a smoother and more accurate measurement but, as will be seen, the classical controller structure imposes a serious restriction. The state-space controller structure incorporating an observer (Chaps. 5 and 8), however, overcomes this. For any high-precision control applications entailing position control with negligible mechanical friction forces, it is essential to employ a derivative term in the controller, as has already been demonstrated using the quarter-vehicle suspension analogy in Sect. 1.3.1. Differentiating a noisy signal will unfortunately yield a derivative estimate that is contaminated by even more noise. Let $z(t)$ be the quantity to be measured, such as the attitude angle of a space telescope about one of the three control axes, $y(t)$ be the measurement and $n_m(t)$ be the measurement noise. Then

$$y(t) = K_s z(t) + n_m(t), \qquad (1.43)$$

where K_s is the sensor scaling constant, assuming a linear sensor transfer characteristic. The derivative terms of the traditional controllers introduced above would employ software differentiation (Chap. 6) which, with a sufficiently short sampling time, would produce a very close approximation to pure continuous differentiation, yielding

$$\dot{y}(t) = K_s \dot{z}(t) + \dot{n}_m(t), \qquad (1.44)$$

which may be written as

$$v(t) = v_{\text{ideal}}(t) + \dot{n}_m(t), \qquad (1.45)$$

where $v(t) = \dot{y}(t)$ is the derivative estimate formed in the control algorithm (Chap. 5) before multiplying by K_D. By its very nature, $n_m(t)$ has relatively high peak values of its instantaneous slope, $\dot{n}_m(t)$, which is the noise contaminating the measurement derivative, $\dot{y}(t)$. It is enlightening to view this situation in the frequency domain. Taking Laplace transforms of (1.45) yields

$$V(s) = V_{\text{ideal}}(s) + s N_m(s). \qquad (1.46)$$

The measurement error is therefore

$$\varepsilon(s) = V(s) - V_{\text{ideal}}(s) = s N_m(s) \qquad (1.47)$$

and the error in the frequency domain is given by setting $s = j\omega$ to yield

$$\varepsilon(j\omega) = j\omega N_m(j\omega) \Rightarrow |\varepsilon(j\omega)| = \omega |N_m(j\omega)|, \qquad (1.48)$$

where ω is the angular frequency of the sinusoidal component of the signal propagating through the system. This means that higher-frequency components of the measurement noise spectrum are amplified far more than lower-frequency components due to the differentiation process, the amplification factor being the frequency itself. The technique sometimes employed in traditional controllers and certainly introduced during the era of analogue electronic implementation to help to overcome this problem is to introduce a first-order low-pass filter together with the differentiator by replacing the transfer function, s, of the pure differentiator by the transfer function, $s/(1 + sT_f)$, where T_f is the filtering time constant with a cut-off (-3 dB) frequency of $1/T_f$[rad/s]. This attenuates the high-frequency components of the noise contaminating the measurement derivative. This, however, is at the expense of introducing a dynamic element into the control loop that will affect the transient performance by an amount that increases with the time constant, T_f. If T_f is relatively small in the sense that it is much smaller than the timescale on which the closed-loop system is intended to operate, then its effect on the transient response will be insignificant and it will considerably reduce the undesirable control actuator activity. For the high-precision applications, however, the propagation of the noise signal through to the controlled variable, $z(t)$, will still be too much and more filtering will be needed by increasing T_f. The effects of the measurement noise and the low-pass filter will now be demonstrated by carrying out simulations of a spacecraft attitude control system using a DP controller with and without the low-pass filter and adding measurement noise, as shown in Fig. 1.5.

The spacecraft parameters are set as previously, i.e., a moment of inertia of $J = 200$ [Kg m^2] and a reaction wheel torque constant of $K_w = 0.05$ [Nm/V], yielding $b_0 = K_w/J = 2.5 \times 10^{-4}$ [rad/s/s/V]. Initially, K_P and K_D will be set to yield a realistic settling time of a $T_s = 45$ [s] with critical damping and no overshoot in the step response based on Fig. 1.12a and therefore ignoring the measurement noise filter. The measurement noise will be taken as 5×10^{-6} [rad] $\cong$ $\widehat{1"}$ [arcsecond] $= 1/3,600$ [deg], rms with a white spectrum and a Gaussian distribution. The results are shown in Fig. 1.13. In Fig. 1.13a–c, a step attitude reference of $z_r(t) = \widehat{5"} \cong 2.4 \times 10^{-5}$ [rad] is applied. The attitude angle being controlled, $z(t)$, is plotted rather than its measurement, $y(t)$. The step reference input is denoted $z_r(t)$ for consistency. To explain the notation, $z_{d,T_f}(t)$ is the *deterministic* response that would occur without measurement noise, while $z_{s,T_f}(t)$ is the *stochastic* (meaning pertaining to a random process) response with the measurement noise, both having measurement rate filtering with a time constant of T_f [s].

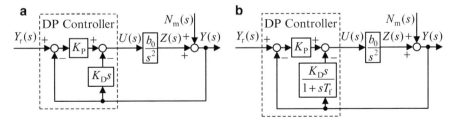

Fig. 1.12 Single-axis, rigid-body spacecraft attitude control system with significant measurement noise. (**a**) without derivative noise filtering and (**b**) with derivative noise filtering

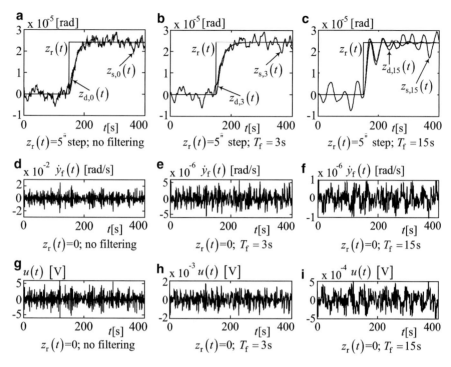

Fig. 1.13 Step responses, measurement derivative noise together with steady-state control activity for spacecraft attitude control

The random variations in the actual attitude angle, $z_{s,T_f}(t)$, are due to the response of the controller and the actuator to the measurement noise. The filtered output derivative, $\dot{y}_f(t)$, is defined by $\dot{Y}_f(s) = [s/(1 + sT_f)] Y(s)$ (Fig. 1.12).

Figure 1.13a, d, g show the effects of the measurement noise without any filtering (Fig. 1.12a), the worst of which is the random control activity in Fig. 1.13g that corresponds to torque variations peaking at about 0.25 [Nm] which is near the maximum value for moderately sized reaction wheels. Figure 1.13e, h show the drastic reduction in the random variations of $\dot{y}_f(t)$ and $u(t)$ brought about by the introduction of the filtering (Fig. 1.12b) with $T_f = 3$ [s], which does not have a significant impact on the transient performance, as can be seen by comparing $z_{d,0}(t)$ of Fig. 1.13a with $z_{d,3}(t)$ of Fig. 1.13b, which are nearly the same. Increasing T_f to 15 s reduces $\dot{y}_f(t)$ and $u(t)$ further but causes a very oscillatory step response as can be seen in $z_{d,15}(t)$ of Fig. 1.13c, indicating that two of the closed-loop poles have become complex conjugates with a small damping ratio.

Unfortunately, as is evident in Fig. 1.13a–c, increasing T_f from zero does not improve the control accuracy. Since the noise spectrum is white, then it will include low-frequency components down to zero frequency (DC) that cannot be filtered, but the problem is exacerbated in Fig. 1.13c by the lightly damped oscillatory mode

1.3 Review of Traditional PID Controllers and Their Variants

being continually excited by the random noise. This situation can be improved, however, by employing a different approach to the gain determination, accepting that the system is of third order. Controller adjustment by trial and error would be time consuming and a more orderly approach is as follows. Essentially, T_f is first chosen. Then K_P and K_D are calculated to yield three coincident real, negative poles at $s_{1,2,3} = r$ which eliminate any possibility of an oscillatory closed-loop mode (Chap. 3). The starting point is the closed-loop transfer function for Fig. 1.12b. First setting $T_f = 1/p$ and then applying Mason's formula with $N_m(s) = 0$ (as stochastic inputs are not needed for this) yields

$$\frac{Y(s)}{Y_r(s)} = \frac{\frac{K_P b_0}{s^2}}{1 - \left[-\frac{b_0}{s^2} \left(\frac{K_D p s}{s+p} + K_P \right) \right]} = \frac{K_P b_0 (s+p)}{s^2(s+p) + K_D p s + b_0 K_P (s+p)}. \tag{1.49}$$

Equating the denominator of (1.49) to the desired one then yields

$$s^3 + ps^2 + b_0(K_D p + K_P) + b_0 K_P p = (s+r)^3 = s^3 + 3rs^2 + 3r^2 s + r^3 \Rightarrow$$
$$r = p/3, \quad K_P = r^3/(b_0 p) \quad \text{and} \quad K_D = (3r^2/b_0 - K_P)/p. \tag{1.50}$$

It is proven in Sect. 1.4.4.3 that if the zero of a transfer function is considerably greater in magnitude than the poles, then its effect on the system response will be negligible. Since $p = 3r$, this is the case here and the dynamic response will be almost the same as the system with the closed-loop transfer function

$$G_{cl}(s) = \frac{r^3}{(s+r)^3}. \tag{1.51}$$

It is well known that as the magnitude of the closed-loop poles reduces, the speed of response of the system reduces and therefore the settling time increases. The author's 5 % settling time formula developed in Chap. 4 yields

$$T_s = 6/r \tag{1.52}$$

and in view of the first of Eq. (1.50),

$$T_s = 6(3/p) = 18T_f. \tag{1.53}$$

The penalty for improving the control system accuracy in the presence of measurement noise by increasing the filtering time constant is therefore a drastic increase in the settling time of the control loop. Figure 1.14 shows the results. The $\pm \widehat{1}^{\prime\prime}$ error limits are included in Fig. 1.14 d, e, f as these are typical of the accuracy

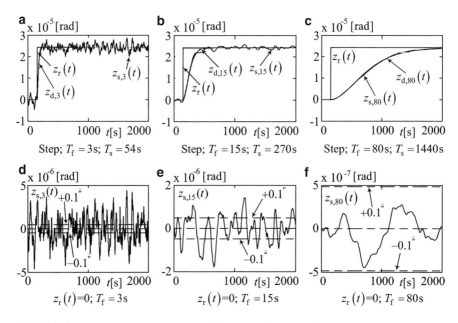

Fig. 1.14 Step responses and stochastic performance of spacecraft attitude control system with triple pole placement

achieved in a space telescope. It is evident that setting $T_f = 80$ [s] would enable this accuracy to be achieved but this would need $T_s = 1,440$ s. Long settling times are common for spacecraft attitude control due to the limited actuator torque and power, but the state-space control structure with the observer introduced in Chap. 3 avoids settling times as long as in Fig. 1.14. The principles learnt through this example are relevant to many other applications but, of course, with different parameter values, such as T_s, which could be as little as a few tens of milliseconds for an earthbound electric drive.

By now, the reader will wonder why the measurement noise filtering was not applied to $y(t)$ as well as $\dot{y}(t)$. The primary reason is that early practitioners of PID and PD controllers experiencing problems with actuator activity and control saturation (the analogue output of the controller being driven to the voltage limits) found that filtering $\dot{y}(t)$ alone to be sufficient. The justification for this is that $\dot{y}(t)$ contains a higher proportion of noise at high frequencies than $y(t)$. Before leaving this topic, however, filtering of $y(t)$ too will be considered, as this leads to another form of traditional controller, i.e. the *lead compensator*. This controller is defined by the following transfer function relationships:

$$E(s) = Y_r(s) - Y(s)$$
$$U(s) = G_c(s)E(s) = K\left(\frac{1+\alpha sT}{1+sT}\right)E(s) \quad (1.54)$$

where K is the controller gain, T is the time constant of the built-in low-pass filter and α is a constant set to a value greater than unity to obtain a controller action

1.3 Review of Traditional PID Controllers and Their Variants

similar to derivative action. The term, compensator, comes from the frequency domain as it is used to shape the amplitude and phase responses of the open-loop system with transfer function, $Y(s)/E(s) = G_c(s)G_p(s)$, where $G_p(s) = Y(s)/U(s)$ is the plant transfer function, in a way that is guaranteed to improve the closed-loop performance. So a plant with a poor amplitude and phase response, $|G_p(j\omega)| \angle \phi_p(\omega)$, in the sense of having inadequate gain and phase margins [briefly discussed in Chap. 3] is cascaded with the compensator having an amplitude and phase response, $|G_c(j\omega)| \angle \phi_c(\omega)$, to yield a more satisfactory combined open-loop amplitude and phase response, $|G_p(j\omega)||G_c(j\omega)| \angle \phi_p(\omega) + \phi_c(\omega)$. This approach to control system design that entails elements of trial and error is traditional and widely documented. It is not dealt with in detail here as the other methods presented in this book are not only capable of producing equally satisfactory performance but can guarantee that the closed-loop system is closely approximated by a known differential equation relating $y_r(t)$ to $y(t)$, which is very useful in certain applications, such as motion control (Chap. 10).

If the traditional PD controller with the transfer function relationships,

$$E(s) = Y_r(s) - Y(s), \quad \dot{E}(s) = \dot{Y}_r(s) - \dot{Y}(s) = sY_r(s) - sY(s)$$
$$U'(s) = K_P E(s) + K_D \dot{E}(s) \qquad , \qquad (1.55)$$

is taken and the previous low-pass filter applied to $Y(s)$ as well as $\dot{Y}(s)$, also to $Y_r(s)$ and $\dot{Y}_r(s)$ to compare like with like when forming $E(s)$ and $\dot{e}(s)$, then the transfer function relationships of the resulting modified controller are

$$E(s) = Y_r(s) - Y(s), \quad E_f(s) = \tfrac{1}{1+sT_f} E(s), \quad \dot{E}(s) = \dot{Y}_r(s) - \dot{Y}(s), \quad \dot{E}_f(s) = \tfrac{1}{1+sT_f} \dot{E}(s)$$
$$U(s) = K_P E_f(s) + K_D \dot{E}_f(s) = K_P E_f(s) + K_D s E_f(s) = \tfrac{K_P + K_D s}{1 + s T_f} E(s),$$

i.e.

$$E(s) = Y_r(s) - Y(s)$$

and

$$U(s) = \left(\frac{K_P + K_D s}{1 + s T_f} \right) E(s). \qquad (1.56)$$

Then direct comparison with (1.54) reveals that this combined filter and PD controller are the same as the phase-lead compensator if

$$T_f = T, \quad K_P = K \text{ and } \alpha = \frac{K_D}{K_P T_f}. \qquad (1.57)$$

For completeness, the spacecraft attitude control example will now be continued by applying the triple pole assignment method using the modified controller, the control system block diagram being shown in Fig. 1.15.

Fig. 1.15 Modified single-axis, rigid-body spacecraft attitude control system with significant measurement noise

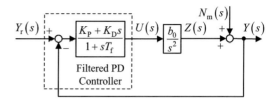

Setting $N_m(s) = 0$ as this external input is not needed for the pole assignment and setting $T_f = 1/p$ as previously, the closed-loop transfer function is

$$\frac{Y(s)}{Y_r(s)} = \frac{\left(\frac{(K_P + K_D s)p}{s+p}\right) \cdot \frac{b_0}{s^2}}{1 + \left(\frac{(K_P + K_D s)p}{s+p}\right) \cdot \frac{b_0}{s^2}} = \frac{(K_P p + K_D ps)b_0}{s^3 + ps^2 + K_D p b_0 s + K_P p b_0}. \quad (1.58)$$

Equating the denominator of (1.58) to the closed-loop characteristic polynomial for placement of the closed-loop poles at $s_{1,2,3} = r$ as previously yields

$$s^3 + ps^2 + K_D p b_0 s + K_P p b_0 = (s+r)^3 = s^3 + 3rs^2 + 3r^2 s + r^3. \quad (1.59)$$

First the filtering time constant, T_f, is chosen according to the cut-off frequency, $1/T_f$ [rad/s], required. According to (1.59), this fixes the triple closed-loop pole location. Then the equations for the controller gains follow. Thus,

$$r = \frac{p}{3}, \quad K_D = \frac{3r^2}{pb_0} \quad \text{and} \quad K_P = \frac{r^3}{pb_0}. \quad (1.60)$$

In this case there is a zero introduced by the controller with magnitude, $K_P/K_D = r/3$ (in contrast to $3r$ in the system of Fig. 1.12b), so the zero has a smaller magnitude than the triple pole, and according to the theory to be presented in Sect. 1.4.4.3, this will cause a significant overshoot in the step response. Figure 1.16 shows the results corresponding to those of Fig. 1.14 obtained previously with a DP controller and output derivative filtering.

Figure 1.16a–c confirm the overshoot and since the ratio between the zero and pole magnitudes remains constant at 1 : 3, the percentage overshoot is the same in each case (Chap. 4). Comparing the stochastic performances displayed in Figs. 1.16 and 1.14d, e, f reveals that the introduction of the filtering of the measurement as well as its derivative does not produce a visible reduction in the random errors, confirming that output derivative filtering alone is sufficient.

1.3.5 Anti-windup Loop for Integral Term

Every control actuator has finite upper and lower limits on its output due to the hardware limitations. For example, the power electronics feeding a motor in an

1.3 Review of Traditional PID Controllers and Their Variants

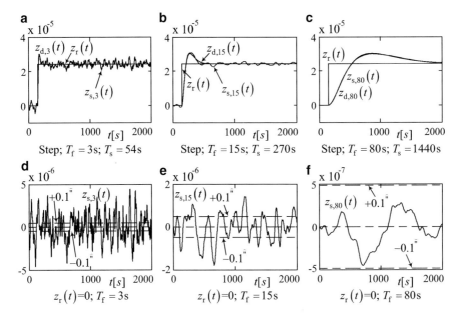

Fig. 1.16 Step responses and stochastic performance of modified spacecraft attitude control system with triple pole placement

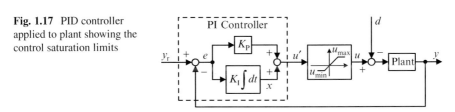

Fig. 1.17 PID controller applied to plant showing the control saturation limits

electric drive is operated with finite power supply voltages and if the controller demands a higher voltage, then the actual voltage supplied to the motor will fall short of that demanded and saturate at one of the limits. In other applications such as various forms of heating (Chap. 2), the control variable is only positive to yield a positive temperature of the heating element, no provision being made for cooling, so the lower limit of the control variable is zero.

Due to the control variable constraints, *any* control loop containing an integral term to avoid a steady-state error with a constant reference input can experience problems during the transient operation following the initial loop closure, step changes in the reference input or suddenly applied external disturbances. Figure 1.17 shows the simplest and most familiar of these, i.e. PI control loop with the nonlinear saturation element inserted between the control input, u', demanded by the controller and the physically realisable control input, u.

The central segment of the transfer characteristic between u' and u has unity slope and passes through the origin as $u = u'$ in absence of the control saturation.

By convention, the abscissa is the input and the ordinate is the output. In many applications where the actuator is able to deliver positive or negative outputs, $u_{\min} = -u_{\max}$ and therefore the origin lies at the midpoint of the segment with unity slope, but in others, such as the heating applications already mentioned, $u_{\min} = 0$ and the origin lies at the lower saturation point.

With reference to Fig. 1.17, consider first the control system operation with y_r and d piecewise constant assuming that the controller gains, K_P and K_I, are adjusted to yield a specified transient performance. If any change in y_r or d takes place, then the ensuing transient should include changes in $u(t)$ that act through the plant to bring the integrator output, $x(t)$, to the constant value required to maintain zero steady-state error, i.e. $e(t) \to e_{ss} = 0$ (Sect. 1.3.1). For a sufficiently large change in y_r or d, $u(t)$ will just reach one of the saturation limits and the system operation remains linear. For any change in y_r or d larger than this, $u'(t)$ will exceed the saturation limit and $u(t)$ will saturate at $u_{\max}$ or $u_{\min}$, depending on the sign of y_r or d and therefore the system enters nonlinear operation. It should be noted that unlike a linear system, the dynamic character and even the stability of a nonlinear system depends upon the initial values of the system variables and the values of its inputs (Chap. 2). Prediction of such behaviour analytically is often very difficult and therefore a practicable approach to this task is simulation. During this saturation, the system cannot drive $e(t)$ to zero as fast as it would without the saturation, thereby increasing the settling time beyond that specified. Unfortunately, the system as it stands in Fig. 1.17 can suffer from worse problems for even larger changes in y_r or d. It should be stressed that the system behaviour under control saturation depends upon the type of the plant, i.e. the number of pure integrators it has in the forward path. Suppose first that the plant is of type '0' and that the steady-state value, u_{ss}, of $u(t)$, needed to maintain zero steady-state error, lies outside the control saturation limits. Then it will be impossible for the steady-state error to be zero. u will then remain at the saturation limit, while $y(t)$ settles to a steady-state value of $y_{ss} = K_{DC}u_{\max}$ or $y_{ss} = K_{DC}u_{\min}$, depending on the sign of y_r or d, where K_{DC} is the DC gain of the plant. This will differ from y_r, resulting in a constant non-zero steady-state error, $e_{ss} = y_r - y_{ss}$, at the input of the integrator. The integrator usually resides in the software of the digital processor implementing the controller (Chap. 5) and its output will therefore ramp indefinitely and $u'(t) = x(t) + K_P e_{ss}$ will do likewise. The plant will remain in steady state under unidirectional control saturation, while $x(t)$ and $u(t)$ continue to ramp. If eventually the cause of the problem is removed, i.e. y_r or d changes to a value for which $u'_{ss} = u_{ss} \in (u_{\min}, u_{\max})$ leading to $e_{ss} = 0$, then the system will come out of saturation and settle correctly but will take a very long time to do so because the integrator output, $x(t)$, will have reached a relatively high value and the system will undergo a transient of long duration to drive it towards the correct steady-state value, x_{ss}. The integrator ramping effect has been described as *integral windup*, the analogy being the storage of potential energy in a clock spring as it is wound up. In an analogue controller, the charge on the feedback capacitor of the integrator's operational amplifier ramps up (but saturates at one of the power supply voltages) and also causes stored *physical* energy that eventually must be

1.3 Review of Traditional PID Controllers and Their Variants

dissipated. In the modern digital implementation, however, the system *mimics* this energy storage but only if, as in the case of the traditional controllers, the control algorithm is based on the linearised equations of the analogue electronic controller (Chap. 5).

If the plant is of type '1' or greater, then if y_r or d changes to a value for which $u_{ss} \notin (u_{min}, u_{max})$, the control will saturate initially but the plant output cannot reach a steady state as occurs for type '0' plants. For a constant saturated control input, the output must be changing due to the presence of at least one pure integrator in the forward path of the plant. Depending upon the particular plant and the controller gain settings, the control may remain in saturation, while the plant output grows indefinitely in one direction, indicating monotonic instability. It may subsequently come out of saturation but be driven into the opposite saturation limit, this process being repeated so that the control variable oscillates between the two saturation limits while the other variables oscillate, usually with increasing amplitude, indicating oscillatory instability. At best, the system might enter a stable limit cycle (Chap. 7) in which all the variables oscillate, being periodic functions. So the integral windup as defined above cannot take place unless the plant is of type '1', but the integrator output, $x(t)$, could undergo much larger excursions than it would without the control saturation and, as will be seen by example, the measures that can be taken to prevent integral windup with type '0' plants can improve the control system behaviour with plants of type '1' or greater under control saturation.

The traditional way of preventing integral windup is to close an additional control loop around the output of the integrator to keep the *demanded* control, $u'(t)$, at the control saturation limit it would otherwise exceed. This indirectly keeps the integrator output, $x(t)$, to much smaller values than it would otherwise reach. The general block diagram of this scheme is shown in Fig. 1.18. Since it contains a nonlinear element, formally it should not contain transfer functions and Laplace transformed signals, as these are only applicable to linear systems, but this is a commonly accepted notation, as can be seen in the MATLAB®–SIMULINK® block diagrams. It should be understood, however, that for any block diagram containing nonlinear elements, linear analysis such as the application of Mason's formula (Appendix A4) can only be applied to linear subsystems within the diagram.

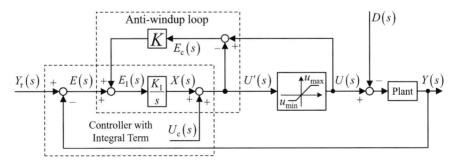

Fig. 1.18 Feedback loop method for integral anti-windup

An example is the anti-windup loop of Fig. 1.18. Linear analysis can only be applied to a system if it is assumed to operate in a linear regime. For example, the system of Fig. 1.18 would have to be operating within the control saturation limits.

The control component, $U_c(s)$, is the proportional term output for the PI or IP controllers and the sum of the proportional and derivative term outputs for the PID or IPD controllers. In the time domain, for unsaturated operation, $u(t) = u'(t) \Rightarrow e_c(t) = 0$ and the anti-windup loop is inactive. Whenever $u'(t) > u_{max}$, however, $u(t)$ saturates at u_{max}. Consequently the anti-windup loop actuation error, $e_c(t) = u(t) - u'(t) = u_{max} - u'(t)$, becomes negative and through the gain, K, controls the integrator output, $x(t)$, to drive $e_c(t)$ back towards zero and hence drive $u'(t)$ back towards u_{max}. Similarly, whenever $u'(t) < u_{min}$, the anti-windup loop controls the integrator output, $x(t)$, to drive $u'(t)$ back towards u_{min}. Thus, the effect of the anti-windup loop is to keep $u'(t) = u_{min} - \varepsilon$ or $u_{max} + \varepsilon$, where $|\varepsilon| << u_{max} - u_{min}$ whenever $u(t) = u_{min}$ or u_{max}, provided the gain, K, is sufficiently large. To determine a suitable value of K, some simple analysis may be carried out on the first-order anti-windup loop. With reference to Fig. 1.18, when the loop is active, $u(t) = u_{max}$ or u_{min} is regarded as the reference input, while $u'(t)$ is the controlled output and $u_c(t)$ is regarded as an external disturbance. The transfer function relationship of the loop is then

$$U'(s) = \frac{\frac{KK_I}{s}U(s) + 1.U_c(s)}{1 - \left(-\frac{KK_I}{s}\right)} = \frac{KK_I U(s) + sU_c(s)}{s + KK_I}. \tag{1.61}$$

If K is increased indefinitely, then

$$\lim_{K \to \infty} U'(s) = U(s), \tag{1.62}$$

indicating complete rejection of $u_c(t)$ and ideal following of $u(t)$. In practice, however, K is limited by the discrete implementation. According to the theory of Chap. 5, if the sampling period is h seconds, then

$$K \in (0, \ 2/(K_I h)). \tag{1.63}$$

Next, the performances of two control systems embodying integral terms in their controllers will be compared by simulations with and without anti-integral windup. The first example is the PID control of an electric kiln with two dominant first-order modes having time constants, T_1 for the refractory bricks and T_2 for the work-piece (Chap. 2), and DC gain, K_{DC}, as shown in Fig. 1.19.

This is a type '0' plant for which the integral anti-windup is the most beneficial.

First, rather than the attempt to tune the controller by trial and error, the pole placement method generalised in Chap. 3 will be applied to calculate the controller gains that yield an acceptable step response assuming that the system is operating within the saturation limits and therefore linear. In this case, $u(s) = u'(s)$ and

1.3 Review of Traditional PID Controllers and Their Variants

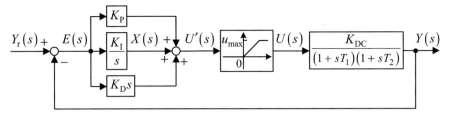

Fig. 1.19 PID control of an electric kiln showing the control saturation limits

therefore the nonlinear element can be removed. Before deriving the closed-loop transfer function, the work will be simplified by setting

$$\frac{K_{DC}}{(1+sT_1)(1+sT_2)} = \frac{K_{DC}}{T_1T_2s^2 + (T_1+T_2)s + 1} = \frac{\frac{K_{DC}}{T_1T_2}}{s^2 + \left(\frac{1}{T_1}+\frac{1}{T_2}\right)s + \frac{1}{T_1T_2}}$$

$$= \frac{b_0}{s^2 + a_1s + a_0}. \quad (1.64)$$

Then the closed-loop system block diagram becomes that of Fig. 1.7 and the closed-loop transfer function has already been derived as (1.30) which is

$$\frac{y(s)}{y_r(s)} = \frac{\left(K_D s^2 + K_P s + K_1\right) b_0}{s^3 + (a_1 + K_D b_0) s^2 + (a_0 + K_P b_0) s + K_1 b_0}. \quad (1.65)$$

As in the previous section, the author's 5 % settling time formula for linear systems (Chap. 3) will be applied to achieve a non-oscillatory step response with a settling time of T_s seconds to determine the desired characteristic polynomial and the gains determined by choosing them to realise this polynomial. Thus,

$$s^3 + (a_1 + K_D b_0) s^2 + (a_0 + K_P b_0) s + K_1 b_0 = (s+p)^3 = s^3 + 3ps^2$$
$$+ 3p^2 s + p^3 \Rightarrow$$
$$K_D = (3p - a_1)/b_0, \quad K_P = \left(3p^2 - a_0\right)/b_0 \text{ and } K_1 = p^3/b_0, \quad (1.66)$$

where $p = 6/T_s$. The workpiece temperature transducer output can be converted numerically to °C and so this will be taken as the units of y. To determine typical plant parameters, full power will be produced by $u = u_{max}$ and applying this permanently will yield a steady-state temperature of $y_{ss} = 1,000$ °C. Taking u as an analogue voltage with $u_{max} = 10$ [V] yields the DC gain as $K_{dc} = y_{ss}/u_{max} = 100$ °C/V. The time constants will be taken as $T_1 = 200$ s and $T_2 = 50$ s. It should be ensured that $u'(t)$ does not go negative, i.e., below the lower saturation limit.

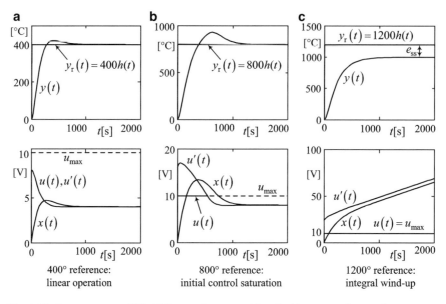

Fig. 1.20 Step responses of kiln PID control system with and without control saturation

To achieve this, a relatively long settling time has to be chosen to enable linear operation and this is $T_s = 600$ s. As will be seen in Chaps. 6 and 7, much shorter settling times can be achieved for such plants with nonlinear time-optimal controllers.

Figure 1.20 shows step response simulations for the control system of Fig. 1.19 without integral anti-windup. In Fig. 1.20a, the small overshoot in $y(t)$ is due to the zeros introduced by the PID controller (Sect. 1.4.4).

This example illustrates all that has been stated earlier in this subsection about the effects of control saturation with type '0' plants containing integral terms. In Fig. 1.20b, c, $u'(t)$ and $x(t)$ exceed the maximum voltage limit of 10 V but they are not physical voltages (in contrast to $u(t)$) but are variables within the digital processor implementing the controller and are not subject to this constraint.

Figure 1.23 compares the performances of the control system of Fig. 1.19 with and without integral anti-windup.

A $1,200\,[^\circ C]$ step reference input is first applied and followed by a step change in the reference input to 800 °C at $t = 3,000$ [s], which is within the system capability. Integral windup occurs without the preventive measure since the system cannot meet the demand of $y_r(t) = 1,200$ [°C] and therefore saturates at $y_{ss} = 1,000$ [°C]. Figure 1.21a shows the integral windup taking place, causing $x(t)$ and hence $u'(t)$ to ramp up. After the reduction of $y_r(t)$ to 800 °C, there is a long delay before the control comes out of saturation due to the time taken for $x(t)$ and hence $u'(t)$ to ramp down again. The system is seen to recover after $t = 7,000$ s. In Fig. 1.21b the response to the reduction in $y_r(t)$ at $t = 3,000$ s is almost immediate thanks to the integral anti-windup preventing $x(t)$ increasing beyond the value for

1.3 Review of Traditional PID Controllers and Their Variants

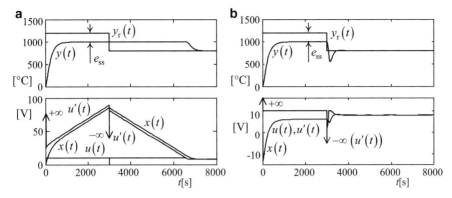

Fig. 1.21 Comparison of performances of kiln PID control system. (**a**) without integral anti-windup and (**b**) with integral anti-windup

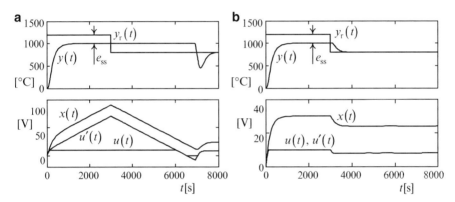

Fig. 1.22 Comparison of performances of kiln IPD control system. (**a**) without integral anti-windup and (**b**) with integral anti-windup

which $u(t)$ saturates. The infinite spikes in $u'(t)$, indicated by the arrows, are due to the action of the derivative term on the step changes in $y(t)$. The jumps in $u'(t)$ are caused by the steps in $y_r(t)$ being fed forward via the proportional term. These discontinuities are the cause of the overshoot in Fig. 1.21b in the response to the step reduction in $y_r(t)$ at $t = 3,000$ s.

An interesting further comparison is the control of the same kiln with the same controller gain settlings but using an IPD controller. The simulation results of Fig. 1.22 correspond with those of Fig. 1.21 and may be compared directly.

Now the effect of the integral anti-windup loop on the performance of a control system including a plant of type '2' will be considered. The example taken is a vacuum air-bearing linear motor-actuated position control where the integral term has been included to counteract constant external disturbance forces to yield a zero steady-state error. In this case, an IPD controller is used.

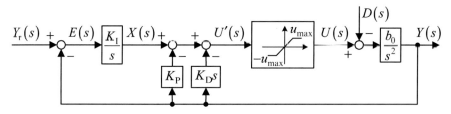

Fig. 1.23 IPD control of mass on linear motor-actuated vacuum air bearing

This achieves a non-overshooting step response with coincident pole placement, as in this application it is often essential to approach a new position monotonically from one side only. A relevant application is numerical machine tool control in which overshooting would cause the material to be erroneously removed from a workpiece that could not be replaced. The system block diagram is shown in Fig. 1.23.

Using Mason's formula (Appendix A4), without control saturation, the closed-loop transfer function is

$$\frac{Y(s)}{Y_r(s)} = \frac{\frac{K_1 b_0}{s^3}}{1 - \left[-\frac{b_0}{s^2}\left(K_D s + K_P + \frac{K_1}{s}\right)\right]} = \frac{K_1 b_0}{s^3 + b_0\left(K_D s^2 + K_P s + K_1\right)}. \quad (1.67)$$

Again utilising the author's 5 % settling time formula (Chap. 4) for coincident closed-loop poles yields the desired characteristic polynomial, which is equated to the denominator of (1.67). Thus, with $a = 6/T_s$, where T_s is the settling time,

$$(s+a)^3 = s^3 + 3as^2 + 3a^2 s + a^3 = s^3 + b_0\left(K_D s^2 + K_P s + K_1\right) \Rightarrow \quad (1.68)$$
$$K_D = 3a/b_0, \quad K_P = 3a^2/b_0 \text{ and } K_1 = a^3/b_0.$$

The plant parameters are taken from an experimental rig at the Mechatronics Research Institute of the Bern Fachhochschule, Switzerland, and are mass, $M = 3.3$ [kg]; linear motor force constant $K_m = 11.1$ [N/A] and transconductance amplifier constant $K_a = 0.8$ [A/V]. This yields $b_0 = K_m K_a/M = 2.69$ [N/V]. In the simulations of Fig. 1.24, T_s is set to 0.1 [s]. Note the contrast of the practicable timescale of this application with the previous spacecraft attitude control and heating applications. This is possible due to the relatively high power ratings of the motors of electric drives.

Figure 1.24a shows a response to a step reference position of 1 [μm] ≡ 10^{-6} [m], which is on a scale appropriate to integrated circuit manufacturing operations for which this type of air bearing is suited. This response is without control saturation and without any external disturbance. In Fig. 1.24b–f, the reference input is held at zero to attempt to hold the mass in a fixed position, while a step disturbance force is applied at $t = 0$ s and held constant until $t = 0.4$ s when it is removed. The control force acting on the mass is proportional to $u(t)$ so in Fig. 1.24a, the initial positive-going portion of $u(t)$ accelerates the mass.

1.3 Review of Traditional PID Controllers and Their Variants

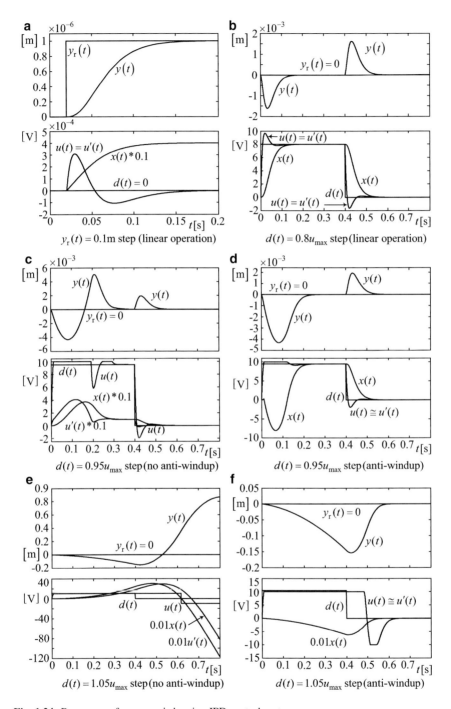

Fig. 1.24 Responses of vacuum air-bearing IPD control system

Also in Fig. 1.24a, the negative-going portion of $u(t)$ decelerates it and brings it to rest in the steady state at the required position. It is important to note that the output, $x(t)$, of the integral term would go to zero in a PID controller used in the same situation (due to the plant being of type '2'), but it rises to a constant value in the IPD controller because, with reference to Fig. 1.23, it is the only controller term acting on the error and the proportional term only acts on the controlled output, $y(t)$.

In Fig. 1.24b, the application and removal of the disturbance force step are followed by equal and opposite transient position errors that decay to zero in the steady state, as expected with linear operation. It should be noted that although the control system is not driven into saturation, a disturbance force of 80 % of the maximum available control force magnitude is relatively high.

In Fig. 1.24c, d, the very high level of disturbance force (95 % of the maximum control force magnitude) causes initial control saturation, and therefore without the integral anti-windup (Fig. 1.24c), a relatively large excursion of the integrator output, $x(t)$, and therefore the demanded control variable, $u'(t)$, occurs, keeping $u(t)$ saturated for too long, causing the first relatively large positive position error peak. This does not occur in Fig. 1.24d due the integral anti-windup loop keeping $u'(t) \cong u_{max}$ as soon as the system first enters control saturation, indirectly keeping $x(t)$ to smaller proportions. Hence, $u(t)$ comes out of saturation sooner than in Fig. 1.24d and the positive position error peak does not occur while the disturbance force is still being applied. The overall result with integral anti-windup is similar to that of Fig. 1.24b but the negative error transient is larger than the positive one due to the initial control saturation.

In Fig. 1.24e, f, the disturbance force is greater in magnitude than the available control force and therefore the system cannot possibly control the mass position during application of this disturbance. In Fig. 1.24e, the prolonged initial control saturation due to the large positive excursion of $x(t)$ causes the first positive position error to be so large that the following negative-going control that attempts to correct it also saturates resulting in a large negative excursion of $x(t)$ and the process repeats. Even after the disturbance force is removed, the system has entered an oscillatory instability from which it cannot recover. Fortunately, the introduction of the integral anti-windup in Fig. 1.24f keeps $x(t)$ to relatively small proportions and the system is able to immediately recover from the uncontrollable situation once the disturbance force is removed. Such a disturbance force, of course, would not be expected in normal operation, but under mechanical fault conditions, it might occur. Also, the position error excursions are larger than would be possible in most practical applications, as the controlled mass would be constrained by end stops. In the system simulated, this imposes position error limits of ± 0.01 [m]. Figure 1.24e, f indicate the system behaviour without them.

It is important to note that modern digital implementation media renders practicable more sophisticated nonlinear control techniques for plants operating with control saturation that can enable satisfactory operation where linear controllers would yield excessive overshooting or instability.

In overall conclusion, anti-windup is highly recommended in any application using traditional controllers with integral terms. In such applications where the

1.4 Dominance in the Pole–Zero Distribution

1.4.1 Background

When characterising a linear control system by means of its set of closed-loop poles and zeros, it is sometimes possible to identify a subset that has more influence on the transient response than the others. This enables a *simplified transfer function* to be formed of a system having nearly the same transient performance. The poles and zeros of this transfer function will be referred to as *dominant* poles and zeros. As will be seen, with appropriate control techniques, the control system designer can exercise a great deal of influence over the closed-loop pole locations and can therefore take advantage of pole dominance. Plant zeros, on the other hand, are invariant with respect to the loop closure, but sometimes it is practicable to design the controller to place some of the closed-loop poles to cancel these zeros, or near them to reduce their unwanted effects. In general, the degree of *dominance* of the plant zeros will influence the choice of the closed-loop poles.

This section is concerned principally with *closed-loop* linear control systems. It is, however, possible to apply the principles of pole and zero dominance to linear plant models in order to arrive at simplified models for control system design, but *this approach must be taken with extreme caution* because the loop closure via any linear controller can shift the poles that were apparently insignificant in the plant model to significant closed-loop positions, thereby causing departure from the intended dynamic performance or even *instability*.

The following sections quantify pole and zero dominance by means of two parameters (due to author) that can aid control system design. One is the *pole-to-pole dominance ratio* that enables the influence of one pole or group of poles on the closed loop dynamic performance to be compared with that of another pole or group of poles. The other is the *pole-to-zero dominance ratio* that enables assessment of the effect of a pole or group of poles closer to the origin of the s-plane than a zero or group of zeros in *reducing* the influence of that zero or group of zeros on the closed-loop dynamic performance. An understanding of the system *modes* defined in the following subsection is needed to develop criteria for selection of a set of poles and zeros for a simplified transfer function using the dominance ratios.

1.4.2 Modes of Linear Systems

Consider a closed-loop control system whose poles, s_i, $i = 1, 2, \ldots, n$, have negative real parts, which is a prerequisite for stability. Then, without reference to the control system structure, its transfer function may be expressed as

$$\frac{Y(s)}{Y_\mathrm{r}(s)} = K_\mathrm{DCL} \frac{\prod_{i=1}^{n} p_i \prod_{k=1}^{m}(s+z_k)}{\prod_{k=1}^{m} z_k \prod_{i=1}^{n}(s+p_i)} = \underbrace{\sum_{i=1}^{n_\mathrm{d}} \frac{A_i}{s+p_i}}_{\text{Terms from distinct poles}}$$

$$+ \underbrace{\sum_{j=1}^{n_m}\sum_{k=1}^{m_j} \frac{A_{jk}}{(s+p_j)^k}}_{\text{Terms from repeated real poles}}, \quad \begin{array}{l} n = n_\mathrm{d} + \sum_{j=1}^{n_m} m_j, \\ m < n \end{array} \tag{1.69}$$

where K_DCL is the closed-loop DC gain. It should be noted that (1.69) is simplified by including any complex conjugate pairs of poles in the set of distinct poles, which do not have to be real. This is elaborated in Sect. 1.4.2.2. Repeated closed-loop real poles are included as they are often a design goal, a simple example being a critically damped second-order system.

The presence of any finite zeros in the transfer function is not immediately evident from the partial fraction expansion but is reflected in the relative weightings of the coefficients, A_i, $i = 1, 2, \ldots, n_\mathrm{d}$ and A_{jk}, $j = 1, 2, \ldots, n_m$, $k = 1, 2, \ldots, m_j$.

The reference input, $y_\mathrm{r}(t)$, to be considered to assess the dynamic character of a system is the Dirac delta impulse function, $\delta(t)$, that is an infinite impulse of infinitesimal duration commencing at $t = 0$ and of unit strength, i.e. $\int_0^\infty \delta(t)\,\mathrm{d}t = 1$. Henceforth, this will be referred to as the unit impulse function. This, of course, is impracticable as a test signal for a physical system. It is useful, however, for mathematical analysis because it excites the system only at $t = 0^+$ and the ensuing transient will not be further influenced by the input and therefore reflect the dynamic character of the system. Three different types of contribution to the impulse response, $y_\delta(t)$, will be identified according to the type of behaviour they represent, associated with (a) real distinct poles, (b) complex conjugate pole pairs and (c) repeated poles. These contributions are referred to as *modes*.

1.4.2.1 Exponential Modes

After taking the inverse Laplace transform of the distinct pole part of the partial fraction expansion (1.69), it is evident that each *real* pole at $s_k = -p_k$, $p_k > 0$, will contribute a component of $y_\delta(t)$ equal to

$$A_k e^{-p_k t} = A_k e^{-t/p_k} \tag{1.70}$$

1.4 Dominance in the Pole–Zero Distribution

where T_{pk} will be called the time constant of the kth pole. The system behaviour represented by a term such as $e^{-t/T_{pk}}$ is called an *exponential mode* of the system.

1.4.2.2 Oscillatory Modes

Each oscillatory mode is associated with a pair of complex conjugate poles and is therefore represented by two terms in the partial fraction expansion (1.69). Let these two terms be adjacent in the sequence, meaning that $\frac{A_l}{s+p_l}$ and $\frac{A_{l+1}}{s+p_{l+1}}$ are terms corresponding to poles at $s_l = -p_l$ and $s_{l+1} = -p_{l+1} = -\overline{p}_l$, where $\overline{p}_l$ represents the complex conjugate of p_l. It should also be noted that $A_{l+1} = \overline{A}_l$. These two terms are then combined to determine the contributing component of the impulse response, which involves only real coefficients. Thus,

$$\frac{A_l}{s+p_l} + \frac{A_{l+1}}{s+p_{l+1}} = \frac{B_l + jC_l}{s+r_l+jq_l} + \frac{B_l - jC_l}{s+r_l-jq_l} = \frac{2(B_l s + B_l r_l + C_l q_l)}{(s+r_l)^2 + q_l^2}. \tag{1.71}$$

Taking the inverse Laplace transform (Table 1 in Tables) then yields

$$2e^{-r_l t}\left[C_l \sin(q_l t) + B_l \cos(q_l t)\right] = 2e^{-t/T_{pl}}\left[C_l \sin(q_l t) + B_l \cos(q_l t)\right]$$

$$= 2\sqrt{C_l^2 + B_l^2}\, e^{-t/T_{pl}} \sin(q_l t + \phi_l), \quad \phi_l = \tan^{-1}(B_l/C_l). \tag{1.72}$$

where T_{pl} is the time constant of the complex conjugate pole pair. The system behaviour represented by a term such as $e^{-t/T_{pl}} \sin(q_l t + \phi_l)$ is an *oscillatory mode*.

1.4.2.3 Polynomial Exponential Modes

With reference to the repeated pole part of the partial fraction expansion in (1.69) and again using a table of Laplace transforms (Table 1 in Tables), each *real* pole of multiplicity, m_j, at $s_k = -p_j$, $p_j > 0$, will contribute a component to $y_\delta(t)$ given by

$$z_j(t) = e^{-p_j t}\sum_{k=1}^{m_j} C_{jk} t^{k-1} = e^{-t/T_{pj}}\sum_{k=1}^{m_j} C_{jk} t^{k-1}, \text{ where } C_{jk} = \frac{A_{jk}}{(k-1)!}. \tag{1.73}$$

The term, *mode*, is usually used in connection with the analysis of uncontrolled plants or other dynamical systems, such as a vibrating structure of a building subject to seismic disturbances. Since it would be rare to find instances of repeated poles in the mathematical models of such systems, literature elsewhere does not contain a name for the associated mode. The partial fraction expansion of

(1.69), however, caters for repeated closed-loop poles since they may be chosen in the model-based approach to control system design to achieve a smooth, non-overshooting step response. Hence, for completeness, the author has given the name *polynomial exponential mode* to the system behaviour represented by a term such as $e^{-t/T_{Pj}} \sum_{k=1}^{m_j} C_{jk} t^{k-1}$. This type of mode is characterised by the dynamic behaviour consisting of one or more stationary points, determined by the degree of the polynomial factor and ultimate decay towards zero caused by the exponential factor since the closed-loop system is assumed to be stable. Although the polynomial factor, $\sum_{k=1}^{m_j} C_{jk} t^{k-1}$, becomes infinite as t increases indefinitely, being dominated by the highest degree term, the exponential factor, $e^{-t/T_{Pj}}$, will ensure that $z_j(t) \to 0$ as $t \to \infty$. This may be proven as follows.

$$\lim_{t \to \infty} e^{-t/T_{Pj}} \sum_{k=1}^{m_j} C_{jk} t^{k-1} = \lim_{t \to \infty} \frac{\sum_{k=1}^{m_j} C_{jk} t^{k-1}}{e^{-t/T_{Pj}}}$$

$$= \lim_{t \to \infty} \frac{\sum_{k=1}^{m_j} C_{jk} t^{k-1}}{\sum_{i=0}^{\infty} \left[(t/T_{Pj})^i / i! \right]} = 0. \quad (1.74)$$

1.4.2.4 Modal Decomposition

The process of forming the partial fraction expansion (1.69) of the transfer function, arranging the terms in groups according to the three mode categories and then taking the inverse Laplace transform, results in

$$y_\delta(t) = \underbrace{\sum_{k=1}^{n_e} A_k e^{-t/T_{Pk}}}_{\text{Exponential modes}} + \underbrace{\sum_{l=1}^{n_o} D_l e^{-t/T_{Pl}} \sin(q_l t + \phi_l)}_{\text{Oscillatory modes}}$$

$$+ \sum_{j=1}^{n_r} \underbrace{e^{-p_j t} \sum_{k=1}^{m_j} C_{jk} t^{k-1}}_{\text{Polynomial exponential modes}}. \quad (1.75)$$

This is referred to as *modal decomposition* since the mathematical expressions defining each mode appear separately as summed terms. In Chap. 2, this topic will be revisited but in the state space and in connection with plant modelling.

Example 1.3 Modal decomposition of a 7th-order system

To demonstrate typical forms of the three types of mode defined above, modal decomposition of a closed-loop system with a simple pole, a complex conjugate pair of poles and a pole of multiplicity 4 will now be carried out. The poles are chosen to

1.4 Dominance in the Pole–Zero Distribution

emphasise the typical features of the different types of mode. A simulation will be carried out to plot the system impulse response together with the three modes that comprise this impulse response. The transfer function taken is

$$\frac{Y(s)}{Y_r(s)} = \frac{16}{(s+2)(s^2+s+8)(s+1)^4}. \qquad (1.76)$$

Carrying out the partial fraction expansion yields

$$\frac{Y(s)}{Y_r(s)} = \underbrace{\frac{A_1}{s+2}}_{Z_3(s)} + \underbrace{\frac{B_1 s + B_2}{s^2+s+8}}_{Z_2(s)} + \underbrace{\frac{C_1}{s+1} + \frac{C_2}{(s+1)^2} + \frac{C_3}{(s+1)^3} + \frac{C_4}{(s+1)^4}}_{Z_1(s)}$$
$$(1.77)$$

where $A_1 = 0.4444$, $B_1 = 0.2086$, $B_2 = -0.2018$, $C_1 = -0.6531$, $C_2 = 0.4898$, $C_3 = 0$ and $C_4 = 1.1429$. The impulse response, $y_\delta(t)$ [i.e. $y(t)$ when $y_r(t) = \delta(t)$], of the complete system is the sum of the modal impulse responses, $z_1(t)$, $z_2(t)$ and $z_3(t)$, which is the sum of the inverse Laplace transforms of the partial fraction terms of the transfer function indicated in (1.77). Thus,

$$y_\delta(t) = \underbrace{A_1 e^{-2t}}_{\substack{\text{Exponential} \\ \text{mode} \\ z_3(t)}} + \underbrace{B_1 \frac{(b-a)^2 + \omega^2}{\omega} e^{-at} \sin(\omega t + \phi)}_{\substack{\text{Oscillatory mode} \\ z_2(t)}}$$

$$+ \underbrace{e^{-t}\left(C_1 + C_2 t + \frac{1}{2}C_3 t^2 + \frac{1}{6}C_4 t^3\right)}_{\substack{\text{Polynomial exponential mode} \\ z_1(t)}}, \qquad (1.78)$$

where $\omega = \sqrt{15}/4$, $b = B_2/B_1$, $a = 1/8$ and $\phi = \tan^{-1}[\omega/(b-a)]$. Figure 1.25 shows state-variable block diagrams (Chap. 2) that would generate $\ddot{z}_1(t)$, $\ddot{z}_2(t)$, $\ddot{z}_3(t)$ and $\dot{y}_\delta(t)$ if $\delta(t)$, which is not realisable, were to be applied in a MATLAB® SIMULINK® simulation. In this case, applying the unit step function, $h(t) = \int_0^t \delta(\tau) \, d\tau$, to the same system generates $\int_0^t \dot{z}_i(\tau) \, d\tau = z_i(t)$, $i = 1, 2, 3$, and hence $y_\delta(t) = z_1(t) + z_2(t) + z_3(t)$, as required. If block diagrams directly realising the transfer functions of (1.77) were to be generated, then the result would be similar to Fig. 1.25 but with the output trees containing the partial fraction coefficients connected to the integrator *outputs* rather than their inputs as shown. This is done to obtain the required impulse responses as the first derivatives of the unit step responses, the reasoning being as follows. In the time domain, the input of an integrator is the first derivative of its output.

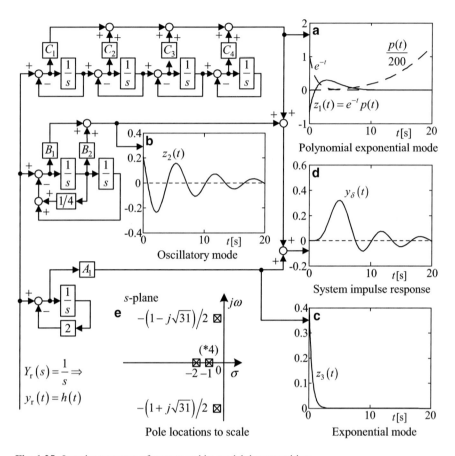

Fig. 1.25 Impulse response of system and its modal decomposition

In Fig. 1.25a, the polynomial function, $p(t) = C_1 + C_2 t + \frac{1}{2}C_3 t^2 + \frac{1}{6}C_4 t^3$, is plotted separately together with the function, e^{-t}, that multiplies it to produce the modal impulse response, $z_1(t)$, which decays despite $p(t)$ growing rapidly.

It is evident that the oscillatory mode, $z_2(t)$, of Fig. 1.25b greatly influences the output $y_\delta(t)$ of Fig. 1.25d but the relatively rapidly decaying exponential mode of Fig. 1.25 has very little influence.

In Fig. 1.25e, the polynomial exponential mode is associated with the pole at -1 with multiplicity, 4 (indicated by (*4) above the pole location), the oscillatory mode is associated with the complex conjugate pole pair at $-0.5 \pm 0.5j\sqrt{31}$ and the exponential mode is associated with the pole at -2. It is evident that the modes associated with the poles closest to the imaginary axis of the s-plane have the most influence on $y_\delta(t)$. This is shown to be generally true in the following sections.

It should be noted that the closed-loop poles are indicated by the symbol ⊠, while open-loop or plant poles are indicated by the symbol ×.

1.4 Dominance in the Pole–Zero Distribution

1.4.3 Dominance in Pole Distributions

1.4.3.1 Pole Dominance Sequence

It is evident that all the three types of mode discussed above will decay at a rate determined by the exponential factor, $e^{-t/T_{pi}}$, $i = 1, 2, ..., n$, where $T_{pi} = 1/\text{Re}(s_i)$, s_i being the pole value. It is important to realise that according to the impulse response (1.69), the modes with the slowest exponential decay are the most dominant and the poles associated with these slowest decaying modes are said to be the most dominant poles. In the notation adopted here, a time constant will be associated with *every* pole. For each real distinct pole, the time constant is simply $T_{pi} = 1/\text{Re}(s_i) = 1/|s_i|$. For an oscillatory exponential mode, the time constants linked to the associated complex conjugate poles, s_j and $s_{j+1} = s_j^*$, are both equal to the time constant of the exponential decay factor, i.e. $T_{pj} = T_{pj+1} = 1/\text{Re}(s_i) = 1/\text{Re}(s_{i+1})$. Similarly, the time constant associated with a pole of multiplicity, r, with value, s_k, is $T_{pk} = T_{pk+1} = \cdots = T_{pk+r-1} = 1/\text{Re}(s_k)$ where $s_k = s_{k+1} = \cdots = s_{k+r-1}$. Suppose now that the poles are placed in a sequence with reducing time constants, i.e. with increasing magnitudes of their real parts, those with equal real parts or members of a multiple pole being placed together.

Then,

$$|\text{Re}(s_1)| \leq |\text{Re}(s_2)| \leq \cdots \leq |\text{Re}(s_n)|. \tag{1.79}$$

and

$$T_{p1} \geq T_{p2} \geq \cdots \geq T_{pn}. \tag{1.80}$$

The poles are now ordered in decreasing degree of dominance. A procedure is devised in the following section, using this sequence, to select a subset of the closed-loop poles to form a reduced order, and therefore simplified, transfer function representing almost the same behaviour as the original system.

1.4.3.2 Pole-to-Pole Dominance Ratio

Having arranged the poles in the sequence according to (1.79) and (1.80), the following procedure selects the set of poles for the reduced-order model.

$$\begin{cases} \text{For } i = 1, 2, ..., n : r_{pp\,i} = \text{Re}(s_{i+1})/\text{Re}(s_i) = T_{pi}/T_{pi+1}. \\ \text{If } r_{pp\,i} \leq r_{pp\,\min}, \text{ keep } s_{i+1}, \text{ else the selection is complete.} \end{cases} \tag{1.81}$$

where $r_{pp\,i}$ is the *pole-to-pole dominance ratio* of pole, s_i, with respect to pole, s_{i+1}, and $r_{pp\,\min}$ is the minimum value of this ratio below which the degree of dominance is deemed to be insufficient to simplify the transfer function.

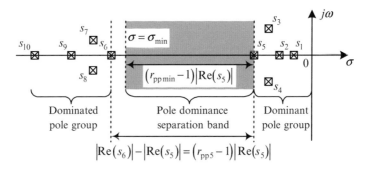

Fig. 1.26 Illustration of pole dominance in the s-plane

The determination of $r_{pp\,min}$ will be addressed shortly. The greater $r_{pp\,i}$, the greater is the dominance of pole, s_i, over poles, s_{i+1}, s_{i+2}, $\cdots$ and s_n.

The criterion for determination of $r_{pp\,min}$ is that the impulse response corresponding to the simplified transfer function closely approximates the impulse response of the original system. For each system, however, the smallest suitable value of $r_{pp\,min}$, which yields the lowest-order simplified transfer function, depends on the order of the system and the distribution of its poles. The task may be simplified, however, by finding a worst-case pole distribution that enables the minimum value of $r_{pp\,min}$ to be found that will be suitable for all systems of a given order. $r_{pp\,min}$ can be regarded as a measure of the minimum allowed separation between two groups of poles in the direction of the real axis of the s-plane, in order for the group to the right be sufficiently dominant to form the approximating transfer function, as illustrated in Fig. 1.26.

In the dominated closed-loop pole group of this illustration, poles s_7, s_8, s_9 and s_{10} have less influence on the dynamic behaviour of the system than pole s_6, and therefore, in the worst case, the dominated poles would all have equal real parts and lie on the vertical line, $\sigma = \sigma_{min}$. Also, the dominated poles have less influence on poles, s_1, s_2, s_3 and s_4 of the dominant pole group than they have on dominant pole, s_5. In view of this, the worst-case situation requiring the maximum width of the pole dominance separation band is for the dominant pole group to have equal real parts and the dominated pole group also to have equal real parts. This situation could only occur in a control system design in which the closed-loop poles could all be placed in desired locations. As will be seen in Chap. 8, such is the case in robust pole placement but here there are no complex conjugate poles and therefore there are only two multiple pole locations to consider. The very worst case demanding the greatest pole dominance separation band for a given system order, n, would be just one dominant pole and $n - 1$ colocated dominated poles. In order for the first pole to be dominant, the group of dominated poles would have to be well separated and by an amount increasing with n.

To simplify the process of determining suitable values of $r_{pp\,min}$ for a given system, a set of generic worst-case systems will be considered of order n having

1.4 Dominance in the Pole–Zero Distribution

Fig. 1.27 Worst-case systems for determination of minimum pole-to-pole dominance ratios

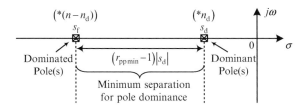

n_d dominant poles, as shown in Fig. 1.27. The symbol, $(*m)$, above a pole location indicates that it has a multiplicity of m. The generic closed-loop system transfer function,

$$\frac{Y(s)}{Y_r(s)} = \frac{p_d^{n_d} p_f^{n-n_d}}{(s+p_d)^{n_d}(s+p_f)^{n-n_d}}, \quad (1.82)$$

will be considered where $p_d = -s_d$ and $p_f = -s_f$. This has unity DC gain, which is usually required. Since $p_f = r_{pp} p_d$, there will be no loss of generality in normalising with respect to the dominant pole location, i.e. setting $p_d = 1$. Then (1.82) becomes

$$\frac{Y(s)}{Y_r(s)} = \frac{r_{pp}^{n-n_d}}{(s+1)^{n_d}(s+r_{pp})^{n-n_d}}. \quad (1.83)$$

The unit step responses of this system are compared with those of the ideal closed-loop system,

$$\frac{Y_{\text{ideal}}(s)}{Y_r(s)} = \frac{1}{(s+1)^{n_d}}, \quad (1.84)$$

that they are intended to approximate. Figure 1.28 shows the results.

It remains to define a condition common to all the systems (generated by the different combinations of n and n_d), for which $r_{pp} = r_{pp\,\text{min}}$. Rather than compare the unit impulse responses of systems (1.83) and (1.84), it is more convenient to compare the unit step responses because the error, $e_h(t) = y_{\text{ideal}}(t) - y(t)$, must satisfy $e_h(0) = e_h(\infty) = 0$ and therefore has a maximum magnitude, $|e_h|_{\text{max}}$, for a finite value of t. Then $r_{pp\,\text{min}}$ is defined as the value of r_{pp} for which $|e_h|_{\text{max}} = 0.05$. This means that the maximum allowed deviation of $y(t)$ from $y_{\text{ideal}}(t)$ is 5 % of the step reference input magnitude.

The choice of 5 % is a matter of engineering judgement. An attempt at analytical determination of $r_{pp\,\text{min}}$ would lead to transcendental equations that would require numerical solution. Instead, the approach adopted here is to run a computer simulation of the unit step responses of systems (1.83) and (1.84) for selected combinations of n and n_d. Then $r_{pp\,\text{min}}$ is adjusted until the maximum value of $|e_h|(t)$ is $|e_h|_{\text{max}} = 0.05$. This has been done for n ranging between 2 and 6 and $n_d = 1, 2, \ldots, n-1$, as this will cover most systems to be dealt with in practice.

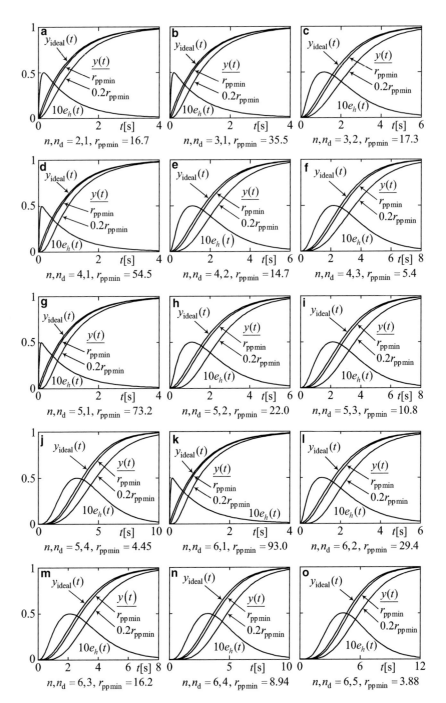

Fig. 1.28 Unit step responses of generic worst-case systems for pole-to-pole dominance equal to and below the minimum value

1.4 Dominance in the Pole–Zero Distribution

The error responses, $e_h(t)$, are plotted for $r_{pp} = r_{pp\,min}$. As expected, the step responses for $r_{pp} = r_{pp\,min}$ are fairly close approximations to the ideal ones. For illustrative purposes, these are each accompanied by an additional step response for $r_{pp} = 0.2 r_{pp\,min}$, which is far too small for the dominance to be effective, the pole separation being insufficient for the response of system (1.83) to be a good approximation to that of system (1.84). The values of $r_{pp\,min}$ given in Fig. 1.28 will be referred to as a design aid for pole placement in Chaps. 3, 4 and 8.

1.4.4 Dominance in Systems with Zeros

1.4.4.1 Plant Zero Invariance

As already stated in Sect. 1.4.1, plant zeros are invariant with respect to the loop closure unless cancelled by closed-loop poles under exceptional circumstances. It is therefore imperative to investigate their effects. This zero invariance will now be proven. Let the general plant transfer function be represented as

$$\frac{Y(s)}{U(s)} = G(s) = \frac{N(s)}{D(s)}, \qquad (1.85)$$

where $N(s)$ and $D(s)$ are, respectively, the numerator and denominator polynomials. Any linear controller for a single input, single output plant can be formulated with two inputs, $y_r(s)$ and $y(s)$, producing a single output, $u(s)$, defined by the general transfer function relationship,

$$U(s) = G_{\mathrm{ff}}(s) Y_r(s) - G_{\mathrm{fb}}(s) Y(s), \qquad (1.86)$$

where $G_{\mathrm{ff}}(s)$ and $G_{\mathrm{fb}}(s)$ are, respectively, the feedforward and feedback transfer functions. Applying controller (1.86) to plant (1.85) yields the closed-loop system block diagram of Fig. 1.29.

Elementary block diagram algebra then yields the closed-loop transfer function.

$$\frac{Y(s)}{Y_r(s)} = G_{\mathrm{cl}}(s) = G_{\mathrm{ff}}(s) \cdot \frac{\frac{N(s)}{D(s)}}{1 + \frac{N(s)}{D(s)} G_{\mathrm{fb}}(s)} = \frac{G_{\mathrm{ff}}(s) N(s)}{D(s) + G_{\mathrm{fb}}(s) N(s)}. \qquad (1.87)$$

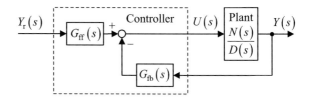

Fig. 1.29 A general linear SISO control system block diagram

Thus, $N(s)$ is a factor of both $G(s)$ and $G_{cl}(s)$. Hence, the zeros are invariant with respect to the loop closure, i.e. they cannot be altered by the feedback process. They can, however, be cancelled by the poles of $G_{ff}(s)$ or by the closed-loop poles, i.e. the roots of $D(s) + G_{fb}(s)N(s) = 0$, provided the roots of $N(s) = 0$ all have negative real parts (Chap. 3). As already seen in Sect. 1.3.1, however, the PID controller and its relatives, whose control actions all operate on the control error, introduce zeros whose values depend on the controller gains and the plant transfer function. These zeros therefore depend upon the closed-loop poles but can, under certain conditions, be cancelled by a subset of the closed-loop poles (Chap. 3).

1.4.4.2 Derivative Effect of Zeros

In Sect. 1.3, attention has already been drawn to the zeros introduced by some of the traditional controllers and their effects on the control system performance due to their derivative action. Further to this, the effects of zeros, regardless of their origin, on the dynamic behaviour of a closed-loop control system will be considered in more detail, in preparation for defining the *pole-to-zero dominance ratio*, r_{pz}, which is analogous to the pole-to-pole dominance ratio, r_{pp}, of Sect. 1.4.3.2.

The general closed-loop transfer function (1.69) may be written as

$$\frac{Y(s)}{Y_r(s)} = K_{\text{DCL}} \frac{\prod_{i=1}^{n} p_i \prod_{k=1}^{m}(s+z_k)}{\prod_{k=1}^{m} z_k \prod_{i=1}^{n}(s+p_i)} = K_{\text{DCL}} \frac{a_0}{b_0} \cdot \frac{\sum_{j=0}^{m} b_j s^j}{s^n + \sum_{i=0}^{n-1} a_i s^i}. \tag{1.88}$$

It is clear from (1.88) that the degree of the numerator polynomial is equal to the number of finite zeros. The effects of these zeros on the dynamic response will now be determined, by comparison of the response with that of the associated system,

$$\frac{X(s)}{Y_r(s)} = K_{\text{DCL}} \frac{\prod_{i=1}^{n} p_i}{\prod_{i=1}^{n}(s+p_i)} = \frac{K_{\text{DCL}} a_0}{s^n + \sum_{i=0}^{n-1} a_i s^i}, \tag{1.89}$$

subject to the same input. This has no finite zeros but the same poles and DC gain. The effect of the finite zeros may be revealed clearly if the step response of system (1.89) is non-overshooting. This is assured if all the closed-loop poles are real and negative, now being proven for coincident poles in which case (1.89) becomes

$$\frac{X(s)}{Y_r(s)} = K_{\text{DCL}} \left(\frac{p}{s+p}\right)^n. \tag{1.90}$$

1.4 Dominance in the Pole–Zero Distribution

The response to a unit impulse, $y_r(t) = \delta(t)$, (Table 1 in Tables) is then

$$x_\delta(t) = K_{\text{DCL}} p^n \frac{1}{(n-1)!} e^{-pt} t^{n-1}. \quad (1.91)$$

This is the first time derivative of the response, $x_h(t)$, to a unit step reference input, $y_r(t) = h(t)$. If $x_h(t)$ had any overshoots or undershoots, then the times at which these occur would be the roots of $x_\delta(t) = \dot{x}_h(t) = 0$ for $t \in (0, \infty)$. With $x_\delta(t)$ given by (1.91), the only values of t satisfying $x_\delta(t) = 0$ are $t = 0$ and $t = \infty$. Hence, there cannot be any overshoots and undershoots of $x_h(t)$. This is illustrated by the block diagram of Fig. 1.30 in which the system with transfer function (1.90) is represented by a set of cascaded first-order systems with transfer functions, $p_i/(s + p_i), i = 1, 2, \ldots, n$, and covers the cases for $n = 1, 2, 3$ and 4. Setting a unity DC gain and normalising with respect to the pole magnitude loses no essential information.

If the poles are chosen according to (1.90), then the original system transfer function (1.88) becomes

$$\frac{Y(s)}{Y_r(s)} = \frac{K_{\text{DCL}} \frac{p^n}{b_0} \sum_{j=0}^{m} b_j s^j}{(s+p)^n}. \quad (1.92)$$

Dividing (1.92) by (1.90) then yields

$$\frac{Y(s)}{Y_r(s)} \bigg/ \frac{X(s)}{Y_r(s)} = \frac{Y(s)}{X(s)} = \frac{1}{b_0} \sum_{j=0}^{m} b_j s^j. \quad (1.93)$$

The following transfer function relationship may then be written.

$$Y(s) = \left(1 + \sum_{i=1}^{m} c_i s^i\right) X(s), \quad (1.94)$$

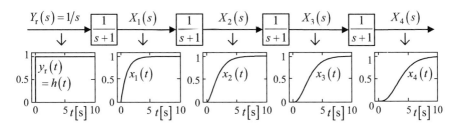

Fig. 1.30 First-order cascaded block representation of linear systems without finite zeros together with step responses

where

$$c_i = b_i/b_0, \quad i = 1, 2, \ldots, m. \tag{1.95}$$

In the time domain, (1.94) may be written

$$y(t) = x(t) + \sum_{i=1}^{m} c_i \frac{d^i}{dt^i} x(t). \tag{1.96}$$

Thus, the step response, $y_h(t)$, of the original system is the step response, $x_h(t)$, of the corresponding zero-less system plus a weighted sum of the derivatives of $x_h(t)$ up to an order equal to the number of zeros. The zeros therefore have a *differentiating* effect. Figure 1.31 shows the first three derivatives computed for $n = 4$, and normalisation with respect to the pole position by setting $p = 1$, which is sufficient for the present purposes.

The block diagram realisation of the transfer function,

$$\frac{X(s)}{Y_r(s)} = \frac{1}{(s+1)^4} = \frac{1}{s^4 + 4s^3 + 6s^2 + 4s + 1}, \tag{1.97}$$

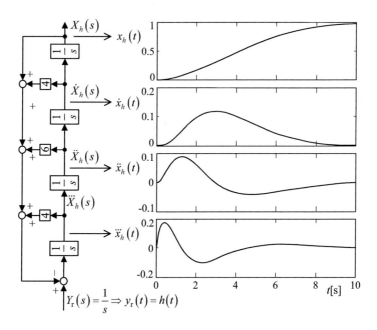

Fig. 1.31 Output of a zero-less fourth-order system with a multiple pole together with its first three derivatives

1.4 Dominance in the Pole–Zero Distribution

is in the control canonical form (Chap. 3) containing four cascaded integrators whose inputs are the required derivatives. That this block diagram has transfer function (1.97) is easily proven by means of Mason's formula (Appendix A4). Since, $\dot{x}_h(t) = x_\delta(t)$, it has just one stationary point and a single maximum at this point, as shown in Fig. 1.31. It is evident in Fig. 1.31 that $\ddot{x}_h(t)$ has a single maximum and a single minimum and that $\dddot{x}_h(t)$ has two maxima and a single minimum.

The notation, $x^{(k)}(t) \equiv d^k x/dt^k$, will be adopted. As already shown, with a unit step input, the output of the system with transfer function (1.90) has the first derivative given by

$$x_h^{(1)}(t) = Ce^{-pt}t^{n-1} \tag{1.98}$$

where $C = K_{\text{DCL}} p^n / (n-1)!$. There are no real roots of $x_h^{(1)}(t) = 0$ in the interval, $t \in (0, \infty)$, confirming that the step response, $x_h^{(0)}(t) \equiv x_h(t)$, has no maxima or minima in this interval, i.e. it is monotonically increasing. Differentiating (1.98) yields

$$x_h^{(2)}(t) = Ce^{-pt}\left[(n-1)t^{n-2} - pt^{n-1}\right] = Ce^{-pt}t^{n-2}\left[(n-1) - pt\right] \tag{1.99}$$

In this case, there is one root of $x_h^{(2)}(t) = 0$ at

$$t_{1\,(1)} = (n-1)/p, \tag{1.100}$$

This indicates a stationary point of $x_h^{(1)}(t)$ at $t = t_{1\,(1)}$. The subscript in parenthesis indicates the order of the derivative having the stationary point at the time indicated. This stationary point has to be at a maximum because elementary analysis of the function (1.98) shows that it initially increases from zero due to the term, t^{n-1}, and eventually reduces and tends to zero from positive values as $t \to \infty$ due to the term, e^{-pt}. Figure 1.31 confirms this for $n = 4$ and gives $t_{1\,(1)} = 3$ s, which can be seen. Differentiating (1.99) yields

$$\begin{aligned} x_h^{(3)}(t) &= Ce^{-pt}\left[(n-2)(n-1)t^{n-3} - 2p(n-1)t^{n-2} + p^2 t^{n-1}\right] \\ &= Ce^{-pt}t^{n-3}\left[(n-2)(n-1) - 2p(n-1)t + p^2 t^2\right]. \end{aligned} \tag{1.101}$$

In this case, there are two stationary points at

$$t_{1,\,2\,(2)} = \left[(n-1) \pm \sqrt{(n-1)}\right]/p, \tag{1.102}$$

Since the stationary point $x_h^{(1)}(t_{1\,(1)})$ is at a maximum, the derivative, $x_h^{(2)}(t)$, must be positive for $t < t_{1\,(1)}$ and negative for $t > t_{1\,(1)}$ so $x_h^{(2)}(t)$ must have a maximum at $t_{1\,(2)} = \left[(n-1) - \sqrt{(n-1)}\right]/p$ and a minimum at $t_{2\,(2)} = \left[(n-1) - \sqrt{(n-1)}\right]/p$. For the system of Fig. 1.31, $t_{1\,(2)} = 3 - \sqrt{3} = 1.2679$ s and $t_{2\,(2)} = 3 + \sqrt{3} = 4.7321$ s, which again can be seen. Continuing in this fashion, it is evident that at each step, the degree of the polynomial, whose real roots in the interval $t \in (0, \infty)$ are the times at which the stationary points occur, increases by 1. Differentiating (1.101) yields

$$\begin{aligned}x_h^{(4)}(t) &= Ce^{-pt}\left[(n-1)(n-2)(n-3)t^{n-4} - 3p(n-1)(n-2)t^{n-3}\right.\\ &\quad \left.+ 3p^2(n-1)t^{n-2} - p^3 t^{n-1}\right]\\ &= Ce^{-pt}t^{n-4}\left[(n-1)(n-2)(n-3) - 3p(n-1)(n-2)t\right.\\ &\quad \left.+ 3p^2(n-1)t^2 - p^3 t^3\right].\end{aligned}$$

(1.103)

Analytical solutions exist for cubic and quartic equations with arbitrary coefficients but the Abel–Ruffini theorem states, remarkably, that no such analytical solution exists for polynomial equations of fifth degree or greater [1]. Also the formulae for the solutions to cubic and quartic equations are relatively complex. It is only necessary, however, to determine how many overshoots and undershoots of each step response derivative occur that lie in the interval, $t \in (0, \infty)$. Proceeding with a similar argument as applied above, since $x_h^{(2)}(t)$ has a maximum at $t = t_{1\,(2)}$ and a minimum at $t = t_{2\,(2)}$, with $t_{2\,(2)} > t_{1\,(2)}$, $x_h^{(3)}(t) > 0$ for $0 < t < t_{1\,(2)}$, $x_h^{(3)}(t) < 0$ for $t_{1\,(2)} < t < t_{2\,(2)}$ and $x_h^{(3)}(t) > 0$ for $t > t_{2\,(2)}$. It follows that $x_h^{(3)}(t)$ has a maximum at $t_{1\,(3)} \in (0, t_{1\,(2)})$, a minimum at $t_{2\,(3)} \in (t_{1\,(2)}, t_{2\,(2)})$ and another maximum at $t_{3\,(3)} \in (t_{2\,(2)}, \infty)$. The times, $t_{1,2,3\,(3)}$, are the roots of

$$(n-1)(n-2)(n-3) - 3p(n-1)(n-2)t + 3p^2(n-1)t^2 - p^3 t^3 = 0,$$

(1.104)

from (1.103). This argument may be continued leading to the conclusion that for a system with n coincident poles and no finite zeros that if n is even, the derivative, $x_h^{(k-1)}(t)$, where $k \le n$, has $k/2$ maxima and $(k/2) - 1$ minima, and if k is odd, it has $(k-1)/2$ maxima and $(k-1)/2$ minima and that as t increases from zero, the sequence of stationary points are alternately at maxima and minima, commencing with a maximum. The total number of overshoots and undershoots is $k - 1$. This may be seen in Fig. 1.31. Thus, the qth derivative of the step response of a system with n coincident real poles and no finite zeros, with $q \le n - 1$, has a total of q alternate overshoots and undershoots and therefore can be seen to *oscillate* for a finite number of half cycles. This is the basis for determining the effect of the finite zeros in (1.92). Using the compact derivative notation, (1.96) can be written

$$y(t) = x(t) + \sum_{i=1}^{m} c_i x^{(i)}(t).$$

(1.105)

1.4 Dominance in the Pole–Zero Distribution

The terms, $c_i x^{(i)}(t)$, $i = 1, 2, .., m$, due to the combined effect of the finite zeros individually have a total of i maxima and minima and these can cause overshoots and undershoots in $y(t)$, but according to the analysis above, they will be finite in number, in contrast to the overshoots and undershoots due to complex conjugate poles, which are infinite in number, as evident in the previous section. Whether or not *all* the individual maxima and minima of $c_i x^{(i)}(t)$ produce corresponding overshoots and undershoots in $y(t)$, however, depends on the weighting coefficients, c_i. In fact, if these coefficients are sufficiently small, there will be *no* overshoots and/or undershoots in $y(t)$.

Having established that the zeros have a differentiating effect, the relationship between this effect and the zero positions in the s-plane will be explored. In view of (1.95), the general system transfer function (1.88) may be written

$$\frac{Y(s)}{Y_r(s)} = K_{DCL} \frac{\prod_{i=1}^{n} p_i}{\prod_{k=1}^{m} z_k} \cdot \frac{\prod_{k=1}^{m}(s + z_k)}{\prod_{i=1}^{n}(s + p_i)} = K_{DCL} a_0 \frac{\sum_{j=0}^{m} c_j s^j}{s^n + \sum_{i=0}^{n-1} a_i s^i} \qquad (1.106)$$

from which

$$\prod_{k=1}^{m}(1 + s/z_k) = \sum_{j=0}^{m} c_j s^j \qquad (1.107)$$

where $c_0 = 1$, and with reference to (1.105), it is clear that the derivative weighting coefficients, c_i, $i = 1, 2, .., m$, increase as the zero magnitudes, $|z_k|$, decrease, indicating that *the closer the zeros are located to the origin of the s-plane, the more effect they will have on the dynamic performance of the system* for fixed pole locations.

Consider now the effect of a single zero on the dynamic response of a closed-loop control system to its reference inputs, where the closed-loop poles have been made coincident. The closed-loop transfer function is then the particular case of (1.92) with $m = 1$, and if $K_{DCL} = 1$, which will not affect the relative weightings of the partial fraction coefficients, may be written

$$\frac{Y(s)}{Y_r(s)} = \frac{p^n}{z} \cdot \frac{s + z}{(s + p)^n} = \frac{p^n}{(s + p)^n} + \frac{p^n}{z} \frac{s}{(s + p)^n}. \qquad (1.108)$$

The transfer function of the zero-less system is

$$\frac{X(s)}{Y_r(s)} = \frac{p^n}{(s + p)^n}. \qquad (1.109)$$

It then follows from (1.108) and (1.109) that

$$Y(s) = X(s) + \frac{1}{z}sX(s) \Rightarrow y(t) = x(t) + \frac{1}{z}\frac{d}{dt}x(t). \quad (1.110)$$

As only the first derivative is present, there is a possibility of a single overshoot in the step response, $y_h(t)$, if $z > 0$ or a single undershoot if $z < 0$, the first derivative of which is the unit impulse response, $y_\delta(t)$, which may change sign in the interval, $t \in (0, \infty)$. Using the table of Laplace transforms (Table 1 in Tables) yields

$$\begin{aligned} y_\delta(t) &= \frac{p^n}{(n-1)!}e^{-pt}t^{n-1} + \frac{p^n}{z(n-1)!}\frac{d}{dt}\left(e^{-pt}t^{n-1}\right) \\ &= \frac{p^n}{(n-1)!}e^{-pt}t^{n-1} + \frac{p^n}{z(n-1)!}e^{-pt}\left[(n-1)t^{n-2} - pt^{n-1}\right] \\ &= \frac{p^n}{(n-1)!}e^{-pt}t^{n-2}\left[\left(1 - \frac{p}{z}\right)t + \frac{(n-1)}{z}\right]. \end{aligned} \quad (1.111)$$

This crosses zero at

$$t_1 = \frac{n-1}{p-z} \quad (1.112)$$

and an overshoot or undershoot can only occur for $t_1 \in (0, \infty)$. Hence, for $z > 0$ (meaning the zero is in the left half of the s-plane at $z_1 = -z$), a single overshoot of the unit step response, $y_h(t)$, exists only if $p > z$. On the other hand, for $z < 0$ (meaning the zero is in the right half of the s-plane at $z_1 = -z$), a single undershoot exists in any case since $p > 0$ for the poles to be located in the stable region at $s_1 = \cdots = s_n = -p$, i.e. the left half of the s-plane. It is important to note that, in general, zeros can have positive real parts without affecting the closed-loop stability, which depends only on the pole values.

Plants with zeros lying in the right half of the s-plane are *non-minimum-phase* plants (Chap. 2). This term comes from the frequency domain and refers to the phase response, $\phi(\omega)$, of the frequency response, $G(j\omega) = |G(j\omega)|e^{j\phi(\omega)}$, of a plant with transfer function, $G(s)$, where the phase angle function, $\phi(\omega)$, cannot have a minimum if the zeros have positive real parts.

1.4.4.3 Pole-to-Zero Dominance Ratio

It is evident from the above that, like the real parts of the poles, the smaller the *magnitudes* of the zeros, the greater their effect on the control system performance, for given pole locations. Hence, following similar lines to Sect. 1.4.3.2, in which the poles are ordered in increasing magnitude of their real parts, let the zeros be ordered in increasing magnitude. Thus,

$$|z_1| \leq |z_2| \leq \cdots \leq |z_m|. \quad (1.113)$$

1.4 Dominance in the Pole–Zero Distribution

Since (a) the timescale on which a linear system operates depends upon the magnitudes of the real parts of its poles and (b) the zeros exercise their influence through the time derivatives of the zero-less version of the system with weightings proportional to the reciprocals of the zero magnitudes, the pole-to-zero dominance ratio,

$$r_{pz} = |z_i| / |\text{Re}(p_n)|, i = 1, 2, ..., m, \quad (1.114)$$

is defined, p_n being the pole with the largest real part and therefore yielding the smallest value of r_{pz} for each zero. Following similar lines to Sect. 1.4.3.2, the minimum pole-to-zero dominance ratio, $r_{pz\,min}$, will be determined, such that the zeros for which $r_{pz} \leq r_{pz\,min}$ will be deemed dominant and will have to be taken into account in the control system design. Qualitatively, the further the pole from the imaginary axis of the s-plane, the faster will be the response to changes in $y_r(t)$ and therefore the greater will be the peak values of the derivatives of $y(t)$ caused by the zeros. It remains to determine a suitable value for $r_{pz\,min}$.

Returning to study of the system with transfer function (1.108),

$$r_{pz} = |z|/p. \quad (1.115)$$

Noting that $p > 0$ for stability and z can be of either sign. Then, the system transfer function (1.108) and its step response become, respectively,

$$\frac{Y(s)}{Y_r(s)} = \frac{p^n}{(s+p)^n} + \text{sgn}(z) \frac{s}{(s+p)^n} \cdot \frac{p^{n-1}}{r_{pz}} \quad (1.116)$$

and

$$y_h(t) = \mathcal{L}^{-1}\left\{ \frac{1}{s}\left[\frac{p^n}{(s+p)^n} + \text{sgn}(z)\frac{s}{(s+p)^n} \cdot \frac{p^{n-1}}{r_{pz}} \right] \right\}$$
$$= \underbrace{1 - \sum_{i=0}^{n-1} \frac{1}{i!}(pt)^i e^{-pt}}_{x_h(t)} + \text{sgn}(z)\frac{p^{n-1}}{r_{pz}} \frac{1}{(n-1)!} t^{n-1} e^{-pt}, \quad (1.117)$$

where $\text{sgn}(x) \triangleq -1$ for $x < 0$, 0 for $x = 0$, $+1$ for $x > 0$. The error between the unit step response of the system and that of the corresponding zero-less system is then

$$e_h(t) = y_h(t) - x_h(t) = \text{sgn}(z) \frac{p^{n-1}}{r_{pz}} \frac{1}{(n-1)!} t^{n-1} e^{-pt}. \quad (1.118)$$

The extreme (meaning global maximum or minimum) error occurs at $t = t_e$ when $\dot{e}_h(t) = 0$, i.e. when

$$\text{sgn}(z) \frac{p^{n-1}}{r_{pz}} \cdot \frac{1}{(n-1)!} t_e^{n-2} [(n-1) - pt_e] e^{-pt_e} = 0 \Rightarrow t_e = \frac{n-1}{p}. \quad (1.119)$$

It is a maximum for $z > 0$ and a minimum for $z < 0$. So the extreme error magnitude is given by

$$|e_{he}| = \frac{1}{r_{pz}} \cdot \frac{1}{(n-2)!} (n-1)^{n-2} e^{-(n-1)}. \quad (1.120)$$

Since this is for a unit step reference input, i.e. $y_r(t) = h(t)$ rather than $y_r(t) = Y_r h(t)$, then $e_h(t)$ of (1.118) and hence e_{he} of (1.120) are already normalised with respect to the step reference value, Y_r. Then it remains to decide a maximum per unit extreme error magnitude, $|e_{he\,max}|$, from which the corresponding value of $r_{pz\,min}$ can be calculated, using (1.120). As for $r_{pp\,min}$ of section 1.4.3.2, this is a matter of engineering judgement and a value of

$$|e_{h\,e\,max}| = 0.05 \quad (1.121)$$

is chosen. As expected, the value of $r_{pz\,min}$ yielding (1.121) depends upon the system order, n, and Table 1.3 gives the values obtained using up to $n = 6$.

The effect of the zero and its degree of dominance is illustrated in Fig. 1.32 by the simulations of the systems with transfer functions (1.116) and (1.109) for different orders and for zeros in the right and left halves of the s-plane.

The subscript, k, in y_{hk} and x_{hk} indicates the system order. The pole location is fixed at $s_1 = \cdots = s_n = -1 \Rightarrow p = 1$ and r_{pz} is set by the value of z. The upper three families of step responses in Fig. 1.32 are for zero locations, z_1, in the left half of the s-plane ($z > 0 \Rightarrow z_1 < 0$) for which it has already been predicted theoretically that a single overshoot occurs only if $p > z$ and in view of (1.115) this implies $r_{pz} < 1$, which can be seen. In contrast to the overshooting due to complex conjugate poles, no oscillations follow the first overshoot. The lower three families of step responses in Fig. 1.32 are for zero locations in the right half of the s-plane ($z < 0 \Rightarrow z_1 > 0$ yielding a non-minimum phase system). Also it has been theoretically predicted that a single undershoot occurs $\forall z < 0$, with no oscillations following and this is visible. The initial movement of the controlled variable, $y(t)$, in the opposite direction to the step change in the reference input, $y_r(t)$, which can be seen clearly, is an easily recognised characteristic of non-minimum

Table 1.3 Variation of minimum pole-to-zero dominance ratio with system order

System order, n	2	3	4	5	6
Minimum pole-to-zero dominance ratio, $r_{pz\,min}$	7.37	5.41	4.48	3.91	3.57

1.4 Dominance in the Pole–Zero Distribution

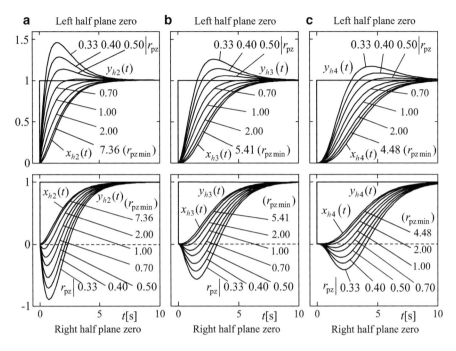

Fig. 1.32 Unit step responses of linear systems with coincident poles, a single zero and various pole-to-zero dominance ratios. (**a**) $n = 2$ (**b**) $n = 3$ (**c**) $n = 4$

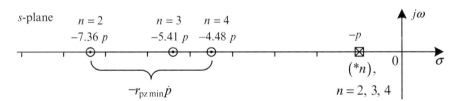

Fig. 1.33 Minimum separations of multiple pole and zero for negligible effect of zero

phase systems. For clarification, the values of r_{pz} indicated in Fig. 1.32 as $r_{pz\,min}$ are the maximum ones shown on each family of step responses but are the *minimum* values for which the zero is judged not to have a significant influence, in evidence through $y_{hn}(t) \cong x_{hn}(t)$. For the lower values of r_{pz}, the difference between $y_{hn}(t)$ and $x_{hn}(t)$ is significant.

An important observation is that the zero has to be considerably larger in magnitude than the poles in order for the effect of the zero on the step response to be negligible and Fig. 1.33 shows the minimum separations needed, to scale, for the system whose step responses are shown in Fig. 1.32. As expected, the smaller the number of poles, the more effect a zero in a given location will have. Conversely, for a given maximum error magnitude between the step responses, $y_{hn}(t)$ and $x_{hn}(t)$, as n increases, the minimum separation between the poles and the zero decreases.

The combined effect of several zeros depends in a complex fashion on their relative locations to one another and to the pole locations. Attempting to derive a general mathematical relationship to cover all cases would yield unwieldy expressions. It is therefore recommended that individual control systems involving zeros are analysed individually. Complex conjugate zero pairs, however, often arise in the control of mechanical objects containing vibration modes, as will be seen in Chap. 2. For this reason, the effect of such a pair of zeros will be considered for various locations in the s-plane. On the basis that it is possible to design linear controllers for linear plants with a free choice of the closed-loop poles, a system with transfer function

$$\frac{Y(s)}{Y_r(s)} = \frac{p^n (s+z)(s+z^*)}{zz^*(s+p)^n} = \frac{p^n \left(s^2 + 2\eta v_n s + v_n^2\right)}{v_n^2 (s+p)^n} = \frac{p^n \left(1 + c_1 s + c_2 s^2\right)}{(s+p)^n} \quad (1.122)$$

where

$$c_1 = 2\eta/v_n \text{ and } c_2 = v_n^2 \quad (1.123)$$

will be investigated. Here, the parameters, η and v_n correspond, respectively, to the damping ratio, ζ, and the undamped natural frequency, ω_n, for a complex conjugate pole pair but do not indicate the presence of continuous oscillations that decay in amplitude exponentially. It is well known, however, that the root loci obtained with a classical control structure containing a single proportional gain, K, terminate on the zeros as $K \to \infty$. For relatively high gains, therefore, there are two complex conjugate closed-loop poles close to each complex conjugate pair of zeros in which case η and v_n become *nearly* the damping ratio and undamped natural frequency of an oscillatory mode. This does not apply, however, to the system with closed-loop transfer function (1.122) because it is assumed at the outset that a control structure is employed enabling all the closed-loop poles to be placed at $s_1 = s_2 = \cdots = s_n = -p$. The associated zero-less system transfer function is

$$\frac{X(s)}{Y_r(s)} = \frac{p^n}{(s+p)^n}. \quad (1.124)$$

Dividing (1.122) by (1.124) then yields

$$Y(s) = \left(1 + c_1 s + c_2 s^2\right) X(s) \Rightarrow y(t) = x(t) + c_1 \dot{x}(t) + c_2 \ddot{x}(t). \quad (1.125)$$

If $y_r(t)$ is the unit step, $h(t)$, then it has already been established that $\dot{x}(t)$ has a single maximum and $\ddot{x}(t)$ has a single maximum followed by a single minimum. In view of

1.4 Dominance in the Pole–Zero Distribution

(1.125), therefore, $y(t)$ may or may not exhibit overshoots or undershoots, depending upon the coefficients, c_1 and c_2. The zeros are the roots of $s^2 + 2\eta v_n s + v_n^2 = 0$, which are

$$s_{1,2} = \left(-2\eta v_n \pm j\sqrt{4v_n^2 - 4\eta^2 v_n^2}\right)/2 = -\eta v_n \pm j v_n \sqrt{1 - \eta^2}.$$

The zero magnitude is therefore $\sqrt{(-\eta v_n)^2 + v_n^2(1 - \eta^2)} = v_n$. The same pole-to-zero dominance ratio applies to both zeros since they have the same magnitude of

$$r_{pz} = v_n/p. \tag{1.126}$$

Then substituting for c_1 and c_2 in (1.122) using (1.123) yields

$$\frac{Y(s)}{Y_r(s)} = \frac{p^n \left[1 + (2\eta/v_n)s + (1/v_n^2)s^2\right]}{(s+p)^n}. \tag{1.127}$$

This may be expressed in terms of r_{pz} by substituting for v_n in (1.127) using $v_n = r_{pz}p$ from (1.126). Thus,

$$\frac{Y(s)}{Y_r(s)} = \frac{p^n \left[1 + \left[2\eta/(r_{pz}p)\right]s + \left[1/(r_{pz}^2 p^2)\right]s^2\right]}{(s+p)^n}. \tag{1.128}$$

Since r_{pz} is a measure of how much dominance the poles have over the zeros regarding their influence on the dynamic performance, there is an inverse relationship between this parameter and each of the derivative coefficients:

$$c_1 = 2\eta/(r_{pz}p) \text{ and } c_2 = 1/(r_{pz}^2 p^2). \tag{1.129}$$

To explore the variations in the shapes of the step responses with r_{pz} and η, Fig. 1.34 shows the pole location in the s-plane for $n = 3$ and ten different locations of the complex conjugate zero pair together with the corresponding step responses normalised with respect to the pole location, i.e. with $p = 1$. It should be noted that no information is lost through this normalisation, since linearly expanding or contracting the pole–zero pattern of any linear system, respectively, contracts or expands the timescale of the step response without changing its basic form. This is proven in Chap. 3. Five of the pairs of complex conjugate zero locations lie on the circle for $r_{pz} = 1/3$, where the zeros would be expected to be dominant. The extreme overshooting and undershooting in the step responses confirm this. For Fig. 1.34a, e, c_1 has the maximum magnitude and $\dot{x}(t)$ results in a large overshoot in Fig. 1.34a and a large undershoot in Fig. 1.34e.

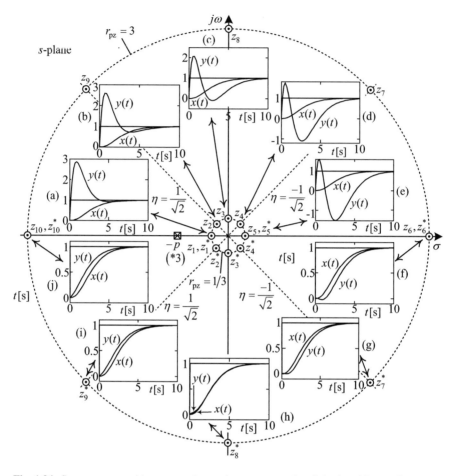

Fig. 1.34 Step responses with two complex conjugate zeros and a triple closed-loop pole

In Fig. 1.34b, d, c_1 has a smaller magnitude, while c_2 remains unaltered so the undershoot and overshoot in $y(t)$ contributed by $\ddot{x}(t)$ is more significant. In Fig. 1.34c, $c_1 = 0$ and the large overshoot and large undershoot in $y(t)$ due to $\ddot{x}(t)$ alone are clearly visible.

The large undershoot characteristic of non-minimum-phase systems may be seen in Fig. 1.34d, e. For the remaining five complex conjugate zero locations, lying on the circle for $r_{pz} = 3$, the poles dominate over the zeros and therefore in Fig. 1.34f–j, $y(t)$ resembles $x(t)$ in form. In Fig. 1.34h, the approximation of $y(t)$ to $x(t)$ is closer than in the other cases because $c_1 = 0$ and therefore $\dot{x}(t)$ has no influence, while it has more influence than $\ddot{x}(t)$ in Fig. 1.34f, g, i, j, due to $c_1 \gg c_2$. Regarding the control system design, the step responses for $r_{pz} = 3$ and greater would be acceptable in form but not so for considerably smaller values of r_{pz}.

1.4 Dominance in the Pole–Zero Distribution

Example 1.4 Influence of zeros on the attitude control of a flexible spacecraft

A single attitude control axis of a flexible spacecraft consisting of a rigid centre body with flexible appendages will be considered in which the inter-axis coupling is negligible and only one vibration mode in the flexible appendages is significant (Chap. 2) and this has negligible natural damping. Then the system is of fourth order with two complex conjugate zeros on the imaginary axis of the s-plane. Assuming that a controller is employed enabling the closed-loop poles to be placed at $s_{1,2,3,4} = -p$, to avoid oscillatory modes, the closed-loop transfer function is

$$\frac{Y(s)}{Y_r(s)} = \frac{p^4}{v^2} \cdot \frac{s^2 + v^2}{(s+p)^4} = \left(1 + \frac{s^2}{v^2}\right) \cdot \frac{p^4}{(s+p)^4} \quad (1.130)$$

where v is the encastre natural frequency, i.e. the frequency of vibration of the flexible appendages if the centre body were to be held fixed with respect to inertial space. As with most control systems, the closed-loop DC gain is unity. In this case the spacecraft attitude angle has to reach a constant reference attitude angle with zero steady-state error. It is required to find the largest value of p, giving the shortest possible settling time, for which the step response moves towards the reference value monotonically, i.e. it has no local or global maxima and minima.

As quoted in some previous cases, if $y_r(t)$ is the unit step function, $h(t)$, then $\dot{y}(t)$ is the unit impulse response which is the inverse Laplace transform of transfer function (1.130). An overshoot or undershoot in the step response will occur when $\dot{y}(t) = 0$ for $t \in (0, \infty)$. Using the table of Laplace transforms (Table 1 of Tables),

$$\begin{aligned}\dot{y}(t) &= \tfrac{1}{3!} t^3 e^{-pt} + \tfrac{1}{v^2} \tfrac{d^2}{dt^2}\left(\tfrac{1}{3!} t^3 e^{-pt}\right) = \tfrac{1}{6} t^3 e^{-pt} + \tfrac{1}{v^2} \tfrac{d}{dt}\left[\tfrac{1}{2} t^2 e^{-pt} - \tfrac{1}{6} p t^3 e^{-pt}\right] \\&= \tfrac{1}{6} t^3 e^{-pt} + \tfrac{1}{v^2}\left[te^{-pt} - \tfrac{1}{2} p t^2 e^{-pt} - \tfrac{1}{2} p t^2 e^{-pt} + \tfrac{1}{6} p^2 t^3 e^{-pt}\right] \\&= te^{-pt}\left[\tfrac{1}{6}\left(1 + p^2/v^2\right) t^2 - \left(p/v^2\right) t + 1/v^2\right].\end{aligned} \quad (1.131)$$

Any roots of $\dot{y}(t) = 0$ satisfying $t \in (0, \infty)$ will be roots of

$$\tfrac{1}{6}\left(1 + p^2/v^2\right) t^2 - \left(p/v^2\right) t + 1/v^2 = 0, \quad (1.132)$$

which are

$$\begin{aligned}t_{1,2} &= \left[p/v^2 \pm \sqrt{p^2/v^4 - \tfrac{2}{3}\left(1 + p^2/v^2\right)(1/v^2)}\right] / \left[\tfrac{1}{3}\left(1 + p^2/v^2\right)\right] \\&= \left[p/v \pm \sqrt{\tfrac{1}{3}\left[(p^2/v^2) - 2\right]}\right] / \left[\left(1 + p^2/v^2\right)(v/3)\right].\end{aligned} \quad (1.133)$$

Hence, if $p^2/v^2 < 2$, the roots are imaginary and no stationary points and therefore no overshoots or undershoots in $y(t)$ due to the zeros can occur. On the other hand, if $p^2/v^2 > 2$, $t_{1,2} \in (0, \infty)$ and therefore there are two stationary points of $y(t)$, a maximum occurring at $t = t_1$ and a minimum at $t = t_2$, with $t_1 < t_2$. Hence, the maximum value of p is given by

$$p_{\max}^2/v^2 = 2 \Rightarrow p_{\max} = \sqrt{2}\, v. \tag{1.134}$$

The pole–zero dominance ratio for this system is

$$r_{pz} = v/p. \tag{1.135}$$

Hence, (1.134) is tantamount to stating the minimum pole–zero dominance ratio as

$$r_{pz\ \min} = v/p_{\max} = 1/\sqrt{2}. \tag{1.136}$$

Some step responses, $y(t)$, of this system are shown in Fig. 1.35, together with the corresponding step responses, $x(t)$, of the associated system without the finite zeros, for $v = 0.1$ rad/s, which is realistic for some communications satellites with long, flexible solar panels.

The settling times of several tens of seconds are also realistic for this application in view of the limited control torques from the control actuators which limit the peak acceleration and deceleration magnitudes. It may be proven using Mason's formula (Appendix A4) that the block diagram shown has transfer function (1.130). The reason for employing this block diagram is that $x(t)$ is the output of a chain of pure integrators from which $\ddot{x}(t)$ is readily attainable to form $y(t)$ as shown. This is for five multiple pole locations, including the case for $r_{pz} = r_{pz\ \min} = 1/\sqrt{2}$ where a stationary point is visible in $y(t)$ at $t = t_1 = t_2$ according to (1.133). For $r_{pz} < 1/\sqrt{2}$, the double root, $t_1 = t_2$, splits into two real roots, t_1 and t_2, $t_1 < t_2$, at which a local minimum and a local maximum occur in $y(t)$, which are clearly visible for $r_{pz} = 1/\sqrt{2} - 0.2$. When r_{pz} is reduced further to $1/\sqrt{2} - 0.4$, the roots become more separated and the local maximum becomes a global maximum, indicating an overshoot.

Thus, as r_{pz} is reduced by moving the multiple pole location further into the left half on the s-plane, the complex conjugate zeros (only one of which is shown in Fig. 1.35) become more dominant. As r_{pz} is increased by reducing the multiple pole magnitude, the poles become more dominant, the stationary point vanishes as the roots, t_1 and t_2 become complex and $y(t)$ becomes closer to $x(t)$.

It is important to note that this example is intended to demonstrate pole dominance and the effects of finite zeros. It is not necessary to place all the closed-loop poles at one location in a real control system design, and recommended approaches for such applications are given in Chaps. 5, 8 and 10.

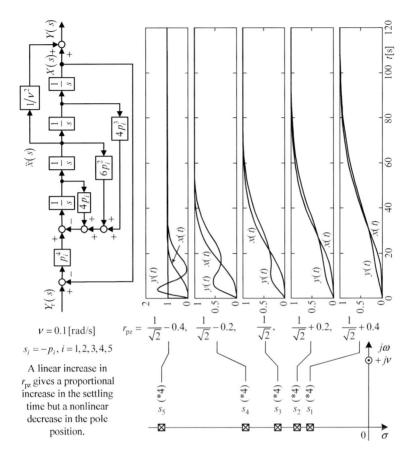

Fig. 1.35 Step responses for attitude control of flexible spacecraft for various pole–zero dominance ratios

1.5 The Steps of Control System Design

This book focuses on the theoretical and technical background needed to undertake control system design, meaning the devising of a system that meets a given performance specification containing information such as the accuracy, the settling time and the required operational envelope of the variables. The sequence of steps needed are presented in the flow chart of Fig. 1.36.

Sometimes the performance specification is incomplete. It is then necessary for the control system designer to liaise with the originator of the specification to agree on the further information needed for the control system design to proceed.

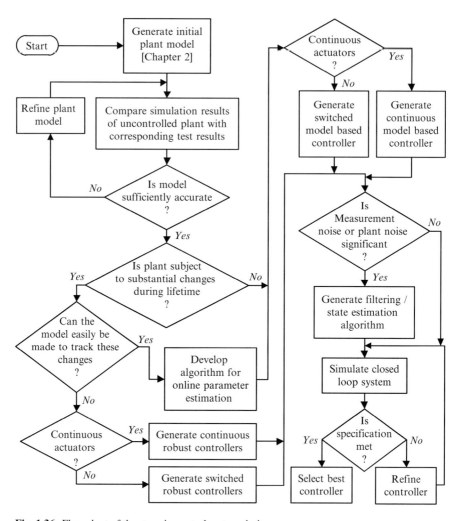

Fig. 1.36 Flow chart of the steps in control system design

The control system designer must sometimes be prepared to undertake decisions regarding the hardware, such as the sizing of motors for electric drives in terms of their maximum torque and power rating or the selection of suitable measurement instrumentation, such as star sensors and rate-integrating gyroscopes for spacecraft attitude control.

In an industrial scenario, the most economic design should be produced that satisfies the performance specification. This implies first considering a standard traditional controller. With the advent of the modern digital processor, however, the most economical approach could be a more flexible generic controller such as the polynomial controller of Chap. 5.

The domain of the control system designer is one of creativity in which ingenuity is required to devise a system that will solve a sufficiently challenging control problem, necessitating a venture beyond the procurement of standard controllers. This textbook is intended to help the reader acquire the relevant theoretical and technical expertise.

1.6 The Flexibility of Digital Implementation

In the era of discrete-component electronic controller implementation, the control system designer working in industry had to aim for the simplest possible controller for economic reasons. This also maximised the reliability. Nowadays, however, the computing power of a modern digital processor, in terms of memory and computational speed, by far exceeds that needed to implement a PID controller. Most importantly, the functionality of a controller based upon such a processor is entirely contained in programmable software. Digital implementation therefore offers flexibility in that a relatively inexpensive mass-produced digital processor-based controller can be programmed to control a very wide range of plants. A sufficient number of analogue-to-digital input interfaces and digital-to-analogue output interfaces, however, are needed.

Flexible digital processor-based controllers can also rapidly realise the results of control system research and development. In addition to implementing new, improved control techniques for existing applications, new control techniques can be very quickly brought into the laboratory and refined to a point of commercial viability.

Reference

1. Jacobson N (2012) Basic algebra I, 2nd edn. Courier Corporation

Chapter 2
Plant Modelling

2.1 Introduction

2.1.1 Overview

A plant model is a mathematical model consisting of one or more equations whose solutions replicate the behaviour of the physical plant. Its purpose is to provide a means by which the behaviour of the plant may be studied to enable the determination of a suitable control technique and the subsequent design of a controller. The plant model also forms the basis of computer simulations for control system development.

This chapter develops the background theory and gives the knowledge needed to generate plant models. After an introduction to the basic character of plants and their components, a subsection on physical modelling is presented. This is based on the underlying science of the various applications. A comprehensive coverage of this, however, would occupy several volumes and therefore, within the space limitations, the main emphasis is on mechanical systems and electric motors as actuators to cater for a large proportion of control systems. Some introductory material on thermal and fluid systems is also given.

A case study of plant modelling in industry is presented in Appendix A2.

2.1.2 Dynamical and Non-Dynamical Systems

First a system can broadly be defined as a collection of interconnected objects that fulfil a specified function. In this context, a plant or its model may be described as a system. A part of a plant or its model will be termed a subsystem. A dynamical system is one whose outputs depend upon the past history of the inputs. Plants to which feedback control is applied are dynamical systems and most are continuous, meaning that their variables are continuous functions of time. These are systems that obey differential equations in which time is the independent variable. The differential equation is the basic form of mathematical model of a plant from which other forms of model may be derived. The physical plant is usually separable into a number of constituent parts, or subsystems, each of which can be modelled separately. This is the physical modelling approach of Sect. 2.2.

The term 'dynamics' has two meanings in the field of control engineering. The first is the way in which the output of a dynamical system responds to its inputs. A system with fast (or high) dynamics is one whose outputs respond quickly to changes in its inputs. The second meaning is related to the modelling of mechanical systems and is defined in Sect. 2.2.2.

A non-dynamical system is one in which the present output depends only on the present input. These are subsystems such as a measurement device. A single input, single output (SISO) non-dynamical system can be represented by an equation defining the relationship between the input, $x(t)$ and the output, $y(t)$, in the form

$$y(t) = f[x(t)], \qquad (2.1)$$

where $f(\bullet)$ is a single-valued and continuous function of its argument. Similarly, a multiple input, multiple output (MIMO or multivariable) non-dynamical subsystem can be represented by the set of equations

$$y_i(t) = f_i[x_1(t), x_2(t), \ldots, x_r(t)], \quad i = 1, 2, \ldots, m. \qquad (2.2)$$

where x_j, $j = 1, 2, \ldots, r$, are the inputs and y_i, $i = 1, 2, \ldots, m$, are the outputs.

A dynamical system is one in which the present outputs depend on past values of the inputs and usually past values of the outputs. It is important to note that the present output of a controlled plant in the real world cannot respond instantaneously to a step change in the control input. The mathematical model therefore cannot contain algebraic dependence of the output on the input. To take a practical example, a step change in the torque demand of an electric drive on a train locomotive cannot cause a step change in the speed of the train, which is the controlled output.

Initially, consider a first-order SISO plant with input, $u(t)$, and output, $y(t)$. This obeys a first-order differential equation of the generic form

$$\dot{y}(t) = f[y(t), u(t)], \qquad (2.3)$$

where $f(\bullet)$ is a single-valued and continuous function of its arguments. If (2.3) is written as an integral equation,

$$y(t) = \int_0^t f[y(\tau), u(\tau)] d\tau, \qquad (2.4)$$

then it may be readily seen that the present output, $y(t)$, depends upon the continuum of past values of the input $u(\tau)$ and the output, $y(\tau)$, for $\tau \in [0, t]$.

2.1 Introduction

An SISO plant of nth order can be modelled by an ordinary differential equation of the generic form,

$$y^{(n)} = f\left(y^{(n-1)}, \ldots, y^{(1)}, y, u^{(m)}, u^{(m-1)}, \ldots, u^{(1)}, u\right), \tag{2.5}$$

where terms such as $x^{(q)}$ mean dx^q/dt^q. A fundamental restriction on this model is $0 \le m < n$. If $m = n$ then there would be a direct dependence of $y^{(n)}$ on $u^{(n)}$, implying a direct dependence of y on u. This, in turn, would imply that a step change in $u(t)$ would cause a step change in $y(t)$. Such behaviour is not found in physical plants. On the other hand, a step change in $u(t)$ will cause a step change in an output derivative of a certain order. To return to the train traction example, a step change in the drive torque demand will produce a step change in the acceleration of the train, which is the first derivative of the controlled speed. Considering the generic model (2.5), a step change in $u(t)$ will cause a step change in $y^{(r)}$, where $r = n - m$, but cannot cause a step change in any of the lower derivatives of y. As can be seen, $y^{(n)}$ depends algebraically on $u^{(m)}$ implying that a step change in $u^{(m)}$ will cause a step change in $y^{(n)}$, in turn implying that a step change in $u^{(0)}$, i.e. $u(t)$, will cause a step change in $y^{(n-m)}$. An important parameter in control system design that is related to this plant property is the *relative degree*, defined as

$$r = n - m. \tag{2.6}$$

This will be met in Chaps. 8 and 10. The term originates from its application to the transfer function model of a linear time-invariant plant, in which it is defined as the difference in degree between the denominator polynomial and numerator polynomial.

Since

$$n > m, \tag{2.7}$$

then, in view of (2.6),

$$r > 0. \tag{2.8}$$

The most general model of a multivariable plant of nth order is a set of interconnected ordinary differential equations of the generic form,

$$y_i^{(n_i)} = f_i\left(y_j^{(n_j-1)}, \ldots, y_j^{(1)}, y_j, u_k^{(m_{k,i})}, u_k^{(m_{k,i}-1)}, \ldots, u_k^{(1)}, u_k\right), \tag{2.9}$$
$$i = 1, 2, \ldots, m, \; j = 1, 2, \ldots, m, \; k = 1, 2, \ldots, p$$

where

$$\sum_{l=1}^{m} n_i = n \tag{2.10}$$

and

$$0 \leq m_{k,i} < n_i. \qquad (2.11)$$

Equation (2.10) merely states that the total order of the system is equal to the sums of the orders of the subsystems modelled by the individual ordinary differential equations of (2.9). Inequality (2.11) is analogous to (2.7) for SISO plants and represents similar practical limitations of real plants.

The relative degree of the plant with respect to the ith output is defined as

$$r_i = \min_{k} (n_i - m_{k,i}). \qquad (2.12)$$

It is the minimum order of the derivative of the output, y_i, that depends algebraically on *any* control input. This parameter is important in control system design, particularly when applying the techniques of sliding mode control, feedback linearisation or forced dynamic control.

2.1.3 Linearity and Nonlinearity

All systems are classified as linear or nonlinear. A linear system may be readily recognised through every mathematical expression being of the general form,

$$\sum_i C_i v_i(t) + B_i. \qquad (2.13)$$

The scalar coefficients, C_i, are usually constant, in which case the term linear time-invariant (LTI) system applies. Some of the variables, $v_i(t)$, are derivatives in the case of dynamical systems. The constant bias, B_i, is included for generality but is not present in many cases. Occasionally, one or more of the coefficients are time varying, in which case the term linear time-varying (LTV) system applies. This provides a straightforward means of recognising a linear model. A nonlinear system is readily recognised as it contains at least one expression not of the general form (2.13). A simple example of a linear non-dynamic LTI subsystem is the model of a liquid level transducer that gives a voltage, y, proportional to the height, h, of the liquid in a cylindrical vessel, given by the linear equation,

$$y = K_h h \qquad (2.14)$$

where K_h is the height measurement constant. This subsystem is an SISO one with input, h, and output, y. An example of a nonlinear non-dynamic subsystem is the model of the process in an electromagnet that produces the force, f, given the current, i, which may be written

2.1 Introduction

$$f = K_f \frac{i^2}{(x-x_0)^2}, \qquad (2.15)$$

where K_f is the electromagnetic force constant and x is the length of the air gap and x_0 is a positive constant. In this case the subsystem is a multiple input, single output (MISO) one with i and x as inputs and output f.

An LTI system exhibits the *scaling property* and the *superposition property*. It is important to note that if bias terms such as in (2.13) are present, then the equations must be reformulated in terms of *changes* in the variables to test for these properties. The scaling property is as follows. If the inputs of a system are multiplied by a constant, then the outputs will be multiplied by the same constant. The superposition property is as follows. Let a sequence of inputs be applied to a linear system one after the other and the corresponding outputs recorded. Then if a single input is applied that is the sum of the inputs previously applied, the output will be the sum of the previously recorded outputs. A nonlinear plant model will exhibit neither of these properties.

If the non-dynamical SISO subsystem modelled by (2.1) is linear, then it has the scaling property,

$$f\left[\lambda x(t)\right] = \lambda f\left[x(t)\right], \qquad (2.16)$$

where λ is a scalar. It will also have the superposition property as follows. Let $x_k(t)$, $k = 1, 2, \ldots, p$, be a set of inputs applied separately and let $y_k(t)$ be the corresponding outputs. After this, let a single input, $x(t) = \sum_{k=1}^{p} x_k(t)$, be applied. Then the resulting output is $y(t) = \sum_{k=1}^{p} y_k(t)$. If the system is linear, then (2.1) is of the form,

$$y(t) = f\left[x(t)\right] = Cx(t), \qquad (2.17)$$

where C is a coefficient that is usually constant. It is straightforward to confirm that the non-dynamical system (2.17) has the scaling and superposition properties. If, as is occasionally the case, C is time varying, the scaling property still holds.

Consider now the non-dynamical multivariable subsystem modelled by (2.2). If the sets of inputs and outputs are expressed as input and output vectors, defined as $\mathbf{x}(t) \triangleq [x_1(t)\, x_2(t) \ldots x_r(t)]^T$ and $\mathbf{y}(t) \triangleq [x_1(t)\, x_2(t) \ldots x_m(t)]^T$, then (2.2) becomes

$$\mathbf{y}(t) = \mathbf{f}\left[\mathbf{x}(t)\right]. \qquad (2.18)$$

The expression of the scaling property is then

$$\mathbf{f}\left[\lambda \mathbf{x}(t)\right] = \lambda \mathbf{f}\left[\mathbf{x}(t)\right]. \qquad (2.19)$$

If the same subsystem obeys the superposition property, then the following is true.

Let $\mathbf{x}_k(t)$, $k = 1, 2, \ldots, q$, be a set of input vectors applied separately and let $\mathbf{y}_k(t)$ be the corresponding output vectors. After this, let a single input vector, $\mathbf{x}(t) = \sum_{k=1}^{q} \mathbf{x}_k(t)$, be applied. Then if the system exhibits the superposition property, the resulting output vector is $\mathbf{y}(t) = \sum_{k=1}^{q} \mathbf{y}_k(t)$. If the system is linear, then (2.18) has the form,

$$\mathbf{y}(t) = \mathbf{f}[\mathbf{x}(t)] = \mathbf{C}\mathbf{x}(t), \qquad (2.20)$$

where $\mathbf{C}$ is a matrix of coefficients. In view of (2.20), the LHS of (2.19) is

$$\mathbf{f}[\lambda \mathbf{x}(t)] = \mathbf{C}\lambda \mathbf{x}(t) \qquad (2.21)$$

and the RHS of (2.19) is

$$\lambda \mathbf{f}[\mathbf{x}(t)] = \lambda \mathbf{C}\mathbf{x}(t). \qquad (2.22)$$

Since the RHS of (2.21) and (2.22) are equal, the system has the scaling property. If the input vectors, $\mathbf{x}_k(t)$, $k = 1, 2, \ldots, q$, are applied, one at a time, the output vectors will be $\mathbf{y}_k(t) = \mathbf{C}\mathbf{x}_k(t)$. Then let the single input vector, $\mathbf{x}(t) = \sum_{k=1}^{q} \mathbf{x}_k(t)$, be applied. This yields the output vector $y(t) = \mathbf{C}\sum_{k=1}^{q}\mathbf{x}_k(t)$. This may be written as $\sum_{k=1}^{q} \mathbf{C}\mathbf{x}_k(t) = \sum_{k=1}^{q} \mathbf{y}_k(t)$. The system therefore has the superposition property.

Considering now the SISO dynamical system modelled by (2.5), if it is an LTI system, it will be of the form

$$y^{(n)} = a_{n-1} y^{(n-1)} + \cdots + a_1 y^{(1)} + a_0 y + b_m u^{(m)}$$
$$+ b_{m-1} u^{(m-1)} + \cdots + b_1 u^{(1)} + b_0 u^{(0)}. \qquad (2.23)$$

where a_i, $i = 0, 1, \ldots, n-1$, and b_j, $j = 0, 1, \ldots, m$, are constant coefficients. Its scaling property may be stated as follows. If $y_1(t)$ is the output for given initial conditions, $y_1^{(i)}(0)$, $i = 0, 1, \ldots, n-1$ and a given input, $u_1(t)$, with finite derivatives, $u_1^{(j)}$, $j = 1, 2, \ldots, m$, then if the scaled input, $\lambda u_1(t)$, whose derivatives are $\lambda u_1^{(j)}$, $j = 1, 2, \ldots, m$, and the initial conditions are similarly scaled, i.e. $\lambda y_1^{(i)}(0)$, $i = 0, 1, \ldots, n-1$, then the output will be scaled by the same factor, i.e. it will be $\lambda y_1(t)$. This is tantamount to stating that scaling the input and its derivatives by λ scales the output and its derivatives by λ. That this is the case for system (2.23) is immediately apparent when both sides are multiplied by λ.

It will now be shown that system (2.23) has the superposition property. Let $u_k(t)$, $k = 1, 2, \ldots, q$, be a set of inputs applied separately and let $y_k(t)$ be the corresponding outputs with initial conditions, $y_k^{(i)}(0)$, $i = 0, 1, \ldots, n-1$. After this, let a single input,

2.1 Introduction

$$u(t) = \sum_{k=1}^{q} u_k(t), \qquad (2.24)$$

be applied with initial conditions, $\sum_{k=1}^{q} y_k^{(i)}(0)$, $i = 0, 1, \ldots, n-1$. Then the output is

$$y(t) = \sum_{k=1}^{n} y_k(t). \qquad (2.25)$$

To show that system (2.23) has this property, first it is observed that each of the input-output pairs, $u_k(t)$ and $y_k(t)$, $k = 1, 2, \ldots, q$, satisfies (2.23), noting that $y_k(t) = y_k^{(0)}(t)$. If all q equations are added, while grouping the corresponding terms, then the result is as follows.

$$\sum_{k=1}^{q} y_k^{(n)}(t) = a_{n-1} \sum_{k=1}^{q} y_k^{(n-1)}(t) + \cdots + a_1 \sum_{k=1}^{q} y_k^{(1)}(t) + a_0 \sum_{k=1}^{q} y_k^{(0)}(t)$$
$$+ b_m \sum_{k=1}^{q} u_k^{(m)}(t) + b_{m-1} \sum_{k=1}^{q} u_k^{(m-1)}(t) + \cdots + b_1 \sum_{k=1}^{q} u_k^{(1)}(t) + b_0 \sum_{k=1}^{q} u_k^{(0)}(t). \qquad (2.26)$$

The summation terms in (2.23) may then be expressed in terms of $u(t)$ using (2.24) and $y(t)$ using (2.25). Thus

$$y^{(n)}(t) = a_{n-1} y^{(n-1)}(t) + \cdots + a_1 y^{(1)}(t) + a_0 y^{(0)}(t)$$
$$+ b_m u^{(m)}(t) + b_{m-1} u^{(m-1)}(t) + \cdots + b_1 u^{(1)}(t) + b_0 u^{(0)}(t). \qquad (2.27)$$

Since this is the differential equation of system (2.23), then the system has the superposition property.

It can be similarly shown that the multivariable linear system (2.9) has the scaling and superposition properties if it is in the LTI form,

$$y_i^{(n_i)} = \sum_{j=1}^{m} \sum_{k=0}^{n_j - 1} a_{ijk} y_j^{(k)} + \sum_{j=1}^{p} \sum_{k=0}^{m_{j,i}} b_{ijk} u_j^{(k)}, \quad i = 1, 2, \ldots, m. \qquad (2.28)$$

where a_{ijk} and b_{ijk} are constant coefficients.

The scaling and superposition properties of LTI systems enable several different forms of plant model to be derived that are useful in control system design. These will be met in Chap. 3. For a nonlinear plant, the only alternative to (2.5) or (2.9) is the state-space model of Chap. 3. The same is true for LTV dynamical systems.

The coil of an electromagnet with resistance, R, and inductance, L, is a linear dynamical subsystem in which the current, i, is related to the applied voltage, v, by

$$Li^{(1)} + Ri = v, \qquad (2.29)$$

where $i^{(1)} = di/dt$. This is an SISO subsystem with input, v, and output, i.

An example of a nonlinear dynamical subsystem is the dynamics model of a rigid body in free fall subject to externally applied torque components. This could be part of a spacecraft model. In this case, the subsystem is a multivariable one with three inputs, the torque components, γ_x, γ_y and γ_z, along the three mutually perpendicular principal axes of inertia, x, y and z, and three outputs, the body angular velocity components, ω_x, ω_y and ω_z. Thus

$$\begin{aligned} \dot{\omega}_x &= k_x \, \omega_y \omega_z + b_x \gamma_x \\ \dot{\omega}_y &= k_y \, \omega_z \omega_x + b_y \, \gamma_y \\ \dot{\omega}_z &= k_z \, \omega_x \omega_y + b_z \, \gamma_z \end{aligned} \qquad (2.30)$$

where $k_x = (J_{yy} - J_{zz})/J_{xx}$, $k_y = (J_{zz} - J_{xx})/J_{yy}$, $k_z = (J_{yy} - J_{xx})/J_{zz}$, $b_x = 1/J_{xx}$, $b_y = 1/J_{yy}$ and $b_z = 1/J_{zz}$ where J_{xx}, J_{yy} and J_{zz} are the principal axis moments of inertia.

2.1.4 Modelling Categories and Basic Forms of Model

2.1.4.1 White-Box Modelling and Differential Equations

White-box modelling, occasionally referred to as glass-box modelling, is a general term used to describe the process of studying each internal component of a system, forming a mathematical model of each and then assembling the whole using shared inputs and outputs to form an overall mathematical model. In control engineering, white-box modelling applies to plants for which control systems are to be created. This is the physical modelling of Sect. 2.2. All plants are continuous-time dynamical systems and therefore the fundamental form of model emerging from the physical modelling is the differential equation, many examples of which will be found throughout this chapter. All other forms of plant model stem from the differential equation.

2.1.4.2 Black Box Modelling and Transfer Functions

Black box modelling is a general term used to describe the process of creating a mathematical model of a system by collecting information from observations of the responses of its outputs to given inputs, without the study of its internal components. This approach is often taken in industry as it is less time consuming and therefore more cost effective than white-box modelling but is restricted to linear models. In control engineering, various control inputs are applied to the plant and its measurement variables observed. There are three basic approaches to the processing of these variables, covered in Sect. 2.2. One is the step response method of Sect. 2.3.2, applicable to first and second-order plants and leads to Laplace transfer function models. The second is the frequency domain method of Sect. 2.3.3, which gives Laplace transfer function models and the third is the time domain method of Sect. 2.3.4, which gives z-transfer function models. The second and third methods are applicable to plants of arbitrary order.

2.2 Physical Modelling

2.2.1 Introduction

In physical plant modelling, the underlying science of each hardware component is applied to derive a mathematical component model in the form of algebraic and/or differential equations. The plant model is then the set of component models interconnected by means of common input and output variables. In the following subsections, some basic components are modelled that may be used to construct various plant models, starting with the most elementary. The equations of these basic elements may be used to build models of specific plants.

2.2.2 Mechanical Modeling Principles

2.2.2.1 Dynamic and Kinematic Subsystems

The *dynamic* subsystem is defined as the part of a mechanical system that relates translational and/or rotational velocity components to applied forces and/or torques. The *kinematic* subsystem is defined as the part of a mechanical system that relates the translational and/or rotational displacements to the translational and/or rotational velocities. The dynamic subsystem involves the inertial parameters of mass and/or moment of inertia, while the kinematic subsystem does not.

2.2.2.2 Degrees of Freedom of Motion

The number of degrees of freedom (d.o.f.) of movement of a mechanical system is the number of coordinates needed to define its position, as illustrated in Fig. 2.1. Many mechanisms such as the linear and rotational actuators of Fig. 2.1a, b, only have a single degree of freedom. The *x-y* positioning drive of Fig. 2.1c has many applications in industry. The two directions of movement are referred to as *axes* and for this reason the generic name for such mechanisms is the *multi-axis mechanism.*

Many multi-axis mechanisms have combinations of rotational and translational degrees of freedom. The axis for a rotational degree of freedom is that about which rotation of one component takes place relative to another component. The number of axes is equal to the number of degrees of freedom of movement.

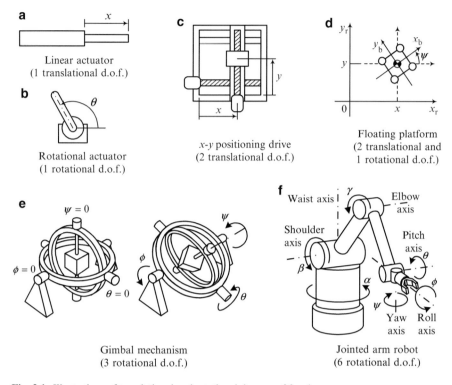

Fig. 2.1 Illustrations of translational and rotational degrees of freedom

Also, the number of degrees of freedom is equal to the number of control actuators, two worm drives being employed in the example of Fig. 2.1c.

Applications at sea such as oil rigs and wind turbines have to be positioned on platforms that are floated out to locations with translational coordinates, x and y, in a frame of reference, (x_r, y_r), and orientated about the yaw axis by an angle, ψ, as shown in Fig. 2.1d. prior to anchoring. In this case, there is no mechanism connecting the platform to the frame of reference so that movement between moving parts can be measured. Instead the two translational coordinates are measured using the global positioning system (GPS) and the rotational one by a compass.

The gimbal mechanism of Fig. 2.1e consists of frames mounted one within the other. The relative rotation axes and associated motors and angle sensors are arranged so that the object in the centre (shown as a cube) can be brought to any orientation by means of three control loops.

The universal multi-axis machine is the jointed-arm robot in which a workpiece held by the gripper is positioned in the finite three-dimensional work space with any orientation. Such robots can have various configurations. The one of Fig. 2.1f has six rotational degrees of freedom controlled using motors and joint angle sensors. Importantly, through the arm configuration, the purely rotational degrees of freedom of the joints are used to control the three *translational* and three rotational degrees of

2.2 Physical Modelling

freedom of the workpiece referred to a fixed frame of reference. Many mechanisms exist in which a given degree of freedom of motion entails both rotation and translation, an example being the crank mechanism.

Three of the rotational movements in Fig. 2.1f are named *roll*, *pitch* and *yaw*. This terminology originated in nautical applications, where the rotation of a ship about its longitudinal (x) axis is called *roll* motion, rotation about the vertical (z) axis perpendicular to the roll axis is called *yaw* motion and rotation about the horizontal (y) axis, perpendicular to the x and z axes, is called *pitch* motion. The same terminology is used for aircraft, underwater vehicles, spacecraft and automobiles.

In applications such as spacecraft and underwater vehicles it is necessary to position and orientate the vehicle body (considered to be rigid here) in the absence of a mechanism connecting the body to the frame of reference. Then it is possible to either select a *fictitious* mechanism connecting the body with the frame of reference, such as the gimbal mechanism of Fig. 2.1e, whose kinematic differential equations (KDEs) can be used as a basis for generating position and attitude coordinates, or select another set of KDEs not associated with any mechanism (Sect. 2.2.4.3).

For illustrative purposes, the gimbal mechanism-based roll, pitch and yaw attitude angles for a vehicle (represented by a cube) are shown in Fig. 2.2.

Starting with the body-fixed frame (x_b, y_b, z_b) aligned with the reference frame (x_r, y_r, z_r) as shown in Fig. 2.2a, a body rotation is made through an angle, ϕ, about the roll axis, bringing it to the orientation shown in Fig. 2.2b. Then a further body rotation is made through an angle, θ, about the newly orientated pitch axis, y_b, bringing it to the orientation shown in Fig. 2.2c. Then a final rotation of the body is made through an angle, ψ, about the newly orientated yaw axis, z_b, bringing it to the orientation shown in Fig. 2.2d. There are, in fact, twelve such attitude representations corresponding to the twelve permutations of the axis rotation orders, the one of Fig. 2.2 being r-p-y (roll–pitch–yaw), the remaining ones being r-y-p, p-r-y, p-y-r, y-r-p, y-p-r, r-p-r, r-y-r, p-r-p, p-y-p, y-r-y and y-p-y. The first six permutations have equivalent gimbal mechanisms. The last six permutations do not as they entail the first and last rotations about the same body-fixed axis. They do produce valid attitude representations, however, because this body-fixed axis has different orientations with respect to the reference frame for the first and third rotations.

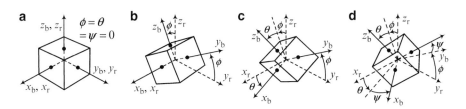

Fig. 2.2 A gimbal mechanism-based roll–pitch–yaw attitude representation for a vehicle. (**a**) Home attitude. (**b**) Roll rotation. (**c**) Pitch rotation. (**d**) Yaw rotation

The rotational dynamics and kinematics for three (x_i, y_i, z_i) degrees of freedom, which are much more complex than those for a single degree of freedom, are important as they are relevant to many applications in vehicle control and robotics.

2.2.2.3 Rigid Body with One Degree of Freedom

Figure 2.3a represents a rigid body of mass, M, constrained to move in a straight line to which is applied a control force, f_c, acting through the centre of mass (indicated by the standard symbol, ⊙) opposed by a force, f_o, due to friction, drag (i.e. air or fluid resistance) or retention spring or a combination thereof, and subject to an external disturbance force, f_d, giving rise to a velocity, v, and displacement, x, relative to an inertial frame of reference.

An inertial frame of reference refers to a set of axes that are not undergoing either translational or rotational acceleration relative to inertial space. Hence Newton's laws of motion would hold in a laboratory fixed with respect to an inertial frame. The frame of reference is shown as three dimensional, which is usual, but in this case, only the x_i axis is relevant. Figure 2.3b similarly represents a rigid body with moment of inertia, J, about an axis parallel to the y_i axis, constrained to rotate about this axis subject to an applied torque, γ_a, opposed by a torque, γ_o, giving rise to an angular velocity, ω, and angular displacement, θ. The forces, torques, linear velocity and angular velocity are vector quantities as indicated by the arrows in Fig. 2.3 but in these simple cases are co-linear.

The dynamics equations are obtained for Fig. 2.3a by equating the net force to the first derivative of the linear momentum and for Fig. 2.3b by equating the net torque to the first derivative of the angular momentum. The KDEs for simple single-degree-of-freedom mechanical components are simply statements that the velocity is the first derivative of the displacement. Assuming that the mass is constant, the model for Fig. 2.3a is as follows.

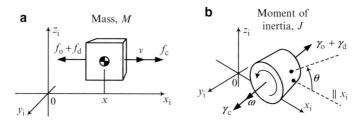

Fig. 2.3 Rigid body constrained to move with one degree of freedom. (**a**) Translational. (**b**) Rotational

2.2 Physical Modelling

$$\text{Dynamic subsystem}: \frac{d}{dt}(Mv) = f_a - f_o \Rightarrow \dot{v} = \frac{1}{M}(f_a - f_o). \tag{2.31}$$

$$\text{Kinematic subsystem}: \dot{x} = v. \tag{2.32}$$

The corresponding model for Fig. 2.3b is as follows.

$$\text{Dynamic subsystem}: \frac{d}{dt}(J\omega) = \gamma_a - \gamma_o \Rightarrow \dot{\omega} = \frac{1}{J}(\gamma_a - \gamma_o). \tag{2.33}$$

$$\text{Kinematic subsystem}: \dot{\theta} = \omega. \tag{2.34}$$

The opposing force of the translational model may be written as

$$f_o = f_g + f_f + f_s + f_l, \tag{2.35}$$

where f_g is the drag force, f_f is the friction force, f_s is the spring force and f_l is the inertial force of any other connected masses. For specific applications, some or all of these may be zero. For the rotational model, the equivalent breakdown of the opposing torque is

$$\gamma_o = \gamma_g + \gamma_f + \gamma_s + \gamma_l. \tag{2.36}$$

The use of (2.31) or (2.33) in conjunction with the material of the following three subsections is referred to, respectively, as the *force or torque balance methods*.

2.2.2.4 Drag Forces and Torques

This topic is relevant to applications such as aircraft, surface ships and underwater vehicles. With reference to Fig. 2.3a, Raleigh's equation for fluid drag [1] is

$$f_g = \frac{1}{2}\rho v^2 C_d A, \tag{2.37}$$

where ρ is the density of the fluid in which the body is immersed, A is the area of the orthographic projection onto a plane perpendicular to the direction of motion and C_d is the dimensionless drag coefficient of the fluid. To simplify any plant model of which this is part and at the same time to ensure that the drag force opposes the motion, (2.37) may be replaced by

$$f_g = K_d v^2 \text{sgn}(v) = K_d |v| v, \tag{2.38}$$

where $\mathrm{sgn}(v) \triangleq \begin{cases} +1, v > 0 \\ 0, v = 0 \\ -1, v < 0 \end{cases}$ and $K_\mathrm{d} = \frac{1}{2}\rho C_\mathrm{d} A$.

For the rotational case of Fig. 2.3, the equivalent relationship,

$$\gamma_\mathrm{g} = K_\mathrm{g}\omega^2 \mathrm{sgn}(v) = K_\mathrm{d}|\omega|\omega, \qquad (2.39)$$

may be used but in this case, the expression of K_g in terms of the parameters of Rayleigh's equation (2.37) is not straightforward as the relative velocity between a rotating body of arbitrary shape and the fluid in which it is immersed is a function of the position on its surface. In practice, this problem would be circumvented by determination of K_g experimentally.

2.2.2.5 Friction Forces and Torques

Many controlled plants embody mechanisms with relatively moving surfaces. A friction force is broadly divided into three components. Thus

$$f_\mathrm{f} = f_\mathrm{fs} + f_\mathrm{fc} + f_\mathrm{fv}, \qquad (2.40)$$

where f_fs is the static friction component, f_fc is the Coulomb friction component and f_fv is the viscous friction component.

Detailed information on the underlying physical processes of friction, including explanations at the molecular level, may be found in works on mechanical engineering [2] but the models given here should be understandable from common experience with the behaviour of relatively moving objects in contact.

Suppose that the object of Fig. 2.3a has a flat base and is resting on a flat surface. Then starting from rest, if the applied force, f_a, is gradually increased from zero, at first there will be no relative movement, but above a certain level, $f_\mathrm{a} = f_\mathrm{fs}$, the object will suddenly move. This is the static friction force given by

$$f_\mathrm{fs} = \mu f_\mathrm{n} \mathrm{sgn}(f_\mathrm{a}) = F_\mathrm{fs} \mathrm{sgn}(f_\mathrm{a}), \qquad (2.41)$$

where f_n is the normal force keeping the surfaces in contact and μ is the coefficient of static friction, determined experimentally. Sometimes, static friction is referred to as *stiction* or *stick–slip friction* due to the sticking effect for $|f_\mathrm{a}| < F_\mathrm{fs}$. For rotating objects in contact such as bearings, the identification of the normal force is less obvious but a similar phenomenon exists in which the static friction torque is

$$\gamma_\mathrm{fs} = \Gamma_\mathrm{fs} \mathrm{sgn}(\gamma_\mathrm{a}), \qquad (2.42)$$

where Γ_fs is the constant static friction torque magnitude.

2.2 Physical Modelling

In the following, v is the relative velocity between moving parts rather than the velocity of one part with respect to an inertial frame of reference.

For the translational case, once f_{fs} of (2.41) is exceeded, the physics of the friction changes as the relative movement begins and a model of the opposing force is

$$f_f = f_{fc} + f_{fv} \qquad (2.43)$$

where

$$f_{fc} = F_{fc}\text{sgn}(v), \qquad (2.44)$$

F_{fc} being constant and also $F_{fc} < F_{fs}$. This is the Coulomb friction force. Also

$$f_{fv} = K_v v. \qquad (2.45)$$

This is the viscous friction force, with coefficient, K_v. Viscous friction is sometimes referred to as kinetic friction or dynamic friction, as it is a continuous function of v, in contrast to the other components whose magnitudes are constant.

Figure 2.4a shows an example of the transfer characteristic, i.e. the graph of f_f against v that results when the effects of static, coulomb and viscous friction described above are combined.

When experiments are conducted to measure the friction transfer characteristic, however, a similar result is obtained but with a continuous transition between the static friction and the combined Coulomb and viscous friction, as shown in Fig. 2.4b. The dot–dashed straight line segments are those of Fig. 2.4a for comparison. This form of transfer characteristic is preferred in view of its being more realistic and also better behaved regarding the accuracy of the numerical integration in simulations. The author has devised the following convenient function for this.

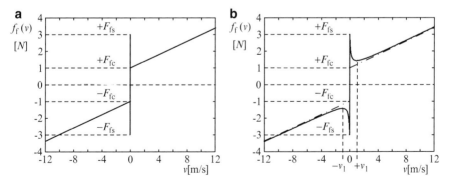

Fig. 2.4 Friction transfer characteristics. (**a**) Theoretical form. (**b**) Realistic form

$$f_\mathrm{f}(v) = K_\mathrm{v} v + \left(\frac{F_\mathrm{fs} + F_\mathrm{fc} K |v|}{1 + K |v|}\right) \mathrm{sgn}(v). \tag{2.46}$$

Here, K is a parameter that may be adjusted to obtain the required sharpness of transition, this being increased as K is increased. If v_1 is specified, then since the transfer characteristic is reflected in the origin, the required value of K is obtained by setting $\frac{\mathrm{d}}{\mathrm{d}v} f_\mathrm{f}(v) = 0$ for $v = v_1 > 0$ for which (2.46) becomes

$$f_\mathrm{f}(v) = K_\mathrm{v} v + \frac{F_\mathrm{fs} + F_\mathrm{fc} K v}{1 + K v} \tag{2.47}$$

Then

$$\frac{\mathrm{d}}{\mathrm{d}v}\left[K_\mathrm{v} v + \frac{F_\mathrm{fs} + F_\mathrm{fc} K v}{1 + K v}\right] = K_\mathrm{v} + \frac{(1 + K v) F_\mathrm{fc} K - (F_\mathrm{fs} + F_\mathrm{fc} K v) K}{(1 + K v)^2}$$

$$= \frac{\left(1 + 2 K v + K^2 v^2\right) K_\mathrm{v} - (F_\mathrm{fs} - F_\mathrm{fc}) K}{(1 + K v)^2} \Rightarrow$$

$$K_\mathrm{v} v_1^2 K^2 + (2 K_\mathrm{v} v_1 - F_\mathrm{fs} + F_\mathrm{fc}) K + K_\mathrm{v} = 0 \Rightarrow$$

$$K = \frac{F_\mathrm{fs} - F_\mathrm{fc} - 2 K_\mathrm{v} v_1 \pm \sqrt{(2 K_\mathrm{v} v_1 - F_\mathrm{fs} + F_\mathrm{fc})^2 - 4 K_\mathrm{v}^2 v_1^2}}{2 K_\mathrm{v} v_1^2}. \tag{2.48}$$

Since, $\sqrt{(2 K_\mathrm{v} v_1 - F_\mathrm{fs} + F_\mathrm{fc})^2 - 4 K_\mathrm{v}^2 v_1^2} < |F_\mathrm{fs} - F_\mathrm{fc} - 2 K_\mathrm{v} v_1|$, a necessary condition for $K > 0$ is

$$F_\mathrm{fs} - F_\mathrm{fc} - 2 K_\mathrm{v} v_1 > 0. \tag{2.49}$$

Then both roots are positive and therefore valid but the largest one will be chosen as this yields the sharpest transition between the static friction and the combined Coulomb and viscous friction. Hence

$$K = \frac{F_\mathrm{fs} - F_\mathrm{fc} - 2 K_\mathrm{v} v_1 + \sqrt{(2 K_\mathrm{v} v_1 - F_\mathrm{fs} + F_\mathrm{fc})^2 - 4 K_\mathrm{v}^2 v_1^2}}{2 K_\mathrm{v} v_1^2}. \tag{2.50}$$

In fact, the transfer characteristic of Fig. 2.1b was produced using (2.46) and (2.50) with $F_\mathrm{fs} = 6$ [N], $F_\mathrm{fc} = 1$ [N], $K_\mathrm{v} = 0.4$ [N/ (m/s)] and $v_1 = 2$ [m/s]. For the rotational case, F_fs, F_fc, K_v and v_1 are replaced, respectively, by Γ_fs, Γ_fc, K_ω and ω_1.

Static friction can cause steady-state errors in traditional control loops without an integral term and limit cycling with integral terms. The reader should be aware, however, that in most real applications the parameters of the friction model are highly dependent upon environmental conditions, particularly temperature, and therefore a model-based controller with inbuilt friction compensation would be

2.2 Physical Modelling

difficult to implement. The recommended approach is to use the model presented in this section with typical parameters for the application in hand in a simulation to predict its effect on a traditional control loop and change over to a robust control technique (Chaps. 9 and 10) if unsatisfactory performance is predicted.

2.2.2.6 Spring Force and Torque

In some mechanical systems, a rigid body such as illustrated in Fig. 2.3 is either physically connected to one or more components via springs or is part of a so-called lumped parameter model of a flexible structure consisting of rigid bodies interconnected by linear springs or torsion springs. In either case, with reference to (2.35) and (2.36),

$$f_s = \sum_{i=1}^{m} f_{si} \quad \text{or} \quad \gamma_s = \sum_{i=1}^{m} \gamma_{si} \qquad (2.51)$$

where f_{si} or γ_{si}, $i = 1, 2, \ldots, m$, is the individual spring forces or torques. Figure 2.5 shows the translational and rotational single-degree-of-freedom cases in which all the displacements and velocities shown are with respect to the inertial frame of reference. The object could be another movable rigid body or a fixed point with respect to the inertial frame of reference.

In Fig. 2.5a, x_i is the displacement of the point on the ith object to which one end of the spring is attached and x_{0i} is the displacement of the same end of the spring if detached from the object and at rest with zero spring force. In these models, the springs are regarded as having zero mass. Hence, $x(t) - x_{0i}(t) = \text{const}$. Since, however, the physical structure does not have to be preserved in a mathematical model used only for control system design, the model is simplified by setting $x_{0i} = x$. Similar statements may be made for Fig. 2.5b, so $\theta_{0i} = \theta$. Assuming Hooke's law holds, the spring forces and torques are respectively,

$$f_{si} = K_{si}(x_i - x) \quad \text{and} \quad \gamma_{si} = K'_{si}(\theta_i - \theta), \quad i = 1, 2, \ldots, n. \qquad (2.52)$$

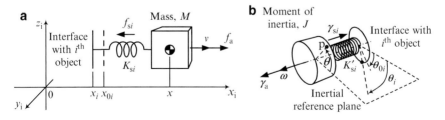

Fig. 2.5 Interaction of a mass with other objects via springs. (**a**) Translational. (**b**) Rotational

2.2.2.7 Inertial Force and Torque

The component of the opposing force or torque due to an *additional* rigid body fixed to either rigid body of Fig. 2.3 is given by

$$f_1 = M_1 \dot{v} \quad \text{or} \quad \gamma_1 = J_1 \dot{\omega} \tag{2.53}$$

where M_1 and J_1 are, respectively, the mass and moment of inertia of the additional rigid body. An example is an inertial mechanical load bolted to an electric motor.

2.2.2.8 Lagrangian Mechanics

Controlled mechanisms sometimes have several degrees of freedom of motion and, for modelling purposes, can be considered as a set of interconnected rigid bodies, possibly including springs, to which are applied actuator forces and torques. An example is the jointed-arm robotic manipulator introduced in Chap. 12. Derivation of the differential equations of motion of such systems using the force and torque balance methods introduced in Sect. 2.2.2.3 can be very laborious and time consuming, particularly if the number of chosen translational and rotational position coordinates exceeds the number of degrees of freedom, in which case the mechanical constraints have to be carefully incorporated. This task may be greatly eased, however, by applying Lagrangian mechanics [3]. First, the Lagrangian is defined as

$$L = T - V \tag{2.54}$$

where T is the total kinetic energy and V is the total potential energy. Then, if the ith mechanical displacement (either translational or rotational) corresponding to the ith degree of freedom is denoted by q_i, the equations of motion are given by:

$$\frac{d}{dt}\left(\frac{\partial L}{\partial \dot{q}_i}\right) - \frac{\partial L}{\partial q_i} + f(\dot{q}_i) = \tau_i, \quad i = 1, 2, \ldots, d, \tag{2.55}$$

where τ_i is the external force or torque and $f(\dot{q}_i)$ is the friction force or torque associated with the ith degree of freedom.

As an example, the equations of motion will be derived for the single-degree-of-freedom translational and rotational examples of Fig. 2.5. First, consider Fig. 2.5a. Let $x = q_1$. Then $\dot{q}_1 = v$. Also $\tau_1 = f_a$. The kinetic energy is then

$$T = \frac{1}{2} M v^2 = \frac{1}{2} M \dot{q}_1^2 \tag{2.56}$$

2.2 Physical Modelling

and the potential energy is

$$V = \frac{1}{2}K_s[x - (x_i - x_{0i})]^2 = \frac{1}{2}K_s[q_1 - (x_i - x_{0i})]^2. \quad (2.57)$$

The Lagrangian is therefore

$$L = T - V = \frac{1}{2}\left\{M\dot{q}_1^2 - K_s[q_1 - (x_i - x_{0i})]^2\right\}. \quad (2.58)$$

To derive the equation of motion from (2.55), in this example, there is no friction, so

$$\frac{\partial L}{\partial \dot{q}_1} = M\dot{q}_1 \Rightarrow \frac{d}{dt}\left(\frac{\partial L}{\partial \dot{q}_1}\right) = M\ddot{q}_1 \quad \text{and} \quad \frac{\partial L}{\partial q_1} = -K_s[q_1 - (x_i - x_{0i})]. \quad (2.59)$$

Hence (2.55) yields

$$M\ddot{q}_1 + K_s[q_1 - (x_i - x_{0i})] = \tau_1 \quad (2.60)$$

Noting that $\dot{x} = v$, (2.60) may be written

$$M\ddot{x} + K_s[x - (x_i - x_{0i})] = f_a. \quad (2.61)$$

The rotational example of Fig. 2.5b is similar. Letting $\theta = q_1$, $\tau_1 = \gamma_a$ and noting $\dot{\theta} = \omega$, the kinetic and potential energies are $T = \frac{1}{2}J\omega^2 = \frac{1}{2}J\dot{q}_1^2$ and $V = \frac{1}{2}K_s[\theta - (\theta_i - \theta_{0i})]^2 = \frac{1}{2}K_s[q_1 - (\theta_i - \theta_{0i})]^2$. Applying the Lagrangian method then yields

$$J\ddot{\theta} + K_s[\theta - (\theta_i - \theta_{0i})] = \gamma_a. \quad (2.62)$$

Another appropriate example is the cart and inverted pendulum mechanism sometimes used to demonstrate control techniques, shown in Fig. 2.6.

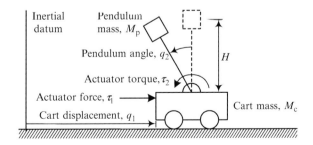

Fig. 2.6 Cart and pole mechanism

In this case,

$$L = T - V = \frac{1}{2}\left\{M_c \dot{q}_1^2 + M_p\left[(\dot{q}_1 - \dot{q}_2 H \cos(q_2))^2 + (\dot{q}_2 H \sin(q_2))^2\right]\right\} - M_p g H \cos(q_2). \tag{2.63}$$

Again, the friction will be assumed negligible. Then applying (2.55) to obtain the equations of motion yields the following.

$$\frac{d}{dt}\left(\frac{\partial L}{\partial \dot{q}_1}\right) - \frac{\partial L}{\partial q_1} = \tau_1 \Rightarrow \frac{d}{dt}\left[M_c \dot{q}_1 + M_p(\dot{q}_1 - \dot{q}_2 H \cos(q_2))\right] - 0 = \tau_1 \Rightarrow$$
$$(M_c + M_p)\ddot{q}_1 - M_p H \ddot{q}_2 \cos(q_2) + M_p H \dot{q}_2^2 \sin(q_2) = \tau_1 \tag{2.64}$$

and

$$\frac{d}{dt}\left(\frac{\partial L}{\partial \dot{q}_2}\right) - \frac{\partial L}{\partial q_2} = \tau_2 \Rightarrow$$
$$\frac{d}{dt}\left\{-M_p\left[(\dot{q}_1 - \dot{q}_2 H \cos(q_2)) H \cos(q_2) - \dot{q}_2 H^2 \sin^2(q_2)\right]\right\}$$
$$+ M_p g H \sin(q_2) = \tau_2 \Rightarrow \tag{2.65}$$
$$\frac{d}{dt}\left\{M_p\left[\dot{q}_2 H^2 - \dot{q}_1 H \cos(q_2)\right]\right\} + M_p g H \sin(q_2) = \tau_2 \Rightarrow$$
$$M_p\left[H^2 \ddot{q}_2 + H \dot{q}_1 \dot{q}_2 \sin(q_2) - H \ddot{q}_1 \cos(q_2) + g H \sin(q_2)\right] = \tau_2$$

Many of the terms in these equations would be difficult to deduce by inspection of Fig. 2.6 and application of the force and torque balance method.

2.2.3 Two Basic Mechanical Components

2.2.3.1 Gear Trains and Referred Mechanical Parameters

Many controlled mechanisms such as robot joint actuators employ gear trains to match the motor output to the mechanical load regarding the torque and speed requirements. A gear train is a set of several toothed wheels, i.e. gear wheels, that mesh with one another. For incorporation in a plant model for control system design, it is sufficient to represent a gear train comprising two or more gear wheels using an equivalent train of just two wheels as shown in Fig. 2.7.

Here, R_1 and R_2 are, respectively, the radii of the input and output wheels, γ_1 and γ_2 are the input and output torques, ω_1 and ω_2 are the angular velocities of the input and output shafts, while θ_1 and θ_2 are the corresponding angles of rotation. Starting with points, p_1 and p_2, on the wheel peripheries that are coincident with the point of

2.2 Physical Modelling

Fig. 2.7 Basic two-wheel representation of a gear system

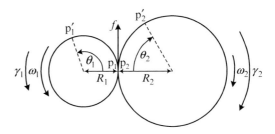

contact, as the wheels rotate, these points move to new positions, p_1' and p_2', on arcs through the same distance, d, requiring

$$d = R_1 \theta_1 = R_2 \theta_2. \tag{2.66}$$

Differentiating (2.66) and letting $\dot{\theta}_1 = \omega_1$ and $\dot{\theta}_2 = \omega_2$ then yields

$$\omega_1 R_1 = \omega_2 R_2 \tag{2.67}$$

The *gear ratio* is defined as

$$G = \frac{\omega_1}{\omega_2} = \frac{R_2}{R_1}. \tag{2.68}$$

This is a basic parameter in modelling a mechanical system containing a gear train.

The tangential force, f, at the wheel interface produced by the applied torque, γ_1, satisfies $\gamma_1 = f R_1$. This also gives rise to an output torque of $\gamma_2 = f R_2$. Hence

$$f = \frac{\gamma_1}{R_1} = \frac{\gamma_2}{R_2} \Rightarrow \gamma_2 = \frac{R_2}{R_1} \gamma_1.$$

i.e.

$$\gamma_2 = G \gamma_1. \tag{2.69}$$

So if $R > 1$, the gear train achieves torque amplification. The mechanical input power, $\gamma_1 \omega_1$, may be expressed in terms of γ_2 and ω_2 using (2.69) and (2.68) as follows.

$$\gamma_1 \omega_1 = \frac{1}{G} \gamma_2 G \omega_2 = \gamma_2 \omega_2. \tag{2.70}$$

This means that the gear system transmits mechanical power with zero loss, indicating that the model is of an ideal gear system. Power losses in a real system, however, may be modelled using viscous friction parameters, as shown below.

Fig. 2.8 Mechanical system with viscous friction and inertia containing a gear system

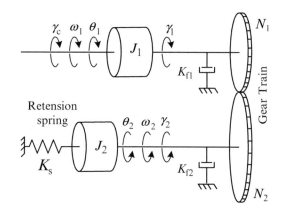

It may be observed that the gear system model described above is similar to that of an ideal electrical transformer, in that ω_1 and ω_2 are analogous to the primary and secondary currents, γ_1 and γ_2 are analogous to the primary and secondary voltages, and R_1 and R_2 are analogous to the numbers of primary and secondary turns. Hence, a gear system may be regarded as a mechanical transformer. Indeed, the viscous friction and inertial components of a mechanism of which the gear system is a part may be referred to either the input shaft side or the output shaft side to simplify the model, in a similar way to referring the inductive and resistive components of an electrical circuit to either the primary side or the secondary side of a transformer. Consider the mechanical system shown in Fig. 2.8.

This is a single-degree-of-freedom mechanism consisting of two balanced masses with moments of inertia, J_1 and J_2, connected by a gear train. A control torque, γ_1, is applied to mass 1 (the torque actuator not being included in this example) and the movement of the system is restrained by a torsion spring attached to mass 2. Also N_1 and N_2 are the numbers of gear teeth. Then the gear ratio is

$$G = N_2/N_1. \tag{2.71}$$

The torque balance equations for sides 1 and 2 of the system are

$$J_1\ddot{\theta}_1 + K_{f1}\dot{\theta}_1 = \gamma_c - \gamma_1 \tag{2.72}$$

$$J_2\ddot{\theta}_2 + K_{f2}\dot{\theta}_2 + K_s\theta_2 = \gamma_2 \tag{2.73}$$

It follows from (2.66) and (2.68) that

$$\theta_1 = G\theta_2 \tag{2.74}$$

2.2 Physical Modelling

and this, together with (2.69). completes the model by connecting (2.72) and (2.73). It also enables the model to be simplified by referring all the quantities to side 1 or side 2 of the gear train.

For side 1, it is necessary to substitute for θ_2 and γ_2 in (2.73) using, respectively, (2.74) and (2.69). Thus

$$J_2 \frac{1}{R} \ddot{\theta}_1 + K_{f2} \frac{1}{R} \dot{\theta}_1 + K_s \frac{1}{R} \theta_1 = G\gamma_1. \tag{2.75}$$

Then substituting for γ_1 in (2.72) using (2.75) yields

$$J_1 \ddot{\theta}_1 + K_{f1} \dot{\theta}_1 = \gamma_c - \left(J_2 \frac{1}{G^2} \ddot{\theta}_1 + K_{f2} \frac{1}{G^2} \dot{\theta}_1 + K_s \frac{1}{G^2} \theta_1 \right) \tag{2.76}$$

which may be written

$$J_{r1} \ddot{\theta}_1 + K_{fr1} \dot{\theta}_1 + K_{sr1} \theta_1 = \gamma_c, \tag{2.77}$$

where

$$J_{r1} = J_1 + \frac{1}{G^2} J_2, \ K_{fr1} = K_{f1} + \frac{1}{G^2} K_{f2} \text{ and } K_{sr1} = \frac{1}{G^2} K_s. \tag{2.78}$$

Then (2.77) is the simplified model, which is equivalent to a single mass moving against viscous friction and a torsion spring without a gear train, whose parameters, J_{r1}, K_{fr1} and K_{s1}, are, respectively, the moment of inertia, the viscous friction coefficient and the spring constant referred to side 1 of the gear train.

For the alternative of referring all the parameters to side 2 of the gear train, it is possible to start with (2.76) and substitute for θ_1 using (2.74), which gives

$$J_1 G \ddot{\theta}_2 + K_{f1} G \dot{\theta}_2 = \gamma_c - \left(J_2 \frac{1}{G} \ddot{\theta}_2 + K_{f2} \frac{1}{G} \dot{\theta}_2 + K_s \frac{1}{G} \theta_2 \right) \tag{2.79}$$

which may be written

$$J_{r2} \ddot{\theta}_2 + K_{fr2} \dot{\theta}_2 + K_{sr2} \theta_2 = G\gamma_c, \tag{2.80}$$

where

$$J_{r2} = J_2 + G^2 J_1, \ K_{fr2} = K_{f2} + G^2 K_{f1} \text{ and } K_{sr2} = K_s. \tag{2.81}$$

Equations (2.78) and (2.81) are similar to those referring reactive and resistive components to the primary or secondary circuits of an electrical transformer.

2.2.3.2 Hard Stops

Controlled mechanisms often have mechanical hard stops, limiting the range of movement to lie between maximum and minimum values. It is necessary to model these hard stops for simulation purposes if there is any likelihood of them being reached in the application under study. It will be assumed that the part of the mechanism constrained by the hard stops may be modelled as a rigid body when not in contact with these stops and has one degree of freedom of movement, which may be rotational or translational. Let the differential equation of motion of this body be

$$\ddot{x} = b(u_b + u_h) \qquad (2.82)$$

where x is either the translational or rotational displacement, $b = 1/M$ or $b = 1/J$, where M is the body mass, J is the body moment of inertia about the axis of rotation, u_b is the total torque or force applied to the body without contacting either hard stop and u_h is the additional force or torque acting on the body as a result of contact with either of the two hard stops. Upon contacting a hard stop, small elastic deformations will occur in the body and the hard stop material. This will be modelled as a very stiff spring. Also a small proportion of the kinetic energy of the body will be lost as heat dissipated in the body due to the nature of its material or energy transfer to the structure upon which the hard stops are mounted. This energy dissipation, if considered significant, will be modelled as viscous damping. Thus

$$e_h = \begin{cases} x_{max} - x, & x > x_{max} \\ 0, & x_{min} \leq x \leq x_{max} \\ x_{min} - x, & x < x_{min} \end{cases}, \quad u_h = K_{sh} e_h + K_{vh} \dot{e}_h, \qquad (2.83)$$

where K_{sh} and K_{vh} are, respectively, the spring constant and viscous damping coefficient representing the elastic deformation and the energy loss during the stop contact. Figure 2.9 shows a block diagram of the model based on (2.82) and (2.83).

The 'max' and 'min' functions shown are as in SIMULINK® and are defined as

$$\max(p, q) \triangleq \begin{cases} p, & p \geq q \\ q, & p < q \end{cases} \quad \text{and} \quad \min(p, q) \triangleq \begin{cases} p, & p \leq q \\ q, & p > q \end{cases}. \qquad (2.84)$$

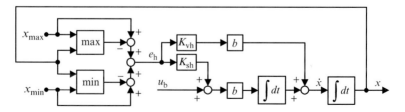

Fig. 2.9 Hard stop model for simulation

2.2 Physical Modelling

Also, the injection of the signal, $bK_{vh}e_h$, between the two integrators realises the term, $K_{vh}\dot{e}_h$, in (2.83), thereby avoiding the implementation of the differentiation.

2.2.4 Modelling for Vehicle Attitude and Position Control

2.2.4.1 Three-Axis Rotational Dynamics

An unconstrained rigid body, usually representing a vehicle such as a spacecraft or underwater vehicle has three rotational degrees of freedom as discussed in Sect. 2.2.2.2. It will be assumed that attitude control actuators are mounted on the body that produce torque components, γ_{bx}, γ_{by} and γ_{bz}, about the body-fixed axes, x_b, y_b and z_b. The dynamic subsystem is obtained by equating the net applied torque to the first derivative of the body angular momentum vector. This is far more involved than the single-degree-of-freedom case due to the derivative of a vector having two parts, the rate of change of magnitude and the rate of change of direction. For the single-degree-of-freedom case, only the rate of change of magnitude of the angular momentum vector is needed, since it does not change direction, yielding (2.33). For an unconstrained rigid body, however, the motion in all three rotational degrees of freedom occurs and the theory of dynamics is needed [4]. Let the instantaneous angular velocity vector be

$$\boldsymbol{\omega} = \omega_x \hat{i} + \omega_y \hat{j} + \omega_z \hat{k}, \tag{2.85}$$

where $\hat{i}, \hat{j}$ and $\hat{k}$ are the unit vectors directed along the mutually orthogonal body-fixed axes, x_b, y_b and z_b and ω_x, ω_y and ω_z are the angular velocity components along these axes. Let the body angular momentum be similarly represented as

$$\mathbf{L}_b = l_{bx}\hat{i} + l_{by}\hat{j} + l_{bz}\hat{k}, \tag{2.86}$$

Using the same notation, the net torque vector acting on the body is

$$\boldsymbol{\gamma} = \boldsymbol{\gamma}_a - \boldsymbol{\gamma}_o, \tag{2.87}$$

where

$$\boldsymbol{\gamma}_a = \gamma_{ax}\hat{i} + \gamma_{ay}\hat{j} + \gamma_{az}\hat{k} \tag{2.88}$$

is the actuator torque vector and

$$\boldsymbol{\gamma}_o = \gamma_{ox}\hat{i} + \gamma_{oy}\hat{j} + \gamma_{oz}\hat{k} \tag{2.89}$$

is the opposing torque vector that could be due, for example, to hydrodynamic drag in an underwater vehicle or solar radiation pressure in the case of a space satellite. The dynamics equation can then be written as

$$\dot{\mathbf{L}}_b + \boldsymbol{\omega} \wedge \mathbf{L}_b = \boldsymbol{\gamma}_a - \boldsymbol{\gamma}_0 \qquad (2.90)$$

where

$$\dot{\mathbf{L}}_b \triangleq \dot{l}_{bx}\widehat{i} + \dot{l}_{by}\widehat{j} + \dot{l}_{bz}\widehat{k} \qquad (2.91)$$

is the rate of change of the magnitude of the angular momentum vector and the second term on the LHS of (2.90) is the rate of change of direction of the angular momentum vector. This is called the gyroscopic torque component since it is responsible for the behaviour of a gyroscope. A familiar example of gyroscopic torques at work is the prevention of a bicycle in motion falling over due to the wheel angular momentum vectors.

It is important to note that (2.90) is valid only for actuators such as thrusters that do not have angular momentums affecting the motion of the body. In spacecraft, however, reaction wheels or control moment gyros are commonly employed. These actuators accrue their own angular momentums through their principle of operation.

A reaction wheel consists of an electric motor with the stator bolted to the spacecraft body, directly driving a balanced flywheel. When the motor develops torque, the wheel undergoes an angular acceleration and its angular momentum magnitude therefore changes. The equal and opposite reaction torque acts on the spacecraft body via the stator, controlling the attitude as required. This also changes the angular momentum of the spacecraft body by an equal and opposite amount to the change in the wheel angular momentum. This is due to the principle of conservation of angular momentum, which states that the total angular momentum of a mechanical system is constant if no external torque acts on it. This is also called a conservative system because the angular momentum is conserved. Since a set of reaction wheels on a spacecraft effects momentum exchange between the spacecraft body and the wheels, these actuators are called momentum exchange actuators. At least three reaction wheels are needed to control the three rotational degrees of freedom, usually a set of four with their spin axes arranged so that any combination of three wheels can be used to maintain the mission with one wheel failure.

The control moment gyro (CMG) consists of a flywheel running at constant speed mounted in a single or a two-axis gimbal system, each gimbal axis equipped with an electromagnetic transducer that provides an attitude control torque component. Various configurations of control moment gyros can be employed to achieve controllability of the three rotational degrees of freedom. In this case, the wheel angular momentum magnitudes are constant but attitude control torques from the electromagnetic transducers produce equal and opposite torques acting at right angles to the wheel spin axes which change the direction of the angular momentum

2.2 Physical Modelling

vector of each CMG. Again, during attitude control manoeuvres, angular momentum is exchanged between the spacecraft body and the set of CMGs, the total angular momentum remaining constant if no external torques are acting.

If external torques do act on a spacecraft equipped with momentum exchange actuators, then the fundamental equation of rotational dynamics states that this is equal to the rate of change of the total angular momentum. If the attitude control maintains the spacecraft body stationary with respect to inertial space, then the actuators absorb the angular momentum until they reach a saturation condition. Either one of the reaction wheels reaches a maximum speed limit or one of the CMG gimbals reaches an angular limit. Then the stored angular momentum has to be removed by transferral to molecules of exhaust emission of a set of thrusters.

A model for control system simulation and design can be formed that is independent of the type of momentum exchange actuator and the configuration of the actuator set. This is done by expressing the actuator angular momentum vector as

$$\mathbf{L}_a = l_{ax}\widehat{i} + l_{ay}\widehat{j} + l_{az}\widehat{k} \tag{2.92}$$

and then replacing the dynamic subsystem (2.90) with the following.

$$\dot{\mathbf{L}}_b + \boldsymbol{\omega} \wedge (\mathbf{L}_b + \mathbf{L}_a) = \boldsymbol{\gamma}_a - \boldsymbol{\gamma}_0, \quad \dot{\mathbf{L}}_a = -\boldsymbol{\gamma}_a, \tag{2.93}$$

Next, to render the dynamic subsystems (2.90) and (2.93) useful for attitude control system design, they should be reformulated in the matrix–vector form with the vectors represented as 3×1 column vectors. First, however, the cross products will be expanded, noting that

$$\widetilde{i} \wedge \widetilde{i} = \widetilde{j} \wedge \widetilde{j} = \widetilde{k} \wedge \widetilde{k} = 0; \; \widetilde{i} \wedge \widetilde{j} = \widetilde{k}; \; \widetilde{j} \wedge \widetilde{i} = -\widetilde{k};$$

$$\widetilde{j} \wedge \widetilde{k} = \widetilde{i}; \; \widetilde{k} \wedge \widetilde{j} = -\widetilde{i}; \; \widetilde{k} \wedge \widetilde{i} = \widetilde{j}; \; \widetilde{i} \wedge \widetilde{k} = -\widetilde{j}$$

Then in (2.90)

$$\boldsymbol{\omega} \wedge \mathbf{L}_b = \left(\omega_x\widehat{i} + \omega_y\widehat{j} + \omega_z\widehat{k}\right) \wedge \left(l_{bx}\widehat{i} + l_{by}\widehat{j} + l_{bz}\widehat{k}\right)$$
$$= \left(\omega_y l_{bz} - \omega_z l_{by}\right)\widehat{i} + \left(\omega_z l_{bx} - \omega_x l_{bz}\right)\widehat{j} + \left(\omega_x l_{by} - \omega_y l_{bx}\right)\widehat{k} \tag{2.94}$$

Then the matrix–vector form of (2.90) follows.

$$\begin{bmatrix} \dot{l}_{bx} \\ \dot{l}_{by} \\ \dot{l}_{bz} \end{bmatrix} + \begin{bmatrix} 0 & -\omega_z & \omega_y \\ \omega_z & 0 & -\omega_x \\ -\omega_y & \omega_x & 0 \end{bmatrix} \begin{bmatrix} l_{bx} \\ l_{by} \\ l_{bz} \end{bmatrix} = \begin{bmatrix} \gamma_{ax} \\ \gamma_{ay} \\ \gamma_{az} \end{bmatrix} - \begin{bmatrix} \gamma_{ox} \\ \gamma_{oy} \\ \gamma_{oz} \end{bmatrix}, \tag{2.95}$$

which, in compact form, may be written

$$\dot{\mathbf{L}}_b + \boldsymbol{\Omega}_3 \mathbf{L}_b = \boldsymbol{\gamma}_a - \boldsymbol{\gamma}_o \tag{2.96}$$

Similarly, (2.93) becomes

$$\dot{\mathbf{L}}_b + \mathbf{\Omega}_3 [\mathbf{L}_b + \mathbf{L}_a] = \boldsymbol{\gamma}_a - \boldsymbol{\gamma}_o, \quad \dot{\mathbf{L}}_a = -\boldsymbol{\gamma}_a \qquad (2.97)$$

To obtain the dynamic subsystem (2.97) in a form equivalent to (2.33), the moment of inertia matrix, often called the moment of inertia tensor [4], is needed. This is

$$\mathbf{J} = \begin{bmatrix} J_{xx} & J_{xy} & J_{xz} \\ J_{yx} & J_{yy} & J_{yz} \\ J_{zx} & J_{zy} & J_{zz} \end{bmatrix} = \lim_{\substack{n \to \infty \\ \delta m_i \to 0}} \sum_{i=1}^{n} \delta m_i \begin{bmatrix} y_i^2 + z_i^2 & x_i y_i & x_i z_i \\ y_i x_i & x_i^2 + z_i^2 & y_i z_i \\ z_i x_i & z_i y_i & x_i^2 + y_i^2 \end{bmatrix}. \qquad (2.98)$$

The diagonal terms are the moments of inertia of the body about the axes, x_b, y_b and z_b, while the off-diagonal terms are the products of inertia. The rigid body is divided into a large number of elements of mass, m_i with coordinates, x_i, y_i and z_i. Then the number of elements is allowed to become infinitely large, each of infinitesimal mass. The angular momentum vector is the product of the moment of inertia matrix and the angular velocity vector. Thus

$$\mathbf{L}_b = \mathbf{J}\boldsymbol{\omega}. \qquad (2.99)$$

Then (2.95), applicable when using thruster-based actuators, becomes

$$\mathbf{J}\dot{\boldsymbol{\omega}} + \mathbf{\Omega}_3 \mathbf{J}\boldsymbol{\omega} = \boldsymbol{\gamma}_a - \boldsymbol{\gamma}_o \qquad (2.100)$$

and (2.97), applicable when using momentum exchange actuators, becomes

$$\mathbf{J}\dot{\boldsymbol{\omega}} + \mathbf{\Omega}_3 [\mathbf{J}\boldsymbol{\omega} + \mathbf{L}_a] = \boldsymbol{\gamma}_a - \boldsymbol{\gamma}_o, \quad \dot{\mathbf{L}}_a = -\boldsymbol{\gamma}_a. \qquad (2.101)$$

2.2.4.2 Basic Three-Axis Rotational Kinematics

As already pointed out in Sect. 2.2.2.2, there are several sets of attitude coordinates that can be chosen for a rigid body. These are attitude representations [5]. For each attitude representation, there is a set of kinematic differential equations [KDEs] relating the derivatives of the attitude coordinates to themselves and the body angular velocity components, ω_x, ω_y and ω_z, defined *i*th in subsection 2.2.4.1. One of the twelve different attitude representations introduced in Sect. 2.2.2.2 based on three successive rotations about the body-fixed axes could be chosen, each with a different set of KDEs. These, however, have a common drawback. Consider, for example, the gimbal mechanism of Fig. 2.1e for orientations of the central cube requiring the two gimbal frames to be coplanar. Then the mechanism loses one degree of freedom of motion. This condition is referred to as gimbal lock. As will be seen, this manifests as a singularity in the corresponding set of KDEs. Fortunately,

2.2 Physical Modelling

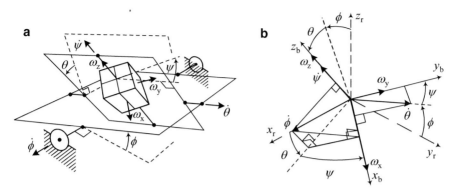

Fig. 2.10 Diagrams for derivation of kinematic differential equations for Fig. 2.2. (**a**) Equivalent gimbal system. (**b**) Angular velocity vector diagram

in addition to the 12 attitude representations already mentioned, there exist others whose sets of rotational KDEs do not exhibit singularities. One of these, based on the quaternion, is presented in the following subsection as it is suited to control system design.

Before introducing the quaternion-based attitude representation, the KDEs for the attitude representation of Fig. 2.2 will be derived to demonstrate the gimbal lock condition and the singularity. Figure 2.10a shows the fictitious gimbal system for this attitude representation that enables it to be visualised.

Figure 2.10b shows the geometry of the body angular velocity vectors and the gimbal joint angular velocity vectors. The gimbal joint angles are the attitude coordinates. The equations for ω_x, ω_y and ω_z, produced by given gimbal angular velocities, $\dot{\phi}$, $\dot{\theta}$ and $\dot{\psi}$, are obtained by studying Fig. 2.10b and are as follows.

$$\omega_x = \dot{\phi} \cos(\theta) \cos(\psi) + \dot{\theta} \sin(\psi) \tag{2.102}$$

$$\omega_y = \dot{\theta} \cos(\psi) - \dot{\phi} \cos(\theta) \sin(\psi) \tag{2.103}$$

$$\omega_z = \dot{\psi} + \dot{\phi} \sin(\theta) \tag{2.104}$$

The required KDEs are then obtained by solving (2.102), (2.103) and (2.104) for $\dot{\phi}$, $\dot{\theta}$ and $\dot{\psi}$. Hence (2.102) × cos(ψ) − (2.103) × sin(ψ) yields

$$\dot{\phi} \cos(\theta) \cos^2(\psi) + \dot{\theta} \sin(\psi) \cos(\psi) - \dot{\theta} \cos(\psi) \sin(\psi) + \dot{\phi} \cos(\theta) \sin^2(\psi)$$
$$= \omega_x \cos(\psi) - \omega_y \sin(\psi) \Rightarrow \dot{\phi} \cos(\theta) = \omega_x \cos(\psi) - \omega_y \sin(\psi) \Rightarrow$$
$$\dot{\phi} = \tfrac{1}{\cos(\theta)} \left[\omega_x \cos(\psi) - \omega_y \sin(\psi) \right] \tag{2.105}$$

Similarly, $(2.102) \times \sin(\psi) + (2.103) \times \cos(\psi)$ yields

$$\dot{\phi}\cos(\theta)\cos(\psi)\sin(\psi) + \dot{\theta}\sin^2(\psi) + \dot{\theta}\cos^2(\psi) - \dot{\phi}\cos(\theta)\sin(\psi)\cos(\psi)$$
$$= \omega_x \sin(\psi) + \omega_y \cos(\psi) \Rightarrow \quad (2.106)$$
$$\dot{\theta} = \omega_x \sin(\psi) + \omega_y \cos(\psi)$$

Finally, substituting for in (2.104) using (2.105) and making $\dot{\psi}$ the subject of the resulting equation yields

$$\dot{\psi} = \omega_z - \left[\omega_x \cos(\psi) - \omega_y \sin(\psi)\right]\tan(\theta). \quad (2.107)$$

The set of three KDEs are (2.105), (2.106) and (2.107), and they constitute a *possible* kinematic subsystem for the rigid-body rotational model. The singularity is evident on the RHS of (2.105) and (2.107), as $\dot{\phi} \to \infty$ and $\dot{\psi} \to \infty$ as $\theta \to \pi/2$ provided $\omega_x \cos(\psi) - \omega_y \sin(\psi) \neq 0$, which will usually be true. Observing Fig. 2.1a, the inner gimbal is in the same plane as the outer gimbal for $\theta = \pi/2$ and the body cannot be rotated about an axis perpendicular to the gimbal plane for this condition. This is gimbal lock. For applications such as surface ships, civil airliners and mobile robots, however, this situation is tolerable as the attitude of the vehicle is limited. For applications, such as spacecraft in which the attitude is unlimited, singularity-free attitude representations exist, two of which are presented in the following subsection.

2.2.4.3 Singularity-Free Three-Axis Rotational Kinematics

Two sets of kinematic differential equations are derived from the first principles in Appendix A2 that, in contrast to the basic kinematic differential equations of Sect. 2.2.4.2, are free of singularities and trigonometric functions. The first is the set of direction cosine-based kinematic differential equations, as follows.

$$\begin{bmatrix} \dot{c}_{xx} & \dot{c}_{xy} & \dot{c}_{xz} \\ \dot{c}_{yx} & \dot{c}_{yy} & \dot{c}_{yz} \\ \dot{c}_{zx} & \dot{c}_{zy} & \dot{c}_{zz} \end{bmatrix} = \begin{bmatrix} 0 & \omega_z & -\omega_y \\ -\omega_z & 0 & \omega_x \\ \omega_y & -\omega_x & 0 \end{bmatrix} \begin{bmatrix} c_{xx} & c_{xy} & c_{xz} \\ c_{yx} & c_{yy} & c_{yz} \\ c_{zx} & c_{zy} & c_{zz} \end{bmatrix} \quad (2.108)$$

Here, c_{ij}, are the set of direction cosines of three mutually orthogonal unit vectors fixed in the vehicle body with respect to a set of three mutually orthogonal unit vectors fixed in the frame of reference. Although these constitute nine attitude coordinates, (2.108) obeys six constraint equations that reduce the total number of rotational degrees of freedom to three. In this case, (2.108) can be numerically integrated in a vehicle application and the three attitude coordinates taken as a suitable subset of three direction cosines for control purposes [5].

2.2 Physical Modelling

The second set of singularity-free kinematic differential equations is based on the quaternion. This interesting mathematical entity is introduced and fully discussed in Appendix A2, together with the derivation of the equations as follows.

$$\begin{bmatrix} \dot{q}_0 \\ \dot{q}_1 \\ \dot{q}_2 \\ \dot{q}_3 \end{bmatrix} = \frac{1}{2} \begin{bmatrix} 0 & -\omega_x & -\omega_y & -\omega_z \\ \omega_x & 0 & \omega_z & -\omega_y \\ \omega_y & -\omega_z & 0 & \omega_x \\ \omega_z & \omega_y & -\omega_x & 0 \end{bmatrix} \begin{bmatrix} q_0 \\ q_1 \\ q_2 \\ q_3 \end{bmatrix} \quad (2.109)$$

Here, the quaternion components, q_i, are the attitude coordinates which obey the single constraint equation,

$$q_0^2 + q_1^2 + q_2^2 + q_3^2 = 1, \quad (2.110)$$

thereby reducing the total number of rotational degrees of freedom to three, so a suitable subset of these coordinates, usually q_1, q_2 and q_3, can be used for control.

2.2.4.4 Translational Dynamics and Kinematics

In applications such as spacecraft and underwater vehicles, the translational dynamic subsystem is the differential equation relating the control force vector, $\mathbf{f}_c$, the opposing force vector, $\mathbf{f}_o$, and the external disturbance force vector, $\mathbf{f}_d$, to the velocity vector, $\mathbf{v}_r$, of the centre of mass. The translational kinematic subsystem is the differential equation relating the velocity and position vectors of the centre of mass. These are straightforward but the force vectors are usually formulated in the vehicle body-fixed frame, (x_b, y_b, z_b), because the actuators are mounted on the body, while the position and velocity vectors are formulated in the reference frame (x_r, y_r, z_r). In this case, the direction cosine matrix, $\mathbf{C}$, discussed in Appendix A2, is needed in the dynamic subsystem to convert the given force components along the body-fixed axes to components along the reference frame axes. Thus, if $\mathbf{f}_r$ is the net force vector acting on the body centre of mass with components along the reference axes, then

$$\mathbf{f}_r = \mathbf{C}^T \left[\mathbf{f}_c - \mathbf{f}_o - \mathbf{f}_d \right]. \quad (2.111)$$

The basic dynamics equation is then obtained by equating the rate of change of linear momentum to $\mathbf{f}_r$. Thus,

$$\frac{d}{dt}(M \mathbf{v}_r) = \mathbf{f}_r \quad (2.112)$$

where M is the vehicle mass. If M is constant, then the dynamic subsystem equation is obtained by combining (2.111) and (2.112). In the component form, this is

$$M \begin{bmatrix} \dot{v}_{rx} \\ \dot{v}_{ry} \\ \dot{v}_{rz} \end{bmatrix} = \begin{bmatrix} c_{xx} & c_{yx} & c_{zx} \\ c_{xy} & c_{yy} & c_{zy} \\ c_{xz} & c_{yz} & c_{zz} \end{bmatrix} \left(\begin{bmatrix} f_{cx} \\ f_{cy} \\ f_{cz} \end{bmatrix} - \begin{bmatrix} f_{ox} \\ f_{oy} \\ f_{oz} \end{bmatrix} - \begin{bmatrix} f_{dx} \\ f_{dy} \\ f_{dz} \end{bmatrix} \right). \quad (2.113)$$

The elements, $c_{ij}, i = x, y, z, j = x, y, z$, would have to be given in terms of the coordinates of the attitude representation for the rotational kinematic subsystem [5].

2.2.5 Electric Motors

2.2.5.1 Introduction

Many systems for controlling the position or velocity of a mechanical object employ electric motors as actuators. This trend is increasing. For example, electric drives, comprising motors, associated power electronics and digital processors, are replacing the internal combustion engine for vehicle propulsion and also replacing the hydraulic actuators on some aircraft control surfaces. Motors are therefore important electrical components to understand and model to be able to create plant models. It is important to note that these machines can operate as generators to return kinetic energy stored in controlled mechanisms to the power supply during deceleration, referred to as regenerative operation, but suitably designed power electronic circuits are needed for this purpose. This is an important practical feature of any system employing electric motors, such as an electric vehicle, designed to recycle energy that would otherwise be lost in the form of heat.

The following subsections present models of three electric motor types found in industry, i.e. the DC motor, and the two basic types of AC motor, i.e. the synchronous motor and the induction motor. The DC motor is relatively straightforward to model and incorporate in a plant model. On the other hand, AC motor models used by power systems engineers are usually equivalent circuits based on a sinusoidal power supply voltage at fixed amplitude and frequency, stemming from the era in which variable speed or position control was quite primitive and entailed varying the supply voltage amplitude only. For the effective use as actuators in feedback control systems, both the amplitude and frequency have to be variable. A plant model in which the motor supply voltage amplitude and frequency are input variables would, however, be nonlinear, also requiring the phasing to be varied for bidirectional control. These problems are circumvented in modern electric drives by the method of vector control [6] described in Sect. 2.2.6 that applies software-implemented transformations through which the AC motor appears similar to a DC motor. The induction and synchronous motor models presented in Sects. 2.2.6.4 and 2.2.6.6 have the DC motor-like input and output variables and therefore include the transformations.

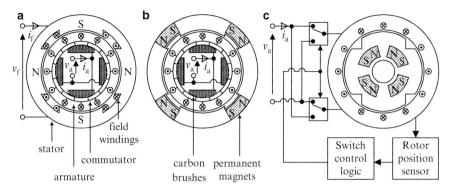

Fig. 2.11 DC motor types. (**a**) Separately excited. (**b**) Permanent magnet. (**c**) Brushless

Models of rotary motors are given, those for linear motors being similar. The models comprise a common mechanical subsystem given in Sect. 2.2.5.3 and specific electrical subsystems given in Sects. 2.2.5.4, 2.2.6.4 and 2.2.6.6.

2.2.5.2 The Three Basic Motor Types

Basic motor descriptions are given in this subsection in sufficient detail to enable the unfamiliar reader to understand the models presented subsequently. It should be noted that the detailed configurations, proportions and practical features are not given here but are available in specialist texts such as [8] and design aspects together with the underlying electromagnetic theory are covered by texts such as [9].

The basic forms of DC motor shown in Fig. 2.11a, b comprise a stator of magnetic material in which a cylindrical armature, also of magnetic material rotates, separated by a small air gap. The stators of large DC motors are configured to produce a magnetic field pattern with alternate North and South poles around the cylindrical air gap as illustrated in Fig. 2.11a, b. In relatively large motors rated in the Megawatt region, such as employed in steel rolling mills, the magnetic field is produced by applying a voltage, v_f, to drive a current, i_f, through field windings in the stator, as shown in Fig. 2.11a. In much smaller DC motors rated in the Kilowatt region and below, such as used in small positioning mechanisms, the magnetic field is produced by permanent magnets as shown in Fig. 2.11b. The crosses on the conductor sections indicate current direction away from the observer while the dots indicate current direction towards the observer.

The armature contains a set of conductors through which a controlled current, i_a, is passed to produce a tangential force, and hence torque, through interaction with the magnetic field. As the armature rotates, when its conductors move from South to North poles and vice versa, the current direction in those conductors is reversed by means of a commutator mounted on the armature shaft, to maintain the torque in the required direction despite the change of direction of the magnetic field as 'seen' by the moving conductors. The commutator consists of a number of copper segments arranged on an insulated cylinder, on which carbon brushes delivering the armature current are kept in good contact by springs (not shown). The commutator is shown 'inside out' in Fig. 2.11a, b for clarity of illustration. In practice the brushes are placed on the outside of the cylindrical set of commutator segments. In brushless DC motors, which are rated in the Kilowatt region and below, the permanent magnets are mounted in the rotor instead of the stator and the torque producing conductors are mounted in the stator as shown in Fig. 2.11c. The commutation is electronic, being carried out with the aid of a rotor position sensor and switching logic driving power electronic switches. The control variable is the armature voltage, v_a. In the separately excited DC motor, v_f is normally kept constant so that the constant steady-state i_f maintains a constant magnetic flux, but this can be reduced as the speed increases to allow a higher maximum speed by reducing the armature back e.m.f. which, since it opposes the armature voltage, would cause loss of control of the armature current due to the voltage saturation limits.

Synchronous motors and induction motors, sometimes called asynchronous motors, have similar stators of magnetic material as illustrated in Fig. 2.12. As in the DC motor, the stator is configured to produce a field pattern with alternate North and South poles around the cylindrical air gap, but in contrast, this field pattern can also be made to rotate. This is achieved by means of a number of coils whose conductors are distributed in slots around the inside cylindrical surface of the stator. Each coil is referred to as a phase. There are at least two phases, usually three in AC motors used as control actuators.

The distribution of the conductors of each phase is such that if a constant direct current is passed through the coil, a magnetic field pattern results with an air gap flux density that has a nearly sinusoidal variation with angular position around the air gap. There are p cycles of flux density per 360° of angular position variation, the overall field pattern being equivalent to that produced by p bar magnets, each with a North pole and a South pole, as if they were buried in the stator in a symmetrical pattern instead of the coil. The integer, p, is therefore referred to as the number

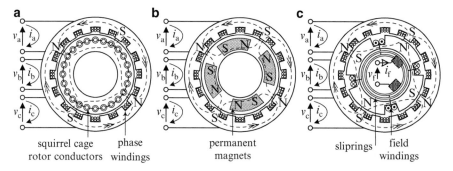

Fig. 2.12 Basic AC motor types. (**a**) Induction (asynchronous motor). (**b**) Permanent magnet synchronous motor (PMSM). (**c**) Separately excited synchronous motor

of pole pairs. The coil of each phase is similar but angularly separated from its neighbour. For a three-phase motor, the mechanical separation is $120°/p$. Then if sinusoidal currents at a frequency of ω_e [rad/s] are driven through the coils, the current in each phase being separated by $120°$ in electrical angle, then the magnetic field pattern will rotate about the rotor centre at an angular velocity of $\omega_s = \omega_e/p$ [rad/s], called the mechanical synchronous angular frequency, while maintaining its shape.

For all three AC motors, a rotor of magnetic material is placed inside the stator separated by a small air gap. The rotors of induction motors usually contain a set of conductors placed axially in a cylindrical configuration near the periphery. The ends of these conductors are electrically connected as shown. This type of rotor is called a squirrel-cage rotor. The induction motor is therefore similar to a transformer with a short-circuited secondary winding. Indeed, if the rotor is locked, the rotating field pattern generates e.m.f.s in the rotor conductors that give rise to relatively high circulating currents. With the rotor free to move, however, the torque generated by the interaction of the rotor currents with the rotating magnetic field causes the rotor to accelerate until it reaches a constant speed if the supply voltage amplitude and frequency are constant. If the mechanical load is purely inertial with zero friction (hypothetical in practice) then the rotor would reach ω_s [rad/s] for which there would be no relative movement between the magnetic field and the rotor conductors. In a real situation, however, a steady torque would be required to maintain a constant rotor speed, which would be ω_m, where $|\omega_m| < |\omega_s|$, as there has to be relative movement between the magnetic field and rotor conductors, called rotor slip, for there to be e.m.f.s driving currents through the rotor to produce the necessary torque.

In the permanent magnet synchronous motor (PMSM) illustrated in Fig. 2.12b, permanent magnets are buried in the rotor to produce a magnetic field with a set of poles equal in number to the set of poles of the rotating magnetic field. Then unlike poles of the rotating magnetic field attract like poles of the rotor and it is 'dragged' round at an angular velocity of ω_s, so there is no rotor slip. The magnetic field, however, is distorted under a mechanical load resulting in an angular displacement between the stator magnetic field and rotor called the load angle. The PMSMs are rated in the Kilowatt region or below. The separately excited synchronous motors are illustrated in Fig. 2.12c.

2.2.5.3 Mechanical Subsystem

The mechanical part is common to all motor types and effectively comprises the single-degree-of-freedom rotational dynamic and kinematic models of Sect. 2.2.2.3, which are as follows:

$$\begin{aligned} \text{Dynamical subsystem} &: \dot{\omega}_r = \tfrac{1}{J_r}(\gamma_e - \gamma_L) \\ \text{Kinematic subsystem} &: \dot{\theta}_r = \omega_r \end{aligned} \quad (2.114)$$

Here, J_r is the rotor moment of inertia, ω_r is the rotor angular velocity, θ_r is the rotor angle (relative to an inertial frame of reference), γ_e is the electromagnetic torque developed by the motor and γ_L is the load torque given by

$$\gamma_L = \gamma_d + \gamma_o, \qquad (2.115)$$

where γ_d is the external disturbance torque and γ_o is the opposing torque defined in Sect. 2.2.2.3.

2.2.5.4 DC Motor Electrical Subsystem

The basic model of a DC motor may be developed by first considering a conductor of length of l metres, carrying a current of i Amperes at right angles to a uniform magnetic field of flux density B Tesla situated on a cylindrical armature at a mean radius of r metres. The torque developed is then

$$\gamma = rBli \quad [\text{Nm}]. \qquad (2.116)$$

The back e.m.f., which enables the motor to operate as a generator when needed, is

$$e = Blr\omega_r [\text{V}] \qquad (2.117)$$

It is evident from (2.116) and (2.117) that the torque and back e.m.f. constants are both equal. For a complete DC motor, similar equations hold that are written as

$$\gamma_e = C\Phi i_a \qquad (2.118)$$

and

$$e_b = C\Phi\omega_r \qquad (2.119)$$

where $K_m = C\Phi$, Φ is the magnetic flux interacting with the armature conductors and C is a constant determined by the configuration of the motor.

In addition, the inductive armature circuit has to be modelled. Also the magnetic field of relatively large DC motors, such as in steel rolling mills, is produced by another inductive circuit but by a permanent magnet in smaller DC motors such as those used in the reaction wheels of spacecraft. Both of these circuits are shown in Fig. 2.13.

Applying Kirchhoff's second law to the field and armature circuits then yields the following differential equations.

$$v_a = R_a i_a + L_a \frac{di_a}{dt} + e_b \Rightarrow \frac{di_a}{dt} = \frac{1}{L_a}(v_a - e_b - R_a i_a) \qquad (2.120)$$

2.2 Physical Modelling

Fig. 2.13 Representation of DC motor including its equivalent circuit

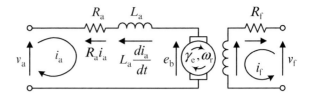

and

$$v_f = R_f i_f + L_f \frac{di_f}{dt} \Rightarrow \frac{di_f}{dt} = \frac{1}{L_f}(v_f - R_f i_f). \qquad (2.121)$$

The DC motor model is then given by (2.114), (2.118), (2.119), (2.120) and (2.121).

2.2.6 Vector-Controlled AC Motors as Control Actuators

2.2.6.1 Concept of Vector Control

In a DC motor, the armature current, the magnetic flux linkage and the torque produced by the Lorenz force may be regarded as vectors, $\hat{\mathbf{i}}$, $\boldsymbol{\psi}$ and $\boldsymbol{\gamma}$. Then the torque equation (2.118) becomes

$$\boldsymbol{\gamma} = C' \boldsymbol{\psi} \wedge \mathbf{i} \qquad (2.122)$$

noting that the constant, C' is not the same as C due to $\boldsymbol{\psi}$ being the flux linkage rather than the total flux. In literature on vector control, however, $\boldsymbol{\psi}$ is usually referred to simply as the magnetic flux. Since, in the vector cross product, $|\boldsymbol{\gamma}| = C'|\boldsymbol{\psi}||\mathbf{i}|\sin(\alpha)$, where α is the angle between the vectors, $\boldsymbol{\psi}$ and $\mathbf{i}$, $|\boldsymbol{\gamma}|$ is maximised by maintaining $\alpha = \pi/2$, i.e. mutual orthogonality between these vectors. This is achieved in a DC motor through its physical design, but for AC motors it is achieved by vector control. For a synchronous motor, the rotor position is determined by measurement so that the orientation of $\boldsymbol{\psi}$ is known. Then the components of $\mathbf{i}$ are controlled to (a) keep $\mathbf{i}$ changing direction relative to the stator to follow rotor so that it is perpendicular to $\boldsymbol{\psi}$ and (b) its magnitude is set to produce the required torque. For an induction motor, the position of the rotor is measured (or estimated using a mathematical model of the motor in the so-called sensorless control). In this case, however, the magnetic flux vector, $\boldsymbol{\psi}$, results from the induced rotor currents and has to be estimated from a mathematical model of the motor. Then the components of $\mathbf{i}$ are determined, as for the synchronous motor, to maintain mutual orthogonality with $\boldsymbol{\psi}$ and produce the required torque.

2.2.6.2 The Transformations of Vector Control

The vectors representing the alternating voltages, currents and magnetic fluxes in an AC machine are expressed with respect to certain frames of reference, usually fixed to the stator or rotor of the motor. The stator currents and voltages are components of their vectors directed along stator-fixed axes and therefore alternate as the vectors rotate relative to the stator in a plane perpendicular to the rotation axis of the motor. The stator current vectors are usually controlled in the frame rotating with the rotor and their components along the axes of the rotating frame do not alternate in the same way. The same is true of the magnetic flux vector as this rotates with the rotor. The calculation of, for example, the stator current vector components in the stator-fixed frame is achieved by a rotational transformation similar to a two-dimensional version of the rotation matrix (direction cosine matrix) **C**, of Sect. 2.2.4.3. In the rotating frame, the components of the vectors appear as variable DC quantities and it is these that are controlled. The motor models needed to achieve this are in the form of differential equations and those available to the control system designer already incorporate the rotational transformations so that the input and output variables are the variable DC ones.

For a multiphase AC motor there exists an equivalent two-phase motor model and it is this that is used in vector control. Since most multiphase motors are three-phase motors, these are assumed in the following description. As shown in Fig. 2.12, the standard subscripts denoting the three phases are a, b and c. The corresponding phases of the two-phase equivalent motor model are denoted α and β. Let $\hat{\mathbf{x}}$ be a vector, corresponding to the applied stator voltage or the stator current, that is rotating at ω [rad/s] in a plane with components, $x_a(t)$, $x_b(t)$ and $x_c(t)$, along three axes, a, b and c equally separated in angle by $2\pi/3$ [rad], as shown in Fig. 2.14a. These are fixed with respect to the stator of the motor.

By convention, if the component of the vector along an axis is towards the arrow, then it is positive: otherwise, it is negative. If ω is constant, then $x_a(t)$, $x_b(t)$ and

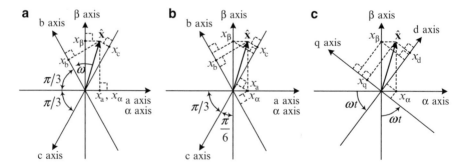

Fig. 2.14 Equivalent three and two-phase alternating variables generated by rotating vector. (**a**) Generation of 2 and 3 phase variables from the same vector. (**b**) Construction for derivation of Clarke transformation. (**c**) Construction for derivation of Park transformation

2.2 Physical Modelling

$x_c(t)$ are sinusoidal with angular frequency ω [rad/s] with amplitude $|\widehat{x}|$ separated in phase by $2\pi/3$ [rad], i.e. balanced three-phase variables. Since the α and β axes are perpendicular, then the components, $x_\alpha(t)$ and $x_\beta(t)$ are sinusoidal variables, also at an angular frequency of ω [rad/s] with amplitudes equal to $|\widehat{x}|$ separated in phase by $\pi/2$ [rad]. Figure 2.14b may then be used to write down equations for $x_a(t)$, $x_b(t)$ and $x_c(t)$ in terms of $x_\alpha(t)$ and $x_\beta(t)$, as follows:

$$x_a(t) = x_\alpha(t)$$
$$x_b(t) = -x_\alpha(t)\sin(\pi/6) + x_\beta(t)\cos(\pi/6) = -\frac{1}{2}x_\alpha(t) + \frac{\sqrt{3}}{2}x_\beta(t) \quad (2.123)$$
$$x_c(t) = -x_\alpha(t)\sin(\pi/6) - x_\beta(t)\cos(\pi/6) = -\frac{1}{2}x_\alpha(t) - \frac{\sqrt{3}}{2}x_\beta(t)$$

i.e.

$$\begin{bmatrix} x_a(t) \\ x_b(t) \\ x_c(t) \end{bmatrix} = \begin{bmatrix} 1 & 0 \\ -1/2 & \sqrt{3}/2 \\ -1/2 & -\sqrt{3}/2 \end{bmatrix} \begin{bmatrix} x_\alpha(t) \\ x_\beta(t) \end{bmatrix}. \quad (2.124)$$

The left pseudo inverse of the matrix on the RHS may then be used to obtain the two-phase variables in terms of the three-phase variables. Thus

$$\begin{bmatrix} x_\alpha(t) \\ x_\beta(t) \end{bmatrix} = \left(\begin{bmatrix} 1 & -\frac{1}{2} & -\frac{1}{2} \\ 0 & \frac{\sqrt{3}}{2} & -\frac{\sqrt{3}}{2} \end{bmatrix} \begin{bmatrix} 1 & 0 \\ -\frac{1}{2} & \frac{\sqrt{3}}{2} \\ -\frac{1}{2} & -\frac{\sqrt{3}}{2} \end{bmatrix} \right)^{-1} \begin{bmatrix} 1 & -\frac{1}{2} & -\frac{1}{2} \\ 0 & \frac{\sqrt{3}}{2} & -\frac{\sqrt{3}}{2} \end{bmatrix} \begin{bmatrix} x_a(t) \\ x_b(t) \\ x_c(t) \end{bmatrix}$$

$$= \begin{bmatrix} \frac{3}{2} & 0 \\ 0 & \frac{3}{2} \end{bmatrix}^{-1} \begin{bmatrix} 1 & -\frac{1}{2} & -\frac{1}{2} \\ 0 & \frac{\sqrt{3}}{2} & -\frac{\sqrt{3}}{2} \end{bmatrix} \begin{bmatrix} x_a(t) \\ x_b(t) \\ x_c(t) \end{bmatrix}$$

i.e.

$$\begin{bmatrix} x_\alpha(t) \\ x_\beta(t) \end{bmatrix} = \begin{bmatrix} \frac{2}{3} & -\frac{1}{3} & -\frac{1}{3} \\ 0 & \frac{1}{\sqrt{3}} & -\frac{1}{\sqrt{3}} \end{bmatrix} \begin{bmatrix} x_a(t) \\ x_b(t) \\ x_c(t) \end{bmatrix}. \quad (2.125)$$

Transformation (2.125) is the Clarke transformation and (2.124) is called the inverse Clarke transformation.

Figure 2.14c represents a single-degree-of-freedom rotational transformation in which the components, $x_d(t)$ and $x_q(t)$, in a new frame of reference with axes, d and q, are expressed in terms of the components, $x_\alpha(t)$ and $x_\beta(t)$, in the frame of reference with axes, α and β, already introduced. If the d - q frame rotates with the vector, $\widehat{x}$, and $|\widehat{x}|$ is constant, then x_d and x_q are constant. The transformation equations follow from the figure and may be written as

$$\begin{bmatrix} x_d(t) \\ x_q(t) \end{bmatrix} = \begin{bmatrix} \cos(\omega t) & \sin(\omega t) \\ -\sin(\omega t) & \cos(\omega t) \end{bmatrix} \begin{bmatrix} x_\alpha(t) \\ x_\beta(t) \end{bmatrix}. \qquad (2.126)$$

This is called the Park transformation. As this is a rotational transformation, it is orthogonal and therefore the matrix of the inverse Park transformation is the transpose of that of the RHS of (2.126). Thus

$$\begin{bmatrix} x_\alpha(t) \\ x_\beta(t) \end{bmatrix} = \begin{bmatrix} \cos(\omega t) & -\sin(\omega t) \\ \sin(\omega t) & \cos(\omega t) \end{bmatrix} \begin{bmatrix} x_d(t) \\ x_q(t) \end{bmatrix}. \qquad (2.127)$$

2.2.6.3 Vector Control Implementation

In vector control, the transformations, (2.124), (2.125), (2.126) and (2.127) are implemented on a digital processor interfaced with the motor according to Fig. 2.15. The d-q frame is fixed in the rotor and 'd' denotes the direct axis along which the magnetic flux vector should be directed and 'q' denotes the quadrature axis along which the current component producing the torque is directed. The purpose of vector control is to keep the current and flux vectors mutually perpendicular, which produces the maximum torque for given vector magnitudes, as in a DC motor in which the armature current direction is perpendicular to the magnetic flux direction. It is important to realise that the current and flux vectors referred to in vector control are independent of the machine geometry and correspond to, rather than equal, the physical fluxes and currents [6].

A few practical features in Fig. 2.15 require explanation. First, nearly all electric drive applications employ *switched mode* power electronics to minimise the energy loss in the physical devices used to control the motor by regulating its electrical power input. The inverter is a set of six electronic switches that are controlled by a pulse modulator to apply physical stator voltages, $v_{as}(t)$, $v_{bs}(t)$ and $v_{cs}(t)$, that rapidly switch between $\pm V_{DC}$ with continuously varying mark space ratios such that the short-term mean values equal $v_a(t)$, $v_b(t)$ and $v_c(t)$, the frequency being high enough for the system performance to be indistinguishable from a hypothetical one in which these continuously varying stator voltages were to be directly applied.

Pulse modulation is covered in some detail in Chap. 8. Second, it is usual for shaft encoders to be employed as speed and position sensors on the shafts of motors used in controlled electric drives. These provide digital outputs with pulse patterns enabling direction of motion to be detected. The frequency of the pulse trains can be determined by pulse timing and this yields an angular velocity measurement. The pulse count yields the angle of rotation. Software-implemented signal processing provides the position and velocity measurements. Third, to minimise instrumentation, it is usual to measure only two stator-phase currents, such as i_b and i_c and calculate the third using the well-known constraint equation of a balanced three-phase load, $i_a + i_b + i_c = 0$, yielding $i_a = -(i_b + i_c)$. Specialist texts such as [7] may be read for more details.

2.2 Physical Modelling

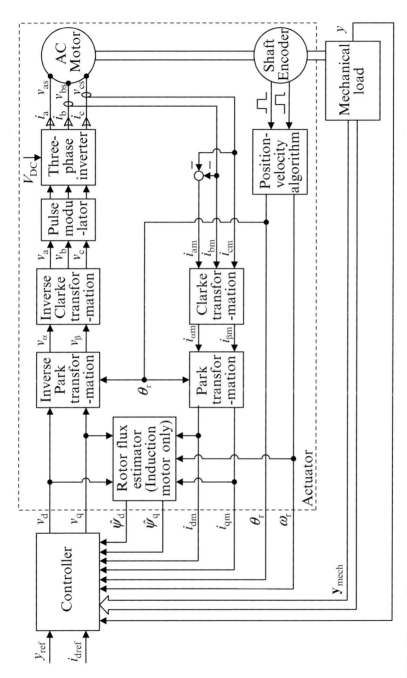

Fig. 2.15 AC motor with vector control transformations for use as a control actuator

Table 2.1 Variables used in vector control of AC motors and their models

Variables	Units	Description
v_a, v_b, v_c	[V]	Continuous three-phase stator voltage demands
V_{DC}	[V]	DC power supply voltage
v_{as}, v_{bs}, v_{cs}	[V]	Switched physical stator voltages with mean values, v_a, v_b, v_c
i_a, i_b, i_c	[A]	Physical stator-phase currents
i_{am}, i_{bm}, i_{cm}	[A]	Measured stator-phase currents
$i_{\alpha m}, i_{\beta m}$	[A]	Equivalent two-phase stator current measurements (Fig. 2.14b)
i_{dm}, i_{qm}	[A]	Measured stator current vector direct and quadrature axis components (Fig. 2.14c)
v_d, v_q	[V]	Continuous stator voltage vector direct and quadrature axis components to be applied
v_α, v_β	[V]	Continuous two-phase stator voltage demands
i_{dref}	[A]	Reference input value of direct-axis stator current vector component.
y, y_{ref}	a	Controlled plant output and corresponding reference input
$\mathbf{y}_{mech}$	a	Optional measurements, $y_i, i = 1, 2, \ldots$, from controlled mechanism

[a]Units dependent upon application

Many vector control schemes in industrial electric drives employ more than one of the traditional PI controllers. One controls the direct component, i_{dm}, of the transformed measured stator current in an induction motor to control the magnetic field, using the direct component, v_d, of the transformed stator voltage. In a PMSM, this PI controller is used to keep i_{dm} as close to zero as possible by setting the reference current to $i_{dref} = 0$. Exceptionally, i_{dref} is made a function of ω_r to reduce the magnetic field at high speeds to extend the speed range by reducing the stator back e.m.f. for a given speed to avoid stator voltage saturation due to the finite DC power supply voltage, V_{DC}, but this technique, known as flux weakening, is carried out with extreme care to avoid demagnetising the permanent magnets. Another PI controller is employed to control the rotor speed, ω_r, using the quadrature component, v_q, of the transformed stator voltage. Typically, if the rotor position is to be controlled, a third PI controller is added using the reference input of the speed control loop as its control variable.

The variables of Fig. 2.15 are described in Table 2.1.

This traditional arrangement of PI controllers, however, often requires much time-consuming tuning at commissioning time and retuning during the lifetime of an electric drive. Also, the traditional philosophy is to control the position or the speed of the motor with the additional torque due to the mechanical load regarded as external disturbance torque. For some applications, such as those entailing mechanical vibration modes, acceptable control is difficult to attain in this way, certainly not a specified dynamic response of the closed-loop system to the reference inputs. The more general control structure of Fig. 2.15 can overcome these problems with a suitable choice of the single controller shown, the freedom of choice being wide with modern digital implementation. In this spirit, the block arrow signal, $\mathbf{y}_{mech}$, represents the additional measurements, such as flexural deflections in

2.2 Physical Modelling

mechanical structures, that enable the best control to be attained within the hardware limitations. As far as this chapter is concerned, the commonly found PI-based vector-controlled electric drives are not included as complete actuators. Instead, the actuator is regarded as just the motor and the transformations, as shown in Fig. 2.15, which can be accurately modelled as part of the plant to be controlled. The controller choice is left open so the mechanical load, the motor and the transformations together constitute the plant. Control techniques other than the traditional ones are advantageous in electric drives [10].

When the motor is viewed through the input and output signals of the 'actuator' block in Fig. 2.15, its behaviour resembles that of the DC motor, at least in that the variables are not required to oscillate, and this demonstrates the great advantage of vector control in enabling AC motors to be used as actuators with relatively sophisticated controllers. The alternating voltages required by the motor are automatically produced by the inverse Park transformation, due to its time-varying elements, $\sin(\omega t)$ and $\cos(\omega t)$. As the motor accelerates and decelerates, however, the frequencies and amplitudes of its alternating voltages, currents and magnetic fluxes will change, in contrast with such motors used directly with AC power supplies. Also, the time-varying Park transformation removes the oscillations of the alternating variables of the motor from the measured current components, i_{dm} and i_{dm}. The oscillations of the AC variables in the motor are therefore 'invisible' to the control engineer in the d-q models in the following subsections, which are in the form of differential equations that may be used directly for control system design. Detailed derivations of these models may be found in specialist texts [11].

2.2.6.4 Induction Motor d-q Model

The complete d-q induction model comprises the following set of first-order differential equations.

$$
\begin{aligned}
\frac{di_d}{dt} &= -Ai_d + B\psi_d + C\omega_r \psi_q + Dv_d + p\omega_r i_q \\
\frac{di_q}{dt} &= -Ai_q + B\psi_q - C\omega_r \psi_d + Dv_q - p\omega_r i_d \\
\frac{d\psi_d}{dt} &= -E\psi_q + Fi_d \\
\frac{d\psi_q}{dt} &= -E\psi_q + Fi_q \\
\frac{d\omega_r}{dt} &= G\left(\Psi_d i_q - \Psi_q i_d\right) - H\gamma_L, \quad \frac{d\theta_r}{dt} = \omega_r
\end{aligned} \quad (2.128)
$$

where

$$
\begin{aligned}
D &= \frac{L_r}{L_s L_r - L_m^2}; \ A = D \cdot \left(R_s + \frac{L_m^2}{L_r^2} R_r\right) \ B = D \cdot \frac{L_m R_r}{L_r^2} \ C = D \cdot \frac{L_m}{L_r} \cdot p \\
E &= \frac{R_r}{L_r} \qquad F = \frac{L_m}{L_r} R_r \qquad\qquad G = \frac{1}{J_r} \cdot \frac{3p}{2} \cdot \frac{L_m}{L_r} \ H = \frac{1}{J_r}
\end{aligned} \quad (2.129)
$$

Here, ψ_d and ψ_q are the rotor magnetic flux vector components, and γ_L is the load torque defined in Sects. 2.2.5.3 and 2.2.2.3. L_s, L_r and L_m are, respectively, the stator, rotor and mutual inductances. R_s and R_r are the stator and rotor resistances, p is the number of stator pole pairs and J_r is the rotor moment of inertia.

2.2.6.5 Induction Motor α-β Model

The induction motor model equivalent to that of Sect. 2.2.6.4 but formulated in terms of the vector components along the α and β axes fixed with respect to the stator are presented here as they could be useful in simulations to display the alternating variables of the motor. Also, it is possible to create a controller based directly on this model with an internal oscillatory mode that automatically creates the alternating variables of the machine without the aid of the time-varying Park and inverse Park transformations, only the Clarke and inverse Clarke transformations being necessary in Fig. 2.15 [10]. Thus,

$$\begin{aligned}
\frac{di_\alpha}{dt} &= -Ai_\alpha + B\psi_\alpha + C\omega_r\psi_\beta + Dv_\alpha \\
\frac{di_\beta}{dt} &= -Ai_\beta + B\psi_\beta - C\omega_r\psi_\alpha + Du_\beta \\
\frac{d\psi_\alpha}{dt} &= -E\psi_\alpha - p\omega_r\psi_\beta + Fi_\alpha \\
\frac{d\psi_\beta}{dt} &= -E\psi_\beta + p\omega_r\psi_\alpha + Fi_\beta \\
\frac{d\omega_r}{dt} &= G\left(\Psi_\alpha i_\beta - \Psi_\beta i_\alpha\right) - H\gamma_L, \quad \frac{d\theta_r}{dt} = \omega_r
\end{aligned} \quad (2.130)$$

The constants are as defined in (2.129).

2.2.6.6 Synchronous Motor d-q Model

The complete d-q permanent magnet synchronous motor model comprises the following set of first-order differential equations:

$$\begin{aligned}
\frac{di_d}{dt} &= -\frac{R_s}{L_d}i_d + p\omega_r\frac{L_q}{L_d}i_q + \frac{1}{L_d}v_d \\
\frac{di_q}{dt} &= -p\omega_r\frac{L_d}{L_q}i_d - \frac{R_s}{L_q}i_q - \frac{p\omega_r}{L_q}\Psi_{PM} + \frac{1}{L_q}v_q \\
\frac{d\omega_r}{dt} &= \frac{1}{J_r}\left(\gamma_e - \gamma_L\right) \\
\frac{d\theta_r}{dt} &= \omega_r
\end{aligned} \quad (2.131)$$

where

$$\gamma_e = \frac{3p}{2}\left[\Psi_{PM}i_q + \left(L_d - L_q\right)i_d i_q\right]. \quad (2.132)$$

2.2 Physical Modelling

Here, Ψ_{PM} is the permanent magnet flux, R_s is the stator resistance, L_d and L_q are the direct and quadrature axis inductances and p is the number of pole pairs.

2.2.7 Fluid and Thermal Subsystems

2.2.7.1 Introduction

Some plants involve heat flow and/or fluid flow and this subsection presents some relevant models. Specialist texts such as [1, 12] may be consulted for a comprehensive coverage.

2.2.7.2 Coupled-Tank Systems

Many industrial processes involve one or more interconnected tanks through which liquid is passed and it is necessary to control the liquid heights in the tanks and the flow rates. In such cases, the liquid may be regarded incompressible. The general coupled-tank system of Fig. 2.16 covers several specific examples.

Pumps, P_1 and P_2, supply the liquid at controlled volume flow rates of q_1 and q_2 via the control variables, u_1 and u_2. Assuming that these pumps and their electric drives are linear and the dynamical effects of these drives are negligible, then

$$q_i = b_i u_i, \quad i = 1, 2. \tag{2.133}$$

The fluid pressures at the bases of the tanks are

$$p_i = \rho g h_i, \quad i = 1, 2. \tag{2.134}$$

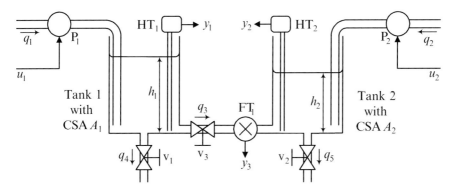

Fig. 2.16 A coupled-tank system

Here, ρ is the liquid density and $g = 9.81$ [m/s/s] is the acceleration due to gravity. The transducers, HT$_1$ and HT$_2$, measure these pressures but in view of (2.134) they are calibrated to measure h_1 and h_2, which are also referred to as the liquid heads. Then, assuming linearity of the transducers, the liquid height measurements are

$$y_i = K_\text{h} h_i, \quad i = 1, 2 \tag{2.135}$$

where K_h is the height measurement constant. The volume flow rate, q_3, between the tanks, which can be positive or negative, is measured by the transducer, FT$_1$, and assuming this is linear, the measurement is

$$y_3 = K_\text{f} q_3. \tag{2.136}$$

The valves, V_1, V_2 and V_3, can be preset to yield different flow rates for given values of h_1 and h_2. According to the theory of fluid dynamics [1], the Reynolds numbers of the valve orifices are dimensionless parameters given by

$$N_{\text{Re}i} = \frac{\rho v_i L_i}{\mu}, \quad i = 1, 2, 3, \tag{2.137}$$

where μ is the fluid dynamic viscosity, v_i is the fluid velocity and L_i is a characteristic linear dimension dependent on the valve setting. So the Reynolds numbers vary with the flow rates and the valve settings, but if they remain sufficiently small ($N_{\text{Re}i} < 2,000$, $i = 1, 2, 3$), then the flow is laminar and the relationship between the pressure drop across each valve and the flow rate through it is linear, yielding

$$q_4 = h_1 / R_{\text{f}1}, \tag{2.138}$$

$$q_5 = h_2 / R_{\text{f}2} \tag{2.139}$$

and

$$q_3 = (h_1 - h_2) / R_{\text{f}3}. \tag{2.140}$$

where $R_{\text{f}i}$, $i = 1, 2, 3$, are defined as the fluid resistances of the orifices. An electrical analogy is immediately apparent in which volume flow rate is equivalent to electric current and the liquid heads are equivalent to voltages. If, on the other hand, the Reynolds numbers of the valve orifices are relatively large ($N_{\text{Re}i} > 4,000$, $i = 1, 2, 3$), then the flow is turbulent and the relationship between the pressure drop across each valve and the flow rate through it becomes nonlinear. Thus

$$q_4 = K_{\text{V}1} \sqrt{h_1}, \tag{2.141}$$

2.2 Physical Modelling

$$q_5 = K_{V2}\sqrt{h_2} \tag{2.142}$$

and

$$q_3 = K_{V3}\sqrt{|h_1 - h_2|}\,\text{sgn}\,(h_1 - h_2), \tag{2.143}$$

where $\text{sgn}(x) \triangleq \{+1, x > 0; 0, x = 0; = 1, x < 0\}$. In this case, the fluid resistances are defined as the changes in the liquid heads divided by the changes in the volume flow rates, i.e.

$$R_1 = 1 / \frac{dq_4}{dh_1} = \frac{2}{K_{V1}}\sqrt{h_1}, \tag{2.144}$$

$$R_2 = 1 / \frac{dq_5}{dh_2} = \frac{2}{K_{V2}}\sqrt{h_2} \tag{2.145}$$

and

$$R_3 = 1 / \left.\frac{dq_3}{dh}\right|_{h=|h_1-h_2|} = \frac{2}{K_{V3}}\sqrt{|h_1 - h_2|}. \tag{2.146}$$

The model is completed by relating the rates of change of the liquid heights to the rates of change of liquid volume in the tanks, using the cross-sectional areas. Thus,

$$\dot{h}_1 = [q_1 - (q_3 + q_4)] / A_1 \tag{2.147}$$

and

$$\dot{h}_2 = (q_3 + q_2 - q_5) / A_2. \tag{2.148}$$

2.2.7.3 Thermal Systems

Plants involving heat flow have continuous spatial temperature distributions for which partial differential equations would be needed to form a precise mathematical model [12]. Such models are referred to as distributed parameter models. While they may be numerically integrated on a computer to predict the system behaviour, they are not convenient for control system design. For this purpose, it is usual to replace a partial differential equation with a finite set of ordinary differential equations whose solutions are accurate at a number of discrete points. Fortunately many thermal systems may be divided into subsystems in which the temperature is nearly uniform, substantial temperature gradients being restricted to the interfaces between the subsystems. Then the number of ordinary differential equations required can be quite small, just one for each subsystem. The complete model is then referred to as a lumped parameter model.

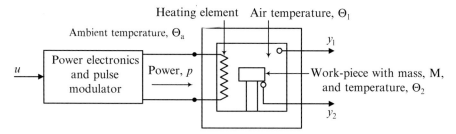

Fig. 2.17 Electric kiln

An example of a controllable heating system with two subsystems is an electric kiln such as illustrated in Fig. 2.17. It is assumed that the pulse modulator, present to operate the power electronics in a switched mode to minimise the energy loss, is designed so that the mean power dissipation in the heating element is directly proportional to the control variable, u. Thus

$$p = K_e u \tag{2.149}$$

where K_e is the heating element power constant. It will be assumed that the convection currents ensure a uniform air temperature within the kiln.

Let q_a be the total amount of heat contained in the air within the kiln, $\dot{q}_s$ be the rate of supply of heat from the heating element and $\dot{q}_w$ be the rate of flow of heat from the kiln wall, which will be negative since heat is actually being lost through the wall due to its imperfect insulation. Then

$$\dot{q}_s = p \tag{2.150}$$

and Fourier's law of heat conduction gives

$$\dot{q}_w = -k_w A_w \frac{d\Theta}{dx}, \tag{2.151}$$

where k_w is the wall conductivity, A_w is the inside area of the kiln wall and $\frac{d\Theta}{dx}$ is the temperature gradient in the wall. Assuming this is constant, then

$$\frac{d\Theta}{dx} = \frac{\Theta_1 - \Theta_a}{D}, \tag{2.152}$$

where D is the wall thickness. The differential equation governing the air temperature is then obtained as follows. First

$$\dot{q}_a = \dot{q}_s + \dot{q}_w = p - k_w A_w \frac{d\Theta}{dx} = K_e u - \frac{k_w A_w}{D}(\Theta_1 - \Theta_a). \tag{2.153}$$

2.3 Identification of LTI Plants from Measurements

Then, if C_a is the specific heat capacity of air and M_a is the mass of air contained in the kiln,

$$q_a = M_a C_a \Theta_1 \Rightarrow \Theta_1 = \frac{1}{M_a C_a} q_a \Rightarrow \dot{\Theta}_1 = \frac{1}{M_a C_a} \dot{q}_a. \quad (2.154)$$

This enables (2.153) to be written as

$$\dot{\Theta}_1 = \frac{1}{T_1}[bu - \Theta_1 + \Theta_a], \quad (2.155)$$

where $T_1 = \frac{M_a C_a D}{k_w A_w}$ is the air heating time constant and $b = \frac{K_e D}{k_w A_w}$ is the aiming temperature constant. The first subsystem of the plant model is given by (2.155).

Let the total amount of heat in the workpiece be q_p, the heat transfer coefficient between the surrounding air and the work-piece be h_p and the surface area of the workpiece be A_p. Then Newton's Law of heating yields

$$\dot{q}_p = h_p A_p (\Theta_1 - \Theta_2). \quad (2.156)$$

If C_p is the specific heat capacity of the workpiece and M_p is its mass, then

$$q_p = M_p C_p \Theta_2 \Rightarrow \Theta_2 = \frac{1}{M_p C_p} q_p \Rightarrow \dot{\Theta}_2 = \frac{1}{M_p C_p} \dot{q}_p, \quad (2.157)$$

enabling (2.156) to be written as

$$\dot{\Theta}_2 = \frac{1}{T_2}(\Theta_1 - \Theta_2) \quad (2.158)$$

where $T_2 = \frac{M_p C_p}{h_p A_p}$ is the workpiece time constant. The second subsystem of the plant model is (2.158).

Finally, the temperature measurement transducers are usually linear so that $y_1 = K_T \Theta_1$ and $y_2 = K_T \Theta_2$, where K_T is the temperature measurement constant.

2.3 Identification of LTI Plants from Measurements

2.3.1 Overview

Plant (or system) identification is the determination of a mathematical model of the plant using measured data. In contrast to the physical modelling of Sect. 2.1.4, the plant is thought of as a 'black box' with input and output signals. The approach is then to determine a model that fits the observed output responses to given inputs.

Here, time domain and frequency domain methods are introduced for obtaining such models in the form of transfer functions.

For some identification methods, an input of known form is applied to the plant, which implies that the operation has to be on an open-loop basis. In these cases, the plant has to be known in advance to be stable. On the other hand, it is possible to identify an unstable plant if a feedback controller can be applied yielding closed-loop stability. Then given reference inputs are applied and the resulting control and measurement variables are observed to perform the identification.

The following subsections commence with elementary methods for the simplest SISO LTI plants, assuming open-loop stability, and progress to more sophisticated methods for the identification of general LTI plants.

2.3.2 Plant Model Determination from Step Response

2.3.2.1 First-Order Plant

Consider a first-order plant characterised by its DC gain, K_{dc}, and time constant, T. The Laplace transfer function model is then

$$\frac{Y(s)}{U(s)} = \frac{K_{dc}}{1 + sT}. \tag{2.159}$$

Let a step input, $u(t) = Ah(t)$, be applied, where A is a constant and $h(t)$ is the unit step function. Then $U(s) = A/s$ and

$$Y(s) = \frac{K_{dc}}{1 + sT} \cdot \frac{A}{s}. \tag{2.160}$$

Then using the table of Laplace transforms and their inverses [Table 1],

$$y(t) = \mathcal{L}^{-1}\left\{\frac{K_{dc}}{1+sT} \cdot \frac{A}{s}\right\} = K_{dc} A \left(1 - e^{-t/T}\right). \tag{2.161}$$

The steady-state value of the response is then

$$y_{ss} = \lim_{t \to \infty} y(t) = K_{dc} A. \tag{2.162}$$

Then (2.161) may be written as

$$y(t) = y_{ss}\left(1 - e^{-t/T}\right). \tag{2.163}$$

Figure 2.18 illustrates an experimental step response.

2.3 Identification of LTI Plants from Measurements

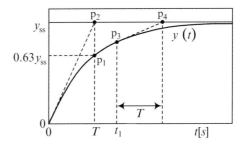

Fig. 2.18 Step response and parameters for estimation of first-order transfer function

The measured steady-state value of $y(t)$ may be used to obtain K_{dc} from (2.162) as

$$K_{dc} = \frac{y_{ss}}{A} \qquad (2.164)$$

The time constant, T, may be determined by three methods for cross-checking:

Method 1 From (2.163),

$$y(T) = y_{ss}\left(1 - e^{-1}\right) = 0.6321 y_{ss} \cong 0.63 y_{ss}. \qquad (2.165)$$

The time constant, T, may therefore be estimated as the time at which the graph of $y(t)$ intersects the line, $y = 0.63 y_{ss}$, at the point, p_1, as shown in Fig. 2.18.

Method 2 With reference to Fig. 2.18, the time constant may be estimated as the time at which the tangent to the graph of $y(t)$ at $t = 0$ intersects the line, $y = y_{ss}$, at the point, p_2. This relationship may be proven as follows. Differentiating (2.163) yields

$$\dot{y}(t) = \frac{y_{ss}}{T} e^{-t/T}. \qquad (2.166)$$

The slope of the tangent, 0-p_2, is therefore $\dot{y}(0) = y_{ss}/T$ and its equation is

$$f_0(t) = \frac{y_{ss}}{T} t. \qquad (2.167)$$

This straight line intersects the horizontal straight line, $y = y_{ss}$, when $f_0(t) = y_{ss}$ and by inspection of (2.167) this is when $t = T$.

Method 3 This method is a generalisation of Method 2 and, with reference to Fig. 2.18, is based on the fact that the tangent to the graph of $y(t)$ at an arbitrary point, p_3, and time, t_1, intersects the horizontal straight line, $y = y_{ss}$, at a point, p_4, where $t = t_1 + T$. This enables T to be estimated at any point on the graph of $y(t)$. The proof of this relationship is as follows. The equation of the tangent, $p_3 - p_4$, is

$$f_{t_1}(t - t_1) = \dot{y}(t_1)(t - t_1) + y(t_1). \qquad (2.168)$$

From (2.163) and (2.166),

$$y(t_1) = y_{ss}\left(1 - e^{-t_1/T}\right) \quad \text{and} \quad \dot{y}(t_1) = \frac{y_{ss}}{T}e^{-t_1/T} \quad (2.169)$$

Substituting for $\dot{y}(t_1)$ and $y(t_1)$ in (2.168) using (2.169) then yields

$$f_{t_1}(t - t_1) = \frac{y_{ss}}{T}e^{-t_1/T}(t - t_1) + y_{ss}\left(1 - e^{-t_1/T}\right) \quad (2.170)$$

By inspection of (2.170), $f_{t_1}(t - t_1) = y_{ss}$ when $t - t_1 = T \Rightarrow t = t_1 + T$.

It is recommended that the three methods are applied, Method 3 for several points, and the set of time constant estimates averaged.

If the plant transfer function model is required in the standard form,

$$\frac{Y(s)}{U(s)} = \frac{b_0}{s + a_0} \quad (2.171)$$

and the model has been obtained in the form of (2.159), then (2.171) can be manipulated into the same form to obtain b_0 and a_0 in terms of K_{dc} and T, as follows.

$$\frac{Y(s)}{U(s)} = \frac{b_0/a_0}{s/a_0 + 1} = \frac{K_{dc}}{1 + sT} \Rightarrow a_0 = \frac{1}{T} \text{ and } \frac{b_0}{a_0} = K_{dc} \Rightarrow b_0 = a_0 K_{dc} = \frac{K_{dc}}{T}. \quad (2.172)$$

2.3.2.2 Underdamped Second-Order Plant

In this case, the plant is characterised by the undamped natural frequency, ω_n, the damping ratio, ζ, where $0 < \zeta < 1$ and the DC gain, K_{dc}, the transfer function being

$$\frac{Y(s)}{U(s)} = \frac{K_{dc}\omega_n^2}{s^2 + 2\zeta\omega_n s + \omega_n^2}. \quad (2.173)$$

If the input is $u(t) = Ah(t)$, where A is a constant and $h(t)$ is the unit step function, then $U(s) = A/s$ and the Laplace transform of the output is

$$Y(s) = \frac{K_{dc}\omega_n^2}{s^2 + 2\zeta\omega_n s + \omega_n^2} \cdot \frac{A}{s} \quad (2.174)$$

Using the final value theorem, the steady-state output is

$$y_{ss} = \lim_{t \to \infty} y(t) = \lim_{s \to 0} sY(s) = K_{dc}A \quad (2.175)$$

Then using the table of Laplace transforms and their inverses (Table 1 in Tables),

2.3 Identification of LTI Plants from Measurements

$$y(t) = \mathcal{L}^{-1}\left\{\frac{K_{dc}\omega_n^2}{s^2 + 2\zeta\omega_n s + \omega_n^2} \cdot \frac{A}{s}\right\} = y_{ss}\left[1 - \frac{1}{\sqrt{1-\zeta^2}}e^{-\zeta\omega_n t}\sin(\omega_d t + \phi)\right] \quad (2.176)$$

where $\omega_d = \omega_n\sqrt{1-\zeta^2}$ is the damped natural frequency and $\phi = \cos^{-1}(\zeta)$.

The parameters, K_{dc}, ω_n and ζ, may be estimated from an experimentally obtained step response of the form shown in Fig. 2.19.

The DC gain can be estimated using y_{ss} obtained from Fig. 2.19 and the known step input level, A, using the following equation from (2.175).

$$K_{dc} = \frac{y_{ss}}{A}. \quad (2.177)$$

An expression for the peak output, y_p, will now be derived using the step response of (2.176). This will first be converted to a more convenient form as follows.

$$\sin(\omega_d t + \phi) = \sin(\omega_d t)\cos(\phi) + \cos(\omega_d t)\sin(\phi)$$
$$= \zeta \sin(\omega_d t) + \sqrt{1-\zeta^2}\cos(\omega_d t), \quad (2.178)$$

recalling that $\cos(\phi) = \zeta$ and therefore $\sin(\phi) = \sqrt{1-\cos^2(\phi)} = \sqrt{1-\zeta^2}$. Then (2.176) may be rewritten as

$$y(t) = y_{ss}\left[1 - e^{-\zeta\omega_n t}\left(\frac{\zeta}{\sqrt{1-\zeta^2}}\sin(\omega_d t) + \cos(\omega_d t)\right)\right] \quad (2.179)$$

The first peak occurs at $t = T_p$, which is the smallest value of $t > 0$ for which $\dot{y}(t) = 0$. Again with the aid of (2.174) and the Laplace transform tables (Table 1 in Tables),

$$\dot{y}(t) = \mathcal{L}^{-1}\{sY(s)\} = \mathcal{L}^{-1}\left\{\frac{y_{ss}\omega_n^2}{s^2 + 2\zeta\omega_n s + \omega_n^2}\right\} = \frac{y_{ss}\omega_n}{\sqrt{1-\zeta^2}}\sin(\omega_d t). \quad (2.180)$$

Fig. 2.19 Step response and parameters for estimation of second-order transfer function

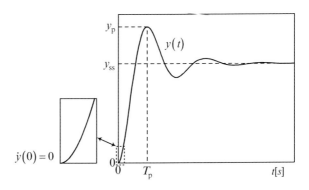

It is clear from (2.180) that the peak time is half the oscillation period. Thus

$$T_p = \frac{1}{2} \cdot \frac{2\pi}{\omega_d} = \frac{\pi}{\omega_d} = \frac{\pi}{\omega_n\sqrt{1-\zeta^2}}. \tag{2.181}$$

Then from (2.179),

$$y(T_p) = y_p = y_{ss}\left[1 + e^{-\frac{\zeta\pi}{\sqrt{1-\zeta^2}}}\right] \Rightarrow -\frac{\zeta\pi}{\sqrt{1-\zeta^2}} = \ln\left(\frac{y_p}{y_{ss}} - 1\right) \Rightarrow$$

$$\zeta^2\pi^2 = (1-\zeta^2)\ln^2\left(\frac{y_p}{y_{ss}} - 1\right) \Rightarrow \zeta^2\left[\pi^2 + \ln^2\left(\frac{y_p}{y_{ss}} - 1\right)\right] = \ln^2\left(\frac{y_p}{y_{ss}} - 1\right) \Rightarrow$$

$$\zeta = -\ln\left(\frac{y_p}{y_{ss}} - 1\right)/\sqrt{\pi^2 + \ln^2\left(\frac{y_p}{y_{ss}} - 1\right)}, \tag{2.182}$$

noting that $0 < y_p/y_{ss} - 1 < 1 \Rightarrow \ln(y_p/y_{ss} - 1) < 0$ and ζ must be positive. This enables the damping ratio, ζ, to be estimated from y_p and y_{ss} obtained from Fig. 2.19. Then the undamped natural frequency can be estimated using T_p from Fig. 2.19 and ζ from (2.182) using the following equation from (2.181).

$$\omega_n = \frac{\pi}{T_p\sqrt{1-\zeta^2}}. \tag{2.183}$$

In many cases, the form of the transfer function will have been established by physical modelling as covered in section 2.1.4. If only the experimental step response is available, then it is important to examine it to check that it is suitable for fitting the model of (2.173). Apart from the need for the step response to be oscillatory, as shown, it is wise to check that it commences with zero slope, i.e. $\dot{y}(0) = 0$, as shown in the insert of Fig. 2.19. This is an indication that the plant satisfies the requirement of having no finite zeros. That fact that the initial slope is non-zero if the plant has a finite zero is proven as follows. Let the plant transfer function have the following transfer function with a finite zero at $s = -1/T_z$.

$$\frac{Y'(s)}{U(s)} = \frac{K_{dc}\omega_n^2(1 + sT_z)}{s^2 + 2\zeta\omega_n s + \omega_n^2} \tag{2.184}$$

The Laplace transform of the step response is therefore

$$Y'(s) = \frac{K_{dc}\omega_n^2(1 + sT_z)}{s^2 + 2\zeta\omega_n s + \omega_n^2} \cdot \frac{A}{s} \tag{2.185}$$

and that of the first derivative is

$$\mathcal{L}\{\dot{y}(t)\} = sY(s) = \frac{AK_{dc}\omega_n^2(1 + sT_z)}{s^2 + 2\zeta\omega_n s + \omega_n^2}. \tag{2.186}$$

2.3 Identification of LTI Plants from Measurements

Applying the initial value theorem then yields

$$\dot{y}(0) = \lim_{s \to \infty} s\mathcal{L}\{\dot{y}(t)\} = \lim_{s \to \infty} \frac{AK_{dc}\omega_n^2 \left(s + s^2 T_z\right)}{s^2 + 2\zeta\omega_n s + \omega_n^2} = AK_{dc}\omega_n^2 T_z. \quad (2.187)$$

Hence $\dot{y}(0) \neq 0$ for $T_z \neq 0$. If transfer function (2.182) does not have a finite zero, then $T_z = 0$. In this case (2.187) yields $\dot{y}(0) = 0$, which is correct.

2.3.3 Plant Model Determination from Frequency Response

2.3.3.1 Introduction

Experimental plant data is often available in the frequency domain. The aim here is to use this data to estimate a linear SISO plant model in the form,

$$\frac{Y(s)}{U(s)} = G(s) = K_{dc} \frac{\prod_i \left(1 + \frac{s}{v_i}\right) \prod_i \left(1 + \frac{2\eta_i}{v_{ni}}s + \frac{1}{v_{ni}^2}s^2\right)}{s^q \prod_i \left(1 + \frac{s}{\omega_i}\right) \prod_i \left(1 + \frac{2\zeta_i}{\omega_{ni}}s + \frac{1}{\omega_{ni}^2}s^2\right)}, \quad (2.188)$$

where K_{dc} is the DC gain, ω_i, ω_{ni}, v_i and v_{ni} are the corner frequencies, ζ_i are the damping ratios of the complex conjugate poles and η_i are equivalent parameters for the complex conjugate zeros. All these parameters, except K_{dc}, are always positive and therefore all the poles and zeros are assumed to lie in the left half of the s-plane. Plants with poles and/or zeros in the right half of the s-plane, however, are dealt with in Sect. 2.3.3.10.

The identification method used depends upon the relative positions of the poles and zeros. These, of course, are not known in advance, but an initial examination of the measured data reveals features that enable an appropriate method to be chosen.

Some background theory in the frequency domain will now be reviewed in preparation for developing the methods of transfer function model determination. First consider the frequency domain transfer function,

$$G(j\omega) = |G(j\omega)| \angle G(j\omega) = M(\omega) e^{\phi(\omega)}. \quad (2.189)$$

This can be displayed graphically in the form of the magnitude, $M(\omega)$, and the phase angle, $\phi(\omega)$. The simplest way to obtain this data is to carry out tests using sinusoidal plant excitation over the frequency range, $\omega \in (0, \omega_b)$, where ω_b is the specified bandwidth for the control system to be designed. Assuming plant linearity, once the initial transients have decayed to negligible proportions the output is a sinusoid, as shown in Fig. 2.20. This is the steady-state sinusoidal response.

Fig. 2.20 Illustration of steady-state sinusoidal response of a linear SISO plant

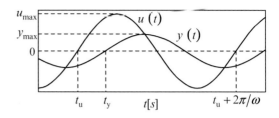

Then

$$M(\omega) = \frac{y_{\max}}{u_{\max}} \quad \text{and} \quad \phi(\omega) = \omega(t_y - t_u) \tag{2.190}$$

Applying sinusoids at many different frequencies, however, could be laborious and time consuming. A faster computer aided alternative is therefore introduced below.

It will be recalled that, in theory, $G(s)$ is the Laplace transform of the unit impulse response of the plant. The Laplace transform becomes the Fourier transform if $s = j\omega$. Then $G(j\omega)$ is the Fourier transform of the unit impulse response of the plant. The Fourier transform of the unit impulse, $\delta(t)$, itself is

$$\int_0^\infty \delta(t) e^{-j\omega t} dt = 1 \tag{2.191}$$

indicating that $\delta(t)$ has a flat Fourier spectrum, meaning that it is composed of an infinite continuum of sinusoidal components having the same magnitude over an infinite frequency range. The unit impulse function, however, cannot be applied to a plant in practice, because it is infinite in magnitude for an infinitesimal time, but an alternative method is possible using a realisable input. Let the Fourier transforms of $y(t)$ and $u(t)$ be, respectively, $Y(j\omega)$ and $U(j\omega)$. Then by analogy with the Laplace transfer function, the frequency domain transfer function is

$$G(j\omega) = \frac{Y(j\omega)}{U(j\omega)}. \tag{2.192}$$

In principle, any $u(t)$ could be used provided it has sufficiently rich frequency content over the frequency range, $\omega \in (0, \omega_b)$, to adequately excite the plant. Then $u(t)$ would be applied in real time, while both $y(t)$ and $u(t)$ are data logged. The Fourier transforms, $Y(j\omega)$ and $U(j\omega)$, would then be computed numerically to yield points on the corresponding magnitude and phase functions.

$$M_u(\omega) = |U(j\omega)|, \phi_u(\omega) = \angle M_u(\omega) \tag{2.193}$$

and

$$M_y(\omega) = |Y(j\omega)|, \phi_y(\omega) = \angle Y_u(\omega) \tag{2.194}$$

2.3 Identification of LTI Plants from Measurements

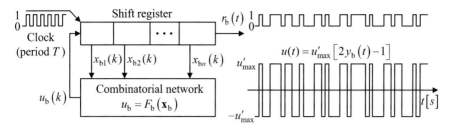

Fig. 2.21 Pseudo-random binary sequence (PRBS) generator

Then the required magnitude and phase of $G(j\omega)$ would be obtained as

$$M(\omega) = \frac{M_y(\omega)}{M_u(\omega)} \quad \text{and} \quad \phi(\omega) = \phi_y(\omega) - \phi_u(\omega). \qquad (2.195)$$

It remains to find a realisable form of $u(t)$ for which $|U(j\omega)|$ is nearly constant for $\omega \in (0, \omega_b)$. White noise has a perfectly flat Fourier spectrum for $\omega \in (0, \infty)$ and therefore a realisable signal approximating this would be suitable. Such a signal is the pseudo-random binary sequence (PRBS). This is a discrete binary signal, $r_b(t)$, produced by an algorithm emulating a shift register with a combinatorial network in the feedback loop [13]. The input of this network is the binary state, $\mathbf{x}_b = [x_{b1}\; x_{b2} \ldots x_{bn}]^T$, of the register and its binary output, u_b, is the register input, as shown in Fig. 2.21. The integer, k, increases by 1 every clock pulse.

The term, pseudo-random, applies because the sequence of register states repeats every n_c clock pulses and is strictly deterministic. The Boolean function, $F_b(\mathbf{x}_b)$, is chosen so that $n_c = 2^n - 1$, which is the maximal length sequence to achieve the best approximation to randomness for a given shift register length. Importantly, for a good approximation to the continuous $G(j\omega)$ to be obtained for the highest angular frequencies approaching, ω_b, the clock frequency, $1/T$, must be chosen several times greater than $f_b = \omega_b/(2\pi)$. The recommendation is $1/T > 5\omega_b/(2\pi) \Rightarrow$

$$T < \frac{2\pi}{5\omega_b}. \qquad (2.196)$$

The amplitude, u'_{max}, of the excitation signal, $u(t)$, sent out to the plant should be less than the physical control saturation limit, u_{max}, to ensure nominally linear operation. A safe value would be

$$u'_{max} = 0.5 u_{max}. \qquad (2.197)$$

The computationally efficient discrete fast Fourier transform [14] is employed for the numerical computation of $M(\omega)$ and $\phi(\omega)$.

The frequency domain transfer function, $G(j\omega)$, will be expressed in the form of Bode plots. These can be produced by real time hardware 'in the loop' systems whose software implements the PRBS and Fourier transform based method such as MATLAB® used with dSPACE® for on-line identification. This consists of two plots, one displaying $M(\omega)$ and the other $\phi(\omega)$. In both plots, the abscissa is ω plotted on a logarithmic scale. $M(\omega)$ is plotted in decibels, meaning

$$M_{\text{dB}}(\omega) \triangleq 20\log_{10}[M(\omega)]. \quad (2.198)$$

This is called the Bode magnitude plot and this alone enables the parameters of transfer function (2.188) to be estimated using the methods presented in the following subsection. As will be seen in Sect. 2.3.3.10, all these methods are useful for modeling plants with poles and/or zeros in the right half of the s-plane but only give the magnitudes of the real parts of the poles and zeros. To determine which poles and/or zeros lie in the right half of the s-plane, more information is required and this is obtained from the graph of $\phi(\omega)$.

The frequency domain transfer function corresponding to (2.188) is

$$G(j\omega) = K_{\text{dc}} \frac{\prod_i \left(1 + j\frac{\omega}{v_i}\right) \prod_i \left(1 - \frac{\omega^2}{v_{ni}^2} + j\frac{2\eta_i \omega}{v_{ni}}\right)}{(j\omega)^q \prod_i \left(1 + j\frac{\omega}{\omega_i}\right) \prod_i \left(1 - \frac{\omega^2}{\omega_{ni}^2} + j\frac{2\zeta_i \omega}{\omega_{ni}}\right)}. \quad (2.199)$$

The corresponding magnitude function is therefore

$$M(\omega) = K_{\text{dc}} \frac{\prod_i \sqrt{1 + \frac{\omega^2}{v_i^2}} \prod_i \sqrt{\left(1 - \frac{\omega^2}{v_{ni}^2}\right)^2 + 4\frac{\eta_i^2 \omega^2}{v_{ni}^2}}}{\omega^q \prod_i \sqrt{1 + \frac{\omega^2}{\omega_i^2}} \prod_i \sqrt{\left(1 - \frac{\omega^2}{\omega_{ni}^2}\right)^2 + 4\frac{\eta_i^2 \omega^2}{\omega_{ni}^2}}}. \quad (2.200)$$

The expression for the Bode magnitude plot of this general transfer function, using definition (2.198), is as follows.

$$M_{\text{dB}}(\omega) = 20\log_{10}(K_{\text{dc}}) - 20q\log_{10}(\omega) + \sum_i 10\log_{10}\left(1 + \frac{\omega^2}{v_i^2}\right)$$

$$- \sum_i 10\log_{10}\left(1 + \frac{\omega^2}{\omega_i^2}\right) + \sum_i 10\log_{10}\left[\left(1 - \frac{\omega^2}{v_{ni}^2}\right)^2 + 4\frac{\eta_i^2 \omega^2}{v_{ni}^2}\right]$$

$$- \sum_i 10\log_{10}\left[\left(1 - \frac{\omega^2}{\omega_{ni}^2}\right)^2 + 4\frac{\eta_i^2 \omega^2}{\omega_{ni}^2}\right]. \quad (2.201)$$

2.3 Identification of LTI Plants from Measurements

The following subsections present methods for estimating the plant parameters from experimentally obtained Bode magnitude plots.

2.3.3.2 DC Gain of a Plant with No Pure Integrators

In this case, $q = 0$ in (2.201), which then satisfies

$$M_{\text{dB}}(0) = 20\log_{10}(K_{\text{dc}}). \tag{2.202}$$

If $M_{\text{dB}}(0)$ could be measured, K_{dc} could be estimated but the logarithmic frequency scale of the Bode plot does not permit this to be shown. If, however, the minimum frequency, $\omega_{\min}$, of the Bode magnitude plot satisfies $\omega_{\min} < \omega_a$ where

$$\omega_a \ll \nu_i, \omega_a \ll \nu_{ni}, \omega_a \ll \omega_i \text{ and } \omega_a \ll \omega_{ni}, \forall i, \tag{2.203}$$

then for $\omega \in (\omega_{\min}, \omega_a)$, (2.201) is approximated by

$$M_{\text{dB}}(\omega) = 20\log_{10}(K_{\text{dc}}) \tag{2.204}$$

which is the equation of the low-frequency asymptote of the Bode magnitude plot. This enables K_{dc} to be approximated by reading $M_{\text{dB}}(\omega_{\min})$ and then calculating

$$K_{\text{dc}} = 10^{M_{\text{db}}(\omega_{\min})/20}. \tag{2.205}$$

If necessary, $\omega_{\min}$ should be reduced to ensure that (2.205) is accurate.

The method is illustrated in Fig. 2.22 for the simple plant model,

$$\frac{Y(s)}{U(s)} = \frac{K_{\text{dc}}}{1 + s/\omega_1} = \frac{10}{1 + s/2}. \tag{2.206}$$

Model (2.206) is initially unknown and its DC gain has to be determined from the Bode magnitude plot. Both asymptotes are shown, one of which is the straight line with a slope of -6 [dB/octave] (or -20 [dB/decade]) passing through the point,

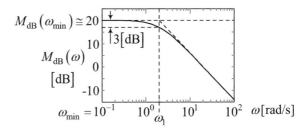

Fig. 2.22 Bode magnitude plot of first-order plant

(ω_1, $M_{dB}(0)$), which the plot approaches as $\omega \to \infty$. The other asymptote is the horizontal straight line given by (2.204). Provided the lowest frequency, ω_{min}, on the horizontal axis (0.1 [rad/s] in Fig. 2.22) is an order of magnitude less than the corner frequency, ω_1, or lower, the plot is so close to the asymptote as $\omega \to \omega_{min}$ that reading $M_{dB}(\omega_{min})$ from the plot yields a close approximation to the required $M_{dB}(0)$. In Fig. 2.22, $M_{dB}(\omega_{min}) \cong 20$ [dB]. Assuming $M_{dB}(0) = M_{dB}(\omega_{min})$, (2.205) gives the correct value of

$$K_{dc} = 10^{20/20} = 10. \tag{2.207}$$

2.3.3.3 The Number of Pure Integrators

Consider now a plant containing q pure integrators, where $q \geq 1$. As ω is reduced to values below ω_a of (2.203), then (2.201) is approximated by

$$M_{dB}(\omega) = 20\log_{10}(K_{dc}) - 20q\log_{10}(\omega). \tag{2.208}$$

If necessary, ω_{min} should be reduced until a nearly straight line segment of the Bode magnitude plot is visible at the low-frequency end. The corresponding asymptote is the straight line of (2.208) with a slope of $-20q$ [dB/decade] (or $-6q$ [dB/octave]). Then if the slope of this line is estimated as S_d [dB/decade] or S_o [dB/octave], the number of integrators is the nearest integer to $S_d/20$ or $S_o/6$.

2.3.3.4 DC Gain of a Plant with Pure Integrators

Once the Bode magnitude plot of Sect. 2.3.3.3 has been obtained then (2.208) applies. With $\omega = \omega_{min}$ this yields

$$K_{dc} = 10^{[M_{dB}(\omega_{min}) + 20q\log_{10}(\omega_{min})]/20}. \tag{2.209}$$

The method will be illustrated for the plant model

$$\frac{Y(s)}{U(s)} = \frac{K_{dc}}{s^2(1+s/\omega_1)} = \frac{10}{s^2(1+s/2)}. \tag{2.210}$$

Model (2.210) is initially unknown. The Bode magnitude plot from which the DC gain is to be found is shown in Fig. 2.23. The slope, S_d, of the plot measured between $\omega = 10^{-2}$ [rad/s] and $\omega = 10^{-1}$ [rad/s] is -40 dB/decade. The number of pure integrators, q, is then the nearest integer to $S_d/20 = 40/20 = 2$.

Thus $q = 2$, agreeing with (2.210). Then $M_{dB}(\omega_{min}) = 100$ dB and $\omega_{min} = 10^{-2}$ [rad/s]. Equation (2.209) then yields $K_{dc} = 10^{[100+40\times(-2)]/20} = 10$, also agreeing with (2.210).

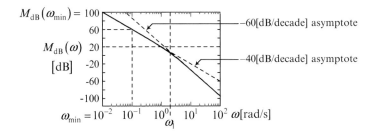

Fig. 2.23 Bode magnitude plot of third order plant containing two pure integrators

2.3.3.5 Asymptotes and the Asymptotic Approximation

Each term under the summation signs in (2.201) is a function having two asymptotes, one for $\omega \to 0$ and the other for $\omega \to \infty$. First consider the terms,

$$M_{\mathrm{dB}i}(\omega) = 10Q\log_{10}\left(1 + \frac{\omega^2}{q_i^2}\right), \qquad (2.211)$$

where $Q = 1$ and $q = \nu$ for real negative zeros while $Q = -1$ and $q = \omega$ for real negative poles. As $\omega \to 0$, (2.211) approaches the asymptote,

$$M_{\mathrm{dB}i}^0(\omega) = 10Q\log_{10}(1) = 0, \qquad (2.212)$$

which is a straight line with zero slope. As $\omega \to \infty$, $\omega \gg q_i$ and therefore (2.211) approaches the asymptote,

$$M_{\mathrm{dB}i}^\infty(\omega) = 10Q\log_{10}\left(\frac{\omega^2}{q_i^2}\right) = 20Q\left[\log_{10}(\omega) - \log_{10}(q_i)\right], \qquad (2.213)$$

which is a straight line satisfying $M_{\mathrm{dB}i}^\infty(q_i) = 0$ with slope $20Q$ [dB/decade], i.e. $6Q$ [dB/octave], noting that the abscissa is $\log_{10}(\omega)$ on the Bode magnitude plot. The Bode magnitude plot and asymptotes for these terms are shown in Fig. 2.24. Importantly, the asymptotes intersect at the corner frequencies, at $\omega = \omega_i$ for the pole and at $\omega = \nu_i$ for the zero. At these frequencies, $M_{\mathrm{dB}i}(\omega)$ has fallen by 3 [dB] for the pole and increased by 3 [dB] for the zero. These features enable the pole magnitudes, which are the corner frequencies, to be estimated from the Bode plot.

An important feature is that $M_{\mathrm{dB}i}(\omega)$ nearly coincides with the asymptote, $M_{\mathrm{dB}i}^0(\omega)$, for $\omega < \omega_{ni}/10$ and $\omega < \nu_{ni}/10$. Similarly $M_{\mathrm{dB}i}(\omega)$ nearly coincides with the asymptote, $M_{\mathrm{dB}i}^\infty(\omega)$, for $\omega > 10\omega_{ni}$ and $\omega > 10\nu_{ni}$. As will be seen in the following subsections, the parameter estimation process is simplified if the corner frequencies, ω_i, ν_i, ω_{ni} and ν_{ni}, are separated by at least two orders of magnitude.

The piecewise linear function, $A_i(\omega)$, is the concatenation of two segments of the asymptote functions, $M_{\mathrm{dB}i}^0(\omega)$ and $M_{\mathrm{dB}i}^\infty(\omega)$, and is defined as follows.

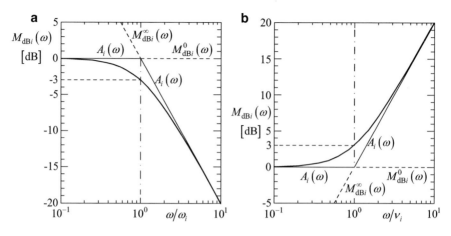

Fig. 2.24 Bode magnitude contributions of pole and zero and their asymptotes. (**a**) Real negative pole. (**b**) Real negative zero

$$A_i(\omega) = \begin{cases} M_{dBi}^0(\omega), & 0 < \omega < q_i \\ M_{dBi}^\infty(\omega), & \omega \geq q_i \end{cases}. \tag{2.214}$$

It is evident that this is an approximation to the Bode magnitude plot, $M_{dBi}(\omega)$. It will be called a *concatenated asymptote function*. The approximation appears better when viewed on larger frequency and amplitude scales due to the closer approach of $M_{dBi}(\omega)$ to the asymptotes at frequencies far removed from the corner frequency.

Next, consider the terms,

$$M_{dBi}(\omega) = 10Q\log_{10}\left[\left(1 - \frac{\omega^2}{q_{ni}^2}\right)^2 + 4\frac{d_i^2\omega^2}{q_{ni}^2}\right], \tag{2.215}$$

of (2.201) where $Q = 1$, $q = \nu$ and $d = \eta$ for the complex conjugate zeros and $Q = -1$, $q = \omega$ and $d = \zeta$ for the complex conjugate poles. As $\omega \to 0$, (2.215) approaches the asymptote,

$$M_{dBi}^0(\omega) = 10Q\log_{10}(1) = 0, \tag{2.216}$$

which is a straight line with zero slope. As $\omega \to \infty$, $\omega \gg q_{ni}$ and therefore (2.215), which can be expanded as

$$M_{dBi}(\omega) = 10Q\log_{10}\left(1 - 2\frac{\omega^2}{q_{ni}^2} + \frac{\omega^4}{q_{ni}^4} + 4\frac{d_i^2\omega^2}{q_{ni}^2}\right),$$

approaches the asymptote,

2.3 Identification of LTI Plants from Measurements

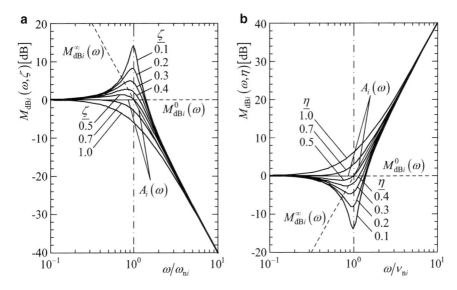

Fig. 2.25 Bode magnitude contributions of complex conjugate pole and zero pairs in the left half of the s-plane together with the asymptotes. (**a**) Complex conjugate pole pair. (**b**) Complex conjugate zero pair

$$M_{\mathrm{dB}i}^{\infty}(\omega) = 10Q\log_{10}\left(\omega^4/q_{\mathrm{n}i}^4\right) = 40Q\left[\log_{10}(\omega) - \log_{10}(q_{\mathrm{n}i})\right], \quad (2.217)$$

which is a straight line satisfying $M_{\mathrm{dB}i}^{\infty}(q_{\mathrm{n}i}) = 0$ with slope $40Q$ [dB/decade].

The Bode magnitude plot for a range of damping ratios is shown in Fig. 2.25 together with the two asymptotes and the concatenated asymptote function, $A_i(\omega)$.

In this case, the asymptotes intersect at the undamped natural frequency, $\omega_{\mathrm{n}i}$, for the poles and at the frequency, $v_{\mathrm{n}i}$, for the zeros.

As in Fig. 2.24, the Bode magnitude plots of Fig. 2.25 closely approach $A_i(\omega)$ at frequencies more than an order of magnitude different from the corner frequency.

The peaks visible in Fig. 2.25a are referred to as resonance peaks, while the dips in Fig. 2.25b are referred to as the anti-resonance dips. The parameters, ζ_i and η_i, can be estimated by determining the amplitudes of the resonance peaks and the anti-resonance dips from the graph of $M_{\mathrm{dB}i}(\omega)$.

Another feature that should be observed is that the frequencies at which the maxima of the peaks in Fig. 2.25a occur, which are the resonance frequencies, $\omega_{\mathrm{r}i}$, are lower than $\omega_{\mathrm{n}i}$. Similarly, the frequencies at which the minima of the dips in Fig. 2.25b occur, which are the anti-resonance frequencies, $v_{\mathrm{r}i}$, are lower than $v_{\mathrm{n}i}$. This effect is more pronounced as ζ_i and η_i increase and must be taken into account when determining the asymptote intersections by examination of the Bode magnitude plot for estimation of $\omega_{\mathrm{n}i}$ and $v_{\mathrm{n}i}$. This is addressed in Sect. 2.3.3.8. For lightly damped cases, however, where $0 < \zeta_i \leq 0.1$ and $0 < \eta_i \leq 0.1$,

$\omega_{ri} \cong \omega_{ni}$ and $\nu_{ri} \cong \nu_{ni}$. Then the asymptote intersections occur approximately at the frequencies of the maxima and minima in Fig. 2.25, rendering the parameter estimation from the Bode magnitude plot more straightforward.

The concatenated asymptote function, $A_i(\omega)$, may be considered to be a good approximation to M_{dBi} for ζ_i and η_i between 0.5 and 0.7. For lower damping ratios the approximation is not so good but (Appendix A2) the error, $P_i(\omega) = M_{dBi}(\omega) - A_i(\omega)$, referred to as the resonance peak function, is useful in the parameter estimation process when the complex conjugate pole or zero pairs are close enough for the resonance peak functions to overlap.

In view of the close approach of $M_{dBi}(\omega)$ to $A_i(\omega)$ at frequencies removed from the associated corner frequency by more than an order of magnitude, if the corner frequencies, ω_i, ν_i, ω_{ni} and ν_{ni}, of a complete plant model are separated by at least two orders of magnitude, a piecewise linear asymptotic approximation, $L_{dB}(\omega)$ to $M_{dB}(\omega)$ can be formed by summing the two-segment concatenated asymptote functions, $A_i(\omega)$, from all of the terms in (2.201) together with the DC gain term and the linear term contributed by any pure integrators. Thus,

$$L_{dB}(\omega) = 20\log_{10}(K_{dc}) - 20q\log_{10}(\omega) + \sum_i A_i(\omega). \quad (2.218)$$

The vertices of $L_{dB}(\omega)$ occur at the corner frequencies, ω_i, ν_i, ω_{ni} and ν_{ni}. Since these are parameters of the required transfer function, finding an estimate, $\widehat{L}_{dB}(\omega)$, of $L_{dB}(\omega)$, using the graph of $M_{dB}(\omega)$ is the first step of the parameter estimation.

The closeness of approach of $\widehat{L}_{dB}(\omega)$ to $L_{dB}(\omega)$, is of paramount importance to ensure an accurate transfer function model. The task is aided by the knowledge that each of the straight line segments of $L_{dB}(\omega)$ has a slope that is an integral multiple of -20 [dB/decade], i.e. -6 [dB/octave]. It is usual for all the segment slopes to be zero or negative due to the domination of the poles of the transfer function. Bode plots may be found, however, containing segments of $L_{dB}(\omega)$ with positive slopes but these are usually those of the open loop transfer function, $G(s)G_p(s)$, of a control system designed with the aid of classical methods including a compensator with transfer function, $G_p(s)$, having the corner frequencies of its zeros lower than those of its poles.

If the corner frequencies are separated by at least two orders of magnitude, it is evident from the foregoing that $M_{dB}(\omega)$ will have almost straight portions with slopes of nearly $-20n$ [dB/decade], where n is an integer, enabling $\widehat{L}_{dB}(\omega)$ to be found by simply fitting tangents to $M_{dB}(\omega)$. If the corner frequencies are closer, however, the nearly straight portions of $M_{dB}(\omega)$ are less well defined. Although $\widehat{L}_{dB}(\omega)$ cannot be a good approximation to $M_{dB}(\omega)$ in these cases, $L_{dB}(\omega)$ still exists whose corner frequencies are the required plant parameters. It is therefore still important to find $\widehat{L}_{dB}(\omega)$. Means of calculating the required corner frequencies, using the graph of $M_{dB}(\omega)$, are developed in Appendix A2.

As all parameter estimation methods are subject to errors, it is recommended, in any case, to generate a Bode magnitude plot, $\widehat{M}_{dB}(\omega)$, from the estimated transfer function and compare this with $M_{dB}(\omega)$. If necessary, adjustments may be made to the transfer function parameters to bring $\widehat{M}_{dB}(\omega)$ closer to $M_{dB}(\omega)$.

2.3.3.6 Well-Separated Real Poles and Zeros

As shown in Sect. 2.3.3.5, the magnitudes of the real poles and zeros are the corner frequencies, ω_i and ν_i, of the Bode magnitude plot. If these are separated by at least two orders of magnitude, then the Bode magnitude plot contains segments that are nearly linear with relatively sharp changes of slope between them, which enable the piecewise linear approximation, $\widehat{L}_{dB}(\omega)$, to be fitted easily and accurately to the graph of $M_{dB}(\omega)$. Also, the vertices of $\widehat{L}_{dB}(\omega)$ may be assumed to be displaced vertically from $M_{dB}(\omega)$ by ± 3 [dB]. Moving from left to right along the Bode magnitude plot, a positive change of slope of $\widehat{L}_{dB}(\omega)$ at a vertex indicates a pole while a negative change of slope indicates a zero. Single poles or zeros cause, respectively, a decrease or increase in slope by 20 [dB/decade] (or 6 [dB/octave]). Occasionally a plant has a real pole of multiplicity, m, which can be detected by a decrease in slope of 20 m [dB/decade] (or 6 m [dB/octave]).

To demonstrate the method, suppose that the Bode magnitude plot, $M_{dB}(\omega)$, of the plant with transfer function,

$$\frac{Y(s)}{U(s)} = \frac{100\,(1+s/10)}{(1+s/0.1)^2\,(1+s/1{,}000)}, \qquad (2.219)$$

is given and it is required to find the corner frequencies by estimating $L_{dB}(\omega)$ using the tangent fitting. The Bode magnitude plot is shown in Fig. 2.26. The asymptotes required to form $\widehat{L}_{dB}(\omega)$ are denoted A_1, A_2, A_3 and A_4. These are drawn tangential to the nearly linear portions of $M_{dB}(\omega)$. The change of slope from asymptote, A_1, to asymptote, A_2, is $S_{d2} - S_{d1} = -40 - 0 = -40$ [dB/decade]. This therefore indicates a double pole. The corresponding corner frequency is $\omega_1 = 0.1$ [rad/s].

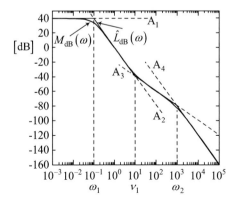

Fig. 2.26 Bode magnitude plot of third-order plant with a double pole, single pole and a zero

The change of slope from asymptote, A_2, to asymptote, A_3, is $S_{d3} - S_{d2} = -20 - (-40) = +20$ [dB/decade].

This therefore indicates a distinct zero. The corresponding corner frequency is $v_1 = 10$ [rad/s]. The change of slope from asymptote, A_3, to asymptote, A_4, is $S_{d4} - S_{d3} = -40 - (-20) = -20$, indicating a distinct pole. The corresponding corner frequency is $\omega_1 = 1,000$ [rad/s]. To complete the determination of the transfer function, the DC gain may be calculated using (2.205). Thus $K_{dc} = 10^{M_{db}(10^{-3})/20} = 10^{40/20} = 100$. The transfer function is therefore of the form,

$$\frac{Y(s)}{U(s)} = \frac{K_{dc}(1 + s/v_1)}{(1 + s/\omega_1)^2 (1 + s/\omega_2)}. \tag{2.220}$$

Inserting the parameter values calculated above yields transfer function (2.219).

2.3.3.7 Plants with Complex Conjugate Poles and Zeros

The following subsections develop methods for estimation of the parameters of transfer function factors corresponding to complex conjugate poles and zeros, by extracting information from the Bode magnitude plot. The next subsection focuses on second-order plants with one complex conjugate pair of poles and no finite zeros. The procedure developed for this model is applicable, without modification, to higher-order plants provided all the corner frequencies are separated by at least an order of magnitude, as shown in Sect. 2.3.3.9. In Appendix A2, this procedure is extended to cater for plants in which the corner frequencies corresponding to the complex conjugate poles and zeros may be made arbitrarily close.

Regarding notation, with reference to (2.199) the general term representing ω_{ni} or v_{ni} is υ_{ni} and that representing ζ_i or η_i is d_i.

2.3.3.8 Second Order Underdamped Model

Plants that can be modelled by the second-order underdamped plant model,

$$\frac{Y(s)}{U(s)} = \frac{K_{dc}}{1 + \frac{2\zeta}{\omega_{ni}} s + \frac{1}{\omega_{ni}^2} s^2}, \tag{2.221}$$

are considered here. The corresponding frequency domain transfer function is

$$\frac{Y(j\omega)}{U(j\omega)} = \frac{K_{dc}}{1 - \frac{\omega^2}{\omega_n^2} + \frac{2\zeta}{\omega_n} j\omega} \tag{2.222}$$

The magnitude function is therefore

2.3 Identification of LTI Plants from Measurements

$$M(\omega) = \frac{K_{dc}}{\sqrt{\left(1-\omega^2/\omega_n^2\right)^2 + 4\zeta^2\omega^2/\omega_n^2}}. \quad (2.223)$$

It will now be shown that a resonance peak occurs for $0 < \zeta < 1/\sqrt{2}$, an expression for which is derived below, enabling ζ to be determined from the Bode magnitude plot. For simplification, if any maximum of (2.223) exists then a minimum of

$$x(\omega) = \left(1-\omega^2/\omega_n^2\right)^2 + 4\zeta^2\omega^2/\omega_n^2 \quad (2.224)$$

also exists for the same value of ω. This satisfies

$$\frac{dx(\omega)}{d\omega} = \frac{8\zeta^2\omega}{\omega_n^2} - 4\left(1-\frac{\omega^2}{\omega_n^2}\right)\frac{\omega}{\omega_n^2} = 0 \Rightarrow 2\zeta^2 - 1 + \frac{\omega^2}{\omega_n^2} = 0$$
$$\Rightarrow \omega = \pm\omega_n\sqrt{1-2\zeta^2}. \quad (2.225)$$

The positive root is the required value of ω, which will be referred to as the *resonance frequency*, ω_r. This has to be real and non-zero for the resonance peak to exist, thereby restricting the damping ratio to the range,

$$0 < \zeta < 1/\sqrt{2}. \quad (2.226)$$

Then

$$\omega_r = \omega_n\sqrt{1-2\zeta^2}. \quad (2.227)$$

The resonance peak magnitude is given by (2.223) with $\omega = \omega_r$ of (2.227). Thus

$$M_p = \frac{K_{dc}}{\sqrt{(1-(1-2\zeta^2))^2 + 4\zeta^2(1-2\zeta^2)}} = \frac{K_{dc}}{2\zeta\sqrt{1-\zeta^2}}. \quad (2.228)$$

Note that for $\zeta = 1/\sqrt{2}$, (2.228) yields $M_p = K_{dc}$ and according to (2.227) this occurs at $\omega_r = 0$ so that $M(\omega)$ monotonically decreases with ω for $\omega > 0$ and therefore, in this case, there is no resonance peak.

On the Bode magnitude plot, the peak of (2.228) for $0 < \zeta < 1/\sqrt{2}$ is

$$M_{pdB} = 20\log_{10}(M_p) = 20\log_{10}(K_{dc}) + 20\log_{10}\left(\frac{1}{2\zeta\sqrt{1-\zeta^2}}\right). \quad (2.229)$$

The resonance peak w.r.t. the DC level (for $\omega \to 0$) is therefore as follows.

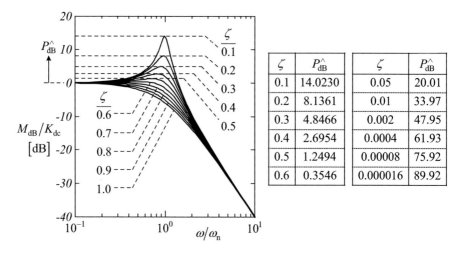

Fig. 2.27 Normalised Bode magnitude plots for underdamped second-order plant models

$$P_{dB}^{\wedge} = 20\log_{10}\left(\frac{1}{2\zeta\sqrt{1-\zeta^2}}\right). \quad (2.230)$$

Figure 2.27 shows a family of normalised Bode magnitude plots and the resonance peak values for different damping ratios.

Plots for $\zeta > 1/\sqrt{2} \cong 0.7071$ are included to demonstrate the lack of the resonance peak for these cases. Furthermore it is evident that the method is only practicable for $\zeta \leq 0.5$, since P_{dB} could not be accurately read for higher damping ratios. The first inset table of Fig. 2.27 indicates the values of P_{dB} calculated using (2.230) for the Bode magnitude plots shown. The second inset table indicates the values of P_{dB} for much lower damping ratios, which could be relevant to mechanical structures requiring vibration control, space satellites with flexible appendages being a case in point.

It remains to derive an equation for ζ using (2.230) to enable its estimation, given a reading from the Bode magnitude plot. First let

$$P = 10^{P_{dB}^{\wedge}/20}. \quad (2.231)$$

Then (2.230) becomes

$$P = \frac{1}{2\zeta\sqrt{1-\zeta^2}} \Rightarrow 4P^2\zeta^2\left(1-\zeta^2\right) = 1 \Rightarrow \zeta^4 - \zeta^2 + \frac{1}{4P^2} = 0. \quad (2.232)$$

Solving (2.232) for ζ^2 then yields

2.3 Identification of LTI Plants from Measurements

$$\zeta^2 = \frac{1 \pm \sqrt{1 - 1/P^2}}{2} \tag{2.233}$$

Since $P_{dB} \geq 0$, then according to (2.231), $P \geq 1$. If $P = 1$ then (2.233) yields $\zeta^2 = 1/2$. As P increases from 1, then ζ^2 must *reduce* from 1/2.

Hence the negative root is taken in (2.233). Finally the positive root is taken for ζ, resulting in

$$\zeta = \sqrt{\frac{1 - \sqrt{1 - 1/P^2}}{2}}. \tag{2.234}$$

Next, the resonance frequency, ω_r, is read from the Bode magnitude plot and ω_n is calculated using (2.227) with the estimate of ζ from (2.234). Thus

$$\omega_n = \frac{\omega_r}{\sqrt{1 - 2\zeta^2}}. \tag{2.235}$$

To complete the estimation of the transfer function, K_{dc} is determined using the method of Sect. 2.3.3.2.

2.3.3.9 Well-Separated Complex Conjugate Poles and Zeros

The procedure of Sect. 2.3.3.8 can be applied to estimate the parameters of complex conjugate pole factors, $1/\left(1 + \frac{2\zeta_i}{\omega_{ni}}s + \frac{1}{\omega_{ni}^2}s^2\right)$, or complex conjugate zero factors, $1 + \frac{2\eta_i}{\nu_{ni}}s + \frac{1}{\nu_{ni}^2}s^2$, of the transfer function if the corner frequencies are separated by at least an order of magnitude. First the amplitude, $P_{dBi}^{\wedge}$, of the measured resonance peak or anti-resonance dip is determined from the Bode magnitude plot, $M_{dB}(\omega)$. From (2.231) and (2.234),

$$d_i = \sqrt{\frac{1 - \sqrt{1 - \frac{1}{10^{P_{dBi}^{\wedge}/10}}}}{2}}. \tag{2.236}$$

where $d_i = \zeta_i$ for the pole factors and $d_i = \eta_i$ for the zero factors. Figure 2.28 shows graphs of this function that can be used to determine d_i directly from $P_{dBi}^{\wedge}$.

Figure 2.29 shows a Bode magnitude plot for the transfer function model,

$$\frac{Y(s)}{U(s)} = K_{dc} \left(\frac{1}{1 + \frac{2\zeta_1}{\omega_{n1}}s + \frac{1}{\omega_{n1}^2}s^2}\right)\left(\frac{1}{1 + \frac{2\zeta_2}{\omega_{n2}}s + \frac{1}{\omega_{n2}^2}s^2}\right). \tag{2.237}$$

The parameters are $\zeta_1 = 0.2$, $\zeta_2 = 0.1$, $\omega_{n1} = 1$ [rad/s], $\omega_{n2} = 100$ [rad/s] and $K_{dc} = 1$. The peak (or dip) magnitude may be determined by measuring the vertical distance of the tangent from the peak (or the dip) on $M_{dB}(\omega)$ parallel to one of the two asymptotes intersecting at the vertex of that peak (or dip). There are two alternatives for measuring each peak or dip magnitude. For each resonance peak (or anti-resonance dip) the tangent parallel to the left-hand asymptote meets $M_{dB}(\omega)$ at the resonance frequency while the tangent parallel to the right-hand asymptote meets $M_{dB}(\omega)$ at the mirrored resonance frequency. Using the left-hand tangents, on the vertical scale of this figure, the two resonance peak magnitudes are seen to be $\hat{P}_{dB1} \cong 8$ [dB] and $\hat{P}_{dB2} \cong 14$ [dB]. The left-hand graph of Fig. 2.28 then gives the correct values of $\zeta_1 = 0.2$ and $\zeta_2 = 0.1$. Reading the two resonance frequencies from Fig. 2.29 yields $\omega_{r1} \cong 0.96$ [rad/s] and $\omega_{r2} \cong 98$ [rad/s].

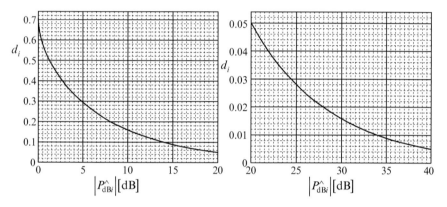

Fig. 2.28 Peak to damping ratio graphs

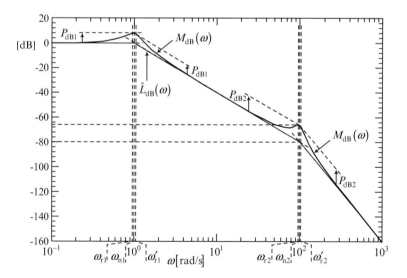

Fig. 2.29 Bode magnitude plot of plant with two underdamped well-separated modes

2.3 Identification of LTI Plants from Measurements

The undamped natural frequencies are then given by (2.235) as $\omega_{n1} = 1.001$ [rad/s] and $\omega_{n2} = 98.99$ [rad/s]. This precision is acceptable for this graphical method.

With reference to (2.201), the contributions of any complex conjugate zeros to the Bode magnitude plot are of the same form as the contributions from the complex conjugate poles but are opposite in sign. Complex conjugate zeros may therefore be recognised from dips in the plot, their contributions being reflections of those shown in Fig. 2.27 about the horizontal line, $P_{dB} = 0$. The frequencies at which the dips occur will be called *anti-resonance frequencies*, v_{ri}. It follows that the parameters, η_i and v_{ni}, may be estimated by a similar method to that above for ζ and ω_n.

Plant modelling with relatively close real poles and zeros or complex conjugate pole or pole–zero pairs is addressed in Appendix A2.

2.3.3.10 Zeros in the Right Half of the s-Plane

The transfer function model (2.188) may be modified with a few additions to cater for poles or zeros in the right or left halves of the s-plane, as follows.

$$\frac{Y(s)}{U(s)} = G(s) = K_{dc} \frac{\prod_i \left(1 + B_i \frac{s}{v_i}\right) \prod_i \left(1 + D_i \frac{2\eta_i}{v_{ni}} s + \frac{1}{v_{ni}^2} s^2\right)}{s^q \prod_i \left(1 + A_i \frac{s}{\omega_i}\right) \prod_i \left(1 + C_i \frac{2\zeta_i}{\omega_{ni}} s + \frac{1}{\omega_{ni}^2} s^2\right)} \quad (2.238)$$

Setting either A_i or B_i, to -1 then indicates that the ith real pole or zero is a right half plane [RHP] pole at $s_i = \omega_i$ or an RHP zero at $s_i = v_i$. Similarly setting C_i or D_i to -1 yields complex conjugate RHP pole pairs at $s_{i,i+1} = \omega_{ni} \left(\zeta_i \pm j\sqrt{1 - \zeta_i^2}\right)$ or RHP zero pairs at $s_{i,i+1} = v_{ni} \left(\eta_i \pm j\sqrt{1 - \eta_i^2}\right)$. Of course, if any coefficient is set to $+1$ the associated poles or zeros are left half plane (LHP) poles or zeros.

In the frequency domain, the magnitude of (2.238) is given by (2.200) since $|A_i| = |B_i| = |C_i| = |D_i| = 1$. Since $|G(j\omega)|$ is independent of A_i, B_i, C_i and D_i, the Bode magnitude plot is insufficient alone to determine the transfer function, unless the poles and zeros are all known to lie in the left half plane of the s-plane, which is often the case. Otherwise, the phase information may be used to determine in which half of the s-plane every pole and zero lies. Assuming $K_{dc} > 0$, the phase angle of transfer function (2.238) is given by

$$\phi(\omega) = \sum_i \phi_{rzi}(\omega) + \sum_i \phi_{czi}(\omega) + \sum_i \phi_{rpi}(\omega) + \sum_i \phi_{cpi}(\omega) \quad (2.239)$$

where

$$\phi_{rzi}(\omega) = \tan^{-1}\left(B_i \frac{\omega}{v_i}\right), \quad \phi_{czi}(\omega) = \tan^{-1}\left(\frac{2\eta_i D_i \frac{\omega}{v_{ni}}}{1-\frac{\omega^2}{v_{ni}^2}}\right)$$
$$\phi_{rpi}(\omega) = -\tan^{-1}\left(A_i \frac{\omega}{\omega_i}\right), \quad \phi_{cpi}(\omega) = -\tan^{-1}\left(\frac{2\zeta_i C_i \frac{\omega}{\omega_{ni}}}{1-\frac{\omega^2}{\omega_{ni}^2}}\right)$$
(2.240)

It is evident that as ω goes from 0 to ∞, the changes in the phase angle contributed by the various poles and zeros are as indicated in Table 2.2. Thus, if any real pole or zero, or any complex conjugate pole or zero pair is transferred to its mirror image location, reflected in the $j\omega$ axis, then the Bode magnitude plot will not change but the phase angle contribution of the transferred poles or zeros will change sign.

Figure 2.30 shows the graphs of the individual phase angle contributions of real poles and zeros and complex conjugate poles and zeros.

Four features may be observed in these phase functions as follows.

1. Each of the phase functions of (2.240) reaches half the maximum contribution indicated in Table 2.2 at the corner frequency. This is confirmed by substitution. Thus

$$\phi_{rpi}(\omega_i) = -A_i \frac{\pi}{4}, \phi_{rzi}(v_i) = B_i \frac{\pi}{4}, \phi_{cpi}(\omega_{ni}) = -C_i \frac{\pi}{2} \text{ and } \phi_{czi}(v_{ni}) = D_i \frac{\pi}{2}.$$

2. It appears that the slopes of the phase angle contributions on the RHS of (2.240) have maximum magnitudes at the corner frequencies, ω_i, v_i, ω_{ni} and v_{ni}, due to ω being on a logarithmic scale. This will now be proven. Let $x = \log(\omega)$. Then

$$\frac{dx}{d\omega} = \frac{1}{\omega} \Rightarrow \frac{d\omega}{dx} = \omega.$$
(2.241)

First consider the real zeros. Let the ith contribution be denoted

$$\phi_{rzi}(\omega) = \tan^{-1}(B_i \omega/v_i).$$
(2.242)

Then, noting that $B_i^2 = 1$,

$$\frac{d\phi_{rzi}(x)}{dx} = \frac{d\phi_{rzi}(\omega)}{d\omega} \cdot \frac{d\omega}{dx} = \frac{\omega}{1+\omega^2/v_i^2} \cdot \frac{B_i}{v_i}$$
(2.243)

Using (2.241), a maximum or minimum value of $d\phi_{rzi}(x)/dx$ occurs if

$$\frac{d^2\phi_{rzi}(x)}{dx^2} = 0 \Rightarrow \frac{d}{dx}\left(\frac{\omega}{1+\frac{\omega^2}{v_i^2}}\right) = 0 \Rightarrow \left[\frac{d}{d\omega}\left(\frac{\omega}{1+\frac{\omega^2}{v_i^2}}\right)\right]\frac{d\omega}{dx} = 0 \Rightarrow$$
$$\omega \frac{d}{d\omega}\left(\frac{\omega}{1+\frac{\omega^2}{v_i^2}}\right) = 0 \Rightarrow \frac{d}{d\omega}\left(\frac{\omega}{1+\frac{\omega^2}{v_i^2}}\right) = 0 \Rightarrow 1+\frac{\omega^2}{v_i^2} - \omega \cdot \frac{2\omega}{v_i^2} = 0 \Rightarrow \omega = v_i.$$
(2.244)

2.3 Identification of LTI Plants from Measurements

Table 2.2 Phase angle contributions of poles and zeros as ω goes from 0 to ∞

Key: LHP $\equiv$ Left Half Plane RHP $\equiv$ Right Half Plane	Real poles		Real zeros		Complex conjugate poles		Complex conjugate zeros	
	LHP: $A_i = +1$	RHP: $A_i = -1$	LHP: $B_i = +1$	RHP: $B_i = -1$	LHP: $C_i = +1$	RHP: $C_i = -1$	LHP: $D_i = +1$	RHP: $D_i = -1$
Contribution [deg]	0 to $-90°$	0 to $+90°$	0 to $+90°$	0 to $-90°$	0 to $-180°$	0 to $+180°$	0 to $+180°$	0 to $-180°$

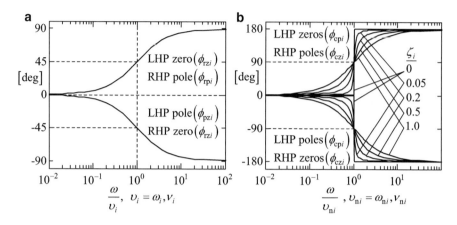

Fig. 2.30 Phase angle contributions of poles and zeros. (**a**) Real pole/zero. (**b**) Complex conjugate pole/zero pair

Since each contribution from the real poles is identical in form to (2.242), it follows that its slope has a maximum magnitude at $\omega = \omega_i$.

Now consider the complex conjugate zeros. Let each phase angle contribution from (2.239) be denoted

$$\phi_{czi}(\omega) = \tan^{-1}\left(\frac{2\eta_i D_i \omega/v_{ni}}{1 - \omega^2/v_{ni}^2}\right) \quad (2.245)$$

Then using (2.241) and noting that $D_i^2 = 1$,

$$\frac{d\phi_{czi}(\omega)}{dx} = \frac{d\phi_{czi}(\omega)}{d\omega} \cdot \frac{d\omega}{dx}$$

$$= \omega \cdot \left[\frac{1}{1+\left(\frac{2\eta_i D_i \omega/v_{ni}}{1-\omega^2/v_{ni}^2}\right)^2}\right] \cdot \frac{\left[(1-\omega^2/v_{ni}^2) - \omega \cdot (-2\omega/v_{ni}^2)\right]}{\left(1-\omega^2/v_{ni}^2\right)^2} \cdot \frac{2\eta_i}{v_{ni}} D_i .$$

Hence

$$\frac{d\phi_{czi}(\omega)}{dx} = \omega \cdot \frac{1 + \omega^2/v_{ni}^2}{\left(1-\omega^2/v_{ni}^2\right)^2 + (2\eta_i \omega/v_{ni})^2} \cdot \frac{2\eta_i}{v_{ni}} D_i . \quad (2.246)$$

A maximum or minimum of $\frac{d\phi_{czi}(\omega)}{dx}$ occurs if $\frac{d^2\phi_{czi}(x)}{dx^2} = 0 \Rightarrow$

2.3 Identification of LTI Plants from Measurements

$$\left[1 - 2\left(1 - 2\eta_i^2\right)\frac{\omega^2}{v_{ni}^2} + \frac{\omega^4}{v_{ni}^4}\right]\left(1 + \frac{3\omega^2}{v_{ni}^2}\right)$$
$$- \left(\omega + \frac{\omega^3}{v_{ni}^2}\right) \cdot 4\left(\frac{\omega^3}{v_{ni}^4} - \left(1 - 2\eta_i^2\right)\frac{\omega}{v_{ni}^2}\right) = 0. \quad (2.247)$$

Inserting $\omega = v_{ni}$ in the LHS of (2.247) yields $16\eta_i^2 - 16\eta_i^2 = 0$. Hence (2.247) is satisfied. The maximum slope magnitude of this phase contribution therefore occurs at $\omega = v_{ni}$. Since the phase contribution of each complex conjugate pole pair is of the same form as (2.245), then by inspection of (2.239), this contribution has the maximum slope magnitude at $\omega = \omega_{ni}$. The sign of the derivative in any case is equal to the sign of the corresponding contribution given in Table 2.2.

Setting $\omega = v_{ni}$ in (2.246) yields the derivative of maximum magnitude as

$$\left.\frac{d\phi_{czi}(v_{ni})}{dx}\right|_{max} = \frac{1}{\eta_i} D_i. \quad (2.248)$$

This explains the increase in the sharpness of the transition between 0 and the extreme value, $D_i\pi/2$, as the damping ratio, η_i, is reduced, which is visible in Fig. 2.30b. In the limit, as $\eta_i \to 0$, the maximum derivative of (2.248) becomes infinite and the phase function becomes a step function of ω, switching between 0 and $D_i\pi/2$ at $\omega = v_{ni}$. Similar relationships hold for $\phi_{cpi}(\omega)$.

3. With reference to Fig. 2.30, each phase function appears to be an odd function, meaning $\phi(-q) = -\phi(q)$, for all real q, if the origin were to be moved to the point where half the extreme value is reached at the corner frequency. This will now be proven. Consider first the real-zero phase function of (2.240). Thus

$$\phi_{rzi}(\omega) = \tan^{-1}\left(B_i \frac{\omega}{v_i}\right) \quad (2.249)$$

The origin has to be shifted to the point, $[\omega, \phi_{rzi}(\omega)] = [v_i, (\pi/4) B_i]$, with ω on a logarithmic scale. To achieve this, the variables are changed to

$$q = \log(\omega) - \log(v_i) = \log(\omega/v_i) \quad (2.250)$$

and

$$\theta_{rzi}(\omega) = \phi_{rzi}(\omega) - (\pi/4) B_i \quad (2.251)$$

From (2.250),

$$\omega/v_i = e^q. \quad (2.252)$$

Substituting for ω/v_i in (2.249) using (2.252) and then inserting the resulting expression for $\phi_{rzi}(\omega)$ in (2.251) then yields the following.

Fig. 2.31 A simple geometric aid

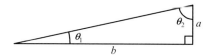

$$\theta_{rzi}(q) = \tan^{-1}(B_i e^q) - (\pi/4) \, B_i = B_i \left[\tan^{-1}(e^q) - \pi/4\right]. \quad (2.253)$$

If $\theta_{rzi}(q)$ is an odd function, then

$$\theta_{rzi}(q) + \theta_{rzi}(-q) = 0 \Rightarrow \tan^{-1}(e^q) + \tan^{-1}(e^{-q}) - \pi/2 = 0. \quad (2.254)$$

This can be proven geometrically using the right-angled triangle of Fig. 2.31.

Let the side lengths, a and b, be chosen such that

$$\frac{a}{b} = e^q \Rightarrow \frac{b}{a} = e^{-q} \quad (2.255)$$

Then using (2.255) and the geometry of Fig. 2.31,

$$\theta_1 + \theta_2 = \frac{\pi}{2} \Rightarrow \tan^{-1}\left(\frac{a}{b}\right) + \tan^{-1}\left(\frac{b}{a}\right) = \frac{\pi}{2} \Rightarrow \tan^{-1}(e^q) + \tan^{-1}(e^{-q}) = \frac{\pi}{2} \quad (2.256)$$

which satisfies (2.254).

Q.E.D.

This also applies to the real-pole function of (2.240) since it is of the same form as (2.249).

Similar analysis of the complex conjugate zero function of (2.240) will now be carried out. Thus

$$\phi_{czi}(\omega) = \tan^{-1}\left(2\eta_i D_i \frac{\omega}{\nu_{ni}} \Big/ \left(1 - \frac{\omega^2}{\nu_{ni}^2}\right)\right). \quad (2.257)$$

In this case, the origin has to be shifted to the point, $[\omega, \phi_{rzi}(\omega)] = [\nu_{ni}, (\pi/2) D_i]$.

The variables are therefore changed to

$$q = \log(\omega) - \log(\nu_{ni}) = \log(\omega/\nu_{ni}) \quad (2.258)$$

and

$$\theta_{czi}(\omega) = \phi_{czi}(\omega) - (\pi/2) D_i \quad (2.259)$$

From (2.258),

$$\omega/\nu_{ni} = e^q \quad (2.260)$$

2.3 Identification of LTI Plants from Measurements

Substituting for ω/ν_{ni} in (2.257) using (2.260) and then inserting the resulting expression for $\phi_{czi}(\omega)$ in (2.259) then yield

$$\theta_{czi}(q) = D_i \left[\tan^{-1}\left(\frac{2\eta_i e^q}{1-e^{2q}}\right) - \frac{\pi}{2} \right] = -D_i \cot^{-1}\left(\frac{2\eta_i e^q}{1-e^{2q}}\right) \quad (2.261)$$

If $\theta_{czi}(q)$ is an odd function, then so would be

$$f(q) = -D_i \cot[\theta_{czi}(q)] = \frac{2\eta_i e^q}{1-e^{2q}} \quad (2.262)$$

Then

$$f(q) + f(-q) = 0 \Rightarrow \frac{e^q}{1-e^{2q}} + \frac{e^{-q}}{1-e^{-2q}} = 0. \quad (2.263)$$

The LHS of (2.263) is $\dfrac{(1-e^{-2q})e^q + (1-e^{2q})e^{-q}}{(1-e^{2q})(1-e^{-2q})} = \dfrac{e^q - e^{-q} + e^{-q} - e^q}{2 - e^{2q} - e^{-2q}}$

$$= 0.$$

Q.E.D.

4. The observation of Fig. 2.30 indicates that if ω is lower than the corner frequency by at least an order of magnitude, the phase angle contribution is negligible but if ω is greater than the corner frequency by at least an order of magnitude, the phase angle contribution is near its maximum value. Consequently, if the corner frequencies of a transfer function are separated by an order of magnitude or more, the changes in the phase angle due to the individual real poles and zeros and the individual complex conjugate pole and zero pairs will be clearly visible, thereby identifying which half of the s-plane the poles and zeros lie.

The following demonstration stresses that the phase information is essential in frequency domain-based plant identification if it is not known in advance which half of the s-plane the poles and/or zeros lie. It also shows how the LHP and RHP allocations are made. The following three different plant transfer functions are taken since they have the same Bode magnitude functions.

$$\left.\begin{array}{c} G_1(s) = \dfrac{1}{s\left(1+\dfrac{s}{\omega_1}\right)} \cdot \dfrac{1+\dfrac{2\eta}{\nu_n}s+\dfrac{1}{\nu_n^2}s^2}{1+\dfrac{2\zeta}{\omega_n}s+!\dfrac{1}{\omega_n^2}s^2}, \quad G_2(s) = \dfrac{1}{s\left(1+\dfrac{s}{\omega_1}\right)} \cdot \dfrac{1+\dfrac{2\eta}{\nu_n}s+\dfrac{1}{\nu_n^2}s^2}{1-\dfrac{2\zeta}{\omega_n}s+\dfrac{1}{\omega_n^2}s^2} \\[2em] G_3(s) = \dfrac{1}{s\left(1+\dfrac{s}{\omega_1}\right)} \cdot \dfrac{1-\dfrac{2\eta}{\nu_n}s+\dfrac{1}{\nu_n^2}s^2}{1+\dfrac{2\zeta}{\omega_n}s+\dfrac{1}{\omega_n^2}s^2} \end{array}\right\}.$$

$$(2.264)$$

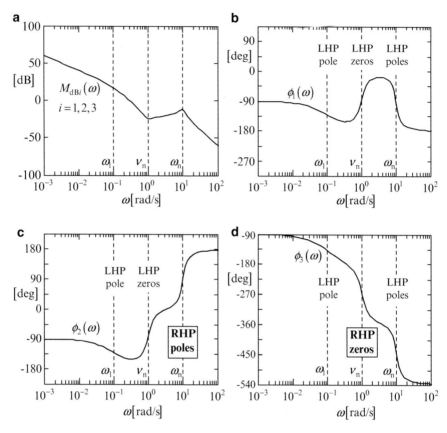

Fig. 2.32 Bode plots of plants with poles or zeros in the RHP. (**a**) Common Bode magnitude plot. (**b**) Phase angle: all LHP poles and zero. (**c**) Phase angle: RHP complex conj. poles. (**d**) Phase angle: RHP complex conj. zero

Here, $\omega_1 = 0.1$ [rad/s], $\omega_n = 10$ [rad/s], $\zeta = 0.2$, $\nu_n = 1$ [rad/s] and $\eta = 0.3$. Figure 2.32 shows the Bode plots.

First, the break frequencies are determined from the Bode magnitude plot of Fig. 2.32a using the methodology of the previous subsections. This enables the identification of a real pole with a corner frequency of ω_1, a complex conjugate zero pair with a corner frequency of ν_n and a complex conjugate pole pair with a corner frequency of ω_n. The damping ratios, ζ and η, can also be estimated using the methods of the previous sections.

The differences between the phase plots are very pronounced due to the separation of the corner frequencies by an order of magnitude. Essentially, the slope of the phase plot for each pole or zero at its corner frequency indicates the half of the s-plane in which the poles or zeros lie. LHP poles have phase plots with negative slopes and LHP zeros have phase plots with positive slopes at the corner frequency.

RHP poles have phase plots with positive slopes and LHP zeros have phase plots with negative slopes at the corner frequency.

2.3.4 Recursive Parameter Estimation: An Introduction

2.3.4.1 The z-Transfer Function Model

A digital controller provides a piecewise constant control input, usually with a constant update period of h seconds. Then a sample/hold unit is placed at the output of an LTI plant with the samples taken at the same instants as the control updates. It is then found that the plant, when observed through the piecewise constant input and sample/hold output obeys a linear difference equation. If the corresponding Laplace transform relationship between these signals is formed, then it is found that the complex variable, s, appears in the form, e^{sh}, throughout the transfer function. This is then simplified by defining a new complex transform variable,

$$z = e^{sh}. \qquad (2.265)$$

This gives the z-transfer function of the plant as the ratio of two polynomials in z, the general form of which is given at the beginning of the following subsection. This basic knowledge regarding the z-transfer function is all that is needed here but the underlying theory is fully developed in Chap. 3, Sect. 3.4.3.

2.3.4.2 Statement of the Problem and Approach

In this subsection, the problem of determining the z-transfer function model of a nominally linear time invariant plant by correlating sampled inputs and outputs is addressed. The z-transfer function model is in the general form,

$$\mathbf{Y}(z) = \frac{\mathbf{B}_1 z^{-1} + \mathbf{B}_2 z^{-2} + \cdots + \mathbf{B}_n z^{-n}}{1 + a_1 z^{-1} + a_2 z^{-2} + \cdots + a_n z^{-n}} \mathbf{U}(z), \qquad (2.266)$$

where $a_i, i = 1, 2, \ldots, n$, are constant characteristic polynomial coefficients, $\mathbf{B}_i \in \Re^{m \times r}, i = 1, 2, \ldots, n$, are constant matrices and n is the plant order. This models SISO plants by setting $m = r = 1$.

Recursive parameter estimation schemes exist that instead estimate the parameters of a discrete state space model once the state representation has been chosen. This, however, requires the theory covered in Chaps. 3 and 5, and is less straightforward than working with (2.266). In any case, a state space model (Chap. 3) can be formed from (2.266).

It is not possible to offer a universal solution to the recursive parameter estimation problem, which is rendered particularly difficult in practice by (a) the presence of plant and measurement noise, the latter being particularly troublesome, and (b) the uncertainty of the model order in some cases. Many different algorithms may be devised, all of which work 'perfectly' in an ideal noise-free environment and if the order of the chosen model is correct. In practice, the best approach depends upon the particular application. An exhaustive treatment of the subject warrants a dedicated work, such as that by Ljung [16]. The purpose of this subsection is to introduce the subject by developing a particular approach and presenting simulation-based demonstrations of applications. There is actually much room for invention here and the reader is encouraged to devise schemes and try them out.

The approach taken here is to form a difference equation corresponding to (2.266), which consists of one or more linear equations relating the regularly sampled input and output measurements to the transfer function coefficients. These equations are repeated for past input and output measurements to form a completely determined set that may be expressed in the matrix form,

$$\mathbf{M}(k)\mathbf{p} = \mathbf{v}(k), \qquad (2.267)$$

where $\mathbf{M}(k)$ is a matrix, which will be referred to as the *solution matrix*, and $\mathbf{v}(k)$ is a column vector, whose elements are the input and output measurements, k is the sample number and $\mathbf{p}$ is a column vector consisting of the plant parameters to be estimated. The estimate of $\mathbf{p}$ will be denoted by $\widehat{\mathbf{p}}$.

2.3.4.3 The Condition Number

Let the solution matrix, $\mathbf{M}(k)$ be square and non-singular. Consider the solution,

$$\widehat{\mathbf{p}}(k) = \mathbf{M}^{-1}(k)\mathbf{v}(k). \qquad (2.268)$$

Although, in theory, $\widehat{\mathbf{p}}$ is constant, in practice it will vary due to measurement noise (already introduced in Chap. 1) and plant noise (originating in the actuators and additive to the control inputs). Hence it is shown as a function of k. Under certain circumstances even in the hypothetical case of zero measurement and plant noise sources, $\widehat{\mathbf{p}}$ will vary significantly due to imperfect calculations imposed by the finite wordlength of any digital processor. This effect is marked if $\mathbf{M}(k)$ is *ill conditioned*. Conditioning is a term in numerical analysis pertaining to the accuracy of calculations. In general it is the proportional change in the result of a computation due to an erroneous proportional change of an input parameter due, for example, to rounding errors. Thus, (2.268) may or *may not* produce an accurate result and the conditioning is quantified by the *condition number* of the matrix, $\mathbf{M}(k)$, given by

$$\operatorname{cond}(\mathbf{M}(k)) = \frac{|\lambda_{\max}(k)|}{|\lambda_{\min}(k)|}, \qquad (2.269)$$

where $\lambda_{\max}(k)$ and $\lambda_{\min}(k)$ are, respectively, the eigenvalues of $\mathbf{M}(k)$ with the maximum and minimum magnitude [15]. The condition number therefore varies between 1 and ∞, the smaller being the better. For this reason, constant or slowly varying inputs are unsuitable. If the control input was constant and the plant stable, then all the variables would settle to constant values, resulting in singularity of $\mathbf{M}(k)$ and an infinite condition number due to at least one of the eigenvalues being zero. It is therefore necessary to excite the plant, preferably by random signals. These must be within the control saturation limits to ensure linear plant operation.

2.3.4.4 The Simultaneous Equations

The z-transfer function model of (2.266) is taken rather than the version in terms of positive powers of z, in order that the corresponding difference equation is in terms of accessible present and past sampled values of $\mathbf{y}(t)$ and $\mathbf{u}(t)$. Thus

$$\left(1 + a_1 z^{-1} + a_2 z^{-2} + \cdots + a_n z^{-n}\right) \mathbf{Y}(z) = \left(\mathbf{B}_1 z^{-1} + \mathbf{B}_2 z^{-2} + \cdots + \mathbf{B}_n z^{-n}\right) \mathbf{U}(z). \tag{2.270}$$

In the discrete domain, this becomes

$$\mathbf{y}(k) + a_1 \mathbf{y}(k-1) + a_2 \mathbf{y}(k-2) + \cdots + a_n \mathbf{y}(k-n)$$
$$= \mathbf{B}_1 \mathbf{u}(k-1) + \mathbf{B}_2 \mathbf{u}(k-2) + \cdots + \mathbf{B}_n \mathbf{u}(k-n) \Rightarrow$$
$$\mathbf{y}(k) = -a_1 \mathbf{y}(k-1) - a_2 \mathbf{y}(k-2) - \cdots - a_n \mathbf{y}(k-n)$$
$$+ \mathbf{B}_1 \mathbf{u}(k-1) + \mathbf{B}_2 \mathbf{u}(k-2) + \cdots + \mathbf{B}_n \mathbf{u}(k-n). \tag{2.271}$$

The basic approach will be to form sets of linear simultaneous equations in the plant parameters that are not underdetermined by taking sufficient input–output samples and solving them for the plant parameters. They will be expressed in the matrix–vector form and the first step towards this is to write the component equations of (2.271) as

$$y_i(k) = \begin{bmatrix} a_1 & \mathbf{b}_{i\,1}^{\mathrm{T}} & a_2 & \mathbf{b}_{i\,2}^{\mathrm{T}} & \cdots & a_n & \mathbf{b}_{i\,n}^{\mathrm{T}} \end{bmatrix} \begin{bmatrix} -y_i(k-1) \\ \mathbf{u}(k-1) \\ -y_i(k-2) \\ \mathbf{u}(k-2) \\ \vdots \\ -y_i(k-n) \\ \mathbf{u}(k-n) \end{bmatrix}, \quad i = 1, 2, \ldots, m, \tag{2.272}$$

where $\mathbf{b}_{ij}^T$ is the ith row of $\mathbf{B}_j$. Alternatively, (2.272) will be written as

$$y_i(k) = \underbrace{\left[-y_i(k-1)\ \mathbf{u}^T(k-1)\ -y_i(k-2)\ \mathbf{u}^T(k-2)\ \cdots\ -y_i(k-n)\ \mathbf{u}^T(k-n) \right]}_{\mathbf{w}_i^T}$$

$$\times \underbrace{\begin{bmatrix} a_1 \\ \mathbf{b}_{i1} \\ a_2 \\ \mathbf{b}_{i2} \\ \vdots \\ a_n \\ \mathbf{b}_{in} \end{bmatrix}}_{\mathbf{p}_i}. \qquad (2.273)$$

Then this may be written compactly as

$$y_i(k) = \mathbf{w}_i^T(k)\mathbf{p}_i, \quad i = 1, 2, \ldots, m. \qquad (2.274)$$

Since there are $N = (r+1)n$ elements in the plant parameter vector, $\mathbf{p}_i$, N linearly independent equations are needed to calculate $\mathbf{p}_i$. The first equation is (2.274) and a further $N-1$ equations are obtained by repeatedly forming (2.274) for past values of y_i, as follows.

$$\underbrace{\begin{bmatrix} y_i(k) \\ y_i(k-1) \\ \vdots \\ y_i(k-N+1) \end{bmatrix}}_{\mathbf{y}_i(k)} = \underbrace{\begin{bmatrix} \mathbf{w}_i^T(k) \\ \mathbf{w}_i^T(k-1) \\ \vdots \\ \mathbf{w}_i^T(k-N+1) \end{bmatrix}}_{\mathbf{W}_i(k)} \mathbf{p}_i, \quad i = 1, 2, \ldots, m \qquad (2.275)$$

The complete set of equations in the plant parameters is therefore

$$\mathbf{W}_i(k)\mathbf{p}_i = \mathbf{y}_i(k), \quad i = 1, 2, \ldots, m. \qquad (2.276)$$

For multivariable plants, the coefficients, a_i, $i = 1, 2, \ldots, n$, are estimated m times but advantage can be taken of this by averaging to obtain more accurate results.

2.3.4.5 Algorithm Avoiding Matrix Inversion

An apparently elegant method of solution of (2.276) that avoids the inversion of the solution matrix, $\mathbf{W}_i(k)$, in the directly calculated estimate,

2.3 Identification of LTI Plants from Measurements

$$\widehat{\mathbf{p}}_i = \mathbf{W}_i^{-1}(k)\mathbf{y}_i(k), \quad i = 1, 2, \ldots, m, \tag{2.277}$$

is to premultiply both sides of (2.276) by $\mathbf{W}_i^T(k)$ and form a matrix-vector differential equation, as follows:

$$\dot{\widehat{\mathbf{p}}}_i + \mathbf{W}_i^T(k)\mathbf{W}_i(k)\widehat{\mathbf{p}}_i = \mathbf{W}_i^T(k)\mathbf{y}_i(k). \tag{2.278}$$

Since the eigenvalues of $\mathbf{W}_i^T(k)\mathbf{W}_i(k)$ are positive and real, then if they are non-zero, $\dot{\widehat{\mathbf{p}}}(t) \to \mathbf{0}$ as $t \to \infty$ and (2.278) approaches $\mathbf{W}_i(k)\widehat{\mathbf{p}}_i = \mathbf{y}_i(k)$. It then follows by comparison with (2.276) that $\widehat{\mathbf{p}}_i(t) \to \mathbf{p}_i$ as required. If the plant is well excited then, in general, the smallest eigenvalues of $\mathbf{M}_i(k) = \mathbf{W}_i^T(k)\mathbf{W}_i(k)$ increase in value and the convergence rate increases. Conversely, if the plant approaches a steady-state condition, then the matrix, $\mathbf{M}_i(k)$, approaches singularity and, as already mentioned, at least one of its eigenvalues approaches zero. In this case the system 'gracefully fails' through its convergence rate reducing to zero instead of a numerical overflow upon attempting to compute (2.277).

The author has found this algorithm works for every plant he has tried but only in simulations in which noise contamination of the measured signals is absent.

2.3.4.6 Noise Contamination

The approach of Sect. 2.3.4.5 could only be successful in practice in cases where the matrix, $\mathbf{W}_i(k)$, is sufficiently well conditioned for plant noise, and particularly measurement noise, to have little effect. Suppose the plant is approaching a settled condition. Then in the presence of noise contamination, the minimum eigenvalue of the matrix, $\mathbf{M}_i(k)$, will tend to increase. In the extreme, if the plant states were constant, $\mathbf{M}_i(k)$ would, in theory, be singular but not so in practice, causing the plant parameter estimates to 'wander' towards incorrect values when using algorithm (2.278). Such random errors in the parameter estimates could be reduced, however, by taking more than the minimum of N component equations in (2.275). This would give $\mathbf{W}_i(k)$ more rows than columns, but $\mathbf{M}_i(k) = \mathbf{W}_i^T(k)\mathbf{W}_i(k)$ would still be square and of dimension, $N \times N$, rendering (2.278) workable. The redundant measurement samples would effectively be filtered in the process of forming $\widehat{\mathbf{p}}_i$, the algorithm performing a type of moving window averaging. In addition, passing $\widehat{\mathbf{p}}_i(k)$ through a low-pass filter with output, $\widehat{\mathbf{p}}_{fi}(k)$, will result in significant attenuation of the fluctuations in $\widehat{\mathbf{p}}_{fi}(k)$ compared with those of $\widehat{\mathbf{p}}_i(k)$. A suitable filter would be the IIR (infinite impulse response) filter with unity DC gain defined by

$$\widehat{\mathbf{p}}_{fi}(k+1) = \alpha\widehat{\mathbf{p}}_{fi}(k) + (1-\alpha)\widehat{\mathbf{p}}_i(k), \tag{2.279}$$

with $\alpha = e^{-h/T_f}$, where T_f is the time constant of the equivalent continuous low-pass filter.

There is, however, another problem that the filtering cannot remove alone. Errors in $\hat{\mathbf{p}}(k)$ with *non-zero mean values* can occur due to propagation of the noise contamination through squared elements of $\mathbf{W}_i(k)$ within $\mathbf{M}_i(k) = \mathbf{W}_i^T(k)\mathbf{W}_i(k)$. This phenomenon is referred to as *biasing*.

As a simple illustration, consider a single element, $w = \overline{w} + n$, where $\overline{w}$ is the signal that would have occurred without the noise and n is a noise signal with zero mean value. Incidentally, if the noise contamination is biased, then so will be the parameter estimation. Hence zero mean value of all noise signals is a mandatory condition. The squared element then gives

$$(\overline{w} + n)^2 = \overline{w}^2 + 2\overline{w}n + n^2, \tag{2.280}$$

in which the term, n^2, is responsible for the biasing.

Biasing is not a real problem in some applications but is very significant in others. Hence the following subsection is included.

2.3.4.7 Biasing Minimisation

The diagonal elements of $\mathbf{W}_i^T(k)\mathbf{W}_i(k)$ are w_{jji}^2, $j = 1, 2, \ldots, n$. Biasing can therefore never be eliminated in algorithm (2.278). Hence reverting to (2.277) is considered but with the protection of monitoring the condition number using (2.269), for which practicable real-time algorithms are available. This protection would be given by an algorithm of the following basic form.

$$\left\{ \begin{array}{l} \text{If cond}(\mathbf{W}_i(k)) < C_{\max}, \text{ then compute } \mathbf{p}_i = \mathbf{W}_i^{-1}(k)\mathbf{y}_i(k) \\ \text{else set } \mathbf{p}_i \text{ to the last computed value and skip the inversion.} \end{array} \right\}, \quad i = 1, 2, \ldots, m, \tag{2.281}$$

where $C_{\max} > 1$ is a selected threshold that can be very large, even of the order of 10^8 with floating processors.

Next, the question of biasing in (2.281) must be addressed. For this purpose, (2.277) can be written as

$$\mathbf{p}_i = \frac{\text{adj}\,[\mathbf{W}_i(k)]}{\det[\mathbf{W}_i(k)]}\mathbf{y}_i(k). \tag{2.282}$$

As will be recalled from (2.273) and (2.275), each row of $\mathbf{W}_i(k)$ is formed by shifting the row below two elements to the right, losing the last two elements and reforming the first two elements with new data samples. There are therefore $N - 2$ common elements between every adjacent row. As an example, consider an SISO third-order plant, for (2.273) becomes (2.283) below. The common elements between the rows of $\mathbf{W}_1(k)$ and between $\mathbf{y}_1$ and $\mathbf{W}_1(k)$ can clearly be seen. This, as it stands, would give rise to biasing in (2.282), partially due to the product, $\text{adj}\,[\mathbf{W}_1(k)]\mathbf{y}_1(k)$.

2.3 Identification of LTI Plants from Measurements

$$\underbrace{\begin{bmatrix} y(k) \\ y(k-1) \\ y(k-2) \\ y(k-3) \\ y(k-4) \\ y(k-5) \end{bmatrix}}_{\mathbf{y}_1} = \underbrace{\begin{bmatrix} -y(k-1) & u(k-1) & -y(k-2) & u(k-2) & -y(k-3) & u(k-3) \\ -y(k-2) & u(k-2) & -y(k-3) & u(k-3) & -y(k-4) & u(k-4) \\ -y(k-3) & u(k-3) & -y(k-4) & u(k-4) & -y(k-5) & u(k-5) \\ -y(k-4) & u(k-4) & -y(k-5) & u(k-5) & -y(k-6) & u(k-6) \\ -y(k-5) & u(k-5) & -y(k-6) & u(k-6) & -y(k-7) & u(k-7) \\ -y(k-6) & u(k-6) & -y(k-7) & u(k-7) & -y(k-8) & u(k-8) \end{bmatrix}}_{\mathbf{W}_1(k)} \underbrace{\begin{bmatrix} a_1 \\ b_1 \\ a_2 \\ b_2 \\ a_3 \\ b_3 \end{bmatrix}}_{\mathbf{p}_1}.$$

(2.283)

Biasing due to this product, however, could be eliminated by forming the successive rows from the component equations (2.273) of (2.275) separated by a delay of $n+1$ sampling periods instead of just one sampling period. Then $\mathbf{y}_i(k)$ and $\mathbf{W}_i(k)$ would be redefined and in the $n=3$ example, (2.283) would be replaced by

$$\underbrace{\begin{bmatrix} y(k) \\ y(k-4) \\ y(k-8) \\ y(k-12) \\ y(k-16) \\ y(k-20) \end{bmatrix}}_{\mathbf{y}_1} = \underbrace{\begin{bmatrix} -y(k-1) & u(k-1) & -y(k-2) & u(k-2) & -y(k-3) & u(k-3) \\ -y(k-5) & u(k-5) & -y(k-6) & u(k-6) & -y(k-7) & u(k-7) \\ -y(k-9) & u(k-9) & -y(k-10) & u(k-10) & -y(k-11) & u(k-11) \\ -y(k-13) & u(k-13) & -y(k-14) & u(k-14) & -y(k-15) & u(k-15) \\ -y(k-17) & u(k-17) & -y(k-18) & u(k-18) & -y(k-19) & u(k-19) \\ -y(k-21) & u(k-21) & -y(k-22) & u(k-22) & -y(k-23) & u(k-23) \end{bmatrix}}_{\mathbf{W}_1(k)} \underbrace{\begin{bmatrix} a_1 \\ b_1 \\ a_2 \\ b_2 \\ a_3 \\ b_3 \end{bmatrix}}_{\mathbf{p}_1}.$$

(2.284)

Now no common elements reside in the rows of $\mathbf{W}_1(k)$ or the vector $\mathbf{y}_1$. By analogy with (2.284) and with reference to (2.275) and (2.273), the general set of equations is as follows.

$$\underbrace{\begin{bmatrix} y_i(k) \\ y_i(k-(n+1)) \\ y_i(k-2(n+1)) \\ \vdots \\ y_i(k-(N-1)(n+1)) \end{bmatrix}}_{\mathbf{y}_i(k)}$$

$$= \underbrace{\begin{bmatrix} -y_i(k-1) & \cdots & \mathbf{u}^\mathrm{T}(k-n) \\ -y_i(n-(n+1)-1) & \cdots & \mathbf{u}^\mathrm{T}(k-(n+1)-n) \\ y_i(k-2(n+1)-1) & \cdots & \mathbf{u}^\mathrm{T}(k-2(n+1)-n) \\ \vdots & \cdots & \vdots \\ y_i(k-(N-1)(n+1)-1) & \cdots & \mathbf{u}^\mathrm{T}(k-(N-1)(n+1)-n) \end{bmatrix}}_{\mathbf{W}_i(k)} \underbrace{\begin{bmatrix} a_1 \\ b_1 \\ \vdots \\ a_n \\ b_n \end{bmatrix}}_{\mathbf{p}_i}$$

(2.285)

To consider the impact of noise contamination, in the same vein as (2.280), $\det[\mathbf{W}_i(k)]$ will contain products of the form, $(\overline{w}_i + n_i)(\overline{w}_j + n_j)\ldots, i \neq j \neq \ldots$, where $n_i, n_j, \ldots$ are uncorrelated noise signals with zero mean values. Then it is reasonable to suppose that each of the resulting noise signal product combinations, $n_i, n_j, \ldots$, will have zero mean values. Hence $\det[\mathbf{W}_i(k)]$ alone will not contribute any biasing. This is also true of the product, $\operatorname{adj}[\mathbf{W}_1(k)]\mathbf{y}_1(k)$, since $\operatorname{adj}[\mathbf{W}_1(k)]$ is made up of determinants, each of which will not produce biasing and the post-multiplication by $\mathbf{y}_1(k)$ just generates more product terms of the form already discussed, each of which will have zero mean errors due to the noise signals. The errors in $\det[\mathbf{W}_i(k)]$ and $\operatorname{adj}[\mathbf{W}_1(k)]\mathbf{y}_1(k)$ are, however, correlated, since they are formed from the same noisy signal samples. Unfortunately, therefore, it cannot be concluded that algorithm (2.282) is entirely free of biasing, but the author has not yet encountered an example in which this has been troublesome.

2.3.4.8 Filtering

In view of the foregoing, a practicable algorithm could be based on (2.281) with first-order low-pass filtering operating on the parameter estimates defined by (2.279). The filter (2.279) will be transformed into the z-domain (Chap. 3) in preparation for the system block diagram to be presented. Thus

$$z\widehat{\mathbf{P}}_{fi}(z) = \alpha \widehat{\mathbf{P}}_{fi}(z) + (1-\alpha)\widehat{\mathbf{P}}_i(z) \Rightarrow \widehat{\mathbf{P}}_{fi}(z) = \left(\frac{1-\alpha}{z-\alpha}\right)\mathbf{I}_N \widehat{\mathbf{P}}_i(z), i = 1, 2, \ldots, m,$$

(2.286)

where $\mathbf{I}_N$ is the unit matrix of dimension $N \times N$, which is introduced to remind the reader that in the physical implementation there are N identical first-order filters.

2.3 Identification of LTI Plants from Measurements

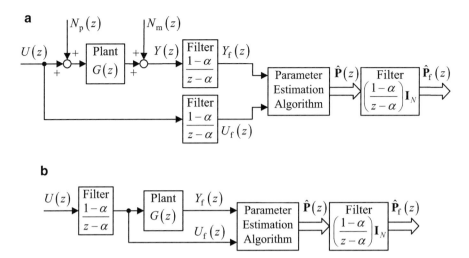

Fig. 2.33 Block diagram manipulation illustrating validity of input–output signal filtering. (**a**) Parameter estimation system with filtering. (**b**) Equivalent simplified deterministic system

In addition, it would be highly desirable to apply similar filtering to the noisy signals before they are processed by the parameter estimation algorithm (2.281), but not with an excessive time constant, T_f; otherwise, there will be insufficient short-term plant excitation to yield a well-conditioned data matrix, $\mathbf{W}_i(k)$. This form of filtering is valid only if identical filtering is applied to the control signals and the measurement signals, as illustrated in Fig. 2.33 for an SISO plant.

In Fig. 2.33a, $N_p(z)$ is the plant noise originating in the control actuators, referred to the plant input and $N_m(z)$ is the measurement noise originating in the measurement instrumentation and referred to the plant output. It is important to realise that since the parameter estimation algorithm is nonlinear, a z-transfer function relationship for it does not exist. The transformed variables are only valid mathematically for the transfer function relationships of the linear parts of the system, i.e. the plant and the filters. To show the filtering arrangement of Fig. 2.33a consider the hypothetical system operation in the absence of the plant and measurement noise signals. It may then be seen that Fig. 2.33b is obtained from Fig. 2.33a by block diagram manipulation for linear systems. In Fig. 2.33b, the plant input, $U_f(z)$, and the plant output, $Y_f(z)$ are fed directly to the parameter estimation algorithm, which will yield the correct estimates. Since these are the same signals as would be applied to the parameter estimator in the physical system of Fig. 2.33a, the correct estimates will be obtained by identically filtering the plant input and output separately, as shown.

2.3.4.9 Sampling Interval and Modal Timescale Ratio

In order for the set of input and output samples to contain sufficient information about the plant for accurate estimation of the transfer function coefficients, not only must the signal levels be sufficiently; high to mask the effects of any noise contamination and the limited number representation in the digital processor; but they must also capture the changes in the plant variables that result from its dynamical behaviour. This behaviour is characterised by the plant modes. Oscillatory and exponential modes have already been introduced in Chap. 1 as describing the behaviour of feedback control systems and are treated more mathematically in Chap. 3. The term, however, also applies to other dynamical systems, including plants taken in isolation without feedback control, whose modelling is the subject of this chapter. Two examples are briefly discussed in the following paragraphs.

In civil engineering, vibration modes occur in building structures. These are particular types of motion due to the combined elasticity and mass of connected elements in the structure, occurring at specific frequencies, called the eigenfrequencies. These are oscillatory modes.

In thermal systems, if the electrical power supply to an electric kiln is switched on, the temperature of the heating element will rise exponentially with a relatively short time constant towards a steady-state value. On the overall timescale of the kiln operation, the heating element time constant is usually negligible compared with the heating time constant of the workpiece. Then the workpiece temperature will rise nearly exponentially. In any case, the workpiece temperature rise has two exponential components. These are exponential modes.

The estimation window period, over which the recursive parameter estimation algorithm acquires a complete set of input–output samples is, with reference to (2.285), given by

$$T_w = [1 + (N-1)\ (n+1) - n]h = (n+1)Nh \qquad (2.287)$$

where h is the sampling period. In order to collect sufficient information about the dynamic response of the plant to the changing input, T_w, should not be very much less than the longest period of the oscillatory modes or the longest time constant of the exponential modes. If T_w is too small, in absence of noise, there would be insufficient changes of the variables over the window duration for the solution matrix, $\mathbf{W}_i(k)$, to be sufficiently well conditioned and the noise sources would cause unacceptable errors in the parameter estimates. On the other hand, T_w must not be very much greater than the shortest period of the oscillatory modes or the shortest time constant of the exponential modes; otherwise, the input–output samples will be too infrequent to capture sufficient information about the plant behaviour, again yielding inaccurate parameter estimation.

The period of an oscillatory mode or the time constant of an exponential mode are of the same order of magnitude as the reciprocals of the associated poles. These will be defined as the modal timescales. It will be recalled from Chap. 1 that (a) an exponential mode is a first-order mode and therefore only a single pole is associated

2.3 Identification of LTI Plants from Measurements

with the mode and (b) an oscillatory mode is a second-order mode associated with a complex conjugate pole pair, both poles sharing the same magnitude, given by the undamped natural frequency. The modal timescale of an exponential mode is therefore equal to its time constant, while the modal timescale of an oscillatory mode is $T/(2\pi)$, where T is the period of the oscillation at the undamped natural frequency.

As a general guideline, the estimation window period should satisfy

$$1 < T_{\text{w}}/T < 100 \tag{2.288}$$

for each exponential or oscillatory mode.

For a linear plant model, the modal timescale ratio may be defined as

$$R_{\text{m}} \triangleq \frac{|T_{\text{max}}|}{|T_{\text{min}}|} = \frac{|s_{\text{max}}|}{|s_{\text{min}}|}, \tag{2.289}$$

where $T_{\text{min}} = 1/s_{\text{max}}$ and $T_{\text{max}} = 1/s_{\text{min}}$, s_{max} and s_{min} being the modal poles with the largest and smallest *non-zero* magnitudes. As a general guideline,

$$1 < R_{\text{m}} < 100, \tag{2.290}$$

It must be born in mind, however, that the most appropriate limits of (2.288) and (2.290) could vary significantly from one plant to another.

Pure integrators, either distinct or multiple, are associated with particular forms of polynomial exponential modes (Chap. 1), which do not have modal impulse responses that decay on a finite timescale and are not found to pose problems. They are therefore not considered when determining the estimation window period, T_{w}.

If there is only one oscillatory mode or one exponential mode with a modal timescale of T, then only (2.288) has to be satisfied.

The modal timescale ratio is related to another similarly defined quantity called the stiffness ratio, which is the ratio between the largest and smallest real parts of the poles. Most readers will be familiar with the term stiffness in connection with elastic elements of mechanical systems and, by analogy, with control loops in which it is defined as $(de/du_{\text{d}})_{\text{ss}}$, where e is the error between the reference input and the controlled output, u_{d} is a constant external disturbance referred to the control input and the suffix, ss, refers to the steady-state condition. Such stiffness is generally brought about in control loops by means of high gain values. As will be seen in Chap. 10, this is often associated with a closed-loop pole with a large negative real value in the s-domain relative to a group of dominant poles. This gives a large stiffness ratio as defined above. This form of stiffness, not always associated with high control-loop gains, is a property of some dynamical systems that poses a challenge in obtaining an accurate numerical solution to the ordinary differential equations that model them [17]. The reader may have observed that some SIMULINK® simulations run slowly with systems containing fast and slow modes. In fact, various numerical integration algorithms are provided that are specifically designed to simulate stiff systems. If changing to one of these substantially speeds up the simulation, this is an indication that the system under study may be stiff. In view of this, it is not surprising that stiff plants pose problems when attempting recursive parameter estimation.

Unfortunately, for many plants, $R_m \gg 100$, but in some cases, model order reduction may be possible since the modes associated with poles having relatively large negative real parts will contribute far less to the dynamic response of the plant than modes associated with poles with much smaller real parts, which are dominant poles (Chap. 1). In these cases, recursive parameter estimation is possible with a reduced-order model, but it must be realised that the accuracy of the derived model will deteriorate to some extent for values of R_m on the borderline between having to estimate with a full or reduced-order model, i.e. $R_m \cong 100$.

Of the following two examples, the first is a plant for which the order of the estimated z-transfer function can be the same as the known plant order. The second is an example requiring model order reduction.

Example 2.1 Recursive parameter estimation for attitude control of a flexible spacecraft

The Laplace transfer function between the reaction wheel drive input, $U(s)$, and the rate gyro output, $Y(s)$, for attitude control of a flexible spacecraft is

$$\frac{Y(s)}{U(s)} = K_{\text{DC}} \frac{1 + s^2/v^2}{s(1 + s^2/\omega^2)}, \qquad (2.291)$$

where the free natural frequency is $\omega = 1$ [rad/s], the encastre natural frequency is $v = 0.8$ [rad/s] and the DC gain is $K_{\text{DC}} = 1$.

To determine a suitable estimation window duration, T_w, since there is just one oscillatory mode, the integrator not being considered, the modal time-scale ratio of (2.289) does not apply.

Since the plant order is $n = 3$, the number of plant parameters is $N = 2n = 6$. The algorithm is therefore given by (2.284). Evaluating (2.287) with $h = 0.5$ [s] then yields $T_w = (n+1)Nh = 12$ [s]. The modal timescale of the oscillatory mode is $T = 1/\omega = 1$ [s], which is of the same order as T_w. So according to (2.288), the choice of h is suitable for the application. Figure 2.34 shows some simulation results.

The model of (2.291) is implemented in MATLAB®–SIMULINK® with zero-order sample and hold unit placed at the output with a sampling period of $h = 0.5$ [s] and a simulation run with the parameter estimation algorithm of Sect. 2.3.4.7, shown in Fig. 2.33a, with $C_{\max} = 10^4$ and the filtering of Sect. 2.3.4.8 with $T_f = 40$ [s]. This value was arrived at by repeated estimation runs, increasing T_f in steps until the random variations of the filtered parameter estimates were reduced to acceptable proportions. A total estimation time of 400 [s] is taken to allow the filtered parameter estimates to reach steady-state (practically). The z-transfer function model obtained is then simulated alongside the sampled continuous model to check that the parameter estimation has been successful.

2.3 Identification of LTI Plants from Measurements

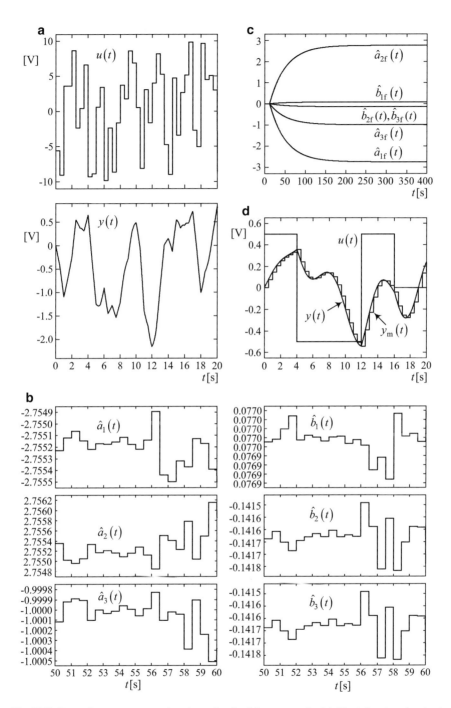

Fig. 2.34 Recursive parameter estimation of a flexible spacecraft. (**a**) Plant input and output variables. (**b**) Unfiltered parameter estimates. (**c**) Filtered parameter estimates. (**d**) Responses of plant and model

The stimulation signal is a piecewise constant random sequence input voltage with a uniform statistical distribution between the reaction wheel drive saturation limits of ±10 [V], synchronised with the 0.5 [s] sampling. This is shown in Fig. 2.34a.

The reaction wheel noise added to this signal has a Gaussian distribution with a 3σ value of 1 [mV]. The measurement noise signal added to the sensor output is also Gaussian at 1.5×10^{-4} [V] $\equiv 0.54$ [arc sec /s], 3σ, referred to the attitude measurement, noting that 1 [arcsec] $= 1/3600$ [deg], which is realistic with a rate gyro of moderate accuracy. Figure 2.34b shows the unfiltered estimates of the z-transfer function coefficients, updated at $h = 0.5$ [s]. A ten second window is shown so that the piecewise constant estimates can be seen. The fluctuations are, of course, due to the measurement and plant noise inputs. The filter initial conditions are zero, resulting in the exponential convergence of the filtered parameter estimates towards the required constant values, as shown in Fig. 2.34c.

These values were inserted in a SIMULINK® z-transfer function and a piecewise constant input, $u(t)$, applied over a 20 [s] period, resulting in a discrete model output, $y_m(t)$. This is shown together with the output, $y(t)$, of the continuous-time plant simulation in Fig. 2.34d, indicating no visible errors on the scale of the graph.

Example 2.2 Recursive parameter estimation for plate angle control in a throttle valve

There are many examples of electromagnetic control actuators throughout industry that have very small time constants associated with the electrical part of the model and much longer time constants associated with the mechanical part of the model. In such cases model order reduction effectively ignores the electrical time constant. This example is a case in point.

The air-to-fuel ratio of an internal combustion engine is controlled by means of a throttle valve consisting of a pivoted plate mounted in the air intake tube, driven by a DC motor through a gear system. A pre-windup coil spring applies a residual torque ensuring the valve is open in case of an electrical failure. The plate position is measured by a position sensor attached to the plate. The linearised continuous-time model supplied courtesy of Delphi Diesel Systems Ltd has the transfer function

$$\frac{Y(s)}{U(s)} = \frac{K_{dc}}{(1+sT_1)(1+sT_2)(1+sT_3)}, \qquad (2.292)$$

where $K_{dc} = 1.8136$, $T_1 = 0.5464$ [s], $T_2 = 0.0266$ [s] and $T_3 = 3.03 \times 10^{-4}$ [s]. The time constant, T_3, is associated with the armature circuit of the DC motor, the time constant, T_2, is associated with the inertia and viscous friction of the moving plate and the time constant, T_3, results from the coil spring, whose spring constant is insufficient to give the system a complex conjugate pole pair. As it stands, this model has a modal time-scale ratio of $R_m = T_1/T_3 = 1803.3$, which by far violates (2.290). In fact, attempts at estimating a third-order model (not displayed) proved to be totally intolerant of the realistic noise levels. A reduced-order model, however,

2.3 Identification of LTI Plants from Measurements

would ignore T_3 and yield a new modal time-scale ratio of $R_m = T_1/T_3 = 1803.3$, which satisfies (2.290). The z-transfer function to be estimated is then

$$\frac{Y(z)}{U(z)} = \frac{b_1 z^{-1} + b_2 z^{-2}}{1 + a_1 z^{-1} + a_2 z^{-2}} \quad (2.293)$$

For $n = 2$, the number of plant parameters is $N = 2n = 4$. Then taking $h = 0.1$ [s], (2.287) gives the estimation window duration as $T_w = (n+1)Nh = 1.2$ [s]. Then $T_w/T_1 = 2.196$ and $T_w/T_2 = 45.11$, both of which satisfy (2.288).

The model of (2.292) will be implemented in MATLAB®–SIMULINK® with a zero-order sample and hold unit c sampling period of $h = 0.5$ [s]. A simulation is then run with the parameter estimation algorithm of Sect. 2.3.4.7, shown in Fig. 2.33a, with $C_{max} = 10^4$ and the filtering of Sect. 2.3.4.8 with $T_f = 50$ [s]. This value was arrived at by repeated estimation runs, increasing T_f in steps until the random variations of the filtered parameter estimates were reduced to acceptable proportions. A total estimation time of 300 [s] is taken to allow the filtered parameter estimates to reach a sufficiently close approximation to their steady-state values. The stimulation signal is a piecewise constant random input voltage sequence with levels having a uniform statistical distribution between limits of 0 [V] and 5 [V], to keep the plate between its end stops corresponding to an output voltage range of $0 \text{ [V]} \leq y \leq 10 \text{ [V]}$.

This is updated to synchronise with the 0.1 [s] sampling, as shown in Fig. 2.35a. The plant noise added to this signal has a Gaussian distribution with a 3σ value of 1 [mV]. Similarly, the measurement noise signal added to the plate angle sensor output is also Gaussian with a 3σ value of 1 [mV]. Figure 2.35b shows the unfiltered estimates of the z-transfer function coefficients, updated at $h = 0.1$ [s]. A five second window is shown so that the piecewise constant estimates can be seen. The filter initial conditions are zero, resulting in the exponential convergence of the filtered parameter estimates towards the required constant values, as shown in Fig. 2.35c. These values were inserted in a SIMULINK® z-transfer function with a ramped input, $u(t)$, which is often used in this application, over a 3 [s] period, resulting in a discrete model output, $y_m(t)$. This is shown together with the output, $y(t)$, of the continuous-time plant simulation in Fig. 2.35d. In this case, a small error between the original plant output, $y_m(t)$, and the estimated discrete model output, $y_m(t)$, is visible, which is attributed to the model order reduction rather than biasing. This has been confirmed by carrying out the whole of the above parameter estimation simulation again but with zero noise sources, no change of the error relative to that of Fig. 2.35d being visible.

In some applications, the sampling periods required for recursive parameter estimation may be too long for satisfactory discrete control to be attainable. To solve this problem, means of changing the coefficients of a z-transfer function plant model to cater for a reduced sampling period are given in Chap. 3, Sect. 3.4.4.

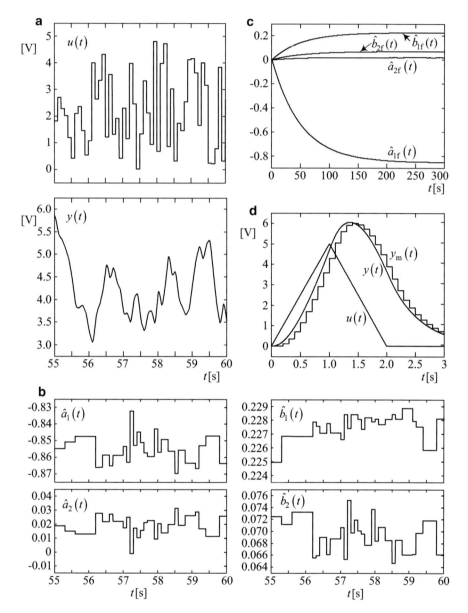

Fig. 2.35 Recursive parameter estimation of a throttle valve for Diesel engines. (**a**) Plant input and output variables. (**b**) Unfiltered parameter estimates. (**c**) Filtered parameter estimates. (**d**) Responses of plant and model

References

1. Batchelor GK (2000) An introduction to fluid dynamics, 2nd edn. Cambridge University Press, Cambridge
2. Popov VL (2010) Contact mechanics and friction. Springer-Verlag, Berlin Heidelberg
3. Taylor JR (2005) Classical mechanics. University Science Books, Herndon, USA. ISBN 1-891389-22
4. Kasdin NJ, Paley DA (2011) Engineering dynamics: a comprehensive introduction. Princeton University Press, Princeton, New Jersey
5. Junkins J, Turner JD (1986) Optimal spacecraft attitude maneuvers. Elsevier, Amsterdam, The Netherlands
6. Quang NP, Dittrich JA (2008) Vector control of three-phase AC machines. Springer, Berlin Heidelberg
7. Haitham AR et al (2012) High performance control of AC drives with Matlab/Simulink models. Wiley, Hoboken, New Jersey
8. Kothari DP, Nagrath IJ (2004) Electric machines. Tata McGraw Hill, New Delhi
9. Pyrhonen J et al (2013) Design of rotating electrical machines. Wiley, Chichester, England
10. Vittek J, Dodds SJ (2003) Forced dynamics control of electric drives. University of Zilina Press, Zilina, Slovakia. ISBN 80-8070-087-7
11. Chiasson JN (2005) Modelling and high performance control of electric machines. Wiley, Hoboken, New Jersey
12. Jaluria Y (2007) Design and optimization of thermal systems, 2nd edn. CRC Press/Taylor and Francis, Boca Raton, Florida
13. Landau ID, Zito G (2006) Digital control systems: design, identification and implementation. Springer-Verlag, London. ISBN 1846280567
14. James JF (2011) A students guide to fourier transforms with applications in physics and engineering. Cambridge University Press, New York. ISBN 0 521 80826/00428
15. Cheney W, Kincaid D (2008) Numerical mathematics and computing. Thomson Brooks/Cole, Monterey. ISBN 978-0-521-17683-5
16. Ljung L (1999) System identification: theory for the user. Prentice Hall, New Jersey
17. Iserles A (2009) A first course in the numerical analysis of differential equations. Cambridge University Press, New York

Chapter 3
Plant Model Manipulation and Analysis

3.1 Introduction

In Chap. 2, two means of obtaining plant models were presented. First, the physical modelling led to differential equations parameterised by coefficients relating directly to the physical components of the plant. Second, the identification from input and output measurements led to transfer functions either in the continuous Laplace domain or in the discrete z-domain. While these models can be used directly for control system design, other forms of plant model exist, which are presented in this chapter. Some can be more convenient for specific control techniques and others lend themselves more readily for analysis to determine the nature of the plant, which will influence the choice of the control technique. This chapter presents these other forms of plant model and the means of converting from one to another, as needed for the application of specific control techniques..

Controllability and observability analyses are included. Controllability, as its name suggests is a property that a plant must have in order for it to be possible to design an effective controller. In many cases the plant model is sufficiently simple for this matter to be resolved by inspection. In less obvious cases, however, it is advisable to form a plant model as early as possible, preferably prior to the physical construction of the plant. Then, if the controllability analysis proves negative, the plant design can be amended to avoid having to implement expensive modifications at a later stage. Observability is concerned with the ability to extract sufficient information about the plant behaviour from the measured output response to given inputs, to enable the plant to be controlled. In many cases, it is known by experience, especially for relatively simple plants, that an effective controller may be readily designed, but again, for less obvious cases it is advisable to carry out this analysis before the final plant construction to enable any additional measurement instrumentation to be included if necessary.

3.2 The State Space Model

3.2.1 Introduction

3.2.1.1 The Concept of State

The state is a set of variables of a dynamical system equal in number to its order. Its understanding is of fundamental importance in the modelling and control of plants. Essentially, the state of a plant is a minimal set of its variables that enables its future behaviour to be predicted under a known control input. It may be regarded as a minimal set of variables that define the present plant behaviour. These are referred to as state variables. It might be argued that creating a controller that uses these state variables enables the best control to be attained within the physical limitations of the hardware such as control saturation limits. This is found to be true in most cases but there can be exceptions necessitating hardware changes to the plant, as highlighted in Sects. 3.2.5, 3.2.6, and 3.2.7.

As an introduction, a set of state variables will now be identified for the following general model of an SISO linear plant of order, n, with constant coefficients.

$$\frac{d^n y}{dt^n} + a_{n-1}\frac{d^{n-1} y}{dt^{n-1}} + \cdots a_1\frac{dy}{dt} + a_0 y = c_{n-1}\frac{d^{n-1} u}{dt^{n-1}} + \cdots + c_1\frac{du}{dt} + c_0 u, \quad (3.1)$$

where $a_i, i = 0, 2, \ldots, n-1$ and $c_i, i = 0, 2, \ldots, n-1$ are constant plant parameters, y is the measurement variable to be controlled and u is the control variable. To carry out the task in hand, (3.1) will be separated into two parts as follows. Consider the following associated system without derivatives on the RHS.

$$\frac{d^n x}{dt^n} + a_{n-1}\frac{d^{n-1} x}{dt^{n-1}} + \cdots a_1\frac{dx}{dt} + a_0 x = u. \quad (3.2)$$

Also, suppose that the initial conditions are zero for (3.1) and (3.2). Since the model is linear, the principle of superposition applies. Also since $x(t)$ is the output with input $u(t)$, then $\frac{d^i x}{dt^i}$ is the output with input, $\frac{d^i u}{dt^i}$, $i = i = 0, 1, \ldots, n-1$. Hence if u is replaced by the weighted sum of u and its derivatives on the RHS of (3.1), x is replaced by the same weighted sum of x and its derivatives. This must also be y. Thus

$$y = c_{n-1}\frac{d^{n-1} x}{dt^{n-1}} + \cdots + c_1\frac{dx}{dt} + c_0 x. \quad (3.3)$$

Then (3.2) and (3.3) are, together, equivalent to plant model (3.1) with $x(t)$ as an intermediate variable.

Recalling the theory of ordinary differential equations with constant coefficients, the general solution, $x(t)$, to (3.2) is given by

3.2 The State Space Model

$$x(t) = x_c(t) + x_p(t), \qquad (3.4)$$

where $x_c(t)$ is the complementary function that depends on the initial conditions, $x^{(i)}(0)$, $i = 0, 1, \ldots, n - 1$, where $x^{(i)} \triangleq d^i x/dt^i$, and $x_p(t)$ is the particular integral that depends upon the forcing function which, in this case, is the control variable, $u(t)$. The initial conditions therefore constitute all the dynamical information about the plant that must be known in order to be able to predict its future behaviour under $u(t)$. These initial conditions could be described as the *initial state* of the plant. Now, suppose the solution is 'frozen' at an arbitrary time, $t = t_1$. Then $x^{(i)}(t_1)$, $i = 0, 1, \ldots, n - 1$, may be considered as a new set of initial conditions enabling the behaviour of the plant for $t > t_1$ to be predicted under the influence of a given control variable, $u(t)$. This set of 'initial' conditions may therefore be described as the *state* of the plant at time, $t = t_1$. Thus, the set of plant variables, $x^{(i)}(t)$, $i = 0, 1, \ldots, n - 1$, are individually referred to as *state variables* and collectively may be described as the *state* of the plant at time, t. The concept of state applies to any dynamical system, including nonlinear ones and general means of determining state variables will be introduced. The following definition of state is widely accepted.

Definition 3.1 *The state of a dynamical system is the minimal set of its variables, whose knowledge enables its future behaviour to be predicted under a known input stimuli.*

3.2.1.2 State Space and State Trajectories

The term 'state space' is used since the set of n state variables of a plant of nth order, $(x_1, x_2, \ldots, x_n)$, may be regarded as the coordinates of a point in a Euclidian space of n dimensions, called the *state point*. Over a period of time, the state variables usually change and consequently the state point moves on a path through the state space, called a *state trajectory*. This may be readily visualised and even displayed graphically for $n = 1, 2$, and 3. Indeed, this will prove to be very useful in Chaps. 8 and 9.

3.2.2 Forming a State-Space Model

Any state-space model consists of a set of first-order differential equations some of whose input and output variables are common, referred to as *state differential equations*. In fact, the numerous sets of first-order differential equations derived in Chap. 2 for the subsystems of the physical plant models are state differential equations of those plant models.

Continuing with the linear SISO plant model defined by (3.2) and (3.3), a set of state variables, $x^{(i)}(0)$, $i = 0, 1, \ldots, n-1$, has been established. In the standard notation, these state variables are

$$x_i(t) = x^{(i-1)}(t), \quad i = 1, 2, \ldots, n. \tag{3.5}$$

The first $n-1$ state differential equations are obtained by differentiating (3.5). Thus, $\dot{x}_i(t) = x^{(i)}(t)$, $i = 1, 2, \ldots, n-1$, which may be written as

$$\begin{aligned} \dot{x}_1 &= x_2 \\ \dot{x}_2 &= x_3 \\ &\vdots \\ \dot{x}_{n-1} &= x_n \end{aligned} \tag{3.6}$$

The final equation is obtained from (3.2), which may be written as

$$x^{(n)} + a_{n-1}x^{(n-1)} + \cdots + a_1 x^{(1)} + a_0 x^{(0)} = u. \tag{3.7}$$

From (3.5), $x_n = x^{(n-1)} \Rightarrow \dot{x}_n = x^{(n)}$. So using (3.7) and (3.5), the final state differential equation is

$$\dot{x}_n = u - (a_0 x_1 + a_1 x_2 + \cdots + a_{n-1} x_n). \tag{3.8}$$

The plant state-space model is completed by writing (3.3) as

$$y = c_0 x_1 + c_1 x_2 + \cdots + c_{n-1} x_n. \tag{3.9}$$

This an example of a *measurement* equation.

3.2.3 The General State-Space Model

The most general plant state-space model may be written as

$$\begin{aligned} \dot{\mathbf{x}} &= \mathbf{f}(\mathbf{x}, \mathbf{u}, \mathbf{d}, t) \text{ (state differential equation)} \\ \mathbf{y} &= \mathbf{h}(\mathbf{x}, \mathbf{u}, t) \text{ (measurement or output equation)} \end{aligned}, \tag{3.10}$$

where $\mathbf{x} \in \Re^n$ represents the set of state variables, $(x_1, x_2, \ldots, x_n)$; $\mathbf{u} \in \Re^r$ represents the set of control variables, $(u_1, u_2, \ldots, u_r)$; $\mathbf{d} \in \Re^r$ represents the set of external disturbances, $(d_1, d_2, \ldots, d_r)$, referred to the control inputs and $\mathbf{y} \in \Re^m$ represents the set of measurement (or output) variables, $(y_1, y_2, \ldots, y_m)$, where usually $m = r$. The individual equations of (3.10) are

3.2 The State Space Model

$$\dot{x}_i = f_i\left[(x_1, x_2, \ldots, x_n), (u_1, u_2, \ldots, u_r), (d_1, d_2, \ldots, d_p), t\right], \quad i = 1, 2, \ldots, n$$
$$y_j = h_j\left[(x_1, x_2, \ldots, x_n), (u_1, u_2, \ldots, u_r), t\right], \quad j = 1, 2, \ldots, m.$$
(3.11)

In contrast with all the examples presented in Chap. 2, **y** is a function of **u**, which is called *input feedforward*. This is included for generality and it is commonly found in the literature, but has never been present in any real plant encountered by the author, to date. No controlled output of a real physical plant can respond instantaneously to a step change in any control input, unless the plant manufacturers purposely created a direct path from the control variable to the measurement instrumentation, which is unlikely! Other dynamical systems do exist, however, with input feedforward, that are not controlled plants, an example being a crossover filter in an audio reproduction system including a high-pass filter section, so the term is left in the general model and the linear model of the following section.

If *any* of the component functions on the right-hand side of (3.11) are nonlinear, then the plant is nonlinear. Specific examples of nonlinear plants have already been met in Chap. 2 and more will appear in Chaps. 7, 8, 9 and 10. If, on the other hand, all the component functions on the right hand side of (3.11) are linear functions of their arguments, then the plant is linear and can be formulated as in the following section.

If the argument, t, is absent from every term in (3.11), then it is defined as *time invariant*, or *stationary*; otherwise, it is defined as *time varying* or *nonstationary*.

It is most important to mention at this point that no state-space plant model is unique. In theory, there exist infinitely many different models of a given plant because there are infinitely many sets of state variables that may be chosen. This subject is addressed at length in Sect. 3.3.

3.2.4 The General LTI State-Space Model

If the state-space model (3.11) is linear and time invariant (LTI), then it becomes

$$\dot{x}_i = \sum_{j=1}^{n} a_{ij} x_j + \sum_{k=1}^{r} b_{ik} u_k - \sum_{l=1}^{p} e_{il} d_l, \quad i = 1, 2, \ldots, n$$
$$y_i = \sum_{j=1}^{h} c_{ij} x_j + \sum_{k=1}^{r} d_{ik} u_k, \quad i = 1, 2, \ldots, m,$$
(3.12)

which may be expressed in the matrix form

$$\begin{bmatrix} \dot{x}_1 \\ \dot{x}_2 \\ \vdots \\ \dot{x}_n \end{bmatrix} = \begin{bmatrix} a_{11} & a_{12} & \cdots & a_{1n} \\ a_{21} & a_{22} & & \vdots \\ \vdots & & \ddots & \\ a_{n1} & \cdots & & a_{nn} \end{bmatrix} \begin{bmatrix} x_1 \\ x_2 \\ \vdots \\ x_n \end{bmatrix} + \begin{bmatrix} b_{11} & \cdots & b_{1r} \\ b_{21} & & b_{2r} \\ \vdots & & \vdots \\ b_{n1} & \cdots & b_{nr} \end{bmatrix} \begin{bmatrix} u_1 \\ \vdots \\ u_r \end{bmatrix}$$

$$- \begin{bmatrix} e_{11} & \cdots & e_{1p} \\ e_{21} & & e_{2p} \\ \vdots & & \vdots \\ e_{n1} & \cdots & e_{np} \end{bmatrix} \begin{bmatrix} d_1 \\ \vdots \\ d_p \end{bmatrix}$$
(3.13)

$$\begin{bmatrix} y_1 \\ y_2 \\ \vdots \\ y_m \end{bmatrix} = \begin{bmatrix} c_{11} & c_{12} & \cdots & c_{1n} \\ c_{21} & c_{22} & \cdots & c_{2n} \\ \vdots & & & \vdots \\ c_{m1} & \cdots & & c_{mn} \end{bmatrix} \begin{bmatrix} x_1 \\ x_2 \\ \vdots \\ x_n \end{bmatrix} + \begin{bmatrix} d_{11} & \cdots & d_{1r} \\ d_{21} & & d_{2r} \\ \vdots & & \vdots \\ d_{m1} & \cdots & d_{mr} \end{bmatrix} \begin{bmatrix} u_1 \\ \vdots \\ u_r \end{bmatrix}, \quad (3.14)$$

i.e.

$$\begin{aligned} \dot{\mathbf{x}} &= \mathbf{A}\mathbf{x} + \mathbf{B}\mathbf{u} - \mathbf{E}\mathbf{d} \\ \mathbf{y} &= \mathbf{C}\mathbf{x} + \mathbf{D}\mathbf{u} \end{aligned} \quad (3.15)$$

where $\mathbf{A} \in \Re^{n \times n}$ is often referred to as the *system matrix* but is referred to here as the *plant matrix* as all the state-space models are of plants to be controlled, $\mathbf{B} \in \Re^{n \times r}$ is the *input matrix*, $\mathbf{E} \in \Re^{n \times p}$ is the external disturbance input matrix, $\mathbf{C} \in \Re^{m \times n}$ is referred to either as the *output matrix* or the *measurement matrix* and $\mathbf{D} \in \Re^{m \times r}$ is the *feedforward matrix*, which would be a null matrix in a controlled plant but is retained so that the reader is aware of its existence in other dynamical systems. The reasons are explained at the end of Sect. 3.2.3.

Linear time invariant (LTI) means that the matrices of (3.15) are constant.

The component equations of (3.13) are the *state differential equations*, but it is usual to refer to the first of equations (3.15) simply as the *state differential equation*.

The minus sign of the external disturbance term is merely a matter of convention, originating in the field of electric drives in which, *by convention*, a positive external load torque slows down the motor output shaft to which it is applied if the motor angular velocity is positive.

Since all variables in (3.14) are arranged in column vectors, $\mathbf{x}$, $\mathbf{u}$, $\mathbf{d}$ and $\mathbf{y}$ are often referred to, respectively, as the *state vector*, the *control vector*, the *external disturbance vector* and the *measurement vector*. If $r = m = 1$, the plant is referred to as *a single input, single output* (SISO) plant; otherwise, it is a *multivariable* or *multiple input, multiple output* (MIMO) plant. This terminology only refers to the inputs and outputs that are electrical signals, which therefore excludes the external disturbance inputs.

3.2.5 Some Preliminary Control Theory

3.2.5.1 Introduction

Once an accurate plant model has been established, an important step in the process of creating a control system is to analyse the plant model to check that it is even possible to produce a controller. The plant state variables represent all the dynamical information about the plant behaviour, and on a few occasions, it proves impossible to control the plant even if all the state variables are available for use in the controller. The plant is then said to be *uncontrollable*. The only course of action is then to modify the plant hardware to render the plant controllable. In other cases, in which it is known that good control satisfying the performance specification could be obtained if the state variables were all available, only a few measurement variables are available and all attempts at producing estimates of the state variables by processing the measurements, such as presented in Chap. 5, fail. This means that it is impossible to deduce the current state by observing the measurements over a finite period of time. The plant is then said to be *unobservable*. Then the only course of action is to add measurement instrumentation to render the plant observable. The sections on controllability and observability to come provide the means of analysing a linear plant model to answer the above questions. The two following sections provide the necessary control theory needed to derive the conditions of controllability and observability. This theory also provides preparation for Sect. 3.4 on discrete linear plant models.

The general theory of controllability and observability is available for nonlinear plant models [1] but a more straightforward and practical approach is taken in this book. Controllability is tested in the process of deriving a nonlinear model-based control law for a specific plant. In this way the total amount of work required from the control system designer is minimised by combining the controllability test with the control law derivation. Similarly, the observability test is combined with the design of a state estimator. This is addressed in Chap. 6.

3.2.5.2 General Solution to LTI State Differential Equation

First, the external disturbance input is not required for the controllability and observability analysis and the plant has control feedforward. The plant state-space model is therefore (3.15) with $\mathbf{d} = \mathbf{0}$ and $\mathbf{D} = \mathbf{0}$. Thus

$$\dot{\mathbf{x}} = \mathbf{A}\mathbf{x} + \mathbf{B}\mathbf{u} \qquad (3.16)$$

$$\mathbf{y} = \mathbf{C}\mathbf{x}. \qquad (3.17)$$

For the controllability analysis, only the state differential equation (3.16) is needed. Since this is of the same form as a linear first-order scalar differential equation, i.e. $\dot{x} = ax + bu$, which has a complementary function, $e^{at}x(0)$, as part of the general solution, this suggests that the general solution of (3.16) should have a complementary function, $e^{\mathbf{A}t}\mathbf{x}(0)$. As will be seen, this turns out to be the case. The exponent here is a matrix and before proceeding further, *matrix functions of matrices* will be briefly discussed. Functions of square matrices may be expanded in an infinite series, as for scalar functions. The Maclaurin series for the exponential matrix function is then

$$e^{\mathbf{A}t} = \mathbf{I}_n + \mathbf{A}\frac{t}{1!} + \mathbf{A}^2\frac{t^2}{2!} + \cdots. \tag{3.18}$$

Note that the first term of the series is the unit matrix, $\mathbf{I}_n \in \Re^{n \times n}$. The remaining terms are of the same dimension, as required for dimensional compatibility.

The general *analytical solution* of (3.16) will first be derived for an arbitrary control vector, $\mathbf{u}(t)$. In the following, the correct order of multiplication has to be maintained as matrix multiplication is noncommutative. The approach will be to suppose a solution of a given form and then substitute it into both sides of (3.16) to determine whether it is valid. Suppose that a solution of the form

$$\mathbf{x}(t) = e^{\mathbf{A}(t-t_0)}\mathbf{f}(t) \tag{3.19}$$

exists, where $\mathbf{f}(t)$ is a vector function to be determined and t_0 is the initial time. Differentiating both sides of (3.19) then yields

$$\dot{\mathbf{x}}(t) = \mathbf{A}e^{\mathbf{A}(t-t_0)}\mathbf{f}(t) + e^{\mathbf{A}(t-t_0)}\dot{\mathbf{f}}(t). \tag{3.20}$$

Substituting for $\mathbf{x}(t)$ and $\dot{\mathbf{x}}(t)$ in (3.16) using (3.19) and (3.20) then yields

$$\mathbf{A}e^{\mathbf{A}(t-t_0)}\mathbf{f}(t) + e^{\mathbf{A}(t-t_0)}\dot{\mathbf{f}}(t) = \mathbf{A}e^{\mathbf{A}(t-t_0)}\mathbf{f}(t) + \mathbf{B}\mathbf{u}(t) \tag{3.21}$$

from which

$$\dot{\mathbf{f}}(t) = e^{-\mathbf{A}(t-t_0)}\mathbf{B}\mathbf{u}(t). \tag{3.22}$$

Integrating from the initial time, t_0, to the present time, t, then yields

$$\mathbf{f}(t) = \int_{t_0}^{t} e^{-\mathbf{A}(\tau-t_0)}\mathbf{B}\mathbf{u}(\tau)\, d\tau + \mathbf{f}(t_0) \tag{3.23}$$

It follows from (3.19) that $\mathbf{f}(t_0) = \mathbf{x}(t_0)$. Then, substituting for $\mathbf{f}(t)$ in (3.19) using (3.23) yields the required general solution of (3.16). Thus

$$\mathbf{x}(t) = e^{\mathbf{A}(t-t_0)}\mathbf{x}(t_0) + \int_{t_0}^{t} e^{\mathbf{A}(t-\tau)}\mathbf{B}\mathbf{u}(\tau)\, d\tau. \tag{3.24}$$

The first term on the RHS of (3.24) is the complementary function and the second term is the particular integral.

3.2 The State Space Model

3.2.5.3 The Cayley–Hamilton Theorem

The Cayley–Hamilton theorem states that a square matrix satisfies its own characteristic equation. The plant matrix, $\mathbf{A} \in \mathfrak{R}^{n \times n}$, has the characteristic equation,

$$|s\mathbf{I}_n - \mathbf{A}| = s^n + a_{n-1}s^{n-1} + \cdots + a_1 s^1 + a_0 s^0 = 0 \quad (3.25)$$

The theorem therefore permits s to be replaced by $\mathbf{A}$ in (3.25), yielding

$$\mathbf{A}^n + a_{n-1}\mathbf{A}^{n-1} + \cdots + a_1 \mathbf{A} + a_0 \mathbf{I}_n = 0. \quad (3.26)$$

The theorem may be proven as follows [2]. The characteristic polynomial of $\mathbf{A}$ is

$$D(s) = |s\mathbf{I}_n - \mathbf{A}| = s^n + a_{n-1}s^{n-1} + \cdots + a_1 s^1 + a_0 s^0. \quad (3.27)$$

Let $\mathbf{B}(s) = \text{adj}\,[s\mathbf{I}_n - \mathbf{A}]$. Then every element of $\mathbf{B}(s)$ is a cofactor of $s\mathbf{I}_n - \mathbf{A}$ and therefore a polynomial in s of maximum degree, $n - 1$. This permits the adjoint matrix to be written as

$$\mathbf{B}(s) = \mathbf{B}_{n-1}s^{n-1} + \mathbf{B}_{n-2}s^{n-2} + \cdots + \mathbf{B}_1 s^1 + \mathbf{B}_0 s^0, \quad (3.28)$$

where $\mathbf{B}_i \in \mathfrak{R}^{n \times n}$, $i = 0, 1, \ldots, n - 1$, are constant matrices. From basic matrix theory, it will be recalled that

$$[s\mathbf{I}_n - \mathbf{A}]^{-1} = \frac{\text{adj}\,[s\mathbf{I}_n - \mathbf{A}]}{|s\mathbf{I}_n - \mathbf{A}|} = \frac{\mathbf{B}(s)}{|s\mathbf{I}_n - \mathbf{A}|} \Rightarrow [s\mathbf{I}_n - \mathbf{A}]\mathbf{B}(s) = |s\mathbf{I}_n - \mathbf{A}|\mathbf{I}_n. \quad (3.29)$$

Substituting for $\mathbf{B}(s)$ and $|s\mathbf{I}_n - \mathbf{A}|$ in (3.29) using (3.28) and (3.25) then yield

$$[s\mathbf{I}_n - \mathbf{A}] \left[\mathbf{B}_{n-1}s^{n-1} + \mathbf{B}_{n-2}s^{n-2} + \cdots + \mathbf{B}_1 s^1 + \mathbf{B}_0 s^0 \right] \\ = \left(s^n + a_{n-1}s^{n-1} + \cdots + a_1 s^1 + a_0 s^0 \right) \mathbf{I}_n. \quad (3.30)$$

Equating the matrix coefficients of like powers of s then gives

$$\begin{aligned} \mathbf{B}_{n-1} &= \mathbf{I}_n \\ \mathbf{B}_{n-2} - \mathbf{A}\mathbf{B}_{n-1} &= a_{n-1}\mathbf{I}_n \\ \mathbf{B}_{n-3} - \mathbf{A}\mathbf{B}_{n-2} &= a_{n-1}\mathbf{I}_n \\ &\vdots \\ \mathbf{B}_0 - \mathbf{A}\mathbf{B}_1 &= a_1 \mathbf{I}_n \\ -\mathbf{A}\mathbf{B}_0 &= a_0 \mathbf{I}_n. \end{aligned} \quad (3.31)$$

Finally, multiplying the sequence of equations in (3.31) by, respectively, $\mathbf{A}^n, \mathbf{A}^{n-1}, \mathbf{A}^{n-2}, \ldots, \mathbf{A}^1$ and $\mathbf{A}^0 = \mathbf{I}_n$ and adding yields

$$\mathbf{A}^n + a_{n-1}\mathbf{A}^{n-1} + \cdots + a_1\mathbf{A} + a_0\mathbf{I}_n = 0. \tag{3.32}$$

3.2.5.4 Matrix Function of a Square Matrix

The general matrix function of a matrix, $\mathbf{f}(\mathbf{A})$, will be *defined* by first expanding the associated scalar function, $f(x)$, in a Maclaurin series as

$$f(x) = \sum_{q=0}^{\infty} \frac{f^{(q)}(0)}{q!} x^q = \sum_{q=0}^{\infty} c_q x^q, \tag{3.33}$$

where $f^{(q)}(x) \triangleq \frac{d^q}{dx^q} f(x)$. Then, the scalar, x, is replaced by $\mathbf{A}$ to yield

$$\mathbf{f}(\mathbf{A}) = \sum_{q=0}^{\infty} c_q \mathbf{A}^q. \tag{3.34}$$

This matrix function is therefore of the same dimension as its argument, $\mathbf{A}$, i.e. $n \times n$. Applying the Cayley–Hamilton theorem then reveals the remarkable result that such a function may, in contrast with a scalar function of a scalar, be expressed *precisely* as a series with only n terms. From (3.32),

$$\mathbf{A}^n = -\left[a_{n-1}\mathbf{A}^{n-1} + \cdots + a_1\mathbf{A} + a_0\mathbf{I}_n\right]. \tag{3.35}$$

The purpose of this step is to express $\mathbf{A}^n$ in terms of lower powers of $\mathbf{A}$. Multiplying (3.35) by $\mathbf{A}$ and then substituting for $\mathbf{A}^n$ using (3.35) yield

$$\mathbf{A}^{n+1} = -\left[a_{n-1}\mathbf{A}^n + a_{n-2}\mathbf{A}^{n-1} \cdots + a_1\mathbf{A}^2 + a_0\mathbf{A}\right]$$
$$= a_{n-1}\left[a_{n-1}\mathbf{A}^{n-1} + \cdots + a_1\mathbf{A} + a_0\mathbf{I}_n\right] - \left[a_{n-2}\mathbf{A}^{n-1} + \cdots + a_1\mathbf{A}^2 + a_0\mathbf{A}\right] \Rightarrow$$
$$\mathbf{A}^{n+1} = b_{n-1}\mathbf{A}^{n-1} + \cdots + b_1\mathbf{A} + b_0\mathbf{I}_n, \tag{3.36}$$

where $b_{n-1} = a_{n-1}^2 - a_{n-2}, \ldots, b_2 = a_{n-1}a_2 - a_1$ and $b_0 = a_{n-1}a_1 - a_0$.

Repeating this procedure an arbitrary number of times then reveals

$$\mathbf{A}^q = \beta_{q,n-1}\mathbf{A}^{n-1} + \cdots + \beta_{q,1}\mathbf{A} + \beta_{q,0}\mathbf{I}_n, q > n, \tag{3.37}$$

where $\beta_{q,i}$, $i = 0, 1, \ldots, n-1$, are constant coefficients. If $\mathbf{A}^q$ in (3.34) is then replaced by the RHS of (3.37) for all $q > n - 1$, the result is a polynomial function of degree, $n - 1$. Thus

$$\mathbf{f}(\mathbf{A}) = \alpha_{n-1}\mathbf{A}^{n-1} + \cdots + \alpha_1\mathbf{A} + \alpha_0\mathbf{I}_n \tag{3.38}$$

where α_i, $i = 0, 1, \ldots, n-1$, are constant coefficients.

3.2 The State Space Model

3.2.6 *Controllability Analysis of Continuous LTI Plant Models*

Once a plant model has been established, it is important to analyse this to check that the plant can actually be controlled. In the rare event of the plant being uncontrollable, the only course of action is to change the plant hardware to remove the issue. In many cases, a mere glance will confirm that there is no problem. In case this is not so obvious, however, this section presents a formal method of analysis that can be applied to LTI plant models. For nonlinear or time-varying linear plant models, the most economic method in time and effort to be recommended is to attempt a model-based controller design. Then if the equations to be solved for the controller parameters or to derive the control algorithm are soluble, then the plant is controllable.

A plant is defined to be controllable if it is possible, in theory, to change its state from an arbitrary initial value, $\mathbf{x}(t_0)$, to an arbitrary final value, $\mathbf{x}(t_f)$, in a finite time interval, $t_f - t_0$, without saturation constraints on any of the variables. Then, assuming the plant is LTI with state differential (3.16), if the plant is controllable, (3.24) will read

$$\mathbf{x}(t_f) = e^{\mathbf{A}(t_f - t_0)} \mathbf{x}(t_0) + \int_{t_0}^{t_f} e^{\mathbf{A}(t_f - \tau)} \mathbf{B} \mathbf{u}(\tau) \, d\tau. \tag{3.39}$$

There is no loss of generality in setting $\mathbf{x}(t_0) = \mathbf{0}$ and $t_0 = 0$. Then (3.39) becomes

$$\mathbf{x}(t_f) = \int_0^{t_f} e^{\mathbf{A}(t_f - \tau)} \mathbf{B} \mathbf{u}(\tau) \, d\tau = e^{\mathbf{A} t_f} \int_0^{t_f} e^{-\mathbf{A}\tau} \mathbf{B} \mathbf{u}(\tau) \, d\tau \tag{3.40}$$

This yields

$$e^{-\mathbf{A} t_f} \mathbf{x}(t_f) = \int_0^{t_f} e^{-\mathbf{A}\tau} \mathbf{B} \mathbf{u}(\tau) \, d\tau. \tag{3.41}$$

Applying (3.38) to the function, $e^{-\mathbf{A}\tau}$, the coefficients are functions of τ. Hence,

$$e^{-\mathbf{A}\tau} = \alpha_{n-1}(\tau) \mathbf{A}^{n-1} + \cdots + \alpha_1(\tau) \mathbf{A} + \alpha_0(\tau) \mathbf{I}_n. \tag{3.42}$$

Then (3.41) may be written as

$$\begin{aligned} e^{-\mathbf{A} t_f} \mathbf{x}(t_f) &= \int_0^{t_f} \left[\alpha_{n-1}(\tau) \mathbf{A}^{n-1} + \cdots + \alpha_1(\tau) \mathbf{A} + \alpha_0(\tau) \mathbf{I}_n \right] \mathbf{B} \mathbf{u}(\tau) \, d\tau \\ &= \mathbf{A}^{n-1} \mathbf{B} \int_0^{t_f} \alpha_{n-1}(\tau) \mathbf{u}(\tau) \, d\tau + \cdots + \mathbf{A} \mathbf{B} \int_0^{t_f} \alpha_1(\tau) \mathbf{u}(\tau) \, d\tau \\ &\quad + \mathbf{B} \int_0^{t_f} \alpha_0(\tau) \mathbf{u}(\tau) \, d\tau \\ &\mathbf{A}^{n-1} \mathbf{B} \mathbf{v}_{n-1}(t_f) + \cdots + \mathbf{A} \mathbf{B} \mathbf{v}_1(t_f) + \mathbf{B} \mathbf{v}_0(t_f). \end{aligned} \tag{3.43}$$

This may be expressed as a partitioned matrix product as follows:

$$\underbrace{e^{-\mathbf{A}t_f}\mathbf{x}(t_f)}_{n\times 1} = \underbrace{\begin{bmatrix} \mathbf{B} & | & \mathbf{AB} & | & \cdots & | & \mathbf{A}^{n-1}\mathbf{B} \end{bmatrix}}_{\mathbf{M}_c(\mathbf{A},\mathbf{B}),\ n\times nr} \overbrace{\begin{bmatrix} \mathbf{v}_0(t_f) \\ \mathbf{v}_1(t_f) \\ \vdots \\ \mathbf{v}_{n-1}(t_f) \end{bmatrix}}^{nr\times 1}. \quad (3.44)$$

In principle, the equations,

$$\int_0^{t_f} \alpha_i(\tau)\mathbf{u}(\tau)\,d\tau = \mathbf{v}_i(t_f), \quad i = 0, 1, \ldots, n-1, \quad (3.45)$$

could be solved for $\mathbf{u}(\tau)$, if $\mathbf{v}_i(t_f)$, $i = 0, 1, \ldots, n-1$, can been found from (3.44). The dimensions of the matrices are indicated above in (3.44), from which it is evident that if the plant is multivariable ($r > 1$), the system of equations is underdetermined. Many solutions of (3.44) are then possible, but only if the matrix, $\mathbf{M}_c(\mathbf{A}, \mathbf{B})$, which is called the *controllability matrix*, has at least n linearly independent columns, meaning that the condition for controllability of the plant is

$$\text{rank}\left[\mathbf{M}_c(\mathbf{A},\mathbf{B})\right] = \text{rank}\begin{bmatrix} \mathbf{B} & | & \mathbf{AB} & | & \cdots & | & \mathbf{A}^{n-1}\mathbf{B} \end{bmatrix} \geq n. \quad (3.46)$$

For an SISO plant, $r = 1$, the input matrix becomes a column vector, denoted, $\mathbf{b}$, and $\mathbf{M}_c(\mathbf{A}, \mathbf{b})$ is square. Then (3.46) is equivalent to

$$\det\left[\mathbf{M}_c(\mathbf{A},\mathbf{b})\right] = \left|\mathbf{b} \ \middle| \ \mathbf{Ab} \ \middle| \ \cdots \ \middle| \ \mathbf{A}^{n-1}\mathbf{b}\right| \neq 0, \quad (3.47)$$

meaning $\mathbf{M}_c(\mathbf{A}, \mathbf{b})$ has to be non-singular. Conditions (3.46) and (3.47) are referred to as the *controllability criteria*.

It is important to note that for any controllable plant (which most will be), (3.45) has infinitely many solutions for $\mathbf{u}(\tau)$ with a multivariable plant and also infinitely many solutions for $u(\tau)$ with an SISO plant, although there is a unique solution of (3.44) in this case. Obtaining actual control functions as solutions of (3.44) followed by (3.45) is difficult and this method does not lend itself to feedback control. Instead the many feedback control techniques presented in elsewhere in this book, many of which are continuously adjustable, provide these solutions in a practical way. Interestingly, however, the equivalent controllability derivation for discrete state-space models does yield a useful feedback control technique. The only practical purpose of this section is to derive the conditions (3.46) and (3.47) for controllability.

3.2 The State Space Model

Example 3.1 Flexible spacecraft with two vibration modes

The state-space model is given for one attitude control axis of a space satellite consisting of a rigid centre body having flexible appendages with two significant vibration modes and negligible natural damping, the interaction from the other two axes being regarded negligible. Thus,

$$\begin{bmatrix}\dot{x}_1\\\dot{x}_2\\\dot{x}_3\\\dot{x}_4\\\dot{x}_5\\\dot{x}_6\end{bmatrix} = \underbrace{\begin{bmatrix}0 & 1 & 0 & 0 & 0 & 0\\0 & 0 & 0 & 0 & 0 & 0\\0 & 0 & 0 & 1 & 0 & 0\\0 & 0 & -\omega_1^2 & 0 & 0 & 0\\0 & 0 & 0 & 0 & 0 & 1\\0 & 0 & 0 & 0 & -\omega_2^2 & 0\end{bmatrix}}_{\mathbf{A}}\begin{bmatrix}x_1\\x_2\\x_3\\x_4\\x_5\\x_6\end{bmatrix} + \underbrace{\begin{bmatrix}0\\1\\0\\1\\0\\1\end{bmatrix}}_{\mathbf{b}} u, \quad y = \underbrace{\begin{bmatrix}c_0 & 0 & c_1 & 0 & c_2 & 0\end{bmatrix}}_{\mathbf{c}^T}\begin{bmatrix}x_1\\x_2\\x_3\\x_4\\x_5\\x_6\end{bmatrix}$$

(3.48)

In this model, which is in the block diagonal modal form (Sect. 3.3.4), the second-order subsystem involving the leading 2×2 diagonal submatrix of $\mathbf{A}$ is the so-called rigid-body mode, where x_1 and x_2 are, respectively, the attitude angle and angular velocity of the equivalent rigid-body spacecraft having the same moment of inertia. The remaining two second-order subsystems associated with the second and third 2×2 diagonal submatrices are, respectively, associated with the vibration mode with angular frequency, ω_1 [rad/s], and the vibration mode with angular frequency, ω_2 [rad/s]. In this case, the plant has only one input and therefore the controllability criterion (3.47) applies. The controllability matrix is then

$$\mathbf{M}_c(\mathbf{A},\mathbf{b}) = \begin{bmatrix}\mathbf{b} & \mathbf{A}\mathbf{b} & \mathbf{A}^2\mathbf{b} & \mathbf{A}^3\mathbf{b} & \mathbf{A}^4\mathbf{b} & \mathbf{A}^5\mathbf{b}\end{bmatrix} = \begin{bmatrix}0 & 1 & 0 & 0 & 0 & 0\\1 & 0 & 0 & 0 & 0 & 0\\0 & 1 & 0 & -\omega_1^2 & 0 & \omega_1^4\\1 & 0 & -\omega_1^2 & 0 & \omega_1^4 & 0\\0 & 1 & 0 & -\omega_2^2 & 0 & \omega_2^4\\1 & 0 & -\omega_2^2 & 0 & \omega_2^4 & 0\end{bmatrix}.$$

(3.49)

Then

$$|\mathbf{M}_c(\mathbf{A},\mathbf{b})| = -\begin{vmatrix}1 & 0 & 0 & 0 & 0\\0 & 0 & -\omega_1^2 & 0 & \omega_1^4\\1 & -\omega_1^2 & 0 & \omega_1^4 & 0\\0 & 0 & -\omega_2^2 & 0 & \omega_2^4\\1 & -\omega_2^2 & 0 & \omega_2^4 & 0\end{vmatrix} = -\begin{vmatrix}0 & -\omega_1^2 & 0 & \omega_1^4\\-\omega_1^2 & 0 & \omega_1^4 & 0\\0 & -\omega_2^2 & 0 & \omega_2^4\\-\omega_2^2 & 0 & \omega_2^4 & 0\end{vmatrix}$$

$$= -\omega_1^2\begin{vmatrix}-\omega_1^2 & \omega_1^4 & 0\\0 & 0 & \omega_2^4\\-\omega_2^2 & \omega_2^4 & 0\end{vmatrix} + \omega_1^4\begin{vmatrix}-\omega_1^2 & 0 & \omega_1^4\\0 & -\omega_2^2 & 0\\-\omega_2^2 & 0 & \omega_2^4\end{vmatrix}$$

$$= \omega_1^4\omega_2^6\left(\omega_1^2 - \omega_2^2\right) - \omega_1^6\omega_2^4\left(\omega_2^2 - \omega_1^2\right)$$

$$= \omega_1^4\omega_2^4\left(\omega_1^2 + \omega_2^2\right)\left(\omega_1^2 - \omega_2^2\right) = \omega_1^4\omega_2^4\left(\omega_1^4 - \omega_2^4\right).$$

(3.50)

Hence, if $\omega_1 \neq \omega_2$, the plant is controllable since $|\mathbf{M}_c(\mathbf{A}, \mathbf{b})| \neq 0$, meaning that it is possible to take the rigid-body substate (x_1, x_2) together with the vibration mode substates, (x_3, x_4) and (x_5, x_6) from an arbitrary initial state to an arbitrary final state in a finite time by an appropriate control function. On the other hand, if $\omega_1 = \omega_2$, then $|\mathbf{M}_c(\mathbf{A}, \mathbf{b})| = 0$, and therefore the plant is uncontrollable. Even if the frequencies of vibration modes in such applications become close together, but not equal, active damping using feedback control is difficult, requiring excessive control peaks causing saturation in practice. In fact, flexible spacecraft structures are designed to have well-separated vibration mode frequencies to avoid controllability issues.

Example 3.2 Roll and heave degrees of freedom of maglev vehicle

Figure 3.1 illustrates a cross section of a vehicle designed to follow a track with electromagnetic levitation (maglev).

This shows two controlled electromagnets for the control of the heave displacement, y, and the roll angle, ϕ. This could be a single vehicle or one of the several vehicles forming a train. There are, in total, five degrees of freedom of motion to be controlled relative to the track on such a vehicle, but only two are considered for this controllability demonstration. The roll angle is exaggerated for clarity. In practice, the magnetic gaps would only be a few millimetres and differ by small proportions.

The electromagnets are highly nonlinear, the attractive forces being given by

$$f_k = K_e \frac{i_k^2}{(g_i + g_0)^2}, k = 1, 2. \tag{3.51}$$

where K_e and g_0 are positive constants, g_1 and g_2 are the variable electromagnet air gaps, g_{1m} and g_{2m} are the corresponding gap measurements and i_1 and i_2 are the electromagnet coil currents. It is assumed, however, that the coil currents are calculated using a linearisation algorithm to realise the linear equation

$$\begin{aligned} f_k - Mg &= K_f u_k \Rightarrow \\ f_k &= K_f u_k + Mg, k = 1, 2, \end{aligned} \tag{3.52}$$

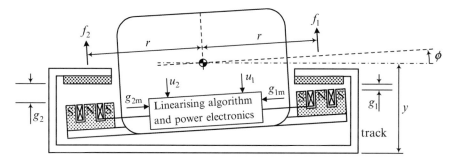

Fig. 3.1 Cross section through electromagnetically levitated vehicle

3.2 The State Space Model

where K_f is the force constant, g is the acceleration due to gravity and u_k, $k = 1, 2$, are the control variables. The linearisation algorithm is obtained by equating the right-hand sides of (3.51) and (3.52) and solving for the currents. Thus,

$$i_k = (g_i + g_0)\sqrt{\frac{K_f u_k + Mg}{K_e}}, \quad k = 1, 2. \tag{3.53}$$

This enables the following LTI state-space model to be used.

$$\begin{bmatrix} \dot{x}_1 \\ \dot{x}_2 \\ \dot{x}_3 \\ \dot{x}_4 \end{bmatrix} = \underbrace{\begin{bmatrix} 0 & 1 & 0 & 0 \\ 0 & 0 & 0 & 0 \\ 0 & 0 & 0 & 1 \\ 0 & 0 & 0 & 0 \end{bmatrix}}_{\mathbf{A}} \begin{bmatrix} x_1 \\ x_2 \\ x_3 \\ x_4 \end{bmatrix} + \underbrace{\begin{bmatrix} 0 & 0 \\ b_1 & b_1 \\ 0 & 0 \\ b_2 & -b_2 \end{bmatrix}}_{\mathbf{B}} \begin{bmatrix} u_1 \\ u_2 \end{bmatrix}. \tag{3.54}$$

The state variables are $x_1 = y$, $x_2 = \dot{y}$, $x_3 = \phi$ and $x_4 = \dot{\phi}$ and the input constants are $b_1 = K_f/M$ and $b_2 = K_f r/J$, M is the vehicle mass and J is the wagon moment of inertia about the roll axis. In this case, the plant is a multivariable one and therefore the controllability criterion (3.46) applies. Thus

$$\mathbf{M}_c(\mathbf{A}, \mathbf{B}) = \begin{bmatrix} \mathbf{B} & | & \mathbf{A}\mathbf{B} & | & \mathbf{A}^3\mathbf{B} & | & \mathbf{A}^3\mathbf{B} \end{bmatrix} = \begin{bmatrix} 0 & 0 & | & b_1 & b_1 & | & 0 & 0 & | & 0 & 0 \\ b_1 & b_1 & | & 0 & 0 & | & 0 & 0 & | & 0 & 0 \\ 0 & 0 & | & b_2 & -b_2 & | & 0 & 0 & | & 0 & 0 \\ b_2 & -b_2 & | & 0 & 0 & | & 0 & 0 & | & 0 & 0 \end{bmatrix}. \tag{3.55}$$

The first four columns of $\mathbf{M}_c$ are the only linearly independent ones and therefore rank $[\mathbf{M}_c] = 4$, which is the dimension, n, of $\mathbf{A}$ and therefore this is sufficient for the plant to be controllable.

3.2.7 Observability Analysis of Continuous LTI Plant Models

The motivation behind this section is the fact that if a complete set of state variables is made available, then the controller may be designed to directly use them to attain the best possible control within the hardware limitations. It is very often the case that not all the state variables are available as measurements, in which case the control system designer wishing to pursue this approach has to devise a means of estimating them (Chap. 5). The purpose of this section is to provide a formal means of analysing the plant model to establish whether it is possible to estimate the unmeasured state variables. If not, then the only course of action would be to modify the plant hardware, usually entailing repositioning or adding measurement instrumentation, to remove the issue.

A plant is defined to be observable if it is possible, in theory, to determine the state variables by using the plant state-space model together with observations of the inputs and the measured outputs over a finite time interval. This section presents a formal analysis method applicable to any LTI plant model of the form of (3.16) and (3.17), which is repeated here for convenience.

$$\dot{\mathbf{x}} = \mathbf{A}\mathbf{x} + \mathbf{B}\mathbf{u} \tag{3.56}$$

$$\mathbf{y} = \mathbf{C}\mathbf{x} \tag{3.57}$$

Let an arbitrary input, $\mathbf{u}(t)$, be applied with the restriction that all of its derivatives are finite up to order $n - 1$, where n is the plant model order. Information about the plant behaviour can be gained by evaluating the derivatives of $\mathbf{y}(t)$ as follows. Differentiating (3.57) and then substituting for $\dot{\mathbf{x}}$ using (3.56) yield

$$\dot{\mathbf{y}} = \mathbf{C}\dot{\mathbf{x}} = \mathbf{C}\mathbf{A}\mathbf{x} + \mathbf{C}\mathbf{B}\mathbf{u}. \tag{3.58}$$

Repeating this for (3.58) gives

$$\ddot{\mathbf{y}} = \mathbf{C}\mathbf{A}\dot{\mathbf{x}} + \mathbf{C}\mathbf{B}\dot{\mathbf{u}} = \mathbf{C}\mathbf{A}^2\mathbf{x} + \mathbf{C}\mathbf{A}\mathbf{B}\mathbf{u} + \mathbf{C}\mathbf{B}\dot{\mathbf{u}} \tag{3.59}$$

Continuing this process up to the derivative of order, $n - 1$, then yields

$$\left.\begin{aligned}\dddot{\mathbf{y}} &= \mathbf{C}\mathbf{A}^3\mathbf{x} + \mathbf{C}\mathbf{A}^2\mathbf{B}\mathbf{u} + \mathbf{C}\mathbf{A}\mathbf{B}\dot{\mathbf{u}} + \mathbf{C}\mathbf{B}\ddot{\mathbf{u}} \\ &\vdots \\ \mathbf{y}^{(n-1)} &= \mathbf{C}\mathbf{A}^{n-1}\mathbf{x} + \mathbf{C}\mathbf{A}^{n-2}\mathbf{B}\mathbf{u} + \mathbf{C}\mathbf{A}^{n-3}\mathbf{B}\dot{\mathbf{u}} + \cdots + \mathbf{C}\mathbf{A}\mathbf{B}\mathbf{u}^{(n-3)} + \mathbf{C}\mathbf{B}\mathbf{u}^{(n-2)}\end{aligned}\right\}. \tag{3.60}$$

Equations (3.57), (3.58), (3.59) and (3.60) may then be assembled into one equation using matrix partitioning as follows:

$$\underbrace{\begin{bmatrix}\mathbf{y} \\ \dot{\mathbf{y}} \\ \ddot{\mathbf{y}} \\ \vdots \\ \mathbf{y}^{(n-1)}\end{bmatrix}}_{nm \times 1} = \underbrace{\begin{bmatrix}\mathbf{C} \\ \mathbf{C}\mathbf{A} \\ \mathbf{C}\mathbf{A}^2 \\ \vdots \\ \mathbf{C}\mathbf{A}^{n-1}\end{bmatrix}}_{\mathbf{M}_o(\mathbf{C},\mathbf{A})}^{nm \times n} \underbrace{\mathbf{x}}_{n \times 1} + \underbrace{\begin{bmatrix}\mathbf{0} \\ \mathbf{C}\mathbf{B}\mathbf{u} \\ \mathbf{C}\mathbf{A}\mathbf{B}\mathbf{u} + \mathbf{C}\mathbf{B}\dot{\mathbf{u}} \\ \vdots \\ \mathbf{C}\mathbf{A}^{n-2}\mathbf{B}\mathbf{u} + \ldots + \mathbf{C}\mathbf{A}\mathbf{B}\mathbf{u}^{(n-3)} + \mathbf{C}\mathbf{B}\mathbf{u}^{(n-2)}\end{bmatrix}}_{nm \times 1}. \tag{3.61}$$

This set of nm linear equations for determining the n state variables is overdetermined for $m > 1$ and in this case, a solution exists provided

3.2 The State Space Model

$$\text{rank}\left[\mathbf{M}_\text{o}\left(\mathbf{C},\mathbf{A}\right)\right]=\text{rank}\begin{bmatrix}\mathbf{C}\\ \hline \mathbf{CA}\\ \hline \mathbf{CA}^2\\ \hline \vdots\\ \hline \mathbf{CA}^{n-1}\end{bmatrix}\geq n, \quad (3.62)$$

where $\mathbf{M}_\text{o}(\mathbf{C},\mathbf{A})$ is the *observability matrix*. If $m = 1$, then $\mathbf{C}$ becomes a row vector, $\mathbf{c}^\text{T}$, and (3.62) becomes

$$\det\left[\mathbf{M}_\text{o}\left(\mathbf{c}^\text{T},\mathbf{A}\right)\right]=\begin{vmatrix}\mathbf{c}^\text{T}\\ \hline \mathbf{c}^\text{T}\mathbf{A}\\ \hline \mathbf{c}^\text{T}\mathbf{A}^2\\ \hline \vdots\\ \hline \mathbf{c}^\text{T}\mathbf{A}^{n-1}\end{vmatrix}\neq 0. \quad (3.63)$$

Conditions (3.62) and (3.63) are referred to as the *observability criteria*.

Example 3.3 Positioning mechanism

A positioning mechanism consists of a DC motor coupled directly to an inertial load subject to vicious friction. The only measurement available is the load angular velocity. The plant state-space model is then as follows:

$$\begin{bmatrix}\dot{x}_1\\ \dot{x}_2\\ \dot{x}_3\end{bmatrix}=\underbrace{\begin{bmatrix}0 & 1 & 0\\ 0 & -a_1 & a_2\\ 0 & -a_3 & -a_4\end{bmatrix}}_{\mathbf{A}}\begin{bmatrix}x_1\\ x_2\\ x_3\end{bmatrix}+\underbrace{\begin{bmatrix}0\\ 0\\ b\end{bmatrix}}_{\mathbf{b}}u, \quad y=\underbrace{\begin{bmatrix}0 & 1 & 0\end{bmatrix}}_{\mathbf{c}^\text{T}}\begin{bmatrix}x_1\\ x_2\\ x_3\end{bmatrix}, \quad (3.64)$$

where x_1 is angular position of the load, x_2 is the angular velocity and x_3 is the armature current. Regarding the plant parameters, $a_1 = B/J$, where B is the viscous friction coefficient and J is the combined moment of inertia of the load and the motor armature; a_2 is the motor torque/back e.m.f. constant; $a_3 = a_2/L_\text{a}$, where L_a is the motor armature inductance; $a_4 = R_\text{a}/L_\text{a}$, where R_a is the motor armature resistance; and $b = K_\text{v}/L_\text{a}$, where K_v is the voltage gain of the power amplifier driving the motor. Since there is only one measurement variable, criterion (3.63) applies. So

$$\mathbf{M}_\text{o}\left(\mathbf{c}^\text{T},\mathbf{A}\right)=\begin{bmatrix}\mathbf{c}^\text{T}\\ \hline \mathbf{c}^\text{T}\mathbf{A}\\ \hline \mathbf{c}^\text{T}\mathbf{A}^2\end{bmatrix}=\begin{bmatrix}0 & 1 & 0\\ 0 & -a_1 & a_2\\ 0 & a_1^2-a_2a_3 & -(a_1a_2+a_2a_4)\end{bmatrix}. \quad (3.65)$$

It is clear from the first column that $\left|\mathbf{M}_\text{o}\left(\mathbf{c}^\text{T},\mathbf{A}\right)\right| = 0$, and therefore, the plant is unobservable as it stands. This situation could be rectified by adding a position sensor, but a potentially less expensive solution would be to add a retention spring. In this case, one more entry would be made in the matrix, **A**, and state differential equation of (3.64) would become

$$\begin{bmatrix} \dot{x}_1 \\ \dot{x}_2 \\ \dot{x}_3 \end{bmatrix} = \begin{bmatrix} 0 & 1 & 0 \\ -a_5 & -a_1 & a_2 \\ 0 & -a_3 & -a_4 \end{bmatrix} \begin{bmatrix} x_1 \\ x_2 \\ x_3 \end{bmatrix} + \begin{bmatrix} 0 \\ 0 \\ b \end{bmatrix} u. \tag{3.66}$$

Here, $a_5 = K_s/J$, where K_s is the retention spring constant. In this case,

$$\mathbf{M}_o\left(\mathbf{c}^T, \mathbf{A}\right) = \begin{bmatrix} \mathbf{c}^T \\ \mathbf{c}^T \mathbf{A} \\ \mathbf{c}^T \mathbf{A}^2 \end{bmatrix} = \begin{bmatrix} 0 & 1 & 0 \\ -a_5 & -a_1 & a_2 \\ a_5 a_1 & -a_5 + a_1^2 - a_2 a_3 & -(a_1 a_2 + a_2 a_4) \end{bmatrix}. \tag{3.67}$$

$\left|\mathbf{M}_o\left(\mathbf{c}^T, \mathbf{A}\right)\right| = a_2 a_5 a_1 - a_5\left(a_1 a_2 + a_2 a_4\right) = -a_5 a_2 a_4 \neq 0$. Hence the plant is observable.

Example 3.4 Spacecraft rendezvous and docking

Consider two spacecraft whose dynamics can be approximated by those of rigid body, i.e. the vibration modes are insignificant. Suppose that spacecraft 1 wishes to rendezvous with spacecraft 2 without exercising any control over it. It is therefore necessary for the controller of spacecraft 1 to match its state with that of spacecraft 2. Both spacecraft have six degrees of freedom with respect to an inertial frame of reference (Chap. 2), but to simplify this example, which only concerns observability, it will be supposed that the roll axes have been aligned and it is simply required to match the roll attitude angle, ϕ_1, of spacecraft 1 with the roll angle, ϕ_2, of spacecraft 2 as part of the rendezvous and docking manoeuvre. It will be assumed that the two measurement variables on spacecraft 1 are $y_1 = \phi_1$ from a star sensor and $y_2 = \phi_2 - \phi_1$ from a camera and image recognition software on board spacecraft 1. Then the plant will consist of the roll dynamics and kinematics of both spacecraft and the measurement hardware. The state-space model of the plant is then

$$\begin{bmatrix} \dot{x}_1 \\ \dot{x}_2 \\ \dot{x}_3 \\ \dot{x}_4 \end{bmatrix} = \underbrace{\begin{bmatrix} 0 & 1 & 0 & 0 \\ 0 & 0 & 0 & 0 \\ 0 & 0 & 0 & 1 \\ 0 & 0 & 0 & 0 \end{bmatrix}}_{\mathbf{A}} \begin{bmatrix} x_1 \\ x_2 \\ x_3 \\ x_4 \end{bmatrix} + \underbrace{\begin{bmatrix} 0 & 0 \\ b_1 & 0 \\ 0 & 0 \\ 0 & b_2 \end{bmatrix}}_{\mathbf{B}} \begin{bmatrix} u_1 \\ u_2 \end{bmatrix}, \quad \begin{bmatrix} y_1 \\ y_2 \end{bmatrix} = \underbrace{\begin{bmatrix} 1 & 0 & 0 & 0 \\ -1 & 0 & 1 & 0 \end{bmatrix}}_{\mathbf{C}} \begin{bmatrix} x_1 \\ x_2 \\ x_3 \\ x_4 \end{bmatrix}, \tag{3.68}$$

where $x_1 = \phi_1$, $x_2 = \dot{\phi}_1$, $x_3 = \phi_2$ and $x_4 = \dot{\phi}_2$. The plant parameters are $b_i = K_{wi}/J_{xxi}$ where K_{wi} and J_{xxi}, $i = 1, 2$, are, respectively, the reaction wheel torque constants and roll axis moments of inertia of spacecrafts 1 and 2.

3.2 The State Space Model

The observability matrix is

$$[\mathbf{M}_o(\mathbf{C},\mathbf{A})] = \begin{bmatrix} \mathbf{C} \\ \mathbf{CA} \\ \mathbf{CA}^2 \\ \mathbf{CA}^3 \end{bmatrix} = \begin{bmatrix} 1 & 0 & 0 & 0 \\ -1 & 0 & 1 & 0 \\ \hline 0 & 1 & 0 & 0 \\ 0 & -1 & 0 & 1 \\ \hline 0 & 0 & 0 & 0 \\ 0 & 0 & 0 & 0 \\ \hline 0 & 0 & 0 & 0 \\ 0 & 0 & 0 & 0 \end{bmatrix}.$$

(3.69)

Evaluating the determinant of the upper 4×4 submatrix yields $(1)(-1)(1) = -1 \neq 0$. The first four rows are therefore linearly independent. Hence rank $[\mathbf{M}_o(\mathbf{C},\mathbf{A})] = 4$.
$= n$. The plant is therefore observable.

3.2.8 The State-Variable Block Diagram

The state-variable block diagram is another way of expressing the state-space model of a plant. For presentation purposes, it shows the model structure and the signal flow within it. It is particularly useful for programming simulation software with a graphical interface such as MATLAB®–SIMULINK®.

The only dynamical element of a state-variable block diagram is the *pure integrator*, all other elements being non-dynamic, such as gains, functions and summing junctions. The pure integrator is the key to forming a state-variable block diagram from a given state-space model. Each state variable, x_i, is the output of a pure integrator, and therefore, the input of the integrator is $\dot{x}_i$ as shown in Fig. 3.2a.

Although the state-space model is based in the time domain, it is common to express a state-variable block diagram in the Laplace domain, often called the s-domain, by showing the integrators as in Fig. 3.2b. Note that $\mathcal{L}\{\dot{x}_i(t)\} = sx_i(s) - x_i(0)$ and therefore the initial condition, $x_i(0)$, is set to zero. This is in keeping with the transfer function definition, which is the ratio of the output Laplace transform to the input Laplace transform with the initial conditions set to zero. It is understood, however, that a plant being represented by a state-variable block diagram in the Laplace domain can have non-zero initial conditions. The integrators of SIMULINK® are shown in this way but they have facilities for setting the initial

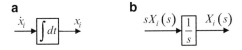

Fig. 3.2 The basic dynamical element of a state-variable block diagram: the pure integrator. (**a**) In the time domain. (**b**) In the Laplace (s) domain

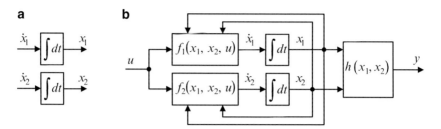

Fig. 3.3 Formation of a state variable block diagram. (**a**) Setting up pure integrators. (**b**) Completion using equations of state-space model

conditions to arbitrary values. If the plant model is LTI, then this notation is useful for the derivation of transfer functions. If the integrators of a *nonlinear* plant model are shown as in Fig. 3.2b, then they must be interpreted in the time domain, the notion of the transfer function being inappropriate.

To demonstrate the method for creating a state-variable block diagram from the state-space model, consider the arbitrary second-order plant model defined by (3.11) with no external inputs, no input feedforward, $n = 2$ and $r = m = 1$.

$$\left. \begin{array}{l} \dot{x}_1 = f_1(x_1, x_2, u) \\ \dot{x}_2 = f_2(x_1, x_2, u) \\ y = h(x_1, x_2) \end{array} \right\} \quad (3.70)$$

The number of integrators is equal to the plant order, n, so two are required in this case. The starting point is to draw the integrators showing their inputs and outputs as in Fig. 3.2. Then it is a straightforward matter to complete the remainder of the block diagram to agree with the equations of the state-space model, as shown in Fig. 3.3.

Example 3.5 Permanent magnet synchronous motor

It is required to form a SIMULINK® block diagram for simulation of a permanent magnet synchronous motor (PMSM) given its state differential equations as follows:

$$\frac{d}{dt}\begin{bmatrix} i_d \\ i_q \end{bmatrix} = \begin{bmatrix} -A & B\omega_r \\ -C\omega_r & -D \end{bmatrix}\begin{bmatrix} i_d \\ i_q \end{bmatrix} - \omega_r \begin{bmatrix} 0 \\ E \end{bmatrix} + \begin{bmatrix} F & 0 \\ 0 & G \end{bmatrix}\begin{bmatrix} u_d \\ u_q \end{bmatrix} \quad (3.71)$$

$$\frac{d\omega_r}{dt} = M(\gamma_e - \gamma_{Le}) = (H + Ki_d)i_q - M\gamma_{Le},$$

where i_d, i_q and u_d, u_q are the stator current and voltage vector components along the direct and quadrature axes (Chap. 2), ω_r is the rotor angular velocity, γ_e is the electromagnetic torque and γ_{Le} is the external load torque. Note that the model is not LTI since the plant contains product nonlinearities, i.e. products of state variables, through the plant matrix of the first of equations (3.71) being a function of ω_r and the second equation containing the term $i_d i_q$. The constant plant parameters are

3.2 The State Space Model

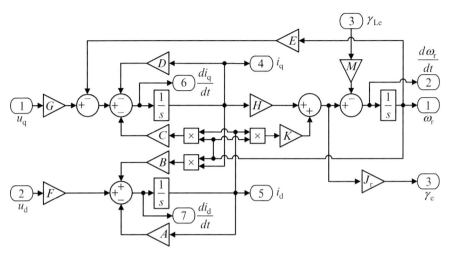

Fig. 3.4 SIMULINK® state-variable block diagram of PMSM

$$\left.\begin{array}{l} A = R_s/L_d;\ B = pL_q/L_d;\ C = pL_d/L_q;\ D = R_s/L_q;\ E = p\Psi_{PM}/L_q \\ F = 1/L_d;\ G = 1/L_q;\ H = 3p\Psi_{PM}/(2J_r);\ K = 3p(L_d - L_q)/(2J_r); \\ M = 1/J_r \end{array}\right\},$$
(3.72)

where Ψ_{PM} is the permanent magnet flux, R_s is the stator resistance, L_d and L_q are the direct and quadrature axis inductances and p is the number of pole pairs.

Since there are three state variables in the PMSM state-space model, then the starting point is three integrators. The remainder of the diagram follows from (3.71), as shown in Fig. 3.4.

As such a motor model will be only part of a control system simulation it is shown as a SIMULINK® subsystem with numbered input and output ports.

Some remarks should be made here. First, the notation of plant models received by control engineers from specialists is often specific to the application rather than adhering to the conventions of, for example, (3.70). In (3.71), the state variables are i_d, i_q and ω_r rather than x_1, x_2 and x_3. Which to use is a matter of preference but it might be argued that the notation in universal use for the physical quantities would be preferable regarding ease of communication between team members of a project. The notation of u, rather than v, to represent a voltage is standard in most countries outside the USA and the UK but is retained in this example since u is a standard symbol to represent control variables. Lastly, the simplification obtained through the constant coefficient definitions, such as (3.72), is recommended as good practice as it minimises human errors in the model preparation and reduces the computational load. For example, if the term $3p\Psi_{PM}/(2J_r)$ were to be implemented in the SIMULINK® diagram, instead of H, then three multiplications and one division would be executed upon every iteration of the numerical integration algorithm, instead of just one, completely unnecessary for constant parameters.

3.2.9 Transfer Function from Continuous LTI State Space Model

The Laplace transfer function of an LTI plant can be determined from its continuous state space model by taking Laplace transforms with zero initial conditions, since such a model comprises first-order ordinary differential equations and linear algebraic equations. This procedure carries over to equations in the matrix–vector form provided the order of matrix multiplication is preserved through the manipulations.

The starting point is the LTI state-space model (3.15), reproduced here.

$$\begin{aligned}\dot{\mathbf{x}}(t) &= \mathbf{A}\mathbf{x}(t) + \mathbf{B}\mathbf{u}(t) - \mathbf{E}\mathbf{d}(t)\\ \mathbf{y}(t) &= \mathbf{C}\mathbf{x}(t) + \mathbf{D}'\mathbf{u}(t).\end{aligned} \quad (3.73)$$

Although, as pointed out in Sect. 3.2.3, $\mathbf{D}' = \mathbf{0}$ in the plant models of real applications, it is deliberately included here to (a) demonstrate the consequences of including it by examining the transfer function relationships and (b) provide the reader the means of modelling and simulating a high-pass filter if needed. Taking Laplace transforms with zero initial conditions yields

$$s\mathbf{X}(s) = \mathbf{A}\mathbf{X}(s) + \mathbf{B}\mathbf{U}(s) - \mathbf{E}\mathbf{D}(s) \quad (3.74)$$

$$\mathbf{Y}(s) = \mathbf{C}\mathbf{X}(s) + \mathbf{D}'\mathbf{U}(s) \quad (3.75)$$

$\mathbf{D}'$ is used for the feedforward matrix instead of $\mathbf{D}$ to avoid notational confusion with $\mathbf{D}(s)$. A transfer function relationship of the form,

$$\mathbf{Y}(s) = \mathbf{G}_u(s)\mathbf{U}(s) - \mathbf{G}_d(s)\mathbf{D}(s), \quad (3.76)$$

is required, and this is achieved by eliminating $\mathbf{X}(s)$ between (3.74) and (3.75).

First, $s\mathbf{X}(s)$ in (3.74) is rewritten as $s\mathbf{I}_n\mathbf{X}(s)$ for dimensional compatibility to enable the following manipulations:

$$[s\mathbf{I}_n - \mathbf{A}]\mathbf{X}(s) = \mathbf{B}\mathbf{U}(s) - \mathbf{E}\mathbf{D}(s) \Rightarrow \mathbf{X}(s) = [s\mathbf{I}_n - \mathbf{A}]^{-1}[\mathbf{B}\mathbf{U}(s) - \mathbf{E}\mathbf{D}(s)]. \quad (3.77)$$

Then, substituting for $\mathbf{X}(s)$ in (3.75) using (3.77) yields

$$\begin{aligned}\mathbf{Y}(s) &= \mathbf{C}[s\mathbf{I}_n - \mathbf{A}]^{-1}[\mathbf{B}\mathbf{U}(s) - \mathbf{E}\mathbf{D}(s)] + \mathbf{D}'\mathbf{U}(s)\\ &= \mathbf{C}\frac{\text{adj}[s\mathbf{I}_n - \mathbf{A}]}{|s\mathbf{I}_n - \mathbf{A}|}[\mathbf{B}\mathbf{U}(s) - \mathbf{E}\mathbf{D}(s)] + \mathbf{D}'\mathbf{U}(s)\\ &= \frac{\mathbf{C}\,\text{adj}[s\mathbf{I}_n - \mathbf{A}][\mathbf{B}\mathbf{U}(s) - \mathbf{E}\mathbf{D}(s)] + |s\mathbf{I}_n - \mathbf{A}|\mathbf{D}'\mathbf{U}(s)}{|s\mathbf{I}_n - \mathbf{A}|}\\ &= \frac{\{\mathbf{C}\,\text{adj}[s\mathbf{I}_n - \mathbf{A}]\mathbf{B} + |s\mathbf{I}_n - \mathbf{A}|\mathbf{D}'\}\mathbf{U}(s) - \mathbf{C}\,\text{adj}[s\mathbf{I}_n - \mathbf{A}]\mathbf{E}\mathbf{D}(s)}{|s\mathbf{I}_n - \mathbf{A}|}\end{aligned} \quad (3.78)$$

3.2 The State Space Model

Comparing (3.78) with (3.76) gives the following:

$$\mathbf{G_u}(s) = \frac{\overset{m \times n}{\mathbf{C}} \overset{n \times n}{\operatorname{adj}[s\mathbf{I}_n - \mathbf{A}]} \overset{n \times r}{\mathbf{B}} + |s\mathbf{I}_n - \mathbf{A}| \overset{m \times r}{\mathbf{D'}}}{|s\mathbf{I}_n - \mathbf{A}|} \quad \text{and}$$

$$\mathbf{G_d}(s) = \frac{\overset{m \times n}{\mathbf{C}} \overset{n \times n}{\operatorname{adj}[s\mathbf{I}_n - \mathbf{A}]} \overset{n \times p}{\mathbf{E}}}{|s\mathbf{I}_n - \mathbf{A}|}, \quad (3.79)$$

in which the dimensions are indicated above the matrices. The matrices, $\mathbf{G_u}(s)$ and $\mathbf{G_d}(s)$, are referred to as *transfer function matrices*, with the exception of an SISO plant in which case $m = r = 1$ and therefore $\mathbf{G_u}(s) = G_u(s)$ is a scalar and simply a transfer function. Similarly, if only a single disturbance source is modelled, then $p = 1$ and $\mathbf{G_d}(s) = G_d(s)$ is also a scalar and simply a transfer function.

The determinant is a polynomial of degree, n. Since this is the denominator polynomial of both transfer functions, then the roots of

$$|s\mathbf{I}_n - \mathbf{A}| = 0 \quad (3.80)$$

are the plant poles. This is also the characteristic equation of the plant matrix, $\mathbf{A}$, often written as $|\lambda \mathbf{I}_n - \mathbf{A}| = 0$, (the notation used in Sect. 3.3.4), whose roots are the *eigenvalues* of $\mathbf{A}$. Hence, the *poles of the plant transfer function model are equal to the eigenvalues of the plant matrix* in the state-space model.

It follows from the matrix theory that the elements of adj $[s\mathbf{I}_n - \mathbf{A}]$ are polynomials of degree, $n - 1$, or less. Hence, if $\mathbf{D'} = \mathbf{0}$, then

$$\mathbf{G_u}(s) = \frac{\mathbf{C} \operatorname{adj}[s\mathbf{I}_n - \mathbf{A}] \mathbf{B}}{|s\mathbf{I}_n - \mathbf{A}|} = \frac{\mathbf{N}(s)}{D(s)}. \quad (3.81)$$

Each element of the numerator matrix, $\mathbf{N}(s)$, is of degree $n - 1$, or less, which must be the case for any real plant, since none of its outputs can respond instantaneously to a step change in any of its inputs. This may be understood by considering the plant in the frequency domain, in which $s = j\omega$. Since $\deg[n_{ij}(s)] \leq n - 1, i = 1, 2, \ldots, m, j = 1, 2, \ldots, r$ and $\deg[D(s)] = n$, then $\lim_{\omega \to \infty} \mathbf{G_u}(j\omega) = \mathbf{0}$. Since any input with a step change has infinite frequency components, then the output cannot have infinite frequency components and cannot respond instantaneously. If, on the other hand, $\mathbf{D'} \neq \mathbf{0}$, from (3.79), $\mathbf{N}(s) = \mathbf{C}\operatorname{adj}[s\mathbf{I}_n - \mathbf{A}]\mathbf{B} + |s\mathbf{I}_n - \mathbf{A}|\mathbf{D'}$ and therefore $\deg[n_{ij}(s)] = n$ for all i and j for which $d'_{ij} \neq 0$. In this case $\mathbf{G_u}(j\infty)$ is finite meaning that some of the outputs will respond instantaneously to a step change in an input. This is confirmed by the feed-forward term, $\mathbf{D'u}(t)$, in (3.73). As stated previously, however, this situation will never be found in a real plant, and therefore, $\mathbf{D'} = \mathbf{0}$ in the state space models unless the application is different, such as a high-pass filter.

3.2.10 Relative Degree

3.2.10.1 Motivation

The relative degree is a property of a plant model that is important in the design of feedback linearising and forced dynamic controllers (Chap. 7) and sliding mode controllers (Chap. 10).

3.2.10.2 SISO Plant

For an LTI SISO plant model with transfer function,

$$\frac{Y(s)}{U(s)} = \frac{\sum_{j=0}^{m} b_j s^j}{s^n + \sum_{i=0}^{n-1} b_i s^i} \tag{3.82}$$

the relative degree is defined as

$$R = n - m. \tag{3.83}$$

There is another definition based on the general state-space model that also applies to nonlinear plants. Consider first the time invariant version of plant model (3.10) for an SISO plant without control feedforward (which never occurs in a real plant), ignoring external disturbances. This is as follows.

$$\dot{\mathbf{x}} = \mathbf{f}(\mathbf{x}, u) \tag{3.84}$$

$$y = h(\mathbf{x}) \tag{3.85}$$

where $\mathbf{x} \in \Re^n$ is the state vector, $u \in \Re$ is control input and $y \in \Re$ is the output. Then the relative degree, R, is determined by the following procedure. First, (3.85) is differentiated w.r.t. time, using the chain rule, to obtain

$$\dot{y} = \sum_{i=1}^{n} \frac{\partial h(\mathbf{x})}{\partial x_i} \dot{x}_i = \begin{bmatrix} \frac{\partial h(\mathbf{x})}{\partial x_1} & \frac{\partial h(\mathbf{x})}{\partial x_2} & \cdots & \frac{\partial h(\mathbf{x})}{\partial x_n} \end{bmatrix} \begin{bmatrix} \dot{x}_1 \\ \dot{x}_2 \\ \vdots \\ \dot{x}_n \end{bmatrix} = \frac{dh(\mathbf{x})}{d\mathbf{x}} \dot{\mathbf{x}}$$

which, in view of (3.84), may be written as

$$\dot{y} = \frac{dh(\mathbf{x})}{d\mathbf{x}} \mathbf{f}(\mathbf{x}, u) \tag{3.86}$$

3.2 The State Space Model

This is defined as the Lie derivative of $h(\mathbf{x})$ along $\mathbf{f}(\mathbf{x}, u)$ and may be expressed using the following brief notation:

$$\dot{y} = L_f h(\mathbf{x}, u) \tag{3.87}$$

This may not directly depend on u, meaning that u does not appear explicitly on the RHS, in many cases. Then it is written as

$$\dot{y} = L_f h(\mathbf{x}) \tag{3.88}$$

and differentiated again to yield

$$\ddot{y} = \frac{dL_f h(\mathbf{x})}{d\mathbf{x}} \mathbf{f}(\mathbf{x}, u) \triangleq L_f^2 h(\mathbf{x}, u). \tag{3.89}$$

If this again does not directly depend on u, then the differentiation is repeated. The direct dependence on u, however, will occur for a finite order of derivative and this is the relative degree, R. Thus the relative degree is the lowest order of output derivative that directly depends upon u and is determined by the following finite sequence of Lie derivatives:

$$\begin{aligned} y &= h(\mathbf{x}) \\ \dot{y} &= L_f h(\mathbf{x}) \\ \ddot{y} &= L_f^2 h(\mathbf{x}) \\ &\vdots \\ y^{(R-1)} &= L_f^{R-1} h(\mathbf{x}) \\ y^{(R)} &= L_f^R h(\mathbf{x}, u) \end{aligned} \tag{3.90}$$

3.2.10.3 Multivariable Plant

Consider now the multivariable plant model

$$\dot{\mathbf{x}} = \mathbf{f}(\mathbf{x}, \mathbf{u}) \tag{3.91}$$

$$\mathbf{y} = \mathbf{h}(\mathbf{x}) \tag{3.92}$$

where $\mathbf{x} \in \Re^n$ is the state vector, $\mathbf{u} \in \Re^r$ is the control vector and $\mathbf{y} \in \Re^m$ is the output vector. First, (3.92) may be written as

$$y_i = h_i(\mathbf{x}), \quad i = 1, 2, \ldots, m. \tag{3.93}$$

Then the relative degree of the plant relative to the output, y_i, is defined in a similar way to that based on (3.90) for an SISO plant. The following sequence of derivatives of (3.93) is formed:

$$\begin{aligned} y_i &= h_i(\mathbf{x}) \\ \dot{y}_i &= L_\mathbf{f} h_i(\mathbf{x}) \\ \ddot{y}_i &= L_\mathbf{f}^2 h_i(\mathbf{x}) \\ &\vdots \\ y_i^{(R_i-1)} &= L_\mathbf{f}^{R_i-1} h_i(\mathbf{x}) \\ y_i^{(R_i)} &= L_\mathbf{f}^{R_i} h_i(\mathbf{x}, \mathbf{u}) \end{aligned} \qquad , i = 1, 2, \ldots, m \qquad (3.94)$$

Thus a multivariable plant has m relative degrees, each equal to the order of the lowest output vector component derivative that directly depends on $\mathbf{u}$, meaning one or more control vector components appear on the RHS.

3.3 State Representation

3.3.1 Introduction

3.3.1.1 Nonuniqueness of State Variables

Consider the following set of state differential equations of a plant model.

$$\dot{x}_i = f_i[(x_1, x_2, \ldots, x_n), (u_1, u_2, \ldots, u_r)], \quad i = 1, 2, \ldots, n. \qquad (3.95)$$

The external disturbances included in (3.11) are ignored as they are not really important in this discussion. Let another variable, z, be identified in the same plant, not in $(x_1, x_2, \ldots, x_n)$. Then if an equation of the same form as (3.95), i.e.

$$\dot{z} = f_z[(x_1, x_2, \ldots, x_n, z), (u_1, u_2, \ldots, u_r)], \qquad (3.96)$$

can be derived, then *by definition*, z is a state variable. If derivatives of the control variables appear on the right-hand side of the equation for $\dot{z}$, then z is *not* a state variable. This means that z cannot have direct functional dependence on any of the control variables. Hence, if

$$z = z(x_1, x_2, \ldots, x_n, u_q), \quad q = 1, 2, \ldots, \text{ and/or } r, \qquad (3.97)$$

then z is not a state variable. It should be noted that a set of state variables can be used directly for feedback control, meaning

$$u_j = g_j(x_1, x_2, \ldots, x_n, y_{r1}, y_{r2}, \ldots, y_{rm}), \quad j = 1, 2, \ldots, r, \qquad (3.98)$$

3.3 State Representation

where y_{rk}, $k = 1, 2, \ldots, m$, are the measurement variables. Then if (3.97) holds and z were to be included in the variables fed back, u_j would be a function of itself as well as the other variables. This is called an *algebraic loop*, which could lead to implementation difficulties and should therefore be avoided. In general, all the plant variables fed back in a control system are state variables.

The following two examples illustrate the nonuniqueness of the state variables chosen for linear and nonlinear plant models and introduce the state transformation.

Example 3.6 State-space model of a low-frequency infinite baffle loudspeaker

The state-space model of a subwoofer (low-frequency) loudspeaker (Fig. 3.5) will now be used to demonstrate the existence of more than one set of state variables.

The loudspeaker consists of a drive unit mounted in an infinite baffle enclosure (closed box). The drive unit consists of a rigid cone constrained to move parallel to its axis of symmetry by means of a flexible suspension system and actuated by a coil mounted on a former at the centre of the cone, moving through the radial field of a permanent magnet. The control variable, u, is the input to the power amplifier, whose output voltage, v_c, is applied to the voicecoil. The power amplifier is DC coupled to facilitate feedback control, whose purpose is to flatten and extend the frequency response of the loudspeaker, essential for serious music reproduction.

The mathematical model is similar to that of a DC motor driving a mechanical load with viscous damping and a retention spring. Figure 3.6 shows the relevant state-variable block diagram. Here, the state variables, x_1, x_2 and x_3, are, respectively, the cone displacement, x, the cone velocity, $\dot{x}$ and the voice-coil current i_c. The alternating force applied to the cone is f and the back e.m.f. due to the voicecoil moving through the magnetic field is e_b.

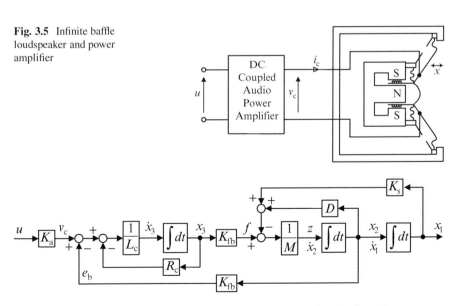

Fig. 3.5 Infinite baffle loudspeaker and power amplifier

Fig. 3.6 State-variable block diagram of a subwoofer driver in an infinite baffle enclosure

The plant parameters are the power amplifier voltage gain, K_a, the voicecoil inductance and resistance, L_c and R_c, the force/back e.m.f. constant, K_{fb}; the mass, M, of the moving part of the drive unit; the damping coefficient, D, due mainly to the flexible cone surround; and the voice-coil drive circuit (dependent on the output impedance of the power amplifier) and the stiffness coefficient, K_s, due to the inner and outer suspension components combined with the air contained in the enclosure.

The state differential equations may be written down by inspection of Fig. 3.6 using the reverse procedure to that described in Sect. 3.2.8 to form the state-variable block diagram. Thus,

$$\dot{x}_1 = x_2, \tag{3.99}$$

$$\dot{x}_2 = \frac{1}{M}(K_{fb}x_3 - Dx_2 - K_sx_1), \tag{3.100}$$

and

$$\dot{x}_3 = \frac{1}{L_c}(K_a u - R_c x_3 - K_{fb} x_2) \tag{3.101}$$

Now consider the variable, z, in Fig. 3.6. This is the cone acceleration. This may be written in terms of the other variables by inspection of the figure. Thus,

$$z = \frac{1}{M}(K_{fb}x_3 - Dx_2 - K_sx_1). \tag{3.102}$$

To carry out the formal test, differentiating (3.102) and then substituting for $\dot{x}_1$, $\dot{x}_2$ and $\dot{x}_3$ using (3.99), (3.100) and (3.101) then yield

$$\dot{z} = \frac{1}{M}\left[\frac{K_{fb}}{L_c}(K_a u - R_c x_3 - K_{fb} x_2) - \frac{D}{M}(K_{fb}x_3 - Dx_2 - K_sx_1) - K_s x_2\right], \tag{3.103}$$

Since this is in the form of (3.96), z is a state variable.

In view of the above, it would be possible to define state variables as follows:

$$\begin{aligned} z_1 &= x_1 \\ z_2 &= x_2 \\ z_3 &= z = \tfrac{1}{M}(K_{fb}x_3 - Dx_2 - K_sx_1). \end{aligned} \tag{3.104}$$

In fact another choice could have been $z_1 = x_1$, $z_2 = x_3$ and $z_3 = z$. A further choice could be $z_1 = x_3$, $z_2 = x_2$ and $z_3 = z$. Many others are possible. Important

3.3 State Representation

observations are that equations (3.104) are linear (due to the plant model being linear in this instance) and that the equations are linearly independent. This means that (z_1, z_2, z_3) contains the same information about the plant behaviour as (x_1, x_2, x_3). In fact, (3.104) is a *state transformation equation* of the form

$$\mathbf{z} = \mathbf{P}\mathbf{x}. \qquad (3.105)$$

Specifically,

$$\begin{bmatrix} z_1 \\ z_2 \\ z_3 \end{bmatrix} = \begin{bmatrix} 1 & 0 & 0 \\ 0 & 1 & 0 \\ -K_s/M & -D/M & K_{fb}/M \end{bmatrix} \begin{bmatrix} x_1 \\ x_2 \\ x_3 \end{bmatrix} \qquad (3.106)$$

In general, if conversely, the *reverse* state transformation,

$$\mathbf{x} = \mathbf{Q}\mathbf{z}, \qquad (3.107)$$

could be written down, where $\mathbf{Q} = \mathbf{P}^{-1}$. Solving (3.104) for $\mathbf{x}$ yields

$$x_1 = z_1$$
$$x_2 = z_2$$
$$x_3 = \frac{1}{K_{fb}}(Mz_3 + Dz_2 + K_s z_1),$$

which may be written as

$$\begin{bmatrix} x_1 \\ x_2 \\ x_3 \end{bmatrix} = \begin{bmatrix} 1 & 0 & 0 \\ 0 & 1 & 0 \\ K_s/K_{fb} & D/K_{fb} & M/K_{fb} \end{bmatrix} \begin{bmatrix} z_1 \\ z_2 \\ z_3 \end{bmatrix}. \qquad (3.108)$$

That $\mathbf{QP} = \mathbf{I}_3$ may be confirmed from (3.106) and (3.108) as follows:

$$\mathbf{QP} = \begin{bmatrix} 1 & 0 & 0 \\ 0 & 1 & 0 \\ K_s/K_{fb} & D/K_{fb} & M/K_{fb} \end{bmatrix} \begin{bmatrix} 1 & 0 & 0 \\ 0 & 1 & 0 \\ -K_s/M & -D/M & K_{fb}/M \end{bmatrix} = \begin{bmatrix} 1 & 0 & 0 \\ 0 & 1 & 0 \\ 0 & 0 & 1 \end{bmatrix} = \mathbf{I}_3.$$
$$(3.109)$$

In general, if $\mathbf{x}$ is the state vector of an established plant model and it is desired to form a second model with a different choice state vector, $\mathbf{z}$, then $\mathbf{z} = \mathbf{P}\mathbf{x}$ is defined as the *forward transformation equation* and $\mathbf{x} = \mathbf{Q}\mathbf{z}$ is defined as the *reverse transformation equation*.

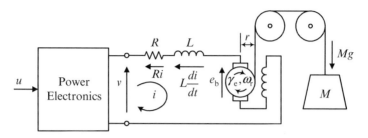

Fig. 3.7 Series wound DC motor lifting load on crane

Example 3.7 Series wound DC motor lifting load on crane

Series wound DC motors are sometimes used to operate cranes due to their high accelerating torques but the torque is proportional to the square of the current since the field current and armature current are equal, as shown in Fig. 3.7. The power electronics produces a voltage, v, with a short-term mean value proportional to the control variable, u, noting that to minimise the power losses, v is rapidly switched by a modulator with a continuously varying mark space ratio, the modulation frequency being sufficiently high for the ripple in the current and torque to have a negligible effect on the plant due to its low-pass filtering action.

Regarding the plant parameters, K_{pe} is the voltage gain of the power electronics, R and L are the combined resistance and inductance of the armature and field windings, K_b is the motor back e.m.f. constant, K_t is the motor torque constant, J is the sum of the armature moment of inertia and the equivalent rotor moment of inertia of the mechanical load referred to the motor output shaft, r is the radius of the pulley wheel driven directly by the motor, M is the load mass and g is the acceleration due to gravity.

The state variables are initially chosen as the output shaft angle, $x_1 = \theta$ the motor speed, $x_2 = \omega_r$; and the motor current, $x_3 = i$, yielding the state-space model

$$\dot{x}_1 = x_2, \quad \dot{x}_2 = \frac{1}{J}\left(K_t x_3^2 - Mgr\right), \quad \dot{x}_3 = \frac{1}{L}\left(K_{pe}u - Rx_3 - K_b x_2\right). \quad (3.110)$$

Now suppose that the motor acceleration, $z = \dot{x}_2$, is to be considered instead of the state variable, x_3. To check that z is valid as a state variable, from (3.110),

$$z = \frac{1}{J}\left(K_t x_3^2 - Mgr\right). \quad (3.111)$$

Differentiating and then substituting for $\dot{x}_3$ using (3.110) yield

$$\dot{z} = 2\frac{K_t}{J}x_3 \dot{x}_3 = 2\frac{K_t}{JL}x_3\left(K_{pe}u - Rx_3 - K_b x_2\right). \quad (3.112)$$

This is in the form of (3.96) without any derivative of u on the RHS. Hence z is a state variable. Let the new set of state variables be

3.3 State Representation

$$z_1 = x_1$$
$$z_2 = x_2 \qquad (3.113)$$
$$z_3 = \dot{z} = \tfrac{1}{J}\left(K_t x_3^2 - Mgr\right),$$

using (3.111). Then, (3.113) is the set of transformation equations. This transformation is nonlinear due to the last equation. The reverse transformation is then obtained by solving (3.113) for x_1, x_2 and x_3. Thus,

$$x_1 = z_1$$
$$x_2 = z_2 \qquad (3.114)$$
$$x_3 = \sqrt{\frac{Jz_3 + Mgr}{K_t}}.$$

Note that making x_3 positive is not restrictive because the series wound DC motor produces a unidirectional torque regardless of the direction of the current, the force applied to the load being bidirectional due to the gravitational force, Mg. To obtain the state differential equations using the new state variables, the first two are

$$\dot{z}_1 = z_2$$
$$\dot{z}_2 = z_3. \qquad (3.115)$$

The third equation is then obtained as follows. Since $\dot{z}_3 = \ddot{z}$, (3.112) becomes

$$\dot{z}_3 = 2\frac{K_t}{JL} x_3 \left(K_{pe} u - R x_3 - K_b x_2\right).$$

Then, substituting for x_2 and x_3 using (3.114) yields

$$\dot{z}_3 = 2\frac{K_t}{JL} \sqrt{\frac{Jz_3 + Mgr}{K_t}} \left(K_{pe} u - R\sqrt{\frac{Jz_3 + Mgr}{K_t}} - K_b z_2\right). \qquad (3.116)$$

3.3.1.2 State Transformations

The plant state can be thought of as an entity independent of the particular choice of the set of state variables. It is the information about the plant behaviour. In the two examples above, the set of state variables forming **x** is one way of representing this information, while the alternative set of variables forming **z** is another way of representing the same information. A particular set of state variables is therefore referred to as a *state representation*. Until now, nearly all the plant models or models of their subsystems derived in Chap. 2 have state variables that are physical quantities such as displacements, velocities and temperatures. In all these cases, the term *physical state representation* is used.

Examples 3.6 and 3.7 have introduced linear and nonlinear forward and reverse state transformations. Transformations between an initially established plant model with state $\mathbf{x} \in \Re^n$ and a second model of the same plant with state $\mathbf{z} \in \Re^n$ will now be considered. Linear transformations are written as

$$\left.\begin{array}{l}\mathbf{z} = \mathbf{Px} \\ \mathbf{x} = \mathbf{Qz}\end{array}\right\} \Rightarrow \mathbf{Q} = \mathbf{P}^{-1}, \qquad (3.117)$$

where $\mathbf{P}$ is the forward transformation matrix and $\mathbf{Q}$ is the reverse transformation matrix. Given that $\mathbf{x}$ is a complete state vector, a necessary condition for $\mathbf{z}$ to be a complete state vector is that all the rows of $\mathbf{P}$ are linearly independent. Care must be taken to ensure that this condition is maintained when choosing new state variables from a linear plant model. For example, in Fig. 3.6, the set of variables, $z_1 = x_1$, $z_2 = f$ and $z_3 = x_3$, are *not* a complete set of state variables because $f = K_{fb}x_3$ and therefore z_2 and z_3 contain the same information about the plant behaviour. The matrix, $\mathbf{P}$, in this case would be

$$\mathbf{P} = \begin{bmatrix} 1 & 0 & 0 \\ 0 & 0 & K_{fb} \\ 0 & 0 & 1 \end{bmatrix}, \qquad (3.118)$$

in which the last two rows are linearly dependent. The reverse transformation, $\mathbf{Q}$, does not exist in this case.

General, including nonlinear, forward and reverse transformations are written as

$$\mathbf{z} = \mathbf{p}(\mathbf{x}) \qquad (3.119)$$

and

$$\mathbf{x} = \mathbf{q}(\mathbf{z}). \qquad (3.120)$$

where $\mathbf{p}(\mathbf{x})$ and $\mathbf{q}(\mathbf{z})$ are vector functions. The test on $\mathbf{z}$ as a state vector is the solubility of (3.119) to give the solution, (3.120) for $\mathbf{x}$, as in Example 3.7.

3.3.2 LTI SISO State-Space Models from Transfer Functions

3.3.2.1 Motivation

Insight into the topic of state representation may be gained by studying transfer function block diagrams for LTI SISO plants and their relationships with the state-space models. The basic canonical forms of state representation are introduced in this section before the general theory of state representation is given in the following sections, which is cast in the matrix–vector notation for the inclusion of multivariable plant models. Transfer function block diagrams, however, are the basis of MATLAB®–SIMULINK® simulations and these may be formed from the matrix–vector state-space models of multivariable plants, if desired. This can be useful if the effects of nonlinear elements, such as control saturation, are to be investigated by simulation, as they may easily be incorporated in SIMULINK® diagrams.

It is important to note that similar relationships to those presented in this section exist between z-transfer function block diagrams and the discrete state-space models to be introduced in Sect. 3.4.

3.3.2.2 State Representations from Block Diagrams

It is well known that more than one block diagram can have the same transfer function. It has also been pointed out in Sect. 3.2.8 that the state variables may be chosen as the outputs of the integrators of state-variable block diagrams (SVBD). Any transfer function block diagram model of a given plant can be converted to a state-variable block diagram by first separating the inherent integrators from other elements of the block diagram by suitable manipulations. Different state representations may therefore be found by first forming different but equivalent transfer function block diagrams. For illustration, numerous state representations will be found for the general LTI SISO third-order plant model transfer function

$$\frac{Y(s)}{U(s)} = \frac{r_2 s^2 + r_1 s + r_0}{s^3 + a_2 s^2 + a_1 s + a_0}. \tag{3.121}$$

where r_i and a_i are constant coefficients. The state representations introduced in the following sections are standard ones using a minimal set of parameters to model a given plant, referred to broadly as *canonical* state representations, or more briefly as *canonical forms*.

First suppose that the plant has distinct real poles located at $s_i = -p_i$, $i = 1, 2, 3$. Then, (3.121) may be written as

$$\frac{Y(s)}{U(s)} = \frac{r_2 s^2 + r_1 s + r_0}{(s + p_1)(s + p_2)(s + p_3)}. \tag{3.122}$$

A block diagram may be formed based on a chain of the three first order subsystems with transfer function, $1/(s + p_i)$, $i = 1, 2, 3$, if the numerator is rewritten in terms of the three denominator factors as follows:

$$\frac{Y(s)}{U(s)} = \frac{C_3 (s + p_1)(s + p_2) + C_2 (s + p_1) + C_1}{(s + p_1)(s + p_2)(s + p_3)}. \tag{3.123}$$

where $C_3 = r_2$, $C_2 = r_1 - C_3 (p_1 + p_2)$ and $C_1 = q_0 - C_2 p_1 - C_3 p_1 p_2$. The corresponding block diagram is shown in Fig. 3.8.

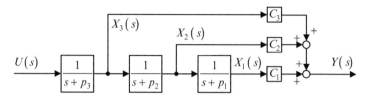

Fig. 3.8 SVBD of a third-order plant based on a first order subsystem chain

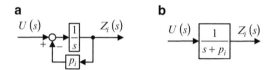

Fig. 3.9 Equivalence between first order subsystem and integrator with feedback. (a) Pure integrator with feedback. (b) First-order subsystem

In fact, five more state representations are possible with the block diagram structure of Fig. 3.8 obtained by reordering the first-order subsystem blocks in the chain.

The justification for Fig. 3.9 being referred to as a state-variable block diagram despite pure integrators not being shown as separate elements is the equivalence of each first-order subsystem to a pure integrator with a feedback loop via a gain equal to minus the pole value, as shown in Fig. 3.9.

Elementary block diagram algebra confirms that the transfer function, $Z_i(s)/U(s)$, of Fig. 3.9a is the same as that shown in the block of Fig. 3.9b. In view of this, the formation of state-variable block diagrams could be simplified by using first-order subsystem blocks rather than integrators where appropriate.

The key to forming the state differential equations of the state-space model from a state-variable block diagram is converting the transfer function relationship of each first-order subsystem to the time domain as follows. Taking the subsystem of Fig. 3.9b as an example yields

$$Z_i(s) = \frac{1}{s + p_i} U(s) \Rightarrow (s + p_i) Z_i(s) = U(s) \Rightarrow \dot{z}_i + p_i z_i = u \Rightarrow \dot{z}_i \quad (3.124)$$
$$= -p_i z_i + u.$$

The last equation of (3.124) is in the form of a state differential equation, as required. Once all the state differential equations of the first order blocks in a plant state-variable block diagram have been formed in this way, the remaining equations that connect the blocks to one another and to the inputs and outputs are self-evident by inspection. Employing this approach for the state-variable block diagram of Fig. 3.9 yields the following state-space model:

3.3 State Representation

Fig. 3.10 SVBD of a third-order plant with distinct real poles, in the modal form

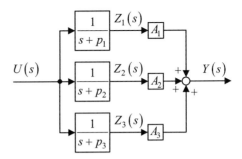

$$\begin{bmatrix} \dot{x}_1 \\ \dot{x}_2 \\ \dot{x}_3 \end{bmatrix} = \begin{bmatrix} -p_1 & 1 & 0 \\ 0 & -p_2 & 1 \\ 0 & 0 & -p_3 \end{bmatrix} \begin{bmatrix} x_1 \\ x_2 \\ x_3 \end{bmatrix} + \begin{bmatrix} 0 \\ 0 \\ 1 \end{bmatrix} u, \quad y = \begin{bmatrix} C_1 & C_2 & C_3 \end{bmatrix} \begin{bmatrix} x_1 \\ x_2 \\ x_3 \end{bmatrix}. \quad (3.125)$$

The unity elements on the superdiagonal of the subsystem matrix occur due to the chain structure of the state-variable block diagram.

The following sections introduce the standard canonical forms.

3.3.2.3 Modal Form with Distinct Poles

A state-variable block diagram with a completely different structure from that of Fig. 3.8 may be formed from transfer function (3.121), comprising three first-order subsystems driven by the control input, using the partial fraction expansion. Thus,

$$\frac{Y(s)}{U(s)} = \frac{A_1}{s+p_1} + \frac{A_2}{s+p_2} + \frac{A_3}{s+p_3}, \quad (3.126)$$

where $A_1 = \frac{r_2 p_1^2 - r_1 p_1 + r_0}{(p_2 - p_1)(p_3 - p_1)}$, $A_2 = \frac{r_2 p_2^2 - r_1 p_2 + r_0}{(p_1 - p_2)(p_3 - p_2)}$ and $A_3 = \frac{r_2 p_3^2 - r_1 p_3 + r_0}{(p_1 - p_3)(p_2 - p_3)}$.

This is shown in Fig. 3.10.

The state-space model for Fig. 3.10 is as follows:

$$\begin{bmatrix} \dot{z}_1 \\ \dot{z}_2 \\ \dot{z}_3 \end{bmatrix} = \begin{bmatrix} -p_1 & 0 & 0 \\ 0 & -p_2 & 0 \\ 0 & 0 & -p_3 \end{bmatrix} \begin{bmatrix} z_1 \\ z_2 \\ z_3 \end{bmatrix} + \begin{bmatrix} 1 \\ 1 \\ 1 \end{bmatrix} u, \quad y = \begin{bmatrix} A_1 & A_2 & A_3 \end{bmatrix} \begin{bmatrix} z_1 \\ z_2 \\ z_3 \end{bmatrix}. \quad (3.127)$$

This is an example of the *modal* state representation in which the plant matrix is diagonal, meaning that there is no interaction between the first-order subsystems via their state variables, z_1, z_2 and z_3. Each of these is associated with an *exponential mode*, characterised by the response of the unforced plant to arbitrary initial states. Thus, the unforced state differential equations may be written separately as follows.

Fig. 3.11 SVBD of a third-order plant with one distinct pole and a double pole, in the modal form

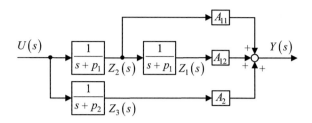

$$\dot{x}_1 = -p_1 x_1, \dot{x}_2 = -p_2 x_2 \text{ and } \dot{x}_3 = -p_3 x_3. \tag{3.128}$$

The solutions to these equations are obtained using Laplace transforms as follows.

$$\dot{x}_i = -p_i x_i \Rightarrow s X_i(s) - x_i(0) = -p_i X_i(s) \Rightarrow (s + p_i) X_i(s) = x_i(0) \Rightarrow$$
$$X_i(s) = \frac{1}{s+p_i} x_i(0), \; i = 1, 2, 3 \Rightarrow$$
$$x_1(t) = e^{-p_1 t} x_1(0), x_2(t) = e^{-p_2 t} x_2(0) \text{ and } x_3(t) = e^{-p_3 t} x_3(0). \tag{3.129}$$

This modal state representation with distinct real poles is generalised to include multivariable plants of any order in Sect. 3.3.4.3.

In contrast to (3.127), the state representation of (3.125), which is one of many that are not named, has interaction between the state variables. In this case, x_1 depends upon x_2 which, in turn, depends upon x_3.

3.3.2.4 Modal Form Including Multiple Poles

Suppose now that two of the real poles are repeated so that

$$\frac{Y(s)}{U(s)} = \frac{q_2 s^2 + q_1 s + q_0}{(s + p_1)^2 (s + p_2)} = \frac{A_{11}}{s + p_1} + \frac{A_{12}}{(s + p_1)^2} + \frac{A_2}{s + p_2}, \tag{3.130}$$

where $A_2 = \frac{r_2 p_2^2 - r_1 p_2 + r_0}{(p_1 - p_2)^2}$, $A_{12} = \frac{r_2 p_1^2 - r_1 p_1 + r_0}{p_2 - p_1}$ and $A_{11} = r_2 - A_2$. In this case, the partial fraction expansion cannot separate the plant model into three independent first-order subsystems. This becomes apparent by inspection of the state-variable block diagram corresponding to (3.130), shown in Fig. 3.11.

The state-space model is as follows:

$$\begin{bmatrix} \dot{z}_1 \\ \dot{z}_2 \\ \dot{z}_3 \end{bmatrix} = \begin{bmatrix} -p_1 & 1 & 0 \\ 0 & -p_1 & 0 \\ 0 & 0 & -p_2 \end{bmatrix} \begin{bmatrix} z_1 \\ z_2 \\ z_3 \end{bmatrix} + \begin{bmatrix} 0 \\ 1 \\ 1 \end{bmatrix} u, \; y = \begin{bmatrix} A_{12} & A_{11} & A_2 \end{bmatrix} \begin{bmatrix} z_1 \\ z_2 \\ z_3 \end{bmatrix}$$
$$\tag{3.131}$$

3.3 State Representation

The two interconnected first-order blocks with poles at $s_{1,2} = -p_1$ in Fig. 3.11, correspond to the second-order subsystem whose subsystem matrix is boxed in (3.131), the corresponding substate differential equation being

$$\begin{bmatrix} \dot{z}_1 \\ \dot{z}_2 \end{bmatrix} = \begin{bmatrix} -p_1 & 1 \\ 0 & -p_1 \end{bmatrix} \begin{bmatrix} z_1 \\ z_2 \end{bmatrix} + \begin{bmatrix} 0 \\ 1 \end{bmatrix} u. \tag{3.132}$$

The state-variable, z_1 depends on the state-variable z_2, due to the unity element on the superdiagonal of the subsystem matrix. In general, repeated poles produce chains of identical first-order subsystems in the state-variable block diagram corresponding to the partial fraction expansion of the transfer function. The number of first-order subsystems in each chain is equal to the multiplicity of the associated pole. So if the pole with value, $-p_1$, in (3.130), had multiplicity 3, instead of 2, the associated subsystem would comprise a chain of three identical first-order blocks with the substate differential equation

$$\begin{bmatrix} \dot{z}_1 \\ \dot{z}_2 \\ \dot{z}_3 \end{bmatrix} = \begin{bmatrix} -p_1 & 1 & 0 \\ 0 & -p_1 & 1 \\ 0 & 0 & -p_1 \end{bmatrix} \begin{bmatrix} z_1 \\ z_2 \\ z_3 \end{bmatrix} + \begin{bmatrix} 0 \\ 0 \\ 1 \end{bmatrix} u \tag{3.133}$$

The two unity elements on the superdiagonal of the subsystem matrix correspond to the two interconnections of the three first-order blocks in the state-variable block diagram. In general, the state-space model derived from the partial fraction expansion of a plant transfer function with multiple poles has a plant matrix with a block diagonal form, simple poles contributing diagonal elements equal to the pole values and multiple poles contributing blocks centred on the main diagonal with unity elements on the superdiagonal, such as the boxed one in (3.131). These are *Jordan blocks* named after the French mathematician. The plant matrix is said to be in the *Jordan canonical form* and the associated modal state representation is classified as the Jordan canonical form. The subsystem associated with a Jordan block defines a mode characterised by its unforced response to arbitrary initial states. Returning to the example of Fig. 3.11, solutions to the three component equations of (3.131) with $u = 0$ will be found to characterise the modes. Thus

$$\dot{z}_1 = -p_1 z_1 + z_2, \tag{3.134}$$

$$\dot{z}_2 = -p_1 z_2 \tag{3.135}$$

and

$$\dot{z}_3 = -p_2 z_3. \tag{3.136}$$

Taking Laplace transforms yields

$$sZ_1(s) - z_1(0) = -p_1 Z_1(s) + Z_2(s) \Rightarrow Z_1(s) = \frac{1}{s + p_1}[z_1(0) + Z_2(s)], \tag{3.137}$$

$$sZ_2(s) - z_2(0) = -p_1 Z_2(s) \Rightarrow Z_2(s) = \frac{1}{s + p_1} z_2(0) \tag{3.138}$$

and

$$sZ_3(s) - z_3(0) = -p_1 Z_3(s) \Rightarrow Z_3(s) = \frac{1}{s + p_1} z_3(0). \tag{3.139}$$

Substituting for $Z_2(s)$ in (3.137) using (3.138) yields

$$Z_1(s) = \frac{1}{s + p_1}\left[z_1(0) + \frac{1}{s + p_1} z_2(0)\right] = \frac{1}{s + p_1} z_1(0) + \frac{1}{(s + p_1)^2} z_2(0). \tag{3.140}$$

The second-order mode is then characterised by the solutions for the state variables, $z_1(t)$ and $z_2(t)$, of the second-order subsystem given by the inverse Laplace transforms of (3.137) and (3.140). Thus,

$$\begin{aligned} z_1(t) &= e^{-p_1 t}[z_1(0) + t z_2(0)] \\ z_2(t) &= e^{-p_1 t} z_2(0) \end{aligned} \tag{3.141}$$

The factor of $z_1(t)$ that is linear in t occurs due to the double pole. As will be seen in the more general treatment of Sect. 3.3.4.5, a pole of multiplicity, m, will give rise to factors that are polynomials in t of degree $m - 1, m - 2, \ldots, 2, 1$ and 0 (i.e. a constant factor), for the individual state variables. This is understakable as the state variables are the outputs of each of the first-order blocks that are connected in a chain. Since the common factor is an exponential function, the modes associated with multiple poles are referred to as *polynomial exponential modes*. To complete this case, $z_3(t)$ is associated with the simple pole at $s_3 = -p_2$ and the corresponding simple exponential mode, being given by the inverse Laplace transform of (3.139) as follows:

$$z_3(t) = e^{-p_2 t} z_3(0). \tag{3.142}$$

3.3.2.5 Modal Form with Complex Conjugate Poles

Next, suppose that two of the poles are complex conjugates. Then the plant transfer function may be written

3.3 State Representation

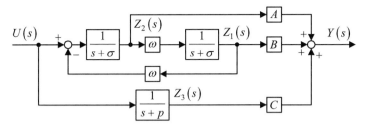

Fig. 3.12 State-variable block diagram of a third-order plant with complex conjugate poles

$$\frac{Y(s)}{U(s)} = \frac{r_2 s^2 + r_1 s + r_0}{\left[(s+\sigma)^2 + \omega^2\right](s+p)}. \tag{3.143}$$

so that there are complex conjugate poles at $s_{1,2} = -\sigma \pm j\omega$. The partial fraction expansion leading to the required state-variable block diagram is

$$\frac{Y(s)}{U(s)} = \frac{A(s+\sigma) + B\omega}{(s+\sigma)^2 + \omega^2} + \frac{C}{s+p}, \tag{3.144}$$

where $C = \frac{r_2 p^2 - r_1 p + r_0}{(\sigma-p)^2 + \omega^2}$, $A = r_2 - C$ and $B = \frac{1}{\omega}\left[r_0 - \omega a p B - (\sigma^2 + \omega^2) C\right]$.

Since the complex conjugate poles are distinct, it would have been possible, in principle, to form a partial fraction expansion in the form of (3.126), but the state variables and the coefficients would have been complex and therefore not easily realised, except by separating the imaginary and real parts of the model, each of which would be in a 'real' state representation similar to the following. The state-variable block diagram corresponding to (3.144) is shown in Fig. 3.12.

The state-space model is

$$\begin{bmatrix} \dot{z}_1 \\ \dot{z}_2 \\ \dot{z}_3 \end{bmatrix} = \begin{bmatrix} -\sigma & \omega & 0 \\ -\omega & -\sigma & 0 \\ 0 & 0 & -p \end{bmatrix} \begin{bmatrix} z_1 \\ z_2 \\ z_3 \end{bmatrix} + \begin{bmatrix} 0 \\ \omega \\ 1 \end{bmatrix} u, \quad y = \begin{bmatrix} B & A & C \end{bmatrix} \begin{bmatrix} z_1 \\ z_2 \\ z_3 \end{bmatrix}. \tag{3.145}$$

As in the double pole case, the state variables z_1 and z_2 are interdependent and therefore the plant model has just two separate subsystems, one of second order involving z_1 and z_2 and the other of first order involving z_3. The mode produced by the second-order subsystem is characterised by the unforced solutions to the first two state differential equations with arbitrary initial states. Thus, with $u = 0$,

$$\dot{z}_1 = -\sigma z_1 + \omega z_2 \tag{3.146}$$

and

$$\dot{z}_2 = -\sigma z_2 - \omega z_1. \tag{3.147}$$

Taking Laplace transforms of (3.146) and (3.147) then yields

$$sZ_1(s) - z_1(0) = -\sigma Z_1(s) + \omega Z_2(s) \Rightarrow (s+\sigma)Z_1(s) = z_1(0) + \omega Z_2(s) \tag{3.148}$$

$$sZ_2(s) - z_2(0) = -\sigma Z_2(s) - \omega Z_1(s) \Rightarrow (s+\sigma)Z_2(s) = z_2(0) - \omega Z_1(s) \tag{3.149}$$

The solution of (3.148) and (3.149) for $Z_1(s)$ and $Z_2(s)$ is then given by

$$\begin{bmatrix} (s+\sigma) & -\omega \\ \omega & (s+\sigma) \end{bmatrix} \begin{bmatrix} Z_1(s) \\ Z_2(s) \end{bmatrix} = \begin{bmatrix} z_1(0) \\ z_2(0) \end{bmatrix} \Rightarrow$$
$$\begin{bmatrix} Z_1(s) \\ Z_2(s) \end{bmatrix} = \frac{1}{(s+\sigma)^2 + \omega^2} \begin{bmatrix} (s+\sigma) & \omega \\ -\omega & (s+\sigma) \end{bmatrix} \begin{bmatrix} z_1(0) \\ z_2(0) \end{bmatrix}. \tag{3.150}$$

Taking inverse Laplace transforms then yields the required solution as

$$\begin{aligned} z_1(t) &= e^{-\sigma t} \left[z_1(0) \cos(\omega t) + z_2(0) \sin(\omega t) \right] \\ z_2(t) &= e^{-\sigma t} \left[z_2(0) \cos(\omega t) - z_1(0) \sin(\omega t) \right] \end{aligned} \tag{3.151}$$

In view of the sinusoidal terms, this type of mode is called an *oscillatory mode*.

As in the previous cases, the third mode associated with the simple pole at $s = -p$ is a simple exponential mode characterised by the solution to the unforced state differential equation, $\dot{z}_3 = -pz_3$, i.e.

$$z_3(t) = e^{-pt} z_3(0). \tag{3.152}$$

It is evident from the above that the state-space model of a plant containing oscillatory modes in the modal state representation has a plant matrix in the block diagonal form with oscillatory mode submatrices of dimension 2×2 taking the form of the blocked submatrix of (3.145). It should be mentioned, however, that other structures of substate-variable block diagrams could be chosen for the oscillatory second-order subsystems. The one of Fig. 3.12 is chosen since it is similar to that used for the more general treatment in Sect. 3.3.4.4.

3.3.2.6 The Controller Canonical Form

The controller canonical form is given its name because the model-based control system design method of pole assignment applied to the linear state feedback control

3.3 State Representation

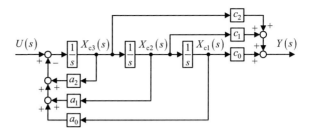

Fig. 3.13 State-variable block diagram of a third-order plant in the controller canonical form

laws yields the most straightforward gain calculations when the plant model has this state representation. Also, any real and/or complex conjugate pole locations are catered for.

The starting point is a chain of integrators equal in number to the plant order, whose outputs are the state variables. Then a set of cascaded feedback loops is formed, each loop connected between the output of each integrator and the input of the first integrator via a gain equal to one of the transfer function denominator coefficients. Working from the innermost loop outwards, the gains are the coefficients of descending powers of s. The output is formed as a weighted sum of the integrator outputs. Starting in sequence from the first integrator in the chain, the weighting coefficients are the coefficients of descending powers of s in the numerator of the transfer function. The state variable block diagram is shown in Fig. 3.13.

The transfer function of this block diagram may be quickly derived using Mason's formula (Appendix A4). Thus,

$$\frac{Y(s)}{U(s)} = \frac{\frac{c_2}{s} + \frac{c_1}{s^2} + \frac{c_0}{s^3}}{1 - \left(-\frac{a_2}{s} + \frac{a_1}{s^2} + \frac{a_0}{s^3}\right)} = \frac{c_2 s^2 + c_1 s + c_0}{s^3 + a_2 s^2 + a_1 s + a_0}. \tag{3.153}$$

This is the general third-order transfer function (3.121). The corresponding state-space model may be derived, observing that the input of a pure integrator is the derivative of its output. Thus for Fig. 3.13,

$$\underbrace{\begin{bmatrix} \dot{x}_{c1} \\ \dot{x}_{c2} \\ \dot{x}_{c3} \end{bmatrix}}_{\dot{\mathbf{x}}_c} = \underbrace{\begin{bmatrix} 0 & 1 & 0 \\ 0 & 0 & 1 \\ -a_0 & -a_1 & -a_2 \end{bmatrix}}_{\mathbf{A}_c} \underbrace{\begin{bmatrix} x_{c1} \\ x_{c2} \\ x_{c3} \end{bmatrix}}_{\mathbf{x}_c} + \underbrace{\begin{bmatrix} 0 \\ 0 \\ 1 \end{bmatrix}}_{\mathbf{b}_c} u, \quad y = \underbrace{\begin{bmatrix} c_0 & c_1 & c_2 \end{bmatrix}}_{\mathbf{c}_c^T} \underbrace{\begin{bmatrix} x_{c1} \\ x_{c2} \\ x_{c3} \end{bmatrix}}_{\mathbf{x}_c}.$$

(3.154)

The general structure is self-evident and should enable a similar model for a linear plant of any order to be formed. A generalisation to multivariable plant models is given in Sect. 3.3.5.

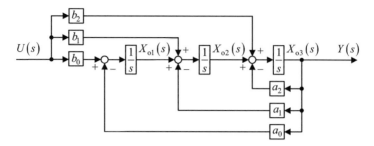

Fig. 3.14 State variable block diagram of a third-order plant in the observer canonical form

3.3.2.7 The Observer Canonical Form

If not all the state-variables are available as measurements, then they are estimated for use in a linear state feedback control law. The purpose of the observer introduced in Chap. 5 is to provide the state estimates and its design is the most straightforward if the plant model is in the observer canonical form. As for the control canonical form, this state representation caters for any real and/or complex conjugate plant poles.

The starting point in forming the state variable block diagram is a chain of integrators equal in number to the plant order, as in Fig. 3.13, but also a summing junction is provided at the input of every integrator, as shown in Fig. 3.14.

Then a cascaded set of feedback loops is formed. A loop connects the output of the last integrator to the input of every integrator via a gain equal to one of the transfer function denominator coefficients, a_i, $i = 1, 2, 3$. Working from the innermost loop outwards, the gains are the coefficients of descending powers of s. The input is fed forward to every integrator input via a coefficient, b_i, $i = 1, 2, 3$. Starting in sequence from the first integrator in the chain, these coefficients are those of ascending powers of s in the numerator of the transfer function. This can be determined easily using Mason's formula (Appendix A4) and the result is (3.153), with $b_i = c_i$, $i = 1, 2, 3$, i.e. the general third order transfer function.

The state-space model may be deduced readily from Fig. 3.14 and is as follows.

$$\underbrace{\begin{bmatrix} \dot{x}_{o1} \\ \dot{x}_{o2} \\ \dot{x}_{o3} \end{bmatrix}}_{\dot{\mathbf{x}}_o} = \underbrace{\begin{bmatrix} 0 & 0 & -a_0 \\ 1 & 0 & -a_1 \\ 0 & 1 & -a_2 \end{bmatrix}}_{\mathbf{A}_o} \underbrace{\begin{bmatrix} x_{o1} \\ x_{o2} \\ x_{o3} \end{bmatrix}}_{\mathbf{x}_o} + \underbrace{\begin{bmatrix} b_0 \\ b_1 \\ b_2 \end{bmatrix}}_{\mathbf{b}_o} u, \quad y = \underbrace{\begin{bmatrix} 0 & 0 & 1 \end{bmatrix}}_{\mathbf{c}_o^T} \underbrace{\begin{bmatrix} x_{o1} \\ x_{o2} \\ x_{o3} \end{bmatrix}}_{\mathbf{x}_o}. \quad (3.155)$$

Comparison with the controller canonical state-space model (3.154) reveals that

$$\mathbf{A}_o = \mathbf{A}_c^T, \, \mathbf{b}_o = \mathbf{c}_c \text{ and } \mathbf{c}_o = \mathbf{b}_c. \quad (3.156)$$

3.3 State Representation

This enables the observer canonical model to be formed once the controller canonical model is known. Its general structure is clear from this third-order example, and studying this should enable a similar model to be formed for any other linear plant.

A generalisation to multivariable plant models is given in Sect. 3.3.5.

3.3.3 Transformation Matrices Connecting Linear Models

It is often the case that two state-space models of the same plant have been derived with different state representations and it is necessary to derive the forward transformation matrix, **P**, or the reverse transformation matrix, **Q**, that connects the two state vectors. Certain control system designs need such a transformation. The two state-space models are

Model 1

$$\dot{\mathbf{x}} = \mathbf{A}\mathbf{x} + \mathbf{B}\mathbf{u} \quad (3.157)$$

and

$$\mathbf{y} = \mathbf{C}\mathbf{x}, \quad (3.158)$$

where $\mathbf{x} \in \Re^n$, $\mathbf{u} \in \Re^r$ and $\mathbf{y} \in \Re^m$, and

Model 2

$$\dot{\mathbf{z}} = \mathbf{F}\mathbf{z} + \mathbf{G}\mathbf{u} \quad (3.159)$$

and

$$\mathbf{y} = \mathbf{H}\mathbf{z}, \quad (3.160)$$

where $\mathbf{z} \in \Re^n$. Usually, $m \geq r$. The transformation from Model 1 to Model 2 will be defined as the forward transformation, the state vectors being linked by the forward transformation equation

$$\mathbf{z} = \mathbf{P}\mathbf{x}. \quad (3.161)$$

Substituting for **z** in Model 2 using (3.161) yields

$$\mathbf{P}\dot{\mathbf{x}} = \mathbf{F}\mathbf{P}\mathbf{x} + \mathbf{G}\mathbf{u} \quad (3.162)$$

and

$$\mathbf{y} = \mathbf{H}\mathbf{P}\mathbf{x}. \quad (3.163)$$

Pre multiplying (3.157) by **P** gives

$$\mathbf{P\dot{x} = PAx + PBu}. \tag{3.164}$$

Comparing (3.162) with (3.164) and (3.158) with (3.163) then reveals the following state independent equations that could be solved for **P**:

$$\mathbf{PA = FP}, \tag{3.165}$$

$$\mathbf{PB = G} \tag{3.166}$$

and

$$\mathbf{HP = C}. \tag{3.167}$$

There are n^2 component equations of (3.165) in the n^2 elements of **P**, but this is a homogeneous set of equations. The trivial solution of $\mathbf{P} = 0$ cannot apply as this would imply $\mathbf{G} = \mathbf{0}$ from (3.166), meaning that the control variables would have no effect on the state of Model 2, and $\mathbf{C} = \mathbf{0}$ from (3.167), meaning that there would be no outputs from Model 1, neither of which would be useful. If (3.165) is taken alone, therefore, there are an infinity of solutions, but only one of these solutions satisfies (3.166) and (3.167). As will be seen, (3.165) can be taken with (3.166) to yield a solution for **P** (which will be called solution (a)). Alternatively, (3.165) can be taken with (3.167) to yield, apparently, another solution for **P** (which will be called solution (b)). The complete set of equations therefore appears to be overdetermined but solutions (a) and (b) cannot conflict: otherwise, $\mathbf{z} = \mathbf{Px}$ would yield two different state vectors for the same plant state differential equation (3.159), which cannot be true.

First (3.165) and (3.166) will be solved to yield an equation for **P** in terms of **A**, **B**, **F** and **G**. The approach is to form a set of n^2 linear simultaneous equations that are linearly independent with respect to the elements of **P** and then solve them for **P** algebraically to obtain a formula rather than a numerical solution. First, (3.166) provides *nr* equations as it stands. A second set of *nr* equations is obtained by postmultiplying both sides of (3.165) by **B** and substituting for **PB** on the RHS using (3.166) to yield

$$\mathbf{PAB = FPB = FG}. \tag{3.168}$$

A third set of *nr* equations is then obtained by postmultiplying both sides of (3.165) by **AB**, substituting for **PA** on the RHS using (3.165) and again substituting for **PB** in the term so formed using (3.166) to yield

$$\mathbf{PA^2B = FPAB = F^2PB = F^2G}. \tag{3.169}$$

3.3 State Representation

A pattern may now be seen to develop in the sequence of equations (3.166), (3.168) and (3.169). The submatrices **B**, **AB** and **A**^{2}**B** on the LHS are the first three submatrices of the controllability matrix derived in Sect. 3.2.6, based on the state representation with the state vector, **x**, which will be denoted by $\mathbf{M}_{cx}(\mathbf{A}, \mathbf{B})$. Similarly, the submatrices **G**, **FG** and **F**^{2}**G** on the RHS are the first three submatrices of the controllability matrix based on the state representation with the state vector, **z**, which will be denoted $\mathbf{M}_{cz}(\mathbf{F}, \mathbf{G})$. Repetition of the process yielding (3.168) and (3.169) produces the remaining submatrices of the controllability matrices. Postmultiplying (3.165) by $\mathbf{A}^{k-1}\mathbf{B}$ and then repeatedly substituting for **PA** in the expressions generated on the RHS using (3.165) until **A** is eliminated and then finally substituting for **PB** using (3.166) yield

$$\mathbf{PA}^k\mathbf{B} = \mathbf{FPA}^{k-1}\mathbf{B} = \mathbf{F}^2\mathbf{PA}^{k-2}\mathbf{B} = \mathbf{F}^3\mathbf{PA}^{k-3}\mathbf{B} = \cdots = \mathbf{F}^k\mathbf{PB} = \mathbf{F}^k\mathbf{G}. \quad (3.170)$$

Taking k from 0 to $n-1$ yields all the equations required, which may be written in the following partitioned matrix form:

$$\mathbf{P}\left[\mathbf{B} \mid \mathbf{AB} \mid \mathbf{A}^2\mathbf{B} \mid \ldots \mid \mathbf{A}^{n-1}\mathbf{B}\right] = \left[\mathbf{G} \mid \mathbf{FG} \mid \mathbf{F}^2\mathbf{G} \mid \ldots \mid \mathbf{F}^{n-1}\mathbf{G}\right],$$

i.e.

$$\underbrace{\mathbf{P}}_{n \times n} \underbrace{\mathbf{M}_{cx}(\mathbf{A},\ \mathbf{B})}_{n \times nr} = \underbrace{\mathbf{M}_{cz}(\mathbf{F},\ \mathbf{G})}_{n \times nr}, \quad (3.171)$$

As the plant has to be controllable, the rank of the controllability matrix is at least n and therefore $\mathbf{M}_{cx}$ is pseudo-invertible. As shown by the dimensions above the matrices, for a multivariable plant, the component equations are overdetermined but cannot produce conflicting solutions, as already stated. This is analogous to three simultaneous equations in two unknowns whose graphs pass through the same point. The $n \times nr$ simultaneous equations, however, may be combined to form n^2 equations conveniently by postmultiplying both sides of (3.171) by $\mathbf{M}_{cx}^T$ to yield

$$\underbrace{\mathbf{P}}_{n \times n} \underbrace{[\mathbf{M}_{cx}(\mathbf{A},\ \mathbf{B})][\mathbf{M}_{cx}(\mathbf{A},\ \mathbf{B})]^T}_{n \times n} = \underbrace{[\mathbf{M}_{cz}(\mathbf{F},\ \mathbf{G})][\mathbf{M}_{cx}(\mathbf{A},\ \mathbf{B})]^T}_{n \times n}. \quad (3.172)$$

A theorem [3] states that if a matrix **M** of dimension $n \times N$, where $N > n$, has rank $\geq n$, then $\mathbf{MM}^T$ is nonsingular. This allows the following formula for the transformation matrix:

$$\mathbf{P} = [\mathbf{M}_{cz}(\mathbf{F},\ \mathbf{G})][\mathbf{M}_{cx}(\mathbf{A},\ \mathbf{B})]^T \left\{ [\mathbf{M}_{cx}(\mathbf{A},\ \mathbf{B})][\mathbf{M}_{cx}(\mathbf{A},\ \mathbf{B})]^T \right\}^{-1}. \quad (3.173)$$

A formula for the reverse transformation matrix immediately follows.

$$\mathbf{Q} = \mathbf{P}^{-1} = [\mathbf{M}_{cx}(\mathbf{A},\mathbf{B})][\mathbf{M}_{cx}(\mathbf{A},\mathbf{B})]^T \left\{ [\mathbf{M}_{cz}(\mathbf{F},\mathbf{G})][\mathbf{M}_{cx}(\mathbf{A},\mathbf{B})]^T \right\}^{-1}. \tag{3.174}$$

There are, however, two more alternative equations, one for $\mathbf{P}$ and one for $\mathbf{Q}$, using the controllability matrices. To obtain these, let (3.171) be expressed instead in terms of $\mathbf{Q}$. Thus,

$$\mathbf{M}_{cx}(\mathbf{A},\mathbf{B}) = \mathbf{Q}\mathbf{M}_{cz}(\mathbf{F},\mathbf{G}). \tag{3.175}$$

Then the solution is through postmultiplication by $[\mathbf{M}_{cz}(\mathbf{F},\mathbf{G})]^T$, yielding

$$[\mathbf{M}_{cx}(\mathbf{A},\mathbf{B})][\mathbf{M}_{cz}(\mathbf{F},\mathbf{G})]^T = \mathbf{Q}[\mathbf{M}_{cz}(\mathbf{F},\mathbf{G})][\mathbf{M}_{cz}(\mathbf{F},\mathbf{G})]^T \Rightarrow$$

$$\mathbf{Q} = [\mathbf{M}_{cx}(\mathbf{A},\mathbf{B})][\mathbf{M}_{cz}(\mathbf{F},\mathbf{G})]^T \left\{ [\mathbf{M}_{cz}(\mathbf{F},\mathbf{G})][\mathbf{M}_{cz}(\mathbf{F},\mathbf{G})]^T \right\}^{-1}. \tag{3.176}$$

This leads directly to the alternative equation for $\mathbf{P}$ as follows:

$$\mathbf{P} = \mathbf{Q}^{-1} = [\mathbf{M}_{cz}(\mathbf{F},\mathbf{G})][\mathbf{M}_{cz}(\mathbf{F},\mathbf{G})]^T \left\{ [\mathbf{M}_{cx}(\mathbf{A},\mathbf{B})][\mathbf{M}_{cz}(\mathbf{F},\mathbf{G})]^T \right\}^{-1}. \tag{3.177}$$

In some cases, an advantageous choice can be made that minimises the amount of computation needed, depending upon the simplicity of the plant matrices in particular state representations.

For an SISO plant, the controllability matrices are of dimension $n \times n$, and therefore the steps leading to (3.173), (3.174), (3.175), (3.176) and (3.177) are unnecessary, a simpler formula being derived directly from (3.171) as

$$\mathbf{P} = \mathbf{M}_{cz}(\mathbf{F},\mathbf{g})[\mathbf{M}_{cx}(\mathbf{A},\mathbf{b})]^{-1}. \tag{3.178}$$

but this is the only option using the controllability matrices. The formula for the corresponding reverse transformation matrix follows directly from (3.178) as

$$\mathbf{Q} = \mathbf{P}^{-1} = \mathbf{M}_{cx}(\mathbf{A},\mathbf{b})[\mathbf{M}_{cz}(\mathbf{F},\mathbf{g})]^{-1}. \tag{3.179}$$

Similar formulae to (3.173), (3.174), (3.175), (3.176), (3.177), (3.178) and (3.179) may be derived in terms of the *observability matrices* by instead solving (3.165) and (3.167) for $\mathbf{P}$, which are repeated here with exchanged RHS and LHS for convenience. Thus,

$$\mathbf{FP} = \mathbf{PA} \tag{3.180}$$

3.3 State Representation

and

$$\mathbf{C} = \mathbf{HP}. \tag{3.181}$$

Premultiplying (3.180) by $\mathbf{HF}^{k-1}$ and then repeatedly substituting for $\mathbf{FP}$ in the expressions generated on the RHS using (3.180) until $\mathbf{F}$ is eliminated and then finally substituting for $\mathbf{HP}$ using (3.181) yield

$$\mathbf{HF}^k\mathbf{P} = \mathbf{HF}^{k-1}\mathbf{PA} = \mathbf{HF}^{k-2}\mathbf{PA}^2 = \mathbf{HF}^{k-3}\mathbf{PA}^3 = \cdots = \mathbf{HPA}^k = \mathbf{CA}^k. \tag{3.182}$$

Taking k from 0 to $n-1$ then yields n matrix equations in which the terms on the RHS are the $m \times n$ submatrices of the observability matrix derived in Sect. 3.2.7. Since this is formulated in the state representation using the state vector, $\mathbf{x}$, it will be denoted by $\mathbf{M}_{ox}(\mathbf{C}, \mathbf{A})$. Similarly, the terms postmultiplying $\mathbf{H}$ on the LHS are the $m \times n$ submatrices of the observability matrix formulated in the state representation using the state vector, $\mathbf{z}$, which will be denoted by $\mathbf{M}_{oz}(\mathbf{H}, \mathbf{F})$. The set of n equations may be expressed as a single equation with partitioning as follows:

$$\begin{bmatrix} \mathbf{H} \\ \hline \mathbf{HF} \\ \hline \mathbf{H}^2\mathbf{F} \\ \hline \vdots \\ \hline \mathbf{H}^{n-1}\mathbf{F} \end{bmatrix} \mathbf{P} = \begin{bmatrix} \mathbf{C} \\ \hline \mathbf{CA} \\ \hline \mathbf{C}^2\mathbf{A} \\ \hline \vdots \\ \hline \mathbf{C}^{n-1}\mathbf{A} \end{bmatrix}, \text{ i.e., } \overbrace{\left[\mathbf{M}_{oz}(\mathbf{H},\mathbf{F})\right]}^{nm \times n} \overbrace{\mathbf{P}}^{n \times n} = \overbrace{\mathbf{M}_{ox}(\mathbf{C},\mathbf{A})}^{nm \times n}. \tag{3.183}$$

This represents $nm \times n$ linear simultaneous equations in the n^2 elements of $\mathbf{P}$, which can be converted to n^2 equations through premultiplication by $\mathbf{M}_{oz}^\mathrm{T}$. Thus

$$\overbrace{[\mathbf{M}_{oz}(\mathbf{H},\mathbf{F})]^\mathrm{T}[\mathbf{M}_{oz}(\mathbf{H},\mathbf{F})]}^{n \times n} \overbrace{\mathbf{P}}^{n \times n} = \overbrace{[\mathbf{M}_{oz}(\mathbf{H},\mathbf{F})]^\mathrm{T}[\mathbf{M}_{ox}(\mathbf{C},\mathbf{A})]}^{n \times n}. \tag{3.184}$$

A similar matrix theorem to that applied in (3.172) may be used here [3], assuming that the plant is observable. This states that if a matrix, $\mathbf{M}$, of dimension $N \times n$, where $N > n$, has rank $\geq n$, then $\mathbf{M}^\mathrm{T}\mathbf{M}$ is nonsingular. This validates the following formula for $\mathbf{P}$ derived from (3.184):

$$\mathbf{P} = \left\{[\mathbf{M}_{oz}(\mathbf{H},\mathbf{F})]^\mathrm{T}[\mathbf{M}_{oz}(\mathbf{H},\mathbf{F})]\right\}^{-1}[\mathbf{M}_{oz}(\mathbf{H},\mathbf{F})]^\mathrm{T}[\mathbf{M}_{ox}(\mathbf{C},\mathbf{A})]. \tag{3.185}$$

This directly yields a formula for the reverse transformation matrix as

$$\mathbf{Q} = \mathbf{P}^{-1} = \left\{[\mathbf{M}_{oz}(\mathbf{H},\mathbf{F})]^\mathrm{T}[\mathbf{M}_{ox}(\mathbf{C},\mathbf{A})]\right\}^{-1}[\mathbf{M}_{oz}(\mathbf{H},\mathbf{F})]^\mathrm{T}[\mathbf{M}_{oz}(\mathbf{H},\mathbf{F})]. \tag{3.186}$$

Alternative formulas for **P** and **Q** are found by expressing (3.183) in terms of **Q** as follows:

$$\mathbf{M}_{oz}(\mathbf{H}, \mathbf{F}) = [\mathbf{M}_{ox}(\mathbf{C}, \mathbf{A})]\mathbf{Q}. \quad (3.187)$$

The formulae are then obtained through premultiplication by $\mathbf{M}_{ox}(\mathbf{C},\mathbf{A})$, yielding

$$[\mathbf{M}_{ox}(\mathbf{C},\mathbf{A})]^T [\mathbf{M}_{oz}(\mathbf{H},\mathbf{F})] = [\mathbf{M}_{ox}(\mathbf{C},\mathbf{A})]^T [\mathbf{M}_{ox}(\mathbf{C},\mathbf{A})]\mathbf{Q} \Rightarrow$$

$$\mathbf{Q} = \left\{[\mathbf{M}_{ox}(\mathbf{C},\mathbf{A})]^T [\mathbf{M}_{ox}(\mathbf{C},\mathbf{A})]\right\}^{-1} [\mathbf{M}_{ox}(\mathbf{C},\mathbf{A})]^T [\mathbf{M}_{oz}(\mathbf{H},\mathbf{F})] \text{ and } \quad (3.188)$$

$$\mathbf{P} = \mathbf{Q}^{-1} = \left\{[\mathbf{M}_{ox}(\mathbf{C},\mathbf{A})]^T [\mathbf{M}_{oz}(\mathbf{H},\mathbf{F})]\right\}^{-1} [\mathbf{M}_{ox}(\mathbf{C},\mathbf{A})]^T [\mathbf{M}_{ox}(\mathbf{C},\mathbf{A})]. \quad (3.189)$$

For an SISO plant, the observability matrices are of dimension $n \times n$, and therefore a simpler formula follows directly from (3.183). Thus,

$$\mathbf{P} = \left[\mathbf{M}_{oz}(\mathbf{h}^T, \mathbf{F})\right]^{-1} \mathbf{M}_{ox}(\mathbf{c}^T, \mathbf{A}), \quad (3.190)$$

which is the only option in terms of the observability matrices, together with

$$\mathbf{Q} = \mathbf{P}^{-1} = \left[\mathbf{M}_{ox}(\mathbf{c}^T, \mathbf{A})\right]^{-1} \mathbf{M}_{oz}(\mathbf{h}^T, \mathbf{F}). \quad (3.191)$$

Although analytical formulas for **P** and **Q** have been derived, their application algebraically would be onerous in most cases. It is therefore recommended that the formulae are programmed in computer software for calculation of **P** and/or **Q** numerically, given numerical values of the plant parameters.

When the first of equations (3.180) is rearranged as

$$\mathbf{F} = \mathbf{P}\mathbf{A}\mathbf{P}^{-1}, \quad (3.192)$$

it is referred to as a *similarity transformation*. This is a term of linear algebra meaning that since the eigenvalues of the transformed matrix, **F**, are the same as those of **A**, then these matrices are defined as being *similar*. Since **A** and **F** are the system matrices of two state-space models of the same plant, their transfer function matrices must be the same. Thus, from (3.81),

$$\mathbf{G}_u(s) = \frac{\mathbf{C}\operatorname{adj}[s\mathbf{I}_n - \mathbf{A}]\mathbf{B}}{|s\mathbf{I}_n - \mathbf{A}|} = \frac{\mathbf{H}\operatorname{adj}[s\mathbf{I}_n - \mathbf{F}]\mathbf{G}}{|s\mathbf{I}_n - \mathbf{F}|}, \quad (3.193)$$

The roots of $|s\mathbf{I}_n - \mathbf{A}| = 0$ and of $|s\mathbf{I}_n - \mathbf{F}| = 0$ are the plant poles, and therefore, the eigenvalues of **F** are the same as those of **A**. Alternatively, using (3.192),

$$|\lambda \mathbf{I}_n - \mathbf{F}| = |\mathbf{P}| |\lambda \mathbf{I}_n - \mathbf{F}| |\mathbf{P}^{-1}| = |\mathbf{P}[\lambda \mathbf{I}_n - \mathbf{F}] \mathbf{P}^{-1}| = |\mathbf{P}\lambda \mathbf{I}_n \mathbf{P}^{-1} - \mathbf{PFP}^{-1}|$$
$$= |\lambda \mathbf{I}_n - \mathbf{A}|.$$

3.3.4 Modal Forms for Multivariable LTI Plants: Transformations

3.3.4.1 Introduction

In the previous section, the forward and reverse transformation matrices between two given LTI state-space models of the same plant with different state representations were determined. In contrast, the purpose of this section is to find a transformation matrix, $\mathbf{P}$, that converts a given LTI state-space model

$$\dot{\mathbf{x}} = \mathbf{A}\mathbf{x} + \mathbf{B}\mathbf{u} \tag{3.194}$$

$$\mathbf{y} = \mathbf{C}\mathbf{x}, \tag{3.195}$$

with an arbitrary state representation using the transformation equation

$$\mathbf{z} = \mathbf{P}\mathbf{x} \tag{3.196}$$

and the reverse transformation equation

$$\mathbf{x} = \mathbf{Q}\mathbf{z}, \tag{3.197}$$

to and from another state-space model

$$\dot{\mathbf{z}} = \mathbf{F}\mathbf{z} + \mathbf{G}\mathbf{u} \tag{3.198}$$

$$\mathbf{y} = \mathbf{H}\mathbf{z}, \tag{3.199}$$

in which (3.198) comprises a number of subsystems whose state variables are not affected by those of the other subsystems, where $\mathbf{x} \in \mathfrak{R}^n$, $\mathbf{z} \in \mathfrak{R}^n$, $\mathbf{u} \in \mathfrak{R}^r$ and $\mathbf{y} \in \mathfrak{R}^m$. The dynamic character of each of these subsystems is captured by their responses to arbitrary initial states, referred to as *modes*. These have already been introduced for SISO plants in Sect. 3.3.2 but the extension to multivariable plants necessitates the matrix–vector approach. All the theory developed, however, applies to SISO plants by setting $r = m = 1$.

The main uses of the modal form are (a) assessing the effectiveness of a controller in keeping individual modes under control by simulation and (b) assessing the degree of excitation of modes that the controller is not designed to keep under control, by simulation. It can also be used for model-based control system design where the objective is to control specific modes.

3.3.4.2 Eigenvalues and Eigenvectors

Determination of the transformation matrix, **P**, of the previous section requires the theory of eigenvalues and eigenvectors of the plant matrix, **A**. The starting point is the definition of an eigenvector, $\mathbf{v} \in \Re^n$. This obeys the equation

$$\mathbf{A}\mathbf{v} = \lambda \mathbf{v}, \tag{3.200}$$

where λ is a scalar.

Definition 3.2 *An eigenvector of a square matrix, $\mathbf{A} \in \Re^{n \times n}$, is any non-zero vector, $\mathbf{v} = [v_1 \ v_2 \ldots v_n]^T$, whose direction in the n-dimensional Euclidean space with coordinates, v_i, $i = 1, 2, \ldots, n$, is unaltered when premultiplied by* **A**.

Equation (3.200) may be written as

$$[\lambda \mathbf{I}_n - \mathbf{A}] \mathbf{v} = \mathbf{0}. \tag{3.201}$$

This represents a homogeneous set of n linear simultaneous equations in v_i, $i = 1, 2, \ldots, n$, that has a nontrivial solution, i.e., $\mathbf{v} \neq \mathbf{0}$, only if

$$|\lambda \mathbf{I}_n - \mathbf{A}| = 0, \tag{3.202}$$

where

$$|\lambda \mathbf{I}_n - \mathbf{A}| = \lambda^n + a_{n-1}\lambda^{n-1} + \cdots + a_1\lambda + a_0 = (\lambda - \lambda_1)(\lambda - \lambda_2)\ldots(\lambda - \lambda_n) \tag{3.203}$$

is the *characteristic polynomial* of **A** and the roots, λ_i, $i = 1, 2, \ldots, n$, of (3.202) are the *eigenvalues* of **A**. In general, substituting the different eigenvalues into (3.201) yields different solutions. Hence, for each eigenvalue, there is an associated eigenvector. Once the eigenvalues have been determined, the complete set of equations to be solved for the eigenvectors is

$$[\lambda_i \mathbf{I}_n - \mathbf{A}] \mathbf{v}_i = \mathbf{0}, \quad i = 1, 2, \ldots, n. \tag{3.204}$$

It is important to note that since each of these is a homogeneous set of equations, there are an infinite number of solutions, $\mathbf{v}_i$, for each i. By inspection of (3.204), if $\mathbf{v}_i = \mathbf{w}_i$ is a solution, then so is $r_i \mathbf{w}_i$, where r_i is an arbitrary scaling constant. The infinite set of solutions, $r_i \mathbf{w}_i$, for each λ_i, with $r_i \in (-\infty, \infty)$, may be written as

$$\begin{bmatrix} v_{1i} \\ v_{2i} \\ \vdots \\ v_{ni} \end{bmatrix} = r_i \begin{bmatrix} w_{1i} \\ w_{2i} \\ \vdots \\ w_{ni} \end{bmatrix}, \quad i = 1, 2, \ldots, n. \tag{3.205}$$

3.3 State Representation 219

It is clear from (3.205) that the direction of each vector is the same for all r_i, as required. The vector $\mathbf{w}_i$, which is any valid eigenvector, is referred to as a basis vector from which all the other possible eigenvectors, $\mathbf{v}_i$, can be generated. The MATLAB® 'eigs' routine accepts a matrix input numerically and returns eigenvalues and eigenvectors normalised to have unity magnitude by choosing

$$r_i = \frac{1}{\|\mathbf{w}_i\|} = \frac{1}{\sqrt{w_{1i}^2 + w_{2i}^2 + \cdots + w_{ni}^2}}, \quad i = 1, 2, \ldots, n. \tag{3.206}$$

Various algorithms could be used to calculate the basis vectors. To give some insight, a possible algorithm will be described here. Let (3.204) be written with $\mathbf{w}_i$ instead of $\mathbf{v}_i$ as

$$\mathbf{M}_i \mathbf{w}_i = \mathbf{0}. \tag{3.207}$$

A linear constraint equation,

$$\mathbf{m}_i^T \mathbf{w}_i = c, \tag{3.208}$$

can be formed and solved with (3.207) to produce a particular solution that can be used as the basis vector, where $\mathbf{m}_i \in \Re^n$ is a column vector and c is a constant scalar. Then (3.207) and (3.208) can be written together as

$$\overbrace{\begin{bmatrix} \overbrace{\mathbf{M}_i}^{n \times n} \\ \hline \underbrace{\mathbf{m}_i^T}_{1 \times n} \end{bmatrix}}^{(n+1) \times n} \mathbf{w}_i = \overbrace{\begin{bmatrix} \overbrace{\mathbf{0}}^{n \times n} \\ \hline \underbrace{c}_{1 \times 1} \end{bmatrix}}^{(n+1) \times n}, \tag{3.209}$$

where the dimensions are indicated above the matrices for clarification. Since this must be an overdetermined set of simultaneous equations with nonconflicting solutions, the same technique can be employed as used in Sect. 3.3.3. Thus

$$\overbrace{\begin{bmatrix} \mathbf{M}_i^T & | & \mathbf{m}_i \end{bmatrix}}^{n \times (n+1)} \overbrace{\begin{bmatrix} \mathbf{M}_i \\ \hline \mathbf{m}_i^T \end{bmatrix}}^{(n+1) \times n} \mathbf{w}_i = \overbrace{\begin{bmatrix} \mathbf{M}_i^T & | & \mathbf{m}_i \end{bmatrix}}^{n \times (n+1)} \overbrace{\begin{bmatrix} \mathbf{0} \\ \hline c \end{bmatrix}}^{(n+1) \times n} \Rightarrow$$

$$\mathbf{w}_i = \left\{ \begin{bmatrix} \mathbf{M}_i^T & | & \mathbf{m}_i \end{bmatrix} \begin{bmatrix} \mathbf{M}_i \\ \hline \mathbf{m}_i^T \end{bmatrix} \right\}^{-1} \begin{bmatrix} \mathbf{M}_i^T & | & \mathbf{m}_i \end{bmatrix} \begin{bmatrix} \mathbf{0} \\ \hline c \end{bmatrix} \tag{3.210}$$

Finally

$$\mathbf{v}_i = \mathbf{w}_i / \|\mathbf{w}_i\|, \quad i = 1, 2, \ldots, n. \tag{3.211}$$

The following sections present three different classes of subsystem revealed by a similarity transformation in which the plant matrix eigenvalues and eigenvectors are used to form the transformation matrix. The unforced responses of these subsystems to arbitrary initial states define three corresponding classes of mode.

3.3.4.3 Model with Distinct Real Eigenvalues

In this case, there are n *eigenvectors* of $\mathbf{A}$ that are linearly independent. Let the columns of the reverse transformation matrix, $\mathbf{Q}$, of (3.197) be formed using these eigenvectors as follows:

$$\mathbf{Q} = [\mathbf{v}_1 \mid \mathbf{v}_2 \mid \ldots \mid \mathbf{v}_n]. \tag{3.212}$$

Substituting $\mathbf{x} = \mathbf{Q}\mathbf{z}$ in (3.194) yields

$$\mathbf{Q}\dot{\mathbf{z}} = \mathbf{A}\mathbf{Q}\mathbf{z} + \mathbf{B}\mathbf{u}. \tag{3.213}$$

Here,

$$\mathbf{A}\mathbf{Q} = \mathbf{A}[\mathbf{v}_1 \mid \mathbf{v}_2 \mid \ldots \mid \mathbf{v}_n] = [\mathbf{A}\mathbf{v}_1 \mid \mathbf{A}\mathbf{v}_2 \mid \ldots \mid \mathbf{A}\mathbf{v}_n]. \tag{3.214}$$

In view of (3.200),

$$\mathbf{A}\mathbf{v}_i = \lambda_i \mathbf{v}_i, \quad i = 1, 2, \ldots, n. \tag{3.215}$$

and therefore (3.214) may be written as

$$\mathbf{A}\mathbf{Q} = [\lambda_1 \mathbf{v}_1 \mid \lambda_2 \mathbf{v}_2 \mid \ldots \mid \lambda_n \mathbf{v}_n] = [\mathbf{v}_1 \mid \mathbf{v}_2 \mid \ldots \mid \mathbf{v}_n] \begin{bmatrix} \lambda_1 & 0 & \cdots & 0 \\ 0 & \lambda_2 & \ddots & \vdots \\ \vdots & \ddots & \ddots & 0 \\ 0 & \cdots & 0 & \lambda_n \end{bmatrix} = \mathbf{Q}\mathbf{\Lambda}. \tag{3.216}$$

Then (3.213) may be written as

$$\mathbf{Q}\dot{\mathbf{z}} = \mathbf{Q}\mathbf{\Lambda}\mathbf{z} + \mathbf{B}\mathbf{u}. \tag{3.217}$$

Premultiplying (3.217) by $\mathbf{P} = \mathbf{Q}^{-1}$ yields

$$\dot{\mathbf{z}} = \mathbf{\Lambda}\mathbf{z} + \mathbf{P}\mathbf{B}\mathbf{u}. \tag{3.218}$$

From (3.195) and (3.197), the corresponding measurement equation is

$$\mathbf{y} = \mathbf{C}\mathbf{Q}\mathbf{z}. \tag{3.219}$$

3.3 State Representation

Since the plant matrix, $\boldsymbol{\Lambda}$, is diagonal, the transformed plant model consists of n first-order subsystems whose state variables, z_i, $i = 1, 2, \ldots, n$, do not influence one another. It is important to recall, however, that the scaling of each eigenvector forming the columns of $\mathbf{Q}$ can be arbitrary. There are therefore an infinite number of state representations that could result from the transformation, but all of them share the property of their state variables not influencing one another due to the common diagonal plant matrix, $\boldsymbol{\Lambda}$. These state representations, therefore, can only differ from one another in the trivial manner of having different scalings of their state variables. To show this, let the eigenvectors forming the columns of $\mathbf{Q}$ be scaled by arbitrary constant factors, r_i, $i = 1, 2, \ldots, n$, to form a new transformation matrix, $\mathbf{Q}_s$, and let the resulting state vector be $\mathbf{z}_s$. Then,

$$\mathbf{Q}_s = [r_1\mathbf{v}_1 \mid r_2\mathbf{v}_2 \mid \cdots \mid r_n\mathbf{v}_n] = [\mathbf{v}_1 \mid \mathbf{v}_2 \mid \cdots \mid \mathbf{v}_n] \begin{bmatrix} r_1 & 0 & \cdots & 0 \\ 0 & r_2 & \ddots & \vdots \\ \vdots & \ddots & \ddots & 0 \\ 0 & \cdots & 0 & r_n \end{bmatrix} = \mathbf{QR} \quad (3.220)$$

$$\text{Hence } \mathbf{x} = \mathbf{Q}_s \mathbf{z}_s = \mathbf{QR}\mathbf{z}_s. \quad (3.221)$$

Since $\mathbf{x} = \mathbf{Qz}$, it follows from (3.221) that $\mathbf{z} = \mathbf{Rz}_s \Rightarrow \mathbf{z}_s = \mathbf{R}^{-1}\mathbf{z}$. Since $\mathbf{R}$ is diagonal, then so is $\mathbf{R}^{-1}$. It follows that the components of $\mathbf{z}_s$ are just differently scaled components of $\mathbf{z}$, i.e. $z_{si} = z_i/r_i$, $i = 1, 2, \ldots, n$.

The dynamic character of the plant is contained in the modes. In order that this may be viewed free of influence by the input, $\mathbf{u}(t)$, the state differential equation with zero input but non-zero initial state may be considered. The output, $\mathbf{y}(t)$, in general has output components that are linear weighted sums of the modes through (3.219). To examine the character of the individual modes, therefore, the general solution to the unforced state differential equation,

$$\dot{\mathbf{z}} = \boldsymbol{\Lambda}\mathbf{z}, \quad (3.222)$$

with $\mathbf{z}(0) \neq 0$ is determined. This comprises the n separate responses as follows:

$$\mathbf{z}(t) = e^{\boldsymbol{\Lambda}t}\mathbf{z}(0) \Rightarrow z_i(t) = e^{\lambda_i t}z_i(0) = e^{(t/T_i)\,\mathrm{sgn}(\lambda_i)}z_i(0), \quad i = 1, 2, \ldots, n. \quad (3.223)$$

These are *exponential modes* already introduced in Chap. 2 for closed-loop systems. They have time constants, $T_i = 1/|\lambda_i|$.

3.3.4.4 Model Including Complex Conjugate Eigenvalues

As a first step, the transformation can be carried out as in Sect. 3.3.4.3 to yield a diagonalised plant matrix, $\boldsymbol{\Lambda}$, but the state variables associated with the complex conjugate eigenvalues will be complex and so will be the associated eigenvectors and elements of $\mathbf{\Lambda}$. Such a plant model would be mathematically correct but not practicable for direct implementation in a simulation, or for derivation of a model-based controller. A practicable version of this model may be formed, however, by taking each pair of state differential equations associated with the complex conjugate eigenvalues and converting them to an equivalent pair of state differential equations with real state variables and coefficients. For this purpose, the complex conjugate eigenvalues will be placed consecutively so that the state differential equations to be converted are together. The real state variables of the converted equations, however, will depend upon one another, resulting in off-diagonal elements in the plant matrix, which is therefore in the block diagonal form rather than the simple diagonal form of Sect. 3.3.4.3.

Once the state-space model has been transformed to the form of (3.218) and (3.219), then it will be helpful to take Laplace transforms with zero initial conditions as in Sect. 3.2.9, giving

$$\mathbf{Y}(s) = \mathbf{H}\mathbf{Z}(s) \tag{3.224}$$

and

$$\mathbf{Z}(s) = [s\mathbf{I}_n - \mathbf{\Lambda}]^{-1}\mathbf{G}\mathbf{U}(s) \tag{3.225}$$

where $\mathbf{H} = \mathbf{CQ}$ and $\mathbf{G} = \mathbf{PB}$. Let $\mathbf{H}$ be partitioned into column vectors and $\mathbf{G}$ be partitioned into row vectors so that (3.224) and (3.225) are written as

$$\mathbf{Y}(s) = [\mathbf{h}_1 \mid \mathbf{h}_2 \mid \cdots \mid \mathbf{h}_n]\mathbf{Z}(s) \tag{3.226}$$

and

$$\mathbf{Z}(s) = [s\mathbf{I}_n - \mathbf{\Lambda}]^{-1} \begin{bmatrix} \mathbf{g}_1^T \\ \mathbf{g}_2^T \\ \vdots \\ \mathbf{g}_n^T \end{bmatrix} \mathbf{U}(s). \tag{3.227}$$

Since the matrix, $s\mathbf{I}_n - \mathbf{\Lambda}$, is diagonal, then (3.227) may be written as follows:

$$\mathbf{Z}(s) = \begin{bmatrix} (1/(s-\lambda_1))\mathbf{g}_1^T \\ (1/(s-\lambda_2))\mathbf{g}_2^T \\ \vdots \\ (1/(s-\lambda_n))\mathbf{g}_2^T \end{bmatrix} \mathbf{U}(s) \tag{3.228}$$

Eliminating $\mathbf{Z}(s)$ between (3.226) and (3.228) then yields

3.3 State Representation

$$Y(s)=[\mathbf{h}_1 \mid \mathbf{h}_2 \mid \cdots \mid \mathbf{h}_n]\begin{bmatrix}(1/(s-\lambda_1))\mathbf{g}_1^T \\ \hline (1/(s-\lambda_2))\mathbf{g}_2^T \\ \hline \vdots \\ \hline (1/(s-\lambda_n))\mathbf{g}_2^T\end{bmatrix}\mathbf{U}(s)=\sum_{k=1}^{n}\frac{\mathbf{h}_k\mathbf{g}_k^T}{s-\lambda_k}\mathbf{U}(s)=\sum_{k=1}^{n}\frac{\mathbf{M}_k}{s-\lambda_k}\mathbf{U}(s).$$
(3.229)

The expression premultiplying $\mathbf{U}(s)$ is a partial fraction expansion of the plant transfer function matrix, of dimension, $m \times r$.

Let λ_i and $\lambda_{i+1} = \bar{\lambda}_i$ be a complex conjugate pair of eigenvalues. Then, the transfer function relationship of the corresponding subsystem of (3.229) is

$$\mathbf{Y}_i(s) = \left[\frac{\mathbf{M}_i}{s - \lambda_i} + \frac{\mathbf{M}_{i+1}}{s - \bar{\lambda}_i}\right]\mathbf{U}(s), \tag{3.230}$$

where $\mathbf{y}_i(s)$ is the contribution to $\mathbf{y}(s)$. This may be written as

$$\begin{aligned}\mathbf{Y}_i(s) &= \left[\frac{\mathbf{M}_i}{s + \sigma_i + j\omega} + \frac{\mathbf{M}_{i+1}}{s + \sigma_i - j\omega}\right]\mathbf{U}(s) \\ &= \frac{\mathbf{M}_i(s + \sigma_i - j\omega) + \mathbf{M}_{i+1}(s + \sigma_i + j\omega)}{(s + \sigma_i)^2 + \omega_i^2}\mathbf{U}(s).\end{aligned} \tag{3.231}$$

Since $\mathbf{u}(s)$ and $\mathbf{y}_i(s)$ are real then so must be all the numerator coefficient matrices. This would be satisfied by $\mathbf{M}_{i+1} = \mathbf{M}_i \in \Re^{m \times r}$ but this would yield

$$\mathbf{Y}_i(s) = \frac{2\mathbf{M}_i(s + \sigma_i)}{(s + \sigma_i)^2 + \omega_i^2}\mathbf{U}(s) \tag{3.232}$$

and therefore impose the constraint that the transfer function zero is fixed at $-\sigma_i$, which is also the real part of the complex conjugate poles. To remove this constraint, $\mathbf{M}_i$ and $\mathbf{M}_{i+1}$ are allowed to be complex but they are conjugates. Thus,

$$\mathbf{M}_i = \mathbf{M}_{Ri} + j\mathbf{M}_{Ii} \tag{3.233}$$

and

$$\mathbf{M}_{i+1} = \mathbf{M}_{Ri} - j\mathbf{M}_{Ii}. \tag{3.234}$$

Then, (3.231) becomes

$$\mathbf{Y}_i(s) = 2\frac{\mathbf{M}_{Ri}(s + \sigma_i) + \mathbf{M}_{Ii}\omega_i}{(s + \sigma_i)^2 + \omega_i^2}\mathbf{U}(s). \tag{3.235}$$

A constraint on $\mathbf{M}_{Ri}$ and $\mathbf{M}_{Ii}$ is needed to ensure that subsystem (3.235) can be represented by a state-space model with only two state variables. In order to find this, first $\mathbf{M}_{Ri}$ and $\mathbf{M}_{Ii}$ will be expressed in terms of the real and imaginary parts of the column vectors $\mathbf{h}_i$ and $\mathbf{h}_{i+1} = \overline{\mathbf{h}}_i$ and the row vectors $\mathbf{g}_i^T$ and $\mathbf{g}_i^T = \mathbf{g}_i^{-T}$. Thus,

$$\mathbf{h}_i = \mathbf{h}_{Ri} - j\mathbf{h}_{Ii}, \quad \mathbf{h}_{i+1} = \mathbf{h}_{Ri} + j\mathbf{h}_{Ii} \qquad (3.236)$$

and

$$\mathbf{g}_i^T = \mathbf{g}_{Ri}^T - j\mathbf{g}_{Ii}^T, \quad \mathbf{g}_{i+1}^T = \mathbf{g}_{Ri}^T - j\mathbf{g}_{Ii}^T. \qquad (3.237)$$

Then

$$\mathbf{M}_i = \mathbf{h}_i \mathbf{g}_i^T = [\mathbf{h}_{Ri} - j\mathbf{h}_{Ii}][\mathbf{g}_{Ri}^T - j\mathbf{g}_{Ii}^T] = [\mathbf{h}_{Ri}\mathbf{g}_{Ri}^T - \mathbf{h}_{Ii}\mathbf{g}_{Ii}^T] - j[\mathbf{h}_{Ri}\mathbf{g}_{Ii}^T + \mathbf{h}_{Ii}\mathbf{g}_{Ri}^T]. \qquad (3.238)$$

Hence

$$\mathbf{M}_{Ri} = \mathbf{h}_{Ri}\mathbf{g}_{Ri}^T - \mathbf{h}_{Ii}\mathbf{g}_{Ii}^T \qquad (3.239)$$

and

$$\mathbf{M}_{Ii} = \mathbf{h}_{Ri}\mathbf{g}_{Ii}^T + \mathbf{h}_{Ii}\mathbf{g}_{Ri}^T \qquad (3.240)$$

In view of (3.235), consider an associated subsystem with state variables x and y and a scalar input u, satisfying the following transfer function relationship.

$$\begin{bmatrix} X(s) \\ Y(s) \end{bmatrix} = \frac{1}{(s+\sigma_i)^2 + \omega_i^2} \begin{bmatrix} s+\sigma_i \\ \omega_i \end{bmatrix} U(s). \qquad (3.241)$$

A state-variable block diagram in the Laplace domain realising this is shown in Fig. 3.15. This may be verified easily by block diagram reduction.

This may be recognised as the second-order SISO subsystem arising in Sect. 3.3.2.5. Substituting for $\mathbf{M}_{Ri}$ and $\mathbf{M}_{Ii}$ in (3.235) using (3.239) and (3.240) yields

$$\mathbf{y}_i(s) = 2\frac{[\mathbf{h}_{Ri}\mathbf{g}_{Ri}^T - \mathbf{h}_{Ii}\mathbf{g}_{Ii}^T](s+\sigma_i) + [\mathbf{h}_{Ri}\mathbf{g}_{Ii}^T + \mathbf{h}_{Ii}\mathbf{g}_{Ri}^T]\omega_i}{(s+\sigma_i)^2 + \omega_i^2}\mathbf{u}(s) \qquad (3.242)$$

Fig. 3.15 State variable block diagram of a second-order linear element

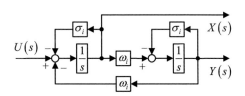

3.3 State Representation

In view of (3.241), this can be realised with the second-order element of Fig. 3.15 if

$$\mathbf{g}_{li}^T = \mathbf{g}_{Ri}^T, \qquad (3.243)$$

$$u(s) = \mathbf{g}_{Ri}^T \mathbf{u}(s), \qquad (3.244)$$

and

$$\mathbf{y}_i(s) = 2\left[\mathbf{h}_{Ri} - \mathbf{h}_{li}\right] x(s) + 2\left[\mathbf{h}_{Ri} + \mathbf{h}_{li}\right] y(s). \qquad (3.245)$$

To comply with the notation of (3.226) and (3.227), Fig. 3.16 shows the time domain state-variable block diagram in which the state variables are redesignated as $x = z_i$ and $y = z_{i+1}$.

It should be noted that z_i and z_{i+1} were originally complex and are now the state variables of the second-order subsystem. Similarly, the column vectors of (3.245) have been redesignated as

$$\mathbf{h}_{Ri} - \mathbf{h}_{li} = \mathbf{h}_i \qquad (3.246)$$

and

$$\mathbf{h}_{Ri} + \mathbf{h}_{li} = \mathbf{h}_{i+1}, \qquad (3.247)$$

and the row vector of (3.244) has been redesignated as

$$\mathbf{g}_{Ri}^T = \mathbf{g}_i^T. \qquad (3.248)$$

After the formation of the second subsystem,

$$\mathbf{g}_{i+1}^T = \mathbf{0}, \qquad (3.249)$$

since there is no direct algebraic dependence of $\dot{z}_{i+1}$ on $\mathbf{u}$, which is evident in Fig. 3.16 through there being no direct input of $\mathbf{u}$ to the input of the second integrator.

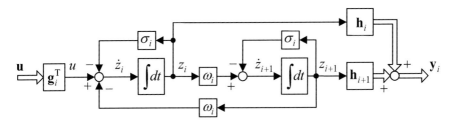

Fig. 3.16 State variable block diagram of a second-order subsystem of modal model

The state-space equations of the second-order subsystem may be written down by inspection of Fig. 3.16. Thus

$$\begin{bmatrix} \dot{z}_i \\ \dot{z}_{i+1} \end{bmatrix} = \begin{bmatrix} -\sigma_i & -\omega_i \\ \omega_i & -\sigma_i \end{bmatrix} \begin{bmatrix} z_i \\ z_{i+1} \end{bmatrix} + \begin{bmatrix} \mathbf{g}_i^T \\ \mathbf{0} \end{bmatrix} \mathbf{u}, \quad y_i = [\mathbf{h}_i \mid \mathbf{h}_{i+1}] \begin{bmatrix} z_i \\ z_{i+1} \end{bmatrix}$$

(3.250)

In the complete state-space model,

$$\begin{aligned} \dot{\mathbf{z}} &= \mathbf{\Lambda}_b \mathbf{z} + \mathbf{G}\mathbf{u} \\ \mathbf{y} &= \mathbf{H}\mathbf{z} \end{aligned}$$

(3.251)

every second-order subsystem such as (3.250) contributes a 2×2 submatrix of the *block diagonal* plant matrix, $\mathbf{\Lambda}_b$, with common diagonal elements as follows:

$$\mathbf{\Lambda}_b = \begin{bmatrix} \lambda_1 & 0 & 0 & \cdots & & \cdots & & \cdots & 0 \\ 0 & \ddots & & \ddots & & \ddots & & & \vdots \\ 0 & \ddots & \lambda_{i-1} & 0 & & \ddots & & & \vdots \\ \vdots & \ddots & 0 & -\sigma_i & -\omega_i & & \ddots & & \vdots \\ & & \ddots & \omega_i & -\sigma_i & 0 & \ddots & & \vdots \\ \vdots & & & \ddots & 0 & \lambda_{i+2} & \ddots & 0 \\ \vdots & & & & \ddots & & \ddots & \ddots & 0 \\ 0 & \cdots & & \cdots & & \cdots & 0 & 0 & \lambda_n \end{bmatrix}$$

(3.252)

According to the unforced solution (3.223), the two contributions due to the complex conjugate eigenvalues are

$$z_i(t) = e^{(-\sigma_i - j\omega_i)t} z_i(0) = e^{-\sigma_i t} [\cos(\omega_i t) - j \sin(\omega_i t)] z_i(0) \quad (3.253)$$

and

$$z_{i+1}(t) = e^{(-\sigma_i + j\omega_i)t} z_{i+1}(0) = e^{-\sigma_i t} [\cos(\omega_i t) + j \sin(\omega_i t)] z_{i+1}(0). \quad (3.254)$$

These give an indication of the dynamic character of the second-order subsystem, i.e. sinusoidal oscillations with exponentially varying amplitude, but the variables are complex, which was the reason for converting this subsystem to (3.250). The corresponding solution to the state differential equation of (3.250) with $\mathbf{u} = \mathbf{0}$ is

$$\begin{bmatrix} \dot{z}_i(t) \\ \dot{z}_{i+1}(t) \end{bmatrix} = \left\{ \exp \begin{bmatrix} -\sigma_i & -\omega_i \\ \omega_i & -\sigma_i \end{bmatrix} t \right\} \begin{bmatrix} z_i(0) \\ z_{i+1}(0) \end{bmatrix}. \quad (3.255)$$

3.3 State Representation

This is an example of finding the solution of the general unforced linear system,

$$\dot{\mathbf{x}} = \mathbf{A}\mathbf{x}, \quad (3.256)$$

and is given by (3.24) with $\mathbf{u}(\tau) = \mathbf{0}$. Taking $t_0 = 0$ yields

$$\mathbf{x}(t) = e^{\mathbf{A}t}\mathbf{x}(0). \quad (3.257)$$

The component equations are needed, however, to investigate the dynamic behaviour. These can be obtained using the transformation matrix, $\mathbf{Q} = [\mathbf{v}_1 \mid \mathbf{v}_2 \mid \cdots \mid \mathbf{v}_n]$, of Sect. 3.3.4.3, where $\mathbf{v}_i$, $i = 1, 2, \ldots, n$, are eigenvectors of $\mathbf{A}$.

The solution of the state differential equation,

$$\dot{\mathbf{z}} = \mathbf{\Lambda}\mathbf{z}, \quad (3.258)$$

of the plant transformed by

$$\mathbf{x} = \mathbf{Q}\mathbf{z} \quad (3.259)$$

is

$$\mathbf{z}(t) = e^{\mathbf{\Lambda}t}\mathbf{z}(0) = \exp\left\{\begin{bmatrix} \lambda_1 & 0 & \cdots & 0 \\ 0 & \lambda_2 & \ddots & \vdots \\ \vdots & \ddots & \ddots & 0 \\ 0 & \cdots & 0 & \lambda_n \end{bmatrix} t\right\} \mathbf{z}(0) = \begin{bmatrix} e^{\lambda_1 t} & 0 & \cdots & 0 \\ 0 & e^{\lambda_2 t} & \ddots & \vdots \\ \vdots & \ddots & \ddots & 0 \\ 0 & \cdots & 0 & e^{\lambda_n t} \end{bmatrix} \mathbf{z}(0). \quad (3.260)$$

where λ_i, $i = 1, 2, \ldots, n$, are the plant eigenvalues. From (3.256), (3.258) and (3.259),

$$\mathbf{\Lambda} = \mathbf{Q}^{-1}\mathbf{A}\mathbf{Q}. \quad (3.261)$$

Hence,

$$\mathbf{\Lambda}^n = \underbrace{\mathbf{Q}^{-1}\mathbf{A}\mathbf{Q}}_{1}\underbrace{\mathbf{Q}^{-1}\mathbf{A}\mathbf{Q}}_{2}\cdots\underbrace{\mathbf{Q}^{-1}\mathbf{A}\mathbf{Q}}_{n} = \mathbf{Q}^{-1}\mathbf{A}^n\mathbf{Q}. \quad (3.262)$$

Since $e^{\mathbf{\Lambda}t}$ and $e^{\mathbf{A}t}$ can, respectively, be expanded as infinite power series in $\mathbf{\Lambda}$ and $\mathbf{A}$, it follows from (3.262) that

$$e^{\mathbf{\Lambda}t} = \mathbf{Q}^{-1}e^{\mathbf{A}t}\mathbf{Q}. \quad (3.263)$$

From (3.263) and (3.259), which is valid for $\mathbf{x}(0)$ and $\mathbf{z}(0)$,

$$e^{\Lambda t}\mathbf{z}(0) = \mathbf{Q}^{-1}e^{\mathbf{A}t}\mathbf{Q}\mathbf{z}(0) = \mathbf{Q}^{-1}e^{\mathbf{A}t}\mathbf{x}(0) \Rightarrow$$

$$e^{\mathbf{A}t}\mathbf{x}(0) = \mathbf{Q}e^{\Lambda t}\mathbf{z}(0) = \begin{bmatrix} \mathbf{v}_1 & | & \mathbf{v}_2 & | & \cdots & | & \mathbf{v}_n \end{bmatrix} \begin{bmatrix} e^{\lambda_1 t} & 0 & \cdots & 0 \\ 0 & e^{\lambda_2 t} & \ddots & \vdots \\ \vdots & \ddots & \ddots & 0 \\ 0 & \cdots & 0 & e^{\lambda_n t} \end{bmatrix} \mathbf{z}(0)$$

So the solution is

$$\begin{aligned}\mathbf{x}(t) = e^{\mathbf{A}t}\mathbf{x}(0) &= \begin{bmatrix} e^{\lambda_1 t}\mathbf{v}_1 & | & e^{\lambda_2 t}\mathbf{v}_2 & | & \cdots & | & e^{\lambda_n t}\mathbf{v}_n \end{bmatrix} \mathbf{z}(0) \\ &= \begin{bmatrix} e^{\lambda_1 t}\mathbf{v}_1 & | & e^{\lambda_2 t}\mathbf{v}_2 & | & \cdots & | & e^{\lambda_n t}\mathbf{v}_n \end{bmatrix} \mathbf{Q}^{-1}\mathbf{x}(0),\end{aligned} \quad (3.264)$$

Returning to (3.250), the solution to the unforced state differential equation,

$$\begin{bmatrix} \dot{x}_1 \\ \dot{x}_2 \end{bmatrix} = \begin{bmatrix} -\sigma_i & -\omega_i \\ \omega_i & -\sigma_i \end{bmatrix} \begin{bmatrix} x_1 \\ x_2 \end{bmatrix}, \quad (3.265)$$

is required, where $x_1 = z_i$ and $x_2 = z_{i+1}$. The eigenvalues of the system matrix are known as $\lambda_i = -\sigma_i - j\omega_i$ and $\lambda_{i+1} = -\sigma_i + j\omega_i$. The eigenvector equation is therefore

$$\mathbf{A}_i \begin{bmatrix} \mathbf{v}_1 & | & \mathbf{v}_2 \end{bmatrix} = \begin{bmatrix} \lambda_i \mathbf{v}_1 & | & \lambda_{i+1} \mathbf{v}_2 \end{bmatrix}. \quad (3.266)$$

The equation for $\mathbf{v}_1$ is then

$$\begin{bmatrix} -\sigma_i & -\omega_i \\ \omega_i & -\sigma_i \end{bmatrix} \begin{bmatrix} v_{11} \\ v_{21} \end{bmatrix} = (-\sigma_i - j\omega_i) \begin{bmatrix} v_{11} \\ v_{21} \end{bmatrix} \Rightarrow \quad (3.267)$$
$$-\sigma_i v_{11} - \omega_i v_{21} = (-\sigma_i - j\omega_i) v_{11} \Rightarrow \omega_i v_{21} = j\omega_i v_{11}$$

and

$$\omega_i v_{11} - \sigma_i v_{21} = (-\sigma_i - j\omega_i) v_{21} \Rightarrow \omega_i v_{11} = -j\omega_i v_{21} \Rightarrow j\omega_i v_{11} = \omega_i v_{21}. \quad (3.268)$$

As would be expected, (3.267) and (3.268) are linearly dependent, allowing an infinite number of solutions, since an eigenvector can have any scaling factor, only its direction being preserved. By inspection, a suitable eigenvector is

$$\begin{bmatrix} v_{11} \\ v_{21} \end{bmatrix} = c_i \begin{bmatrix} 1 \\ j \end{bmatrix}. \quad (3.269)$$

where c_i is an arbitrary scaling factor. Similarly, the equation for $\mathbf{v}_2$ is

$$\begin{bmatrix} -\sigma_i & -\omega_i \\ \omega_i & -\sigma_i \end{bmatrix} \begin{bmatrix} v_{12} \\ v_{22} \end{bmatrix} = (-\sigma_i + j\omega_i) \begin{bmatrix} v_{12} \\ v_{22} \end{bmatrix} \Rightarrow, \quad (3.270)$$

3.3 State Representation

$$-\sigma_i v_{12} - \omega_i v_{22} = (-\sigma_i + j\omega_i) v_{12} \Rightarrow -\omega_i v_{22} = j\omega_i v_{12} \quad (3.271)$$

and

$$\omega_i v_{12} - \sigma_i v_{22} = (-\sigma_i + j\omega_i) v_{22} \Rightarrow \omega_i v_{12} = j\omega_i v_{22} \Rightarrow j\omega_i v_{12} = -\omega_i v_{22}. \quad (3.272)$$

Again by inspection, a suitable eigenvector is

$$\begin{bmatrix} v_{12} \\ v_{22} \end{bmatrix} = c_i \begin{bmatrix} 1 \\ -j \end{bmatrix}. \quad (3.273)$$

where c_i is the same arbitrary scaling factor as in (3.269) to ensure that $\mathbf{v}_1$ and $\mathbf{v}_2$ are complex conjugate eigenvectors corresponding to the complex conjugate eigenvalues, as needed to arrive at the required real coefficients in the solution.
The solution of (3.265) will now be obtained from (3.264). First

$$\mathbf{Q}^{-1} = [\mathbf{v}_1 \mid \mathbf{v}_2]^{-1} = \begin{bmatrix} 1 & 1 \\ j & -j \end{bmatrix}^{-1} = \frac{1}{2j}\begin{bmatrix} -j & -1 \\ -j & 1 \end{bmatrix} = \frac{1}{2}\begin{bmatrix} 1 & -j \\ 1 & j \end{bmatrix} \quad (3.274)$$

Then,

$$\begin{bmatrix} x_1(t) \\ x_2(t) \end{bmatrix} = \begin{bmatrix} e^{-(\sigma_i+j\omega_i)t}\begin{bmatrix} 1 \\ j \end{bmatrix} c_i \mid e^{-(\sigma_i-j\omega_i)t}\begin{bmatrix} 1 \\ -j \end{bmatrix} c_i \end{bmatrix}\begin{bmatrix} 1 & 1 \\ j & -j \end{bmatrix}^{-1}\begin{bmatrix} x_1(0) \\ x_2(0) \end{bmatrix}$$

$$= c_i e^{-\sigma_i t}\begin{bmatrix} \cos(\omega_i t) - j\sin(\omega_i t) & \cos(\omega_i t) + j\sin(\omega_i t) \\ \sin(\omega_i t) + j\cos(\omega_i t) & \sin(\omega_i t) - j\cos(\omega_i t) \end{bmatrix}\frac{1}{2}\begin{bmatrix} 1 & -j \\ 1 & j \end{bmatrix}\begin{bmatrix} x_1(0) \\ x_2(0) \end{bmatrix}$$

$$= \frac{1}{2}c_i e^{-\sigma_i t}\begin{bmatrix} 2\cos(\omega_i t) & -2\sin(\omega_i t) \\ 2\sin(\omega_i t) & 2\cos(\omega_i t) \end{bmatrix}\begin{bmatrix} x_1(0) \\ x_2(0) \end{bmatrix}$$

Setting $t = 0$ reveals that the correct eigenvector scaling factor is $c_i = 1$. Then

$$\begin{aligned} x_1(t) &= e^{-\sigma_i t}[x_1(0)\cos(\omega_i t) - x_2(0)\sin(\omega_i t)] \\ x_2(t) &= e^{-\sigma_i t}[x_1(0)\sin(\omega_i t) + x_2(0)\cos(\omega_i t)] \end{aligned}. \quad (3.275)$$

Since it has been necessary to combine two first-order subsystems with complex conjugate eigenvalues and eigenvectors to form a second-order subsystem with real coefficients and variables, the unforced state response given by (3.275) is considered to characterise a single mode. Since the state variables oscillate it is referred to as an *oscillatory mode*. It is the oscillatory mode introduced in Chap. 1 for closed-loop systems, but is viewed in the state space. A vibrating member of a mechanical structure or a resonant circuit containing inductance and capacitance would be examples of physical systems containing such modes.

3.3.4.5 Model Including Repeated Eigenvalues

Kinematic integrators in motion control applications yield double eigenvalues of zero value, equivalent to double poles at the origin of the s-plane. An example is spacecraft attitude control in which three rotational degrees of freedom require control. Other numbers and values of repeated eigenvalues are only infrequently found in uncontrolled plants but this section is sufficiently general to enable any plant model with repeated eigenvalues to be dealt with. Also, the material should be useful preparation for the chapters on model-based control that contain many examples in which the closed-loop system is *designed* to have repeated eigenvalues.

As has been seen, if the eigenvalues of **A** are distinct, then the transformation

$$\mathbf{F} = \mathbf{Q}^{-1}\mathbf{A}\mathbf{Q} \tag{3.276}$$

with

$$\mathbf{Q} = [\mathbf{v}_1 \mid \mathbf{v}_2 \mid \ldots \mid \mathbf{v}_n] \tag{3.277}$$

where $\mathbf{v}_i$, $i = 1, 2, \ldots, n$, are linearly independent eigenvectors of **A**, diagonalises **A** in the sense that **F** is a diagonal matrix with the eigenvalues of **A** as elements. If any of the eigenvalues are repeated, then such diagonalisation *may or may not* be possible. In every case, however, transformation to a form close to the diagonal form is always possible with a transformation matrix whose columns are eigenvectors as in (3.277). This is the *Jordan canonical form*

$$\mathbf{J} = \begin{bmatrix} \mathbf{J}_1 & 0 & \cdots & 0 \\ 0 & \mathbf{J}_2 & & \vdots \\ \vdots & & \ddots & 0 \\ 0 & \cdots & 0 & \mathbf{J}_{n_J} \end{bmatrix}, \tag{3.278}$$

which is in the block diagonal form with submatrices, $\mathbf{J}_i$, $i = 1, 2, \ldots, n_J$, each of these being referred to as a *Jordan block*. Each Jordan block has repeated eigenvalues on the leading diagonal but unity elements on the superdiagonal as follows.

$$\mathbf{J}_i = \begin{bmatrix} \lambda_i & 1 & 0 & \cdots & 0 \\ 0 & \lambda_i & 1 & \ddots & \vdots \\ \vdots & \ddots & \lambda_i & \ddots & 0 \\ \vdots & & \ddots & \ddots & 1 \\ 0 & \cdots & \cdots & 0 & \lambda_i \end{bmatrix}. \tag{3.279}$$

More than one Jordan block may share a single eigenvalue. For example, consider an eigenvalue, λ_q, with algebraic multiplicity, 4, meaning that it contributes a factor,

$(\lambda - \lambda_q)^4$, to the characteristic polynomial of **A**. The Jordan block combinations given by (3.280) below are then possible. Each Jordan block is boxed in this illustration. Thus, in case 1 the eigenvalue, λ_q, is associated with only one Jordan block of dimension, 4×4: in case 2 it is associated with two Jordan blocks, one of dimension, 3×3 and the other of dimension, 1×1: in case 3 it is associated again with two Jordan blocks but this time both are of dimension, 2×2: in case 4 it is associated with three Jordan blocks, one of dimension 2×2 and the other two both of dimension, 1×1: and finally, in case 5 it is associated with four Jordan blocks, each of dimension 1×1. Some other cases have been omitted that are trivial variations on the cases illustrated, entailing only exchanging the subscripts of the associated state variables. It is now evident that there can be very many different Jordan canonical forms for a given set of eigenvalues of **A**, each instance of a repeated eigenvalue 'splitting' into one of several Jordan blocks as illustrated in (3.280). An important observation is that some Jordan blocks are only of dimension 1×1 and have no unity elements above on the superdiagonal of **J**. The number of Jordan blocks associated with a given repeated eigenvalue is defined as the *geometric multiplicity* of the eigenvalue. It is evident from the above example that if the geometric multiplicity is equal to the algebraic multiplicity for every repeated eigenvalue, then no unity elements appear on the superdiagonal of **J** and therefore **A** can be diagonalised, **J** being equivalent to **Λ** in Sect. 3.3.4.3. This also means that each eigenvalue, λ_i, of algebraic multiplicity, n_i, associated with n_i Jordan blocks of dimension 1×1 is also associated with n_i linearly independent eigenvectors.

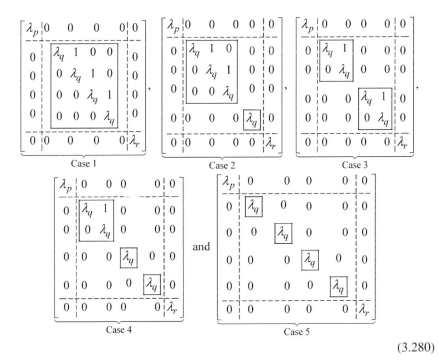

(3.280)

If the geometric multiplicity, m_i, is less than the algebraic multiplicity, n_i, of an eigenvalue, λ_i, then m_i linearly independent eigenvectors can be found to build up the columns of the transformation matrix, **Q**, but $n_i - m_i$ additional linearly independent vectors are needed to complete this process, for every repeated eigenvalue. Fortunately, this is possible, and, using the standard terminology of linear algebra, these additional vectors are called *generalised eigenvectors*. The means of determining these will be presented shortly.

The mathematical procedure for finding the way each repeated eigenvalue of a given matrix, **A**, splits into Jordan blocks, as illustrated by the cases of (3.280), is not straightforward but fortunately the plants usually met in real control engineering have only a few multiple eigenvalues and, most importantly, the split of the eigenvalues should be self-evident by examining the physical structure of the plant. Here, it is important to realise that the unity elements of the superdiagonal of **J** indicate sets of state variables that cannot be separated in the sense of Sect. 3.3.4.3, each Jordan block being the plant submatrix of a subsystem that cannot be further divided. Consider, for example, a double integrator plant model with the state differential equation

$$\underbrace{\begin{bmatrix} \dot{x}_1 \\ \dot{x}_2 \end{bmatrix}}_{\dot{\mathbf{x}}} = \underbrace{\begin{bmatrix} 0 & 1 \\ 0 & 0 \end{bmatrix}}_{\mathbf{A}} \underbrace{\begin{bmatrix} x_1 \\ x_2 \end{bmatrix}}_{\mathbf{x}} + \underbrace{\begin{bmatrix} 0 \\ b \end{bmatrix}}_{\mathbf{b}} u, \quad (3.281)$$

which could, for example, represent one degree of freedom of an air-bearing positioner with negligible friction. It is well known that this is *not* separable into two first-order subsystems by transforming to a state space model of the form,

$$\underbrace{\begin{bmatrix} \dot{z}_1 \\ \dot{z}_2 \end{bmatrix}}_{\dot{\mathbf{z}}} = \underbrace{\begin{bmatrix} \lambda_1 & 0 \\ 0 & \lambda_2 \end{bmatrix}}_{\Lambda} \underbrace{\begin{bmatrix} z_1 \\ z_2 \end{bmatrix}}_{\mathbf{z}} + \underbrace{\begin{bmatrix} g_1 \\ g_2 \end{bmatrix}}_{\mathbf{g}} u. \quad (3.282)$$

This can be shown by inserting the plant eigenvalues in (3.282), which from (3.281) are the roots of the characteristic equation

$$|\lambda \mathbf{I}_2 - \mathbf{A}| = 0 \Rightarrow \begin{vmatrix} \lambda & -1 \\ 0 & \lambda \end{vmatrix} = 0 \Rightarrow \lambda^2 = 0 \Rightarrow \lambda_{1,2} = 0. \quad (3.283)$$

Then, applying the transformation equation, $\mathbf{x} = \mathbf{Qz}$, would yield $\mathbf{AQ} = \mathbf{Q\Lambda} = \mathbf{0} \Rightarrow$

$$\begin{bmatrix} 0 & 1 \\ 0 & 0 \end{bmatrix} \begin{bmatrix} q_{11} & q_{12} \\ q_{21} & q_{22} \end{bmatrix} = \begin{bmatrix} 0 & 0 \\ 0 & 0 \end{bmatrix} \Rightarrow q_{21} = q_{22} = 0 \Rightarrow x_2 = q_{21}z_1 + q_{22}z_2 = 0,$$

(3.284)

3.3 State Representation

which is false, indicating diagonalisation is not feasible. In fact, **A** is already in the Jordan canonical form, which is the closest achievable approach to diagonalisation.

An example of a state-space plant model that is not already in the Jordan canonical form is the model of a three-axis stabilised rigid-body spacecraft with roll, pitch and yaw attitude angles, x_1, x_2 and x_3, and the corresponding body rates, x_4, x_5 and x_6, sufficiently small for the kinematic equations and the dynamic equations to be approximately linear (Chap. 2). The linear state-space model that follows does allow inter-axis coupling due to the mutually orthogonal principal axes of inertia being misaligned with respect to the mutually orthogonal control axes.

$$\begin{bmatrix} \dot{x}_1 \\ \dot{x}_2 \\ \dot{x}_3 \\ \dot{x}_4 \\ \dot{x}_5 \\ \dot{x}_6 \end{bmatrix} = \begin{bmatrix} 0 & 0 & 0 & 1 & 0 & 0 \\ 0 & 0 & 0 & 0 & 1 & 0 \\ 0 & 0 & 0 & 0 & 0 & 1 \\ 0 & 0 & 0 & 0 & 0 & 0 \\ 0 & 0 & 0 & 0 & 0 & 0 \\ 0 & 0 & 0 & 0 & 0 & 0 \end{bmatrix} \begin{bmatrix} x_1 \\ x_2 \\ x_3 \\ x_4 \\ x_5 \\ x_6 \end{bmatrix} + \begin{bmatrix} 0 & 0 & 0 \\ 0 & 0 & 0 \\ 0 & 0 & 0 \\ b_{41} & b_{42} & b_{43} \\ b_{51} & b_{52} & b_{53} \\ b_{61} & b_{62} & b_{63} \end{bmatrix} \begin{bmatrix} u_1 \\ u_2 \\ u_3 \end{bmatrix} \quad (3.285)$$

In this example, it is known that the model must embody three separated double integrators, none of which can be separated into first-order subsystems. The eigenvalues are therefore all zero and the Jordan canonical form of the plant matrix contains three identical Jordan blocks, one for each double integrator. Thus

$$\mathbf{J} = \begin{bmatrix} \boxed{\begin{matrix} \lambda_1 & 1 \\ 0 & \lambda_1 \end{matrix}} & \begin{matrix} 0 & 0 \\ 0 & 0 \end{matrix} & \begin{matrix} 0 & 0 \\ 0 & 0 \end{matrix} \\ \begin{matrix} 0 & 0 \\ 0 & 0 \end{matrix} & \boxed{\begin{matrix} \lambda_2 & 1 \\ 0 & \lambda_2 \end{matrix}} & \begin{matrix} 0 & 0 \\ 0 & 0 \end{matrix} \\ \begin{matrix} 0 & 0 \\ 0 & 0 \end{matrix} & \begin{matrix} 0 & 0 \\ 0 & 0 \end{matrix} & \boxed{\begin{matrix} \lambda_3 & 1 \\ 0 & \lambda_3 \end{matrix}} \end{bmatrix} = \begin{bmatrix} \boxed{\begin{matrix} 0 & 1 \\ 0 & 0 \end{matrix}} & \begin{matrix} 0 & 0 \\ 0 & 0 \end{matrix} & \begin{matrix} 0 & 0 \\ 0 & 0 \end{matrix} \\ \begin{matrix} 0 & 0 \\ 0 & 0 \end{matrix} & \boxed{\begin{matrix} 0 & 1 \\ 0 & 0 \end{matrix}} & \begin{matrix} 0 & 0 \\ 0 & 0 \end{matrix} \\ \begin{matrix} 0 & 0 \\ 0 & 0 \end{matrix} & \begin{matrix} 0 & 0 \\ 0 & 0 \end{matrix} & \boxed{\begin{matrix} 0 & 1 \\ 0 & 0 \end{matrix}} \end{bmatrix}. \quad (3.286)$$

The three double eigenvalues, λ_1, λ_2 and λ_3, one for each Jordan block, are shown to display the structure of **J**, but are all zero in this example. It remains to find the transformation matrix that enables the complete state-space model to be formed in the Jordan canonical form. This will now be examined for a general Jordon block of dimension $n_i \times n_i$. The starting point is the given plant model in an arbitrary state representation

$$\dot{\mathbf{x}} = \mathbf{Ax} + \mathbf{Bu} \quad (3.287)$$

$$\mathbf{y} = \mathbf{Cx} \quad (3.288)$$

the desired state differential equation in the Jordan canonical form

$$\dot{\mathbf{z}} = \mathbf{Jz} + \mathbf{Gu} \quad (3.289)$$

$$\mathbf{y} = \mathbf{Hz} \tag{3.290}$$

and the transformation equation

$$\mathbf{x} = \mathbf{Qz}. \tag{3.291}$$

The plant matrix, $\mathbf{J}$, is already known together with $\mathbf{A}$, $\mathbf{B}$ and $\mathbf{C}$. The task is to find $\mathbf{Q}$ to enable the transformed input and output matrices to be calculated as

$$\mathbf{G} = \mathbf{Q}^{-1}\mathbf{B} \text{ and } \mathbf{H} = \mathbf{CQ}. \tag{3.292}$$

Substituting for $\mathbf{x}$ in (3.287) using (3.291), premultiplying (3.289) by $\mathbf{Q}$ and comparing the resulting equations yield

$$\mathbf{AQ} = \mathbf{QJ}. \tag{3.293}$$

Next, (3.293) is expressed in a partitioned form that enables the columns of $\mathbf{Q}$ corresponding to the Jordon block, $\mathbf{J}_i$, to be determined.

$$\mathbf{A} \begin{bmatrix} \cdots \mathbf{v}_{k+1} & \mathbf{v}_{k+2} & \cdots & \mathbf{v}_{k+n_i-1} & \cdots \end{bmatrix}$$
$$= \begin{bmatrix} \cdots \mathbf{v}_{k+1} & \mathbf{v}_{k+2} & \cdots & \mathbf{v}_{k+n_i} & \cdots \end{bmatrix} \begin{bmatrix} \lambda_{i-1} & 0 & 0 & \cdots & \cdots & 0 \\ 0 & \lambda_i & 1 & \ddots & & \vdots \\ \vdots & \ddots & \lambda_i & \ddots & \ddots & \vdots \\ \vdots & & \ddots & \ddots & 1 & 0 \\ \vdots & & & \ddots & \lambda_i & 0 \\ 0 & \cdots & \cdots & \cdots & 0 & \lambda_{i+1} \end{bmatrix}, \tag{3.294}$$

where $k = \sum_{j=1}^{i-1} n_j$. The component equations of (3.294) involving $\mathbf{J}_i$ are then

$$\mathbf{A}\mathbf{v}_{k+1} = \lambda_i \mathbf{v}_{k+1} \tag{3.295}$$

$$\mathbf{A}\mathbf{v}_{k+2} = \mathbf{v}_{k+1} + \lambda_i \mathbf{v}_{k+2}, \; \mathbf{A}\mathbf{v}_{k+3} = \mathbf{v}_{k+2} + \lambda_i \mathbf{v}_{k+3}, \; \ldots, \; \mathbf{A}\mathbf{v}_{k+n_i}$$
$$= \mathbf{v}_{k+n_i-1} + \lambda_i \mathbf{v}_{k+n_i}. \tag{3.296}$$

According to (3.295), $\mathbf{v}_{k+1}$ is an eigenvector of $\mathbf{A}$ corresponding to the eigenvalue, λ_i. Once this has been determined, then the remaining vectors, $\mathbf{v}_{k+l}$, $l = 2, 3, \ldots, n_i$, can be determined from Eq. (3.296) in order. This is referred to as a *generalised eigenvector chain*. Once this task is completed, normalisation can be carried out as described at the end of Sect. 3.3.4.2.

Returning to the three-axis stabilised spacecraft example, the above method could be applied to obtain the plant model in the Jordan canonical form, but it is

3.3 State Representation

possible to achieve this more quickly in this case by just reordering the individual state differential equations of the given model (3.285) as follows.

$$\begin{bmatrix} \dot{x}_1 \\ \dot{x}_4 \\ \dot{x}_2 \\ \dot{x}_5 \\ \dot{x}_3 \\ \dot{x}_6 \end{bmatrix} = \begin{bmatrix} 0 & 1 & 0 & 0 & 0 & 0 \\ 0 & 0 & 0 & 0 & 0 & 0 \\ 0 & 0 & 0 & 1 & 0 & 0 \\ 0 & 0 & 0 & 0 & 0 & 0 \\ 0 & 0 & 0 & 0 & 0 & 1 \\ 0 & 0 & 0 & 0 & 0 & 0 \end{bmatrix} \begin{bmatrix} x_1 \\ x_4 \\ x_2 \\ x_5 \\ x_3 \\ x_6 \end{bmatrix} + \begin{bmatrix} 0 & 0 & 0 \\ b_{41} & b_{42} & b_{43} \\ 0 & 0 & 0 \\ b_{51} & b_{52} & b_{53} \\ 0 & 0 & 0 \\ b_{61} & b_{62} & b_{63} \end{bmatrix} \begin{bmatrix} u_1 \\ u_2 \\ u_3 \end{bmatrix} \quad (3.297)$$

The state vector of the transformed plant model is therefore

$$\begin{bmatrix} z_1 & z_2 & z_3 & z_4 & z_5 & z_6 \end{bmatrix}^T = \begin{bmatrix} x_1 & x_4 & x_2 & x_5 & x_3 & x_6 \end{bmatrix}^T. \quad (3.298)$$

The corresponding transformation equations, $\mathbf{z} = \mathbf{P}\mathbf{x}$ and $\mathbf{x} = \mathbf{Q}\mathbf{z}$, are

$$\begin{bmatrix} z_1 \\ z_2 \\ z_3 \\ z_4 \\ z_5 \\ z_6 \end{bmatrix} = \underbrace{\begin{bmatrix} 1 & 0 & 0 & 0 & 0 & 0 \\ 0 & 0 & 0 & 1 & 0 & 0 \\ 0 & 1 & 0 & 0 & 0 & 0 \\ 0 & 0 & 0 & 0 & 1 & 0 \\ 0 & 0 & 1 & 0 & 0 & 0 \\ 0 & 0 & 0 & 0 & 0 & 1 \end{bmatrix}}_{\mathbf{P}} \begin{bmatrix} x_1 \\ x_2 \\ x_3 \\ x_4 \\ x_5 \\ x_6 \end{bmatrix} \quad \text{and} \quad \begin{bmatrix} x_1 \\ x_2 \\ x_3 \\ x_4 \\ x_5 \\ x_6 \end{bmatrix} = \underbrace{\begin{bmatrix} 1 & 0 & 0 & 0 & 0 & 0 \\ 0 & 0 & 1 & 0 & 0 & 0 \\ 0 & 0 & 0 & 0 & 1 & 0 \\ 0 & 1 & 0 & 0 & 0 & 0 \\ 0 & 0 & 0 & 1 & 0 & 0 \\ 0 & 0 & 0 & 0 & 0 & 1 \end{bmatrix}}_{\mathbf{Q}} \begin{bmatrix} z_1 \\ z_2 \\ z_3 \\ z_4 \\ z_5 \\ z_6 \end{bmatrix} \quad (3.299)$$

The odd columns of $\mathbf{Q}$ in (3.299) are the eigenvectors of $\mathbf{A}$ corresponding to the double eigenvalues of the three Jordan blocks, and the even columns are the generalised eigenvectors. Similarly, the columns of $\mathbf{P}$ are the eigenvectors and generalised eigenvectors of $\mathbf{J}$. An observation is that $\mathbf{P} = \mathbf{Q}^T$ and since $\mathbf{P} = \mathbf{Q}^{-1}$, then $\mathbf{P}$ and $\mathbf{Q}$ are orthogonal matrices.

It will be recalled that in Sect. 3.3.4.4, every oscillatory mode is represented by the unforced response of a second-order subsystem of the plant model to arbitrary initial state variables, since the two associated state variables are inseparable. Similarly, for a plant with repeated eigenvalues, the state variables associated with every Jordan block in $\mathbf{J}$ are inseparable and therefore, this is the system matrix of a subsystem whose unforced response to an arbitrary initial state represents a mode. The state differential equation for the ith subsystem of the plant model is

$$\dot{\mathbf{z}}_i = \mathbf{J}_i \mathbf{z}_i + \mathbf{H}_i \mathbf{u}, \quad (3.300)$$

where $\mathbf{z}_i = \begin{bmatrix} z_{k+1} & z_{k+2} & \cdots & z_{k+n_i} \end{bmatrix}^T$, $\mathbf{H}_i = \begin{bmatrix} \mathbf{h}_{k+1}^T & \mathbf{h}_{k+2}^T & \cdots & \mathbf{h}_{k+n_i}^T \end{bmatrix}^T$ and $k = \sum_{j=1}^{i-1} n_j$.

For the unforced case, the solution to $\dot{\mathbf{z}}_i = \mathbf{J}_i \mathbf{z}_i$ is

$$\mathbf{z}_i(t) = e^{\mathbf{J}_i t} \mathbf{z}_i(0) \qquad (3.301)$$

To find the component equations of this solution, the Taylor series shows that

$$f\begin{bmatrix} \lambda_i & 1 & 0 & \cdots & 0 \\ 0 & \lambda_i & 1 & & \vdots \\ \vdots & 0 & \lambda_i & \ddots & 0 \\ \vdots & & \ddots & \ddots & 1 \\ 0 & \cdots & \cdots & 0 & \lambda_i \end{bmatrix} = \begin{bmatrix} \frac{1}{0!} f(\lambda_i) & \frac{1}{1!} f'(\lambda_i) & \frac{1}{2!} f''(\lambda_i) & \cdots & \frac{1}{n!} f^{(n)}(\lambda_i) \\ 0 & \frac{1}{0!} f(\lambda_i) & \frac{1}{1!} f'(\lambda_i) & & \frac{1}{n-1!} f^{(n-1)}(\lambda_i) \\ \vdots & & 0 & \frac{1}{0!} f(\lambda_i) & \ddots & \vdots \\ \vdots & & & \ddots & \ddots & \frac{1}{1!} f'(\lambda_i) \\ 0 & & \cdots & & 0 & \frac{1}{0!} f(\lambda_i) \end{bmatrix}. \qquad (3.302)$$

Applying this to (3.302) then yields

$$\mathbf{z}_i(t) = \begin{bmatrix} e^{\lambda_i t} & te^{\lambda_i t} & \cdots & & \frac{t^{n-1}}{n-1!} e^{\lambda_i t} & \frac{t^n}{n!} e^{\lambda_i t} \\ 0 & e^{\lambda_i t} & te^{\lambda_i t} & & & \frac{t^{n-1}}{n-1!} e^{\lambda_i t} \\ 0 & 0 & e^{\lambda_i t} & \ddots & & \vdots \\ \vdots & & \ddots & \ddots & \ddots & \vdots \\ \vdots & & & \ddots & e^{\lambda_i t} & te^{\lambda_i t} \\ 0 & \cdots & \cdots & \cdots & 0 & e^{\lambda_i t} \end{bmatrix} \mathbf{z}_i(0). \qquad (3.303)$$

The component equations are therefore as follows:

$$\begin{aligned} z_{k+1}(t) &= e^{\lambda_i t} z_{k+1}(0) + t e^{\lambda_i t} z_{k+2}(0) + \tfrac{1}{2!} t^2 e^{\lambda_i t} z_{k+3}(0) + \cdots + \tfrac{1}{n!} t^n e^{\lambda_i t} z_{k+n_i}(0) \\ z_{k+2}(t) &= e^{\lambda_i t} z_{k+2}(0) + t e^{\lambda_i t} z_{k+3}(0) + \tfrac{1}{2!} t^2 e^{\lambda_i t} z_{k+4}(0) + \cdots + \tfrac{1}{n-1!} t^{n-1} e^{\lambda_i t} z_{k+n_i}(0) \\ &\vdots \\ z_{k+n_i-1}(t) &= e^{\lambda_i t} z_{k+n_i-1}(0) + t e^{\lambda_i t} z_{k+n_i}(0) \\ z_{k+n_i}(t) &= e^{\lambda_i t} z_{k+n_i}(0) \end{aligned} \qquad (3.304)$$

which may be written as

$$\begin{aligned} z_{k+1}(t) &= e^{\lambda_i t} \left[\tfrac{1}{n!} z_{k+n_i}(0) t^n + \tfrac{1}{n-1!} z_{k+n_i-1}(0) t^{n-1} + \cdots + z_{k+2}(0) t + z_{k+1}(0) \right] \\ z_{k+2}(t) &= e^{\lambda_i t} \left[\tfrac{1}{n-1!} z_{k+n_i}(0) t^{n-1} + \tfrac{1}{n-2!} z_{k+n_i-1}(0) t^{n-2} + \cdots + z_{k+3}(0) t + z_{k+2}(0) \right] \\ &\vdots \\ z_{k+n_i-1}(t) &= e^{\lambda_i t} \left[z_{k+n_i}(0) t + z_{k+n_i-1}(0) \right] \\ z_{k+n_i}(t) &= e^{\lambda_i t} z_{k+n_i}(0) \end{aligned} \qquad (3.305)$$

Each of these variables is the product of an exponential and a polynomial in t, and they constitute the *polynomial exponential mode* introduced in Chap. 1 for closed-loop systems, viewed in the state space.

3.3.5 SISO Controller and Observer Canonical Forms

3.3.5.1 Introduction

The controller canonical form of plant model simplifies the design of linear state feedback controllers (Chap. 4) and the observer canonical form of plant model simplifies the design of the observer, which is a form of state estimator (Chap. 5). Further to their introduction for SISO LTI plants in Sects. 3.3.2.6 and 3.3.2.7 this section provides a generalisation to cater for SISO plants of any order.

For SISO plant models, the approach taken is to start with the plant model in the form of a rationalised transfer function, $Y(s)/U(s)$. This enables the state-variable block diagrams and the corresponding state-space models to be obtained in a straightforward manner without the need for transformation matrices. These are needed, however, to transform the state vector in a control system. For example, the state estimate from an observer formulated with the observer canonical form would have to be transformed for use in a linear state feedback control law formulated for use with the controller canonical form. Hence all the possible transformation matrices that might be needed are derived in this section in terms of the known plant parameters, based on the method of Sect. 3.3.3.

Multivariable plant models of arbitrary order are developed by starting with the plant model in the form of a transfer function matrix whose elements are rationalised with common denominators in the columns for the controller canonical form and in the rows for the observer canonical form. The state-space models follow readily from this transfer function matrix. Then the transformations between the controller and observer state representations and to and from any other state representation are obtained using the method of Sect. 3.3.3.

3.3.5.2 Models

The state-space model with the controller canonical form is given by

$$\dot{\mathbf{x}}_c = \mathbf{A}_c \mathbf{x}_c + \mathbf{b}_c u \qquad (3.306)$$

$$y = \mathbf{c}_c^T \mathbf{x}_c \qquad (3.307)$$

where

$$\mathbf{A}_c = \begin{bmatrix} 0 & 1 & 0 & \cdots & 0 \\ 0 & 0 & 1 & \ddots & \vdots \\ \vdots & & \ddots & \ddots & 0 \\ 0 & 0 & \cdots & 0 & 1 \\ -a_0 & -a_1 & -a_2 & \cdots & -a_{n-1} \end{bmatrix}, \mathbf{b}_c = \begin{bmatrix} 0 \\ 0 \\ \vdots \\ 0 \\ 1 \end{bmatrix} \text{ and } \mathbf{c}_c^T = \begin{bmatrix} c_0 & c_1 & \cdots & c_{n-1} \end{bmatrix}.$$

(3.308)

The state-space model with the observer canonical form is given by

$$\dot{\mathbf{x}}_o = \mathbf{A}_o \mathbf{x}_o + \mathbf{b}_o u \quad (3.309)$$

$$y = \mathbf{c}_o^T \mathbf{x}_o \quad (3.310)$$

where

$$\mathbf{A}_o = \begin{bmatrix} 0 & 0 & \cdots & 0 & -a_0 \\ 1 & 0 & & \vdots & -a_1 \\ 0 & 1 & \ddots & 0 & -a_2 \\ \vdots & \ddots & \ddots & 0 & \vdots \\ 0 & \cdots & 0 & 1 & -a_{n-1} \end{bmatrix}, \mathbf{b}_o = \begin{bmatrix} b_0 \\ b_1 \\ \vdots \\ \vdots \\ b_{n-1} \end{bmatrix} \text{ and } \mathbf{c}_o^T = \begin{bmatrix} 0 & 0 & \cdots & 0 & 1 \end{bmatrix}. \quad (3.311)$$

By analogy with the third-order plant of Sects. 3.3.2.6 and 3.3.2.7, the transfer function of the general SISO plant represented by the above canonical models is

$$\frac{Y(s)}{U(s)} = \frac{r_{n-1}s^{n-1} + \cdots + r_1 s + r_0}{s^n + a_{n-1}s^{n-1} + \cdots + a_1 s + a_0}, \quad (3.312)$$

where $b_i = c_i = r_i$, $i = 1, 2, \ldots, n$. Then comparison of (3.308) with (3.311) reveals

$$\mathbf{A}_o = \mathbf{A}_c^T, \mathbf{b}_o = \mathbf{c}_c \text{ and } \mathbf{c}_o = \mathbf{b}_c, \quad (3.313)$$

The collective term used to describe plant matrices in the form of $\mathbf{A}_c$ or $\mathbf{A}_o$ in which one row or column is composed of the coefficients of the characteristic polynomial is the *companion form*.

3.3 State Representation

3.3.5.3 Transformations

The forward transformation matrices, $\mathbf{P}_c$ and $\mathbf{P}_o$, leading from the plant model,

$$\dot{\mathbf{x}} = \mathbf{A}\mathbf{x} + \mathbf{b}u \tag{3.314}$$

$$y = \mathbf{c}^T\mathbf{x}, \tag{3.315}$$

to the controller canonical model, (3.306) and (3.307), and the observer canonical model, (3.309) and (3.310), are now determined in terms of the plant model parameters, together with the reverse transformation matrices, $\mathbf{Q}_c$ and $\mathbf{Q}_o$. The forward transformation equations are

$$\mathbf{x}_c = \mathbf{P}_c\mathbf{x} \tag{3.316}$$

and

$$\mathbf{x}_o = \mathbf{P}_o\mathbf{x} \tag{3.317}$$

while the reverse transformation equations are

$$\mathbf{x} = \mathbf{Q}_c\mathbf{x}_c \tag{3.318}$$

and

$$\mathbf{x} = \mathbf{Q}_o\mathbf{x}_o. \tag{3.319}$$

Once the transfer function (3.312) of the given plant model of (3.314) and (3.315) is known, then the matrix sets, (3.308) and (3.311), of the controller and observer canonical state-space models are immediately known. The method of Sect. 3.3.3 can then be used to determine $\mathbf{P}_c$, $\mathbf{P}_o$, $\mathbf{Q}_c$ and $\mathbf{Q}_o$. Each of these can be determined using the controllability matrices based on (3.178) and (3.179) or using the observability matrices based on (3.190) and (3.191), as follows:

$$\mathbf{P}_c = \mathbf{M}_{cx_c}(\mathbf{A}_c, \mathbf{b}_c)[\mathbf{M}_{cx}(\mathbf{A}, \mathbf{b})]^{-1} \tag{3.320}$$

or

$$\mathbf{P}_c = [\mathbf{M}_{ox_c}(\mathbf{c}_c^T, \mathbf{A}_c)]^{-1}\mathbf{M}_{ox}(\mathbf{c}^T, \mathbf{A}), \tag{3.321}$$

$$\mathbf{P}_o = \mathbf{M}_{cx_o}(\mathbf{A}_o, \mathbf{b}_o)[\mathbf{M}_{cx}(\mathbf{A}, \mathbf{b})]^{-1} \tag{3.322}$$

or

$$\mathbf{P}_o = [\mathbf{M}_{ox_o}(\mathbf{c}_o^T, \mathbf{A}_o)]^{-1}\mathbf{M}_{ox}(\mathbf{c}^T, \mathbf{A}), \tag{3.323}$$

$$\mathbf{Q}_c = \mathbf{P}_c^{-1} = \mathbf{M}_{cx}(\mathbf{A}, \mathbf{b})[\mathbf{M}_{cx_c}(\mathbf{A}_c, \mathbf{b}_c)]^{-1} \tag{3.324}$$

or

$$Q_c = P_c^{-1} = [M_{ox}(c^T, A)]^{-1} M_{ox_c}(c_c^T, A_c), \quad (3.325)$$

$$Q_o = P_o^{-1} = M_{cx}(A, b)[M_{cx_o}(A_o, b_o)]^{-1} \quad (3.326)$$

or

$$Q_o = P_o^{-1} = [M_{ox}(c^T, A)]^{-1} M_{ox_o}(c_o^T, A_o). \quad (3.327)$$

Equations (3.320), (3.321), (3.322), (3.323), (3.324), (3.325), (3.326), and (3.327) provide any transformation needed for an SISO plant model, but it is interesting to note that two of the matrices on the RHS take on a simple form that avoids numerical evaluation and therefore reduces the computation burden. Specifically,

$$[M_{cx_c}(A_c, b_c)]^{-1} = [M_{ox_o}(c_o^T, A_o)]^{-1} = \begin{bmatrix} a_1 & a_2 & a_3 & \cdots & a_{n-1} & 1 \\ a_2 & a_3 & & a_{n-1} & 1 & 0 \\ a_3 & & \cdot\cdot\cdot & \cdot\cdot\cdot & 0 & 0 \\ \vdots & a_{n-1} & \cdot\cdot\cdot & \cdot\cdot\cdot & & \vdots \\ a_{n-1} & 1 & 0 & & & 0 \\ 1 & 0 & 0 & \cdots & 0 & 0 \end{bmatrix}. \quad (3.328)$$

This matrix will be denoted by M_n.

The elements of $M_{cx_c}(A_c, b_c)$ and $M_{ox_o}(c_o^T, A_o)$ increase considerably in unwieldiness with $i + j$, where i and j are, respectively, the row and column number of an element but this problem can be circumvented by computing M_n^{-1}, which avoids setting up these controllability and observability matrices in the usual partitioned form. This enables (3.320), (3.323), (3.324) and (3.327) to be selected as preferable options, which may be written in terms of M_n as follows:

$$P_c = M_n^{-1} \big[b \mid Ab \mid \ldots \mid A^{n-1}b \big]^{-1} = \Big[\big[b \mid Ab \mid \ldots \mid A^{n-1}b \big] M_n \Big]^{-1}, \quad (3.329)$$

$$P_o = M_n \begin{bmatrix} c^T \\ \hline c^T A \\ \vdots \\ \hline c^T A^{n-1} \end{bmatrix}, \quad (3.330)$$

$$Q_c = \big[b \mid Ab \mid \ldots \mid A^{n-1}b \big] M_n \quad (3.331)$$

3.3 State Representation

and

$$Q_o = \left[\begin{array}{c} c^T \\ \hline c^T A \\ \hline \vdots \\ \hline c^T A^{n-1} \end{array}\right]^{-1} M_n^{-1} = \left[M_n \left[\begin{array}{c} c^T \\ \hline c^T A \\ \hline \vdots \\ \hline c^T A^{n-1} \end{array}\right]\right]^{-1}. \quad (3.332)$$

The validity of M_n as given by (3.328) will now be demonstrated for $n = 4$.

Here, $A_c = \begin{bmatrix} 0 & 1 & 0 & 0 \\ 0 & 0 & 1 & 0 \\ 0 & 0 & 0 & 1 \\ -a_0 & -a_1 & -a_2 & -a_3 \end{bmatrix}$ and $b_c = \begin{bmatrix} 0 \\ 0 \\ 0 \\ 1 \end{bmatrix}$. Then,

$$M_{cx_c}(A_c, b_c) = \left[b_c \mid A_c b_c \mid A_c^2 b_c \mid A_c^3 b_c\right] = \begin{bmatrix} 0 & 0 & 0 & 1 \\ 0 & 0 & 1 & -a_3 \\ 0 & 1 & -a_3 & a_3^2 - a_2 \\ 1 & -a_3 & a_3^2 - a_2 & -a_1 + 2a_2 a_3 - a_3^3 \end{bmatrix}. \quad (3.333)$$

Either postmultiplication or premultiplication of this symmetric matrix by M_4 should yield I_4. Thus,

$$\begin{bmatrix} 0 & 0 & 0 & 1 \\ 0 & 0 & 1 & -a_3 \\ 0 & 1 & -a_3 & a_3^2 - a_2 \\ 1 & -a_3 & a_3^2 - a_2 & -a_1 + 2a_2 a_3 - a_3^3 \end{bmatrix} \begin{bmatrix} a_1 & a_2 & a_3 & 1 \\ a_2 & a_3 & 1 & 0 \\ a_3 & 1 & 0 & 0 \\ 1 & 0 & 0 & 0 \end{bmatrix} = \begin{bmatrix} 1 & 0 & 0 & 0 \\ 0 & 1 & 0 & 0 \\ 0 & 0 & 1 & 0 \\ 0 & 0 & 0 & 1 \end{bmatrix}. \quad (3.334)$$

which is correct. This also holds for $M_{ox_o}(c_o^T, A_o)$.

Example 3.8 Different state representations for one control axis of a flexible spacecraft

A space satellite consists of a rigid centre body to which flexible solar panels are attached. For one control axis, the transfer function between the reaction wheel drive input, $U(s)$, and a measurement, $Y(s)$, numerically equal to the attitude angle, is

$$\frac{Y(s)}{U(s)} = \frac{b(s^2 + v^2)}{s^2(s^2 + \omega^2)}, \quad (3.335)$$

where ω is the free natural frequency of the single significant vibration mode and v is the encastre natural frequency, i.e. the frequency at which the solar panels would vibrate if the centre body were to be fixed with respect to inertial space. Also,

$$b = \frac{K_w \omega^2}{Jv^2}, \tag{3.336}$$

where J is the total spacecraft moment of inertia about the control axis and K_w is the reaction wheel torque constant.

The validity of the transformations, (3.329) and (3.330), will be demonstrated by applying them to the state vector of a model in the modal form and running a step response simulation of this model together with models in the controller canonical and observer canonical forms to check the outputs of the transformation matrices.

The rationalised transfer function for the controller and observer canonical models follows directly from (3.335) and is

$$\frac{Y(s)}{U(s)} = \frac{bs^2 + bv^2}{s^4 + \omega^2 s^2}. \tag{3.337}$$

For the modal state representation, the partial fraction expansion is

$$\frac{Y(s)}{U(s)} = \frac{A}{s^2} + \frac{B}{s^2 + \omega^2}, \tag{3.338}$$

where $A = bv^2/\omega^2$ and $B = b - A$.

The corresponding state-variable block diagrams are shown in Fig. 3.17.

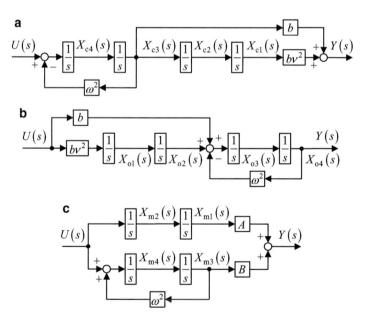

Fig. 3.17 State variable block diagram models of flexible spacecraft. (**a**) Controler canonical form. (**b**) Observer canonical form. (**c**) Modal form

3.3 State Representation

The state-space models may then be formed by inspection. Thus

$$\begin{bmatrix} \dot{x}_{c1} \\ \dot{x}_{c2} \\ \dot{x}_{c3} \\ \dot{x}_{c4} \end{bmatrix} = \underbrace{\begin{bmatrix} 0 & 1 & 0 & 0 \\ 0 & 0 & 1 & 0 \\ 0 & 0 & 0 & 1 \\ 0 & 0 & -\omega^2 & 0 \end{bmatrix}}_{\mathbf{A}_c} \begin{bmatrix} x_{c1} \\ x_{c2} \\ x_{c3} \\ x_{c4} \end{bmatrix} + \underbrace{\begin{bmatrix} 0 \\ 0 \\ 0 \\ 1 \end{bmatrix}}_{\mathbf{b}_c} u, \quad y = \underbrace{\begin{bmatrix} bv^2 & 0 & b & 0 \end{bmatrix}}_{\mathbf{c}_c^T} \begin{bmatrix} x_{c1} \\ x_{c2} \\ x_{c3} \\ x_{c4} \end{bmatrix},$$

(3.339)

$$\begin{bmatrix} \dot{x}_{o1} \\ \dot{x}_{o2} \\ \dot{x}_{o3} \\ \dot{x}_{o4} \end{bmatrix} = \underbrace{\begin{bmatrix} 0 & 0 & 0 & 0 \\ 1 & 0 & 0 & 0 \\ 0 & 1 & 0 & -\omega^2 \\ 0 & 0 & 1 & 0 \end{bmatrix}}_{\mathbf{A}_o} \begin{bmatrix} x_{o1} \\ x_{o2} \\ x_{o3} \\ x_{o4} \end{bmatrix} + \underbrace{\begin{bmatrix} bv^2 \\ 0 \\ b \\ 0 \end{bmatrix}}_{\mathbf{b}_o} u, \quad y = \underbrace{\begin{bmatrix} 0 & 0 & 0 & 1 \end{bmatrix}}_{\mathbf{c}_o^T} \begin{bmatrix} x_{o1} \\ x_{o2} \\ x_{o3} \\ x_{o4} \end{bmatrix},$$

(3.340)

and

$$\text{and} \quad \begin{bmatrix} \dot{x}_{m1} \\ \dot{x}_{m2} \\ \dot{x}_{m3} \\ \dot{x}_{m4} \end{bmatrix} = \underbrace{\begin{bmatrix} 0 & 1 & 0 & 0 \\ 0 & 0 & 0 & 0 \\ 0 & 0 & 0 & 1 \\ 0 & 0 & -\omega^2 & 0 \end{bmatrix}}_{\mathbf{A}_m} \begin{bmatrix} x_{m1} \\ x_{m2} \\ x_{m3} \\ x_{m4} \end{bmatrix} + \underbrace{\begin{bmatrix} 0 \\ 1 \\ 0 \\ 1 \end{bmatrix}}_{\mathbf{b}_m} u, \quad y = \underbrace{\begin{bmatrix} A & 0 & B & 0 \end{bmatrix}}_{\mathbf{c}_m^T} \begin{bmatrix} x_{m1} \\ x_{m2} \\ x_{m3} \\ x_{m4} \end{bmatrix}.$$

(3.341)

This is an example in which the blocks centred on the diagonal of the plant matrix, $\mathbf{A}_m$, are of different modal forms. The first is a 2×2 Jordan block for the two eigenvalues (poles) at $s_{1,2} = 0, 0$. This second-order mode is called the *rigid-body mode* as it is the only mode that would be present if the satellite body was entirely rigid. The second 2×2 block is the oscillatory mode block. This is in a different form to the symmetrical one introduced in Sect. 3.3.2.5 and results from formulating the oscillatory subsystem in the control canonical form, so that the block is in a companion form. If the previous state representation were to be used, then the state-space model (3.341) would be replaced by the following one:

$$\begin{bmatrix} \dot{z}_{m1} \\ \dot{z}_{m2} \\ \dot{z}_{m3} \\ \dot{z}_{m4} \end{bmatrix} = \begin{bmatrix} 0 & 1 & 0 & 0 \\ 0 & 0 & 0 & 0 \\ 0 & 0 & 0 & \omega \\ 0 & 0 & -\omega & 0 \end{bmatrix} \begin{bmatrix} z_{m1} \\ z_{m2} \\ z_{m3} \\ z_{m4} \end{bmatrix} + \begin{bmatrix} 0 \\ 1 \\ 0 \\ 1 \end{bmatrix} u, \quad y = \begin{bmatrix} A & 0 & C & 0 \end{bmatrix} \begin{bmatrix} z_{m1} \\ z_{m2} \\ z_{m3} \\ z_{m4} \end{bmatrix}.$$

(3.342)

where $C = \omega B$. The required transformations are

$$\mathbf{x}_c = \mathbf{P}_c \mathbf{x}_m \text{ and } \mathbf{x}_o = \mathbf{P}_o \mathbf{x}_m, \tag{3.343}$$

where

$$\mathbf{P}_c = \left[\begin{array}{c|c|c|c} \mathbf{b}_m & \mathbf{A}_m \mathbf{b}_m & \mathbf{A}_m^2 \mathbf{b}_m & \mathbf{A}_m^3 \mathbf{b}_m \end{array} \right] \mathbf{M}_4 \Big]^{-1} \tag{3.344}$$

and

$$\mathbf{P}_o = \mathbf{M}_4 \begin{bmatrix} \mathbf{c}_m^T \\ \hline \mathbf{c}_m^T \mathbf{A}_m \\ \hline \mathbf{c}_m^T \mathbf{A}_m^2 \\ \hline \mathbf{c}_m^T \mathbf{A}_m^3 \end{bmatrix}. \tag{3.345}$$

with

$$\mathbf{M}_4 = \begin{bmatrix} a_1 & a_2 & a_3 & 1 \\ a_2 & a_3 & 1 & 0 \\ a_3 & 1 & 0 & 0 \\ 1 & 0 & 0 & 0 \end{bmatrix} = \begin{bmatrix} 0 & \omega^2 & 0 & 1 \\ \omega^2 & 0 & 1 & 0 \\ 0 & 1 & 0 & 0 \\ 1 & 0 & 0 & 0 \end{bmatrix}. \tag{3.346}$$

Simulations of step responses with zero initial states are presented in Fig. 3.18 for $J = 200$ [Kg m^2], $K_w = 0.01$ [Nm/V], $\omega = 0.6$ [rad/s] and $\nu = 0.2$ [rad/s].

As expected, the graphs of $y(t)$ are identical in all three state representations although the state variables are different. The state variables of Fig. 3.18b, c are obtained from simulations of (3.339) and (3.340). These are precisely overlaid by the state variables obtained from (3.341) via transformations (3.343), thereby demonstrating the validity of these transformations.

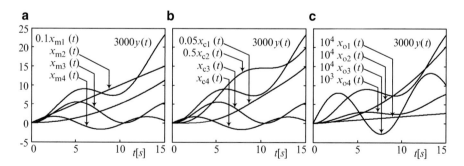

Fig. 3.18 Step responses of modal, controller canonical and observer canonical models. (**a**) Modal form. (**b**) Controller canonical form (**c**) Observer canonical form

3.3 State Representation

3.3.6 Multivariable Controller and Observer Canonical Forms

3.3.6.1 Introduction

As for the SISO plant models, the purpose of developing the controller and observer canonical forms of multivariable state-space plant models is to simplify the control system design. The implementation then necessitates a transformation matrix from the observer to the controller canonical state representations and this is given in Sect. 3.3.6.5.

3.3.6.2 Starting with the Transfer Function Matrix

As demonstrated in Sect. 3.3.2, SISO state-space models in the controller or observer canonical forms may be derived from a rationalised transfer function model of the plant via state-variable block diagrams. A similar approach will be taken here for forming multivariable observer and controller canonical forms. Usually, the numbers of controlled outputs and control variables are equal, so the transfer function matrix, $\mathbf{G}(s)$, is square. The plant input–output transfer function relationship may then be written as

$$\underbrace{\begin{bmatrix} Y_1(s) \\ Y_2(s) \\ \vdots \\ Y_m(s) \end{bmatrix}}_{\mathbf{Y}(s)} = \underbrace{\begin{bmatrix} G_{11}(s) & G_{12}(s) & \cdots & G_{3m}(s) \\ G_{21}(s) & G_{22}(s) & & G_{2m}(s) \\ \vdots & & \ddots & \vdots \\ G_{m1}(s) & G_{m2}(s) & \cdots & G_{mm}(s) \end{bmatrix}}_{\mathbf{G}(s)} \underbrace{\begin{bmatrix} U_1(s) \\ U_2(s) \\ \vdots \\ U_m(s) \end{bmatrix}}_{\mathbf{U}(s)}. \quad (3.347)$$

Before proceeding further, it is important to discuss the order of this plant model. This depends on how it is manipulated. First suppose that the transfer function matrix has been derived from a state-space model,

$$\begin{aligned} \dot{\mathbf{x}} &= \mathbf{A}\mathbf{x} + \mathbf{B}\mathbf{u} \\ \mathbf{y} &= \mathbf{C}\mathbf{x}, \end{aligned} \quad (3.348)$$

in an arbitrary state representation, where $\mathbf{x} \in \mathfrak{R}^n$, $\mathbf{u} \in \mathfrak{R}^m$ and $\mathbf{y} \in \mathfrak{R}^m$. Following a similar procedure to that of Sect. 3.2.9, the transfer function relationship is

$$\mathbf{Y}(s) = \mathbf{C}[s\mathbf{I}_n - \mathbf{A}]^{-1}\mathbf{B}\mathbf{U}(s). \quad (3.349)$$

Then

$$\mathbf{G}(s) = \mathbf{C}[s\mathbf{I}_n - \mathbf{A}]^{-1}\mathbf{B} = \frac{\text{Cadj}\,[s\mathbf{I}_n - \mathbf{A}]\,\mathbf{B}}{|s\mathbf{I}_n - \mathbf{A}|} = \frac{\mathbf{Q}(s)}{P(s)}, \quad (3.350)$$

where the degree of every element of $\mathbf{Q}(s)$ is less than $\deg(P(s)) = n$. The order of the state-space model is equal to the number of state variables, n. Since the characteristic polynomial, $P(s)$, is brought out as a common denominator of the transfer function matrix of (3.350) the order of this transfer function is also n. Each of the n^2 constituent subsystems,

$$Y_i(s) = \frac{Q_{ij}(s)}{P(s)} U_j(s), \quad i = 1, 2, \ldots, n, \quad j = 1, 2, \ldots, n, \qquad (3.351)$$

of (3.347), however, would contribute n state variables when implemented in the SISO state-space form and therefore be of order n. The total number of state variables would then be n^3. Then, the complete model would contain $n^3 - n$ redundant state variables, since only n state variables are contained in model (3.348).

Importantly, the model based on (3.351) would be *uncontrollable* since the n subsystems sharing each control variable also share the same characteristic polynomial, $P(s)$. This may be seen by forming n state differential equations for each subsystem in the SISO control canonical form using only the factor, $1/P(s)$, and implementing the factor, $Q_{ij}(s)$, as part of the output equation. There are then n instances of n identical dynamical subsystems driven by a common control variable. A plant model consisting of just two identical dynamical systems driven by the same control variable is uncontrollable. For example, an overhead gantry crane cannot properly control two identical loads hanging on the same truck. If they are made to swing with arbitrary initial angular displacements and velocities, they cannot be brought to rest together at the same time by moving the truck.

The model based on (3.351) is also *unobservable* since any tests carried out by viewing only the output responses to the inputs would only enable n of the n^3 state variables to be estimated, because, as a black box, the model would be indistinguishable from model (3.348).

The order of the transfer function matrix model may be reduced by cancelling common factors between the numerator and denominator polynomials of (3.351), but this would be insufficient to reduce the order of the model to n. The procedures developed in Sects. 3.3.6.3 and 3.3.6.4, however, seek to achieve this. For simplicity, the following 3×3 transfer function matrix model will be studied, assuming the cancellations referred to above have been carried out. Applying the method to transfer function matrices of different dimensions should be straightforward.

$$\underbrace{\begin{bmatrix} Y_1(s) \\ Y_2(s) \\ Y_3(s) \end{bmatrix}}_{\mathbf{Y}(s)} = \underbrace{\begin{bmatrix} \dfrac{M_{11}(s)}{D_{11}(s)} & \dfrac{M_{12}(s)}{D_{12}(s)} & \dfrac{M_{13}(s)}{D_{13}(s)} \\ \dfrac{M_{21}(s)}{D_{21}(s)} & \dfrac{M_{22}(s)}{D_{22}(s)} & \dfrac{M_{23}(s)}{D_{23}(s)} \\ \dfrac{M_{31}(s)}{D_{31}(s)} & \dfrac{M_{32}(s)}{D_{32}(s)} & \dfrac{M_{33}(s)}{D_{33}(s)} \end{bmatrix}}_{\mathbf{G}(s)} \underbrace{\begin{bmatrix} U_1(s) \\ U_2(s) \\ U_3(s) \end{bmatrix}}_{\mathbf{U}(s)} \qquad (3.352)$$

3.3 State Representation

3.3.6.3 Multivariable Controller Canonical Form

Since each column of **G**(s) in (3.347) is associated with one control input, a single input state-space subsystem can be formed for each by expressing every element of the column with a common denominator polynomial. Then

$$\begin{bmatrix} Y_1(s) \\ Y_2(s) \\ Y_3(s) \end{bmatrix} = \begin{bmatrix} \dfrac{N_{11}(s)}{D_1(s)} & \dfrac{N_{12}(s)}{D_2(s)} & \dfrac{N_{13}(s)}{D_3(s)} \\ \dfrac{N_{21}(s)}{D_1(s)} & \dfrac{N_{22}(s)}{D_2(s)} & \dfrac{N_{23}(s)}{D_3(s)} \\ \dfrac{N_{31}(s)}{D_1(s)} & \dfrac{N_{32}(s)}{D_2(s)} & \dfrac{N_{33}(s)}{D_3(s)} \end{bmatrix} \begin{bmatrix} U_1(s) \\ U_2(s) \\ U_3(s) \end{bmatrix} \quad (3.353)$$

which can be expressed as

$$\begin{bmatrix} Y_{1i}(s) \\ Y_{2i}(s) \\ Y_{3i}(s) \end{bmatrix} = \underbrace{\dfrac{\begin{bmatrix} N_{1i}(s) \\ N_{2i}(s) \\ N_{3i}(s) \end{bmatrix}}{D_i(s)} U_i(s)}_{\text{Subsystem } i}, \quad i = 1, 2, 3,$$

$$\begin{bmatrix} Y_1(s) \\ Y_2(s) \\ Y_3(s) \end{bmatrix} = \begin{bmatrix} Y_{11}(s) \\ Y_{21}(s) \\ Y_{31}(s) \end{bmatrix} + \begin{bmatrix} Y_{12}(s) \\ Y_{22}(s) \\ Y_{32}(s) \end{bmatrix} + \begin{bmatrix} Y_{13}(s) \\ Y_{23}(s) \\ Y_{33}(s) \end{bmatrix}. \quad (3.354)$$

Let each of the constituent rationalised transfer functions be written as

$$\dfrac{N_{ki}(s)}{D_i(s)} = \dfrac{n_{kin_i-1}s^{n_i-1} \cdots + n_{ki1}s + n_{ki0}}{s^{n_i} + d_{in_i-1}s^{n_i-1} \cdots + d_{i1}s + d_{i0}}, \quad (3.355)$$

Suppose subsystems 1, 2 and 3 are, respectively, of orders $n_1 = 2$, $n_2 = 3$ and $n_3 = 4$. Then the subsystem transfer function relationships will be

$$\begin{bmatrix} Y_{11}(s) \\ Y_{21}(s) \\ Y_{31}(s) \end{bmatrix} = \dfrac{\begin{bmatrix} n_{111}s + n_{110} \\ n_{211}s + n_{210} \\ n_{311}s + n_{310} \end{bmatrix}}{s^2 + d_{11}s + d_{10}} U_1(s), \quad \begin{bmatrix} Y_{12}(s) \\ Y_{22}(s) \\ Y_{32}(s) \end{bmatrix} = \dfrac{\begin{bmatrix} n_{122}s^2 + n_{121}s + n_{120} \\ n_{222}s^2 + n_{221}s + n_{220} \\ n_{322}s^2 + n_{321}s + n_{320} \end{bmatrix}}{s^3 + d_{22}s^2 + d_{21}s + d_{20}} U_2(s)$$

and

$$\begin{bmatrix} Y_{13}(s) \\ Y_{23}(s) \\ Y_{33}(s) \end{bmatrix} = \dfrac{\begin{bmatrix} n_{133}s^3 + n_{132}s^2 + n_{131}s + n_{130} \\ n_{233}s^3 + n_{232}s^2 + n_{231}s + n_{230} \\ n_{333}s^3 + n_{332}s^2 + n_{331}s + n_{330} \end{bmatrix}}{s^4 + d_{33}s^3 + d_{32}s^2 + d_{31}s + d_{30}} U_3(s).$$

(3.356)

Then, if the substate vectors of subsystems 1, 2 and 3 are, respectively, chosen as $\begin{bmatrix} x_{c1} & x_{c2} \end{bmatrix}^T$, $\begin{bmatrix} x_{c3} & x_{c4} & x_{c5} \end{bmatrix}^T$ and $\begin{bmatrix} x_{c6} & x_{c7} & x_{c8} & x_{c9} \end{bmatrix}^T$, the individual subsystem state-space models are as follows:

$$\begin{bmatrix} \dot{x}_{c1} \\ \dot{x}_{c2} \end{bmatrix} = \begin{bmatrix} 0 & 1 \\ -d_{10} & -d_{11} \end{bmatrix} \begin{bmatrix} x_{c1} \\ x_{c2} \end{bmatrix} + \begin{bmatrix} 0 \\ 1 \end{bmatrix} u_1, \quad \begin{bmatrix} y_{11} \\ y_{21} \\ y_{31} \end{bmatrix} = \begin{bmatrix} n_{110} & n_{111} \\ n_{210} & n_{211} \\ n_{310} & n_{311} \end{bmatrix} \begin{bmatrix} x_{c1} \\ x_{c2} \end{bmatrix}$$

(3.357)

$$\begin{bmatrix} \dot{x}_{c3} \\ \dot{x}_{c4} \\ \dot{x}_{c5} \end{bmatrix} = \begin{bmatrix} 0 & 1 & 0 \\ 0 & 0 & 1 \\ -d_{20} & -d_{21} & -d_{22} \end{bmatrix} \begin{bmatrix} x_{c3} \\ x_{c4} \\ x_{c5} \end{bmatrix} + \begin{bmatrix} 0 \\ 0 \\ 1 \end{bmatrix} u_2, \quad \begin{bmatrix} y_{12} \\ y_{22} \\ y_{32} \end{bmatrix}$$

$$= \begin{bmatrix} n_{120} & n_{121} & n_{122} \\ n_{220} & n_{221} & n_{222} \\ n_{320} & n_{321} & n_{322} \end{bmatrix} \begin{bmatrix} x_{c3} \\ x_{c4} \\ x_{c5} \end{bmatrix}$$

$$\begin{bmatrix} \dot{x}_{c6} \\ \dot{x}_{c7} \\ \dot{x}_{c8} \\ \dot{x}_{c9} \end{bmatrix} = \begin{bmatrix} 0 & 1 & 0 & 0 \\ 0 & 0 & 1 & 0 \\ 0 & 0 & 0 & 1 \\ -d_{30} & -d_{31} & -d_{32} & -d_{33} \end{bmatrix} \begin{bmatrix} x_{c6} \\ x_{c7} \\ x_{c8} \\ x_{c9} \end{bmatrix} + \begin{bmatrix} 0 \\ 0 \\ 0 \\ 1 \end{bmatrix} u_3, \quad \begin{bmatrix} y_{13} \\ y_{23} \\ y_{33} \end{bmatrix}$$

$$= \begin{bmatrix} n_{130} & n_{131} & n_{132} & n_{133} \\ n_{230} & n_{231} & n_{232} & n_{233} \\ n_{330} & n_{331} & n_{332} & n_{333} \end{bmatrix} \begin{bmatrix} x_{c6} \\ x_{c7} \\ x_{c8} \\ x_{c9} \end{bmatrix}$$

The complete state-space model, $\dot{\mathbf{x}}_c = \mathbf{A}_c \mathbf{x}_c + \mathbf{B}_c \mathbf{u}$, $\mathbf{y} = \mathbf{C}_c \mathbf{x}_c$, is then

$$\begin{bmatrix} \dot{x}_{c1} \\ \dot{x}_{c2} \\ \dot{x}_{c3} \\ \dot{x}_{c4} \\ \dot{x}_{c5} \\ \dot{x}_{c6} \\ \dot{x}_{c7} \\ \dot{x}_{c8} \\ \dot{x}_{c9} \end{bmatrix} = \begin{bmatrix} 0 & 1 & 0 & 0 & 0 & 0 & 0 & 0 & 0 \\ d_{10} & d_{11} & 0 & 0 & 0 & 0 & 0 & 0 & 0 \\ 0 & 0 & 0 & 1 & 0 & 0 & 0 & 0 & 0 \\ 0 & 0 & 0 & 0 & 1 & 0 & 0 & 0 & 0 \\ 0 & 0 & d_{20} & d_{21} & d_{22} & 0 & 0 & 0 & 0 \\ 0 & 0 & 0 & 0 & 0 & 0 & 1 & 0 & 0 \\ 0 & 0 & 0 & 0 & 0 & 0 & 0 & 1 & 0 \\ 0 & 0 & 0 & 0 & 0 & 0 & 0 & 0 & 1 \\ 0 & 0 & 0 & 0 & 0 & d_{30} & d_{31} & d_{32} & d_{33} \end{bmatrix} \begin{bmatrix} x_{c1} \\ x_{c2} \\ x_{c3} \\ x_{c4} \\ x_{c5} \\ x_{c6} \\ x_{c7} \\ x_{c8} \\ x_{c9} \end{bmatrix} + \begin{bmatrix} 0 & 0 & 0 \\ 1 & 0 & 0 \\ 0 & 0 & 0 \\ 0 & 0 & 0 \\ 0 & 1 & 0 \\ 0 & 0 & 0 \\ 0 & 0 & 0 \\ 0 & 0 & 0 \\ 0 & 0 & 1 \end{bmatrix} \begin{bmatrix} u_1 \\ u_2 \\ u_3 \end{bmatrix}$$

$$\begin{bmatrix} y_1 \\ y_2 \\ y_3 \end{bmatrix} = \begin{bmatrix} n_{110} & n_{111} & n_{120} & n_{121} & n_{122} & n_{130} & n_{131} & n_{132} & n_{133} \\ n_{210} & n_{211} & n_{220} & n_{221} & n_{222} & n_{230} & n_{231} & n_{232} & n_{233} \\ n_{310} & n_{311} & n_{320} & n_{321} & n_{322} & n_{330} & n_{331} & n_{332} & n_{333} \end{bmatrix} \begin{bmatrix} x_{c1} \\ x_{c2} \\ \vdots \\ x_{c9} \end{bmatrix}.$$

(3.358)

3.3 State Representation

The important feature of this state representation is that the state differential equation is completely decoupled into three independent subsystems and any interaction between u_i and y_j, $i \neq j$, is entirely contained in the measurement matrix. This renders the design of a multivariable linear state feedback control law by eigenvalue assignment (Chap. 7) straightforward.

There are 'trivial' variants on this plant model that could be obtained by interchanging the columns of $\mathbf{G}(s)$ in (3.347), which would just reorder the three subsystems, or by interchanging the rows of $\mathbf{G}(s)$, which would reorder the measurement variables. In most applications, the number of measurement variables would equal the number of control variables and the rows and columns of $\mathbf{G}(s)$ would be arranged such that u_i has the most influence on y_i for $i = 1, 2, \ldots, r$.

3.3.6.4 Multivariable Observer Canonical Form

The approach is similarly to that adopted for the controller canonical form in Sect. 3.3.6.3 but through forming a single subsystem from each row of the transfer function matrix, $\mathbf{G}(s)$, rather than from each column. Every element in each row is expressed with a common denominator polynomial. Thus, (3.347) becomes

$$\begin{bmatrix} Y_1(s) \\ Y_2(s) \\ Y_3(s) \end{bmatrix} = \begin{bmatrix} \frac{N'_{11}(s)}{D'_1(s)} & \frac{N'_{12}(s)}{D'_1(s)} & \frac{N'_{13}(s)}{D'_1(s)} \\ \frac{N'_{21}(s)}{D'_2(s)} & \frac{N'_{22}(s)}{D'_2(s)} & \frac{N'_{23}(s)}{D'_2(s)} \\ \frac{N'_{31}(s)}{D'_3(s)} & \frac{N'_{32}(s)}{D'_3(s)} & \frac{N'_{33}(s)}{D'_3(s)} \end{bmatrix} \begin{bmatrix} U_1(s) \\ U_2(s) \\ U_3(s) \end{bmatrix}. \quad (3.359)$$

which can be expressed as individual subsystems,

$$Y_i(s) = \frac{\begin{bmatrix} N'_{i1}(s) & N'_{i2}(s) & N'_{i3}(s) \end{bmatrix}}{D'_i(s)} \begin{bmatrix} U_1(s) \\ U_2(s) \\ U_3(s) \end{bmatrix}, \quad i = 1, 2, 3. \quad (3.360)$$

The rationalised transfer function elements are given by

$$\frac{N'_{ki}(s)}{D'_i(s)} = \frac{n'_{ki n_i - 1} s^{n_i - 1} \cdots + n'_{ki1} s + n'_{ki0}}{s^{n_i} + d'_{i n_i - 1} s^{n_i - 1} \cdots + d'_{i1} s + d'_{i0}}, \quad (3.361)$$

As previously, the orders of subsystems 1, 2 and 3 are taken, respectively, as $n_1 = 2$, $n_2 = 3$ and $n_3 = 4$. The transfer function relationships of the subsystems are then as follows:

$$Y_1(s) = \frac{\left[\left(n'_{111}s + n'_{110}\right) \left(n'_{211}s + n'_{210}\right) \left(n'_{311}s + n'_{310}\right)\right] \begin{bmatrix} U_1(s) \\ U_2(s) \\ U_3(s) \end{bmatrix}}{s^2 + d'_{11}s + d'_{10}}$$

$$Y_2(s) = \frac{\left[\begin{pmatrix} n'_{122}s^2 \\ +n'_{121}s+n'_{120} \end{pmatrix} \begin{pmatrix} n'_{222}s^2 \\ +n'_{221}s+n'_{220} \end{pmatrix} \begin{pmatrix} n'_{322}s^2 \\ +n'_{321}s+n'_{320} \end{pmatrix}\right] \begin{bmatrix} U_1(s) \\ U_2(s) \\ U_3(s) \end{bmatrix}}{s^3 + d'_{22}s^2 + d'_{21}s + d'_{20}}$$

$$Y_3(s) = \frac{\left[\begin{pmatrix} n'_{133}s^3 + n'_{132}s^2 \\ +n'_{131}s+n'_{130} \end{pmatrix} \begin{pmatrix} n'_{233}s^3 + n'_{232}s^2 \\ +n'_{231}s + n'_{230} \end{pmatrix} \begin{pmatrix} n'_{333}s^3 + n'_{332}s^2 \\ +n'_{331}s + n'_{330} \end{pmatrix}\right] \begin{bmatrix} U_1(s) \\ U_2(s) \\ U_3(s) \end{bmatrix}}{s^4 + d'_{33}s^3 + d'_{32}s^2 + d'_{31}s + d'_{30}}$$

(3.362)

Then if the substate vectors of subsystems 1, 2 and 3 are, respectively, chosen as $\begin{bmatrix} x_{o1} & x_{o2} \end{bmatrix}^T$, $\begin{bmatrix} x_{o3} & x_{o4} & x_{o5} \end{bmatrix}^T$ and $\begin{bmatrix} x_{o6} & x_{o7} & x_{o8} & x_{o9} \end{bmatrix}^T$, the substate-space models are

$$\begin{bmatrix} \dot{x}_{o1} \\ \dot{x}_{o2} \end{bmatrix} = \begin{bmatrix} 0 & -d'_{10} \\ 1 & -d'_{11} \end{bmatrix} \begin{bmatrix} x_{o1} \\ x_{o2} \end{bmatrix} + \begin{bmatrix} n'_{110} & n'_{210} & n'_{310} \\ n'_{111} & n'_{211} & n'_{311} \end{bmatrix} \begin{bmatrix} u_1 \\ u_2 \\ u_3 \end{bmatrix}, \quad y_1 = \begin{bmatrix} 0 & 1 \end{bmatrix} \begin{bmatrix} x_{o1} \\ x_{o2} \end{bmatrix}$$

$$\begin{bmatrix} \dot{x}_{o3} \\ \dot{x}_{o4} \\ \dot{x}_{o5} \end{bmatrix} = \begin{bmatrix} 0 & 0 & -d'_{20} \\ 1 & 0 & -d'_{21} \\ 0 & 1 & -d'_{22} \end{bmatrix} \begin{bmatrix} x_{o3} \\ x_{o4} \\ x_{o5} \end{bmatrix} + \begin{bmatrix} n'_{120} & n'_{220} & n'_{320} \\ n'_{121} & n'_{221} & n'_{321} \\ n'_{122} & n'_{222} & n'_{322} \end{bmatrix} \begin{bmatrix} u_1 \\ u_2 \\ u_3 \end{bmatrix},$$

$$y_2 = \begin{bmatrix} 0 & 0 & 1 \end{bmatrix} \begin{bmatrix} x_{o3} \\ x_{o4} \\ x_{o5} \end{bmatrix}$$

$$\begin{bmatrix} \dot{x}_{o6} \\ \dot{x}_{o7} \\ \dot{x}_{o8} \\ \dot{x}_{o9} \end{bmatrix} = \begin{bmatrix} 0 & 0 & 0 & -d'_{30} \\ 1 & 0 & 0 & -d'_{31} \\ 0 & 1 & 0 & -d'_{32} \\ 0 & 0 & 1 & -d'_{33} \end{bmatrix} \begin{bmatrix} x_{o6} \\ x_{o7} \\ x_{o8} \\ x_{o9} \end{bmatrix} + \begin{bmatrix} n'_{130} & n'_{230} & n'_{330} \\ n'_{131} & n'_{231} & n'_{331} \\ n'_{132} & n'_{232} & n'_{332} \\ n'_{133} & n'_{233} & n'_{333} \end{bmatrix} \begin{bmatrix} u_1 \\ u_2 \\ u_3 \end{bmatrix},$$

$$y_3 = \begin{bmatrix} 0 & 0 & 0 & 1 \end{bmatrix} \begin{bmatrix} x_{o6} \\ x_{o7} \\ x_{o8} \\ x_{o9} \end{bmatrix}.$$

(3.363)

The complete state-space model, $\dot{\mathbf{x}}_o = \mathbf{A}\mathbf{x}_o + \mathbf{B}_o\mathbf{u}$, $\mathbf{y} = \mathbf{C}_o\mathbf{x}_o$, is then

3.3 State Representation

$$\begin{bmatrix} \dot{x}_{o1} \\ \dot{x}_{o2} \\ \dot{x}_{o3} \\ \dot{x}_{o4} \\ \dot{x}_{o5} \\ \dot{x}_{o6} \\ \dot{x}_{o7} \\ \dot{x}_{o8} \\ \dot{x}_{o9} \end{bmatrix} = \begin{bmatrix} 0 & -d'_{10} & 0 & 0 & 0 & 0 & 0 & 0 & 0 \\ 1 & -d'_{11} & 0 & 0 & 0 & 0 & 0 & 0 & 0 \\ 0 & 0 & 0 & 0 & -d'_{20} & 0 & 0 & 0 & 0 \\ 0 & 0 & 1 & 0 & -d'_{21} & 0 & 0 & 0 & 0 \\ 0 & 0 & 0 & 1 & -d'_{22} & 0 & 0 & 0 & 0 \\ 0 & 0 & 0 & 0 & 0 & 0 & 0 & 0 & -d'_{30} \\ 0 & 0 & 0 & 0 & 0 & 1 & 0 & 0 & -d'_{31} \\ 0 & 0 & 0 & 0 & 0 & 0 & 1 & 0 & -d'_{32} \\ 0 & 0 & 0 & 0 & 0 & 0 & 0 & 1 & -d'_{33} \end{bmatrix} \begin{bmatrix} x_{o1} \\ x_{o2} \\ x_{o3} \\ x_{o4} \\ x_{o5} \\ x_{o6} \\ x_{o7} \\ x_{o8} \\ x_{o9} \end{bmatrix} + \begin{bmatrix} n'_{110} & n'_{210} & n'_{310} \\ n'_{111} & n'_{211} & n'_{311} \\ n'_{120} & n'_{220} & n'_{320} \\ n'_{121} & n'_{221} & n'_{321} \\ n'_{122} & n'_{222} & n'_{322} \\ n'_{130} & n'_{230} & n'_{330} \\ n'_{131} & n'_{231} & n'_{331} \\ n'_{132} & n'_{232} & n'_{332} \\ n'_{133} & n'_{233} & n'_{333} \end{bmatrix} \begin{bmatrix} u_1 \\ u_2 \\ u_3 \end{bmatrix}$$

$$\begin{bmatrix} y_1 \\ y_2 \\ y_3 \end{bmatrix} = \begin{bmatrix} 0 & 1 & 0 & 0 & 0 & 0 & 0 & 0 & 0 \\ 0 & 0 & 0 & 0 & 1 & 0 & 0 & 0 & 0 \\ 0 & 0 & 0 & 0 & 0 & 0 & 0 & 0 & 1 \end{bmatrix} \begin{bmatrix} x_{o1} \\ x_{o2} \\ \vdots \\ x_{o9} \end{bmatrix}.$$

(3.364)

3.3.6.5 Transformations

The general transformation matrices of Sect. 3.3.3 may be used to calculate the state vectors of either the controller or observer canonical plant models given the state vector of a plant model with an arbitrary state representation having matrices **A**, **B** and **C**. As will be recalled, there are *four* options in every case and this allows a choice to be made that minimises the amount of computation needed by exploiting the simple forms of $\mathbf{B}_c$ in (3.358) and $\mathbf{C}_o$ in (3.364). For the controller canonical form,

$$\mathbf{x}_c = \mathbf{P}_c \mathbf{x}, \qquad (3.365)$$

and the four options based on (3.173), (3.177), (3.185) and (3.189) are as follows:

$$\mathbf{P}_c = [\mathbf{M}_{cx_c}(\mathbf{A}_c, \mathbf{B}_c)][\mathbf{M}_{cx}(\mathbf{A}, \mathbf{B})]^T \left\{ [\mathbf{M}_{cx}(\mathbf{A}, \mathbf{B})][\mathbf{M}_{cx}(\mathbf{A}, \mathbf{B})]^T \right\}^{-1}, \quad (3.366)$$

$$\mathbf{P}_c = [\mathbf{M}_{cx_c}(\mathbf{A}_c, \mathbf{B}_c)][\mathbf{M}_{cx_c}(\mathbf{A}_c, \mathbf{B}_c)]^T \left\{ [\mathbf{M}_{cx}(\mathbf{A}, \mathbf{B})][\mathbf{M}_{cx_c}(\mathbf{A}_c, \mathbf{B}_c)]^T \right\}^{-1},$$

(3.367)

$$\mathbf{P}_c = \left\{ [\mathbf{M}_{ox_c}(\mathbf{C}_c, \mathbf{A}_c)]^T [\mathbf{M}_{ox_c}(\mathbf{C}_c, \mathbf{A}_c)] \right\}^{-1} [\mathbf{M}_{ox_c}(\mathbf{C}_c, \mathbf{A}_c)]^T [\mathbf{M}_{ox}(\mathbf{C}, \mathbf{A})],$$

(3.368)

$$\mathbf{P}_c = \left\{ [\mathbf{M}_{ox}(\mathbf{C}, \mathbf{A})]^T [\mathbf{M}_{ox_c}(\mathbf{C}_c, \mathbf{A}_c)] \right\}^{-1} [\mathbf{M}_{ox}(\mathbf{C}, \mathbf{A})]^T [\mathbf{M}_{ox}(\mathbf{C}, \mathbf{A})]. \quad (3.369)$$

The simplest of the matrices on the RHS to numerically evaluate is $\mathbf{M}_{cx_c}(\mathbf{A}_c, \mathbf{B}_c)$. This appears three times in (3.367), once only in (3.366) and not at all in (3.368) or (3.369). Hence (3.367) is the best choice.

For the observer canonical form,

$$\mathbf{x}_o = \mathbf{P}_o \mathbf{x}, \qquad (3.370)$$

where the choices of equation for the transformation matrix are

$$\mathbf{P}_o = [\mathbf{M}_{cx_o}(\mathbf{A}_o, \mathbf{B}_o)][\mathbf{M}_{cx}(\mathbf{A}, \mathbf{B})]^T \{[\mathbf{M}_{cx}(\mathbf{A}, \mathbf{B})][\mathbf{M}_{cx}(\mathbf{A}, \mathbf{B})]^T\}^{-1}, \qquad (3.371)$$

$$\mathbf{P}_o = [\mathbf{M}_{cx_o}(\mathbf{A}_o, \mathbf{B}_o)][\mathbf{M}_{cx_o}(\mathbf{A}_o, \mathbf{B}_o)]^T \{[\mathbf{M}_{cx}(\mathbf{A}, \mathbf{B})][\mathbf{M}_{cx_o}(\mathbf{A}_o, \mathbf{B}_o)]^T\}^{-1}, \qquad (3.372)$$

$$\mathbf{P}_o = \{[\mathbf{M}_{ox_o}(\mathbf{C}_o, \mathbf{A}_o)]^T [\mathbf{M}_{ox_o}(\mathbf{C}_o, \mathbf{A}_o)]\}^{-1} [\mathbf{M}_{ox_o}(\mathbf{C}_o, \mathbf{A}_o)]^T [\mathbf{M}_{ox}(\mathbf{C}, \mathbf{A})], \qquad (3.373)$$

$$\mathbf{P}_o = \{[\mathbf{M}_{ox}(\mathbf{C}, \mathbf{A})]^T [\mathbf{M}_{ox_o}(\mathbf{C}_o, \mathbf{A}_o)]\}^{-1} [\mathbf{M}_{ox}(\mathbf{C}, \mathbf{A})]^T [\mathbf{M}_{ox}(\mathbf{C}, \mathbf{A})]. \qquad (3.374)$$

In this case, the matrix on the RHS demanding the least computation in the numerical evaluation is $\mathbf{M}_{ox_o}(\mathbf{C}_o, \mathbf{A}_o)$. This appears three times in (3.373), only once in (3.374) and not in either (3.371) or (3.372). The best choice is therefore (3.373).

David G Luenberger, who originated the state observer (Chap. 5), which the observer canonical form of plant model is designed to serve, established an approach [4] to finding the transformation matrices for canonical forms similar to but not identical to those of this section. The transformation matrices are simpler, one being based on selections of n linearly independent columns of the controllability matrix, $\mathbf{M}_{cx}(\mathbf{A}, \mathbf{B})$. Remarkably, this yields the same companion form submatrices as those of the observer canonical model of (3.364) centred on the leading diagonal of the plant matrix, $\mathbf{A}_o$. Some interaction between the subsystems is present, however, due to non-zero elements outside the block diagonal structure. Also, interestingly, the simple input matrix, $\mathbf{B}_c$, of the control canonical model (3.358) appears with this plant matrix. A modification to the transformation matrix, however, similar to the introduction of the matrix, $\mathbf{M}_n$, in Sect. 3.3.5.3, yields a plant matrix similar to that of the control canonical form, $\mathbf{A}_c$, but with the additional elements outside the block diagonal structure.

Example 3.9 State-space models for three-axis attitude control of rigid-body spacecraft

The attitude dynamics and kinematics model of a three-axis stabilised rigid-body spacecraft may be combined to form the transfer function matrix model

3.3 State Representation

$$\begin{bmatrix} Y_1(s) \\ Y_2(s) \\ Y_3(s) \end{bmatrix} = \begin{bmatrix} \frac{b_{11}}{s^2} & \frac{b_{12}}{s^2} & \frac{b_{13}}{s^2} \\ \frac{b_{21}}{s^2} & \frac{b_{22}}{s^2} & \frac{b_{23}}{s^2} \\ \frac{b_{31}}{s^2} & \frac{b_{32}}{s^2} & \frac{b_{33}}{s^2} \end{bmatrix} \begin{bmatrix} U_1(s) \\ U_2(s) \\ U_3(s) \end{bmatrix} \tag{3.375}$$

where $b_{ij} = K_w[\mathbf{J}^{-1}]_{ij}$, $i = 1, 2, 3$, $j = 1, 2, 3$, where $\mathbf{J}$ is the body moment of inertia matrix and K_w is the reaction wheel torque constant. State-space models are required in the controller and observer canonical forms together with the transformation matrix yielding the controller canonical state variables given the observer canonical state variables.

For the controller canonical form, the three transfer function matrix-based subsystems are of the same form and given by

$$\begin{bmatrix} Y_{1i}(s) \\ Y_{2i}(s) \\ Y_{3i}(s) \end{bmatrix} = \frac{\begin{bmatrix} b_{1i}(s) \\ b_{2i}(s) \\ b_{3i}(s) \end{bmatrix}}{s^2} U_i(s), \quad i = 1, 2, 3, \quad \begin{bmatrix} Y_1(s) \\ Y_2(s) \\ Y_3(s) \end{bmatrix} \tag{3.376}$$

$$= \begin{bmatrix} Y_{11}(s) \\ Y_{21}(s) \\ Y_{31}(s) \end{bmatrix} + \begin{bmatrix} Y_{12}(s) \\ Y_{22}(s) \\ Y_{32}(s) \end{bmatrix} + \begin{bmatrix} Y_{13}(s) \\ Y_{23}(s) \\ Y_{33}(s) \end{bmatrix}.$$

The three state-space subsystems in the SISO controller canonical form are then

$$\begin{bmatrix} \dot{x}_{c1} \\ \dot{x}_{c2} \end{bmatrix} = \begin{bmatrix} 0 & 1 \\ 0 & 0 \end{bmatrix} \begin{bmatrix} x_{c1} \\ x_{c2} \end{bmatrix} + \begin{bmatrix} 0 \\ 1 \end{bmatrix} u_1, \quad \begin{bmatrix} y_{11} \\ y_{21} \\ y_{31} \end{bmatrix} = \begin{bmatrix} b_{11} & 0 \\ b_{21} & 0 \\ b_{31} & 0 \end{bmatrix} \begin{bmatrix} x_{c1} \\ x_{c2} \end{bmatrix}$$

$$\begin{bmatrix} \dot{x}_{c3} \\ \dot{x}_{c4} \end{bmatrix} = \begin{bmatrix} 0 & 1 \\ 0 & 0 \end{bmatrix} \begin{bmatrix} x_{c3} \\ x_{c4} \end{bmatrix} + \begin{bmatrix} 0 \\ 1 \end{bmatrix} u_2, \quad \begin{bmatrix} y_{12} \\ y_{22} \\ y_{32} \end{bmatrix} = \begin{bmatrix} b_{12} & 0 \\ b_{22} & 0 \\ b_{32} & 0 \end{bmatrix} \begin{bmatrix} x_{c3} \\ x_{c4} \end{bmatrix} \tag{3.377}$$

$$\begin{bmatrix} \dot{x}_{c5} \\ \dot{x}_{c6} \end{bmatrix} = \begin{bmatrix} 0 & 1 \\ 0 & 0 \end{bmatrix} \begin{bmatrix} x_{c5} \\ x_{c6} \end{bmatrix} + \begin{bmatrix} 0 \\ 1 \end{bmatrix} u_3, \quad \begin{bmatrix} y_{12} \\ y_{22} \\ y_{32} \end{bmatrix} = \begin{bmatrix} b_{13} & 0 \\ b_{23} & 0 \\ b_{33} & 0 \end{bmatrix} \begin{bmatrix} x_{c5} \\ x_{c6} \end{bmatrix}$$

and the overall output is given by

$$\begin{bmatrix} Y_1(s) \\ Y_2(s) \\ Y_3(s) \end{bmatrix} = \begin{bmatrix} Y_{11}(s) \\ Y_{21}(s) \\ Y_{31}(s) \end{bmatrix} + \begin{bmatrix} Y_{12}(s) \\ Y_{22}(s) \\ Y_{32}(s) \end{bmatrix} + \begin{bmatrix} Y_{13}(s) \\ Y_{23}(s) \\ Y_{33}(s) \end{bmatrix} \tag{3.378}$$

Assembling (3.377) and (3.378) to form the required multivariable controller canonical state-space model then yields

$$\begin{bmatrix} \dot{x}_{c1} \\ \dot{x}_{c2} \\ \dot{x}_{c3} \\ \dot{x}_{c4} \\ \dot{x}_{c5} \\ \dot{x}_{c6} \end{bmatrix} = \underbrace{\begin{bmatrix} 0 & 1 & 0 & 0 & 0 & 0 \\ 0 & 0 & 0 & 0 & 0 & 0 \\ \hline 0 & 0 & 0 & 1 & 0 & 0 \\ 0 & 0 & 0 & 0 & 0 & 0 \\ \hline 0 & 0 & 0 & 0 & 0 & 1 \\ 0 & 0 & 0 & 0 & 0 & 0 \end{bmatrix}}_{\mathbf{A}_c} \begin{bmatrix} x_{c1} \\ x_{c2} \\ x_{c3} \\ x_{c4} \\ x_{c5} \\ x_{c6} \end{bmatrix} + \underbrace{\begin{bmatrix} 0 & 0 & 0 \\ 1 & 0 & 0 \\ \hline 0 & 0 & 0 \\ 0 & 1 & 0 \\ \hline 0 & 0 & 0 \\ 0 & 0 & 1 \end{bmatrix}}_{\mathbf{B}_c} \begin{bmatrix} u_1 \\ u_2 \\ u_3 \end{bmatrix}, \begin{bmatrix} y_1 \\ y_2 \\ y_3 \end{bmatrix} = \underbrace{\begin{bmatrix} b_{11} & 0 & b_{12} & 0 & b_{13} & 0 \\ b_{21} & 0 & b_{22} & 0 & b_{23} & 0 \\ b_{31} & 0 & b_{32} & 0 & b_{33} & 0 \end{bmatrix}}_{\mathbf{C}_c} \begin{bmatrix} x_{c1} \\ x_{c2} \\ x_{c3} \\ x_{c4} \\ x_{c5} \\ x_{c6} \end{bmatrix}. \tag{3.379}$$

For the observer canonical form, the three transfer function matrix based subsystems are also of the same form and given by

$$Y_i(s) = \frac{\begin{bmatrix} b_{i1} & b_{i2} & b_{i3} \end{bmatrix}}{s^2} \begin{bmatrix} U_1(s) \\ U_2(s) \\ U_3(s) \end{bmatrix}, \quad i = 1, 2, 3 \tag{3.380}$$

The corresponding three state-space subsystems are

$$\begin{bmatrix} \dot{x}_{o1} \\ \dot{x}_{o2} \end{bmatrix} = \begin{bmatrix} 0 & 0 \\ 1 & 0 \end{bmatrix} \begin{bmatrix} x_{o1} \\ x_{o2} \end{bmatrix} + \begin{bmatrix} b_{11} & b_{12} & b_{13} \\ 0 & 0 & 0 \end{bmatrix} \begin{bmatrix} u_1 \\ u_2 \\ u_3 \end{bmatrix}, \quad y_1 = \begin{bmatrix} 0 & 1 \end{bmatrix} \begin{bmatrix} x_{o1} \\ x_{o2} \end{bmatrix}$$

$$\begin{bmatrix} \dot{x}_{o3} \\ \dot{x}_{o4} \end{bmatrix} = \begin{bmatrix} 0 & 0 \\ 1 & 0 \end{bmatrix} \begin{bmatrix} x_{o3} \\ x_{o4} \end{bmatrix} + \begin{bmatrix} b_{21} & b_{22} & b_{23} \\ 0 & 0 & 0 \end{bmatrix} \begin{bmatrix} u_1 \\ u_2 \\ u_3 \end{bmatrix}, \quad y_2 = \begin{bmatrix} 0 & 1 \end{bmatrix} \begin{bmatrix} x_{o3} \\ x_{o4} \end{bmatrix} \tag{3.381}$$

$$\begin{bmatrix} \dot{x}_{o5} \\ \dot{x}_{o6} \end{bmatrix} = \begin{bmatrix} 0 & 0 \\ 1 & 0 \end{bmatrix} \begin{bmatrix} x_{o5} \\ x_{o6} \end{bmatrix} + \begin{bmatrix} b_{31} & b_{32} & b_{33} \\ 0 & 0 & 0 \end{bmatrix} \begin{bmatrix} u_1 \\ u_2 \\ u_3 \end{bmatrix}, \quad y_3 = \begin{bmatrix} 0 & 1 \end{bmatrix} \begin{bmatrix} x_{o5} \\ x_{o6} \end{bmatrix}$$

These are then assembled to create the required state-space model in the observer canonical form as follows:

$$\begin{bmatrix} \dot{x}_{o1} \\ \dot{x}_{o2} \\ \dot{x}_{o3} \\ \dot{x}_{o4} \\ \dot{x}_{o5} \\ \dot{x}_{o6} \end{bmatrix} = \underbrace{\begin{bmatrix} 0 & 0 & 0 & 0 & 0 & 0 \\ 1 & 0 & 0 & 0 & 0 & 0 \\ \hline 0 & 0 & 0 & 0 & 0 & 0 \\ 0 & 0 & 1 & 0 & 0 & 0 \\ \hline 0 & 0 & 0 & 0 & 0 & 0 \\ 0 & 0 & 0 & 0 & 1 & 0 \end{bmatrix}}_{\mathbf{A}_o} \begin{bmatrix} x_{o1} \\ x_{o2} \\ x_{o3} \\ x_{o4} \\ x_{o5} \\ x_{o6} \end{bmatrix} + \underbrace{\begin{bmatrix} b_{11} & b_{12} & b_{13} \\ 0 & 0 & 0 \\ b_{21} & b_{22} & b_{32} \\ 0 & 0 & 0 \\ b_{13} & b_{23} & b_{33} \\ 0 & 0 & 0 \end{bmatrix}}_{\mathbf{B}_o} \begin{bmatrix} u_1 \\ u_2 \\ u_3 \end{bmatrix}, \begin{bmatrix} y_1 \\ y_2 \\ y_3 \end{bmatrix} = \underbrace{\begin{bmatrix} 0 & 1 & 0 & 0 & 0 & 0 \\ 0 & 0 & 0 & 1 & 0 & 0 \\ 0 & 0 & 0 & 0 & 0 & 1 \end{bmatrix}}_{\mathbf{C}_o} \begin{bmatrix} x_{o1} \\ x_{o2} \\ x_{o3} \\ x_{o4} \\ x_{o5} \\ x_{o6} \end{bmatrix}. \tag{3.382}$$

In this example, $\mathbf{A}_o = \mathbf{A}_c^T$, $\mathbf{C}_o = \mathbf{B}_c^T$ and since the moment of inertia matrix is symmetrical, $\mathbf{B}_o = \mathbf{C}_c^T$. These are similar to the relationships that hold between the SISO observer and controller canonical state-space models.

The transformation equation using (3.367) is

$$\mathbf{x}_c = \mathbf{P}_c \mathbf{x}_o, \tag{3.383}$$

3.4 Discrete LTI Plant Models

where

$$\mathbf{P}_c = [\mathbf{M}_{cx_c}(\mathbf{A}_c, \mathbf{B}_c)][\mathbf{M}_{cx_c}(\mathbf{A}_c, \mathbf{B}_c)]^T \{[\mathbf{M}_{cx_o}(\mathbf{A}_o, \mathbf{B}_o)][\mathbf{M}_{cx_c}(\mathbf{A}_c, \mathbf{B}_c)]^T\}^{-1}.$$
(3.384)

In this case, using (3.379) and (3.382) yields

$$\mathbf{M}_{cx_c}(\mathbf{A}_c, \mathbf{B}_c) = \begin{bmatrix} \mathbf{B}_c & | & \mathbf{A}_c \mathbf{B}_c & | & \mathbf{A}_c^2 \mathbf{B}_c & | & \mathbf{A}_c^3 \mathbf{B}_c & | & \mathbf{A}_c^4 \mathbf{B}_c & | & \mathbf{A}_c^5 \mathbf{B}_c \end{bmatrix},$$
$$\mathbf{M}_{cx_o}(\mathbf{A}_o, \mathbf{B}_o) = \begin{bmatrix} \mathbf{B}_o & | & \mathbf{A}_o \mathbf{B}_o & | & \mathbf{A}_o^2 \mathbf{B}_o & | & \mathbf{A}_o^3 \mathbf{B}_o & | & \mathbf{A}_o^4 \mathbf{B}_o & | & \mathbf{A}_o^5 \mathbf{B}_o \end{bmatrix}$$
(3.385)

Simulations will now be presented that demonstrate the validity of the controller canonical and observer canonical state-space models (3.379) and (3.382) together with the transformation (3.383). The plant parameters are as follows:

$$\mathbf{J} = \begin{bmatrix} 600 & 100 & -160 \\ 100 & 500 & 120 \\ -160 & 120 & 300 \end{bmatrix} [\text{Kg m}^2]; \, K_w = 0.1 \, [\text{Nm/V}].$$
(3.386)

The off-diagonal elements of $\mathbf{J}$ are larger in proportion to the diagonal elements than usual. With negligible off-diagonal elements, the state transformation would be trivial: just a proportional scaling. It is assumed that the attitude measurements are scaled in the digital processor to be numerically equal to the attitude angles in radians. Figure 3.19 shows the results.

The initial state variables are set to zero and then a step reaction wheel control voltage vector of $[u_1(t) \; u_2(t) \; u_3(t)]^T = [4 \; -5 \; 3]^T h(t)$ [V] is applied to both models. Figure 3.19a, b shows the state variables of the separate models. Such large differences of scale between the state variables of different state representations are to be expected for many applications. Figure 3.18c shows the transformed state variables according to $\mathbf{x}'_c = \mathbf{P}_c \mathbf{x}_o$ and indicates, as expected, that $\mathbf{x}'_c = \mathbf{x}_c$.

3.4 Discrete LTI Plant Models

3.4.1 Formation of the Discrete State Space Model

The software implementation of controllers using digital processors and the digital simulation of control systems require discrete models of the plants to be controlled, if the iteration interval, h, i.e. the period required to execute one cycle of calculations of the control algorithm, normally assumed to be the same as the signal sampling time, is not to be limited.

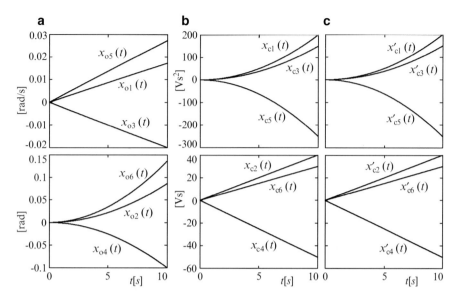

Fig. 3.19 Open-loop step responses of rigid-body satellite models: state variables. (**a**) Controller canonical form. (**b**) Observer canonical form. (**c**) Transformed from observer canonical form

Each of the forms of continuous plant model already dealt with, i.e. the differential equation, Laplace transfer function, transfer function block diagram, continuous state-space model and state-variable block diagram, has discrete equivalents. These are developed in the following sections in preparation for the discrete linear control system design approach of Chap. 6.

The starting point is taken as the continuous state space model of (3.16) and (3.17), reproduced here for convenience.

$$\dot{\mathbf{x}} = \mathbf{A}\mathbf{x} + \mathbf{B}\mathbf{u} \quad (3.387)$$

$$\mathbf{y} = \mathbf{C}\mathbf{x} \quad (3.388)$$

where $\mathbf{x} \in \Re^n$, $\mathbf{u} \in \Re^r$ and $\mathbf{y} \in \Re^m$. A computer implementing a controller produces a piecewise constant control input, $\mathbf{u}(t)$, updated at the end of every iteration interval. Then, as will be shown shortly, it is possible to obtain an *exact* solution to (3.387) analytically. The solution at the end of each iteration interval depends upon the state at the beginning of each iteration interval and the control value over each iteration interval. This yields an alternative model to the continuous state-space model that may be *iterated* to yield precisely the same values of $\mathbf{x}(t)$ and $\mathbf{y}(t)$ as the continuous model at the instants between the iteration intervals when the control variable is

3.4 Discrete LTI Plant Models

updated. This generic model will now be derived. The general solution to (3.387) was derived in Sect. 3.2.5.2 as

$$\mathbf{x}(t) = e^{\mathbf{A}(t-t_0)}\mathbf{x}(t_0) + \int_{t_0}^{t} e^{\mathbf{A}(t-\tau')}\mathbf{B}\mathbf{u}(\tau')\,d\tau'. \tag{3.389}$$

It is clear from this that once the initial state, $\mathbf{x}(t_0)$, is known, the behaviour of the plant subject to a known control input may be predicted. The matrix, $e^{\mathbf{A}(t-t_0)}$, which will be denoted by $\mathbf{\Phi}(t, t_0)$, is referred to as the *state transition matrix* (sometimes called the *fundamental matrix*) since it determines the transition from the initial state to the current state with zero control input. It will be recalled from Sects. 3.3.2 and 3.3.4 that the unforced plant state response with arbitrary initial state, which can now be written as

$$\mathbf{x}(t) = \mathbf{\Phi}(t, t_0)\mathbf{x}(t_0), \tag{3.390}$$

exhibits the individual plant modes when in the modal state representation. In any state representation, $\mathbf{x}(t)$ contains linear weighted sums of these plant modes.

Thus dynamic character of the plant is encaptured in the matrix, $\mathbf{\Phi}(t, t_0)$.

For an arbitrary control variable, $\mathbf{u}(\tau')$, solution (3.389) is mathematically exact but cannot be computed numerically with infinite precision. As already stated, however, in a digital control system, the control variable is piecewise constant. In this case it is possible to obtain an analytical solution computed to an accuracy limited only by the word-length and number representation in the digital processor. Also, if $\mathbf{u}$ is constant, (3.389) may be simplified slightly by changing the variable of integration from τ' to τ such that $\tau' = t - \tau$. This yields

$$\mathbf{x}(t) = e^{\mathbf{A}(t-t_0)}\mathbf{x}(t_0) + \int_{t-t_0}^{0} e^{\mathbf{A}\tau}\mathbf{B}\mathbf{u}(-d\tau) = e^{\mathbf{A}(t-t_0)}\mathbf{x}(t_0) + \int_{0}^{t-t_0} e^{\mathbf{A}\tau}\mathbf{B}\mathbf{u}\,d\tau. \tag{3.391}$$

In order to apply (3.391) to model a general LTI plant with $\mathbf{u}(t)$ supplied by a digitally implemented controller the initial time is $t_0 = t_k$ at the beginning of the kth iteration interval and the final time is $t = t_{k+1}$ at the end of this interval. The constant control value applied over this interval is updated by the controller at $t = t_k$ and therefore denoted by $\mathbf{u}(t_k)$. Solution (3.391) then becomes

$$\begin{aligned}\mathbf{x}(t_{k+1}) &= e^{\mathbf{A}(t_{k+1}-t_k)}\mathbf{x}(t_k) + \int_0^{t_{k+1}-t_k} e^{\mathbf{A}\tau}\mathbf{B}\mathbf{u}(t_k)\,d\tau \\ &= e^{\mathbf{A}(t_{k+1}-t_k)}\mathbf{x}(t_k) + \left[\int_0^{t_{k+1}-t_k} e^{\mathbf{A}\tau}\mathbf{B}\,d\tau\right]\mathbf{u}(t_k).\end{aligned} \tag{3.392}$$

The notation may now be simplified. First, the iteration interval is usually constant and has already been denoted by h. Thus $t_{k+1} - t_k = h$. Then $\mathbf{x}(t_k)$ and $\mathbf{u}(t_k)$ may be denoted, respectively, by $\mathbf{x}(k)$ and $\mathbf{u}(k)$. Equation (3.392) then becomes

$$\mathbf{x}(k+1) = e^{\mathbf{A}h}\mathbf{x}(k) + \left[\int_0^h e^{\mathbf{A}\tau}\mathbf{B}d\tau\right]\mathbf{u}(k). \tag{3.393}$$

Furthermore, the state transition matrix,

$$\mathbf{\Phi}(h) = e^{\mathbf{A}h}, \tag{3.394}$$

is constant. This will be called the *discrete plant matrix* which, like $\mathbf{A}$, is of dimension $n \times n$. In view of the finite integration limits, the matrix,

$$\mathbf{\Psi}(h) = \int_0^h e^{\mathbf{A}\tau}\mathbf{B}d\tau = \int_0^h \mathbf{\Phi}(\tau)\mathbf{B}d\tau, \tag{3.395}$$

is also constant and will be called the *discrete input matrix*. This is of dimension $n \times r$, as is the input matrix $\mathbf{B}$. The generic discrete state-space model of an LTI plant is then

$$\begin{aligned}\mathbf{x}(k+1) &= \mathbf{\Phi}(h)\mathbf{x}(k) + \mathbf{\Psi}(h)\mathbf{u}(k) \\ \mathbf{y}(k) &= \mathbf{C}\mathbf{x}(k).\end{aligned} \tag{3.396}$$

The analogy with the continuous state-space model of (3.387) and (3.388) is self-evident. An important distinction between the two models is that while the continuous model plant matrices, $\mathbf{A}$ and $\mathbf{B}$, are fixed, the corresponding discrete plant matrices, $\mathbf{\Phi}(h)$ and $\mathbf{\Psi}(h)$, depend on the iteration interval, h. The measurement matrix, $\mathbf{C}$, is, of course, the same in both models.

The discrete SISO plant model is a particular case of (3.396), but by analogy with its continuous counterpart,

$$\begin{aligned}\dot{\mathbf{x}} &= \mathbf{A}\mathbf{x} + \mathbf{b}u \\ y &= \mathbf{c}^T\mathbf{x},\end{aligned} \tag{3.397}$$

it will be written as

$$\begin{aligned}\mathbf{x}(k+1) &= \mathbf{\Phi}(h)\mathbf{x}(k) + \boldsymbol{\psi}(h)u(k) \\ y(k) &= \mathbf{c}^T\mathbf{x}(k).\end{aligned} \tag{3.398}$$

In some publications, the symbols, $\mathbf{A}$, $\mathbf{B}$ (or $\mathbf{b}$) and $\mathbf{C}$ (or $\mathbf{c}^T$) are retained for discrete state space models. The reader should be able to understand these from the context, but the different notation of (3.396) and (3.398) is preferred here to make the distinction absolutely clear.

3.4.2 State Space Model Derivation from Modal Basis Functions

3.4.2.1 Basic Principle

Attention is now turned to the determination of $\mathbf{\Phi}(h)$ and $\mathbf{\Psi}(h)$ (or $\mathbf{\psi}(h)$) for particular plants. Numerical approximations may be obtained by truncating the finite series for $e^{\mathbf{A}h}$ but since analytical determination is possible and yields an accuracy limited only by the number representation and wordlength of the digital processor, only this will be considered henceforth. A straightforward method is presented here based on the plant modal responses. These were derived in Sects. 3.3.2 and 3.3.4 as unforced responses to arbitrary initial states. Instead, a simpler approach will be adopted here, using the impulse response of the *associated* subsystem with transfer function,

$$\frac{X(s)}{V(s)} = \frac{1}{D(s)}, \qquad (3.399)$$

where $D(s)$ is the plant characteristic polynomial, $V(s)$ is the notional input and $X(s)$ is the notional output. The justification for this is that the relative scalings of the modal responses are unimportant at this stage, since the subsequent steps described below automatically establish the correct relative weightings of the modes in $\mathbf{\Phi}(t)$ (ultimately to become $\mathbf{\Phi}(h)$), using the given continuous state-space model of the plant. So the purpose of (3.399) is to deliver the 'essential ingredients for the final recipe'. These consist of the simplest possible mathematical functions that express the essential dynamic character of the plant obtained from the inverse Laplace transform of the partial fraction expansion of (3.399), defined as the *modal basis functions*, $f_i(t)$, $i = 1, 2, \ldots, n$, where n is the plant order. Such partial fraction expansions will be recalled from Sects. 3.3.2.3, 3.3.2.4, and 3.3.2.5.

Each element of $\mathbf{\Phi}(t)$ is a linear weighted sum of these functions. Hence,

$$\mathbf{\Phi}(t) = \sum_{i=1}^{n} \mathbf{M}_i f_i(t), \qquad (3.400)$$

where $\mathbf{M}_i$, $i = 1, 2, \ldots, n$, are weighting matrices that can be determined as follows.

The following properties of the state transition matrix are needed. These can be determined by means of the following matrix Maclaurin expansion.

$$\mathbf{\Phi}(t) = e^{\mathbf{A}t} = \mathbf{I}_n + \mathbf{A}t + \tfrac{1}{2!}\mathbf{A}^2 t^2 + \cdots \qquad (3.401)$$

It immediately follows that

$$\mathbf{\Phi}(0) = \mathbf{I}_n. \qquad (3.402)$$

Successive differentiations of (3.401) and setting $t = 0$ then yields

$$\begin{aligned}\dot{\boldsymbol{\Phi}}(t) &= \mathbf{A} + \mathbf{A}^2 t + \tfrac{1}{2!}\mathbf{A}^3 t^2 + \cdots \Rightarrow \dot{\boldsymbol{\Phi}}(0) = \mathbf{A}, \\ \ddot{\boldsymbol{\Phi}}(t) &= \mathbf{A}^2 + \mathbf{A}^3 t + \tfrac{1}{2!}\mathbf{A}^4 t^2 + \cdots \Rightarrow \ddot{\boldsymbol{\Phi}}(0) = \mathbf{A}^2, \\ &\vdots \\ \boldsymbol{\Phi}^{(n-1)}(t) &= \mathbf{A}^{n-1} + \mathbf{A}^{n-2} t + \cdots \Rightarrow \boldsymbol{\Phi}^{(n-1)}(0) = \mathbf{A}^{n-1}.\end{aligned} \quad (3.403)$$

Then (3.402) and (3.403) can be written as

$$\boldsymbol{\Phi}^{(j)}(t) = \mathbf{A}^j, \quad j = 0, 1, \ldots, n-1. \quad (3.404)$$

The same operations on (3.400) then yield

$$\boldsymbol{\Phi}^{(j)}(t) = \sum_{i=1}^{n} \mathbf{M}_i f_i^{(j)}(t) \Rightarrow \boldsymbol{\Phi}^{(j)}(0) = \sum_{i=1}^{n} \mathbf{M}_i f_i^{(j)}(0), \quad j = 0, 1, \ldots, n-1. \quad (3.405)$$

Equating the RHS of (3.404) and the RHS of (3.405) then yields n linear simultaneous matrix equations from which the weighting matrices can be determined. Thus,

$$\sum_{i=1}^{n} \mathbf{M}_i f_i^{(j)}(0) = \mathbf{A}^j, \quad j = 0, 1, \ldots, n-1. \quad (3.406)$$

Once $\mathbf{M}_i$, $i = 1, 2, \ldots, n$, have been determined, the discrete plant matrix is given by (3.400) with $t = h$. Thus

$$\boldsymbol{\Phi}(h) = \sum_{i=1}^{n} \mathbf{M}_i f_i(h), \quad (3.407)$$

and the discrete drive matrix is given by (3.395) with $e^{\mathbf{A}\tau} = \boldsymbol{\Phi}(\tau)$ given by (3.400) with $t = \tau$. Thus,

$$\boldsymbol{\Psi}(h) = \left[\int_0^h \sum_{i=1}^{n} \mathbf{M}_i f_i(\tau)\, d\tau\right] \mathbf{B}. \quad (3.408)$$

3.4.2.2 Plants with Real Distinct Poles

In this case, the plant characteristic polynomial is

$$D(s) = \prod_{i=1}^{n} (s + a_i), \; a_k \neq a_j \, \forall k \neq j, \; k \in [1, n] \text{ and } j \in [1, n] \quad (3.409)$$

3.4 Discrete LTI Plant Models

and the transfer function of the associated subsystem, expanded in a partial fraction expansion for generation of the modal basis functions, is

$$\frac{X(s)}{V(s)} = \frac{1}{D(s)} = \sum_{i=1}^{n} \frac{A_i}{s + a_i}, \quad i = 1, 2, \ldots, n. \tag{3.410}$$

The impulse response of this system is then

$$x(t) = \mathcal{L}^{-1}\left(\sum_{i=1}^{n} \frac{A_i}{s + a_i}\right) = \sum_{i=1}^{n} A_i e^{-a_i t}, \quad i = 1, 2, \ldots, n. \tag{3.411}$$

The modal basis functions may then be extracted without regard to the weighting functions, A_i (which therefore do not require evaluation). Thus,

$$f_i(t) = e^{-a_i t}, i = 1, 2, \ldots, n. \tag{3.412}$$

The discrete plant matrix is then

$$\mathbf{\Phi}(h) = \sum_{i=1}^{n} \mathbf{M}_i e^{-a_i h} \tag{3.413}$$

and the discrete input matrix is

$$\mathbf{\Psi}(h) = \left[\int_0^h \sum_{i=1}^{n} \mathbf{M}_i e^{-a_i \tau} d\tau\right] \mathbf{B} = \left[\sum_{i=1}^{n} \mathbf{M}_i [-a_i e^{-a_i \tau}]_0^h\right] \mathbf{B}$$

$$= \left[\sum_{i=1}^{n} \mathbf{M}_i a_i \left(1 - e^{-a_i h}\right)\right] \mathbf{B}. \tag{3.414}$$

Example 3.10 Two zone kiln

A kiln consists of two zones with heating time constants, T_1 and T_2, controlled by separate electric heaters with control variables, u_1 and u_2, the zone temperatures being x_1 and x_2. The continuous state-space model, $\dot{\mathbf{x}} = \mathbf{A}\mathbf{x} + \mathbf{B}\mathbf{u}$, $\mathbf{y} = \mathbf{C}\mathbf{x}$, is

$$\begin{bmatrix} \dot{x}_1 \\ \dot{x}_2 \end{bmatrix} = \begin{bmatrix} -1/T_1 & 0 \\ 0 & -1/T_2 \end{bmatrix} \begin{bmatrix} x_1 \\ x_2 \end{bmatrix} + \begin{bmatrix} 1 & \lambda \\ \lambda & 1 \end{bmatrix} \begin{bmatrix} u_1 \\ u_2 \end{bmatrix}, \quad \begin{bmatrix} y_1 \\ y_2 \end{bmatrix} = \begin{bmatrix} c & 0 \\ 0 & c \end{bmatrix} \begin{bmatrix} x_1 \\ x_2 \end{bmatrix}, \tag{3.415}$$

where $\lambda \in (0, 1)$ is an interaction factor representing the influence of each heater on the zone it is not intended to control and c is the temperature measurement constant. It is required to determine the corresponding discrete state-space model for an iteration/sampling period of h seconds.

First, letting $1/T_1 = p_1$ and $1/T_2 = p_2$, the plant characteristic polynomial is

$$|s\mathbf{I}_2 - \mathbf{A}| = \begin{vmatrix} s + p_1 & 0 \\ 0 & s + p_2 \end{vmatrix} = (s + p_1)(s + p_2). \tag{3.416}$$

The associated subsystem transfer function expanded in partial fractions is therefore

$$\frac{X(s)}{V(s)} = \frac{1}{D(s)} = \frac{1}{(s + p_1)(s + p_2)} = \frac{A_1}{s + p_1} + \frac{A_2}{s + p_2} \tag{3.417}$$

and the corresponding impulse response is

$$x(t) = A_1 e^{-p_1 t} + A_2 e^{-p_2 t}. \tag{3.418}$$

Disregarding the weighting coefficients, A_1 and A_2, the modal basis functions are extracted as $f_1(t) = e^{-p_1 t}$ and $f_2(t) = e^{-p_2 t}$, yielding the state transition matrix

$$\mathbf{\Phi}(t) = \mathbf{M}_1 e^{-p_1 t} + \mathbf{M}_2 e^{-p_2 t}. \tag{3.419}$$

where $\mathbf{M}_1$ and $\mathbf{M}_2$ are the weighting matrices.

As the plant is of second order, condition (3.402) and the first condition of (3.403) are needed to form the simultaneous equations needed to determine $\mathbf{M}_1$ and $\mathbf{M}_2$. Thus,

$$\mathbf{\Phi}(0) = \mathbf{I}_2 \Rightarrow \mathbf{M}_1 + \mathbf{M}_2 = \mathbf{I}_2 \tag{3.420}$$

and

$$\dot{\mathbf{\Phi}}(0) = \mathbf{A} \Rightarrow -p_1 \mathbf{M}_1 - p_2 \mathbf{M}_2 = \mathbf{A}. \tag{3.421}$$

Then, $p_2 \times$ (3.420) + (3.421) yields

$$\mathbf{M}_1 = \frac{1}{p_2 - p_1} [p_2 \mathbf{I}_2 + \mathbf{A}] = \frac{1}{p_2 - p_1} \left[\begin{bmatrix} p_2 & 0 \\ 0 & p_2 \end{bmatrix} + \begin{bmatrix} -p_1 & 0 \\ 0 & -p_2 \end{bmatrix} \right] = \begin{bmatrix} 1 & 0 \\ 0 & 0 \end{bmatrix} \tag{3.422}$$

and from (3.420),

$$\mathbf{M}_2 = \mathbf{I}_2 - \mathbf{M}_1 = \begin{bmatrix} 0 & 0 \\ 0 & 1 \end{bmatrix}. \tag{3.423}$$

Then, substituting in (3.419) for $\mathbf{M}_1$ and $\mathbf{M}_2$ using (3.422) and (3.423) yields

$$\mathbf{\Phi}(t) = \mathbf{M}_1 e^{-p_1 t} + \mathbf{M}_2 e^{-p_2 t} = \begin{bmatrix} e^{-p_1 t} & 0 \\ 0 & e^{-p_2 t} \end{bmatrix}. \tag{3.424}$$

3.4 Discrete LTI Plant Models

Then,

$$\Psi(h) = \int_0^h \Phi(\tau) \mathbf{B} d\tau = \int_0^h \begin{bmatrix} e^{-p_1\tau} & 0 \\ 0 & e^{-p_2\tau} \end{bmatrix} \begin{bmatrix} 1 & \lambda \\ \lambda & 1 \end{bmatrix} d\tau$$

$$= \begin{bmatrix} \left[\frac{-1}{p_1}e^{-p_1\tau}\right]_0^h & 0 \\ 0 & \left[\frac{-1}{p_2}e^{-p_2\tau}\right]_0^h \end{bmatrix} \begin{bmatrix} 1 & \lambda \\ \lambda & 1 \end{bmatrix} \Rightarrow$$

$$\Psi(h) = \begin{bmatrix} \frac{1}{p_1}\left(1-e^{-p_1h}\right) & 0 \\ 0 & \frac{1}{p_2}\left(1-e^{-p_2h}\right) \end{bmatrix} \begin{bmatrix} 1 & \lambda \\ \lambda & 1 \end{bmatrix}$$

$$= \begin{bmatrix} \frac{1}{p_1}\left(1-e^{-p_1h}\right) & \frac{\lambda}{p_1}\left(1-e^{-p_1h}\right) \\ \frac{\lambda}{p_2}\left(1-e^{-p_2h}\right) & \frac{1}{p_2}\left(1-e^{-p_2h}\right) \end{bmatrix}.$$

(3.425)

The discrete state-space model is therefore as follows.

$$\begin{bmatrix} x_1(k+1) \\ x_2(k+1) \end{bmatrix} = \begin{bmatrix} e^{-p_1h} & 0 \\ 0 & e^{-p_2h} \end{bmatrix} \begin{bmatrix} x_1(k) \\ x_2(k) \end{bmatrix} + \begin{bmatrix} \frac{1}{p_1}\left(1-e^{-p_1h}\right) & \frac{\lambda}{p_1}\left(1-e^{-p_1h}\right) \\ \frac{\lambda}{p_2}\left(1-e^{-p_2h}\right) & \frac{1}{p_2}\left(1-e^{-p_2h}\right) \end{bmatrix} \begin{bmatrix} u_1(k) \\ u_2(k) \end{bmatrix}, \begin{bmatrix} y_1(k) \\ y_2(k) \end{bmatrix} = \begin{bmatrix} c & 0 \\ 0 & c \end{bmatrix} \begin{bmatrix} x_1(k) \\ x_2(k) \end{bmatrix}.$$

(3.426)

3.4.2.3 Plants with Oscillatory Modes

Since oscillatory modes are produced by complex conjugate pairs of poles, which are distinct, the recommended approach is the same as for plants with real distinct poles. This will lead to complex terms in the discrete state-space model, but these will occur in complex conjugate pairs and may be combined to arrive at a useful model with only real terms. In this case, if the complex conjugate pair of poles are at $s_{i,i+1} = -\sigma \pm j\omega$, then the corresponding modal basis functions extracted from the associated subsystem will be

$$f_i(t) = e^{-(\sigma-j\omega)t} = e^{-\sigma t}\left[\cos(\omega t) + j\sin(\omega t)\right]$$

(3.427)

and

$$f_{i+1}(t) = e^{-(\sigma+j\omega)t} = e^{-\sigma t}\left[\cos(\omega t) - j\sin(\omega t)\right].$$

(3.428)

An alternative, however, is to eliminate the imaginary terms at the outset by choosing different modal basis functions that are judiciously chosen linearly independent weighted sums of the complex conjugate modal functions already extracted. Thus,

$$f_i'(t) = \tfrac{1}{2}\left[f_i(t) + f_{i+1}(t)\right] = e^{-\sigma t}\cos(\omega t)$$

(3.429)

$$f'_{i+1}(t) = \tfrac{1}{2j}[f_i(t) - f_{i+1}(t)] = e^{-\sigma t} \sin(\omega t) \tag{3.430}$$

Both methods will be demonstrated in the following example.

Example 3.11 Underdamped second-order system with electric drive applications

Consider first an underdamped second-order SISO plant whose continuous state differential equation, $\dot{\mathbf{x}} = \mathbf{A}\mathbf{x} + \mathbf{b}u$, is

$$\begin{bmatrix} \dot{x}_1 \\ \dot{x}_2 \end{bmatrix} = \begin{bmatrix} -\sigma & \omega \\ -\omega & -\sigma \end{bmatrix} \begin{bmatrix} x_1 \\ x_2 \end{bmatrix} + \begin{bmatrix} 0 \\ b \end{bmatrix} u \tag{3.431}$$

This could be, for example, the underdamped model of an automobile suspension system (quarter vehicle only) provided with an electromagnetic actuator for active control. In this case, the plant characteristic polynomial is

$$\begin{vmatrix} s+\sigma & -\omega \\ \omega & s+\sigma \end{vmatrix} = s^2 + 2\sigma s + \sigma^2 + \omega^2 = (s + \sigma + j\omega)(s + \sigma - j\omega) \tag{3.432}$$

and therefore the associated subsystem, once expanded in partial fractions, is

$$\frac{X(s)}{V(s)} = \frac{A_1}{s+\sigma+j\omega} + \frac{A_2}{s+\sigma-j\omega} \tag{3.433}$$

and the impulse response is

$$x(t) = A_1 e^{-(\sigma+j\omega)t} + A_2 e^{-(\sigma-j\omega)t} \tag{3.434}$$

Disregarding the coefficients, A_1 and A_2, the modal basis functions are extracted as

$$f_1(t) = e^{-(\sigma+j\omega)t} \text{ and } f_2(t) = e^{-(\sigma-j\omega)t}. \tag{3.435}$$

The state transition matrix is then

$$\mathbf{\Phi}(t) = \mathbf{M}_1 e^{-(\sigma+j\omega)t} + \mathbf{M}_2 e^{-(\sigma-j\omega)t} \tag{3.436}$$

The simultaneous equations for determination of $\mathbf{M}_1$ and $\mathbf{M}_2$ are then

$$\mathbf{\Phi}(0) = \mathbf{I}_2 \Rightarrow \mathbf{M}_1 + \mathbf{M}_2 = \mathbf{I}_2 \tag{3.437}$$

$$\dot{\mathbf{\Phi}}(0) = \mathbf{A} \Rightarrow -(\sigma+j\omega)\mathbf{M}_1 - (\sigma-j\omega)\mathbf{M}_2 = \mathbf{A} \tag{3.438}$$

Then $(\sigma - j\omega) \times$ (3.437) + (3.438) yields

3.4 Discrete LTI Plant Models

$$\mathbf{M}_1 = \frac{-1}{2j\omega}[(\sigma - j\omega)\mathbf{I}_2 + \mathbf{A}] = \tfrac{1}{2}\begin{bmatrix} 1 + j\frac{\sigma}{\omega} & 0 \\ 0 & 1 + j\frac{\sigma}{\omega} \end{bmatrix} + \frac{j}{2\omega}\begin{bmatrix} -\sigma & \omega \\ -\omega & -\sigma \end{bmatrix}$$

$$= \tfrac{1}{2}\begin{bmatrix} 1 & j \\ -j & 1 \end{bmatrix}. \tag{3.439}$$

and from (3.437),

$$\mathbf{M}_2 = \mathbf{I}_2 - \mathbf{M}_1 = \tfrac{1}{2}\begin{bmatrix} 1 & -j \\ j & 1 \end{bmatrix}. \tag{3.440}$$

Then, (3.436) becomes

$$\begin{aligned}
\mathbf{\Phi}(t) &= \tfrac{1}{2}\begin{bmatrix} 1 & j \\ -j & 1 \end{bmatrix} e^{-(\sigma+j\omega)t} + \tfrac{1}{2}\begin{bmatrix} 1 & -j \\ j & 1 \end{bmatrix} e^{-(\sigma-j\omega)t} \\
&= \tfrac{e^{-\sigma t}}{2}\begin{bmatrix} 1 & j \\ -j & 1 \end{bmatrix}[\cos(\omega t) - j\sin(\omega t)] + \tfrac{e^{-\sigma t}}{2}\begin{bmatrix} 1 & -j \\ j & 1 \end{bmatrix}[\cos(\omega t) + j\sin(\omega t)] \\
&= e^{-\sigma t}\begin{bmatrix} \cos(\omega t) & \sin(\omega t) \\ -\sin(\omega t) & \cos(\omega t) \end{bmatrix}.
\end{aligned} \tag{3.441}$$

The behaviour of an underdamped second-order system is visible in the elements of $\mathbf{\Phi}(t)$. The discrete drive matrix is then obtained by integration by parts as

$$\begin{aligned}
\mathbf{\psi}(h) &= \int_0^h \mathbf{\Phi}(\tau)\mathbf{b}\,d\tau = \int_0^h e^{-\sigma t}\begin{bmatrix} \cos(\omega t) & \sin(\omega t) \\ -\sin(\omega t) & \cos(\omega t) \end{bmatrix}\begin{bmatrix} 0 \\ b \end{bmatrix} d\tau \\
&= b\int_0^h e^{-\sigma t}\begin{bmatrix} \sin(\omega t) \\ \cos(\omega t) \end{bmatrix} d\tau \\
&= \left[\frac{be^{-\sigma t}}{\sigma^2 + \omega^2}\begin{bmatrix} -\omega\cos(\omega t) - \sigma\sin(\omega t) \\ \omega\sin(\omega t) - \sigma\cos(\omega t) \end{bmatrix}\right]_0^h \Rightarrow
\end{aligned} \tag{3.442}$$

$$\mathbf{\psi}(h) = \frac{b}{\sigma^2 + \omega^2}\begin{bmatrix} \omega\left[1 - e^{-\sigma h}\cos(\omega h)\right] - \sigma e^{-\sigma h}\sin(\omega h) \\ \omega e^{-\sigma h}\sin(\omega h) + \sigma\left[1 - e^{-\sigma h}\cos(\omega h)\right] \end{bmatrix}.$$

The required model $\mathbf{x}(k+1) = \mathbf{\Phi}(h)\mathbf{x}(k) + \mathbf{\psi}(h)u(k)$ is then

$$\begin{aligned}
\begin{bmatrix} x_1(k+1) \\ x_2(k+1) \end{bmatrix} &= e^{-\sigma h}\begin{bmatrix} \cos(\omega h) & \sin(\omega h) \\ -\sin(\omega h) & \cos(\omega h) \end{bmatrix}\begin{bmatrix} x_1(k) \\ x_2(k) \end{bmatrix} \\
&+ \tfrac{b}{\sigma^2+\omega^2}\begin{bmatrix} \omega\left[1 - e^{-\sigma h}\cos(\omega h)\right] - \sigma e^{-\sigma h}\sin(\omega h) \\ \omega e^{-\sigma h}\sin(\omega h) + \sigma\left[1 - e^{-\sigma h}\cos(\omega h)\right] \end{bmatrix} u(k).
\end{aligned} \tag{3.443}$$

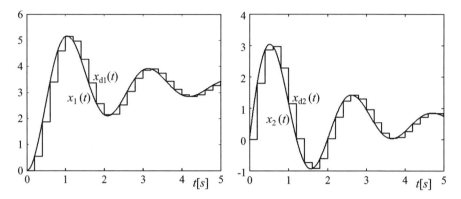

Fig. 3.20 Step response of continuous and discrete state-space models of underdamped second-order plant

Figure 3.20 shows a simulation of step responses for illustration.

The continuous plant model (3.431) has $\omega = 3$ [rad/s], $\sigma = 0.5$ [s^{-1}] and $b = 10$ and the equivalent discrete plant model of (3.443) has the iteration/sampling period set to a relatively large value of $h = 0.2$ [s] so that the essential differences between the continuous model states, $x_1(t)$ and $x_2(t)$, and the corresponding discrete states, $x_{d1}(t)$ and $x_{d2}(t)$, are clearly visible. As intended, the state variables of the discrete model precisely match those of the continuous model at the iteration update times of $t = kh$, $k = 0, 1, 2, \ldots$, but are constant between these times. It should be noted, however, that the plots of the continuous model state variables are also discrete variables, as they are produced by a digital computer, but are very close approximations to the theoretical continuous variables using a variable step algorithm in SIMULINK® with far shorter intervals between the updates.

Next, if $\sigma = 0$ and $b = 0$, the continuous plant model (3.431) becomes an unforced and undamped dynamical system with state differential equation,

$$\begin{bmatrix} \dot{x}_1 \\ \dot{x}_2 \end{bmatrix} = \begin{bmatrix} 0 & \omega \\ -\omega & 0 \end{bmatrix} \begin{bmatrix} x_1 \\ x_2 \end{bmatrix}. \tag{3.444}$$

It is then a two-phase oscillator that could be used to generate the elements of the Park transformation matrix in vector control of AC motors (Chap. 2), i.e.

$$\mathbf{M}_{\text{Pk}}(t) = \begin{bmatrix} \cos(\omega t) & \sin(\omega t) \\ -\sin(\omega t) & \cos(\omega t) \end{bmatrix}. \tag{3.445}$$

The state variables of (3.444) are the required time-varying elements of (3.445) if the initial state is $[x_1 \ x_2]^{\text{T}} = [0 \ 1]^{\text{T}}$, giving

$$x_1(t) = \sin(\omega t) \text{ and } x_2(t) = \cos(\omega t). \tag{3.446}$$

3.4 Discrete LTI Plant Models

This may be confirmed by substitution in (3.444). This two-phase oscillator can be used to produce three-phase variables, $x_a(t)$, $x_b(t)$ and $x_c(t)$, for an AC motor or to control the power electronics of an autonomous power supply using the inverse Clarke transformation (Chap. 2) as the output matrix with (3.444) as follows.

$$\begin{bmatrix} x_a \\ x_b \\ x_c \end{bmatrix} = \begin{bmatrix} 1 & 0 \\ -1/2 & \sqrt{3}/2 \\ -1/2 & -\sqrt{3}/2 \end{bmatrix} \begin{bmatrix} x_1 \\ x_2 \end{bmatrix}. \qquad (3.447)$$

It is possible to implement such a system on a digital processor using software for numerical integration such as available for simulation of continuous systems in MATLAB®–SIMULINK®, but this can only be approximate. With most numerical integration algorithms, the result of attempting to implement the two-phase oscillator (3.444) would be a gradual change in the oscillation amplitude. The use of the discrete state-space model would solve this problem as it is based on the continuous solution evaluated precisely each iteration. The required discrete two-phase oscillator algorithm can be obtained simply by setting $\sigma = 0$ and $b = 0$ in (3.443) but it will be derived again to demonstrate the use of the alternative real modal basis functions. These are given by (3.429) and (3.430) with $\sigma = 0$ so that the required state transition matrix is of the form

$$\mathbf{\Phi}(t) = \mathbf{M}_1 \cos(\omega t) + \mathbf{M}_2 \sin(\omega t). \qquad (3.448)$$

The simultaneous equations for determination of $\mathbf{M}_1$ and $\mathbf{M}_2$ are then

$$\mathbf{\Phi}(0) = \mathbf{I}_2 \Rightarrow \mathbf{M}_1 = \mathbf{I}_2 = \begin{bmatrix} 1 & 0 \\ 0 & 1 \end{bmatrix} \qquad (3.449)$$

and

$$\dot{\mathbf{\Phi}}(0) = \mathbf{A} \Rightarrow [-\mathbf{M}_1 \omega \sin(\omega t) + \mathbf{M}_2 \omega \cos(\omega t)]_{t=0} = \mathbf{A} \Rightarrow$$
$$\mathbf{M}_2 = \frac{\mathbf{A}}{\omega} = \frac{1}{\omega} \begin{bmatrix} 0 & \omega \\ -\omega & 0 \end{bmatrix} = \begin{bmatrix} 0 & 1 \\ -1 & 0 \end{bmatrix} \qquad (3.450)$$

Then (3.448) gives

$$\mathbf{\Phi}(t) = \begin{bmatrix} 1 & 0 \\ 0 & 1 \end{bmatrix} \cos(\omega t) + \begin{bmatrix} 0 & 1 \\ -1 & 0 \end{bmatrix} \sin(\omega t) = \begin{bmatrix} \cos(\omega t) & \sin(\omega t) \\ -\sin(\omega t) & \cos(\omega t) \end{bmatrix}. \qquad (3.451)$$

The required discrete two-phase oscillator algorithm is therefore

$$\begin{bmatrix} x_1(k+1) \\ x_2(k+1) \end{bmatrix} = \begin{bmatrix} \cos(\omega h) & \sin(\omega h) \\ -\sin(\omega h) & \cos(\omega h) \end{bmatrix} \begin{bmatrix} x_1(k) \\ x_2(k) \end{bmatrix}. \qquad (3.452)$$

3.4.2.4 Plants with Multiple Poles

The most common plant with multiple poles is the double integrator and this will be one of the examples given. Applications include rigid-body spacecraft and air-bearing-based positioning systems operating in a contactless fashion and therefore without friction. For completeness, however, and to explain the theory, a more general case will be discussed. Let a plant contain repeated real poles at $s_{1,2,\ldots,q} = -p$. Then, the plant characteristic polynomial will be $(s+p)^q D'(s)$ where all the roots of $D'(s) = 0$ differ from $-p$. In this case, the partial fraction expansion of the associated dynamical system will be of the form

$$\frac{X(s)}{V(s)} = \frac{1}{(s+p)^q D'(s)} = \frac{A_{r1}}{s+p} + \frac{A_{r2}}{(s+p)^2} + \cdots + \frac{A_{rq}}{(s+p)^q} + P'(s). \tag{3.453}$$

where $P'(s)$ is the part of the partial fraction expansion corresponding to $D'(s)$. The impulse response with zero initial conditions will therefore be of the form

$$x(t) = A_{r1} e^{-pt} + \frac{A_{r2}}{1!} t e^{-pt} + \cdots + \frac{A_{rq}}{(q-1)!} t^{q-1} e^{-pt} + \mathcal{L}^{-1}\{P'(s)\}. \tag{3.454}$$

The time-varying terms, $t^i e^{-pt}$, $i = 0, 1, \ldots, q-1$, capture the essential dynamic character of the plant, and therefore, the coefficients, A_{ri}, $i = 1, 2, \ldots, q$, and the factorial terms can be disregarded. The modal basis functions are then extracted as

$$f_1 = e^{-pt}, \quad f_2 = t e^{-pt}, \quad \ldots, \quad f_q = t^{q-1} e^{-pt}. \tag{3.455}$$

As in the distinct pole cases, they will automatically appear in the state transition matrix correctly weighted by the matrices, $\mathbf{M}_i$, $i = 1, 2, \ldots, n$, of (3.400). One case to be discussed, however, is that of the plant only having multiple poles at one location, the associated dynamical system transfer function being

$$\frac{X(s)}{V(s)} = \frac{1}{(s+p)^n}. \tag{3.456}$$

An apparent issue is that the partial fraction expansion (3.453) does not exist. There are still, however, n modal basis functions given by (3.455) with $q = n$. This is shown as follows. Consider a state-variable block diagram of (3.456) consisting of a chain of n identical first-order blocks with transfer function, $1/(s+p)$. Then, the state variables, z_i, $i = 1, 2, \ldots, n$, chosen as the block outputs satisfy

$$Z_1(s) = \frac{1}{s+p} V(s), \quad Z_2(s) = \frac{1}{(s+p)^2} V(s), \quad \ldots, \quad Z_n(s) = \frac{1}{(s+p)^n} V(s). \tag{3.457}$$

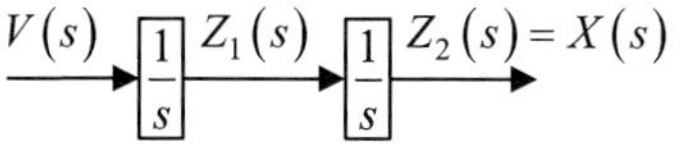

3.4 Discrete LTI Plant Models

$$\xrightarrow{V(s)} \boxed{\dfrac{1}{s}} \xrightarrow{Z_1(s)} \boxed{\dfrac{1}{s}} \xrightarrow{Z_2(s) = X(s)}$$

Fig. 3.21 State-variable block diagram of double integrator associated dynamical system

If a unit impulse, $v(t) = \delta(t)$, is applied, then $V(s) = 1$ and in the time domain,

$$z_i(t) = \mathcal{L}^{-1}\left(\frac{1}{s+p}\right)^i = \frac{1}{(i-1)!} t^{i-1} e^{-pt}, \quad i = 1, 2, \ldots, n \quad (3.458)$$

It is then evident that, with $q = n$, the modal basis functions of (3.455) are extracted from (3.458) by disregarding the factorial terms.

Example 3.12 Double integrator plant

The state-space model of a double integrator plant, $\mathbf{x} = \mathbf{Ax} + \mathbf{b}u$, $y = \mathbf{c}^T\mathbf{x}$, is

$$\begin{bmatrix} \dot{x}_1 \\ \dot{x}_2 \end{bmatrix} = \begin{bmatrix} 0 & 1 \\ 0 & 0 \end{bmatrix} \begin{bmatrix} x_1 \\ x_2 \end{bmatrix} + \begin{bmatrix} 0 \\ b \end{bmatrix} u, \quad y = \begin{bmatrix} 1 & 0 \end{bmatrix} \begin{bmatrix} x_1 \\ x_2 \end{bmatrix}. \quad (3.459)$$

where b is the lumped plant parameter. The plant characteristic polynomial is

$$|s\mathbf{I}_2 - \mathbf{A}| = \left\| \begin{bmatrix} s & 0 \\ 0 & s \end{bmatrix} - \begin{bmatrix} 0 & 1 \\ 0 & 0 \end{bmatrix} \right\| = \begin{vmatrix} s & -1 \\ 0 & s \end{vmatrix} = s^2 \quad (3.460)$$

and therefore the transfer function of the associated dynamical system is

$$\frac{X(s)}{V(s)} = \frac{1}{s^2}. \quad (3.461)$$

The state-variable block diagram is shown in Fig. 3.21.

The state variables, z_1 and z_2, yield the modal basis functions if the input, $v(t)$, is a unit impulse function, implying $V(s) = 1$. Thus, assuming $f_i(0) = 0$, $i = 1, 2$,

$$z_1(t) = \mathcal{L}^{-1}\left(\frac{1}{s}\right) = h(t) = 1 \forall t > 0 \text{ and } z_2(t) = \mathcal{L}^{-1}\left(\frac{1}{s^2}\right) = th(t) = t, \forall t > 0, \quad (3.462)$$

where $h(t)$ is the unit step function. In this case, there are no coefficients to disregard, and the modal basis functions are

$$f_1(t) = 1 \text{ and } f_2(t) = t, \quad (3.463)$$

giving the state transition matrix as

$$\mathbf{\Phi}(t) = \mathbf{M}_1 + \mathbf{M}_2 t. \quad (3.464)$$

The equations for determination of $\mathbf{M}_1$ and $\mathbf{M}_2$ are then

$$\Phi(0) = \mathbf{I}_2 \Rightarrow \mathbf{M}_1 = \mathbf{I}_2 = \begin{bmatrix} 1 & 0 \\ 0 & 1 \end{bmatrix} \tag{3.465}$$

and

$$\dot{\Phi}(0) = \mathbf{A} \Rightarrow \mathbf{M}_2 = \mathbf{A} = \begin{bmatrix} 0 & 1 \\ 0 & 0 \end{bmatrix}. \tag{3.466}$$

Then, the state transition matrix (3.464) becomes

$$\Phi(t) = \begin{bmatrix} 1 & 0 \\ 0 & 1 \end{bmatrix} + \begin{bmatrix} 0 & 1 \\ 0 & 0 \end{bmatrix} t = \begin{bmatrix} 1 & t \\ 0 & 1 \end{bmatrix}. \tag{3.467}$$

giving the discrete drive matrix,

$$\boldsymbol{\psi}(h) = \int_0^h \Phi(\tau) \mathbf{B} d\tau = \int_0^h \begin{bmatrix} 1 & \tau \\ 0 & 1 \end{bmatrix} \begin{bmatrix} 0 \\ b \end{bmatrix} d\tau = \int_0^h \begin{bmatrix} b\tau \\ b \end{bmatrix} d\tau = \begin{bmatrix} \frac{1}{2}bh^2 \\ bh \end{bmatrix}. \tag{3.468}$$

Then the discrete state-space model, $\mathbf{x}_{k+1} = \Phi(h)\mathbf{x}_k + \boldsymbol{\psi}(h)u_k$, $y_k = \mathbf{c}\mathbf{x}_k$, is

$$\begin{bmatrix} x_1(k+1) \\ x_2(k+1) \end{bmatrix} = \begin{bmatrix} 1 & h \\ 0 & 1 \end{bmatrix} \begin{bmatrix} x_1(k) \\ x_2(k) \end{bmatrix} + \begin{bmatrix} \frac{1}{2}bh^2 \\ bh \end{bmatrix} u(k), \quad y(k) = \begin{bmatrix} 1 & 0 \end{bmatrix} \begin{bmatrix} x_1(k) \\ x_2(k) \end{bmatrix}. \tag{3.469}$$

Example 3.13 Triple integrator plant

A relevant application is spacecraft attitude control using variable geometry solar panels (Chap. 8). The continuous state-space model, $\mathbf{x} = \mathbf{A}\mathbf{x} + \mathbf{b}u$, $y = \mathbf{c}^T\mathbf{x}$, is

$$\begin{bmatrix} \dot{x}_1 \\ \dot{x}_2 \\ \dot{x}_3 \end{bmatrix} = \begin{bmatrix} 0 & 1 & 0 \\ 0 & 0 & 1 \\ 0 & 0 & 0 \end{bmatrix} \begin{bmatrix} x_1 \\ x_2 \\ x_3 \end{bmatrix} + \begin{bmatrix} 0 \\ 0 \\ b \end{bmatrix} u, \quad y = \begin{bmatrix} 1 & 0 & 0 \end{bmatrix} \begin{bmatrix} x_1 \\ x_2 \\ x_3 \end{bmatrix}, \tag{3.470}$$

and therefore, the plant characteristic polynomial is

$$|s\mathbf{I}_3 - \mathbf{A}| = \left| \begin{bmatrix} s & 0 & 0 \\ 0 & s & 0 \\ 0 & 0 & s \end{bmatrix} - \begin{bmatrix} 0 & 1 & 0 \\ 0 & 0 & 1 \\ 0 & 0 & 0 \end{bmatrix} \right| = \begin{vmatrix} s & -1 & 0 \\ 0 & s & -1 \\ 0 & 0 & s \end{vmatrix} = s^3. \tag{3.471}$$

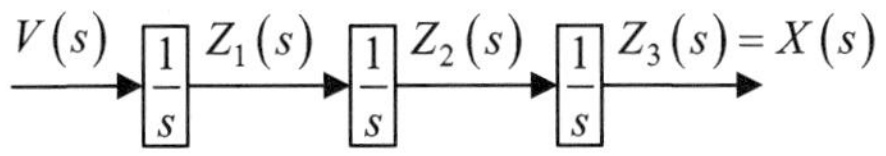

3.4 Discrete LTI Plant Models

$$\xrightarrow{V(s)} \boxed{\tfrac{1}{s}} \xrightarrow{Z_1(s)} \boxed{\tfrac{1}{s}} \xrightarrow{Z_2(s)} \boxed{\tfrac{1}{s}} \xrightarrow{Z_3(s)=X(s)}$$

Fig. 3.22 State-variable block diagram of triple integrator associated dynamical system

The transfer function of the associated dynamical system is therefore

$$\frac{X(s)}{V(s)} = \frac{1}{s^3} \tag{3.472}$$

and the corresponding state-variable block diagram is shown in Fig. 3.22.

The responses of the state variables, z_1, z_2 and z_3, to a unit impulse input, assuming $z_i(0) = 0$, $i = 1, 2, 3$, are then the modal basis functions

$$z_1(t) = 1, \quad z_2(t) = t, \quad z_3(t) = \tfrac{1}{2}t^2, \forall\ t > 0. \tag{3.473}$$

In this case, the simplest set of modal basis functions that can be formed is

$$f_1(t) = 1, \quad f_2(t) = t, \quad f_3(t) = t^2 \tag{3.474}$$

giving the state transition matrix

$$\mathbf{\Phi}(t) = \mathbf{M}_1 + \mathbf{M}_2 t + \mathbf{M}_3 t^2. \tag{3.475}$$

The equations for determination of $\mathbf{M}_1$, $\mathbf{M}_2$ and $\mathbf{M}_3$ are then

$$\mathbf{\Phi}(0) = \mathbf{I}_3 \Rightarrow \mathbf{M}_1 = \mathbf{I}_3 = \begin{bmatrix} 1 & 0 & 0 \\ 0 & 1 & 0 \\ 0 & 0 & 1 \end{bmatrix}, \tag{3.476}$$

$$\dot{\mathbf{\Phi}}(t) = \mathbf{M}_2 + 2\mathbf{M}_3 t \Rightarrow \dot{\mathbf{\Phi}}(0) = \mathbf{A} \Rightarrow \mathbf{M}_2 = \mathbf{A} = \begin{bmatrix} 0 & 1 & 0 \\ 0 & 0 & 1 \\ 0 & 0 & 0 \end{bmatrix} \tag{3.477}$$

and

$$\ddot{\mathbf{\Phi}}(t) = 2\mathbf{M}_3 \Rightarrow \ddot{\mathbf{\Phi}}(0) = 2\mathbf{M}_3 = \mathbf{A}^2 \Rightarrow \mathbf{M}_3 = \tfrac{1}{2}\mathbf{A}^2 = \begin{bmatrix} 0 & 0 & \tfrac{1}{2} \\ 0 & 0 & 0 \\ 0 & 0 & 0 \end{bmatrix} \tag{3.478}$$

Then the state transition matrix (3.475) becomes

$$\mathbf{\Phi}(t) = \begin{bmatrix} 1 & 0 & 0 \\ 0 & 1 & 0 \\ 0 & 0 & 1 \end{bmatrix} + \begin{bmatrix} 0 & 1 & 0 \\ 0 & 0 & 1 \\ 0 & 0 & 0 \end{bmatrix} t + \begin{bmatrix} 0 & 0 & \tfrac{1}{2} \\ 0 & 0 & 0 \\ 0 & 0 & 0 \end{bmatrix} t^2 = \begin{bmatrix} 1 & t & \tfrac{1}{2}t^2 \\ 0 & 1 & t \\ 0 & 0 & 1 \end{bmatrix} \tag{3.479}$$

and the discrete input matrix is

$$\Psi(h) = \int_0^h \Phi(\tau)\mathbf{B}\,d\tau = \int_0^h \begin{bmatrix} 1 & \tau & \frac{1}{2}\tau^2 \\ 0 & 1 & \tau \\ 0 & 0 & 1 \end{bmatrix} \begin{bmatrix} 0 \\ 0 \\ b \end{bmatrix} d\tau = \int_0^h \begin{bmatrix} \frac{1}{2}b\tau^2 \\ b\tau \\ b \end{bmatrix} d\tau = \begin{bmatrix} \frac{1}{6}bh^3 \\ \frac{1}{2}bh^2 \\ bh \end{bmatrix}$$

(3.480)

Then the discrete state-space model, $\mathbf{x}_{k+1} = \Phi(h)\mathbf{x}_k + \Psi(h)u_k$, $y_k = \mathbf{c}\mathbf{x}_k$, is

$$\begin{bmatrix} x_1(k+1) \\ x_2(k+1) \\ x_3(k+1) \end{bmatrix} = \begin{bmatrix} 1 & h & \frac{1}{2}h^2 \\ 0 & 1 & h \\ 0 & 0 & 1 \end{bmatrix} \begin{bmatrix} x_1(k) \\ x_2(k) \\ x_3(k) \end{bmatrix} + \begin{bmatrix} \frac{1}{6}bh^3 \\ \frac{1}{2}bh^2 \\ bh \end{bmatrix} u(k),$$

(3.481)

$$y(k) = \begin{bmatrix} 1 & 0 & 0 \end{bmatrix} \begin{bmatrix} x_1(k) \\ x_2(k) \\ x_3(k) \end{bmatrix}.$$

3.4.2.5 Relationships for Checking Derivations

The following relationships may be used to check the correctness of the matrices, $\Phi(h)$ and $\Psi(h)$ (or $\psi(h)$ for SISO plant models), once they have been derived:

$$\frac{d}{dh}\Phi(h) = \mathbf{A}\Phi(h) \qquad (3.482)$$

$$\frac{d}{dh}\Psi(h) = \Phi(h)\mathbf{B} \qquad (3.483)$$

For SISO plants, the matrices, $\Psi(h)$ and $\mathbf{B}$, are, respectively, replaced by the column vectors, $\psi(h)$ and $\mathbf{b}$. Relationship (3.482) follows directly by differentiating

$$\Phi(h) = e^{\mathbf{A}h}. \qquad (3.484)$$

The second relationship is proven, using the matrix Maclaurin series, as follows:

$$\Psi(h) = \int_0^h e^{\mathbf{A}\tau}\mathbf{B}\,d\tau = \int_0^h \left[\mathbf{I}_n + \mathbf{A}\tau + \tfrac{1}{2!}\mathbf{A}^2\tau^2 + \cdots\right]\mathbf{B}\,d\tau$$
$$= \left[\mathbf{I}_n h + \tfrac{1}{2!}\mathbf{A}h^2 + \tfrac{1}{3!}\mathbf{A}^2 h^3 + \cdots\right]\mathbf{B}$$

(3.485)

Differentiating (3.485) w.r.t. h then yields

$$\frac{d}{dh}\Psi(h) = \left[\mathbf{I}_n + \mathbf{A}h + \tfrac{1}{2!}\mathbf{A}^2 h^2 + \cdots\right]\mathbf{B} = \Phi(h)\mathbf{B}. \qquad (3.486)$$

3.4 Discrete LTI Plant Models

Applying these checks to Example 3.9, the discrete plant matrix is derived as

$$\Phi(h) = e^{-\sigma h} \begin{bmatrix} \cos(\omega h) & \sin(\omega h) \\ -\sin(\omega h) & \cos(\omega h) \end{bmatrix}, \qquad (3.487)$$

Then

$$\begin{aligned}
\mathbf{A}\Phi(h) &= e^{-\sigma h} \begin{bmatrix} -\sigma & \omega \\ -\omega & -\sigma \end{bmatrix} \begin{bmatrix} \cos(\omega h) & \sin(\omega h) \\ -\sin(\omega h) & \cos(\omega h) \end{bmatrix} \\
&= e^{-\sigma h} \begin{bmatrix} -\sigma \cos(\omega h) - \omega \sin(\omega h) & -\sigma \sin(\omega h) + \omega \cos(\omega h) \\ -\omega \cos(\omega h) + \sigma \sin(\omega h) & -\omega \sin(\omega h) - \sigma \cos(\omega h) \end{bmatrix}
\end{aligned} \qquad (3.488)$$

and

$$\frac{d}{dh}\Phi(h) = e^{-\sigma h} \begin{bmatrix} -\omega \sin(\omega h) - \sigma \cos(\omega h) & +\omega \cos(\omega h) - \sigma \sin(\omega h) \\ -\omega \cos(\omega h) + \sigma \sin(\omega h) & -\omega \sin(\omega h) - \sigma \cos(\omega h) \end{bmatrix}, \qquad (3.489)$$

which agrees with the RHS of (3.488).

The discrete input matrix is derived as

$$\Psi(h) = \frac{b}{\sigma^2 + \omega^2} \begin{bmatrix} \omega\left[1 - e^{-\sigma h} \cos(\omega h)\right] - \sigma e^{-\sigma h} \sin(\omega h) \\ \omega e^{-\sigma h} \sin(\omega h) + \sigma\left[1 - e^{-\sigma h} \cos(\omega h)\right] \end{bmatrix}. \qquad (3.490)$$

Then

$$\begin{aligned}
\frac{d}{dh}\Psi(h) &= \frac{be^{-\sigma h}}{\sigma^2 + \omega^2} \begin{bmatrix} \omega\sigma \cos(\omega h) + \omega^2 \sin(\omega h) + \sigma^2 \sin(\omega h) - \omega\sigma \cos(\omega h) \\ -\omega\sigma \sin(\omega h) + \omega^2 \cos(\omega h) + \sigma^2 \cos(\omega h) + \omega\sigma \sin(\omega h) \end{bmatrix} \\
&= be^{-\sigma h} \begin{bmatrix} \sin(\omega h) \\ \cos(\omega h) \end{bmatrix}
\end{aligned} \qquad (3.491)$$

and

$$\Phi(h)\mathbf{b} = e^{-\sigma h} \begin{bmatrix} \cos(\omega h) & \sin(\omega h) \\ -\sin(\omega h) & \cos(\omega h) \end{bmatrix} \begin{bmatrix} 0 \\ b \end{bmatrix} = be^{-\sigma h} \begin{bmatrix} \sin(\omega h) \\ \cos(\omega h) \end{bmatrix} \qquad (3.492)$$

which agrees with the RHS of (3.491).

3.4.3 Plant z-Transfer Function Model

3.4.3.1 Introduction

It will be recalled from Sect. 3.2.9 that Laplace transfer function models can be derived from continuous state-space models. Similarly, it is possible to derive z-transfer function models from the discrete state-space models, and this section provides the necessary theory. The z-transform is related to linear difference equations with constant coefficients as the Laplace transform is related to linear differential equations with constant coefficients.

The less common notation of $Q(z)$ for the plant z-transfer function is used rather than $G(z)$ since the Laplace transfer function of the plant is denoted by $G(s)$.

3.4.3.2 Sampling Process and Definition of the z-Transform

It will be recalled that the Laplace transform of a variable, $y(t)$, is defined as

$$\mathcal{L}\{y(t)\} \triangleq \int_0^\infty y(t)e^{-st}\,dt = Y(s). \tag{3.493}$$

An equivalent expression for the z-transform will now be derived. Figure 3.23 shows a continuous variable, $x(t)$, its sampled version and a corresponding train of infinite impulses representing the samples, introduced for mathematical convenience.

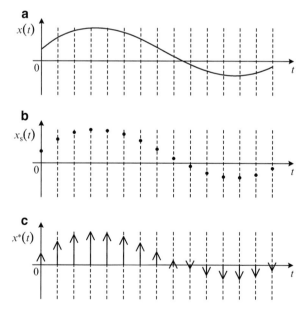

Fig. 3.23 A variable and representations of its sampled values. (**a**) Original continuous variables. (**b**) Sequence of sampled values. (**c**) Sequence of infinite impulses

3.4 Discrete LTI Plant Models

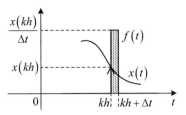

Fig. 3.24 Definition of $k$impulse representing the th sampled value of a variable, $x(t)$

In Fig. 3.23c, the infinite impulses that are represented by arrows with lengths equal to the magnitudes of the samples in Fig. 3.23b have *strengths* equal to the values of these samples, as defined in Fig. 3.24.

The shaded rectangular impulse, $f(t)$, has a strength equal to

$$\int_0^\infty f(t)\,dt = x(kh). \tag{3.494}$$

As $\Delta t \to 0$, the pulse becomes of infinite magnitude and,

$$\lim_{\Delta t \to 0} f(t) = x(kh)\delta(t - kh) \tag{3.495}$$

where $\delta(t - kh)$ is a Dirac delta impulse function, which has a strength of unity and occurs at $t = kh$.

Hence

$$\int_0^\infty x(t)\delta(t - kh)\,dt = x(kh). \tag{3.496}$$

The integrand of equation (3.496) is the kth impulse of $x^*(t)$. It follows that $x^*(t)$ may be expressed as

$$x^*(t) = \sum_{k=0}^\infty [x(t)\,\delta(t - kh)]. \tag{3.497}$$

The Laplace transform of $x^*(t)$ is therefore

$$X^*(s) = \int_0^\infty x^*(t)\,e^{-st}\,dt = \int_0^\infty \left(\sum_{k=0}^\infty (x(t)\,\delta(t - kh))\right) e^{-st}\,dt$$

$$= \sum_{k=0}^\infty \left(\int_0^\infty (x(t)\,\delta(t - kh))\,e^{-st}\,dt\right) = \sum_{k=0}^\infty x(kh)\,e^{-skh} = \sum_{k=0}^\infty x(kh)\left(e^{-sh}\right)^k.$$

$$\tag{3.498}$$

Since $s = \sigma + j\omega$ is a complex variable, then so is e^{-sh} since

$$e^{sh} = e^{-(\sigma + j\omega)h} = e^{-\sigma h} e^{-j\omega h} = e^{-\sigma h} [\cos(\omega h) + j \sin(\omega h)] \quad (3.499)$$

This enables a considerable simplification to be made by introducing a new variable,

$$z = e^{sh}. \quad (3.500)$$

Making this substitution in equation (3.498) then yields

$$X(z) = \sum_{k=0}^{\infty} x(kh) z^{-k}. \quad (3.501)$$

The asterisk has been removed as $X^*(s)$ and $X(z)$ are different functions of their arguments although $X^*(s) = X(z)$. In the functional notation, $X^*(s) = X^*(z)$ would imply $s = z$, which is incorrect. It is usual to write (3.501) as

$$X(z) = \sum_{k=0}^{\infty} x(k) z^{-k}, \quad (3.502)$$

understanding that the samples are separated by intervals of h seconds. Although (3.502) has been derived using the Laplace transform, it is often quoted as the definition of the z-transform of a sequence of sampled values of $x(t)$. The notation used for taking z-transforms is similar to that used for Laplace transforms. Thus, (3.502) is written as

$$X(z) = \mathcal{Z}\{x(k)\}. \quad (3.503)$$

3.4.3.3 Time Shifting Property

The time-shifting property of the z-transform is the key to deriving z-transfer function models of LTI plants starting with difference equation models in the discrete domain. There are two cases to consider. The first is the z-transform of $x(k)$ delayed by q iteration steps, with $q > 0$. Thus

$$\mathcal{Z}\{x(k-q)\} = \sum_{k=0}^{\infty} x(k-q) z^{-k} = \sum_{k=0}^{\infty} x(k-q) z^{-k} z^q z^{-q}$$

$$= z^{-q} \sum_{k=0}^{\infty} x(k-q) z^{-(k-q)} = z^{-q} \sum_{k-q=-q}^{\infty} x(k-q) z^{-(k-q)}$$

$$(3.504)$$

3.4 Discrete LTI Plant Models

This may be split into two parts, one of which includes $X(z)$, as follows:

$$\mathcal{Z}\{x(k-q)\} = z^{-q} \sum_{k-q=-q}^{-1} x(k-q) z^{-(k-q)} + z^{-q} \sum_{k-q=0}^{\infty} x(k-q) z^{-(k-q)}$$

$$= z^{-q} \sum_{k-q=-q}^{-1} x(k-q) z^{-(k-q)} + z^{-q} X(z).$$

(3.505)

If $x(j) = 0$ for $j < 0$, then the first part vanishes since $j = k - q < 0$ and (3.505) reduces to

$$\mathcal{Z}\{x(k-q)\} = z^{-q} X(z).$$

(3.506)

The second case is the z-transform of $x(k)$ projected q iteration steps into the future, with $q > 0$. This result is obtained by reversing the sign in front of q (as opposed to changing the sign of q) in (3.504). Thus

$$\mathcal{Z}\{x(k+q)\} = \sum_{k+q=q}^{\infty} x(k+q) z^{-(k+q)} z^q.$$

(3.507)

To again obtain a term involving $X(z)$, it is necessary to add terms that take $k + q$ from zero to $q - 1$ and then subtract them to make the equation correct. Thus,

$$\mathcal{Z}\{x(k+q)\} = z^q \sum_{k+q=0}^{\infty} x(k+q) z^{-(k+q)} - z^q \sum_{k+q=0}^{q-1} x(k+q) z^{-(k+q)}$$

$$= z^q \left[X(z) - x(0) - z^{-1} x(1) - \cdots - z^{q-1} x(q-1) \right]$$

(3.508)

In this case, the simpler relationship equivalent to (3.506) is valid only if $x(i) = 0$, $i = 1, 2, \ldots, q - 1$.

3.4.3.4 Transfer Function from Discrete State-Space Model

The use of the time-shifting property to obtain a z-transfer function from a discrete LTI state-space plant model will now be demonstrated. This is similar to the determination of the Laplace transfer function from the continuous state-space model given in Sect. 3.2.9. The z-transform is valid for vector variables as well as scalar variables. So taking z-transforms of (3.396) with zero initial conditions, i.e. $\mathbf{x}(0) = \mathbf{0}$ with $q = 1$ in (3.508), yields

$$\mathcal{Z}\{\mathbf{x}(k+1)\} = \mathcal{Z}\{\mathbf{\Phi}(h)\mathbf{x}(k) + \mathbf{\Psi}(h)\mathbf{u}(k)\} \tag{3.509}$$

and

$$\mathcal{Z}\{\mathbf{y}(k)\} = \mathcal{Z}\{\mathbf{Cx}(k)\} \tag{3.510}$$

From (3.509) and (3.510),

$$z\mathbf{X}(z) = \mathbf{\Phi}(h)\mathbf{X}(z) + \mathbf{\Psi}(h)\mathbf{U}(z) \Rightarrow [z\mathbf{I}_n - \mathbf{\Phi}(h)]\mathbf{X}(z) = \mathbf{\Psi}(h)\mathbf{U}(z) \Rightarrow$$
$$\mathbf{X}(z) = [z\mathbf{I}_n - \mathbf{\Phi}(h)]^{-1}\mathbf{\Psi}(h)\mathbf{U}(z) \tag{3.511}$$

and

$$\mathbf{Y}(z) = \mathbf{CX}(z) \tag{3.512}$$

Eliminating $\mathbf{X}(z)$ between (3.511) and (3.512) gives

$$\mathbf{Y}(z) = \mathbf{Q}(z)\mathbf{U}(z), \tag{3.513}$$

where

$$\mathbf{Q}(z) = \mathbf{C}[z\mathbf{I}_n - \mathbf{\Phi}(h)]^{-1}\mathbf{\Psi}(h) \tag{3.514}$$

is the plant transfer function matrix. For SISO plants modelled by (3.398), the result is the scalar transfer function,

$$Q(z) = \mathbf{c}^T[z\mathbf{I}_n - \mathbf{\Phi}(h)]^{-1}\boldsymbol{\psi}(h). \tag{3.515}$$

The transfer function matrix of (3.514) can be expanded as

$$\mathbf{Q}(z) = \mathbf{C}\frac{\operatorname{adj}[z\mathbf{I}_n - \mathbf{\Phi}(h)]}{\det[z\mathbf{I}_n - \mathbf{\Phi}(h)]}\mathbf{\Psi}(h) = \frac{\mathbf{B}_{n-1}(h)z^{n-1} + \cdots + \mathbf{B}_1(h)z + \mathbf{B}_0(h)}{z^n + a_{n-1}(h)z^{n-1} + \cdots + a_1(h)z + a_0(h)}, \tag{3.516}$$

where the numerator coefficient matrices, $\mathbf{B}_i(h)$, $i = 0, 1, \ldots, n-1$, are of dimension $m \times r$. For a SISO plant, the numerator matrices are replaced by scalar coefficients, $b_i(h)$, $i = 0, 1, \ldots, n-1$. Note that $\mathbf{Q}(z)$ is used rather than $\mathbf{G}(z)$ because the plant transfer function matrix is $\mathbf{G}(s)$ in the continuous domain and $\mathbf{Q}(z)$ is not the same function of z that $\mathbf{G}(s)$ is of s, for the same plant.

The dependence of the z-transfer function coefficients on the iteration/sampling interval, h, means that the z-transfer function of a given LTI plant is not unique. It follows that the plant poles and zeros in the z-plane also depend upon h. The functional notation used for the coefficients is a reminder of this.

3.4 Discrete LTI Plant Models

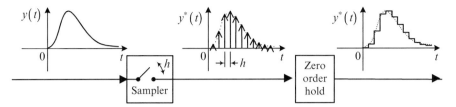

Fig. 3.25 Traditional model of a sample and hold circuit at the input of an A/D converter

3.4.3.5 Pulse Transfer Functions and z-Transfer Functions

The z-transfer function plant models derived from the discrete-state space models in Sect. 3.4.3.4 are ready for direct use in linear discrete control system designs. Tables are published widely, however, containing Laplace domain transfer functions and corresponding z-transfer functions that are *not* directly suitable for linear discrete control system design. These common tables of z-transforms can, however, be used with a traditional procedure concerning the *sample and hold* unit to derive z-transfer functions identical to those obtained from the state-space models. This section is provided to highlight the difference between the two sets of different z-transfer functions by explaining the traditional procedure. It is actually unnecessary to undertake this procedure if a table containing the directly useful plant z-transfer functions is available. This is provided in Table 3 in the Tables section preceding the appendix and contains both sets of z-transfer functions for comparison.

First, the traditional modelling of the *sample and hold* unit will be described. Let $y(t)$ be the impulse response of the plant to be modelled. Then a sampled version, $y^*(t)$, is produced similar to that illustrated in Fig. 3.23c. The traditional symbol for a sampler producing this signal is shown in Fig. 3.25.

The switch implementing the sampler closes for an infinitesimal duration at the sampling instants. A zero-order hold circuit keeps the last value sampled by the switch and is updated with each new sample. It should be noted that an *r*th-order hold fits an *r*th-order polynomial to the last sample and the previous *r* samples and uses this to predict $y(t)$ between the last sample and the next sample. So the zero-order hold fits a zero-order polynomial to the last sample, which is simply a constant equal to the value of the last sample. The zero-order hold is the most common.

The train of infinite impulses produced by the switch in the mathematical model cannot be realised. It is introduced in this form only to aid the derivation of the traditional model of the sample and hold as a Laplace transfer function. So the signal, $y(t)$, and its sampled and held version, $y_s(t)$, exist in the physical system but $y^*(t)$ does not. The ideal zero-order hold block operates in the following way. It is effectively a pure integrator whose output is reset to zero an infinitesimal time before each impulse of $y^*(t)$ occurs. Thus, each impulse of $y^*(t)$ is integrated with zero initial conditions so that the sampled value of $y(t)$ corresponding to the impulse is produced at the integrator output in an infinitesimal time. This process continues so that the combined sampler and zero order hold produces $y_s(t)$.

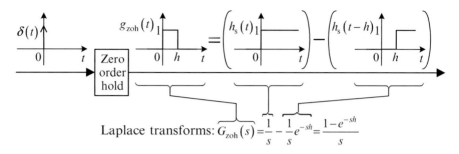

Fig. 3.26 Derivation of Laplace transfer function of zero-order hold

The zero-order hold unit may be modelled as a Laplace transfer function. The fundamental definition of the Laplace transfer function is used here, i.e. the Laplace transfer function of a linear dynamical system is the Laplace transform of its output when the input is a unit impulse function. Figure 3.26 shows the unit impulse response, $g_{zoh}(t)$, of the zero-order hold unit together with its Laplace transform, $G_{zoh}(s)$, which is therefore the required Laplace transfer function.

Thus, $g_{zoh}(t)$ is a square pulse with unity height and a duration of h seconds. This is expressed as the difference between the unit step function and another unit step function delayed by h seconds. The Laplace transform of the delayed step function follows from the shifting property, $\mathcal{L}\{f(t-\tau)\} = e^{-s\tau}\mathcal{L}\{f(t)\}$. Thus,

$$\frac{X_s(s)}{X^*(s)} = G_{zoh}(s) = \frac{1-e^{-sh}}{s} \quad (3.517)$$

It is this transfer function that is used together with the z-transforms from commonly available tables to obtain the plant model as the z-transfer function between the control variable from the computer and the measurement variable from the plant.

Let $G(s)$ be the Laplace transfer function of the plant. The corresponding z-transfer function is to be found that can be used for discrete control system design, including the sample and hold. This will be denoted by $Q(z)$ rather than the usual $G(z)$ since it is a different function of z than $G(s)$ is of s. Then, $G(s)$ is the inverse Laplace transform of the impulse response. On the same basis, the corresponding z-transfer function, $P(z)$, commonly found in tables, is the z-transform of the sampled impulse response. It is the z-transfer function between the sampled input, comprising a train of impulses equal in strength to the sampled input values, *without a sample and hold*, and the sampled output. For this reason, these z-transfer functions will be referred to as *pulse transfer functions*. Figure 3.27 shows a continuous control signal, $u(t)$, applied to a linear plant and its output $y_1(t)$. For comparison, it shows the input, $u^*(t)$, that would have to be applied by a controller if the z-transfer functions obtained from standard tables were to be used directly as plant models. The signal $u^*(t)$ is not of the piecewise constant form of the control

3.4 Discrete LTI Plant Models

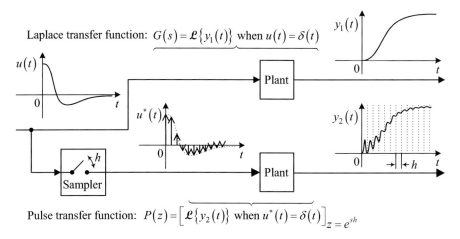

Fig. 3.27 Pulse transfer function corresponding to Laplace transfer function for the same plant

input of a discrete control system and it is not realisable. So a sample and hold is needed in this model to convert $u^*(t)$ to the required piecewise constant $u_s(t)$.

Figure 3.28 shows pictorially the process of determining $Q(z)$, given $P(z)$ from standard tables. To continue in the same vein as Fig. 3.27, the starting point is a fictitious continuous control variable, $u'(t)$, which when sampled yields the values of the piecewise control, $u(t)$, applied to the plant and these are represented by a sequence of impulses with strengths equal to these control values.

This step is necessary to be able to develop the method of using standard tables to obtain $Q(z)$. Then the zero-order hold (ZOH) unit is inserted to obtain the piecewise constant $u(t)$. The pure integrator in the ZOH transfer function is then transferred to the plant block to form the augmented plant with transfer function, $G(s)/s$. It is the augmented plant transfer function that is looked up in the standard tables to obtain the corresponding $P(z)$. The remaining part of the ZOH transfer function is easily transferred to the z-domain using (3.500), i.e. $z = e^{sh}$, to yield $(z-1)/z$. Finally, the required plant z-transfer function is

$$Q(z) = \frac{z-1}{z} P(z). \tag{3.518}$$

The above procedure is the traditional one for obtaining the plant z-transfer function model from the Laplace transfer function model for discrete domain design of linear controllers, but to save time, the reader may refer to Table 3 of the Tables section preceding the appendix, which covers commonly encountered transfer functions.

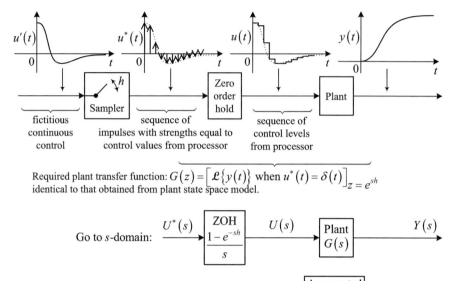

Fig. 3.28 Plant z-transfer function derivation for discrete control system synthesis

3.4.4 Change of Sampling Period for z-Transfer Function Models

3.4.4.1 Introduction

With reference to Chap. 2, Sect. 2.3.4, z-transfer function models generated by recursive parameter estimation may have longer iteration periods than those required for satisfactory discrete control. This section therefore provides means of converting a given z-transfer function model, $Q_1(z)$, for sampling period, h_1, to a corresponding transfer function model, $Q_2(z)$, for a different sampling period, h_2.

To begin, suppose the Laplace transfer function model, $G(s)$, of the plant is available. Then the corresponding continuous state-space model in the controller canonical form follows directly. This can be converted to a discrete state-space model with any chosen sampling period (Sect. 3.4.1) and then to a corresponding z-transfer function model (Sect. 3.4.3.4). This process is shown in Table 3.1, steps 2, 3 and 4, and leads to $Q_2(z)$ for iteration period, h_2. Similarly, another z-transfer

3.4 Discrete LTI Plant Models

Table 3.1 Hypothetical steps for conversion to a different sampling period

Starting point:	z-Transfer function model for sampling period, h_1	$\frac{Y(z)}{U(z)} = Q_1(z)$
	Connecting diagram:	Discrete state-variable block diagram in the controller canonical form
Step 1:	Form discrete state-space model for sampling period, h_1:	$\mathbf{x}(k+1) = \boldsymbol{\Phi}(h_1)\mathbf{x}(k) + \boldsymbol{\psi}(h_1)u(k) \quad y(k) = \mathbf{c}^T\mathbf{x}(k)$
	Connecting equations:	$\boldsymbol{\Phi}(h_1) = e^{\mathbf{A}h_1}, \ \boldsymbol{\psi}(h_1) = \int_0^{h_1} e^{\mathbf{A}\tau}\mathbf{b}d\tau$
Step 2:	Form continuous state-space model:	$\dot{\mathbf{x}} = \mathbf{A}\mathbf{x} + \mathbf{b}u \quad y = \mathbf{c}^T\mathbf{x}$
	Connecting equations:	$\boldsymbol{\Phi}(h_2) = e^{\mathbf{A}h_2}, \ \boldsymbol{\psi}(h_2) = \int_0^{h_2} e^{\mathbf{A}\tau}\mathbf{b}d\tau$
Step 3:	Form discrete state-space model for sampling period, h_2:	$\mathbf{x}(k+1) = \boldsymbol{\Phi}(h_2)\mathbf{x}(k) + \boldsymbol{\psi}(h_2)u(k) \quad y(k) = \mathbf{c}^T\mathbf{x}(k)$
Step 4:	Form z-transfer function model for sampling period, h_2	$\frac{Y(z)}{U(z)} = Q_2(z)$

function model, $Q_1(z)$, for iteration period, h_1, could be obtained by starting at step 2 in Table 3.1 and working *backwards* through step 1 to the starting point. In theory, these last steps could be taken in the correct order starting with $Q_1(z)$ and deriving the continuous state-space model, from which $Q_2(z)$ could be derived.

All the steps may be carried out in the order of Table 3.1 with relative ease except step 2. This would require solving the connecting equations between steps 1 and 2 to obtain $\mathbf{A}$ and $\mathbf{b}$ in terms of $\boldsymbol{\Phi}(h_1)$ and $\boldsymbol{\psi}(h_1)$. Unfortunately this problem is not generally tractable but instead, the conversion between the discrete models can be carried out via an *approximate* continuous linear plant model whose transfer function will be denoted by $F(s)$, instead of the continuous state space model.

3.4.4.2 A Variation of Tustin's Transformation

Let $Q(z)$ be the z-transfer function model of the plant for sampling period, h. The aim is then to find a functional relationship and its inverse, i.e.,

$$z = f(s, h) \quad \text{and} \quad s = f^{-1}(z, h), \tag{3.519}$$

where s is the Laplace transform variable, such that $Q(f(s, h)) = F(s, h)$ is the transfer function of a system having the same order as $Q(z)$ and that approximates the true Laplace transfer function, $G(s)$, of the plant. Since the transformation,

$$z = e^{sh}, \tag{3.520}$$

is the basis of forming the z-transfer function starting with the Laplace transform of a sampled signal [Sect. 3.4.3.2), one might be tempted to use this for (3.519). Unfortunately, however, $Q(e^{sh})$ is of infinite order since this transfer function can only be expressed as the ratio of two polynomials in s by expanding e^{sh} as an infinite power series in sh. To circumvent this problem, however, an approximation to e^{sh} may be used that yields a transfer function, $F(s)$, that is of the same order as $Q(z)$. An approximation satisfying this requirement is

$$z = e^{sh} = \frac{e^{s\lambda h}}{e^{-s(1-\lambda)h}} = \frac{1 + s\lambda h + \frac{1}{2!}(s\lambda h)^2 + \cdots}{1 - s(1-\lambda)h + \frac{1}{2!}(-s(1-\lambda)h)^2 + \cdots} \cong \frac{1 + s\lambda h}{1 - s(1-\lambda)h}. \tag{3.521}$$

where λ has been introduced as a tuning parameter. Then, if

$$z = f(s, h) = \frac{1 + s\lambda h}{1 + s(\lambda - 1)h}, \tag{3.522}$$

the transfer function, $F(s) = Q(f(s))$, is an approximate Laplace transfer function model of the plant in the continuous domain.

The inverse of transformation (3.522) is

$$s = \frac{1}{(1-\lambda)h} \cdot \frac{z-1}{z + \frac{\lambda}{1-\lambda}}. \tag{3.523}$$

Let

$$Q_1(z) = \frac{\sum_{j=0}^{n-1} b_{1j} z^j}{z^n + \sum_{i=0}^{n-1} a_{1i} z^{-i}}. \tag{3.524}$$

Then the first step is to convert to the s-domain using (3.522) with $h = h_1$. Thus,

$$z = \frac{1 + s\lambda h_1}{1 + s(\lambda - 1)h_1}. \tag{3.525}$$

The approximate continuous transfer function model, $F_1(s)$, would then be obtained by substituting for z in (3.524) using (3.525), but since this is not needed for later use, direct conversion back to the z-domain may be carried out by substituting for s in (3.525) using (3.523) with $h = h_2$. Then the conversion relationship is

3.4 Discrete LTI Plant Models

$$z := \frac{1 + \dfrac{\lambda h_1}{(1-\lambda) h_2} \cdot \dfrac{z-1}{z + \dfrac{\lambda}{1-\lambda}}}{1 - \dfrac{h_1}{h_2} \cdot \dfrac{z-1}{z + \dfrac{\lambda}{1-\lambda}}} = \frac{1 + \alpha \beta \dfrac{z-1}{z+\beta}}{1 - \alpha \dfrac{z-1}{z+\beta}} = \frac{z + \beta + \alpha\beta(z-1)}{z + \beta - \alpha(z-1)}$$

$$= \frac{(1+\alpha\beta)z + \beta(1-\alpha)}{(1-\alpha)z + \alpha + \beta} = \frac{z + \beta \left(\dfrac{1-\alpha}{1+\alpha\beta} \right)}{\left(\dfrac{1-\alpha}{1+\alpha\beta} \right) z + \dfrac{\alpha+\beta}{1+\alpha\beta}} = \frac{z+q}{pz+r}, \tag{3.526}$$

where $\alpha = \frac{h_1}{h_2}$, $\beta = \frac{\lambda}{1-\lambda}$, $p = \frac{1-\alpha}{1+\alpha\beta}$, $q = \beta p$ and $r = \frac{\alpha+\beta}{1+\alpha\beta}$.

Note that the symbol, :=, indicates replacement in contrast to equality. Replacing z in (3.524) using (3.526) then yields

$$P_2(z) = \frac{\sum_{j=0}^{n-1} b_{1j} \left(\dfrac{z+q}{pz+r} \right)^j}{\left(\dfrac{z+q}{pz+r} \right)^n + \sum_{i=0}^{n-1} a_{1i} \left(\dfrac{z+q}{pz+r} \right)^i} = \frac{\sum_{j=0}^{n-1} b_{1j} (z+q)^j (pz+r)^{n-j}}{(z+q)^n + \sum_{i=0}^{n-1} a_{1i} (z+q)^i (pz+r)^{n-i}}. \tag{3.527}$$

While the order of $P_2(z)$ is the same as that of $Q_1(z)$, i.e. n, the number of zeros of $P_2(z)$ is also n, while the number of zeros of $G_1(z)$ is $n-1$ or less. This is a consequence of approximation (3.521) being a transfer function with zero relative degree. To obtain a model with the same relative degree as $Q_1(z)$ but with the poles of $P_2(z)$ to preserve its dynamic character, the numerator factor, $(pz+r)^{n-j}$, of (3.527) is replaced by its DC gain (by setting $z = 1$) to obtain

$$Q_2(z) = \frac{\sum_{j=0}^{n-1} b_{1j} (z+q^j)(p+r)^{n-j}}{(z+q)^n + \sum_{i=0}^{n-1} a_{1i}(z+q)^i (pz+r)^{n-i}}. \tag{3.528}$$

If $\lambda = 0.5$, (3.522) becomes the bilinear transformation of Tustin's method, which maps all points in the left half of the complex s-plane to the unit disc in the z-plane. Since this is also true of the transformation, $z = e^{sh}$, each pole of $G(s)$ has a corresponding pole of $F_1(s)$ such that both poles lie within the left half of the s-plane or outside it. It also follows that each pole of $Q_1(z)$ is associated with a pole of $Q_2(z)$ through transformation (3.526) such that both these poles lie outside the unit disc in the z-plane or outside it. Since this is also true of the pair of transfer functions $Q_1(z)$ and $Q_2(z)$ connected by the steps of the exact transformation of Table 3.1, the new transfer function (3.528) would first be simulated together with transfer function

(3.524) first with $\lambda = 0.5$ with a common step control input, to yield two outputs, $y_1(t)$ and $y_2(t)$. Then λ would be adjusted to obtain a better fit of the outputs at the sample points.

Another simple adjustment is provided by setting the actual sampling period to $\tilde{h}_2$ in the new discrete model and varying this to change the timescale of $y_2(t)$. On the other hand, if h_2 is specified, the converse scheme may be employed of replacing h_2 by $\tilde{h}_2$ in (3.526) and varying this instead, starting with $\tilde{h}_2 = h_2$.

Example 3.14 Reduction of sampling period for a throttle valve model

The second-order discrete transfer function model,

$$\frac{Y(z)}{U(z)} = \frac{b_1 z^{-1} + b_2 z^{-2}}{1 + a_1 z^{-1} + a_2 z^{-2}} \qquad (3.529)$$

for a throttle valve of an internal combustion engine has been obtained by recursive parameter estimation with a sampling period of $h_1 = 0.1$ [s], where $b_1 = 0.2271$, $b_2 = 0.0689$, $a_1 = -0.8531$ and $a_2 = 0.0189$. For control system implementation, since a control-loop settling time of $T_s = 0.1$ [s] is required, a shorter sampling period of $h_2 = 0.02$ [s] is needed. With $\lambda = 0.5$, the conversion equation (3.526) then becomes

$$z := \frac{6z - 4}{-4z + 6} \Rightarrow z^{-1} := -\frac{4z - 6}{6z - 4} \qquad (3.530)$$

The original plant model is shown in Fig. 3.29a and the converted plant model in Fig. 3.29b, in block diagram form suitable for simulation in SIMULINK® and ultimately implementation using dSPACE®.

It is straightforward to confirm that Fig. 3.29a has transfer function (3.529) by means of Mason's formula (Appendix A4). Figure 3.30 shows the responses of the original discrete model, $y_1(t)$: the continuous model, $y_c(t)$, obtained by replacing z in (3.529) by w of (3.525); and the required discrete model, $y_1(t)$.

Despite the approximation of (3.521), the errors between $y_2(t)$, $y_1(t)$ and $y_c(t)$ at the sample points are invisible even on the magnified inset of Fig. 3.30, indicating that the new discrete model is satisfactory.

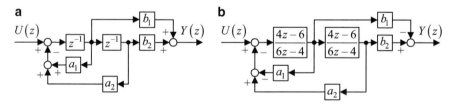

Fig. 3.29 Second-order discrete models for a throttle valve in block diagram form. (**a**) Original model with $h = 0.1$ [s]. (**b**) Model converted to $h = 0.02$ [s]

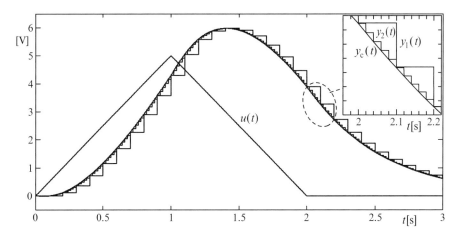

Fig. 3.30 Simulations of throttle valve models

3.4.5 Controllability Analysis of Discrete LTI Plant Models

3.4.5.1 SISO Plants

The starting point is the discrete state-space LTI SISO plant model (3.398), i.e.,

$$\mathbf{x}(k+1) = \mathbf{\Phi}(h)\mathbf{x}(k) + \mathbf{\psi}(h)u(k) \tag{3.531}$$

$$y(k) = \mathbf{c}^{\mathrm{T}}\mathbf{x}(k). \tag{3.532}$$

This plant model is controllable if it is possible to change the state from an arbitrary initial value, $\mathbf{x}(0)$, to a different arbitrary value in a finite number of iterations with a sequence of finite control values. To determine a condition for controllability, let a sequence of q control values, $u(0), u(1), \ldots u(q-1)$, be applied and the possibility of reaching an arbitrary state be investigated. Thus, with $\mathbf{\Phi}(h)$ and $\mathbf{\psi}(h)$ being abbreviated, respectively, to $\mathbf{\Phi}$ and $\mathbf{\psi}$,

$$\mathbf{x}(1) = \mathbf{\Phi}\mathbf{x}(0) + \mathbf{\psi}u(0)$$

$$\mathbf{x}(2) = \mathbf{\Phi}\mathbf{x}(1) + \mathbf{\psi}u(1) = \mathbf{\Phi}\left[\mathbf{\Phi}\mathbf{x}(0) + \mathbf{\psi}u(0)\right] + \mathbf{\psi}u(1)$$

$$= \mathbf{\Phi}^2\mathbf{x}(0) + \mathbf{\Phi}\mathbf{\psi}u(0) + \mathbf{\psi}u(1)$$

$$\mathbf{x}(3) = \mathbf{\Phi}\mathbf{x}(2) + \mathbf{\psi}u(2) = \mathbf{\Phi}\left[\mathbf{\Phi}^2\mathbf{x}(0) + \mathbf{\Phi}\mathbf{\psi}u(0) + \mathbf{\psi}u(1)\right] + \mathbf{\psi}u(2)$$

$$= \mathbf{\Phi}^3\mathbf{x}(0) + \mathbf{\Phi}^2\mathbf{\psi}u(0) + \mathbf{\Phi}\mathbf{\psi}u(1) + \mathbf{\psi}u(2)$$

$$\vdots$$

$$\mathbf{x}(q) = \mathbf{\Phi}^q\mathbf{x}(0) + \mathbf{\Phi}^{q-1}\mathbf{\psi}u(0) + \mathbf{\Phi}^{q-2}\mathbf{\psi}u(1) + \ldots + \mathbf{\Phi}\mathbf{\psi}u(q-2) + \mathbf{\psi}u(q-1). \tag{3.533}$$

The last equation may be written as

$$\mathbf{x}(q) = \mathbf{\Phi}^q \mathbf{x}(0) + \left[\mathbf{\psi} \mid \mathbf{\Phi}\mathbf{\psi} \mid \ldots \mid \mathbf{\Phi}^{q-2}\mathbf{\psi} \mid \mathbf{\Phi}^{q-1}\mathbf{\psi} \right] \begin{bmatrix} u(q-1) \\ u(q-2) \\ \vdots \\ u(1) \\ u(0) \end{bmatrix}. \qquad (3.534)$$

This is a set of n linear simultaneous equations to solve for the q control values. If $q < n$, it is overdetermined and no solution exists that simultaneously satisfies every equation. If $q > n$, it is underdetermined, and there are an infinite number of possible solutions. The minimum number of control values in the sequence for which there could be a unique solution is therefore $q = n$. The square matrix

$$\mathbf{M}_c(\mathbf{\Phi}, \mathbf{\psi}) = \left[\mathbf{\psi} \mid \mathbf{\Phi}\mathbf{\psi} \mid \ldots \mid \mathbf{\Phi}^{n-2}\mathbf{\psi} \mid \mathbf{\Phi}^{n-1}\mathbf{\psi} \right], \qquad (3.535)$$

therefore has to be nonsingular for the plant to be controllable and is called the discrete *controllability matrix*. The controllability condition is

$$\det[\mathbf{M}_c(\mathbf{\Phi}, \mathbf{\psi})] \neq 0. \qquad (3.536)$$

This corresponds to the controllability matrix, $\mathbf{M}_c(\mathbf{A}, \mathbf{b})$, derived in Sect. 3.2.7 for the continuous state-space plant model. For a controllable plant model, (3.535) and (3.534) with $q = n$ yield

$$\begin{bmatrix} u(n-1) \\ u(n-2) \\ \vdots \\ u(1) \\ u(0) \end{bmatrix} = \mathbf{M}_c^{-1}(\mathbf{\Phi}, \mathbf{\psi}) [\mathbf{x}_n - \mathbf{\Phi}^q \mathbf{x}_0]. \qquad (3.537)$$

As will be seen in Chap. 6, a discrete linear state feedback control law can be formed using (3.537) having a step response that settles *precisely* in n iterations, which is called a dead-beat response.

3.4.5.2 Multivariable Plants

In this case, the plant model to consider is (3.396), i.e.,

$$\begin{aligned} \mathbf{x}(k+1) &= \mathbf{\Phi}(h)\mathbf{x}(k) + \mathbf{\Psi}(h)\mathbf{u}(k) \\ \mathbf{y}(k) &= \mathbf{C}\mathbf{x}(k). \end{aligned} \qquad (3.538)$$

3.4 Discrete LTI Plant Models

The plant model is controllable if a sequence of unbounded control vector values, $\mathbf{u}(0), \mathbf{u}(1), \ldots, \mathbf{u}(q)$, can be applied to change the plant state from an arbitrary initial value, $\mathbf{x}(0)$, to an arbitrary final value, $\mathbf{x}(q)$. Abbreviating $\mathbf{\Phi}(h)$ and $\mathbf{\Psi}(h)$, respectively, to $\mathbf{\Phi}$ and $\mathbf{\Psi}$, a sequence of states similar to that of (3.533) may be produced leading to

$$\mathbf{x}(q) = \mathbf{\Phi}^q \mathbf{x}(0) + \mathbf{\Phi}^{q-1}\mathbf{\Psi}\mathbf{u}(0) + \mathbf{\Phi}^{q-2}\mathbf{\Psi}\mathbf{u}(1) + \cdots + \mathbf{\Phi}\mathbf{\Psi}\mathbf{u}(q-2) + \mathbf{\Psi}\mathbf{u}(q-1), \tag{3.539}$$

which may be written as

$$\mathbf{x}(q) = \mathbf{\Phi}^q \mathbf{x}(0) + \left[\mathbf{\Psi} \mid \mathbf{\Phi}\mathbf{\Psi} \mid \ldots \mid \mathbf{\Phi}^{q-2}\mathbf{\Psi} \mid \mathbf{\Phi}^{q-1}\mathbf{\Psi} \right] \begin{bmatrix} \mathbf{u}(q-1) \\ \mathbf{u}(q-2) \\ \vdots \\ \mathbf{u}(1) \\ \mathbf{u}(0) \end{bmatrix}. \tag{3.540}$$

This is a set of n simultaneous equations to be solved for qr control vector components. If $qr > n$, they are underdetermined and there could be an infinite number of solutions. If $qr < n$, they are overdetermined and no solution exists. Since q is an integer, it is only possible to obtain a completely determined set of equations if n/r is an integer, thereby permitting $qr = n$. If not, then the minimum possible number of iterations is $q_m = \lceil n/r \rceil$, i.e. the smallest integer greater than n/r. A set of equations containing more than $q_m r$ control vector components, however, is permissible and it is usual to take $q = n$ iterations as in the SISO case of Sect. 3.4.5.1. Then the matrix,

$$\mathbf{M}_c(\mathbf{\Phi}, \mathbf{\Psi}) = \left[\mathbf{\Psi} \mid \mathbf{\Phi}\mathbf{\Psi} \mid \ldots \mid \mathbf{\Phi}^{n-2}\mathbf{\Psi} \mid \mathbf{\Phi}^{n-1}\mathbf{\Psi} \right], \tag{3.541}$$

in (3.540) is the discrete *controllability matrix*. It has dimension, $n \times nr$, and therefore the condition for controllability is

$$\text{rank } [\mathbf{M}_c(\mathbf{\Phi}, \mathbf{\Psi})] \geq n. \tag{3.542}$$

This corresponds to the controllability matrix, $\mathbf{M}_c(\mathbf{A}, \mathbf{B})$, derived in Sect. 3.2.7 for the continuous state-space plant model.

For the minimal number of iterations, the solution for the sequence of control vectors may be obtained as follows. First $\mathbf{M}_c(\mathbf{\Phi}, \mathbf{\Psi})$ is abbreviated to $\mathbf{M}_c$. Then premultiplying both sides of (3.540) by $\mathbf{M}_c^T$ yields

$$\mathbf{M}_c^T\mathbf{x}(q_m) = \mathbf{M}_c^T\mathbf{\Phi}^{q_m}\mathbf{x}_0 + \overbrace{\mathbf{M}_c^T}^{rq_m \times n}\overbrace{\mathbf{M}_c}^{n \times rq_m} \begin{bmatrix} \mathbf{u}(q_m-1) \\ \mathbf{u}(q_m-2) \\ \vdots \\ \mathbf{u}(1) \\ \mathbf{u}(0) \end{bmatrix} \Rightarrow$$

$$\begin{bmatrix} \mathbf{u}(q_m-1) \\ \mathbf{u}(q_m-2) \\ \vdots \\ \mathbf{u}(1) \\ \mathbf{u}(0) \end{bmatrix} = \left[\mathbf{M}_c^T\mathbf{M}_c\right]^{-1}\mathbf{M}_c^T\left[\mathbf{x}_{q_m} - \mathbf{\Phi}^{q_m}\mathbf{x}_0\right]$$

(3.543)

Note that the matrix, $\left[\mathbf{M}_c^T\mathbf{M}_c\right]^{-1}\mathbf{M}_c^T$, is the left pseudo inverse of the non-square matrix, $\mathbf{M}_c$, of dimension, $n \times rq_m$. As for the SISO case, a discrete linear state feedback control law can be formed using (3.543). This, however, settles precisely in q_m iterations, where $q_m < n$, which results from there being more 'degree of control' of a multivariable plant than an SISO plant of the same order, thanks to more than one control variable working together to perform the task.

3.4.6 Analysis of Discrete LTI Plant Models

3.4.6.1 SISO Plants

The starting point is the plant model of (3.531) and (3.532), which is

$$\mathbf{x}(k+1) = \mathbf{\Phi}(h)\mathbf{x}(k) + \mathbf{\psi}(h)u(k) \tag{3.544}$$

$$y(k) = \mathbf{c}^T\mathbf{x}(k). \tag{3.545}$$

The state, $\mathbf{x}(k)$, contains all the information about the dynamic behaviour of the plant at time, t_k, but $y(k)$ cannot since it is only a scalar quantity. Clearly more information can be gained by taking more output samples at different times. The question then arises of whether *all* the information can be gained about the dynamical behaviour of the plant in this way, aided by the mathematical model of the plant. This is the property of observability. In the discrete domain, a mathematical condition for observability can be derived by posing the question of whether or not the initial plant state, $\mathbf{x}_0$, assumed to be initially unknown, can be determined using only sets of input and output sequences, $u(0), u(1), \ldots u(q-1)$ and $y(0), y(1), \ldots, y(q-1)$. If this can be achieved then the state at any other time can be determined by means of the state difference equation (3.544). With $\mathbf{\Phi}(h)$ and $\mathbf{\psi}(h)$ abbreviated, respectively,

3.4 Discrete LTI Plant Models

to $\boldsymbol{\Phi}$ and $\boldsymbol{\psi}$, (3.545) and (3.544) yield

$$y(0) = \mathbf{c}^T\mathbf{x}(0)$$
$$y(1) = \mathbf{c}^T\mathbf{x}(1) = \mathbf{c}^T[\boldsymbol{\Phi}\mathbf{x}(0) + \boldsymbol{\psi}u(0)] = \mathbf{c}^T\boldsymbol{\Phi}\mathbf{x}(0) + \mathbf{c}^T\boldsymbol{\psi}u(0)$$
$$y(2) = \mathbf{c}^T\mathbf{x}(2) = \mathbf{c}^T[\boldsymbol{\Phi}\mathbf{x}(1) + \boldsymbol{\psi}u(1)] = \mathbf{c}^T[\boldsymbol{\Phi}[\boldsymbol{\Phi}\mathbf{x}(0) + \boldsymbol{\psi}u(0)] + \boldsymbol{\psi}u(1)]$$
$$= \mathbf{c}^T\boldsymbol{\Phi}^2\mathbf{x}(0) + \mathbf{c}^T[\boldsymbol{\Phi}\boldsymbol{\psi}u(0) + \boldsymbol{\psi}u(1)]$$
$$\vdots$$
$$y(q-1) = \mathbf{c}^T\boldsymbol{\Phi}^{q-1}\mathbf{x}(0) + \mathbf{c}^T\left[\boldsymbol{\Phi}^{q-2}\boldsymbol{\psi}u(0) + \cdots + \boldsymbol{\Phi}\boldsymbol{\psi}u(q-3)\right] + \boldsymbol{\psi}u(q-2)$$
(3.546)

which may be written as

$$\begin{bmatrix} y(0) \\ y(1) \\ y(2) \\ \vdots \\ y(q-1) \end{bmatrix} = \begin{bmatrix} \mathbf{c}^T \\ \hline \mathbf{c}^T\boldsymbol{\Phi} \\ \hline \mathbf{c}^T\boldsymbol{\Phi}^2 \\ \hline \vdots \\ \hline \mathbf{c}^T\boldsymbol{\Phi}^{q-1} \end{bmatrix} \mathbf{x}(0) + \begin{bmatrix} 0 \\ \hline \mathbf{c}^T\boldsymbol{\psi}u(0) \\ \hline \mathbf{c}^T\left[\boldsymbol{\Phi}\boldsymbol{\psi}u(0) + \boldsymbol{\psi}u(1)\right] \\ \hline \vdots \\ \hline \mathbf{c}^T\left[\boldsymbol{\Phi}^{q-2}\boldsymbol{\psi}u(0) + \ldots + \boldsymbol{\Phi}\boldsymbol{\psi}u(q-3) + \boldsymbol{\psi}u(q-2)\right] \end{bmatrix}.$$
(3.547)

This is a set of q linear equations in the n components of $\mathbf{x}_0$ to be found. If $q < n$, it is underdetermined in which case there is an infinite number of solutions. This, however, has a different implication than in Sect. 3.4.5 concerning controllability in which all the different solutions were valid control sequences that produced the required change of the state. In this case, there is only one valid solution, i.e. the state, $\mathbf{x}_0$, so an infinite number of solutions implies that $\mathbf{x}_0$ cannot be found. If $q > n$, the set of equations is overdetermined. Again the situation is different from that concerning controllability where this implies an insufficient choice of control values to achieve the required change of state. In this case, there are more input–output samples than necessary and $\mathbf{x}_0$ can possibly be calculated from any subset of n input–output samples taken from the complete set. It would be possible to obtain the solution with $q > n$ using all the input–output samples, and this would lead to an observability condition similar to that for multivariable plants to be derived in Sect. 3.4.6.2. A simpler condition, however, is obtained with the minimum number, $q = n$, of input–output samples. In this case, with reference to (3.547), the matrix,

$$\mathbf{M}_o\left(\mathbf{c}^T, \boldsymbol{\Phi}\right) = \begin{bmatrix} \mathbf{c}^T \\ \hline \mathbf{c}^T\boldsymbol{\Phi} \\ \hline \mathbf{c}^T\boldsymbol{\Phi}^2 \\ \hline \vdots \\ \hline \mathbf{c}^T\boldsymbol{\Phi}^{n-1} \end{bmatrix},$$
(3.548)

has to be nonsingular for the solution to exist. This is the discrete *observability matrix*. The condition for observability is therefore

$$\det \left[\mathbf{M}_o \left(\mathbf{c}^T, \mathbf{\Phi} \right) \right] \neq 0. \tag{3.549}$$

3.4.6.2 Multivariable Plants

The relevant plant model is now that of (3.538), which is

$$\begin{aligned} \mathbf{x}(k+1) &= \mathbf{\Phi}(h)\mathbf{x}(k) + \mathbf{\Psi}(h)\mathbf{u}(k) \\ \mathbf{y}(k) &= \mathbf{C}\mathbf{x}(k). \end{aligned} \tag{3.550}$$

A sequence of input–output pairs may be sampled similar to that of Sect. 3.4.6.1 to yield

$$\begin{bmatrix} \mathbf{y}(0) \\ \mathbf{y}(1) \\ \mathbf{y}(2) \\ \vdots \\ \mathbf{y}(q-1) \end{bmatrix} = \begin{bmatrix} \mathbf{C} \\ \mathbf{C}\mathbf{\Phi} \\ \mathbf{C}\mathbf{\Phi}^2 \\ \vdots \\ \mathbf{C}\mathbf{\Phi}^{q-1} \end{bmatrix} \mathbf{x}(0) + \begin{bmatrix} 0 \\ \mathbf{C}\mathbf{\Psi}\mathbf{u}(0) \\ \mathbf{C}\left[\mathbf{\Phi}\mathbf{\Psi}\mathbf{u}(0) + \mathbf{\Psi}u(1)\right] \\ \vdots \\ \mathbf{C}\left[\mathbf{\Phi}^{q-2}\mathbf{\Psi}\mathbf{u}(0) + \ldots + \mathbf{\Phi}\mathbf{\Psi}\mathbf{u}(q-3) + \mathbf{\Psi}\mathbf{u}(q-2)\right] \end{bmatrix}. \tag{3.551}$$

This is a set of mq linear equations in the n components of $\mathbf{x}_0$ to be found. If $mq < n$, then it is underdetermined, giving an infinite number of solutions, meaning that the number of input–output pairs is insufficient for $\mathbf{x}_0$ to be found. If $mq > n$, the set of equations is overdetermined meaning that more than one combination of n output vector components and the corresponding input terms may be used to determine $\mathbf{x}_0$. The set of equations would be completely determined if $mq = n$, but since q is an integer then so should be n/m. If not, then the minimum possible number of input–output samples is $q_{\min} = \lceil n/m \rceil$, i.e. $q_{\min}$ is the smallest integer satisfying $mq > n$. Since, however, taking in more than the minimum number of input–output samples is valid, it is usual to take $q = n$ as for the SISO case. Then the matrix

$$\mathbf{M}_o(\mathbf{C}, \mathbf{\Phi}) = \begin{bmatrix} \mathbf{C} \\ \mathbf{C}\mathbf{\Phi} \\ \mathbf{C}\mathbf{\Phi}^2 \\ \vdots \\ \mathbf{C}\mathbf{\Phi}^{n-1} \end{bmatrix}, \tag{3.552}$$

in (3.551) is the discrete *observability matrix*. It has dimension $nm \times n$, and therefore, the condition for observability is

$$\text{rank}\,[\mathbf{M}_o\,(\mathbf{C}, \mathbf{\Phi})] \geq n. \quad (3.553)$$

This corresponds to the observability matrix, $\mathbf{M}_o(\mathbf{C}, \mathbf{A})$, derived in Sect. 3.2.7 for the continuous state-space model.

For the minimal number of input–output samples, $q_{\min}$, the solution for $\mathbf{x}_0$ is obtained as follows. First, $\mathbf{M}_o(\mathbf{C}, \mathbf{\Phi})$ is abbreviated to $\mathbf{M}_o$. Then, premultiplying both sides of (3.551) by $\mathbf{M}_o^T$ yields

$$\mathbf{M}_o^T \begin{bmatrix} \mathbf{y}(0) \\ \mathbf{y}(1) \\ \mathbf{y}(2) \\ \vdots \\ \mathbf{y}(q_m-1) \end{bmatrix} = \underbrace{\mathbf{M}_o^T}_{n \times mq_m} \underbrace{\mathbf{M}_o}_{mq_m \times n} \mathbf{x}(0) + \mathbf{M}_o^T \begin{bmatrix} \mathbf{0} \\ \hline \mathbf{C}\mathbf{\Psi}\mathbf{u}(0) \\ \hline \mathbf{C}\left[\mathbf{\Phi}\mathbf{\Psi}\mathbf{u}(0)+\mathbf{\Psi}\mathbf{u}(1)\right] \\ \hline \vdots \\ \hline \mathbf{C}\left[\mathbf{\Phi}^{q-2}\mathbf{\Psi}\mathbf{u}(0)+\ldots+\mathbf{\Phi}\mathbf{\Psi}\mathbf{u}(q_m-3)+\mathbf{\Psi}\mathbf{u}(q_m-2)\right] \end{bmatrix} \Rightarrow$$

$$\mathbf{x}(0) = \left[\mathbf{M}_o^T \mathbf{M}_o\right]^{-1} \mathbf{M}_o^T \left(\begin{bmatrix} \mathbf{y}(0) \\ \mathbf{y}(1) \\ \mathbf{y}(2) \\ \vdots \\ \mathbf{y}(q_m-1) \end{bmatrix} - \begin{bmatrix} \mathbf{0} \\ \hline \mathbf{C}\mathbf{\Psi}\mathbf{u}(0) \\ \hline \mathbf{C}\left[\mathbf{\Phi}\mathbf{\Psi}\mathbf{u}(0)+\mathbf{\Psi}\mathbf{u}(1)\right] \\ \hline \vdots \\ \hline \mathbf{C}\left[\mathbf{\Phi}^{q-2}\mathbf{\Psi}\mathbf{u}(0)+\ldots+\mathbf{\Phi}\mathbf{\Psi}\mathbf{u}(q_m-3)+\mathbf{\Psi}\mathbf{u}(q_m-2)\right] \end{bmatrix} \right)$$

$$(3.554)$$

The matrix, $\left[\mathbf{M}_o^T \mathbf{M}_o\right]^{-1} \mathbf{M}_o^T$, is the left pseudo inverse of the non-square observability matrix, $\mathbf{M}_o$, of dimension $mq_m \times n$, which reduces to $\mathbf{M}_o^{-1}$ in the SISO case for which $m = 1$ and $q_m = n$, and therefore, $\mathbf{M}_o$ is of dimension $n \times n$. As will be seen in Chap. 8, this provides the basis of a discrete state estimation algorithm that can estimate the current state using q_m past input–output samples.

References

1. Cheng D et al (2010) Analysis and design of nonlinear control systems. Science Press, Beijing and Springer, Berlin Heidelberg
2. Rukmangadachari E (2009) Mathematical methods. Dorling Kindersley (India) Pvt. Ltd., licensees of Pearson Education in South Asia
3. Gantmakher FR, Brenner JL (2005) The application of the theory of matrices. Dover Publications, Mineola/New York
4. Luenberger DG (1967) Canonical forms for multivariable systems. IEEE Trans Autom Control AC-12(3), 290–293

Chapter 4
Traditional Controllers: Model Based Design

4.1 Approach

This chapter commences with the simplest feedback control systems to ensure continuity and provide some revision for readers who have only undertaken 1 year of undergraduate study of linear control systems. As the chapter progresses, various performance demands are introduced together with increases in the plant order. Controllers are developed through the needs of application examples. At each stage, features, either in the control structure or design methodology, are introduced that meet the specification. With this approach, the reader will fully understand the features and be able to design the simplest controller to meet a given performance specification for any linear SISO plant.

Following the introduction of closed loop control using analogue electronic implementation in the 1940s, the classical design procedures based mainly in the frequency domain evolved. These were influenced by the need to minimise the complexity of the analogue electronics implementing the controllers, thereby minimising cost and maximising reliability but restricting the attainable performance for many plants. This tradition has continued over the years with the consequence that, at the time of writing this book, most industries have not ventured much further than the PID controller and its relatives presented in Chap. 1. Digital processors, however, are now available at a reasonable cost and with such computational powers that the above restrictions need no longer apply. More complex but more effective controllers can be created merely by software changes without any increase in the hardware complexity.

The sampling frequencies of digital processors in many applications are so high that continuous time theory is applicable, as in this chapter, but in certain applications the time scale of operation is so short that discrete control theory is needed (Chap. 6) using the discrete plant models of Chaps. 2 and 3.

The approach to linear control system design taken in this book is to create controllers that have a number of adjustable parameters *at least equal in number* to the total order of the closed-loop system. This enables *any* linear closed loop dynamics to be attained for a given order but within the limitations of the hardware such as control saturation limits. It also enables a *particularly simple design approach* to be adopted that minimises trial and error. First the closed-loop pole locations are chosen that produce a desirable dynamic behaviour. Then the controller parameters are chosen that yield those pole locations. This process, which has already been briefly introduced in Chap. 1, is referred to as *pole assignment*. The control techniques of this chapter and Chap. 5 are *model based* in that a reasonably accurate plant model is needed but Chaps. 8 and 9 present control techniques that enable pole assignment despite relatively severe plant modelling uncertainties and external disturbances.

This chapter contains sufficient material for the design of traditional linear controllers for linear time invariant (LTI), single input, single output (SISO) plants of first and second order, that fall into the general class of plants that can be modelled by the general transfer function relationship,

$$Y(s) = \frac{\sum_{j=0}^{m} b_j s^j}{s^n + \sum_{i=0}^{n-1} a_i s^i} [U(s) - D(s)], \quad m < n, \tag{4.1}$$

where b_j, $j = 1, 2, ..., m$ and a_i, $i = 1, 2, ..., n$ are constant, real coefficients and $D(s)$ is the external disturbance referred to the control input, $U(s)$, equivalent to all the physical external disturbances acting on the plant by having precisely the same effect on the controlled output, y (Chap. 2). This disturbance is often significant and the controller must be designed to adequately counteract it. The minus sign with $D(s)$ is the convention adopted in electric drive applications, originating from the fact that an externally applied load torque to an electric motor reduces its speed. Since no generality is lost using this convention and many control applications have electric drives as the actuators, it will be adopted throughout this book.

For plants of third or higher order, the more sophisticated controllers introduced in Chap. 5 are needed.

The degree to which the closed loop performance differs from that specified due to plant modelling uncertainties and external disturbances is defined here as *sensitivity*. Conversely, the ability of the control system to maintain the closed loop performance near to that specified despite these imperfections is defined as *robustness*. Quantitative definitions of robustness and sensitivity are given in Sect. 4.6. By 'specified closed loop performance' is meant a precisely defined relationship between the reference input and the controlled output, such as a step response specification in terms of settling time and percentage overshoot. Achieving closed loop stability is only part of the requirement of a real control system design. To illustrate this point, a *stable* control system may have a step response that is unacceptable through being very oscillatory and/or having a settling time that is too long.

The importance of simulation in the control system design process has already been emphasised in Chap. 1 and this must always be carried out to check that the design is correct.

Control system performance specifications are often in the frequency domain and Sect. 4.6 includes their conversion to time domain specifications that meet them, enabling the design methods of this chapter to be applied.

Block diagram algebra and reduction can be used to simplify the block diagrams of linear control systems leading to the closed loop transfer function. To achieve this for some control loop structures more complex than the simplest ones consisting of forward and feedback transfer functions in a single loop, however, the application of Mason's formula is simpler and is therefore utilised frequently in this book. The procedure is shortened by eliminating the traditional step of forming a signal flow graph [1, 2] since the block diagram contains the same information. Another useful bi-product of this method frequently used throughout the book is the derivation of the characteristic polynomial by equating the determinant of Mason's formula to zero (Appendix A4).

4.2 Pole Assignment

The general principle of pole assignment (also referred to as pole placement) in linear single input, single output control systems will now be presented. As will be seen in the numerous examples following this section, the characteristic equation of any linear closed loop system may be expressed in the general form,

$$s^n + c_{n-1}(\mathbf{k}) s^{n-1} + \cdots + c_1(\mathbf{k}) s + c_0(\mathbf{k}) = 0, \tag{4.2}$$

where n is the order and the constant coefficients, $C_i(\mathbf{k})$, $i = 1, 2, \ldots, n$, are functions of the controller gains represented by $\mathbf{k} = [k_1, k_2, \ldots, k_r]^T$, where $r \leq n$. This column vector representation is commonly used and is part of the matrix-vector notation introduced in Chap. 4, but here it just represents the set of controller gains. The coefficients, $C_i(\mathbf{k})$, are, of course, also functions of the plant parameters but these are not shown explicitly in (4.2) as they are not generally adjustable.

If the control system designer is free to implement any controller, the recommendation here is to formulate one for which $r = n$. Applying the traditional controllers in cases where $r < n$, however, will be considered and the limitations discussed.

Having arrived at a suitable controller structure for the application in hand, then (4.2) is established by analysing the closed loop system using the available plant model. It is important to realise that the characteristic equation must be normalised with respect to a selected coefficient if it is not already unity. This is achieved simply by dividing every coefficient by the coefficient selected for normalisation. Normalisation with respect to the coefficient of s^n will be followed in most cases throughout this book.

The next step is to formulate the characteristic equation of the same order as (4.2) that yields the required closed loop dynamics, also normalised with respect to the coefficient of s^n. If s_i, $i = 1, 2, .., n$, are the required closed loop pole values, then the corresponding desired closed loop characteristic equation is

$$(s - s_1)(s - s_2)\ldots(s - s_n) = s^n + d_{n-1}s^{n-1} + \ldots d_1 s + d_0 = 0, \qquad (4.3)$$

where $d_{n-1} = -(s_1 + s_2 + \cdots + s_n), \ldots, d_0 = (-1)^n s_1 s_2 \ldots s_n$.

The set of poles to be selected depends very much on the application in hand, including the hardware limitations and many examples are provided throughout the text. The task, however, is greatly eased by the author's settling time formulae presented subsequently in this chapter.

The next step is to adjust **k** to force (4.2) to be identical to (4.3). This is achieved by equating the characteristic polynomials as follows.

$$s^n + c_{n-1}(\mathbf{k})s^{n-1} + \cdots + c_1(\mathbf{k})s + c_0(\mathbf{k}) = s^n + d_{n-1}s^{n-1} + \ldots sd_1 + d_0. \qquad (4.4)$$

Equation (4.4) must be satisfied for all s and this requires the coefficients of like powers of s to be equated, yielding the following set of simultaneous equations.

$$c_i(k_1, k_2, \ldots, k_r) = d_i, \quad i = 0, 1, .., n - 1 \qquad (4.5)$$

If $r = n$, then the simultaneous equations are completely determined and the required gains, $k_1, k_2, \ldots$, and k_n, are obtained directly as the solution of (4.5), but provided the equations are independent of one another in the sense that one equation may not be made equivalent to any of the others by algebraic manipulation. This is a generalisation of linear independence of a set of linear simultaneous equations. It should be noted that these equations provide an alternative and very simple means of assessing the *controllability* of a plant, addressed formally in Chap. 3. If the plant is controllable, then there exists a solution to (4.5). If not, then modification of the plant, such as changing the actuator location, to make it controllable would have to be considered. Most real plants, however, are controllable.

If $r < n$, then partial pole placement is possible (Appendix A4).

It is important to note that pole assignment can also be used to aid the design of *observers* for estimating the plant state that are themselves linear subsystems of control systems containing gains to be determined that are equal in number to the order of the observer (Chap. 8).

In many cases, simultaneous Eq. (4.5) are linear with respect to k_i, $i = 1, 2, .., r$, and to avoid lengthy derivations of formulae for the gains in the more complex control systems, this task may be carried out by computer aided design, the procedure of which is given in Chap. 11.

The method of pole assignment is a common thread running through the remaining sections of this chapter, commencing with the simplest examples.

4.3 Definition of Settling Time

If a step reference input is applied to a feedback control system, then assuming it is stable, the controlled output, y(t), will undergo a transient period followed by settling to a constant steady state value, y_{ss}. A useful design criterion for feedback control systems, very simple regardless of the order is the settling time. Since the step responses of linear continuous control systems take, in theory, an infinite time to settle, a practicable settling criterion is used so that a finite settling time can be specified. To achieve this, the settling time is defined as follows.

Definition 4.1 *The settling time, T_s, according to the x% criterion is the time for the step response, y(t), to reach and stay within a band having limits of $y_{ss}(1 \pm x/100)$.*

This is illustrated in Fig. 4.1. The question then arises of how wide to make the band. This is not a matter of optimisation and the standard values used have just been agreed in the engineering community as being reasonable. The most commonly used value of a_1 is 5 but occasionally 2 is used. The settling criteria based on these percentages are referred to, respectively, as the 5 % and the 2 % criteria.

Standard formulae for the settling times of first and second order linear systems using the 5 % criterion are in common use and these will be derived in Sects. 4.4.1 and 4.4.2. Until recently, these formulae were the only ones available. For this reason, some designers try to make higher order control systems have one or two dominant poles (Chap. 1) so that the well known theory of first or second order systems can be applied. This approach also has the advantage of increasing the robustness against plant modelling uncertainties and external disturbances (Sect. 8.5.2) through placing the dominated poles with larger magnitudes, but usually at the expense of increased control actuator activity in the presence of plant noise (Chap. 5). Also the standard settling time formula for second order linear systems is restricted to under-damped systems with the damping ratio in the range, $0 < \zeta < 1$. These restrictions, however, are removed by the author's own settling time formulae, which are derived in Sect. 4.5.4 for systems of arbitrary order using the 5 % and 2 % criteria. These are, however, restricted to linear systems with multiple poles but this yields a well behaved closed loop dynamics.

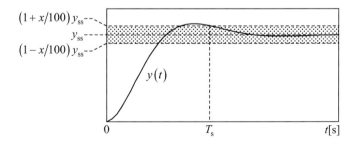

Fig. 4.1 Control system step response and illustration of the settling time

Also, if a small overshoot in the step response is desired, this can be catered for by separating the closed loop poles into complex conjugate pairs and applying the scaling law introduced in Sect. 4.5.2 to retain the specified settling time.

In Sect. 4.4, control system design is introduced using the traditional controllers, commencing with the simplest and includes the standard settling time formulae.

4.4 PID Controllers and Their Variants

4.4.1 First Order Systems

This section is restricted to plants modelled by (4.1) with $n = 1$. Thus

$$Y(s) = \left(\frac{b_0}{s + a_0}\right)[U(s) + D(s)] \qquad (4.6)$$

The simplest linear closed loop controller is the proportional controller introduced in Chap. 1. This has a single adjustable parameter, K, i.e., the proportional gain. Figure 4.2 shows this applied to plant (4.6). First, the external disturbance input will be set to zero but introduced later in an application example. This will serve to review some basic relationships between the controller parameters, pole locations, dynamic (i.e., transient) responses and steady state responses.

The determination of K that achieves an acceptable performance will now be considered. The starting point is usually the closed-loop transfer function. Figure 4.2 has the classical feedback structure of Fig. 4.3a for which the familiar basic closed loop and error transfer function formulae are given in Fig. 4.3b, c.

Applying this to obtain the closed loop transfer function for Fig. 4.2 yields

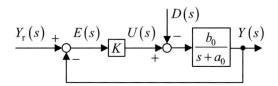

Fig. 4.2 Proportional controller applied to a first order plant

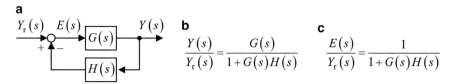

Fig. 4.3 Basic feedback loop and transfer functions. (a) Block diagram (b) Closed loop transfer function (c) Error transfer function

4.4 PID Controllers and Their Variants

$$\frac{Y(s)}{Y_r(s)} = \frac{Kb_0/(s+a_0)}{1 + Kb_0/(s+a_0)} = \frac{Kb_0}{s + (a_0 + Kb_0)} \quad (4.7)$$

which may also be written

$$\frac{Y(s)}{Y_r(s)} = \frac{K_{\text{DCL}}}{1 + sT_c} \quad (4.8)$$

where K_{DCL} is the *closed-loop* DC gain and T_c is the *closed-loop* time constant, given, respectively, by

$$K_{\text{DCL}} = \frac{Kb_0}{a_0 + Kb_0} \quad (4.9)$$

and

$$T_c = \frac{1}{a_0 + Kb_0} \quad (4.10)$$

As K is increased, T_c reduces, indicating a faster dynamic response to changes in $y_t(t)$ and $K_{\text{DCL}} \to 1$ as $K \to \infty$, indicating a reduced steady state error. These observations will be considered in more detail in the following example.

Example 4.1 Control of the refractory wall temperature of an electric kiln

The plant considered here is an electric kiln (Ref., Chap. 2) consisting of a variable power electric element embedded in the refractory brick enclosure containing a work-piece. The complete plant model is of second order, but in this example, a control loop is closed just on the refractory brick temperature so that the plant is only of first order. In practice, a second loop could be closed for control of the work-piece temperature forming a cascade control structure (Sect. 4.4.3). The block diagram of the refractory wall temperature sub-plant is shown in Fig. 4.4a whose parameters are defined in Chap. 2.

The work-piece temperature is typically several thousand °C, and therefore the ambient temperature, even within an extreme range of $(-50, +50)$ °C, would have a negligible effect and is therefore not included in the plant model.

By inspection of Fig. 4.4a, the plant DC gain is $K_{\text{DC}} = K_t K_h$. The closed loop system using the simplest linear controller, i.e., the proportional controller is shown in Fig. 4.4b. The closed loop transfer function is as follows.

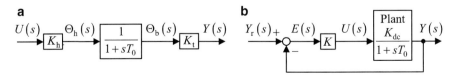

Fig. 4.4 Closed loop proportional control of a first order plant. (**a**) Plant (**b**) Closed loop system

$$\frac{Y(s)}{Y_r(s)} = \frac{\frac{KK_{DC}}{1+sT_0}}{1+\frac{KK_{DC}}{1+sT_0}} = \frac{KK_{DC}}{1+KK_{DC}+sT_0} = \frac{\frac{KK_{DC}}{1+KK_{DC}}}{1+\frac{sT_0}{1+KK_{DC}}} \quad (4.11)$$

i.e.,

$$\frac{Y(s)}{Y_r(s)} = \frac{K_{DCL}}{1+sT_c}. \quad (4.12)$$

where the closed-loop DC gain and time constant are, respectively,

$$K_{dcl} = \frac{KK_{DC}}{1+KK_{DC}} \quad (4.13)$$

and

$$T_c = \frac{T_0}{1+KK_{dc}}. \quad (4.14)$$

The settling time (5 % criterion) is three closed loop time constants. Thus

$$T_s = 3T_c. \quad (4.15)$$

and in view of (4.14) this is

$$T_s = \frac{3T_0}{1+KK_{DC}}. \quad (4.16)$$

This becomes smaller as K increases. The closed-loop pole, according to (4.12), is given by the single root of the first degree closed-loop characteristic equation,

$$1+sT_c = 0, \quad (4.17)$$

which is

$$s_1 = -\frac{1}{T_c} = -\frac{1+KK_{DC}}{T_0}. \quad (4.18)$$

Thus, if K is increased from zero, the closed loop pole starts at $-1/T_0$, which is the open loop pole location, i.e., the pole of the plant transfer function in Fig. 4.4, and becomes more negative. The locus of the closed loop pole position in the s-plane as K increases, i.e., the root locus, is shown in Fig. 4.5.

As can be seen from (4.14) and (4.15), T_c and T_s both decrease with K. Hence the speed of response of the control system increases with K. Importantly, the *settling*

4.4 PID Controllers and Their Variants

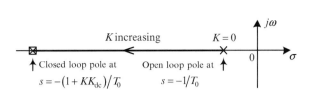

Fig. 4.5 Root locus with respect to K for the electric kiln control system

time is inversely proportional to the pole magnitude, as can be seen by substituting $T_c = T_s/3$ in (4.15), which yields

$$T_s = -\frac{3}{s_1} \qquad (4.19)$$

Another important effect occurring in this particular system is the variation of the *steady state error* to a step reference input with K. Applying the error transfer function of Fig. 4.3 to the control system of Fig. 4.4b yields

$$\frac{E(s)}{Y_r(s)} = \frac{1}{1 + \dfrac{KK_{DC}}{1+sT_0}} = \frac{1+sT_0}{1+sT_0+KK_{DC}} = \frac{\dfrac{1+sT_0}{1+KK_{dc}}}{1 + \dfrac{sT_0}{1+KK_{dc}}} \qquad (4.20)$$

Let a step reference input of $y_r(t) = Y_s h(t)$ be applied, where Y_s is the value of the step and $h(t)$ is the unit step function. Then $y_r(s) = Y_s/s$ and the steady state error obtained from (4.20) using the Final Value Theorem is

$$e_{ss} = \lim_{t \to \infty} e(t) = \lim_{s \to 0} sE(s) = \lim_{s \to 0} s \frac{\dfrac{1+sT_0}{1+KK_{DC}}}{1 + \dfrac{sT_0}{1+KK_{dc}}} \frac{Y_s}{s} = \frac{Y_s}{1+KK_{dc}}. \qquad (4.21)$$

This well known result shows that increasing K reduces the steady-state error.

The equation for the step response is obtained from (4.12) with the aid of a table of Laplace transforms and their inverses (Table 4.1). Thus,

$$y(t) = \mathcal{L}^{-1}\left\{\frac{K_{DCL}}{1+sT_c} \cdot \frac{Y_s}{s}\right\} = K_{DCL} Y_s \left[1 - e^{-t/T_c}\right]. \qquad (4.22)$$

and substituting for K_{DCL} and T_c using (4.9) and (4.10) yields

$$y(t) = \frac{KK_{DC}}{1+KK_{DC}} Y_s \left(1 - e^{-\frac{1+KK_{dc}}{T_0}t}\right). \qquad (4.23)$$

Figure 4.6 shows step response simulations with $T_0 = 100$ [s], $K_h = 200$ [°C/V] and $K_t = 0.005$ [V/°C] giving $K_{dc} = 1$.

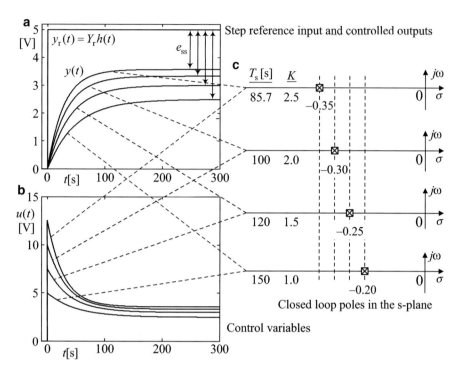

Fig. 4.6 Variation of the step response and the closed loop pole location with the proportional gain for the refractory wall temperature control system

The family of step responses is for $Y_r = 5$ V corresponding to $1{,}000\,^\circ\text{C}$ and shows the variation with the proportional gain, K, the corresponding settling time and the closed-loop pole locations. It is evident that the settling time and the steady state error may be reduced by increasing the proportional gain, K. This also follows from elementary analysis of (4.23).

Following the approach adopted for control system design in this chapter, a *design formula* can be derived that enables the proportional gain, K, to be calculated to yield a specified settling time, T_s, for given plant parameters. This is obtained by making K the subject of (4.16), yielding

$$K = \frac{3\,(T_0/T_s) - 1}{K_{\text{dc}}}. \tag{4.24}$$

In theory, any desired settling time can be attained by calculating the required value of K, *but in practice,* for a given step reference input, increasing K beyond a certain limit will cause the control to saturate, which sets a lower limit on the settling time. This is due to the physical limitations of the control actuator and/or the electronic drive circuit. It should be mentioned that such control saturation limits determine the *minimum* attainable settling time in the nonlinear control technique of *time optimal*

4.4 PID Controllers and Their Variants

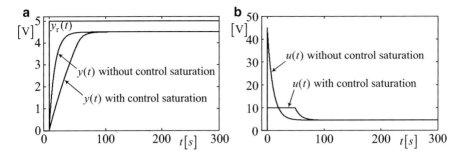

Fig. 4.7 Effect of control saturation on the electric kiln control system. (**a**) Controlled output and reference input (**b**) Control variable

Fig. 4.8 Proportional control of the electric kiln including a reference input scaling coefficient to reduce the steady state error

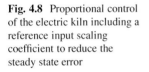 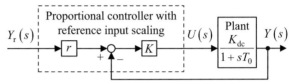

control (Chap. 8). Consequently the minimum settling time attainable by adjustment of any *linear* controller with the same control saturation constraints will be longer than optimal. In the heating process, the lower limit is zero (with the heating element off) and the upper limit is set by the maximum power input to the heating element. This corresponds to a maximum value, u_{max}, of u. In Fig. 4.6, where the control saturation limits have not been applied in the simulation, it is evident that if $u_{max} = 10$ V, the settling time of $T_s = 40$ s cannot be attained with this particular controller. For clarity of illustration, this situation is exaggerated in the simulations of Fig. 4.7, where the proportional gain is set to $K = 9$ yielding $T_s = 30$ s, in theory.

This is an instance of the plant hardware limitations preventing the desired control system performance being attained. In this application, the steady state value of $u(t)$ needed to raise the temperature sufficiently for nominally zero steady state error for temperature setpoints (i.e., constant reference inputs) beyond a certain level would exceed the upper saturation limit, causing wind-up of the integral term with a PI or IP controller (Chap. 1), in which case modification of the plant to provide more heating power would have to be considered.

A simple alternative to the PI or IP controller that brings the steady state error of the step response to nearly zero regardless of the settling time is the introduction of a *reference input scaling coefficient*, r, as shown in Fig. 4.8.

This would be set to the reciprocal of the closed loop DC gain of (3.3), yielding

$$r = \frac{1}{K_{DCL}} = \frac{1 + KK_{DC}}{KK_{DC}} \qquad (4.25)$$

The *nominal* closed loop DC gain would then be unity, yielding zero steady state error for a step reference input, *in theory*. One potential problem with this approach is the dependence upon accurate plant parameter estimates: in this case the plant DC gain, K_{dc}. Let the imprecise estimate of K_{dc} be $\tilde{K}_{dc}$. Then the DC gain of the system of Fig. 4.8 is

$$K'_{DCL} = rK_{DCL} = \frac{(1 + K\tilde{K}_{DC})(KK_{DC})}{(1 + KK_{DC})(K\tilde{K}_{DC})}. \quad (4.26)$$

There will therefore be a residual steady state error given by

$$e'_{ss} = Y_s - y_{ss} = (1 - K'_{DCL}) Y_s$$

$$= \frac{(1 + KK_{DC})(K\tilde{K}_{DC}) - (1 + K\tilde{K}_{DC})(KK_{DC})}{(1 + KK_{DC})(K\tilde{K}_{DC})} Y_s = \frac{K(\tilde{K}_{DC} - K_{DC})}{(1 + KK_{DC})(K\tilde{K}_{DC})} Y_s. \quad (4.27)$$

Note if $\tilde{K}_{DC} < K_{DC}$, the residual steady state error is negative meaning $y_{ss} > Y_s$.

Example 4.2 Control of greenhouse temperature.

In this application, the external disturbance, $D(s)$, shown in Fig. 4.9 is significant.

Steady state error analysis will first be carried out. It is most important to realise that with a non-unity r, the control error, $E(s)$, is *not* the input to the proportional gain, K, but as shown. The Principle of Superposition for linear systems may be applied to derive the error transfer function relationship from Fig. 4.9, which is

$$E(s) = Y_r(s) - r \cdot \frac{\frac{KK_{DC}}{1 + sT_0}}{1 + \frac{KK_{DC}}{1 + sT_0}} y_r(s) + \frac{\frac{K_{DC}}{1 + sT_0}}{1 + \frac{KK_{DC}}{1 + sT_0}} D(s) \quad (4.28)$$

and if $y_r(t) - Y_s h(t) \Rightarrow Y_r(s) - Y_s/s$ and $d(t) = D_s h(t) \Rightarrow D(s) = D_s/s$ then the steady state error is

$$e_{ss} = \lim_{s \to 0} s \left(\frac{Y_s}{s} - r \cdot \frac{\frac{KK_{DC}}{1 + sT_0}}{1 + \frac{KK_{DC}}{1 + sT_0}} \cdot \frac{Y_s}{s} + \frac{\frac{K_{DC}}{1 + sT_0}}{1 + \frac{KK_{DC}}{1 + sT_0}} \cdot \frac{D}{s} \right) \quad (4.29)$$

$$= \left[1 - r \left(\frac{KK_{DC}}{1 + KK_{DC}} \right) \right] Y_s + \left(\frac{K_{DC}}{1 + KK_{DC}} \right) D_s$$

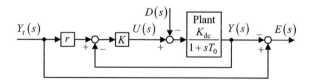

Fig. 4.9 Steady state error in system with reference input scaling coefficient

4.4 PID Controllers and Their Variants

Substituting for r in (4.29) using (4.25) then yields

$$e_{ss} = \left(\frac{K_{DC}}{1 + KK_{DC}}\right) D_s. \tag{4.30}$$

A non-zero steady state error therefore exists for finite K caused by a constant external disturbance. Only the component (4.27) is absent through assuming a perfectly known plant but will still contribute a small steady state error in practice. If *zero* steady state error is essential, a controller containing an integral term is needed. The IP or PI controllers introduced in Chap. 1 would suffice, yielding a second order closed loop system, as exemplified in the following section.

4.4.2 Second Order Systems

The first control system design objective taken is for the closed loop transfer function to be the following second order one without finite zeros and unity DC gain.

$$\frac{Y(s)}{Y_r(s)} = \frac{\omega_n^2}{s^2 + 2\zeta\omega_n s + \omega_n^2}. \tag{4.31}$$

Here, the undamped natural frequency, ω_n, and the damping ratio, ζ, are chosen to yield an acceptable step response for the application in hand. Figure 4.10 shows a family of responses to a step reference input, $y_r(t) = Y_s h(t)$, where Y_s is a constant, for different values of the damping ration, ζ. This can be used as a design aid.

A derivation of the mathematical expression of these responses will follow shortly. Note that they are normalised with respect to Y_r and the time is scaled by a factor of ω_n. Although it is most unlikely that the oscillatory responses for $\zeta < 0.7$ will be chosen, they are included for completeness. They also illustrate

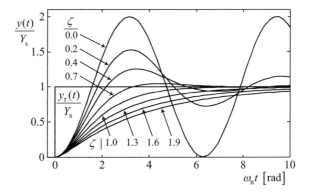

Fig. 4.10 Step responses of a second order linear system without finite zeros for a range of different damping ratios and constant undamped natural frequency

Fig. 4.11 Proportional controller applied to second order plant

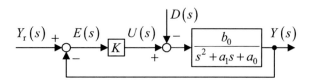

possible responses with incorrect controller gains or fault conditions. Importantly, such errors can sometimes cause instability.

Ideally the designer would be able to determine the controller parameters that produce the value of ζ yielding the selected step response shape and ω_n for the required time scale. As will be seen, not all the traditional controllers enable ζ and ω_n to be chosen independently.

The control of the general class of second order linear plants without finite zeros having transfer function relationship (4.1) with $n = 2$ and $b_1 = 0$, i.e.,

$$Y(s) = \left(\frac{b_0}{s^2 + a_1 s + a_0} \right) [U(s) + D(s)] \qquad (4.32)$$

will now be considered. This covers many applications, some of which are included in Chap. 2. First, the simple proportional controller will again be applied. Again the often significant external disturbance, $D(s)$, referred to the control input, is included. The control system block diagram is then as shown in Fig. 4.11.

Applying the Principle of Superposition for linear systems, the closed-loop transfer function relationship is

$$Y(s) = \frac{[KY_r(s) - D(s)] \frac{b_0}{s^2 + a_1 s + a_0}}{1 - \left[-K \frac{b_0}{s^2 + a_1 s + a_0} \right]} = \frac{b_0 [KY_r(s) - D(s)]}{s^2 + a_1 s + a_0 + K b_0} \qquad (4.33)$$

This system has two closed-loop poles but it is not possible to obtain *any* desired second order closed loop dynamics. This is because the one and only adjustable controller parameter, K, is insufficient to obtain a completely determined set of simultaneous equations for pole assignment such as (4.5) with $r = n$. In this case, $r = 1$ and $n = 2$. The system will now be analysed to determine the resulting limitations on the attainable dynamic performance and the steady state performance. First, let (4.33) be expressed in the standard form,

$$Y(s) = \frac{K_{\text{DCL}} \omega_n^2 \left[Y_r(s) - \frac{1}{K} D(s) \right]}{s^2 + 2\zeta \omega_n s + \omega_n^2}, \qquad (4.34)$$

where K_{DCL} is the closed loop DC gain, *which may not be unity*. The closed loop transfer function (4.31) with unity DC gain would yield zero steady state error of the step response but only applies with proportional control in cases where the plant has at least one pure integrator in the forward path, meaning that its transfer

4.4 PID Controllers and Their Variants

function contains $1/s$ as a factor, requiring $a_0 = 0$. Other than this, keeping the steady state error down to acceptable proportions might require the introduction of a reference input scaling coefficient or another controller containing an integral term. The closed-loop DC gain is defined as

$$K_{\text{DCL}} \triangleq \lim_{s \to 0} \left[\frac{Y(s)}{Y_r(s)} \right] \bigg|_{D(s)=0}. \qquad (4.35)$$

For the system of Fig. 4.11 this yields

$$K_{\text{DCL}} = \frac{K b_0}{a_0 + K b_0}. \qquad (4.36)$$

With $y_r(t) = Y_s h(t) \Rightarrow y_r(s) = Y_s/s$ where Y_s is the step reference input value and $d(t) = D_s h(t) \Rightarrow d(s) = D_s/s$, where D_s is the step disturbance value, applying the Final Value Theorem yields the steady state error,

$$\begin{aligned}
e_{ss} &= \lim_{t \to \infty} e(t) = \lim_{s \to 0} s E(s) = \lim_{s \to 0} s \left[Y_r(s) - Y(s) \right] \\
&= \lim_{s \to 0} s \left[y_r(s) - \frac{b_0 \left[K Y_r(s) - D(s) \right]}{s^2 + a_1 s + a_0 + K b_0} \right] \\
&= \lim_{s \to 0} s \left[\frac{(s^2 + a_1 s + a_0) \frac{Y_s}{s} + b_0 \frac{D_s}{s}}{s^2 + a_1 s + a_0 + K b_0} \right] = \frac{a_0 Y_s + b_0 D_s}{a_0 + K b_0}.
\end{aligned} \qquad (4.37)$$

If $a_0 = 0$, (4.36) indicates a unity DC gain and according to (4.37) there would be zero steady state error for a step reference input, *but only in cases where the external disturbance is not present.* A similar result will be recalled in Sect. 4.4.1 for $n = 1$, In fact, these restrictions apply for any plant order.

Selecting the two closed-loop poles independently (with, of course, the restriction that if one closed-loop pole is complex, the other closed-loop pole must be its complex conjugate) is equivalent to selecting the undamped natural frequency, ω_n and the damping ratio, ζ, independently. If this could be done, then the denominators of (4.33) and (4.34) would be identical for *any* selected combination of ζ and ω_n, resulting in the following pair of simultaneous equations.

$$a_0 + K b_0 = \omega_n^2 \qquad (4.38)$$

$$a_1 = 2\zeta \omega_n \qquad (4.39)$$

It is clear from (4.38) that ω_n can be selected and the required K calculated using

$$K = \left(\omega_n^2 - a_0 \right) / b_0 \qquad (4.40)$$

but (4.39) then fixes the value of ζ which may or may not be acceptable. Conversely, if ζ is selected to yield acceptable damping, then according to (4.39),

$$\omega_n = \frac{a_1}{2\zeta}, \quad (4.41)$$

which determines the time scale of the step response (Fig. 4.10). The value of the proportional gain yielding the selected value of ζ would then be given by (4.40) with ω_n according to (4.41). Thus

$$K = \frac{a_1^2}{4\zeta^2} - a_0. \quad (4.42)$$

For the analysis of system (4.34) to determine the effects constraint (4.41) on the step response, the standard settling time formula for the under-damped case, i.e., with $0 < \zeta < 1$, is needed. This will now be derived. If $y_r(t) = Y_s h(t) \Rightarrow y_r(s) = Y_s/s$, with zero initial conditions and $D(s) = 0$, (4.34) becomes

$$Y(s) = \frac{K_{\text{DCL}} \omega_n^2}{s^2 + 2\zeta\omega_n s + \omega_n^2} \cdot \frac{Y_s}{s}. \quad (4.43)$$

Then using the Final Value Theorem, the steady state value of $y(t)$ is $y_{ss} = \lim_{s \to 0}[sy(s)] = K_{\text{dcl}} Y_s$ and therefore (4.43) can be expressed as

$$y(s) = \frac{\omega_n^2}{s^2 + 2\zeta\omega_n s + \omega_n^2} \cdot \frac{y_{ss}}{s} \quad (4.44)$$

Taking inverse Laplace transforms of (4.44) yields

$$y(t) = \left[1 - \frac{1}{\sqrt{1-\zeta^2}} e^{-\zeta\omega_n t} \sin\left(\omega_n\sqrt{1-\zeta^2}\, t + \cos^{-1}(\zeta)\right)\right] y_{ss}. \quad (4.45)$$

Definition 4.1 for the settling time (5 % criterion) will now be applied to (4.45). Thus

$$1 - \frac{1}{\sqrt{1-\zeta^2}} e^{-\zeta\omega_n T_s} \sin\left(\omega_n\sqrt{1-\zeta^2}\, T_s + \cos^{-1}(\zeta)\right) = 0.95 \text{ or } 1.05 \Rightarrow$$

$$\frac{1}{\sqrt{1-\zeta^2}} e^{-\zeta\omega_n T_s} \sin\left(\omega_n\sqrt{1-\zeta^2}\, T_s + \cos^{-1}(\zeta)\right) = \pm 0.05. \quad (4.46)$$

The required value of the settling time is the smallest solution of (4.46) for T_s. Unfortunately, this is a transcendental equation without an analytical solution. The problem, however, can be made mathematically tractable by redefining the settling time for the undamped linear second order system, as the time taken

4.4 PID Controllers and Their Variants

for the exponential envelope function, $\left(1/\sqrt{1-\zeta^2}\right)e^{-\zeta\omega_n t}$, to reach 0.05. This is practicable because $y(T_s)$ would lie *within* the - band rather than at its boundary. According to the redefined settling time,

$$\frac{1}{\sqrt{1-\zeta^2}}e^{-\zeta\omega_n T_s} = 0.05 \Rightarrow e^{-\zeta\omega_n T_s} = 0.05\sqrt{1-\zeta^2} \Rightarrow -\zeta\omega_n T_s$$

$$-\ln\left(0.05\sqrt{1-\zeta^2}\right) \Rightarrow \zeta\omega_n T_s = \ln\left[1/\left(0.05\sqrt{1-\zeta^2}\right)\right] = \underbrace{\ln(20)}_{2.995732} - 0.5\ln\left(1-\zeta^2\right)$$

$$\cong 3 - 0.5\ln\left(1-\zeta^2\right) \Rightarrow T_s = \frac{1}{\zeta\omega_n}\left[3 - 0.5\ln\left(1-\zeta^2\right)\right]. \tag{4.47}$$

Since the oscillatory term in (4.45) does not exist for critically or over-damped systems having $\zeta \geq 1$ the exponential envelope function also cannot exist and therefore (4.47) is invalid. This is reflected in the term, $\ln\left(1-\zeta^2\right) = j\pi \ln\left(\zeta^2 - 1\right)$, resulting in T_s complex according to (3.31), which is inadmissible. The author's settling time formulae derived in Sect. 4.5.4, however, are applicable to critically damped second order linear systems *and higher order systems with multiple closed loop poles*.

Returning to the limitations of the attainable dynamic performance of the closed loop system of Fig. 4.11, (4.39) yields $\zeta\omega_n = a_1/2$ and therefore the settling time according to (4.47) becomes

$$T_s = \frac{2}{a_1}\left[3 - 0.5\ln\left(1-\zeta^2\right)\right]. \tag{4.48}$$

Hence if ζ has been chosen to yield an acceptable form of step response, the resulting settling time would have either to be accepted or another controller selected that is free of the restriction. An attempt to substantially reduce the settling time by increasing K in the system of Fig. 4.11 would fail since from (4.42),

$$\zeta^2 = \frac{a_1^2}{4(K+a_0)} \tag{4.49}$$

and therefore increasing K will reduce ζ and cause the step response to be more oscillatory. Furthermore substituting for ζ^2 in (4.48) using (4.49) yields

$$T_s = \frac{2}{a_1}\left\{3 - 0.5\ln\left[1 - \frac{a_1^2}{4(K+a_0)}\right]\right\}. \tag{4.50}$$

and therefore while increasing K does reduce T_s, it is limited to $T_{s\,\mathrm{min}} = \lim_{K\to\infty} T_s = 6/a_1$. These effects may be observed graphically in Fig. 4.12. This applies to two different plants modelled by (4.33) with $D(s) = 0$, both with $a_1 = 2$ to yield the same negative real part of the complex branches of the root loci.

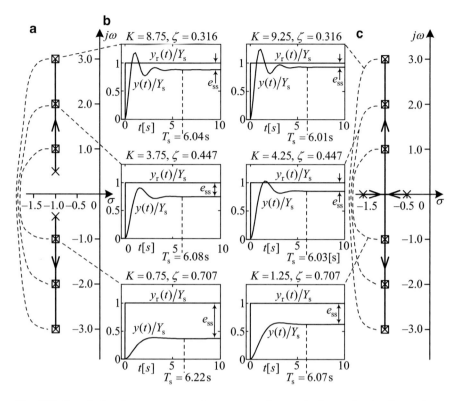

Fig. 4.12 Root loci and step responses for proportional control of second order linear plants. (a) Root locus for $a_0 = 1.25$ (b) Step responses (c) Root locus for $a_0 = 0.75$

One plant has $a_0 = 1.25$ yielding complex conjugate open loop poles. A practical example of this would be the position control of a mechanism containing a retention spring to return the mechanism to a safe position in case of an actuator failure, giving the open loop system an oscillatory mode. The other plant has $a_0 = 0.75$ yielding two real and negative open loop poles, a practical example being the position control of a mechanism without a retention spring. It is also evident by comparing the coefficients of s in the denominators of (4.33) and (4.34) that the coefficient, a_1, gives the system 'natural' damping. This is brought about by the energy dissipation process of viscous friction in the mechanism being controlled. The settings of the gain, K, yielding $\zeta = 0.707$ would yield an acceptable shape of the step response but an important observation in Fig. 4.12 is the relatively large steady state errors due to $a_0 \neq 0$ in (4.37) which would be unacceptable.

It is clear that increasing K to reduce them would result in highly oscillatory step responses that would also be unacceptable. In some other cases, however, a_0 could be sufficiently small for the steady state error to be acceptable or even negligible and if in addition, the settling time is sufficiently short, then the simple proportional controller would suffice. Otherwise, another controller would have to be considered either containing a reference input scaling coefficient or an integral term.

4.4 PID Controllers and Their Variants

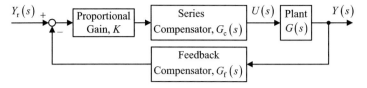

Fig. 4.13 General control structure for SISO plant with compensators

The classical approach to control system design aimed at overcoming constraints on the attainable performance with simple proportional control is the introduction of *compensators* and Fig. 4.13 shows the general control structure with series and feedback compensators for a SISO plant. Usually either one or the other are applied. The compensators are linear subsystems designed in the frequency domain with the aid of Bode amplitude and phase frequency response diagrams, Nyquist diagrams or Nichols diagrams. If simple proportional control alone is applied, the constraints on the attainable performance discussed above would be manifest as poor performance in the frequency domain indicated by inadequate stability margins (gain and phase margins) revealed in these diagrams. The classical control system design approach would be to design the transfer function, $G_c(s)$ or $G_f(s)$, to improve the performance margins. These frequency domain design methods evolved during the era of analogue electronic implementation of the 1940s and 1950s and the complexity of $G_c(s)$ or $G_c(s)$ was constrained by the economics of increased production costs of larger circuit boards and reliability in service. These traditional methods usually lead to an acceptable form of step response in the time domain but achieving a precisely defined closed loop dynamic performance, such as a prescribed settling time and maximum overshoot in the step response by this means involves trial and error. This is avoided in the control techniques and design methods presented in this book. For those wishing to study the classical design methods in depth, however, excellent treatises are available [1–3].

It is appropriate to remark that *any* SISO linear control system can be shown in the form of Fig. 4.13 and therefore all the linear control system design methods presented in this book can be regarded as alternative methods of compensator design. A special case is the polynomial control method of Chap. 4 that permits design by complete pole assignment and has a control structure similar to that of Fig. 4.13.

The control techniques still to be presented are free of constraints such as that suffered by the system of Fig. 4.11, even for systems of third and higher order.

Continuing with the control of the second order linear plant (4.32), the DP controller introduced in Chap. 2 can solve the problem of limited attainable dynamic performance experienced with the simple proportional controller. The block diagram of this control system is shown in Fig. 4.14. In practice, the derivative term, $K_D s$, would be implemented either by direct measurement of $\dot{y}$, software differentiation or an observer (Chap. 5).

Fig. 4.14 DP controller applied to second order linear plant

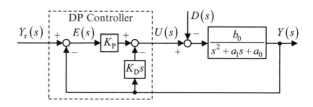

It should be noted that this controller can be regarded as an example of the more general state feedback controller introduced in Chap. 4.

Using the Principle of Superposition, the transfer function relationship between $Y_r(s)$, $D(s)$ and $E(s)$ is as follows.

$$E(s) = \frac{\left[1 - \left(-\frac{b_0}{s^2 + a_1 s + a_0} K_D s\right)\right] Y_r(s) + \frac{b_0}{s^2 + a_1 s + a_0} D(s)}{1 - \left\{-\left(-\frac{b_0}{s^2 + a_1 s + a_0}\right)(K_D s + K_P)\right\}} \quad (4.51)$$

$$= \frac{\left[s^2 + (a_1 + b_0 K_D) s + a_0\right] Y_r(s) + b_0 D(s)}{s^2 + (a_1 + b_0 K_D) s + (a_0 + b_0 K_P)}.$$

With step inputs, $y_r(t) = Y_s h(t) \Rightarrow Y_r(s) = Y_s/s$ and $d(t) = D_s h(t) \Rightarrow D(s) = D_s/s$, the steady state error is

$$e_{ss} = \lim_{s \to 0} s \frac{\left[s^2 + (a_1 + b_0 K_D) s + a_0\right] \frac{Y_s}{s} + b_0 \frac{D_s}{s}}{s^2 + (a_1 + b_0 K_D) s + (a_0 + b_0 K_P)} = \frac{a_0 Y_s + b_0 D_s}{a_0 + b_0 K_P}. \quad (4.52)$$

So the result is similar to (4.37) obtained with the simple proportional controller but in this case, the introduction of the derivative feedback coefficient, K_D, enables the proportional gain, K_P, to be increased to reduce e_{ss} to acceptable proportions for some applications without sacrificing damping. At this point the method of pole assignment (Sect. 4.2) can be applied. The denominator of (4.51), is the closed loop characteristic polynomial in terms of the controller gains while the desired characteristic polynomial can be the standard one, $s^2 + 2\zeta\omega_n s + \omega_n^2$, enabling ζ and ω_n to be chosen directly. Thus

$$s^2 + (a_1 + b_0 K_D) s + (a_0 + b_0 K_P) = s^2 + 2\zeta\omega_n s + \omega_n^2. \quad (4.53)$$

Equating the corresponding coefficients yields the simultaneous equations,

$$a_0 + b_0 K_P = \omega_n^2 \quad (4.54)$$

$$a_1 + b_0 K_D = 2\zeta\omega_n. \quad (4.55)$$

4.4 PID Controllers and Their Variants

Since K_P and K_D may be freely chosen, it is possible to first determine the values of ζ and ω_n that yield the required dynamic performance using, for example, a family of step responses such as in Fig. 4.10. Then the values of K_P and K_D that achieve this may be determined just by making these gains the subjects of (4.54) and (4.55), resulting in the *design formulae*,

$$
\begin{aligned}
K_p &= \frac{\omega_n^2 - a_0}{b_0} \\
K_d &= \frac{2\zeta\omega_n - a_1}{b_0}.
\end{aligned}
\quad (4.56)
$$

The steady state error of (4.52) could therefore be reduced by increasing ω_n while retaining the same damping ratio, ζ.

Example 4.3 Single axis, rigid body spacecraft attitude control

An informative application is the attitude control about a single axis of a spacecraft whose dynamics can be closely represented by that of a rigid body. Figure 4.15 shows a DP controller applied to this plant.

Here, J is the spacecraft moment of inertia about the control axis, K_w is the torque constant of the reaction wheel K_s is the optical attitude sensor constant and K_g is the rate gyro scaling constant. Before carrying out the steady state analysis and deriving the design formulae, however, the plant model will be converted to the standard form exemplified in Fig. 4.14 to simplify the algebra. The transfer function relationship of the plant in Fig. 4.15 is

$$
Y(s) = \frac{K_s}{Js^2} [K_w U(s) - \Gamma_d(s)]. \quad (4.57)
$$

The corresponding transfer function relationship of the general second order plant without finite zeros shown in Fig. 4.14 is

$$
Y(s) = \frac{b_0}{s^2 + a_1 s + a_0} [U(s) - D(s)]. \quad (4.58)
$$

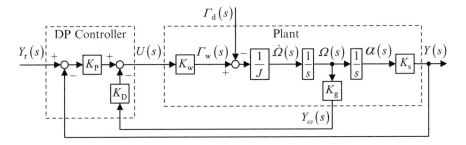

Fig. 4.15 Single axis, rigid body spacecraft attitude control system

Fig. 4.16 Simplified block diagram of single axis rigid body spacecraft attitude control system for analysis and design

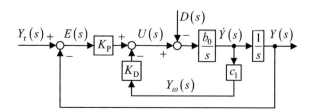

Then making (4.58) equivalent to (4.57) requires

$$a_0 = a_1 = 0, \quad b_0 = \frac{K_s K_w}{J} \quad \text{and} \quad D(s) = \frac{1}{K_w}\Gamma_d(s) \quad (4.59)$$

Also the measurement, $Y_\omega(s)$ in Fig. 4.15 has to be included. This is given by

$$Y_\omega(s) = K_g\Omega(s) = K_g s\alpha(s) = (K_g/K_s) sY(s). \quad (4.60)$$

Let the corresponding equation in the simplified plant model be

$$Y_\omega(s) = c_1 s Y(s) \quad (4.61)$$

where c_1 is a constant to be determined. Then comparing (4.61) with (4.60) yields

$$c_1 = \frac{K_g}{K_s}. \quad (4.62)$$

Figure 4.16 shows the control system block diagram with the simplified plant model.

The pole placement design may be carried out as follows. First, the characteristic equation is obtained by equating the determinant of Mason's formula to zero. Thus

$$1 - \left[-\left(K_D c_1 + \frac{K_P}{s}\right)\frac{b_0}{s}\right] = 0 \Rightarrow s^2 + K_D c_1 b_0 s + K_P b_0 = 0. \quad (4.63)$$

This is in the standard form where the coefficient of the highest power of s is unity. Equating the characteristic polynomial to the corresponding polynomial expressed in terms of the damping ratio, ζ, and the undamped natural frequency, ω_n, yields

$$s^2 + K_D c_1 b_0 s + K_P b_0 = s^2 + 2\zeta\omega_n s + \omega_n^2 \quad (4.64)$$

This places the closed loop poles in locations determined by the chosen values of ζ and ω_n. The gain formulae are then obtained by equating the coefficients of like degree terms in s as follows.

4.4 PID Controllers and Their Variants

$$K_P = \frac{\omega_n^2}{b_0}, \quad K_D = \frac{2\zeta\omega_n}{(c_1 b_0)}. \tag{4.65}$$

It should be observed that setting $K_D = 0$ reduces the DP controller to the simple proportional controller and this would be entirely unsuited to this application because (4.65) would imply that $\zeta = 0$ resulting in the undamped sinusoidal oscillatory step response of Fig. 4.10. In other applications such as the position control of mechanisms in factories, the viscous component of bearing friction provides some natural damping but in the spacecraft it is entirely provided by the controller. *Like* the viscous bearing friction, The controller produces a component of the reaction wheel torque that is proportional to the body angular velocity and opposes the motion, as would viscous friction if it was present. This is an example of *active damping*, a technique used also to provide damping of a mechanical structure using actuators instead of passive dampers. This will be addressed later in this chapter.

Space satellites often operate with piecewise constant reference attitude inputs and constant (or slowly varying) disturbance torques, so it would be appropriate to carry out the steady state error analysis with $y_r(t) = Y_s h(t) \Rightarrow y_r(s) = Y_s/s$ and $d(t) = D_s h(t) \Rightarrow d(s) = D_s/s$. Thus

$$E(s) = \frac{1 \cdot \left\{1 - \left[-(K_D c_1)\frac{b_0}{s}\right]\right\} Y_r(s) - \frac{b_0}{s^2} D(s)}{1 - \left[-(K_D c_1 + \frac{K_P}{s})\frac{b_0}{s}\right]} = \frac{(s^2 + K_D c_1 b_0) Y_r(s) + b_0 D(s)}{s^2 + K_D c_1 b_0 s + K_P b_0}.$$

Hence

$$e_{ss} = \lim_{s \to 0} s \cdot \frac{(s^2 + K_D c_1 b_0)\frac{Y_s}{s} + b_0 \frac{D_s}{s}}{s^2 + K_D c_1 b_0 s + K_P b_0} = \frac{D}{K_P}. \tag{4.66}$$

This is an example of an application in which the plant contains pure integrators and therefore has no steady state error component due to the step reference input but suffers from a steady state error due to a constant disturbance. This could be reduced by increasing K_P via increase of ω_n according to (4.65). It could only be entirely eliminated by means of another controller containing an integral term.

Suppose that the settling time (5 % criterion) of a large angle slewing manoeuvre is specified together with the damping ratio, ζ. Then the required undamped natural frequency, ω_n, could be found from (4.47). Thus

$$\omega_n = \frac{3 - 0.5 \ln(1 - \zeta^2)}{\zeta T_s}. \tag{4.67}$$

Figure 4.17a shows the simulation results for a rigid body spacecraft attitude control system designed with the aid of (4.67).

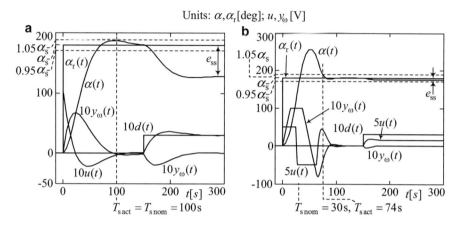

Fig. 4.17 Slew manoeuvre of spacecraft attitude control system using a DP controller. (**a**) Linear operation $T_s = 100$ s. (**b**) Reaction wheel and rate gyro saturation $T_s = 50$ s

The step reference input is $\alpha_r(t) = \alpha_s h(t)$, where $\alpha_s = 180°$ (the largest possible slew demand), designed for $\zeta = 1/\sqrt{2} \cong 0.7071$, which gives a small overshoot of about 4 %, and $T_s = 100$ [s]. The spacecraft parameters are $J = [150 \text{ kg m}^2]$; $K_w = 0.1$ [Nm/V]; $K_g = 100$ [V/(rad/s)]; $K_s = 3$ [V/rad]; control saturation limits $\pm u_{max} = \pm 10$ V; rate gyro output saturation limits $\pm y_{\omega \text{ max}} = \pm 10$ V. Also, a step disturbance torque of $\Gamma_d(t) = 0.3h(t - 150)$ [Nm] $\equiv d(t) = 3h(t - 150)$ [V] is applied, which is quite large but realistic in the case of a disturbance torque component due to the force vector of an orbit-change thruster not being directed precisely through the spacecraft centre of mass (Chap. 2).

For simulations supporting the design of control systems, it is important to plot $u(t)$ and also other variables, such as $y_\omega(t)$ in this example, to check that they remain within their saturation limits. This ensures *linear* operation of the control system since the intention is to employ a *linear* control system design procedure.

It is also important to include imperfections in the simulation such as the control and rate gyro output saturations in this example, to be able to predict the results of driving the control system against such limits, which can occur occasionally in practice. The electrical power for the attitude control subsystem of a spacecraft is generally limited to a few tens of Watts and this severely limits the maximum torque output from the reaction wheel and the maximum body angular velocity. With this in mind, $T_s = 100$ s is realistic for this application. Figure 4.17b, predicts the results of attempting to design the system for $T_s = 30$ s. Saturation occurs in the rate gyro and also the reaction wheel, resulting in the settling time being more than double that specified. The reaction wheel cannot produce high enough angular acceleration peaks due to its limited torque and the gyro is not designed to measure angular velocities outside the operational envelope of the spacecraft. A large overshoot occurs, and the reason for such behaviour under control saturation is given in the nonlinear phase-plane analysis of Chap. 8.

4.4 PID Controllers and Their Variants

In Fig. 4.17a, the steady state error, e_{ss}, of about 50° due to the disturbance torque is unacceptable but after the system comes out of saturation in Fig. 4.17b the steady state error has been brought down to a much smaller level within the settling band of ±9°, but this would be unacceptable with such a large overshoot during the slew manoeuvre. If a zero steady state performance is required, then a controller containing an integral term, such as the IPD controller, would have to be considered. This will be addressed in a further example in Sect. 4.5.

As already pointed out in Chap. 1, the commonly used PI, PD and PID controllers introduce zeros in the closed loop transfer function, which can cause a finite number of overshoots and undershoots in the step response even with negative real closed loop poles. It is sometimes possible, however, but under restrictions on the attainable closed loop dynamics, to alleviate the effects of the zeros where only the PI, PD and PID controllers are available, by cancelling them with closed loop poles and this is addressed in Appendix A4. If the control system designer is free to programme any algorithm on the control processor, however, then the IP, DP or IPD controllers can be employed so as not to introduce the problem.

4.4.3 Cascade Control Structure

Plants are often structured as a number of sub-plants with the interconnecting variables available as measurements. The technique of *cascade control* is often practiced in these cases, which entails several controllers, one for each sub-plant arranged as a number of nested, or cascaded, loops as shown in Fig. 4.18.

The general cascade control structure for an arbitrary number of p sub-plants is readily deduced from Fig. 4.18. Sometimes, however, each of the measurement variables is shown entering a summing junction at the input of each controller, but this restricts the system to the use of traditional PID, P, PI or PD controllers.

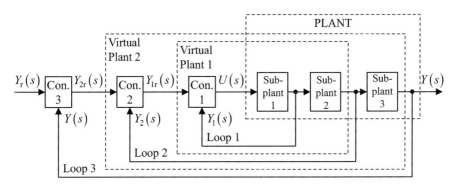

Fig. 4.18 Cascade control structure for SISO plant

These traditional controllers have just one input, i.e., the control error, i.e., the controller inputs would be $E_i(s) = Y_{ri}(s) - Y_i(s)$, $i = 1, 2, \ldots, p$. Figure 4.18 is more general, catering not only for the traditional controllers but also for other controllers with different structures for which $Y_{ri}(s)$ and $Y_i(s)$ have to be separate inputs. For example the IPD controller in Fig. 4.14 has $Y(s)$ and $Y_r(s)$ as separate inputs because the derivative term acts on $Y(s)$ alone.

The cascade control approach can ease the overall control system design task by separating it into a number of simpler tasks. In Fig. 4.18, Subplant 1 and Controller 1 form the innermost Loop 1 with $U(s)$ as the control variable. The control loops are designed individually, starting with Loop 1 and working outwards. Loop 1 and Subplant 2 form Virtual Plant 1 whose control variable is $Y_{1r}(s)$ provided by Controller 2. Similarly, Loop 2 and Subplant 3 form Virtual Plant 2 whose control variable is $Y_{2r}(s)$ provided by Controller 3. The design process is greatly simplified by designing the system such that $y_{1r}(t)$ is sufficiently slowly varying for the tracking error, $y_{1r}(t) - y_1(t)$, of Loop 1 to be negligible. Then Loop 2 can be designed ignoring the dynamics of Loop 1, by assuming that $y_{1r}(t)$ is the control variable of Plant 2. Similarly $y_{2r}(t)$ will be made sufficiently slowly varying for the tracking error, $y_{2r}(t) - y_2(t)$, of Loop 2 to be negligible. Then Loop 3 can be designed ignoring the dynamics of Loop 2, assuming that $y_{2r}(t)$ is the control variable of Plant 3. This is achieved as follows. The settling time of each loop is made at least an order of magnitude less than the next loop, working outwards, the settling time of each loop being that of its response to a step reference input with the loop considered in isolation from the rest of the system. This approach succeeds for the following reason. As the settling time of a loop is reduced, the gain or gains of its controller increases, thereby tightening the loop and reducing the tracking error for a given continuously varying reference input. Referring again to Fig. 4.18, a rule of thumb for design is that the settling time of Loop 1 is one tenth of the settling time of Loop 2, or less and the settling time of Loop 2 is one tenth of the settling time of Loop 3, or less. It must be realised, however, that the response of the controlled output, $y(t)$, to the reference input, $y_r(t)$, will not be precisely the same as in the ideal case of the direct control of Subplant 3 assuming $y_2(t) = y_{2r}(t)$, since in the real system, $y_2(t) \cong y_{2r}(t)$. In view of this uncertainty, a simulation is recommended to check that the overall control system performance meets the given specification for specific cases.

It would be possible to design a cascade control system taking the dynamics of Loops 2 and 3 into account, thereby removing the uncertainty in performance referred to in the last paragraph. This, however, would increase the complexity of the design task to a level comparable to that needed for the design of a single controller using all the available measurements.

Example 4.4 Cascade control of a throttle valve for internal combustion engines

Figure 4.19 shows a position control system for a throttle valve using a DC actuator, having the cascade control structure. Here, J is the moment of inertia, F is the viscous friction coefficient and K_s is the constant of the retension spring provided to ensure that the throttle valve automatically closes if the drive fails.

4.4 PID Controllers and Their Variants

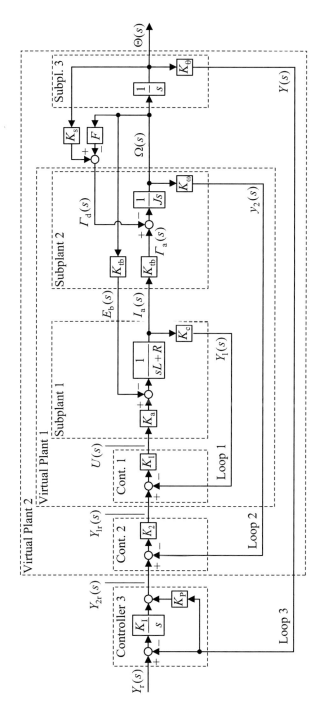

Fig. 4.19 Cascade position control of throttle valve for internal combustion engines

L is the armature inductance, R is the armature resistance and K_a is the drive power amplifier voltage gain. K_θ, K_ω and K_c are, respectively, the measurement constants for the vane angle, $\theta(t)$, the vane angular velocity, $\omega(t)$ and armature current, $i_a(t)$. Since sub-plants 1, 2 and 3 are each of first order, a proportional controller for each of the three loops would suffice to achieve closed loop stability. The retension spring, however, renders the plant of type '0', and would cause a steady state error in $\theta(t)$ for a non-zero constant reference angle, θ_r. To avoid this, Controller 3 is an IP controller. Then the output of the integral term will counteract the retension spring torque for a non-zero vane angle with an equal component of the actuator torque, $\Gamma_a(t)$, indirectly via Controller 1 and Controller 2.

In practice, the arbitrary use of PI controllers is sometimes found in cascade control solutions and they are usually tuned by trial and error, which can be time consuming. In the example of Fig. 4.19, this would introduce three integral terms where just one would be sufficient, the total system order being raised to six. Following the approach introduced in Sect. 4.5, however, it would still be possible to design the system to achieve a prescribed sixth order closed loop dynamics by the method of pole placement, but equally good results could be achieved with the fourth order dynamics of the system of Fig. 4.19. The alternative approach of simplification through the choice of the loop settling times will be followed here.

Commencing with the current control Loop 1, the back e.m.f. is regarded as an external disturbance and the transfer function relationship is

$$Y_1(s) = \frac{\dfrac{K_1 K_a Y_{1r}(s)}{sL+R} - \dfrac{E_b(s)}{sL+R}}{1 + \dfrac{K_1 K_a K_c}{sL+R}}$$

$$= \frac{K_1 K_a K_c Y_{1r}(s) - K_c E_b(s)}{sL + R + K_1 K_a K_c} = \frac{K_{\text{DCL1}} Y_{1r}(s) - K_{\text{DCE}} E_b(s)}{1 + sT_{c1}} \quad (4.68)$$

where

$$T_{c1} = \frac{L}{R + K_1 K_a K_c}, \quad K_{\text{DCL1}} = \frac{K_1 K_a K_c}{R + K_1 K_a K_c} \quad \text{and} \quad K_{\text{DCE}} = \frac{K_c}{R + K_1 K_a K_c} \quad (4.69)$$

As the gain, K_1, is increased, it is evident from (4.69) that the closed loop time constant, T_{c1}, reduces, the closed loop DC gain, K_{DCL1}, approaches the ideal value of unity and that the disturbance input DC gain, K_{DCE}, approaches zero, meaning that Loop 1 can be made very insensitive to the back e.m.f. The actuator electrical time constant, L/R, is of the order of milliseconds and therefore T_{c1} will be of this order, or less if $K_1 > 0$. To formalise the design of Loop 1, its settling time, T_{s1}, will be specified. Since, using the 5 % criterion, $T_{s1} = 3T_{c1}$, the value of the proportional gain, K_1, needed to realise it is given by (4.69) as

4.4 PID Controllers and Their Variants

$$K_1 = \frac{1}{K_a K_c}\left(\frac{3L}{T_{s1}} - R\right). \tag{4.70}$$

As observed previously, reducing the settling time increases the controller gains. Substituting for K_1 in K_{DCL1} and K_{DCE} given by (4.69) using (4.70) yields

$$K_{\text{DCL1}} = \frac{\frac{1}{K_a K_c}\left(\frac{3L}{T_{s1}} - R\right)K_a K_c}{R + \frac{1}{K_a K_c}\left(\frac{3L}{T_{s1}} - R\right)K_a K_c} = \frac{3L - RT_{s1}}{3L} \tag{4.71}$$

and

$$K_{\text{DCE}} = \frac{K_c}{R + \frac{1}{K_a K_c}\left(\frac{3L}{T_{s1}} - R\right)K_a K_c} = \frac{K_c}{3L}T_{s1}. \tag{4.72}$$

According to (4.71), reducing T_{s1} causes K_{DCL1} to approach the ideal value of unity and from (4.72) K_{DCE} approaches the ideal value of zero.

Speed control Loop 2 operates on a much longer time scale than Loop 1, the mechanical time constant, J/F, being typically of the order of 0.1 [s] so $y_{1r}(t)$ should be sufficiently slowly time varying to ensure $y_1(t) \cong y_{1r}(t)$, the assumption, $y_1(t) = y_{1r}(t)$, being made to simplify this loop to that of Fig. 4.20.

In this case, the torque, $\Gamma_{\text{Ld}}(s)$, is treated as an external disturbance. The transfer function relationship is

$$Y_2(s) = \frac{\frac{K_2 K_{\text{tb}} K_\omega}{K_c}\cdot\frac{1}{Js}Y_{2r}(s) - \frac{K_\omega}{Js}\Gamma_{\text{Ld}}(s)}{1 + \frac{K_2 K_{\text{tb}} K_\omega}{K_c}\cdot\frac{1}{Js}} = \frac{Y_{2r}(s) - \frac{K_c}{K_2 K_{\text{tb}}}\Gamma_{\text{Le}}(s)}{\frac{K_c J}{K_2 K_{\text{tb}} K_\omega}s + 1}$$

$$= \frac{K_{\text{DCL2}}Y_{2r}(s) - K_{\text{DCF}}\Gamma_{\text{Le}}(s)}{1 + sT_{c2}} = \frac{K_{\text{DCL2}}Y_{2r}(s) - K_{\text{DCF}}\Gamma_{\text{Le}}(s)}{1 + s\frac{T_{s2}}{3}},$$

$$\tag{4.73}$$

where T_{s2} is the settling time of Loop 2 (5 % criterion). If T_{s2} is specified, then the formula for calculating the required gain, K_2, is obtained from (4.73) as follows.

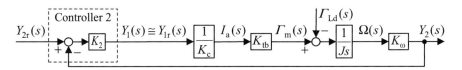

Fig. 4.20 Block diagram of simplified speed control Loop 2

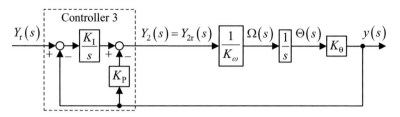

Fig. 4.21 Block diagram of simplified position control Loop 3

$$\frac{T_{s2}}{3} = \frac{K_c J}{K_2 K_{tb} K_\omega} \Rightarrow K_2 = \frac{3 K_c J}{K_{tb} K_\omega T_{s2}} \tag{4.74}$$

Inspection of (4.73) shows that the closed loop DC gain is $K_{DCL2} = 1$. An expression for the disturbance input DC gain, $K_{DC\Gamma}$, may also be obtained from (4.73) and then expressed in terms of T_{s2} by substituting for K_2 using (4.74). Thus

$$K_{DC\Gamma} = \frac{K_c}{K_2 K_{tb}} = \frac{K_c K_\omega}{\frac{3 K_c J}{K_{tb} K_\omega T_{s2}} K_{tb} K_\omega} = \frac{K_\omega}{3J} T_{s2}. \tag{4.75}$$

Hence reducing T_{s2} causes $K_{DC\Gamma}$ to approach the ideal value of zero.

If the position control Loop 3 is designed to have a settling time, T_{s3}, satisfying $T_{s3} \gg T_{s2}$, then $y_{2r}(t)$ should be sufficiently slowly varying for for $y_2(t) \cong y_{2r}(t)$, the assumption, $y_2(t) = y_{2r}(t)$, being made to simplify this loop to that of Fig. 4.21.

The closed loop transfer function is

$$\frac{Y(s)}{Y_r(s)} = \frac{\frac{K_\theta K_I}{K_\omega s^2}}{1 - \left[-\frac{K_\theta}{K_\omega s} \left(\frac{K_I}{s} + K_P \right) \right]} = \frac{\frac{K_\theta K_I}{K_\omega}}{s^2 + \frac{K_\theta K_P}{K_\omega} s + \frac{K_\theta K_I}{K_\omega}}. \tag{4.76}$$

As expected, this has a DC gain of unity due to the integral term of Controller 3. The settling time formula derived in Sect. 4.5.4 for the 5 % criterion applied to this system with critical damping yields

$$T_{s3} = 4.5 T_{c3} \tag{4.77}$$

where T_{c3} is the time constant of the double closed loop pole, meaning the reciprocal of its magnitude. The characteristic polynomial of (4.76) is then

$$s^2 + \frac{K_\theta K_P}{K_\omega} s + \frac{K_I K_\theta}{K_\omega} = \left(s + \frac{1}{T_{c3}}\right)^2 = \left(s + \frac{4.5}{T_{s3}}\right)^2 = s^2 + \frac{9}{T_{s3}} s + \frac{81}{4 T_{s3}^2} \tag{4.78}$$

from which the controller gains are given by

$$K_P = \frac{9K_\omega}{T_{s3}K_\theta} \text{ and } K_I = \frac{81K_\omega}{4T_{s3}^2 K_\theta}. \tag{4.79}$$

Next, step response simulations of the system of Fig. 4.19 will be compared with that of the ideal system having transfer function (4.76) with the gains of (4.79)]. The drive parameters are as follows. $R = 1.25$ [Ω]; $L = 0.02$ [H]; $K_{tb} = 0.026$ [Nm/A]; $J = 0.003$ [Kg m^2]; $F = 2 \times 10^{-3}$ [Nm/(rad/s)]; $K_s = 0.093$ [Nm/rad]; $K_c = 10$; $K_\omega = 1$; $K_\theta = 1$. The angular velocity and angular position constants are set to unity as they are derived from an encoder output and scaled in the control computer to be numerically in radians and radians per second. The specified settling time is fixed at $T_{s3} = 0.3$ [s]. The settling times of Loop 2 and Loop 1 are, respectively, set to $T_{s2} = \lambda T_{s3}$ and $T_{s1} = \lambda T_{s2}$, where $0 < \lambda < 1$. Two simulations are presented in Fig. 4.22. In the first, $\lambda = 0.1$ to satisfy the requirements for the simple loop by loop design approach. In the second, $\lambda = 0.2$. As shown in Fig. 4.22a, $\lambda = 0.1$ enables $y(t)$ to follow the ideal step response, $y_{\text{ideal}}(t)$, fairly closely and, as intended, the control errors, $y_{r1}(t) - y_1(t)$, and $y_{r2}(t) - y_2(t)$, are kept to relatively small proportions. The corresponding simulation for $\lambda = 0.2$ is shown in Fig. 4.22b on the same scales as Fig. 4.22a so that the increase in the errors may be seen. In this case, a significant difference between $y(t)$ and $y_{\text{ideal}}(t)$ is visible, the actual settling time, $T_{s\,\text{actual}}$, exceeding T_{s3}. Remarkably, the performance of Fig. 4.22a is achieved without the need for values of the plant parameters, K_s and F. This is an example of robustness introduced in Sect. 4.6.3 and achieved by the control techniques of Chaps. 9 and 10.

4.5 Systems of Third and Higher Order

4.5.1 Attainable Closed Loop Dynamics

Consider a linear control system whose transfer function has no finite zeros. Then if there is a sufficient number of independently adjustable controller parameters that can be chosen to attain any desired set of characteristic polynomial coefficients, the pole assignment procedure introduced in Sect. 4.2 can be applied to achieve any specified closed loop dynamics (settling time, percentage overshoot, and so forth), within the limitations set by the hardware. Then it is clear that the order of a control system that can be designed by pole assignment is considerably limited when employing traditional controllers due to their small numbers of gains. So a proportional controller is limited to first order plants as it only has one gain. An IP controller is also limited to first order plants although the closed loop system is of second order, since the integrator in the controller contributes 1 to the order.

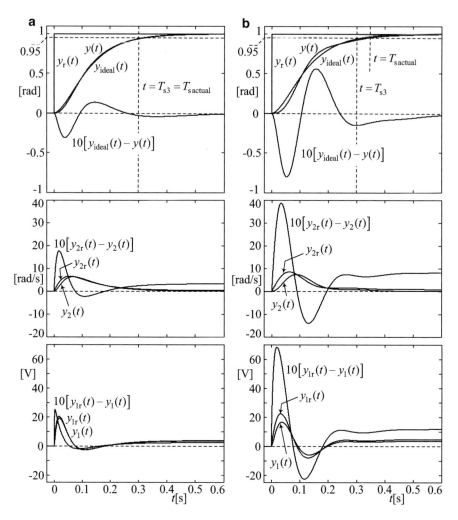

Fig. 4.22 Step responses of cascade position control of throttle valve. (**a**) $\lambda = 0.1$. (**b**) $\lambda = 0.2$

A DP controller is limited to second order plants as it has two gains. An IPD controller is also limited to second order plants although the closed loop system is of third order, again since the integrator in the controller contributes 1 to the order.

It is, however, possible to attain a satisfactory performance for some applications where it is not necessary to design to achieve a precisely specified closed loop dynamics. The classical frequency domain design methods accommodate this. Also, it is possible to apply *partial* pole assignment, in which a limited subset of the closed loop poles are realised but the attainable closed loop dynamics is limited due to the unassigned 'free' poles having to be dominated (Chap. 1) by the assigned poles.

4.5 Systems of Third and Higher Order

This is addressed in Appendix A4. The remaining sections of this chapter provide preparatory material for studying the control techniques presented in Chap. 5, which are free of the above restrictions.

4.5.2 The Laplace to Time Domain Inverse Scaling Law

It has already been established in Chap. 1 that the shapes of the impulse and step responses of any linear system, i.e., its dynamic character, depend on the *relative* locations of its poles and zeros, which will be referred to as the *pole-zero pattern*. For a third order system, the range of different pole-zero patterns and corresponding variations in the dynamic character is considerably larger than that of a second order system (Sect. 4.4.2). As the order increases further, the range of these variations of pole-zero pattern and dynamic character grows enormously. This potentially presents a challenge to the control system designer but the task of determining suitable closed loop pole locations for linear control systems is eased by the existence of a law relating the scales of the pole-zero patterns to the time scales of the step responses without altering their dynamic character. Linearly changing the scale of the pole-zero pattern in the s-plane by a factor of λ causes the time scale of the impulse or step response to be changed by a factor of $1/\lambda$ while preserving its shape. This inverse scaling law is proven as follows. The transfer function of a linear system with unit impulse response, $g_1(t)$, is given by its Laplace transform,

$$G_1(s) = \int_0^\infty e^{-st} g_1(t) dt \tag{4.80}$$

Consider another system with an impulse response, $g_2(t)$, of precisely the same shape and amplitude scale as $g_1(t)$, but on a different time scale such that

$$g_2(t) = g_1(\lambda t) \tag{4.81}$$

where λ is an arbitrary time scaling factor. Then its transfer function is

$$G_2(s) = \int_0^\infty e^{-st} g_2(t) dt = \int_0^\infty e^{-st} g_1(\lambda t) dt$$
$$= \frac{1}{\lambda} \int_0^\infty e^{-\frac{s}{\lambda} \lambda t} g_1(\lambda t) d(\lambda t) = \frac{1}{\lambda} G_1\left(\frac{s}{\lambda}\right). \tag{4.82}$$

Let p_i, $1 = 1, 2, \ldots, n$ and z_i, $1 = 1, 2, \ldots, m$ be, respectively, the poles and zeros of $G_1(s)$. Then according to (4.82), the poles and zeros of $G_2(s)$ are, respectively p_i/λ, $1 = 1, 2, \ldots, n$ and z_i/λ, $1 = 1, 2, \ldots, m$. This proves the scaling law. The coefficient, $1/\lambda$, multiplying $G_1(s/\lambda)$ in (4.82) is a result of the amplitude scales of the two systems being identical.

Example 4.5 Demonstration of the s to time domain scaling law

Here, the scaling law is demonstrated for a third order system. Preserving a DC gain of unity, the 'slow' version has transfer function,

$$\frac{y_s(s)}{y_r(s)} = \frac{1}{s+1} \cdot \frac{(0.5+j4)(0.5-j4)}{(s+0.5+j4)(s+0.5-j4)} = \frac{16.25}{(s+1)(s^2+s+16.25)}. \quad (4.83)$$

The 'fast' version has the pole pattern increased in scale by a factor of $\lambda = 2$, yielding

$$\frac{y_f(s)}{y_r(s)} = \frac{1}{(s/2)+1} \cdot \frac{16.25}{(s/2)^2+(s/2)+16.25} = \frac{2}{s+2} \cdot \frac{65}{s^2+2s+65}. \quad (4.84)$$

Figure 4.23a shows the pole patterns of the 'slow' and 'fast' versions and Fig. 4.23b shows the corresponding unit step responses. Figure 4.23b also exemplifies the step response shape, untypical of those of the familiar first and second order systems step responses, that exhibits oscillatory peaks, the first of which is *less* than the steady state value, followed by peaks at higher values. In this example, it is due to the influence of the exponential mode. For the two arbitrary values, y_1 and y_2, of the output marked on Fig. 4.23b, it is evident that the slow system takes precisely twice as long as the fast system to reach the same value.

If the controller has a set of adjustable parameters (gains and/or other adjustable coefficients) that permit the closed loop poles to be placed in any desired locations, i.e., pole assignment is possible, then the scaling law presented above can be used as a control system design tool. First, a system without finite zeros will be discussed. Suppose that a set of closed loop poles, $(s'_{c1}, s'_{c2}, \ldots, s'_{cn})$, has been found that produces the required form of step response but the settling time, T'_s,

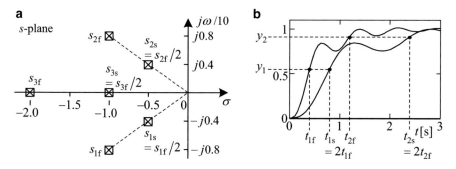

Fig. 4.23 Demonstration of the s to time domain inverse scaling law for a 3rd order system. (**a**) s domain. (**b**) Time domain

4.5 Systems of Third and Higher Order

is different from the specified value of T_s. Then the scaling factor, $\lambda = T_s/T'_s$, is formed. Calculating the controller parameters to produce a new set of poles, $(s_{c1}, s_{c2}, \ldots, s_{cn}) = (s'_{c1}/\lambda, s'_{c2}/\lambda, \ldots, s'_{cn}/\lambda)$ then yields the required step response.

If the plant transfer function has finite zeros, they are also zeros of the closed loop system that cannot be adjusted as part of the time scaling process. If they cause unacceptable overshoots and/or undershoots and are in the left half of the s-plane, then an external pre-compensator (Chap. 4) should be used to cancel them.

It is possible to create controllers that introduce independently adjustable zeros, but they would normally be used to cancel all the closed loop poles to achieve zero dynamic lag (Chap. 12) in which case the step response scaling law is irrelevant.

4.5.3 Step Responses with Coincident Closed Loop Poles

It will be recalled from Sect. 4.2 that a closed loop system in which complete pole assignment is possible contains at least n parameters that may be adjusted to yield a specified set of coefficients of the characteristic polynomial. The flexibility of modern digital implementation renders this possible for all controllable plants (nonlinear ones to which feedback linearisation is applied being included in Chap. 7). If the plant is linear and contains transport delays, then similar closed loop dynamics can be obtained using the discrete modelling of Chap. 2 and the discrete control techniques of Chap. 6. It has already been established in Chap. 1 that a closed loop system having real negative poles in the s-plane and no finite zeros has a step response that monotically increases and therefore does not overshoot. If the system has coincident closed loop poles, then the problem of completing the control system design hinges on the determination of just one parameter, i.e., the multiple pole location. Recalling the s to time domain scaling law of Sect. 4.5.2, the settling time would be a natural choice of design criterion. A formula relating the settling time of the step response to the order as well as the pole location would therefore be a valuable design aid. Hence the step responses of a set of closed loop linear systems of increasing order with coincident poles will be studied. The generic transfer function is:

$$\frac{Y(s)}{Y_r(s)} = \left(\frac{1}{1 + sT_c}\right)^n, \quad n = 1, 2, 3\ldots \quad (4.85)$$

where T_c is the time constant of the exponential decay of the polynomial exponential mode (Sect. 1.5.2). If $y_r(t) = Y_s h(t)$, where $h(t)$ is the unit step function and Y_s is the reference input level, then with the aid of Laplace (Table 1 in Tables),

$$y(t) = Y_s\left[1 - \sum_{i=0}^{n-1} \frac{1}{i!}(t/T_c)^i e^{-t/T_c}\right], \quad n = 1, 2, 3\ldots \quad (4.86)$$

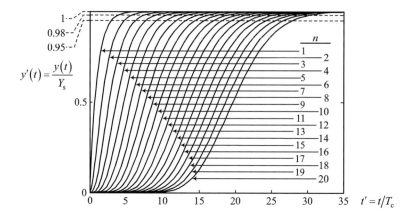

Fig. 4.24 Family of normalised step responses of linear system with multiple poles

Figure 4.24 shows a family of step responses for orders ranging between 1 and 20. The outputs and the time are normalised, respectively, with respect to Y_r and T_c to cover all linear systems. Then (4.86) becomes

$$y'(t) = 1 - \sum_{i=1}^{n-1} \frac{1}{i!}(t')^i e^{-t'}. \qquad (4.87)$$

The normalised settling times for the 5 % and 2 % criteria are, respectively, the times, $T_{s5\%}$ and $T_{s2\%}$, at which the normalised step responses cross the horizontal straight lines, $y' = 0.95$ and $y' = 0.98$. For the $x\%$ criterion, (4.87) therefore yields $y'(T_{sx\%}) = 1 - \sum_{i=1}^{n-1} \frac{1}{i!}(T_{sx\%})^i e^{-T'_{sx\%}}$ and since $x = 100\,[1 - y'(T_{sx\%})]$, then

$$0.01x = \sum_{i=1}^{n-1} \frac{1}{i!}(T'_{sx\%})^i e^{-T'_{sx\%}}. \qquad (4.88)$$

The solutions for $x = 5$ and $x = 2$ would be the required settling time formulae but they do not exist in the closed form,

$$T'_{sx\%} = f(x, n), \qquad (4.89)$$

as (4.88) is a transcendental equation. Numerical approximations, however, can yield settling time formulae that are sufficiently accurate for practical purposes.

4.5.4 Derivation of the Settling Time Formulae

4.5.4.1 Numerical Approach

Figure 4.24 reveals that the differences between the settling times of the responses of systems differing in order by 1 are roughly equal. Approximations to (4.88) of the form,

$$\tilde{T}'_{sx\%} = C(x) + M(x)n, \qquad (4.90)$$

should therefore exist that would serve as easily applied settling time formulas. To find these, the normalised settling times have been precisely computed for $n = 1, 2, \ldots, 20$ using a MATLAB®–SIMULINK® variable step simulation and the results are given in Table 4.1 and plotted in Fig. 4.25 as points indicated by ⊙. The question now arises of the choice of the linear approximation method. The classical approach would be to find the least squares fit using all 20 points of each plot. This, however, is more appropriate for data subject to random errors while the data of Table 4.1 is deterministic. Instead, two *fixing points* have been chosen (indicated by) on each plot, through which the straight lines will pass, to provide better approximations for the most common system orders.

With reference to Table 4.1, the formula should yield the well known result of $T_{s5\%} = 2.9957 T_c \cong 3 T_c$ for $n = 1$, given by (4.15). To guarantee this, the fixing point, $(n, T_{s5\%}) = (1, \overline{3})$, is selected for the straight line fit of the 5 % criterion. Also, for $n=1$, $T_{s2\%}=3.9118 T_c \cong 4 T_c$. The fixing point, $(n, T_{s2\%}) = (1, \overline{4})$, is therefore chosen for the straight line fit of the 2 % criterion. The slopes of the continuous curves satisfying (4.88) that would pass through the computed points shown in Fig. 4.25 decrease slightly with n. It follows that if a second fixing point is chosen on each of these curves for a selected value, $n_1 > 1$, of n, through which the straight lines will pass, then their equations will be exact at the fixing points and only slightly under-estimate the settling times between the points.

Table 4.1 Precise normalised settling times for multiple pole systems of order, n

n	$T'_{s5\%} = T_{s5\%}/T_c$	$T'_{s2\%} = T_{s2\%}/T_c$	n	$T'_{s5\%} = T_{s5\%}/T_c$	$T'_{s2\%} = T_{s2\%}/T_c$
1	2.9957	3.9118	11	16.9623	18.8298
2	4.7438	5.8338	12	18.2075	20.1352
3	6.2958	7.5165	13	19.4426	21.4279
4	7.7537	9.0841	14	20.6686	22.7094
5	9.1536	10.5804	15	21.8865	23.9809
6	10.5131	12.0270	16	23.0972	25.2434
7	11.8424	13.4364	17	24.3012	26.4976
8	13.1482	14.8166	18	25.4992	27.7444
9	14.4347	16.1731	19	26.6918	28.9844
10	15.7053	17.5098	20	27.8793	30.2181

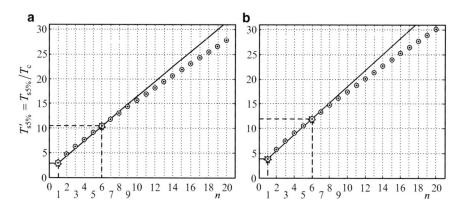

Fig. 4.25 Normalised settling time against system order and straight line fits. (**a**) 5 % criterion. (**b**) 2 % criterion

This is evident in Fig. 4.25. The second point has been chosen at $n = 6$ since commonly $n \in [1, 6]$. For $n = 6$, $T_{s5\%} = 10.513 T_c \cong 10.5 T_c$ giving the point, $(n, T_{s5\%}) = (6, 10.5)$ and $T_{s2\%} = 12.027 T_c \cong 12 T_c$ giving the point, $(n, T_{s2\%}) = (6, 12)$. The straight line fit (4.90) will be written as

$$\tilde{T}'_{sx\%} = C_x + M_x n \qquad (4.91)$$

as the functions, $C(x)$ and $M(x)$, are not needed analytically. Then the coefficients, C_x and M_x, will be determined using the fixing points, for $x = 5$ and $x = 2$.

4.5.4.2 The 5 % Formula

Using the fixing points of Fig. 4.25a, indicated by, yields

$$\begin{cases} C_5 + M_5 = 3 \\ C_5 + 6 M_5 = 10.5 \end{cases} \Rightarrow \begin{cases} 5 M_5 = 7.5 \Rightarrow M_5 = 1.5 \\ C_5 = 3 - M_5 = 1.5 \end{cases} \Rightarrow \tilde{T}'_{s5\%} = 1.5 (1 + n). \qquad (4.92)$$

For simplicity of notation, the settling time, using either approximation (4.92) or the one to be derived for the 2 % criterion, will be denoted T_s, the criterion being applied in a particular application being stated. Hence $\tilde{T}'_{s5\%} = T_s / T_c$ and therefore the settling time formula for the 5 % criterion follows from (4.92) as

$$\underline{T_s = 1.5 (1 + n) T_c}. \qquad (4.93)$$

The 5 % criterion will be used frequently in this book, as is common practice.

4.5 Systems of Third and Higher Order

During a control system design by pole placement, the working can be shortened by writing the characteristic polynomial directly as

$$\left(s + \frac{1}{T_c}\right)^n = \left[s + \frac{1.5(1+n)}{T_s}\right]^n. \tag{4.94}$$

4.5.4.3 The 2 % formula

Using the fixing points of Fig. 4.25b, indicated by, yields

$$\begin{Bmatrix} C_2 + M_2 = 4 \\ C_2 + 6M_2 = 12 \end{Bmatrix} \Rightarrow \begin{Bmatrix} 5M_2 = 8 \Rightarrow M_2 = 1.6 \\ C_2 = 4 - M_2 = 2.4 \end{Bmatrix} \Rightarrow \tilde{T}'_{s2\%} = 1.6(1.5+n) \tag{4.95}$$

where $\tilde{T}'_{s2\%} = T_s/T_c$. The settling time formula for the 2 % criterion is therefore

$$T_s = 1.6(1.5+n)T_c. \tag{4.96}$$

Again, the characteristic polynomial may be written directly as

$$\left(s + \frac{1}{T_c}\right)^n = \left[s + \frac{1.6(1+n)}{T_s}\right]^n. \tag{4.97}$$

4.5.5 Settling Time Formula Error Determination and Correction

Let the *actual* settling time obtained after application of (4.93) or (4.96) with a nominal settling time of T_s be denoted by T_{sa}. Then the normalised actual settling time will be defined as

$$T''_{sa} \triangleq \frac{T_{sa}}{T_s} \tag{4.98}$$

and would therefore be unity with zero error. The percentage error is then

$$e_\% = ((T_{sa} - T_s)/T_s) * 100\% = (T''_{sa} - 1) * 100 \ \%. \tag{4.99}$$

Table 4.2 shows the errors for systems up to 20th order. For $n = 1, 2, \ldots 10$, the errors are within ± 5 %. This is considered acceptable for most control system designs, as illustrated by the families of step responses in Fig. 4.26, which all nearly pass through the point for which $t = T_s$.

Table 4.2 Percentage errors of settling time formulae

n	5 % Criterion		2 % Criterion	
	T''_{sa}	$e_\%$	T''_{sa}	$e_\%$
1	0.9986	−0.14	0.9780	−2.20
2	1.0542	+5.42	1.0418	+4.18
3	1.0493	+4.93	1.0440	+4.40
4	1.0338	+3.38	1.0323	+3.23
5	1.0171	+1.71	1.0173	+1.71
6	1.0012	+0.12	1.0022	+0.22
7	0.9869	−1.31	0.9880	−1.20
8	0.9739	−2.61	0.9748	−2.52
9	0.9623	−3.77	0.9627	−3.73
10	0.9518	−4.82	0.9516	−4.84
11	0.9423	−5.77	0.9415	−5.85
12	0.9337	−6.63	0.9322	−6.78
13	0.9258	−7.42	0.9236	−7.64
14	0.9186	−8.14	0.9157	−8.43
15	0.9119	−8.81	0.9083	−9.17
16	0.9058	−9.42	0.9015	−9.85
17	0.9000	−10.00	0.8951	−10.9
18	0.8947	−10.53	0.8892	−11.08
19	0.8897	−11.03	0.8836	−11.64
20	0.8850	−11.50	0.8784	−12.16

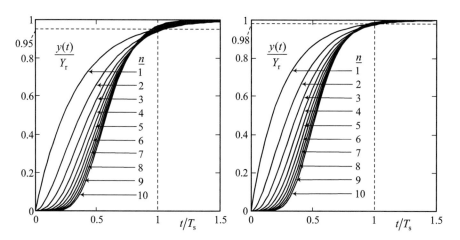

Fig. 4.26 Normalised step responses of systems designed using the settling time formulae

If the specified settling time has to be realised more accurately, then using the scaling law of Sect. 4.5.2, T''_{sa} can be looked up in Table 4.2 and the desired characteristic polynomial formed using the compensated settling time, T_{sc}, as follows.

4.5 Systems of Third and Higher Order

$$\left[s + \frac{1.5(1+n)}{T_{\text{sc}}}\right]^n \text{ (5 \% criterion) or } \left[s + \frac{1.6(1.5+n)}{T_{\text{sc}}}\right]^n \text{ (2 \% criterion)}, \tag{4.100}$$

where T_{sc} is calculated using the demanded settling time, T_{sd}, and T''_{sa} as follows.

$$\frac{T_{\text{sa}}}{T_{\text{s}}} = \frac{T_{\text{sd}}}{T_{\text{sc}}} \Rightarrow T_{\text{sc}} = \frac{T_{\text{s}}}{T_{\text{sa}}}.T_{\text{sd}} \Rightarrow T_{\text{sc}} = \frac{T_{\text{sd}}}{T''_{\text{sa}}}. \tag{4.101}$$

4.5.6 Closed Loop Poles for Given Overshoot and Settling Time

For some applications it may be desirable to design the control system to exhibit a small overshoot in the step response. For example, it has already been mentioned in Sect. 1.4.1 that the derivative kick produced by some of the traditional controllers due to the zeros introduced can reduce the steady state errors due to the nonlinear stick slip friction in position controlled mechanisms. Most of the controllers introduced in Chap. 4, however, do not introduce zeros, but complex conjugate pole placement may be used for the same purpose.

The procedure adopted here is as follows. Starting with multiple closed loop poles at $s_{1, 2,..., n} = -1/T_{\text{c}}$ according to one of the settling time formulae, i.e.,

$$T_{\text{c}} = \left\{ \frac{T_{\text{s}}}{1.5(1+n)} \text{ (5\% criterion)}, \text{ or } \frac{T_{\text{s}}}{1.6(1.5+n)} \text{ (2\% criterion)} \right\}, \tag{4.102}$$

the correct settling time would result but without overshooting. Then, if the order, n, of the closed loop system is even, all of the closed loop poles are moved to complex conjugate locations, $-(1 \pm jb)/T_{\text{c}}$. Then b is increased from zero until the required % overshoot is obtained. This, however, reduces the settling time so the final step is to apply the correction procedure of Sect. 4.5.5 based on the inverse scaling law of Sect. 4.5.2 to linearly shrink the closed loop pole pattern to increase the settling time to that specified without altering the percentage overshoot. The closed loop pole values then become

$$s_{1, 2,..., n} = -\frac{\lambda(1 \pm jb)}{T_{\text{c}}} \tag{4.103}$$

where λ is the scaling constant, which is reduced from unity until the required settling time is obtained. If n is odd, then the procedure is similar, but one pole has to remain real while the remaining poles are initially given values, $-(1 \pm jb)/T_{\text{c}}$, as before. If the real pole remained at $-1/T_{\text{c}}$, then its influence would prevent an overshoot occurring, as illustrated in Fig. 4.23 for a third order system. To avoid this, the magnitude of the real pole can be increased by an amount equal to the imaginary part magnitude, yielding a value of $-(1+b)/T_{\text{c}}$.

After increasing b until the specified percentage overshoot is obtained, the scaling is applied to increase the settling time to the specified value, as before, without altering the percentage overshoot, the closed loop pole locations finally being

$$\left\{ s_1 = -\frac{\lambda(1+b)}{T_c}, \quad s_{2,3,\ldots,n} = -\frac{\lambda(1 \pm jb)}{T_c} \right\}. \quad (4.104)$$

The pole locations are shown in Fig. 4.27, (*m) indicating multiplicity, m.

The desired characteristic polynomials will now be derived for $n \in [2, 6]$, as this should cater for most cases. The reader may use the same approach for higher orders if needed. Each complex conjugate pole pair will contribute a quadratic factor,

$$[s + \lambda(1+jb)/T_c][s + \lambda(1-jb)/T_c] = s^2 + 2\lambda/T_c s + \lambda^2(1+b^2)/T_c^2. \quad (4.105)$$

To simplify the polynomial expansions, let $\lambda/T_c = a$, $1 + b^2 = c$ and $1 + b = e$. Then the desired characteristic polynomials are as follows.

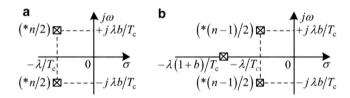

Fig. 4.27 Closed loop pole patterns for specified settling time and overshoot. (**a**) n even. (**b**) n odd

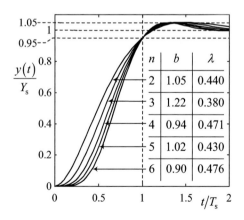

Fig. 4.28 Normalised step responses for 5 % overshoot and 5 % settling criterion

n	b	λ
2	1.05	0.440
3	1.22	0.380
4	0.94	0.471
5	1.02	0.430
6	0.90	0.476

$$\left.\begin{array}{l}\underline{n=2}: d(s)=s^2+2as+ca^2; \quad \underline{n=3}: d(s)=\left(s^2+2as+ca^2\right)(s+ea);\\ \underline{n=4}: d(s)=\left(s^2+2as+ca^2\right)^2; \quad \underline{n=5}: d(s)=\left(s^2+2as+ca^2\right)^2(s+ea);\\ \underline{n=6}: d(s)=\left(s^2+2as+ca^2\right)^3.\end{array}\right\}.$$
(4.106)

The corresponding normalised step responses are shown in Fig. 4.28.

4.6 Performance Specifications in the Frequency Domain

4.6.1 Background

Performance specifications are often given in the frequency domain. This is largely traditional and originates from the era of analogue control systems (Chap. 1) in which frequency domain based design methods prevailed. This, however, has continued through the modern branch of robust control theory known as H_∞ optimisation, which minimises the sensitivity (Sect. 4.6.3), equivalent to maximising the robustness to minimise the effects of plant modelling inaccuracies and external disturbances on the closed loop performance. This material is beyond the scope of this book but the reader may refer to authoritative texts on H_∞ (4, 5). Different approaches, however, to achieving robustness are given in Chaps. 8 and 9.

This section covers the few frequency domain performance parameters that are sometimes found in specifications and, where appropriate, relates them to the time domain in which most of the material of this book resides.

4.6.2 Closed Loop System Bandwidth

Given the closed loop transfer function, $G_{cl}(s)$, the closed loop system bandwidth, ω_b, is defined as the maximum angular frequency for which

$$|G_{cl}(j\omega_b)| = \frac{|G_{cl}(j0)|}{\sqrt{2}} \Rightarrow \frac{|G_{cl}(j\omega_b)|}{|G_{cl}(j0)|} = \frac{1}{\sqrt{2}}. \tag{4.107}$$

Expressed in Decibels, (4.107) becomes

$$20\log_{10}|G_{cl}(j\omega_b)| - 20\log_{10}|G_{cl}(j0)| = -3 \text{ dB}. \tag{4.108}$$

The more rapidly the reference input, $y_r(t)$, changes, the higher the proportion of high frequency components in its Fourier spectrum. The closed loop system must respond to these if it is to force the controlled output, $y(t)$, to follow $y_r(t)$ accurately. It follows that the higher the closed loop system bandwidth, the more accurately it follows rapid changes of $y_r(t)$ and there is an inverse relationship between the settling time, T_s, and the bandwidth, ω_b. This can be demonstrated quantitatively through the settling time formulae (4.93) or (4.96), which presumes that the control system structure permits complete pole assignment. Assuming unity closed loop DC gain and taking the 5 % criterion, from (4.93),

$$T_c = \frac{T_s}{1.5(1+n)} \quad (4.109)$$

Then $\frac{Y(s)}{Y_r(s)} = G_{cl}(s) = \frac{1}{(1+sT_c)^n} \Rightarrow \frac{1}{|1+j\omega_b T_c|^n} = \frac{1}{\sqrt{2}} \Rightarrow |1+j\omega_b T_c|^n = \sqrt{2} \Rightarrow$

$$\left(1+\omega_b^2 T_c^2\right)^{\frac{n}{2}} = \sqrt{2} \Rightarrow \left(1+\omega_b^2 T_c^2\right)^n = 2 \Rightarrow \omega_b^2 T_c^2 = 2^{1/n} - 1 \Rightarrow$$

$$T_c = \frac{\sqrt{2^{1/n}-1}}{\omega_b} \quad (4.110)$$

Finally, eliminating T_c between (4.109) and (4.110) yields the formula that can be used to find the settling time corresponding to the closed loop system bandwidth. Thus

$$T_s = \frac{1.5(1+n)\sqrt{2^{1/n}-1}}{\omega_b}. \quad (4.111)$$

This confirms the inverse relationship between T_s and ω_b. Corresponding families of Bode magnitude plots and step responses are shown in Fig. 4.29 for systems ranging in order between 1 and 6.

Each Bode plot passes through the -3 dB point at $\omega = \omega_b$ rad/s. For a given value of ω_b, the settling time is almost independent of the order for $n = 1$, 2 and 3 due to the term, $\sqrt{2^{1/n}-1}$ reducing with n while the term, $1+n$, increases with n

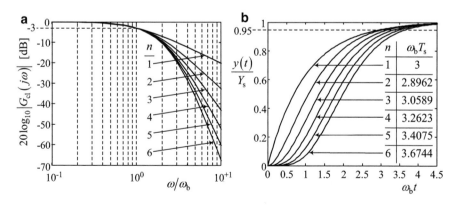

Fig. 4.29 Normalised closed loop plots for systems designed using the 5 % settling time formula. (a) Bode magnitude plot, (b) step responses

4.6 Performance Specifications in the Frequency Domain

by almost the same factor. For $n > 4$, however, the term, $\sqrt{2^{1/n} - 1}$, does not reduce sufficiently with n, resulting in T_s increasing with n, but not by huge proportions.

If the control system structure does not permit complete pole assignment, then invariably the closed loop system will not have multiple poles, but an estimate of the settling time, given the closed loop system bandwidth, could be obtained from (4.111) but replacing n by the number of dominant poles, minus the number of dominant zeros (Chap. 1).

4.6.3 Sensitivity and Robustness

4.6.3.1 Sensitivity

Definition 4.2 *The sensitivity of a control system is a measure of how much the dynamic response to the reference input differs from the specified one in the presence of plant modelling uncertainties and external disturbances.*

For a linear control system, an expression for the sensitivity can be derived using the closed loop transfer function since this uniquely corresponds to the specified step response. A linear continuous control system with any structure can be transformed to a transfer function block diagram of the classical form shown in Fig. 4.30.

Here, $G(s)$ is the transfer function of the physical plant. Let the transfer function of its model be $\tilde{G}(s)$ such that

$$G(s) = \tilde{G}(s) + \delta G(s) \tag{4.112}$$

where $\delta G(s)$ is the plant modelling error. Similarly, let the closed loop transfer function of the physical system (with the link, L, connected) be

$$\left. \frac{Y(s)}{Y_r(s)} \right|_{D(s)=0} = G_{cl}(s) \tag{4.113}$$

and let the closed loop transfer function obtained by applying the same controller to the plant model, $\tilde{G}(s)$, be $\tilde{G}_{cl}(s)$. Then the following relationship may be written.

$$G_{cl}(s) = \tilde{G}_{cl}(s) + \delta G_{cl}(s), \tag{4.114}$$

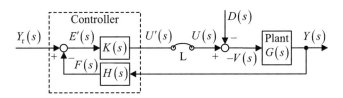

Fig. 4.30 Classical representation of linear continuous SISO control system

where $\delta G_{cl}(s)$ is the amount by which the closed loop transfer function changes if the plant transfer function changes from $\tilde{G}(s)$ to $G(s)$. Then the *sensitivity* of the closed loop transfer function with respect to the plant transfer function is defined as

$$S_P^C(s) = \lim_{\delta G(s) \to 0} \frac{\delta G_{cl}(s)/G_{cl}(s)}{\delta G(s)/G(s)} = \frac{G(s)}{G_{cl}(s)} \lim_{\delta G(s) \to 0} \frac{\delta G_{cl}(s)}{\delta G(s)} = \frac{G(s)}{G_{cl}(s)} \frac{dG_{cl}(s)}{dG(s)}. \tag{4.115}$$

It is a measure of the per unit change in the closed loop transfer function caused by a per unit change in the plant transfer function (due to the combined effect of the individual errors introduced when forming each part of the plant model).

Applying (4.115) to Fig. 4.30 but using the known model transfer function, $\tilde{G}(s)$, in place of $G(s)$ yields the well known expression for the sensitivity,

$$\begin{aligned}
S_P^C(s) &= \frac{1 + K(s)\tilde{G}(s)H(s)}{K(s)} \cdot \frac{d}{d\tilde{G}(s)} \left(\frac{K(s)\tilde{G}(s)}{1 + K(s)\tilde{G}(s)H(s)} \right) \\
&= \frac{1 + K(s)\tilde{G}(s)H(s)}{K(s)} \cdot \frac{\left[1 + K(s)\tilde{G}(s)H(s)\right] K(s) - K(s)\tilde{G}(s)K(s)H(s)}{\left[1 + K(s)\tilde{G}(s)H(s)\right]^2} \\
&= \frac{1}{1 + K(s)\tilde{G}(s)H(s)} = \frac{1}{1 + \tilde{G}_L(s)}.
\end{aligned} \tag{4.116}$$

where $\tilde{G}_L(s)$ is the loop gain, since removing the link, L, in Fig. 4.30 and replacing $G(s)$ by $\tilde{G}(s)$ yields

$$-\frac{U'(s)}{U(s)}\bigg|_{Y_r(s)=D(s)=0} = K(s)\tilde{G}(s)H(s) = \tilde{G}_L(s). \tag{4.117}$$

The smaller the sensitivity of (4.116), the better is the control system performance with respect to the modelling uncertainties and external disturbances. In the frequency domain with $s = j\omega$, for a SISO plant, H_∞ optimisation [4, 5] may be used to find the controller transfer functions that minimise the peak value of the sensitivity function, $|S_P^C(j\omega)|$, over the range of frequencies that the control system is expected to operate. This method is readily generalised for multiple input, multiple output (MIMO), i.e., multivariable, linear plants and can achieve closed loop stability over a wide range of uncertain plant parameters but is not suited to directly satisfying a given transient response specification. It is, however, excellent

4.6 Performance Specifications in the Frequency Domain

in applications where the closed loop dynamics is not critical. This book, however, focuses mainly on control techniques that yield specified closed loop dynamics. The sensitivity function will be used, where appropriate, to assess the performance when applying these control techniques, some of which (Chap. 9 and 10) are deliberately created to yield low sensitivity.

With reference to Fig. 4.30,

$$\left.\frac{V(s)}{D(s)}\right|_{Y_r(s)=0} = \frac{1}{1 + K(s)\tilde{G}(s)H(s)} = S_P^C(s). \qquad (4.118)$$

Although the sensitivity has been defined in terms of the plant transfer function uncertainties, (4.118) indicates that it is also a measure of how effectively the controller rejects external disturbances, $V(s) = D(s) - U(s)$ being the residual plant input due to the external disturbance, which would be zero with an ideal controller that would perfectly counteract the external disturbance by applying $U(s) = D(s)$.

Again with reference to Fig. 4.30,

$$\left.\frac{E'(s)}{Y_r(s)}\right|_{d(s)=0} = \frac{1}{1 + K(s)\tilde{G}(s)H(s)} = S_P^C(s) \qquad (4.119)$$

The control system analysis toolbox in MATLAB®–SIMULINK® may be used to display sensitivity plots on the Decibel scale in the form,

$$\left|S_P^C(j\omega)\right|_{dB} \triangleq 20\log_{10}\left|S_P^C(j\omega)\right| \qquad (4.120)$$

from a block diagram with input, $Y_r(s)$, and output, $E'(s)$.

It is common for unity feedback control systems to be referred to when carrying out sensitivity analysis. For this reason, readers may be tempted to convert any control system to the unity feedback control structure before carrying out the sensitivity analysis, regardless of its original structure, but it will now be proven that this will lead to incorrect sensitivity assessment. Suppose that the generic system of Fig. 4.30 is converted to the unity feedback structure of Fig. 4.31.

In order to force the system of Fig. 4.31 to be equivalent to that of Fig. 4.30, it must have the same closed loop transfer function, as follows.

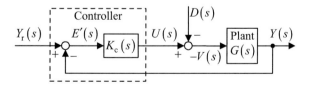

Fig. 4.31 Generic linear control system block diagram with the unity feedback structure

$$\left.\frac{Y(s)}{Y_r(s)}\right|_{D(s)=0} = \frac{K_c(s)G(s)}{1+K_c(s)G(s)} = \frac{K(s)G(s)}{1+K(s)G(s)H(s)} \Rightarrow$$
$$[1+K(s)G(s)H(s)]K_c(s) = [1+K_c(s)G(s)]K(s) \Rightarrow$$
$$\{1+K(s)G(s)[H(s)-1]\}K_c(s) = K(s) \Rightarrow$$

$$K_c(s) = \frac{K(s)}{1+K(s)G(s)[H(s)-1]}. \tag{4.121}$$

For a system already having the unit feedback structure, $H(s) = 1$, and as would be expected, $K_c(s) = K(s)$. The sensitivity function for the system of Fig. 4.31 is

$$S_P^{C'}(s) = \frac{1}{1+K_c(s)\tilde{G}(s)}. \tag{4.122}$$

Expressing this in terms of the original controller transfer functions, $K(s)$ and $H(s)$, using (4.121) then yields

$$S_P^{C'}(s) = \frac{1}{1+\frac{K(s)\tilde{G}(s)}{1+K(s)\tilde{G}(s)[H(s)-1]}} = \frac{1+K(s)\tilde{G}(s)[H(s)-1]}{1+K(s)\tilde{G}(s)H(s)}. \tag{4.123}$$

By comparison with (4.116), $S_P^{C'}(s) \neq S_P^C(s)$ for $H(s) \neq 1$.

In order to establish a numerical scale by means of which the goodness of a control system may be judged using sensitivity evaluation, the upper end of the scale will be the worst case, which will be taken as the sensitivity of an open loop system with transfer function,

$$\frac{Y(s)}{Y_r(s)} = P(s)G(s) = G_{ol}(s), \tag{4.124}$$

where $P(s)$ is the pre-compensator transfer function designed to yield the required *nominal* dynamic response of $y(t)$ to $y_r(t)$, since such a control system cannot take any corrective action, as is the case with feedback, to compensate for plant modelling errors or external disturbances. In this case, following similar lines to the derivation of (4.115) and using (4.124), the sensitivity of the open loop transfer function with respect to the plant transfer function is

$$S_P^O(s) = \frac{G(s)}{G_{ol}(s)} \frac{dG_{ol}(s)}{dG(s)} = \frac{G(s)}{P(s)G(s)} \frac{d[P(s)G(s)]}{dG(s)} = 1. \tag{4.125}$$

Thus the sensitivity scale between 'ideal' and very poor for a closed loop system is simply $S_P^C \in [0, 1]$. Exceptionally, however, it is possible for S_P^C to exceed unity. On the Decibel scale of (4.120), all control systems should have negative sensitivities, the worst case being 0 dB. Those considered robust would be expected to exhibit sensitivity plots satisfying $\left|S_P^C(j\omega)\right|_{dB} < -10$ dB.

4.6 Performance Specifications in the Frequency Domain

4.6.3.2 Robustness

Definition 4.3 *The robustness of a control system is a measure of how well the specified response to the reference input is maintained in the presence of plant modelling uncertainties and external disturbances.*

For quantification, an expression is formed that evaluates robustness on the same scale as sensitivity, such that a control system with the minimum possible sensitivity has the maximum possible robustness. Recalling (4.116), this is

$$R_P^C(s) = 1 - S_P^C(s) = \frac{K(s)\tilde{G}(s)H(s)}{1 + K(s)\tilde{G}(s)H(s)} = \frac{\tilde{G}_L(s)}{1 + \tilde{G}_L(s)}. \quad (4.126)$$

Since robustness is complementary to sensitivity, $R_P^C(s)$ is often called the complementary sensitivity function.

4.6.3.3 Sensitivity and Robustness as Specifications

In the frequency domain,

$$\left|S_P^C(j\omega)\right| = \frac{1}{\left|1 + \tilde{G}_L(j\omega)\right|} \quad \text{and} \quad \left|R_P^C(j\omega)\right| = \frac{\left|\tilde{G}_L(j\omega)\right|}{\left|1 + \tilde{G}_L(j\omega)\right|}. \quad (4.127)$$

The control system design specification is sometimes given in the form of a boundary below which $|S_P^C(j\omega)|_{dB}$, given by (4.120), must lie for all $\omega \in (0, \omega_b)$.

Alternatively, the specification can be a boundary above which

$$\left|R_P^C(j\omega)\right|_{dB} \triangleq 20\log_{10}\left|R_P^C(j\omega)\right| \quad (4.128)$$

must lie for all $\omega \in (0, \omega_b)$.

It is important to note that sensitivity and robustness are separate performance specifications from the closed loop bandwidth defined in Sect. 4.6.2, or transient response specifications such as settling time and percentage overshoot. By varying the control system structure or the controller parameters, it is possible even to keep the closed loop transfer function the same while varying the robustness or sensitivity, as demonstrated in the following example.

Example 4.6 Hybrid IPD/IDP control of a double integrator plant

Figure 4.32 shows a control system with two poles and one finite zero that can all be placed independently.

The double integrator plant could represent applications such as single axis rigid body spacecraft attitude control or an air bearing based positioning mechanism.

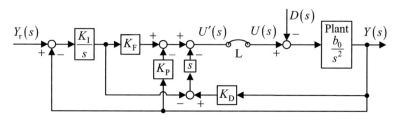

Fig. 4.32 Hybrid IPD/IDP controller applied to a double integrator plant

It will be subsequently demonstrated that reducing the sensitivity (or increasing the robustness) reduces the deviation of the step response from the ideal step response due to the external disturbance, $D(s)$, and any plant modelling errors.

The closed loop transfer function with the link, L, connected is as follows.

$$\left.\frac{Y(s)}{Y_r(s)}\right|_{D(s)=0} = \frac{b_0 K_I (s + K_F)}{s^3 + b_0 [K_D s^2 + (K_P + K_I) s + K_F K_I]}. \quad (4.129)$$

Let the closed loop system have two poles at $s_{1,2} = -p_d$ and the other pole at $s_3 = -p_f$. Then the zero could be placed to cancel this pole by setting

$$K_f = p_f, \quad (4.130)$$

the closed loop transfer function then becoming

$$\frac{Y(s)}{Y_r(s)} = \frac{b_0 K_I (s + K_F)}{(s + p_d)^2 (s + p_f)} = \frac{b_0 K_I}{(s + p_d)^2}, \quad (4.131)$$

producing a critically damped second order step response with zero overshoot. The 5 % settling time formula (4.93) could then be used with $n = 2$ by setting

$$p_d = \frac{4.5}{T_s}. \quad (4.132)$$

To determine the remaining controller parameters, the desired closed loop characteristic polynomial is

$$(s + p_d)^2 (s + p_f) = s^3 + (p_f + 3p_d) s^2 + p_d (p_d + 2p_f) s + p_d p_f^2. \quad (4.133)$$

Equating this to the denominator of (4.128) and using (4.129) then yields

$$K_d = (p_f + 2p_d)/b_0, \quad K_I = p_d^2/b_0 \quad \text{and} \quad K_p = 2 p_d p_f / b_0. \quad (4.134)$$

4.6 Performance Specifications in the Frequency Domain

It is clear that closed loop system response to the reference input is independent of the pole magnitude, p_f. The robustness and sensitivity, however, will vary with p_f, and this will now be investigated. With the link, L, removed, the loop transfer function is

$$G_L(s) = -\frac{U'(s)}{U(s)}\bigg|_{Y_r(s)=D(s)=0} = \frac{b_0}{s^3}\left[K_D s^2 + (K_P + K_I)s + K_F K_I\right]. \quad (4.135)$$

The sensitivity is therefore

$$S_P^C(s, p_f) = \frac{1}{1 + G_L(s)} = \frac{s^3}{s^3 + b_0[K_D s^2 + (K_P + K_I)s + K_F K_I]}$$

$$= \frac{s^3}{(s + p_d)^2 (s + p_f)} \quad (4.136)$$

and the robustness is

$$R_P^C(s, p_f) = 1 - \frac{s^3}{(s + p_d)^2 (s + p_f)}. \quad (4.137)$$

It is evident by inspection of (4.136) and (4.137) that

$$\lim_{p_f \to \infty} S_P^C(s, p_f) = 0 \quad \text{and} \quad \lim_{p_f \to \infty} R_P^C(s, p_f) = 1. \quad (4.138)$$

It follows that there exists a finite value of p_f above which plant parameteric uncertainties and external disturbances have negligible effects on the control system performance. If s were to be real and finite, it would be clear by inspection of (4.136) and (4.137) that as p_f is increased from zero, with $p_d > 0$, $S_P^C(s, p_f)$ would commence at a finite positive value less than unity and reduce monotonically towards zero, while $R_P^C(s, p_f)$ would commence at a finite positive value less than unity and increase monotonically towards unity. Since s is a complex variable, however, this conclusion cannot be made without further analysis. From (4.136), the magnitude of the sensitivity in the frequency domain is

$$\left|S_P^C(j\omega, p_f)\right| = \frac{\omega^3}{(\omega^2 + p_d^2)\sqrt{\omega^2 + p_f^2}}. \quad (4.139)$$

This shows that as p_f is varied from 0 to ∞, $|S_P^C(j\omega, p_f)|$ commences at $\omega^2/(\omega^2 + p_d^2)$ and monotonically reduces to zero.

From (4.137) the robustness is

$$R_P^C(s, p_f) = 1 - S_P^C(s, p_f) = \frac{(2p_d + p_f)s^2 + (p_d^2 + 2p_d p_f)s + p_d^2 p_f}{s^3 + (2p_d + p_f)s^2 + (p_d^2 + 2p_d p_f)s + p_d^2 p_f}. \quad (4.140)$$

Then the magnitude of the robustness in the frequency domain is

$$\left|R_P^C(j\omega, p_f)\right| = \sqrt{\frac{\left[p_d^2 p_f - (2p_d + p_f)\omega^2\right]^2 + \left[(p_d^2 + 2p_d p_f)\omega\right]^2}{\left[p_d^2 p_f - (2p_d + p_f)\omega^2\right]^2 + \left[(p_d^2 + 2p_d p_f)\omega - \omega^3\right]^2}}. \quad (4.141)$$

The convergence behaviour is certainly not evident by inspection of (4.141) and therefore some further manipulation will be carried out to enable its determination. Like terms may be recognised in the numerator and denominator of (4.141) and therefore it may be rewritten as

$$\left|R_P^C(j\omega, p_f)\right| = \sqrt{\frac{f(\omega, p_f)}{f(\omega, p_f) + \omega^6 - 2(p_d^2 + 2p_d p_f)\omega^4}}, \quad (4.142)$$

where

$$\begin{aligned}
f(\omega, p_f) &= \left[p_d^2 p_f - (2p_d + p_f)\omega^2\right]^2 + \left[(p_d^2 + 2p_d p_f)\omega\right]^2 \\
&= p_d^4 p_f^2 - 2p_d^2 p_f(2p_d + p_f)\omega^2 + (2p_d + p_f)^2\omega^4 + (p_d^4 + 4p_d^3 p_f + 4p_d^2 p_f^2)\omega^2 \\
&= (4p_d^2 + 4p_d p_f + p_f^2)\omega^4 + (p_d^4 + 2p_d^2 p_f^2)\omega^2 + p_d^4 p_f^2.
\end{aligned} \quad (4.143)$$

Substituting for $f(\omega, p_f)$ in (4.142) using (4.143) yields

$$\left|R_P^C(j\omega, p_f)\right| = \sqrt{\frac{(4p_d^2 + 4p_d p_f + p_f^2)\omega^4 + (p_d^4 + 2p_d^2 p_f^2)\omega^2 + p_d^4 p_f^2}{\omega^6 + (2p_d^2 + p_f^2)\omega^4 + (p_d^4 + 2p_d^2 p_f^2)\omega^2 + p_d^4 p_f^2}},$$

which can be written as

$$\left|R_P^C(j\omega, p_f)\right| = \sqrt{\frac{(4p_d^2 + 4p_d p_f)\omega^4 + p_d^4 \omega^2 + (\omega^4 + 2p_d^2 \omega^2 + p_d^4) p_f^2}{\omega^6 + 2p_d^2 \omega^4 + p_d^4 \omega^2 + (\omega^4 + 2p_d^2 \omega^2 + p_d^4) p_f^2}}. \quad (4.144)$$

By inspection of (4.144), as p_f is increased from zero, $|R_P^C(j\omega, p_f)|$ commences at $\sqrt{\frac{4p_d^2 \omega^2 + p_d^4}{\omega^4 + 2p_d^2 \omega^2 + p_d^4}}$ and varies monotonically towards $\lim_{p_f \to \infty} |R_P^C(j\omega, p_f)| = 1$, since the quadratic terms in p_f, which are identical in the numerator and denominator of the fraction inside the square root, become dominant.

Some simulation results are presented in Fig. 4.33 in which the estimated plant parameter is $\tilde{b}_0 = 1$ and $T_s = 1$ [s]. In Fig. 4.33a–c the third pole is placed at $p_f = p_d = 4.5/T_s$. In Fig. 4.33b–d, p_f is increased to $100 p_d$. The effects of changing the magnitude of the pole, s_3, are presented in Fig. 4.33. In Fig. 4.33b, e, a step external

4.6 Performance Specifications in the Frequency Domain

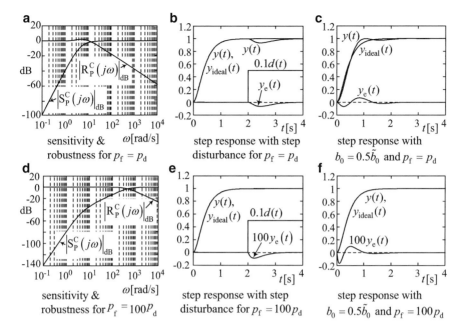

Fig. 4.33 Effects of pole placement on robustness and sensitivity

disturbance, $d(t) = 5h(t-2)$, is applied. In Fig. 4.33c, f, the plant parameter, b_0, is reduced to one half of the value, $\tilde{b}_0$, assumed in the control system design. Comparison of Fig. 4.33a–c with d–f indicates the extreme increase in robustness and reduction in sensitivity brought about by increasing the third pole magnitude by a factor of 100.

This is due to the increases in the controller gains K_p, K_d and K_f which can be seen in (4.130) and (4.134). Comparison of Fig. 4.33b, e for $t \in (0, 2)$ [s], i.e., before the external disturbance is applied, indicates identical transient responses despite the movement of the third pole, due to the pole-zero cancellation.

Example 4.6 is intended to emphasise the fact that the transient response specifications and sensitivity/robustness specifications are essentially separate and can be considered complementary. For example, if the controllers of Chaps. 4 and 10 control the same plant with the same dynamics, much higher robustness will be achieved with the controllers of Chap. 10.

4.6.4 Stability Analysis in the Frequency Domain

4.6.4.1 Background

The stability analysis of linear control systems using Bode, Nyquist or Nichols plots stem from the era of analogue electronic controllers in which the main adjustable parameter was the gain, K, of a power amplifier. This analysis enabled control systems to be designed that tolerated changes (usually reductions) of the gain, K, due to component aging and was able to take into account the potentially destabilising effects of decoupling capacitors and noise filters built into the circuit design.

The basic stability margins, described in the following sections, are often quoted as performance measures and the designer of a nominally linear control system may be asked to quote them, or work with them as specifications, regardless of the control system structure. This section therefore includes definitions of the stability margins that apply to any linear control system. The traditional control system structure for which this analysis was originally developed is that of Fig. 4.30, containing a series compensator, $K(s)$, and/or a feedback compensator, $H(s)$, for which frequency domain design methods are widely published [1–3]. The analysis is based on the steady state sinusoidal response of the open loop system, which is determined by the relationship,

$$U'(j\omega) = G_L(j\omega) U(j\omega). \quad (4.145)$$

Also the external disturbance is not considered in this analysis and set to zero as in a linear system it will not affect the stability. A restriction is that the open loop transfer function, $G_L(j\omega)$, (defined as $-U'(s)/U(s)$ with the link, L, removed) has a low pass filter characteristic, requiring the degree of the numerator polynomial to be less than that of the denominator polynomial, otherwise the loop closure would create an algebraic loop. As ω increases, the phase angle,

$$\phi(\omega) = \arg[G_L(j\omega)] = \tan^{-1}\{\operatorname{Im}[G_L(j\omega)]/\operatorname{Re}[G_L(j\omega)]\} \quad (4.146)$$

becomes negative as $\omega \to \infty$, so that the sinusoid, $u'(t)$, lags behind the sinusoid, $u(t)$.

4.6.4.2 Nyquist Stability Criterion

A useful means of determining the stability of a linear control system in the frequency domain is the Nyquist Stability Criterion [1, 2, 6], based on the plot of $G_L(j\omega) = G_{re}(\omega) + jG_{im}(\omega)$, which is also the polar plot, of $|G_L(j\omega)|$ against $\phi(\omega)$. This is the Nyquist plot. The Nyquist Stability Theorem is as follows. Let n_{ce} be the number of clockwise encirclements of the critical point, $(G_{re}, G_{im}) = (-1, 0)$, as ω goes from $-\infty$ to ∞, counting -1 for any anti-clockwise encirclement. Let n_{cp} be the number of closed loop poles in the right half of the s-plane and let n_{op} be the number of poles of $G_L(s)$, in the right half of the s-plane. Then

$$n_{cp} = n_{op} + n_{ce}. \quad (4.147)$$

4.6 Performance Specifications in the Frequency Domain

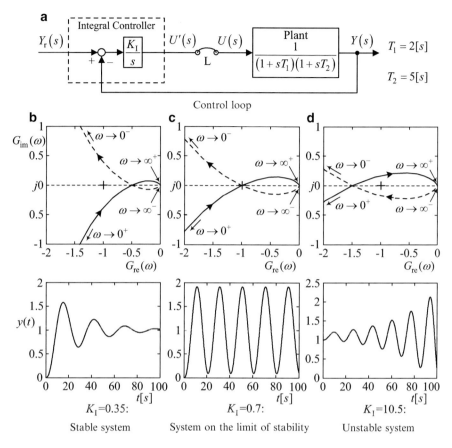

Fig. 4.34 Nyquist plots for an integral control loop

The Nyquist Stability Criterion then states that for the closed loop system to be stable, $n_{cp} \leq 0$. Figure 4.34 shows three Nyquist plots of an integral control system for different gain settings.

$G_L(j\omega)$ is given by (4.145) with the link, L, removed. The controller here is a relatively rare variant on the traditional PID controller that contains only the integral term. If the plant has negative real poles, then it is possible for an integral control loop to be stable for a certain range of gain settings, as shown in this example, which could be a heating process with two dominant time constants. In Fig. 4.34b, the Nyquist plot does not encircle the critical point (marked + in all the Nyquist plots to follow) and the system is stable. In Fig. 4.34c, it passes through the critical point twice without encircling it and the system oscillates at a constant amplitude, on the borderline between stability and instability. In Fig. 4.34c, the Nyquist plot executes one clockwise encirclement of the critical point and the system is unstable.

4.6.4.3 Relative Stability

The Nyquist stability criterion establishes whether or not a system is stable but it is useful for the control system designer to also have a measure of how close the system is to instability. Specifically, let

$$G_L(s) = K_{DC} \frac{G_1(s)}{s^q} \qquad (4.148)$$

where K_{DC} is the open loop DC gain and $q \geq 0$ is an integer, then the controller gains have to be set to values such that there is some margin by which K_{DC} can increase, possibly due to any changes in the plant, without exceeding the stability limit and encountering a situation such as demonstrated in Fig. 4.34. The stability margins of the following sections fulfil this purpose.

4.6.4.4 Gain Margin

Suppose that there exists a finite frequency, referred to as the phase crossover frequency, ω_{pc}, at which the phase angle plot, $\phi(\omega) = -180°$, or the phase plot accompanying a Bode plot crosses the $-180°$ line. Then It follows from a study of the Nyquist plot that if $|G_L(j\omega_{pc})| < 1$, the closed loop system would be stable but if $|G_L(j\omega_{pc})| > 1$, the closed loop system would be unstable. Figure 4.35 illustrates this using the Nyquist plot for the stable case of Fig. 4.34b.

The *gain margin* is the factor, G_m, by which the open loop DC gain, K_{DC}, would have to be increased to result in $|G_m G_L(j\omega_{pc})| = 1 \Rightarrow G_m |G_L(j\omega_{pc})| = 1$, thereby bringing the system to the stability limit. Hence

$$G_m = 1/|G_L(j\omega_{pc})| \quad \text{or} \quad G_{mdB} = -20\log_{10}|G_L(j\omega_{pc})|. \qquad (4.149)$$

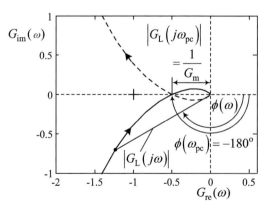

Fig. 4.35 Gain margin shown on a Nyquist plot

4.6 Performance Specifications in the Frequency Domain

Hence if $G_m > 1 \Rightarrow G_{mdB} > 0$, the closed loop system is stable, otherwise unstable. If $G_m = 1 \Rightarrow G_{mdB} = 0$ and the system is on the limit of stability. This can be illustrated by applying the construction of Fig. 4.35 to the Nyquist plot of Fig. 4.34c. If, on the other hand, $0 < G_m < 1 \Rightarrow G_{mdB} < 0$, then the system is unstable. This can be illustrated by applying the construction of Fig. 4.35 to the Nyquist plot of Fig. 4.34d.

The gain margin is interpreted differently for conditionally stable systems, as in Example 4.7 below. These systems are typically unstable for K_{DC} *below* a critical value, K_{DCmin}. Then the Nyquist plot executes at least one clockwise encirclement of the critical point, $-1+j0$, for all $K_{DC} \in (0, K_{DC\,min})$, above which the system is stable. Assuming that the controller gains have been set so that $K_{DC} \in (0, K_{DC\,min})$, then the gain margin is the reduction factor, G_m, such that $G_m K_{DC} = K_{DC\,min}$. Then according to (4.149), if $G_m < 1 \Rightarrow G_{mdB} < 0$, the closed loop system is stable, otherwise it is unstable.

4.6.4.5 Phase Margin

This is an alternative to the gain margin. Suppose that there exists a finite frequency, referred to as the gain crossover frequency, ω_{gc}, at which $|G_L(j\omega_{gc})| = 1$ or the Bode magnitude plot, $|G_L(j\omega)|_{dB}$ crosses the horizontal straight line at 0 dB. Then it follows from a study of the Nyquist plot that if $\phi(\omega_{gc}) > -180°$, loop closure would result in stability but if $\phi(\omega_{gc}) < -180°$, the closed loop system would be unstable.

The *phase margin*, ϕ_m, is the angle by which $\phi(\omega_{gc})$ would have to be reduced to make the net phase angle of the open loop system $-180°$, as shown in Fig. 4.36. Thus

$$\phi(\omega_{gc}) - \phi_m = -180° \Rightarrow \phi_m = \phi(\omega_{gc}) + 180°. \quad (4.150)$$

It follows that if $\phi_m > 0$, the closed loop system is stable, otherwise it is unstable.

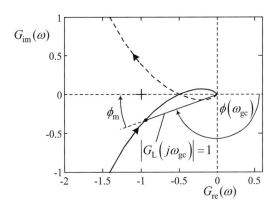

Fig. 4.36 Phase margin shown on a Nyquist plot

4.6.4.6 Delay Margin

Significant pure time delay elements can occur in plants, typically in the process industry (Chap. 2) also in communications links of remotely controlled objects. It will be assumed that just one such element with delay, τ, is present such that its transfer function, $e^{-s\tau}$, is a factor of $G_L(s)$. Then

$$G_L(s) = e^{-s\tau} G'_L(s) \qquad (4.151)$$

where $G'_L(s)$ is the non-delay factor of the open loop transfer function. Then in the frequency domain,

$$\arg[G_L(j\omega)] = \arg[G'_L(j\omega)] + \arg[e^{-j\omega\tau}] = \arg[G'_L(j\omega)] - \omega\tau \qquad (4.152)$$

and since $|e^{-j\omega\tau}| = 1$,

$$|G_L(j\omega)| = |G'_L(j\omega)| \qquad (4.153)$$

The phase margin is then given by (4.150) as

$$\phi_m = \arg[G'_L(j\omega_{gc})] - \omega_{gc}\tau + 180°. \qquad (4.154)$$

where $|G'_L(j\omega_{gc})| = 1$. The question then arises of how much additional time delay could be tolerated while maintaining stability. Thus the *delay margin* will be defined as the additional time delay, τ_m, that will bring the system to the limit of stability. This brings the phase margin to zero. Then (4.154) becomes

$$\begin{aligned} 0 &= \arg[G'_L(j\omega_{gc})] - \omega_{gc}(\tau + \tau_m) + 180° \Rightarrow \\ \tau_m &= \frac{1}{\omega_{gc}} \{\arg[G'_L(j\omega_{gc})] + 180°\} - \tau \end{aligned} \qquad (4.155)$$

In process control, this margin would be useful in determining the minimum tolerable flow rate of a fluid in a pipe responsible for the time delay of, for example, heat supply to a vessel, or the maximum distance of a remotely controlled object if the feedback controller is at the control station rather than in the object. It should be remarked that the time delays of the transmitted signals between spacecraft and the ground based mission control station can be very significant, ranging between seconds to hours depending upon the distance, but it is usual for the feedback controller to operate autonomously on board the spacecraft. The reference inputs, i.e., attitude and translational coordinate demands, however, are subject to these delays.

4.6 Performance Specifications in the Frequency Domain

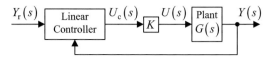

Fig. 4.37 Introduction of proportional gain, K, into a linear control system for classical stability analysis in the frequency domain

4.6.4.7 Stability Margin Assessment Via a Proportional Gain

In order to apply the traditional methods of analysis in the frequency domain to *any* linear control system, including those presented in Chap. 5, a proportional gain, K, may be introduced in the forward path of the block diagram, as in Fig. 4.37.

The control system would be implemented without this additional gain, equivalent to $K = 1 \Rightarrow U(s) = U_c(s)$, since the parameters of the controller will have been set to yield the required performance. The classical methods of analysis, however, could be useful in providing a measure of robustness as variation of the gain, K, could be regarded as being equivalent to a variation of or uncertainty in the open loop DC gain due to plant modelling errors and drift of parameters due to ageing.

Example 4.7 Classical stability analysis of the IPD/IDP control loop of Example 4.6.

In this case the open loop transfer function, $G_L(s)$, is that of (4.135) but multiplied by the proportional gain, K, as described above. Thus

$$G_L(s) = K \frac{b_0 \left[K_D s^2 + (K_P + K_I) s + K_F K_I \right]}{s^3}. \tag{4.156}$$

Figure 4.38 shows the root locus, Nyquist and Bode plots of the system of Example 4.6 with open loop transfer function (4.135), for $p_f = p_d = 4.5/T_s$, where $T_s = 1$ s.

First, it may be observed from the root locus that the system is conditionally stable. It is unstable for $K \in (0, K_{min}]$ and stable for $K \in (K_{min}, \infty]$. The three loci can be seen to pass simultaneously through the point, $s_{1,2,3} = -4.5/T_s = -4.5$ for $K = 1$, as the controller has been designed with this pole placement. The conditional stability can also be observed in the Nyquist plot.

It should be noted that the '−1' point is encircled in an anti-clockwise sense as ω goes from $-\infty$ to $+\infty$ indicating system stability. This situation holds for all $K > 1$, because increasing K linearly expands the plot and therefore the system remains stable, in keeping with the root locus. Reducing K linearly contracts the plot and the system remains stable until $K = K_{min}$ when the point where the plot crosses itself at $\omega = \pm \omega_{pc}$ coincides with the critical point, $-1 + j0$. For $K < K_{min}$, the '$\omega = \pm \omega_{pc}$' point lies to the right of the '−1' point. This point is now encircled

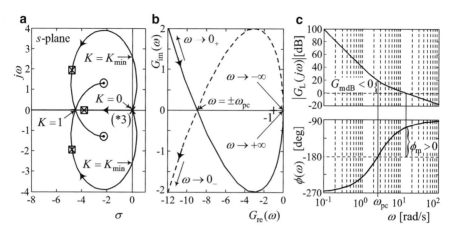

Fig. 4.38 Bode plots of the open loop system of Fig. 4.32. (**a**) root locus. (**b**) Nyquist plot for $K = 1$. (**c**) Bode plot for $K = 1$

in the clockwise sense as ω goes from $-\infty$ to $+\infty$, (noting that the locus is closed at $\omega = 0$ an infinite distance from the origin) indicating instability, as also indicated by the root locus.

In keeping with Sect. 4.6.4.4, Fig. 4.38c indicates a negative gain margin and therefore *stability*, as the system is conditionally stable, in agreement with the root locus and the Nyquist plot for $K = 1$. The phase margin is about $+70°$, which is generally considered to be good.

References

1. Nise NS (2010) Control systems engineering, 6th edn. Wiley, New York
2. D'Azzo JJ et al (2003) Linear control system analysis and design, 5th edn. Marcel Dekker, New York
3. Ogata K (2010) Modern control engineering, 5th edn. Pearson, Englewood Cliffs
4. Skogestad S, Postlethwaite I (2006) Multivariable feedback control: analysis and design, 2nd edn. Wiley, New York
5. McFarlane DC, Glover K (1990) Robust controller design using normalized coprime factor plant descriptions, 2nd edn. Springer, Berlin/New York
6. Dorf RC, Bishop RH (2010) Modern control systems, 12th edn. Pearson, Upper Saddle River

Chapter 5
Linear Controllers for LTI SISO Plants of Arbitrary Order: Model-Based Design

5.1 Overview

Every linear SISO controller applied to an LTI plant realises a closed-loop differential equation of the form

$$y^{(N)} + p_{N-1} y^{(N-1)} + \cdots + p_1 \dot{y} + p_0 y = q_M y_r^M + \cdots + q_1 \dot{y}_r + q_0 y_r, \quad M < N. \tag{5.1}$$

The traditional controllers of Chap. 3 that may be designed to achieve complete pole assignment have N independently adjustable controller gains that enable any desired set of the constant coefficients, p_i, $i = 0, 1, \ldots, N - 1$, to be realised. Following similar lines to the development of the state space model from the plant differential equation in Chap. 2, a similar state space representation of (5.1) may be formed, as

$$\begin{cases} x^{(N)} + p_{N-1} x^{(N-1)} + \cdots + p_1 \dot{x} + p_0 x = y_r \\ y = q_M x^{(M)} + \cdots + q_1 \dot{x} + q_0 x \end{cases} \tag{5.2}$$

in which the natural choice of state variables is $x_i = x^{(i)}$, $i = 0, 1, \ldots, N-1$. Each of these state variables is associated with a mode of the closed-loop system, each of which is made stable and determined by the control system designer via complete pole placement to achieve the desired behaviour. In this sense, complete control of the system state is achieved. To afford this luxury, however, the traditional controllers are limited to plants of up to second order only and contain at most one integrator, imposing an upper limit of $N_{max} = 3$. Of course, it is possible to control plants of higher than second order with traditional controllers, and the frequency domain design methods based on stability margins (gain and phase margins) enable this to be done, but the attainable closed-loop performance is limited by the inability to realise any desired set of closed-loop poles. In the spirit

© Springer-Verlag London 2015
S.J. Dodds, *Feedback Control*, Advanced Textbooks in Control
and Signal Processing, DOI 10.1007/978-1-4471-6675-7_5

of this book, which is to fully exploit the digital implementation medium, the two control techniques developed in this chapter remove this restriction, enabling complete pole assignment, implying complete control of the system state of (5.2).

The first part of this chapter is devoted to linear state feedback control. It is important to note, however, that a state feedback controller designed to operate with only a single measurement variable, y, and the corresponding reference input, y_r, comprises the following two parts.

1. The linear state feedback (LSF) control law, when receiving estimates of the state variables, calculates the control variable, u, to be applied to the plant. This is developed in the following section.
2. The state estimator or observer, when presented with the measurement variable, y, and the control variable, u, produces estimates of the state variables needed by the control law. The whole of Chap. 8 is devoted to this topic.

Many examples of the pole assignment method are included in this chapter, developing the reader's understanding through the derivation of controller gain formulae. Once familiarity has been gained, however, the reader may use one of two numerical methods to achieve the same goal without the sometimes lengthy algebraic manipulations. The first is Ackermann's gain formula (Appendix A5) and the second is linear characteristic polynomial interpolation (Appendix A5).

The last section of this chapter is devoted to polynomial control. It is included because it is not only capable of delivering the same performance as linear state feedback control but, with appropriate settings of its parameters, can offer similar performance to the robust control techniques (Chap. 10). Furthermore, a polynomial controller has a simpler structure than an LSF controller including (b) above and is relatively straightforward to design.

The two continuous linear control techniques of this chapter can be applied to any LTI SISO plant but excluding those containing transport delays, which are catered for in the discrete domain (Chap. 6).

5.2 Linear Continuous State Feedback Control

5.2.1 Introduction

The important concept of state has already been discussed in Chap. 3 in which it was highlighted that the state of a plant contains all the information about its behaviour at every instant of time. It follows that if this information is available to the controller, then it can be designed to achieve the desired closed-loop dynamic behaviour but within the hardware limitations. This section develops a generic linear state feedback controller for LTI SISO plants. Each of the state variables is either measured or estimated (Chap. 5). The general block diagram is shown in Fig. 5.1.

5.2 Linear Continuous State Feedback Control

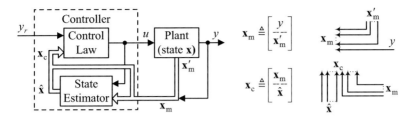

Fig. 5.1 General structure of SISO linear state feedback control system

The block arrows represent several signal lines. As indicated by the definitions of $\mathbf{x}_m$ and $\mathbf{x}_c$ given in Fig. 5.1, merged block arrows or a merged block arrow and single arrow carry all of the signals in the two merging paths.

Sometimes not all the state variables are measured either due to equipment cost constraints or impracticability. Then a state estimator is included as shown. This may require the control input to the plant as one of its inputs together with the set of measured state variables represented by $\mathbf{x}_m$, which is a subset of the complete set of state variables represented by $\mathbf{x}$. This is dealt with fully in Chap. 8, but the reason for introducing the state estimator here is to explain the difference between a *controller* and a *control law*. The *control law* is that part of a controller that has a complete set of state variables as its inputs, represented by $\mathbf{x}_c$ in Fig. 5.1, together with the reference input, y_r, and uses these to calculate the control variable, u. In contrast, the *controller* is that part of a control system that accepts y_r together with the *measured* state variables, denoted by $\mathbf{x}_m$, including the controlled output, y (and which may consist of only y), and produces the control variable, u. Thus, the control law is part of a state feedback controller, as shown. For the remainder of this chapter, the state estimator will be assumed ideal so that the system is simplified by setting $\mathbf{x}_c = \mathbf{x}$.

No integral term is included in Fig. 5.1 as it is not needed for the introduction of state feedback control, but it is subsequently added to combine its benefits, already demonstrated in Chap. 1 for the traditional controllers, with those of state feedback control. For the same reason, no external disturbance input is shown initially.

5.2.2 Linear State Feedback Control Law

The linear state feedback (LSF) control law for LTI SISO plants with constant coefficients is straightforward and has the same form for any plant order or relative degree. The control variable, u, is made equal to a linear weighted sum of the state variables, $x_1, x_2, ..., x_n$, and the reference input, y_r, as follows:

$$u = r y_r - (k_1 x_1 + k_2 x_2 + \cdots + k_n x_n). \tag{5.3}$$

Fig. 5.2 State variable block diagram of generic linear state feedback control system

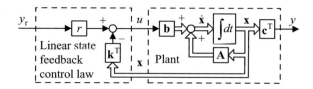

Here, r is the reference input scaling coefficient that can be adjusted, if necessary, to yield a closed-loop DC gain of unity and $k_1, \ldots, k_n$ are constant state feedback gains that may be chosen to place the closed-loop poles in the locations yielding the desired closed-loop dynamics. The minus sign is present merely as a matter of convention. Control law (5.3) may be written in the matrix–vector form as

$$u = ry_r - [k_1 k_2 \ldots k_n] \begin{bmatrix} x_1 \\ x_2 \\ \vdots \\ x_n \end{bmatrix}, \quad \text{i.e.,} \quad u = ry_r - \mathbf{k}^T \mathbf{x}. \tag{5.4}$$

Figure 5.2 shows this applied to a SISO plant represented by its state space model,

$$\dot{\mathbf{x}} = \mathbf{A}\mathbf{x} + \mathbf{b}u, \quad y = \mathbf{c}^T \mathbf{x}, \tag{5.5}$$

introduced in Chap. 3.

5.2.3 Matrix–Vector Formulation

The approach to the design of SISO linear state feedback control systems is similar to that made in Chap. 4 for the traditional control systems that can be designed by pole assignment. The characteristic polynomial of the closed-loop system is first found, which is in terms of the linear state feedback gains and the plant parameters. Then this is equated to the desired closed-loop characteristic polynomial that gives the desired dynamic behaviour. The resulting simultaneous equations are then solved for the linear state feedback gains. These steps will now be applied to the generic linear state feedback control system using the matrix–vector formulation (5.5) of the plant state space model and (5.4) of the linear state feedback control law.

The *closed-loop* state space equations are obtained by substituting for u in (5.5) using (5.4). Hence,

$$\dot{\mathbf{x}} = \mathbf{A}\mathbf{x} + \mathbf{b}\left[ry_r - \mathbf{k}^T \mathbf{x}\right] \Rightarrow \dot{\mathbf{x}} = \left[\mathbf{A} - \mathbf{b}\mathbf{k}^T\right]\mathbf{x} + \mathbf{b}ry_r, \quad \text{or}$$
$$\dot{\mathbf{x}} = \mathbf{A}_{cl}\mathbf{x} + \mathbf{b}_{cl} y_r \tag{5.6}$$

5.2 Linear Continuous State Feedback Control

This is in the same form as (5.5), the plant matrix, **A**, being replaced by the closed-loop system matrix,

$$\mathbf{A}_{cl} = \mathbf{A} - \mathbf{bk}^T, \tag{5.7}$$

and the control input matrix, **b**, being replaced by the closed-loop system reference input matrix, $\mathbf{b}_{cl} = \mathbf{b}r$. As will be seen, the elements of $\mathbf{A}_{cl}$ can be altered via the adjustable elements of the state feedback gain matrix, $\mathbf{k}^T$, to obtain the desired closed-loop dynamics. It was proven in Chap. 3 that the eigenvalues of **A** are equal to the plant poles. It follows that the eigenvalues of $\mathbf{A}_{cl}$ are equal to the closed-loop poles. Assuming controllability of the plant (Chap. 3), it will now be proven that $\mathbf{k}^T$ may be adjusted to yield any desired set of closed-loop poles.

The eigenvalues of $\mathbf{A}_{cl}$ are the roots of its characteristic equation,

$$|s\mathbf{I} - \mathbf{A}_{cl}| = |s\mathbf{I} - \mathbf{A} + \mathbf{bk}^T| = 0, \tag{5.8}$$

where **I** is the unit matrix with the same dimension as $\mathbf{A}_{cl}$. Let the state space model be transformed to the control canonical form (Chap. 3) through

$$\mathbf{x}_c = \mathbf{P}_c \mathbf{x} \Rightarrow \mathbf{x} = \mathbf{P}_c^{-1} \mathbf{x}_c \tag{5.9}$$

where $\mathbf{P}_c$ is the transformation matrix. Since $\mathbf{P}_c$ requires the inverse of the controllability matrix for its formation, the plant must be controllable. Then (5.6) becomes $\mathbf{P}_c^{-1} \dot{\mathbf{x}}_c = \mathbf{A}_{cl} \mathbf{P}_c^{-1} \mathbf{x}_c + \mathbf{b}_{cl} y_r \Rightarrow$

$$\dot{\mathbf{x}}_c = \mathbf{P}_c \mathbf{A}_{cl} \mathbf{P}_c^{-1} \mathbf{x}_c + \mathbf{P}_c \mathbf{b}_{cl} y_r \tag{5.10}$$

and from (5.7) this may be written

$$\dot{\mathbf{x}}_c = \mathbf{P}_c \left[\mathbf{A} - \mathbf{bg}^T\right] \mathbf{P}_c^{-1} \mathbf{x}_c + \mathbf{P}_c \mathbf{b}_{cl} y_r = \left[\mathbf{P}_c \mathbf{A} \mathbf{P}_c^{-1} - \mathbf{P}_c \mathbf{bg}^T \mathbf{P}_c^{-1}\right] \mathbf{x}_c + \mathbf{P}_c \mathbf{b}_{cl} y_r \tag{5.11}$$

$\mathbf{P}_c \mathbf{A} \mathbf{P}_c^{-1}$ is the plant matrix, $\mathbf{A}_c$, in the control canonical form. Then $\mathbf{P}_c \mathbf{b}$, $\mathbf{k}^T \mathbf{P}_c^{-1}$ and $\mathbf{P}_c \mathbf{b}_{cl}$ are, respectively, the plant input matrix, $\mathbf{b}_c$; the linear state feedback gain matrix, $\mathbf{k}_c^T$; and the closed-loop system reference input matrix, $\mathbf{b}_{clc}$, for the plant in the control canonical state representation. Then (5.11) may be written

$$\dot{\mathbf{x}}_c = \left[\mathbf{A}_c - \mathbf{b}_c \mathbf{k}_c^T\right] \mathbf{x}_c + \mathbf{b}_{clc} y_r \tag{5.12}$$

With reference to Chap. 3,

$$\mathbf{A}_c = \begin{bmatrix} 0 & 1 & 0 & 0 \\ \vdots & \ddots & \ddots & 0 \\ 0 & \cdots & 0 & 1 \\ a_0 & a_1 & \cdots & a_{n-1} \end{bmatrix} \quad \text{and} \quad \mathbf{b}_c = \begin{bmatrix} 0 \\ \vdots \\ 0 \\ 1 \end{bmatrix}. \tag{5.13}$$

The corresponding closed-loop system matrix is then

$$\mathbf{A}_{clc} = \begin{bmatrix} 0 & 1 & 0 & & 0 \\ \vdots & \ddots & \ddots & & 0 \\ 0 & \cdots & 0 & & 1 \\ a_0 & a_1 & \cdots & & a_{n-1} \end{bmatrix} - \begin{bmatrix} 0 \\ \vdots \\ 0 \\ 1 \end{bmatrix} \begin{bmatrix} k_{c0} & k_{c1} & \cdots & k_{c1n-1} \end{bmatrix}$$

$$= \begin{bmatrix} 0 & 1 & 0 & & 0 \\ \vdots & & \ddots & \ddots & 0 \\ 0 & & \cdots & 0 & 1 \\ a_0 - k_{c0} & a_1 - k_{c1} & \cdots & a_{n-1} - k_{cn-1} \end{bmatrix}$$

(5.14)

The closed-loop characteristic equation is then $|s\mathbf{I} - \mathbf{A}_{clc}| = 0$, i.e.

$$\begin{vmatrix} s & -1 & 0 & & 0 \\ \vdots & \ddots & \ddots & & 0 \\ 0 & & \cdots & s & -1 \\ k_{c0} - a_0 & k_{c1} - a_1 & \cdots & s + k_{cn-1} - a_{n-1} \end{vmatrix} = 0 \Rightarrow$$

$$s^n + (k_{cn-1} - a_{n-1})s^{n-1} + \cdots + (k_{c1} - a_1)s + (k_{c0} - a_0) = 0$$

(5.15)

It is evident that the n coefficients of (5.15) may be independently adjusted to yield any desired closed-loop poles. Recall that $\mathbf{k}_c^T = \mathbf{k}^T \mathbf{P}_c^{-1}$. The corresponding linear state feedback gains of the untransformed system are therefore given by

$$\mathbf{k}^T = \mathbf{k}_c^T \mathbf{P}_c.$$

(5.16)

The above proves that complete pole placement is possible for the system comprising plant (5.5) with control law (5.4), provided the plant is controllable.

5.2.4 Closed-Loop Transfer Function

Taking Laplace transforms of (5.6) with zero initial conditions yields

$$s\mathbf{X}(s) = \mathbf{A}_{cl}\mathbf{X}(s) + \mathbf{b}_{cl}Y_r(s)$$

(5.17)

Similarly taking Laplace transforms of the measurement equation in (5.5) yields

$$Y(s) = \mathbf{c}^T \mathbf{X}(s)$$

(5.18)

The closed-loop transfer function is then obtained by eliminating $\mathbf{X}(s)$ between (5.17) and (5.18). First, $\mathbf{X}(s)$ is made the subject of (5.17). To maintain dimensional

5.2 Linear Continuous State Feedback Control

compatibility after $\mathbf{X}(s)$ has been taken out as a factor, recalling from elementary matrix algebra that $\mathbf{IX}(s) = \mathbf{X}(s)$, where $\mathbf{I}$ is the appropriately dimensioned unit matrix, (5.17) is first written as

$$s\mathbf{IX}(s) = \mathbf{A}_{cl}\mathbf{X}(s) + \mathbf{b}_{cl}Y_r(s) \Rightarrow \quad (5.19)$$

$$[s\mathbf{I} - \mathbf{A}_{cl}]\mathbf{X}(s) = \mathbf{b}_{cl}Y_r(s) \Rightarrow \mathbf{X}(s) = [s\mathbf{I} - \mathbf{A}_{cl}]^{-1}\mathbf{b}_{cl}Y_r(s) \quad (5.20)$$

which can be written as

$$\mathbf{X}(s) = \frac{\text{adj}\,[s\mathbf{I} - \mathbf{A}_{cl}]}{|s\mathbf{I} - \mathbf{A}_{cl}|}\mathbf{b}_{cl}Y_r(s). \quad (5.21)$$

Substituting for $\mathbf{X}(s)$ in (5.18) using (5.21) yields $Y(s) = \mathbf{c}^T \frac{\text{adj}[s\mathbf{I}-\mathbf{A}_{cl}]}{|s\mathbf{I}-\mathbf{A}_{cl}|}\mathbf{b}_{cl}Y_r(s) \Rightarrow$

$$\frac{Y(s)}{Y_r(s)} = \mathbf{c}^T \frac{\text{adj}\,[s\mathbf{I} - \mathbf{A}_{cl}]}{|s\mathbf{I} - \mathbf{A}_{cl}|}\mathbf{b}_{cl} \quad (5.22)$$

which is the required closed-loop transfer function. This last step is possible due to the matrix product on the right-hand side being a scalar, as the following dimensional compatibility analysis confirms:

$$\frac{Y(s)}{Y_r(s)} = \frac{\overbrace{\underbrace{\mathbf{c}^T}_{1\times n}\,\underbrace{\text{adj}\,[s\mathbf{I} - \mathbf{A}_{cl}]}_{n\times n}\,\underbrace{\mathbf{b}_{cl}}_{n\times 1}}^{1\times 1:\text{ scalar polynomial of degree }\leq n-1}}{\underbrace{|s\mathbf{I} - \mathbf{A}_{cl}|}_{\text{scalar polynomial of degree }n}} \equiv \frac{\sum_{i=0}^{n-1} b_{cl\,i}\,s^i}{s^n + \sum_{j=0}^{n-1} a_{cl\,j}\,s^j} \quad (5.23)$$

This proves that the denominator polynomial of the closed-loop transfer function is equal to the characteristic polynomial of the closed-loop system matrix and therefore that the closed-loop system poles are equal to the eigenvalues of the closed-loop system matrix.

The determinant, $|s\mathbf{I} - \mathbf{A}_{cl}|$, is an nth degree polynomial in s. Since adj $[s\mathbf{I} - \mathbf{A}_{cl}]$ is the transpose of the matrix of cofactors, each of which is a determinant of dimension, $(n-1) \times (n-1)$, then every element of adj $[s\mathbf{I} - \mathbf{A}_{cl}]$ must be a polynomial of *maximum* degree equal to $n-1$.

5.2.5 Pole Assignment Using the Matrix–Vector Formulation

As far as the control system design is concerned, it remains to determine the locations of the closed-loop poles that yield an acceptable or specified closed-loop dynamics. Any means at the reader's disposal may be used to achieve this, but the straightforward method using the 5 % settling time formula of Chap. 4 will be applied in the various examples in the remainder of this chapter.

To proceed with the control system design, note that the closed-loop system matrix, $\mathbf{A}_{cl}$, is a function of the linear state feedback gain vector, $\mathbf{k}$, as well as the plant parameters, in which case (5.23) can be written as

$$\frac{Y(s)}{Y_r(s)} = \frac{\mathbf{c}^T \mathrm{adj}\,[s\mathbf{I} - \mathbf{A}_{cl}(\mathbf{k})]\,\mathbf{b}_{cl}}{|s\mathbf{I} - \mathbf{A}_{cl}(\mathbf{k})|} = \frac{\sum_{i=0}^{n-1} b_{cli}(\mathbf{k})\,s^i}{s^n + \sum_{j=0}^{n-1} a_{clj}(\mathbf{k})\,s^j}. \tag{5.24}$$

To complete the control system design using the 5 % settling time formula, the desired closed-loop characteristic polynomial is equated to the closed-loop characteristic polynomial in (5.24) and the linear state feedback gains determined by equating like degree terms in s to form n simultaneous equations. Thus,

$$[s + 1.5\,(1+n)/T_s]^n = s^n + \sum_{j=0}^{n-1} a_{clj}(\mathbf{k})\,s^j \tag{5.25}$$

from which equations for the gains as functions of the plant parameters (matrices $\mathbf{A}$ and $\mathbf{B}$ only) and the settling time follow, which can be written as

$$k_i = f_i\,(\mathbf{A}, \mathbf{B}, T_s), \quad i = 1, 2, \ldots, n. \tag{5.26}$$

Alternatively the desired closed-loop transfer function can be equated to (5.24) and the linear state feedback gains determined as above together with the reference input scaling coefficient, r, which is also a function of the plant parameters (matrix $\mathbf{C}$ as well as matrices $\mathbf{A}$ and $\mathbf{B}$) and the settling time. Thus,

$$r = f\,(\mathbf{A}, \mathbf{B}, \mathbf{C}, T_s) \tag{5.27}$$

Example 5.1 Control of a heating process with two dominant time constants: version 1.

A heating process has the following state space model:

$$\dot{x}_1 = \frac{1}{T_1}(bu - x_1); \quad \dot{x}_2 = \frac{1}{T_2}(x_1 - x_2); \quad y = K_T x_2. \tag{5.28}$$

It is required to design a linear state feedback control law that yields critical damping of the closed-loop system, a settling time of T_s seconds (5 % criterion) and a closed-loop DC gain of unity, using the matrix–vector formulation. First, the plant model (5.28) is converted to the matrix vector form

5.2 Linear Continuous State Feedback Control

$$\begin{bmatrix} \dot{x}_1 \\ \dot{x}_2 \end{bmatrix} = \underbrace{\begin{bmatrix} -1/T_1 & 0 \\ 1/T_2 & -1/T_2 \end{bmatrix}}_{\mathbf{A}} \begin{bmatrix} x_1 \\ x_2 \end{bmatrix} + \underbrace{\begin{bmatrix} b/T_1 \\ 0 \end{bmatrix}}_{\mathbf{b}} u, \quad y = \underbrace{\begin{bmatrix} 0 & K_T \end{bmatrix}}_{\mathbf{c}^T} \begin{bmatrix} x_1 \\ x_2 \end{bmatrix}. \tag{5.29}$$

Since the plant is of second order, the state feedback gain matrix is $\mathbf{k}^T = \begin{bmatrix} k_1 & k_2 \end{bmatrix}$. The closed-loop system matrix is $\mathbf{A}_{cl} = \mathbf{A} - \mathbf{b}\mathbf{k}^T$, where

$$\mathbf{b}\mathbf{k}^T = \begin{bmatrix} b/T_1 \\ 0 \end{bmatrix} \begin{bmatrix} k_1 & k_2 \end{bmatrix} = \begin{bmatrix} bk_1/T_1 & bk_2/T_1 \\ 0 & 0 \end{bmatrix}$$

and hence

$$\mathbf{A}_{cl} = \begin{bmatrix} -1/T_1 & 0 \\ 1/T_2 & -1/T_2 \end{bmatrix} - \begin{bmatrix} bk_1/T_1 & bk_2/T_1 \\ 0 & 0 \end{bmatrix} = \begin{bmatrix} -(1+bk_1)/T_1 & -bk_2/T_1 \\ 1/T_2 & -1/T_2 \end{bmatrix} \tag{5.30}$$

The closed-loop transfer function is given by (5.22) for which is needed

$$s\mathbf{I} - \mathbf{A}_{cl} = \begin{bmatrix} s & 0 \\ 0 & s \end{bmatrix} - \begin{bmatrix} -(1+bk_1)/T_1 & -bk_2/T_1 \\ 1/T_2 & -1/T_2 \end{bmatrix}$$

$$= \begin{bmatrix} s + (1+bk_1)/T_1 & bk_2/T_1 \\ -1/T_2 & s + 1/T_2 \end{bmatrix}. \tag{5.31}$$

Then

$$\mathrm{adj}\,[s\mathbf{I} - \mathbf{A}_{cl}] = \begin{bmatrix} s + 1/T_2 & 1/T_2 \\ -bk_2/T_1 & s + (1+bk_1)/T_1 \end{bmatrix}^T$$

$$= \begin{bmatrix} s + 1/T_2 & -bk_2/T_1 \\ 1/T_2 & s + (1+bk_1)/T_1 \end{bmatrix} \tag{5.32}$$

and

$$|s\mathbf{I} - \mathbf{A}_{cl}| = \begin{vmatrix} s + (1+bk_1)/T_1 & bk_2/T_1 \\ -1/T_2 & s + 1/T_2 \end{vmatrix}$$

$$= s^2 + \left(\frac{1+bk_1}{T_1} + \frac{1}{T_2} \right) s + \frac{1 + b(k_1 + k_2)}{T_1 T_2} \tag{5.33}$$

Then the closed-loop transfer function is given by (5.22), yielding

$$\frac{Y(s)}{Y_r(s)} = \frac{\begin{bmatrix} 0 & K_T \end{bmatrix} \begin{bmatrix} s+1/T_2 & -bk_2/T_1 \\ 1/T_2 & s+(1+bk_1)/T_1 \end{bmatrix} \begin{bmatrix} rb/T_1 \\ 0 \end{bmatrix}}{|s\mathbf{I} - \mathbf{A}_{cl}|}$$

$$= \frac{\begin{bmatrix} 0 & K_T \end{bmatrix} \begin{bmatrix} (s+1/T_2)\, rb/T_1 \\ rb/(T_1 T_2) \end{bmatrix}}{|s\mathbf{I} - \mathbf{A}_{cl}|} \quad (5.34)$$

$$\Rightarrow \frac{Y(s)}{Y_r(s)} = \frac{rbK_T/(T_1 T_2)}{s^2 + \left(\dfrac{1+bk_1}{T_1} + \dfrac{1}{T_2}\right)s + \dfrac{1+b(k_1+k_2)}{T_1 T_2}}$$

Using the 5 % settling time formula, the required closed-loop transfer function is

$$\frac{y(s)}{y_r(s)} = \left[\frac{1.5(1+n)/T_s}{s+1.5(1+n)/T_s}\right]^n\bigg|_{n=2} = \left(\frac{9/(2T_s)}{s+9/(2T_s)}\right)^2$$

$$= \frac{81/(4T_s^2)}{s^2 + (9/T_s)s + 81/(4T_s^2)}. \quad (5.35)$$

Equating the corresponding terms in (5.34) and (5.35) then yields

$$k_1 = \left[\left(\frac{9}{T_s} - \frac{1}{T_2}\right)T_1 - 1\right]\frac{1}{b}, \quad k_2 = \left(\frac{81 T_1 T_2}{4T_s^2} - 1\right)\frac{1}{b} - k_1 \text{ and } r = \frac{81 T_1 T_2}{4T_s^2 b K_T}.$$

$$(5.36)$$

While the relatively simple foregoing example introduces the reader to the detailed workings of the matrix–vector method, it would be tedious for systems of third or higher order, and therefore, computer-aided design packages are recommended. The alternative method of the following section, however, enables the linear state feedback gains to be derived analytically with relative ease in many cases.

5.2.6 Pole Assignment Using Mason's Formula

A straightforward method of algebraic determination of the closed-loop characteristic polynomial or transfer function analytically for the derivation of design formulae is to apply Mason's formula to the state variable block diagram of the closed-loop system. This method (Appendix A4) yields the transfer function in the general form:

$$\frac{Y(s)}{Y_r(s)} = \frac{\sum_k p_k(s)\Delta_k(s)}{\Delta(s)}, \quad (5.37)$$

5.2 Linear Continuous State Feedback Control

where $\Delta(s)$, $p_k(s)$ and $\Delta_k(s)$ are, respectively, the system determinant, the transmittance of the kth forward path and the associated cofactor. It is necessary to convert this transfer function to the standard form:

$$\frac{Y(s)}{Y_r(s)} = \frac{\sum_{j=0}^{n-1} b_j s^j}{s^n + \sum_{i=0}^{n-1} a_i s^i}. \tag{5.38}$$

It is possible to normalise with respect to other terms such as the constant term of the characteristic polynomial, but in general, using (5.38) will involve the least algebraic effort. This standard form can be obtained from (5.37) by multiplying the numerator and denominator by a normalising polynomial, $q(s)$. Thus,

$$\frac{Y(s)}{Y_r(s)} = \frac{q(s) \sum_k p_k(s) \Delta_k(s)}{q(s) \Delta(s)} \tag{5.39}$$

The normalised characteristic polynomial is then $q(s)\Delta(s)$. Comparing (5.39) with (5.22) and noting that $|s\mathbf{I} - \mathbf{A}_{cl}|$ is already in the standard form then yield

$$|s\mathbf{I} - \mathbf{A}_{cl}| = q(s)\Delta(s) \quad \text{and} \quad \mathbf{c}^T \text{adj}\,[s\mathbf{I} - \mathbf{A}_{cl}]\,\mathbf{b}_{cl} = q(s) \sum_k p_k(s)\Delta_k(s). \tag{5.40}$$

In general, it is much easier to expand $q(s)\Delta(s)$ and $q(s)\sum_k p_k(s)\Delta_k(s)$ than their matrix–vector equivalents, as the following examples demonstrate.

The desired closed-loop transfer function using the 5 % settling time formula is

$$\frac{Y(s)}{Y_r(s)} = \left(\frac{1.5\,(1+n)/T_s}{s + 1.5\,(1+n)/T_s}\right)^n \tag{5.41}$$

It is important to ensure that the closed-loop transfer function derived from the state variable block diagram and the desired closed-loop transfer function are both in the standard form (5.38); otherwise, incorrect controller parameters will result.

Example 5.2 Control of a heating process with two dominant time constants: version 2.

The design formulae for Example 5.2 will again be derived, with the same specification, but using the state variable block diagram of Fig. 5.3.

Applying Mason's formula to obtain the closed-loop transfer function yields

$$\frac{Y(s)}{Y_r(s)} = \frac{rbK_T \left(\frac{1}{1+sT_1}\right)\left(\frac{1}{1+sT_2}\right)}{1 - \left[-\frac{b}{1+sT_1}\left(k_1 + \frac{k_2}{1+sT_2}\right)\right]} = \frac{\frac{rbK_T}{T_1 T_2}\left(\frac{1}{s+1/T_1}\right)\left(\frac{1}{s+1/T_2}\right)}{1 + \frac{1}{T_1 T_2}\left[\frac{b}{s+1/T_1}\left(T_2 k_1 + \frac{k_2}{s+1/T_2}\right)\right]}$$

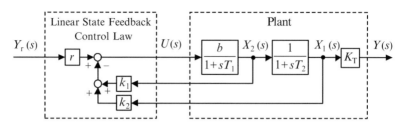

Fig. 5.3 State variable block diagram of linear state feedback control of a heating process

By inspection, $q(s) = (s + 1/T_1)(s + 1/T_2)$. Then

$$\frac{Y(s)}{Y_r(s)} = \frac{rbK_T/(T_1 T_2)}{\left(s + \frac{1}{T_1}\right)\left(s + \frac{1}{T_2}\right) + \frac{b}{T_1 T_2}\left[T_2 k_1 \left(s + \frac{1}{T_2}\right) + k_2\right]}$$

$$= \frac{rbK_T/(T_1 T_2)}{s^2 + \left(\frac{1+bk_1}{T_1} + \frac{1}{T_2}\right)s + \frac{1+b(k_1+k_2)}{T_1 T_2}}. \quad (5.42)$$

This is identical to (5.34). The remainder of the working is as in Example 5.1.

The next example illustrates the ease with which a linear state feedback control law may be designed for a fourth-order plant using the state variable block diagram.

Example 5.3 Position control of an electric drive with a flexible coupling.

Figure 5.4 shows the state variable block diagram of a linear state feedback controller applied to an electric drive with a flexible coupling.

It should be noted that in this 'per unit' model, the control variable, u, and the measurement variable, y, are normalised and are therefore proportional to but not numerically equal to the voltages in the physical system. The time constants, T_1, T_2 and T_{sp}, are functions of the moments of inertia and the spring constant.

It is required to derive design formulae for the controller parameters that yield a settling time of T_s seconds (5 % criterion) and a closed-loop DC gain of unity.

To ease the algebra, let $1/T_1 = q_1$, $1/T_{sp} = q_s$ and $1/T_2 = q_2$. Then from Fig. 5.4,

$$\frac{Y(s)}{Y_r(s)} = \frac{rq_1 q_s q_2 / s^4}{1 - \left[-\frac{q_1 k_4}{s} - \frac{q_1 q_s}{s^2} - \frac{q_s q_2}{s^2} - \frac{q_1 q_s k_3}{s^2} - \frac{q_1 q_s q_2 k_2}{s^3} - \frac{q_1 q_s q_2 k_1}{s^4}\right] + \left(-\frac{q_1 k_4}{s}\right)\left(-\frac{q_s q_2}{s^2}\right)}$$

$$= \frac{rq_1 q_s q_2}{s^4 + q_1 k_4 s^3 + q_s (q_1 + q_2 + q_1 k_3) s^2 + q_1 q_s q_2 (k_2 + k_4) s + q_1 q_s q_2 k_1}. \quad (5.43)$$

5.2 Linear Continuous State Feedback Control

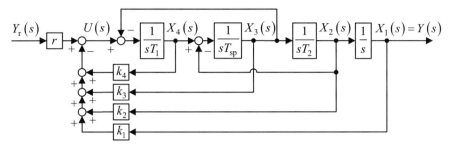

Fig. 5.4 State variable block diagram of linear state feedback control of a flexible drive

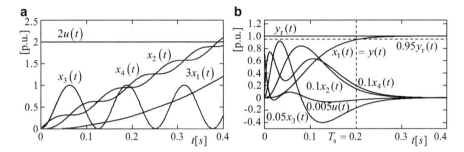

Fig. 5.5 Step responses of uncontrolled and controlled flexible drive. (**a**) Uncontrolled plant. (**b**) Controlled plant

The desired closed-loop transfer function is

$$\frac{Y(s)}{Y_r(s)} = \left(\frac{1.5(1+n)/T_s}{s + 1.5(1+n)/T_s}\right)^n \bigg|_{n=4} = \left(\frac{p}{s+p}\right)^4$$

$$= \frac{p^4}{s^4 + 4ps^3 + 6p^2 s^2 + 4p^3 s + p^4}, \qquad (5.44)$$

where $p = 15/(2T_s)$. Comparing (5.43) with (5.44) then yields

$$k_4 = \frac{4p}{q_1}, \quad k_3 = \frac{(6p^2/q_s) - (q_1 + q_2)}{q_1}, \quad k_2 = \frac{4p^3}{q_1 q_s q_2} - k_4 \text{ and } k_1 = \frac{p^4}{q_1 q_s q_2}. \qquad (5.45)$$

For a unity DC gain, from (5.43), $y(0)/y_r(0) = r/g_1 = 1 \Rightarrow$

$$r = g_1. \qquad (5.46)$$

The following plant parameters are given: $T_1 = T_2 = 0.1$ s and $T_{sp} = 0.008$ s. Figure 5.5a shows the oscillatory step response of the uncontrolled plant, and Fig. 5.5b shows the response of the closed-loop system to a step reference angle input, demonstrating how the plant state is brought under control.

5.2.7 Pole Assignment for Plants with Significant Zeros

5.2.7.1 Determination of Zero Significance

The degree of influence of any finite zeros on the step response of a linear SISO control system can be assessed using the pole-to-zero dominance ratios (Chap. 1). Suppose that in the absence of the zeros, the required settling time, T_s, is achieved by coincident negative real closed-loop pole placement at $s_{1,2,...,n} = p_c$ using one of the settling time formulae (Chap. 4). Let the set of zeros be located at z_i, $i = 1, 2, .., m$. The pole-to-zero dominance ratio for each zero is then

$$r_{pz\,i} = \frac{|z_i|}{|p_c|}. \tag{5.47}$$

The ith zero will have a significant effect if

$$r_{pz\,i} \leq r_{pz\,min}, \tag{5.48}$$

where $r_{pz\,min}$ is the minimum value of r_{pzi} for which the effect of the zero is negligible. This reduces with the system order but is greater than unity (Table 1.3).

During the development of a control system, it is advisable to run two simulations with all the closed-loop poles placed at p_c to achieve the specified settling time, one including the zeros and a fictitious one without the zeros. A comparison will then indicate whether or not measures have to be taken in the control system design to reduce or eliminate the influence of the zeros.

Since the control system designer should be free to use a controller that does not introduce zeros, only plant zeros will be considered henceforth.

5.2.7.2 Symbols for Pole Categories

To clarify the pole–zero plots displayed in further examples, three classifications of pole are distinguished by the symbols of Table 5.1.

Table 5.1 Three categories of system poles and their symbols

Symbol	Category
×	Plant or open-loop pole
⊠	Closed-loop pole
✳	Pole of pre-compensator external to feedback loop

5.2.7.3 Pole–Zero Cancellation

It may be possible to cancel the effects of significant zeros with poles at the same locations, but, of course, they must lie in the left half of the s-plane for system stability. There are two methods for implementing pole–zero cancellation:

(i) A dynamic pre-compensator whose poles are coincident with the zeros is inserted in the reference input channel.
(ii) A subset of the closed-loop poles are made coincident with the zeros.

If $r_{pz\,i} \leq 1$ according to (5.47), then method (i) is recommended, otherwise method (ii). The reason for this is that in general, the sensitivity (Chap. 4) increases as the magnitude of the closed-loop poles reduces due to the reduction in the controller gains. Method (i) will not influence the sensitivity since the pre-compensator is outside the feedback loop. Before considering method (ii), however, it must be realised that the number of closed-loop poles influencing the response to the reference input reduces by an amount equal to the number of zeros, and for a given settling time, they will have smaller magnitudes than the poles of method (i) and therefore contribute to an increase in sensitivity. Therefore, if $r_{pz\,i} > 1$ but is close to unity, the sensitivities obtained with both methods will have to be calculated and compared and the method yielding the lower sensitivity selected. To demonstrate the points made above, both methods will be applied in the following example, although $r_{pz\,i} < 1$. If a system has some zeros with $r_{pz\,i} > 1$ and others with $r_{pz\,i} < 1$, it may be appropriate to use a combination of methods (i) and (ii).

Example 5.4 Heading angle (yaw attitude) control of a surface ship

The yaw dynamics and kinematics of a surface ship have the following transfer function relationship between the rudder servo input, $U(s)$; a disturbance input, $U_d(s)$ (referred to the control input and representing an external disturbance caused by wave motion); and the yaw gyrocompass measurement, $Y(s)$.

$$Y(s) = \frac{K_f(s + 1/T_3)[U(s) + U_d(s)]}{(s + 1/T_r)(s^2 + (1/T_1 + 1/T_2)s + c/(T_1 T_2))s}$$

$$= \frac{K_f(s + b_0)[U(s) + U_d(s)]}{s^4 + a_2 s^3 + a_1 s^2 + a_0 s}, \qquad (5.49)$$

where $b_0 = 1/T_3$, $a_2 = 1/T_r + 1/T_1 + 1/T_2$, $a_1 = (1/T_r)(1/T_1 + 1/T_2) + c/(T_1 T_2)$ and $a_0 = c/(T_r T_1 T_2)$. The plant parameters are given as $1/T_1 = 0.2$ s^{-1}, $1/T_2 = 0.4$ s^{-1}, $1/T_3 = 0.17$ s^{-1}, $1/T_r = 0.5$ s^{-1}, $c = 1.2$ and $K_f = 0.15$.

It is required to design a linear state feedback control law that yields a closed-loop step response with a settling time of $T_s = 15$ s and with zero overshoot. The rudder servo input is the demanded rudder angle that is limited to ± 1.3 rad.

Ignoring the zero, the three poles would be placed at $p_c = -1.5(1+n)/T_s|_{n=4} = -15/(2T_s) = -0.5$ s^{-1}. The plant zero is at $q_p = -0.17$ s^{-1}. The pole-to-zero dominance ratio is therefore $r_{pz} = 0.17/0.5 = 0.34$. Table 1.4 gives $r_{pz\,min} = 5.41$.

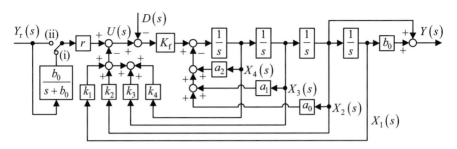

Fig. 5.6 Linear state feedback control of the yaw attitude angle of a surface ship

Since $r_{pz} < r_{pz\ min}$, the zero is significant. The control law will first be designed ignoring the zero and the resulting system simulated to assess its effect. Then method (i) will be applied. Finally, the system will be designed using method (ii) and again simulated. For both methods, the effects of a step external disturbance and mismatches in the plant parameter, c, which depends upon the ship's speed will be assessed and the sensitivities compared.

Figure 5.6 shows the state variable block diagram with the plant in the control canonical form to minimise the algebra, including the options of method (i) and method (ii) selected via the switch shown.

With the switch in position (ii), the closed-loop transfer function relationship between $Y_r(s)$, $U_d(s)$ and $Y(s)$ is

$$Y(s) = \frac{\left[rK_f Y_r(s) + K_f U_d(s)\right](s+b_0)/s^4}{1-(-a_2/s - a_1/s^2 - a_0/s^3) - K_f(-k_4/s - k_3/s^2 - k_2/s^3 - k_1/s^4)} \quad (5.50)$$

$$= \frac{\left[rK_f Y_r(s) + K_f U_d(s)\right](s+b_0)}{s^4 + (a_2 + K_f k_4)s^3 + (a_1 + K_f k_3)s^2 + (a_0 + K_f k_2)s + K_f k_1}.$$

Ignoring the zero, the closed-loop characteristic polynomial would be given by

$$(s+p)^4 = s^4 + 4ps^3 + 6p^2 s^2 + 4p^3 s + p^4, \quad (5.51)$$

where $p = 1.5(1+n)/T_s|_{n=4} = (15/2)T_s$. This also applies for method (i). Equating (5.51) to the denominator of (5.50) and setting r to give a unity closed-loop DC gain then yields

$$k_1 = \frac{p^4}{K_f}, \quad k_2 = \frac{4p^3 - a_0}{K_f}, \quad k_3 = \frac{6p^2 - a_1}{K_f}, \quad k_4 = \frac{4p - a_2}{K_f}, \quad \text{and} \quad r = \frac{g_1}{b_0}. \quad (5.52)$$

For method (ii), also with a unity closed-loop DC gain, $Y(s)/Y_r(s)|_{U_d(s)=0}$ yields

5.2 Linear Continuous State Feedback Control

$$\frac{[rK_{\mathrm{f}}y_{\mathrm{r}}(s) + K_{\mathrm{f}}d(s)](s + b_0)}{s^4 + (a_2 + K_{\mathrm{f}}k_4)s^3 + (a_1 + K_{\mathrm{f}}k_3)s^2 + (a_0 + K_{\mathrm{f}}k_2)s + K_{\mathrm{f}}g_1}$$

$$= \frac{q^3(s + b_0)}{(s + q)^3(s + b_0)}. \tag{5.53}$$

where $q = 1.5(1 + n)/T_\mathrm{s}|_{n=3} = 6/T_\mathrm{s}$. Hence,

$$s^4 + (a_2 + K_{\mathrm{f}}g_4)s^3 + (a_1 + K_{\mathrm{f}}g_3)s^2 + (a_0 + K_{\mathrm{f}}g_2)s + K_{\mathrm{f}}g_1$$
$$= s^4 + (3q + b_0)s^3 + 3q(b_0 + q)s^2 + q^2(3b_0 + q)s + q^3 b_0 \Rightarrow$$

$$k_1 = \frac{q^3 b_0}{K_{\mathrm{f}}}, \quad k_2 = \frac{q^2(3b_0 + q) - a_0}{K_{\mathrm{f}}}, \quad k_3 = \frac{3q(b_0 + q) - a_1}{K_{\mathrm{f}}}, \quad k_4 = \frac{3q + b_0 - a_2}{K_{\mathrm{f}}}, \quad r = \frac{g_1}{b_0}. \tag{5.54}$$

The considerable overshoot induced by the zero in Fig. 5.7a violates the design specification, demonstrating the need for zero cancellation. Figure 5.7c shows the pole–zero plot of the plant together with the closed-loop poles for both methods and the pre-compensator pole for method (ii). Both methods yield the correct settling time as is evident in Fig. 5.7d, g, but the disturbance causes a much larger steady-state error with method (ii) due to the higher sensitivity, which is displayed in Fig. 5.7i. The plant parameter mismatch causes a larger departure of the step response from the ideal one using method (ii) than using method (i), for the same reason, as confirmed by Fig. 5.7e, h.

It should be noted that an integral term would usually be employed in the controller for this application to reduce the steady-state error to zero for a constant external disturbance component, but transient heading errors of a similar magnitude to the steady-state errors in Fig. 5.7d, g would occur with a step external disturbance for similar pole placements.

Another important point is the increased control activity resulting from ignoring significant zeros, as is evident in Fig. 5.7f. This is caused by the larger angular acceleration in Fig. 5.7a associated with the larger angular excursion, leading to the overshoot, in a shorter time, in contrast with the smaller angular excursions in longer times associated with the non-overshooting responses of Fig. 5.7d, g. This is crucial in this application since increased control activity increases the fuel consumption of the engine powering the hydraulic rudder servo. If in another ship stabilisation application a short settling time is not essential and overshooting is allowed, then it is advisable to position two of the closed-loop poles as close as possible to the complex conjugate plant poles, but with a higher damping ratio, so that the main function of the controller is to perform active damping, which will save fuel.

5.2.7.4 Complex Conjugate s-Plane Zero Compensation

Some plants exhibiting oscillatory modes have zeros in their transfer functions. This is not the case in the flexible electric drive of Example 5.3. Here, the actuator, i.e.

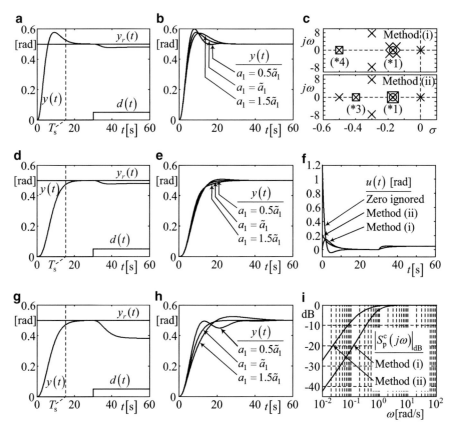

Fig. 5.7 Step and disturbance responses of ship yaw rate control system. (**a**) Heading with disturbance (zero ignored). (**b**) Heading with plant parameter mismatching (zero ignored). (**c**) Pole–zero plots. (**d**) Heading with disturbance [Method (i)]. (**e**) Heading with plant parameter mismatching [Method (i)]. (**f**) Control activity. (**g**) Heading with disturbance [Method (ii)]. (**h**) Heading with plant parameter mismatching [Method (ii)]. (**i**) System sensitivity

the motor, applies torque to the first mass to control the position or speed of the second mass by mounting the sensor on this mass. If, on the other hand, the sensor is colocated with the actuator on the first mass to control the motor speed or position, a complex conjugate pair of zeros is present with the imaginary part magnitude equal to the encastre natural frequency, i.e. the frequency of oscillation of the second mass with the motor rotor locked. Such complex conjugate zeros can also occur in other mechanical structures even if the sensors and actuators are not co-located and certainly will do so if any of them *are* co-located.

Let the *i*th pair of complex conjugate zeros associated with an oscillatory mode in the plant be the roots of

$$s^2 + 2\eta_i v_i s + v_i^2 = 0. \tag{5.55}$$

5.2 Linear Continuous State Feedback Control

They are regarded near the jw axis if:

(a) They are significant according to Sect. 5.2.7.1.
(b) If the zeros were to be cancelled by the poles of an external pre-compensator, the damping ratio of the resulting oscillatory mode satisfies $\eta_i < \eta_{\min}$, where $\eta_{\min}$ is a minimum damping ratio below which the system dynamics would be considered unsatisfactory. A value of $\eta_{\min} = 0.2$ will suffice for most applications.

Although an external pre-compensator would not be excited by external disturbances, sudden changes in the reference input would certainly excite any lightly damped oscillatory modes that it contains. Suppose the pre-compensator cancels the zeros and the closed-loop poles are real and negative. Then a step reference input will cause the output of the pre-compensator to oscillate, but the controlled output responds monotonically *without any oscillations*. This, however, could be unsatisfactory because a certain subsystem of the plant, referred to as subsystem 1, is forced by the pre-compensator to oscillate. The remainder of the plant constitutes subsystem 2, whose state variables, including the controlled output, do not oscillate. A compromise must therefore be made in which the closed-loop poles are placed as close as possible to the complex conjugate zeros but in locations giving sufficient damping of the oscillations. These, however, will now influence subsystem 2 including the controlled output, due to the imperfect pole–zero cancellation, requiring a large amount of active damping. An example will follow shortly.

Figure 5.8 shows one pair of complex conjugate zeros at point z in the upper half of the s-plane together with two candidate positions of the corresponding compensator pole pair, at points p and q on a circle of radius, r_i, centred on the point z.

The optimal pole position on this circle is at point p, where the oscillatory mode in the step response decays fastest, i.e. with the minimum time constant, T_{pi}, of the exponentially decaying envelope function. It is important to realise that the location, q, giving the maximum damping ratio for a given r_i does not give the maximum rate of decay of the oscillatory mode. Hence the ith pre-compensator pole pair has the

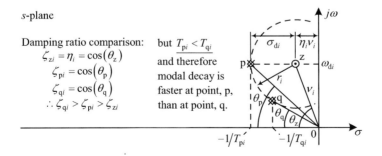

Fig. 5.8 Geometric construction for optimal pre-compensator pole placement

same imaginary part, ω_{di} (i.e. damped natural frequency), as the corresponding ith complex conjugate zero. Since

$$\omega_{di} = v_i \sqrt{1 - \eta_i^2}, \tag{5.56}$$

the required pre-compensator transfer function is

$$\frac{Y_{rp}(s)}{Y_r(s)} = \frac{r}{(s + \sigma_i)^2 + \omega_{di}^2}, \tag{5.57}$$

where

$$\sigma_i = \eta_i v_i + \sigma_{di} \tag{5.58}$$

and r is the reference input scaling coefficient. This increases the modal damping by moving the pre-compensator poles σ_{di} units to the left of the ith pair of zeros without altering the imaginary part, yielding pole location, p, in Fig. 5.8. The additional damping term, σ_{di}, is increased with the aim of achieving adequate damping while retaining an acceptable step response of the controlled output.

Example 5.5 Linear state feedback attitude control of a large flexible satellite.

Figure 5.9 shows the state variable block diagram of one attitude control axis of a space satellite with a dominant flexure mode in the solar panels, to which is applied a linear state feedback control law together with an external pre-compensator for minimising the effect of the two imaginary complex conjugate zeros.

No damping is included in the model as the solar panel damping ratios are typically 10^{-3} and are therefore negligible. Just one axis is considered in this example as the cross coupling torques due to flexure modes of the other axes are

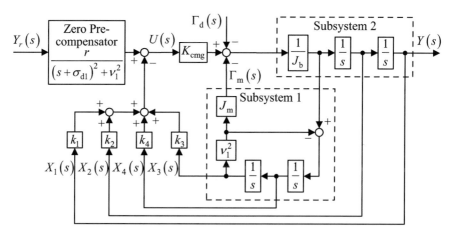

Fig. 5.9 Linear state feedback single-axis attitude control of large flexible satellite

5.2 Linear Continuous State Feedback Control

assumed negligible and two control moment gyros per control axis are employed with equal and opposite variable gimbal angles [1]. The plant is shown in the feedback modal form as the two state variables associated with the flexure mode are then the physical modal deflection, x_3, of the equivalent modal mass of the lumped parameter model and its first derivative, x_4, and the other two state variables are the centre-body attitude, x_1, which also is the output variable, y, to be controlled, and the angular velocity, x_2, both of which may be measured. γ_d is the external disturbance torque acting on the main body of the spacecraft. The plant transfer function is

$$\left.\frac{y(s)}{u(s)}\right|_{\Gamma_d=0} = \frac{K_{\text{cmg}}}{J_b s^2} \cdot \frac{\{1-[-v_1^2/s^2]\}}{1-[-v_1^2(1+J_m/J_b)/s^2]} = \frac{b}{s^2} \cdot \frac{s^2 + v_1^2}{s^2 + \omega_1^2} \tag{5.59}$$

where J_b is the centre-body moment of inertia; J_m is the moment of inertia of the mass of the equivalent mass-spring system representing the vibration mode; v_1 is the encastre natural frequency, i.e. the frequency at which the structure would vibrate if the centre body were to be held fixed with respect to inertial space; $\omega_1 = v_1\sqrt{1+J_m/J_b}$ is the free natural frequency of the vibration mode; and $b = K_{\text{cmg}}/J_b$, where K_{cmg} is the torque constant of the control moment gyro sct. Since $\eta_1 = 0$ in (5.56) and (5.58), $\omega_{d1} = v_1$ and $\sigma_1 = \sigma_{d1}$. Then according to (5.57), the characteristic polynomial of the external pre-compensator required to reduce but not eliminate the effect of the two imaginary conjugate zeros is $(s + \sigma_{d1})^2 + v_1^2$, as shown in Fig. 5.9. The closed-loop transfer function with the pre-compensator is

$$\left.\frac{Y(s)}{Y_r(s)}\right|_{\Gamma_d(s)=0}$$

$$= \frac{\dfrac{r}{(s+\sigma_{d1})^2 + v_1^2} \cdot \dfrac{b}{s^2}\left\{1 - \left[-\dfrac{v_1^2}{s^2}\right]\right\}}{1 - \left[-\dfrac{\omega_1^2}{s^2} - b\left(\dfrac{k_4}{s} + \dfrac{k_3}{s^2} + \dfrac{k_2}{s} + \dfrac{k_1}{s^2}\right)\right] + \left(-\dfrac{v_1^2}{s^2}\right)\left[-b\left(\dfrac{k_2}{s} + \dfrac{k_1}{s^2}\right)\right]}$$

$$= \frac{r}{(s+\sigma_{d1})^2 + v_1^2} \cdot \frac{b(s^2 + v_1^2)}{s^4 + b(k_4+k_2)s^3 + \left[b(k_3+k_1) + \omega_1^2\right]s^2 + bv_1^2 k_2 s + bv_1^2 k_1}. \tag{5.60}$$

With $\sigma_{d1} = 0$, complete pole–zero cancellation would occur and then the system could be designed to achieve a specified step response settling time for large angle slewing manoeuvres using one of the settling time formulae. For the 5 % criterion, the desired characteristic polynomial would be

$$(s+p)^4 = s^4 + 4ps^3 + 6p^2s^2 + 4p^3s + p^4, \tag{5.61}$$

where $p = 1.5\,(1+n)/T_s|_{n=4} = (15/2T_s)$. Equating this with the closed-loop characteristic polynomial of (5.60) and setting r for unity closed-loop DC gain yield

$$k_1 = \frac{p^4}{bv_1^2},\ k_2 = \frac{4p^3}{bv_1^2},\ k_3 = \frac{6p^2 - \omega_1^2}{b} - k_1,\ k_4 = \frac{4p}{b} - k_2,\ r = \left(\sigma_{d1}^2 + v_1^2\right)k_1.$$
(5.62)

Typical plant parameters for a large spacecraft will be taken as follows: $J_b = 500$ kg m^2, $J_m = 1{,}500$ kg m^2, $v_1 = 0.1$ rad/s and $K_{cmg} = 5$ Nm/V. The control law is to be designed to have the minimum settling time for which $|u(t)| \le u_{max}$ where $u_{max} = 10$ V, for a 180° slew manoeuvre. This is achieved by running a simulation and adjusting T_s and σ_{d1} until the specification is satisfied.

Figure 5.10 shows some simulation results for a 180° slew manoeuvre. This includes the application of a step external disturbance torque of 25 Nm applied at $t = 100$ s, which could be the side effect of the firing of an orbit change thruster (Chap. 2). Here, T_s is the nominal settling time as calculated using the 5 % settling time formula while T_{sa} is the actual settling time achieved.

In this example, T_{sa} differs considerably from T_s because the effect of the zeros on the imaginary axis of the s-plane cannot be entirely eliminated.

To demonstrate the need for the pre-compensator, Fig. 5.10g shows the unacceptable overshoot and a large undershoot induced by the zeros without a pre-compensator when $T_s = 23.5$ s, the optimal value found with the pre-compensator. This could not be realised because the peak value of $|u(t)|$ by far exceeds the limit of 10 V. With control saturation, the closed-loop system becomes nonlinear and oscillates. Means of avoiding such behaviour are presented in Chap. 8 as in practice, nominally linear control systems must be designed to accommodate control saturation.

Since this chapter is restricted to linear systems, the system of this example is designed to avoid control saturation. The large excursions of $y(t)$ and the control saturation can be avoided by increasing T_s to 58 s, as shown in Fig. 5.10b, e, but at the expense of reduced sensitivity as indicated by Fig. 5.10i, one consequence being the unacceptable steady-state error due to the disturbance torque shown in Fig. 5.10b.

The introduction of the external pre-compensator enables a much shorter settling time and lower sensitivity to be obtained within the specified control saturation limits, as indicated in Fig. 5.10a, d. The local maxima and minima of $y(t)$ during the transient are a diminished version of the large excursions in Fig. 5.10g, brought about by the pre-compensator. Note the much smaller steady-state error obtained due to the reduced T_s increasing the state feedback gains of (5.62).

If complete pole–zero cancellation was to be attempted by setting $\sigma_{d1} = 0$, then an ideal output step response would be obtained as shown in Fig. 5.10c, but oscillations occur in the state variables, $x_3(t)$ and $x_4(t)$, manifested as oscillations in $\Gamma_m(t)$ in Fig. 5.10f. With reference to Fig. 5.9, the oscillations of $\Gamma_m(t)$ are counteracted by oscillations in $u(t)$ originating from the undamped pre-compensator output. In the physical spacecraft, even after the centre-body attitude reached

5.2 Linear Continuous State Feedback Control

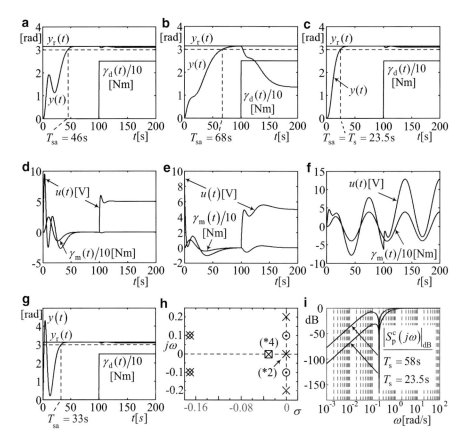

Fig. 5.10 Single-axis attitude control of flexible spacecraft. (a) $\sigma_{dl} = 0.17$, $T_s = 23.5$ s. (b) No pre-comp., $T_s = 58$ s. (c) Pole–zero cancellation. (d) $\sigma_{dl} = 0.17$, $T_s = 23.5$ s. (e) No pre-comp., $T_s = 58$ s. (f) Pole–zero cancellation. (g) No pre-comp., $T_s = 23.5$ s. (h) Pole–zero plot. (i) System sensitivity

steady state, the solar panel would be continuously flexing in the first s-shaped bending mode (Chap. 2), while the controller creates a control torque component counteracting the oscillating spring torque acting on the centre-body, thereby decoupling the solar panel motion from the attitude control. This uncontrolled motion of the solar panels, which could cause mechanical damage, is avoided by introducing a non-zero value of σ_{dl} in the pre-compensator.

5.2.7.5 Right Half s-Plane Zero Compensation

Cancellation of zeros in the right half of the *s*-plane (referred to as RHP zeros) with poles is impracticable due to the unstable modes created. A suggested approach is to 'mirror' the zeros in the *jw* axis of the *s*-plane with a set of poles. The underlying reasoning is as follows. The resulting mirroring factors of the closed-loop transfer function, which are of the form

$$f_r(s) = -\frac{s-a}{s+a} \tag{5.63}$$

for real right half plane (RHP) zeros, where $a > 0$ and

$$f_c(s) = \frac{s^2 - 2\zeta\omega_n s + \omega_n^2}{s^2 + 2\zeta\omega_n s + \omega_n^2} \tag{5.64}$$

for complex conjugate RHP zeros, can be regarded as approximations to pure time delays. That this is the case can be shown by forming Padé's qth order approximation to the transfer function of a pure time delay of τ seconds. This is

$$e^{-s\tau_d} = e^{-s\tau_d/2} \cdot e^{-s\tau_d/2} = \frac{e^{-s\tau_d/2}}{e^{+s\tau_d/2}} \cong \frac{\sum_{k=0}^{q} s^k (-\tau_d/2)^k / k!}{\sum_{k=0}^{q} s^k (\tau_d/2)^k / k!}. \tag{5.65}$$

For $q = 1$, this is

$$e^{-s\tau_d} \cong \frac{1 - s\tau_d/2}{1 + s\tau_d/2} = -\frac{s-a}{s+a}, \tag{5.66}$$

where $\tau_d = 2/a$. The first-order mirroring factor (5.63) therefore approximates a pure time delay of $2/a$ seconds. Replacing this factor with a pure time delay, of course, would eliminate the large initial undershoot (Chap. 1) at would otherwise occur. It follows that if (5.66) is a good approximation, the mirroring process should at least reduce the undershoot, possibly to acceptable proportions. To investigate this, the step responses of the non-mirrored closed-loop system,

$$\frac{Y(s)}{Y_r(s)} = \frac{1 - sT_z}{(1 + sT_{cn})^n}, \tag{5.67}$$

will now be compared with those of the closed-loop system,

$$\frac{Y'(s)}{Y_r(s)} = \frac{1 - sT_z}{(1 + sT_z)(1 + sT_{cm})^n}, \tag{5.68}$$

mirrored using an external pre-compensator, for the same pole-to-zero dominance ratios, with significant zeros satisfying

$$r_{pz} = T_c / T_z < r_{pz\ min} \quad \text{[Table 1.4]}. \tag{5.69}$$

where, for fair comparison, $T_c = T_{cn} = T_{cm}$. It is found that the undershoot is reduced but certainly not eliminated, as evident in Fig. 5.11 for $n = 2$ and $r_{pz\ min} = 7.37$.

5.2 Linear Continuous State Feedback Control

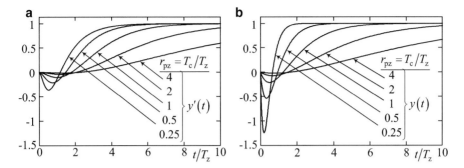

Fig. 5.11 Comparison of step responses for the same pole-to-zero dominance ratios. (**a**) Step responses with zero mirroring. (**b**) Step responses without zero mirroring

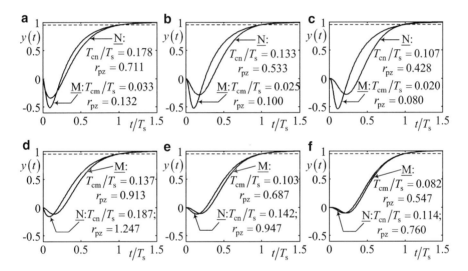

Fig. 5.12 Step responses with and without zero mirroring for the same settling times. (**a**) $T_z/T_s = 0.25$; $n = 2$. (**b**) $T_z/T_s = 0.25$; $n = 3$. (**c**) $T_z/T_s = 0.25$; $n = 4$. (**d**) $T_z/T_s = 0.15$; $n = 2$. (**e**) $T_z/T_s = 0.15$; $n = 3$. (**g**) $T_z/T_s = 0.15$; $n = 4$. Key: M Mirroring, N No-Mirroring; $rpx = T_c/T_z$

Similar results may be obtained for higher orders. Mirroring might be recommended on the basis of the reduction in overshoot, but it should be observed that for small values of r_{pz}, which indicate the highest influence of the zero, the settling time is significantly increased. For the design of control systems in which the settling time is specified, however, it would be more appropriate to adjust T_{cn} and T_{cm} in (5.67) and (5.68) to different values to achieve the same settling time, T_s, for a given value of T_z, and *then* compare the magnitudes of the undershoots. This is done in Fig. 5.12 for two different values of the ratio: T_z/T_s. With this constant settling time constraint, it is evident that on the whole, the zero mirroring does not offer any advantage in reducing the initial undershoots and, at best, only marginally reduces them. As indicated in Fig. 5.12a–c, as the zero becomes more significant by reducing the settling time, the zero mirroring results in a *larger* undershoot. A possible justification for the use of the mirroring method, however, is the reduction of the control system sensitivity that occurs for the following reason. To achieve a given overall settling time, accommodating the delay introduced by the mirroring factor, $(1 - sT_z)/(1 + sT_z)$, in (5.68) necessitates *reduction* of the settling time of the remaining subsystem, $1/(1 + sT_{cm})^n$ by reducing T_{cm} to a smaller value than T_{cn} which makes the gains of the controller yielding (5.68) larger than those of the controller yielding (5.67). Hence, the sensitivity with the mirroring is less than without, for the same settling time. On the other hand, similar step responses to those without the mirroring can be obtained using the robust pole assignment method of Chap. 9, thereby reducing the sensitivity to lower levels than attainable with the mirroring and with smaller undershoots.

In conclusion, the recommendation is to simply carry out multiple pole assignment yielding (5.67) to achieve, if possible, a settling time and an undershoot magnitude below the maximum specified values. Then, if necessary, the robust pole assignment method of Chap. 10 can be used to reduce the system sensitivity below specified levels. If the undershoot is too large for the application in hand, then the only course of action is to increase the settling time until the step response is acceptable.

The following example demonstrates that state feedback enables effective control of an unstable plant as well as dealing with right half plane zeros.

Example 5.6 Pitch attitude control of a forward swept wing aircraft by state feedback.

The pitch attitude of a highly manoeuvrable, low drag aircraft, the Grumman X29, is controlled by a combination of two sets of control surfaces consisting of flaperons and canard wings. The simplified transfer function model,

$$\frac{Y(s)}{A(s)} = \underbrace{\frac{(s+d)(s-c)}{(s+a)(s-b)(s+c)}}_{\text{Dynamics+kinematics}}, \quad \frac{A(s)}{u(s)} = \underbrace{\frac{e}{(s+e)}}_{\text{Equivalent actuator}}, \quad (5.70)$$

is given where the constant parameters are $a = 10\ \text{s}^{-1}$, $b = 6\ \text{s}^{-1}$, $c = 26\ \text{s}^{-1}$, $d = 3\ \text{s}^{-1}$ and $e = 20\ \text{s}^{-1}$. The equivalent actuator is a fictitious control surface with angle, $A(s)$, equivalent to the two physical actuators, together with a position control servomechanism, where $U(s)$ is the demanded angle as well as the control variable. The range of equivalent control surface angles that cannot be exceeded due to mechanical constraints is $-0.5 \le \alpha \le +1$ [rad].

It so happens that the plant model contains a mirroring factor, $(s-c)/(s+c)$, and this is not connected with the control system design.

The control system design specification is (a) settling time $T_s = 0.3$ [s] (5 % criterion) for step reference input of $Y_{rs} = 0.02$ [rad], (b) sensitivity $\left|S_p^c(j\omega)\right|_{\text{dB}} <$

5.2 Linear Continuous State Feedback Control

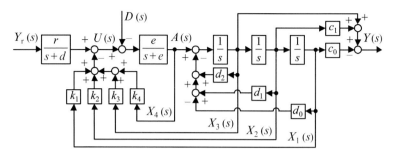

Fig. 5.13 Linear state feedback control law applied to forward swept wing aircraft

-20 dB for $0.01 < \omega < 1$ [rad/s] and (c) steady-state error $\leq 0.1\ Y_{rs}$ due to external disturbance (wind gust) equivalent to a 0.1[rad] step, $u_d(t)$, applied at the control input.

First, the zero at $s = -3[\text{s}^{-1}]$ is cancelled by an external pre-compensator rather than by closed-loop pole placement because, using the 5 % settling time formula (Chap. 3) for $n = 4$ and for $T_s = 0.3$ [s] would require pole placement at $s = -0.04\ \text{s}^{-1}$, ignoring the zeros, and a 'slower' cancelling pole at $s = -3[\text{s}^{-1}]$ would increase the sensitivity.

To simplify the derivation of the linear state feedback gains, let

$$\frac{(s+d)(s-c)}{(s+a)(s-b)(s+c)} = \frac{s^2 + c_1 s - c_0}{s^3 + d_2 s^2 + d_1 s - d_0} \tag{5.71}$$

where $c_1 = d - c$, $c_0 = dc$, $d_2 = a + c - b$, $d_1 = ac - ab - bc$ and $d_0 = abc$. Figure 5.13 shows the control system block diagram.

The closed-loop transfer function is

$$\frac{Y(s)}{Y_r(s)}$$

$$= \frac{\frac{r}{s+d} \cdot \frac{e}{s+e} \cdot \left(\frac{1}{s} + \frac{c_1}{s^2} - \frac{c_0}{s^3}\right)}{1 - \left[-\frac{d_2}{s} - \frac{d_1}{s^2} + \frac{d_0}{s^3} - \left(\frac{e}{s+e}\right)\left(g_4 + \frac{g_3}{s} + \frac{g_2}{s^2} + \frac{g_1}{s^3}\right)\right] + \left(-\frac{g_4 e}{s+e}\right)\left(-\frac{d_2}{s} - \frac{d_1}{s^2} + \frac{d_0}{s^3}\right)}$$

$$= \frac{re(s-c)}{(s+e)\left(s^3 + d_2 s^2 + d_1 s - d_0\right) + e\left[g_4 s^3 + (g_3 + g_4 d_2) s^2\right.}$$
$$\left. + (g_2 + g_4 d_1) s + (g_1 - g_4 d_0)\right]$$

$$= \frac{re(s-c)}{s^4 + [d_2 + e(1 + g_4)]s^3 + [d_1 + e(d_2 + g_3 + g_4 d_2)]s^2} \tag{5.72}$$
$$+ [-d_0 + e(d_1 + g_2 + g_4 d_1)]s + e[g_1 - d_0(1 + g_4)]$$

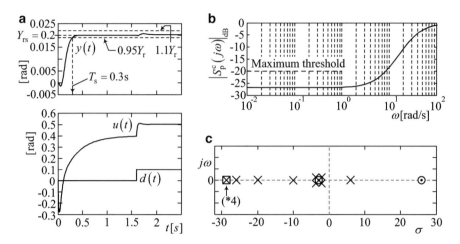

Fig. 5.14 Simulation results for X29 aircraft pitch attitude control. (**a**) Step and disturbance responses. (**b**) Sensitivity. (**c**) Pole–zero map

The desired closed-loop characteristic polynomial will first be set according to the settling time formula (Chap. 3) and then adjustments made to the multiple pole value if necessary to allow for the effect of the zero at $s = +26 \; [\text{s}^{-1}]$. In this case, the desired characteristic polynomial is

$$(s+p)^4 = s^4 + 4ps^3 + 6p^2s^2 + 4p^3s + p^4 \tag{5.73}$$

where $p = 1.5(1+n)/T_s|_{n=4} = (15/2T_s)$. Equating this with the closed-loop characteristic polynomial of (5.72) and setting r for unity closed-loop DC gain yield

$$\left. \begin{array}{ll} g_4 = \frac{(4p-d_2)}{e-1}, & g_3 = \frac{(6p^2-d_1)}{e-d_2(1+g_4)}, \\ g_2 = \frac{(4p^3+d_0)}{e-d_1(1+g_4)}, & g_1 = \frac{p^4}{e+d_0(1+g_4)}, \end{array} \quad r = \frac{d_0(1+g_4)-g_1}{c} \right\}. \tag{5.74}$$

The simulations of Fig. 5.14 indicate that the design specifications have been met. The response of the control system to the external disturbance applied at $t = 6$ s can be seen in Fig. 5.14a when $u(t)$ rapidly increases to counteract it. Interestingly the specified step response is attained despite the relatively slow exponential mode due to the external pre-compensator that is visible in $u(t)$. This is due to the derivative action of the plant zero being compensated.

It should be mentioned that in this application, the small initial undershoot of the step response is not a desirable feature. If, for example, the pilot requires to rapidly gain altitude, the pitch attitude angle must be increased. The necessary backward movement of the control stick produces a momentary negative pitch movement, i.e. 'nose down' before the required 'nose up'. This would give the plane an uncomfortable handling quality and render it difficult to control. It would be possible to eliminate the initial undershoot by sufficiently increasing T_s, but at the expense of a sluggish feel experienced by the pilot. It is possible, however, that nonlinear reference input processing techniques could eliminate the overshoot with an acceptable value of T_s.

5.2.7.6 Plants with Zeros at the Origin of the s-Plane

It is rare for a controlled plant to have a transfer function with a zero at the origin of the *s*-plane, which is a differentiator in the time domain. This case is included, however, in view of an important application, i.e. the atomic fusion-reaction based Tokomak currently approaching maturity for large-scale electrical power generation.

The class of plant to be considered has a transfer function of the form

$$\frac{Y(s)}{U(s)} = G(s) = sG_0(s) = s \cdot \frac{\sum_{i=0}^{n-1} b_i s^i}{s^n + \sum_{i=0}^{n-1} a_i s^i}. \quad (5.75)$$

where $G_0(s)$ is of type '0', meaning that $G_0(0)$ is finite and non-zero. Despite the zero DC gain of the plant, however, it is possible to design a control system that enables $y(t)$ to reach a constant set point, Y_r, or indeed any other reference input, $y_r(t)$, with a non-zero long-term average, but for a limited time that is dependent on the restrictions of the plant hardware. The approach is pole–zero cancellation as in Sect. 5.2.7.3, but in this case, the pre-compensator is a pure integrator and is inserted directly in the plant input. Then the problem reduces, mathematically, to that of controlling $G_0(s)$ which can be done with the aid of any suitable linear control technique. It is important to realise, however, that the pole–zero cancellation means that the mode associated with the integrator, which is on the verge of instability, is uncontrollable. Figure 5.15 shows the general control system block diagram, using linear state feedback to enable pole assignment, to consider the operation of the integral pre-compensator.

The input of the integrator, $u'(t)$, becomes the control variable generated by the linear state feedback control law, and this will be called the primary control input. It is assumed, as is usual, that a unity closed loop DC gain is required.

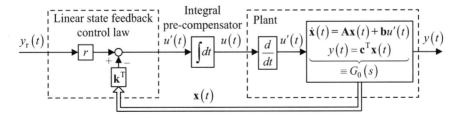

Fig. 5.15 Linear state feedback control of type '−1' plant using an integral pre-compensator

Then the steady-state value of $y(t)$ will, in theory, be $y_{ss} = Y_r$, requiring a constant steady-state primary control input, $u'_{ss} = Y_r/G_0(0)$, and therefore, the *physical* control variable satisfies $u(t) \to [Y_r/G_0(0)]t + C$ as $t \to \infty$, where C is an arbitrary constant of integration. In theory, $u(t)$ will have reached an infinite value and would continue to increase at a constant rate of $Y_r/G_0(0)$ units per second in the 'steady state' leading to the conclusion that the method is impracticable. It is possible, however, that such a system could operate on a cyclic basis, such as would be found in repeating operations on production lines or batch processes. The example to follow is a case in point. In practice, with zero initial values of all the variables, with a step reference input, all the variables undergo a transient phase and approach nearly constant values as t exceeds the settling time, T_s, except $u(t)$, which approaches a linear ramp function. In an application where it is sufficient for $y(t)$ to stay close to Y_r until a finite time, T_f, while $u(t)$ stays within the saturation limits determined by the hardware, the method is practicable. After $t = T_f$, the system can be reset and the cycle repeated when required.

Example 5.7 Linear state feedback plasma current control in a Tokomak fusion reactor.

Once the plasma has formed within the Toroidal vacuum vessel of a Tokomak fusion reactor, it effectively forms a single turn short circuited secondary coil of a transformer, the primary coil being the central poloidal coil [2]. The plasma current is one of the heating methods needed to maintain a temperature of about $10^8[°C]$ to maintain a fusion reaction and is controlled at a set point of several mega-amperes [MA] for each repeated Tokomak cycle lasting several tens of seconds. The control variable, $u(t)$, typically operates pulse-modulated multilevel power electronics, a form of power amplifier, already introduced in Chap. 2, and effectively produces a proportional poloidal coil voltage. The plasma current is inductively controlled via this voltage. The circumferential electric field required to drive electric current through the plasma is produced by an increasing magnetic field flux and is proportional to the first derivative of the poloidal coil current. This is reflected as a zero of the plant transfer function at the origin of the s-plane. This transfer function is

$$\frac{Y(s)}{U(s)} = \frac{b_1 s}{s^2 + a_1 s + a_0}, \tag{5.76}$$

where $b_1 = -K_a K_m M/\lambda$, $a_1 = (R_c L_p + R_p L_c)/\lambda$, $a_0 = R_c R_p/\lambda$ and $\lambda = L_c L_p - M^2$, where K_a is the power amplifier voltage gain, K_m is the plasma current measurement constant, R_c and L_c are the poloidal coil resistance and inductance, R_p and L_p are the plasma 'single coil turn' resistance and induction, M is the mutual inductance and λ is the electromagnetic coupling coefficient. For this example, the values are $R_p = 5 \times 10^{-8}$ [Ω], $L_p = 2 \times 10^{-6}$ [H], $R_c = 10^{-4}$ [Ω], $L_c = 10^{-4}$ [H], $M = 4 \times 10^{-6}$ [H], $K_a = 100$ and $K_m = 1$ [V/MA]. The control saturation level is $|u|_{max} = 10$ [V], yielding a maximum primary voltage

5.2 Linear Continuous State Feedback Control

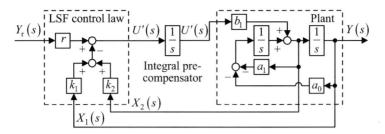

Fig. 5.16 Linear state feedback control law and integral pre-compensator applied for plasma current control of a Tokomak fusion reactor

of 1,000 [V]. It must be noted, however, that the resistance, R_p, is highly dependent on the plasma temperature, and in practice, this would have to be measured and the coefficient, a_1, continuously updated during the transient part of the Tokomak cycle. In this example, however, all the plant parameters will be assumed constant for simplicity of illustration, this example focusing on the operation of the integral pre-compensator.

The control system specification is as follows. The response to a 4 [MA] step reference input has a settling time of $T_\mathrm{s} = 30$ [s] (2 % criterion) and is applied until $t = 80$ [s], when it is stepped down to zero and the plasma current allowed to decay in readiness for the next Tokomak cycle.

Figure 5.16 shows the control system block diagram in which the plant model is in the control canonical form except for $u(s)$ being injected into the second integrator input to realise the zero at the origin of the s-plane.

It may easily be confirmed that the plant has transfer function (5.76). The state variables are $x_1 = y$ and $x_2 = \dot{y}$ which are easily made available. It should be noted that although x_2 depends algebraically on the physical control variable, u (Chap. 2), which would disqualify it as a state variable when considering the plant in isolation, in the complete system, the control variable is u', which renders x_2 a state variable. The closed-loop transfer function is

$$\frac{Y(s)}{Y_\mathrm{r}(s)} = \frac{rb_1/s^2}{1 - \left[-\frac{a_1}{s} - \frac{a_0}{s^2} - \frac{k_2 b_1}{s} - \frac{k_1 b_1}{s^2}\right]} = \frac{rb_1}{s^2 + (a_1 + k_2 b_1)s + (a_0 + k_1 b_1)}. \tag{5.77}$$

The 2 % settling time formula yields the desired closed-loop transfer function

$$\frac{Y(s)}{Y_\mathrm{r}(s)} = \left(\frac{\frac{1.6(1.5+n)}{T_\mathrm{s}}}{s + \frac{1.6(1.5+n)}{T_\mathrm{s}}}\right)^n \bigg|_{n=2} = \left(\frac{\frac{5.6}{T_\mathrm{s}}}{s + \frac{5.6}{T_\mathrm{s}}}\right)^2 = \frac{\frac{31.36}{T_\mathrm{s}^2}}{s^2 + \frac{11.2}{T_\mathrm{s}}s + \frac{31.36}{T_\mathrm{s}^2}}. \tag{5.78}$$

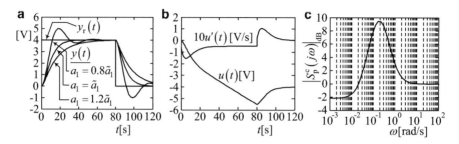

Fig. 5.17 Response of linear state feedback control system for Tokomak plasma current. (**a**) Plasma current responses. (**b**) Control variables. (**c**) Sensitivity

Equating (5.77) and (5.78) then yield

$$k_1 = \left(\frac{31.36}{T_s^2} - a_0\right)/b_1; \quad k_2 = \left(\frac{11.2}{T_s} - a_1\right)/b_1; \quad r = \frac{31.36}{b_1 T_s^2}. \qquad (5.79)$$

Figure 5.17 shows some simulation results. In this example, the performance specification did not refer to robustness or sensitivity. It is always wise, however, to investigate this as the control system designer has to consider robust control techniques (Chaps. 8 and 9) if this reveals problems. The aforementioned plant parameter, a_1, that is subject to variations, has been mismatched by ± 20 % with respect to the nominal value, $\tilde{a}_1$, and Fig. 5.17a shows that this causes considerable departures from the correct, non-overshooting step response. This is reflected in the high sensitivity displayed in Fig. 5.17c.

As a general guideline, a sensitivity of less than -20 dB over a frequency range of $\omega \in (0, \omega_b)$ is regarded as good (Chap. 4). Hence, this control application will be revisited using robust control techniques (Chap. 11).

Turning attention to Fig. 5.17b, the primary control variable, $u'(t)$, has positive and negative excursions, as is often the case but initially goes negative to produce a positive increase in $y(t)$. This is due to the negative value of b_1. Such plants having negative DC gains usually require negative controller gains and the model-based methods presented in this book cater for this automatically. The 'settling' of the physical control variable, $u(t)$, to a ramp function as $y(t)$ approaches the demanded value can clearly be seen. After $y_r(t)$ returns to zero, another transient occurs in which $u'(t) \rightarrow 0$ and $u(t) \rightarrow$ const. Although the timescale of operation of this plant is several tens of seconds, the plant parameters are such that $|u(t)| < |u|_{\max} = 10$ V. Before the next Tokomak cycle, however, $u(t)$ will have to be returned to zero. It ends at a constant non-zero value in Fig. 5.17b but simply produces a constant poloidal coil voltage and a constant magnetic flux, and therefore, zero induced e.m.f. in the plasma. At the end of the cycle, however, the fusion reactions will have taken place and all the gases exhausted in readiness for the new injection of deuterium or tritium vapour in readiness for the next cycle. At this time, u can be safely set

5.2 Linear Continuous State Feedback Control

to zero and the poloidal coil current, I_p, will decay to zero with the time constant, L_p/R_p. Alternatively, I_p could be controlled to zero faster with another control law, directly using u as the control variable.

5.2.8 State Feedback Controllers with Additional Integral Terms

5.2.8.1 Steady-State Analysis of Basic LSF Control System

Let the basic linear state feedback control system of Sect. 5.2.2 formed by applying control law (5.4) to plant (5.5) be subject to a constant external disturbance, d, such that u is replaced by $u - d$. Then the system and output error equations are

$$\dot{\mathbf{x}} = \mathbf{A}\mathbf{x} + \mathbf{b}\left[ry_{\mathrm{r}} - \mathbf{k}^{\mathrm{T}}\mathbf{x} - d\right], \quad y = \mathbf{c}^{\mathrm{T}}\mathbf{x}, \quad e = y_{\mathrm{r}} - \mathbf{c}^{\mathrm{T}}\mathbf{x}. \quad (5.80)$$

If y_{r} and d are constant, in the steady state, $\dot{\mathbf{x}} = \mathbf{0}$. Then if the steady-state values of the variables are denoted by $\mathbf{x}_{\mathrm{ss}}$, y_{ss} and e_{ss}, (5.80) becomes

$$\mathbf{A}\mathbf{x}_{\mathrm{ss}} + \mathbf{b}\left[ry_{\mathrm{r}} - \mathbf{k}^{\mathrm{T}}\mathbf{x}_{\mathrm{ss}} - d\right] = \mathbf{0}, \quad y_{\mathrm{ss}} = \mathbf{c}^{\mathrm{T}}\mathbf{x}_{\mathrm{ss}}, \quad e_{\mathrm{ss}} = y_{\mathrm{r}} - y_{\mathrm{ss}} \quad (5.81)$$

Let the nominal parameter matrices and vectors of the plant model upon which the control law is based be denoted, $\widehat{\mathbf{A}}, \widehat{\mathbf{b}}$ and $\widehat{\mathbf{c}}^{\mathrm{T}}$ and the corresponding steady-state values of the variables be denoted $\widehat{\mathbf{x}}_{\mathrm{ss}}$, $\widehat{y}_{\mathrm{ss}}$ and $\widehat{e}_{\mathrm{ss}}$. The control law is also designed to yield zero steady-state error with $d = 0$, in which case (5.81) becomes

$$\widehat{\mathbf{A}}\widehat{\mathbf{x}}_{\mathrm{ss}} + \widehat{\mathbf{b}}\left[ry_{\mathrm{r}} - \mathbf{k}^{\mathrm{T}}\widehat{\mathbf{x}}_{\mathrm{ss}}\right] = \mathbf{0}, \quad \widehat{y}_{\mathrm{ss}} = \widehat{\mathbf{c}}^{\mathrm{T}}\widehat{\mathbf{x}}_{\mathrm{ss}}, \quad \widehat{e}_{\mathrm{ss}} = y_{\mathrm{r}} - \widehat{y}_{\mathrm{ss}} = 0. \quad (5.82)$$

It is clear from the first equation of (5.81) that $\mathbf{x}_{\mathrm{ss}}$ is a function of d. Then with the nominal plant parameters, if $d \neq 0$, (5.82) becomes

$$\begin{cases} \widehat{\mathbf{A}}\left(\widehat{\mathbf{x}}_{\mathrm{ss}} + \delta\widehat{\mathbf{x}}_{\mathrm{ss}}\right) + \widehat{\mathbf{b}}\left[ry_{\mathrm{r}} - \mathbf{k}^{\mathrm{T}}\left(\widehat{\mathbf{x}}_{\mathrm{ss}} + \delta\widehat{\mathbf{x}}_{\mathrm{ss}}\right) - d\right] = \mathbf{0}, \\ \widehat{y}_{\mathrm{ss}} = \widehat{\mathbf{c}}^{\mathrm{T}}\left(\widehat{\mathbf{x}}_{\mathrm{ss}} + \delta\widehat{\mathbf{x}}_{\mathrm{ss}}\right), \quad \widehat{e}_{\mathrm{ss}} = y_{\mathrm{r}} - \widehat{y}_{\mathrm{ss}} = -\widehat{\mathbf{c}}^{\mathrm{T}}\delta\widehat{\mathbf{x}}_{\mathrm{ss}} \neq 0 \end{cases}. \quad (5.83)$$

It follows from the theorem of superposition for linear systems that if the external disturbance consists of the sum of a constant component and a random component, the error will randomly vary about a non-zero mean value in the steady state.

Also, starting again with $d = 0$ and nominal plant parameters with (5.82) satisfied, if the plant parameters, $\widehat{\mathbf{A}}, \widehat{\mathbf{b}}$ and $\widehat{\mathbf{c}}^{\mathrm{T}}$, are replaced by the actual ones, $\mathbf{A}, \mathbf{b}$ and $\mathbf{c}^{\mathrm{T}}$, the first of equations (5.82) requires the steady-state value of $\mathbf{x}(t)$ to change from $\widehat{\mathbf{x}}_{\mathrm{ss}}$ to, say, $\widehat{\mathbf{x}}_{\mathrm{ss}} + \delta'\widehat{\mathbf{x}}_{\mathrm{ss}}$. Then (5.82) becomes

$$\begin{cases} \mathbf{A}\left(\widehat{\mathbf{x}}_{ss} + \delta'\widehat{\mathbf{x}}_{ss}\right) + \mathbf{b}\left[ry_r - \mathbf{k}^T\left(\widehat{\mathbf{x}}_{ss} + \delta'\widehat{\mathbf{x}}_{ss}\right)\right] = \mathbf{0}, \\ y_{ss} = \mathbf{c}^T\left(\widehat{\mathbf{x}}_{ss} + \delta'\widehat{\mathbf{x}}_{ss}\right), \quad e_{ss} = \widehat{y}_r - \widehat{y}_{ss} \neq 0 \end{cases} \quad (5.84)$$

There is one exception to this, however, in that if the plant is of type 'r' with $r \geq 1$, $e_{ss} = 0$ with $d = 0$. This can be proven by transformation of the plant model to the control canonical form but is unnecessary at this point.

To summarise, a constant external disturbance will give rise to a steady-state error in any basic linear state feedback control system, and parametric errors in the plant modelling will cause steady-state errors in the absence of a constant external disturbance if the plant is of type greater than 0.

5.2.8.2 Linear State Feedback Plus Integral Control

In order to ensure that zero steady-state error is experienced in a linear state feedback control system with a step reference and/or step external disturbance input, the reference input term, ry_r, may be replaced by an integral term similar to that often employed in the traditional controllers (Chap. 1). The standard linear state feedback control law (5.4) is then replaced by

$$u = x_I - \mathbf{k}^T \mathbf{x}, \quad x_I = K_I \int (y_r - y) dt. \quad (5.85)$$

The implementation is based on (5.85), but for the application of Ackermann's gain formula (Chap. 5), the integral term is rewritten as a state differential equation, $\dot{x}_I = K_I(y_r - y) = K_I(y_r - \mathbf{c}^T \mathbf{x})$, and this is appended to the plant state equations to form the augmented plant,

$$\begin{bmatrix} \dot{x}_I \\ \dot{\mathbf{x}} \end{bmatrix} = \begin{bmatrix} 0 & -\mathbf{c}^T \\ \mathbf{0} & \mathbf{A} \end{bmatrix} \begin{bmatrix} x_I \\ \mathbf{x} \end{bmatrix} + \begin{bmatrix} 0 \\ \mathbf{b} \end{bmatrix} u + \begin{bmatrix} K_I \\ \mathbf{0} \end{bmatrix} y_r \equiv \dot{\mathbf{x}}_s = \mathbf{A}_s \mathbf{x}_s + \mathbf{b}_s u + \mathbf{b}_r y_r, \quad (5.86)$$

to which is applied the state feedback control law

$$u = -\begin{bmatrix} -1 & \mathbf{k}^T \end{bmatrix} \begin{bmatrix} x_I \\ \mathbf{x} \end{bmatrix} \equiv u = -\mathbf{k}_s^T \mathbf{x}_s, \quad (5.87)$$

The matrices input to Ackermann's formula (Chap. 12) are $\mathbf{A}_s$, $\mathbf{b}_s$ and $\mathbf{k}_s^T$.

If n is the plant order, then the $n + 1$ eigenvalues of the closed-loop system matrix, $\mathbf{A}_{scl} = \mathbf{A}_s - \mathbf{b}_s \mathbf{k}_s^T$ (i.e. the poles of the closed-loop system), may be placed by independent choice of the n linear state feedback gains, $g_1, g_2, ..., g_n$, which are the components of $\mathbf{k}^T$, and the integral term gain, K_I. Figure 5.18 shows the state

5.2 Linear Continuous State Feedback Control

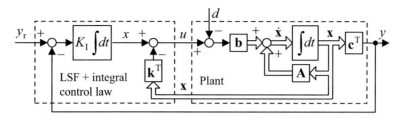

Fig. 5.18 General linear state feedback plus integral control system

variable block diagram of the closed-loop system including the control law in the form of (5.85) for implementation, with the inclusion of an external disturbance input, d, referred to the control input, u.

Example 5.8 Cruise speed control for a road transportation vehicle.

This application is a Diesel driveline for a transportation vehicle that is subject to relatively large external disturbances. As is well known, such vehicles often have to travel on roads with a gradient giving rise to a gravitational force component along the direction of travel that reduces or increases the vehicle speed for a given throttle setting if, respectively, the gradient is positive or negative. An integral term in the cruise controller is needed to counteract this disturbance without changing the vehicle speed. The example is a real application (data obtained with the courtesy Delphi Diesel) in which the measurement variable is the engine speed rather than the road wheel speed, resulting in a pair of complex conjugate zeros corresponding to the lightly damped complex conjugate poles associated with the torsional vibrations of the propeller shaft. The plant therefore falls into the category of Sect. 5.2.7.4, and an external pre-compensator is needed.

The plant transfer function between the engine speed measurement, $Y(s)$ (numerically in [r/min]), and the throttle input, $U(s)$ (electrical input with saturation limit of $u_{max} = 10$ [V]), is obtained from an identification algorithm based on measurements of the response to pseudo-random binary inputs (Chap. 2) and is given as

$$\frac{Y(s)}{U(s)} = \frac{c_2 s^2 + c_1 s + c_0}{s^3 + a_2 s^2 + a_1 s + a_0}. \qquad (5.88)$$

The coefficients depend on the selected gear and the vehicle mileage (due to wear) and are obtained by on-line identification for gear 1 and an intermediate mileage as $a_0 = -0.2454$, $a_1 = 245.4$, $a_2 = 7.594$, $c_0 = 9236$, $c_1 = 285.5$ and $c_2 = 475$.

The specified response to a step reference input of 200[r/min] should have the shortest possible settling time but be monotonically increasing without any stationary points (also implying zero overshoot) and constrained by $0 \le u(t) \le 15$ [V]. A guideline of $T_s = 1$ [s] (5 % criterion) is suggested as a starting point.

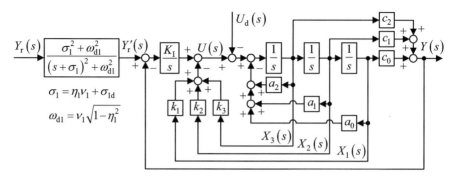

Fig. 5.19 Linear state feedback plus integral control of a Diesel driveline

The numerator polynomial of (5.88) can be written as

$$s^2 + \frac{c_1}{c_2}s + \frac{c_0}{c_2} = s^2 + 2\eta_1 v_1 s + v_1^2 \Rightarrow v_1 = \sqrt{\frac{c_0}{c_2}} \cong 4.4 \text{ [rad/s]}, \tag{5.89}$$

$$\eta_1 = \frac{c_1}{2\sqrt{c_0 c_2}} \cong 0.07.$$

The pre-compensator transfer function is then given by (5.56), (5.57) and (5.58) with $i=1$, and the pre-compensator DC gain is set to unity as the integral term ensures that the main control loop has a unity DC gain. Figure 5.19 shows the control system block diagram.

The state variables would have to be estimated using an observer (Chap. 6). The approach for the pole assignment will be as follows. First, a simulation of the system of Fig. 5.19 will be set up with $T_s = 1$ [s], bearing in mind that the actual settling time, T_{sa}, will differ from this for $\sigma_{1d} > 0$. Then σ_{1d} will be increased from zero until sufficient modal damping is achieved by observation of $x_1(t)$ (which is proportional to the vehicle road speed) and, in particular $u(t)$, does not go negative (as an internal combustion engine cannot produce negative torque in response to a negative throttle input!). Then T_s will be set to the smallest value for which the step response has no stationary points.

Only the closed-loop characteristic equation of Fig. 5.19 is needed for the gain determination as the closed-loop DC gain is unity due to the integral term. Thus,

$$1 - \left[-\frac{a_2 + k_3}{s} - \frac{a_1 + k_2 + c_2 K_I}{s^2} - \frac{a_0 + k_1 + c_1 K_I}{s^3} - \frac{c_0 K_I}{s^4} \right] = 0 \Rightarrow$$

$$s^4 + (a_2 + k_3)s^3 + (a_1 + k_2 + c_2 K_I)s^2 + (a_0 + k_1 + c_1 K_I)s + c_0 K_I = 0 \tag{5.90}$$

The desired characteristic equation is

$$(s+p)^4 = s^4 + 4ps^3 + 6p^2 s^2 + 4p^3 s + p^4 = 0 \tag{5.91}$$

5.2 Linear Continuous State Feedback Control

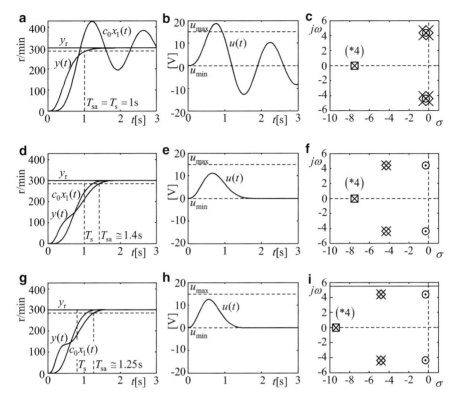

Fig. 5.20 Step reference responses for LSF plus integral control of a Diesel driveline. (a) $T_s = 1$ s, $\sigma_{dl} = 0$. (b) $T_s = 1$ s, $\sigma_{dl} = 0$. (c) $T_s = 1$ s, $\sigma_{dl} = 0$. (d) $T_s = 1$ s, $\sigma_{dl} = 4$ s^{-1}. (e) $T_s = 1$ s, $\sigma_{dl} = 4$ s^{-1}. (f) $T_s = 1$ s, $\sigma_{dl} = 4$ s^{-1}. (g) $T_s = 0.8$ s, $\sigma_{dl} = 4.5$ s^{-1}. (h) $T_s = 0.8$ s, $\sigma_{dl} = 4.5$ s^{-1}. (i) $T_s = 0.8$ s, $\sigma_{dl} = 4.5$ s^{-1}

where $p = 1.5(1+n)/T_s|_{n=4} = 15/(2T_s)$. Equating (5.90) and (5.91) then yields

$$K_I = p^4/c_0, \quad g_1 = 4p^3 - (a_0 + c_1 K_I), \quad g_2 = 6p^2 - (a_1 + c_2 K_I), \quad g_3 = 4p - a_2. \tag{5.92}$$

Figure 5.20 shows simulations with a step engine speed reference of 300[r/min].

Figure 5.20a–c show that exact pole–zero cancellation yields an ideal engine speed response but an oscillating vehicle movement due to the torsional vibrations of the propeller shaft. This unacceptable situation is similar to that experienced in Example 5.5. Also $u(t)$ exceeds the specified limits, $u_{\max}$ and $u_{\min}$, and would saturate in practice. For $T_s = 1$ s, Fig. 5.20d–f show the results for the minimum setting of $\sigma_{ld} = 4$ s^{-1} required to reduce the overshoot in $y(t)$ to zero. In Fig. 5.20g–i, σ_{ld} has been increased and T_s reduced to obtain a shorter actual settling time of $T_{sa} \cong 1.25$ s.

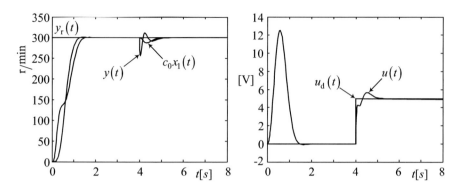

Fig. 5.21 Step disturbance response for LSF plus integral control of a Diesel driveline

It may be observed in Fig. 5.20d, g that $x_1(t)$, and therefore the vehicle road speed, rises more smoothly than $y(t)$ towards its steady-state value due to its not being influenced by the zeros. This can be confirmed by deriving the transfer function, $X_1(s)/Y_r(s)$, which has no finite zeros.

Figure 5.21 shows the engine speed response to a step disturbance, $u_d(t)$, equivalent to an increase in the road gradient to a constant value. The initial transient is the same as that of Fig. 5.20g as the vehicle reaches the demanded speed and is virtually in the steady state when the disturbance is applied at $t = 4$ s.

The integral term action is evident in the transient following this disturbance as $y(t)$ momentarily falls below $y_r(t)$ and then returns with zero steady-state error. Following the initial peak in $u(t)$ that accelerates the vehicle up to the required speed, the rapid action of the controller in counteracting the disturbance is evident in $u(t)$ following $u_d(t)$.

5.2.8.3 Linear State Feedback Plus Multiple Integral Control

The linear state feedback plus integral control law of the previous section may be generalised to yield a zero steady-state error when the reference input is a polynomial of the general form

$$y_r(t) = \sum_{k=0}^{R} R_k t^k h(t) \tag{5.93}$$

where $h(t)$ is the unit step function. This is achieved by introducing $R+1$ integrators in an outer loop around the state feedback loop structured so that the system between $e(s)$ and $y(s)$ is of type '$R+1$', and there are $R+1$ independently adjustable gains introduced so that the closed-loop poles introduced may be freely placed. Figure 5.22 shows the block diagram.

5.2 Linear Continuous State Feedback Control

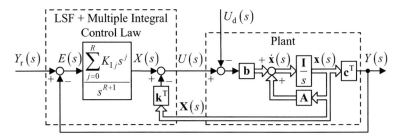

Fig. 5.22 General linear state feedback plus multiple integral control system

The R zeros introduced by this scheme, which are the roots of

$$\sum_{j=0}^{R} K_{1j} s^j = 0 \qquad (5.94)$$

are unavoidable due to the controller structure. Unlike the plant zeros, the controller zeros depend on the closed-loop pole locations and the plant parameters. Under these circumstances, if a settling time criterion has to be satisfied and overshooting in the step response is undesirable or unacceptable, the recommended approach is to first consider allocating R of the closed-loop poles for zero compensation and place the remaining $n-R$ closed-loop poles to yield the required settling time in the absence of the zeros. Then complete pole–zero cancellation should be considered and applied if feasible (i.e. the zeros have sufficient negative real parts). If this approach is not successful in a given application, then all of the n closed-loop poles should be colocated to achieve the specified settling time, and if this causes unacceptable overshooting/undershooting due to the zeros, a longer settling time yielding an acceptable overshoot should be considered. It is important to point out, however, that for applications in which overshooting has to be very limited or entirely eliminated, the polynomial controller is really more suitable as it has a structure in which closed-loop transfer function zeros are not introduced with multiple integral terms, and therefore, no overshooting occurs in the step response if there are no plant zeros and the closed-loop poles are placed on the negative real axis of the s-plane. In fact, the $R+1$ integrators are placed in the plant input, thereby increasing its 'type' by $R+1$. As will be seen, these integrators are treated as part of the plant in the controller design. This might appear to be an attractive approach for linear state feedback control, as no zeros would be introduced, but the need to feed back the integrator states as well as the plant states for the pole placement would introduce a steady state error and therefore defeat the purpose of the integrators.

An external zero cancelling pre-compensator is unsuitable as it would introduce a constant steady-state error for $R=1$ and cause an infinite steady-state error for $R>1$. This is proven by considering a pre-compensator with unity DC gain of the form

$$\frac{Y'_r(s)}{Y_r(s)} = P(s) = \frac{p_0}{\sum_{j=0}^{R} p_j s^j}. \qquad (5.95)$$

In terms of the error, $E_r(s) = Y_r(s) - Y_r'(s)$,

$$\frac{E_r(s)}{Y_r(s)} = 1 - P(s) = \frac{\sum_{j=1}^{R} p_j s^j}{\sum_{j=0}^{R} p_j s^j}. \tag{5.96}$$

If the reference input is the polynomial (5.93), then its Laplace transform is

$$Y_r(s) = \sum_{k=0}^{R} \frac{R_k k!}{s^{k+1}} \tag{5.97}$$

and the steady-state error of the pre-compensator is then

$$e_{rss} = \lim_{s \to 0} sE_r(s) = \lim_{s \to 0} \sum_{k=0}^{R} \frac{R_k k! \sum_{j=1}^{R} p_j s^j}{s^k \sum_{j=0}^{R} p_j s^j} = \begin{cases} 0 \text{ for } R = 0 \\ \frac{r_1 p_1}{p_0} \text{ for } R = 1 \\ \infty \text{ for } R > 1 \end{cases}. \tag{5.98}$$

For the steady-state error analysis of the system of Fig. 5.22, let the transfer function relationship between $X(s)$, $U_d(s)$ and $Y(s)$ be written as

$$Y(s) = G_{sf}(s)[X(s) - U_d(s)]. \tag{5.99}$$

where $G_{sf}(s)$ is the closed-loop transfer function resulting from the linear state feedback applied alone, without the multiple integral outer loop closed. The closed-loop transfer function relationship of the complete system is then

$$E(s) = \frac{Y_r(s) + G_{sf}(s)U_d(s)}{1 + G_{sf}(s)\left[\sum_{j=0}^{R} K_{1j} s^j / s^{R+1}\right]} = \frac{s^{R+1}[Y_r(s) + G_{sf}(s)U_d(s)]}{s^{R+1} + G_{sf}(s)\sum_{j=0}^{R} K_{1j} s^j}. \tag{5.100}$$

On the assumption that $G_{sf}(0)$ is finite, it would appear from (5.100) that if the system produces a zero steady-state error for the polynomial reference input function (5.93), then it will also do so if the external disturbance is a similar polynomial function:

$$u_d(t) = \sum_{k=0}^{R} D_k t^k h(t). \tag{5.101}$$

Then taking Laplace transforms of (5.101) yields

$$U_d(s) = \sum_{k=0}^{R} \frac{D_k k!}{s^{k+1}}. \tag{5.102}$$

5.2 Linear Continuous State Feedback Control

Substituting for $Y_r(s)$ and $U_d(s)$ in (5.100) using (5.102) yields

$$E(s) = \frac{s^{R-k}\sum_{k=0}^{R}[R_k + D_k G_{sf}(s)]\, k!}{s^{R+1} + G_{sf}(s)\sum_{j=0}^{R} K_{1j} s^j}. \tag{5.103}$$

The steady-state error is therefore

$$e_{ss} = \lim_{s\to 0} sE(s) = \lim_{s\to 0}\frac{\sum_{k=0}^{R} s^{R-k+1}[R_k + D_k G_{sf}(s)]\,k!}{s^{R+1} + G_{sf}(s)\sum_{j=0}^{R} K_{1j} s^j} = 0. \tag{5.104}$$

Example 5.9 Slewing control of a rigid-body spacecraft.

Spacecraft are sometimes required to turn about a single axis at a constant angular velocity and reach specified attitude angles at specified times. This requires the attitude control system to follow a ramp reference input with zero steady-state error given by (5.93) with $R = 1$, i.e.

$$y_r(t) = r_0 + r_1 t \tag{5.105}$$

Two integrators are required in the integral outer loop closed around the linear state feedback control loop, as shown in Fig. 5.23.

Here, $b = K_w/J$, where K_w is the reaction wheel torque constant and J is the spacecraft moment of inertia. These plant parameters are given as $J = 100\ [\text{Kg m}^2]$ and $K_w = 0.02\ [\text{Nm/V}]$. The specified settling time (2 % criterion) is $T_s = 120$ s.

The closed-loop characteristic equation of the system of Fig. 5.23 is

$$1 - \left[-\frac{b}{s}\left(k_2 + \frac{k_1}{s} + \frac{K_{11}}{s^2} + \frac{K_{10}}{s^3}\right)\right] = 0$$
$$\Rightarrow s^4 + b\left(k_2 s^3 + k_1 s^2 + K_{11} s + K_{10}\right) = 0. \tag{5.106}$$

At this stage it is assumed that one of the closed-loop poles can be placed to cancel the zero introduced by the double integral term of the controller at $-K_{11}/K_{10}$.

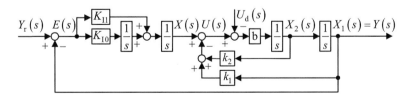

Fig. 5.23 Linear state feedback plus double integral slewing control of a rigid-body spacecraft

Then the desired closed-loop characteristic equation, given by the 2 % settling time formula (Chap. 4)] for $n = 3$, is

$$(s+q)(s+p)^3 = (s+q)\left(s^3 + 3ps^2 + 3p^2 s + p^3\right)$$
$$= s^4 + (3p+q)s^3 + 3p(p+q)s^2 + p^2(p+3q)s + p^3 q = 0, \quad (5.107)$$

where $p = 1.6(1.5+n)/T_s|_{n=3} = 7.2/T_s$ and

$$q = \frac{K_{10}}{K_{11}}. \quad (5.108)$$

Equating the characteristic polynomials of (5.106) and (5.107) then yields:

$$K_{10} = p^3 q/b, \quad (5.109)$$

$$K_{11} = p^2(p+3q)/b, \quad (5.110)$$

$$k_1 = 3p(p+q)/b, \quad (5.111)$$

$$k_2 = (3p+q)/b \quad (5.112)$$

Since q depends on the two integral term gains through (5.108), Eqs. (5.109), (5.110), (5.111) and (5.112) have to be solved for the four controller gains after substituting for q. Then (5.109) yields

$$K_{11} = \frac{p^3}{b}. \quad (5.113)$$

Substituting for K_{11} in (5.108) using (5.113) gives

$$q = \frac{K_{10} b}{p^3}. \quad (5.114)$$

Substituting for K_{11} and q in (5.110) using (5.113) and (5.114) then yields

$$p^3/b = p^2\left(p + 3K_{10}b/p^3\right)/b \Rightarrow K_{10} = 0. \quad (5.115)$$

Unfortunately, this effectively removes the first integrator in Fig. 5.23, thereby reducing the double integral term to a single integral term, which will not remove the steady-state error for a ramp reference input. Pole–zero cancellation is therefore not feasible in this case. The next option is to colocate all four closed-loop poles by setting $q = p$. From (5.109) and (5.110), the resulting zero magnitude is $z = K_{10}/K_{11} = p/4$. Hence, the pole-to-zero dominance ratio is $r_{pz} = z/p = 0.25$.

5.2 Linear Continuous State Feedback Control

From Table 1.4, the minimum value below in which the effect of the zero is significant is $r_{pz\,min} = 4.48$ for $n = 4$. Hence, some noticeable overshooting is expected. Despite this, a simulation of the slew manoeuvre will be carried out to demonstrate the zero steady-state error and examine the initial transient. Since r_{pz} is independent of p, changing T_s (which is achieved by changing p in inverse proportion) will not change the percentage overshoot. Hence, for this particular application, the minimum value of T_s is determined by the control saturation limits, $\pm u_{max}$.

Setting $q = p$ in (5.109), (5.110), (5.111) and (5.112) yields the required controller gains as

$$K_{10} = \frac{p^4}{b}, \quad K_{11} = \frac{4p^3}{b}, \quad k_1 = \frac{6p^2}{b} \quad \text{and} \quad k_2 = \frac{4p}{b} \quad (5.116)$$

where in this case $p = 1.6\,(1.5 + n)\,/\,T_s|_{n=4} = 8.8/T_s$.

Figure 5.24 shows the control system response with $r_0 = -0.5$ rad, $r_1 = 0.01$ rad/s and a step disturbance of $u_d(t) = 0.2u_{max}h\,(1-60)$ s, with $u_{max} = 10$ V, representing a disturbance torque due to the firing of an imperfectly aligned orbit adjustment thruster.

The initial state variables are all zero. Figure 5.24a, b show the error, $e(t) = y(t) - y_r(t)$, converging to zero following the initial transient. The spacecraft angular acceleration is directly proportional to $u(t)$, and hence, Fig. 5.24c indicates the initial acceleration and deceleration needed to bring the spacecraft into line with the attitude demand ramp. The step disturbance, $u_d(t)$, at $t = 200$ s is counteracted by a relatively rapid rise in $u(t)$ to meet it, and the inset shows that the mean value of the resulting transient, $e(t)$, is zero, which is due to the integral terms of the controller.

The step response (not shown) has a large overshoot of about 35 % due to the controller zero. The polynomial control technique of the following section does not introduce zeros with integral terms and therefore produces no such overshoot.

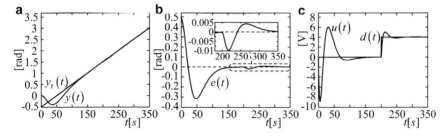

Fig. 5.24 Ramp response of rigid-body spacecraft attitude control system with disturbance. (**a**) Attitude angle and demand. (**b**) Attitude angle error. (**c**) Control and disturbance

5.3 Polynomial Control

5.3.1 Introduction

Polynomial control can be regarded as a generalisation of the RST controller [3] formulated in the discrete *z*-domain for digital implementation. The transfer function polynomials that form the basis of this controller are denoted: $R(z)$, $S(z)$ and $T(z)$. The polynomial controller has the same structure and is formulated in the Laplace domain here. It will be revisited in the z-domain in Chap. 7.

Polynomial control achieves a similar performance to linear state feedback control for a given application, through the process of complete pole assignment, but using only the measured output.

A straightforward route to polynomial control is recognising that every linear continuous SISO controller has two inputs, $Y_r(s)$ and $Y(s)$, and a single output, $U(s)$, that can be represented by the general transfer function relationship:

$$U(s) = G_r(s)Y_r(s) - G_y(s)Y(s) \qquad (5.117)$$

The corresponding block diagram is shown in Fig. 5.25.

For example, consider the linear controller comprising a linear state feedback control law and an observer for state estimation shown in Fig. 5.26. Observers will be dealt with in Chap. 8, but an understanding of them is not necessary for this illustration. All that is needed is to apply the principle of superposition and Mason's formula to derive the transfer function relationship of the form of (5.117) for the controller.

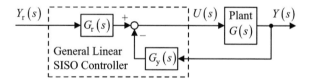

Fig. 5.25 Basic general linear continuous SISO control system representation

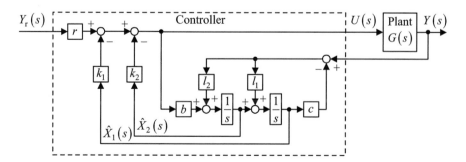

Fig. 5.26 Example of linear state feedback controller incorporating an observer

5.3 Polynomial Control

This is done with the controller considered in isolation from the plant. Thus,

$$U(s) = \frac{r\left\{1-\left[-\frac{cl_1}{s}-\frac{cl_2}{s^2}\right]\right\}Y_r(s)-\left(\frac{l_1k_1}{s}+\frac{l_2k_2}{s}+\frac{l_2k_1}{s^2}\right)Y(s)}{1-\left[-\frac{cl_1}{s}-\frac{cl_2}{s^2}-\frac{bk_2}{s}-\frac{bk_1}{s^2}\right]+\frac{cbl_1k_2}{s^2}} \quad (5.118)$$

$$= \frac{r(s^2+cl_1s+cl_2)Y_r(s)-[(l_1k_1+l_2k_2)s+l_2k_1]Y(s)}{s^2+(cl_1+bk_2)s+(cl_2+bk_1+cbl_1k_2)}$$

Hence in this case,

$$G_r(s) = \frac{r\left(s^2+cl_1s+cl_2\right)}{s^2+(cl_1+bk_2)s+(cl_2+bk_1+cbl_1k_2)} \quad (5.119)$$

and

$$G_y(s) = \frac{(l_1k_1+l_2k_2)s+l_2k_1}{s^2+(cl_1+bk_2)s+(cl_2+bk_1+cbl_1k_2)}. \quad (5.120)$$

This example also illustrates the relatively complex way the adjustable controller parameters (in this case the gains k_1, k_2, l_1 and l_2) can contribute to the coefficients of $G_r(s)$ and $G_y(s)$ for a relatively complicated control structure. It may be observed, however, that these transfer functions share a common denominator and that (5.118) is expressed in terms of polynomials. This, in turn, raises the question of whether it is possible to use the coefficients of the polynomials directly in the design process. As will be seen, pursuing this approach leads to a universal linear controller for SISO plants that is referred to as the *polynomial controller*.

5.3.2 Formulation of Polynomial Controller Structure

Expressing the controller of Fig. 5.25 using polynomials and assuming a common denominator polynomial for $G_r(s)$ and $G_y(s)$ as in the example of Fig. 5.26 yield

$$U(s) = \frac{R(s)Y_r(s) - H(s)Y(s)}{F(s)} \quad (5.121)$$

where $R(s)$ is the reference input polynomial, $H(s)$ is the feedback polynomial and the denominator polynomial, $F(s)$, will be called the filtering polynomial

Since (5.121) can be expressed as

$$U'(s) = R(s)Y_r(s) - H(s)Y(s), \quad U(s) = \frac{1}{F(s)}U'(s) \quad (5.122)$$

in which any amplified high-frequency components of measurement noise (Chap. 1) in $U'(s)$ due to the weighted derivative action of $H(s)$ are attenuated before reaching the controller output, $U(s)$, due to the low-pass filtering action of the transfer function, $1/F(s)$.

Fig. 5.27 General block diagram of SISO polynomial control system

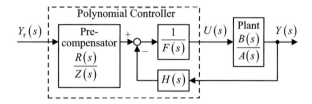

It will be recalled from Sect. 5.2.7 that an external pre-compensator is sometimes desirable whose poles can be chosen independently of the closed-loop system poles in order to effect plant zero compensation without reducing the sensitivity. A further polynomial, $Z(s)$, can be included for this purpose as the denominator of the reference input transfer function, by replacing (5.121) with

$$U(s) = \frac{1}{F(s)}\left[\frac{R(s)}{Z(s)}Y_r(s) - H(s)Y(s)\right]. \tag{5.123}$$

This results in the control system block diagram of Fig. 5.27.

To facilitate the development of the design procedure, the general plant transfer function is expressed in terms of its denominator and numerator polynomials,

$$A(s) = s^{n_a} + \sum_{i=0}^{n_a-1} a_i s^i \quad \text{and} \quad B(s) = \sum_{i=0}^{n_b} b_i s^i, \tag{5.124}$$

noting that normalisation has been carried out with respect to the coefficient of s^{n_a} in $A(s)$. Similarly expressed, the controller polynomials are

$$\begin{cases} H(s) = \sum_{i=0}^{n_h} h_i s^i, & F(s) = \sum_{i=0}^{n_f} f_i s^i \\ Z(s) = \sum_{i=0}^{n_z} z_i s^i, & R(s) = \sum_{i=0}^{n_r} r_i s^i \end{cases}. \tag{5.125}$$

The adjustable controller parameters used for the pole placement are a subset of the coefficients of $H(s)$ and $F(s)$. They are not normalised to yield a matrix–vector equation of simple form for calculation of these coefficients that can be easily implemented in computer-aided design software (Chap. 11). The closed loop DC gain can be set as desired (usually unity).

The purposes of each of the controller polynomials are defined in Table 5.2.

Although the transfer function, $1/F(s)$, increases the order of the closed-loop system by n_f, it also introduces $n_f + 1$ adjustable coefficients and therefore allows complete pole assignment despite its introduction.

By inspection of Fig. 5.27, the closed-loop transfer function is

$$\frac{Y(s)}{Y_r(s)} = \frac{R(s)}{Z(s)} \cdot \frac{\frac{B(s)}{A(s)F(s)}}{1 + \frac{B(s)H(s)}{A(s)F(s)}} = \frac{R(s)}{Z(s)} \cdot \frac{B(s)}{A(s)F(s) + B(s)H(s)}. \tag{5.126}$$

5.3 Polynomial Control

Table 5.2 Purposes of components of polynomial controller

Component	Purpose
$H(s)$	To provide a sufficient number of adjustable coefficients in the controller to enable complete pole placement
$1/F(s)$	To prevent high-frequency components of measurement noise amplified by $H(s)$ appearing in $u(s)$
$Z(s)$	To act as an external zero pre-compensator when needed, otherwise $n_z = 0$ and $z_0 = b_0$
$R(s)$	To cancel the closed-loop poles (for zero dynamic lag (Chap. 13)), otherwise $n_r = 0 \Rightarrow R(s) = r_0$, i.e. the reference input scaling coefficient

For pole placement of the closed-loop system, the characteristic polynomial is

$$A(s)F(s) + B(s)H(s). \tag{5.127}$$

As $R(s)$ and $Z(s)$ are outside the feedback loop, they are determined separately.

5.3.3 Constraints on Controller Polynomial Degrees

With reference to (5.127), the order of the closed-loop part of the system, i.e. excluding the pre-compensator, is

$$N = \max\{(n_a + n_f), (n_b + n_h)\}. \tag{5.128}$$

Importantly, the closed-loop system cannot contain an algebraic loop. This is guaranteed if the relative degree of the transfer function is positive. Thus,

$$\mathrm{rdg}\left(\frac{B(s)H(s)}{A(s)F(s)}\right) = n_a + n_f - (n_b + n_h) > 0. \tag{5.129}$$

It then follows from (5.128) and (5.129) that the total system order is

$$N = n_a + n_f. \tag{5.130}$$

It has already been stated that $F(s)$ is present to avoid high-frequency measurement noise components amplified by the differentiating action of $H(s)$ being transmitted to $U(s)$, but in order for this to be successful, the transfer function, $H(s)/F(s)$, must not exhibit any differentiating action, meaning that it must have a non-negative relative degree. Thus,

$$n_f \geq n_h. \tag{5.131}$$

The minimum number of adjustable controller parameters needed to achieve complete pole placement is equal to the closed-loop system order, N, excluding the pre-compensator. Furthermore, the N equations relating the adjustable controller parameters to the N coefficients of the closed-loop characteristic equation must be linearly independent in order for a solution to exist. In terms of (5.124) and (5.125), the closed-loop characteristic polynomial (5.127) may be expressed as

$$\left(s^{n_a} + \sum_{i=0}^{n_a-1} a_i s^i\right)\sum_{i=0}^{n_f} f_i s^i + \left(\sum_{i=0}^{n_b} b_i s^i\right)\left(\sum_{i=0}^{n_h} h_i s^i\right). \quad (5.132)$$

In view of (5.130), inspection of this polynomial reveals that it may be normalised with respect to the highest degree term via division by f_{n_f}, without reducing the number of closed-loop poles that can be independently placed. The result is

$$\left(s^{n_a} + \sum_{i=0}^{n_a-1} a_i s^i\right)\left(s^{n_f} + \sum_{i=0}^{n_f-1} f_i' s^i\right) + \left(\sum_{i=0}^{n_b} b_i s^i\right)\left(\sum_{i=0}^{n_h} h_i' s^i\right) \quad (5.133)$$

where $f_i' = f_i/f_{n_f}, i = 0, 1, .., n_f - 1$, $h_i' = h_i/f_{n_f}$ and $i = 0, 1, .., n_h$. There are therefore $n_f + n_h + 1$ adjustable coefficients in (5.133) for the pole assignment but $n_f + n_h + 2$ adjustable coefficients in (5.132), one of which is redundant. This is f_{n_f}, which becomes unity when the desired closed-loop characteristic equation is normalised with respect to the highest degree term. As will be seen, the solution of the pole placement equations to be derived automatically performs this normalisation.

In order for complete pole placement to be possible, the minimum number of adjustable parameters that may be adjusted must be equal to the order of the closed-loop system, excluding the pre-compensator. Thus,

$$n_h + n_f + 1 = N. \quad (5.134)$$

In view of (5.130) and (5.134),

$$n_h = n_a - 1. \quad (5.135)$$

Finally, (5.131) and (5.135) may be combined to form the following single expression for the polynomial degree constraints that can be used alone as the first step in the control system design.

$$n_f \geq n_h = n_a - 1 \quad (5.136)$$

It would be advantageous to increase n_f beyond the minimum to give $H(s)/F(s)$ low-pass filtering properties in applications where even the raw measurement noise would cause too much control activity. This degree of freedom for the controller design is not afforded by other control techniques.

5.3.4 Determination of the Controller Parameters

5.3.4.1 Closed-Loop DC gain

First, the value of r_0 yielding the specified closed-loop DC gain, assumed to be unity, is obtained by replacing the polynomials in (5.126) by their constant terms.

$$K_{\text{dcl}} = \frac{y(0)}{y_r(0)} = \frac{r_0}{z_0} \cdot \frac{b_0}{a_0 f_0 + b_0 h_0} = 1 \quad \Rightarrow \quad r_0 = \frac{z_0 (a_0 f_0 + b_0 h_0)}{b_0}. \quad (5.137)$$

5.3.4.2 Pole Placement for the Feedback Loop

A general algorithm for calculating the coefficients of $H(s)$ and $F(s)$ for the pole assignment with a plant of arbitrary order may be deduced from the solution for the general third-order plant with transfer function:

$$\frac{y(s)}{u(s)} = \frac{b_2 s^2 + b_1 s + b_0}{s^3 + a_2 s^2 + a_1 s + a_0}. \quad (5.138)$$

Equating the characteristic polynomial (5.127) to the desired one yields

$$\begin{aligned}
&\left(s^3 + a_2 s^2 + a_1 s + a_0\right)\left(f_2 s^2 + f_1 s + f_0\right) + \left(b_2 s^2 + b_1 s + b_0\right)\left(h_2 s^2 + h_1 s + h_0\right) \\
&= f_2 s^5 + (a_2 f_2 + f_1) s^4 + (a_1 f_2 + a_2 f_1 + f_0) s^3 \\
&\quad + (a_0 f_2 + a_1 f_1 + a_2 f_0) s^2 + (a_0 f_1 + a_1 f_0) s + a_0 f_0 \\
&\quad + b_2 h_2 s^4 + (b_1 h_2 + b_2 h_1) s^3 + (b_0 h_2 + b_1 h_1 + b_2 h_0) s^2 + (b_0 h_1 + b_1 h_0) s + b_0 h_0 \\
&= s^5 + d_4 s^4 + d_3 s^3 + d_2 s^2 + d_1 s + d_0
\end{aligned} \quad (5.139)$$

The specific structure of the polynomial controller yields an ordered structure in the equations needed to achieve the pole placement, thereby facilitating computer-aided design. This avoids much algebraic work that would be entailed in deriving the controller parameters from first principles in individual cases. The simultaneous equations for the required coefficients of $H(s)$ and $F(s)$ can be expressed by a single matrix–vector equation, and for (5.139), it is

$$\underbrace{\begin{bmatrix} 1 & 0 & 0 & 0 & 0 & 0 \\ a_2 & 1 & 0 & b_2 & 0 & 0 \\ a_1 & a_2 & 1 & b_1 & b_2 & 0 \\ a_0 & a_1 & a_2 & b_0 & b_1 & b_2 \\ 0 & a_0 & a_1 & 0 & b_0 & b_1 \\ 0 & 0 & a_0 & 0 & 0 & b_0 \end{bmatrix}}_{\mathbf{P}} \underbrace{\begin{bmatrix} f_2 \\ f_1 \\ f_0 \\ h_2 \\ h_1 \\ h_0 \end{bmatrix}}_{\mathbf{k}} = \underbrace{\begin{bmatrix} 1 \\ d_4 \\ d_3 \\ d_2 \\ d_1 \\ d_0 \end{bmatrix}}_{\mathbf{d}} \Rightarrow \mathbf{k} = \mathbf{P}^{-1}\mathbf{d}$$

Plant denominator partition   Plant numerator partition

$$(5.140)$$

The 1's on the leading diagonal of the plant denominator partition of **P** are the repeated coefficient of $s^{N_a} = s^3$ in $A(s)$, i.e. $a_{N_a} = a_3 = 1$.

In contrast with the pole placement procedure for other linear control techniques, the number of simultaneous equations is $N + 1$ rather than N due to the redundant coefficient, $f_{n_f} = f_2$, which is unity according to the first component equation of (5.140), resulting from equating the coefficients of $s^N = s^5$ in (5.139). If this coefficient was set to unity in the first place, then the number of simultaneous equations would be reduced to N, but the form of the matrix in the equation corresponding to (5.140) would be less convenient.

The general matrix–vector equation for pole placement may be deduced from the above and is as follows.

$$\underbrace{\begin{bmatrix} 1 & 0 & \cdots & 0 & 0 & \cdots & \cdots & 0 \\ a_{n_a-1} & 1 & \ddots & \vdots & b_{n_b} & \ddots & & \vdots \\ a_{n_a-2} & a_{n_a-1} & \ddots & 0 & b_{n_b-1} & b_{n_b} & \ddots & \vdots \\ \vdots & a_{n_a-2} & \ddots & 1 & \vdots & b_{n_b-1} & \ddots & 0 \\ a_0 & \vdots & \ddots & a_{n_a-1} & b_0 & \vdots & \ddots & b_{n_b} \\ 0 & a_0 & \ddots & a_{n_a-2} & 0 & b_0 & \ddots & b_{n_b-1} \\ \vdots & \ddots & \ddots & \vdots & \vdots & \ddots & \ddots & \vdots \\ 0 & \cdots & 0 & a_0 & 0 & \cdots & 0 & b_0 \end{bmatrix}}_{\mathbf{P}} \underbrace{\begin{bmatrix} f_{n_f} \\ f_{n_f-1} \\ \vdots \\ f_0 \\ \hline h_{n_h} \\ h_{n_h-1} \\ \vdots \\ h_0 \end{bmatrix}}_{\mathbf{k}} = \underbrace{\begin{bmatrix} 1 \\ d_{N-1} \\ d_{N-2} \\ \vdots \\ \vdots \\ \vdots \\ d_1 \\ d_0 \end{bmatrix}}_{\mathbf{d}}$$

Example 5.10 Polynomial speed control of an electric drive with a flexible coupling.

The plant is as in Example 5.3. With reference to Fig. 5.4, the transfer function is

$$\frac{Y(s)}{U(s)} = \frac{\frac{1}{T_1 T_2 T_{sp} s^3}}{1 - \left[-\frac{1}{T_1 T_{sp} s^2} - \frac{1}{T_2 T_{sp} s^2}\right]} = \frac{\frac{1}{T_1 T_2 T_{sp}}}{s^3 + \frac{T_1+T_2}{T_1 T_2 T_{sp}} s} = \frac{b_0}{s^3 + a_1 s}. \quad (5.141)$$

As in Example 5.3, the control variable, u, and the measurement variable, y, are normalised in this 'per unit' model and are proportional to but not numerically equal to the voltages in the physical system. The time constants, T_1, T_2 and T_{sp}, are functions of the of the moments of inertia and the spring constant.

In this example, a minimal order polynomial controller will be designed to yield a step response with a specified settling time (5 % criterion) of T_s seconds, using a zero-order pre-compensator without pole cancellation just to achieve a closed-loop DC gain of unity.

In this case, $n_a = 3$. Then according to (5.136), the minimal order controller is obtained by setting $n_f = n_h = n_a - 1 = 2$. According to (5.130), the order of

5.3 Polynomial Control

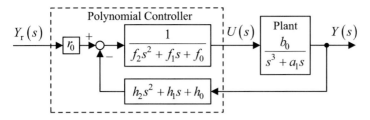

Fig. 5.28 Polynomial speed control of an electric drive with a flexible coupling

the closed-loop system without any dynamic pre-compensator is $N = n_a + n_f = 3 + 2 = 5$. Then to achieve the specified T_s, the desired closed-loop characteristic polynomial is

$$s^5 + d_4 s^4 + d_3 s^3 + d_2 s^2 + d_1 s + d_0$$
$$= (s+d)^5 = s^5 + 5ds^4 + 10d^2 s^3 + 10d^3 s^2 + 5d^4 s + d^5 \tag{5.142}$$

where $d = 1.5(1+N)/T_s|_{N=5} = 9/T_s$.

Figure 5.28 shows the control system block diagram.

Here, the plant has no finite zeros, then **P** is lower triangular, (5.140) becoming

$$\begin{bmatrix} 1 & 0 & 0 & 0 & 0 & 0 \\ 0 & 1 & 0 & 0 & 0 & 0 \\ a_1 & 0 & 1 & 0 & 0 & 0 \\ 0 & a_1 & 0 & b_0 & 0 & 0 \\ 0 & 0 & a_1 & 0 & b_0 & 0 \\ 0 & 0 & 0 & 0 & 0 & b_0 \end{bmatrix} \begin{bmatrix} f_2 \\ f_1 \\ f_0 \\ h_2 \\ h_1 \\ h_0 \end{bmatrix} = \begin{bmatrix} 1 \\ d_4 \\ d_3 \\ d_2 \\ d_1 \\ d_0 \end{bmatrix}. \tag{5.143}$$

The solution for the controller parameters is then obtained by back substitution as

$$\begin{cases} f_2 = 1, \quad f_1 = d_4, \quad a_1 f_2 + f_0 = d_3 \Rightarrow f_0 = d_3 - a_1 f_2 \\ a_1 f_1 + b_0 h_2 = d_2 \Rightarrow h_2 = (d_2 - a_1 f_1)/b_0, \\ a_1 f_0 + b_0 h_1 = d_1 \Rightarrow h_1 = (d_1 - a_1 f_0)/b_0, \quad b_0 h_0 = d_0 \Rightarrow h_0 = d_0/b_0 \end{cases} \tag{5.144}$$

Since there is no zero compensation in this example, $Z(s) = z_0 = 1$ and with $a_0 = 0$, (5.137) yields the reference input scaling coefficient as

$$r_0 = \frac{a_0 f_0 + b_0 h_0}{b_0} = h_0. \tag{5.145}$$

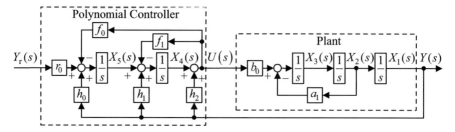

Fig. 5.29 Polynomial speed control of an electric drive with a flexible coupling showing controller implementation block diagram and plant state space representation

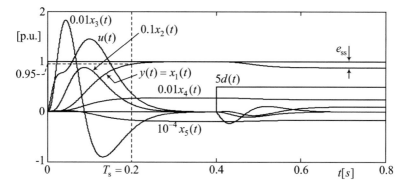

Fig. 5.30 All variables for step response of basic polynomial control of flexible drive

Figure 5.29 shows an implementation block diagram of the polynomial controller, which is its state variable block diagram in the observer canonical form (Chap. 3). State space representations are ideal for the formation of algorithms for such dynamic controllers. The plant is also shown as a state variable block diagram, to emphasise the point made at the beginning of this chapter that polynomial controllers exercise complete control of the system state.

Here, this comprises the plant state, $[x_1, x_2, x_3]^T$, and the controller state, $[x_4, x_5]^T$.

Figure 5.30 shows a simulation based on Fig. 5.29 in which the plant parameters are as specified in Example 5.3 and $T_s = 0.2$ [s] (5 % criterion). It is clear that the state variables are under control and the step response of $y(t)$ is as specified.

A step disturbance torque referred to the control input, equal to $0.1 u_{max}$, is applied at $t = 0.4$ [s]. This results in a considerable steady-state error, e_{ss}, as shown, but this can be reduced to much smaller proportions with the aid of the robust pole placement of Chap. 10 or entirely eliminated by means of an integral term, such as introduced in the following section.

5.3 Polynomial Control

5.3.4.3 Pre-compensators

Zero compensation is fully explained in Sect. 5.2.7. The denominator, $Z(s)$, of the pre-compensator in Fig. 5.27 performs this function.

Examples of the use of $R(s)$ to cancel a subset of the closed-loop poles not needed for their contribution to the response of the controlled output, $y(t)$, to the reference input, $y_r(t)$, are given in Chap. 11 and its use to achieve zero dynamic lag between $y_r(t)$ and $y(t)$, as needed for some motion control applications.

5.3.5 The Polynomial Integral Controller

To enable a polynomial control system to have a zero steady-state error with a reference input and/or an external disturbance input that are polynomials in t (step inputs being particular cases), the plant may be augmented by a multiple integrator of order, m, where $m \geq 1$, placed at the plant input as shown in Fig. 5.31.

Although the multiple integrator is technically part of the controller, it is considered part of the plant for the purpose of the control system design, forming the augmented plant with control input, $U'(s)$. Any pure integral action in the plant, which can reduce the number, m, of additional integrators needed, is shown by the factor, s^r, in the denominator of its transfer function, where $r \geq 0$ is the plant type. Then $s^r A'(s) = A(s)$ of (5.124). The controller polynomials, $H(s)$ and $F(s)$, can then be determined as in Sect. 5.3.4.

A dynamic pre-compensator polynomial, $R(s)$, is included as this will be needed to obtain a zero steady-state error.

The closed-loop transfer function relationship of the system of Fig. 5.31 is

$$Y(s) = \frac{\frac{B(s)R(s)}{s^{m+r}A'(s)F(s)}Y_r(s) - \frac{B(s)}{s^r A'(s)}U_d(s)}{1 - \left[-\frac{B(s)H(s)}{s^{m+r}A'(s)F(s)}\right]} = \frac{B(s)\left[R(s)Y_r(s) - s^m F(s)U_d(s)\right]}{s^{m+r}A'(s)F(s) + B(s)H(s)}.$$

(5.146)

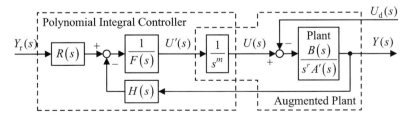

Fig. 5.31 SISO polynomial control system with multiple integrator plant augmentation

Let the polynomial reference and disturbance inputs

$$y_r(t) = \sum_{j=0}^{p} R_j t^j h(t), \quad p \geq 0 \tag{5.147}$$

and

$$u_d(t) = \sum_{k=0}^{q} D_k t^k h(t), \quad q \geq 0 \tag{5.148}$$

be applied, where $h(t)$ is the unit step function. With zero initial conditions, the Laplace transforms of (5.147) and (5.148) are

$$Y_r(s) = \sum_{j=0}^{p} R_j \frac{j!}{s^{j+1}} \tag{5.149}$$

and

$$U_d(s) = \sum_{k=0}^{q} D_k \frac{k!}{s^{k+1}}. \tag{5.150}$$

The minimum number, m, of integrators and the polynomial, $R(s)$, that are needed to ensure zero steady-state error will now be determined as a function of the input polynomial degrees, p and q, and the plant type, r.

Using the transfer function relationship of (5.146), the corresponding relationship for the error, $E(s) = Y_r(s) - Y(s)$, may be derived as

$$E(s) = \frac{\{s^{m+r} A'(s) F(s) + B(s)[H(s) - R(s)]\} Y_r(s) + s^m B(s) F(s) U_d(s)}{s^{m+r} A'(s) F(s) + B(s) H(s)}. \tag{5.151}$$

Substituting for $Y_r(s)$ and $U_d(s)$ using (5.149) and (5.150) then yields

$$E(s) = \frac{\left[\{s^{m+r} A'(s) F(s) + B(s)[H(s) - R(s)]\} \sum_{j=0}^{p} R_j \frac{j!}{s^{j+1}} + s^m B(s) F(s) \sum_{k=0}^{q} D_k \frac{k!}{s^{k+1}} \right]}{s^{m+r} A'(s) F(s) + B(s) H(s)} \tag{5.152}$$

The steady-state error is then given by $e_{ss} = \lim_{s \to 0} s E(s)$. For the disturbance input, (5.152) reveals that a necessary condition for $e_{ss} = 0$ is

$$m \geq q + 1. \tag{5.153}$$

5.3 Polynomial Control

For the reference input, a necessary condition is $m + r \geq p + 1 \Rightarrow$

$$m \geq p - r + 1. \tag{5.154}$$

Conditions (5.153) and (5.154) may be combined by writing

$$m \geq \max(\overset{a}{p} - r + 1, \, q + 1). \tag{5.155}$$

To complete the set of conditions, if (5.153) and (5.154) are satisfied, then by inspection of (5.152), $e_{ss} = 0$ if

$$\lim_{s \to 0} B(s) [H(s) - R(s)] \sum_{j=0}^{p} R_j \frac{j!}{s^j} = 0. \tag{5.156}$$

Since $B(0) \neq 0$, (5.156) can be simplified to

$$\lim_{s \to 0} [H(s) - R(s)] \sum_{j=0}^{p} R_j \frac{j!}{s^j} = 0 \Rightarrow \lim_{s \to 0} [H(s) - R(s)] \frac{1}{s^p} = 0. \tag{5.157}$$

Expanding (5.157) then enables the lowest degree $R(s)$ to be determined, which will minimise the possible number of overshoots and undershoots caused by the associated zeros. Thus,

$$\lim_{s \to 0} \left[(h_0 - r_0) + (h_1 - r_1)s + \ldots (h_{n_h} - r_{n_h}) s^{n_h} - r_{n_h+1} s^{n_h+1} - \ldots \right] \frac{1}{s^p} = 0. \tag{5.158}$$

All the terms in the square parentheses of (5.158) with degree greater than p vanish in the limit and are therefore not of concern.

The polynomial, $R(s)$, of minimum degree can then be obtained by setting

$$r_i = 0, \quad i = p+1, \, p+2, \ldots, \tag{5.159}$$

All the terms of degree p or lower have to be made zero by setting

$$r_i = h_i, \quad i = 0, 1, \ldots, p. \tag{5.160}$$

Hence, the pre-compensator polynomial of minimum degree satisfying (5.158) is

$$R(s) = \sum_{i=0}^{p} h_i s^i. \tag{5.161}$$

Regarding the degrees of the controller polynomials, the plant order, n_a, is replaced by the augmented plant order, $n_a + m$, in (5.136) to yield

$$n_f \geq n_h = n_a + m - 1. \quad (5.162)$$

In cases where $p < n_h$, the pre-compensator, $R(s)$, is a truncated version, $H_L(s)$, of the feedback polynomial, $H(s)$, in which the coefficients of $R(s)$ are the coefficients of $H(s)$, from h_0 to h_p. Then the control error, $E(s) = Y_r(s) - Y(s)$, is input to the controller, rather than $Y(s)$ alone, via $H_L(s)$. Also, the output, $Y(s)$, is input to the controller via the remaining fragment,

$$H_H(s) = \sum_{i=p+1}^{n_h} h_i s^i \quad (5.163)$$

of $H(s)$, such that

$$H(s) = H_H(s) + H_L(s). \quad (5.164)$$

Fig. 5.32 shows the resulting controller structure.

This may be seen in the following example.

For a step reference input, $p = 0$ and $H_L(s) = h_0$, which is only a reference input scaling coefficient, and no controller zeros are introduced. According to (5.154), only one integrator is needed to augment the plant, which is sufficient for many applications. In contrast, the PID controller introduces two zeros. If the degree, q, of the disturbance input is increased sufficiently for the number, m, of the augmenting integrators, given by (5.155), to increase, then the degree of $H_L(s)$ remains unaltered and no more controller zeros are introduced.

Example 5.11 Polynomial integral position control of a flexible electric drive.

The plant is the same as that of Example 5.10 but with an additional kinematic integrator to change from speed control to position control, the transfer function being

$$\frac{Y(s)}{U(s)} = \frac{\frac{1}{T_1 T_2 T_{sp}}}{s^4 + \frac{T_1+T_2}{T_1 T_2 T_{sp}} s^2} = \frac{b_0}{s^4 + a_2 s^2}. \quad (5.165)$$

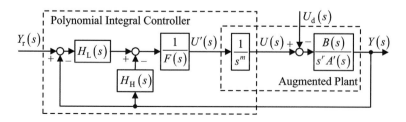

Fig. 5.32 Structure of polynomial integral controller with split feedback polynomial

5.3 Polynomial Control

The position measurement, $y(t)$, has to follow each segment of a cubic spline (Chap. 13), subject to a step external disturbance, with zero steady-state error. The specified settling time is $T_s = 0.2$ [s].

The disturbance is a step, i.e. a polynomial of degree $q = 0$. Then (5.153) gives $m \geq 1$. By inspection of (5.165), the number of pure integrators in the plant is $r = 2$. Since the reference input is a cubic polynomial, $p = 3$. (5.154) gives $m \geq p - r + 1 = 2$. Hence, the minimum number of pure integrators needed to augment the plant is $m = 2$. The augmented plant transfer function is therefore

$$\frac{Y(s)}{U'(s)} = \frac{b_0}{s^6 + a_4 s^4}, \qquad (5.166)$$

where $a_4 = a_2$ in (5.165). According to (5.162), the minimal order controller is obtained by setting $n_f = n_h = n_a + m - 1 = 4 + 2 - 1 = 5$. Then the total system order is $N = n_a + m + n_f = 4 + 2 + 5 = 11$. Then assuming coincident closed-loop pole placement using the 5 % settling time formula, the desired characteristic polynomial is

$$\begin{aligned}(s+d)^{11} &= s^{11} + 11ds^{10} + 55d^2 s^9 + 165d^3 s^8 + 330d^4 s^7 + 462d^5 s^6 \\ &\quad + 462d^6 s^5 + 330d^7 s^4 + 165d^8 s^3 + 55d^9 s^2 + 11d^{10}s + d^{11} \\ &= s^{11} + d_{10}s^{10} + d_9 s^9 + d_8 s^8 + d_7 s^7 + d_6 s^6 + d_5 s^5 + d_4 s^4 \\ &\quad + d_3 s^3 + d_2 s^2 + d_1 s + d_0,\end{aligned} \qquad (5.167)$$

where $d = 1.5(1+N)/T_s|_{N=11} = 18/T_s$. The controller parameter equation is then

$$\begin{bmatrix} 1 & 0 & 0 & 0 & 0 & 0 & 0 & 0 & 0 & 0 & 0 & 0 \\ 0 & 1 & 0 & 0 & 0 & 0 & 0 & 0 & 0 & 0 & 0 & 0 \\ a_4 & 0 & 1 & 0 & 0 & 0 & 0 & 0 & 0 & 0 & 0 & 0 \\ 0 & a_4 & 0 & 1 & 0 & 0 & 0 & 0 & 0 & 0 & 0 & 0 \\ 0 & 0 & a_4 & 0 & 1 & 0 & 0 & 0 & 0 & 0 & 0 & 0 \\ 0 & 0 & 0 & a_4 & 0 & 1 & 0 & 0 & 0 & 0 & 0 & 0 \\ 0 & 0 & 0 & 0 & a_4 & 0 & b_0 & 0 & 0 & 0 & 0 & 0 \\ 0 & 0 & 0 & 0 & 0 & a_4 & 0 & b_0 & 0 & 0 & 0 & 0 \\ 0 & 0 & 0 & 0 & 0 & 0 & 0 & 0 & b_0 & 0 & 0 & 0 \\ 0 & 0 & 0 & 0 & 0 & 0 & 0 & 0 & 0 & b_0 & 0 & 0 \\ 0 & 0 & 0 & 0 & 0 & 0 & 0 & 0 & 0 & 0 & b_0 & 0 \\ 0 & 0 & 0 & 0 & 0 & 0 & 0 & 0 & 0 & 0 & 0 & b_0 \end{bmatrix} \begin{bmatrix} f_5 \\ f_4 \\ f_3 \\ f_2 \\ f_1 \\ f_0 \\ h_5 \\ h_4 \\ h_3 \\ h_2 \\ h_1 \\ h_0 \end{bmatrix} = \begin{bmatrix} 1 \\ d_{10} \\ d_9 \\ d_8 \\ d_7 \\ d_6 \\ d_5 \\ d_4 \\ d_3 \\ d_2 \\ d_1 \\ d_0 \end{bmatrix}. \qquad (5.168)$$

As the plant has no finite zeros, the plant parameter matrix, **P**, has the lower triangular form, and therefore, solution by back substitution is possible. From (5.168),

$$f_5 = 1, \quad f_4 = d_{10}, \quad a_4 f_5 + f_3 = d_9 \Rightarrow f_3 = d_9 - a_4 f_5, \quad a_4 f_4 + f_2 = d_8 \Rightarrow$$
$$f_2 = d_8 - a_4 f_4,$$
$$a_4 f_3 + f_1 = d_7 \Rightarrow f_1 = d_7 - a_4 f_3, \quad a_4 f_2 + f_0 = d_6 \Rightarrow f_0 = d_6 - a_4 f_2,$$
$$a_4 f_1 + b_0 h_5 = d_5 \Rightarrow h_5 = (d_5 - a_4 f_1)/b_0, \quad a_4 f_0 + b_0 h_4 = d_4 \Rightarrow$$
$$h_4 = (d_4 - a_4 f_0)/b_0$$
$$b_0 h_3 = d_3 \Rightarrow h_3 = d_3/b_0, \quad b_0 h_2 = d_2 \Rightarrow h_2 = d_2/b_0,$$
$$b_0 h_1 = d_1 \Rightarrow h_1 = d_1/b_0, \quad b_0 h_0 = d_0 \Rightarrow h_0 = d_0/b_0 \tag{5.169}$$

Finally the reference input pre-compensator is given by (5.161) with $p = 3$. Thus,

$$R(s) = h_0 + h_1 s + h_2 s^2 + h_3 s^3. \tag{5.170}$$

It should be noted that such pole placement in higher order systems can require very large controller parameter values. This may cause numerical problems, particularly with fixed point digital processors, but the increasing computational capability of these processors should alleviate future occurrences of this problem.

The control system implementation block diagram is shown in Fig. 5.33.

This state space form of block diagram (in which the state variables are the integrator outputs) is ideal for producing a discrete algorithm for digital implementation and may also be used directly in the z-domain (Chap. 6).

Figure 5.34 shows the responses of the system to cubic reference inputs with zero initial conditions on the six integrators of the controller and zero initial plant state. Figure 5.1a clearly shows the error asymptotically approaching zero for $t > T_s$, as required. The oscillations of the acquisition transient are due to the zeros introduced

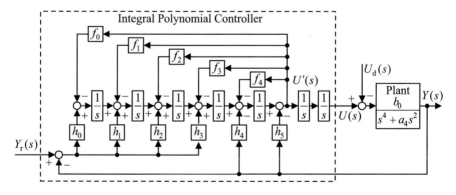

Fig. 5.33 Implementation block diagram of integral polynomial controller for position control of an electric drive with a flexible coupling

5.3 Polynomial Control

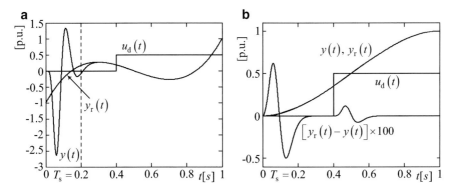

Fig. 5.34 Responses of flexible drive control system to cubic reference inputs. (**a**) $y_r(t) = 16t^3 - 24t^2 + 10t - 1$. (**b**) $y_r(t) = 3t^2 - 2t^3$

by the pre-compensator (5.170), which are the roots of $H_L(s) = 0$. It should be noted that the oscillations are finite in number, in contrast with oscillations due to complex conjugate pole pairs (Chap. 1 and Sect. 5.2.7). They are relatively large in this case since $y(0) \neq y_r(0)$, but in applications where $y_r(t)$ is a cubic spline (Appendix A11) comprising contiguous cubic segments, $y(t) \cong y_r(t)$ and $\dot{y}(t) \cong \dot{y}_r(t)$ at the beginning of each segment.

This is simulated in Fig. 5.1b where $y(0) = y_r(0) = 0$ and $\dot{y}(0) = \dot{y}_r(0) = 0$, but there are differences between $\ddot{y}(0)$ and $\ddot{y}_r(0)$, similarly between $\dddot{y}(0)$ and $\dddot{y}_r(0)$, which give rise to a transient error, $e(t) = y_r(t) - y(t)$. This, however, is hardly visible in the superimposed graphs of $y(t)$ and $y_r(t)$. It is therefore shown separately and magnified.

A step external disturbance of 50 % of the maximum control magnitude is applied at $t = 0.4$ [s]. A small transient control error with zero steady-state value occurs, which is too small to be visible in Fig. 5.34a and hence magnified in Fig. 5.34b.

In the implementation of any controller containing integral terms, it is advisable to include anti-windup measures (Chap. 1). Although potential instabilities due to integral windup will have been removed, it is essential to carry out an investigation to ensure that the control saturation, which can occur in any case, will not cause instability or other undesirable behaviours. Simulation is the most practicable tool for this.

Finally, it is important to again raise the general issue of plant model uncertainty. Although any model-based control system will produce the intended responses to the reference inputs, investigations should be carried out to ascertain the effects of the most extreme plant model mismatching perceived to be possible in the particular application. For the simpler nominally linear systems, Routh's stability criterion (Appendix A5) could be useful, but simulation would be a more practicable approach for more complex control systems.

References

1. Hughes PC (2004) Spacecraft attitude dynamics. Dover Publications, Mineola, New York
2. Ariola M, Pironti A (2008) Magnetic control of Tokamak Plasmas. Springer, London
3. Landau ID, Zito G (2006) Digital control systems, design, identification and implementation. Springer, Berlin/New York

Chapter 6
Discrete Control of LTI SISO Plants

6.1 Introduction

Controllers implemented by means of digital processors are referred to as discrete (or digital) controllers, since this entails a sequence of internal binary states between which discrete transitions occur. These sequences of states are determined by discrete control algorithms, which comprise a set of instructions repeatedly carried out. Each execution of this set of instructions is called an *iteration*. The iterations are separated by h seconds, called the iteration period or the step length. It is also often called the sampling period because the measurement variables from the plant are sampled and held in the processor at the beginning of each iteration.

The control systems for linear time-invariant (LTI) plants of the previous chapters have been formulated in the continuous domain. Their simulations, however, are in the discrete domain, implemented by software on digital computers. The simulation results *appear* continuous due to the iteration period being very small compared with the duration of the transient being simulated. So the controllers of the simulations are already implemented in a discrete form. This implies that it is possible to implement controllers formulated in the continuous domain to control real plants using the gains and any other parameters determined in the continuous domain. As will be seen, this is possible with a sufficiently short iteration period.

Real-time control and simulation software is available in which a controller developed in the simulation environment is disconnected from the simulated plant and interfaced with the real plant via digital-to-analogue and analogue-to-digital converters, dSPACE® being a prime example. It must be realised, however, that simulation software includes a choice of different computing engines, i.e. numerical methods for computing the solutions to differential equations, some of which are 'variable step', meaning that h varies from one iteration to the next to attain high accuracy. Only fixed step computing engines are used with simulation and real-time control software, because the plant–controller interface has a fixed sampling period with a minimum limit of h_{min}. There is also an upper limit, h_{max}, above which the transient response of a closed-loop system will differ significantly from that predicted in the continuous domain. This is the criterion presented in Sect. 6.4. If $h_{max} > h_{min}$, then a realisable step length satisfying $h_{max} > h > h_{min}$ exists and the straightforward use of the continuous domain controller parameters in the digital implementation described in Sect. 6.5 is possible. If, on the other hand, $h_{max} < h_{min}$, then the controller can be designed in the discrete domain with the aid of the z-transform, the theory of which has been introduced in Chap. 3 and is continued in Sect. 6.3. In this approach, the continuous dynamic elements of controllers are replaced by their discrete equivalents as explained in Sect. 6.5 and the model-based design carried out according to Sect. 6.6, plants containing pure time delays also being catered for. Some practitioners may prefer to develop real-time control software independently using a high-level computing language, and Sect. 6.5 provides a straightforward route to this goal.

6.2 Real-Time Operation of Digital Controllers

Figure 6.1 shows how a digital processor operating in discrete time to implement a controller interfaces with the continuous-time world outside.

The continuously varying measurement variable, $y(t)$, enters the digital processor via an analogue-to-digital (A/D) converter which takes a sample at the beginning of each iteration period and carries out a conversion. The converter output is held at the value of the last sample and updated using every subsequent sample. The result is a *sampled and held* (S/H) version of $y(t)$, denoted by $y_s(t)$. This is shown graphically in Fig. 6.1. The integer, k, is referred to as the iteration number or the sample number. The reference input, $y_r(t)$, is either S/H, as shown, or is directly produced in the software. The control algorithm uses $y_s(t)$ and $y_r(t)$ to calculate the control signal, $u(t)$, that is piecewise constant and delivered to the plant via

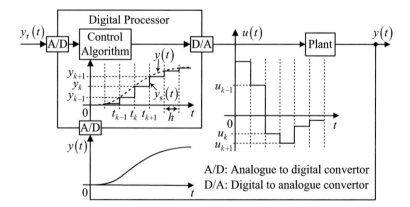

Fig. 6.1 Basic input and output signals of digital controller and the sampling process

6.3 Dynamics of Discrete Linear Systems

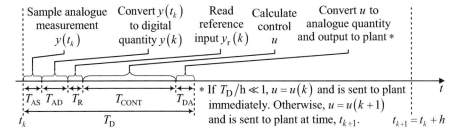

Fig. 6.2 Sequence of events during one iteration interval of a digital controller

a digital-to-analogue (D/A) converter. Since the plant is a continuous dynamical system acting as a low-pass filter, its output, $y(t)$, is continuous despite $u(t)$ being discontinuous.

If the effect of the sampling is predicted to be significant with the minimum value of h, then it is of interest to consider the sequence of events shown in Fig. 6.2. The periods, T_{AS}, T_{AD}, T_R and T_{DA}, depend on the processor hardware. The period, T_{CONT}, also depends on the control algorithm. In some cases, $T_D/h \approx 0$.

Then the controller design can be carried out using a plant model that implicitly assumes that once the sample, $y(t_k)$, is taken, the corresponding control value, $u(k)$, is calculated and output to the plant at time, t_k, *with no delay*. This theoretical ideal cannot be precisely implemented but is approximated if $T_D/h \ll 1$. This simplifies the control system design using z-transforms. If, on the other hand, $0 < T_D/h \leq 1$, then the delay might cause the closed-loop performance to differ considerably from that predicted with zero delay. Model-based control system design to overcome this problem would be difficult, particularly if T_D varies from one iteration to the next. In such cases, however, the control variable can be designated $u(k+1)$ and its output delayed until time, t_{k+1}, which introduces a fixed time delay of h seconds in the plant model. This can easily be accommodated in the control system design using the control techniques of Sect. 6.6.3.

Note that k represents the present discrete time as t represents the present time.

Although processor computational power is increasing, the trade-off between the hardware cost and the computational speed is important in the processor choice for mass-produced applications, where increasing the design effort to produce a controller capable of longer iteration periods using a cheaper processor is more economical.

6.3 Dynamics of Discrete Linear Systems

6.3.1 Stability Analysis in the z-Plane

If a control system is designed as a continuous-time system and then the controller is software implemented on a digital processor as in Fig. 6.1, then the iteration interval, h, cannot be arbitrarily long. Beyond a certain limit, h_{max}, established in Sect. 6.4, the dynamic performance will differ considerably from that predicted in the continuous domain and beyond a higher and *critical* threshold, h_{crit}, discussed in this subsection, instability will result.

It will be recalled from Sect. 3.4.3 that the z-transform is based on the Laplace transform through setting

$$z = e^{sh}, \quad (6.1)$$

where s and z are, respectively, the complex variables of the Laplace transform and the z-transform. Equation (6.1) enables the stability boundary and region in the z-plane corresponding to the stability boundary and region in the s-plane to be determined as follows.

$$z = e^{sh} = e^{(\sigma + j\omega)h} = e^{\sigma h} e^{j\omega h} = e^{\sigma h} \left[\cos(\omega h) + j \sin(\omega h) \right]. \quad (6.2)$$

Let

$$z = x + jy. \quad (6.3)$$

Then equating (6.2) and (6.3) yields

$$x = e^{\sigma h} \cos(\omega h). \quad (6.4)$$

and

$$y = e^{\sigma h} \sin(\omega h). \quad (6.5)$$

Squaring and adding (6.4) and (6.5) gives

$$x^2 + y^2 = e^{2\sigma h} \left[\cos^2(\omega h) + \sin^2(\omega h) \right] \Rightarrow x^2 + y^2 = r^2. \quad (6.6)$$

which is a circle with radius, r, and centre at the origin of the z-plane, where

$$r = e^{\sigma h}. \quad (6.7)$$

The stability boundary in the s-plane is the imaginary axis defined by

$$\sigma = 0, \quad (6.8)$$

which, according to (6.7), gives $r = 1$. Then (6.6) is the equation of the stability boundary in the z-plane, which is the unit circle.

The stability region in the s-plane is the left half plane, defined by $\sigma \in (0, \infty)$, as shown shaded in Fig. 6.3a. Then in the z-plane, according to (6.7),

$$r \in (0, 1). \quad (6.9)$$

6.3 Dynamics of Discrete Linear Systems

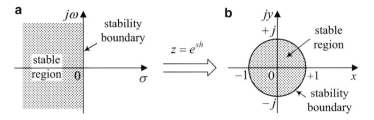

Fig. 6.3 Stability regions. (**a**) s-plane. (**b**) z-plane

Then (6.6) gives the unit disc centred at the origin as shown in Fig. 6.3b.

To summarise, (6.1) is a transformation that maps the left half of the s-plane onto the unit disc centred on the origin in the z-plane.

The dependence of the dynamic behaviour and stability of linear discrete systems on the locations of their poles in the z-plane will now be illustrated by a simple example. Consider the first-order system with transfer function,

$$\frac{Y(z)}{V(z)} = \frac{1-p}{z-p}. \tag{6.10}$$

This is not intended to specifically represent an uncontrolled plant or a closed-loop system, just an arbitrary linear system. Recall that the DC gain of a continuous LTI SISO system, containing no pure integrators, with transfer function, $G(s)$, is $G_{\mathrm{DC}} = \lim_{s \to 0} G(s)$. From (6.1), as $s \to 0$, $z \to 1$, so the DC gain of a discrete LTI SISO system with transfer function, $Q(z)$, also containing no pure integrators, meaning that the transfer function denominator cannot have $z-1$ as a factor (Table 3 in Tables), then the DC gain is

$$Q_{\mathrm{DC}} = \lim_{z \to 1} Q(z). \tag{6.11}$$

Hence, the system with transfer function (6.10) has a unity DC gain. A set of unit step responses for different pole locations, $z = p$, is shown in Fig. 6.4.

The difference equation corresponding to (6.10) is obtained by first cross multiplying, giving $(z-p)\,Y(z) = (1-p)\,V(z)$. Then taking inverse z-transforms yields

$$\mathcal{Z}^{-1}\{(z-p)\,Y(z)\} = \mathcal{Z}^{-1}\{(1-p)\,V(z)\} \Rightarrow \mathcal{Z}^{-1}\{zY(z)\} - p\mathcal{Z}^{-1}\{Y(z)\}$$
$$= (1-p)\,\mathcal{Z}^{-1}\{V(z)\} \Rightarrow$$
$$y(k+1) - py(k) = (1-p)\,v(k) \Rightarrow y(k+1) = py(k) + (1-p)\,v(k) \tag{6.12}$$

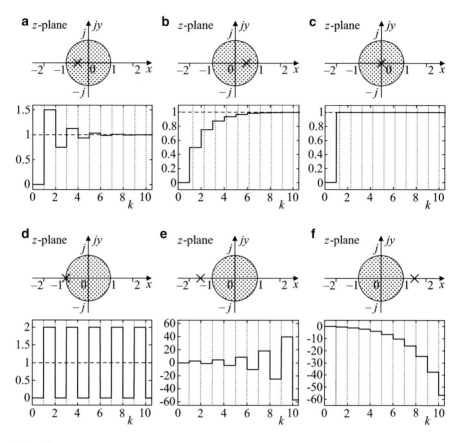

Fig. 6.4 Pole locations and step responses of a discrete LTI SISO system. (**a**) $p=-0.5$ (**b**) $p=0.5$ (**c**) $p=0$ (**d**) $p=-1$ (**e**) $p=-1.5$ (**f**) $p=1.5$

The responses are obtained with the discrete step input, $v(k) = \{0 \text{ for } k < 0, 1 \text{ for } k \geq 0\}$, and zero initial state, $y(0) = 0$. This simple example demonstrates the system stability for the pole within the unit circle in the z-plane and the instability resulting when the pole is without this region. It also highlights several features that are counter-intuitive at first sight when already familiar with the continuous domain. First, when the pole is within the left half of the z-plane, exemplified by Fig. 6.4a, d, e, the response is oscillatory. Oscillations can never occur in a first-order continuous LTI SISO system. These oscillations are peculiar to discrete systems and occur at a fixed frequency of $1/(2h)$ [Hz]. Second, if the pole is at the origin of the z-plane as in Fig. 6.4c, the system settles precisely with zero steady-state error in just one iteration period. This is an example of a *dead-beat* response. If the pole is in the right half of the z-plane, as in Fig. 6.4b, f, the response bears a closer resemblance to the

6.3 Dynamics of Discrete Linear Systems

familiar exponential step response of a continuous first-order LTI system. As will be seen, if a discrete control system is created that closely emulates the behaviour of the equivalent continuous system, its poles lie in the right half of the unit disc in the z-plane.

The stability of linear discrete systems may be checked by examining their characteristic polynomials for unstable roots, as in the continuous domain, but these have to lie outside the unit circle in the z-plane. For systems up to fourth order, the roots can be found analytically. For third- and fourth-order systems, however, the formulae for the roots are very long and time consuming to apply [1]. No analytical formulae for the roots of polynomial equations of greater than 4th degree exist but they can be found numerically with the aid of a computer.

An alternative to finding the roots of the characteristic polynomial is to determine whether any roots lie outside the unit circle, rendering the system unstable. Two methods are available that are similar to Routh's stability test in the continuous domain. The first is the bilinear transformation,

$$z = \frac{1+w}{1-w}, \tag{6.13}$$

which maps the unit disc in the z-plane to the left half of the complex w-plane. This yields a polynomial, $p(w)$, whereas (6.1) does not yield a polynomial in s. In this case, Routh's test may be applied directly to $p(w)$ for detecting roots in the right half of the w-plane, implying the detection of roots outside the unit circle in the z-plane. The second method is Jury's method, in which a Jury array is formed that is analogous to the Routh array, without the need for the bilinear transformation [2]. Both methods, however, are more time-consuming than Routh's method in the continuous domain.

An easily applied approach to stability analysis of a discrete LTI control system of any order is as follows. After deriving the closed-loop characteristic polynomial,

$$z^N + n_{N-1}(h)z^{N-1} + \cdots + n_1(h)z + n_0(h), \tag{6.14}$$

the root locus with respect to the iteration period, h, is computed and displayed. This will immediately find the upper limit, h_{crit}, on h beyond which instability results. Software such as MATLAB© with SIMULINK© can be adapted to generate such a locus using a numerical root finder. It is important to realise that the standard root locus software is inapplicable since the variable parameter is a gain that is a factor of the open-loop transfer function, while the variable parameter in this case is h, of which each coefficient in (6.14) is a different function.

Example 6.1 Stability analysis of a digitally implemented IPD control system

Figure 6.5 shows the z-transfer function block diagram of one axis of a three-axis spacecraft attitude control system in which the inter-axis coupling is negligible.

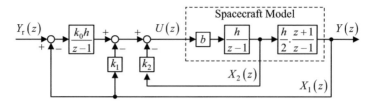

Fig. 6.5 Spacecraft attitude control system (single axis) with digital LSF + integral controller

The linear state feedback plus integral controller has been designed in the continuous domain to achieve a settling time of $T_s = 40$ [s] with zero overshoot using the 5 % settling time formula, the controller gains being given by

$$k_0 = \frac{p^3}{b}, \quad k_1 = \frac{3p^2}{b} \quad \text{and} \quad k_2 = \frac{3p}{b}, \tag{6.15}$$

where $p = 6/T_s$, $b = K_w/J$, where K_w is the reaction wheel torque constant and J is the moment of inertia about the control axis. The parameter values are $J = 200$ [Kg m^2], $K_w = 0.1$ [Nm/V] and $T_s = 30$ [s].

It is required to plot a root locus with respect to the iteration period, h, and use this to find the upper limit of h, beyond which the system will be unstable.

The algebra is simplified by setting

$$q = z - 1. \tag{6.16}$$

Then the characteristic polynomial of the system of Fig. 6.5 may be obtained as

$$p(q) = r(q)\Delta(q) \tag{6.17}$$

where $\Delta(q)$ is the determinant of Mason's formula and $r(q)$ is a polynomial of minimal order such that $p(q)$ is a polynomial. Hence,

$$\Delta(q) = 1 - \left[-\frac{bh}{q} \left(k_2 + \frac{k_1 h}{2} \cdot \frac{q+2}{q} + \frac{k_0 h^2}{2} \cdot \frac{q+2}{q^2} \right) \right] \tag{6.18}$$

Choosing $r(q) = q^3$ then yields

$$r(q) = q^3 + bh \left(k_2 q^2 + \frac{k_1 h}{2} \cdot (q^2 + 2q) + \frac{k_0 h^2}{2} \cdot (q+2) \right)$$
$$= q^3 + b\left(k_2 h + \tfrac{1}{2} k_1 h^2\right) q^2 + b\left(k_1 h^2 + \tfrac{1}{2} k_0 h^3\right) q + b k_0 h^3 \tag{6.19}$$

According to (6.16), the required closed-loop poles of the system in the z-plane are

$$z_{1,2,3} = q_{1,2,3} + 1 \tag{6.20}$$

where $q_{1,2,3}$ are the roots of $r(q) = 0$.

6.3 Dynamics of Discrete Linear Systems

In order to find the critical value, h_{crit}, of the upper stability limit, h is increased from $0\,[\text{s}]$ until one of the branches of the root locus just reaches the stability boundary. Figure 6.6a shows the root locus up to $h = 3.1\,[\text{s}]$.

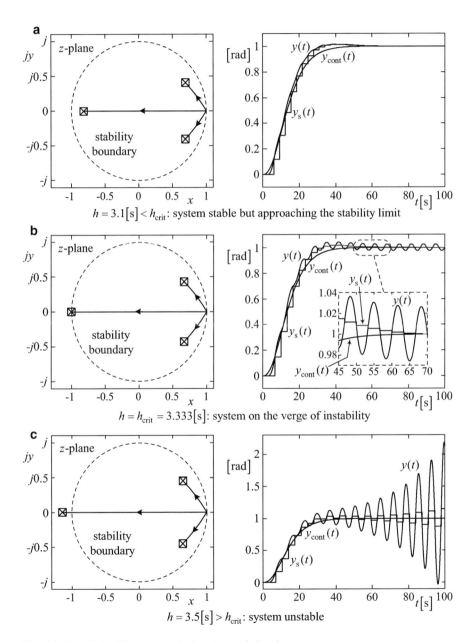

Fig. 6.6 Root loci with respect to the iteration period and step responses

The real branch reaches the boundary at $h = 3.333$ [s] as shown in Fig. 6.6b and therefore this is the required maximum value. The corresponding step responses are also shown so that the system behaviour may be correlated with the pole locations. Here $y_{cont}(t)$ is the step response of the equivalent continuous system upon which the controller design is based, $y(t)$ is the step response of the digitally controlled plant and $y_s(t)$ is the sampled and held version of $y(t)$, which is also the solution to the linear difference equation obeyed by the closed-loop system corresponding to the closed-loop z-transfer function relationship, $Y(z)/Y_r(z)$. In Fig. 6.6b, $y(t)$ is seen to 'settle' to a constant amplitude oscillation although the successive steps of $y_s(t)$ reduce with time. This illustrates the importance of predicting the behaviour of the continuous plant between the iteration instants. Figure 6.6c shows the result of increasing h to 3.5 [s], where the real pole lies outside the stability boundary. In this case, the system suffers from oscillatory instability.

When applying this method, the root locus commences on the stability boundary at $z = 1 + j0$, as in Example 6.1. The reasoning is as follows. It may be seen, by examples taken from Table 3 in the section, Tables, preceding the appendix, that if the general plant transfer function is denoted

$$Q(z) = \frac{B(z, h)}{A(z, h)}, \tag{6.21}$$

where

$$B(z) = \sum_{i=0}^{m} b_i(h) z^i \quad \text{and} \quad A(z) = \sum_{j=0}^{n} a_j(h) z^j, \tag{6.22}$$

then

$$\lim_{h \to 0} B(z, h) = 0 \quad \text{and} \quad \lim_{h \to 0} A(z, h) = a(z - 1)^n. \tag{6.23}$$

where a is a constant. Next, any linear discrete controller can be represented by the general transfer function relationship,

$$U(z) = \frac{1}{F(z, h)} [R(z, h) Y_r(z) - H(z, h) Y(z)]. \tag{6.24}$$

where $F(z, h)$ and $H(z, h)$ are, respectively, the denominator and feedback polynomials and $R(z, h)$ is the reference input transfer function. Since the controller under study here is a continuous controller that has been converted to the discrete domain by using a table such as Table 3 of Tables preceding the appendix, then, similarly to (6.23),

$$\lim_{h \to 0} F(z, h) = f(z - 1)^p. \tag{6.25}$$

Figure 6.7 shows the transfer function block diagram of controller (6.24) applied to plant (6.21).

6.3 Dynamics of Discrete Linear Systems

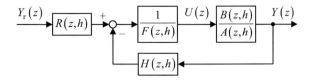

Fig. 6.7 General representation of SISO LTI discrete control system

By inspection of this diagram, the characteristic equation is

$$1 - \left[-\frac{B(z,h)}{A(z,h)} \cdot \frac{H(z,h)}{F(z,h)} \right] = 0 \Rightarrow A(z,h) F(z,h) + B(z,h) H(z,h) = 0. \tag{6.26}$$

It then follows from (6.23) and (6.25) that as $h \to 0$, (6.26) becomes

$$(z - 1)^N = 0, \tag{6.27}$$

where $N = n + p$. Hence, the root locus with respect to h starts with $h = 0$ at the point, $1 + j0$, in the z-plane. It is evident from this that if h tends to zero, which yields a performance approaching that of the equivalent ideal continuous control system, then the z-plane poles approach coincidence and also become closer to the stability boundary when viewed on the scale of Fig. 6.6c. This does not indicate an approach to instability, however, as the equivalent ideal continuous system is stable with its closed loop poles chosen appropriately in the left half of the s-plane.

6.3.2 Connection Between Dynamic Behaviour and the z-Plane Pole Locations

6.3.2.1 General Considerations

The purpose of this section is to enable the dynamic behaviour of an SISO LTI discrete system to be assessed by inspection of its z-plane poles. The dynamic behaviour of a system may be described as its natural behaviour with non-zero initial states and zero inputs. It may be evaluated by examining the impulse response, because this input is applied only at $t = 0$ to disturb the states and for $t > 0$ it is zero.

Consider a continuous SISO LTI feedback control system with input, $y_r(t)$, and output, $y(t)$. It should be noted that this is a fictitious control system created in theory for the sole purpose of enabling the dynamic characteristics of a discrete system to be recognised by viewing the pole locations in the z-plane. Let a sampled and held output, $y_s(t)$, be formed with a sampling period of h [s]. If $y_r(t)$ is similarly sampled and held, then a linear difference equation relating $y_s(t)$ to $y_r(t)$ may be formed, giving a closed-loop z-transfer function,

$$\frac{Y_s(z)}{Y_r(z)} = Q_{cl}(z) \qquad (6.28)$$

The fictitious linear continuous control system with Laplace transfer function,

$$\frac{Y(s)}{Y_r(s)} = G_{cl}(s), \qquad (6.29)$$

is then the *equivalent continuous system* of the discrete system. The relationship between the systems defined by (6.28) and (6.29) is the same as that between the continuous and discrete models of the same plant established in Chap. 3, Sect. 3.4.

It follows that the poles of $G_{cl}(s)$ are related to the poles of $Q_{cl}(z)$ by

$$z = e^{sh}. \qquad (6.30)$$

It is important to understand, however, that the output of a continuous plant controlled by a discrete controller to have z-transfer function (6.28) will not be precisely the same as the output of the fictitious equivalent continuous system with s-transfer function (6.29). This is the plane to see when considering that the output of a discrete controller is piecewise constant while that of the fictitious equivalent continuous controller is continuous. Furthermore, several different plants may be controlled by a discrete controller to have precisely the same closed-loop z-transfer function but the outputs of each of these plants will be different between the sample points. The good news is that the outputs of all these discrete control systems will be precisely the same as that of the equivalent continuous system at the sample points, which is why the familiar parameters characterising the dynamic behaviour of the equivalent continuous system are useful to indicate the general behaviour of the discrete system. The deviations of the output of a discrete control system from the equivalent continuous system between the sample points are only short term and small for relatively small h but increase with h and this issue is addressed in Sect. 6.6.

The dynamic character of a discrete LTI system is similar to that of the equivalent continuous LTI system, since the impulse response of the discrete system is the sampled and held version of the impulse response of the continuous system. The only difference is the stepping effect of the sample and hold process. Hence

$$\lim_{h \to 0} y_s(t) = y(t). \qquad (6.31)$$

Hence simple real poles in the right half of the z-plane may be associated with the time constants of the corresponding continuous exponential modes of the equivalent continuous system. Similarly, complex conjugate poles in the z-plane may be associated with the damping ratios and undamped natural frequencies of the continuous oscillatory modes of the equivalent continuous system. A set of multiple

6.3 Dynamics of Discrete Linear Systems

poles in the right half of the z-plane may be similarly associated with the time constant of the polynomial exponential mode of the equivalent continuous system. In theory, however, there exist LTI discrete systems that are not associated with plant models. These are dealt with separately in Sect. 6.3.2.5.

The following three subsections present the process of determining the contributions of the three basic types of mode to the dynamic character of a discrete LTI system, given the z-plane pole locations.

6.3.2.2 Exponential Modes

Suppose a discrete system is known to have an isolated exponential mode with a pole at $z = z_1$ in the right half of the z-plane and within the unit circle (assuming a stable closed-loop system). Then the equivalent continuous system will have a corresponding exponential mode with a pole at $s = s_1 < 0$ satisfying (6.30). Thus,

$$z_1 = e^{s_1 h} \tag{6.32}$$

In this case,

$$s_1 = \frac{1}{h} \log_e (z_1) \tag{6.33}$$

with time constant,

$$T_1 = -\frac{1}{s_1} = -\frac{h}{\log_e (z_1)}. \tag{6.34}$$

If the system is subject to a transient disturbance, then once this has subsided, the contribution of this mode will take approximately

$$T_s = 3T_1 \tag{6.35}$$

seconds to decay to negligible proportions, which is the settling time of a first-order LTI system. Also, the number of iteration steps in this settling time may be taken as a measure of the closeness of the discrete exponential mode to the equivalent continuous exponential mode and is given by

$$n_1 = \frac{3T_1}{h} = -\frac{3}{s_1 h} = -\frac{3}{\log_e (z_1)}. \tag{6.36}$$

Figure 6.8 shows the z-plane pole locations and the corresponding s-plane pole locations together with the impulse responses for two different cases.

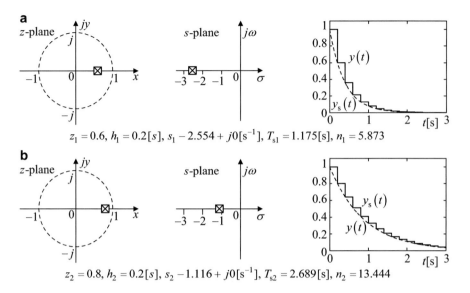

Fig. 6.8 Pole locations and corresponding impulse responses for two different discrete exponential modes with a common iteration period

6.3.2.3 Oscillatory Modes

A discrete oscillatory mode is now considered having a known pair of complex conjugate poles, $z_{1,2} = x \pm jy$. The equivalent continuous oscillatory mode has a pair of complex conjugate poles, $s_{1,2} = \sigma \pm j\omega_d$, where

$$\omega_d = \omega_n \sqrt{1 - \zeta^2} \qquad (6.37)$$

and $\sigma = -\zeta\omega_n$. Here, ζ is the damping ratio, ω_n is the undamped natural frequency and ω_d is the damped natural frequency, i.e. the frequency of oscillation. The two pairs of poles are connected by (6.30). Thus,

$$z_{1,2} = e^{s_{1,2}h} \Rightarrow x \pm jy = e^{(\sigma \pm j\omega_d)h} = e^{-\zeta\omega_n h}[\cos(\omega_d h) + j\sin(\omega_d h)]. \quad (6.38)$$

The z-plane poles therefore lie on a circle of radius,

$$r = e^{-\zeta\omega_n h}, \qquad (6.39)$$

and make angles, $\pm\theta$, with the real axis, where

$$\theta = \omega_d h, \qquad (6.40)$$

as shown in Fig. 6.9. As h increases, the angle, $\omega_d h$, increases, which can bring the z-plane poles into the left half plane. The value, h_1, at which the z-plane poles lie on

6.3 Dynamics of Discrete Linear Systems

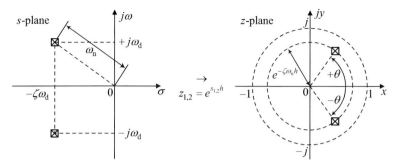

Fig. 6.9 Mapping of oscillatory mode poles from s-plane to z-plane

the imaginary axis, is given by $\omega_d h_1 = \pi/2 \Rightarrow h_1 = \pi/(2\omega_d)$. Since the period of the modal oscillation is $T = 2\pi/\omega_d$, then $h_1 = T/4$. The iteration period, however, does not normally exceed one quarter of the modal oscillation period. The z-plane poles are, therefore, usually in the right half plane.

Correlation of the dynamic behaviour with the complex conjugate pole locations on the z-plane may be aided by a family of curves with equal damping ratio, ζ. These enable ζ for a given pair of z-plane poles be readily estimated. The frequency of oscillation can then be determined by dividing the measured angle, $\theta = \omega_d h$, by h. The equations for generating the equal ζ contours are obtained as follows. Eliminating ω_n between (6.37) and (6.39) together with (6.40) gives

$$r = e^{-\frac{\zeta}{\sqrt{1-\zeta^2}}\theta}, \quad \theta \in (0, \pi). \tag{6.41}$$

Converting to Cartesian coordinates then yields

$$x = r\cos(\theta), \quad y = r\sin(\theta). \tag{6.42}$$

By analogy with the approach of Sect. 6.3.2.2, if the system undergoes a transient disturbance, then once this has subsided, the contribution of the oscillatory mode will take approximately the time for the exponential envelope function, $e^{\zeta\omega_n t}$, (Chap. 4, Sect. 4.4.4.2) to decay to negligible proportions. In this case, the settling time formula for first-order systems with exponential impulse responses will be applied using the envelope function time constant, $1/\zeta\omega_n$, yielding

$$T_s = \frac{3}{\zeta\omega_n}. \tag{6.43}$$

The closeness of the discrete mode to the equivalent continuous mode is then the number of iteration periods in this settling time. Thus, for the ith mode,

$$n_i = \frac{3}{\zeta_i \omega_{ni} h}. \tag{6.44}$$

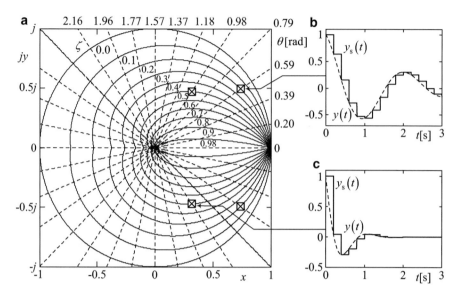

Fig. 6.10 ζ and θ contours with poles and impulse responses for two discrete modes. (a) Pole locations with ζ and θ contours in z-plane (b) $\zeta_1 = 0.2$, $\omega_{n1} = 5.67$[rad/s] (c) $\zeta_2 = 0.5$, $\omega_{n2} = 3.01$[rad/s]

Figure 6.10a shows a family of contours of constant ζ and θ. Figure 6.10b, c show the impulse responses of two different discrete oscillatory modes, together with the impulse responses of the equivalent continuous modes. The complex conjugate pole pairs of these discrete modes are shown in Fig. 6.10a, both with an iteration interval of $h = 0.2$ [s].

Regarding (6.44), for Fig. 6.10b, c, $n_1 = \frac{3}{\zeta_1 \omega_{n1} h_1} = \frac{3}{0.2 \times 5.67 \times 0.2} = 13.23$ and $n_2 = \frac{3}{\zeta_2 \omega_{n2} h_2} = \frac{3}{0.5 \times 3.01 \times 0.2} = 9.97$. In keeping with this, $y_s(t)$ can be seen to be a slightly better approximation to $y(t)$ in Fig. 6.10b than it is in Fig. 6.10c.

6.3.2.4 Polynomial Exponential Modes

Consider now an isolated discrete polynomial exponential mode with a pole of multiplicity, r, at $z_{1,2,\ldots,r} = z_1$ in the z-plane. This is associated with an equivalent continuous polynomial exponential mode with a pole at $s_{1,2,\ldots,r} = s_1$ in the s-plane. Then

$$z_1 = e^{s_1 h} \qquad (6.45)$$

The s-plane pole of the equivalent continuous mode is obtained from (6.45) as

$$s_1 = \frac{1}{h} \log_e (z_1) \qquad (6.46)$$

6.3 Dynamics of Discrete Linear Systems

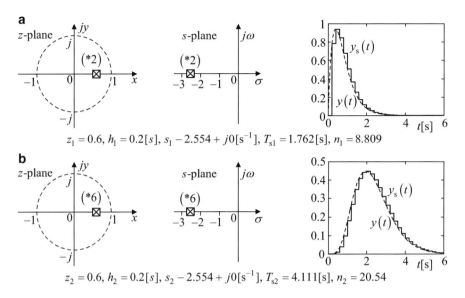

Fig. 6.11 Pole locations and impulse responses for two polynomial exponential modes

which constrains the pole to lie in the right half of the z-plane. The time constant is

$$T_1 = -\frac{1}{s_1} = -\frac{h}{\log_e(z_1)} \tag{6.47}$$

The duration of a polynomial exponential mode is longer than that of a simple exponential mode and is given approximately by the 5 % settling time formula. Thus,

$$T_{s1} = 1.5(r+1)T_1 = -\frac{1.5(r+1)h}{\log_e(z_1)}. \tag{6.48}$$

The number of iteration steps in this settling time is a measure of the closeness of the discrete mode to the equivalent continuous mode and is

$$n_1 = \frac{T_{s1}}{h} = -\frac{1.5(r+1)}{\log_e(z_1)}. \tag{6.49}$$

Figure 6.11 shows the impulse responses of two different exponential modes with different orders, r, but the same pole locations and iteration periods.

6.3.2.5 Discrete LTI Systems Not Used for Plant Models

Suppose a stable discrete LTI system is formulated whose poles, $z_i, i = 1, 2, \ldots, n$, lie within the unit circle in the z-plane, not necessarily with the intention of this modelling a specific plant. Then if the equivalent continuous system exists, its poles, s_i, $i = 1, 2, \ldots, n$, would have to satisfy

$$z_i = e^{s_i h} \Rightarrow x_i + j y_i = e^{(\sigma_i + j \omega_i) h} = e^{\sigma_i h} \left[\cos(\omega_i h) + j \sin(\omega_i h) \right]. \quad (6.50)$$

As already evident from Sects. 6.3.2.1, 6.3.2.2, 6.3.2.3, and 6.3.2.4, if the z-plane poles lie in the right half plane, $x_i > 0$ and values of σ_i and ω_i satisfying (6.50) exist implying that the equivalent continuous system exists. On the other hand, if any poles lie in the left half of the z-plane, $x_i < 0$ for some values of i and since $e^{\sigma_i h} > 0$ for all i, (6.50) would require that

$$\cos(\omega_i h) < 0 \Rightarrow \frac{\pi}{2} < \omega_i h - 2k\pi < \frac{3\pi}{2}, \quad k = 0, 1, 2, \ldots \quad (6.51)$$

Then for the equivalent continuous system to exist with an arbitrary value of h,

(a) The poles for which $x_i < 0$ would have to occur in complex conjugate pairs since, in general, $\sin(\omega_i h) \neq 0$
(b) The minimum value of h would be $h_{\min} = \pi/(2\omega_i)$, where ω_i is the frequency of oscillation of an oscillatory mode, having a period of $T_i = 2\pi/\omega_i$, giving $h_{\min} = T_i/4$, which is too long to be practicable for plant modelling.

For real negative z-plane poles, (6.50) would require that

$$\omega_i h = \frac{\pi}{2} + 2k\pi, \quad k = 0, 1, 2, \ldots, \quad (6.52)$$

In this case, $T_i = 1/\omega_i$ would be the time constant of an exponential mode of the equivalent continuous system and according to (6.52) the minimum iteration period would be $h_{\min} = (\pi/2) T_i \cong 1.5 T_i$. This is far too long to be practicable.

In view of the above, discrete LTI systems with poles in the left half of the z-plane are not of practical value in forming discrete models of continuous plants but the reader should be aware of their existence. Their dynamic character has already been demonstrated among the arbitrary first-order discrete systems considered in Sect. 6.3.1, those of Fig. 6.4a, d being relevant. Thus, poles in the left half of the z-plane are associated with oscillations at a fixed frequency of $1/(2h)$ [Hz]. The system of Fig. 6.4 might be useful as a square wave generator.

6.3.3 The Effects of Zeros in the z-Plane

6.3.3.1 Introduction

Consider an arbitrary LTI discrete system with the general z-transfer function,

$$\frac{Y(z)}{V(z)} = \frac{\sum_{i=0}^{n-1} b_i z^i}{D(z)}, \quad (6.53)$$

6.3 Dynamics of Discrete Linear Systems

where $D(z) = z^n + \sum_{i=0}^{n-1} d_i z^i$. The zeros are the roots of $\sum_{i=0}^{n-1} b_i z^i = 0$.
An associated system with no zeros will now be formed with transfer function,

$$\frac{X(z)}{V(z)} = \frac{1}{D(z)}. \tag{6.54}$$

Then

$$Y(z) = \left(\sum_{i=0}^{n-1} b_i z^i\right) X(z) \tag{6.55}$$

Introducing a factor, z^r, in a transfer function causes a time shift of rh seconds in the time domain. The difference equation corresponding to (6.55) is therefore

$$y(k) = \sum_{i=0}^{n-1} b_i x(k+i). \tag{6.56}$$

The effect of zeros in the z-domain is therefore to produce a weighted sum of time-shifted versions of the output that would occur in the absence of the zeros. It will be recalled from Chap. 1, Sect. 1.6, that for an LTI continuous system, the effect of zeros in the Laplace transfer function is to produce a weighted sum of the derivatives of the output that would occur in absence of the zeros. This is quite different from the discrete case and therefore poses the question of how zeros in the Laplace transfer function, $G(s)$, of the equivalent continuous system manifest in the z-transfer function, $Q(z)$, of the discrete system.

6.3.3.2 Discretisation Zeros and Plant Zeros

It may often be observed that zeros appear in $Q(z)$ but not in $G(s)$. Since these zeros appear only as a result of the process of transferring from the continuous domain to the discrete domain, they will be referred to as *discretisation zeros*. These may be seen in Plant Models: Laplace and z-Transfer Functions [Tables]. For example, the double integrator plant has Laplace transfer function,

$$\frac{Y(s)}{U(s)} = G(s) = \frac{1}{s^2} \tag{6.57}$$

and z-transfer function,

$$\frac{Y(z)}{U(z)} = Q(z) = \frac{\frac{1}{2}h^2(z+1)}{(z-1)^2}. \tag{6.58}$$

To assess the effect of the discretisation zero at $-1 + j0$, let $X(z)$ be the output of the associated zero-less system with transfer function,

$$\frac{X(z)}{U(z)} = \frac{h^2}{(z-1)^2}. \tag{6.59}$$

Then

$$Y(z) = \tfrac{1}{2}(z+1)X(z) \Rightarrow y(k) = \tfrac{1}{2}[x(k+1) + x(k)] \tag{6.60}$$

which is a moving-window averaging filter operating on only two successive values of x. This will create a difference between $x(t)$ and $y(t)$ but, noting that $t = kh$, this difference will approach zero as $h \to 0$. This is typical of the effect of discretisation zeros in other cases.

The s-plane zeros of the equivalent continuous closed-loop system, which are also zeros of the plant Laplace transfer function, can cause overshooting in the step response (Chap. 1, Sect. 1.6) and this effect also occurs in the associated discrete system through corresponding z-plane zeros. Since these zeros originate as plant zeros in the s-domain, they will be referred to simply as *plant zeros*.

To exemplify the distinction between plant zeros and discretisation zeros, suppose that the plant is a single attitude control axis of a spacecraft comprising a rigid centre body with a flexible appendage causing one dominant vibration mode. The Laplace transfer function is

$$\frac{Y(s)}{U(s)} = \frac{b\left(s^2 + v^2\right)}{s^2\left(s^2 + \omega^2\right)}, \tag{6.61}$$

where ω is the free natural frequency, i.e. the frequency of vibration of the uncontrolled spacecraft, v is the encastre natural frequency, i.e. the frequency of vibration of the appendage that would occur with the centre body held stationary with respect to inertial space and the constant, b, depends on the spacecraft moment of inertia and the torque actuator constant. The corresponding z-transfer function may be found from Table 3 in Tables but after partial fraction expansion since transfer function (6.61) is not listed. Thus,

$$\frac{Y(s)}{U(s)} = b\left[\frac{v^2}{\omega^2} \cdot \frac{1}{s^2} + \left(1 - \frac{v^2}{\omega^2}\right) \cdot \frac{1}{s^2 + \omega^2}\right] \tag{6.62}$$

The corresponding z-transfer function is then

$$\frac{Y(z)}{U(z)} = b\left[\frac{v^2}{\omega^2} \cdot \frac{\tfrac{1}{2}h^2(z+1)}{(z-1)^2} + \left(1 - \frac{v^2}{\omega^2}\right) \cdot \frac{1}{\omega^2} \cdot \frac{[1 - \cos(\omega h)](z+1)}{z^2 - 2z\cos(\omega h) + 1}\right]. \tag{6.63}$$

Thus, each part of the partial fraction expansion has a discretisation zero at $-1 + j0$. Expressing transfer function (6.63) in the factored form then yields

$$\frac{Y(z)}{U(z)} = \frac{(z+1)\{Az^2 + Bz + A\}}{\omega^2(z-1)^2 [z^2 - 2z\cos(\omega h) + 1]}, \qquad (6.64)$$

where $A = b\left(\frac{1}{2}v^2h^2 + C\right)$, $B = b\left(v^2h^2\cos(\omega h) + 2C\right)$ and $C = \left(1 - \frac{v^2}{\omega^2}\right)[1 - \cos(\omega h)]$. Unless the controller embodies pole–zero cancellation, which is only feasible if the zeros lie within the stable region in the z-plane, i.e. the unit circle, the numerator of (6.64) is invariant with respect to the loop closure via a linear controller. Hence, in this case the plant zeros that can cause overshooting in the step response are the roots of $Az^2 + Bz + A = 0$. To be able to determine whether the control system will exhibit this behaviour by direct examination of the z-plane poles and zeros would require a theory of pole–zero dominance in the z-plane to be developed paralleling that of Chap. 1 for the s-plane in Sect. 1.6. As this is not available at present, the recommended approach is to analyse the equivalent continuous system using the theory of pole–zero dominance in the s-plane.

6.4 Criterion for Applicability of Continuous LTI System Theory

A heuristic criterion will now be developed that can be used to identify applications in which the continuous controllers of the previous chapters can be directly converted to control algorithms without changing their gains and any other parameters. This is not a stability criterion but a means of ensuring the attainment of a specified dynamic performance.

It is clear from Fig. 6.1 and many of the subsequent examples that if h is small enough, $y_s(t)$ is a good approximation to $y(t)$. Then, it would be reasonable to suppose that a controller designed in the continuous domain could be implemented digitally as in Fig. 6.1 and give a similar closed-loop performance. It would be possible to investigate the effect of the sampling process in particular cases by simulation but a criterion can be devised that enables the control system designer to select a successful approach before embarking on a complete design and test by simulation.

It will be recalled from Chap. 1 that each mode of a linear system is characterised by the impulse response of the modal transfer function that appears in the partial fraction expansion of the transfer function of the overall system, which can either be an uncontrolled plant or a closed-loop system. Let the impulse response of the ith modal transfer function be $y_{mi}(t)$ and the corresponding sampled and held response be $y_{msi}(t)$. Then a measure of the accuracy of $y_{msi}(t)$ could be taken as the maximum magnitude of the error, $y_{mi}(t) - y_{msi}(t)$, as a proportion of the maximum magnitude, $|y_{mi}|_{max}$, i.e.

$$E_s \triangleq \frac{|y_{mi}(t) - y_{msi}(t)|_{max}}{|y_{mi}|_{max}} \qquad (6.65)$$

This will be defined as the *peak sampling error* to ensure that the accuracy assessment is independent of the scaling of the impulse response. Since the system is linear, changing the scale of $y_{mi}(t)$ by an arbitrary factor will change the scale of $y_{msi}(t)$ and the maximum magnitude, $|y_{mi\,max}|$, of $y_{mi}(t)$ by the same factor. The maximum error magnitude occurs where $y_{mi}(t)$ has the maximum *derivative* magnitude, $|\dot{y}_{mi}|_{max}$. Note that the error at each sample point is zero and reaches a local maximum at the end of each iteration period. The numerator term of (6.65) is the largest of these local maxima over the duration of the impulse response. Then (6.65) may be replaced by

$$E_s \cong \frac{|\dot{y}_{mi}|_{max} h}{|y_{mi}|_{max}}. \tag{6.66}$$

Then a maximum value of this peak relative error must be chosen below which $y_{msi}(t)$ is considered a good approximation to $y_{mi}(t)$. There is no optimal choice but 1/5 is considered reasonable. Then the required criterion can be formed as

$$E_s \leq \tfrac{1}{5} \Rightarrow h \leq \frac{|y_{mi}|_{max}}{5|\dot{y}_{mi}|_{max}}. \tag{6.67}$$

Two modes will now be studied in order to determine (6.67) in terms of the pole magnitude. The first is a stable exponential mode with modal transfer function,

$$\frac{Y_{m1}(s)}{U(s)} = \frac{1}{s+a}, \quad a > 0 \tag{6.68}$$

and the corresponding impulse response is

$$y_{m1}(t) = e^{-at}. \tag{6.69}$$

Then, by inspection of (6.69), $|y_{m1\,max}| = y_{m1}(0) = 1$. Since

$$\dot{y}_{m1}(t) = -a e^{-at}, \tag{6.70}$$

then $|\dot{y}_{m1}|_{max} = |\dot{y}_{m1}(0)| = a$ and, with $i = 1$, (6.67) becomes

$$h \leq \frac{1}{5a} \tag{6.71}$$

This is in terms of the pole magnitude, which is a.

The second mode is oscillatory with transfer function,

$$\frac{Y_{m2}(s)}{U(s)} = \frac{\omega_n}{s^2 + 2\zeta\omega_n s + \omega_n^2}, \quad \omega_n > 0, \quad 0 \leq \zeta < 1 \tag{6.72}$$

6.4 Criterion for Applicability of Continuous LTI System Theory

and the corresponding impulse response is

$$y_{m2}(t) = A e^{-\zeta \omega_n t} \sin(\omega_d t), \quad A = 1/\sqrt{1-\zeta^2}, \quad \omega_d = \omega_n \sqrt{1-\zeta^2}. \qquad (6.73)$$

Then

$$\dot{y}_{m2}(t) = A \omega_d e^{-\zeta \omega_n t} \cos(\omega_d t) - \zeta \omega_n A e^{-\zeta \omega_n t} \sin(\omega_d t) \qquad (6.74)$$

and

$$|\dot{y}_{m2}|_{max} = \dot{y}_{m2}(0) = A \omega_d. \qquad (6.75)$$

Next, $|y_{m2}|_{max} = y_{m2}(t_1)$, where $\dot{y}_{m2}(t_1) = 0 \Rightarrow$

$$\omega_d \cos(\omega_d t_1) - \zeta \omega_n \sin(\omega_d t_1) = 0 \Rightarrow$$

$$\omega_d t_1 = \tan^{-1}\left(\frac{\omega_d}{\zeta \omega_n}\right) = \tan^{-1}\left(\frac{\sqrt{1-\zeta^2}}{\zeta}\right) = \sin^{-1}\left(\sqrt{1-\zeta^2}\right). \qquad (6.76)$$

Then substituting for $\omega_d t_1$ in (6.73) with $t = t_1$ using (6.76) yields

$$|y_{m2}|_{max} = y_{m2}(t_1) = e^{-\frac{\zeta}{\sqrt{1-\zeta^2}} \omega_d t_1} = e^{-\frac{\zeta}{\sqrt{1-\zeta^2}} \sin^{-1}\left(\sqrt{1-\zeta^2}\right)}. \qquad (6.77)$$

With $i = 2$, criterion (6.67) then becomes

$$h \leq \frac{e^{-\frac{\zeta}{\sqrt{1-\zeta^2}} \sin^{-1}\left(\sqrt{1-\zeta^2}\right)}}{5 \omega_n}. \qquad (6.78)$$

For $\zeta = 0$, which gives an oscillatory undamped mode, this becomes

$$h \leq \frac{1}{5 \omega_n} \qquad (6.79)$$

which is equivalent to (6.71) as ω_n is the pole magnitude. For other values of ζ, the numerator of (6.78) differs from unity but is of the same order.

Considering a complete set of poles of a sampled system, it is clear from the above that the most critical poles are those with the maximum magnitudes.

The foregoing leads to the following criterion, the usefulness of which will be demonstrated by examples in Sect. 6.5.

Criterion 6.1 The iteration interval, h, is sufficiently small for a digitally implemented controller to be designed in the continuous domain if

$$h \leq \frac{1}{5\,|s_{max}|}, \qquad (6.80)$$

where s_{max} is the pole with the largest magnitude taken from the combined set of plant poles and the equivalent continuous closed-loop system poles.

The criterion has to be applied to the plant poles but upon loop closure they cease to exist. If the controller is linear and discrete with $h = \text{const.}$, then the poles are in the z-domain. If, on the other hand, with the linear continuous equivalent controller satisfying the performance specification for the application in hand, the poles are shifted to new locations in the s-plane. It is this hypothetical case that is relevant to Criterion 6.1. The argument is really as follows. Suppose this equivalent continuous linear closed-loop system is available. Then the discrete control algorithm to be created in reality could be approximated by inserting a S/H unit in the output of the continuous equivalent controller. It is then clear that the modes of the continuous equivalent closed-loop system will determine whether or not h is sufficiently small for the insertion of the S/H unit to have a negligible effect on the closed-loop performance. Hence, Criterion 6.1 applies to both sets of poles. This will be demonstrated by a forthcoming example.

Figure 6.12 shows the modal impulse responses considered above plotted together with their sampled versions for $h = h_{max} = 1/(5\,|s_{max}|)$.

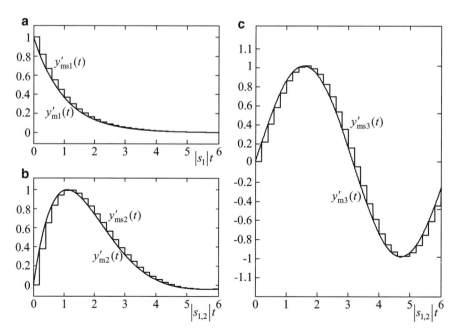

Fig. 6.12 Normalised impulse responses together with their sampled and held versions. (**a**) Exponential mode (**b**) Oscillatory mode, $\zeta = 1/\sqrt{2}$ (**c**) Oscillatory mode, $\zeta = 0$

Here, $y'_{mi}(t) = y_{mi}(t)/|y_{mi}|_{max}$ and $y'_{msi}(t) = y_{msi}(t)/|y_{mi}|_{max}$, $i = 1, 2, 3$. They are plotted on the same scales so that the sawtooth-shaped errors between the impulse responses and their sampled and held versions may be seen to be of similar magnitudes. Arguably, they are on the borderline of acceptability of $y_{msi}(t)$ being a good approximation to $y_{mi}(t)$.

It is important to realise that (a) how much h can be increased beyond h_{max} before the performance differs considerably from that of the equivalent continuous system and (b) the margin of stability will both vary from one system to the next.

6.5 Discrete Control for Small Iteration Intervals

6.5.1 Introduction

This section provides the means of converting any continuous linear controller, including those presented in the previous chapters, to discrete control algorithms, assuming that the sampling period, h, is sufficiently small for the discrete control system performance to be similar to the original continuous control system performance using the same values of the gains and any other adjustable parameters. This requires Criterion 6.1 of the previous section to be satisfied.

A convenient starting point is a block diagram of the controller taken in isolation. Any dynamic elements should then be converted to state-variable block diagrams. Then the complete controller block diagram will be in the form of interconnected basic elements, i.e. summing junctions, constant gains, differentiators and integrators. The input and output variables of each element will then be labelled. Then it is a straightforward matter to write down the continuous equation of each element. These equations will then be converted to a discrete form. With reference to Fig. 6.2, if $T_D/h \ll 1$, the computed control is $u(t_k)$; otherwise, it is $u(t_{k+1})$. Finally, the discrete equations are converted to algorithms in the form of flow charts from which software can be prepared in any high-level language.

6.5.2 Discrete Equations of the Basic Elements

Once all the dynamic elements of the block diagram of a linear controller have been changed to the state-space form, then the whole block is composed of all or a subset of the four basic elements shown in Fig. 6.13.

For the summing junction, in the continuous domain,

$$z(t) = x_1(t) + \cdots + x_n(t) - y_1(t) - \cdots - y_m(t).$$

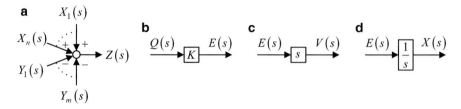

Fig. 6.13 The basic elements of a linear controller formulated in the continuous domain. (**a**) summing junction (**b**) constant gain (**c**) differentiator (**d**) integrator

At the sampling instants, this becomes

$$z(t_k) = x_1(t_k) + \cdots + x_n(t_k) - y_1(t_k) - \cdots - y_m(t_k). \tag{6.81}$$

By convention, t_k is the *present time* at the instant of the kth iteration update. The iteration updates are assumed to be separated by a constant period of h seconds. In converting to the discrete form for the software implementation, a slightly simpler notation will be used so (6.81) becomes

$$z(k) = x_1(k) + \cdots + x_n(k) - y_1(k) - \cdots - y_m(k). \tag{6.82}$$

Thus, $z(k)$ represents a quantity in the computer software equal numerically to $z(t_k)$, similarly for the other terms.

For the constant gain, in the continuous domain, $e(t) = Kq(t) \Rightarrow e(t_k) = Kq(t_k)$, which is written as

$$e(k) = Kq(k). \tag{6.83}$$

For the differentiator, in the continuous domain,

$$v(t) = \frac{\mathrm{d}}{\mathrm{d}t}e(t). \tag{6.84}$$

For this, a numerical approximation is needed since at time, t_k, only the samples, $e(t_k)$, $e(t_{k-1})$, ..., are available. Several numerical methods are available to estimate the derivative, $v(t_k) = \dot{e}(t_k)$, using the available samples of $e(t)$. All of these are based on polynomial fitting to the sample points on the graph of $e(t_q)$ against t_q, $q = k, k-1, \ldots$, and evaluation of this polynomial at time, t_k. This is termed *software differentiation*. The simplest possible version will be taken here, which fits a polynomial of first degree, i.e. a straight line, to the two points, $[t_{k-1}, e(t_{k-1})]$ and $[t_k, e(t_k)]$, on the graph of $e(t)$, as shown in Fig. 6.14.

6.5 Discrete Control for Small Iteration Intervals

Fig. 6.14 Linear derivative approximation

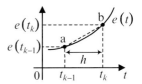

Fig. 6.15 Trapezoidal integral approximation

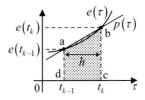

The derivative approximation is the slope of the chord, ab, which is

$$\dot{e}(t_k) \cong \frac{e(t_k) - e(t_{k-1})}{t_k - t_{k-1}} = \frac{e(t_k) - e(t_{k-1})}{h}. \qquad (6.85)$$

For the software implementation, this is written as

$$v(k) = \frac{1}{h}[e(k) - e(k-1)]. \qquad (6.86)$$

The sign, $\cong$, has been replaced by the sign, $=$, with the understanding that (6.86) is an approximate differentiation algorithm implemented in the software that yields a sequence of values, $v(k)$, $k = 0, 1, 2, \ldots$, that approximate the continuous derivative, $\dot{y}(t)$, at the sampling times, t_k, $k = 0, 1, 2, \ldots$.

For the integrator, in the continuous domain,

$$x(t) = \int_0^t e(\tau) d\tau \qquad (6.87)$$

Again a numerical approximation is needed using the available samples, $e(t_k)$, $e(t_{k-1})$, ..., and several methods are available based on a polynomial fit, $p(\tau)$, to the points. The integral of the polynomial function is evaluated analytically from $\tau = t_0$ to $\tau = t_1$, then from $\tau = t_1$ to $\tau = t_2$ and so on. In this way, the approximate integral is updated recursively as

$$x(k) = x(k-1) + \int_{t_{k-1}}^{t_k} p(\tau) d\tau. \qquad (6.88)$$

In this case, only the points, $e(t_k)$ and $e(t_k - 1)$, are utilised. As for the differentiator, the polynomial is of first degree, i.e. a straight line fit, as shown in Fig. 6.15.

Fig. 6.16 Euler integral approximation

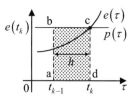

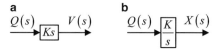

Fig. 6.17 Combined elements of control system block diagrams. (**a**) Differentiator with gain. (**b**) Integrator with gain

The integral in (6.88) is the area of the trapezoid, abcd, in Fig. 6.15. Hence (6.88) may be written as $x(k) = x(k-1) + \frac{1}{2}[e(t_k) + e(t_{k-1})]h$. In the notation used for the software implementation, this is

$$x(k) = x(k-1) + \tfrac{1}{2}[e(k) + e(k-1)]h \qquad (6.89)$$

The simpler but less accurate Euler integration algorithm is included as it is commonly found. This uses only the current sample, $e(k)$, to calculate the next contribution to the integral, as the area of a rectangular strip as shown in Fig. 6.16.

In this case, $p(\tau)$ is a polynomial of degree zero, i.e. a constant, $p(\tau) = e(t_k)$ and the integral of (6.88) becomes the area of the rectangle, abcd. Then

$$x(k) = x(k-1) + e(k)h. \qquad (6.90)$$

Since differentiators and integrators are commonly combined with constant gains in control system block diagrams, they can be treated as basic elements to shorten the derivations of control algorithms. First, the common variable, $E(s)$, may be eliminated between Fig. 6.13b, c to form the block of a differentiator with gain as shown in Fig. 6.17a.

The corresponding algorithm is then obtained by eliminating $e(k)$ and $e(k-1)$ between (6.83), (6.89) and the equation,

$$e(k-1) = Kq(k-1), \qquad (6.91)$$

which follows from (6.83). Then substituting for $e(k)$ and $e(k-1)$ in (6.86) using (6.83) and (6.91) yields

$$v(k) = \frac{K}{h}[q(k) - q(k-1)] \qquad (6.92)$$

6.5 Discrete Control for Small Iteration Intervals

Second, the common variable, $E(s)$, may be eliminated between Fig. 6.13 (b) and (d) to form the block of an integrator with gain as shown in Fig. 6.17b. The corresponding algorithm is then obtained by eliminating $e(k)$ and $e(k-1)$ between (6.83), (6.91) and (6.89), taking the trapezoidal integration algorithm. Thus

$$x(k) = x(k-1) + \tfrac{1}{2}[q(k) + q(k-1)]Kh. \tag{6.93}$$

6.5.3 Discrete Controller Block Diagrams for Simulation

This subsection provides the means of setting up simulations with discrete elements using software such as SIMULINK®. This requires block diagrams of the control system with dynamic elements represented using z-transforms. Consider initially the following Laplace transforms of the differentiation and integration operations with zero initial conditions.

$$\mathcal{L}\left\{\frac{\mathrm{d}x(t)}{\mathrm{d}t}\right\} = s\mathcal{L}\{x(t)\} = sX(s) \tag{6.94}$$

and

$$\mathcal{L}\left\{\int_0^t x(\tau)\,\mathrm{d}\tau\right\} = \frac{1}{s}\mathcal{L}\{x(t)\} = \frac{1}{s}X(s) \tag{6.95}$$

The analogous relationships between the discrete domain and the z-domain are

$$\mathcal{Z}\{x(k+1)\} = z\mathcal{Z}\{x(k)\} = zX(z) \tag{6.96}$$

and

$$\mathcal{Z}\{x(k-1)\} = z^{-1}\mathcal{Z}\{x(k)\} = \frac{1}{z}X(z) \tag{6.97}$$

The background theory of the z-transform is given in Sect. 6.6. For the purpose of this subsection, however, it is only necessary to be aware of relationship (6.97), since the block diagram of any linear discrete dynamical system can be formed from just three types of basic element, i.e. summing junctions, constant gains and blocks with transfer function, $1/z$, which correspond to (6.97) and are elements with fixed iteration periods of h seconds, as shown in Fig. 6.18.

Fig. 6.18 Delay element and the associated notation

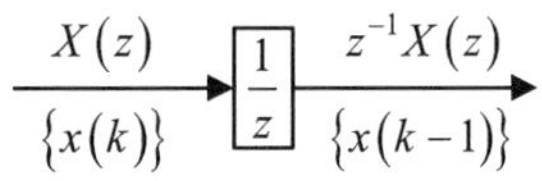

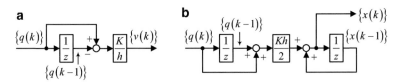

Fig. 6.19 Discrete approximations of continuous dynamic elements. (**a**) Approximate differentiator with gain. (**b**) Approximate integrator (trapezoidal) with gain

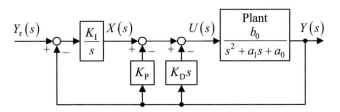

Fig. 6.20 Equivalent continuous control system block diagram

The signal notation shown above the arrows is correct as the block diagram is strictly in the z-domain. The simulation software, however, contains blocks labelled with z-transfer functions although it does not operate in the z-domain. The time domain variables used in simulations are shown in curly parentheses below.

Figure 6.19 shows block diagrams of the differentiator with gain defined by (6.92) and the integrator with gain defined by (6.93).

Example 6.2 Position control of a throttle valve for an internal combustion engine

The plant model taken here is the linearised and reduced-order version of the detailed throttle valve model developed in Chap. 2, Example 2.1. A discrete IPD controller is to be developed from the continuous one shown in Fig. 6.20.

The controller will be designed for a specified settling time, T_s, with zero overshoot using the 5 % settling time formula, following the pole placement method of Chap. 4, Sect. 4.5. In this case, the closed-loop characteristic polynomial is

$$\left(s^3 + a_1 s^2 + a_0 s\right)\left[1 - \tfrac{b_0}{s^2 + a_1 s + a_0}\left(-K_D s - K_P - \tfrac{K_I}{s}\right)\right] \\ = s^3 + (a_1 + b_0 K_D) s^2 + (a_0 + b_0 K_P) s + b_0 K_I, \quad (6.98)$$

which is also required to be

$$\left[s + \frac{1.5(1+n)}{T_s}\right]^n \bigg|_{n=3} = (s+p)^3 = s^3 + 3ps^2 + 3p^2 s + p^3, \quad (6.99)$$

where $p = 6/T_s$. Equating (6.98) and (6.99) then yields the controller gains as

$$K_I = \frac{p^3}{b_0}, \quad K_P = \frac{3p^2 - a_0}{b_0} \quad \text{and} \quad K_D = \frac{3p - a_1}{b_0}. \quad (6.100)$$

6.5 Discrete Control for Small Iteration Intervals

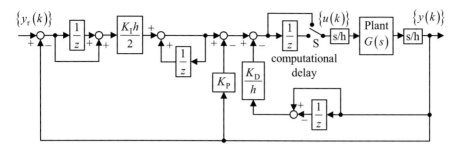

Fig. 6.21 Simulation block diagram for discrete controller applied to continuous plant

Figure 6.21 shows a simulation block diagram corresponding to Fig. 6.20 in which the integral and differential terms have been replaced by the equivalent discrete elements of Fig. 6.19, with the appropriate gains, and $G(s) = b_0/\left(s^2 + a_1 s + a_0\right)$.

The two options of either allowing a whole iteration period for the computational delay or assuming that the computational delay is negligible are catered for by the switch, S. The mixture of Laplace and z-transfer functions is not strictly correct but is shown as it follows the notation used in software such as ©MATLAB-SIMULINK, which really represents the real-time elements being simulated. Hence some of the discrete variables are indicated in parenthesis. The blocks marked, s/h, represent sample and hold functions that are included in the software.

The author thanks Delphi Diesel for the data that gives the following plant parameters. $a_0 = 68.1845$, $a_1 = 104.8611$ and $b_0 = 124.0962$. The plant poles are the roots of $s^2 + a_1 s + a_0 = 0$, i.e. $s_{1,2} = -104.2068\ [\text{s}^{-1}]$ and $-0.6543\ [\text{s}^{-1}]$. The specification for settling time given by Delphi is $T_s = 0.1\ [\text{s}]$. In view of (6.99), the magnitude of the triple closed-loop pole of the equivalent continuous system is $p = 6/T_s = 6/0.1 = 60\ [\text{s}^{-1}]$. In this case, $|s_{\max}| = \max\{104.2, 0.6543, 60\} = 104.2\ [\text{s}^{-1}]$. In this case, the plant is critical in determining the maximum recommended iteration interval, which is $h_{\max} = 1/(5|s_{\max}|) = 0.002\ [\text{s}]$ [Criterion 6.1]. This is well above the value of $h = 0.001\ [\text{s}]$ used in this application.

Figure 6.22 shows simulated step responses assuming negligible computational delay for the control algorithm, for which the switch, S, in Fig. 6.21 is in the upper position.

Here, $y(t)$ is the controlled plant output, $y_s(t)$ is the sampled and held version of $y(t)$ and $y_{ec}(t)$ is the output of the equivalent continuous control system upon which the discrete control system design is based.

In Fig. 6.22a, b, c, $T_s = 0.1\ [\text{s}]$. In Fig. 6.22d, e, f, the prescribed settling time has been reduced to $T_s = 0.025\ [\text{s}]$, yielding $p = 6/T_s = 240\ [\text{s}^{-1}]$, $|s_{\max}| = \max\{104.2068, 0.6543, 240\} = 240\ [\text{s}^{-1}]$. Now the closed-loop poles of the equivalent closed-loop system required to produce the shorter settling time determine $h_{\max}$, yielding $h_{\max} = 1/(5|s_{\max}|) = 0.0008\ [\text{s}]$.

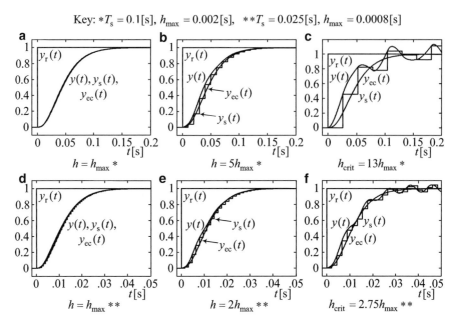

Fig. 6.22 Step responses with discrete IPD controller for different iteration intervals assuming negligible computational delay

In Fig. 6.22a, d, $h = h_{max}$ and, as required, $y(t) \cong y_{ec}(t)$, the difference not being visible on the scales of the complete transient. Figure 6.22b, e show the effect of increasing h considerably beyond h_{max}, where the difference between $y(t)$ and $y_{ec}(t)$ becomes visible, The settling time specification is still satisfied and this may be sufficient for some applications. It would, however, cause unacceptable errors in high-precision motion control applications with a continuously varying reference input, $y_r(t)$, using a zero dynamic lag pre-compensator (Chap. 13). In Fig. 6.22c, f, h has been taken up to the critical value, h_{crit}, bringing the system to the verge of instability. Increasing h further would result in oscillatory instability similar to that of Fig. 6.6c but not necessarily at a frequency of $1/(2h)$, which would be caused by a real negative pole outside the unit circle. The constant amplitude oscillations in Fig. 6.23c, f are at a lower frequency, indicative of complex conjugate poles on the unit circle.

The simulations presented in Fig. 6.23 correspond to those of Fig. 6.22, but with a time delay of h seconds by moving the switch, S, to the lower position in Fig. 6.21, representing time allowed for the control computations in the real application.

The marked contrast with Fig. 6.22 is the much reduced stability margins, $h_{crit} - h_{max}$, due to the destabilising effect of the time delay.

6.5 Discrete Control for Small Iteration Intervals

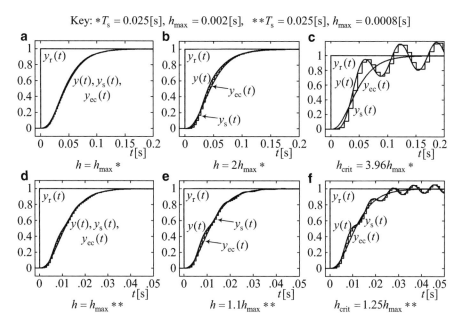

Fig. 6.23 Step responses with discrete IPD controller for different iteration intervals allowing one iteration interval for control computations

6.5.4 Control Algorithms and Flow Charts

This subsection is provided for those who wish to generate their own real-time control software rather than employ a system such as ©dSPACE. Once a control system block diagram has been formed, such as that of Fig. 6.21, the controller may be separated from the diagram and converted to a control algorithm. This is accomplished as follows. First, the controller usually contains dynamic elements such as an integral term, a differential term, or noise filters. In the discrete domain, the basic building block for dynamic elements is the delay block with transfer function, $1/z$. These blocks are in some ways analogous to the pure integrator in the continuous domain, so similarly, their outputs are the state variables of the controller and are labelled, $x_{c1}(k), x_{c2}(k), \ldots$. Then the set of discrete equations relating these state variables to one another and to the reference input, $y_r(k)$, the control output, $u(k)$, the controlled output, $y(k)$, and any other measured state variables from the plant are written down. In a few cases, such as a linear state feedback controller in which all the state variables, $x_i(k), i = 1, 2, \ldots, n$, are measured, if the digital processor is sufficiently fast to ignore the computational delay, the discrete equation of the controller is simply

$$u(k) = R y_r(k) - k_1 x_1(k) - \cdots - k_n x_n(k) \qquad (6.101)$$

where R is the reference input scaling coefficient, usually set to give a unity closed-loop DC gain and k_i, $i = 1, 2, \ldots, n$, are the state feedback gains. If the computational delay cannot be ignored and a whole iteration interval is allowed for the computations, then a delay element is introduced such as by moving the switch, S, to the lower position in Fig. 6.21, in which case $u(k)$ becomes a state variable. Let this be $x_{c1}(k)$, then (6.101) is replaced by

$$x_{c1}(k+1) = R y_r(k) - k_1 x_1(k) - \cdots - k_n x_n(k) \qquad (6.102)$$

$$u(k) = x_{c1}(k) \qquad (6.103)$$

In this case, the controller is a dynamic one of first order, comprising the first-order difference equation (6.102) and the output equation (6.103).

The iteration index (sample number), k, is a notation in the above discrete equations for indicating at which point in time variables are updated. It is not used in the software implementing the controller. Instead, the order in which the instructions of the programme are written, and the instructions themselves determine such timing. For this an alternative notation will be developed in which k is eliminated and the discrete equations of the controller become a set of *instructions* in a computer programme, i.e. a *control algorithm*. This will be expressed in the form of a flow chart enabling implementation with any chosen programming language.

The general set of equations for a discrete linear SISO controller may be written,

$$\mathbf{x}_c(k+1) = \mathbf{A}_c \mathbf{x}_c(k) + \mathbf{b}_r y_r(k)$$
$$u(k) = \mathbf{c}_c^T \mathbf{x}_c(k) + \mathbf{c}_p^T \mathbf{y}(k) + R y_r(k) \qquad (6.104)$$

where $\mathbf{x}_c(k) \in \Re^{n_c}$ is the controller state; $\mathbf{y} \in \Re^m$ is a set of measured plant state variables including the controlled output, which may be chosen as $y_1(k)$; $y_r(k)$ is the reference input; and $\mathbf{A}_c \in \Re^{n_c \times n_c}$, $\mathbf{b}_r \in \Re^{n_c \times 1}$, $\mathbf{c}_c^T \in \Re^{1 \times n_c}$, $\mathbf{c}_p^T \in \Re^{1 \times m}$ and R are the controller constants. As already indicated in Sect. 6.2, k represents the present discrete time, so that all the discrete variables appended by (k) are the current ones. Variables appended by $(k+1)$ are 'new'. To change to the notation of the control algorithm, (k) is dropped from the current variables and $(k+1)$ is replaced by a subscript 'new'. Then (6.104) is replaced by

$$\mathbf{x}_{c\,new} = \mathbf{A}_c \mathbf{x}_c + \mathbf{b}_r y_r \qquad (6.105)$$

$$u = \mathbf{c}_c^T \mathbf{x}_c + \mathbf{c}_p^T \mathbf{y} + R y_r. \qquad (6.106)$$

Figure 6.24 shows the corresponding flowchart.

In the real-time control software, the RHS of (6.105) is evaluated, followed by the RHS of (6.106). For the initial iteration of the controller the question has to be

6.5 Discrete Control for Small Iteration Intervals

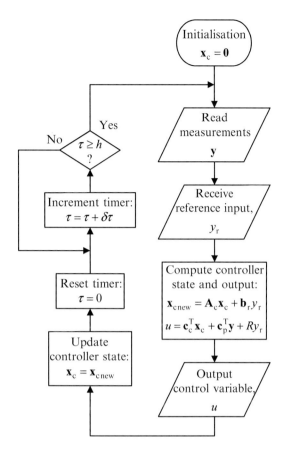

Fig. 6.24 Flow chart for general discrete SISO linear control algorithm

asked of whether or not the information is available for the evaluation. If not, then an initial value has to be chosen and set up before loop closure, i.e. the starting of the controller once connected to the plant. This is *initialisation*. Taking each term one at a time, the evaluation of $\mathbf{A}_c \mathbf{x}_c$ requires an initial value of the controller state, $\mathbf{x}_c$. Usually, this is set to zero. The term, $\mathbf{b}_r y_r$, requires y_r, which may assumed to be available from a reference input generator, or a user. The term, $\mathbf{c}_c^T \mathbf{x}_c$, may now be computed as the controller state, $\mathbf{x}_c$, has been initialised. The term, $\mathbf{c}_p^T \mathbf{y}$, can be evaluated, assuming that the plant instrumentation providing the measurement vector, $\mathbf{y}$, has been turned on. Finally, the term, $R y_r$, can be evaluated since the reference input, y_r, has already been stated as being available. So the only variables of a control algorithm requiring initialisation are its internal state variables.

It is important to note the elements in the return path of the flow chart, i.e. the controller state update and the timer that keeps the sampling and controller iterations repeating at the correct intervals of h seconds. The timing increment, τ, is chosen such that h/τ is an integer so that this timing is precise.

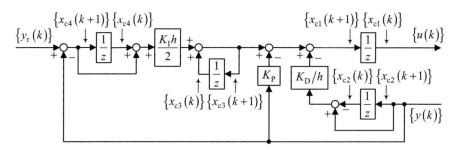

Fig. 6.25 Discrete state-space block diagram of IPD controller

Example 6.3 Algorithm and flow chart for an IPD controller

The starting point here is the discrete state-space block diagram of the IPD controller contained in Fig. 6.21. The version including the computational delay allowance is taken. This is shown in Fig. 6.25 together with all the variables needed. The discrete equations of the controller may be written by inspection of this diagram. First the state difference equations are as follows.

$$\begin{aligned}
x_{c1}(k+1) &= \tfrac{K_1 h}{2}[x_{c4}(k) + y_r(k) - y(k)] - K_P y(k) - \tfrac{K_D}{h}[y(k) - x_{c2}(k)] \\
x_{c2}(k+1) &= y(k) \\
x_{c3}(k+1) &= x_{c3}(k) + \tfrac{K_1 h}{2}[x_{c4}(k) + y_r(k) - y(k)] \\
x_{c4}(k+1) &= y_r(k) - y(k)
\end{aligned}$$

$$u(k) = x_{c1}(k) \tag{6.107}$$

The corresponding equations for forming the control algorithm are

$$\left. \begin{aligned}
x_{c1\text{new}} &= \tfrac{K_1 h}{2}[x_{c4} + y_r - y] - K_P y - \tfrac{K_D}{h}[y - x_{c2}] \\
x_{c2\text{new}} &= y(k) \\
x_{c3\text{new}} &= x_{c3}(k) + \tfrac{K_1 h}{2}[x_{c4} + y_r - y] \\
x_{c4\text{new}} &= y_r - y \\
u &= x_{c1}
\end{aligned} \right\} \tag{6.108}$$

Figure 6.26 shows the flow chart for the control algorithm, including the introduction of two intermediate variables to avoid repeated terms for efficient computation.

The discrete control systems described so far are *synchronous* and are in common use with microprocessors and DSPs. It is possible, however, to have an

6.5 Discrete Control for Small Iteration Intervals 451

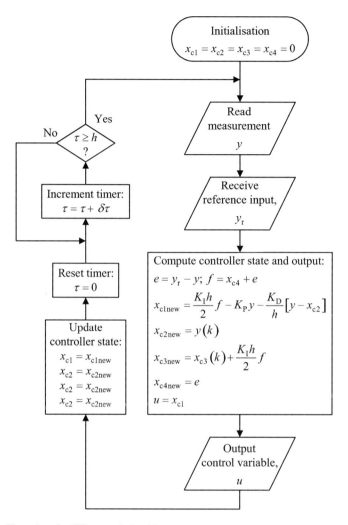

Fig. 6.26 Flow chart for IPD control algorithm

asynchronous system in which the input data reading and control output are executed as soon as the algorithm computations are completed. In this case, the increment timer of Figs. 6.24 and 6.26 is not needed. For example, a system taking full advantage of field-programmable gate array (FPGA) implementation will be asynchronous and will execute several parts of the algorithm simultaneously by means of *independently operating accumulators and adders*. Under these circumstances, h, will be highly variable for each of the functions being implemented in parallel but typically much shorter than that set by Criterion 6.1.

6.6 Discrete Control with Unlimited Iteration Intervals

6.6.1 Pole Placement Design with the Settling Time Formulae

6.6.1.1 Introduction

In Sect. 6.5, the pole placement design was carried out in the s-plane and controllers implemented in the discrete domain using approximations to the continuous dynamic elements, i.e. integrators and differentiators. This approach resulted in satisfactory performance with the iteration interval, h, no greater than a maximum value, $h_{\max}$, according to Criterion 6.1 but could yield oscillatory behaviour for higher values of h and would definitely suffer from instability beyond a higher critical threshold, h_{crit}. In contrast, the method presented in this section entails pole placement design in the z-plane. This is entirely free of these restrictions, in theory imposing no upper limit on h.

It follows from Sects. 6.3 and 6.4 that given a continuous-time linear control system with s-plane poles, s_i, $i = 1, 2, \ldots, n$, if the transformation formula, $z = e^{sh}$, is applied to yield a set of corresponding z-plane poles, z_i, $i = 1, 2, \ldots, n$, satisfying

$$z_i = e^{s_i h}, \quad 1, 2, \ldots, n, \tag{6.109}$$

then provided $h < h_{\max}$ according to Criterion 6.1, any discrete linear control system having these poles will exhibit similar dynamic behaviour to the continuous-time linear control system, referred to as the equivalent continuous linear control system. This provides the basis for pole placement design of linear discrete control systems in the z-plane.

At the sample points, the step response, $y_s(t)$, of the discrete system will coincide with the step response, $y_{\mathrm{ce}}(t)$, of the equivalent continuous system if the z-transfer function of the discrete system contains no zeros introduced by the discretisation as opposed to zeros corresponding to those in the Laplace transfer function of the continuous plant model [Sect. 6.3.3]. In presence of these zeros, $y_s(t)$ will approach, but not equal, $y_{\mathrm{ce}}(t)$ at the sample points. This can be understood by considering that these zeros depend on the plant parameters while the ideal closed-loop system has been formulated independently of these plant parameters.

Since the control variable of the discrete system is constant over each iteration period, the continuous plant output, $y(t)$, will be according to the natural dynamics of the plant and therefore be different to $y_{\mathrm{ce}}(t)$ during these periods. There remains, therefore, the question of how the discrete system behaves for $h > h_{\max}$. As will be seen in the following subsections, the dynamic behaviour of $y_s(t)$ is non-oscillatory if the equivalent continuous system behaviour is non-oscillatory and stability is guaranteed with an accurate plant model.

6.6.1.2 The Dead-Beat Step Responses

The s-plane poles of the equivalent continuous control system must have negative real parts and finite or zero imaginary parts. Thus,

$$s_k h = \sigma_k h, \sigma_k < 0, k = 1, 2, \ldots, \quad \text{for real poles}$$
$$s_l h = (\sigma_l \pm j\omega_l) h, \sigma_l < 0, l = 1, 2, \ldots, \quad \text{for complex conjugate poles,}$$
(6.110)

For a given value of h, if the continuous control system is required to respond faster with the same step response shape, then the s-plane poles are increased in magnitude by a scaling factor, $\lambda > 1$, (Chap. 4, Sect. 4.5.1). Then the corresponding z-plane poles are given by (6.109) and (6.110) as

$$z_k = e^{\lambda \sigma_k h}, \sigma_k < 0, k = 1, 2, \ldots, \quad \text{for real poles}$$
$$z_l = e^{\lambda(\sigma_l \pm j\omega_l)h} = e^{\lambda \sigma_l h}[\cos(\omega_l h) \pm j \sin(\omega_l h)],$$
$$\sigma_l < 0, \quad l = 1, 2, \ldots, \quad \text{for complex conjugate poles}$$
(6.111)

Suppose the violation of Criterion 6.1 is taken to an extreme such that $|s_i| h \to \infty$. Then this is equivalent to letting $\lambda \to \infty$ in (6.111), which yields

$$z_i \to 0, \quad i = 1, 2, \ldots, n. \tag{6.112}$$

It will therefore be of interest to investigate the dynamic character of linear discrete control systems with all their poles located at the origin of the z-plane. It should also be noted that it is feasible to design the controller to yield this pole location regardless of any association with continuous-time control systems.

With the pole location of interest, the discrete SISO control system will have the transfer function,

$$\frac{Y(z)}{Y_r(z)} = \frac{N(z)}{z^n}, \tag{6.113}$$

where $N(z)$ is a polynomial that depends upon the specific application but has degree, $n-1$ or less, for the same reason as in the Laplace domain, i.e. a real control system cannot respond instantaneously to a step reference input change. Also, unity DC gain is usually required. The DC gain in the Laplace domain is found by letting $s \to 0$. Hence through (6.109), $z \to 1$ for discrete linear control systems. A unity DC gain in (6.113) therefore requires $N(1) = 1$. Next, (6.113) will be written as

$$\frac{Y(z)}{Y_r(z)} = \frac{b_0 + b_1 z + \cdots + b_{n-1} z^{n-1}}{z^n}, \tag{6.114}$$

where $b_0 + b_1 + \cdots + b_{n-1} = 1$.

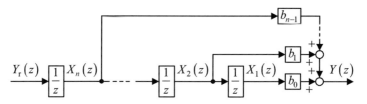

Fig. 6.27 State-variable block diagram of system with poles at the origin of the z-plane

The dynamic nature of the system will now be investigated for different orders. For this, the discrete state-variable block diagram with transfer function (6.114) shown in Fig. 6.27 will be used.

Let $y_r(k) = AH(k) = \begin{cases} 0, & k < 0 \\ A, & k \geq 0 \end{cases}$, $A = $ const. Then since the blocks with transfer function, $1/z$, represent delays of duration, h [s], and k increments by 1 every h [s],

$$x_n(k) = AH(k-1), \, x_{n-1}(k) = AH(k-2), \\ \ldots, x_2(k) = AH[k-(n-1)], \, x_1(k) = AH(k-n). \quad (6.115)$$

With zero initial state, when $k = 1$, $x_n(k)$ steps from zero to A and remains at this value. When $k = 2$, $x_{n-1}(k)$ similarly steps from zero to A and remains at this value. Once $k = n$, $x_1(k)$ steps from zero to A, by which time all the state variables are constant at A. A discrete linear system of nth order with all of its poles at the origin of the z-plane therefore has a step response that settles precisely in n steps, i.e. in a finite period of nh seconds. It is important to realise, however, that this description of the behaviour only applies at the sample points, as it is based on the z-transfer function and difference equations that describe the behaviour only at the sample points. The behaviour of $\mathbf{x}(t)$ and $y(t)$ *between* the sampling points depends on the plant pole locations in the s-plane. In the following subsection, examples of design by pole placement that yield the same closed-loop z-transfer function for different plants will be presented that exemplify this dependence.

6.6.1.3 Application of the Settling Time Formulae

If the settling time formulae of Chap. 4 are used, then the desired closed-loop poles of a continuous linear control system are located at $s_{c1, 2, \ldots, n} = s_c$, where

$$s_c = \begin{cases} -1.5(1+n)/T_{sd} & \text{for the 5 \% criterion} \\ -1.6(1.5+n)/T_{sd} & \text{for the 2 \% criterion} \end{cases} \quad (6.116)$$

and T_{sd} is the demanded settling time. The closed-loop poles of the linear discrete control system in the z-plane are then $z_{c1, 2, \ldots, n} = z_c$, where, using (6.109),

$$z_c = e^{s_c h}. \quad (6.117)$$

6.6 Discrete Control with Unlimited Iteration Intervals

It is evident from (6.117) that $|z_c| < 1$ for $\operatorname{Re}(s_c) < 0$ and $h > 0$. Then as $h \to \infty$, $z_c \to 0$, meaning that the multiple z-plane pole remains within the unit circle, i.e. the region of stability, and approaches the origin of the z-plane, *which is the most stable location*, as h is increased indefinitely. This is in complete contrast with the approach of Sect. 6.5 in which one or two (complex conjugate) closed-loop poles reached the unit circle at a finite critical value of $h = h_{\text{crit}}$ and entered the unstable region outside the unit circle as h increased beyond h_{crit}. In view of the previous subsection, as $h \to \infty$, the corresponding step response approaches a dead-beat response, which settles precisely in a period of nh seconds but this situation is, of course, impracticable as the dead-beat settling time would become infinite. Conversely, however, if h is fixed and the settling time, T_s, is reduced, then as $T_s \to 0$, again $z_c \to 0$, meaning that the step response would approach the dead-beat response, precisely settling in nh seconds, which is the shortest possible settling time of a linear discrete control system for a given value of h.

The demanded settling time, T_{sd} of (6.116), would usually be set to the specified settling time, T_s. If this results in the *actual* settling time, T_{sa}, being substantially greater than T_s, then provided $T_{sd} \geq nh$, T_{sd} can be reduced until $T_{sa} = T_s$.

6.6.1.4 Following of the Equivalent Continuous System

Consider a linear discrete control system without any plant zeros whose equivalent continuous system has the transfer function

$$\frac{Y_{ec}(s)}{Y_r(s)} = \frac{p_c^n}{(s + p_c)^n}. \tag{6.118}$$

where $p_c = -s_c$ and s_c is given by (6.116). Then, using Table 3 in Tables, the z-transfer function of the *ideal* discrete system, whose step response, $y_{id}(t)$, is the sampled and held version of $y_{ec}(t)$, is

$$\frac{Y_{id}(z)}{Y_r(z)} = 1 - \sum_{k=1}^{n} \frac{(-p_c)^{k-1}}{(k-1)!} \frac{\partial^{k-1}}{\partial p_c^{k-1}} \left(\frac{z-1}{z - e^{-p_c h}} \right) = \frac{D_{id}(z)}{(z - e^{-p_c h})^n}. \tag{6.119}$$

The roots of $D_{id}(z) = 0$ are the discretisation zeros. The plant z-transfer function is

$$\frac{Y(z)}{U(z)} = \frac{D_p(z)}{A(z)}. \tag{6.120}$$

If any linear controller is used that does not introduce zeros, then, assuming a DC gain of unity, the closed-loop transfer function will be

$$\frac{Y(z)}{Y_r(z)} = \frac{\left[(1 - e^{-p_c h})^n / D_p(1)\right] D_p(z)}{(z - e^{-p_c h})^n}. \tag{6.121}$$

In general, the discretisation zeros of the ideal and actual systems are not equal because $D_p(z)$ depends upon the plant parameters, while $D_{id}(z)$ does not. Hence, $y_s(t) \neq y_{id}(t)$ meaning that the plant output, $y(t)$, will not follow the output of the equivalent continuous system perfectly at the sample points. The one exception is $n = 1$ since in this case there are no discretisation zeros.

6.6.2 Pole Placement for Negligible Digital Processing Time

6.6.2.1 Introduction

Sometimes a control digital processor is required to share its computational capacity with other tasks, such as data collection and processing for health monitoring, resulting in a relatively large iteration interval, h. Also, the processor could be so fast that T_D [ref., Fig. 6.2] is relatively small. Both situations could result in $T_D \ll h$. Then the controller can be designed as if the control variable, $u(k)$, is applied to the plant at the same time that the measurement sample, $y(k)$, is taken. This is assumed throughout this subsection.

If the controller is formulated so that it contains a number of independent parameters at least equal in number to the order of the control loop, then design by pole assignment may be carried out. It must always be remembered, however, that a perfect plant model does not exist. As for continuous-time control systems, in some cases, the plant parameters must be quite accurate if the real system response is to closely approach the theoretical ideal response. Simulation including mismatch between the real plant and its model is therefore recommended before implementation.

A generic discrete controller designable by pole assignment assuming negligible computational delay is the linear state feedback controller. If the state, $\mathbf{x}(k)$, is available from an LTI SISO plant defined by the discrete state-space model

$$\begin{aligned} \mathbf{x}(k+1) &= \mathbf{\Phi}(h)\mathbf{x}(k) + \mathbf{\psi}(h)u(k) \\ y(k) &= \mathbf{c}^T \mathbf{x}(k), \end{aligned} \tag{6.122}$$

then the control law,

$$u(k) = R y_r(k) - \mathbf{k}\mathbf{x}(k), \tag{6.123}$$

may be designed by pole assignment following similar lines to the design of the continuous linear control systems of Chap. 3. Here $\mathbf{x}(k) \in \Re^n$; the constant plant matrices are $\mathbf{\Phi}(h) \in \Re^{n \times n}$, $\mathbf{\psi}(h) \in \Re^{n \times 1}$ and $\mathbf{c}^T \in \Re^{1 \times n}$; and, in particular, the state feedback gain matrix is $\mathbf{k} \in \Re^{n \times 1}$, which permits complete pole placement. The following subsections consider the design of first-, second- and third-order systems.

Fig. 6.28 Discrete linear control system for first-order LTI plant

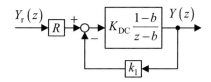

6.6.2.2 First-Order Plants

With reference to Table 3 in Tables, the plant with Laplace transfer function,

$$\frac{Y(s)}{U(s)} = K_{\text{DC}} \frac{a}{s+a} \tag{6.124}$$

has the corresponding z-transfer function,

$$\frac{Y(z)}{U(z)} = K_{\text{DC}} \frac{1-b}{z-b}, \tag{6.125}$$

where $b = e^{-ah}$. The linear state feedback control system is the proportional control system shown in Fig. 6.28.

The closed-loop transfer function is

$$\frac{Y(z)}{Y_r(z)} = \frac{RK_{\text{DC}} \frac{1-b}{z-b}}{1 + k_1 K_{\text{DC}} \frac{1-b}{z-b}} = \frac{RK_{\text{DC}}(1-b)}{z - b + k_1 K_{\text{DC}}(1-b)} \tag{6.126}$$

For a demanded settling time of T_{sd} seconds using the 5 % settling time formula, the closed-loop pole of the equivalent continuous system is $s_{\text{c1}} = -1.5(1+N)/T_{\text{sd}}|_{N=1} = -3/T_s$. The corresponding z-plane pole is therefore

$$z_{\text{c1}} = e^{s_{\text{c1}} h} = e^{-3h/T_{\text{sd}}} \tag{6.127}$$

The desired characteristic polynomial is then

$$z - z_{\text{c1}} \tag{6.128}$$

Equating (6.128) with the denominator of (6.126) then yields

$$k_1 = \frac{b - z_{\text{c1}}}{K_{\text{DC}}(1-b)}. \tag{6.129}$$

For a closed-loop DC gain of unity,

$$\lim_{z \to 1} \frac{Y(z)}{Y_r(z)} = \frac{RK_{DC}}{1 + k_1 K_{DC}} = 1 \Rightarrow R = \frac{1 + k_1 K_{DC}}{K_{DC}}. \qquad (6.130)$$

In the special case of the plant being a pure integrator, then the plant Laplace and z-transfer functions are

$$\frac{Y(s)}{U(s)} = K_{DC}\frac{1}{s} \quad \text{and} \quad \frac{Y(z)}{U(z)} = K_{DC}\frac{h}{z-1} \qquad (6.131)$$

where the DC gain is according to the general definition, $K_{DC} = \lim_{s \to 0} s^r G(s)$, where r is the plant type (Chap. 1). Then the closed-loop transfer function is

$$\frac{Y(z)}{Y_r(z)} = \frac{RK_{DC}\frac{h}{z-1}}{1 + k_1 K_{DC}\frac{h}{z-1}} = \frac{RK_{DC}h}{z - 1 + k_1 K_{DC}h} \qquad (6.132)$$

and equating (6.128) to the denominator of (6.132) yields

$$k_1 = \frac{1 - z_{c1}}{K_{DC}h}. \qquad (6.133)$$

Finally, to achieve a unity closed-loop DC gain, setting $z = 1$ in (6.132) requires

$$R = k_1. \qquad (6.134)$$

Figure 6.29 shows step responses for the same plant parameters, $a = 1$ and $K_{DC} = 1$, a specified settling time of $T_{sd} = 1$ [s] and $h = h_{max}$, $h = 5h_{max}$ and $h = 10h_{max}$, where h_{max} is the maximum iteration period according to Criterion 6.1. In this case, the plant pole is $s_1 = -1$ and the closed-loop pole of the equivalent continuous system is $s_{c1} = -3/T_{sd} = -3$. So $h_{max} = 1/(5|s_{max}|)$ where $|s_{max}| = \max(1, 3) = 3$, giving $h_{max} = 1/15 = 0.067$ [s].

The step response, $y_{ec}(t)$, of the equivalent continuous closed-loop system is the same in each case and, as predicted theoretically, the controlled output, $y(t)$, and its sampled and held version, $y_s(t)$, coincide with $y_{ec}(t)$ at the sample points. The magnitude of the error, $y(t) - y_{ec}(t)$, between the sample points increases with h but the settling time specification is almost kept even for $h = 10h_{max}$. In Fig. 6.30, the plant parameters are as in Fig. 6.29 and the iteration interval is kept at $h = 0.2$ [s].

The demanded settling time is reduced in steps and the system remains well behaved. Even with the unrealisable demand of $T_{sd} = 0$, the system 'does its very best' and produces the response with the shortest possible settling time, i.e. the dead-beat response that settles in a number of iterations equal to the system order. As evident in Fig. 6.30c, $y(t)$ settles in only one iteration. The sampled and held version, $y_s(t)$, jumps to the demanded value at $t = 0.2$ [s] but $y(t)$ rises exponentially to the demanded value under a constant input calculated by the controller.

6.6 Discrete Control with Unlimited Iteration Intervals

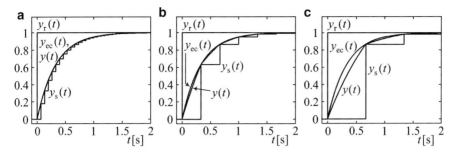

Fig. 6.29 Step responses of first-order discrete LTI control system for different h values. (**a**) $h = h_{\max}$ (**b**) $h = 5h_{\max}$. (**c**) $h = 10h_{\max}$

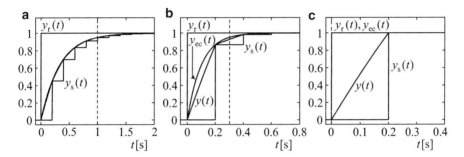

Fig. 6.30 Step responses of first-order discrete LTI control system for different T_{sd} values. (**a**) $T_{sd} = 1[s]$ (**b**) $T_{sd} = 0.3[s]$ (**c**) $T_{sd} = 0[s]$

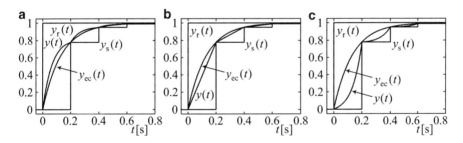

Fig. 6.31 Step responses of first-order discrete LTI control system with different plant poles. (**a**) $s_1 = -10[s^{-1}]$ (**b**) $s_1 = 0[s^{-1}]$ (**c**) $s_1 = +10[s^{-1}]$

Figure 6.31 demonstrates the variations in the behaviour of $y(t)$ between the sample points for different plants controlled to have the same equivalent continuous system step response with a 5 % settling time of $T_{sd} = 0.4$ [s].

In Fig. 6.31a, the plant pole has been chosen larger than the equivalent continuous closed-loop pole at $s_{c1} = -3/0.4 = 7.5 \ [s^{-1}]$ so that the exponential segments of $y(t)$ have a shorter time constant, i.e. the plant time constant, than the time constant of the exponential step response, $y_{ec}(t)$, in contrast with Figs. 6.29 and 6.30. This is the reason for $y(t)$ lying above $y_{cc}(t)$ between the sample points. In Fig. 6.31b, the plant is a pure integrator resulting in $y(t)$ comprising a set of contiguous straight line segments between the sample points. Finally, Fig. 6.31c demonstrates the ability of the control technique to handle unstable plants. The growing exponential segments of $y(t)$ between the sample points where the plant input is constant may be clearly seen. The controller makes these fit the equivalent continuous closed-loop system step response, $y_{ec}(t)$, at the sample points.

As pointed out in Sect. 6.6.1.4, there are no discretisation zeros to cause a mismatch between the closed-loop z-transfer function and the ideal z-transfer function corresponding to the Laplace transfer function of the equivalent continuous system (Table 3 in Tables). This results in $y(t) = y_{ec}(t)$ at the sample points, as shown in Figs. 6.29, 6.30, and 6.31.

On a pragmatic basis, it must be mentioned that the system may be quite sensitive to plant modelling errors for the relatively long iteration intervals used for the above demonstrations. It is therefore recommended to test this sensitivity by simulation with mismatched plant parameters during the control system design process.

6.6.2.3 Second-Order Plants

The following underdamped second-order plant model without finite zeros will be taken, with transfer function,

$$\frac{Y(s)}{U(s)} = \frac{K_{DC}\omega_n^2}{s^2 + 2\zeta\omega_n s + \omega_n^2} = K_{DC}\frac{a^2 + b^2}{(s+a)^2 + b^2}, \quad (6.135)$$

where K_{DC} is the DC gain, ζ is the damping ratio satisfying $0 < \zeta < 1$ and ω_n is the undamped natural frequency. The parameterisation using a and b simplifies the corresponding z-transfer function, which is

$$\frac{Y_{id}(z)}{U(z)} = \frac{b_1 z + b_0}{z^2 + a_1 z + a_0} \quad (6.136)$$

where $b_1 = K_{DC}\left[1 - e^{-ah}\left(c + \frac{a}{b}s\right)\right]$, $b_0 = K_{DC}\left[e^{-2ah} - e^{-ah}\left(c - \frac{a}{b}s\right)\right]$, $c = \cos(bh)$, $s = \sin(bh)$, $a_1 = -2e^{-2ah}\cos(bh)$ and $a_0 = e^{-2ah}$. Also from (6.135),

$$a = \zeta\omega_n \quad (6.137)$$

and

$$a^2 + b^2 = \omega_n^2 \Rightarrow b = \omega_n\sqrt{1-\zeta^2} = \omega_d \quad (6.138)$$

where ω_d is the damped natural frequency, i.e. the actual frequency of oscillation. The zero at $z = -b_0/b_1$ is a discretisation zero as the plant transfer function (6.135) has no finite zeros (Sect. 6.3.3.2).

6.6 Discrete Control with Unlimited Iteration Intervals

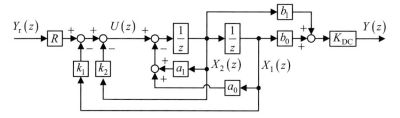

Fig. 6.32 Linear state feedback control law applied to discrete linear second-order SISO plant

Figure 6.32 shows a linear state feedback control law applied to the discrete state-variable block diagram of the plant in the controller canonical form, which renders the gain determination straightforward.

The closed-loop transfer function is

$$\frac{Y(z)}{Y_r(z)} = RK_{DC} \cdot \frac{\frac{b_1}{z} + \frac{b_0}{z^2}}{1 - \left[-(k_2 + a_1)\frac{1}{z} - (k_1 + a_0)\frac{1}{z^2}\right]} = \frac{RK_{DC}(b_1 z + b_0)}{z^2 + (k_2 + a_1)z + k_1 + a_0}.$$

(6.139)

To achieve a non-overshooting step response of the equivalent continuous system with a demanded 5 % settling time of T_{sd}, the closed-loop poles in the s-plane are

$$s_{c1,2} = -\frac{1.5(1+N)}{T_{sd}}\bigg|_{N=2} = -\frac{4.5}{T_{sd}}$$

(6.140)

The corresponding z-plane poles according to (6.117) are

$$z_{c1,2} = e^{s_{c1,2}h} = z_{c1} = e^{-4.5h/T_{sd}}$$

(6.141)

Hence the desired characteristic polynomial of the discrete control system is

$$(z - z_c)^2 = z^2 - 2z_c z + z_c^2.$$

(6.142)

The equations for the controller gains are obtained by equating polynomial (6.142) with the denominator of (6.139). Thus

$$k_1 = z_c^2 - a_0$$

(6.143)

and

$$k_2 = -(a_1 + 2z_c).$$

(6.144)

Assuming a closed-loop DC gain of unity, from (6.139),

$$\lim_{z \to 1} \frac{Y(z)}{Y_r(z)} = 1 = \frac{RK_{DC}(b_1 + b_0)}{1 + k_2 + a_1 + k_1 + a_0} \Rightarrow R = \frac{1 + a_0 + a_1 + k_1 + k_2}{K_{DC}(b_1 + b_0)}. \quad (6.145)$$

Regarding Sect. 6.6.1.4, the ideal discrete control system transfer function corresponding to the equivalent continuous system transfer function,

$$\frac{Y_{ec}(s)}{Y_r(s)} = \left(\frac{4.5/T_{sd}}{s + 4.5/T_{sd}}\right)^2, \quad (6.146)$$

is obtained from Table 3 in Tables as

$$\frac{Y_{id}(z)}{Y_r(z)} = \frac{\left(1 - e^{-4.5h/T_{sd}} - ahe^{-4.5h/T_{sd}}\right)z + e^{-9h/T_{sd}} - e^{-4.5h/T_{sd}} + ahe^{-4.5h/T_{sd}}}{\left(z - e^{-4.5h/T_{sd}}\right)^2}. \quad (6.147)$$

The ideal discretisation zero is therefore located at

$$z_{id1} = \frac{e^{-4.5h/T_{sd}} - e^{-9h/T_{sd}} - (4.5h/T_{sd})e^{-4.5h/T_{sd}}}{1 - e^{-4.5h/T_{sd}} - (4.5h/T_{sd})e^{-4.5h/T_{sd}}}. \quad (6.148)$$

This depends only on h and T_{sd}, while, according to (6.139), the actual discretisation zero is located at

$$z_1 = -\frac{b_0}{b_1} = \frac{e^{-ah}\left(\cos(bh) - \frac{a}{b}\sin(bh)s\right) - e^{-2ah}}{1 - e^{-ah}\left(\cos(bh) + \frac{a}{b}\sin(bh)\right)}, \quad (6.149)$$

and this depends on h and the plant parameters, $a = \zeta\omega_n$ and $b = \omega_n\sqrt{1 - \zeta^2}$.

The denominators of (6.139) and (6.147) are identical but, clearly, $z_1 \neq z_{id1}$ and therefore the controlled output, $y(t)$, cannot equal the continuous equivalent system output, $y_{ec}(t)$, at the sample points. It will now be shown, however, that as $h \to 0$ this following error tends to zero. For (6.148), since

$$e^{-4.5h/T_{sd}} = e^{-\alpha h} = 1 - \alpha h + \tfrac{1}{2!}\alpha^2 h^2 - \ldots \quad \text{and}$$

$$e^{-9h/T_{sd}} = e^{-2\alpha h} = 1 - 2\alpha h + \tfrac{1}{2!}4\alpha^2 h^2 - \ldots,$$

$$\lim_{h \to 0} z_{id1} = \lim_{h \to 0} \frac{1 - \alpha h + \tfrac{1}{2}\alpha^2 h^2 \ldots - \left(1 - 2\alpha h + 2\alpha^2 h^2 - \ldots\right) - \alpha h\left(1 - \alpha h + \tfrac{1}{2}\alpha^2 h^2\right)}{1 - (1 - \alpha h + \tfrac{1}{2}\alpha^2 h^2) - \alpha h(1 - \alpha h + \tfrac{1}{2}\alpha^2 h^2)}$$

$$= \lim_{h \to 0} \frac{-\tfrac{1}{2}\alpha^2 h^2 + \cdots + \text{weighted sum of } h^q, \ q \geq 3}{\tfrac{1}{2}\alpha^2 h^2 - \cdots + \text{weighted sum of } h^q, \ q \geq 3} = -1 \quad (6.150)$$

6.6 Discrete Control with Unlimited Iteration Intervals

For (6.149), since $e^{-ah} = 1 - ah + \frac{1}{2!}a^2h^2 - \ldots$, $e^{-2ah} = 1 - 2ah + \frac{1}{2!}4a^2h^2 - \ldots$,

$\cos(bh) = 1 - \frac{1}{2!}b^2h^2 + \ldots$ and $\sin(bh) = bh - \frac{1}{3!}b^3h^3 + \ldots$,

$$\lim_{h \to 0} z_1 = \lim_{h \to 0} \frac{\left(1 - ah + \frac{1}{2}a^2h^2 - \ldots\right)\left[\left(1 - \frac{1}{2}b^2h^2 + \ldots\right) - \frac{a}{b}\left(bh - \frac{1}{6}b^3h^3 + \ldots\right)\right] - \left(1 - !2ah + 2a^2h^2 - \ldots\right)}{1 - \left(1 - ah + \frac{1}{2}a^2h^2 - \ldots\right)\left[\left(1 - \frac{1}{2}b^2h^2 + \ldots\right) + \frac{a}{b}\left(bh - \frac{1}{6}b^3h^3 + \ldots\right)\right]}$$

$$= \lim_{h \to 0} \frac{-\frac{1}{2}\left(a^2 + b^2\right)h^2 + \text{weighted sum of } h^q, \ q \geq 3}{\frac{1}{2}\left(a^2 + b^2\right)h^2 + \text{weighted sum of } h^q, \ q \geq 3} = -1$$

(6.151)

Hence

$$\lim_{h \to 0} z_1 = \lim_{h \to 0} z_{\text{id}1}, \quad (6.152)$$

indicating that as h is reduced, the discretisation zeros of the closed-loop system become closer to the discretisation zeros of the ideal discrete system whose output, $y_{\text{id}}(t)$, in real time coincides with the output, $y_{\text{ec}}(t)$, of the equivalent continuous system at the sampling points. So $\lim_{h \to 0} y(t) = y_{\text{ec}}(t)$.

Some simulations will now be presented, first to demonstrate the effect of increasing the iteration period, h, starting with the maximum value according to Criterion 6.1, for a fixed demanded settling time of $T_{\text{sd}} = 0.5$ [s]. The plant parameters taken are $\zeta = 0.2$, $\omega_n = 10$ [rad/s] and $K_{\text{DC}} = 1$. The plant pole magnitude is therefore $|s_{1,2}| = \omega_n = 10$ [s^{-1}]. For $T_{\text{sd}} = 0.5$ [s], the closed-loop pole magnitude for the equivalent continuous system is $|s_{c1,2}| = 4.5/T_{\text{sd}} = 9$ [s^{-1}]. According to Criterion 6.1, $|s_{\max}| = \max(|s_{1,2}|, |s_{c1,2}|) = 10$ [s^{-1}] and then the minimum iteration interval guaranteeing a discrete system performance similar to that of the continuous equivalent system is $h_{\max} = 1/(5|s_{\max}|) = 0.02$ [s]. Figure 6.33 shows the results.

These results agree with the foregoing analysis. The departure of $y(t)$ from $y_{\text{ec}}(t)$, however, is negligible for $h = h_{\max}$. Although this error is significant in Fig. 6.33c, h is relatively large since there are only three samples over the transient duration and, in contrast with the approach of Sect. 6.5, the system remains stable.

Figure 6.34 shows step responses with T_{sd} reduced in steps to zero.

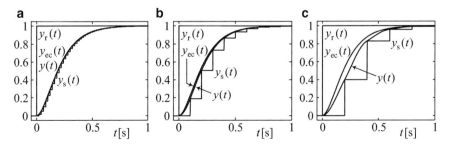

Fig. 6.33 Step responses of second-order discrete LTI control system for different h values. (**a**) $h = h_{\max}$ (**b**) $h = 5h_{\max}$ (**c**) $h = 10h_{\max}$

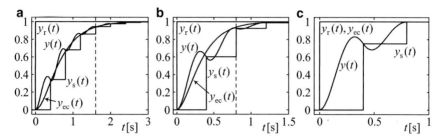

Fig. 6.34 Step responses of second-order discrete LTI control system for different T_{sd} values. (**a**) $T_{sd} = 1.6[s]$ (**b**) $T_{sd} = 0.8[s]$ (**c**) $T_{sd} = 0[s]$

The iteration period is fixed at a relatively high value of $h = 0.4$ [s] to show the behaviour of the underdamped plant during the periods of constant control. The oscillatory behaviour is evident in all three cases but the plant is fully under control as the amplitude of the oscillations is reduced from one iteration period to the next. Figure 6.34c shows the dead-beat response resulting from setting the demanded settling time to zero. Following the theory of Sect. 6.6.1.2, the controlled output is brought to the constant demanded value without any subsequent changes in the state variables in just two iterations.

6.6.2.4 Third- and Higher-Order Plants

If a complete set of state variables is available, then design by pole assignment according to Sect. 6.6.1 may be carried out. Examples of third and higher order are not given here as they are present in the section to follow, the design approach being similar, while the need for all the state variables to be available is removed through the use of polynomial control.

Although polynomial control is used with compensation for the computational delay in the following section, it can also be used without such compensation, resulting in a lower-order system, as will be seen in Example 6.4.

6.6.3 Computational Delay Allowance

6.6.3.1 Introduction

In Chap. 8, the observer will be introduced, the discrete version of which predicts the plant state at the next sampling point, thereby allowing for the computational delay. An alternative that does not require state estimation is presented here.

With reference to Sect. 6.2, if the period, T_D, needed to produce a new control value, is significant compared with h, a complete iteration period can be allowed for the control computations. This effectively introduces a time delay of h seconds in

6.6 Discrete Control with Unlimited Iteration Intervals

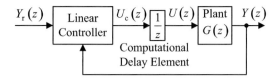

Fig. 6.35 Insertion of delay element in control system to allow computation time

the forward path of the control system, which can be represented on the z-transfer function block diagram of the control system by inserting a block with transfer function, $1/z$, in the control channel, as shown in Fig. 6.35.

This creates an augmented plant with the computed control input, $U_c(z)$, as shown. If the plant order is n, the order of the augmented plant is $n + 1$. If the linear controller contains dynamic elements and is of order, n_c, then the total system order is $N = n_c + n + 1$. If the controller, such as the generic one presented in the following subsection, contains N adjustable parameters that permit design by pole placement, then the approach of Sect. 6.6.1 may be taken.

6.6.3.2 Discrete Polynomial Control

The polynomial controller of Chap. 5 lends itself very well to the discrete domain, easily accommodating the computational delay element of Fig. 6.35. This is essentially the same as the RST controller [3]. The general control system block diagram is shown in Fig. 6.36.

All the algebraic considerations are identical to those presented in Chap. 5 for the equivalent continuous control system but these are summarised in this subsection. The only difference is in the setting up of the desired characteristic polynomial for the pole placement, which follows Sect. 6.6.1. The purposes of the controller polynomials are similar to those in the equivalent continuous control system but since there are some differences, they are given in Table 6.1.

The closed-loop transfer function for Fig. 6.36 is

$$\frac{Y(z)}{Y_r(z)} = \frac{R(z)}{P(z)} \cdot \frac{B(z)}{F(z)A(z) + B(z)H(z)}, \qquad (6.153)$$

The pole placement equation is then

$$F(z)A(z) + B(z)H(z) = D(z) \qquad (6.154)$$

where

$$A(z) = z^{n_a} + \sum_{i=0}^{n_a-1} a_i z^i, \quad B(z) = \sum_{i=0}^{n_b} b_i z^i, \quad F(z) = \sum_{i=0}^{n_f} f_i z^i \text{ and } H(z) = \sum_{i=0}^{n_h} h_i z^i.$$

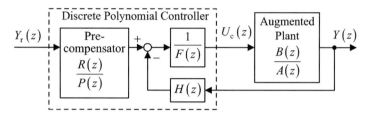

Fig. 6.36 Generic discrete polynomial control system

Table 6.1 Purposes of components of discrete polynomial controller

Component	Purpose
$H(z)$	To provide a minimal number of controller parameters for pole placement
$\frac{1}{F(z)}$	To render the controller causal and therefore realisable while introducing sufficient controller parameters to support design by pole placement despite the resulting increase in the overall system order
$R(z)$	Dynamic lag pre-compensation (Chap. 13) or, in reduced form, as a reference input scaling coefficient, r_0, to obtain the required DC gain
$P(z)$	Compensation for the effects of plant zeros, otherwise set to unity

Also, $D(z) = z^N + \sum_{i=0}^{N-1} d_i z^i$ is the desired closed-loop characteristic polynomial, where the overall system order is

$$N = n_a + n_f \tag{6.155}$$

and the constraint,

$$n_f \geq n_h = n_a - 1 \tag{6.156}$$

on the polynomial degrees applies. As in Chap. 5, once $D(z)$ has been determined, (6.154) can be solved for the coefficients of the controller polynomials, $H(z)$ and $F(z)$ by means of the following matrix–vector equation.

$$\overbrace{\begin{bmatrix} 1 & 0 & \cdots & 0 & 0 & \cdots & & 0 \\ a_{n_a-1} & 1 & \ddots & \vdots & b_{n_b} & \ddots & & \vdots \\ \vdots & a_{n_a-1} & \ddots & 0 & \vdots & b_{n_b} & \ddots & \\ a_1 & \vdots & \ddots & 1 & b_1 & \vdots & \ddots & 0 \\ a_0 & a_1 & & a_{n_a-1} & b_0 & b_1 & & b_{n_b} \\ 0 & a_0 & \ddots & \vdots & 0 & b_0 & \ddots & \vdots \\ \vdots & & \ddots & a_1 & \vdots & & \ddots & b_1 \\ 0 & \cdots & 0 & a_0 & 0 & \cdots & 0 & b_0 \end{bmatrix}}^{\mathbf{P}} \overbrace{\begin{bmatrix} f_{n_f} \\ \vdots \\ f_1 \\ f_0 \\ h_{n_h} \\ \vdots \\ h_1 \\ h_0 \end{bmatrix}}^{\mathbf{k}} = \overbrace{\begin{bmatrix} 1 \\ d_{N-1} \\ \vdots \\ \vdots \\ d_2 \\ d_1 \\ d_0 \end{bmatrix}}^{\mathbf{d}} \Rightarrow \mathbf{k} = \mathbf{P}^{-1}\mathbf{d}.$$

(6.157)

This generic equation can be programmed and solved on a computer to avoid tedious manual derivation of formulae for the controller parameters.

6.6 Discrete Control with Unlimited Iteration Intervals

It may be recalled from Chap. 5 that increasing n_f beyond the minimum value could enhance the filtering of measurement noise. Arguably, this is also true in the discrete domain since the overall dynamics of the discrete polynomial control system will be similar to that of the equivalent continuous polynomial control system.

Once $F(z)$ and $H(z)$ have been determined and, if necessary, $P(z)$ has been determined for the pre-compensation, either $R(z)$ is determined to eliminate the dynamic lag (Chap. 13) or $R(z) = r_0$, where r_0 is set to achieve, usually, a closed-loop DC gain of unity. Assuming the latter case, using (6.153),

$$\lim_{z \to 1} \frac{Y(z)}{Y_r(z)} = 1 \Rightarrow \frac{r_0}{P(1)} \cdot \frac{B(1)}{F(1)A(1) + B(1)H(1)} = 1 \Rightarrow \quad (6.158)$$
$$r_0 = P(1)[F(1)A(1) + B(1)H(1)]/B(1).$$

It should be noted that the element with transfer function, $1/F(z)$ introduces a delay of $n_f h$ [s] in the response to the reference input and if the system is to be designed to a specified settling time of T_s and the iteration interval, h, is a significant proportion of T_s, then the delay can be compensated by setting the *demanded* settling time in the formulae for the controller parameters to

$$T_{sd} = T_s - n_f h. \quad (6.159)$$

Example 6.4 Rigid-body spacecraft attitude control

The z-transfer function for one attitude control axis of a rigid-body spacecraft is

$$\frac{Y(z)}{U(z)} = \frac{\frac{1}{2}bh^2(z+1)}{(z-1)^2}, \quad (6.160)$$

where $b = K_w/J$, K_w is the reaction wheel torque constant and J is the spacecraft moment of inertia about the control axis. If a delay element is inserted in the control channel to allow a whole iteration period for the control computations, then the augmented plant transfer function is

$$\frac{Y(z)}{U_c(z)} = \frac{\frac{1}{2}bh^2(z+1)}{z(z-1)^2} = \frac{b_1 z + b_0}{z^3 + a_2 z^2 + a_1 z + a_0} \quad (6.161)$$

It is required to design a discrete polynomial controller of minimal order, with $h = 1$ [s], to have a non-overshooting step response with a settling time of T_s seconds according to the 2 % criterion. This settling time is to be set to the minimum value such that the control variable magnitude peaks at 10 [V], which is the saturation limit, for the largest slew manoeuvre of 180°. Furthermore, it is required to find the minimum value of h, and hence the settling time, that yields a dead-beat response for the same slew manoeuvre and the same control saturation constraint. The purpose of this is to determine if the slewing time can be shortened using dead-beat control.

The plant parameters are given as $J = 200$ [kg m²] and $K_w = 0.1$ [Nm/V]. Since the controller is of minimal order, (6.156) yields

$$n_f = n_h = n_a - 1 \tag{6.162}$$

Since $n_a = 3$, $n_f = n_h = 2$ and therefore (6.155) yields $N = n_a + n_f = 5$. From (6.161), $b_1 = \frac{1}{2}bh^2$, $b_0 = \frac{1}{2}bh^2$, $a_2 = -2$, $a_1 = 1$ and $a_0 = 0$. Then (6.157) becomes

$$\begin{bmatrix} 1 & 0 & 0 & 0 & 0 & 0 \\ a_2 & 1 & 0 & 0 & 0 & 0 \\ a_1 & a_2 & 1 & b_1 & 0 & 0 \\ a_0 & a_1 & a_2 & b_0 & b_1 & 0 \\ 0 & a_0 & a_1 & 0 & b_0 & b_1 \\ 0 & 0 & a_0 & 0 & 0 & b_0 \end{bmatrix} \begin{bmatrix} f_2 \\ f_1 \\ f_0 \\ h_2 \\ h_1 \\ h_0 \end{bmatrix} = \begin{bmatrix} 1 \\ d_4 \\ d_3 \\ d_2 \\ d_1 \\ d_0 \end{bmatrix}. \tag{6.163}$$

The method of Sect. 6.6.1 will be applied to determine the desired characteristic polynomial coefficients. The pole value of the equivalent continuous system is

$$s_c = -1.6\,(1.5 + N)\,/\,T_{sd} = -10.4/T_{sd}. \tag{6.164}$$

Then the z-plane pole value is

$$z_c = e^{s_c h} \tag{6.165}$$

Let $q_c = -z_c$. Then the desired characteristic polynomial is as follows.

$$\begin{aligned} & z^5 + d_4 z^4 + d_3 z^3 + d_2 z^2 + d_1 z + d_0 \\ & = (z + q_c)^5 = z^5 + 5 q_c z^4 + 10 q_c^2 z^3 + 10 q_c^3 z^2 + 5 q_c^4 z + q_c^5 \Rightarrow \\ & d_4 = 5 q_c,\ d_3 = 10 q_c^2,\ d_2 = 10 q_c^3,\ d_1 = 5 q_c^4 \text{ and } d_0 = q_c^5. \end{aligned} \tag{6.166}$$

Figure 6.37a shows the overall control system block diagram and Fig. 6.37b shows the implementation block diagram of the controller. Note that by inspection of (6.163), $f_2 = 1$ and this immediately simplifies Fig. 6.37.

For a closed-loop DC gain of unity, noting that the zero pre-compensator is not needed, (6.158) yields

$$r_0 = [(1 + f_1 + f_0)(1 + a_2 + a_1 + a_0) + (b_1 + b_0)(h_2 + h_1 + h_0)] / (b_1 + b_0). \tag{6.167}$$

6.6 Discrete Control with Unlimited Iteration Intervals

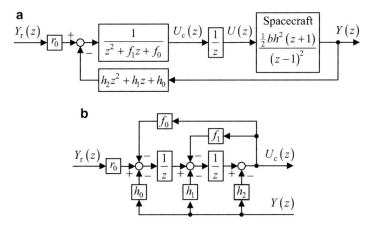

Fig. 6.37 Discrete polynomial control of rigid-body spacecraft. (**a**) Control system block diagram. (**b**) Implementation block diagram of controller

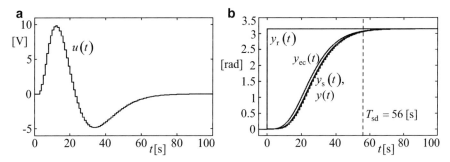

Fig. 6.38 Minimum settling time slew with delay element, reaching control saturation. (**a**) Control input applied to reaction wheel. (**b**) Controlled and e.c. system output

Next, h is fixed at 1 [s] and simulations of a step response with $y_r = \pi$ [rad] $\equiv 180°$ carried out, adjusting T_{sd} until the peak value of $u(t)$ just reaches 10 [V]. The optimal demanded settling time was found to be $T_{sd} = 56$ [s] as shown in Fig. 6.38.

Figure 6.39 shows the dead-beat step response obtained with $T_{sd} = 0$ and adjusting h so that $u(t)$ just reaches saturation. The optimal value is $h = 17.7$ [s].

The sampling times are shown by the vertical dotted lines. Following the theory of Sect. 6.6.1.2, the system settles precisely in a number of iterations equal to the system order, which is five in this case. Remarkably, after the delay of three iterations, the transient is completed in just two further iterations, during which the behaviour is identical to that of the time-optimal control of Chap. 9, the control first saturating to impart the maximum angular acceleration until the spacecraft attitude reaches the halfway point at the end of interval four, when the control changes sign to the opposite saturation limit, imparting the maximum deceleration to bring the spacecraft to rest at the correct attitude by the end of interval five.

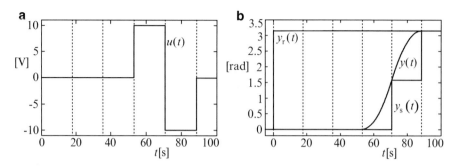

Fig. 6.39 Dead-beat slew without delay element, reaching control saturation. (**a**) Control input applied to reaction wheel. (**b**) Controlled output

The behaviour of the nonlinear time-optimal controller, however, is only mimicked by this linear controller. It must be realised that true time-optimal control from any initial state and for any constant reference input can only be attained by this controller for rest-to-rest manoeuvres and would require a new h value before every manoeuvre.

Comparing Figs. 6.39b with Fig. 6.38b shows that the dead-beat control *increases* the overall settling time but this is due to the initial delay of three iterations. This delay can be explained by considering the closed-loop transfer function, which, after the pole placement, is

$$\frac{Y(z)}{Y_r(z)} = \frac{\frac{1}{2}(z+1)}{z^5} = \frac{1}{2}\left(z^{-4} + z^{-5}\right). \tag{6.168}$$

The corresponding difference equation is

$$y(k+5) = \tfrac{1}{2}\left[y_r(k+1) + y_r(k)\right]. \tag{6.169}$$

Hence if $y_r(k) = AH(k) = \begin{cases} 0, & k < 0 \\ A, & k \geq 0 \end{cases}$ where A is the demanded attitude angle, then $k = -1$ gives $y(4) = A/2$ and $k = 0$ gives $y(5) = A$, agreeing with Fig. 6.39b.

This appears to be a delay of $4h$, which is true for the sampled and held plant output, $y_s(t)$, but the continuous plant output, $y(t)$, begins to respond at $t = 3h$.

The foregoing discussion has been useful in demonstrating the tolerance of arbitrarily large iteration intervals when employing the discrete control system design method using pole placement in the z-plane, but when applying this for dead-beat control in the spacecraft application, it is really unnecessary to include the delay element for the computational time allowance, as a modern digital processor could carry out the control computations in a fraction of a millisecond, which is negligible compared with 17.7 [s]. So it will be interesting to repeat the exercise

6.6 Discrete Control with Unlimited Iteration Intervals

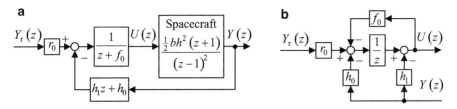

Fig. 6.40 Discrete polynomial spacecraft attitude control without delay element. (**a**) Control system block diagram. (**b**) Controller implementation

using a discrete polynomial controller without the delay element. In this case, the plant (6.160) replaces the augmented plant (6.161), whose z-transfer is

$$\frac{Y(z)}{U_c(z)} = \frac{\frac{1}{2}bh^2(z+1)}{(z-1)^2} = \frac{b_1 z + b_0}{z^2 + a_1 z + a_0}. \tag{6.170}$$

Then with $n_a = 2$, again assuming a minimal order controller, (6.162) yields $n_f = n_h = 1$. The resulting control system block diagram is shown in Fig. 6.40.

The total system order is now $N = n_a + n_f = 3$. From (6.170), $b_1 = \frac{1}{2}bh^2$, $b_0 = \frac{1}{2}bh^2$, $a_1 = -2$ and $a_0 = 1$. Then (6.157) becomes

$$\begin{bmatrix} 1 & 0 & 0 & 0 \\ a_1 & 1 & b_1 & 0 \\ a_0 & a_1 & b_0 & b_1 \\ 0 & a_0 & 0 & b_0 \end{bmatrix} \begin{bmatrix} f_1 \\ f_0 \\ h_1 \\ h_0 \end{bmatrix} = \begin{bmatrix} 1 \\ d_2 \\ d_1 \\ d_0 \end{bmatrix} \tag{6.171}$$

Again using the 2 % settling time formula, the equivalent continuous system pole is

$$s_c = -1.6(1.5 + N)/T_{sd} = -7.2/T_{sd}. \tag{6.172}$$

The corresponding z-plane pole, z_c, is again given by (6.165). With $q_c = -z_c$, the desired closed-loop characteristic equation is

$$\begin{aligned} z^3 + d_2 z^2 + d_1 z + d_0 &= (z + q_c)^3 = z^3 + 3q_c z^2 + 3q_c^2 z + q_c^3 \Rightarrow \\ d_2 &= 3q_c, \quad d_1 = 3q_c^2 \quad \text{and} \quad d_0 = q_c^3. \end{aligned} \tag{6.173}$$

For a closed-loop DC gain of unity, (6.158) yields

$$r_0 = [(1 + f_0)(1 + a_1 + a_0) + (b_1 + b_0)(h_1 + h_0)]/(b_1 + b_0) \tag{6.174}$$

The minimum settling time for which $u(t)$ just reaches one of the saturation limits of ± 10 [V] is $T_{sd} = 61.5$ [s], a slightly worse result than the system including the delay element, which was $T_{sd} = 61.5$ [s]. The simulation is shown in Fig. 6.41a.

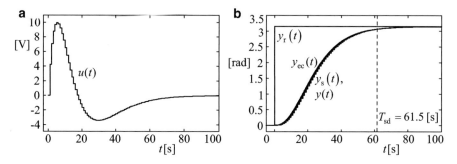

Fig. 6.41 Minimum settling time slew without delay element, reaching control saturation. (**a**) Control input applied to reaction wheel. (**b**) Controlled and e.c. system outputs

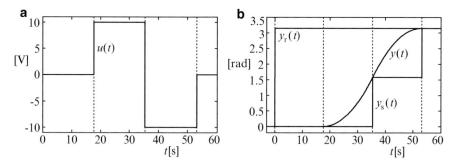

Fig. 6.42 Dead-beat slew without delay element, reaching control saturation. (**a**) Control input applied to reaction wheel. (**b**) Controlled output

The reasoning is as follows. For a given settling time, the peak acceleration of the step response, $y(t)$, reduces as the order of the control loop increases. This requires a larger peak in $u(t)$. It follows that for a given peak value of $u(t)$, the settling time reduces as the order of the control loop increases.

Next, the dead-beat response is again obtained by setting $T_{sd} = 0$ and h is adjusted so that $u(t)$ reaches a control saturation limit. Again $h = 17.7$ [s] and a behaviour similar to the time-optimal control of Chap. 9 results but delayed by only one iteration period as the system order has been reduced from five to three and the transient is therefore of duration, $3h$, as shown in Fig. 6.42.

6.6.4 Discrete Integral Polynomial Control

Next, the inclusion of integral terms in the controller will be considered to eliminate steady-state errors due to reference inputs and/or external disturbances (referred to the control input and denoted d_u) that are polynomial functions of time, constant reference inputs or external disturbances being particular cases. The resulting

6.6 Discrete Control with Unlimited Iteration Intervals

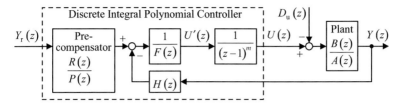

Fig. 6.43 Control system incorporating a polynomial integral controller

controller will be referred to as a discrete integral polynomial controller. A similar controller has already been formed in the continuous domain in Chap. 5, Sect. 5.3.5, and this entailed inserting a multiple integrator of order, m, in the control channel to cater for reference input and external disturbance polynomials not exceeding $m - 1$ in degree. This may be done in the discrete domain and the simplest approach is to insert m blocks with z-transfer function, $1/(z-1)$, which are discrete integrators each with a gain of $1/h$, that will be taken into account in the controller parameter determination. The control system block diagram is shown in Fig. 6.43.

The procedure for calculating the parameters of this controller are the same as for the basic polynomial controller. All that is needed is to regard the multiple integral block as part of an augmented version of the plant with control input, $U'(z)$.

Example 6.5 Discrete integral polynomial control of a combustion engine throttle valve

The plant is the same as in Example 6.2. The armature inductance of the DC motor is ignored, giving the following Laplace transfer function of the plant.

$$\frac{Y(s)}{U(s)} = \frac{K_{dc}}{(1 + sT_1)(1 + sT_2)}. \tag{6.175}$$

The plant parameters (courtesy, Delphi Diesel Systems Ltd) are $K_{dc} = 1.8136$, $T_1 = 0.5464$ [s] and $T_2 = 0.0266$ [s]. The discrete model will be determined for $h = 0.01$ [s]. This is compatible with the specified settling time, $T_s = 0.1$ [s] (5 % criterion). Referring to Table 3 in Tables yields

$$\frac{Y(z)}{U(z)} = \frac{K_{dc}}{b - a}\left(b\frac{1 - e^{-ah}}{z - e^{-ah}} - a\frac{1 - e^{-bh}}{z - e^{-bh}}\right), \tag{6.176}$$

where $a = 1/T_1$ and $b = 1/T_2$. A single integral term is required to avoid any steady-state errors due to variations in the retention spring constant brought about by temperature changes and ageing. Hence, the augmented plant transfer function is obtained by multiplying (6.176) by $1/(z-1)$. A further multiplication by $1/z$ will be made to allow for the computational delay. The resulting augmented plant transfer function also requires manipulation into the polynomial form for the controller parameter determination. Thus

$$\frac{Y(z)}{U_c(z)}$$
$$= \frac{K_{dc}}{b-a} \cdot \frac{[b(1-e^{-ah})-a(1-e^{-bh})]z + a(1-e^{-bh})e^{-ah} - b(1-e^{-ah})e^{-bh}}{z(z-1)(z-e^{-ah})(z-e^{-bh})}$$

$$= \frac{b_1 z + b_0}{z^4 + a_3 z^3 + a_2 z^2 + a_1 z + a_0} \tag{6.177}$$

where $b_1 = \frac{K_{dc}}{b-a}[b(1-e^{-ah}) - a(1-e^{-bh})]$, $b_0 = \frac{K_{dc}}{b-a}[a(1-e^{-bh})e^{-ah} - b(1-e^{-ah})e^{-bh}]$, $a_3 = -(1 + e^{-ah} + e^{-bh})$, $a_2 = e^{-ah} + e^{-bh} + e^{-(a+b)h}$, $a_1 = -e^{-(a+b)h}$ and $a_0 = 0$.

From (6.177), the augmented plant order is $n_a = 4$. Then, choosing the minimal order polynomial controller, (6.156) yields $n_f = n_h = n_a - 1 = 3$. The overall system order is $N = n_a + n_f = 7$, and therefore the equivalent continuous system pole, using the 5 % settling time formula with a demanded settling time of T_{sd}, is

$$s_c = -1.5(1+N)/T_{sd} = -12/T_{sd}. \tag{6.178}$$

The corresponding z-plane pole, z_c, is given by (6.165). With $q_c = -z_c$, the desired closed-loop characteristic polynomial is

$$z^7 + d_6 z^6 + d_5 z^5 + d_4 z^4 + d_3 z^3 + d_2 z^2 + d_1 z + d_0 = (z + q_c)^7$$
$$= z^7 + 7q_c z^6 + 21 q_c^2 z^5 + 35 q_c^3 z^4 + 35 q_c^4 z^3 + 21 q_c^5 z^2 + 7 q_c^6 z + q_c^7 \Rightarrow$$
$$d_6 = 7 q_c,\ d_5 = 21 q_c^2,\ d_4 = 35 q_c^3,\ d_3 = 35 q_c^4,\ d_2 = 21 q_c^5,\ d_1 = 7 q_c^6 \text{ and } d_0 = q_c^7. \tag{6.179}$$

The matrix–vector equation for the controller parameters is

$$\begin{bmatrix} 1 & 0 & 0 & 0 & 0 & 0 & 0 & 0 \\ a_3 & 1 & 0 & 0 & 0 & 0 & 0 & 0 \\ a_2 & a_3 & 1 & 0 & 0 & 0 & 0 & 0 \\ a_1 & a_2 & a_3 & 1 & b_1 & 0 & 0 & 0 \\ a_0 & a_1 & a_2 & a_3 & b_0 & b_1 & 0 & 0 \\ 0 & a_0 & a_1 & a_2 & 0 & b_0 & b_1 & 0 \\ 0 & 0 & a_0 & a_1 & 0 & 0 & b_0 & b_1 \\ 0 & 0 & 0 & a_0 & 0 & 0 & 0 & b_0 \end{bmatrix} \begin{bmatrix} f_3 \\ f_2 \\ f_1 \\ f_0 \\ h_3 \\ h_2 \\ h_1 \\ h_0 \end{bmatrix} = \begin{bmatrix} 1 \\ d_6 \\ d_5 \\ d_4 \\ d_3 \\ d_2 \\ d_1 \\ d_0 \end{bmatrix}. \tag{6.180}$$

The control system block diagram is shown in Fig. 6.44a. The pre-compensator consists of only the reference input scaling coefficient, r_0, in this case. Then setting $z = 1$ in the closed-loop transfer function derived using Fig. 6.44a yields the following expression for the reference input scaling coefficient:

$$r_0 = \frac{(1 + f_2 + f_1 + f_0)(1 + a_3 + a_2 + a_1 + a_0) + (b_1 + b_0)(h_3 + h_2 + h_1 + h_0)}{b_1 + b_0}. \tag{6.181}$$

6.6 Discrete Control with Unlimited Iteration Intervals

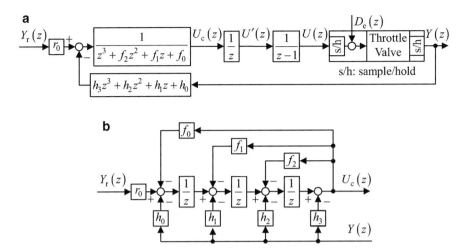

Fig. 6.44 Discrete integral polynomial control of throttle valve. (**a**) Control system block diagram. (**b**) Implementation block diagram of controller

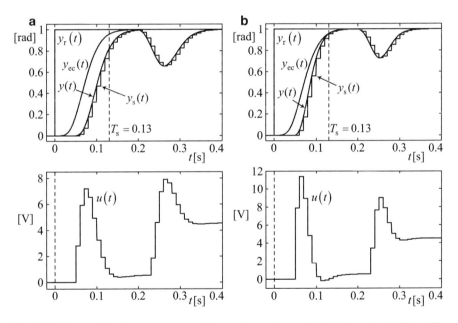

Fig. 6.45 Step and disturbance response of discrete integral control of throttle valve. (**a**) $T_{sd} = T_s$. (**b**) $T_{sd} = T_s - 3h$

The external disturbance, $D_e(z)$, has been included to demonstrate the correct operation of the integral term.

Figure 6.45a shows a step response for $T_{sd} = T_s = 0.13$ [s] with a step disturbance, $d_e(t) = 4h(t - 0.2)$ [V], which is one third of the control saturation limit. In this case, the delay of $n_f h = 3 \times 0.01 = 0.03$ [s] referred to at the end of Sect. 6.6.3.2 is a significant proportion of T_s giving, according to (6.159), an actual settling time of $T_{sa} = T_s + 3h = 0.16$ [s].

The delay of 0.03 [s] between the equivalent continuous system response, $y_{ec}(t)$, and $y(t)$ is clearly visible. The actual settling time, however, is made equal to the specified value in (b) by using (6.159) setting $T_{sd} = T_s - 3h$. The transient dip in the position response caused by the step disturbance is reduced slightly in Fig. 6.45b due to the increase in the controller gains brought about by reducing T_{sd}. The action of the integral term in ensuring zero steady-state error is evident in both runs.

6.6.5 Control of Plants Containing Pure Time Delays

Pure time delays, sometimes called transport delays or dead times, arise in plants through physical hardware limitations that fall into one of two basic categories. These are described in the terms of an LTI plant model as follows.

(a) Sensor delay: The measurement which, for a linear system, is ideally $y(t) = \mathbf{c}^T\mathbf{x}(t)$ is instead $y(t) = \mathbf{c}^T\mathbf{x}(t - \tau_d)$, where $\mathbf{x}$ is the plant state, $\mathbf{c}^T$ is the measurement matrix and τ_d is the sensor delay.
(b) Actuator delay: The effect of the control variable, $u(t)$, on the plant state which, for an LTI system, ideally obeys $\dot{\mathbf{x}}(t) = \mathbf{A}\mathbf{x}(t) + \mathbf{b}u(t)$, where $\mathbf{A}$ is the plant matrix and $\mathbf{b}$ is the input matrix, but instead obeys $\dot{\mathbf{x}}(t) = \mathbf{A}\mathbf{x}(t) + \mathbf{b}u(t - \tau_d)$, where τ_d is the actuator delay.

An example of category (a) is a steel rolling mill in which the roller force is varied to regulate the thickness of the emerging strip of steel. Physical restrictions prevent the thickness measurement transducer from being mounted at the same location as the roller contact line on the strip. Instead it has to be mounted a distance, D, from the contact line, resulting in a measurement time delay of $\tau_d = D/v$, where v is the velocity of the emerging strip. An example of category (b) is a hot water cylinder containing a heat exchanger. The cylinder has to be situated remotely from the controlled boiler. Assuming that the water in the heat exchange circuit is maintained at a velocity, v, in the pipework by a circulating pump and the distance the water has to travel from the controlled heat source to the heat exchanger is D, then changes of temperature at the boiler flow pipe are delayed by $\tau_d = D/v$ before reaching the heat exchanger.

Based on the well-known Laplace transform relationship,

$$\mathcal{L}\{f(t - \tau_d)\} = e^{-s\tau_d}\mathcal{L}\{f(t)\} \tag{6.182}$$

a SISO LTI plant with a pure time delay can be modelled by the transfer function,

$$\frac{Y(s)}{U(s)} = G_p(s)e^{-s\tau_d}, \tag{6.183}$$

6.6 Discrete Control with Unlimited Iteration Intervals

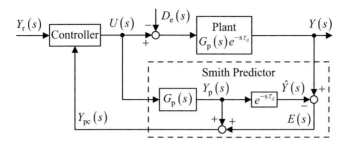

Fig. 6.46 Control of a plant with a pure time delay aided by Smith predictor

where $G_p(s)$ is the plant transfer function that would model the plant in absence of the pure time delay and $e^{-s\tau_d}$ is the transfer function of the pure time delay. It is understandable that if a feedback control system was designed ignoring the time delay, i.e. assuming the plant transfer function to be just $G_p(s)$, then the closed-loop system dynamics with plant transfer function (6.183) would deteriorate with respect to that predicted without the time delay. In the extreme, time delays can cause instability if ignored. The theoretical difficulty arising is that transfer function (6.183) has an infinite number of poles and therefore renders the plant of infinite order. Historically, the problem has been tackled starting with a finite-order approximation such as the rth-order Padé approximation,

$$e^{-s\tau_d} = \frac{e^{-s\tau_d/2}}{e^{s\tau_d/2}} \cong \frac{1 - s\frac{\tau_d}{2} + \frac{1}{2!}s^2\left(\frac{\tau_d}{2}\right)^2 + \cdots + \frac{1}{r!}s^r\left(\frac{\tau_d}{2}\right)^r}{1 + s\frac{\tau_d}{2} + \frac{1}{2!}s^2\left(\frac{\tau_d}{2}\right)^2 + \cdots + (-1)^r\frac{1}{r!}s^r\left(\frac{\tau_d}{2}\right)^r}, \quad (6.184)$$

and then the control system design can proceed with the resulting finite-order plant model using any desired control technique, r being chosen such that, in the frequency domain, the magnitude and phase of $G_d(j\omega)$ are a good approximation to the magnitude and phase of $e^{-j\omega\tau_d}$ over the intended bandwidth of the control system. This, however, can lead to a complicated, high-order solution, still with a degree of uncertainty due to the approximation. O J M Smith [4] overcame this with a different approach which could be regarded as a forerunner of model predictive control [5]. This control scheme is shown in Fig. 6.46.

A real-time model of the non-delay part of the plant is used to produce a prediction, $Y_p(s)$, of the plant output that will occur in real time as a measurement τ_d seconds ahead in time. Then $Y_p(s)$ is fed back to the controller instead of $Y(s)$ so that the controller design can be based on $G_p(s)$ alone. In addition, the time delay is modelled to produce an estimate, $\hat{Y}(s)$, of $Y(s)$ that is used to form an error, $E(s) = Y(s) - \hat{Y}(s)$ that is added to the feedback signal, $Y_p(s)$, to form a predicted and corrected output, $Y_{pc}(s)$. With a perfect plant model and zero external disturbance, then $E(s) = 0$ and the outer loop of Fig. 6.46 would have no effect but it is important to note that in real time the plant output, $y(s)$, is delayed by τ_d [s] relative to the directly controlled $y_p(s)$. With a disturbance, $D_e(s)$, referred to the plant input, however, $E(s) \neq 0$, so the outer loop is present to reduce the control error, $Y(s) - Y_r(s)$, due to the disturbance and also to attempt compensation for plant modelling errors. The method, however, can be sensitive to external disturbances and modelling errors, particularly the delay time [6]. A discrete version of the Smith predictor will be demonstrated in the following example.

In the discrete domain, applying the transformation, $z = e^{sh}$, to the transfer function, $e^{s\tau_d}$, of the pure time delay, yields

$$e^{s\tau_d} = e^{sh(\tau_d/h)} = z^{\tau_d/h} \tag{6.185}$$

as the z-transfer function of a pure time delay of τ_d seconds. Then if h is chosen so that τ_d/h is an integer, q, which should be feasible, the z-transfer function of the pure time delay becomes

$$Q_d(z) = z^{-q}. \tag{6.186}$$

This renders the control system design in the discrete domain more tractable than in the continuous domain, as the pure time delay contributes only a finite amount, q, to the order of the plant model upon which the control system design is based. An approach similar to but not identical to the Smith predictor will be taken using the state observer in Chap. 8 but in the discrete domain.

Example 6.6 Discrete polynomial control of a heat exchanger aided by a Smith predictor

A liquid is heated by passing it at a constant flow rate through a heat exchange coil mounted within a steam jacket. The liquid temperature is measured at a point remote from the heat exchanger and is controlled by varying the steam flow rate. The Laplace transfer function of the heat exchanger is given as

$$\frac{Y(s)}{U(s)} = \frac{K_{dc} e^{-s\tau_d}}{(1 + sT_1)(1 + sT_2)} \tag{6.187}$$

where $U(s)$ varies the steam flow rate, $Y(s)$ is the temperature measurement, $K_{dc} = 1$ is the plant DC gain, $T_1 = 30$ [s] is the time constant associated with the steam jacket, $T_2 = 10$ [s] is the time constant associated with the heat exchange coil and τ_d is the pure time delay due to the distance between the heat exchanger and the temperature measurement point with an estimate of $\tilde{\tau}_d = 5$ [s].

A discrete integral polynomial controller with an additional delay for computational allowance, aided by a discrete Smith predictor, will now be designed to a specification of zero overshoot and a settling time of $T_s = 40$ [s]. The sensitivity with respect to errors in the estimate, $\tilde{\tau}_d$, of τ_d will be assessed.

First, the second-order part of the plant model with Laplace transfer function

$$\frac{Y_p(s)}{U(s)} = \frac{K_{dc}}{(1 + sT_1)(1 + sT_2)} \tag{6.188}$$

6.6 Discrete Control with Unlimited Iteration Intervals

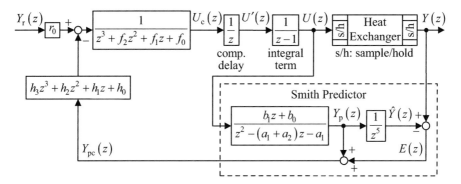

Fig. 6.47 Control of a heat exchanger with a pure time delay aided by a Smith predictor

will be converted to a discrete model with iteration period, h. Then the integral term and computational delay element will be included. This results in an augmented model of precisely the same form as in Example 6.5 but, of course, different values of the coefficients. Thus

$$\frac{Y_\mathrm{p}(z)}{U_\mathrm{c}(z)} = G_\mathrm{p}(z) = \frac{b_1 z + b_0}{z^4 + a_3 z^3 + a_2 z^2 + a_1 z + a_0} \qquad (6.189)$$

where the coefficients are given by the expressions of Example 6.5. This is the transfer function used in the Smith predictor of Fig. 6.47. This will be assumed to have no parametric estimation errors in this example. The iteration period will be chosen as $h = 1$ [s] so $q = \tilde{\tau}_\mathrm{d}/h = 5$. The pure time delay in the Smith predictor therefore has a z-transfer function of $1/z^5$.

The design of the polynomial integral controller with the computational delay allowance will be as in Example 6.5 except for the demanded settling time. There are two components of the output response delay. The first is the delay of $n_\mathrm{f} h$ due to the controller accounted for in (6.159) and the second is the pure time delay, τ_d, itself. In this example, the estimated time delay is $\tau_\mathrm{d} = 5h$. Hence the demanded settling time is $T_\mathrm{sd} = T_\mathrm{s} - n_\mathrm{f} h - \widehat{\tau}_\mathrm{d} = 40 - 3 * 1 - 5 * 1 = 32$ [s]. Figure 6.48a shows the response, with zero initial conditions, to step reference input of half the maximum temperature limit of 5 [V].

This complies with the performance specification. Figure 6.48b, c show, respectively, the responses obtained with errors in the pure time delay estimate of $+15$ % and -15 %. These mismatches both result in oscillations about the set point but with a small degree of damping, rendering the system stable. Increasing the mismatches to about $+30$ % and -40 %, however, results in undamped oscillations of constant amplitude. Further increases of the mismatches result in oscillatory instability. Theoretical stability analysis of such systems [6] is a specialist research topic and is not straightforward as arbitrary mismatches create time delays with transfer functions of the form of (6.185) which render the system of infinite order for τ_d/h non-integer.

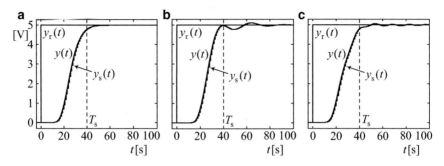

Fig. 6.48 Step responses of Smith predictor-aided discrete polynomial control system. (a) $\tau_d = \tilde{\tau}_d$ (b) $\tau_d = 1.15\tilde{\tau}_d$ (c) $\tau_d = 0.85\tilde{\tau}_d$

Creating new control techniques that yield reduced sensitivity to uncertainties or variations in the pure time delays remains a challenge. One might be tempted to model the time delay, $1/z^5$, in the above example, along with the non-delay part of the plant model, resulting in a single z-transfer function model for the plant and then design a polynomial controller. This would appear an elegant solution and the author has already attempted a simulation for the above example. The result is perfect for $\tau_d = \hat{\tau}_d$ but mismatches as little as $\pm 0.02\%$ cause instability. This approach therefore cannot be recommended due to the extreme sensitivity.

On a positive note, there are many possibilities for *adaptive control* in which an algorithm for on-line estimation of the dead time using observations of the control input and real plant output could be created. An interesting approach would be to create a discrete plant model of finite order but variable dead-time estimate, $\hat{\tau}_d$, by varying the iteration period, h, continuously such that $\hat{\tau}_d/h = q$, where q is a constant integer. The algorithm would then automatically vary h, and hence $\hat{\tau}_d$, until a real-time plant model output tracks the real plant output when a common control input is applied. Effective control would then be obtained using a controller whose parameters adapt to $\hat{\tau}_d(t)$. The challenge is to produce a stable and convergent adaptation algorithm.

References

1. Irving R (2013) Beyond the quadratic formula. The Mathematical Association of America, Washington, DC. ISBN 978-0-88365-783-0
2. Fadali MS, Visioli FA (2013) Digital control engineering: analysis and design. Academic, Amsterdam/Boston. ISBN 978-0-12-394391-0
3. Landau ID, Zito G (2006) Digital control systems, design, identification and implementation. Springer-Verlag London
4. Smith OJM (1959) A controller to overcome dead-time. ISA Trans 6(2):28–33
5. Maciejowski JM (2002) Predictive control: with constraints. Pearson Education, London. ISBN 0-201-39823-0
6. Loiseau JJ et al (2009) Topics in time delay system: analysis, algorithms and control, Lecture notes in control and information sciences. Springer, Berlin. ISBN 978-3-642-02897-7

Chapter 7
Model Based Control of Nonlinear and Linear Plants

7.1 Introduction

During the previous era of analogue circuit implementation of controllers, the realisation of nonlinear functions to deal with nonlinear plants was discouraged due to increased hardware complexity and the resulting increase in costs and decrease in reliability. This led to the traditional approach of linearisation about an operating point for obtaining an approximate linear plant model upon which a traditional linear controller design could be based. Since many control systems of this type are established in the industry, this technique is presented in Sect. 7.2. It suffers, however, from the restriction of the plant state having to be kept in a close neighbourhood of the operating point.

The advent of digital processors for control applications enabled analogue controllers to be replaced with control algorithms that can contain nonlinear terms free of implementation issues. This led to the *feedback linearisation* method in which a nonlinear plant is transformed to a linear plant by means of a nonlinear state feedback control law with an external input that becomes the new control input. Thus, the linear model obtained by linearisation about the operating point is replaced by an augmented *physical* plant that can also be controlled by a traditional linear controller. Most importantly, the operational restriction of the method of linearisation about an operating point is removed by feedback linearisation.

Since feedback linearisation is now becoming established in industry, it is described at the beginning of Sect. 7.3. This, however, is very closely related to *feedback linearising control* (FLC), which goes one step further by implementing a controller directly with all the flexibility allowed by digital implementation, avoiding the need to employ an additional traditional controller. The main emphasis is therefore on FLC, the study of which will still enable the reader to do feedback linearisation if preferred.

It is sometimes advantageous for the specified closed-loop dynamics to be nonlinear. A simple extension of FLC can achieve this for both linear and nonlinear plants. Also linear state feedback control laws for linear plants based on models with specific state representations can be derived more quickly via FLC than using the traditional methods of Chap. 4. In both these cases, however, the term, feedback linearising control, is inappropriate. Instead the more general term, *forced dynamic control* [1], is used here as, in every case, the state feedback control law can be regarded as forcing the closed-loop system to have the specified dynamics.

The control techniques falling under the general category of forced dynamic control may be applied to multivariable plants in a straightforward manner, and for this reason, Sect. 7.3 includes a subsection introducing multivariable control. These control techniques can also be formulated in the discrete domain if a discrete plant model is available. This topic is also covered.

7.2 Linearisation About an Operating Point

7.2.1 Basic Principle

Before presenting the method for developing a linear plant model commencing with a nonlinear plant model, linearisation of the model of a non-dynamic nonlinear plant element will be considered alone as a simple introduction. The example taken is the square-law transfer characteristic between the current, i_m, of a series-connected DC motor and the torque, γ_m, described by the equation

$$\gamma_m = K_t i_m^2, \tag{7.1}$$

where K_t is the torque constant. It is assumed that the motor is used in a position control application. Suppose that the motor shaft is subject to a load torque of Γ_L. Then, taking the sign convention of electric drives, the net torque applied to the mechanical load is

$$\gamma_n = \gamma_m - \gamma_L. \tag{7.2}$$

If the mechanical load is stationary in the steady state, then $\gamma_n = 0$, requiring that the operating value of the motor torque is

$$\gamma_m = \gamma_L \tag{7.3}$$

If, in addition, γ_L is constant, then the operating value of the motor torque is

$$\overline{\gamma}_m = \gamma_L. \tag{7.4}$$

In view of (7.1), the corresponding constant operating value of the motor current is

7.2 Linearisation About an Operating Point

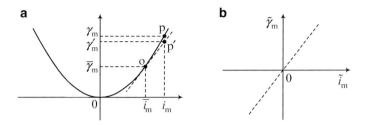

Fig. 7.1 Illustration of the linearisation process. (**a**) Transfer characteristic of nonlinear element. (**b**) Transfer characteristic of linear model

$$\bar{i}_m = \frac{1}{\sqrt{K_t}} \sqrt{\bar{\gamma}_m}, \quad (7.5)$$

the positive root being taken arbitrarily, assuming the current is unidirectional. The *operating point* is defined as the point, $(\bar{i}_m, \bar{\gamma}_m)$, on the graph of γ_m against i_m, which is point, o, in Fig. 7.1a. During normal operation of the control system, $\gamma_m(t)$ and $i_m(t)$ vary while the point, p, moves on the transfer characteristic.

The straight line tangent, op′, to the operating point is the transfer characteristic of the linear model shown in Fig. 7.1b given by

$$\tilde{\gamma}_m = K'_m \tilde{i}_m, \quad (7.6)$$

where

$$\tilde{\gamma}_m = \gamma'_m - \bar{\gamma}_m, \quad (7.7)$$

$$\tilde{i}_m = i_m - \bar{i}_m \quad (7.8)$$

and

$$K'_m = \frac{d}{di_m} \gamma_m \bigg|_{i_m = \bar{i}_m} = \frac{d}{di_m} K_m i_m^2 \bigg|_{i_m = \bar{i}_m} = 2 K_m \bar{i}_m. \quad (7.9)$$

It is important to realise that this linearised model has the origin at $(\tilde{i}_m, \tilde{\gamma}_m) = (0, 0)$, since it is based on *changes* of the variables, i_m and γ_m, about the operating point. Hence, when the control system design is first validated by simulation using the linearised model, the operating point values must be added to the variables.

It is clear from Fig. 7.1 that as $|i_m - \bar{i}_m|$ increases from zero, the points, p and p′, become further apart, indicating that the linear model becomes less accurate. This demonstrates the basic limitation of the method.

7.2.2 Linear State-Space Model

Now the general theory of linearisation about an operating point will be presented. This follows the same approach as in Sect. 7.2.1. The starting point is the general continuous-time state-space plant model

$$\begin{aligned} \dot{\mathbf{x}} &= \mathbf{F}(\mathbf{x}, \mathbf{u}) \\ \mathbf{y} &= \mathbf{H}(\mathbf{x}) \end{aligned} \quad (7.10)$$

where the right-hand sides of both equations, together with their derivatives, are continuous functions. The external disturbance input is absent for simplicity, since it can be easily incorporated in specific examples, particularly if it is referred to the control input.

It is assumed that a steady state is reached with a constant control input vector, $\mathbf{u} = \bar{\mathbf{u}}$, and this defines the required operating point, which is the complete set of nominally constant state variables, control variables and measurement variables, $(\bar{\mathbf{x}}, \bar{\mathbf{u}}, \bar{\mathbf{y}})$, when the closed-loop system is operating in the steady state. This must satisfy (7.10) with $\dot{\mathbf{x}} = \mathbf{0}$. Thus,

$$\mathbf{F}(\bar{\mathbf{x}}, \bar{\mathbf{u}}) = \mathbf{0}, \quad (7.11)$$

$$\bar{\mathbf{y}} = \mathbf{H}(\bar{\mathbf{x}}). \quad (7.12)$$

The nominally constant desired state, $\bar{\mathbf{x}}$, of the closed-loop system is first determined. Then (7.11) is solved for $\bar{\mathbf{u}}$ and (7.12) gives $\bar{\mathbf{y}}$.

The plant DC gain, $\mathbf{K}_{\text{DC}}(\bar{\mathbf{u}}) \in \Re^{n \times m}$, is defined by the equation

$$\bar{\mathbf{y}} = \mathbf{K}_{\text{DC}}(\bar{\mathbf{u}})\,\bar{\mathbf{u}}. \quad (7.13)$$

For a linear plant, this is constant. It should be noted that in plants containing pure integrators, some or all of the components of $\bar{\mathbf{u}}$ will be zero, since the inputs of these integrators must be zero for their outputs to be constant. Consequently some or all of the DC gain elements may be infinite. An extreme example is a three-axis stabilised spacecraft with nominally rigid-body dynamics whose model contains a chain of two pure integrators connected to each of the three control inputs.

As the first step in the formation of the linear state-space model from (7.10), let $\dot{\mathbf{x}} = \mathbf{v}$. Then (7.10) becomes

$$\begin{aligned} \mathbf{v} &= \mathbf{F}(\mathbf{x}, \mathbf{u}) \\ \mathbf{y} &= \mathbf{H}(\mathbf{x}) \Rightarrow \\ v_i &= f_i(x_1, x_2, \ldots, x_n, u_1, u_2, \ldots, u_r), \quad i = 1, 2, \ldots, n \\ y_j &= h_j(x_1, x_2, \ldots, x_n), \quad j = 1, 2, \ldots, m \end{aligned} \quad (7.14)$$

7.2 Linearisation About an Operating Point

and small changes of all the variables about the operating point are given by the total differentials

$$\begin{aligned}\delta v_i &= \left.\frac{\partial f_i}{\partial x_1}\right|_{**}\delta x_1 + \left.\frac{\partial f_i}{\partial x_2}\right|_{**}\delta x_2 + \ldots \left.\frac{\partial f_i}{\partial x_n}\right|_{**}\delta x_n + \\ &\quad + \left.\frac{\partial f_i}{\partial u_1}\right|_{**}\delta u_1 + \left.\frac{\partial f_i}{\partial u_2}\right|_{**}\delta u_2 + \ldots \left.\frac{\partial f_i}{\partial u_r}\right|_{**}\delta u_r, \quad i = 1, 2, \ldots, n, \\ \delta y_j &= \left.\frac{\partial h_j}{\partial x_1}\right|_{*}\delta x_1 + \left.\frac{\partial h_j}{\partial x_2}\right|_{*}\delta x_2 + \ldots \left.\frac{\partial h_j}{\partial x_n}\right|_{*}\delta x_n, \quad j = 1, 2, \ldots, m \\ &** \ (\mathbf{x}, \mathbf{u}) = (\overline{\mathbf{x}}, \overline{\mathbf{u}}), \quad * \ \mathbf{x} = \overline{\mathbf{x}}\end{aligned} \quad (7.15)$$

Sometimes (7.15) is written in the following matrix form:

$$\begin{bmatrix}\delta v_1 \\ \delta v_2 \\ \vdots \\ \delta v_n\end{bmatrix} = \begin{bmatrix}\frac{\partial f_1}{\partial x_1} & \frac{\partial f_1}{\partial x_2} & \cdots & \frac{\partial f_1}{\partial x_n} \\ \frac{\partial f_2}{\partial x_1} & \frac{\partial f_2}{\partial x_2} & & \frac{\partial f_2}{\partial x_n} \\ \vdots & & \ddots & \\ \frac{\partial f_n}{\partial x_1} & \frac{\partial f_n}{\partial x_2} & & \frac{\partial f_n}{\partial x_n}\end{bmatrix}_{**}\begin{bmatrix}\delta x_1 \\ \delta x_2 \\ \vdots \\ \delta x_n\end{bmatrix} + \begin{bmatrix}\frac{\partial f_1}{\partial u_1} & \frac{\partial f_1}{\partial u_2} & \cdots & \frac{\partial f_1}{\partial u_r} \\ \frac{\partial f_2}{\partial u_1} & \frac{\partial f_2}{\partial u_2} & & \frac{\partial f_2}{\partial u_r} \\ \vdots & & \ddots & \\ \frac{\partial f_n}{\partial u_1} & \frac{\partial f_n}{\partial u_2} & & \frac{\partial f_n}{\partial u_r}\end{bmatrix}_{**}\begin{bmatrix}\delta u_1 \\ \delta u_2 \\ \vdots \\ \delta u_r\end{bmatrix}$$

$$\begin{bmatrix}\delta y_1 \\ \delta y_2 \\ \vdots \\ \delta y_m\end{bmatrix} = \begin{bmatrix}\frac{\partial h_1}{\partial x_1} & \frac{\partial h_1}{\partial x_2} & \cdots & \frac{\partial h_1}{\partial x_n} \\ \frac{\partial h_2}{\partial x_1} & \frac{\partial h_2}{\partial x_2} & & \frac{\partial h_2}{\partial x_n} \\ \vdots & & \ddots & \\ \frac{\partial h_m}{\partial x_1} & \frac{\partial h_m}{\partial x_2} & & \frac{\partial h_m}{\partial x_n}\end{bmatrix}_{*}\begin{bmatrix}\delta x_1 \\ \delta x_2 \\ \vdots \\ \delta x_n\end{bmatrix} \quad \begin{array}{l}** \ (\mathbf{x}, \mathbf{u}) = (\overline{\mathbf{x}}, \overline{\mathbf{u}}) \\ * \ \mathbf{x} = \overline{\mathbf{x}}\end{array} \quad (7.16)$$

This may be written in a short-hand notation as

$$\begin{aligned}\delta \mathbf{v} &= \left.\frac{\partial \mathbf{f}}{\partial \mathbf{x}}\right|_{**}\delta \mathbf{x} + \left.\frac{\partial \mathbf{f}}{\partial \mathbf{u}}\right|_{**}\delta \mathbf{u} \\ \delta \mathbf{y} &= \left.\frac{\partial \mathbf{h}}{\partial \mathbf{x}}\right|_{*}\delta \mathbf{x}\end{aligned} \quad \begin{array}{l}** \ (\mathbf{x}, \mathbf{u}) = (\overline{\mathbf{x}}, \overline{\mathbf{u}}) \\ * \ \mathbf{x} = \overline{\mathbf{x}}\end{array}, \quad (7.17)$$

where $\frac{\partial \mathbf{v}}{\partial \mathbf{x}}$, $\frac{\partial \mathbf{v}}{\partial \mathbf{u}}$ and $\frac{\partial \mathbf{y}}{\partial \mathbf{x}}$ are called the *Jacobean matrices* or simply the *Jacobeans*.

Next, the notation will be changed to follow that of Sect. 7.2.1. Thus, let $\delta v_i = \tilde{v}_i$, $\delta x_i = \tilde{x}_i$, $i = 1, 2, \ldots, n$, $\delta u_j = \tilde{u}_j$, $j = 1, 2, \ldots, r$, $\delta y_k = \tilde{y}_k$, $k = 1, 2, \ldots, m$. To simplify the notation further, let

$$\begin{aligned}\left.\frac{\partial f_i}{\partial x_j}\right|_{**} &= \overline{a}_{ij}, & i &= 1, 2, \ldots, n, & & \\ \left.\frac{\partial f_i}{\partial u_j}\right|_{**} &= \overline{b}_{ik}, & j &= 1, 2, \ldots, n, & ** \ (\mathbf{x}, \mathbf{u}) &= (\overline{\mathbf{x}}, \overline{\mathbf{u}}) \\ & & k &= 1, 2, \ldots, r, & * \ \mathbf{x} &= \overline{\mathbf{x}} \\ \left.\frac{\partial h_l}{\partial x_i}\right|_{*} &= \overline{c}_{li}, & l &= 1, 2, \ldots, m, & &\end{aligned} \quad (7.18)$$

Finally, $\tilde{x}_i \cong x_i - \overline{x}_i$ in a close neighbourhood of the operating point then, approximately, $\dot{\tilde{x}}_i = \dot{x}_i = v_i$, $i = 1, 2, \ldots, n$. Then (7.16) becomes

$$\begin{bmatrix} \dot{\tilde{x}}_1 \\ \dot{\tilde{x}}_2 \\ \vdots \\ \dot{\tilde{x}}_n \end{bmatrix} = \begin{bmatrix} \bar{a}_{11} & \bar{a}_{12} & \cdots & \bar{a}_{1n} \\ \bar{a}_{21} & \bar{a}_{22} & & \bar{a}_{2n} \\ \vdots & & \ddots & \vdots \\ \bar{a}_{n1} & \bar{a}_{n2} & \cdots & \bar{a}_{nn} \end{bmatrix} \begin{bmatrix} \tilde{x}_1 \\ \tilde{x}_2 \\ \vdots \\ \tilde{x}_n \end{bmatrix} + \begin{bmatrix} \bar{b}_{11} & \bar{b}_{12} & \cdots & \bar{b}_{1r} \\ \bar{b}_{21} & \bar{b}_{22} & & \bar{b}_{2r} \\ \vdots & & & \vdots \\ \bar{b}_{n1} & \bar{b}_{n2} & \cdots & \bar{b}_{nr} \end{bmatrix} \begin{bmatrix} \tilde{u}_1 \\ \tilde{u}_2 \\ \vdots \\ \tilde{u}_r \end{bmatrix},$$

$$\begin{bmatrix} \tilde{y}_1 \\ \tilde{y}_2 \\ \vdots \\ \tilde{y}_m \end{bmatrix} + \begin{bmatrix} \bar{c}_{11} & \bar{c}_{12} & \cdots & \bar{c}_{1n} \\ \bar{c}_{21} & \bar{c}_{22} & & \bar{c}_{2n} \\ \vdots & & & \vdots \\ \bar{c}_{m1} & \bar{c}_{m2} & \cdots & \bar{c}_{mn} \end{bmatrix} \begin{bmatrix} \tilde{x}_1 \\ \tilde{x}_2 \\ \vdots \\ \tilde{x}_n \end{bmatrix}$$

(7.19)

which may be written in the matrix–vector notation as

$$\dot{\tilde{\mathbf{x}}} = \bar{\mathbf{A}}\tilde{\mathbf{x}} + \bar{\mathbf{B}}\tilde{\mathbf{u}} \\ \tilde{\mathbf{y}} = \bar{\mathbf{C}}\tilde{\mathbf{x}}$$

(7.20)

This is the required linear state-space model of the nonlinear plant, in which the matrices, $\bar{\mathbf{A}}$, $\bar{\mathbf{B}}$ and $\bar{\mathbf{C}}$, are the Jacobian matrices of (7.17). It should be born in mind that the vectors, $\tilde{\mathbf{x}}$, $\tilde{\mathbf{u}}$ and $\tilde{\mathbf{y}}$, do not approximate $\mathbf{x}$, $\mathbf{u}$ and $\mathbf{y}$ but approximate *changes* of the variables with respect to the fixed operating point values. It follows that the DC gain of the linearised model may be quite different from the DC gain of the nonlinear plant. Hence, *it is usual to include an integral term in the linear controller to ensure zero steady-state error* with a constant reference input. This is needed in any case to maintain zero steady-state error with constant external disturbances.

For the reasons explained in Sect. 7.2.1, (7.20) will only accurately replicate relatively small changes in the variables of the nonlinear plant model with respect to the operating point. This implies that a closed-loop system using a controller based on the linear plant model will yield the specified closed-loop dynamics only if the system variables remain close to the operating point. Otherwise the transient behaviour will not follow a specified dynamics accurately.

The reader is advised to check the correctness of the controller design by simulating the step response of the closed-loop system comprising the controller applied to the linearised model. This should be followed by a comparison of two closed-loop system step response simulations of, one with the nonlinear model and the other with the linearised model, both commencing at the operating point. This should be repeated with increased reference input magnitudes to enable the degree of deviation of the nonlinear system response from the ideal linear one to be determined.

Example 7.1 Series wound DC motor lifting load on boom crane

A complete description of this SISO plant together with definitions of its constant parameters and variables is given in Example 2.7, Sect. 2.4.1. The plant state differential equations are

7.2 Linearisation About an Operating Point

$$\begin{aligned}\dot{x}_1 &= f_1(x_2) = x_2 \\ \dot{x}_2 &= f_2(x_3) = \tfrac{1}{J}\left(K_t x_3^2 - Mgr_p\right) \\ \dot{x}_3 &= f_3(x_2, x_3, u) = \tfrac{1}{L}\left(K_{pe}u - Rx_3 - K_b x_2 x_3\right)\end{aligned} \quad (7.21)$$

The measurement equation $y = x_1$ is not included in this case because it is assumed that y is scaled in the computer software to be numerically equal to x_1.

One nonlinear term in (7.21) is the motor torque, $K_t x_3^2$, where x_3 is the current, already described in Sect. 7.2.1. The other nonlinear term, $K_b x_2 x_3$, is the back e.m.f., where x_2 is the motor shaft angular velocity. The task is to form a linear state-space model with an operating point defined by a constant value of the mechanical load position, $\bar{x}_1$, and then design a linear state feedback control law to yield a settling time of T_s with zero overshoot using the 5 % settling time formula.

In this case, the operating point is $(\bar{x}_1, \bar{x}_2, \bar{x}_3, \bar{u})$. From (7.21), the operating point coordinates satisfy

$$\bar{x}_2 = 0, \qquad (7.22)$$

$$K_t \bar{x}_3^2 - Mgr_p = 0, \qquad (7.23)$$

$$K_{pe}\bar{u} - R\bar{x}_3 - K_b \bar{x}_2 \bar{x}_3 = 0 \qquad (7.24)$$

Equation (7.22) gives $\bar{x}_2$ directly, (7.23) gives

$$\bar{x}_3 = \sqrt{\frac{Mgr_p}{K_t}} \qquad (7.25)$$

and finally (7.24), (7.22) and (7.25) give

$$\bar{u} = \frac{R}{K_{pe}}\sqrt{\frac{Mgr_p}{K_t}}. \qquad (7.26)$$

The operating point is therefore

$$(\bar{x}_1, \bar{x}_2, \bar{x}_3, \bar{u}) = \left(\bar{x}_1,\ 0,\ \sqrt{\frac{Mgr_p}{K_t}},\ \frac{R}{K_{pe}}\sqrt{\frac{Mgr_p}{K_t}}\right). \qquad (7.27)$$

It should be noted that the operating point value, $\bar{x}_1$, is arbitrary because, ignoring the mass of the suspension cable, the load can be at any height for the constant value of u given by (7.26) to just balance the effect of gravity to maintain the load stationary.

To derive the linear model, from (7.21), first

$$\begin{aligned} v_1 &= f_1(x_2) = x_2 \\ v_2 &= f_2(x_3) = \tfrac{1}{J}\left(K_t x_3^2 - Mgr_p\right) \\ v_3 &= f_3(x_2, x_3, u) = \tfrac{1}{L}\left(K_{pe}u - Rx_3 - K_b x_2 x_3\right) \end{aligned} \qquad (7.28)$$

The total differentials of (7.28) are then

$$\tilde{v}_1 = \frac{\partial}{\partial x_2}[f_1(x_2)]\,\tilde{x}_2 = \tilde{x}_2$$

$$\tilde{v}_2 = \frac{\partial}{\partial x_3}[f_2(x_3)]\,\tilde{x}_3\bigg|_{x_3=\bar{x}_3} = 2\frac{K_t}{J}\bar{x}_3\tilde{x}_3$$

$$\tilde{v}_3 = \frac{\partial}{\partial x_2}[f_3(x_2, x_3, u)]\,\tilde{x}_2\bigg|_{x_3=\bar{x}_3} + \frac{\partial}{\partial x_3}[f_3(x_2, x_3, u)]\,\tilde{x}_3\bigg|_{x_2=\bar{x}_2} + \frac{\partial}{\partial u}[f_3(x_2, x_3, u)]\,\tilde{u}$$

$$= -\frac{K_b}{L}\bar{x}_3\tilde{x}_2 - \frac{R}{L}\tilde{x}_3 + \frac{K_{pe}}{L}\tilde{u}. \qquad (7.29)$$

which yields the following linear state-space model:

$$\begin{aligned} \dot{\tilde{x}}_1 &= \tilde{x}_2 \\ \dot{\tilde{x}}_2 &= 2\tfrac{K_t}{J}\bar{x}_3\tilde{x}_3 \\ \dot{\tilde{x}}_3 &= -\tfrac{K_b}{L}\bar{x}_3\tilde{x}_2 - \tfrac{R}{L}\tilde{x}_3 + \tfrac{K_{pe}}{L}\tilde{u} \end{aligned} \qquad (7.30)$$

A simple way of introducing an integral term into the controller is employing a cascade control structure with a linear state feedback control law forming the inner loop and an integral controller forming the outer loop. First, (7.30) may be expressed in the matrix–vector form as follows:

$$\begin{bmatrix}\dot{\tilde{x}}_1\\\dot{\tilde{x}}_2\\\dot{\tilde{x}}_3\end{bmatrix} = \begin{bmatrix}0 & 1 & 0 \\ 0 & 0 & 2\bar{x}_3 K_t/J \\ 0 & -\bar{x}_3 K_b/L & -R/L\end{bmatrix}\begin{bmatrix}\tilde{x}_1\\\tilde{x}_2\\\tilde{x}_3\end{bmatrix} + \begin{bmatrix}0\\0\\K_{pe}/L\end{bmatrix}\tilde{u}. \qquad (7.31)$$

The required linear state feedback control law is then

$$u = u' - \begin{bmatrix}k_1 & k_2 & k_3\end{bmatrix}\begin{bmatrix}\tilde{x}_1\\\tilde{x}_2\\\tilde{x}_3\end{bmatrix}, \qquad (7.32)$$

where u' replaces the usual reference input with scaling coefficient and is the output of the integral controller defined by

$$u' = K_I \int (x_{1r} - x_1)\,dt. \qquad (7.33)$$

7.2 Linearisation About an Operating Point

The approach will be to first find the inner loop transfer function, $\tilde{x}_1(s)/u'(s)$, in terms of the linear state feedback gains, k_1, k_2 and k_3, following the methodology presented in Chap. 4, and then use this to find the overall closed-loop transfer function, $\tilde{x}_1(s)/\tilde{x}_{1r}(s)$, in terms of k_1, k_2, k_3 and K_I, which enables these four parameters to be determined by pole placement. First, the closed-loop system matrix is

$$\mathbf{A}_{cl} = \begin{bmatrix} 0 & 1 & 0 \\ 0 & 0 & a_{23} \\ 0 & a_{32} & a_{33} \end{bmatrix} - \begin{bmatrix} 0 \\ 0 \\ b \end{bmatrix} \begin{bmatrix} k_1 & k_2 & k_3 \end{bmatrix} = \begin{bmatrix} 0 & 1 & 0 \\ 0 & 0 & a_{23} \\ -bk_1 & a_{32} - bk_2 & a_{33} - bk_3 \end{bmatrix}. \tag{7.34}$$

where

$$a_{23} = 2\bar{x}_3 K_t/J, a_{32} = -\bar{x}_3 K_b/L, a_{33} = -R/L \text{ and } b = K_{pe}/L.$$

The closed-loop characteristic polynomial is therefore

$$|s\mathbf{I}_3 - \mathbf{A}_{cl}| = \begin{vmatrix} s & -1 & 0 \\ 0 & s & -a_{23} \\ bk_1 & bk_2 - a_{32} & s + bk_3 - a_{33} \end{vmatrix}$$
$$= s\left[s^2 + (bk_3 - a_{33})s + a_{23}(bk_2 - a_{32})\right] + a_{23}bk_1 \tag{7.35}$$
$$= s^3 + (bk_3 - a_{33})s^2 + a_{23}(bk_2 - a_{32})s + a_{23}bk_1.$$

Let the inner loop DC gain be K_0. Since this is $\tilde{X}_1(0)/U'(0)$, then the inner loop transfer function is

$$\frac{\tilde{X}_1(s)}{U'(s)} = \frac{K_0 a_{23} bk_1}{s^3 + (bk_3 - a_{33})s^2 + a_{23}(bk_2 - a_{32})s + a_{23}bk_1}. \tag{7.36}$$

To find K_0 in terms of the linear state feedback gains and the plant parameters, the state differential equation of the inner loop is

$$\begin{bmatrix} \dot{\tilde{x}}_1 \\ \dot{\tilde{x}}_2 \\ \dot{\tilde{x}}_3 \end{bmatrix} = \begin{bmatrix} 0 & 1 & 0 \\ 0 & 0 & a_{23} \\ -bk_1 & a_{32} - bk_2 & a_{33} - bk_3 \end{bmatrix} \begin{bmatrix} \tilde{x}_1 \\ \tilde{x}_2 \\ \tilde{x}_3 \end{bmatrix} + \begin{bmatrix} 0 \\ 0 \\ b \end{bmatrix} u'. \tag{7.37}$$

In the steady state, $\dot{\tilde{x}}_1 = \dot{\tilde{x}}_2 = \dot{\tilde{x}}_3 = 0$. Hence, from (7.37), $\tilde{x}_2 = 0$, $a_{23}\tilde{x}_3 = 0 \Rightarrow \tilde{x}_3 = 0$ and $-bk_1\tilde{x}_1 + bu' = 0 \Rightarrow k_1\tilde{x}_1 = u' \Rightarrow$

$$K_0 = \frac{\tilde{x}_1}{u'} = \frac{1}{k_1}. \tag{7.38}$$

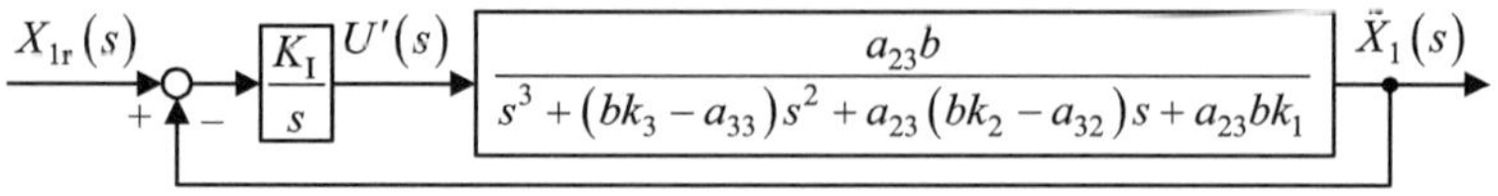

Fig. 7.2 Block diagram of linearised closed-loop system

The inner loop transfer function (7.36) then becomes

$$\frac{\tilde{X}_1(s)}{U'(s)} = \frac{a_{23}b}{s^3 + (bk_3 - a_{33})s^2 + a_{23}(bk_2 - a_{32})s + a_{23}bk_1}. \qquad (7.39)$$

A block diagram of the complete closed-loop system, based on (7.39) and (7.33), is shown in Fig. 7.2.

The overall characteristic polynomial is therefore given by

$$s^4 + (bk_3 - a_{33})s^3 + a_{23}(bk_2 - a_{32})s^2 + a_{23}bk_1 s + a_{23}bK_I \qquad (7.40)$$

For the pole placement, the desired closed-loop characteristic polynomial is

$$\left(s + \frac{1.5(1+n)}{T_s}\right)^n \bigg|_{n=4} = (s+a)^4 = s^4 + 4as^3 + 6a^2 s^2 + 4a^3 s + a^4, \qquad (7.41)$$

where $a = 7.5/T_s$. Equating (7.40) to (7.41) then yields the linear state feedback gains and the integral gain as

$$k_1 = \frac{4a^3}{a_{23}b}, \quad k_2 = \left(\frac{6a^2}{a_{23}} + a_{32}\right)/b, \quad k_3 = \frac{4a + a_{33}}{b} \quad \text{and} \quad K_I = \frac{a^4}{a_{23}b}. \qquad (7.42)$$

The simulation results shown in Fig. 7.3 compare the step response of the system comprising the linear state feedback controller applied to the nonlinear plant model with that of the linearised system of Fig. 7.2. The plant parameters are taken as $K_{pe} = 50$, $R = 1$ [Ω], $L = 0.1$ [H], $K_b = 0.006$ [V/A/(rad/s)], $K_T = 1$ [Nm/A], $r = 0.2$ [m], $M = 200$ [Kg] and $J = 100$ [Kg m^2]. Both will be commenced at the operating point. As the load is moved from one position to another, once the load is brought to rest, the system moves to the operating point again. During acceleration and deceleration, the state of the plant moves away from the operating point to an extent determined by the reference input and the demanded settling time.

In the notation of Fig. 7.3, $x_{1L} = \tilde{x}_1/|x_{1r}|$, which is the step response of the fictitious ideal system comprising the linear controller applied to the linearised plant model, normalised with respect to the magnitude of the reference input magnitude. Similarly, $x_{1N} = x_1/|x_{1r}|$ is the normalised step response predicted from the real system having the same linear controller applied to the nonlinear plant model. This enables the per unit errors between the ideal and real responses to be compared.

In this example, if the step reference input is $x_{1r} = 0.5$ [m] or less, the step response is very close to the ideal one. When, however, x_{1r} is increased to 0.7 [m], the oscillatory error between the real and ideal systems becomes noticeable. The

7.3 Feedback Linearising and Forced Dynamic Control

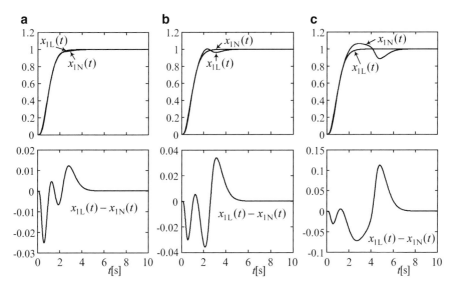

Fig. 7.3 Normalised step responses of crane position control system with linear controller. (**a**) $x_{1r} = 0.5$ [m]. (**b**) $x_{1r} = 0.7$ [m]. (**c**) $x_{1r} = 0.730482$ [m]

error rapidly becomes unacceptable with x_{1r} beyond 0.7 [m], as shown in Fig. 7.3c. In fact the system becomes unstable at $x_{1r} = 0.730493$ [m].

7.2.3 Limitation

In general, the plant state cannot move very far from the operating point; otherwise the linear model upon which the linear controller is based does not accurately represent the behaviour of the real plant, and therefore, the closed-loop dynamics will not be as specified. The forgoing example serves to emphasise this. The technique is therefore mainly limited to set-point controllers (sometimes referred to as regulators) where the reference input is constant, frequently found in industrial process control applications. Even in these applications, investigations should be carried out to test the effects of further than normal excursions beyond the operating point.

7.3 Feedback Linearising and Forced Dynamic Control

7.3.1 Preliminaries

7.3.1.1 The Plant State-Space Model

The most general form of plant model considered has the state differential equation,

$$\dot{\mathbf{x}} = \mathbf{f}(\mathbf{x}, \mathbf{u} - \mathbf{d}), \tag{7.43}$$

and the measured/controlled output equation

$$y = h(x), \quad (7.44)$$

where $x \in \Re^n$ is the state vector, $u \in \Re^m$ is the control vector, $d \in \Re^m$ is the external disturbance vector referred to the control vector and $y \in \Re^m$ is the controlled output vector. The assumption that the dimensions of the control and output vectors are equal simplifies the problem and is valid in most practical applications.

Any variables needed in the derived algorithms are assumed to be available from an observer (Chap. 8) if not directly measured. An estimate, $\hat{d}$, of d will also be assumed available from an observer. Then an auxiliary control vector, u', is introduced such that $u = u' + \hat{d}$. Assuming $\hat{d} = d$, this enables (7.43) to be simplified to

$$\dot{x} = f(x, u'). \quad (7.45)$$

SISO plants are covered by setting $r = m = 1$.

7.3.1.2 Feedback Linearisation

There are two categories of *feedback linearisation* [2]. The first is *state feedback linearisation*, applicable if (7.44) is linear, which is often the case, being equivalent to

$$y = Cx, \quad (7.46)$$

where C is a constant output matrix. Here a linear state-space model with the same state vector as in (7.45) is formed. Thus,

$$\dot{x} = Ax + Bu''. \quad (7.47)$$

Then the RHS of (7.45) and (7.47) are equated and solved for u', giving

$$f(x, u') = Ax + Bu'' \Rightarrow u' = G(x, u''). \quad (7.48)$$

So (7.48) is a form of state feedback control law with external input, u'', that transforms the nonlinear plant (7.45) to the linear one (7.47). A standard linear controller is then applied to the transformed plant consisting of (7.47) and (7.46). It is important to note, however, that the matrices, A and B have to be chosen not only for controllability of (7.47) but also for solubility of (7.48).

The second category, appropriate if (7.44) is nonlinear (or linear) and the plant is of full relative degree (Chap. 3), is *output feedback linearisation*, in which the state variables of (7.45) are transformed to the outputs and their derivatives using the same procedure as in Chap. 3 for determining the relative degrees with respect to the output vector components. Rather than using the standard Lie derivative notation, a simplified notation is used for the successive derivatives, as follows:

7.3 Feedback Linearising and Forced Dynamic Control

$$y_i = h_{0i}(\mathbf{x}), \quad \dot{y}_i = h_{1i}(\mathbf{x}), \quad \ldots, \quad y_i^{(R_i-1)} = h_{R_i-1,i}(\mathbf{x}), \quad i = 1, 2, \ldots, m \tag{7.49}$$

$$y_i^{(R_i)} = h_{R_i,i}(\mathbf{x}, \mathbf{u}') \tag{7.50}$$

where the subscript on the RHS indicates the order of the derivative. Full relative degree means

$$\sum_{i=1}^{m} R_i = n. \tag{7.51}$$

Then the complete set of output derivatives, $y_i^{(k)}$, $k = 0, 1, \ldots, R_i$, $i = 1, 2, \ldots, m$, are n in number and constitute a complete set of state variables in the generalised controller canonical form corresponding to the linear multivariable controller canonical form presented in Chap. 3, Sect. 3.3.6.3. Let these be denoted

$$\begin{aligned}
x_{c1} &= y_1 & x_{cR_1+1} &= y_2 & \cdots & & x_{c\sum_{j=1}^{m-1} R_j+1} &= y_m \\
x_{c2} &= \dot{y}_1 & x_{cR_1+2} &= \dot{y}_2 & & & x_{c\sum_{j=1}^{m-1} R_j+2} &= \dot{y}_m \\
&\vdots & &\vdots & & & &\vdots \\
x_{cR_1} &= y_1^{(R_1-1)} & x_{cR_1+R_2} &= y_2^{(R_2-1)} & & x_{c\sum_{j=1}^{m} R_j} &= cn & &= y_m^{(R_m-1)}
\end{aligned} \tag{7.52}$$

In view of (7.52) and (7.50), the complete sets of state differential equations and measurement equations are as follows:

$$\dot{x}_{ci} = \dot{x}_{ci+1}, \quad \dot{x}_{ci+1} = \dot{x}_{ci+2}, \quad \ldots, \quad \dot{x}_{cR_i-1} = x_{cR_i}, \quad i = 1, 2, \ldots, m. \tag{7.53}$$

$$\dot{x}_{cR_i} = h_{R_i,i}(\mathbf{x}, \mathbf{u}'), \quad i = 1, 2, \ldots, m \tag{7.54}$$

$$y_1 = x_{c1}, \quad y_2 = x_{cR_1+1}, \quad \ldots, \quad y_m = x_{c\sum_{j=1}^{m-1} R_j+1} \tag{7.55}$$

Then a linear state-space model is formed with the same state variables as in (7.53) and the same measurement equations (7.55). The simplest state differential equations that can be used are those of pure integrator chains, i.e.,

$$\dot{x}_{ci} = \dot{x}_{ci+1}, \quad \dot{x}_{ci+1} = \dot{x}_{ci+2}, \quad \ldots, \quad \dot{x}_{cR_i-1} = x_{cR_i}, \quad i = 1, 2, \ldots, m \tag{7.56}$$

$$\dot{x}_{cR_i} = u_i'', \quad i = 1, 2, \ldots, m. \tag{7.57}$$

Then the RHS of (7.54) and (7.57) is equated and the resulting set of m simultaneous equations solved for u_i', $i = 1, 2, \ldots, m$ to yield

$$u_i' = g_i\left(\mathbf{x},\, u_1'',\, u_2'',\, \ldots,\, u_m''\right), \quad i = 1, 2, \ldots, m. \tag{7.58}$$

This is a state feedback control law with external inputs, $u_1'', u_2'', \ldots, u_m''$, that such that the closed loop system becomes m separate pure integrator chains governed by (7.56), (7.57) and (7.55). This may be regarded as a set of separate SISO plants, each comprising a multiple integrator, to which may be applied linear state feedback control. The linear state feedback controllers may be easily designed by pole assignment.

7.3.1.3 Introduction to Multivariable Control

Multivariable control is the control of a plant with more than one output using more than one input. Numerous approaches to multivariable control have evolved over the years [3], but these are almost entirely restricted to linear plants or linearised plant models with operating points following the method of Sect. 7.2. Due to limited space, it is not possible to cover all the existing multivariable control techniques. The general technique of forced dynamic control (FDC) upon which the remainder of this chapter focuses, however, is applicable to nonlinear or linear *multivariable* plants in a relatively straightforward manner. Hence, multivariable control using this technique is covered, and this subsection is included to provide some preparation.

The block diagram of Fig. 7.4 illustrates the plant of (7.43) and (7.44).

Figure 7.5 shows a multivariable controller applied to this plant.

The purpose of a multivariable controller is twofold.

1. In common with other controllers, it has to achieve the required closed-loop dynamic behaviour and robustness against plant parameter uncertainties and external disturbances.

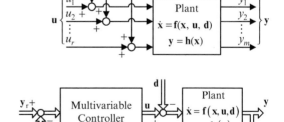

Fig. 7.4 Representation of a multivariable plant

Fig. 7.5 Block diagram representation of a general multivariable control system

7.3 Feedback Linearising and Forced Dynamic Control

2. The multivariable controller must, at least, ensure that any combination of step reference input vector components is responded to with acceptably small or zero steady-state errors due to DC interaction, i.e., steady-state errors in y_i due to constant values of y_j, $j \neq i$. Usually, in addition, dynamic interaction has to be minimised, meaning that a change in $y_{ri}(t)$ results in $y_i(t)$ responding as desired with minimal transient disturbance of $y_j(t)$, $j = 1, 2, \ldots, m$, $j \neq i$.

7.3.1.4 Specified Closed-Loop Differential Equations

In control of a multivariable plant using FLC, the set of desired closed-loop linear differential equations,

$$y_i^{(R_i)} = d_{0i}(y_{ri} - y_i) - d_{1i}\dot{y} - \cdots - d_{R_i-1, i} y^{(R_i-1)}, \quad i = 1, 2, \ldots, m, \quad (7.59)$$

has to be determined, where the degree, R_i, is the relative degree of the plant model with respect to the ith output (Chap. 3). This subsection is concerned with the determination of the constant coefficients, d_{ji}, $j = 0, 1, \ldots, R_i - 1$, $i = 1, 2, \ldots, m$. For an SISO plant, (7.59) is replaced by a single equation without the subscript, i.

The closed-loop transfer function corresponding to (7.59) is

$$\frac{Y_i(s)}{Y_{ri}(s)} = \frac{d_{0i}}{s^{R_i} + d_{R_i-1, i} s^{R_i-1} + \cdots + d_{1i} s + d_{0i}}, \quad i = 1, 2, \ldots, m. \quad (7.60)$$

The required coefficients are those of the denominator polynomials of (7.60) and may be determined by pole assignment. The settling time formulae may be used with the aid of Table 4 or Table 5 in the Table section following Chap. 11.

7.3.2 Feedback Linearising Control of Plants with Full Relative Degree

7.3.2.1 Introduction

Consider a nonlinear SISO plant modelled by a differential equation of the form

$$y^{(n)} = f\left(y^{(n-1)}, y^{(n-2)}, \ldots, \dot{y}, y, u\right), \quad (7.61)$$

where y is the controlled measured output, u is the control input and the RHS is a continuous function of its arguments. Then assuming the derivatives, $\dot{y}$ to $y^{(n-1)}$, are available either as measurements or calculated estimates, it is straightforward to obtain a feedback linearising control (FLC) law. First a linear differential equation is formed that the desired closed-loop system is intended to obey, having the same order, n, as (7.61), which relates y and its derivatives to the reference input y_r.

Thus,

$$y^{(n)} = d_0(y_r - y) - d_1\dot{y} - \cdots - d_{n-1}y^{(n-1)}. \tag{7.62}$$

Next the coefficients, $d_0, d_1, \ldots, d_{n-1}$, are chosen to yield the desired closed-loop dynamics, for which a method is given in Sect. 7.3.1.4. Since the closed-loop system is stable, if $y_r = $ const., then $y^{(k)} \to 0$ as $t \to \infty$, $k = 1, 2, \ldots, n \Rightarrow y \to y_r$ as $t \to \infty$. This indicates a closed-loop DC gain of unity, which is usually required. Finally, the RHS of (7.61) and (7.62) are equated and then solved for u. Thus,

$$f\left(y^{(n-1)}, y^{(n-2)}, \ldots, \dot{y}, y, u\right) = d_0(y_r - y) - d_1\dot{y} - \cdots - d_{n-1}y^{(n-1)}$$

from which

$$u = g\left(y^{(n-1)}, y^{(n-2)}, \ldots, \dot{y}, y - y_r\right) \tag{7.63}$$

This is a nonlinear state feedback control law in which the state variables are $x_1 = y$, $x_2 = \dot{y}$, ..., $y^{(n-1)}$, the state differential equations corresponding to (7.61) and the measurement equation being

$$\begin{aligned}
\dot{x}_1 &= x_2 \\
&\vdots \\
\dot{x}_{n-1} &= x_n \\
\dot{x}_n &= f(x_n, x_{n-1}, \ldots, x_2, x_1, u) \\
y &= x_1
\end{aligned} \tag{7.64}$$

This is in the nonlinear controller canonical state representation that is analogous to that for linear plants given in Chap. 3 but the version with no plant zeros. This means that the relative degree of the plant model (Chap. 3) is equal to the plant order, n.

The following example serves to demonstrate the efficacy of the method.

Example 7.2 Position control of a pendulum with unlimited motion

Figure 7.6 represents a pendulum constrained to move in a vertical plane but through an unlimited angle, α, with measurement, $y = \alpha$, which is to be controlled to follow a reference input, y_r, with a linear second-order dynamic response having damping ratio, ζ, and undamped natural frequency, ω_n. The state differential equations are

$$\begin{aligned}
\dot{x}_1 &= x_2 \\
\dot{x}_2 &= \tfrac{1}{MR^2}(K_T u - MgR\sin(x_1) - Bx_2)
\end{aligned} \tag{7.65}$$

7.3 Feedback Linearising and Forced Dynamic Control

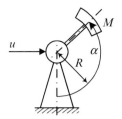

Fig. 7.6 Single degree of freedom pendulum with unlimited motion

where $x_1 = \alpha$, M is the pendulum mass, R is the distance from the centre of rotation to the pendulum centre of mass, g is the acceleration due to gravity, B is the viscous friction coefficient of the drive mechanism and K_T is the torque constant of the electric drive. This state-space model may be recognised as being in the nonlinear controller canonical form, and it is therefore possible to readily write down a second-order differential equation modelling the plant as

$$\ddot{y} = bu - a\sin(y) - c\dot{y} \tag{7.66}$$

where $b = K_T/(MR^2)$, $a = g/R$ and $c = B/(MR^2)$. The general desired second-order closed-loop differential equation corresponding to (7.62) is

$$\ddot{y} = d_0(y_r - y) - d_1\dot{y} \tag{7.67}$$

In terms of ζ and ω_n, (7.67) becomes

$$\ddot{y} = \omega_n^2(y_r - y) - 2\zeta\omega_n\dot{y} \tag{7.68}$$

Equating the RHS of (7.66) and (7.68), then solving the resulting equation for u then yields the required feedback linearising control law as

$$u = \frac{1}{b}\left[\omega_n^2(y_r - y) + (c - 2\zeta\omega_n)\dot{y} + a\sin(y)\right]. \tag{7.69}$$

It is clear that the third term on the RHS of (7.69) cancels the plant nonlinearity and the second term gives the system the required damping factor of $2\zeta\omega_n$, while the first term introduces the control error with the correct weighting.

7.3.2.2 SISO Plants of Full Relative Degree

The plant state-space model is given by (7.45) and (7.44) with $m = 1$. Thus,

$$\dot{\mathbf{x}} = \mathbf{f}(\mathbf{x}, u') \tag{7.70}$$

$$y = h(\mathbf{x}) \tag{7.71}$$

The first step in deriving a FLC law is to determine the relative degree of the plant as in Sect. 7.3.1.2, which yields

$$\left.\begin{array}{l} y = h_0(\mathbf{x}) \\ \dot{y} = h_1(\mathbf{x}) \\ \vdots \\ y^{(R-1)} = h_{R-1}(\mathbf{x}) \end{array}\right\} \tag{7.72}$$

$$y^{(R)} = h_R(\mathbf{x}, u') \tag{7.73}$$

It is assumed that the relative degree, R, of the plant is equal to its order, n. It is then of *full relative degree*. This must be checked in specific cases. Since u' appears on the RHS of (7.73), it is useful for deriving the basic FLC law. This is accomplished simply by equating the RHS of (7.73) to the RHS of the desired closed-loop differential equation (7.62) and making u' the subject of the resulting equation. Thus,

$$\begin{aligned} h_n(\mathbf{x}, u') &= d_0(y_r - y) - d_1 \dot{y} - \cdots - d_{n-1} y^{(n-1)} \Rightarrow \\ u' &= p\left(y_r - y, \dot{y}, \ldots, y^{(n-1)}, \mathbf{x}\right). \end{aligned} \tag{7.74}$$

This can be implemented if each component of $\mathbf{x}$ and each derivative of y is either a measurement or as an estimate from an observer (to be addressed in Chap. 8).

Equations (7.72), however, constitute a nonlinear state transformation. Assuming that the inverse transformation,

$$\mathbf{x} = \mathbf{h}_{\text{inv}}\left(y, \dot{y}, \ldots, y^{(n-1)}\right), \tag{7.75}$$

exists, then substituting for $\mathbf{x}$ in (7.74) using (7.75) yields the alternative FLC law,

$$u' = q\left(y_r - y, \dot{y}, \ldots, y^{(n-1)}\right). \tag{7.76}$$

The state variables, $y, \dot{y}, \ldots, y^{(n-1)}$, are then those of the SISO version of the nonlinear controller canonical state representation introduced in Sect. 7.3.1.2. It is evident that the complete state of the plant is controlled using this method.

Example 7.3 Single degree of freedom electromagnetic levitation system

This example is often used as a laboratory demonstration of the type of contactless suspension used in a maglev (electromagnetically levitated) vehicle. The electromagnetic attractive force acting on a sphere of magnetic material is controlled by means of the voltage applied to an electromagnet to achieve a constant air gap in the steady state. The structure of the complete control system is shown in Fig. 7.7. To simplify the plant model, the current transducer is assumed to deliver an ideal

7.3 Feedback Linearising and Forced Dynamic Control

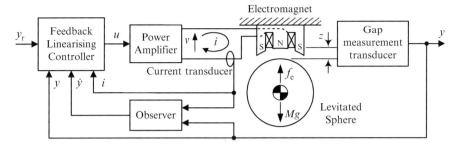

Fig. 7.7 Electromagnetically levitated sphere

current measurement equal numerically to the coil current, i, and the observer is assumed to deliver a perfect estimate of the derivative, $\dot{y}$, of the gap measurement. In this case, the external disturbance is $d = 0 \Rightarrow u' = u$. The equations modelling the plant are as follows. The force balance equation is

$$M\ddot{z} = Mg - f_e \qquad (7.77)$$

where M is the mass of the sphere, z is the gap length, g is the acceleration due to gravity and f_e is the electromagnetic force. The coil inductance is

$$L(z) = \frac{\mu_0 A N^2}{z + l/\mu_r}, \qquad (7.78)$$

where μ_0 is the permeability of free space, μ_r is the relative permeability of the electromagnet core, A is the cross-sectional area of the magnetic circuit, N is the number of coil turns and l is the length of the magnetic circuit within the core. Applying Kirchhoff's second law to the electrical circuit yields

$$v = Ri + \frac{d}{dt}[L(z)i] = Ri + L(z)\frac{di}{dt} + i\frac{dL(z)}{dz}\dot{z} = Ri + \left(\frac{\mu_0 A N^2}{z + l/\mu_r}\right)\frac{di}{dt}$$
$$- i\frac{\mu_0 A N^2}{(z + l/\mu_r)^2}\dot{z}, \qquad (7.79)$$

where L and R are, respectively, the electromagnet coil inductance and resistance. The voltage applied to the coil is

$$v = Bu \qquad (7.80)$$

where B is the voltage gain of the power amplifier. The gap measurement is

$$y = Cz, \qquad (7.81)$$

where C is the measurement constant of the gap length measurement transducer.

Finally, the force is the rate of change of the stored energy in the magnetic field with respect to the gap length. Thus,

$$f_e = -\frac{d}{dz}\tfrac{1}{2}L(z)i^2 = \tfrac{1}{2}\frac{\mu_0 A N^2 i^2}{(z+l/\mu_r)^2}. \tag{7.82}$$

The minus sign is due to the convention of positive forces acting downwards.

Let the state variables be chosen as $x_1 = z$, $x_2 = \dot{z}$ and $x_3 = i$. Then the state-space model of the plant may be formed from the above equations as follows:

$$\dot{x}_1 = x_2 \tag{7.83}$$

$$\dot{x}_2 = g - \frac{E}{(x_1+z_0)^2}x_3^2 \tag{7.84}$$

$$\dot{x}_3 = \frac{x_3 x_2}{x_1+z_0} + \frac{x_1+z_0}{D}(Bu - Rx_3) \tag{7.85}$$

$$y = Cx_1 \tag{7.86}$$

where $D = \mu_0 A N^2$, $E = D/(2M)$ and $z_0 = L/\mu_r$.

First, the relative degree of the plant is determined. Differentiating (7.86) and then substituting for $\dot{x}_1$ using (7.83) yields

$$\dot{y} = C\dot{x}_1 = Cx_2. \tag{7.87}$$

Since u does not appear on the RHS, the process is repeated. Differentiating (7.87) and substituting for $\dot{x}_2$ using (7.84) yields

$$\ddot{y} = C\dot{x}_2 = C\left[g - \frac{E}{(x_1+z_0)^2}x_3^2\right]. \tag{7.88}$$

Again u does not appear on the RHS, and so the process is repeated once more. Differentiating (7.88) and substituting for $\dot{x}_1$ and $\dot{x}_3$ using (7.83) and (7.85) yields

$$\begin{aligned}\dddot{y} &= CE\left[\frac{2}{(x_1+z_0)^3}\dot{x}_1 x_3^2 - \frac{1}{(x_1+z_0)^2}2x_3\dot{x}_3\right] \\ &= \frac{2CEx_3}{(x_1+z_0)^2}\left\{\frac{x_3 x_2}{x_1+z_0} - \left[\frac{x_3 x_2}{x_1+z_0} + \frac{x_1+z_0}{D}(Bu - Rx_3)\right]\right\} \\ &= -\frac{2CEx_3}{D(x_1+z_0)}(Bu - Rx_3) = -\frac{Cx_3}{M(x_1+z_0)}(Bu - Rx_3).\end{aligned} \tag{7.89}$$

7.3 Feedback Linearising and Forced Dynamic Control

Since u appears on the RHS of (7.89), the relative degree of the plant is the order of the derivative of y, which is $R = 3$. This is also the order of the plant since there are three state variables. Next, the desired closed-loop differential equation,

$$\dddot{y} = d_0 (y_r - y) - d_1 \dot{y} - d_2 \ddot{y}, \tag{7.90}$$

is formed, the coefficients of which may be determined using the method of Sect. 7.3.1.4 to achieve a specified transient response. Finally, the FLC law is derived by equating the RHS of (7.89) and (7.90) and then solving for u. Thus,

$$\begin{aligned} -\frac{Cx_3}{M(x_1+z_0)} (Bu - Rx_3) &= d_0 (y_r - y) - d_1 \dot{y} - d_2 \ddot{y} \Rightarrow \\ u &= \frac{1}{B} \left\{ Rx_3 + \frac{M(x_1+z_0)}{Cx_3} [d_0 (y - y_r) + d_1 \dot{y} + d_2 \ddot{y}] \right\} \end{aligned} \tag{7.91}$$

In this example, however, it is necessary to establish a coil current, x_3, before loop closure to avoid a software overflow due to the singularity at $x_3 = 0$, and an infinite control voltage demand, which would be unattainable. This is easily achieved by applying $u = u_{max}$, where u_{max} is the control saturation limit, until $x_3 = x_{3\,ilc}$, where $x_{3\,ilc}$ is the current for initial loop closure. Prior to turning on the levitation system, a maglev vehicle would be resting on safety wheels, which would limit the gap to a maximum value of $x_{1\,max}$. A suitable value for $x_{3\,ilc}$ would be just sufficient to lift the vehicle. In this case, (7.84) should yield zero acceleration in the steady state. Thus,

$$0 = g - \frac{E}{(x_{1\,max} + z_0)^2} x_{3\,ilc}^2 \Rightarrow x_{3\,ilc} = (x_{1\,max} + z_0) \sqrt{\frac{g}{E}} \tag{7.92}$$

It should be noted that the sign of the current makes no difference to the attractive levitation force and it is chosen as positive here.

The plant parameters for the following simulation are $z_0 = 0.01$ [m], $x_{1\,max} = 0.09$ [m], $M = 10$ [Kg], $B = 2$, $R = 1$ [Ω], $E = 0.001$ [m^3/s^2/A^2], $C = 100$ and $u_{max} = 10$ [V]. The desired closed-loop differential equation coefficients are chosen to yield a settling time of $T_s = 5$ [s] (5 % criterion), for which Table 12.1 yields $d_0 = \alpha^3$, $d_1 = 3\alpha^2$ and $d_2 = 3\alpha$, where $\alpha = 6/T_s$.

The gap is first set to $x_{1\,max}$ with all other initial conditions zero and the constant reference gap is set to $x_{1r} = 0.05$ [m]. Figure 7.8 shows the acquisition transient following loop closure. As expected, Fig. 7.8a shows the response of a third-order linear system with coincident closed-loop poles set according to the 5 % settling time formula (Chap. 4). The corresponding control variable, $u(t)$, of Fig. 7.8b is not typical of linear control systems. In this case, it commences close to u_{max} to raise the current, $x_3(t)$, to produce the required lifting force but rapidly falls as less current is required to produce a given force as the gap closes.

It should be noted that although five state variables, x_1, x_3, y, $\dot{y}$ and $\ddot{y}$, are used by control law (7.91), there are only three independent state variables. This is true

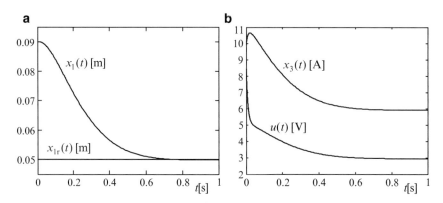

Fig. 7.8 Simulation of acquisition transient of electromagnetic levitation system. (**a**) Levitation gap and constant reference. (**b**) Control voltage and coil current

as y, $\dot{y}$ and $\ddot{y}$ are dependent on x_1, x_2 and x_3 via (7.86), (7.87) and (7.88), which are state transformation equations. Mixed state representations are usable but advantage can be taken of the transformation equations to use the most convenient set of variables in a particular application, depending on which are measurements.

7.3.2.3 Multivariable Plants of Full Relative Degree

Consider a nonlinear multivariable plant modelled by the state differential equation

$$\dot{\mathbf{x}} = \mathbf{f}(\mathbf{x}, \mathbf{u} - \mathbf{d}) \qquad (7.93)$$

and the measured/controlled output equation

$$\mathbf{y} = \mathbf{h}(\mathbf{x}), \qquad (7.94)$$

where $\mathbf{x} \in \Re^n$ is the state vector, $\mathbf{u} \in \Re^r$ is the control vector, $\mathbf{d} \in \Re^r$ is the external disturbance vector referred to the control vector and $\mathbf{y} \in \Re^m$ is the measurement vector. The functions on the RHS of (7.93) and (7.94) are assumed to be continuous. The approach will be similar to that of Sect. 7.3.2.2, so first it will be supposed that an observer is used to produce an estimate, $\widehat{\mathbf{d}}$, of $\mathbf{d}$ and an auxiliary control vector, $\mathbf{u}'$, is introduced such that $\mathbf{u} = \mathbf{u}' + \widehat{\mathbf{d}}$. On the assumption that $\widehat{\mathbf{d}} = \mathbf{d}$, this enables (7.93) to be replaced by

$$\dot{\mathbf{x}} = \mathbf{f}(\mathbf{x}, \mathbf{u}'). \qquad (7.95)$$

The first step in deriving a FLC law is to determine the relative degree of the plant with respect to each output vector component, as in Chap. 3. Using a similar notation to that of (7.72) and (7.73), m sets of output derivatives are generated, each terminating with the derivative that directly depends on *any* of the control vector components, the order of which is the relative degree. Thus,

$$\begin{aligned} y_i &= h_{0i}(\mathbf{x}) \\ \dot{y}_i &= h_{1i}(\mathbf{x}) \\ &\vdots \\ y_i^{(R_i-1)} &= h_{R_i-1,i}(\mathbf{x}) \end{aligned} \quad , i = 1, 2, \ldots, m \quad (7.96)$$

$$y_i^{(R_i)} = h_{R_i,i}(\mathbf{x}, \mathbf{u}) \quad (7.97)$$

As in Sect. 7.3.2.2, the plant will first be assumed to have *full relative degree*. For a multivariable plant, this means

$$\sum_{i=1}^{m} R_i = n, \quad (7.98)$$

where n is the plant order. In this case, the complete set of output derivatives,

$$y_i, \dot{y}_i, \ldots, y_i^{(R_i)}, i = 1, 2, \ldots, m, \quad (7.99)$$

is n in number and constitutes a complete set of state variables in a particular state representation that could be called the general multivariable controller canonical state representation.

Next, for each output, y_i, the desired closed-loop differential equation,

$$y_i^{(R_i)} = d_{0i}(y_{ri} - y_i) - d_{1i}\dot{y}_i - \cdots - d_{R_i-1,i}y_i^{(R_i-1)}, \quad i = 1, 2, \ldots, m, \quad (7.100)$$

is formed, where the constant coefficients, d_{ji}, $j = 0, 1, \ldots, R_i - 1$, $i = 1, 2, \ldots, m$, may be determined using Tables 4 or 5 in Tables. Since the m differential equations of (7.100) have no common terms, forcing the closed-loop system to obey these will achieve elimination of interaction, which is usually a major goal in multivariable control.

Next, the RHS of (7.97) and (7.100) are equated. Thus,

$$h_{R_i,i}(\mathbf{x}, \mathbf{u}) = d_{0i}(y_{ri} - y_i) - d_{1i}\dot{y}_i - \cdots - d_{R_i-1,i}y_i^{(R_i-1)}, \quad i = 1, 2, \ldots, m \quad (7.101)$$

These are then viewed as m simultaneous equations. If a solution exists for the r components of the control vector, $\mathbf{u}$, then this is the required control law. If $r < m$, then elimination of interaction will not be possible, but partial state control may be feasible if the zero dynamics considered in Sect. 7.3.3 is stable.

Example 7.4 Spin control of a rigid-body spacecraft

A rigid-body spacecraft with reaction wheel torque actuators aligned with the principal axes of inertia has the following state-space model (Chap. 2):

$$\begin{bmatrix} J_{xx}\dot{\omega}_x \\ J_{yy}\dot{\omega}_y \\ J_{zz}\dot{\omega}_z \end{bmatrix} = \begin{bmatrix} (J_{yy} - J_{zz})\omega_z\omega_y + \omega_z l_{wy} - \omega_y l_{wz} \\ (J_{zz} - J_{xx})\omega_x\omega_z + \omega_x l_{wz} - \omega_z l_{wx} \\ (J_{xx} - J_{yy})\omega_y\omega_x + \omega_y l_{wx} - \omega_x l_{wy} \end{bmatrix} + K_w \begin{bmatrix} u_x \\ u_y \\ u_z \end{bmatrix} \quad (7.102)$$

together with

$$\begin{bmatrix} \dot{l}_{wx} \\ \dot{l}_{wy} \\ \dot{l}_{wz} \end{bmatrix} = -K_w \begin{bmatrix} u_x \\ u_y \\ u_z \end{bmatrix} \quad (7.103)$$

$$\begin{bmatrix} y_1 & y_2 & y_3 & y_4 & y_5 & y_6 \end{bmatrix}^T = \begin{bmatrix} \omega_x & \omega_y & \omega_z & l_{wx} & l_{wy} & l_{wz} \end{bmatrix}^T. \quad (7.104)$$

Here, J_{xx}, J_{yy} and J_{zz} are the principal axis moments of inertia, K_w is the torque constant of each of the three reaction wheels; ω_x, ω_y and ω_z are the body angular velocity vector components along the principal axes of inertia; and l_{wx}, l_{wy} and l_{wz} are the reaction wheel angular momenta. The plant parameters are given as $J_{xx} = 300$ [Kg m²], $J_{yy} = 600$ [Kg m²], $J_{zz} = 900$ [Kg m²] and $K_w = 0.1$ [Nm/V] with control saturation limits of ±10 [V], which are avoided in feedback linearisation.

Upon release from the launch vehicle, the spacecraft has the initial angular velocities, $\omega_x(0) = 0.07$ [rad/s], $\omega_y(0) = -0.12$ [rad/s] and $\omega_z(0) = -0.08$ [rad/s].

It is required to design a feedback linearising controller that will bring the spacecraft to rest, with non-overshooting linear closed-loop dynamics and a settling time of $T_s = 50$ [s] (5 % criterion), and thereafter be able to respond to changes in the reference inputs without any inter-axis coupling, i.e. a change in one of the reference inputs will only cause the angular velocity component about that control axis to change but not affect the other two angular velocity components. In aerospace applications, the term 'inter-axis coupling' is used, while in other applications such as process control, the term 'interaction' is used: $\omega_{zr} = 0$ [rad/s].

In this example, it is only necessary to control the sub-plant defined by (7.102). So only the state variables, ω_x, ω_y and ω_y, are to be controlled. As will be seen, however, measurements of all the six state variables of (7.104) will be needed. The measurements are shown numerically equal to the state variables as scaling to achieve this can be applied in the on-board control computer.

It is wise to simplify the plant model before carrying out the control law derivation. First let the controlled outputs be denoted $y_1 = x_1 = \omega_x$, $y_2 = x_2 = \omega_y$ and $y_3 = x_3 = \omega_z$. Then, with $u_1 = u_x$, $u_2 = u_y$ and $u_3 = u_z$ (7.102) may be written as

$$\begin{bmatrix} \dot{y}_1 \\ \dot{y}_2 \\ \dot{y}_3 \end{bmatrix} = \begin{bmatrix} c_1 y_3 y_2 + a_1 (y_3 l_{wy} - y_2 l_{wz}) \\ c_2 y_1 y_3 + a_2 (y_1 l_{wz} - y_3 l_{wx}) \\ c_3 y_2 y_1 + a_3 (y_2 l_{wx} - y_1 l_{wy}) \end{bmatrix} + \begin{bmatrix} b_1 u_1 \\ b_2 u_2 \\ b_3 u_3 \end{bmatrix} \quad (7.105)$$

7.3 Feedback Linearising and Forced Dynamic Control

where

$$c_1 = \left(J_{yy} - J_{zz}\right)/J_{xx}, \quad c_2 = (J_{zz} - J_{xx})/J_{yy}, \quad c_3 = \left(J_{xx} - J_{yy}\right)/J_{zz}, \quad (7.106)$$

$a_1 = 1/J_{xx}, a_2 = 1/J_{yy}, a_3 = 1/J_{zz}, b_1 = K_w/J_{xx}, b_2 = K_w/J_{yy}$ and $b_3 = K_w/J_{zz}$.

Conveniently, the three component equations of (7.105) are already in the form of (7.97) with control variables appearing on the RHS, so no differentiation is necessary.

Next, the desired differential equations of the closed-loop system with the same output derivatives as that on the LHS of (7.105) may easily be set up as three decoupled first-order subsystems as follows:

$$\begin{bmatrix} \dot{y}_1 \\ \dot{y}_2 \\ \dot{y}_3 \end{bmatrix} = \begin{bmatrix} 3/T_s & 0 & 0 \\ 0 & 3/T_s & 0 \\ 0 & 0 & 3/T_s \end{bmatrix} \begin{bmatrix} (y_{r1} - y_1) \\ (y_{r2} - y_2) \\ (y_{r3} - y_3) \end{bmatrix} \quad (7.107)$$

where $y_{r1} = \omega_{xr}, y_{r2} = \omega_{yr}$ and $y_{r3} = \omega_{zr}$. That this is correct may easily be seen by deriving the individual transfer functions as

$$\frac{Y_i(s)}{Y_{ri}(s)} = \frac{1}{1 + sT_s/3}, \quad i = 1, 2, 3. \quad (7.108)$$

These each have a unity DC gain, as required, and the closed-loop time constant is

$$T_c = T_s/3 \Rightarrow T_s = 3T_c \quad (7.109)$$

which may be recognised as the 5 % settling time formula, $T_s = 1.5\,(1 + n)\,T_c$ of Chap. 4, with $n = 1$ since each subsystem is only of first order.

The required controllaw is obtained by equating the RHS of (7.105) and (7.107). Then the resulting equations are solved for u_1, u_2 and u_3. Thus,

$$\left.\begin{array}{l} \begin{bmatrix} c_1 y_3 y_2 + a_1 \left(y_3 l_{wy} - y_2 l_{wz}\right) + b_1 u_1 \\ c_2 y_1 y_3 + a_2 (y_1 l_{wz} - y_3 l_{wx}) + b_2 u_2 \\ c_3 y_2 y_1 + a_3 \left(y_2 l_{wx} - y_1 l_{wy}\right) + b_3 u_3 \end{bmatrix} = \begin{bmatrix} (3/T_s)\,(y_{r1} - y_1) \\ (3/T_s)\,(y_{r2} - y_2) \\ (3/T_s)\,(y_{r3} - y_3) \end{bmatrix} \rightarrow \\ u_1 = \left[(3/T_s)\,(y_{r1} - y_1) - c_1 y_3 y_2 - a_1 \left(y_3 l_{wy} - y_2 l_{wz}\right)\right]/b_1 \\ u_2 = \left[(3/T_s)\,(y_{r2} - y_2) - c_2 y_1 y_3 - a_2 (y_1 l_{wz} - y_3 l_{wx})\right]/b_2 \\ u_3 = \left[(3/T_s)\,(y_{r3} - y_3) - c_3 y_2 y_1 - a_3 \left(y_2 l_{wx} - y_1 l_{wy}\right)\right]/b_3 \end{array}\right\} \quad (7.110)$$

Figure 7.9 shows a simulation of the control system with the spacecraft initially allowed to tumble with free motion and zero reaction wheel inputs, loop closure not occurring until $t = 70$ [s]. Before discussing the behaviour of the closed-loop system, the oscillatory behaviour of the angular velocities for $0 < t < 70$ [s] will be explained. This is referred to as *nutation*. This is similar to the wobbling motion

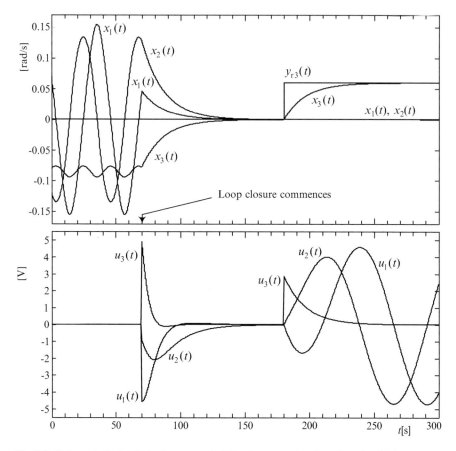

Fig. 7.9 Spin control of rigid-body spacecraft following separation from launch vehicle

of a toy spinning top when disturbed. With $u_1 = u_2 = u_3 = 0$, the solution to the unforced nonlinear differential Eq. (7.105) is non-sinusoidal (in contrast to the spinning top). In the unforced case and with the reaction wheel angular momentums, $l_{wx} = l_{wy} = l_{wz} = 0$, (7.105) reduces to

$$\dot{x}_1 = c_1 x_3 x_2 \tag{7.111}$$

$$\dot{x}_2 = c_2 x_1 x_3 \tag{7.112}$$

$$\dot{x}_3 = c_3 x_2 x_1 \tag{7.113}$$

These are Euler's equations of rigid-body rotational motion derived in Chap. 2. The form of the solutions indicated by the simulation may be understood by finding the solution in the three substate planes, (x_1, x_2), (x_2, x_3) and (x_3, x_1), which is quite

7.3 Feedback Linearising and Forced Dynamic Control

straightforward. Dividing (7.111) by (7.112) conveniently eliminates time and x_3 between the equations. Thus,

$$\frac{\dot{x}_1}{\dot{x}_2} = \frac{c_1 x_3 x_2}{c_2 x_1 x_3} \Rightarrow \frac{dx_1}{dx_2} = \frac{c_1}{c_2} \cdot \frac{x_2}{x_1} \qquad (7.114)$$

The solution by the method of separation of variables is then as follows:

$$\int x_1 dx_1 = \frac{c_1}{c_2} \int x_2 dx_2 \Rightarrow \frac{1}{2} x_1^2 = \frac{c_1}{c_2} \cdot \frac{1}{2} x_2^2 + C \qquad (7.115)$$

where C is an arbitrary constant of integration. If the initial substate is $(x_1(0), x_2(0))$, then the solution may be written as

$$x_1^2 - \frac{c_1}{c_2} \cdot x_2^2 = x_1^2(0) - \frac{c_1}{c_2} \cdot x_2^2(0) \qquad (7.116)$$

By inspection of (7.111), (7.112) and (7.113), the solution in the other two substate planes may be formed from (7.116) by cyclic permutation of the subscripts $(1 \to 2, 2 \to 3, 3 \to 1)$ as follows:

$$x_2^2 - \frac{c_2}{c_3} \cdot x_3^2 = x_2^2(0) - \frac{c_2}{c_3} \cdot x_3^2(0) \qquad (7.117)$$

and

$$x_3^2 - \frac{c_3}{c_1} \cdot x_1^2 = x_3^2(0) - \frac{c_3}{c_1} \cdot x_1^2(0). \qquad (7.118)$$

Since $J_{zz} > J_{yy} > J_{xx}$, in view of (7.106),

$$c_1 < 0, c_2 > 0 \text{ and } c_3 < 0 \Rightarrow \frac{c_1}{c_2} < 0, \frac{c_2}{c_3} < 0 \text{ and } \frac{c_3}{c_1} > 0, \qquad (7.119)$$

meaning that substate trajectories (7.116) and (7.117) are elliptical and substate trajectory (7.118) is hyperbolic. These are projections of the body state trajectory of the spacecraft for $0 \le t < 70$ [s] on the mutually orthogonal (x_1, x_2), (x_2, x_3) and (x_3, x_1) planes, shown in Fig. 7.10a–c.

For the initial state, the substate point repeatedly describes an elliptical path centred on the origin in Fig. 7.10a. In Fig. 7.10b, the substate point also lies on an ellipse centred on the origin (shown dotted) but oscillates between two points, describing an elliptical arc. In Fig. 7.10c, the path of the substate point lies on a hyperbola (both parts shown dotted) and oscillates between two points, thereby describing a hyperbolic arc. The three-dimensional state trajectory is plotted in Fig. 7.10d, and before the loop closure can be seen to describe a path similar to a three-dimensional Lissajous figure in which $x_1(t)$ and $x_2(t)$ oscillate at a fixed frequency but differ in phase by $90°$, while $x_3(t)$ oscillates at twice this frequency.

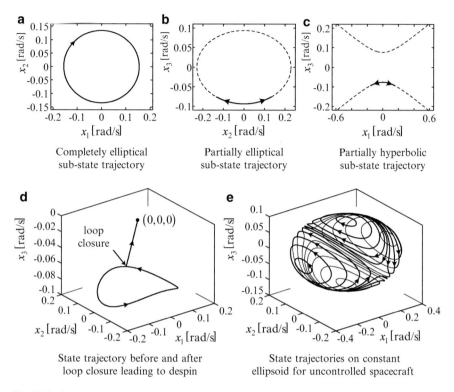

Fig. 7.10 Substate trajectories and state trajectory of uncontrolled plant

It is important to realise, however, that the oscillations are not sinusoidal and the frequencies depend on the initial state.

The accrued angular momentum of any mechanical system is the integral of the *externally* applied torque vector. In the example, this torque vector is zero. Hence, the angular momentum vector is constant. Before loop closure via the reaction wheel torque actuators, all of this angular momentum is contained in the rigid spacecraft body. Let the magnitude of the angular momentum be L. Then,

$$J_{xx}^2 x_1^2 + J_{yy}^2 x_2^2 + J_{zz}^2 x_3^2 = L^2 \qquad (7.120)$$

This is the equation of an ellipsoid. Figure 7.10e shows a family of state trajectories of the tumbling spacecraft body prior to the loop closure commencing with the same initial angular momentum as for the system simulated in Fig. 7.9, but different initial states constrained to satisfy (7.120). According to the law of conservation of angular momentum, as the state evolves, it continues to satisfy (7.120). This pattern of state trajectories, actually representing an infinite continuum of such trajectories, enables the dynamical behaviour of the tumbling spacecraft prior to be understood. The closed trajectory in Fig. 7.10d is one of the trajectories lying on the constant angular momentum ellipsoid of Fig. 7.10e.

7.3 Feedback Linearising and Forced Dynamic Control   509

The drastic change in the dynamic behaviour of the system upon loop closure is evident in Fig. 7.9 where the oscillatory behaviour vanishes and the three body angular velocities decay exponentially to zero with the correct settling time as required. The cyclic behaviour in Fig. 7.10d gives way to *linear* motion towards the origin as the time constant for each control axis is the same.

After the initial spacecraft de-spin, it is common to execute a slow rotation about one control axis to allow a star sensor to detect a star pattern on a swathe of the celestial sphere for initial attitude determination. In the simulation, a step yaw rate demand of 0.06 [rad/s] is commanded at $t = 180$ [s]. The control system is seen to respond with no disturbance of the other angular velocity components, i.e., $x_2(t)$ and $x_3(t)$ occurring. This is the ideal behaviour of this multivariable control system in having no interaction between the control channels.

The oscillatory control variables during the slow spin phase are due to the law of conservation of angular momentum. The spacecraft receives an initial angular momentum due to the imperfect launch vehicle separation mechanism, and this must remain with the spacecraft until external momentum dumping torques are applied using thrusters, which are small rockets: an operation outside the scope of this example. Due to the physics of the spacecraft, it is not possible to control all six state variables, x_1, x_2, x_3, l_x, l_y and l_z, with only the reaction wheel torques. So the angular momentum vector maintains a constant magnitude and direction with respect to inertial space. Once the spacecraft is rotating only about the yaw axis, the angular momentum vector rotates with conical motion with respect to the mutually orthogonal body-fixed control axis set. The angular momentum components along the x and y axes therefore oscillate in quadrature, and these must be produced by the x and y axis reaction wheels. Oscillatory control torques are therefore needed to accelerate and decelerate these wheels.

7.3.3 Feedback Linearising Control of Plants Less Than Full Relative Degree

7.3.3.1 Zero Dynamics

If a plant is less than full relative degree, then $n_c < n$, where $n_c = R$, i.e. the relative degree for an SISO plant or $n_c = \sum_{i=1}^{m} R_i$, where R_i is the relative degree with respect to the ith output for a multivariable plant. Then FLC can still be applied, but only a subsystem of order n_c will be controlled. This means that only n_c of the n state variables are controlled and therefore guaranteed to be associated with stable closed-loop modes. There must therefore be an uncontrolled subsystem of order, $n - n_c$, containing modes that may or may not be stable. These will, in general, be excited by the variables of the controlled subsystem. It is important to realise that the uncontrolled subsystem is not necessarily *uncontrollable*. This fact is highly relevant to the following Sect. 7.3.3.2. As will be recalled from Chap. 3, controllability refers to the ability to achieve any change of the plant state by manipulating the control inputs. If the plant is controllable, then *full state feedback* will enable all n state variables to be controlled and be associated with stable modes.

Let the substate of the plant controlled by FLC be denoted

$$\mathbf{x}_c = \begin{bmatrix} h_1(\mathbf{x}) \\ h_2(\mathbf{x}) \\ \vdots \\ h_{n_c}(\mathbf{x}) \end{bmatrix} = \mathbf{h}(\mathbf{x}). \tag{7.121}$$

Then there exists an uncontrolled substate that may be denoted

$$\mathbf{x}_z = \begin{bmatrix} g_1(\mathbf{x}) \\ g_2(\mathbf{x}) \\ \vdots \\ g_{n-n_c}(\mathbf{x}) \end{bmatrix} = \mathbf{g}(\mathbf{x}) \tag{7.122}$$

with an arbitrary state representation, the functions, $g_i(\mathbf{x})$, $i = 1, 2, \ldots, n - n_c$, being chosen to be independent of one another and of the functions, $h_j(\mathbf{x})$, $j = 1, 2, \ldots, n_c$. This is analogous to the linear independence required between the equations constituting a linear transformation between a given set of state variables of a linear plant and a new set of state variables for the same plant having another state representation. Then (7.121) and (7.122) constitute a transformation from the original state representation with state vector, $\mathbf{x}$, to a new state representation with state vector, $[\mathbf{x}_c \, \mathbf{x}_z]^T$. The plant state differential equations may then be written as

$$\dot{\mathbf{x}}_c = \mathbf{f}_c(\mathbf{x}_c, \mathbf{x}_z, \mathbf{u}). \tag{7.123}$$

and

$$\dot{\mathbf{x}}_z = \mathbf{f}_z(\mathbf{x}_c, \mathbf{x}_z, \mathbf{u}) \tag{7.124}$$

The dynamics of the uncontrolled subsystem defined by (7.124) is the *zero dynamics*. As will be seen in Sect. 7.3.4, for a linear plant, the zero dynamic subsystem (7.124) is linear and has eigenvalues equal to the plant *zeros*. If it is known that the zero dynamic substate, $\mathbf{x}_z$, resides in a close neighbourhood of a fixed operating point, $\mathbf{x}_{zo}$, a fact that is not always easy to establish theoretically (only by practical knowledge of the application or simulation), then eigenvalue analysis is possible after linearisation about this operating point using the method of Sect. 7.2.

Example 7.5 Zero dynamics in feedback linearising control of induction motor drive

In this example, remarkably, oscillatory zero dynamics is exploited to automatically create the rotating magnetic field within an induction motor for speed control. The plant model is taken from Chap. 2 and is the induction motor state-space model

7.3 Feedback Linearising and Forced Dynamic Control

formulated in the stator-fixed α–β frame to avoid the Park transformation. The external load torque is taken as zero in this case. The state differential equations are then

$$\dot{x}_1 = G\left(x_4 x_3 - x_5 x_2\right) \tag{7.125}$$

$$\dot{x}_2 = -A x_2 + B x_4 + C x_1 x_5 + D u_1 \tag{7.126}$$

$$\dot{x}_3 = -A x_3 + B x_5 - C x_1 x_4 + D u_2 \tag{7.127}$$

$$\dot{x}_4 = -E x_4 - p x_1 x_5 + F x_2 \tag{7.128}$$

$$\dot{x}_5 = -E x_5 + p x_1 x_4 + F x_3 \tag{7.129}$$

where $x_1 = \omega_r$, $x_2 = i_\alpha$, $x_3 = i_\beta$, $x_4 = \psi_\alpha$, $x_5 = \psi_\beta$, $u_1 = v_\alpha$ and $u_2 = v_\beta$. The constant parameters are

$$\left. \begin{array}{l} D = \frac{L_r}{L_s L_r - L_m^2}; \; A = D \cdot \left(R_s + \frac{L_m^2}{L_r^2} R_r \right) \quad B = D \cdot \frac{L_m R_r}{L_r^2} \quad C = D \cdot \frac{L_m}{L_r} \cdot p \\ E = \frac{R_r}{L_r} \qquad\qquad\qquad\quad F = \frac{L_m}{L_r} R_r \qquad\qquad G = \frac{1}{J_r} \cdot \frac{3p}{2} \cdot \frac{L_m}{L_r} \end{array} \right\}, \tag{7.130}$$

where the parameters on the RHS of Eq. (7.130) are defined in Chap. 2, Sect. 2.2.9. The two controlled outputs are the motor speed,

$$y_1 = x_1 \tag{7.131}$$

and the magnetic flux norm, i.e., the square of the rotor magnetic flux magnitude,

$$y_2 = x_4^2 + x_5^2. \tag{7.132}$$

First, the relative degree with respect to y_1 is determined. Differentiating (7.131) and substituting for $\dot{x}_1$ using (7.125) yields

$$\dot{y}_1 = \dot{x}_1 = G\left(x_4 x_3 - x_5 x_2\right). \tag{7.133}$$

Neither u_1 nor u_2 appear on the RHS, and therefore, (7.133) is differentiated and substitutions made for $\dot{x}_i$, $i = 2, 3, 4, 5$, using (7.126), (7.127), (7.128) and (7.129).

$$\ddot{y}_1 = G\left(\dot{x}_4 x_3 + \dot{x}_3 x_4 - \dot{x}_5 x_2 - \dot{x}_2 x_5\right)$$

$$= G \left[\begin{array}{l} (-E x_4 - p x_1 x_5 + F x_2) x_3 + (-A x_3 + B x_5 - C x_1 x_4 + D u_2) x_4 \\ - (-E x_5 + p x_1 x_4 + F x_3) x_2 - (-A x_2 + B x_4 + C x_1 x_5 + D u_1) x_5 \end{array} \right]$$

Simplifying and grouping the terms involving the control inputs yields

$$\ddot{y}_1 = G \begin{bmatrix} (-Ex_4 - px_1x_5)\,x_3 + (-Ax_3 - Cx_1x_4)\,x_4 \\ - (-Ex_5 + px_1x_4)\,x_2 - (-Ax_2 + Cx_1x_5)\,x_5 \end{bmatrix} + GD\,(x_4 u_2 - x_5 u_1). \tag{7.134}$$

Since u_1 and u_2 appear on the RHS of (7.134), the relative degree with respect to y_1 is $R_1 = 2$ and therefore, this equation is kept for forming the control law. Rearranging (7.134) in preparation for this yields

$$x_4 u_2 - x_5 u_1 = f_1 \tag{7.135}$$

where

$$f_1 = \frac{1}{GD}\left\{\ddot{y}_1 - G\begin{bmatrix} (-Ex_4 - px_1x_5)\,x_3 + (-Ax_3 - Cx_1x_4)\,x_4 \\ - (-Ex_5 + px_1x_4)\,x_2 - (-Ax_2 + Cx_1x_5)\,x_5 \end{bmatrix}\right\}. \tag{7.136}$$

Next, (7.132) is differentiated and substitutions made for $\dot{x}_4$ and $\dot{x}_5$ using (7.128) and (7.129). Thus,

$$\begin{aligned}\dot{y}_2 &= 2\,(\dot{x}_4 x_4 + \dot{x}_5 x_5) \\ &= 2\,[(-Ex_4 - px_1x_5 + Fx_2)\,x_4 + (-Ex_5 + px_1x_4 + Fx_3)\,x_5] \\ &= 2\,[F\,(x_2 x_4 + x_3 x_5) - E\,(x_4^2 + x_5^2)].\end{aligned} \tag{7.137}$$

Since neither u_1 nor u_2 appear on the RHS, (7.137) is differentiated and substitutions made for $\dot{x}_i$, $i = 2, 3, 4, 5$, using (7.126), (7.127), (7.128) and (7.129). First, it will be advantageous to substitute for $x_4^2 + x_5^2$ in (7.137) using (7.132). Thus,

$$\dot{y}_2 = 2\,[F\,(x_2 x_4 + x_3 x_5) - E y_2]. \tag{7.138}$$

Then,

$$\begin{aligned}\ddot{y}_2 &= 2\,[F\,(\dot{x}_2 x_4 + \dot{x}_4 x_2 + \dot{x}_3 x_5 + \dot{x}_5 x_3) - E\dot{y}_2] \\ &= 2F\begin{bmatrix}(-Ax_2 + Bx_4 + Cx_1x_5 + Du_1)\,x_4 + (-Ex_4 - px_1x_5 + Fx_2)\,x_2 \\ + (-Ax_3 + Bx_5 - Cx_1x_4 + Du_2)\,x_5 + (-Ex_5 + px_1x_4 + Fx_3)\,x_3\end{bmatrix} \\ &\quad - 2E\dot{y}_2 \\ &= 2F\begin{bmatrix}(-Ax_2 + Bx_4)\,x_4 + (-Ex_4 - px_1x_5 + Fx_2)\,x_2 \\ + (-Ax_3 + Bx_5)\,x_5 + (-Ex_5 + px_1x_4 + Fx_3)\,x_3\end{bmatrix} \\ &\quad - 2E\dot{y}_2 + 2FD\,(x_4 u_1 + x_5 u_2).\end{aligned} \tag{7.139}$$

7.3 Feedback Linearising and Forced Dynamic Control

Since u_1 and u_2 appear on the RHS of (7.140), the relative degree with respect to y_2 is $R_2 = 2$. This equation is therefore used together with (7.134) to derive the control law. Rearranging (7.139) in the same form as (7.135) yields

$$x_4 u_1 + x_5 u_2 = f_2. \tag{7.140}$$

where

$$f_2 = \frac{1}{2FD} \left\{ \ddot{y}_2 - 2F \begin{bmatrix} (-Ax_2 + Bx_4)x_4 + (-Ex_4 - px_1x_5 + Fx_2)x_2 \\ + (-Ax_3 + Bx_5)x_5 + (-Ex_5 + px_1x_4 + Fx_3)x_3 \end{bmatrix} \right.$$
$$\left. + 2E\dot{y}_2 \right\}. \tag{7.141}$$

Solving (7.135) and (7.140) for u_1 and u_2 then yields

$$\begin{bmatrix} u_1 \\ u_2 \end{bmatrix} = \begin{bmatrix} -x_5 & x_4 \\ x_4 & x_5 \end{bmatrix}^{-1} \begin{bmatrix} f_1 \\ f_2 \end{bmatrix} = \frac{1}{x_4^2 + x_5^2} \begin{bmatrix} -x_5 & x_4 \\ x_4 & x_5 \end{bmatrix} \begin{bmatrix} f_1 \\ f_2 \end{bmatrix} = \frac{1}{y_2} \begin{bmatrix} -x_5 & x_4 \\ x_4 & x_5 \end{bmatrix} \begin{bmatrix} f_1 \\ f_2 \end{bmatrix}. \tag{7.142}$$

Finally, $\ddot{y}_1$ and $\ddot{y}_2$ are determined according to the desired closed-loop differential equations. Thus,

$$\ddot{y}_1 = d_{01}(y_{1r} - y_1) - d_{11}\dot{y}_1 \tag{7.143}$$

and

$$\ddot{y}_2 = d_{02}(y_{2r} - y_2) - d_{12}\dot{y}_1 \tag{7.144}$$

Then the required control law is (7.142) with, f_1, f_2, $\ddot{y}_1$ and $\ddot{y}_2$ given by (7.136), (7.141), (7.143) and (7.144). If the coefficients of (7.143) and (7.144) are chosen according to Sect. 7.3.1.4, to yield settling times of T_{s1} for y_1 and T_{s2} for y_2, then

$$d_{01} = \alpha^2 \text{ and } d_{11} = 2\alpha \text{ where } \alpha = 4.5/T_{s1} \tag{7.145}$$

$$d_{02} = \beta^2 \text{ and } d_{12} = 2\beta \text{ where } \beta = 4.5/T_{s2} \tag{7.146}$$

It will be observed from (7.142) that loop closure cannot occur with $y_2(0) = 0$. The explanation is that the induction motor must be excited to establish a non-zero rotor flux in order for an accelerating torque to be generated by a finite rotor current.

Figure 7.11 shows the results of a simulation with the following parameter values: $L_s = L_r = 0.02$ [H]; $L_m = 0.018$ [H]; $R_r = 4$ [Ω]; $R_s = 3$ [Ω]; $J_r = 0.05$ [Kg m^2]; $p = 2$. The settling times are set to $T_{s1} = 0.2$ [s] and $T_{s2} = 0.1$ [s]. The motor is initially excited by applying an α-axis stator voltage component of $u_1 = 50$ [V] between $t = 0$ [s] and $t = 0.1$ [s], which is sufficient time for the motor magnetisation to reach approximately steady state as evident in Fig. 7.11b.

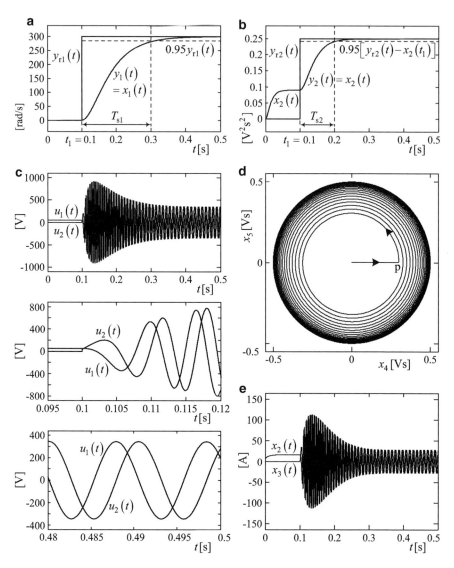

Fig. 7.11 Simulation of feedback linearising control of induction motor drive. (**a**) Rotor speed, (**b**) Rotor flux vector norm. (**c**) Rotor flux vector components. (**d**) Control variables, (**e**) Stator two-phase currents

Then loop closure occurs at $t = 0.1$ [s]. In Fig. 7.11a, b, the system can be seen to respond to the simultaneously applied step reference speed and magnetic flux norm with critically damped second-order linear dynamics and with the correct settling times. This example demonstrates clearly in Fig. 7.11c that the control variables can be of a quite different form from those typical of linear control systems when the plant is highly nonlinear. These commence quasi-sinusoidal with increasing frequency and amplitude as the accelerating torque is being produced and settle to two-phase sinusoidal voltages, as the motor speed approaches its steady state.

The frequency can be seen to be higher during the acceleration phase than in the steady state, where it has synchronous period, $2\pi/300/p = 0.0105$ [s], which indicates the slip needed in the induction motor to produce the accelerating torque. The loop closure occurs at point p in Fig. 7.11d after which the substate trajectory, $[x_4(t), x_5(t)]$, spirals out and converges towards a circle. Since the substate point on this graph is the tip of the magnetic flux vector, this is seen to rotate as in the classical operation of an induction motor with sinusoidal AC supplies. Remarkably, the feedback linearising controller produces the rotating magnetic field and associated alternating internal variables, as in classical induction motor operation, *automatically*. It is the *zero dynamics* that is responsible for this. The closed-loop system, when viewed via the reference inputs and the controlled outputs, is of fourth order and obeys (7.143) and (7.144), while the plant with state differential equations, (7.125), (7.126), (7.127), (7.128) and (7.129), is of fifth order. The zero dynamics is therefore of the first order. It is also remarkable how this can produce the highly oscillatory behaviour of the internal states, $x_i(t)$, $i = 2, 3, 4, 5$, while the four 'transformed' state variables, $y_1(t)$, $\dot{y}_1(t)$, $y_2(t)$ and $\dot{y}_2(t)$, are non-oscillatory, in contrast to a linear system where the zero dynamics would have to be at least of second order.

7.3.3.2 Zero Dynamics Avoidance using Auxiliary Output

If the plant is less than full relative degree and a feedback linearising control law is used based upon the derivatives of the controlled outputs, then the resulting zero dynamics may cause excursions of the physical variables that may be harmful to the plant hardware. This problem may be circumvented, however, by creating an auxiliary non-measured output vector

$$\mathbf{z} = \mathbf{G}(\mathbf{x}) \tag{7.147}$$

that can be produced using an observer, such that the plant with state-space model (7.95) is of full relative degree with respect to this output. In this case, repeated differentiations of the output vector components yield

$$\begin{aligned} z_i &= g_{0i}(\mathbf{x}) \\ \dot{z}_i &= g_{1i}(\mathbf{x}) \\ &\vdots \\ z_i^{(R_i-1)} &= g_{R_i-1,i}(\mathbf{x}) \end{aligned} \quad , i = 1, 2, \ldots, m, \tag{7.148}$$

where

$$\sum_{i=1}^{m} R_i = n \tag{7.149}$$

and therefore the output derivatives,

$$z_i, \dot{z}_i, \ldots, z_i^{(R_i)}, \quad i = 1, 2, \ldots, m, \tag{7.150}$$

constitute a complete set of state variables. FLC based on these variables will then exercise complete control of the plant state. For control of SISO plants, $m = 1$.

A basic requirement is that $\mathbf{G}(\mathbf{x})$ of (7.147) must be chosen such that with a constant reference input vector, $\mathbf{y}_r$, the steady-state output satisfies

$$\mathbf{z}_{ss} = \lim_{t \to \infty} \mathbf{z}(t) = \mathbf{y}_{ss} = \lim_{t \to \infty} \mathbf{y}(t) = \mathbf{y}_r. \tag{7.151}$$

For a multivariable plant, it is usually necessary to minimise or eliminate transient interaction between the control channels. To the knowledge of the author, no general theory is available for choice of $\mathbf{G}(\mathbf{x})$ achieving this with a nonlinear plant, but means can be devised to solve the problem for each application.

Example 7.6 FLC of a flexible electric drive with a hard nonlinear torsion characteristic

Figure 7.12 represents a motor driving an inertial load via a torsional flexible shaft.

The electric drive is assumed to be operated in the torque mode in which the electromagnetic torque, γ_e, responds to its demand with negligible dynamic lag and can therefore be regarded as the control variable for the overall system. With reference to Fig. 7.12a, the electromagnetic torque acts on the rotor with moment of inertia, J_r, which turns through an angle, θ_r, with respect to an inertial datum. The spring torque, γ_s, gives rise to a torsional deflection angle, $\alpha = \theta_L - \theta_r$, where θ_L is the angle of the mechanical load with moment of inertia, J_L, with respect to the inertial datum. In this example, the controlled output is θ_r. The frictional torques are regarded negligible. The torque balance equations are then

$$J_r \ddot{\theta}_r = \gamma_e + \gamma_s \tag{7.152}$$

$$J_L \ddot{\theta}_L = -\gamma_s \tag{7.153}$$

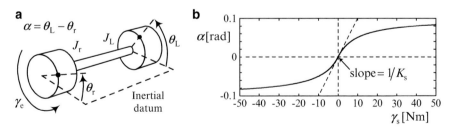

Fig. 7.12 Flexible electric drive. (**a**) Mechanical representation. (**b**) Nonlinear hard torsion spring characteristic

7.3 Feedback Linearising and Forced Dynamic Control

where the torsional flexibility of the coupling shaft is modelled as

$$\gamma_s(\alpha) = \frac{K_s \alpha_{\max} \alpha}{\alpha_{\max} - |\alpha|}, \quad (7.154)$$

where K_s is the spring constant for linear operation with infinitesimal deflections and $\alpha_{\max}$ is the maximum deflection magnitude. This gives rise to the transfer characteristic of Fig. 7.12b shown for $K_s = 100$ [Nm/rad] and $\alpha_{\max} = 0.1$ [rad].

The object is to design a feedback linearising position control system to control θ_r to respond to a step change in the demanded value, $\theta_{r\,\text{dem}}$, with no overshooting and a settling time of T_s seconds (5 % criterion). This will be based on the state-space model of the plant, which is

$$\dot{x}_1 = x_2 \quad (7.155)$$

$$\dot{x}_2 = b\left[u + \gamma_s(x_1, x_3)\right] \quad (7.156)$$

$$\dot{x}_3 = x_4 \quad (7.157)$$

$$\dot{x}_4 = -a\gamma_s(x_1, x_3) \quad (7.158)$$

$$y = x_1 \quad (7.159)$$

where $u = \gamma_e$, $b = 1/J_r$, $a = 1/J_L$ and

$$\gamma_s(x_1, x_3) = \frac{K_s \alpha_{\max}(x_3 - x_1)}{\alpha_{\max} - |x_3 - x_1|}. \quad (7.160)$$

First, a basic FLC design and simulation will be carried out to demonstrate the zero dynamics. Differentiating (7.159) and substituting for $\dot{x}_1$ using (7.155) yields

$$\dot{y} = x_2 \quad (7.161)$$

As u does not appear on the RHS, (7.161) is differentiated and a substitution made for $\dot{x}_2$ using (7.156). Thus,

$$\ddot{y} = b\left[u + \gamma_s(x_1, x_3)\right]. \quad (7.162)$$

As u appears on the RHS, the relative degree of the plant is $R = 2$, which is the order of the closed-loop system that will obey the closed-loop differential equation,

$$\ddot{y} = d_0(y_r - y) - d_1\dot{y}, \quad (7.163)$$

where $y_r = \theta_{r\,\text{dem}}$ and using the method of Sect. 7.3.1.4, $p = 4.5/T_s$, $d_0 = p^2$ and $d_1 = 2p$. Equating the RHS of (7.162) and (7.163) and then solving for u yield the feedback linearising control law

$$u = \frac{1}{b}[d_0(y_r - y) - d_1\dot{y}] - f(x_1, x_3). \tag{7.164}$$

In view of (7.163), (7.159) and (7.161), only the substate, (x_1, x_2), is controlled, and therefore, the sub-plant obeying (7.157) and (7.158) constitutes the zero dynamics. This sub-plant approaches linearity as α and γ_s of Fig. 7.12b approach zero, the linear approximation to the zero dynamic substate equations being

$$\dot{x}_3 = x_4 \tag{7.165}$$

$$\dot{x}_4 = -\omega_n^2(x_3 - x_1) \tag{7.166}$$

where $\omega_n = \sqrt{aK_s} = \sqrt{K_s/J_L}$. It is straightforward to determine that the eigenvalues of this sub-plant are $\pm j\omega_n$ which indicate oscillatory behaviour that is undamped and sinusoidal with constant amplitude once the rotor angle, x_1, has been controlled to a constant value by the FLC law.

Figure 7.13 shows a step response simulation with the following plant parameters: $K_s = 100$ [Nm/rad]; $\alpha_{\max} = 0.5$ [rad] $J_r = 1$ $\left[\text{Kg m}^2\right]$; $J_r = 2$ $\left[\text{Kg m}^2\right]$; control saturation limit $u_{\max} = 400$ [Nm]. The settling time is set to $T_s = 5$ [s]. As is evident in Fig. 7.13a, the rotor angle is controlled as desired with a non-overshooting transient and the correct settling time. The zero dynamics, however, causes undesirable oscillatory behaviour of the load mass angle. The load mass is allowed to oscillate, which creates an oscillatory spring torque acting on the rotor of the motor. The control law generates an equal and opposite control torque component that precisely counteracts this spring torque and a second component that accelerates the motor rotor and decelerates it to bring it to rest at the required angle.

In so doing, however, the control law allows the load mass to oscillate continuously with respect to the motor rotor. This behaviour is evident in Fig. 7.13b–d. The nonlinear hard spring characteristic causes the noticeably non-sinusoidal oscillation of $u(t)$ and $\gamma_s(t)$.

To avoid the zero dynamics, the auxiliary output,

$$z = x_3, \tag{7.167}$$

will be considered, as it is the load mass angle that will equal the rotor angle if, with a step reference input, the system is brought to the desired steady state with zero spring torque according to (7.160). The suitability of this choice will now be established theoretically. If all the state variables are under control, then with a constant reference input, all the system variables will settle to constant values in

7.3 Feedback Linearising and Forced Dynamic Control

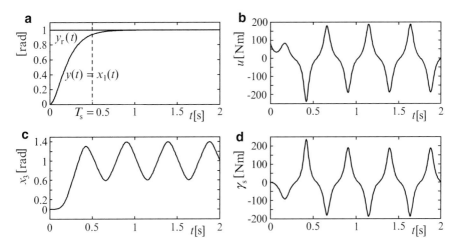

Fig. 7.13 FLC of flexible electric drive with zero dynamics. (**a**) Rotor angle (controlled output). (**b**) Control torque. (**c**) Load mass angle. (**d**) Spring torque

the steady state. Denoting the steady-state values by the subscript, ss, the plant state equations, (7.155), (7.156), (7.157), (7.158) and (7.159), become

$$0 = x_{2ss} \tag{7.168}$$

$$0 = b\left[u_{ss} + \gamma_s\left(x_{1ss}, x_{3ss}\right)\right] \tag{7.169}$$

$$0 = x_{4ss} \tag{7.170}$$

$$0 = -a\gamma_s\left(x_{1ss}, x_{3ss}\right) \tag{7.171}$$

In view of (7.168) and (7.170), the motor rotor and load mass are stationary in the steady state. In view of (7.171) and (7.160), $x_{3ss} = x_{1ss}$. Hence, from (7.167) and (7.159), $z_{ss} = y_{ss}$, which is the condition for $z = x_3$ to be a suitable candidate for an auxiliary output. Next whether or not the plant has full relative degree with respect to z has to be determined. Hence, the candidate auxiliary output (7.167) will be repeatedly differentiated and substitutions made for the state-variable derivatives using the state differential Eqs. (7.155), (7.156), (7.157) and (7.158). Differentiating (7.167) and substituting for $\dot{x}_3$ using (7.157) yields

$$\dot{z} = x_4. \tag{7.172}$$

Differentiating (7.172) and substituting for $\dot{x}_4$ using (7.158) and (7.160) yields

$$\ddot{z} = -a\gamma_s\left(x_1, x_3\right) = -a\frac{K_s\alpha_{\max}\left(x_3 - x_1\right)}{\alpha_{\max} - |x_3 - x_1|}. \tag{7.173}$$

Differentiating (7.173) and substituting for $\dot{x}_1$ and $\dot{x}_3$ using (7.155) and (7.157) gives

$$\dddot{z} = -aK_s\alpha_{max}\frac{(\alpha_{max} - |x_3 - x_1|)(x_4 - x_2) - (x_3 - x_1)[-\text{sgn}(x_3 - x_1)](x_4 - x_2)}{(\alpha_{max} - |x_3 - x_1|)^2}$$

$$= -aK_s\alpha_{max}\frac{(\alpha_{max} - |x_3 - x_1|)(x_4 - x_2) + |x_3 - x_1|(x_4 - x_2)}{(\alpha_{max} - |x_3 - x_1|)^2}$$

$$= -\frac{aK_s\alpha_{max}^2(x_4 - x_2)}{(\alpha_{max} - |x_3 - x_1|)^2}.$$

(7.174)

Finally, differentiating (7.174) and substituting for $\dot{x}_i$, $i = 1, 2, 3, 4$, using (7.155, 7.156, 7.157, and 7.158) yields

$$\ddddot{z} = -aK_s\alpha_{max}^2 \frac{d}{dt}\left[(x_4 - x_2)(\alpha_{max} - |x_3 - x_1|)^{-2}\right]$$

$$= -aK_s\alpha_{max}^2 \left\{\begin{array}{l}(x_4 - x_2)(-2)(\alpha_{max} - |x_3 - x_1|)^{-3}[-\text{sgn}(x_3 - x_1)](x_4 - x_2) \\ + (-a\gamma_s(x_1, x_3) - b[u + \gamma_s(x_1, x_3)])(\alpha_{max} - |x_3 - x_1|)^{-2}\end{array}\right\}$$

$$= \frac{aK_s\alpha_{max}^2}{(\alpha_{max} - |x_3 - x_1|)^3}\left\{\begin{array}{l}-2(x_4 - x_2)^2\,\text{sgn}(x_3 - x_1) \\ + (a\gamma_s(x_1, x_3) + b[u + \gamma_s(x_1, x_3)])(\alpha_{max} - |x_3 - x_1|)\end{array}\right\}$$

Hence

$$\ddddot{z} = \frac{aK_s\alpha_{max}^2}{(\alpha_{max}-|x_3-x_1|)^3}\left\{\begin{array}{l}-2(x_4 - x_2)^2\,\text{sgn}(x_3 - x_1) \\ + \left(a\frac{K_s\alpha_{max}(x_3-x_1)}{\alpha_{max}-|x_3-x_1|} + b\left[u + \frac{K_s\alpha_{max}(x_3-x_1)}{\alpha_{max}-|x_3-x_1|}\right]\right)(\alpha_{max} - |x_3 - x_1|)\end{array}\right\}$$

$$= \frac{aK_s\alpha_{max}^2}{(\alpha_{max}-|x_3-x_1|)^2}\left\{bu + \frac{(a+b)K_s\alpha_{max}(x_3-x_1) - 2(x_4-x_2)^2\,\text{sgn}(x_3-x_1)}{\alpha_{max}-|x_3-x_1|}\right\}.$$

(7.175)

Since u appears on the RHS of (7.175), the relative degree with respect to the auxiliary output, z, is $R = 4$, which is equal to the plant order. There is therefore no zero dynamics, and a feedback linearising control law based on (7.175) will control the complete plant state. In this case, the desired closed-loop differential equation is

$$\ddddot{z} = d_0(z_r - z) - d_1\dot{z} - d_2\ddot{z} - d_3\dddot{z}.$$

(7.176)

Equating the RHS of (7.175) and (7.176) and then solving for u yields the FLC law,

$$u = \frac{1}{b}\left\{\frac{(\alpha_{max}-|x_3-x_1|)^2}{aK_s\alpha_{max}^2}\left[d_0(z_r - z) - d_1\dot{z} - d_2\ddot{z} - d_3\dddot{z}\right] + \frac{2(x_4-x_2)^2\,\text{sgn}(x_3-x_1) - (a+b)K_s\alpha_{max}(x_3-x_1)}{\alpha_{max}-|x_3-x_1|}\right\}.$$

(7.177)

7.3 Feedback Linearising and Forced Dynamic Control

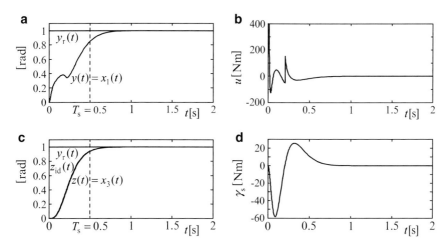

Fig. 7.14 FLC of flexible electric drive using auxiliary output with full relative degree (**a**) Rotor angle (controlled output). (**b**) Control torque. (**c**) Load mass angle (auxilliary output). (**d**) Spring torque

The coefficients, d_i, $i = 1, 2, 3, 4$, can be determined using the method of Sect. 7.3.1.4 to yield a non-overshooting step response with a settling time of T_s seconds. Thus, $p = 1.5\,(1+n)\,/\,T_s|_{n=4} = 7.5/T_s$, $d_1 = p^4$, $d_2 = 4p^3$, $d_3 = 6p^2$ and $d_4 = 4p$.

Figure 7.14 shows a simulation of the step response of the revised control system with the same plant parameters and demanded settling time as previously.

As expected, the oscillatory behaviour shown in Fig. 7.13 due to the zero dynamics is no longer present. A short period of initial control saturation now occurs, but this does not cause a noticeable difference between the step response, $z(t)$, and the ideal step response, $z_{id}(t)$, i.e. the solution to (7.176) with the step input, $z_r(t)$, as evident in Fig. 7.14c. Although the controlled output, $y(t)$, has the correct steady-state value, it has transient behaviour different from that of $z(t)$. This may be analysed as follows. In view of (7.159) and (7.167), (7.173) may be written as

$$\ddot{z} = -a \frac{K_s \alpha_{\max}\,(z-y)}{\alpha_{\max} - |z-y|} \Rightarrow y = z + \frac{\alpha_{\max} - |z-y|}{a K_s \alpha_{\max}} \ddot{z}. \qquad (7.178)$$

So $y(t)$ may be regarded as a state-dependent weighted sum of $z(t)$ and $\ddot{z}(t)$. Considering the smooth monotonically increasing $z(t)$ of Fig. 7.14c, $\ddot{z}(t)$ will have a positive maximum and $\ddot{z}(t)$ will have a positive maximum followed by a negative minimum. Thus, $y(t)$ may have a local maximum followed by a local minimum, and this behaviour is evident in Fig. 7.14a. This behaviour is very similar to that introduced in Chap. 1, Sect. 1.6.4.2, for linear control systems with finite zeros in the transfer functions. In this example, the linearised plant has two complex conjugate zeros which would produce a similar step response. Each specific application would have to be considered to determine whether or not this transient behaviour is acceptable. If not, then the shape of the step response can be improved by means of a pre-compensator as described in Chap. 5, Sect. 5.2.7. Finally, it will be observed that the settling time of $y(t)$ is somewhat longer than that of $z(t)$. If this is critical, then it can be adjusted to the correct value using the settling time correction method described in Chap. 4, Sect. 4.5.5.

7.3.4 Forced Dynamic Control of Continuous LTI Plants

7.3.4.1 The Plant State-Space Models

The most general plant model considered here is for an LTI multivariable plant. Thus,

$$\dot{\mathbf{x}} = \mathbf{A}\mathbf{x} + \mathbf{B}\mathbf{u} \tag{7.179}$$

$$\mathbf{y} = \mathbf{C}\mathbf{x} \tag{7.180}$$

where $\mathbf{x} \in \mathfrak{R}^n$ is the state vector, $\mathbf{u} \in \mathfrak{R}^m$ is the control (or input) vector, $\mathbf{y} \in \mathfrak{R}^m$ is the measurement (or controlled output) vector and $\mathbf{A}$, $\mathbf{B}$ and $\mathbf{C}$ are constant matrices. Since the number of controlled outputs equals the number of inputs, the plant is *square*, this description originating from the fact that the transfer function matrix is square. The LTI SISO plant model is obtained by setting $m = 1$, but the following specific notation will be used:

$$\dot{\mathbf{x}} = \mathbf{A}\mathbf{x} + \mathbf{b}u \tag{7.181}$$

$$y = \mathbf{c}^T\mathbf{x} \tag{7.182}$$

where u and y are the scalar input and output.

The approach in the following four subsections is the same as that of Sects. 7.3.2 and 7.3.3, but the linear plant models enable generic control laws to be derived.

7.3.4.2 SISO LTI Plants of Full Relative Degree

In this case, the relative degree is $R = n$, and therefore all the output derivatives of order up to $n-1$ are linear functions of $\mathbf{x}$ only, the derivative that is a linear function of $\mathbf{x}$ and u having order, n. This sequence of derivatives is as follows:

$$\begin{aligned} \dot{y} &= \mathbf{c}^T\mathbf{A}\mathbf{x} \\ \ddot{y} &= \mathbf{c}^T\mathbf{A}^2\mathbf{x} \\ &\vdots \\ y^{(n-1)} &= \mathbf{c}^T\mathbf{A}^{n-1}\mathbf{x} \end{aligned} \tag{7.183}$$

7.3 Feedback Linearising and Forced Dynamic Control

$$y^{(n)} = \mathbf{c}^T \mathbf{A}^n \mathbf{x} + \mathbf{c}^T \mathbf{A}^{n-1} \mathbf{b} u. \tag{7.184}$$

As for the feedback linearising control, (7.184) is used to form the forced dynamic control law by equating its RHS to that of the desired closed-loop differential equation,

$$y^{(n)} = d_0 (y_r - y) - d_1 \dot{y} - \cdots - d_{n-1} y^{(n-1)}, \tag{7.185}$$

where the constant coefficients may be determined using the method of Sect. 7.3.1.4. Thus,

$$\begin{aligned} \mathbf{c}^T \mathbf{A}^n \mathbf{x} + \mathbf{c}^T \mathbf{A}^{n-1} \mathbf{b} u &= d_0 (y_r - y) - d_1 \dot{y} - \cdots - d_{n-1} y^{(n-1)} \Rightarrow \\ u &= \tfrac{1}{\mathbf{c}^T \mathbf{A}^{n-1} \mathbf{b}} \left[d_0 (y_r - y) - d_1 \dot{y} - \cdots - d_{n-1} y^{(n-1)} - \mathbf{c}^T \mathbf{A}^n \mathbf{x} \right]. \end{aligned} \tag{7.186}$$

This control law is useful as it stands although, as in most of the nonlinear examples of the foregoing subsections, the state representation is mixed, provided, of course, that all the variables are available as either measurements or estimates from an observer. If the original state representation of the plant model is preferred throughout, then (7.183) provides the required state transformation. Then substituting for $y^{(i)}$, $i = 0, 1, \ldots, n - 1$, in (7.186) using (7.183) yields

$$u = r y_r - \mathbf{k}^T \mathbf{x} \tag{7.187}$$

where the reference input scaling coefficient is

$$r = \frac{d_0}{\mathbf{c}^T \mathbf{A}^{n-1} \mathbf{b}} \tag{7.188}$$

and the state feedback gain matrix is

$$\mathbf{k}^T = \frac{1}{\mathbf{c}^T \mathbf{A}^{n-1} \mathbf{b}} \mathbf{c}^T \left(d_0 \mathbf{I}^{[n]} + d_1 \mathbf{A} + \cdots + d_{n-1} \mathbf{A}^{n-1} + \mathbf{A}^n \right). \tag{7.189}$$

Thus, (7.187) is a linear state feedback control law in which (7.188) and (7.189) provide an alternative method of design to that presented in Chap. 5, Sect. 5.2, that is rather more straightforward and lends itself readily to computer-aided design.

7.3.4.3 Multivariable LTI Plants of Full Relative Degree

The derivation of this control law is based on the plant model of (7.179) and (7.180). First, it is convenient to write (7.182) in terms of its component equations as follows:

$$y_i = \mathbf{c}_i^T \mathbf{x}, \quad i = 1, 2, \ldots, m. \tag{7.190}$$

Then the sequence of derivatives required to determine the relative degree with respect to each output is formed as follows:

$$\left.\begin{array}{l}\dot{y}_i = \mathbf{c}_i^T \mathbf{A} \mathbf{x} \\ \ddot{y}_i = \mathbf{c}_i^T \mathbf{A}^2 \mathbf{x} \\ \vdots \\ y_i^{(R_i-1)} = \mathbf{c}_i^T \mathbf{A}^{R_i-1} \mathbf{x}\end{array}\right\}, \quad i = 1, 2, \ldots, m \quad (7.191)$$

$$y_i^{(R_i)} = \mathbf{c}_i^T \mathbf{A}^{R_i} \mathbf{x} + \mathbf{c}_i^T \mathbf{A}^{R_i-1} \mathbf{B} \mathbf{u}, \, i = 1, 2, \ldots, m, \quad (7.192)$$

where

$$\sum_{i=1}^{n} R_i = n \quad (7.193)$$

since the plant is of full relative degree. It is evident from (7.192) that the relative degrees with respect to the individual outputs are determined by evaluating the sequence, $\mathbf{c}_i^T \mathbf{A}^q \mathbf{B}$, $q = 0, 1, \ldots$, until a nonvanishing term is found.

Let this be $\mathbf{c}_i^T \mathbf{A}^{q_{max}} \mathbf{B}$. Then the relative degree with respect to y_i is $R_i = q_{max} + 1$. Then the desired closed-loop differential equation is set up for each output as follows:

$$y_i^{(R_i)} = d_{0i}(y_{ri} - y_i) - d_{1i}\dot{y} - \cdots - d_{R_i-1,i} y^{(R_i-1)}, \quad i = 1, 2, \ldots, m, \quad (7.194)$$

where the constant coefficients, d_{ji}, $j = 0, 1, \ldots, R_i - 1$, $i = 1, 2, \ldots, m$, are determined using the method of Sect. 7.3.1.4. The required FDC law is then obtained by equating the RHS of (7.192) and (7.194) and then solving for $\mathbf{u}$. Thus,

$$\mathbf{c}_i^T \mathbf{A}^{R_i} \mathbf{x} + \mathbf{c}_i^T \mathbf{A}^{R_i-1} \mathbf{B} \mathbf{u} = d_{0i}(y_{ri} - y_i) - d_{1i}\dot{y}_i - \cdots - d_{R_i-1,i} y_i^{(R_i-1)} \Rightarrow$$

$$\mathbf{c}_i^T \mathbf{A}^{R_i-1} \mathbf{B} \mathbf{u} = d_{0i}(y_{ri} - y_i) - d_{1i}\dot{y}_i - \cdots - d_{R_i-1,i} y_i^{(R_i-1)} - \mathbf{c}_i^T \mathbf{A}^{R_i} \mathbf{x} \Rightarrow$$

$$\begin{bmatrix} \mathbf{c}_1^T \mathbf{A}^{R_1-1} \\ \mathbf{c}_2^T \mathbf{A}^{R_2-1} \\ \vdots \\ \mathbf{c}_m^T \mathbf{A}^{R_m-1} \end{bmatrix} \mathbf{B} \mathbf{u} = \begin{bmatrix} d_{01}(y_{r1} - y_1) - d_{11}\dot{y}_1 - \cdots - d_{R_1-1,1} y_1^{(R_1-1)} \\ d_{02}(y_{r2} - y_2) - d_{12}\dot{y} - \cdots - d_{R_2-1,2} y_2^{(R_2-1)} \\ \vdots \\ d_{0m}(y_{rm} - y_m) - d_{1m}\dot{y} - \cdots - d_{R_m-1,m} y_m^{(R_m-1)} \end{bmatrix} - \begin{bmatrix} \mathbf{c}_1^T \mathbf{A}^{R_1} \\ \mathbf{c}_2^T \mathbf{A}^{R_2} \\ \vdots \\ \mathbf{c}_m^T \mathbf{A}^{R_m} \end{bmatrix} \mathbf{x} \Rightarrow$$

$$\mathbf{u} = \left(\begin{bmatrix} \mathbf{c}_1^T \mathbf{A}^{R_1-1} \\ \mathbf{c}_2^T \mathbf{A}^{R_2-1} \\ \vdots \\ \mathbf{c}_m^T \mathbf{A}^{R_m-1} \end{bmatrix} \mathbf{B} \right)^{-1} \left(\begin{bmatrix} d_{01}(y_{r1}-y_1)-d_{11}\dot{y}_1-\cdots-d_{R_1-1,1}y_1^{(R_1-1)} \\ d_{02}(y_{r2}-y_2)-d_{12}\dot{y}_2-\cdots-d_{R_2-1,2}y_2^{(R_2-1)} \\ \vdots \\ d_{0m}(y_{rm}-y_m)-d_{1m}\dot{y}_m-\cdots-d_{R_m-1,m}y_m^{(R_m-1)} \end{bmatrix} - \begin{bmatrix} \mathbf{c}_1^T \mathbf{A}^{R_1} \\ \mathbf{c}_2^T \mathbf{A}^{R_2} \\ \vdots \\ \mathbf{c}_m^T \mathbf{A}^{R_m} \end{bmatrix} \mathbf{x} \right).$$

$$(7.195)$$

7.3 Feedback Linearising and Forced Dynamic Control

This is useful as it stands, although the state representation is mixed between that of the plant model and the output derivatives. On the other hand, if the plant model state representation is preferred throughout, then the required state transformation equations are (7.191). Then substituting for $y_i^{(j)}$, $j = 0, 1, \ldots, R_i - 1$, $i = 1, 2, \ldots, m$, in (7.195) using (7.191) yields

$$\mathbf{u} = \left[\begin{bmatrix} \mathbf{c}_1^T \mathbf{A}^{R_1-1} \\ \mathbf{c}_2^T \mathbf{A}^{R_2-1} \\ \vdots \\ \mathbf{c}_m^T \mathbf{A}^{R_m-1} \end{bmatrix} \mathbf{B} \right]^{-1} \left(\begin{bmatrix} d_{01} y_{r1} \\ d_{02} y_{r2} \\ \vdots \\ d_{0m} y_{rm} \end{bmatrix} - \begin{bmatrix} \mathbf{c}_1^T \left[d_{01} \mathbf{I}^{[n]} + d_{11} \mathbf{A} + \ldots + d_{R_1-1,1} \mathbf{A}^{R_1-1} + \mathbf{A}^{R_1} \right] \\ \mathbf{c}_2^T \left[d_{02} \mathbf{I}^{[n]} + d_{12} \mathbf{A} + \ldots + d_{R_2-1,2} \mathbf{A}^{R_2-1} + \mathbf{A}^{R_2} \right] \\ \vdots \\ \mathbf{c}_m^T \left[d_{0m} \mathbf{I}^{[n]} - d_{1m} \mathbf{A} + \ldots + d_{R_m-1,m} \mathbf{A}^{R_m-1} + \mathbf{A}^{R_m} \right] \end{bmatrix} \mathbf{x} \right),$$

which is in the standard multivariable linear state feedback form,

$$\mathbf{u} = \mathbf{R} \mathbf{y}_r - \mathbf{K} \mathbf{x}, \tag{7.196}$$

where the reference input scaling matrix is

$$\mathbf{R} = \left[\begin{bmatrix} \mathbf{c}_1^T \mathbf{A}^{R_1-1} \\ \mathbf{c}_2^T \mathbf{A}^{R_2-1} \\ \vdots \\ \mathbf{c}_m^T \mathbf{A}^{R_m-1} \end{bmatrix} \mathbf{B} \right]^{-1} \begin{bmatrix} d_{01} & 0 & \cdots & 0 \\ 0 & d_{02} & & \vdots \\ \vdots & & \ddots & 0 \\ 0 & \cdots & 0 & d_{0m} \end{bmatrix} \tag{7.197}$$

and the linear state feedback gain matrix is

$$\mathbf{K} = \left[\begin{bmatrix} \mathbf{c}_1^T \mathbf{A}^{R_1-1} \\ \mathbf{c}_2^T \mathbf{A}^{R_2-1} \\ \vdots \\ \mathbf{c}_m^T \mathbf{A}^{R_m-1} \end{bmatrix} \mathbf{B} \right]^{-1} \begin{bmatrix} \mathbf{c}_1^T \left[d_{01} \mathbf{I}^{[n]} + d_{11} \mathbf{A} + \ldots + d_{R_1-1,1} \mathbf{A}^{R_1-1} + \mathbf{A}^{R_1} \right] \\ \mathbf{c}_2^T \left[d_{02} \mathbf{I}^{[n]} + d_{12} \mathbf{A} + \ldots + d_{R_2-1,2} \mathbf{A}^{R_2-1} + \mathbf{A}^{R_2} \right] \\ \vdots \\ \mathbf{c}_m^T \left[d_{0m} \mathbf{I}^{[n]} - d_{1m} \mathbf{A} + \ldots + d_{R_m-1,m} \mathbf{A}^{R_m-1} + \mathbf{A}^{R_m} \right] \end{bmatrix}. \tag{7.198}$$

Since Eq. (7.194) have no common variables, the control law achieves complete interaction elimination as well as the specified dynamics relating y_i to y_{ri}.

Example 7.7 Lateral motion control of an aircraft

The state-space model governing the lateral motion of an aircraft is given as

$$\begin{bmatrix} \dot{x}_1 \\ \dot{x}_2 \\ \dot{x}_3 \\ \dot{x}_4 \\ \dot{x}_5 \end{bmatrix} = \begin{bmatrix} a_{11} & 0 & a_{13} & 0 & a_{15} \\ 0 & 0 & 1 & 0 & 0 \\ a_{31} & 0 & a_{33} & 0 & a_{35} \\ 0 & 0 & 0 & 0 & 1 \\ a_{51} & 0 & a_{53} & 0 & a_{55} \end{bmatrix} \begin{bmatrix} x_1 \\ x_2 \\ x_3 \\ x_4 \\ x_5 \end{bmatrix} + \begin{bmatrix} b_{11} & b_{12} & b_{13} \\ 0 & 0 & 0 \\ b_{31} & b_{32} & b_{33} \\ 0 & 0 & 0 \\ b_{51} & b_{52} & b_{53} \end{bmatrix} \begin{bmatrix} u_1 \\ u_2 \\ u_3 \end{bmatrix} \quad (7.199)$$

$$\begin{bmatrix} y_1 \\ y_2 \\ y_3 \end{bmatrix} = \begin{bmatrix} 1 & 0 & 0 & 0 & 0 \\ 0 & 1 & 0 & 0 & 0 \\ 0 & 0 & 0 & 1 & 0 \end{bmatrix} \begin{bmatrix} x_1 \\ x_2 \\ x_3 \\ x_4 \\ x_5 \end{bmatrix} \quad (7.200)$$

where x_1 is the sideslip angle, x_2 is the roll attitude angle, x_3 is the roll angular velocity, x_4 is the yaw attitude angle and x_5 is the yaw angular velocity. The plant parameters are given [4] as $\begin{bmatrix} a_{11} & a_{13} & a_{15} \\ a_{31} & a_{33} & a_{35} \\ a_{51} & a_{53} & a_{55} \end{bmatrix} = \begin{bmatrix} -0.163 & 0 & -1.0 \\ 16.6 & -1.08 & -0.13 \\ 15.7 & -0.02 & -0.25 \end{bmatrix}$ and $\begin{bmatrix} b_{11} & b_{12} & b_{13} \\ b_{31} & b_{32} & b_{33} \\ b_{51} & b_{52} & b_{53} \end{bmatrix} = \begin{bmatrix} -0.0054 & 0.05 & -0.025 \\ 42.3 & 6.88 & 0.08 \\ 1.08 & -11.7 & -1.25 \end{bmatrix}$. It is required to design a control system that eliminates interaction and yields non-overshooting step responses with settling times of $T_{s1} = 4$ [s], $T_{s2} = 2$ [s] and $T_{s3} = 3$ [s].

First, (7.200) will be written as

$$y_1 = \underbrace{\begin{bmatrix} 1 & 0 & 0 & 0 & 0 \end{bmatrix}}_{\mathbf{c}_1^T} \begin{bmatrix} x_1 \\ x_2 \\ x_3 \\ x_4 \\ x_5 \end{bmatrix}, \quad y_2 = \underbrace{\begin{bmatrix} 0 & 1 & 0 & 0 & 0 \end{bmatrix}}_{\mathbf{c}_2^T} \begin{bmatrix} x_1 \\ x_2 \\ x_3 \\ x_4 \\ x_5 \end{bmatrix}, \quad y_3 = \underbrace{\begin{bmatrix} 0 & 0 & 0 & 1 & 0 \end{bmatrix}}_{\mathbf{c}_3^T} \begin{bmatrix} x_1 \\ x_2 \\ x_3 \\ x_4 \\ x_5 \end{bmatrix}.$$

(7.201)

The relative degrees with respect to these outputs are determined as follows: $\mathbf{c}_1^T \mathbf{b} = \begin{bmatrix} -0.0054 & 0.05 & -0.025 \end{bmatrix} \neq \mathbf{0}$. Hence, the relative degree w.r.t. y_1 is $R_1 = 1$. $\mathbf{c}_2^T \mathbf{b} = \mathbf{0}$. $\mathbf{c}_2^T \mathbf{Ab} = \begin{bmatrix} 42.3 & 6.88 & 0.08 \end{bmatrix} \neq \mathbf{0}$. Hence, the relative degree w.r.t. y_2 is $R_2 = 2$. $\mathbf{c}_3^T \mathbf{b} = \mathbf{0}$. $\mathbf{c}_3^T \mathbf{Ab} = \begin{bmatrix} 1.08 & -11.7 & -1.25 \end{bmatrix} \neq \mathbf{0}$. Hence, the relative degree w.r.t. y_3 is $R_3 = 2$. Since $R_1 + R_2 + R_3 = 5 = n$, the plant is of full relative degree.

The forced dynamic control law is the linear state feedback control law,

$$\mathbf{u} = \mathbf{R}\mathbf{y}_r - \mathbf{K}\mathbf{x}, \quad (7.202)$$

7.3 Feedback Linearising and Forced Dynamic Control

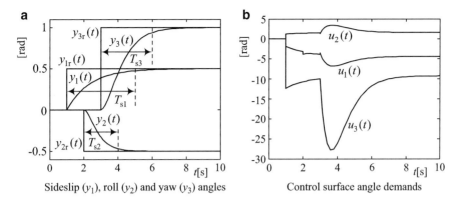

Fig. 7.15 Step responses of forced dynamic lateral motion control of an aircraft

where

$$\mathbf{R} = \left[\begin{bmatrix} \mathbf{c}_1^T \\ \mathbf{c}_2^T \mathbf{A} \\ \mathbf{c}_3^T \mathbf{A} \end{bmatrix} \mathbf{b}\right]^{-1} \begin{bmatrix} d_{01} & 0 & 0 \\ 0 & d_{02} & 0 \\ 0 & 0 & d_{03} \end{bmatrix} \quad \text{and}$$

$$\mathbf{K} = \left[\begin{bmatrix} \mathbf{c}_1^T \\ \mathbf{c}_2^T \mathbf{A} \\ \mathbf{c}_3^T \mathbf{A} \end{bmatrix} \mathbf{b}\right]^{-1} \begin{bmatrix} \mathbf{c}_1^T \left[d_{01}\mathbf{I}^{[5]} + \mathbf{A}\right] \\ \mathbf{c}_2^T \left[d_{02}\mathbf{I}^{[5]} + d_{12}\mathbf{A} + \mathbf{A}^2\right] \\ \mathbf{c}_3^T \left[d_{03}\mathbf{I}^{[5]} + d_{13}\mathbf{A} + \mathbf{A}^2\right] \end{bmatrix}. \quad (7.203)$$

Using the method of Sect. 7.3.1.4, the desired closed-loop differential equation coefficients in terms of the required settling times are given by:

$$p_1 = 3/T_{s1}, d_{01} = p_1. \quad (7.204)$$

$$p_2 = 4.5/T_{s2}, d_{02} = p_2^2, d_{12} = 2p_2. \quad (7.205)$$

$$p_3 = 4.5/T_{s3}, d_{03} = p_3^2, d_{13} = 2p_3. \quad (7.206)$$

Figure 7.15 shows the results of a simulation.

The step reference input components are applied at different times to demonstrate the complete absence of interaction, i.e. $y_{ir}(t)$ influences only $y_i(t)$, $i = 1, 2, 3$. Hence, $y_{1r}(t) = 0.5h(t-1)$ [rad], $y_{2r}(t) = -0.5h(t-2)$ [rad] and $y_{3r}(t) = h(t-3)$ [rad]. As expected, $y_1(t)$, is a first-order step response, while $y_2(t)$ and $y_3(t)$ are critically damped step responses, each with the specified settling time. For $t \in (1, 2)$ [s], $u_1(t)$ acts to produce the required response of $y_1(t)$ while $u_2(t)$ and $u_3(t)$ counteract any interaction that would cause $y_2(t)$ and $y_3(t)$ to change from zero.

7.3.4.4 LTI SISO Plants Less than Full Relative Degree

First, it will be useful to make some comparisons between the FDC of SISO plants and the linear state feedback control of SISO plants introduced in Chap. 5. It has been established in Sect. 7.3.4.2 that an FDC law for a linear SISO plant based on a linear-desired closed-loop dynamics is a linear state feedback control law. The generic FDC law (7.189) for plants of full relative degree could also be applied to a plant having relative degree $R < n$, but with caution. It follows from Sect. 7.3.3.1 that the response of the controlled output to the reference input would have the prescribed dynamics of order R, while the zero dynamics, unobservable by viewing the reference input and the controlled output, is of order $n - R$. For comparison, it will be recalled from Chap. 5, Sect. 5.2.7.3, that if a linear state feedback control system is designed for a plant having a transfer function with finite zeros and the method of pole assignment is used to cancel the zeros, then the closed-loop system is of order $n - m$, where n is the degree of the denominator polynomial of the plant transfer function and m is the degree of the numerator polynomial. This system only *appears* to be of order $n - m$, when viewing only the reference input and output, and is actually still of order n, the m poles that cancel the m zeros constituting an uncontrolled subsystem of order m. It is clear from this comparison that the zero dynamics occurring with FDC of an LTI plant is precisely the same as the dynamics of the uncontrolled subsystem when LSF control is applied with zero cancellation. In conclusion, the *zero dynamics* is a linear uncontrolled subsystem whose poles (eigenvalues) are the transfer function *zeros* and the relative degree of a linear SISO plant is $R = n - m$, i.e. the difference between the degrees of the plant transfer function denominator and numerator polynomials. This explains the choice of the names 'zero dynamics' and 'relative degree' which is not so obvious in the context of FLC of nonlinear plants (Sect. 7.3.3).

It will also be recalled from Chap. 5 that if the poles of a linear state feedback control system are all placed at one location to achieve a specified settling time, then the step response of the controlled output may have a finite number of stationary points, which could be overshoots and undershoots, due to the derivative effect of the plant zeros. In other words, let the transfer function of the closed-loop system be

$$\frac{Y(s)}{Y_r(s)} = \frac{1 + w_1 s + \cdots + w_m s^m}{(1 + sT_c)^n}. \tag{7.207}$$

Then the step response, $z(t)$, of the zero-less system with transfer function

$$\frac{Z(s)}{Y_r(s)} = \frac{1}{(1 + sT_c)^n} \tag{7.208}$$

will monotonically increase towards the steady-state value, while the corresponding response, $y(t)$, of the system with transfer function (7.207) may have a finite number of stationary points due to the derivative terms in

7.3 Feedback Linearising and Forced Dynamic Control

$$y(t) = z(t) + w_1\dot{z}(t) + \cdots + w_m z^{(m)}(t), \quad (7.209)$$

dependent on the values of the derivative weighting coefficients, w_i, $i = 1, 2, \ldots, m$.

If the straightforward FDC of Sect. 7.3.4.2 is applied with relative degree, $R < n$, then it is equivalent to placing m of the closed-loop poles to cancel the zeros so that (7.207) becomes

$$\frac{Y(s)}{Y_r(s)} = \frac{1 + w_1 s + \cdots + w_m s^m}{\underbrace{(1 + w_1 s + \cdots + w_m s^m)(1 + sT_c)^R}_{\text{System characteristic polynomial}}} = \frac{1}{(1 + sT_c)^R}, \quad (7.210)$$

where $R = n - m$, thereby circumventing the derivative effects of the plant zeros. It is evident from (7.210), however, that the zeros of the plant transfer function have to lie in the left half of the s-plane. In any case, the behaviour of the zero dynamic subsystem has to be acceptable regarding the plant hardware. If not, then an approach similar to that of Sect. 7.3.3.2 can be taken. This is equivalent to applying linear state feedback control to place the poles according to (7.207), ignoring the plant zeros, thereby achieving complete control of the plant state, which ensures closed-loop stability but does allow stationary points in the step response due to the plant zeros.

In FDC, complete state control of an LTI plant of less than full relative degree is achieved using the calculated plant output, z, in (7.208) instead of y.

Hence, the model of (7.181) and (7.182) is appended with the auxiliary output,

$$z = \mathbf{g}^T \mathbf{x}, \quad (7.211)$$

where the output matrix, $\mathbf{g}^T$, has to be determined such that:

(a) The plant has full relative degree, $R = n$, with respect to z, where n is the plant order,
(b) If the plant is maintained in the steady state, meaning that (7.181), (7.182) and (7.211) are replaced by

$$\dot{\mathbf{x}}_{ss} = \mathbf{A}\mathbf{x}_{ss} + \mathbf{b}u_{ss} = \mathbf{0}, \quad (7.212)$$

$$y_{ss} = \mathbf{c}^T \mathbf{x}_{ss} \quad (7.213)$$

and

$$z_{ss} = \mathbf{g}^T \mathbf{x}_{ss} \quad (7.214)$$

with all the variables at constant values, then

$$z_{ss} = y_{ss}. \quad (7.215)$$

It follows from (7.213), (7.214) and (7.215) that

$$\mathbf{g}^T \mathbf{x}_{ss} = \mathbf{c}^T \mathbf{x}_{ss}. \tag{7.216}$$

Note that although the plant may be inherently unstable, the above steady-state conditions are those which would be maintained by a controller that ensures closed-loop stability. This condition may be found without having to design the controller first.

To satisfy condition (a), above, $\mathbf{g}^T$ must satisfy a set of equations based on the sequence of output derivatives up to order, $n - 1$, as follows:

$$\dot{z} = \mathbf{g}^T \dot{\mathbf{x}} = \mathbf{g}^T [\mathbf{A}\mathbf{x} + \mathbf{b}u] = \mathbf{g}^T \mathbf{A}\mathbf{x} \Rightarrow \mathbf{g}^T \mathbf{b} = 0 \tag{7.217}$$

$$\ddot{z} = \mathbf{g}^T \mathbf{A}\dot{\mathbf{x}} = \mathbf{g}^T \mathbf{A} [\mathbf{A}\mathbf{x} + \mathbf{b}u] = \mathbf{g}^T \mathbf{A}^2 \mathbf{x} \Rightarrow \mathbf{g}^T \mathbf{A}\mathbf{b} = 0 \tag{7.218}$$

$$\vdots$$

$$z^{(n-1)} = \mathbf{g}^T \mathbf{A}^{n-2} \dot{\mathbf{x}} = \mathbf{g}^T \mathbf{A}^{n-2} [\mathbf{A}\mathbf{x} + \mathbf{b}u] = \mathbf{g}^T \mathbf{A}^{n-1} \mathbf{x} \Rightarrow \mathbf{g}^T \mathbf{A}^{n-2} \mathbf{b} = 0. \tag{7.219}$$

Finally, (7.217), (7.218), (7.219) and (7.216) may be written as

$$\mathbf{g}^T \begin{bmatrix} \mathbf{b} & \mathbf{A}\mathbf{b} & \cdots & \mathbf{A}^{n-2}\mathbf{b} & \mathbf{x}_{ss} \end{bmatrix} = \begin{bmatrix} 0 & 0 & \cdots & 0 & \mathbf{c}^T \mathbf{x}_{ss} \end{bmatrix}. \tag{7.220}$$

The required solution can then be obtained by first finding a suitable steady state, $\mathbf{x}_{ss}$, which is not unique, obtained by applying a constant $u = u_{ss}$. If the plant contains pure integrators, u_{ss} would have to be zero. Then a non-zero $\mathbf{x}_{ss}$ would be obtained by setting non-zero arbitrary initial conditions for these integrators. Once $\mathbf{x}_{ss}$ has been determined, then the required auxiliary output matrix would be calculated from (7.220) as

$$\mathbf{g}^T = \begin{bmatrix} 0 & 0 & \cdots & 0 & \mathbf{c}^T \mathbf{x}_{ss} \end{bmatrix} \begin{bmatrix} \mathbf{b} & \mathbf{A}\mathbf{b} & \cdots & \mathbf{A}^{n-2}\mathbf{b} & \mathbf{x}_{ss} \end{bmatrix}^{-1}. \tag{7.221}$$

The required FDC law would then be (7.186) with $\mathbf{c}^T$ replaced by $\mathbf{g}^T$.

Example 7.8 Forced dynamic cruise speed control of road transportation vehicle

The plant of this example is the Diesel driveline of Chap. 5, Example 5.8. The transfer function model (data courtesy of Delphi Diesel) is given as

$$\frac{Y(s)}{U(s)} = \frac{c_2 s^2 + c_1 s + c_0}{s^3 + a_2 s^2 + a_1 s + a_0} \tag{7.222}$$

7.3 Feedback Linearising and Forced Dynamic Control

where $a_0 = -0.2454$, $a_1 = 245.4$, $a_2 = 7.594$, $c_0 = 9236$, $c_1 = 285.5$ and $c_2 = 475$, $U(s)$ is the throttle input and $Y(s)$ is the measurement of the engine speed [r/min].

The corresponding state-space model in the controller canonical form is

$$\begin{bmatrix} \dot{x}_1 \\ \dot{x}_2 \\ \dot{x}_3 \end{bmatrix} = \begin{bmatrix} 0 & 1 & 0 \\ 0 & 0 & 1 \\ -a_0 & -a_1 & -a_2 \end{bmatrix} \begin{bmatrix} x_1 \\ x_2 \\ x_3 \end{bmatrix} + \begin{bmatrix} 0 \\ 0 \\ 1 \end{bmatrix} u \quad (7.223)$$

$$y = \begin{bmatrix} c_0 & c_1 & c_2 \end{bmatrix} \begin{bmatrix} x_1 \\ x_2 \\ x_3 \end{bmatrix}. \quad (7.224)$$

First, to demonstrate that the plant is less than full relative degree with respect to y,

$$\dot{y} = c_0 \dot{x}_1 + c_1 \dot{x}_2 + c_2 \dot{x}_3 = c_0 x_2 + c_1 x_3 + c_2 u. \quad (7.225)$$

Since this first derivative directly depends on u, the relative degree with respect to y is $R = 1$. This is less than $n = 3$, and therefore an auxiliary output

$$z = \begin{bmatrix} g_0 & g_1 & g_2 \end{bmatrix} \begin{bmatrix} x_1 \\ x_2 \\ x_3 \end{bmatrix} \quad (7.226)$$

is required. To determine g_1, g_2 and g_3, the steady-state solution of (7.223) for x_{1ss}, x_{2ss} and x_{3ss} with a constant input, u_{ss}, is as follows:

$$\begin{bmatrix} \dot{x}_{1ss} \\ \dot{x}_{2ss} \\ \dot{x}_{3ss} \end{bmatrix} = \begin{bmatrix} 0 & 1 & 0 \\ 0 & 0 & 1 \\ -a_0 & -a_1 & -a_2 \end{bmatrix} \begin{bmatrix} x_{1ss} \\ x_{2ss} \\ x_{3ss} \end{bmatrix} + \begin{bmatrix} 0 \\ 0 \\ 1 \end{bmatrix} u_{ss} = \begin{bmatrix} 0 \\ 0 \\ 0 \end{bmatrix} \Rightarrow$$
$$\dot{x}_{1ss} = x_{2ss} = 0 \Rightarrow x_{1ss} = \text{const.}, \ x_{3ss} = 0 \quad (7.227)$$

Then (7.221) becomes

$$\mathbf{g}^T = \begin{bmatrix} 0 & 0 & \mathbf{c}^T \mathbf{x}_{ss} \end{bmatrix} \begin{bmatrix} \mathbf{b} & \mathbf{Ab} & \mathbf{x}_{ss} \end{bmatrix}^{-1} = \begin{bmatrix} 0 & 0 & c_0 x_{1ss} \end{bmatrix} \begin{bmatrix} 0 & 0 & x_{1ss} \\ 0 & 1 & 0 \\ 1 & -a_2 & 0 \end{bmatrix}^{-1}$$

$$= \begin{bmatrix} 0 & 0 & c_0 x_{1ss} \end{bmatrix} \begin{bmatrix} 0 & -a_2 x_{1ss} & -x_{1ss} \\ 0 & -x_{1ss} & 0 \\ -1 & 0 & 0 \end{bmatrix} \frac{1}{-x_{1ss}} = \begin{bmatrix} c_0 & 0 & 0 \end{bmatrix}. \quad (7.228)$$

The required auxiliary output equation is therefore

$$z = \underbrace{\begin{bmatrix} c_0 & 0 & 0 \end{bmatrix}}_{\mathbf{g}^T} \begin{bmatrix} x_1 \\ x_2 \\ x_3 \end{bmatrix}. \qquad (7.229)$$

The relative degree with respect to z is now $R = 3$. The desired closed-loop differential equation of the same order is

$$\dddot{z} = d_0(y_r - z) - d_1\dot{z} - d_2\ddot{z} \qquad (7.230)$$

whose coefficients can be determined by the method of Sect. 7.3.1.4 to achieve a specified settling time, T_s. For the 5 % criterion,

$$p = 6/T_s, d_0 = p^3, d_1 = 3p^2 \text{ and } d_2 = 3p. \qquad (7.231)$$

The control law is then (7.187), (7.188) and (7.189) with $n = 3$ and $\mathbf{c}^T$ replaced by $\mathbf{g}^T$.

Figure 7.16a shows a simulation with a step engine speed reference of 300 [r/min] and a nominal settling time of $T_s = 1$ [s]. As expected, $z(t)$ is monotonic with the correct settling time, but $y(t)$ has two stationary points due to the plant zeros, and this transient effect has increased the actual settling time, T_{sa}. In Fig. 7.16b, the nominal settling time, T_s, has been reduced to bring T_{sa} down to the specified value of 1 [s]. It will be noticed, however, that this has increased the 'kink' in the step response. The reason for this is that reducing T_s increases the magnitude of the triple closed-loop pole while the plant zeros are fixed, these zeros consequently having more influence on the transient response (Chap. 1).

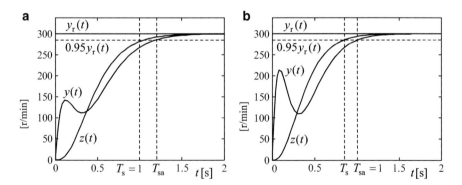

Fig. 7.16 Step response for forced dynamic speed control of a Diesel driveline. (**a**) With nominal settling time. (**b**) With adjusted settling time

7.3 Feedback Linearising and Forced Dynamic Control

7.3.4.5 Multivariable Plants Less than Full Relative Degree

For a multivariable plant, model (7.179) is appended with an auxiliary output vector,

$$\mathbf{z} = \mathbf{G}\mathbf{x}, \tag{7.232}$$

where the output matrix, $\mathbf{G}$, has to be determined such that:

(a) The plant has full relative degree with respect to $\mathbf{z}$, requiring

$$\sum_{i=1}^{m} R_i = n \tag{7.233}$$

where R_i is the relative degree with respect to auxiliary output, z_i.

(b) When the plant is maintained in the steady state, meaning that (7.179), (7.180) and (7.232) are replaced by

$$\dot{\mathbf{x}}_{ss} = \mathbf{A}\mathbf{x}_{ss} + \mathbf{B}\mathbf{u}_{ss} = \mathbf{0}, \tag{7.234}$$

$$\mathbf{y}_{ss} = \mathbf{C}\mathbf{x}_{ss} \tag{7.235}$$

and

$$\mathbf{z}_{ss} = \mathbf{G}\mathbf{x}_{ss}, \tag{7.236}$$

then

$$\mathbf{z}_{ss} = \mathbf{y}_{ss}. \tag{7.237}$$

Each application would have to be studied individually to determine R_i, $i = 1, 2, \ldots, m$.

As already highlighted in Sect. 7.3.4.4 for SISO plants, the steady-state conditions of (b) would be maintained by a controller regardless of whether the plant is inherently stable or unstable and can be derived independently of the controller.

First, the component equations of (7.236) may be written separately as

$$z_i = \mathbf{g}_i^T \mathbf{x}, \quad i = 1, 2, \ldots, m. \tag{7.238}$$

Then the equations that $\mathbf{g}_i^T$, $i = 1, 2, \ldots, m$, must obey to satisfy condition (a) are

$$\left. \begin{array}{l} \dot{z}_i = \mathbf{g}_i^T \dot{\mathbf{x}} = \mathbf{g}_i^T [\mathbf{A}\mathbf{x} + \mathbf{B}\mathbf{u}] = \mathbf{g}_i^T \mathbf{A}\mathbf{x} \Rightarrow \mathbf{g}_i^T \mathbf{B} = \mathbf{0} \\ \ddot{z}_i = \mathbf{g}_i^T \mathbf{A}\dot{\mathbf{x}} = \mathbf{g}_i^T \mathbf{A} [\mathbf{A}\mathbf{x} + \mathbf{B}\mathbf{u}] = \mathbf{g}_i^T \mathbf{A}^2 \mathbf{x} \Rightarrow \mathbf{g}_i^T \mathbf{A}\mathbf{B} = \mathbf{0} \\ \vdots \\ z_i^{(R_i - 1)} = \mathbf{g}_i^T \mathbf{A}^{R_i - 2} \dot{\mathbf{x}} = \mathbf{g}_i^T \mathbf{A}^{R_i - 2} [\mathbf{A}\mathbf{x} + \mathbf{B}\mathbf{u}] \Rightarrow \mathbf{g}_i^T \mathbf{A}^{R_i - 2} \mathbf{B} = \mathbf{0} \end{array} \right\}, \quad i = 1, 2, \ldots, m, \tag{7.239}$$

where R_i is the relative degree with respect to z_i. Further equations are obtained by applying condition (b). Equations (7.236) and (7.236) may be rewritten as

$$y_{iss} = \mathbf{c}_i^T \mathbf{x}_{ss}, \quad i = 1, 2, \ldots, m \tag{7.240}$$

and

$$z_{iss} = \mathbf{g}_i^T \mathbf{x}_{ss}. \quad i = 1, 2, \ldots, m. \tag{7.241}$$

For (7.237) to be satisfied, then in view of (7.240) and (7.241),

$$\mathbf{g}_i^T \mathbf{x}_{ss} = \mathbf{c}_i^T \mathbf{x}_{ss}, \quad i = 1, 2, \ldots, m. \tag{7.242}$$

This enables the determination of a limited number of elements of $\mathbf{g}_i^T$, as was the case in Sect. 7.3.4.4 for SISO plants. Then they could be inserted in Eqs. (7.239) and the remaining elements of $\mathbf{g}_i^T$ calculated, but this would have to be done manually. Instead labour could be saved by including these known elements in a procedure that could be computerised. Let the known elements be denoted by g_{iq}, where q is the column number of $\mathbf{g}_i^T$. Then equations could be written for them in the same form as the equations of (7.239) as follows:

$$\left(\mathbf{g}_i^T \begin{bmatrix} \vdots \\ 1 \\ \vdots \end{bmatrix} \leftarrow \text{Row } q \right) = \mathbf{g}_i^T \mathbf{n}_{iq} = g_{iq}. \tag{7.243}$$

Then (7.239) and (7.243) together constitute a set of simultaneous equations from which $\mathbf{g}_i^T$ can be calculated. These may be written as the following matrix equation:

$$\mathbf{g}_i^T \left[\mathbf{B} \ \mathbf{AB} \ \cdots \ \mathbf{A}^{R_i-2}\mathbf{B} \ \cdots \ \mathbf{n}_{iq} \ \cdots \right] = \left[0 \ 0 \ \cdots \ 0 \ \cdots \ g_{iq} \ \cdots \right]. \tag{7.244}$$

These equations must be at least n in number, meaning that either they are completely determined in which case the solution is

$$\mathbf{g}_i^T = \left[0 \ 0 \ \cdots \ 0 \ \cdots \ g_{iq} \ \cdots \right] \left[\mathbf{B} \ \mathbf{AB} \ \cdots \ \mathbf{A}^{R_i-2}\mathbf{B} \ \cdots \ \mathbf{n}_{iq} \ \cdots \right]^{-1} \tag{7.245}$$

or overdetermined in which case the solution is given by

$$\mathbf{g}_i^T \mathbf{M}\mathbf{M}^T = \left[0 \ 0 \ \cdots \ 0 \ \cdots \ g_{iq} \ \cdots \right] \mathbf{M}^T \Rightarrow$$

$$\mathbf{g}_i^T = \left[0 \ 0 \ \cdots \ 0 \ \cdots \ g_{iq} \ \cdots \right] \mathbf{M}^T \left[\mathbf{M}\mathbf{M}^T \right]^{-1} \tag{7.246}$$

where $\mathbf{M} = \left[\mathbf{B} \ \mathbf{AB} \ \cdots \ \mathbf{A}^{R_i-2}\mathbf{B} \ \cdots \ \mathbf{n}_{iq} \ \cdots \right]$.

7.3 Feedback Linearising and Forced Dynamic Control

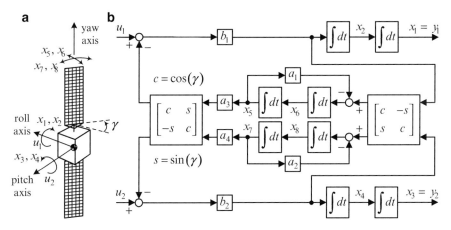

Fig. 7.17 Roll/yaw axis dynamics and kinematics model of a satellite. (**a**) Satellite configuration. (**b**) State-variable block diagram in the inverse dynamic form

Example 7.9 Roll/yaw axes attitude control of satellite with flexible solar panel

Figure 7.17 shows the model of a space satellite for which is required an attitude control system for the roll and pitch axes between which is very significant cross coupling due to s-shaped out-of-plane and in-plane flexural vibration modes in the solar panels. In this model, the two double integrators in the forward path of the state-variable block diagram represent the dynamics and kinematics of the central rigid body about the roll and pitch axes in which the roll and pitch attitude angles are x_1 and x_3, their respective derivatives being x_2 and x_4.

The flexible solar panels are represented by the inverse dynamic model in the feedback path. One vibration mode is modelled for the out-of-plane panel vibrations, by a torsional mass spring system having angular displacement, x_5, and angular velocity, x_6. The in-plane panel vibrations are similarly modelled with angular displacement, x_7, and angular velocity, x_8. In order to maximise the electrical power collected by the solar panels, they are rotated about the yaw axis through an angle, γ, to minimise the angle of incidence of the solar radiation vector. The control variables, u_1 and u_2, are numerically equal to the roll and pitch control torques. The constant coefficients in Fig. 7.17 are expressed in terms of the physical spacecraft parameters as follows:

$$b_1 = 1/J_{xx}, \ b_2 = 1/J_{yy}, \ a_1 = v_O^2, \ a_2 = v_I^2, \ a_3 = J_O a_1 \text{ and } a_4 = J_1 a_2, \tag{7.247}$$

where J_{xx} and J_{yy} are the roll and pitch axis moments of inertia of the rigid centre-body, v_O and J_O are the encastre natural frequency and moment of inertia of the mass-spring system representing the out-of-plane panel vibration mode and v_1 and J_1 are the encastre natural frequency and moment of inertia of the mass-spring system representing the in-plane panel vibration mode. The encastre frequencies are the panel vibration frequencies that would occur with the centre-body fixed in inertial space.

The state-space model corresponding to Fig. 7.17b is as follows:

$$\underbrace{\begin{bmatrix} \dot{x}_1 \\ \dot{x}_2 \\ \dot{x}_3 \\ \dot{x}_4 \\ \dot{x}_5 \\ \dot{x}_6 \\ \dot{x}_7 \\ \dot{x}_8 \end{bmatrix}}_{\dot{\mathbf{x}}} = \underbrace{\begin{bmatrix} 0 & 1 & 0 & 0 & 0 & 0 & 0 & 0 \\ 0 & 0 & 0 & 0 & a_{25} & 0 & a_{27} & 0 \\ 0 & 0 & 0 & 1 & 0 & 0 & 0 & 0 \\ 0 & 0 & 0 & 0 & a_{45} & 0 & a_{47} & 0 \\ 0 & 0 & 0 & 0 & 0 & 1 & 0 & 0 \\ 0 & 0 & 0 & 0 & a_{65} & 0 & a_{67} & 0 \\ 0 & 0 & 0 & 0 & 0 & 0 & 0 & 1 \\ 0 & 0 & 0 & 0 & a_{85} & 0 & a_{87} & 0 \end{bmatrix}}_{A} \underbrace{\begin{bmatrix} x_1 \\ x_2 \\ x_3 \\ x_4 \\ x_5 \\ x_6 \\ x_7 \\ x_8 \end{bmatrix}}_{\mathbf{x}} + \underbrace{\begin{bmatrix} 0 & 0 \\ b_1 & 0 \\ 0 & 0 \\ 0 & b_2 \\ 0 & 0 \\ b_{61} & b_{62} \\ 0 & 0 \\ b_{81} & b_{82} \end{bmatrix}}_{B} \underbrace{\begin{bmatrix} u_1 \\ u_2 \end{bmatrix}}_{\mathbf{u}} \qquad (7.248)$$

$$y_1 = \mathbf{c}_1^T \mathbf{x} = \begin{bmatrix} 1 & 0 & 0 & 0 & 0 & 0 & 0 & 0 \end{bmatrix} \mathbf{x} \qquad (7.249)$$

$$y_2 = \mathbf{c}_2^T \mathbf{x} = \begin{bmatrix} 0 & 0 & 1 & 0 & 0 & 0 & 0 & 0 \end{bmatrix} \mathbf{x} \qquad (7.250)$$

where

$$\begin{aligned} a_{25} &= -b_1 c a_3, \quad a_{27} = -b_1 s a_4, \quad a_{45} = b_2 s a_3, \quad a_{47} = -b_2 c a_4 \\ a_{65} &= -\left[a_1 + \left(c^2 b_1 + s^2 b_2\right) a_3\right], \quad a_{67} = cs\left(b_2 - b_1\right) a_4 \\ a_{85} &= cs\left(b_2 - b_1\right) a_3, \quad a_{87} = -\left[a_2 + \left(c^2 b_2 + s^2 b_1\right) a_4\right]. \\ b_{61} &= cb_1, \quad b_{62} = -sb_2, \quad b_{81} = sb_1, \quad b_{82} = cb_2 \end{aligned} \qquad (7.251)$$

First the relative degrees with respect to y_1 and y_2 are determined to demonstrate that the plant is less than full relative degree. Thus,

$$\dot{y}_1 = \dot{x}_1 = x_2, \quad \ddot{y}_1 = \dot{x}_2 = a_{25}x_5 + a_{27}x_7 + b_1 u_1. \qquad (7.252)$$

Hence, the relative degree with respect to y_1 is $R_1 = 2$. Similarly

$$\dot{y}_2 = \dot{x}_3 = x_4, \quad \ddot{y}_2 = \dot{x}_4 = a_{45}x_5 + a_{47}x_7 + b_2 u_2. \qquad (7.253)$$

Hence, the relative degree with respect to y_2 is $R_2 = 2$. Since $R_1 + R_2 = 4$ and the plant order is $n = 8$, the plant is less than full relative degree. The following auxiliary outputs are therefore needed:

$$z_1 = \mathbf{g}_1^T \mathbf{x} = \begin{bmatrix} g_{11} & g_{12} & g_{13} & g_{14} & g_{15} & g_{16} & g_{17} & g_{18} \end{bmatrix} \mathbf{x} \qquad (7.254)$$

$$z_2 = \mathbf{g}_2^T \mathbf{x} = \begin{bmatrix} g_{21} & g_{22} & g_{23} & g_{24} & g_{25} & g_{26} & g_{27} & g_{28} \end{bmatrix} \mathbf{x}. \qquad (7.255)$$

7.3 Feedback Linearising and Forced Dynamic Control

Next, consider the steady-state equation corresponding to (7.247). This yields

$$\dot{x}_{1ss} = x_{2ss} = 0 \Rightarrow x_{1ss} = \text{const.} \tag{7.256}$$

$$\dot{x}_{2ss} = a_{25}x_{5ss} + a_{27}x_{7ss} + b_1 u_{1ss} = 0 \tag{7.257}$$

$$\dot{x}_{3ss} = x_{4ss} = 0 \Rightarrow x_{3ss} = \text{const.} \tag{7.258}$$

$$\dot{x}_{4ss} = a_{45}x_{5ss} + a_{47}x_{7ss} + b_2 u_{2ss} = 0 \tag{7.259}$$

$$\dot{x}_{5ss} = x_{6ss} = 0 \Rightarrow x_{5ss} = \text{const.} \tag{7.260}$$

$$\dot{x}_{6ss} = -a_1 x_{5ss} + b_{61} u_{1ss} + b_{62} u_{2ss} \Rightarrow x_{6ss} = \text{const.} \tag{7.261}$$

$$\dot{x}_{7ss} = x_{8ss} = 0 \Rightarrow x_{7ss} = \text{const.} \tag{7.262}$$

$$\dot{x}_{8ss} = -a_2 x_{7ss} + b_{81} u_{1ss} + b_{82} u_{2ss} = 0 \Rightarrow x_{8ss} = \text{const.} \tag{7.263}$$

Control torques will cause angular accelerations of the rigid centre-body. This implies that the solution should include $u_{1ss} = u_{2ss} = 0$. Then (7.261) and (7.263) yield

$$x_{5ss} = 0, x_{7ss} = 0. \tag{7.264}$$

The steady-state vector is therefore

$$\mathbf{x}_{ss} = \begin{bmatrix} x_{1ss} & 0 & x_{3ss} & 0 & 0 & 0 & 0 & 0 \end{bmatrix}^T. \tag{7.265}$$

The steady-state auxiliary outputs are therefore

$$z_{1ss} = \mathbf{g}_1^T \mathbf{x}_{ss} = g_{11} x_{1ss} + g_{13} x_{3ss} \tag{7.266}$$

and

$$z_{2ss} = \mathbf{g}_2^T \mathbf{x}_{ss} = g_{21} x_{1ss} + g_{23} x_{3ss}. \tag{7.267}$$

Since

$$z_{1ss} = y_{1ss} = x_{1ss} \text{ and } z_{2ss} = y_{2ss} = x_{2ss}, \tag{7.268}$$

it follows from (7.266) and (7.267) that

$$g_{11} = 1, g_{13} = 0, \quad g_{21} = 0 \text{ and } g_{23} = 1. \tag{7.269}$$

The corresponding equations in the form of (7.243) are then

$$\mathbf{g}_1^T \mathbf{n}_{11} = 1, \ \mathbf{g}_1^T \mathbf{n}_{12} = 0, \ \mathbf{g}_2^T \mathbf{n}_{21} = 0 \text{ and } \mathbf{g}_2^T \mathbf{n}_{22} = 1, \quad (7.270)$$

where $\mathbf{n}_{11} = \mathbf{n}_{21} = \begin{bmatrix} 1 & 0 & 0 & 0 & 0 & 0 & 0 & 0 \end{bmatrix}^T$ and $\mathbf{n}_{12} = \mathbf{n}_{22} = \begin{bmatrix} 0 & 0 & 1 & 0 & 0 & 0 & 0 & 0 \end{bmatrix}^T$. By symmetry of Fig. 7.17, it can be concluded that the relative degrees with respect to z_1 and z_2 are equal and given by

$$R_1 = R_2 = n/2 = 8/2 = 4. \quad (7.271)$$

Then the simultaneous equations (7.244) are

$$\mathbf{g}_1^T \begin{bmatrix} \mathbf{B} & \mathbf{AB} & \mathbf{A}^2\mathbf{B} & \mathbf{n}_{11} & \mathbf{n}_{12} \end{bmatrix} = \begin{bmatrix} 0 & 0 & 0 & 0 & 0 & 0 & 1 & 0 \end{bmatrix} \quad (7.272)$$

and

$$\mathbf{g}_2^T \begin{bmatrix} \mathbf{B} & \mathbf{AB} & \mathbf{A}^2\mathbf{B} & \mathbf{n}_{21} & \mathbf{n}_{22} \end{bmatrix} = \begin{bmatrix} 0 & 0 & 0 & 0 & 0 & 0 & 0 & 1 \end{bmatrix}. \quad (7.273)$$

Since these are completely determined, the solutions are

$$\mathbf{g}_1^T = \begin{bmatrix} 0 & 0 & 0 & 0 & 0 & 0 & 1 & 0 \end{bmatrix} \begin{bmatrix} \mathbf{B} & \mathbf{AB} & \mathbf{A}^2\mathbf{B} & \mathbf{n}_{11} & \mathbf{n}_{12} \end{bmatrix}^{-1} \quad (7.274)$$

and

$$\mathbf{g}_2^T = \begin{bmatrix} 0 & 0 & 0 & 0 & 0 & 0 & 0 & 1 \end{bmatrix} \begin{bmatrix} \mathbf{B} & \mathbf{AB} & \mathbf{A}^2\mathbf{B} & \mathbf{n}_{21} & \mathbf{n}_{22} \end{bmatrix}^{-1}. \quad (7.275)$$

The FDC law based on (7.196) is then

$$\mathbf{u} = \mathbf{R}\mathbf{y}_r - \mathbf{K}\mathbf{x} \quad (7.276)$$

where

$$\mathbf{R} = \left[\begin{bmatrix} \mathbf{g}_1^T \mathbf{A}^3 \\ \mathbf{g}_2^T \mathbf{A}^3 \end{bmatrix} \mathbf{B} \right]^{-1} \begin{bmatrix} d_{01} & 0 \\ 0 & d_{02} \end{bmatrix} \quad (7.277)$$

and

$$\mathbf{K} = \left[\begin{bmatrix} \mathbf{g}_1^T \mathbf{A}^3 \\ \mathbf{g}_2^T \mathbf{A}^3 \end{bmatrix} \mathbf{B} \right]^{-1} \begin{bmatrix} \mathbf{g}_1^T \left[d_{01}\mathbf{I}^{[8]} + d_{11}\mathbf{A} + d_{21}\mathbf{A}^2 + d_{31}\mathbf{A}^3 + \mathbf{A}^4 \right] \\ \mathbf{g}_2^T \left[d_{02}\mathbf{I}^{[8]} + d_{12}\mathbf{A} + d_{22}\mathbf{A}^2 + + d_{32}\mathbf{A}^3 + \mathbf{A}^4 \right] \end{bmatrix}. \quad (7.278)$$

If the method of Sect. 7.3.1.4 is applied to yield non-overshooting step responses of $z_1(t)$ and $z_2(t)$ with specified settling times of T_{s1} and T_{s2} using the 5 % criterion, then if $p_1 = 7.5/T_{s1}$ and $p_2 = 7.5/T_{s2}$, the coefficients of (7.278) are

7.3 Feedback Linearising and Forced Dynamic Control 539

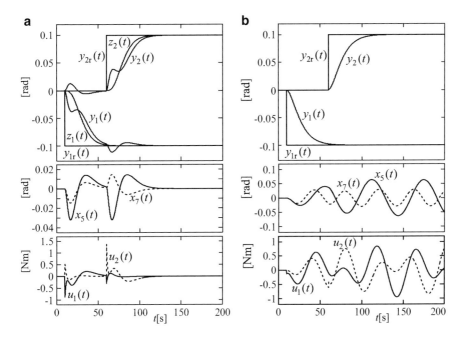

Fig. 7.18 Roll and pitch step responses of flexible satellite attitude control system. (**a**) FDC using auxiliary outputs. (**b**) FDC using controlled outputs

$$d_{0i} = p_i^4, \ d_{1i} = 4p_i^3, \ d_{2i} = 6p_i^2, \ d_{3i} = 4p_i, \quad i = 1, 2. \tag{7.279}$$

Figure 7.18 shows a simulation with $J_{xx} = 300 \ [\text{Kg m}^2]$, $J_{yy} = 250 \ [\text{Kg m}^2]$, $\nu_0 = 0.5 \ [\text{rad/s}]$, $\nu_1 = 1.5 \ [\text{rad/s}]$, $J_0 = 1,000 \ [\text{Kg m}^2]$ and $J_1 = 1,100 \ [\text{Kg m}^2]$. The settling times are set to $T_{s1} = T_{s2} = 40 \ [\text{s}]$. First a step roll attitude demand, $y_{1r}(t)$, of -1 [rad] is applied at $t = 10$ [s]. This is followed by a step pitch attitude demand, $y_{2r}(t)$, of $+1$ [rad] at $t = 60$ [s]. Figure 7.18a shows the response of the system with the above FDC law, while Fig. 7.18b shows the response that would be obtained with an FDC law based directly on the outputs y_1 and y_2.

In Fig. 7.18a, the auxiliary outputs, $z_1(t)$ and $z_2(t)$, have ideal fourth-order step responses with the correct settling times and no interaction. The controlled outputs, $y_1(t)$ and $y_2(t)$, however, exhibit stationary points due to the derivative effect of the plant zeros explained in Sect. 7.3.4.4. There is also some interaction evident. The steady-state reference inputs, however, are reached as intended, without any error. It is clear that all the state variables are under control, since the only state variables not displayed in Fig. 7.18a are derivatives of those shown.

In Fig. 7.18b, the controlled outputs, $y_1(t)$ and $y_2(t)$, behave in an ideal fashion, being second-order step responses with the correct settling time and no interaction, but at the expense of the fourth-order zero dynamics, which manifests as uncontrolled in-plane and out-of-plane vibrations of the solar panels, this being two simple harmonic oscillators. The flexure mode displacements, $x_5(t)$ and $x_7(t)$, show this. The control variables also continue to oscillate after the centre-body attitude has settled in order to precisely cancel the torques acting on the centre-body due to the flexural vibrations. This is a very good practical demonstration of the way zero dynamics can affect the operation of a control system.

7.3.4.6 Elimination of Dynamic Interaction

Dynamic interaction is the effect of reference input y_{ri}, on controlled outputs, y_j, $j = 1, 2, \ldots, m, j \neq i$, which occurs only during transient conditions. The system of Sect. 7.3.4.5 yields zero steady-state errors but is subject to dynamic interaction, as can be seen in the step responses of Example 7.9. It is evident from Sects. 7.3.2.2 and 7.3.4.3 that when applying FLC or FDC, in principle, dynamic interaction is automatically eliminated if the plant has full relative degree. In practice, however, significant dynamic interaction may still occur due to plant modelling errors. If this proves to be the case and it is necessary to reduce the degree of dynamic interaction, then FLC or FDC can be combined with the robust control techniques of Chap. 10. If the plant does not have full relative degree then, for the system presented in Sect. 7.3.4.5, a certain amount of dynamic interaction is inevitable, even with a perfect plant model. It is this case that is catered for here.

Commencing with the state-space model of (7.179) and (7.180), it will be recalled from Chap. 3 that the plant transfer function matrix relationship without external disturbance may be expressed as

$$\mathbf{Y}(s) = \mathbf{G}_u(s)\mathbf{U}(s) = \frac{\mathbf{C}\,\mathrm{adj}\,[s\mathbf{I}_n - \mathbf{A}]\,\mathbf{B}}{|s\mathbf{I}_n - \mathbf{A}|}\mathbf{U}(s). \qquad (7.280)$$

Hence,

$$\mathbf{Y}(s) = \frac{\mathbf{N}(s)}{A(s)}\mathbf{U}(s) \qquad (7.281)$$

where $A(s) = |s\mathbf{I}_n - \mathbf{A}|$ is the common denominator polynomial of degree, n, and $\mathbf{N}(s)$ is a square matrix of dimension, $m \times m$, whose elements are polynomials with degrees varying between 0 and $n-1$. The dynamic interaction may be eliminated by introducing a dynamic de-coupling pre-compensator with square transfer function matrix, $\mathbf{P}(s)$, whose input vector, $\mathbf{U}'(s)$, becomes the new control input, as depicted in Fig. 7.19.

The overall transfer function relationship is then given by

$$\mathbf{Y}(s) = \frac{\mathbf{N}(s)}{A(s)}\mathbf{U}(s) = \frac{\mathbf{N}(s)}{A(s)}\mathbf{P}(s)\mathbf{U}'(s). \qquad (7.282)$$

7.3 Feedback Linearising and Forced Dynamic Control

Fig. 7.19 Introduction of decoupling pre-compensator for elimination of dynamic interaction

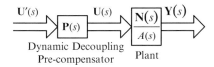

It is necessary to choose **P**(s) such that the overall transfer function matrix,

$$\mathbf{G}(s) = \frac{\mathbf{N}(s)}{A(s)} \mathbf{P}(s), \tag{7.283}$$

is diagonal. Then the task becomes that of designing m SISO control systems with sub-plant transfer functions, $Y_i(s)/U'_i(s) = g_{ii}(s)$, $i = 1, 2, \ldots, m$. Any control technique could then be used including the SISO FDC technique of Sect. 7.3.4.4. First consider setting

$$\mathbf{P}(s) = \mathbf{N}^{-1}(s) = \frac{\text{adj}\,[\mathbf{N}(s)]}{\det\,[\mathbf{N}(s)]}. \tag{7.284}$$

This would certainly accomplish the diagonalisation, yielding m identical sub-plants,

$$\frac{Y_i(s)}{U'_i(s)} = \frac{1}{A(s)}, i = 1, 2, \ldots, m, \tag{7.285}$$

which would appear attractive since not only would the task be reduced to designing one controller and duplicating it for each sub-plant, but the absence of zeros in the sub-plants would enable FDC to be applied without any inflections in the step responses. Implementing (7.284), however, cancels the zeros due to the plant having less than full relative degree. Hence, $\det\,[\mathbf{N}(s)] = 0$ must have roots with negative real parts as the zero dynamics of the system would have poles equal to these roots. If this zero dynamics is unstable or unacceptable, then an alternative approach would be to replace (7.284) with

$$\mathbf{P}(s) - \text{adj}\,[\mathbf{N}(s)]\,\mathbf{Q}(s) \tag{7.286}$$

where $\mathbf{Q}(s)$ is a diagonal matrix to be determined. Then the transfer function relationship (7.282) of the pre-compensated plant would be

$$\mathbf{Y}(s) = \frac{\mathbf{N}(s)}{A(s)} \text{adj}\,[\mathbf{N}(s)]\,\mathbf{Q}(s)\mathbf{U}'(s) \tag{7.287}$$

where $\mathbf{M}(s) = \mathbf{N}(s)\text{adj}\,[\mathbf{N}(s)] = \mathbf{I}_m \det\,(\mathbf{N}(s))$ from the definition of the matrix inverse, $\mathbf{N}^{-1}(s)$. Hence, $\mathbf{M}(s)$ is a diagonal matrix, as required for the decoupling, and has identical elements. Then $\mathbf{Q}(s)$ can also have identical elements to retain the simplification of having identical sub-plants to control. Thus,

$$\mathbf{Q}(s) = \frac{1}{q(s)} \mathbf{I}_m \qquad (7.288)$$

where $q(s)$ is a polynomial chosen to satisfy the following conditions:

(i) It does not cancel any of the plant zeros.
(ii) It is of minimum degree in the interests of simplicity but with the restriction that all the elements of $\mathbf{P}(s)$ have nonnegative relative degree to avoid differentiations in the digital pre-compensator implementation, which could be inaccurate, particularly for higher-order derivatives.

To satisfy condition (ii), it is evident from (7.286) that if adj$[\mathbf{N}(s)]$ is denoted $\mathbf{R}(s)$,

$$\deg(q(s)) = d = \max_{i,j} \{\deg(r_{ij}(s))\} \qquad (7.289)$$

since $1/q(s)$ multiplies every element of $\mathbf{R}(s)$.

Since it is most unlikely that there are plant zeros at the origin of the s-plane, condition (i) above would be satisfied by introducing pure integrators in $\mathbf{Q}(s)$ where necessary. Then, also respecting condition (ii),

$$q(s) = s^d. \qquad (7.290)$$

Example 7.10 Dynamic decoupling pre-compensator for flexible spacecraft

Space telescopes have flexible solar panels that are rotated with respect to the spacecraft centre-body that contains the telescope. In this application, supposing the telescope boresight coincides with the yaw axis, then very precise attitude control is required about the mutually orthogonal roll and pitch axes, pointing accuracies typically being of the order of 0.01 [arc sec] $\cong 3 \times 10^{-6}$ [deg], 3σ, assuming a normal error distribution. The mathematical model is similar to that of Example 7.9, so dynamic decoupling is required to avoid attitude disturbances about one axis, while attitude adjustments are made about the other axis, due to flexural vibrations of the solar panels. The mathematical model in the form of a transfer function matrix relationship is

$$\begin{bmatrix} Y_1(s) \\ Y_2(s) \end{bmatrix} = \frac{\begin{bmatrix} n_{11}(s) & n_{12}(s) \\ n_{21}(s) & n_{22}(s) \end{bmatrix}}{s^2 \left(s^2 + v_0^2\right)\left(s^2 + v_1^2\right)} \begin{bmatrix} U_1(s) \\ U_2(s) \end{bmatrix}, \qquad (7.291)$$

where $n_{11}(s) = \left(b_1 p_1(s) p_2(s) + d_1 C^2 p_2(s) + d_2 S^2 p_1(s)\right)$;

$n_{12}(s) = (d_3 p_1(s) - d_4 p_2(s)) SC, \; n_{21}(s) = \left(e_1 C^2 p_1(s) - e_2 S^2 p_2(s)\right),$

$$n_{22}(s) = \left(b_2 p_1(s) p_2(s) + e_3 C^2 p_1(s) + e_4 S^2 p_2(s)\right), p_1(s) = s^2 + v_O^2,$$

$p_2(s) = s^2 + v_I^2$, v_O and v_I are the fundamental out-of-plane and in-plane encastre vibration frequencies; b_1, b_2, c_i, d_i, $i = 1, 2, 3, 4$ are constants dependent on v_O, v_I and the spacecraft moments of inertia; and $C = \cos(\gamma)$, $S = \sin(\gamma)$ and γ are the solar panel orientation angle. The control inputs, $u_1(t)$ and $u_2(t)$, are numerically equal to the control torques applied by the reaction wheels, and the controlled outputs are numerically equal to the roll and pitch attitude angles.

The transfer function matrix relationship of the pre-compensated plant according to (7.287) and (7.288) is

$$\begin{bmatrix} Y_1(s) \\ Y_2(s) \end{bmatrix} = \frac{\begin{bmatrix} n_{11}(s) & n_{12}(s) \\ n_{21}(s) & n_{22}(s) \end{bmatrix}}{s^2 \left(s^2 + v_O^2\right) \left(s^2 + v_I^2\right)} \cdot \frac{\begin{bmatrix} n_{22}(s) & -n_{12}(s) \\ -n_{21}(s) & n_{11}(s) \end{bmatrix}}{q(s)} \begin{bmatrix} U_1'(s) \\ U_2'(s) \end{bmatrix}. \quad (7.292)$$

According to (7.289) and (7.290), since $\deg(n_{22}(s)) = \deg(n_{11}(s)) = 4$ and $\deg(n_{12}(s)) = \deg(n_{21}(s)) = 2$, then $q = s^4$. The identical pre-compensated sub-plant transfer functions are

$$\frac{Y_1(s)}{U_1'(s)} = \frac{Y_2(s)}{U_2'(s)} = \frac{n_{11}(s) n_{22}(s) - n_{21}(s) n_{22}(s)}{s^6 \left(s^2 + v_O^2\right) \left(s^2 + v_I^2\right)}. \quad (7.293)$$

It is important to mention that the complexity of this approach increases dramatically with the number of control inputs and controlled outputs, m, but it is practicable for the many plants with $m = 2$ and, arguably, for $m = 3$. For higher values of m, the labour involved would be relieved by software capable of algebraic evaluation of the adjoint matrices of square matrices whose elements are polynomials. The relatively high orders of the identical sub-plants to be controlled should be no obstacle with modern digital implementation.

7.3.5 FDC and FLC Using Discrete Plant Models

7.3.5.1 Introduction

Forced dynamic control may be carried out in the discrete domain by replacing the role of a derivative of order, q, in the continuous domain with that of a q-step prediction. This enables control laws to be derived along similar lines to those of Sects. 7.3.2 and 7.3.3 using discrete plant models.

The basic principle of feedback linearising control in the discrete domain is directly analogous to that described in Sect. 7.3.2.1 for the continuous domain. Suppose a nonlinear SISO plant is modelled by the difference equation,

$$y(k+n) = f[y(k), y(k+1), \ldots, y(k+n-1), u(k)], \quad (7.294)$$

where k is the iteration index, which will be assumed to mark updates separated by equal time intervals of h [s]. The integer, n, will be defined as the *order* of the prediction. It is also the order of difference Eq. (7.294). Then a discrete feedback linearising control law can be quickly derived. A linear difference equation, also of order, n, is formed defining the desired closed-loop dynamics as follows:

$$y(k+n) = r_0 y_r(k) - p_0 y(k) - p_1 y(k+1) - \cdots - p_{n-1} y(k+n-1). \tag{7.295}$$

Here, r_0 and p_i, $i = 0, 1, \ldots, n-1$, are constants to be determined (Sect. 7.3.5.3). Then the RHS of (7.294) and (7.295) are equated and the resulting equation solved for $u(k)$. Thus,

$$f[y(k), y(k+1), \ldots, y(k+n-1), u(k)]$$
$$= r_0 y_r(k) - p_0 y(k) - p_1 y(k+1) - \cdots - p_{n-1} y(k+n-1) \Rightarrow$$
$$u(k) = g[y(k), y(k+1), \ldots, y(k+n-1), y_r(k)]. \tag{7.296}$$

Since the variables, $y(k+i)$, $i = 0, 1, \ldots, n-1$, are state variables, (7.296) is a nonlinear state feedback control law. The particular state representation is the discrete SISO controller canonical form in which the state variables are $x_1(k) = y(k), x_2(k) = y(k+1), \ldots, x_n(k) = y(k+n-1)$. The plant state-space model is then

$$\begin{aligned} x_1(k+1) &= x_2(k) \\ x_2(k+1) &= x_3(k) \\ &\vdots \\ x_{n-1}(k+1) &= x_n(k) \\ x_n(k+1) &= f[x_1(k), x_2(k), \ldots, x_n(k), u(k)] \end{aligned} \quad, \quad y(k) = x_1(k). \tag{7.297}$$

State variables not directly available as measurements can be made available as estimates from an observer.

7.3.5.2 Relative Degree in the Discrete Domain

This topic is included as it is fundamental to the derivation of FDC and FLC laws. Consider first an SISO plant with the state-space model,

$$\mathbf{x}(k+1) = \mathbf{f}[\mathbf{x}(k), u(k)] \tag{7.298}$$

$$y(k) = h[\mathbf{x}(k)] \tag{7.299}$$

7.3 Feedback Linearising and Forced Dynamic Control

where $\mathbf{x} \in \Re^n$ is the state vector, u is the control input and y is the measured/controlled output. The vector function, $\mathbf{f}(\cdot)$, and the scalar function, $h(\cdot)$, are smooth.

The relative degree, R, is defined as the lowest-order prediction that is directly dependent on $u(k)$. The process of finding successive predictions of increasing order is similar to the process of finding successive Lie derivatives in the continuous case. Thus, from (7.299),

$$y(k+1) = h[\mathbf{x}(k+1)]. \tag{7.300}$$

Substituting for $\mathbf{x}(k+1)$ using (7.298) may or may not indicate a direct dependence on $u(k)$. If not, then the result may be written as

$$y(k+1) = h_1[\mathbf{x}(k)]. \tag{7.301}$$

This process is continued until a direct dependence on $u(k)$ occurs. Thus,

$$\left. \begin{array}{l} y(k+2) = h_1[\mathbf{x}(k+1)] = h_2[\mathbf{x}(k)] \\ \quad \vdots \\ y(k+R-1) = h_{R-2}[\mathbf{x}(k+1)] = h_{R-1}[\mathbf{x}(k)] \end{array} \right\} \tag{7.302}$$

$$y(k+R) = h_{R-1}[\mathbf{x}(k+1)] = h_R[\mathbf{x}(k), u(k)]. \tag{7.303}$$

So the prediction of order, R, is the lowest-order prediction that is directly dependent on $u(k)$. The relative degree of the plant with respect to the output, $y(k)$, is therefore R. As for the continuous case, if $R = n$, the plant is of full relative degree and if $R < n$, the plant is of less than full degree.

By analogy with the continuous case, for an LTI SISO plant model, the term 'relative degree' is the difference in degree between the denominator and the numerator of the z-transfer function, $Y(z)/U(z)$.

Now a multivariable plant is considered with the state-space model,

$$\mathbf{x}(k+1) = \mathbf{f}[\mathbf{x}(k), \mathbf{u}(k)], \tag{7.304}$$

$$\mathbf{y}(k) = \mathbf{h}[\mathbf{x}(k)], \tag{7.305}$$

where all quantities are as for the SISO plant as above except for the control vector, $\mathbf{u} \in \Re^m$, and the output vector, $\mathbf{y} \in \Re^m$. Let the measurement equation be written in the component form,

$$y_i(k) = h_i[\mathbf{x}(k)], \quad i = 1, 2, \ldots, m. \tag{7.306}$$

Then the relative degree with respect to $y_i(k)$ is the lowest-order prediction that is directly dependent on any element of $\mathbf{u}(k)$. The sequence of predictions of increasing order used to determine the relative degrees are similar to those of (7.301), (7.302), and (7.303) and are as follows. Starting with (7.306),

$$\left.\begin{array}{l} y_i\,(k+1) = h_i\,[\mathbf{x}\,(k+1)] = h_{1i}\,[\mathbf{x}(k)] \\ y_i\,(k+2) = h_{1i}\,[\mathbf{x}\,(k+1)] = h_{2i}\,[\mathbf{x}(k)] \\ \vdots \\ y_i\,(k+R_i-1) = h_{R_i-2i}\,[\mathbf{x}\,(k+R_i-1)] = h_{R_i-1i}\,[\mathbf{x}(k)] \\ y_i\,(k+R_i) = h_{R_i-1i}\,[\mathbf{x}\,(k+R_i)] = h_{R_ii}\,[\mathbf{x}(k),\,\mathbf{u}(k)] \end{array}\right\}. \qquad (7.307)$$

Then the relative degree of the plant with respect to the output, $y_i(k)$, is R_i. Again by analogy with the continuous case, if the plant has full relative degree, then

$$\sum_{i=1}^{n} R_i = n. \qquad (7.308)$$

7.3.5.3 Desired Closed-Loop Difference Equation

In this subsection, the more general multivariable plant will be considered, the SISO plant being catered for by simply setting the number of control inputs and controlled outputs to unity. Usually the closed-loop system is required to have a DC gain of unity for each reference input, output pair, so that if step reference inputs are applied, the steady-state errors are zero. It will be recalled that in the continuous-time case, this is achieved by making the coefficients of y_{ri} and y_i equal to d_{0i} in the desired closed-loop differential equation (7.59),

$$y_i^{(R_i)} = d_{0i}\,(y_{ri} - y_i) - d_{1i}\,\dot{y} - \cdots - d_{R_i-1,\,i}\,y^{(R_i-1)}, \quad i = 1,\,2,\ldots,m, \qquad (7.309)$$

where R_i is the relative degree with respect to the output, y_i. This is because $y_i^{(k)} = 0, k = 1, 2, \ldots, R_i-1$, in the steady state. The desired discrete difference equation equivalent to (7.309) is

$$y_i\,(k+R_i) = r_i\,y_{ri}\,(k) - p_{0i}\,y_i\,(k) - p_{1i}\,y_i\,(k+1) - \cdots - p_{R_i-1i}\,y_i\,(k+R_i-1). \qquad (7.310)$$

In this case, the required steady-state condition for a constant reference input, Y_{ri}, is

$$y_i\,(k+j) = Y_{ri},\,j = 0,\,1,\ldots,\,R_i. \qquad (7.311)$$

Substituting this in (7.310) then yields

$$1 = r_i - p_{0i} - p_{1i} - \cdots - p_{R_i-1i} \Rightarrow r_i = 1 + p_{0i} + p_{1i} + \cdots + p_{R_i-1i}. \qquad (7.312)$$

7.3 Feedback Linearising and Forced Dynamic Control

It remains to determine p_{qi}, $q = 0, 1, \ldots, R_i - 1$. This may be done by revisiting the transformation, $z_j = e^{s_j h}$, between the pole locations, s_j, in the s-plane and the corresponding pole locations, z_j, in the z-plane used when designing a discrete linear control system to approximate the behaviour of the equivalent continuous system designed using one of the settling time formulae (Chap. 6, Sect. 6.4.4). To summarise, the z-transfer functions of the desired closed-loop system corresponding to (7.310) are

$$\frac{Y_i(z)}{Y_{ri}(z)} = \frac{r_i}{z^{R_i} + p_{R_i-1,i}z^{R_i-1} + \cdots + p_{1i}z + p_{0i}}, \quad i = 1, 2, \ldots, m. \quad (7.313)$$

The desired characteristic polynomial of the equivalent continuous system in the s-domain is

$$(s + p_i)^{R_i} \quad (7.314)$$

where

$$p_i = \begin{cases} 1.5(1 + R_i)/T_{si} & \text{for 5\% criterion} \\ 1.6(1.5 + R_i)/T_{si} & \text{for 2\% criterion} \end{cases}. \quad (7.315)$$

Then the desired characteristic polynomial of the discrete system is

$$(z + q_i)^{R_i} \quad (7.316)$$

where

$$q_i = -e^{-p_i h}, i = 1, 2, \ldots, m. \quad (7.317)$$

As pointed out in Chap. 6, Sect. 6.4.4, the demanded settling time, T_{sdi}, will be realised automatically if $hp_i \ll 1$, but the actual settling time, T_{sai}, will approach the dead-beat value of $R_i h$, which is the shortest attainable settling time for a discrete linear system, as $hp_i \to \infty$. Even if $hp_i \ll 1$ is not satisfied, provided $T_{sdi} > R_i h$, $T_{sai} = T_{sdi}$ can be achieved by adjusting T_{si} to a value less than T_{sdi}.

Example 7.11 Discrete feedback linearising position control of an underwater vehicle

Consider one degree of freedom of translational motion of an underwater surveillance vehicle. It will be supposed that the screw drive system has already been designed to yield a propulsion force, f, governed by the first-order differential equation,

$$\dot{f} = \frac{1}{T_d}(u - f), \quad (7.318)$$

where T_d is the screw drive system time constant. If x is the vehicle position coordinate with respect to an inertial datum, the force balance equation is

$$M\ddot{x} + K_{dv}|v|v = f \tag{7.319}$$

where M is the vehicle mass and K_{dv} is the vehicle drag coefficient.

The continuous-time state-space model corresponding to (7.318) and (7.319) is then

$$\begin{aligned} \dot{x}_1 &= x_2 \\ \dot{x}_2 &= Ax_3 - C|x_2|x_2 \\ \dot{x}_3 &= B(u - x_3) \\ y &= x_1 \end{aligned} \tag{7.320}$$

where the state variables are $x_1 = x$, $x_2 = v$ and $x_3 = f$. The constants are $A = 1/M$, $B = 1/T_d$ and $C = K_{dv}/M$. An approximate discrete state-space model may then be formed using the derivative approximation, $\dot{x}_i \cong [x_i(k+1) - x_i(k)]/h$, $i = 1, 2, 3$. Thus,

$$x_1(k+1) = x_1(k) + hx_2(k), \tag{7.321}$$

$$x_2(k+1) = x_2(k) + Ahx_3(k) - Ch|x_2(k)|x_2(k), \tag{7.322}$$

$$x_3(k+1) = x_3(k) + Bh[u(k) - x_3(k)], \tag{7.323}$$

$$y(k) = x_1(k). \tag{7.324}$$

To begin the FLC law derivation, first the relative degree of the plant model with respect to the controlled output, y, is determined. Taking the first-order prediction of (7.324) and then substituting for $x_1(k+1)$ using (7.321) yield

$$y(k+1) = x_1(k) + hx_2(k). \tag{7.325}$$

Since the control input, $u(k)$, does not appear on the RHS of (7.325), a first-order prediction of (7.325) is taken and substitutions made for $x_1(k+1)$ and $x_2(k+1)$ using (7.321) and (7.322). Thus,

$$\begin{aligned} y(k+2) &= x_1(k) + hx_2(k) + h[x_2(k) + Ahx_3(k) - Ch|x_2(k)|x_2(k)] \\ &= x_1(k) + 2hx_2(k) + Ah^2x_3(k) - Ch^2|x_2(k)|x_2(k). \end{aligned} \tag{7.326}$$

Since $u(k)$ does not appear on the RHS of (7.326), a further prediction is taken, and the standard procedure would be to substitute for $x_1(k+1)$, $x_2(k+1)$ and $x_3(k+1)$ using (7.321), (7.322) and (7.323). It can be seen, however, that only

7.3 Feedback Linearising and Forced Dynamic Control

(7.323) yields $u(k)$ on the RHS, so the result will be simpler by leaving $x_1(k+1)$ and $x_2(k+1)$ without the substitution. This yields

$$y(k+3) = x_1(k+1) + \left[2h - Ch^2 |x_2(k+1)|\right] x_2(k+1) \\ + Ah^2 \{x_3(k) + Bh[u(k) - x_3(k)]\}. \tag{7.327}$$

Hence, the plant is of relative degree, $R = 3$, which is also the plant order, n. It is therefore of full relative degree so the FLC control law based on (7.327) will not yield zero dynamics. The desired closed-loop difference equation is therefore

$$y(k+3) = ry_r(k) - p_0 y(k) - p_1 y(k+1) - p_2 y(k+2). \tag{7.328}$$

The discrete FLC law is then obtained by equating the RHS of (7.327) and (7.328) and then solving for $u(k)$. Thus,

$$u(k) = \frac{1}{ABh^3} \left\{ \begin{array}{l} ry_r(k) - p_0 y(k) - p_1 y(k+1) - p_2 y(k+2) - x_1(k+1) \\ - \left[2h - Ch^2 |x_2(k+1)|\right] x_2(k+1) + Ah^2(Bh-1)x_3(k) \end{array} \right\}. \tag{7.329}$$

Since $y(k+1)$, $y(k+2)$, $x_1(k+1)$ and $x_2(k+1)$ are, respectively, given by (7.325), (7.327), (7.321) and (7.322), which are all functions of the state variables, $x_1(k)$, $x_2(k)$ and $x_3(k)$, FLC law (7.329) is a state feedback control law.

Equating the characteristic polynomial of (7.328) in the z-domain to the desired one yielding a nominal settling time of T_s seconds (5 % criterion) and no overshoot, which is desirable in this application, yields

$$z^3 + p_2 z^2 + p_1 z + p_0 = (z+q)^3 = z^3 + 3qz^2 + 3q^2 z + q^3 \Rightarrow \\ p_0 = q^3, \; p_1 = 3q^2, \; p_2 = 3q, \tag{7.330}$$

where

$$q = -e^{-ph}, \quad p = 6/T_s. \tag{7.331}$$

For a unity closed-loop DC gain,

$$r = 1 + p_0 + p_1 + p_2. \tag{7.332}$$

Figure 7.20 shows the simulation results of a manoeuvre in which the reference input position, y_r, is initially stepped to 1 [m] and then stepped to -1 [m] at $t = 20$ [s].

The plant parameters are set to $M = 200$ [Kg], $K_{dv} = 2,500$ $\left[\text{N}/(\text{m/s})^2\right]$ and $T_d = 1$ [s]. The demanded settling time is $T_{sd} = 10$ [s]. The sampling period is set to $h = 0.5$ [s], which is longer than usual to clearly show the discrete control variable.

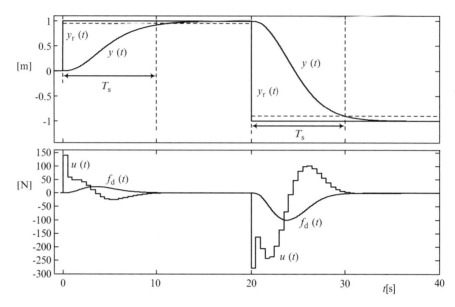

Fig. 7.20 Simulation of discrete FLC of underwater vehicle

It is clear from Fig. 7.20 that the closed-loop system satisfies the specification. Also, the nonlinear drag force, $f_d(t) = K_{dv}|x_2(t)|x_2(t)$, is significant as can be seen by comparing its scale with that of the control variable, $u(t)$, which is numerically equal to the force demand input to the screw drive system. It is important to point out, however, that in this example, the use of the derivative approximation to form the discrete plant model introduces a modelling error that increases as the step-length, h, is increased. For this reason, increasing h significantly beyond 0.5 [s] gives rise to a noticeable overshoot that increases in percentage with the magnitude of the demanded position change. If a more precise discrete model of the nonlinear plant were to be available, then a larger value of h would be tolerated.

7.3.6 Near-Time-Optimal Position Control Through FDC

7.3.6.1 Introduction

In certain mechanical systems requiring position control, it is desired to reach a constant reference position in the minimum time. This situation is often found on production lines where the throughput rate has to be maximised by switching the reference position between constant values. The mechanisms here are either tailored for specific applications or are general-purpose jointed-arm robots. Another relevant application is scientific spacecraft carrying directional instruments such

7.3 Feedback Linearising and Forced Dynamic Control

as telescopes that are required to frequently slew between different objects in the celestial sphere. The relatively low levels of electrical power available for attitude control severely limit the mechanical power that can be delivered from the control actuators during slewing manoeuvres, resulting in rather long slewing times, of the order of several tens of minutes. It is therefore of interest to maximise the proportion of the spacecraft lifetime spent collecting scientific data by minimising the slewing times.

In practice, the manoeuvre time of a mechanism is limited by the maximum accelerating and deceleration force or torque magnitudes that may be applied, and this is reflected by the control saturation limits. The branch of control theory supporting such applications is the *time-optimal control* of Chap. 9, but for the purpose of understanding the material of this section, the following pragmatic explanation will suffice. Suppose a motorist needs to drive a fixed distance from point A to point B in the minimum possible time, commencing and ending with zero velocity. Then it is clear that the maximum possible acceleration must be applied until an intermediate point between A and B is reached. Then maximum braking is applied to bring the vehicle to rest at point B. If the driver is regarded as the controller, then positive control inputs are implemented through the accelerator pedal and negative control inputs are implemented through the brake pedal. In this illustration, the control variable consists of an extreme positive value followed by a switch to an extreme negative value. Time-optimal feedback controllers do this automatically.

7.3.6.2 Plant Model

The most general state-space plant model considered here is as follows:

$$\dot{\mathbf{x}} = \mathbf{f}(\mathbf{x}, \mathbf{u}). \tag{7.333}$$

$$\mathbf{y} = \mathbf{h}(\mathbf{x}). \tag{7.334}$$

The definitions of the terms are as at the beginning of Sect. 7.3.1.1, but y_i, $i = 1, 2, \ldots, m$, are each positions of a mechanism to be controlled. The mechanism could consist of a set of m second-order subsystems behaving approximately as double integrators if the mechanical friction is not too great and the actuator dynamics can be ignored. For example, the actuators of a jointed-arm robot are often synchronous motors with vector-controlled drives containing stator current loops with negligible dynamic lag, designed to accept joint torque demands as control variables. In other mechanisms, motors may be employed in which the electrical time constant, such as the armature time constant of a DC motor, is significant and this raises the order of each subsystem of the plant model to three.

7.3.6.3 Desired Closed-Loop Differential Equations

The desired closed-loop differential equation for each controlled position is based on the near-time-optimal sliding mode control of a double integrator, derived in Chap. 10 and based on the time-optimal control theory of Chap. 9. The solution of this differential equation consists of constant acceleration and deceleration segments between which there is a fast but smooth transition approximating a switch, followed by an exponential convergence to the reference input. The double integrator state equations (not part of the plant model) are

$$\begin{aligned}\dot{x}_{1i} &= x_{2i} \\ \dot{x}_{2i} &= u_{ai}\end{aligned}, \quad i = 1, 2, \ldots, m, \qquad (7.335)$$

where x_{1i} will be the desired mechanism position coordinate, x_{2i} is then the desired velocity and u_{ai} will be called the control acceleration, noting that this is not a plant control component. Then the near-time-optimal control law for this double integrator is defined by the equations

$$x_{1ei} = x_{1i} - x_{1ir} \qquad (7.336)$$

$$x = |x_{2i}| - u_{ai\,max} T_{ci} \qquad (7.337)$$

$$S(x_{1ei}, x_{2i}) = x_{1ei} + \tfrac{1}{4}\left[u_{ai\,max}T_{ci}^2 + \frac{1}{u_{ai\,max}}x_{2i}^2\right]\operatorname{sgn}(x_{2i})[1 + \operatorname{sgn}(x)]$$
$$+ \tfrac{1}{2}T_{ci}x_{2i}[1 - \operatorname{sgn}(x)] \qquad (7.338)$$

and

$$u_{ai} = -\operatorname{sat}[K_i S(x_{1ei}, x_{2i}), -u_{ai\,max}, u_{ai\,max}], \quad i = 1, 2, \ldots, m, \qquad (7.339)$$

where $S(x_{1ei}, x_{2i})$ is the switching function, K_i is a relatively high gain determining the width of the sliding mode boundary layer (Chap. 10), $u_{i\,max}$ is the maximum acceleration magnitude and T_{ci} is the time constant of convergence of $x_{1i}(t)$ towards a constant x_{1ir}. The saturation function is defined as

$$\operatorname{sat}[x, x_{min}, x_{max}] \triangleq \begin{cases} x_{max}, & x > x_{max} \\ x, & x_{min} \le x \le x_{max} \\ x_{min}, & x < x_{max} \end{cases}. \qquad (7.340)$$

Equations (7.335), (7.336), (7.337), (7.338) and (7.339) may be converted to a single differential equation by letting $x_{1i} = y_i$ and $x_{2i} = \dot{y}_i$. Thus,

$$\ddot{y}_i = -\operatorname{sat}[K_i S(y_i - y_{ir}, \dot{y}_i), -u_{ai\,max}, u_{ai\,max}], \quad i = 1, 2, \ldots, m \qquad (7.341)$$

7.3 Feedback Linearising and Forced Dynamic Control

where

$$S(y_i - y_{ir}, \dot{y}_i)$$
$$= y_i - y_{ir} + \frac{1}{4}\left[u_{ai\,max}T_{ci}^2 + \frac{1}{u_{ai\,max}}\dot{y}_i^2\right]\text{sgn}(\dot{y}_i)[1+\text{sgn}(x)] + \tfrac{1}{2}T_{ci}\dot{y}_i[1-\text{sgn}(x)]$$

and

$$x = |\dot{y}_i| - u_{ai\,max}T_{ci}.$$

For plants with a relative degree of $R_i = 2$ with respect to each controlled output, (7.341) will be the desired closed-loop differential for each controlled output used to derive the FDC law.

For plants with a relative degree of $R_i = 3$ with respect to each controlled output, the desired closed-loop differential equations need to be of third order while still yielding accelerations within the given saturation limits of $\pm u_{i\,max}$. This can be achieved by replacing y_i in (7.341) with y'_i and then adding a first-order dynamic lag to redefine y_i. This yields the following coupled first- and second-order differential equations for the desired closed-loop dynamics:

$$\dot{y}_i = \frac{1}{T_{di}}(y'_i - y_i) \tag{7.342}$$

$$\ddot{y}'_i = -\text{sat}\left[K_i S\left(y'_i - y_{ir}, \dot{y}'_i\right), -u_{ai\,max}, u_{ai\,max}\right], \quad i = 1, 2, \ldots, m, \tag{7.343}$$

where $S(y'_i - y_{ir}, \dot{y}'_i) = y'_i - y_{ir}$

$$+ \frac{1}{4}\left[u_{ai\,max}T_{ci}^2 + \frac{1}{u_{ai\,max}}\dot{y}_i'^2\right]\text{sgn}(\dot{y}'_i)[1+\text{sgn}(x')] + \tfrac{1}{2}T_{ci}\dot{y}'_i[1-\text{sgn}(x')]$$

and $x' = |\dot{y}'_i| - u_{ai\,max}T_{ci}$. To convert (7.342) and (7.343) to a third-order differential equation involving only y_i and y_{ir}, as needed for a FDC law derivation, differentiating (7.342) twice yields

$$\dddot{y}_i = \frac{1}{T_{di}}(\ddot{y}'_i - \ddot{y}_i), \quad i = 1, 2, \ldots, m. \tag{7.344}$$

Then $\ddot{y}'_i$ is given by (7.343) where y'_i and $\dot{y}'_i$ are obtained in terms of y_i and $\dot{y}_i$ using (7.342) as follows:

$$y'_i = y_i + T_{di}\dot{y}_i \text{ and } \dot{y}'_i = \dot{y}_i + T_{di}\ddot{y}_i. \tag{7.345}$$

Figure 7.21 shows the step responses of the closed-loop systems governed by (7.341) and (7.344) for two maximum acceleration magnitudes, $u_{ai\,max}$, of 0.6 and 1.5.

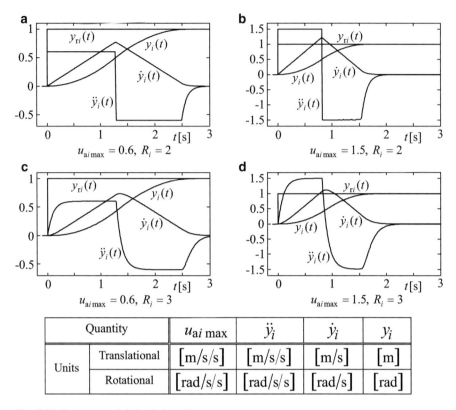

Fig. 7.21 Responses of desired closed-loop systems

The units of the variables depend on whether the position coordinate is rotational or translational, and these are indicated in the table at the foot.

The responses of Fig. 7.21a, b, for use with plants of second relative degree, can be seen to closely resemble those of the double integrator time-optimal control, which has piecewise constant acceleration, piecewise linear velocity and piecewise parabolic position. The approach to this ideal is not quite so close in Fig. 7.21c, d, for use with plants of third relative degree, due to the first-order element with time constant, $T_d = 0.1$ [s], whose effect can be seen by comparison with Fig. 7.21a, b, but the similarity with the double integrator time-optimal control is still evident. This can be made closer by reducing T_d, but a value sufficiently large to demonstrate the effect has been used here.

The settling time reduces automatically as $u_{ai\,max}$ is increased and is very close to the theoretical minimum for a double integrator. The control variables of the FDC law will realise this near-time-optimal double integrator dynamics provided that the actual control variables do not saturate. To obtain the shortest settling time for each controlled position, $u_{ai\,max}$ is adjusted, during the simulation stage of the control system development, until the maximum excursions of the control variable just fall

7.3 Feedback Linearising and Forced Dynamic Control

below the saturation limits. Then, for many mechanisms, the result may be almost time-optimal. The closeness to this optimality may be judged by the closeness of the control function to a piecewise constant function with a single intermediate switch between the control saturation limits. It should be noted that the derivation of truly time optimal feedback control laws for multi-axis mechanisms with significant inter-axis interaction or nonlinear friction would be difficult or intractable. In contrast, the near-time-optimal forced dynamic control system is practicable, though suboptimal. In any case, a shorter settling time than attainable using any linear controller is possible. It is important to mention that since the near-time-optimal double integrator sliding mode controller of Chap. 9 is a robust controller, the FDC control law based on the same dynamics will also exhibit robustness and should tolerate realistic modelling errors and disturbances with minimal deviations from the prescribed dynamics.

Example 7.12 Three-axis slewing control of a rigid-body spacecraft

A rigid-body spacecraft with reaction wheels for attitude control is modelled by the Euler rotational dynamics equations (Chap. 2),

$$\begin{bmatrix} \dot{\omega}_x \\ \dot{\omega}_y \\ \dot{\omega}_z \end{bmatrix} = \begin{bmatrix} J_{xx} & J_{xy} & J_{xz} \\ J_{yx} & J_{yy} & J_{yz} \\ J_{zx} & J_{zy} & J_{zz} \end{bmatrix}^{-1} \left\{ \begin{bmatrix} \Delta J_{yz}\omega_z\omega_y + \omega_z l_{wy} - \omega_y l_{wz} \\ \Delta J_{zx}\omega_x\omega_z + \omega_x l_{wz} - \omega_z l_{wx} \\ \Delta J_{xy}\omega_y\omega_x + \omega_y l_{wx} - \omega_x l_{wy} \end{bmatrix} + K_w \begin{bmatrix} u_x \\ u_y \\ u_z \end{bmatrix} \right\}, \tag{7.346}$$

the reaction wheel momentum differential equations,

$$\begin{bmatrix} \dot{l}_{wx} \\ \dot{l}_{wy} \\ \dot{l}_{wz} \end{bmatrix} = -K_w \begin{bmatrix} u_x \\ u_y \\ u_z \end{bmatrix}, \tag{7.347}$$

and the quaternion-based kinematic differential equations,

$$\begin{bmatrix} \dot{q}_0 \\ \dot{q}_1 \\ \dot{q}_2 \\ \dot{q}_3 \end{bmatrix} = \frac{1}{2} \begin{bmatrix} 0 & -\omega_x & -\omega_y & -\omega_z \\ \omega_x & 0 & \omega_z & -\omega_y \\ \omega_y & -\omega_z & 0 & \omega_x \\ \omega_z & \omega_y & -\omega_x & 0 \end{bmatrix} \begin{bmatrix} q_0 \\ q_1 \\ q_2 \\ q_3 \end{bmatrix}. \tag{7.348}$$

Here, J_{ij}, $i = x, y, z, j = x, y, z$ are the moment of inertia matrix elements; $\Delta J_{yz} = J_{yy} - J_{zz}$, $\Delta J_{zx} = J_{zz} - J_{xx}$ and $\Delta J_{xy} = J_{xx} - J_{yy}$, l_{wi}, $i = x, y, z$ are the reaction wheel angular momenta; K_w is the reaction wheel torque constant; u_x, u_y and u_z are the control inputs; and q_i, $i = 1, 2, 3, 4$, are the quaternion components that represent the spacecraft attitude. It will be recalled from Chap. 2 that since there are only three rotational degrees of rotational freedom, the quaternion components obey a constraint equation,

$$q_0^2 + q_1^2 + q_2^2 + q_3^2 = 1 \tag{7.349}$$

and at $t = 0$, (7.348) is initialised at a zero attitude defined by $\begin{bmatrix} q_0 & q_1 & q_2 & q_3 \end{bmatrix}^T = \begin{bmatrix} 1 & 0 & 0 & 0 \end{bmatrix}^T$. Then for pure rotations about the x (roll) axis, the y (pitch) axis and the z (yaw) axis, (7.348) becomes, respectively,

$$\begin{bmatrix} \dot{q}_0 \\ \dot{q}_1 \end{bmatrix} = \frac{1}{2} \begin{bmatrix} 0 & -\omega_x \\ \omega_x & 0 \end{bmatrix} \begin{bmatrix} q_0 \\ q_1 \end{bmatrix}, \begin{bmatrix} \dot{q}_0 \\ \dot{q}_2 \end{bmatrix} = \frac{1}{2} \begin{bmatrix} 0 & -\omega_y \\ \omega_y & 0 \end{bmatrix} \begin{bmatrix} q_0 \\ q_1 \end{bmatrix}, \begin{bmatrix} \dot{q}_0 \\ \dot{q}_3 \end{bmatrix}$$

$$= \frac{1}{2} \begin{bmatrix} 0 & -\omega_z \\ \omega_z & 0 \end{bmatrix} \begin{bmatrix} q_0 \\ q_3 \end{bmatrix}. \quad (7.350)$$

Then starting at zero attitude, $\dot{q}_1 \cong \frac{1}{2}\omega_x$ for roll axis rotation only, $\dot{q}_2 \cong \frac{1}{2}\omega_y$ for pitch axis rotation only and $\dot{q}_3 \cong \frac{1}{2}\omega_z$ for yaw axis rotation only. Then q_1, q_2 and q_3 are nearly half the physical angles of rotation. For simultaneous three-axis slewing, this simple approximation does not hold, but q_1, q_2 and q_3 do uniquely define the spacecraft attitude, and these will be taken as the measurement variables. Thus,

$$\begin{bmatrix} y_1 \\ y_2 \\ y_3 \end{bmatrix} = \begin{bmatrix} q_1 \\ q_2 \\ q_3 \end{bmatrix}. \quad (7.351)$$

Before setting about the FDC law derivation, the kinematic differential equations will be expressed in an alternative more convenient form. Thus,

$$\begin{bmatrix} \dot{q}_0 \\ \dot{q}_1 \\ \dot{q}_2 \\ \dot{q}_3 \end{bmatrix} = \frac{1}{2} \begin{bmatrix} -q_1 & -q_2 & -q_3 \\ q_0 & -q_3 & q_2 \\ q_3 & q_0 & -q_1 \\ -q_2 & q_1 & q_0 \end{bmatrix} \begin{bmatrix} \omega_x \\ \omega_y \\ \omega_z \end{bmatrix}. \quad (7.352)$$

The relative degrees with respect to all three outputs can now be determined together by working directly with these three equations which, together (7.346), are the relevant plant state differential equations. Thus, differentiating (7.351) and substituting for $\dot{q}_1, \dot{q}_2$ and $\dot{q}_3$ yields

$$\begin{bmatrix} \dot{y}_1 \\ \dot{y}_2 \\ \dot{y}_3 \end{bmatrix} = \frac{1}{2} \begin{bmatrix} q_0 & -q_3 & q_2 \\ q_3 & q_0 & -q_1 \\ -q_2 & q_1 & q_0 \end{bmatrix} \begin{bmatrix} \omega_x \\ \omega_y \\ \omega_z \end{bmatrix}. \quad (7.353)$$

Since no control inputs appear on the RHS, (7.353) is differentiated once more. It is immediately evident through observation of (7.346) that the resulting $\dot{\omega}_x$, $\dot{\omega}_y$ and $\dot{\omega}_z$ will each be directly dependent on the control inputs. In this case, it is not necessary to substitute for $\dot{q}_i, i = 1, 2, 3, 4$, as these are already state variables, not depending directly on the control inputs, as indicated by (7.352). Thus, differentiating (7.353) and substituting for $\dot{\omega}_x$, $\dot{\omega}_y$ and $\dot{\omega}_z$ yields

7.3 Feedback Linearising and Forced Dynamic Control

$$\begin{bmatrix} \ddot{y}_1 \\ \ddot{y}_2 \\ \ddot{y}_3 \end{bmatrix} = \frac{1}{2} \begin{bmatrix} \dot{q}_0 & -\dot{q}_3 & \dot{q}_2 \\ \dot{q}_3 & \dot{q}_0 & -\dot{q}_1 \\ -\dot{q}_2 & \dot{q}_1 & \dot{q}_0 \end{bmatrix} \begin{bmatrix} \omega_x \\ \omega_y \\ \omega_z \end{bmatrix}$$

$$+ \frac{1}{2} \begin{bmatrix} q_0 & -q_3 & q_2 \\ q_3 & q_0 & -q_1 \\ -q_2 & q_1 & q_0 \end{bmatrix} \begin{bmatrix} J_{xx} & J_{xy} & J_{xz} \\ J_{yx} & J_{yy} & J_{yz} \\ J_{zx} & J_{zy} & J_{zz} \end{bmatrix}^{-1} \left\{ \begin{bmatrix} \Delta J_{yz}\omega_z\omega_y + \omega_z l_{wy} - \omega_y l_{wz} \\ \Delta J_{zx}\omega_x\omega_z + \omega_x l_{wz} - \omega_z l_{wx} \\ \Delta J_{xy}\omega_y\omega_x + \omega_y l_{wx} - \omega_x l_{wy} \end{bmatrix} \right.$$

$$\left. + K_w \begin{bmatrix} u_x \\ u_y \\ u_z \end{bmatrix} \right\}. \tag{7.354}$$

The relative degrees are therefore $R_1 = R_2 = R_3 = 2$, and therefore the desired closed-loop differential equations are given by (7.341). Thus,

$$\begin{bmatrix} \ddot{y}_1 \\ \ddot{y}_2 \\ \ddot{y}_3 \end{bmatrix} = \begin{bmatrix} -\operatorname{sat}[K_1 S(y_1 - y_{1r}, \dot{y}_1), -u_{a1\,\max}, u_{a1\,\max}] \\ -\operatorname{sat}[K_2 S(y_2 - y_{2r}, \dot{y}_2), -u_{a2\,\max}, u_{a2\,\max}] \\ -\operatorname{sat}[K_3 S(y_3 - y_{3r}, \dot{y}_3), -u_{a3\,\max}, u_{a3\,\max}] \end{bmatrix}. \tag{7.355}$$

The first part of the required control law is then obtained by first solving (7.354) for the control vector, as follows:

$$\begin{bmatrix} u_x \\ u_y \\ u_z \end{bmatrix} = \frac{1}{K_w} \left\{ \begin{bmatrix} J_{xx} & J_{xy} & J_{xz} \\ J_{yx} & J_{yy} & J_{yz} \\ J_{zx} & J_{zy} & J_{zz} \end{bmatrix} \begin{bmatrix} q_0 & -q_3 & q_2 \\ q_3 & q_0 & -q_1 \\ -q_2 & q_1 & q_0 \end{bmatrix}^{-1} \left\{ 2 \begin{bmatrix} \ddot{y}_1 \\ \ddot{y}_2 \\ \ddot{y}_3 \end{bmatrix} - \begin{bmatrix} \dot{q}_0 & -\dot{q}_3 & \dot{q}_2 \\ \dot{q}_3 & \dot{q}_0 & -\dot{q}_1 \\ -\dot{q}_2 & \dot{q}_1 & \dot{q}_0 \end{bmatrix} \begin{bmatrix} \omega_x \\ \omega_y \\ \omega_z \end{bmatrix} \right\} \right. $$
$$\left. - \begin{bmatrix} \Delta J_{yz}\omega_z\omega_y + \omega_z l_{wy} - \omega_y l_{wz} \\ \Delta J_{zx}\omega_x\omega_z + \omega_x l_{wz} - \omega_z l_{wx} \\ \Delta J_{xy}\omega_y\omega_x + \omega_y l_{wx} - \omega_x l_{wy} \end{bmatrix} \right\}. \tag{7.356}$$

The control law is then completed by (7.355). As a precaution, the nonsingularity of the inverted matrix in (7.356) has to be checked:

$$\begin{vmatrix} q_0 & -q_3 & q_2 \\ q_3 & q_0 & -q_1 \\ -q_2 & q_1 & q_0 \end{vmatrix} = q_0(q_0^2 + q_1^2) + q_3(q_3 q_0 - q_2 q_1) + q_2(q_3 q_1 + q_2 q_0) \tag{7.357}$$

$$= q_0(q_0^2 + q_1^2 + q_2^2 + q_3^2) = q_0.$$

Consider again the single-axis rotations. It is then informative to view the trajectory of $q_i(t)$, $i = 1, 2, 3$, plotted against $q_0(t)$. For this case, it follows from (7.349) that each of these trajectories is a unit circle, as shown in Fig. 7.22. With reference to the table on the right of Fig. 7.22, the solutions to the three subsystems of (7.350), commencing from the zero attitude, are given by

$$q_i(t) = \sin\left[\tfrac{1}{2}\alpha(t)\right], \quad q_0(t) = \cos\left[\tfrac{1}{2}\alpha(t)\right]. \tag{7.358}$$

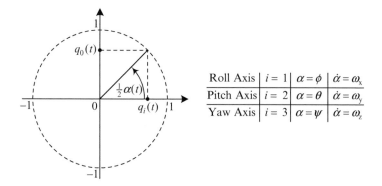

Fig. 7.22 Circular quaternion trajectories for single-axis rotations

Hence, q_0 approaches zero only for the single-axis rotations approaching $\pm\pi$ [rad]. It can be concluded that the inverted matrix referred to above remains nonsingular over a large range of spacecraft attitudes.

Figure 7.23a shows a simulation of the FDC system with the plant parameters
$$\begin{bmatrix} J_{xx} & J_{xy} & J_{xz} \\ J_{yx} & J_{yy} & J_{yz} \\ J_{zx} & J_{zy} & J_{zz} \end{bmatrix} = \begin{bmatrix} 200 & 8 & -15 \\ 8 & 300 & 12 \\ -15 & 12 & 250 \end{bmatrix} [\text{Kg m}^2] \text{ and } K_w = 0.1 \text{ [Nm/V]}.$$ The controller parameters are $K_1 = K_2 = K_3 = 10$ and $T_{c1} = T_{c2} = T_{c3} = 0.4$ [s]. The maximum angular accelerations have been adjusted to $u_{a1\,\text{max}} = 1.8 \times 10^{-3}$ [rad/s/s], $u_{a2\,\text{max}} = 0.9 \times 10^{-3}$ [rad/s/s] and $u_{a3\,\text{max}} = 0.8 \times 10^{-3}$ [rad/s/s] to yield peak control inputs within the maximum magnitude of 10 [V].

It is important to understand that the quaternion components are not physical angles, although together they define the spacecraft attitude. For this reason, units of [rad] are not given. They are dimensionless. To obtain a 'feeling' for the corresponding attitude angle magnitudes, however, the quaternion components are approximately in half radians, which follows from the forgoing analysis of single-axis rotations.

The step reference input, $y_{2r}(t)$, is delayed by 10 [s] with respect to the step reference input, $y_{1r}(t)$, and similarly the step reference input, $y_{3r}(t)$, is delayed with respect to $y_{2r}(t)$ by 10 [s], ensuring that the spacecraft is already in motion when $y_{2r}(t)$ and $y_{3r}(t)$ are applied. This tests the ability of the control system to eliminate interaction under transient conditions and this is successful as demonstrated in Fig. 7.23a. It is evident from Fig. 7.23a that $y_1(t)$, $y_2(t)$ and $y_3(t)$ are of the same double parabolic form as the desired responses of Fig. 7.21a, b. The control inputs required to achieve this in the system simulated in Fig. 7.23a, however, differ considerably from the piecewise constant, single-switch, double integrator time-optimal control required in the system simulated in Fig. 7.21, due to the difference between the interactive and nonlinear plant model given by (7.351) together with (7.346) through (7.349) and three separate double integrators.

7.3 Feedback Linearising and Forced Dynamic Control

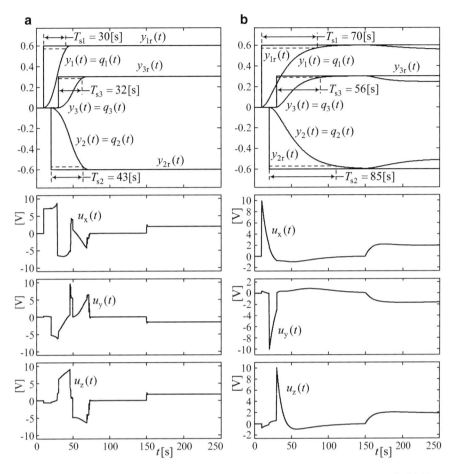

Fig. 7.23 Simulation of large angle three-axis slewing control of rigid-body spacecraft. (**a**) Near time optimal forced dynamic control. (**b**) Feedback linearising control

At $t = 150$ [s], disturbance torque step inputs referred to the control inputs of $u_{dx} = 2$ [V], $u_{dy} = -1.5$ [V] and $u_{dz} = 1.8$ [V] are applied, which are due to orbit change thrusters whose force vectors are not precisely directed through the spacecraft centre of mass. These are seen to have negligible effect on the spacecraft attitude, the control inputs very rapidly counteracting the disturbances, confirming that the control system exhibits extreme robustness.

For comparison, Fig. 7.23b shows a simulation in which the near-time-optimal forced dynamic control law is replaced by a feedback linearising control law for which the desired closed-loop differential equations (7.355) are replaced by the linear differential equations,

$$\begin{bmatrix} \ddot{y}_1 \\ \ddot{y}_2 \\ \ddot{y}_3 \end{bmatrix} = \begin{bmatrix} \left(\frac{4.5}{T_{s1}}\right)^2 (y_{1r} - y_1) - \frac{9}{T_{s1}} \dot{y}_1 \\ \left(\frac{4.5}{T_{s2}}\right)^2 (y_{2r} - y_2) - \frac{9}{T_{s2}} \dot{y}_2 \\ \left(\frac{4.5}{T_{s3}}\right)^2 (y_{3r} - y_3) - \frac{9}{T_{s3}} \dot{y}_3 \end{bmatrix}, \qquad (7.359)$$

which is a set of three critically damped, decoupled, second-order linear subsystems with settling times of T_{s1}, T_{s2} and T_{s3}. These settling times have been adjusted to yield a peak value of reaction wheel drive input at one of the maximum limits of ±10 [V], which occurs at the instant each step reference input is applied, as can be seen in the figure. This form of control is typical of that occurring in other second-order position control loops with nominally linear closed-loop dynamics. As is evident from Fig. 7.23, the near-time-optimal FDC law yields settling times approximately one half of those attainable using FLC.

The FLC system does not have comparable robustness with the near-time-optimal FDC system as is evident from the transient and steady-state errors occurring in Fig. 7.23b. As will be seen in Chap. 11, however, a linear closed-loop system with a dominant group of poles and a much larger dominated pole yields high robustness.

References

1. Vittek J, Dodds SJ (2003) Forced dynamics control of electric drives. University of Zilina Press, Zilina, Slovakia. ISBN 80-8070-087-7
2. Isidori A (1995) Nonlinear control systems, 3rd edn. Springer-Verlag, London. ISBN 3-540-19916-0
3. Albertos P, Sala A (2004) Multivariable control systems. Springer-Verlag, London. ISBN 978-1-85233-843-5
4. McLean D (1990) Automatic flight control systems. Prentice Hall International (UK) Ltd, Hemel Hempstead, Hertfordshire. ISBN 0-13-054008-0

Chapter 8
State Estimation

8.1 Introduction

State feedback control laws sometimes require state variables that cannot be measured or are uneconomical to measure. This chapter is therefore dedicated to providing means of measuring only a minimal set of state variables and estimating the remainder in computer software. The minimal set of variables for measurement is that for which the plant is *observable* (Chap. 3). For a SISO plant, this consists of just a single variable, usually the variable to be controlled. The so-called sensor-less AC electric drive is an exception, where only the stator currents are measured and the rotor speed to be controlled is estimated.

Any device for producing estimates of state variables instead of direct measurements is referred to as a *state estimator*, including the differentiators already introduced in Chap. 1. The observer is a particular class of state estimator distinguished by its principle of operation, and most of this chapter is devoted to this.

The required state variables of a plant are sometimes the derivatives of a measured variable. A simple way to obtain estimates of such derivatives without installing instrumentation for their measurement is to include software differentiation in the real-time control programme of the digital processor implementing the controller. This has already been introduced in Chap. 6. Software differentiation is acceptable provided the measurement is not contaminated by a relatively high level of noise. This topic will be recalled from Chap. 1 where it is emphasised that differentiation of a measurement results in amplification of the measurement noise by a factor proportional to the frequency. The introduction of low-pass filtering can help to overcome this problem but restricts the set of controller parameters to lie within a boundary outside in which instability results, thereby restricting the attainable dynamic performance. Such filtering increases the order of the system. It follows that the pole placement of a basic LSF control system using the derivatives of measurements of state variables is complete without the filtering but is *incomplete* with it unless the filter parameters are included in the pole placement process, which could be mathematically tedious due to the control system structure. The observer circumvents these problems by providing *filtered* derivative estimates without introducing dynamic lag, and the Kalman filter that has a similar structure to the observer minimises the noise contamination of the state estimates under certain assumptions regarding the nature of the measurement noise.

It is important to recall that the continuous polynomial controller introduced in Chap. 5 and carried over to the discrete domain in Chap. 6 achieves complete state control of LTI SISO plants with design by pole assignment without feedback of state variables other than the measured/controlled output. This could therefore be regarded as an alternative to state feedback control of LTI SISO plants aided by state estimators, but at present state estimation is the only option for control of some nonlinear plants by techniques such as the feedback linearising control in Chap. 7.

The state estimation of nonlinear plants is addressed in Appendix A8.

8.2 The Full State Continuous Observer for LTI SISO Plants

8.2.1 Introduction

The purpose of a state observer is to provide estimates of unmeasured state variables needed for the control of a plant and also, where necessary, filtered estimates of measured variables that are contaminated by relatively high levels of noise. The observer is linked to the derivation in Chap. 3 of the observability matrix and the associated condition for observability. It will be recalled that this entails observing the control variables of the plant and its response via the measurement variables until sufficient information has been collected to determine the current plant state. The observer operates in this way to produce the state estimate.

An observer, along with a state feedback control law, is usually implemented by a digital processor. Discrete implementation of controllers is addressed in Chap. 6 which also states a criterion of applicability of continuous time control theory giving a minimum sampling frequency for a given application for which the continuous time approach may be used. The topics will be introduced in the continuous domain, while some of the examples will extend to the discrete domain.

The basic structure of a linear SISO control system employing a full state observer is shown in Fig. 8.1.

To establish some standard terminology, the *controller* is defined as the part of the system that accepts the measurement variable, y, together with the corresponding reference input, y_r, and produces the control variable, u. The *control law* is, in general, only *part* of the controller and forms the complete control system together with the observer (or other state estimator). It is that part of the system which accepts the state vector (or its estimate) together with the reference input, y_r, and produces the control variable, u, as shown.

8.2 The Full State Continuous Observer for LTI SISO Plants

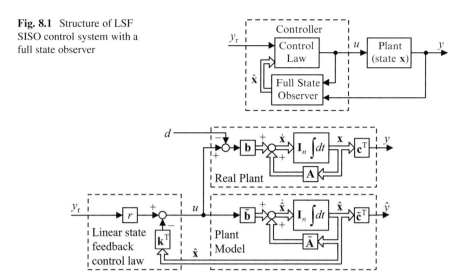

Fig. 8.1 Structure of LSF SISO control system with a full state observer

Fig. 8.2 An attempt at creating an LSF control system using a plant model

The observer will now be developed in a number of steps, which will enable its internal structure and behaviour to be fully appreciated. The starting point is a real-time state-space model of the plant in which the state variables are accessible, many examples of which may be found in Chaps. 2 and 3. A first attempt at creating an LSF control system using a real-time plant model is shown in Fig. 8.2. Here, $\widehat{\mathbf{x}}$ and $\hat{y}$ are, respectively, the estimates of the plant state, $\mathbf{x}$, and the measurement, y. Also, $\tilde{\mathbf{A}}$, $\tilde{\mathbf{b}}$ and $\tilde{\mathbf{c}}^T$ are estimates of the plant matrices, $\mathbf{A}$, $\mathbf{b}$ and $\mathbf{c}^T$.

Suppose that the following conditions are satisfied:

(a) The initial state, $\mathbf{x}(0)$, of the real plant is known and the initial state estimate is set to $\widehat{\mathbf{x}}(0) = \mathbf{x}(0)$.
(b) The external disturbance, d, referred to the plant input is zero.
(c) $\{\mathbf{A}, \mathbf{b}, \mathbf{c}^T\} = \{\tilde{\mathbf{A}}, \tilde{\mathbf{b}}, \tilde{\mathbf{c}}^T\}$.

Then since the real plant and its model are driven by the same control variable, $\widehat{\mathbf{x}}(t) = \mathbf{x}(t)$ and therefore $\widehat{y}(t) = y(t)$, implying that successful control of the plant model yields successful control of the real plant. In practice, however, conditions (a) and (b) are only sometimes satisfied and condition (c) is *never* satisfied. Hence, if condition (a) is satisfied, $\widehat{y}(0) = y(0)$, but an error will build up between $\widehat{\mathbf{x}}(t)$ and $\mathbf{x}(t)$ together with a corresponding error between $\hat{y}(t)$ and $y(t)$. The plant cannot, therefore, be controlled with the scheme of Fig. 8.2 as it stands. This is not surprising since the measurement, y, is not fed back and used in any way. What is needed is some means of correcting the model state, $\widehat{\mathbf{x}}(t)$, using the known error, $e(t) = y(t) - \widehat{y}(t)$, thereby forming a *model correction loop*. This loop is intended to drive the correction loop error, $e(t)$, to proportions regarded negligible for the application in hand. Then if $\{\mathbf{A}, \mathbf{b}, \mathbf{c}^T\} \cong \{\tilde{\mathbf{A}}, \tilde{\mathbf{b}}, \tilde{\mathbf{c}}^T\}$, achieving $e(t) \cong 0$ would at the same time achieve a state estimation error of $\varepsilon(t) \cong \mathbf{0}$, where

$$\varepsilon(t) \triangleq \widehat{\mathbf{x}}(t) - \mathbf{x}(t). \tag{8.1}$$

The model correction loop is formed by feeding $e(t)$ into every integrator of the plant model via an adjustable gain. Provided the plant is observable (Chap. 2), the model correction loop can be designed by pole assignment to achieve the desired convergence of $\varepsilon(t)$ and $e(t)$ to zero, via the adjustable gains that are equal in number to the plant order. These gains form the observer *gain matrix*

$$\mathbf{l} = \begin{bmatrix} l_1 & l_2 & \cdots & l_n \end{bmatrix}^T \tag{8.2}$$

so that the general observer for a SISO plant appears as in Fig. 8.3.

The standard symbol for the observer gain matrix has been established as $\mathbf{l}$ which is a reminder of the inventor of the observer, David Luenberger [1]. Since the observer contains a real-time model of the plant, then its state-space equations are similar to that of the plant, reproduced here for convenience:

$$\dot{\mathbf{x}} = \mathbf{A}\mathbf{x} + \mathbf{b}(u - d), \quad y = \mathbf{c}^T \mathbf{x}. \tag{8.3}$$

The observer equations corresponding to Fig. 8.3 are

$$\dot{\widehat{\mathbf{x}}} = \tilde{\mathbf{A}}\widehat{\mathbf{x}} + \tilde{\mathbf{b}}u + \mathbf{l}(y - \widehat{y}), \quad \widehat{y} = \tilde{\mathbf{c}}^T \widehat{\mathbf{x}}. \tag{8.4}$$

The external disturbance, d, is not modelled here, but observers will be introduced later that are designed to estimate d along with the plant state for use in control laws delivering a control component directly cancelling the disturbance.

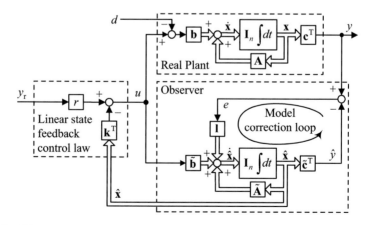

Fig. 8.3 LSF control system employing a full state observer

8.2.2 The Separation Principle and the Transparency Property

In the following, it is assumed that condition (c) is satisfied since this leads to a useful mathematical result that can greatly simplify the control system design by *separating* the observer and linear state feedback controller and designing them independently. This is the separation principle and yields good results in practice provided the plant modelling errors are not too great. A mathematical statement of the separation principle is obtained as follows. The first time derivative of (8.1) is

$$\dot{\varepsilon} = \dot{\hat{x}} - \dot{x}. \tag{8.5}$$

Then substituting for $\dot{x}$ and $\dot{\hat{x}}$ in (8.5) using (8.3) and (8.4) yields

$$\dot{\varepsilon} = \tilde{A}\hat{x} + \tilde{b}u + l\left(c^T x - \tilde{c}^T \hat{x}\right) - [Ax + b(u-d)]. \tag{8.6}$$

If condition (c) is satisfied, the terms involving u on the RHS cancel and (8.6) becomes $\dot{\varepsilon} = \tilde{A}\left(\hat{x} - x\right) - l\,\tilde{c}^T\left(\hat{x} - x\right) + \tilde{b}d = \left[\tilde{A} - l\,\tilde{c}^T\right]\left(\hat{x} - x\right) + \tilde{b}d \Rightarrow$

$$\dot{\varepsilon} = \left[\tilde{A} - l\,\tilde{c}^T\right]\varepsilon + \tilde{b}d. \tag{8.7}$$

If $d = 0$, this becomes the unforced error state differential equation,

$$\dot{\varepsilon} = \left[\tilde{A} - l\,\tilde{c}^T\right]\varepsilon, \tag{8.8}$$

in which the state vector is ε and implies that the determination of l to yield the desired convergence of $\varepsilon(t)$ to zero may be carried out independently of the linear state feedback control law design.

The separation principle will now be substantiated by proving that the characteristic polynomial of the control system has two factors, one being the characteristic polynomial of the observer alone and the other being the characteristic polynomial of the LSF control loop using the plant state variables as if they were all available. The state differential equation of the system of Fig. 8.3 may be written in partitioned form with the state vector, $[x|\varepsilon]^T$, as follows. One component equation is (8.8) and the other is the plant state differential equation with u given by the LSF control law. Thus, noting that $\hat{x} = \varepsilon + x$ from (8.1),

$$\dot{x} = \tilde{A}x + \tilde{b}u = \tilde{A}x + \tilde{b}\left(ry_r - k^T\hat{x}\right) = \tilde{A}x + \tilde{b}\left(ry_r - k^T[\varepsilon + x] - d\right) \Rightarrow$$
$$\dot{x} = \left[A - bk^T\right]x - bk^T\varepsilon + b\left(ry_r - d\right). \tag{8.9}$$

Then (8.9) and (8.8) may be combined as

$$\begin{bmatrix} \dot{\mathbf{x}} \\ \dot{\boldsymbol{\varepsilon}} \end{bmatrix} = \underbrace{\begin{bmatrix} \tilde{\mathbf{A}} - \tilde{\mathbf{b}}\mathbf{k}^T & -\tilde{\mathbf{b}}\mathbf{k}^T \\ 0 & \tilde{\mathbf{A}} - \mathbf{l}\tilde{\mathbf{c}}^T \end{bmatrix}}_{\mathbf{A}_c} \begin{bmatrix} \mathbf{x} \\ \boldsymbol{\varepsilon} \end{bmatrix} + \begin{bmatrix} \tilde{\mathbf{b}}r \\ 0 \end{bmatrix} y_r - \begin{bmatrix} \tilde{\mathbf{b}} \\ 0 \end{bmatrix} d \tag{8.10}$$

If the unit matrix of dimension, n, is denoted by $\mathbf{I}_n$, then since the four matrices comprising $\mathbf{A}_c$ are of dimension, $n \times n$, the characteristic polynomial of $\mathbf{A}_c$ is

$$\left| s\mathbf{I}_{2n} - \mathbf{A}_c \right| = \left| \begin{bmatrix} s\mathbf{I}_n & 0 \\ 0 & s\mathbf{I}_n \end{bmatrix} - \begin{bmatrix} \tilde{\mathbf{A}} - \tilde{\mathbf{b}}\mathbf{k}^T & -\tilde{\mathbf{b}}\mathbf{k}^T \\ 0 & \tilde{\mathbf{A}} - \mathbf{l}\tilde{\mathbf{c}}^T \end{bmatrix} \right| = \left| \begin{bmatrix} s\mathbf{I}_n - \left[\tilde{\mathbf{A}} - \tilde{\mathbf{b}}\mathbf{k}^T\right] & -\tilde{\mathbf{b}}\mathbf{k}^T \\ 0 & s\mathbf{I}_n - \left[\tilde{\mathbf{A}} - \mathbf{l}\tilde{\mathbf{c}}^T\right] \end{bmatrix} \right|.$$

It is known from matrix theory that the determinant of a block triangular matrix comprising square matrices is the product of the determinants of the matrices on the leading diagonal. Hence, the characteristic polynomial is factorised as

$$\left| s\mathbf{I}_{2n} - \tilde{\mathbf{A}}_c \right| = \left| s\mathbf{I}_n - \left[\tilde{\mathbf{A}} - \tilde{\mathbf{b}}\mathbf{k}^T\right] \right| \cdot \left| s\mathbf{I}_n - \left[\tilde{\mathbf{A}} - \mathbf{l}\,\tilde{\mathbf{c}}^T\right] \right|, \tag{8.11}$$

enabling the two characteristic equations

$$\left| s\mathbf{I}_n - \left[\tilde{\mathbf{A}} - \tilde{\mathbf{b}}\mathbf{k}^T\right] \right| = 0 \tag{8.12}$$

and

$$\left| s\mathbf{I}_n - \left[\tilde{\mathbf{A}} - \mathbf{l}\,\tilde{\mathbf{c}}^T\right] \right| = 0, \tag{8.13}$$

to be formed, one for the observer and the other for the LSF control loop. Hence, the pole placement for each can be done separately. This is the separation principle.

It should be noted that since the LHS of (8.13) is the characteristic polynomial of the transfer function between any input–output pair of the observer of Fig. 8.3 considered in isolation, the observer eigenvalues will be referred to as the observer correction loop poles, or simply the observer poles.

Provided conditions (b) and (c) above are satisfied, then if the gain matrix is set so that the observer poles have negative real parts, then

$$\boldsymbol{\varepsilon}(t) \to \mathbf{0} \Rightarrow \widehat{\mathbf{x}}(t) \to \mathbf{x}(t) \text{ as } t \to \infty. \tag{8.14}$$

Then once the correction loop has reached steady state, $\widehat{\mathbf{x}}(t) = \mathbf{x}(t)$, and the system of Fig. 8.3 behaves as if the observer were to be removed, with the true plant state, $\mathbf{x}$, fed directly to the LSF control law. Considering only the response, $y(t)$, to the

8.2 The Full State Continuous Observer for LTI SISO Plants

reference input, $y_r(t)$, it would not be possible to detect the presence of the observer. This is the transparency property of an observer in an LSF control system.

8.2.3 Design of the Real-Time Model Correction Loop

It is evident from Sect. 8.2.2 that the dynamics of the real-time model correction loop depends only on the eigenvalues of the observer correction loop matrix, $\tilde{\mathbf{A}} - \mathbf{l}\,\tilde{\mathbf{c}}^T$, that can be given any desired values via the choice of $\mathbf{l}$, if the plant is observable. A common rule of thumb is that the set of observer correction loop poles, which are the roots of (8.13), should be those of the LSF control loop, which are the roots of (8.12), increased in scale by a factor of at least five. This rule is a matter of engineering judgement, the intention being that at any time, such as the initial loop closure when $\widehat{\mathbf{x}}$ is likely to be very different from $\mathbf{x}$, the model correction loop will drive $\widehat{\mathbf{x}}$ close to $\mathbf{x}$ on a shorter timescale than the LSF control loop dynamics, thereby ensuring that the plant state could not be moved substantially far in the wrong direction due to the initially incorrect value of $\widehat{\mathbf{x}}$. If one of the settling time formulae has been used to design the control system, implying that the closed-loop poles are coincident, then the 'factor of five' rule is implemented by simply setting the nominal observer correction loop settling time, T_{so}, to satisfy

$$T_{so} \le \tfrac{1}{5} T_s \tag{8.15}$$

and carrying out the pole placement of the observer, also using the settling time formulae, by solving the following equation for the gain matrix, $\mathbf{l}$.

$$\left| s\mathbf{I}_n - \left[\tilde{\mathbf{A}} - \mathbf{l}\,\tilde{\mathbf{c}}^T \right] \right| = (s + 1/T_{co})^n, \tag{8.16}$$

where

$$T_{co} = \begin{cases} T_{so}/[1.5(1+n)] & (5\ \%\ \text{criterion}), \\ T_{so}/[1.6(1.5+n)] & (2\ \%\ \text{criterion}). \end{cases} \tag{8.17}$$

In fact the solvability of (8.16) is an alternative to the observability test of Chap. 2.

Some analysis follows to investigate how closely the individual state estimation error components $\varepsilon_i(t) = \widehat{x}_i(t) - x_i(t)$ and $i = 1, 2, \ldots, n$ and the output error $e_o(t) = y(t) - \widehat{y}(t)$ satisfy the settling time specification. It will be recalled from Chap. 4, Sects. 4.5.3 and 4.5.4, that the settling time formulae apply to the step response of a control system with closed-loop transfer function,

$$\frac{Y(s)}{Y_r(s)} = \left(\frac{1}{1 + sT_c} \right)^n, \tag{8.18}$$

given by

$$y(t) = \left[1 - \sum_{k=1}^{n} \frac{1}{(k-1)!}\left(\frac{t}{T_c}\right)^{k-1} e^{-t/T_c}\right] Y_r, \qquad (8.19)$$

where

$$T_c = \begin{cases} T_s/[1.5(1+n)] & (5\ \%\ \text{criterion}) \\ T_s/[1.6(1.5+n)] & (2\ \%\ \text{criterion}) \end{cases}, \qquad (8.20)$$

Y_r is the step reference input level and T_s is the settling time. The control error is

$$e(t) = Y_r - y(t), \qquad (8.21)$$

and therefore, the following alternative definition of the settling time to that given in Chap. 4, Sect. 4.3, applies.

Definition 8.1 *The settling time of a SISO control system step response ($x\%$ criterion) is the time taken for the error, $e(t) = Y_r - y(t)$, to fall to $x\%$ of the initial value, where Y_r is the level of the step reference input, assuming zero initial plant state.*

This definition of settling time applies to the situation in an observer, since some of its state estimation error components start from nonzero values and are brought to zero. Continuing the analysis, however, reveals differences with respect to the control system step response, even if the pole values were made the same. Substituting for $y(t)$ in (8.21) using (8.19) yields the control system step response error as

$$e(t) = \sum_{k=1}^{n} \frac{1}{(k-1)!}\left(\frac{t}{T_c}\right)^{k-1} e^{-t/T_c} Y_r. \qquad (8.22)$$

Since the impulse response of a system with transfer function, $[1/(1+sT_c)]^k$, is a polynomial exponential modal function given by

$$f_k(t) = \frac{1}{(k-1)!}\frac{t^{k-1}}{T_c^k} e^{-t/T_c}, \qquad (8.23)$$

the error, $e(t)$, of (8.22) may be written

$$e(t) = \sum_{k=1}^{n} T_c \frac{1}{(k-1)!}\frac{t^{k-1}}{T_c^k} e^{-t/T_c} Y_r = \sum_{k=1}^{n} m_k f_k(t), \qquad (8.24)$$

8.2 The Full State Continuous Observer for LTI SISO Plants

where

$$m_k = T_c Y_r, \quad k = 1, 2, \ldots, n \tag{8.25}$$

are modal weighting coefficients. Consider now the observer state estimation error differential Eq. (8.8). This is an unforced linear system and is therefore stimulated only by an initial state error, $\boldsymbol{\varepsilon}(0)$, the general solution being

$$\boldsymbol{\varepsilon}(t) = e^{[\tilde{\mathbf{A}} - \mathbf{l}\tilde{\mathbf{c}}^T]t} \boldsymbol{\varepsilon}(0). \tag{8.26}$$

Importantly, in view of the theory presented in Chap. 3, Sect. 3.4.2, if the gain matrix, **l**, has been set according to (8.16), then (8.26) may be expressed as a weighted matrix sum of the modes of (8.23) with T_c replaced by T_{co}. Thus,

$$\boldsymbol{\varepsilon}(t) = \mathbf{M}\mathbf{f}(t) \Rightarrow \varepsilon_i(t) = \sum_{k=1}^{n} m_{ik} f_k(t). \tag{8.27}$$

Considering (8.25), the component, $\varepsilon_i(t)$, can only behave as the step response error (8.24) if $m_{ik} = C_i$, $k = 1, 2, \ldots, n$, where C_i is a constant. This condition cannot be satisfied since it is clear by comparing (8.27) with (8.26) that m_{ik} depends on the initial state error components and that in general $m_{ij} \neq m_{ik}$, $j \neq k$. It should also be pointed out that in view of (8.24), $\varepsilon_i(0) \neq 0$ only if $m_{i1} \neq 0$. If $m_{i1} = 0$, which certainly occurs frequently, then $\varepsilon_i(0) = 0$, but $\varepsilon_i(t)$, $t > 0$, goes through a transient during which it reaches an extreme value, i.e. $|\varepsilon_i(t)|$ reaches a maximum, $|\varepsilon_i|_{\max}$, before ultimately decaying to zero due to the exponential factor, $e^{-t/T_{co}}$. To cater for such components of $\boldsymbol{\varepsilon}(t)$, the definition of settling time will be modified as follows.

Definition 8.2 *For an arbitrary initial state estimation error, $\boldsymbol{\varepsilon}(0)$, the settling time of the component, $\varepsilon_i(t)$ (x% criterion), of an observer is the time taken for $|\varepsilon_i(t)|$ to reduce from the peak value $|e_i|_{\max}$ to $(x/100)|e_i|_{\max}$, measured from the time, t_1, of the peak.*

Assuming condition (c) is satisfied, since $y = \mathbf{c}^T \mathbf{x}(t)$ is another state variable, the output error, $e_o(t) = \mathbf{c}^T \boldsymbol{\varepsilon}(t) = \mathbf{c}^T [\mathbf{x}(t) - \widehat{\mathbf{x}}(t)]$, can be regarded as another state estimation error component. It would be expected that the actual settling times reached by each state estimation error component would be similar in value according to Definition 8.2 since every mode in the system is ultimately reduced to negligible proportions by the factor, $e^{-t/T_{co}}$. These settling times should be close to T_{so} of (8.17). At worst, they will certainly be of the same order of magnitude. Fortunately, it is usually more important to accurately realise the specified settling time, T_s, of the main control loop than the specified observer settling time, T_{so}. So the differences caused by the different modal weightings of the individual state estimation error components should not be too critical. In any case, the settling

Fig. 8.4 Isolated observer state-variable block diagram for gain matrix determination

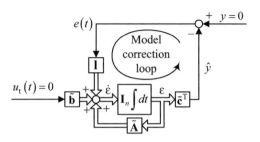

time in (8.17) can be adjusted to a different value, T'_{so}, to reduce any critical state component settling times, identified by simulation, to T_{so}.

For SISO plants, there is an alternative method to the solution of (8.16) for the determination of the gain matrix, $\mathbf{l}$, that would be less time consuming in most cases. In view of the separation principle, it would be possible to disconnect the observer from the control system of Fig. 8.3, as shown in Fig. 8.4a, and design it as a self-contained linear system. This can be confirmed by setting $y = 0$, which is equivalent to setting $\mathbf{x} = \mathbf{0}$, and if $u = 0$, with $\mathbf{A} = \tilde{\mathbf{A}}$ and $\mathbf{c}^T = \tilde{\mathbf{c}}^T$, the observer state differential equation would become

$$\dot{\widehat{\mathbf{x}}} = \left[\mathbf{A} - \mathbf{l}\mathbf{c}^T\right]\widehat{\mathbf{x}}. \qquad (8.28)$$

This is the state estimation error differential Eq. (8.8) if $\widehat{\mathbf{x}}$ is replaced by $\boldsymbol{\varepsilon}$, indicating that the observer may be designed in isolation. Then the straightforward method for SISO plants is to first form the state-variable block diagram, shown in general form in Fig. 8.4.

The state-variable block diagram is drawn in detail for the particular plant, showing the integrators, gains and summing junctions. Then the characteristic polynomial is found using the determinant of Mason's formula, as demonstrated in Chaps. 4 and 5 for basic linear control loops.

Example 8.1 Observer for single axis of rigid-body spacecraft attitude control system

There is a sufficient distribution of stars over the celestial sphere with small enough magnitudes (i.e. large enough brightness) to enable modern three-axis star sensors for spacecraft attitude measurement to operate over 4π steradians, minus, of course, the solid angles occupied by bodies that occlude the stars such as the sun and any nearby planets. Hence, the star sensor is sometimes the only form of sensor provided for the attitude control of modern spacecraft. A state feedback control law requires the angular velocity components as additional state variables, and these may be estimated using an observer. The application considered here is a three-axis stabilised rigid-body spacecraft with negligible inter-axis coupling permitting each axis to be considered in isolation, the plant state-space model being

$$\dot{x}_1 = x_2, \ \dot{x}_2 = b(u - d), \ y = cx_1, \qquad (8.29)$$

8.2 The Full State Continuous Observer for LTI SISO Plants

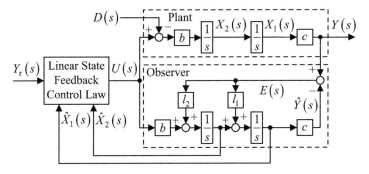

Fig. 8.5 Single-axis rigid-body spacecraft attitude control system employing an observer

where x_1 is the attitude angle, u is the control voltage input to the reaction wheel drive and d is the external disturbance torque referred to the control input. Figure 8.5 shows the control system block diagram.

The plant parameters are $b = K_w/J$ and $c = K_s$, where K_w is the reaction wheel torque constant, J is the moment of inertia of the spacecraft body about the control axis and K_s is the measurement constant of the star sensor. For the simulations to follow, $J = 150$ [Kgm^2], $K_w = 0.1$ [Nm/V] and $K_s = 3$ [V/rad]. The linear state feedback control law yields a settling time of $T_s = 100$ [s].

Condition (c) of Sect. 8.2.1 is assumed to hold as the purpose of this example is to demonstrate the basic operation of the observer free of any mismatch between the plant and its model, except for the external disturbance, d. First, the characteristic polynomial will be derived using the orthodox matrix–vector method. Then, for comparison, it will be derived again using the determinant of Mason's formula.

The state equations for the observer are

$$\begin{bmatrix} \hat{\dot{x}}_1 \\ \hat{\dot{x}}_2 \end{bmatrix} = \underbrace{\begin{bmatrix} 0 & 1 \\ 0 & 0 \end{bmatrix}}_{\mathbf{A}} \begin{bmatrix} \hat{x}_1 \\ \hat{x}_2 \end{bmatrix} + \underbrace{\begin{bmatrix} 0 \\ b \end{bmatrix}}_{\mathbf{b}} u + \underbrace{\begin{bmatrix} l_1 \\ l_2 \end{bmatrix}}_{\mathbf{l}} \left(y - \underbrace{\begin{bmatrix} c & 0 \end{bmatrix}}_{\mathbf{c}^T} \begin{bmatrix} \hat{x}_1 \\ \hat{x}_2 \end{bmatrix} \right). \qquad (8.30)$$

The observer characteristic polynomial is the LHS of (8.13). This example yields

$$\left| s\mathbf{I}_2 - [\mathbf{A} - \mathbf{l}\,\mathbf{c}^T] \right| = \left| s \begin{bmatrix} 1 & 0 \\ 0 & 1 \end{bmatrix} - \left[\begin{bmatrix} 0 & 1 \\ 0 & 0 \end{bmatrix} - \begin{bmatrix} l_1 \\ l_2 \end{bmatrix} \begin{bmatrix} c & 0 \end{bmatrix} \right] \right|$$

$$= \left| \begin{bmatrix} s & -1 \\ 0 & s \end{bmatrix} - \begin{bmatrix} -l_1 c & 1 \\ -l_2 c & 0 \end{bmatrix} \right| = \left| \begin{matrix} s + cl_1 & -1 \\ cl_2 & s \end{matrix} \right| = s^2 + cl_1 s + cl_2. \qquad (8.31)$$

The observer characteristic equation, and hence the polynomial, can be obtained directly by equating to zero the determinant of Mason's formula applied to the observer of Fig. 8.5 as if it was disconnected from the system as in Fig. 8.4. Thus,

$$1 - \left[\frac{cl_1}{s} - \frac{cl_2}{s^2}\right] = 0 \Rightarrow s^2 + cl_1 s + cl_2 = 0. \tag{8.32}$$

Using the settling time formula (5 % criterion), the desired characteristic polynomial of (8.17) is

$$[s + 1.5(1+n)/T_{so}]^n|_{n=2} = (s + 9/(2T_{so}))^2 = s^2 + (9/T_{so})s + 81/(4T_{so}^2). \tag{8.33}$$

Equating (8.33) and (8.31) then yields

$$l_1 = 9/(cT_{so}) \text{ and } l_2 = 81/(4cT_{so}^2). \tag{8.34}$$

Next, a simulation of the initial attitude control loop closure will be carried out. Upon separation from the launch vehicle, the attitude will be incorrect, and there will be an unwanted initial angular velocity due to the imperfect separation devices. The attitude control system will be required to bring the spacecraft angular velocity, x_2, to zero and the attitude, x_1, to the correct value, which will be taken as zero, to begin the mission. In the simulation of Fig. 8.6, the initial attitude angle and angular velocity are, respectively, $x_1(0) = -1.5$ rad and $x_2(0) = 0.05$ rad/s. The corresponding estimates in the observer are set to $\widehat{x}_1(0) = 0$ and $\widehat{x}_2(0) = 0$.

Figure 8.6a shows the results obtained with the observer correction loop settling time set to the maximum recommended value of $T_{so} = T_s/5$, and Fig. 8.6b shows the corresponding results obtained with $T_{so} = T_s/100$.

Figure 8.6a 1 shows the state estimation errors, $\varepsilon_1(t) = \widehat{x}_1 - x_1$ and $\varepsilon_2(t) = \widehat{x}_2 - x_2$. A scale factor of 9.25 is applied to $\varepsilon_2(t)$ so that the peak values of $\varepsilon_1(t)$ and $9.25\varepsilon_2(t)$ are of the same magnitude. This enables the actual settling times of the errors to be assessed using the same ± 5 % levels as shown. These settling times, according to *Definition* 8.2, are seen to be very close to the specified value of $T_{so} = 20$ [s]. This is also true for Fig. 8.6b 1. In this case, however, the scaling factor had to be reduced to 0.596, meaning that reducing T_{so} from 20 [s] to 1 [s] increases the magnitude of the angular velocity error peak by a factor of about 15. This is a consequence of the higher observer gains needed to bring about faster convergence of the errors towards zero.

Figure 8.6a 2 and 3 clearly show $\widehat{x}_1(t)$ and $\widehat{x}_2(t)$ converging towards $x_1(t)$ and $x_2(t)$, as required. This is also true in Fig. 8.6b 2 and 3. Since the convergence is much faster, the plant state has not changed very much over the shorter convergence period compared with Fig. 8.6a 2 and 3. This is the reason for $x_1(t)$ and $x_2(t)$ in Fig. 8.6b 4 following the ideal paths (shown dotted) much more closely than they do in Fig. 8.6a 4. These ideal paths are those followed without any state estimation errors.

It should be noted that in a real space mission, the observer would be turned on and allowed to settle before closing the attitude control loop to avoid the overshooting behaviour evident in Fig. 8.6a 4.

8.2 The Full State Continuous Observer for LTI SISO Plants

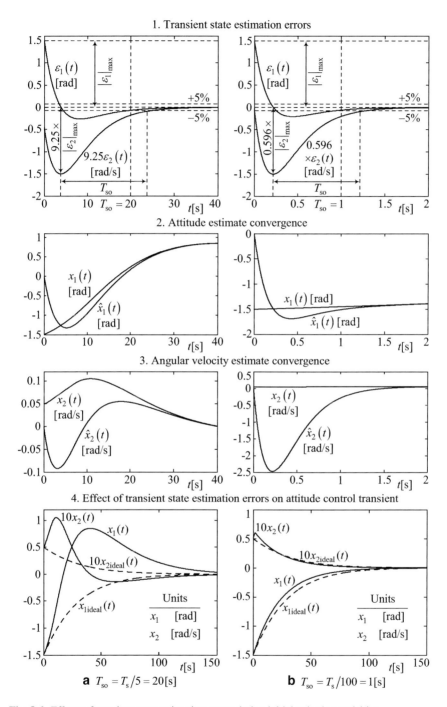

Fig. 8.6 Effects of transient state estimation errors during initial attitude acquisition

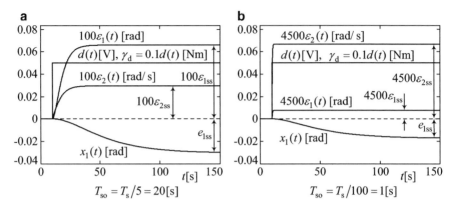

Fig. 8.7 Effect of step external disturbance torque

In this example, the estimated attitude angle, $\hat{x}_1$, is fed to the control law, but it might be argued that using the direct attitude measurement by feeding back $x_1 = y/c$ instead of $\hat{x}_1$ would reduce the effects of the state estimation errors. This could certainly be done, but, as will be seen in later sections, the filtering properties of the observer enable more accurate control in a noisy environment, and this would require only the state estimates to be fed to the control law.

After the initial attitude acquisition, it is usual for orbit adjustment thrusters to be fired that are nominally directed through the spacecraft centre of mass but in practice will be misaligned slightly resulting in a constant disturbance torque. Figure 8.7 shows the steady-state errors that occur in the state estimate as well as the attitude angle, x_1, when a step disturbance torque equal to 5 % of the maximum reaction wheel torque is applied, equivalent to $d = 0.5h\,(t-10)$ [V].

It is evident by comparing Fig. 8.7b with Fig. 8.7a that reducing T_{so} by a factor of 20 considerably reduces the magnitudes of the steady-state estimation errors, ε_{1ss} and ε_{2ss}, but does not substantially reduce the steady-state pointing error, e_{1ss}. It has already been shown in Chap. 5 that a standard linear state feedback control law, such as used here, allows a steady-state error in the controlled variable if the reference input and the external disturbance (referred to the control input) are constant, so the behaviour of $x_1(t)$ in Fig. 8.7 is to be expected. The steady-state errors in the state variables, however, could be eliminated by means of a special observer that includes disturbance estimation as developed in the following subsection, and the disturbance estimate could be used in the control law to counteract the constant disturbance and bring the steady-state pointing error to zero. It should also be mentioned that this could also be achieved using the linear state feedback plus integral control law of Chap. 5, Sect. 5.2.8, regardless of the steady-state errors, ε_{1ss} and ε_{2ss}.

It is important to note that plant model uncertainty has to be taken into account as with the design of controllers in general. The real-time model correction loop of an observer can be considered as a special form of model-based controller applied

8.2 The Full State Continuous Observer for LTI SISO Plants

to the plant model. This will yield the desired decay of the state estimation errors towards zero with a perfect plant model, and simulation may be used to check the correctness of the observer design using a simulation without any deliberate mismatch with respect to the nominal plant. It must also be ensured that the state estimation errors still converge to acceptably small proportions with the perceived worst-case plant modelling errors for the particular application. For relatively simple plant models, Routh's stability criterion (Appendix A5), could be useful: otherwise, simulation is advisable.

8.2.4 Estimation of Disturbances

If the form of the equivalent disturbance, $d(t)$, applied to the plant input is known to be a solution of an unforced linear differential equation with constant coefficients, then this differential equation can be converted to the state-space form (Chap. 3) to create a state-space model of the disturbance. Thus,

$$\dot{\mathbf{x}}_d = \mathbf{A}_d \mathbf{x}_d, \quad d = \mathbf{c}_d^T \mathbf{x}_d, \tag{8.35}$$

where $\mathbf{x}_d \in \mathfrak{R}^{n_d}$. This can be combined with the plant Eq. (8.3) to form augmented plant equations in the partitioned form as follows:

$$\begin{bmatrix} \dot{\mathbf{x}} \\ \dot{\mathbf{x}}_d \end{bmatrix} = \underbrace{\begin{bmatrix} \mathbf{A} & -\mathbf{c}_d^T \\ \mathbf{0} & \mathbf{A}_d \end{bmatrix}}_{\mathbf{A}_a} \underbrace{\begin{bmatrix} \mathbf{x} \\ \mathbf{x}_d \end{bmatrix}}_{\mathbf{x}_a} + \underbrace{\begin{bmatrix} \mathbf{b} \\ \mathbf{0} \end{bmatrix}}_{\mathbf{b}_a} u, \quad \begin{bmatrix} y \\ d \end{bmatrix} = \underbrace{\begin{bmatrix} \mathbf{c}^T & 0 \\ 0 & \mathbf{c}_d^T \end{bmatrix}}_{\mathbf{C}_a} \begin{bmatrix} \mathbf{x} \\ \mathbf{x}_d \end{bmatrix}. \tag{8.36}$$

where subscript, a, indicates 'augmented'. The corresponding observer equation is

$$\begin{bmatrix} \dot{\hat{\mathbf{x}}} \\ \dot{\hat{\mathbf{x}}}_d \end{bmatrix} = \underbrace{\begin{bmatrix} \tilde{\mathbf{A}} & -\tilde{\mathbf{c}}_d^T \\ \mathbf{0} & \tilde{\mathbf{A}}_d \end{bmatrix}}_{\tilde{\mathbf{A}}_a} \underbrace{\begin{bmatrix} \hat{\mathbf{x}} \\ \hat{\mathbf{x}}_d \end{bmatrix}}_{\hat{\mathbf{x}}_a} + \underbrace{\begin{bmatrix} \mathbf{k} \\ \mathbf{k}_d \end{bmatrix}}_{\mathbf{k}_a} (y - \hat{y}) + \underbrace{\begin{bmatrix} \tilde{\mathbf{b}} \\ \mathbf{0} \end{bmatrix}}_{\tilde{\mathbf{b}}_a} u, \quad \begin{bmatrix} \hat{y} \\ \hat{d} \end{bmatrix} = \underbrace{\begin{bmatrix} \tilde{\mathbf{c}}^T & 0 \\ 0 & \tilde{\mathbf{c}}_d^T \end{bmatrix}}_{\tilde{\mathbf{C}}_a} \begin{bmatrix} \hat{\mathbf{x}} \\ \hat{\mathbf{x}}_d \end{bmatrix} \tag{8.37}$$

Since (8.36) and (8.37) are of precisely the same form as (8.3) and (8.4), if $\{\mathbf{A}, \mathbf{A}_d, \mathbf{b}, \mathbf{c}^T\} = \{\tilde{\mathbf{A}}, \tilde{\mathbf{A}}_d, \tilde{\mathbf{b}}, \tilde{\mathbf{c}}^T\}$, the error state equation,

$$\dot{\boldsymbol{\varepsilon}}_a = \left[\tilde{\mathbf{A}}_a - \mathbf{l}_a \tilde{\mathbf{c}}_a^T\right] \boldsymbol{\varepsilon}_a, \text{ i.e., } \begin{bmatrix} \dot{\boldsymbol{\varepsilon}} \\ \dot{\boldsymbol{\varepsilon}}_d \end{bmatrix} = \begin{bmatrix} \tilde{\mathbf{A}} - \mathbf{l} \tilde{\mathbf{c}}^T & -\tilde{\mathbf{c}}_d^T \\ -\mathbf{l}_d \tilde{\mathbf{c}}^T & \tilde{\mathbf{A}}_d \end{bmatrix} \begin{bmatrix} \boldsymbol{\varepsilon} \\ \boldsymbol{\varepsilon}_d \end{bmatrix}, \tag{8.38}$$

follows by analogy with (8.8), where $\varepsilon = \widehat{\mathbf{x}} - \mathbf{x}$, as before, and $\varepsilon_d = \widehat{\mathbf{x}}_d - \mathbf{x}_d$. The observer pole placement equation using the settling time formula is similar to (8.17) and is

$$\left|s\mathbf{I}_{n+n_d} - \left[\widetilde{\mathbf{A}}_a - \mathbf{l}_a\, \widetilde{\mathbf{c}}_a^T\right]\right| = \begin{cases} [s + 1.5\,(1+n+n_d)/T_{so}]^{n+n_d} & (5\%\ \text{criterion}) \\ [s + 1.6\,(1.5+n+n_d)/T_{so}]^{n+n_d} & (2\%\ \text{criterion}) \end{cases} \quad (8.39)$$

In the examples to follow, the notation for the observer gains is

$$\mathbf{l}_a^T = \left[\ \underbrace{l_1\ l_2\ \cdots\ l_n}_{\mathbf{l}}\ \underbrace{l_{n+1}\ l_{n+2}\ \cdots\ l_{n+n_d}}_{\mathbf{l}_d}\ \right]^T \quad (8.40)$$

Example 8.2 LSF control of vacuum air bearing with load torque compensation

This is an example of high-precision position control typical of that needed in the microchip manufacturing industry. The plant is illustrated in Fig. 8.8. The measurement is obtained from an encoder with a resolution of the order of 10 nm, and position changes of the order of 1 μm are demanded.

A disturbance force occurs due to the signal and power leads connected to the payload, but since the payload movements are so small, it can be regarded as constant. This is referred to the control input and represented by $d(t) = D = \text{const}$. Changes in the actuator force induce rotational oscillatory motion through an angle, x_3, about a vertical axis due to the compliance of the lateral bearing air cushion. This contaminates the position measurement, y. Using an observer containing a precise plant model, however, the position, x_1, can be accurately estimated and controlled.

The state differential equation of the disturbance in the form of (8.35) is just $\dot{x}_{d1} = 0$, $d = x_{d1}$, and the corresponding equation in the observer is $\dot{\widehat{x}}_{d1} = k_5 e$,

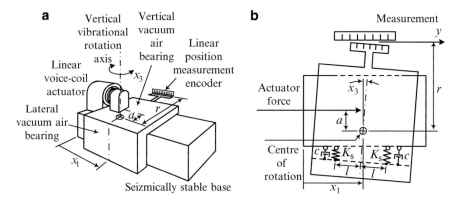

Fig. 8.8 Vacuum air-bearing positioning system. (**a**) Main components. (**b**) Schematic

8.2 The Full State Continuous Observer for LTI SISO Plants

where e is the model correction loop actuation error. The state-space model of the plant is in the modal form comprising the rotational vibration mode and the translational rigid-body mode. It is therefore of fourth order. This, together with the disturbance state differential equation, is as follows, in the form of (8.36):

$$\begin{bmatrix} \dot{x}_1 \\ \dot{x}_2 \\ \dot{x}_3 \\ \dot{x}_4 \\ \hline \dot{x}_{d1} \end{bmatrix} = \begin{bmatrix} 0 & 1 & 0 & 0 & | & 0 \\ 0 & 0 & 0 & 0 & | & 0 \\ 0 & 0 & 0 & 1 & | & 0 \\ 0 & 0 & -a_3 & -a_4 & | & 0 \\ \hline 0 & 0 & 0 & 0 & | & 0 \end{bmatrix} \begin{bmatrix} x_1 \\ x_2 \\ x_3 \\ x_4 \\ \hline x_{d1} \end{bmatrix} + \begin{bmatrix} 0 \\ b_1 \\ 0 \\ b_2 \\ \hline 0 \end{bmatrix} (u-d), \begin{bmatrix} y \\ d \end{bmatrix} \begin{bmatrix} 1 & 0 & c_3 & 0 & | & 0 \\ \hline 0 & 0 & 0 & 0 & | & 1 \end{bmatrix} \begin{bmatrix} x_1 \\ x_2 \\ x_3 \\ x_4 \\ \hline x_{d1} \end{bmatrix}. \tag{8.41}$$

Here, $b_1 = K_a K_f/m$, $b_2 = K_a K_f a/J$, $a_3 = 2K_s l^2/J$, $a_4 = 2cl^2/J$ and $c_3 = R$, where K_a is the trans-conductance of the voice-coil actuator drive amplifier, K_f is the force constant of the voice-coil actuator, m is the payload mass, J is the payload moment of inertia about the vibrational rotation axis, a is the lateral displacement between the actuator force vector and the vibrational rotation axis, l is the distance from the payload centre line of two springs representing the air compliance, as shown in Fig. 8.8b, K_s is the spring constant and c is the associated damping coefficient. The plant parameters are $m = 3.3$ kg, $J = 0.03$ kg m^2, $a = 0.02$ m, $d = 0.01$ m, $K_m = 11.1$ N/A, $K_a = 0.8$ A/V, $K_s = 2.9 \times 10^6$ N/m, $c = 100$ N/(m/s) and $R = 0.13$ m.

The corresponding observer equation is

$$\left. \begin{aligned} \dot{\hat{\mathbf{x}}} &= \begin{bmatrix} 0 & 1 & 0 & 0 & | & 0 \\ 0 & 0 & 0 & 0 & | & 0 \\ 0 & 0 & 0 & 1 & | & 0 \\ 0 & 0 & -\tilde{a}_3 & -\tilde{a}_4 & | & 0 \\ \hline 0 & 0 & 0 & 0 & | & 0 \end{bmatrix} \hat{\mathbf{x}} + \begin{bmatrix} 0 \\ \tilde{b}_1 \\ 0 \\ \tilde{b}_2 \\ \hline 0 \end{bmatrix} (u - \hat{d}) + \begin{bmatrix} l_{a1} \\ l_{a2} \\ l_{a3} \\ l_{a4} \\ l_{a5} \end{bmatrix} (y - \hat{y}), \\ \begin{bmatrix} \hat{y} \\ \hat{d} \end{bmatrix} &= \begin{bmatrix} 1 & 0 & \tilde{c}_3 & 0 & | & 0 \\ \hline 0 & 0 & 0 & 0 & | & 1 \end{bmatrix} \hat{\mathbf{x}}, \text{ where } \hat{\mathbf{x}} = \begin{bmatrix} \hat{x}_1 & \hat{x}_2 & \hat{x}_3 & \hat{x}_4 & | & \hat{x}_{d1} \end{bmatrix}^T \end{aligned} \right\}. \tag{8.42}$$

Figure 8.9 shows a block diagram of the complete control system.

The linear state feedback control law is a standard one designed with multiple pole placement to have a settling time of $T_s = 0.1$ s. The observer settling time is set to $T_{so} = 0.02$ s. A special feature of this control system is that the controller adds the disturbance estimate to the LSF control law output to counteract the disturbance. This is an alternative to including an integral term.

Considering the observer in isolation, the pole placement equation can be obtained by equating the determinant of Mason's formula to zero, as follows.

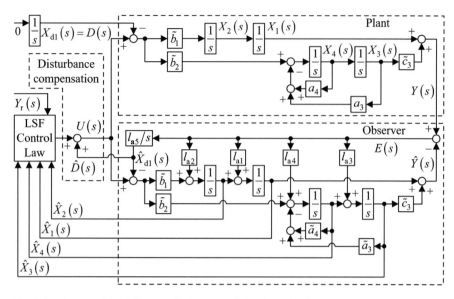

Fig. 8.9 Observer-aided LSF control of vacuum air bearing with disturbance compensation

$$1 - \left[-\frac{l_{a3}\tilde{c}_3}{s} - \frac{l_{a4}\tilde{c}_3}{s^2} - \frac{l_{a1}}{s} - \frac{l_{a2}}{s^2} + \frac{l_{a5}\tilde{b}_1}{s^3} + \frac{l_{a5}\tilde{b}_2\tilde{c}_3}{s^3} - \frac{\tilde{a}_4}{s} - \frac{\tilde{a}_3}{s^2} \right]$$
$$+ \left(-\frac{\tilde{a}_4}{s} - \frac{\tilde{a}_3}{s^2} \right)\left(-\frac{l_{a1}}{s} - \frac{l_{a2}}{s^2} + \frac{l_{a5}\tilde{b}_1}{s^3} \right) = 0 \Rightarrow$$
$$s^5 + (\tilde{c}_3 l_{a3} + l_{a1} + \tilde{a}_4)s^4 + (\tilde{c}_3 l_{a4} + l_{a2} + \tilde{a}_3 + \tilde{a}_4 l_{a1})s^3$$
$$+ \left(\tilde{a}_3 l_{a1} + \tilde{a}_4 l_{a2} - \tilde{b}_1 l_{a5} - \tilde{b}_2 \tilde{c}_3 l_{a5} \right)s^2 + \left(\tilde{a}_3 l_{a2} - \tilde{a}_4 \tilde{b}_1 l_{a5} \right)s - \left(\tilde{a}_3 \tilde{b}_1 l_{a5} \right) = 0.$$
(8.43)

The desired characteristic equation using the 5 % settling time formula is

$$\left[s + \frac{1.5(1+N)}{T_{so}} \right]^N \bigg|_{N=5} = (s+q)^5 = s^5 + 5qs^4 + 10q^2 s^3 + 10q^3 s^2 + 5q^4 s + q^5 = 0$$
(8.44)

where $N = n + n_d$, $n = 4$, $n_d = 1$ and $q = 9/T_{so}$. Equating the RHS of (8.43) and (8.44) then yields the observer gains as

$$\begin{cases} l_{a5} = -\dfrac{q^5}{\tilde{a}_3 \tilde{b}_1},\ l_{a2} = \dfrac{5q^4 + \tilde{a}_4 \tilde{b}_1 l_{a5}}{\tilde{a}_3},\ l_{a1} = \dfrac{10q^3 - \left(\tilde{a}_4 l_{a2} - \tilde{b}_1 l_{a5} - \tilde{b}_2 \tilde{c}_3 l_{a5} \right)}{\tilde{a}_3}, \\ l_{a4} = \dfrac{10q^2 - (l_{a2} + \tilde{a}_3 + \tilde{a}_4 l_{a1})}{\tilde{c}_3} \text{ and } l_{a3} = \dfrac{5q - (l_{a1} + \tilde{a}_4)}{\tilde{c}_3}. \end{cases}$$
(8.45)

8.2 The Full State Continuous Observer for LTI SISO Plants

Before the initial loop closure, the platform is mechanically fixed in the zero position while the vacuum air bearing is pressurised. Then the mechanical lock is removed and the control loop closed with zero position reference input while the observer is allowed to settle. Figure 8.10a, d show the observer settling transients with a residual disturbance force of 10^{-5} N. Such small forces are realistic for this application but very significant with demanded control accuracies with an order of magnitude of nanometres. In Fig. 8.10a, the plant model and the real plant are perfectly matched, but in Fig. 8.10d, all the plant parameters are mismatched by $\pm 5\ \%$ ($m = 1.05\tilde{m}$, $J = 1.05\tilde{J}$, $a = 0.95\tilde{a}$, $d = 0.95\tilde{d}$, $K_m = 0.95\tilde{K}_m$, $K_a = 0.95\tilde{K}_a$, $K_s = 0.95\tilde{K}_s$, $c = 0.95\tilde{c}$ and $R = 1.5\tilde{R}$), this combination being considered the worst case in the sense of reducing the plant forward path gain below the nominal value which will encourage a sluggish and possibly oscillatory response.

In Fig. 8.10a, the disturbance estimate has the correct settling time and the state estimation errors settle in about the same time from their peak values, following *Definition* 8.2. In Fig. 8.10d, the plant mismatching increases the settling times.

Comparing Fig. 8.10b, e reveals that the plant mismatching has no significant detrimental effect on the overall control system performance.

Figure 8.10c shows the steady-state error resulting from the removal of the feedback of $\widehat{d}$ in the disturbance compensator of Fig. 8.9.

The single integrator disturbance model used in this example to estimate a constant external disturbance is the most common since it can also be useful when the disturbance is time varying but of unknown form. The differential equation and the corresponding state-space model of such a disturbance may not exist, and even if it does, it will not be $\dot{x}_{d1} = 0$, $d = x_{d1}$ as above. The argument is as follows. If $d(t)$ is sufficiently slowly varying for any change in its value over a period equal to the correction loop settling time, T_{so}, to be negligible, then $d(t)$ can be regarded constant. Then in practice, the estimate, $\widehat{d}(t)$, will follow $d(t)$ with sufficient accuracy for the disturbance compensation to be effective. Provided $d(t)$ contains no discontinuities, then T_{so} can be reduced, if necessary, to satisfy this requirement.

In the frequency domain, if the bandwidth, ω_{bd}, of the disturbance is known, then Eq. (4.111) developed in Sect. 4.6 can be used to determine a suitable value for T_{so}. This equation gives the settling time of the step response (5 % criterion) of a control system of known bandwidth. If applied to ω_{bd}, this formula gives the minimum period over which significant changes of $d(t)$ should be expected. As a rule of thumb, the maximum value of T_{so} is taken as one fifth of the value yielded by (4.11). Thus,

$$T_{so} \leq 0.3\,(1+N)\,\sqrt{2^{1/N}-1}/\omega_{bd}. \qquad (8.46)$$

If there are discontinuities in $d(t)$, then T_{so} is the settling time of the decay of the error, $\widehat{d}(t) - d(t)$, following a jump in $d(t)$, according to *Definition* 8.2. Then T_{so} should be set to a sufficiently small value to suit the application in hand. The following example demonstrates the use of (8.46).

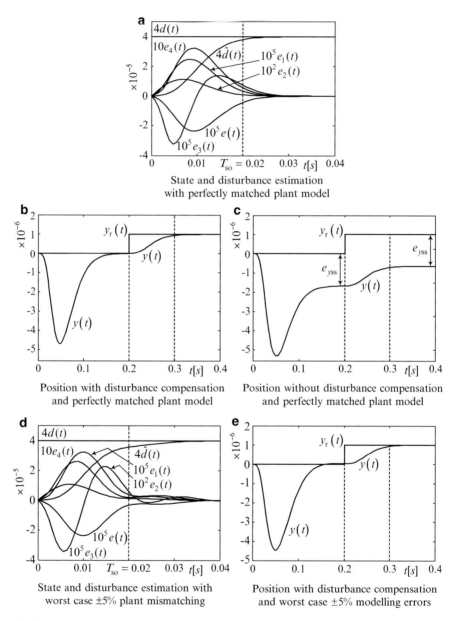

Fig. 8.10 Step response of air-bearing position control system with constant disturbance force

Example 8.3 Ship roll stabilisation with disturbance torque estimation and compensation

The disturbance torque components acting on a surface ship due to wave motion are of a random nature and cannot be modelled by a differential equation. It is

8.2 The Full State Continuous Observer for LTI SISO Plants

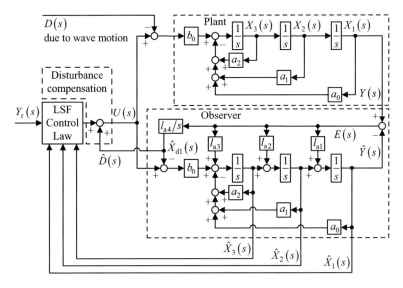

Fig. 8.11 LSF roll control of a ship with disturbance torque estimation and compensation

possible, however, to employ observers with single integrator disturbance models to obtain fairly close approximations to the disturbance components since they are band limited.

Suppose that for the roll axis being considered here, the transfer function relationship between the fin servo reference input, regarded here as the control input, $u(s)$; the disturbance, $d(s)$, referred to the control input; and the roll attitude angle measurement, $y(s)$, is given by

$$y(s) = \frac{K_f\left[u(s) - d(s)\right]}{\left(s^2 + 2\zeta\omega_n s + \omega_n^2\right)(s + 1/T_v)} = \frac{b_0}{s^3 + a_2 s^2 + a_1 s + a_0}\left[u(s) - d(s)\right], \tag{8.47}$$

where the parameters are defined in Chap. 2. Then a linear state feedback control system with a single integrator disturbance model can be designed following similar lines to those of Example 8.2. Figure 8.11 shows the block diagram.

Considering the observer in isolation, the pole placement equation for a correction loop settling time of T_{so} seconds (5 % criterion) is

$$\begin{aligned}
s^4 &\left[1 - \left(-\frac{l_1}{s} - \frac{l_2}{s^2} - \frac{l_3}{s^3} + \frac{b_0 l_4}{s^4} - \frac{a_2}{s} - \frac{a_1}{s^2} - \frac{a_0}{s^3}\right) + \left(-\frac{a_2}{s}\right)\left(-\frac{l_1}{s} - \frac{l_2}{s^2}\right) + \left(-\frac{a_1}{s^2}\right)\left(-\frac{l_1}{s}\right)\right] \\
&= s^4 + (l_1 + a_2)s^3 + (l_2 + a_1 + a_2 l_1)s^2 + (l_3 + a_0 + a_2 l_2 + a_1 l_1)s - b_0 l_4 \\
&= \left[s + 1.5(1+N)/T_{so}\right]^N\bigg|_{N = n + n_d = 4} = (s+q)^4 \\
&= s^4 + 4qs^3 + 6q^2 s^2 + 4q^3 s + q^4 \Rightarrow \\
l_1 &= 4q - a_2, \; l_2 = 6q^2 - (a_1 + a_2 l_1), \\
l_3 &= 4q^3 - (a_0 + a_2 l_2 + a_1 l_1) \text{ and } l_4 = -q^4/b_0
\end{aligned}\right\}. \tag{8.48}$$

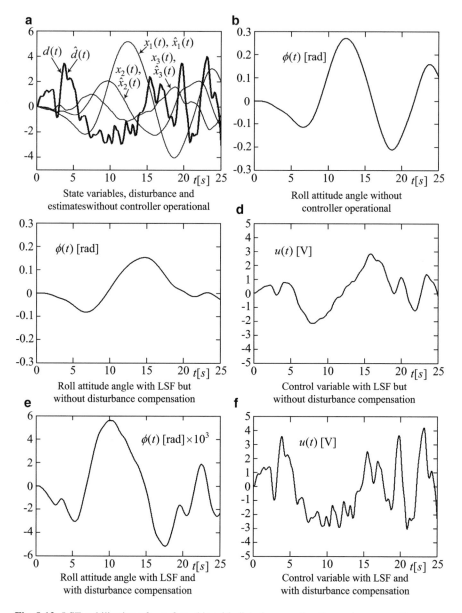

Fig. 8.12 LSF stabilisation of a surface ship with disturbance estimation and compensation

The plant parameters taken are $\omega_n = 0.6$ rad/s, $\zeta = 0.1$, $T_v = 1$ s and constant extreme inputs $u = \pm u_{max} = \pm 10$ V, where u_{max} is the maximum control voltage magnitude, yielding steady-state roll attitude angles of $\pm 30° = \pm 0.5236$ rad $\equiv y_{ss} = \pm 10$ V, requiring $K_f T_v / \omega_n^2 = 1 \Rightarrow K_f = 0.36$. Although not shown in Fig. 8.1, the attitude sensor constant is $K_s = 10/0.5236 = 19.099$ V/rad.

8.2 The Full State Continuous Observer for LTI SISO Plants

The disturbance bandwidth is given as $\omega_{bd} = 3$ rad/s.

The standard LSF control law is designed with $T_s = 5$ s (5 % criterion). For the observer, (8.46) yields the maximum value of the correction loop settling time as $T_{so} = 0.3 \times 5\sqrt{1.189 - 1}/3 = 0.217$ s.

The disturbance is modelled as the output of a low-pass filter with transfer function, $[\omega_{bd}/(s + \omega_{bd})]^3$, where the input is a white, Gaussian, random signal and with the variance set to 1000 in SIMULINK® to yield a filter output peaking at approximately ± 4 V, which is $\pm 0.4 u_{max}$.

Figure 8.12a shows the state variables and disturbance (referred to the control input) together with their estimates from the observer, without the controller operating. The small dynamic lag between $d(t)$ and $\widehat{d}(t)$ is acceptable for this application and in any case can be reduced by making T_{so} smaller, but within the constraints set by the measurement noise level and the finite sampling frequency of the digital processor (Chap. 6). Figure 8.12b shows the roll axis attitude response to the disturbance, again without the controller operating. Figure 8.12c shows the roll attitude fluctuations approximately halved due to the loop closure via the LSF control law but without the disturbance compensation.

Figure 8.12d shows the corresponding control activity. Figure 8.12e shows the drastic reduction in the roll attitude fluctuation with the disturbance compensation added to the LSF control law output, while Fig. 8.12f indicates an increase in the control activity needed to achieve this. It may be necessary, however, to perform a trade-off between the degree of passenger comfort and the fuel consumption by the engines providing the hydraulic or electrical power to operate the fin servos.

The single integrator disturbance model for the observer will suffice for the majority of applications requiring disturbance estimation and compensation, but occasionally it is not practicable to increase the observer eigenvalues sufficiently due to the amplification of measurement noise, addressed in the following sections. The following example is of a high-precision application that benefits from an accurate disturbance model which, if perfect, eliminates any dynamic lag of the estimate after the observer has settled and, in practice, reduces it to negligible proportions.

Example 8.4 Spacecraft attitude control in the presence of an oscillating disturbance torque

A spacecraft contains a three-axis stabilised part containing the propulsion system and a rotating part creating artificial gravity for the crew. The dynamic imbalance causes a sinusoidal oscillating disturbance torque component about the two control axes perpendicular to the spin axis of the rotating part, referred to as the transverse axes. For this relatively simple example, the angular velocities about these two axes are assumed to be small enough for the gyroscopic cross coupling between them to be negligible. In this case, each control axis can be considered separately. Assuming rigid-body dynamics, the attitude measurement, y, about one of the two transverse axes will obey the linear differential equation,

$$\ddot{y}(t) = b[u(t) - d(t)], \tag{8.49}$$

with $b = K_g K_s / J$, where K_g is the torque constant of the control moment gyro, K_s is the attitude sensor constant and J is the moment of inertia about the control axis being considered. The disturbance referred to the control input is given by

$$d(t) = D \sin(\Omega t + \phi), \tag{8.50}$$

where the amplitude, D, and the phase angle, ϕ, depend upon the positions of the crew around the rotating section of the spacecraft. Equation (8.50) is the general solution to the second-order differential equation

$$\ddot{d} + \Omega^2 d = 0. \tag{8.51}$$

State-space models corresponding to (8.49) and (8.51) may be formed with state variables chosen as $x_1 = y$, $x_2 = \dot{y}$, $x_{d1} = d$ and $x_{d2} = \dot{d}$ as follows:

$$\left\{ \begin{array}{l} \dot{x}_1 = x_2 \\ \dot{x}_3 = b(u - d) \\ y = x_1 \end{array} \right\} \text{ plant subsystem,} \quad \left\{ \begin{array}{l} \dot{x}_{d1} = x_{d2} \\ \dot{x}_{d2} = -\Omega^2 x_{d1} \\ d = x_{d1} \end{array} \right\} \text{ disturbance subsystem.} \tag{8.52}$$

In the form of (8.36), these may be written as

$$\begin{bmatrix} \dot{x}_1 \\ \dot{x}_2 \\ \hline \dot{x}_{d1} \\ \dot{x}_{d2} \end{bmatrix} = \begin{bmatrix} 0 & 1 & | & 0 & 0 \\ 0 & 0 & | & -1 & 0 \\ \hline 0 & 0 & | & 0 & 1 \\ 0 & 0 & | & -\Omega^2 & 0 \end{bmatrix} \begin{bmatrix} x_1 \\ x_2 \\ \hline x_{d1} \\ x_{d2} \end{bmatrix} + \begin{bmatrix} 0 \\ b \\ \hline 0 \\ 0 \end{bmatrix} u, \quad \begin{bmatrix} y \\ d \end{bmatrix} = \begin{bmatrix} 1 & 0 & | & 0 & 0 \\ 0 & 0 & | & 1 & 0 \end{bmatrix} \begin{bmatrix} x_1 \\ x_2 \\ \hline x_{d1} \\ x_{d2} \end{bmatrix}. \tag{8.53}$$

In this example, perfectly known plant and disturbance parameters will be assumed so the corresponding observer equation is

$$\begin{bmatrix} \dot{\hat{x}}_1 \\ \dot{\hat{x}}_2 \\ \hline \dot{\hat{x}}_{d1} \\ \dot{\hat{x}}_{d2} \end{bmatrix} = \begin{bmatrix} 0 & 1 & | & 0 & 0 \\ 0 & 0 & | & -1 & 0 \\ \hline 0 & 0 & | & 0 & 1 \\ 0 & 0 & | & -\Omega^2 & 0 \end{bmatrix} \begin{bmatrix} \hat{x}_1 \\ \hat{x}_2 \\ \hline \hat{x}_{d1} \\ \hat{x}_{d2} \end{bmatrix} + \begin{bmatrix} l_{a1} \\ l_{a2} \\ l_{a3} \\ l_{a4} \end{bmatrix} (y - \hat{y}) + \begin{bmatrix} 0 \\ b \\ 0 \\ 0 \end{bmatrix} u, \quad \begin{bmatrix} \hat{y} \\ \hat{d} \end{bmatrix} = \begin{bmatrix} 1 & 0 & | & 0 & 0 \\ 0 & 0 & | & 1 & 0 \end{bmatrix} \begin{bmatrix} \hat{x}_1 \\ \hat{x}_2 \\ \hline \hat{x}_{d1} \\ \hat{x}_{d2} \end{bmatrix}. \tag{8.54}$$

Figure 8.13 shows the block diagram of the complete system including a conventional linear state feedback control law with an additional disturbance compensation term.

The observer pole placement equation for a 5 % settling time of T_{so} may be written

8.2 The Full State Continuous Observer for LTI SISO Plants

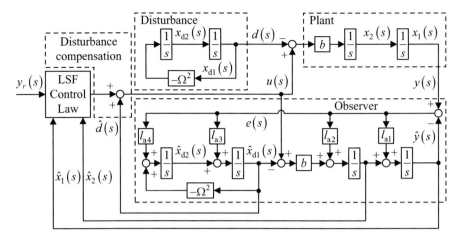

Fig. 8.13 Observer-aided LSF control of spacecraft with oscillating disturbance torque

$$s^4 \left[1 - \left(-\frac{l_{a1}}{s} - \frac{l_{a2}}{s^2} + \frac{bl_{a3}}{s^3} + \frac{bl_{a4}}{s^4} - \frac{\Omega^2}{s^2} \right) + \left(-\frac{\Omega^2}{s^2} \right) \left(-\frac{l_{a1}}{s} - \frac{l_{a2}}{s^2} \right) \right]$$
$$= s^4 + l_{a1}s^3 + \left(l_{a2} + \Omega^2 \right) s^2 + \left(\Omega^2 l_{a1} - bl_{a3} \right) s + \left(\Omega^2 l_{a2} - bl_{a4} \right)$$
$$= [s + 1.5(1+N)/T_{so}]^N \Big|_{N=n+n_d=4} = (s+q)^4$$
$$= s^4 + 4qs^3 + 6q^2 s^2 + 4q^3 s + q^4 \Rightarrow$$
$$l_{a1} = 4q, \ l_{a2} = 6q^2 - \Omega^2, \ l_{a3} = \left(\Omega^2 l_{a1} - 4q^3 \right)/b \text{ and } l_{a4} = \left(\Omega^2 l_{a2} - q^4 \right)/b.$$
(8.55)

The plant parameters taken for the following simulation are $J = 10,000$ [Kg m^2], $K_g = 1$ [Nm/V] with a maximum control voltage magnitude of $u_{max} = 10$ V and $K_s = 10/\pi$ [V/rad]. The crew are situated at a radius of $R = 5$ [m] from the centre of rotation so the spin angular velocity, Ω, required to produce a centrifugal acceleration of $g = 9.8$ [m/s^2] is given by $\Omega^2 R = g \Rightarrow \Omega = \sqrt{g/R} = 1.4$ [rad/s]. The linear state feedback control law is designed to give a settling time of $T_s = 100$ s. Such long settling times are usual for spacecraft applications, in contrast to earthbound motion control systems, due to the large moments of inertia and the control torques being severely limited by the available electrical power of only a few hundred Watts. The observer settling time is set to $T_{so} = 10$ [s].

The simulation starts with zero initial states for the spacecraft transverse control axis and the observer but with the artificial gravity already operating, creating an oscillating disturbance, $d(t) = D \cos(\Omega t)$ with $D = 0.1 u_{max}$. During this initial period, the attitude control loop is not closed so that the effect of the disturbance torque on the spacecraft attitude can be seen, but the observer is operating and settles to follow the disturbance and plant states. At $t = 50$ s, the attitude control loop is closed with zero attitude demand. Finally at $t = 250$ s, a step slew demand of

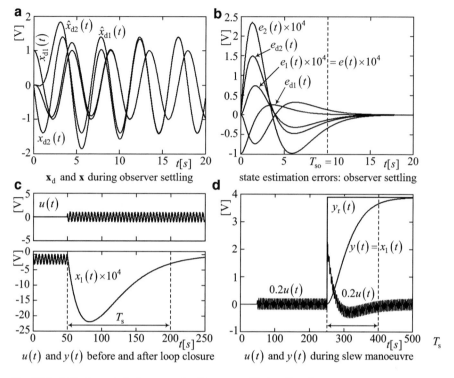

Fig. 8.14 Attitude control of spacecraft subject to oscillating disturbance torque

$70° \equiv 1.22$ rad $\equiv y_r = 1.22 K_s h (t - 250)$ is applied. Figure 8.14 shows the results.

Figure 8.14a shows the oscillatory disturbance states and the plant states together with their estimates before closure of the attitude control loop and the disturbance compensation. The state-variable estimates, $\widehat{x}_1(t)$, $\widehat{x}_2(t)$, $\widehat{x}_{d1}(t)$ and $\widehat{x}_{d2}(t)$, may be seen to almost converge, respectively, to $x_1(t)$, $x_2(t)$, $x_{d1}(t)$ and $x_{d2}(t)$ in the nominal correction loop settling time, T_{so}. This is also evident in Fig. 8.14b that displays the state and disturbance estimation errors that are completely free of the oscillations, due to the separation theorem. In Fig. 8.14c, the effect of the oscillatory disturbance on the spacecraft attitude can be seen to entirely disappear upon the initial loop closure, the oscillation instead appearing in $u(t)$, which is counteracting the continuing disturbance. The transient due to the nonzero states at the instant of the loop closure settles in the nominal settling time of T_s. Finally, Fig. 8.14d shows the slew manoeuvre during which the accelerating and decelerating control variable continues to have an oscillatory component counteracting the continuing disturbance.

8.3 The Full State Discrete Observer for LTI SISO Plants

8.3.1 Introduction

It has already been established in Chap. 6 that linear continuous controllers have discrete equivalents that can be designed by pole placement, in which the closed-loop pole locations, $s_{c1,2,\ldots,n}$, are first determined in the s-plane for the equivalent continuous system and then converted to desired z-plane closed-loop pole locations by means of the transformation, $z_{c1,2,\ldots,n} = \exp(s_{c1,2,\ldots,n}h)$, which will guarantee stability of the discrete closed-loop system for $s_{c1,2,\ldots,n}$ anywhere in the left half of the s-plane. This is also true of the state observer. The basic principles and features already presented in Sect. 8.2 for the continuous observer also apply to the discrete linear observer.

8.3.2 Observer Algorithm and Design Procedure

The generic discrete state space plant model upon which the observer is based is

$$\mathbf{x}(k+1) = \mathbf{\Phi}(h)\mathbf{x}(k) + \mathbf{\psi}(h)[u(k) - d(k)], \ y(k) = \mathbf{c}^T\mathbf{x}(k), \quad (8.56)$$

where $\mathbf{x} \in \Re^n$ is the state vector, u is the control variable, y is the measurement variable, d is the external disturbance referred to the control input, k is the sample number, $\mathbf{\Phi}(h) \in \Re^{n \times n}$ is the discrete plant matrix, $\mathbf{\psi}(h) \in \Re^{n \times 1}$ is the discrete input matrix and h is the sampling/iteration period. For comparison, the generic continuous state-space plant model (8.3) is reproduced here for convenience, together with the corresponding continuous LTI observer equations. Thus,

$$\dot{\mathbf{x}} = \mathbf{A}\mathbf{x} + \mathbf{b}(u - d), \quad y = \mathbf{c}^T\mathbf{x} \quad (8.57)$$

$$\dot{\widehat{\mathbf{x}}} = \tilde{\mathbf{A}}\widehat{\mathbf{x}} + \tilde{\mathbf{b}}u + \mathbf{l}(y - \widehat{y}), \quad \widehat{y} = \tilde{\mathbf{c}}^T\widehat{\mathbf{x}}. \quad (8.58)$$

Comparing (8.56) with (8.57) shows that the operation of differentiation is replaced by prediction one sampling period ahead. This enables the discrete LTI observer equations to be formed by analogy with the continuous LTI observer Eq. (8.58), resulting in

$$\widehat{\mathbf{x}}(k+1) = \tilde{\mathbf{\Phi}}(h)\widehat{\mathbf{x}}(k) + \tilde{\mathbf{\psi}}(h)u(k) + \mathbf{l}[y(k) - \widehat{y}(k)], \ \widehat{y}(k) = \tilde{\mathbf{c}}^T\widehat{\mathbf{x}}(k), \quad (8.59)$$

where $\widehat{\mathbf{x}}(k)$ and $\hat{y}(k)$ are, respectively, estimates of the variables $\mathbf{x}(k)$ and $y(k)$. Similarly, $\tilde{\mathbf{\Phi}}(h)$ and $\tilde{\mathbf{\psi}}(h)$ are, respectively, estimates of the constant parameters $\mathbf{\Phi}(h)$ and $\mathbf{\psi}(h)$. As for the continuous case, $\mathbf{l}$ is the gain matrix through which the model state is corrected using the output error, $y(k) - \widehat{y}(k)$. Since (8.59) is a recursion relation, it may be programmed directly on a digital processor, after eliminating the sample number, k. It may therefore be referred to as an *observer algorithm*. Using the z-transform relationship, $\mathcal{Z}\{q(k+1)\} = z^{-1}\mathcal{Z}\{q(k)\}$, a z-transfer function block diagram may be formed, as shown in Fig. 8.15.

This may be seen to have the same structure as Fig. 8.3 for the continuous case. It should be noted that the plant is shown in discrete form so that with the plant model parameters perfectly matched to the real plant parameters, it may be readily understood that $e(k) \to 0$ and therefore $\hat{\mathbf{x}}(k)$ as $k \to \infty$. The real plant is, of course, continuous, with sampled and held inputs and outputs, enabling the discrete model to be used.

As well as providing visualisation of how the observer works, a transfer function block diagram is particularly useful in simple applications for quickly deriving the correction loop characteristic polynomial for the pole placement, when the individual elements and their connections are shown.

The design procedure is to start with the desired correction loop poles of the equivalent continuous observer. These may be determined by setting the observer settling time to satisfy $T_{\text{so}} \leq T_{\text{s}}/5$, where T_{s} is the main control loop settling time, using one of the settling time formulae to form the desired characteristic polynomial as

$$(s+p)^n, \text{ where } \begin{cases} p = 1.5(1+n)/T_{\text{so}} & (5\ \% \text{ criterion}) \\ p = 1.6(1.5+n)/T_{\text{so}} & (2\ \% \text{ criterion}) \end{cases}. \tag{8.60}$$

Then the corresponding characteristic polynomial of the discrete observer is

$$(z+q)^n, \text{ where } q = -e^{-ph}. \tag{8.61}$$

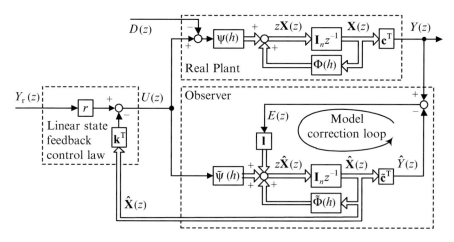

Fig. 8.15 Transfer function block diagram of discrete plant model and observer

8.3 The Full State Discrete Observer for LTI SISO Plants

It follows from the theory presented in Chap. 6 that the actual settling time will be limited to nh and satisfying $nh \leq T_s/5$ may not be possible in some cases. Whether or not this is a problem depends upon the application in hand.

Example 8.5 Observer-aided LSF control of a heating process containing a time delay

The plant is modelled as a first-order transfer function with time constant, T_p; a pure time delay, τ_d; and DC gain, K_{DC}, the transfer function being

$$\frac{Y(s)}{U(s)} = \frac{K_{DC} e^{-s\tau_d}}{1 + sT_p}. \tag{8.62}$$

This example is chosen since the discrete formulation renders linear state feedback control possible while continuous linear state feedback would not be practicable due to the pure time delay rendering the plant of infinite order, since $e^{-s\tau_d} = 1/\sum_{i=0}^{\infty} (\tau_d^i/i!) s^i$. For the discrete plant model, the sampling time, h, is chosen such that $\tau_d = n_d h$, where n_d is an integer. The order of the discrete plant model is then $n_d + 1$, the plant z-transfer function being

$$\frac{Y(z)}{U(z)} = K_{dc} \frac{1 - e^{-h/T_p}}{z - e^{-h/T_p}} \cdot \frac{1}{z^{n_d}}. \tag{8.63}$$

The plant parameters are given as $T_p = 5$ [s], $\tau_d = 8$ [s] and $K_{dc} = 1$. If an observer can be constructed to estimate the unmeasured output, $x(k)$, of the first-order subsystem with transfer function,

$$\frac{X(z)}{U(z)} = K_{dc} \frac{1 - e^{-h/T_p}}{z - e^{-h/T_p}}. \tag{8.64}$$

Then the problem is reduced to that of controlling just this subsystem, the controlled output, $y(k)$, being a delayed version of $x(k)$, i.e. $y(k) = x(k - n_d)$. The specified settling time, T_s, of the overall system must be greater than τ_d, so that the specified settling time of the first-order loop is $T_{s1} = T_s - \tau_d = T_s - n_d h$. Setting $T_{s1} = T_p$ would result in open loop operation so it is necessary to set $0 < T_{s1} \leq T_p$ to obtain the feedback action needed to give the system some robustness. A value of $T_{s1} = T_p/2 = 2.5$ [s] will be taken here. While it is possible to cater for any finite order, the minimum value of n_d will be found for simplification. This depends upon the maximum value, h_{max}, of h, enabling the first-order loop-specified settling time of T_{s1} to be realised. This is the value yielding a dead-beat response (Chap. 6) with an exact settling time of T_{s1}. Since the control loop is only of first order, this is just $h_{max} = T_{s1} = 2.5$ [s]. Then the minimum value of n_d is the smallest integer that is greater than $\tau_d/h_{max} = 3.2$, i.e. $n_d = 4 \Rightarrow h = \tau_d/n_d = 2$ [s].

To enable the first-order control loop to be realised, the plant model of the observer is formulated in the control canonical form. The control system block diagram corresponding to Fig. 8.15 is then shown in Fig. 8.16 for $n_p = 4$, where the model is simplified by setting $K_{dc}\left(1 - e^{-h/T_p}\right) = b$ and $a = e^{-h/T_p}$. Here, $\widehat{x}_5 = \widehat{x}$ and the controller can be regarded as a linear state feedback controller in which all the gains are zero except k that feeds back $\widehat{x}_5$.

In view of the separation principle, the observer characteristic polynomial may be formed as if the observer is separated from the plant and the control law as follows:

$$z^5\left[1 - \left(-\frac{l_1}{z} - \frac{l_2}{z^2} - \frac{l_3}{z^3} - \frac{l_4}{z^4} - \frac{l_5}{z^5} + \frac{a}{z}\right) + \frac{a}{z}\left(-\frac{l_1}{z} - \frac{l_2}{z^2} - \frac{l_3}{z^3} - \frac{l_4}{z^4}\right)\right]$$
$$= z^5 + (l_1 - a) z^4 + (l_2 - al_1) z^3 + (l_3 - al_2) z^2 + (l_4 - al_3) z + (l_5 - al_4). \quad (8.65)$$

The desired characteristic polynomial using the 5 % settling time criterion is then

$$(z + q)^5 = z^5 + 5qz^4 + 10q^2z^3 + 10q^3z^2 + 5q^4z + q^5, \quad (8.66)$$

where $q = -e^{-ph}$ and $p = 1.5(1 + 5) T_{so} = 9/T_{so}$. Then equating (8.65) and (8.66) gives

$$l_1 = 5q + a, \ l_2 = 10q^2 + al_1, \ l_3 = 10q^3 + al_2, \ l_4 = 5q^4 + al_3 \text{ and } l_5 = q^5 + al_4. \quad (8.67)$$

Since u is the control voltage, and y is an analogue measurement voltage, these variables together with x_i and $\widehat{x}_i$, $i = 1, 2, \ldots, 5$, are in volts.

The closed-loop transfer function of the first-order loop is

$$\frac{\widehat{X}_5(z)}{Y_r(z)} = \frac{rbz^{-1}}{1 + (kb - a)z^{-1}} = \frac{rb}{z + kb - a}. \quad (8.68)$$

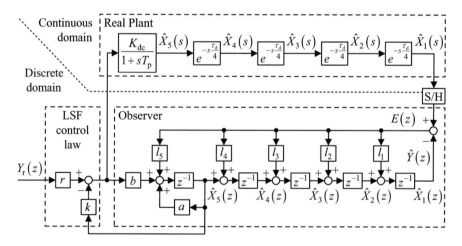

Fig. 8.16 LSF control of plant with pure time delay aided by a discrete state observer

8.3 The Full State Discrete Observer for LTI SISO Plants

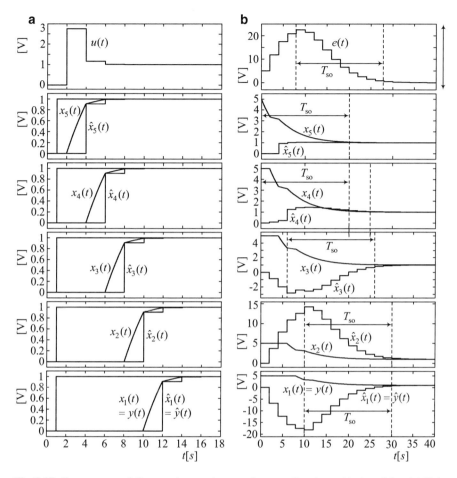

Fig. 8.17 Convergence of discrete observer in control system for plant with time delay. (a) Unit step response. (b) state estimate convergence

For a specified settling time of T_s (5 % criterion) and a unity closed-loop DC gain, the desired closed-loop transfer function is

$$\frac{X_5(z)}{Y_r(z)} = \frac{1 - e^{-3h/T_{s1}}}{z - e^{-3h/T_{s1}}}. \tag{8.69}$$

Equating (8.69) to (8.68) then yields

$$k = \frac{1}{b}\left(a - e^{-3h/T_{s1}}\right) \quad \text{and} \quad r = \frac{1}{b}\left(1 - e^{-3h/T_{s1}}\right). \tag{8.70}$$

The minimum observer settling time is that of the dead-beat response, i.e. $nh = 10$ [s]. Since $T_s = T_{s1} + \tau_d = 10.5$ [s], the criterion, $T_{so} < T_s/5$, for fast observer settling cannot be satisfied, but this is a special application, and it is always possible to run the observer to a settled condition before the main control loop closure. A value of $T_{so} = 20$ [s] is chosen for this example.

Figure 8.17a shows the unit step response of the control system with the state variables of the observer equal to those of the plant at the sample points. Figure 8.17b shows the convergence of the observer state towards the plant state with $x_i(0) = 5$ [V] and $\hat{x}_i(0) = 0$ [V], $i = 1, 2, \ldots, 5$, again with $y_r(t) = h(t)$. The settling time of $y(t)$ to 1 is determined by the observer transient in this case.

8.4 The Full State Observer for Multivariable Plants

8.4.1 Introduction

The design method for LTI SISO plants based on determination of the correction loop characteristic polynomial from the observer transfer function block diagram presented in Sects. 8.2 and 8.3 would generally prove to be tortuous if attempted for LTI multivariable plants. A more practicable method is based directly on the state-space model of the plant. This is first presented for LTI SISO plants to clarify the procedure and then extended for multivariable plants.

8.4.2 Matrix–Vector Design Method for SISO LTI Plants

Starting with the plant model in the form of a transfer function, a model in the observer canonical state representation can be formed as in Chap. 3, Sect. 3.3.5.2, which is reproduced here for convenience. Thus,

$$\dot{\mathbf{x}}_o = \mathbf{A}_o \mathbf{x}_o + \mathbf{b}_o u \tag{8.71}$$

$$y = \mathbf{c}_o^T \mathbf{x}_o, \tag{8.72}$$

where

$$\mathbf{A}_o = \begin{bmatrix} 0 & 0 & \cdots & 0 & -a_0 \\ 1 & 0 & & \vdots & -a_1 \\ 0 & 1 & \ddots & 0 & -a_2 \\ \vdots & \ddots & \ddots & 0 & \vdots \\ 0 & \cdots & 0 & 1 & -a_{n-1} \end{bmatrix}, \quad \mathbf{b}_o = \begin{bmatrix} b_0 \\ b_1 \\ \vdots \\ \vdots \\ b_{n-1} \end{bmatrix} \quad \text{and} \quad \mathbf{c}_o^T = \begin{bmatrix} 0 & 0 & \cdots & 0 & 1 \end{bmatrix}.$$

$$\tag{8.73}$$

8.4 The Full State Observer for Multivariable Plants

Then the corresponding observer equation is

$$\dot{\hat{\mathbf{x}}}_o = \mathbf{A}_o\hat{\mathbf{x}}_o + \mathbf{b}_o u + \mathbf{l}\left(y - \mathbf{c}_o^T\hat{\mathbf{x}}_o\right). \tag{8.74}$$

Subtracting (8.71) from (8.74) then yields the state estimation error equation

$$\dot{\boldsymbol{\varepsilon}}_o = \left[\mathbf{A}_o - \mathbf{l}\mathbf{c}_o^T\right]\boldsymbol{\varepsilon}_o, \tag{8.75}$$

where $\boldsymbol{\varepsilon}_o = \hat{\mathbf{x}}_o - \mathbf{x}_o$ and

$$\mathbf{A}_o - \mathbf{l}\mathbf{c}_o^T = \begin{bmatrix} 0 & 0 & \cdots & 0 & -a_0 \\ 1 & 0 & & \vdots & -a_1 \\ 0 & 1 & \ddots & 0 & -a_2 \\ \vdots & \ddots & \ddots & 0 & \vdots \\ 0 & \cdots & 0 & 1 & -a_{n-1} \end{bmatrix} - \begin{bmatrix} l_0 \\ l_1 \\ l_2 \\ \vdots \\ l_{n-1} \end{bmatrix} \begin{bmatrix} 0\ 0 \cdots 0\ 1 \end{bmatrix}$$

$$= \begin{bmatrix} 0 & 0 & \cdots & 0 & -(a_0 + l_0) \\ 1 & 0 & & \vdots & -(a_1 + l_1) \\ 0 & 1 & \ddots & 0 & -(a_2 + l_2) \\ \vdots & \ddots & \ddots & 0 & \vdots \\ 0 & \cdots & 0 & 1 & -(a_{n-1} + l_{n-1}) \end{bmatrix}. \tag{8.76}$$

The characteristic polynomial of $\mathbf{A}_o - \mathbf{l}\mathbf{c}_o^T$ therefore may be recognised and equated to the desired characteristic polynomial that could be formed as previously using one of the settling time formulae. Thus,

$$s^n + (a_{n-1} + l_{n-1})s^{n-1} + \cdots + (a_1 + l_1)s + (a_0 + l_0) = (s + p)^n, \tag{8.77}$$

where $p = 1.5(1 + n)/T_{so}$ (5 % criterion) or $p = 1.6(1.5 + n)/T_{so}$ (2 % criterion), where $T_{so} \leq T_s/5$, where T_s is the settling time of the main control loop which the observer is supporting. Equating coefficients of s^i, $i = 0, 1, \ldots, n-1$, on the LHS and RHS of (8.77) then immediately gives the required observer gains, l_i, $i = 0, 1, \ldots, n-1$. This demonstrates the convenience of the observer canonical state representation. Any other desired state representation for control purposes can be generated using a transformation matrix as shown in Chap.3, Sect. 3.3.3.

8.4.3 Matrix–Vector Design Method for Multivariable LTI Plants

Once a plant model is available in the form of a transfer function relationship, then a state-space model may be formed in the multivariable observer canonical form as in Chap. 3, Sect. 3.3.6.4, enabling a similar procedure to be followed as presented in Sect. 8.4.2. The general state-space model in the observer canonical form is as follows:

$$\dot{\mathbf{x}}_o = \mathbf{A}_o \mathbf{x}_o + \mathbf{B}_o \mathbf{u} \qquad (8.78)$$

$$\mathbf{y} = \mathbf{C}_o \mathbf{x}_o, \qquad (8.79)$$

where

$$\mathbf{A}_o = \begin{bmatrix} \mathbf{A}_{o1} & 0 & \cdots & 0 \\ 0 & \mathbf{A}_{o2} & & \vdots \\ \vdots & & \ddots & 0 \\ 0 & \cdots & 0 & \mathbf{A}_{om} \end{bmatrix}, \quad \mathbf{B}_o = \begin{bmatrix} b_{o11} & \cdots & b_{o1m} \\ b_{o21} & & b_{o2m} \\ \vdots & & \vdots \\ b_{on1} & \cdots & b_{onm} \end{bmatrix} \text{ and } \mathbf{C}_o = \begin{bmatrix} \mathbf{c}_{o1}^T & 0 & \cdots & 0 \\ 0 & \mathbf{c}_{o2}^T & & \vdots \\ \vdots & & \ddots & 0 \\ 0 & \cdots & 0 & \mathbf{c}_{om}^T \end{bmatrix}, \qquad (8.80)$$

where

$$\mathbf{A}_{oi} = \begin{bmatrix} 0 & \cdots & 0 & -a_0 \\ 1 & \ddots & \vdots & -a_{i1} \\ 0 & \ddots & 0 & \vdots \\ 0 & 0 & 1 & -a_{in_i} \end{bmatrix} \quad \text{and} \quad \mathbf{c}_{oi}^T = \underbrace{\begin{bmatrix} 0 & \cdots & 0 & 1 \end{bmatrix}}_{n_i \text{ elements}}, \quad i = 1, 2, \ldots, m,$$

$$\sum_{i=1}^{m} n_i = n. \qquad (8.81)$$

The corresponding observer equation is then

$$\dot{\widehat{\mathbf{x}}}_o = \mathbf{A}_o \widehat{\mathbf{x}}_o + \mathbf{B}_o \mathbf{u} + \mathbf{L} \left[\mathbf{y} - \mathbf{C}_o \widehat{\mathbf{x}}_o \right] \qquad (8.82)$$

Subtracting (8.78) from (8.82) then yields the state estimation error equation

$$\dot{\boldsymbol{\varepsilon}}_o = \left[\mathbf{A}_o - \mathbf{L} \mathbf{c}_o^T \right] \boldsymbol{\varepsilon}_o. \qquad (8.83)$$

There are a total of nm elements in the gain matrix, $\mathbf{L}$, while only n elements are needed to place the eigenvalues of $\mathbf{A}_o - \mathbf{L} \mathbf{c}_o^T$ as desired. If these elements are arranged in the block diagonal form,

8.4 The Full State Observer for Multivariable Plants

$$\mathbf{L} = \begin{bmatrix} \mathbf{l}_1 & 0 & \cdots & 0 \\ 0 & \mathbf{l}_2 & \ddots & \vdots \\ \vdots & \ddots & \ddots & 0 \\ 0 & \cdots & 0 & \mathbf{l}_m \end{bmatrix} \quad \text{where} \quad \mathbf{l}_i = \begin{bmatrix} l_{i0} \\ l_{i1} \\ \vdots \\ l_{in_i-1} \end{bmatrix}, \tag{8.84}$$

then $\mathbf{A}_o - \mathbf{L}\mathbf{c}_o^T$ is of the same block diagonal form as $\mathbf{A}_o$, each block being of the same form as (8.76). Thus,

$$\mathbf{A}_o - \mathbf{L}\mathbf{c}_o^T = \begin{bmatrix} \mathbf{A}_{o1} & 0 & \cdots & 0 \\ 0 & \mathbf{A}_{o2} & \ddots & \vdots \\ \vdots & \ddots & \ddots & 0 \\ 0 & \cdots & 0 & \mathbf{A}_{om} \end{bmatrix} - \begin{bmatrix} \mathbf{l}_1 & 0 & \cdots & 0 \\ 0 & \mathbf{l}_2 & \ddots & \vdots \\ \vdots & \ddots & \ddots & 0 \\ 0 & \cdots & 0 & \mathbf{l}_m \end{bmatrix} \begin{bmatrix} \mathbf{c}_{o1}^T & 0 & \cdots & 0 \\ 0 & \mathbf{c}_{o2}^T & \ddots & \vdots \\ \vdots & \ddots & \ddots & 0 \\ 0 & \cdots & 0 & \mathbf{c}_{om}^T \end{bmatrix}$$

$$= \begin{bmatrix} \mathbf{A}_{o1} - \mathbf{l}_1 \mathbf{c}_{o1}^T & 0 & \cdots & 0 \\ 0 & \mathbf{A}_{o2} - \mathbf{l}_2 \mathbf{c}_{o2}^T & \ddots & \vdots \\ \vdots & \ddots & \ddots & 0 \\ 0 & \cdots & 0 & \mathbf{A}_{om} - \mathbf{l}_m \mathbf{c}_{om}^T \end{bmatrix} \tag{8.85}$$

where

$$\mathbf{A}_{oi} - \mathbf{l}_i \mathbf{c}_{oi}^T = \begin{bmatrix} 0 & \cdots & 0 & -a_{i0} \\ 1 & \ddots & \vdots & -a_{i1} \\ 0 & \ddots & 0 & \vdots \\ 0 & 0 & 1 & -a_{in_i-1} \end{bmatrix} - \begin{bmatrix} l_{i0} \\ l_{i1} \\ \vdots \\ l_{in_i-1} \end{bmatrix} \begin{bmatrix} 0 & \cdots & 0 & 1 \end{bmatrix}$$

$$= \begin{bmatrix} 0 & \cdots & 0 & -(a_{i0} + l_0) \\ 1 & \ddots & \vdots & -(a_{i1} + l_1) \\ 0 & \ddots & 0 & \vdots \\ 0 & 0 & 1 & -(a_{in_i-1} + l_{n_i-1}) \end{bmatrix}. \tag{8.86}$$

$$i = 1, 2, \ldots, m$$

It follows that the observer has m subsystems whose eigenvalues may be independently placed. If the settling time formulae are again used, the characteristic polynomials of the subsystems may be equated to the desired ones as follows:

$$s^n + (a_{in_i-1} + l_{in_i-1})s^{n-1} + \cdots + (a_{i1} + l_{i1})s + (a_{i0} + l_{i0}) = (s + p_i)^{n_i},$$

$$i = 1, 2, \ldots, m, \tag{8.87}$$

where $p_i = 1.5\,(1+n_i)/T_{soi}$ (5 % criterion) or $p_i = 1.6\,(1.5+n_i)/T_{soi}$ (2 % criterion), where $T_{soi} \le T_s/5$, T_s being the settling time of the main control loop. In most cases, it would be appropriate to make the settling times of the m subsystems the same.

8.5 The Noise Filtering Property of the Observer

8.5.1 Background

Attention is now turned to a benefit brought about by the observer, further to making an estimate of the complete plant state available. This is the reduction of the random state estimation errors due to unwanted noise signals generated by imperfect hardware, by means of the inherent filtering process of the observer. This section leads to an observer pole placement method that seeks to reduce these state estimation errors to acceptable levels in cases where the noise would otherwise prevent the control accuracy specification being met.

First, some relevant terminology will be introduced, in the context of dynamical systems. A *deterministic* system is one in which no random signals affect the development of its future states. In contrast, a *stochastic* system is one whose behaviour is non-deterministic in that its future states are determined not only by its predictable actions but also by random signals. The systems described in this section are stochastic ones. The terminology can also be used to classify signals. A deterministic signal is one containing no random components, and a stochastic signal is one containing random components.

All real plants contain hardware that generates certain levels of noise that may adversely affect the control system performance. These signals fall into two basic categories: plant noise and measurement noise. Instrumentation providing measurements of variables in control systems contains noise generating electronic components that contaminate the measurement variables used by the controller, as already discussed in Chap. 1. This is *measurement noise*. Additionally, the actuators of control systems generate noise since they also contain noise sources. These are not only electronic amplifiers such as those used to drive electromagnetic actuators. For example, the motion of fluid through the spool valve of an electro-hydraulic actuator generates a certain amount of noise in the form of random variations of flow rate which manifest themselves ultimately as random variations in the force output of the actuator. Another example is the cogging torque produced by some synchronous motors. In contrast to measurement noise, the noise generated in actuators directly affects the state variables of the plant and is therefore referred to as *plant noise*.

8.5 The Noise Filtering Property of the Observer

The definitions of plant and measurement noise are as follows.

Definition 8.3 *Measurement noise is any random signal originating in the hardware that contaminates the measurement variables without affecting the state of the uncontrolled plant.*

Definition 8.4 *Plant noise is any random signal originating in the hardware that affects the state of the uncontrolled plant.*

These definitions are made for the uncontrolled plant as *any* noise source will affect both the plant state and the measurements in a closed-loop system. They apply to *all* the plant noise sources including any not within the sensors and actuators.

8.5.2 Lumped Plant Noise and Measurement Noise Sources

To simplify the model upon which a control system design is based, all the physical plant noise sources may be replaced by a single noise source, $n_p(t)$, injected at the same point as $u(t)$, such that the random variations in $y(t)$ are unaltered. Similarly, all the noise sources within the measurement hardware are equivalent to a single noise source, $n_m(t)$, injected at the same point as y. This is shown in Fig. 8.18.

Since this section is not concerned with plant modelling errors, the plant and its model in the observer are shown as identical. Then once any transient part of the state estimation error, $\varepsilon(t) = \widehat{\mathbf{x}}(t) - \mathbf{x}(t)$, has decayed to negligible proportions, the remaining error will be of a random nature and due only to the noise sources.

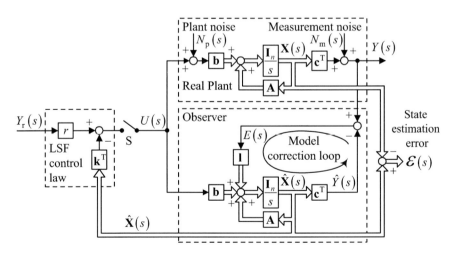

Fig. 8.18 Observer-aided LSF control system for SISO plant with lumped noise sources

8.5.3 State Estimation Error Variation with Observer Gains

The following discussion will be greatly simplified by designing the observer, as previously, with a multiple pole at $s_{1, 2, \ldots, n} = -1/T_{co}$, so that the correction loop settling time (5 % criterion) is $T_{so} = 1.5\,(1 + n)\,T_{co}$. It will also be assumed that, unless otherwise stated, the observer is connected to the real plant but without the LSF control law active, i.e. switch, S, is open in Fig. 8.18.

The model correction loop acts on the error, $e(t)$, and attempts to cause $\hat{y}(t)$ to follow $y(t)$ including any random components it contains, regardless of their cause. As will be seen, this is beneficial for changes in $y(t)$ due to the plant noise but is undesirable for changes in $y(t)$ due to the measurement noise.

First, suppose that the only noise source is measurement noise. Then reducing T_{so} increases the magnitudes of the elements of the gain matrix, **l**, thereby 'tightening' the model correction loop, causing $\hat{y}(t)$ to follow $y(t)$ and the measurement noise it contains more closely. This is *undesirable* because the resulting random variations in $\hat{\mathbf{x}}(t)$ do not occur in $\mathbf{x}(t)$ and therefore constitute state estimation errors. Increasing T_{so}, however, causes the model correction loop to be less effective in following the random variations of $y(t)$, especially the rapid ones, and as T_{so} is increased further, $\hat{y}(t)$ begins to follow a 'middle course' through the random variations and become very close to the deterministic component of $y(t)$. This results in $\hat{\mathbf{x}}(t)$ becoming closer to the deterministic $\mathbf{x}(t)$, thereby reducing the state estimation errors. Hence, provided that T_{so} is not too small, the observer filters out the short-term random variations in $\hat{y}(t)$ and hence $\hat{\mathbf{x}}(t)$ due to the measurement noise. In fact, the transfer function components of $\varepsilon(s)/N_m(s)$ are those of low-pass filters with cut-off frequencies that reduce with T_{so}. In the limit as $T_{so} \to \infty$, for some plants $\mathbf{l} \to \mathbf{0}$, disabling the correction loop so that the measurement noise has no influence at all and in any case may almost completely open the correction loop. If $\mathbf{l} = \mathbf{0}$, $\hat{\mathbf{x}}(0) = \mathbf{x}(0)$ and the plant and its model are identical, then $\hat{\mathbf{x}}(t) = \mathbf{x}(t)$ $\forall t > 0$, but in practice some correction loop action is necessary to compensate for plant modelling errors, allow the initial observer settling transient to take place and compensate for the effects of plant noise, now to be discussed.

Suppose now that the only noise source is plant noise. As already stated in *Definition* 8.4 and evident in Fig. 8.18, the plant noise causes random changes in the state, $\mathbf{x}(t)$, of the real plant, which in turn causes random changes in $y(t)$. As previously, reducing T_{so} 'tightens' the correction loop by increasing the magnitudes of the elements of the gain matrix, **l**, and therefore causes $\hat{y}(t)$ to follow $y(t)$ more closely. This in turn drives $\varepsilon(t) = \hat{\mathbf{x}}(t) - \mathbf{x}(t)$ to smaller proportions, which is desirable. This means that the better the performance of the observer in minimising the state estimation errors due to plant noise, the more noisy will be $\hat{\mathbf{x}}(t)$. At a glance, this may be counter-intuitive, but it must be realised that loop closure via a well-designed linear state feedback control law by closing switch, S, in Fig. 8.18 will nearly eliminate the random variations of $\mathbf{x}(t)$ and $\hat{\mathbf{x}}(t)$ due to $u(t)$ almost counteracting the plant noise, $n_p(t)$.

8.5 The Noise Filtering Property of the Observer

To summarise, tightening the model correction loop by reducing T_{so} reduces the state estimation errors due to plant noise but increases them due to measurement noise. This requires a compromise, and the following two subsections develop a method for finding an intermediate value of T_{so} that approximately minimises the state estimation errors given the measurement and plant noise spectral density functions.

8.5.4 State Estimation Error Transfer Function Relationship

The filtering properties of the observer may be revealed by deriving the transfer function relationships between the plant and measurement noise sources and the state estimation error components. This will also provide the foundation for the design method presented in the following sub-subsection.

With reference to Fig. 8.18, the plant transfer function relationship is given by

$$\mathbf{X}(s) = \frac{\mathbf{I}_n}{s}\{\mathbf{b}[N_p(s) + U(s)] + \mathbf{A}(s)\} \Rightarrow [s\mathbf{I}_n - \mathbf{A}]\mathbf{X}(s) = \mathbf{b}[N_p(s) + U(s)]. \tag{8.88}$$

The reason for leaving (8.88) in this form will become apparent below. Also

$$Y(s) = N_m(s) + \mathbf{c}^T\mathbf{X}(s). \tag{8.89}$$

Similarly, the observer transfer function relationship is given by

$$\widehat{\mathbf{X}}(s) = \frac{\mathbf{I}_n}{s}\{\mathbf{b}U(s) + \mathbf{l}[Y(s) - \mathbf{c}^T\widehat{\mathbf{X}}(s)] + \mathbf{A}\widehat{\mathbf{X}}(s)\} \Rightarrow$$
$$[s\mathbf{I}_n - \mathbf{A}]\widehat{\mathbf{X}}(s) = \mathbf{b}U(s) + \mathbf{l}[Y(s) - \mathbf{c}^T\widehat{\mathbf{X}}(s)]. \tag{8.90}$$

Substituting for $y(s)$ in (8.90) using (8.89) then yields

$$[s\mathbf{I}_n - \mathbf{A}]\widehat{\mathbf{X}}(s) = \mathbf{b}U(s) + \mathbf{l}[N_m(s) + \mathbf{c}^T\mathbf{X}(s) - \mathbf{c}^T\widehat{\mathbf{X}}(s)]. \tag{8.91}$$

Since $\varepsilon(s) = \widehat{\mathbf{x}}(s) - \mathbf{x}(s)$, (8.91) may be written as

$$[s\mathbf{I}_n - \mathbf{A}]\widehat{\mathbf{X}}(s) = \mathbf{b}U(s) + \mathbf{l}[N_m(s) - \mathbf{c}^T\varepsilon(s)], \tag{8.92}$$

and then subtracting (8.88) from (8.92) yields

$$[s\mathbf{I}_n - \mathbf{A}]\varepsilon(s) = \mathbf{l}[N_m(s) - \mathbf{c}^T\varepsilon(s)] - \mathbf{b}N_p(s). \tag{8.93}$$

Finally, making $\varepsilon(s)$ the subject of (8.93) yields

$$\varepsilon(s) = \left[s\mathbf{I}_n - \mathbf{A} + \mathbf{l}\mathbf{c}^T\right]^{-1} \left[\mathbf{l}N_m(s) - \mathbf{b}N_p(s)\right]. \tag{8.94}$$

It is evident from (8.94) that the observer acts as a low-pass filter between $N_m(s)$ and $\varepsilon(s)$ and also between $N_p(s)$ and $\varepsilon(s)$. Reducing the correction loop settling time, T_{so}, in general increases the magnitudes of the elements of $\mathbf{l}$, and it is evident from (8.94) that this reduces the state estimation errors originating from the plant noise, $N_p(s)$, but increases those that originate from the measurement noise, $N_m(s)$, as predicted in Sect. 8.5.3.

Example 8.6 State estimation error transfer function relationships for a rigid-body spacecraft

Many spacecraft have high-precision attitude control systems with pointing accuracy specifications in the arcsecond (1/3600 deg.) or the sub-arc-second region. In these applications, the effects of the plant and measurement noise are often very significant, demanding state estimation techniques that, ideally, minimise the stochastic state estimation errors. Without inter-axis coupling, the plant state-space model for a single control axis is as follows:

$$\dot{x}_1 = x_2, \; \dot{x}_2 = b\left(u + n_m\right), \tag{8.95}$$

where x_1 is the attitude angle and $b = K_w/J$ where K_w is the reaction wheel torque constant and J is the spacecraft body moment of inertia. Figure 8.19 shows a transfer function block diagram of the plant and the observer.

Transfer function relationship (8.94) will now be derived for this example. Thus,

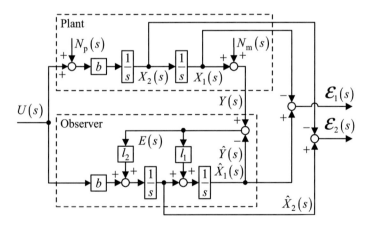

Fig. 8.19 Plant and observer for single axis of rigid-body spacecraft showing noise sources

8.5 The Noise Filtering Property of the Observer

$$\begin{bmatrix} \varepsilon_1(s) \\ \varepsilon_2(s) \end{bmatrix} = \underbrace{\dfrac{\left\{\dfrac{b}{s^2}\left[\dfrac{l_1}{s}+\dfrac{l_2}{s^2}\right]N_p(s) + \left[\dfrac{l_1}{s}+\dfrac{l_2}{s^2}\right]N_m(s)\right\}}{1 - \left[-\dfrac{l_1}{s}-\dfrac{l_2}{s^2}\right]}}_{\text{Contribution via observer plant}}$$

$$-\underbrace{\left\{\begin{bmatrix}\dfrac{b}{s^2}\\ \dfrac{b}{s}\end{bmatrix} N_p(s) - \begin{bmatrix} 0 \\ 0 \end{bmatrix} N_m(s)\right\}}_{\substack{\text{Contribution} \\ \text{directly from plant}}}$$

$$= \dfrac{\left\{b\begin{bmatrix} l_1 s + l_2 \\ l_2 s \end{bmatrix}N_p(s) + \begin{bmatrix} l_1 s^3 + l_2 s^2 \\ l_2 s^3 \end{bmatrix}N_m(s)\right\} - \left\{\begin{bmatrix} b\left(s^2 + l_1 s + l_2\right) \\ b\left(s^3 + l_1 s^2 + l_2 s\right)\end{bmatrix} N_p(s)\right\}}{s^2(s^2 + l_1 s + l_2)}$$

$$= \dfrac{\begin{bmatrix} l_1 s + l_2 \\ l_2 s \end{bmatrix} N_m(s) - b\begin{bmatrix} 1 \\ s + l_1 \end{bmatrix} N_p(s)}{s^2 + l_1 s + l_2}.$$

(8.96)

Designing the observer to have a correction loop settling time of T_{so} seconds, (5 % criterion then requires the pole placement equation,

$$s^2 + l_1 s + l_2 = \left[s + \dfrac{1.5(1+n)}{T_{so}}\right]^n \Big|_{n=2} = \left(s + \dfrac{9}{2T_{so}}\right)^2 = s^2 + \dfrac{9}{T_{so}}s + \dfrac{81}{4T_{so}^2},$$

(8.97)

from which the observer gains are

$$l_1 = 9/T_{so} \quad \text{and} \quad l_2 = 81/\left(4T_{so}^2\right).$$

(8.98)

Substituting for l_1 and l_2 in (8.96) using (8.98) then yields

$$\begin{bmatrix} \varepsilon_1(s) \\ \varepsilon_2(s) \end{bmatrix} = \dfrac{\begin{bmatrix} \dfrac{9}{T_{so}}s + \dfrac{81}{4T_{so}^2} \\ \dfrac{81}{4T_{so}^2}s \end{bmatrix} N_m(s) - b\begin{bmatrix} 1 \\ s + \dfrac{9}{T_{so}} \end{bmatrix} N_p(s)}{s^2 + \dfrac{9}{T_{so}}s + \dfrac{81}{4T_{so}^2}}$$

$$= \dfrac{\begin{bmatrix} \dfrac{4T_{so}}{9}s + 1 \\ s \end{bmatrix} N_m(s) - \dfrac{4T_{so}^2 b}{81}\begin{bmatrix} 1 \\ s + \dfrac{9}{T_{so}} \end{bmatrix} N_p(s)}{1 + \dfrac{4}{9}T_{so}s + \dfrac{4}{81}T_{so}^2 s^2}.$$

(8.99)

Then

$$\lim_{T_{so} \to \infty} \begin{bmatrix} \varepsilon_1(s) \\ \varepsilon_2(s) \end{bmatrix} = \frac{1}{s^2} \begin{bmatrix} 1 \\ s \end{bmatrix} N_p(s) \quad \text{and} \quad \lim_{T_{so} \to 0} \begin{bmatrix} \varepsilon_1(s) \\ \varepsilon_2(s) \end{bmatrix} = \begin{bmatrix} 1 \\ s \end{bmatrix} n_m(s). \tag{8.100}$$

As the model correction loop is tightened by reducing T_{so}, the contribution of the plant noise to the state estimation error diminishes and that of the measurement noise increases. At the extremes, (8.100) indicates that as $T_{so} \to \infty$, the contribution of the measurement noise is eliminated. This is not surprising as (8.98) gives

$$\lim_{T_{so} \to \infty} \begin{bmatrix} l_1 \\ l_2 \end{bmatrix} = \begin{bmatrix} 0 \\ 0 \end{bmatrix}, \tag{8.101}$$

and therefore, the model correction loop is disabled, but this is not practicable.

In theory, as $T_{so} \to 0$, the contribution from the plant noise is eliminated, but $\widehat{x}_1(t)$ is contaminated by the unfiltered measurement noise, and worse still, $\widehat{x}_2(t)$ is contaminated by the unfiltered *derivative* of the measurement noise, which, as already demonstrated in Chap. 1, exaggerates the high-frequency components of the noise. This extreme, however, could not occur in practice as the infinite gains indicated by (8.98) and the associated infinite correction loop poles ($s_{1,2} = -9/(2T_{so})$) could not be realised in practice due to the finite sampling frequency of the digital implementation (Chap. 6), but similar undesirable effects would result by setting T_{so} too small. T_{so} has to be set to a value such that there is sufficient attenuation of the high-frequency components of the measurement noise due to the denominator of (8.99). Setting T_{so} too large would cause excessive state estimation errors due to the plant noise.

8.5.5 Considering Noise Levels in Observer Design

8.5.5.1 Introduction

If the measurement and plant noise signals are quantified, then it is possible to find observer gains which minimise the noise content in a given state estimation error component, ε_i. The *power spectral densities* of the noise signals and the *variance* of the state estimation error are the quantities used for this minimisation, and they are defined in the following subsections. Attempting to carry out the minimisation analytically in the simplest practical applications, such as the single-axis attitude control of a rigid-body spacecraft of Example 8.6, leads to the evaluation of integrals that do not appear even in the most comprehensive tables [3], but constrained minimisation with respect to the observer correction loop settling time, T_{so}, is much simpler and can be tractable, but is suboptimal. This method will be demonstrated for the spacecraft example. For the many cases in which analytical solutions either

8.5 The Noise Filtering Property of the Observer

do not exist or involve an inordinate amount of working, a practicable approach is to carry out the minimisation using a simulation, which will also be demonstrated.

Finally, the Kalman filter will be introduced, which is essentially an observer with the gain matrix calculated using an algorithm based on the variances of the plant and measurement noise signals, instead of pole placement. This is applicable to either time-invariant or time-varying plants and is optimal in the minimisation of the noise contaminating the state estimate provided the noise sources statistically have Gaussian distributions and their spectra in the frequency domain are white.

8.5.5.2 Power Spectral Density and Variance

A random noise signal, $n(t)$, with zero mean value can be characterised by its variance and power spectral density [2]. The variance is a measure of the noise level and is the mean squared value denoted

$$\sigma_{nn}^2 = E\left[n^2(t)\right] = \lim_{T \to \infty} \frac{1}{T} \int_0^T n^2(t) dt. \tag{8.102}$$

Here, σ_{nn} is the standard deviation of statistics, and the notation, $E[n^2(t)]$, means 'the expected value of $n^2(t)$'. In analysing the filtering effect of an observer on the noise signals, the random signal is viewed in the frequency domain by means of its power spectral density (PSD). If a periodic signal is decomposed into a linear weighted sum of its sinusoidal Fourier components, the histogram of the component amplitude at a given frequency, taken from a discrete set comprising the fundamental frequency and its harmonics (i.e. integer multiples of the fundamental frequency), may be referred to as its discrete spectrum. By analogy, the Fourier transform of a more general signal, which is not necessarily periodic, can be plotted as a graph of the amplitude of the sinusoidal signal component as a continuous function of the frequency, which may be referred to as a continuous spectrum. The amplitude of the component is then the spectral density.

The term power originates from analogue communication applications in which the amount of power dissipated in a circuit containing a resistance, R, is $i^2 R$, where i is the current, or alternatively v^2/R, where v is the applied voltage. Of course, in control applications, there is not usually physical power dissipation associated with the noise signal, $n(t)$, but $n^2(t)$ may be referred to as the signal power. The power spectral density of a signal, $n(t)$, is denoted $S_{nn}(j\omega)$, where ω is the angular frequency of the signal component. It is related to the variance as follows:

$$\sigma_{nn}^2 = E\left[n^2(t)\right] = \frac{1}{2\pi} \int_{-\infty}^{\infty} S_{nn}(j\omega) d\omega. \tag{8.103}$$

If $n(t)$ is applied to a linear system with Laplace transfer function, $G(s)$, resulting in an output, $m(t)$. It can then be shown [2] that the variance of $m(t)$ is

$$\sigma_{mm}^2 = E\left[m^2(t)\right] = \frac{1}{2\pi}\int_{-\infty}^{\infty} |G(j\omega)|^2 S_{nn}(j\omega)\,d\omega. \qquad (8.104)$$

8.5.5.3 PSD of the Sum of Two Random Signals

Let

$$n(t) = n_1(t) + n_2(t), \qquad (8.105)$$

where $n_1(t)$ and $n_2(t)$ are zero mean random signals with power spectral densities, $S_{n_1 n_1}(j\omega)$ and $S_{n_2 n_2}(j\omega)$. Then [2], the power spectral density of $n(t)$ is

$$S_{nn}(j\omega) = S_{n_1 n_1}(j\omega) + S_{n_2 n_2}(j\omega) + \tfrac{1}{2}\mathrm{Re}\left[S_{n_1 n_2}(j\omega)\right], \qquad (8.106)$$

where $S_{n_1 n_2}(j\omega)$ is the cross power spectral density, satisfying

$$\sigma_{n_1 n_2}^2 = E\left[n_1(t)n_2(t)\right] = \frac{1}{2\pi}\int_{-\infty}^{\infty} S_{n_1 n_2}(j\omega)\,d\omega \qquad (8.107)$$

which is the covariance between $n_1(t)$ and $n_2(t)$.

8.5.5.4 Band-Limited White Noise

Many real sensors and actuators have substantially constant noise power spectral densities over a *finite* frequency range between zero (DC) and a maximum value, ω_b [rad/s], above which the PSD reduces to negligible proportions. At the frequency, ω_b, the PSD will have fallen by 3 dB, and this frequency is called the bandwidth. This type of noise signal is referred to as band-limited white noise. The term white noise derives from the meaning of white light in which all the wavelengths (or frequencies) of the light components are present. In theory, true white noise is present with a constant PSD over an infinite frequency range. This poses the difficulty of (8.103) yielding an infinite variance. In practice, however, the bandwidth is finite. Taking the PSD as constant, equal to S_{nn} over the frequency range, $\omega \in (0, \omega_b)$, (8.103) becomes finite, i.e.

$$\sigma_{nn}^2 = E\left[n^2(t)\right] = \frac{1}{2\pi}\int_{-\omega_b}^{+\omega_b} S_{nn}\,d\omega = \frac{\omega_b}{\pi} S_{nn}. \qquad (8.108)$$

In a digital simulation, according to the Nyquist sampling theorem, a nominally white noise source with an iteration/sampling period of h [s] will only contain information at frequencies up to $1/(2h)$ [Hz]. In this case, the bandwidth of the sampled signal may be taken as $f_b = 1/(2h)$ [Hz] or $\omega_b = \pi/h$ [rad/s]. In this case, (8.108) may be written

8.5 The Noise Filtering Property of the Observer

$$\sigma_{nn}^2 = E\left[n^2(t)\right] = \frac{1}{h}S_{nn}. \tag{8.109}$$

Any noise source with a frequency-dependent PSD is *coloured noise*, by analogy with coloured light, which contains dominant components at certain wavelengths.

8.5.5.5 Constrained Minimisation of Noise Variance

Consider now any LTI control system with plant noise, $n_p(t)$, having power spectral density, $S_{n_p n_p}(j\omega)$, and measurement noise, $n_m(t)$, having power spectral density, $S_{n_m n_m}(j\omega)$. Let all other inputs to the control system be zero. Then let $e(t)$ be a stochastic error of interest. This would usually be the control error, $e(t) = y(t) - y_r(t)$, where $y(t)$ is the controlled output and $y_r(t)$ is the corresponding reference input, zero in this case. If the system is a linear state feedback control system containing an observer, then other error signals of interest could be $\varepsilon_i(t) = \widehat{x}_i(t) - x_i(t)$, $i = 1, 2, \ldots, n$, where $x_i(t)$ and $\widehat{x}_i(t)$ are, respectively, the ith state variable and its estimate. The transfer function relationship between the selected error and the noise signals may then be written

$$E(s) = G_{en_p}(s)N_p(s) + G_{en_m}(s)N_m(s). \tag{8.110}$$

In view of (8.104) and (8.106), and assuming that $n_p(t)$ and $n_m(t)$ are not statistically correlated, which is reasonable in most cases, then the variance of the selected error signal is

$$\sigma_{ee}^2 = E\left[e^2(t)\right]$$
$$= \frac{1}{2\pi}\int_{-\infty}^{\infty}\left[|G_{en_p}(j\omega)|^2 S_{n_p n_p}(j\omega) + |G_{en_m}(j\omega)|^2 S_{n_m n_m}(j\omega)\right]d\omega. \tag{8.111}$$

If the control system contains an observer designed to have a correction loop settling time of T_{so}, $G_{en_p}(j\omega)$ and $G_{en_m}(j\omega)$ are both functions of T_{so} and therefore so is σ_{ee}^2. In relatively simple cases, the integral of (8.111) could be evaluated analytically and the minimisation carried out by solving the equation,

$$\frac{\mathrm{d}}{\mathrm{d}T_{so}}\sigma_{ee}^2(T_{so}) = 0, \tag{8.112}$$

for T_{so}, giving the optimal value, $T_{so\,opt}$, that yields the minimum variance, $\sigma_{ee\,min}^2$. In many cases, however, the integral evaluation could be very time consuming or even intractable. A recommended alternative practical approach is to run a simulation of the control system excited by the noise sources together with the real-time evaluation of the variance

$$\sigma_{ee}^2(t) = \frac{1}{t+\varepsilon}\int_0^t e^2(\tau)\,\mathrm{d}\tau, \quad t = \int_0^t 1.\mathrm{d}\tau, \tag{8.113}$$

where ε is a small constant to avoid the singularity at $t = 0$. Convergence of $\sigma_{ee}^2(t)$ to a constant, σ_{ee}^2, will be observed. Since the minimisation is only one dimensional, it is a simple matter to repeat the simulation for different values of T_{so} until the minimum, $\sigma_{ee\,\mathrm{min}}^2$, is found. It is a constrained minimisation, however, since the observer gain matrix elements are constrained to yield multiple observer poles. Without this constraint, the minimisation would be n-dimensional since the n gain matrix elements would be independently adjustable except for the condition of the observer correction loop being stable, but this minimisation would be far more complex. The Kalman filter of the following subsection achieves this in an elegant manner but with the constraint of the noise sources being white and also with a Gaussian statistical distribution.

Although the method of this subsection has been presented for control systems containing observers, it may be extended to other control systems configured with adjustable parameters suitable for minimisation of the stochastic control error and is not restricted to white noise sources. Furthermore, if the simulation-based method is employed, then the control system does not have to be linear, in contrast to the analytical method that is based on transfer function relationships.

Example 8.7 Constrained stochastic optimisation of a spacecraft attitude control system

This example continues with the rigid-body spacecraft of Example 8.6. It will be used to demonstrate the analytical approach to minimising the variances of the state estimation errors, $\varepsilon_1(t)$ and $\varepsilon_2(t)$, by variation of the observer correction loop settling time, T_{so}. It will then be used to confirm the theoretical results using the simulation-based method and also to attempt minimisation of the control error, $e(t)$, when the observer is used with a linear state feedback control law with gains set for a settling time of $T_s = 50$ [s] and critical damping.

It will be assumed that the plant noise of the reaction wheel torque actuator and the measurement noise of the star sensor are both band-limited white noise with a sampling time of $h = 0.01$ [s] in the simulation, giving a bandwidth of $f_b = 1/(2h) = 50$ [Hz] $\equiv \omega_b = 314$ [rad/s], the constant power spectral densities being denoted, respectively, by $S_{n_p n_p}$ and $S_{n_m n_m}$. The corresponding variances are given as $\sigma_{n_p n_p}^2 = 0.001$ [V^2] referred to the control voltage input and $\sigma_{n_m n_m}^2 = 10^{-7}$ [rad^2]. Using (8.108), this gives

$$S_{n_m n_m} = \frac{\pi}{\omega_b}\sigma_{n_m n_m}^2 = \frac{10^{-7}}{100} = 10^{-9}\ [\mathrm{rad}^2/\mathrm{Hz}] \tag{8.114}$$

and

$$S_{n_p n_p} = \frac{\pi}{\omega_b}\sigma_{n_p n_p}^2 = \frac{0.001}{100} = 10^{-5}\ [\mathrm{V}^2/\mathrm{Hz}]. \tag{8.115}$$

8.5 The Noise Filtering Property of the Observer

Applying (8.111) using the state estimation error transfer function relationship (8.99) gives the state estimation error variances as

$$\begin{bmatrix} \sigma^2_{\varepsilon_1\varepsilon_1} \\ \sigma^2_{\varepsilon_2\varepsilon_2} \end{bmatrix} = \frac{1}{2\pi} \int_{-\infty}^{\infty} \left\{ \frac{\left[1 + \frac{16T_{so}^2}{81}\omega^2 \right] S_{n_m n_m} + \frac{16T_{so}^4 b^2}{6561} \left[\omega^2 + \frac{1}{\frac{81}{T_{so}^2}} \right] S_{n_p n_p}}{\left(1 + \frac{4T_{so}^2}{81}\omega^2 \right)^2} \right\} d\omega. \tag{8.116}$$

Inspection of (8.116) reveals two integrals upon which the solution can be based, evaluated with the aid of integral tables [3]. Thus,

$$\frac{1}{2\pi} \int_{-\infty}^{\infty} \frac{dx}{(1+a^2x^2)^2} = \frac{1}{2\pi a} \int_{-\infty}^{\infty} \frac{d(ax)}{(1+a^2x^2)^2}$$
$$= \frac{1}{4\pi a} \left[\frac{ax}{(1+a^2x^2)} + \tan^{-1}(ax) \right]_{-\infty}^{\infty} = \frac{1}{4a}, \tag{8.117}$$

where $a = 2T_{so}/9$ and

$$\frac{1}{2\pi} \int_{-\infty}^{\infty} \frac{x^2 dx}{(1+a^2x^2)^2} = \frac{1}{2\pi a^3} \int_{-\infty}^{\infty} \frac{a^2 x^2 d(ax)}{(1+a^2x^2)^2}$$
$$= \frac{1}{4\pi a^3} \left[-\frac{ax}{(1+a^2x^2)} + \tan^{-1}(ax) \right]_{-\infty}^{\infty} = \frac{1}{4a^3}. \tag{8.118}$$

Using (8.117) and (8.118) in (8.116), with $\frac{1}{4a} = \frac{9}{8T_{so}}$ and $\frac{1}{4a^3} = \frac{729}{32T_{so}^3}$, yields

$$\begin{bmatrix} \sigma^2_{\varepsilon_1\varepsilon_1} \\ \sigma^2_{\varepsilon_2\varepsilon_2} \end{bmatrix} = \begin{bmatrix} \frac{9}{8T_{so}} + \frac{9}{2T_{so}} \\ \frac{729}{32T_{so}^3} \end{bmatrix} S_{n_m n_m} + \begin{bmatrix} \frac{2b^2 T_{so}^3}{729} \\ \frac{b^2 T_{so}}{18} + \frac{2b^2 T_{so}}{9} \end{bmatrix} S_{n_p n_p}$$
$$= \begin{bmatrix} \frac{45 S_{n_m n_m}}{8T_{so}} + \frac{2b^2 S_{n_p n_p} T_{so}^3}{729} \\ \frac{729 S_{n_m n_m}}{32T_{so}^3} + \frac{5b^2 S_{n_p n_p} T_{so}}{18} \end{bmatrix}. \tag{8.119}$$

The fact that reducing T_{so} increases the sensitivity to measurement noise and reduces the sensitivity to plant noise is evident in (8.119). To find the optimal correction loop settling time for ε_1, which will be called $T_{so\,opt1}$, equating the derivative of the first component of (8.119) to zero yields

$$\left. \frac{d\sigma^2_{\varepsilon_1\varepsilon_1}}{dT_{so}} \right|_{T_{so}=T_{so\,opt1}} = -\frac{45 S_{n_m n_m}}{8T_{so\,opt1}^2} + \frac{6b^2 S_{n_p n_p} T_{so\,opt1}^2}{729} = 0 \Rightarrow$$

$$T_{\text{so opt1}}^4 = \frac{45 S_{n_m n_m}}{8} \cdot \frac{729}{6b^2 S_{n_p n_p}} \Rightarrow T_{\text{so opt1}} = \frac{5.113}{b^{1/2}} \cdot \left(\frac{S_{n_m n_m}}{S_{n_p n_p}}\right)^{1/4}. \quad (8.120)$$

Substituting $T_{\text{so opt1}}$ for T_{so} in the first equation of (8.119) using (8.120) then yields

$$\sigma_{\varepsilon_1 \varepsilon_1 \text{ min}}^2 = \frac{45 S_{n_m n_m}}{8} \cdot \frac{b^{1/2}}{5.113} \cdot \left(\frac{S_{n_p n_p}}{S_{n_m n_m}}\right)^{1/4} + \frac{2b^2 S_{n_p n_p}}{729} \cdot \frac{133.668}{b^{3/2}} \cdot \left(\frac{S_{n_m n_m}}{S_{n_p n_p}}\right)^{3/4}$$

$$= 1.467 b^{1/2} S_{n_m n_m}^{3/4} S_{n_p n_p}^{1/4}$$

$$(8.121)$$

Repeating this procedure for ε_2, $\left.\dfrac{d\sigma_{\varepsilon_2 \varepsilon_2}^2}{dT_{\text{so}}}\right|_{T_{\text{so}} = T_{\text{so opt2}}} = -3 \cdot \dfrac{729 S_{n_m n_m}}{32 T_{\text{so opt2}}^4} + \dfrac{5b^2 S_{n_p n_p}}{18} = 0 \Rightarrow$

$$T_{\text{so opt2}}^4 = 3 \cdot \frac{729 S_{n_m n_m}}{32} \cdot \frac{18}{5b^2 S_{n_p n_p}} \Rightarrow T_{\text{so opt2}} = \frac{3.961}{b^{1/2}} \cdot \left(\frac{S_{n_m n_m}}{S_{n_p n_p}}\right)^{1/4}. \quad (8.122)$$

Substituting $T_{\text{so opt2}}$ for T_{so} in the second equation of (8.119) using (8.122) yields

$$\sigma_{\varepsilon_2 \varepsilon_2 \text{ min}}^2 = \frac{729 S_{n_m n_m}}{32} \cdot \frac{b^{3/2}}{62.146} \cdot \left(\frac{S_{n_p n_p}}{S_{n_m n_m}}\right)^{3/4} + \frac{5b^2 S_{n_p n_p}}{18} \cdot \frac{3.961}{b^{1/2}} \cdot \left(\frac{S_{n_m n_m}}{S_{n_p n_p}}\right)^{1/4}.$$

$$= 1.467 b^{3/2} S_{n_m n_m}^{1/4} S_{n_p n_p}^{3/4}$$

$$(8.123)$$

Comparing (8.120) and (8.122) shows $T_{\text{so opt1}}$ and $T_{\text{so opt2}}$ to be of the same order of magnitude but sufficiently different to raise the question of which is the best state estimation error is to minimise. In this example, it is arguably ε_1 since x_1 is the spacecraft attitude. This example also illustrates how $T_{\text{so opt1}}$ and $T_{\text{so opt2}}$, hence the observer poles, depend on the ratio of the plant and measurement noise variances.

The plant parameters are taken as $J = 100 \ [\text{Kg m}^2]$ and $K_w = 0.1 \ [\text{Nm/V}]$, giving $b = 10^{-3} \ [\text{rad/s}^2/\text{V}]$. Then (8.121) and (8.123) yield

$$\sigma_{\varepsilon_1 \varepsilon_1 \text{ min}}^2 = 4.6391 \times 10^{-10} \ [\text{rad}^2] \quad \text{and}$$

$$\sigma_{\varepsilon_2 \varepsilon_2}^2 = 4.6391 \times 10^{-11} \ [(\text{rad/s})^2]. \quad (8.124)$$

If the error signals had Gaussian distributions, then the standard deviations would be

$$\sigma_{\varepsilon_1 \varepsilon_1 \text{ min}} = 2.1539 \times 10^{-5} \ [\text{rad}] \equiv 0.0012 \ [\text{deg}] \equiv 4.32 \ [\text{arcsec}] \quad (8.125)$$

8.5 The Noise Filtering Property of the Observer

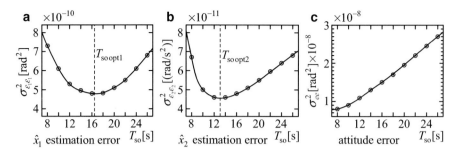

Fig. 8.20 Constrained error variance minimisation for rigid-body spacecraft attitude control

and

$$\sigma_{\varepsilon_2 \varepsilon_2} = 6.8111 \times 10^{-6} \; [\text{rad/s}]. \tag{8.126}$$

The variance of the attitude angle estimate of (8.125) gives an indication of the attainable accuracy of the control system, not considering other factors such as attitude sensor misalignment.

The values of $T_{\text{so opt1}}$ and $T_{\text{so opt2}}$ given by (8.120) and (8.122) are

$$T_{\text{so opt1}} = 16.1687 \; [\text{s}] \quad \text{and} \quad T_{\text{so opt2}} = 12.5258 \; [\text{s}]. \tag{8.127}$$

The linear state feedback control law is designed to yield a main control loop settling time of $T_s = 50$ [s]. Hence, the values of $T_{\text{so opt1}}$ and $T_{\text{so opt2}}$ do not satisfy the fast settling criterion given in Sect. 8.2.3, i.e., $T_{\text{so}} \leq \frac{1}{5} T_s$. As will be seen, however, this is not a problem if the system is to be designed to minimise the variance, σ_{ee}^2, of the stochastic component of the attitude control error, $e = x_{1r} - x_1$. Figure 8.20 shows the results of the simulation-based minimisation. The minima in Fig. 8.20a, b coincide with the theoretically predicted ones. Remarkably, however, there is no minimum in the stochastic attitude error. Further simulations for values of T_{so} outside the range of Fig. 8.20c confirm this. Reducing T_{so} to very small values yields an asymptotic reduction of σ_{ee}^2 towards a constant value of about 4×10^{-9} $[\text{rad}^2]$.

This means that setting $T_{\text{so}} = T_{\text{so opt1}}$ to minimise the variance of the attitude estimation *does not* minimise the stochastic spacecraft attitude control error variance, σ_{ee}^2. This phenomenon is attributed to the effect of the loop closure via the linear state feedback control law in which a weighted sum of the noise signals, ε_1 and ε_2, contaminating $\widehat{x}_1$ and $\widehat{x}_2$, is applied to the plant via u. In this case, T_{so} should be set to as small a value as possible as limited by the sampling period of the digital processor implementing the controller. This will certainly satisfy $T_{\text{so}} \leq \frac{1}{5} T_s$.

Example 8.8 Unconstrained stochastic optimisation of a spacecraft attitude control system

In Example 8.7, the observer gains, l_1 and l_2, were functions of T_{so} and were therefore not varied independently to minimise the state estimation noise variances.

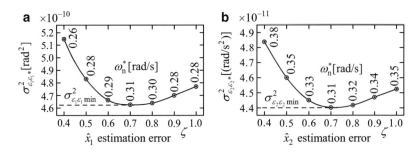

Fig. 8.21 Unconstrained error variance minimisation for rigid-body spacecraft attitude control

This poses the question of whether significant further reductions of the noise variances are achievable by removing this constraint, i.e. by varying l_1 and l_2 independently. In this particular example, removing the pole assignment procedure will not risk instability of the observer provided the minimisation search is carried out only for positive values of l_1 and l_2, since the roots of $s^2 + l_1 s + l_2 = 0$ lie in the left half of the s-plane for $l_1 > 0$ and $l_2 > 0$. Rather than vary l_1 and l_2 directly, the undamped natural frequency, ω_n, and the damping ratio, ζ, of the observer correction loop will be varied. This will ease the minimisation process for the following reason. Since

$$s^2 + l_1 s + l_2 = s^2 + 2\zeta\omega_n s + \omega_n^2, \tag{8.128}$$

$$l_1 = 2\zeta\omega_n \tag{8.129}$$

and

$$l_2 = \omega_n^2 \tag{8.130}$$

For the pole placement in the previous two examples, $\zeta = 1$ and $\omega_n = 9/(2T_{so})$. It has already been established in Example 8.7 that minima of the state estimation noise variances $\sigma_{\varepsilon_1\varepsilon_1}^2$ and $\sigma_{\varepsilon_2\varepsilon_2}^2$ exist with respect to T_{so}. Minima must therefore also exist with respect to ω_n for $\omega_n > 0$ and $\zeta = 1$. It is reasonable to suppose that such minima will also exist for other positive values of ζ, and the purpose of this example will be to find the values of ζ yielding lowest minima of $\sigma_{\varepsilon_1\varepsilon_1}^2$ and $\sigma_{\varepsilon_2\varepsilon_2}^2$. This will effectively find the minima with respect to freely chosen positive values of l_1 and l_2, through (8.129) and (8.130).

In Fig. 8.21, the state estimation error noise variances have been minimised numerically using the simulation-based method with respect to ω_n for constant damping ratios ranging between $\zeta = 0.4$ and $\zeta = 1$, the plant parameters and noise parameters being as for Example 8.7.

These minimum variances are denoted $\sigma_{\varepsilon_i\varepsilon_i*}^2$, $i = 1, 2$, and the values of ω_n at which they occur are denoted, ω_n^*. The minima, $\sigma_{\varepsilon_i\varepsilon_i \text{ min}}^2$, $i = 1, 2$, of these

8.5 The Noise Filtering Property of the Observer

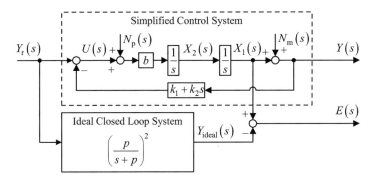

Fig. 8.22 Simplified spacecraft attitude control system showing stochastic control error

w.r.t. ζ are the absolute, unconstrained minima. Importantly, the absolute minima, $\sigma^2_{\varepsilon_1\varepsilon_1\,\text{min}}$ and $\sigma^2_{\varepsilon_2\varepsilon_2\,\text{min}}$, occur for the same values of ζ and ω_n^*, which means that the observer does not have to be optimised for either $\sigma^2_{\varepsilon_1\varepsilon_1}$ or $\sigma^2_{\varepsilon_2\varepsilon_2}$, in contrast with the constrained optimisation of Example 8.7. It is optimal for both state variables. Also, it is known that the Kalman filter to be presented in Sect. 8.6, which yields optimal state estimation in the sense of this subsection, yields a damping ratio of $\zeta = 1/\sqrt{2} = 0.7071$ for a double integrator plant, which agrees with Fig. 8.21.

It should be mentioned that the suboptimal variances of the constrained optimisation based on the settling time formulae with multiple observer poles are the values of $\sigma^2_{\varepsilon_1\varepsilon_1*}$ and $\sigma^2_{\varepsilon_2\varepsilon_2*}$ for $\zeta = 1$ in Fig. 8.21, and these are, respectively, only about 2 % and 3 % greater than the optimal values, $\sigma^2_{\varepsilon_1\varepsilon_1\,\text{min}}$ and $\sigma^2_{\varepsilon_2\varepsilon_2\,\text{min}}$. From a practical viewpoint, therefore, the simple constrained optimisation method is useful.

Before leaving this example, the point made in Example 8.7 that minimising the stochastic state estimation errors does not minimise the overall stochastic control errors is raised once more. In this example, performing the unconstrained minimisation with respect to the observer parameters, no minimum of $\sigma^2_{ee} = E\{e^2\}$, where $e = x_1 - x_{1r}$, could be found. This poses the question of whether this can be minimised by the controller parameters other than those of an observer. This proves to be the case only with controllers not supported by optimally designed observers. An example is the simplified control system obtained by direct measurement and differentiation to obtain y and $\dot{y}$ as estimates of x_1 and x_2 instead of using the observer. This system is shown in Fig. 8.22.

The simplified and ideal closed-loop systems are identical except for the noise sources. Then $E(s)$ is the stochastic control error. Since both systems are linear, there is no loss of generality in setting $Y_r(s) = 0$. Then the transfer function relationship between $E(s)$, $N_p(s)$ and $N_m(s)$ is

$$Y(s) = \frac{\frac{b}{s^2}N_p(s) - (k_2 s + k_1)\frac{b}{s^2}N_m(s)}{1 + (k_2 s + k_1)\frac{b}{s^2}} = \frac{bN_p(s) - (k_2 s + k_1)bN_m(s)}{s^2 + bk_2 s + bk_1}. \quad (8.131)$$

With multiple poles using the 5 % settling time formula, this becomes

$$E(s) = \frac{\frac{4T_s^2 b}{81} N_p(s) - \left(1 + \frac{4T_s}{9}s\right) N_m(s)}{\left(1 + \frac{2T_s}{9}s\right)^2}, \tag{8.132}$$

where T_s is the settling time. The plant parameters and noise sources are as for Example 8.7. Thus, $b = 10^{-3}$ [rad/s^2/V] and the plant and measurement noise signals are band-limited white noise with a sampling time of $h = 0.01$ [s], having constant spectral densities of $S_{n_p n_p} = 10^{-5}$ [V^2/Hz] and $S_{n_m n_m} = 10^{-9}$ [rad^2/Hz]. The variance of the stochastic error of (8.132) is then given by

$$\sigma_{ee}^2 = \frac{1}{2\pi} \int_{-\infty}^{\infty} \left\{ \frac{\frac{16 T_s^4 b^2}{6561} S_{n_p n_p} + \left(1 + \frac{16 T_s^2}{81}\omega^2\right) S_{n_m n_m}}{\left(1 + \frac{4T_s^2}{81}\omega^2\right)^2} \right\} d\omega. \tag{8.133}$$

Since this is identical to the first component of Eq. (8.116) with T_{so} being is replaced by T_s, the minimum variance, $\sigma_{ee\ min}^2$, and the settling time, $T_{s\ opt}$, at which this occurs may be written down by analogy with (8.120) and (8.121). Thus,

$$T_{s\ opt1} = \frac{5.113}{b^{1/2}} \cdot \left(\frac{S_{n_m n_m}}{S_{n_p n_p}}\right)^{1/4} = 4.6391 \times 10^{-10}\ [\text{rad}^2] \tag{8.134}$$

and

$$\sigma_{ee\ min}^2 = 1.467 b^{1/2} S_{n_m n_m}^{3/4} S_{n_p n_p}^{1/4} = 4.6391 \times 10^{-10}\ [\text{rad}^2]. \tag{8.135}$$

It follows that if the minimisation was carried out using the simulation-based method, then the graph of σ_{ee}^2 against T_s would be identical to the graph of $\sigma_{\varepsilon_1 \varepsilon_1}^2$ against T_{so} shown in Fig. 8.20a.

Returning to the system containing the observer optimised stochastically without constraints as shown in Fig. 8.21, attempting stochastic error minimisation by varying the settling time, T_s, will not find a minimum in σ_{ee}^2. Figure 8.23 shows σ_{ee}^2 plotted over a practicable range of settling times for this application.

In fact, below this range, σ_{ee}^2 is found to reduce monotonically with T_s, reducing asymptotically towards the minimum attitude angle estimation error variance, $\sigma_{\varepsilon_1 \varepsilon_1}^2$. The reason for this is that as T_s is made indefinitely small, the LSF control law will tightly control the plant model of the observer and accurately follow the random variations in $\widehat{x}_1$, the control variable at the same time counteracting the disturbance due to the plant noise. The attitude angle, x_1, will then closely follow $\widehat{x}_1$ and therefore have an error variance close to that of $\widehat{x}_1$, i.e. $\sigma_{\varepsilon_1 \varepsilon_1}^2$. If T_s is increased, the effect of the plant noise increases thereby increasing σ_{ee}^2 above $\sigma_{\varepsilon_1 \varepsilon_1}^2$.

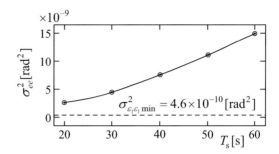

Fig. 8.23 Attempted minimisation of stochastic control error with respect to settling time

8.6 The Kalman Filter

8.6.1 Introduction

The Kalman filter [4] is a state estimator for linear plants that minimises a measure of the stochastic state estimation errors due to plant and measurement noise. It has a similar but not identical structure to an observer and similarly has filtering properties regarding measurement noise and plant noise. This is the reason for its title. The Kalman filter is optimal with respect to the selected error measure when the model is perfectly accurate and the statistics of the noise signals satisfy certain conditions to be given subsequently. The most general plant catered for by the Kalman filter is a linear time-varying multivariable plant modelled by either the continuous state-space model,

$$\dot{\mathbf{x}}(t) = \mathbf{A}(t)\mathbf{x}(t) + \mathbf{B}(t)\left[\mathbf{u}(t) + \mathbf{n}_\mathrm{p}(t)\right] \tag{8.136}$$

$$\mathbf{y}(t) = \mathbf{C}(t)\mathbf{x}(t) + \mathbf{n}_\mathrm{m}(t), \tag{8.137}$$

or the discrete state-space model,

$$\mathbf{x}(k+1) = \mathbf{\Phi}(k,h)\mathbf{x}(k) + \mathbf{\Psi}(k,h)\left[\mathbf{u}(k) + \mathbf{n}_\mathrm{p}(k)\right] \tag{8.138}$$

$$\mathbf{y}(k) = \mathbf{C}(k)\mathbf{x}(k) + \mathbf{n}_\mathrm{m}(k), \tag{8.139}$$

where $\mathbf{x} \in \Re^n$ is the state vector, $\mathbf{u} \in \Re^m$ is the control vector, $\mathbf{y} \in \Re^m$ is the measurement vector, $\mathbf{n}_\mathrm{p} \in \Re^m$ is the plant noise vector, $\mathbf{n}_\mathrm{m} \in \Re^m$ is the measurement noise vector, $\mathbf{A}(k) \in \Re^{n \times n}$ is the plant matrix, $\mathbf{B}(k) \in \Re^{n \times m}$ is the input matrix, $\mathbf{C}(k) \in \Re^{m \times n}$ is the output matrix, $\mathbf{\Phi}(k,h)$ is the discrete system matrix and $\mathbf{\Psi}(k,h)$ is the discrete input matrix and h is the iteration/sampling time. The derivation of the discrete Kalman filter based on the plant model of (8.138) and (8.139) is given here as it is more straightforward than that of the continuous Kalman filter based on the plant model of (8.136) and (8.137) and is directly relevant to digital implementation. Since h is fixed, the notation is simplified by denoting $\mathbf{\Phi}(k,h)$ and $\mathbf{\Psi}(k,h)$, respectively, by $\mathbf{\Phi}(k)$ and $\mathbf{\Psi}(k)$.

In the following subsection, a generalisation of the discrete observer introduced in Sect. 8.3 will be used as a starting point. This will also serve to point out the differences between the observer and the Kalman filter.

8.6.2 The Discrete Observer

The discrete observer defined by (8.59) may be generalised to cater for linear, time-varying multivariable plants as follows:

$$\widehat{\mathbf{x}}(k+1) = \mathbf{\Phi}(k)\widehat{\mathbf{x}}(k) + \mathbf{\Psi}(k)\mathbf{u}(k) + \mathbf{L}(k)\mathbf{e}(k), \qquad (8.140)$$

where

$$\mathbf{e}(k) = \mathbf{y}(k) - \widehat{\mathbf{y}}(k) \qquad (8.141)$$

is the model correction loop actuation error and

$$\widehat{\mathbf{y}}(k) = \mathbf{C}\widehat{\mathbf{x}}(k), \qquad (8.142)$$

since the measurement noise, $\mathbf{n}_m(k)$, is assumed to have zero mean value and is set to zero in the model. The state estimation error is defined as $\boldsymbol{\varepsilon}(k) \triangleq \widehat{\mathbf{x}}(k) - \mathbf{x}(k)$. Then the state estimation error difference equation expressing $\boldsymbol{\varepsilon}(k+1)$ in terms of $\boldsymbol{\varepsilon}(k)$ is derived as follows. Substituting for $\mathbf{e}(k)$ in (8.140) using (8.141) and then for $\mathbf{y}(k)$ and $\widehat{\mathbf{y}}(k)$ using (8.139) and (8.142) yields

$$\widehat{\mathbf{x}}(k+1) = \mathbf{\Phi}(k)\widehat{\mathbf{x}}(k) + \mathbf{\Psi}(k)\mathbf{u}(k) + \mathbf{L}(k)\mathbf{C}(k)\left[\mathbf{x}(k) - \widehat{\mathbf{x}}(k)\right] + \mathbf{L}(k)\mathbf{n}_m(k). \qquad (8.143)$$

Subtracting (8.138) then yields

$$\widehat{\mathbf{x}}(k+1) - \mathbf{x}(k+1) = \mathbf{\Phi}(k)\left[\widehat{\mathbf{x}}(k) - \mathbf{x}(k)\right] + \mathbf{\Psi}(k)\mathbf{u}(k) + \mathbf{L}(k)\mathbf{C}(k)\left[\mathbf{x}(k) - \widehat{\mathbf{x}}(k)\right]$$
$$+ \mathbf{L}(k)\mathbf{n}_m(k) - \mathbf{\Psi}(k)\left[\mathbf{u}(k) + \mathbf{n}_p(k)\right] \Rightarrow$$
$$\boldsymbol{\varepsilon}(k+1) = [\mathbf{\Phi}(k) - \mathbf{L}(k)\mathbf{C}(k)]\boldsymbol{\varepsilon}(k) - \mathbf{\Psi}(k)\mathbf{n}_p(k) + \mathbf{L}(k)\mathbf{n}_m(k). \qquad (8.144)$$

For LTI plants, the gain matrix, $\mathbf{L}$, would usually be chosen to yield eigenvalues of $\mathbf{\Phi} - \mathbf{LC}$ within the unit circle so that if $\mathbf{n}_p(k) = \mathbf{0}$ and $\mathbf{n}_m(k) = \mathbf{0}$, $\boldsymbol{\varepsilon}(k) \to \mathbf{0}$ as $k \to \infty$. This might also be possible for some time-varying plants if $\mathbf{L}(k)$ could be chosen to make $[\mathbf{\Phi}(k) - \mathbf{L}(k)\mathbf{C}(k)]$ constant. The approach of Sect. 8.6.4 for the Kalman filter is entirely different, being based on the noise signal statistics and

8.6 The Kalman Filter

the plant model. It would also be possible to design the observer to minimise the selected measure of the stochastic state estimation error as for the Kalman Filter using information about the noise signals following a method for determining the optimal gain matrix, **L**, similar to that of Sect. 8.6.4. The conventional approach of doing this for the Kalman filter, however, will be followed as this is well established. In preparation for this, the following subsection formulates the state difference equation and derives of the state estimation error difference equation following similar lines to that carried out above for the observer, but it is less straightforward.

8.6.3 The Kalman Filter: State Difference and Error Equations

In the observer of the preceding subsection, $\hat{\mathbf{x}}(k+1)$ is calculated at time, t_k, and is therefore a prediction of the plant state at time, t_{k+1}, allowing h seconds for the calculation of the next control, $\mathbf{u}(k+1)$. Each correction to the plant model state is made using a measurement that was sampled h seconds earlier. In contrast, each correction to the plant model state in the Kalman filter is made without any delay, but requiring the control computation time to be a negligible proportion of h. In preparation for this, the measurement Eq. (8.139) is replaced by

$$\mathbf{y}(k+1) = \mathbf{C}(k)\mathbf{x}(k+1) + \mathbf{n}_m(k+1). \tag{8.145}$$

The state difference equations of the discrete Kalman filter corresponding to the discrete observer of (8.140), (8.141) and (8.142) are as follows:

$$\hat{\mathbf{x}}(k+1|k) = \mathbf{\Phi}(k)\hat{\mathbf{x}}(k|k) + \mathbf{\Psi}(k)\mathbf{u}(k) \quad \text{(prediction)} \tag{8.146}$$

$$\hat{\mathbf{x}}(k+1|k+1) = \hat{\mathbf{x}}(k+1|k) + \mathbf{K}(k)\mathbf{e}(k+1) \quad \text{(correction)}, \tag{8.147}$$

where

$$\mathbf{e}(k+1) = \mathbf{y}(k+1) - \hat{\mathbf{y}}(k+1) \tag{8.148}$$

is the model correction loop actuation error and

$$\hat{\mathbf{y}}(k+1) = \mathbf{C}(k)\hat{\mathbf{x}}(k+1|k). \tag{8.149}$$

Here, $\mathbf{K}(k)$ is the correction loop gain matrix that corresponds to $\mathbf{L}(k)$ in the observer. This is conventional notation and therefore adhered to, but to avoid confusion when dealing with control systems comprising a Kalman filter and a linear state feedback control law, it is recommended to replace the conventional symbol, **K**, for the linear state feedback gain matrix with a different symbol such as **G**.

To explain the standard notation used in (8.146), (8.147), (8.148) and (8.149), considering that the Kalman filter is a recursive algorithm, $\widehat{\mathbf{x}}(k+1|k)$ means an estimate of the state at time, t_{k+1}, using all the past control and measurement samples taken up to and including time t_k, while $\widehat{\mathbf{x}}(k+1|k+1)$ means an estimate of the state at time, t_{k+1}, using all the past control and measurement samples taken up to and including time t_{k+1}. As indicated in (8.146) and (8.147), the Kalman filter embodies the process of prediction and correction. It is, in fact, contained in the general class of *predictor corrector* algorithms of numerical methods [5].

Next, the difference equation governing the state estimation error, defined as

$$\boldsymbol{\varepsilon}(k) = \widehat{\mathbf{x}}(k|k) - \mathbf{x}(k), \qquad (8.150)$$

will be derived. The first step is to use Eqs. (8.146), (8.147), (8.148) and (8.149) and Eq. (8.137) to form a single equation for $\widehat{\mathbf{x}}(k+1|k+1)$ in terms of $\widehat{\mathbf{x}}(k|k)$, $\mathbf{x}(k+1)$ and $\mathbf{u}(k)$. Thus, substituting for $\widehat{\mathbf{y}}(k+1)$ in (8.148) using (8.149), then for $\mathbf{e}(k+1)$ in (8.147) using (8.148) and finally for $\widehat{\mathbf{x}}(k+1|k)$ and $\mathbf{y}(k+1)$ in (8.147) using, respectively, (8.146) and (8.145) yields

$$\begin{aligned}\widehat{\mathbf{x}}(k+1|k+1) &= \boldsymbol{\Phi}(k)\widehat{\mathbf{x}}(k|k) + \boldsymbol{\Psi}(k)\mathbf{u}(k) \\ &+ \mathbf{K}(k)\left[\mathbf{C}(k)\mathbf{x}(k+1) + \mathbf{n}_m(k+1) - \mathbf{C}(k)\left[\boldsymbol{\Phi}(k)\widehat{\mathbf{x}}(k|k) + \boldsymbol{\Psi}(k)\mathbf{u}(k)\right]\right].\end{aligned} \qquad (8.151)$$

The approach will be to subtract (8.138) from (8.151) to yield $\widehat{\mathbf{x}}(k+1|k+1) - \mathbf{x}(k+1) = \boldsymbol{\varepsilon}(k+1)$ on the left-hand side and only terms involving $\widehat{\mathbf{x}}(k|k) - \mathbf{x}(k) = \boldsymbol{\varepsilon}(k)$ on the right-hand side. This first requires substituting for $\mathbf{x}(k+1)$ in (8.151) using (8.138) to yield

$$\begin{aligned}\widehat{\mathbf{x}}(k+1|k+1) &= \boldsymbol{\Phi}(k)\widehat{\mathbf{x}}(k|k) + \boldsymbol{\Psi}(k)\mathbf{u}(k) \\ &+ \mathbf{K}(k)\begin{bmatrix}\left[\mathbf{C}(k)\left[\boldsymbol{\Phi}(k)\mathbf{x}(k) + \boldsymbol{\Psi}(k)\left[\mathbf{u}(k) + \mathbf{n}_p(k)\right]\right] + \mathbf{n}_m(k+1)\right] \\ - \mathbf{C}(k)\left[\boldsymbol{\Phi}(k)\widehat{\mathbf{x}}(k|k) + \boldsymbol{\Psi}(k)\mathbf{u}(k)\right]\end{bmatrix}\end{aligned} \qquad (8.152)$$

which simplifies to

$$\begin{aligned}\widehat{\mathbf{x}}(k+1|k+1) &= \boldsymbol{\Phi}(k)\widehat{\mathbf{x}}(k|k) + \boldsymbol{\Psi}(k)\mathbf{u}(k) \\ &+ \mathbf{K}(k)\left[\mathbf{C}(k)\boldsymbol{\Phi}(k)\left[\mathbf{x}(k) - \widehat{\mathbf{x}}(k|k)\right] + \mathbf{C}(k)\boldsymbol{\Psi}\mathbf{n}_p(k) + \mathbf{n}_m(k+1)\right].\end{aligned} \qquad (8.153)$$

8.6 The Kalman Filter

Subtracting (8.138) from (8.153) then yields

$$\widehat{\mathbf{x}}\left(k+1\middle|k+1\right) - \mathbf{x}(k+1) = \mathbf{\Phi}(k)\left[\widehat{\mathbf{x}}\left(k\middle|k\right) - \mathbf{x}(k)\right] - \mathbf{\Psi}(k)\mathbf{n}_\mathrm{p}(k) \\ + \mathbf{K}(k)\left[\mathbf{C}(k)\mathbf{\Phi}(k)\left[\mathbf{x}(k) - \widehat{\mathbf{x}}\left(k\middle|k\right)\right] + \mathbf{C}(k)\mathbf{\Psi}\mathbf{n}_\mathrm{p}(k) + \mathbf{n}_\mathrm{m}(k+1)\right], \quad (8.154)$$

which gives the required state estimation error difference equation as

$$\boldsymbol{\varepsilon}(k+1) = \mathbf{\Phi}(k)\boldsymbol{\varepsilon}(k) + \mathbf{K}(k)\left[-\mathbf{C}(k)\mathbf{\Phi}(k)\boldsymbol{\varepsilon}(k) + \mathbf{C}(k)\mathbf{\Psi}(k)\mathbf{n}_\mathrm{p}(k) + \mathbf{n}_\mathrm{m}(k+1)\right] \\ - \mathbf{\Psi}(k)\mathbf{n}_\mathrm{p}(k) \Rightarrow$$

$$\boldsymbol{\varepsilon}(k+1) = \left[\mathbf{I}^{(n)} - \mathbf{K}(k)\mathbf{C}(k)\right]\left[\mathbf{\Phi}(k)\boldsymbol{\varepsilon}(k) - \mathbf{\Psi}(k)\mathbf{n}_\mathrm{p}(k)\right] + \mathbf{K}(k)\mathbf{n}_\mathrm{m}(k+1). \quad (8.155)$$

This forms the basis of the optimal gain determination in the following subsection.

8.6.4 Derivation of the Discrete Kalman Gain Algorithm

The assumptions regarding the statistics of the plant and measurement noise signals, $\mathbf{n}_\mathrm{p}(k)$ and $\mathbf{n}_\mathrm{m}(k)$, under which the Kalman filter is optimal are as follows:

1. They have zero mean values.
2. They are band-limited white noise signals, meaning, in practical terms, that they have spectral densities (Sect. 8.5.5) independent of frequency over the frequency range of operation of the control system:

$$\mathrm{cov}\left[\mathbf{n}_\mathrm{p}(k)\right] = E\left\{\mathbf{n}_\mathrm{p}(k)\mathbf{n}_\mathrm{p}^\mathrm{T}(k)\right\} = \mathbf{Q}(k) \quad (8.156)$$

and

$$\mathrm{cov}\left[\mathbf{n}_\mathrm{m}(k)\right] = E\left\{\mathbf{n}_\mathrm{m}(k)\mathbf{n}_\mathrm{m}^\mathrm{T}(k)\right\} = \mathbf{R}(k). \quad (8.157)$$

where $E\{x\}$ is the expected value of x, i.e. its mean value (Sect. 8.5.5) and $q_{ij}(k) = E\left\{n_{\mathrm{p}i}(k)n_{\mathrm{p}j}(k)\right\}$ and $r_{ij}(k) = E\left\{n_{\mathrm{m}i}(k)n_{\mathrm{m}j}(k)\right\}$. Hence the diagonal elements of $\mathbf{Q}(k)$ and $\mathbf{R}(k)$ are, respectively, the variances of the individual plant and measurement noise components (Sect. 8.5.5), and the off-diagonal terms are called the *covariances* between the individual noise signal components, a nonzero value of $q_{ij}(k)$, $i \neq j$, for example, indicating statistical correlation between $n_{\mathrm{p}i}(k)$ and $n_{\mathrm{p}j}(k)$. It is evident that both $\mathbf{Q}(k)$ and $\mathbf{R}(k)$ are symmetrical matrices.

A further condition is often stated that the noise signals should have a Gaussian distribution, but there is nothing in the following derivation that requires this. The dependence of $\mathbf{Q}(k)$ and $\mathbf{R}(k)$ on k means that $\mathbf{n}_p(k)$ and $\mathbf{n}_m(k)$ are white noise signals with a time-varying amplitude, which occurs in a few applications.

The covariance of the state estimation error is

$$\mathbf{P}(k) = \text{cov}\left[\boldsymbol{\varepsilon}(k)\right] = E\left\{\boldsymbol{\varepsilon}(k)\boldsymbol{\varepsilon}^\mathrm{T}(k)\right\}, \qquad (8.158)$$

which is also a symmetric matrix. A recursive relationship giving $\mathbf{P}(k+1)$ as a function of $\mathbf{P}(k)$ will be derived. The Kalman filter sets the gain, $\mathbf{K}(k+1)$ to minimise a scalar function $J[\mathbf{P}(k+1)]$, chosen to be a measure of the magnitude of the stochastic state estimation error. This gives optimal state estimation on the basis that $J[\mathbf{P}(i)], i = k, k-1, k-2, \ldots$, have also been minimised by appropriate values of $\mathbf{K}(i)$. An algorithm that achieves this automatically will be derived. The usual choice of the scalar function for the Kalman filter is

$$J\left[\mathbf{P}(k)\right] = \sum_{i=1}^{n} E\left\{\varepsilon_i^2(k)\right\}, \qquad (8.159)$$

which gives equal emphasis to each state variable being estimated. It should be noted, however, that this can be generalised to $J[\mathbf{P}(k)] = \sum_{i=1}^{n} W_i E\left\{\varepsilon_i^2(k)\right\}$, where W_i are weighting coefficients set to suit the needs of a particular application. The task in hand is to find the gain matrix, $\mathbf{K}(k+1)$, that minimises $\mathbf{P}(k+1)$. First, (8.155) may be written as

$$\boldsymbol{\varepsilon}(k+1) = \mathbf{M}(k)\boldsymbol{\Phi}\boldsymbol{\varepsilon}(k) - \mathbf{M}(k)\boldsymbol{\Psi}\mathbf{n}_p(k) + \mathbf{K}(k)\mathbf{n}_m(k+1), \qquad (8.160)$$

where

$$\mathbf{M}(k) = \mathbf{I}^{(n)} - \mathbf{K}(k)\mathbf{C}(k). \qquad (8.161)$$

Then

$$\begin{aligned}
\boldsymbol{\varepsilon}(k+1)\boldsymbol{\varepsilon}^\mathrm{T}(k+1) &= \left[\mathbf{M}(k)\boldsymbol{\Phi}(k)\boldsymbol{\varepsilon}(k) - \mathbf{M}(k)\boldsymbol{\Psi}(k)\mathbf{n}_p(k) + \mathbf{K}(k)\mathbf{n}_m(k+1)\right] \\
&\quad \times \left[\boldsymbol{\varepsilon}^\mathrm{T}(k)\boldsymbol{\Phi}^\mathrm{T}(k)\mathbf{M}^\mathrm{T}(k) - \mathbf{n}_p^\mathrm{T}(k)\boldsymbol{\Psi}^\mathrm{T}(k)\mathbf{M}^\mathrm{T}(k) + \mathbf{n}_m^\mathrm{T}(k+1)\mathbf{K}^\mathrm{T}(k)\right] \\
&= \mathbf{M}(k)\boldsymbol{\Phi}(k)\boldsymbol{\varepsilon}(k)\boldsymbol{\varepsilon}^\mathrm{T}(k)\boldsymbol{\Phi}^\mathrm{T}(k)\mathbf{M}^\mathrm{T}(k) - \mathbf{M}(k)\boldsymbol{\Phi}(k)\boldsymbol{\varepsilon}(k)\mathbf{n}_p^\mathrm{T}(k)\boldsymbol{\Psi}^\mathrm{T}(k)\mathbf{M}^\mathrm{T}(k) \\
&\quad + \mathbf{M}(k)\boldsymbol{\Phi}(k)\boldsymbol{\varepsilon}(k)\mathbf{n}_m^\mathrm{T}(k+1)\mathbf{K}^\mathrm{T}(k) - \mathbf{M}(k)\boldsymbol{\Psi}(k)\mathbf{n}_p(k)\boldsymbol{\varepsilon}^\mathrm{T}(k)\boldsymbol{\Phi}^\mathrm{T}(k)\mathbf{M}^\mathrm{T}(k) \\
&\quad + \mathbf{M}(k)\boldsymbol{\Psi}(k)\mathbf{n}_p(k)\mathbf{n}_p^\mathrm{T}(k)\boldsymbol{\Psi}^\mathrm{T}(k)\mathbf{M}^\mathrm{T}(k) - \mathbf{M}(k)\boldsymbol{\Psi}(k)\mathbf{n}_p(k)\mathbf{n}_m^\mathrm{T}(k+1)\mathbf{K}^\mathrm{T}(k) \\
&\quad + \mathbf{K}\mathbf{n}_m(k+1)\boldsymbol{\varepsilon}^\mathrm{T}(k)\boldsymbol{\Phi}^\mathrm{T}\mathbf{M}^\mathrm{T} - \mathbf{K}\mathbf{n}_m(k+1)\mathbf{n}_p^\mathrm{T}(k)\boldsymbol{\Psi}^\mathrm{T}\mathbf{M}^\mathrm{T} \\
&\quad + \mathbf{K}\mathbf{n}_m(k+1)\mathbf{n}_m^\mathrm{T}(k+1)\mathbf{K}^\mathrm{T}
\end{aligned}$$
(8.162)

It is reasonable to suppose that $\mathbf{n}_m(k+1)$ is not correlated with $\mathbf{n}_p(k)$ in most applications. Then $E\left\{\mathbf{n}_p(k)\mathbf{n}_m^\mathrm{T}(k+1)\right\} = E\left\{\mathbf{n}_m(k+1)\mathbf{n}_p^\mathrm{T}(k)\right\} = \mathbf{0}$. Since there is dynamic lag between the plant noise input and the state estimation error, $\mathbf{n}_p(k)$

8.6 The Kalman Filter

cannot influence $\boldsymbol{\varepsilon}(k)$, but only $\boldsymbol{\varepsilon}(k+j)$, $j \geq 1$. Hence, $= E\left\{\mathbf{n}_p(k)\boldsymbol{\varepsilon}^T(k)\right\} = \mathbf{0}$ $E\{\boldsymbol{\varepsilon}(k)\mathbf{n}_p^T(k)\}$. Also, $\mathbf{n}_m(k+1)$ cannot possibly influence $\boldsymbol{\varepsilon}(k)$ since $\mathbf{n}_m(k+1)$ occurs after $\boldsymbol{\varepsilon}(k)$ and therefore $E\left\{\boldsymbol{\varepsilon}(k)\mathbf{n}_m^T(k+1)\right\} = E\left\{\mathbf{n}_m(k+1)\boldsymbol{\varepsilon}^T(k)\right\} = \mathbf{0}$.

In this case, taking the expected values of both sides of (8.162) yields

$$E\left\{\boldsymbol{\varepsilon}(k+1)\boldsymbol{\varepsilon}^T(k+1)\right\} = \mathbf{M}(k)\boldsymbol{\Phi}(k)E\left\{\boldsymbol{\varepsilon}(k)\boldsymbol{\varepsilon}^T(k)\right\}\boldsymbol{\Phi}^T(k)\mathbf{M}^T(k)$$
$$+ \mathbf{M}(k)\boldsymbol{\Psi}(k)E\left\{\mathbf{n}_p(k)\mathbf{n}_p^T(k)\right\}\boldsymbol{\Psi}^T(k)\mathbf{M}^T(k)$$
$$+ \mathbf{K}(k)E\left\{\mathbf{n}_m(k+1)\mathbf{n}_m^T(k+1)\right\}\mathbf{K}^T(k)$$

which, using (8.156), (8.157) and (8.158), may be written as the matrix Riccati equation,

$$\begin{aligned}\mathbf{P}(k+1) &= \mathbf{M}(k)\boldsymbol{\Phi}(k)\mathbf{P}(k)\boldsymbol{\Phi}^T(k)\mathbf{M}^T(k) \\ &\quad + \mathbf{M}(k)\boldsymbol{\Psi}(k)\mathbf{Q}(k)\boldsymbol{\Psi}^T(k)\mathbf{M}^T(k) + \mathbf{K}(k)\mathbf{R}(k)\mathbf{K}^T(k) \Rightarrow \\ \mathbf{P}(k+1) &= \mathbf{M}(k)\left[\boldsymbol{\Phi}(k)\mathbf{P}(k)\boldsymbol{\Phi}^T(k) + \boldsymbol{\Psi}(k)\mathbf{Q}(k)\boldsymbol{\Psi}^T(k)\right]\mathbf{M}^T(k) \\ &\quad + \mathbf{K}(k)\mathbf{R}(k)\mathbf{K}^T(k).\end{aligned} \quad (8.163)$$

Let

$$\mathbf{P}^*(k) = \boldsymbol{\Phi}(k)\mathbf{P}(k)\boldsymbol{\Phi}^T(k) + \boldsymbol{\Psi}(k)\mathbf{Q}(k)\boldsymbol{\Psi}^T(k). \quad (8.164)$$

Then (8.163) is simplified to

$$\mathbf{P}(k+1) = \mathbf{M}(k)\mathbf{P}^*(k)\mathbf{M}^T + \mathbf{K}(k)\mathbf{R}(k)\mathbf{K}^T(k). \quad (8.165)$$

Substituting for $\mathbf{M}(k)$ in (8.165) using (8.161) reveals all the terms in $\mathbf{K}(k)$ with respect to which $\mathbf{P}(k+1)$ has to be minimised. Thus,

$$\begin{aligned}\mathbf{P}(k+1) &= \left[\mathbf{I}^{(n)} - \mathbf{K}(k)\mathbf{C}(k)\right]\mathbf{P}^*(k)\left[\mathbf{I}^{(n)} - \mathbf{K}(k)\mathbf{C}(k)\right]^T + \mathbf{K}(k)\mathbf{R}(k)\mathbf{K}^T(k) \\ &= \left[\mathbf{I}^{(n)} - \mathbf{K}(k)\mathbf{C}(k)\right]\mathbf{P}^*(k)\left[\mathbf{I}^{(n)} - \mathbf{C}^T(k)\mathbf{K}^T(k)\right] + \mathbf{K}(k)\mathbf{R}(k)\mathbf{K}^T(k) \\ &= \mathbf{P}^*(k) - \mathbf{K}(k)\mathbf{C}(k)\mathbf{P}^*(k) - \mathbf{P}^*(k)\mathbf{C}^T(k)\mathbf{K}^T(k) \\ &\quad + \mathbf{K}(k)\mathbf{C}(k)\mathbf{P}^*(k)\mathbf{C}^T(k)\mathbf{K}^T(k) + \mathbf{K}(k)\mathbf{R}(k)\mathbf{K}^T(k) \Rightarrow \\ \mathbf{P}(k+1) &= \mathbf{P}^*(k) - \mathbf{K}(k)\mathbf{C}(k)\mathbf{P}^*(k) - \mathbf{P}^*(k)\mathbf{C}^T(k)\mathbf{K}^T(k) \\ &\quad + \mathbf{K}(k)\left[\mathbf{C}\mathbf{P}^*(k)\mathbf{C}^T(k) + \mathbf{R}(k)\right]\mathbf{K}^T(k).\end{aligned} \quad (8.166)$$

In view of (8.158), the scalar function (8.159) to be minimised with respect to $\mathbf{K}(k)$ is the trace of the matrix, $\mathbf{P}(k)$. Thus,

$$J[\mathbf{P}(k)] = \mathrm{tr}[\mathbf{P}(k)]. \qquad (8.167)$$

This enables the matrix differential calculus [6] to be used to minimise $\mathbf{P}(k+1)$ by differentiating the RHS of (8.166) with respect to the matrix, $\mathbf{K}(k)$, and setting the resulting expression to zero. The definition of the derivative of a scalar matrix function, $f[\mathbf{K}]$, with respect to a matrix, $\mathbf{K} \in \Re^{n \times m}$, is

$$\frac{\partial f[\mathbf{K}]}{\partial \mathbf{K}} \triangleq \begin{bmatrix} \frac{\partial f[\mathbf{K}]}{\partial k_{11}} & \cdots & \frac{\partial f[\mathbf{K}]}{\partial k_{1m}} \\ \vdots & & \vdots \\ \frac{\partial f[\mathbf{K}]}{\partial k_{n1}} & \cdots & \frac{\partial f[\mathbf{K}]}{\partial k_{nm}} \end{bmatrix}. \qquad (8.168)$$

This leads to the following relevant mathematical identities. If p is a scalar and not a function of $\mathbf{K}$, then

$$\frac{\partial p}{\partial \mathbf{K}} = \mathbf{0}. \qquad (8.169)$$

If $\mathbf{M}$ is a matrix that is not a function of $\mathbf{K}$, then

$$\frac{\partial}{\partial \mathbf{K}} \mathrm{tr}[\mathbf{KM}] = \mathbf{M}^{\mathrm{T}}, \qquad (8.170)$$

$$\frac{\partial}{\partial \mathbf{K}} \mathrm{tr}\left[\mathbf{MK}^{\mathrm{T}}\right] = \mathbf{M} \qquad (8.171)$$

and

$$\frac{\partial}{\partial \mathbf{K}} \mathrm{tr}\left[\mathbf{KMK}^{\mathrm{T}}\right] = \mathbf{K}\left[\mathbf{M} + \mathbf{M}^{\mathrm{T}}\right]$$
$$= 2\mathbf{KM} \text{ if } \mathbf{M} \text{ is symmetric.} \qquad (8.172)$$

Noting that $\mathbf{P}^*(k)$ given by (8.164) and $\mathbf{CP}^*(k)\mathbf{C}^{\mathrm{T}}(k) + \mathbf{R}(k)$ are symmetrical, applying identities (8.169), (8.170), (8.171) and (8.172) to (8.166) yields

$$\frac{\partial \mathrm{tr}[\mathbf{P}(k+1)]}{\partial \mathbf{K}(k)} = -[\mathbf{C}(k)\mathbf{P}^*(k)]^{\mathrm{T}} - \mathbf{P}^*(k)\mathbf{C}^{\mathrm{T}}(k) + 2\mathbf{K}(k)\left[\mathbf{CP}^*(k)\mathbf{C}^{\mathrm{T}}(k) + \mathbf{R}(k)\right]$$
$$= -\mathbf{P}^{*\mathrm{T}}(k)\mathbf{C}^{\mathrm{T}}(k) - \mathbf{P}^*(k)\mathbf{C}^{\mathrm{T}}(k) + 2\mathbf{K}(k)\left[\mathbf{CP}^*(k)\mathbf{C}^{\mathrm{T}}(k) + \mathbf{R}(k)\right]$$
$$= 2\left[\mathbf{K}(k)\left[\mathbf{C}(k)\mathbf{P}^*(k)\mathbf{C}^{\mathrm{T}}(k) + \mathbf{R}(k)\right] - \mathbf{P}^*(k)\mathbf{C}^{\mathrm{T}}(k)\right]. \qquad (8.173)$$

8.6 The Kalman Filter

The optimal value of $\mathbf{K}(k)$ that minimises $\mathbf{P}(k+1)$ will be called $\mathbf{K}_{opt}$. This makes the RHS of (8.173) zero and therefore

$$\mathbf{K}_{opt}\left[\mathbf{C}(k)\mathbf{P}^*(k)\mathbf{C}^T(k) + \mathbf{R}(k)\right] = \mathbf{P}^*(k)\mathbf{C}^T(k) \tag{8.174}$$

The minimum value of $\mathbf{P}(k+1)$ is then obtained by setting $\mathbf{K}(k) = \mathbf{K}_{opt}$ in (8.166) and substituting for $\mathbf{K}_{opt}\left[\mathbf{C}\mathbf{P}^*(k)\mathbf{C}^T(k) + \mathbf{R}(k)\right]$ using (8.174). Thus,

$$\begin{aligned}\mathbf{P}(k+1) &= \mathbf{P}^*(k) - \mathbf{K}_{opt}\mathbf{C}(k)\mathbf{P}^*(k) - \mathbf{P}^*(k)\mathbf{C}^T(k)\mathbf{K}_{opt}^T + \mathbf{P}^*(k)\mathbf{C}^T(k)\mathbf{K}_{opt}^T \\ &= \left[\mathbf{I}^{(n)} - \mathbf{K}_{opt}\mathbf{C}(k)\right]\mathbf{P}^*(k).\end{aligned} \tag{8.175}$$

Since $\mathbf{P}(k+1)$ is being minimised, the gain that produces the minimum value must occur at the same time and therefore, $\mathbf{K}_{opt}$ is replaced by $\mathbf{K}(k+1)$ in (8.173) and (8.175) to yield, together with (8.164), the following three equations that constitute the Kalman gain algorithm.

$$\mathbf{P}^*(k) = \mathbf{\Phi}(k)\mathbf{P}(k)\mathbf{\Phi}^T(k) + \mathbf{\Psi}(k)\mathbf{Q}(k)\mathbf{\Psi}^T(k) \tag{8.176}$$

$$\mathbf{K}(k+1) = \mathbf{P}^*(k)\mathbf{C}^T(k)\left[\mathbf{C}(k)\mathbf{P}^*(k)\mathbf{C}^T(k) + \mathbf{R}(k)\right]^{-1} \tag{8.177}$$

$$\mathbf{P}(k+1) = \left[\mathbf{I}^{(n)} - \mathbf{K}(k+1)\mathbf{C}(k)\right]\mathbf{P}^*(k). \tag{8.178}$$

8.6.5 The Steady-State Kalman Filter

An LTI plant model in which $\mathbf{\Phi}(k) = \mathbf{\Phi} = \text{const.}$ and $\mathbf{\Psi}(k) = \mathbf{\Psi} = \text{const.}$, for many applications and the noise signal covariance matrices may be assumed constant. In such cases if the Kalman gain algorithm is started with arbitrary initial conditions, $\mathbf{P}(0)$ and $\mathbf{K}(0)$, as $k \to \infty$, $\mathbf{P}(k) \to \mathbf{P} = \text{const.}$ and $\mathbf{K}(k) \to \mathbf{K} = \text{const.}$ where $\mathbf{K}$ is the optimal constant gain matrix that should be used in the discrete Kalman filter state difference Eqs. (8.146), (8.147), (8.148) and (8.149), which become

$$\begin{aligned}\widehat{\mathbf{x}}(k+1|k) &= \mathbf{\Phi}\widehat{\mathbf{x}}(k|k) + \mathbf{\Psi}\mathbf{u}(k) \\ \widehat{\mathbf{y}}(k+1) &= \mathbf{C}\widehat{\mathbf{x}}(k+1|k) \\ \mathbf{e}(k+1) &= \mathbf{y}(k+1) - \widehat{\mathbf{y}}(k+1) \\ \widehat{\mathbf{x}}(k+1|k+1) &= \widehat{\mathbf{x}}(k+1|k) + \mathbf{K}\mathbf{e}(k+1).\end{aligned} \tag{8.179}$$

It is then unnecessary to implement the gain algorithm of (8.176), (8.177) and (8.178), in real time. Instead, it may be run offline until the changes in $\mathbf{K}(k)$ from iteration to iteration are negligible, whence $\mathbf{K}(k) \cong \mathbf{K}$ can be used in (8.179).

8.6.6 The Kalman–Bucy Filter

The continuous equivalent of the discrete Kalman filter is the Kalman–Bucy filter, the derivation of which is given in [7]. The filtering equations are identical in form to the continuous observer of Sect. 8.2 but generalised to cater for multivariable, time-varying plants and is based on the plant model defined by (8.136) and (8.137). It is given by

$$\dot{\widehat{\mathbf{x}}}(t) = \mathbf{A}(t)\widehat{\mathbf{x}}(t) + \mathbf{B}(t)\mathbf{u}(t) + \mathbf{K}(t)\left[\mathbf{y}(t) - \widehat{\mathbf{y}}(t)\right], \quad \widehat{\mathbf{y}}(t) = \mathbf{C}(t)\widehat{\mathbf{x}}(t), \quad (8.180)$$

where $\mathbf{K}(t)$ is given by the Kalman–Bucy gain equations,

$$\begin{aligned}\dot{\mathbf{P}}(t) &= \mathbf{A}(t)\mathbf{P}(t) + \mathbf{P}(t)\mathbf{A}^{\mathrm{T}}(t) + \mathbf{B}(t)\mathbf{Q}(t)\mathbf{B}^{\mathrm{T}}(t) - \mathbf{K}(t)\mathbf{R}(t)\mathbf{K}^{\mathrm{T}}(t), \\ \mathbf{K}(t) &= \mathbf{P}(t)\mathbf{C}^{\mathrm{T}}(t)\mathbf{R}^{-1}(t)\end{aligned} \quad (8.181)$$

where, as for the discrete Kalman filter, $\mathbf{Q}(t)$, $\mathbf{R}(t)$ and $\mathbf{P}(t)$ are the covariance matrices of, respectively, the plant noise, measurement noise and stochastic state estimation error vectors and $\mathbf{K}(t)$ is the model correction loop gain matrix.

As for the discrete Kalman filter, if $\mathbf{Q}$ and $\mathbf{R}$ are constant and an LTI plant model is sufficient, a simpler steady-state Kalman–Bucy filter may be used in which only (8.180) is implemented in real time, the constant gain, $\mathbf{K}$, being determined offline as the steady-state solution to (8.181).

Example 8.9 Kalman–Bucy filter for a double integrator plant
This example demonstrates that the Kalman–Bucy filter for a double integrator plant,

$$\begin{bmatrix}\dot{x}_1 \\ \dot{x}_2\end{bmatrix} = \begin{bmatrix}0 & 1 \\ 0 & 0\end{bmatrix}\begin{bmatrix}x_1 \\ x_2\end{bmatrix} + \begin{bmatrix}0 \\ b\end{bmatrix}(u + n_{\mathrm{p}}), \quad y = \begin{bmatrix}1 & 0\end{bmatrix}\begin{bmatrix}x_1 \\ x_2\end{bmatrix} + n_{\mathrm{m}}, \quad (8.182)$$

where $E\left\{n_{\mathrm{p}}^2\right\} = q = \text{const.}$ and $E\left\{n_{\mathrm{m}}^2\right\} = r = \text{const.}$, has correction loop dynamics with a damping ratio of $\zeta = 1/\sqrt{2}$ for any q or r and a settling time that reduces with the ratio, q/r.

The Kalman–Bucy filter equations are

$$\begin{bmatrix}\dot{\widehat{x}}_1 \\ \dot{\widehat{x}}_2\end{bmatrix} = \begin{bmatrix}0 & 1 \\ 0 & 0\end{bmatrix}\begin{bmatrix}\widehat{x}_1 \\ \widehat{x}_2\end{bmatrix} + \begin{bmatrix}0 \\ b\end{bmatrix}u + \begin{bmatrix}k_1 \\ k_2\end{bmatrix}(y - \widehat{y}), \quad \widehat{y} = \begin{bmatrix}1 & 0\end{bmatrix}\begin{bmatrix}\widehat{x}_1 \\ \widehat{x}_2\end{bmatrix}, \quad (8.183)$$

8.6 The Kalman Filter

where the gain matrix is determined using the solution to the following equations:

$$\begin{bmatrix} \dot{p}_{11} & \dot{p}_{12} \\ \dot{p}_{21} & \dot{p}_{22} \end{bmatrix} = \begin{bmatrix} 0 & 1 \\ 0 & 0 \end{bmatrix} \begin{bmatrix} p_{11} & p_{12} \\ p_{21} & p_{22} \end{bmatrix} + \begin{bmatrix} p_{11} & p_{12} \\ p_{21} & p_{22} \end{bmatrix} \begin{bmatrix} 0 & 0 \\ 1 & 0 \end{bmatrix} + \begin{bmatrix} 0 \\ b \end{bmatrix} q \begin{bmatrix} 0 & b \end{bmatrix}$$

$$- \begin{bmatrix} k_1 \\ k_2 \end{bmatrix} r \begin{bmatrix} k_1 & k_2 \end{bmatrix} \qquad (8.184)$$

and

$$\begin{bmatrix} k_1 \\ k_2 \end{bmatrix} = \begin{bmatrix} p_{11} & p_{12} \\ p_{21} & p_{22} \end{bmatrix} \begin{bmatrix} 1 \\ 0 \end{bmatrix} r^{-1} \Rightarrow \begin{bmatrix} k_1 \\ k_2 \end{bmatrix} = \frac{1}{r} \begin{bmatrix} p_{11} \\ p \end{bmatrix}. \qquad (8.185)$$

Since q and r are constant, the steady-state solution to (8.184) can be found to determine the constant gain matrix for use in (8.183). Usually, an analytical solution is unattainable but is possible in this example. The steady-state solution is the solution to the equation obtained by setting $\dot{\mathbf{P}} = \mathbf{0}$. Thus, from (8.184) and noting that $\mathbf{P}$ is symmetric,

$$\begin{bmatrix} 0 & 0 \\ 0 & 0 \end{bmatrix} = \begin{bmatrix} p & p_{22} \\ 0 & 0 \end{bmatrix} + \begin{bmatrix} p & 0 \\ p_{22} & 0 \end{bmatrix} + \begin{bmatrix} 0 \\ b \end{bmatrix} q \begin{bmatrix} 0 & b \end{bmatrix} - \begin{bmatrix} k_1 \\ k_2 \end{bmatrix} r \begin{bmatrix} k_1 & k_2 \end{bmatrix}, \qquad (8.186)$$

where $p = p_{21} = p_{12}$. Substituting for $\mathbf{K}$ in (8.186) using (8.185) then yields

$$\begin{bmatrix} 2p & p_{22} \\ p_{22} & 0 \end{bmatrix} + \begin{bmatrix} 0 \\ b \end{bmatrix} q \begin{bmatrix} 0 & b \end{bmatrix} = \frac{1}{r} \begin{bmatrix} p_{11} \\ p \end{bmatrix} r \cdot \frac{1}{r} \begin{bmatrix} p_{11} & p \end{bmatrix} \Rightarrow$$

$$\begin{bmatrix} 2p & p_{22} \\ p_{22} & 0 \end{bmatrix} + \begin{bmatrix} 0 & 0 \\ 0 & b^2 q \end{bmatrix} = \frac{1}{r} \begin{bmatrix} p_{11}^2 & p_{11} p \\ p_{11} p & p^2 \end{bmatrix}. \qquad (8.187)$$

The solution is therefore

$$p = b(qr)^{1/2}, \; p_{11} = (2rp)^{1/2} = \sqrt{2} b^{1/2} q^{1/4} r^{3/4},$$
$$p_{22} = \frac{p_{11} p}{r} = \sqrt{2} b^{3/2} q^{3/4} r^{1/4}. \qquad (8.188)$$

Substituting for p and p_{11} in (8.185) using (8.188) gives the following gains:

$$k_1 = \frac{p_{11}}{r} = \sqrt{2} b^{1/2} \left(\frac{q}{r}\right)^{1/4} \quad \text{and} \quad k_2 = \frac{p}{r} = b \left(\frac{q}{r}\right)^{1/2}. \qquad (8.189)$$

This is in keeping with the observation in Sect. 8.5 that increasing the measurement noise variance, r reduces the gains to reduce the sensitivity to measurement noise, while increasing plant noise variance, q, increases the gains to reduce the sensitivity to plant noise. The model correction loop dynamics produced by the gains of (8.189) will now be analysed. From (8.183), the system matrix of the Kalman filter is

$$\begin{bmatrix} 0 & 1 \\ 0 & 0 \end{bmatrix} - \begin{bmatrix} k_1 \\ k_2 \end{bmatrix} \begin{bmatrix} 1 & 0 \end{bmatrix} = \begin{bmatrix} -k_1 & 1 \\ -k_2 & 0 \end{bmatrix}$$

and its characteristic polynomial is therefore

$$\begin{vmatrix} s + k_1 & -1 \\ k_2 & s \end{vmatrix} = s^2 + k_1 s + k_2 = s^2 + 2\zeta\omega_n s + \omega_n^2. \tag{8.190}$$

Then in view of (8.189),

$$\omega_n = b^{1/2} \left(\frac{q}{r}\right)^{1/4} \quad \text{and} \quad \zeta = \frac{k_1}{2\omega_n} = \sqrt{2} b^{1/2} \left(\frac{q}{r}\right)^{1/4} \cdot \frac{1}{2 b^{1/2}} \left(\frac{r}{q}\right)^{1/4} = \frac{1}{\sqrt{2}}. \tag{8.191}$$

The correction loop is therefore well behaved with a constant damping ratio of $\zeta = 0.7071$, independent of the noise levels. The undamped natural frequency, ω_n, increases with the ratio, r/q, meaning that the settling time, $T_s = \frac{1}{\zeta\omega_n}\left[3 - 0.5\ln\left(1-\zeta^2\right)\right]$ (Chap. 4), reduces with this ratio.

References

1. Luenberger DG (1964) Observing the state of a linear system. IEEE Trans Mil Electron 8:74–80
2. Howard RM (2002) Principles of random signal analysis and low noise design: the power spectral density and its applications. Wiley, New York
3. Gradshteyn IS, Ryzhik IM (2007) Table of integrals, series, and products, 7th edn. Academic Press, Elsevier, Burlington Massachusetts, USA
4. Kalman RE (1960) A new approach to linear filtering and prediction problems. Trans ASME J Basic Eng 82:35–45
5. Butcher JC (2003) Numerical methods for ordinary differential equation. Wiley, New York
6. Brookes M (2011) The matrix reference manual, [online]
7. Kalman RE, Bucy RS (1961) New results in linear filtering and prediction theory. Trans ASME J Basic Eng 83:95–108

Chapter 9
Switched and Saturating Control Techniques

9.1 Introduction

9.1.1 Switched Control

Some control actuators are designed for switched rather than continuous operation. Examples are (a) power electronic switches that introduce much smaller electrical power losses than continuously operated power amplifiers [1] and (b) gas-jet thrusters for spacecraft attitude or trajectory control that have to be operated fully on or off to maximise fuel efficiency, measured by the specific impulse, which is the momentum imparted per unit mass of fuel consumed [2]. Each control variable of a switched control system can be implemented by one or more actuators, each of which is turned on or off at every instant of time.

The physical control variable that operates a switched actuator is a logic signal, typically a voltage switching between 0 V and 5 V representing the two logic states, '0' and '1'. The corresponding control variable, $u(t)$, appearing in the plant model used for the control system design is the physical quantity directly producing the changes of state in the plant, such as the voltage applied to an electrical machine or the attitude control torque applied to a spacecraft.

If a single physical actuator is used, then u is switched between the maximum and minimum values, and hence the term *bang–bang control* is often used. If the control is unidirectional, as for a power electronic switch operating an electric heating element of a kiln, where u is switched between a fixed voltage and zero, then the term *on–off control* is used. The term *relay control* may also be found, originating from the early application of electrical relays as an inexpensive form of power amplifier, later to be replaced by power electronic switches. These are forms of *two-level control*. For some applications, such as thruster-based spacecraft attitude control, in which bidirectional control is required but the individual actuators only provide unidirectional forces, two control actuators are provided.

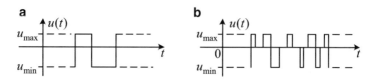

Fig. 9.1 Switched control sequences. (**a**) Two-level control. (**b**) Three-level control

This allows *three* control levels, i.e. u_{max} (actuator 1 'on'), u_{min} (actuator 2 'on'), where usually $u_{min} = -u_{max}$, and 0 (both actuators 'off'). This will be termed *three-level control*. The fourth combined actuator state (both actuators 'on') that could produce a fourth control level, $u_{max} + u_{min}$, if $|u_{max}| \neq |u_{min}|$ is not usually employed in practice as this would be highly inefficient.

Most systems employ only one or two actuators per control variable, but more than two is possible. An example is a multilevel power electronic converter [3] that produces a voltage that follows, as closely as possible, a continuously time-varying voltage demand by switching between a finite number of DC power supply voltages.

The term *switched control* will be used to describe all of the forms of control referred to above. The term *discontinuous* control may also be found in some literature. Typical control sequences for two- and three- level switched control are illustrated in Fig. 9.1.

Switched control techniques fall into one of two basic categories: pulse modulation and switched state feedback. Pulse modulation is a technique whereby the control variable is rapidly switched between the admissible values to yield a short-term mean value that follows a continuous demand from a controller that could be used with continuous actuators. In this way, similar performance to that of continuous actuators can be attained. Section 9.2 is devoted to this. The remainder of this chapter is devoted to switched state feedback control, which entails switching the actuator to the required levels as determined by the current plant state. An example is time-optimal control which minimises the time taken to reach the required state.

Optimal control theory has a large part to play in switched and saturating control but leads to open-loop solutions that are not generally robust with respect to external disturbances and plant modelling errors. A small subset of these solutions, however, may be converted to nonlinear state feedback control laws [4], thereby offering a degree of robustness, some of which will be presented in this chapter.

It is important to note that a linear plant that is controllable according to the criterion given in Chap. 2 may only really be controllable over a finite region of the state space in practice due to the control saturation constraints or if switched control actuators are employed. This controllability region is also disturbance dependent and even *vanishes* if the external disturbance referred to the control input is $d \notin (-u_{max}, +u_{max})$, as then the controller could not counteract it. This, however, is not usually an issue since the hardware should be designed or selected such that a finite controllability region exists that includes the operational envelope of the particular

9.1 Introduction

application. It is important to note that this restriction also applies to control systems employing continuous actuators as these are always subject to upper and lower saturation limits.

Most control systems spend the greater proportion of their lifetimes maintaining the plant state close to that desired, within specified limits. In the case of switched control systems, the actuator on and off times have lower limits imposed by physical constraints, resulting in the real state oscillating about the desired state due to the actuators being switched at a finite frequency. Most systems are designed such that this switching has a regular pattern and the state trajectory is a closed, repeated path. This behaviour is called limit cycling. Several examples of this will be studied in the following sections.

9.1.2 Saturating Control

Finite control saturation limits are present in every control system employing continuous actuators. These are represented in Fig. 9.2 by a saturation element.

Most of these systems are designed to operate continuously within the saturation limits, but it is necessary to ensure stability when non-intended saturation occurs. Some of the material presented in this chapter will be useful in dealing with this. On the other hand, it may be desirable to design the control system to operate intentionally with control saturation for relatively large changes of state. This is especially true for the time-optimal control addressed in Sects. 9.6, 9.7 and 9.8 and also the minimum energy control of Chap. 11.

A layman's example of time-optimal control is the action of the driver of an automobile travelling from position 'a' at rest to position 'b' at rest in the shortest possible time. The required strategy is to apply the maximum acceleration up to an intermediate position between 'a' and 'b' and then apply the brake for maximum deceleration to bring the vehicle to rest at position 'b'. This form of control with a single switch part of the way through the process may be observed in the time-optimal control of second-order plants covered in Sect. 9.8, but more than one switch is required for the time-optimal control of higher-order plants.

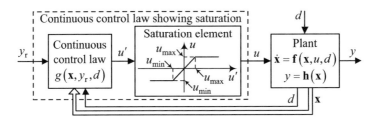

Fig. 9.2 General SISO continuous feedback control system showing the control saturation

Intentionally saturating controllers such as time-optimal controllers will limit cycle in the vicinity of the desired state if operation in the switched mode is continued. This is avoided by changing smoothly from the switched mode to a continuous mode as the desired state is approached. Means for achieving this are developed in Sects. 9.7 and 9.8.

9.2 Pulse Modulation for Use with Continuous Controllers

9.2.1 Basic Concept

Where switched actuators are employed and a closed-loop optimal control system is either unnecessary or unattainable, it is possible to achieve performances similar to the continuous control systems of Chaps. 4, 5 and 7 or the discrete control systems of Chap. 6 by using the controllers suited primarily for continuous actuators to generate a demanded control, $u'(t)$, and then employing a pulse modulator that switches the applied control, $u(t)$, between the allowed control levels so that the short-term mean value, $\bar{u}$, approximates $u'(t)$, meaning

$$\bar{u}(q) = \frac{1}{t_{sq+2} - t_{sq}} \int_{t_{sq}}^{t_{sq+2}} u(t) \mathrm{d}t \cong u'\left(\bar{t}_q\right), \tag{9.1}$$

where t_{sq} is the time of switch q in $u_i(t)$ and $\bar{t}_q = (t_{sq+2} + t_{sq})/2$, as shown in Fig. 9.3.

Let the set of control levels for multilevel switched control be $(u_{s1}, u_{s2}, \ldots, u_{sL})$, where $u_{si} > u_{si+1}$, $i = 1, 2, \ldots, L-1$. The maximum and minimum levels are therefore, respectively, $u_{max} = u_{s1}$ and $u_{min} = u_{sL}$. Then at every instant of time, u is switched between the two levels, $(u_a, u_b) = (u_{sk}, u_{sk+1})$ such that $u_{sk} \geq u' \geq u_{sk+1}$. For two-level control, this reduces to $(u_a, u_b) = (u_{max}, u_{min})$. For three-level control, very often $(u_{s1}, u_{s2}, u_{s3}) = (u_{max}, 0, u_{min})$.

The 'instantaneous' frequency of the waveform of $u(t)$, referred to as the modulation frequency, is

$$f_{mq} = 1/\left(t_{sq+2} - t_{sq}\right) \tag{9.2}$$

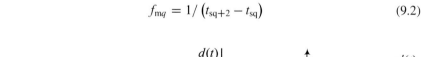

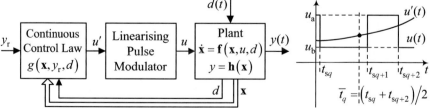

Fig. 9.3 Block diagram of pseudo-linear control system for SISO plant

9.2 Pulse Modulation for Use with Continuous Controllers

and the corresponding modulation period is

$$t_{mq} = t_{sq+2} - t_{sq}. \tag{9.3}$$

Thus, the modulation period contains a single switch at time, t_{sq+1}, such that the mean value, $\bar{u}(q)$, of u over this period is approximately equal to the demanded control, $u'(t_q)$, at the midpoint defined by $t_q = (t_{sq+2} + t_{sq})/2$. The control signal, $u(t)$, can then be considered to comprise two components, $u'(t)$ and $u_s(t)$. Thus,

$$u(t) = u'(t) + u_s(t), \tag{9.4}$$

where $u_s(t)$ is an oscillatory component with zero mean value. According to Fourier analysis, $u_s(t)$ can be further divided into a fundamental component at a frequency of f_{mq} and harmonic components at higher frequencies. This concept applies even with f_{mq} time varying, as it can be regarded as a quasi-periodic function with similarly time-varying harmonic frequencies. The demanded control, $u'(t)$, is not a periodic function but has a Fourier transform that indicates a continuous spectrum in contrast to the multiline spectrum of a periodic function. Since $u'(t)$ varies more slowly than $u_s(t)$, it is dominated by components at frequencies lower than f_{mq}. It follows that since any controlled plant has low-pass filtering properties, provided f_{mq} is higher than the cut-off frequency of the plant, the response of the plant output, $y(t)$, to $u_s(t)$ will be negligible compared with its response to $u'(t)$. Under these circumstances, the closed-loop behaviour of the switched control system will be almost the same as that of the equivalent continuous system employing a continuous actuator, obtained by removing the pulse modulator and setting $u = u'$ in Fig. 9.3.

The term 'linearising pulse modulator' is used to emphasise its purpose in providing a linear relationship between the short-term mean value of $u(t)$ and the demanded $u'(t)$. It is important to note, however, that this linearity can only exist for $u' \in (u_{min}, u_{max})$. The modulator will saturate if $u' \notin (u_{min}, u_{max})$.

9.2.2 Implementation

9.2.2.1 Basic Methods

Some methods of implementation of pulse modulation originate in the era of discrete-component analogue electronic circuits. In some methods, the modulation frequency, defined by (9.2), varies with the magnitude of the modulating signal, u'. The commonly used two-level and three-level methods based on comparators and saw-tooth waveform generators, shown in Fig. 9.4, are often implemented by electronic hardware but can easily be software implemented. They have constant modulation frequencies, enabling the z-domain to be useful in the controller design if the plant is linear and the modulation frequency has to be too small for the criterion of applicability of continuous control theory (Chap. 6) to be met.

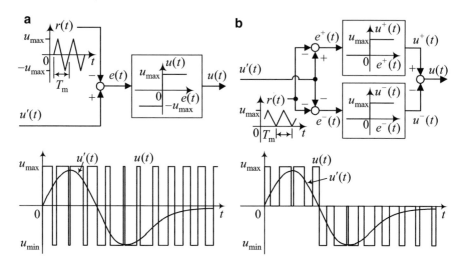

Fig. 9.4 Triangular wave generator-based pulse modulation for switched control. (**a**) Two-level. (**b**) Three-level

It should be noted that the modulation period, T_m, is equivalent to the sampling period, h. In Fig. 9.4a, a triangular wave, $r(t)$, at a constant modulation period of T_m and peak values of $\pm u_{max}$ is subtracted from $u'(t)$ to form an error signal, $e(t)$.

Then, $u(t) = u_{max}\,\text{sgn}\,[e(t)]$. Let t^+ and t^- be, respectively, the proportions of time for which $e(t) > 0$ and $e(t) < 0$ with u' constant. Then due to the linear segments of $r(t)$, $t^+ - t^- \propto u'$ and therefore the mean value of $u(t)$ taken over any period of duration, T_m, is $\bar{u} = u'$, but provided $|u'| < u_{max}$. If $|u'| \geq u_{max}$, then u saturates at a constant value of $u = u_{max}\,\text{sgn}\,(u')$. It must be remembered, however, that u' is often time varying, but $\bar{u} \cong u'$ provided

$$\left|\frac{du'(t)}{dt}\right| \ll \left|\frac{dr(t)}{dt}\right| = \frac{4u_{max}}{T_m}. \tag{9.5}$$

In Fig. 9.4b, separate control signals, $u^+(t)$ and $u^-(t)$, are generated to drive two actuators, one dedicated to producing positive control and the other dedicated to negative control. A pertinent example is spacecraft attitude control using two on-off thrusters per control axis. To achieve this, two error signals, $e^+(t)$ and $e^-(t)$, are generated by subtracting a unidirectional triangular wave at a constant modulation period of T_m and peak values of 0 and u_{max} from, respectively, $u'(t)$ and $-u'(t)$. For $u'(t) > 0$, $e^-(t)$ remains negative and $e^+(t)$ alternates in sign. Similarly for $u'(t) < 0$, $e^+(t)$ remains negative and $e^-(t)$ alternates in sign. Then the unidirectional transfer characteristics between $e^+(t)$ and $u^+(t)$ and between $e^-(t)$ and $u^-(t)$, give

$$u^+(t) = \tfrac{1}{2}\{1 + \text{sgn}\,[e^+(t)]\} \quad \text{and} \quad u^-(t) = \tfrac{1}{2}\{1 + \text{sgn}\,[e^-(t)]\} \tag{9.6}$$

9.2 Pulse Modulation for Use with Continuous Controllers

This yields modulation similar to that of Fig. 9.4a but positive-going control signals to each actuator. The subtraction of $u^-(t)$ from $u^+(t)$ to yield the net control, $u(t)$, would be carried out in a simulation but, of course, does not appear in the control algorithm of a real implementation as the actuator placement and the physical plant perform this function.

The following algorithm is an alternative that caters for both two- and three-level switched control and also has a constant modulation frequency. Let u_a and u_b be the currently selected control levels with $u_a > u_b$. In the following, h is the sampling time and k is the sample number. There will be one control switch between the updates of u'_k. Let τ_{ak} be the period during which $u = u_a$. Then the period for which $u = u_b$ is $\tau_{bk} = h - \tau_{ak}$. The mean value of u during the kth period of duration, h, is set equal to u'_k for $u' \in (u_a, u_b)$; otherwise control saturation occurs. Thus,

$$\begin{cases} \text{For two level control, } u_a = u_{\max} \text{ and } u_b = u_{\min}. \\ \text{For three level control, } (u_a, u_b) = \begin{cases} (u_{\max}, 0) & \text{if } u'_k > 0 \\ (0, u_{\min}) & \text{if } u'_k < 0 \end{cases} \\ \text{If } u' \in (u_a, u_b),\ u'_k = \frac{u_a \tau_{ak} + u_b \tau_{bk}}{h} = \frac{u_a \tau_{ak} + u_b(h - \tau_{ak})}{h} \Rightarrow \\ \qquad \tau_{ak} = \left(\frac{u'_k - u_b}{u_a - u_b}\right) h \text{ and } \tau_{bk} = \left(\frac{u_a - u'_k}{u_a - u_b}\right) h \\ \text{If } u' \notin (u_a, u_b), \text{ then} \\ \text{If } u'_k \geq u_a,\ \tau_{ak} = h \text{ and } \tau_{bk} = 0 \text{ or If } u'_k \leq u_b,\ \tau_{ak} = 0 \text{ and } \tau_{bk} = h \end{cases}$$
(9.7)

In order to implement this algorithm accurately, the digital processor operates with many iterations (such as 100) in real time during the period of h seconds, to ensure a sufficiently high resolution for τ_{ak} and τ_{bk}. This also applies to the digital implementation of the triangular waveform-based schemes shown in Fig. 9.4.

9.2.2.2 Minimum Switching Period

Actuators, like all physical systems, cannot respond in zero time to step demands such as occur in switched control. It is therefore necessary to design a system such that the period between two consecutive control switches, referred to as the switching period, cannot fall below a minimum value, $T_{\min}$, to allow the actuator output to nearly reach a constant steady-state value following a switch. Another reason for this restriction is the need to maximise efficiency. Power electronic switches, for example, ideally present an electrical resistance that switches between zero for 'on' and infinity for 'off'. During the transition from one state to the other, there is electrical power dissipation in the finite resistance of the switching semiconductor. It is therefore desirable to maximise, as far as practicable, the period between these transitions. Another example is on–off thrusters for spacecraft control, in which the specific impulse is known to fall significantly during switching state transitions. The value of $T_{\min}$ depends, of course, on the application.

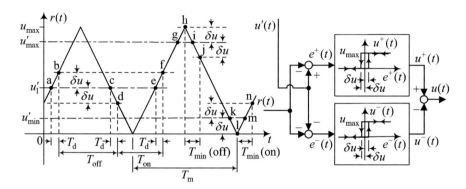

Fig. 9.5 Implementation of minimum switching time using hysteresis switching elements

For power electronics, it is of the order of microseconds, and for spacecraft thrusters, it is of the order of milliseconds.

The minimum switching period can be implemented in the schemes of Fig. 9.4 by replacing the switching elements with hysteresis elements that delay the switching events, as shown in Fig. 9.5 for the three-state switching of Fig. 9.4b. In this system, there are upper and lower limits, u'_{max} and u'_{min}, of the demanded average control level, beyond which no switching can occur and u is held constant at u_{max} if $u' > u'_{max}$ or 0 if $u' < u'_{min}$. First consider a constant intermediate value of u', such as u'_1 in Fig. 9.5. Without the hysteresis, the actuator would be turned off between points 'a' and 'c' and turned on between points 'c' and 'e'. With the hysteresis, the actuator would be turned off between points 'b' and 'd' and turned on between points 'd' and 'f'.

Thus, the switching times, T_{off} and T_{on}, remain unaltered but would be delayed by a constant time, T_d. According to the geometry of the triangular waveform of $r(t)$,

$$\frac{T_d}{\delta u} = \frac{T_m}{2u_{max}} \Rightarrow T_d = \frac{\delta u}{2u_{max}} \cdot T_m \qquad (9.8)$$

Now consider $u' = u'_{max} - \varepsilon$, where ε is infinitesimal and positive. Without the hysteresis, the actuator would turn off at point 'g' for $T_{min} = 2T_d$ seconds and turn on at point 'i'. With the hysteresis, the actuator would be turned off at point 'h' and turned on at point 'j'. Thus, the switching time in the 'off' state would remain unaltered at the minimum value of T_{min} but be delayed by T_d seconds. Similarly, if $u' = u'_{max} + \varepsilon$ without the hysteresis, the actuator would turn on at point 'k' for

$T_{min} = 2T_d$ seconds and turn off at point' 'm'. With the hysteresis, the actuator would be turned on at point 'l' and turned off at point 'n'. Thus, the switching time in the 'on' state would remain unaltered at the minimum value of T_{min} but be delayed by T_d seconds. Since $T_{min} = 2T_d$, in view of (9.8), the hysteresis limit required to realise a given value of T_{min} is given by

9.2 Pulse Modulation for Use with Continuous Controllers

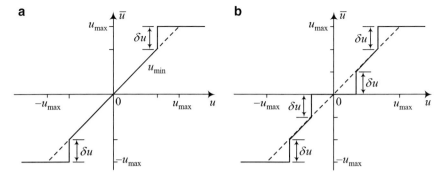

Fig. 9.6 Transfer characteristics of linearising pulse modulators. (**a**) Two-level control. (**b**) Three-level control

$$\delta u = \frac{T_{\min}}{T_{\mathrm{m}}} u_{\max}. \tag{9.9}$$

Figure 9.6 shows the transfer characteristics between the short-term mean value, $\bar{u}$, and u' for different constant values of u' with and without the hysteresis-based minimum switching time scheme described above. The dotted characteristics are without the hysteresis and the solid ones are with the hysteresis. It is important to note that for clarity of presentation, in Figs. 9.5 and 9.6, δu is shown much larger in proportion to $u_{\max}$ than it would be in practice. The transfer characteristics would then appear almost the same as that of a piecewise linear element with saturation limits of $\pm u_{\max}$ and unity slope for $u' \in (-u_{\max}, u_{\max})$.

The introduction of the minimum switching time scheme therefore does not have a significant impact on the performance of the control system. In fact the dead space of width, $2\delta u$, in the transfer characteristic for the three-level control can be beneficial in reducing the frequency of actuator operations once the controller has brought the plant close to the desired state. This will be discussed further in Sect. 9.7.3 and Appendix A9.

Example 9.1 Thruster-based attitude control of flexible spacecraft via pulse modulation

The purpose of this example is to design a linear control system and then carry out simulations, first with continuous control and then with switched control and a linearising pulse modulator to demonstrate that similar performance may be attained.

The plant is one control axis of a spacecraft comprising a rigid centre body with moment of inertia, J_1, and a flexible appendage, with one significant vibration mode, represented by a second mass with moment of inertia, J_2, coupled to the centre body via a torsion spring with stiffness, K_s (Chap. 2). The plant transfer function is then

$$\frac{y(s)}{u(s)} = \frac{1}{J_1 s^2} \cdot \frac{s^2 + v^2}{s^2 + \omega^2}, \tag{9.10}$$

where $v^2 = \frac{K_s}{J_2}$ and $\omega^2 = v^2\left(1 + \frac{J_2}{J_1}\right)$. Here, ω is the free natural frequency of the vibration mode and v is the corresponding encastre natural frequency, i.e. the frequency at which the flexible appendage would vibrate if the centre body were to be held fixed with respect to inertial space. The measurement variable, y, is scaled in the control computer to be numerically equal to the attitude angle in radians. The control variable, u, is scaled in the control computer to be numerically equal to the control torque in Newton metres. The control torque is provided by a pair of on–off thrusters yielding three torque levels of $u = +u_{\max}, -u_{\max}$ or 0. The spacecraft parameters are $v = 0.1$ rad/s, $J_1 = 100$ kg m², $J_2 = 300$ kg m² and $u_{\max} = 1$ Nm.

In this application, the complex conjugate plant zeros can cause an overshoot of the step response, followed by an undershoot, if the closed-loop poles are placed on the real axis of the s-plane according to the settling time formula with too small a pole-to-zero dominance ratio (Chap. 2). Attempting to minimise the settling time exacerbates this situation because the closed-loop poles can be much larger in magnitude than the plant zeros. Hence, a zero pre-compensator is advisable. Setting this to precisely cancel the zeros at $\pm jv$ will yield an ideal monotonically increasing step response, but no active damping of the vibration mode will be provided. The pre-compensator poles are therefore placed to the left of the zeros in the s-plane by an amount that achieves sufficient damping of the vibration mode without introducing undesirable behaviour of the centre-body attitude step response. This modal damping may be assessed by monitoring the angular displacement, β, of the second mass with respect to the first mass. The transfer function via this (Chap. 2) is

$$\frac{\beta(s)}{u(s)} = \frac{-1/J_1}{s^2 + \omega^2}. \tag{9.11}$$

The nominal settling time, T_s, (5 % criterion) is to be minimised within the constraint that the system is to operate without saturation for step reference attitude angle inputs up to $\pm \pi$ rad, and the centre-body attitude step response is monotonic.

The linear controller chosen for this example is a polynomial controller of minimal order. To design this (Chap. 4), (8.3) is first expressed in the standard form,

$$\frac{y(s)}{u(s)} = \frac{b_2 s^2 + b_0}{s^4 + a_2 s^2} \tag{9.12}$$

where $b_0 = v^2/J_1$, $b_2 = 1/J_1$ and $a_2 = \omega^2$. In these terms, (9.11) becomes

$$\frac{\beta(s)}{u(s)} = \frac{-b_2}{s^2 + a_2} \tag{9.13}$$

9.2 Pulse Modulation for Use with Continuous Controllers

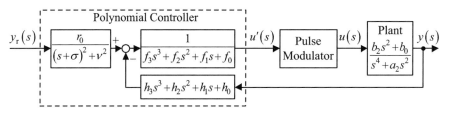

Fig. 9.7 Switched polynomial attitude control of flexible spacecraft for one control axis

Figure 9.7 shows the transfer function block diagram of the control system.

The controller parameters are determined following the procedure in Chap. 4 as if using continuous actuators with $u = u'$. First the desired characteristic polynomial is determined using the 5 % settling time formula. Thus,

$$s^7 + d_6 s^6 + d_5 s^5 + d_4 s^4 + d_3 s^3 + d_2 s^2 + d_1 s + d_0$$
$$= (s+p)^7 = s^7 + 7ps^6 + 21p^2 s^5 + 35p^3 s^4 + 35p^4 s^3 + 21p^5 s^2 + 7p^6 s + p^7. \quad (9.14)$$

where $p = 1.5\,(1+7)/T_s = 12/T_s$. The pole assignment matrix equation is

$$\begin{bmatrix} 1 & 0 & 0 & 0 & 0 & 0 & 0 & 0 \\ 0 & 1 & 0 & 0 & 0 & 0 & 0 & 0 \\ a_2 & 0 & 1 & 0 & b_2 & 0 & 0 & 0 \\ 0 & a_2 & 0 & 1 & 0 & b_2 & 0 & 0 \\ 0 & 0 & a_2 & 0 & b_0 & 0 & b_2 & 0 \\ 0 & 0 & 0 & a_2 & 0 & b_0 & 0 & b_2 \\ 0 & 0 & 0 & 0 & 0 & 0 & b_0 & 0 \\ 0 & 0 & 0 & 0 & 0 & 0 & 0 & b_0 \end{bmatrix} \begin{bmatrix} f_3 \\ f_2 \\ f_1 \\ f_0 \\ h_3 \\ h_2 \\ h_1 \\ h_0 \end{bmatrix} = \begin{bmatrix} 1 \\ d_6 \\ d_5 \\ d_4 \\ d_3 \\ d_2 \\ d_1 \\ d_0 \end{bmatrix}. \quad (9.15)$$

Solving this for the controller parameters then yields:

$$f_3 = 1, \quad f_2 = d_6 \quad (9.16)$$

$$a_2 f_3 + f_1 + b_2 h_3 = d_5 \quad (9.17)$$

$$a_2 f_2 + f_0 + b_2 h_2 = d_4 \quad (9.18)$$

$$a_2 f_1 + b_0 h_3 + b_2 h_1 = d_3 \quad (9.19)$$

$$a_2 f_0 + b_0 h_2 + b_2 h_0 = d_2 \quad (9.20)$$

$$b_0 h_1 = d_1 \Rightarrow \underline{h_1 = d_1/b_0} \tag{9.21}$$

$$b_0 h_0 = d_0 \Rightarrow \underline{h_0 = d_0/b_0} \tag{9.22}$$

Next, all known terms in (9.17), (9.18), (9.19) and (9.20) are moved to the RHS:

$$f_1 + b_2 h_3 = d_5 - a_2 f_3 = c_5 \tag{9.23}$$

$$f_0 + b_2 h_2 = d_4 - a_2 f_2 = c_4 \tag{9.24}$$

$$a_2 f_1 + b_0 h_3 = d_3 - b_2 h_1 = c_3 \tag{9.25}$$

$$a_2 f_0 + b_0 h_2 = d_2 - b_2 h_0 = c_2 \tag{9.26}$$

Then $a_2 \times$ (9.23) $-$ (9.25) gives

$$(a_2 b_2 - b_0) h_3 = a_2 c_5 - c_3 \Rightarrow \underline{h_3 = \frac{a_2 c_5 - c_3}{a_2 b_2 - b_0}} \tag{9.27}$$

Similarly, $a_2 \times$ (9.24) $-$ (9.26) gives

$$(a_2 b_2 - b_0) h_2 = a_2 c_4 - c_2 \Rightarrow \underline{h_2 = \frac{a_2 c_4 - c_2}{a_2 b_2 - b_0}} \tag{9.28}$$

Then (9.23) and (9.24) give

$$\underline{f_1 = c_5 - b_2 h_3} \tag{9.29}$$

$$\underline{f_0 = c_4 - b_2 h_2}. \tag{9.30}$$

Finally, to achieve unity closed-loop DC gain, the pre-compensator gain is set to

$$\underline{r_0 = \frac{v^2 (a_0 f_0 + b_0 h_0)}{b_0}} \tag{9.31}$$

The pulse modulator of Fig. 9.5 is used with $T_{\min} = 5$ ms. Figure 9.8 shows the simulation results. Figure 9.8a shows a simulation of the switched control system, and Fig. 9.8b shows a simulation of the continuous system as a benchmark with which it may be compared.

The benchmark system is set up as follows. For a step reference attitude input of π rad, T_s and the real part magnitude, σ, of the zero pre-compensator are adjusted until $u(t)$ just peaks at $u_{\max}$ and the vibration mode is adequately damped, observed by monitoring $\beta(t)$, while the step response of $y(t)$ is monotonic, as required. The parameter values found to achieve this are $T_s = 85$ s and $\sigma = 0.06$. Figure 9.8a

9.2 Pulse Modulation for Use with Continuous Controllers

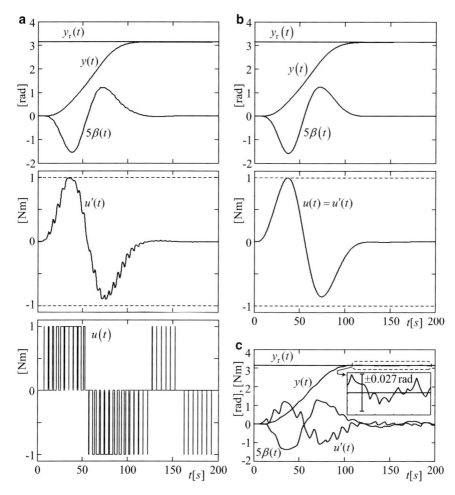

Fig. 9.8 Comparison of step responses with switched and continuous control. (**a**) Switched control: $T_m = 5$ s. (**b**) Continuous control (benchmark): $u = u'$. (**c**) Switched control: $T_m = 10$ s

shows the results obtained with the equivalent switched control system. The pulse modulation period set to $T_m = 5$ s. In fact, setting T_m to the maximum value according to the criterion of applicability of continuous-time control theory given in Chap. 6 yields $T_{m\ min} = 1/(5p) = 60/T_s \cong 0.706$ s (since $p = 12/T_s$ in this example), which is rather small for this application. Using the more practicable value of $T_m = 5$ s (giving a thruster operation every 5 s) does not significantly deteriorate the performance as far as $y(t)$ is concerned. The ripples visible in $u(t)$ and (to a lesser extent) in $\beta(t)$ of Fig. 9.8a are due to the thruster impulses. In Fig. 9.8c, T_m is increased to 10 s, which gives rise to significant thruster-induced variations in $u(t)$ and $\beta(t)$. The overall shape of $y(t)$ is not affected, but the pointing error limits are increased to ± 0.027 rad $\equiv \pm 1.55$ deg, as shown in the inset, which would be too great for many spacecraft. It is important to note that certain modulation periods will excite an oscillatory mode and reduce the control system accuracy. High-precision applications, such as space telescopes demanding pointing accuracies in the region of 0.01 arcseconds $\equiv 2.8 \times 10^{-6}$ deg, would need a shorter modulation period that should also not be an integral multiple of the oscillatory mode period, to minimise the modal excitation.

9.3 Switched State Feedback Control: Basic Concepts

Like the continuous state feedback control of Chaps. 5 and 7, switched state feedback control benefits from the fact that the plant state contains all the information about its dynamic behaviour and therefore enables effective control to be achieved with a suitable control law. At every instant of time, however, the choice of the control is limited to one of a finite set of values, often just two, denoted here as u_{min} and u_{max}. Despite this restriction, continuous control of the plant state is achieved through free choice of the switching times.

Attention is restricted to two- and three-level switched control systems as most practical applications fall into either of these categories. Furthermore, three-level switched control can be regarded as two-level switched control having the control level pair (u_a, u_b) which, at appropriate times, is changed between the three useful combinations allowed by the two actuators, i.e. (u_{max}, u_{min}), $(u_{max}, 0)$ or $(0, u_{min})$. Hence, much of the material that follows concentrates on two-level switched control.

In switched state feedback control, the switching times are not chosen directly but determined by means of a switching boundary, defined by

$$S(\mathbf{x}, y_r, d) = 0, \quad (9.32)$$

where $\mathbf{x}$ is the plant state, y_r is the reference input and d, which is optional, is the external disturbance. This boundary divides the n dimensional state space into two regions. One is designated the 'p' region in which the control is chosen to be $u = u_{max}$ and the other is designated the 'n' region in which $u = u_{min}$. In many cases, $u_{max} > 0$ and $u_{min} < 0$, but in some, $u_{min} = 0$ or both control levels are positive. Switching boundary (9.32) defines the state feedback control law

$$u = \tfrac{1}{2} \{ u_{max} [1 + \text{sgn}[S(\mathbf{x}, y_r, d)]] + u_{min} [1 - \text{sgn}[S(\mathbf{x}, y_r, d)]] \}, \quad (9.33)$$

If $u_{min} = -u_{max}$, this reduces to

$$u = u_{max} \, \text{sgn}[S(\mathbf{x}, y_r, d)]. \quad (9.34)$$

Here, sgn(S) is the signum function defined as

9.3 Switched State Feedback Control: Basic Concepts

$$\text{sgn}(S) \triangleq \begin{cases} +1 \text{ for } S \geq 0 \\ -1 \text{ for } S < 0 \end{cases}. \tag{9.35}$$

It is necessary to use such an asymmetric definition to ensure that u can only be u_{min} or u_{max}. If the symmetric definition, $\text{sgn}(S) \triangleq \{+1 \text{ for } S > 0, 0 \text{ for } S = 0, -1 \text{ for } S < 0\}$, were to be used, (9.33) would return $u = \frac{1}{2}(u_{max} + u_{min})$ for $S = 0$, which is not realisable in the physical system. The alternative asymmetric definition, $\text{sgn}(S) \triangleq \{+1 \text{ for } S > 0, -1 \text{ for } S \leq 0\}$, could be used instead of (9.35) and would yield practically the same performance in a real control system since the occurrence of $S = 0$ is rare.

The function, $S(\mathbf{x}, y_r, d)$, is called the *switching function* as it determines where in the state space the switching occurs. The control switches whenever $S(\mathbf{x}, y_r, d)$ changes sign. Sometimes the switching function is shown simply as $S(\mathbf{x})$, but y_r is included here as most real control systems have to respond to a reference input and d is included as some applications require disturbance estimation and compensation.

The switching boundary (9.32) is an $n-1$- dimensional hypersurface in the n-dimensional state space. This may be demonstrated by first writing (9.32) as

$$S(x_1, x_2, \ldots, x_n, y_r, d) = 0. \tag{9.36}$$

Then, since any one of the state variables may be written as a function of the remaining $n-1$ state variables, such as

$$x_n = f(x_1, x_2, \ldots, x_{n-1}, y_r, d), \tag{9.37}$$

then the switching boundary is of dimension $n-1$ in the n- dimensional state space.

It is usual to use terms such as 'switching boundary' and 'boundary layer' in connection with switched and saturating control of plants of second order or greater, but it is also done in this book for first-order plants to aid understanding of the theory. So first consider the switched state feedback control of the general first-order plant,

$$\dot{x}_1 = f_1(x_1, u, d), \quad y = h(x_1). \tag{9.38}$$

The external disturbance will not be included in the switching function as it is not needed for the switched control of first-order plants. The control system block diagram is shown in Fig. 9.9. The plant is shown in the standard state-space form to emphasise that the plant state, in this case just x_1, is fed back to the control law.

It should be noted, however, that for a first-order plant, the output equation, $y = h(x_1)$, is a trivial state transformation. Hence the measured output, y, is an alternative state variable and is sometimes fed back directly. The control system design could therefore be based on the following simplified plant model.

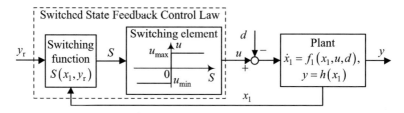

Fig. 9.9 Switched state feedback control of a first-order plant

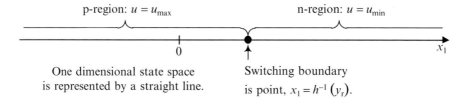

Fig. 9.10 Representation of switching boundary in state space of first-order plant

$$\dot{y} = f(y, u - d). \tag{9.39}$$

Without loss of generality, the external disturbance, d, is referred to the control input, u. The discussion continues, however, using the plant representation (9.38) as this is needed for higher-order plants. As shown in Fig. 9.9, the switched state feedback control law comprises a) the switching function and b) the switching element, which provide signals that turn the actuators on and off. For a first-order plant, since $\dot{x}_1$ and hence $\dot{y}$ is algebraically related to u, y is driven towards y_r just by switching to the value of u that causes the sign of $\dot{y}$ to oppose the error, $y - y_r$. The switching function achieving this is

$$S(x_1, y_r) = y_r - h(x_1). \tag{9.40}$$

It is evident by inspection of the switching element transfer characteristic of Fig. 9.9 that a switch occurs whenever

$$S(x_1, y_r) = 0, \tag{9.41}$$

which defines the switching boundary. The state space is only one dimensional and the switching boundary is of dimension one less than this, i.e. zero dimensional and is just a single point in the state space, as illustrated in Fig. 9.10.

Examples of first-order switched control systems are given in Sect. 9.7.

Figure 9.11 shows the general block diagram for plants of arbitrary order, n. It is necessary for $\mathbf{x} \in \Re^n$, and sometimes, d, to be accessible to achieve satisfactory control. A state estimator or observer (Chap. 8) is needed to provide estimates of

9.4 Switching Function Sign Convention

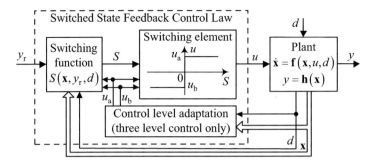

Fig. 9.11 General block diagram of SISO switched state feedback control system

these variables in a real system but, for clarity of explanation, is not shown in most of the systems discussed in this chapter.

Three-level control could be implemented by switching the pair of control levels (u_a, u_b) between the three useful combinations allowed by the two actuators, i.e. ($u_{\max}$, $u_{\min}$), ($u_{\max}$, 0) or (0, $u_{\min}$) as determined by the optional control level adaptation block. The external disturbance, d, is shown as an argument of the switching function for direct disturbance compensation but this is not needed for applications in which state feedback alone provides sufficient compensation.

The task of the control system designer is to determine the switching boundary that achieves the basic objective of driving the control error,

$$\varepsilon = y - y_r, \tag{9.42}$$

to zero and yields an acceptable dynamical closed-loop behaviour. Sections 9.7, 9.8 and 9.9 contain several examples.

9.4 Switching Function Sign Convention

For a given switching boundary, the switching function and corresponding switching boundary equation is not unique. Several different solutions exist. Consider, for example, the double integrator time-optimal switching boundary

$$S(x_{1e}, x_{2e}) = -\left(x_{1e} + \frac{1}{2b_0 u_{\max}} x_{2e} |x_{2e}|\right) = 0 \tag{9.43}$$

where $x_{1e} = x_1 - y_r$ and $x_{2e} = x_2$ are the error state coordinates; $x_1 = y$ is the controlled output, $x_2 = \dot{x}_1$; and b_0 is a constant plant parameter. This is derived in Sect. 9.8.3, but the reader need not refer to it at this stage. The important point emphasised here is that algebraic manipulation of (9.43) yields alternative switching functions for the same switching boundary. A trivial but relevant alternative is obtained by multiplying the switching function by -1 to yield

$$S'(x_{1e}, x_{2e}) = x_{1e} + \frac{1}{2b_0 u_{max}} x_{2e} |x_{2e}| = 0 \tag{9.44}$$

The convention adopted is that if $u_{min} = -u_{max}$, the switching function, $S(\mathbf{x}, y_r, d)$, is chosen such that the required control law is

$$u = u_{max} \operatorname{sgn}[S(\mathbf{x}, y_r, d)]. \tag{9.45}$$

9.5 Boundary Layer for Saturating Control Systems

Certain switched control techniques, such as the time-optimal ones to be studied in Sects. 9.7, 9.8 and 9.9, are beneficial with continuous actuators operated in saturation. Once the plant state is brought to the desired value using switched control, the control law switches the control variable relatively rapidly between u_{min} and u_{max} to maintain the desired state within certain limits, leading to the limit cycling behaviour studied in Sect. 9.7.3 and Appendix A9. This behaviour, however, is unnecessary when using continuous actuators as the control can be continuously varied to maintain the desired state. The boundary layer achieves this.

First consider a switched state feedback control system with the general switching boundary, $S(\mathbf{x}, y_r, d) = 0$, introduced in Sect. 9.3. The boundary layer may then be viewed as a region of the state space between two new boundaries placed either side of and in a close neighbourhood of the boundary, $S(\mathbf{x}, y_r, d) = 0$. These will be called the *saturation boundaries*. The boundary layer replaces the switching boundary and performs the function of effecting a continuous transition between $u = u_{min}$ and $u = u_{max}$ as the state trajectory passes through it. This is sometimes referred to as *soft switching*. The discontinuous switch between $u = u_{min}$ and $u = u_{max}$ that occurs in switched control when a state trajectory crosses the switching boundary, $S(\mathbf{x}, y_r, d) = 0$, is sometimes referred to as *hard switching*. The control system performance during relatively large state changes, in which the 'size' of the state trajectory is much larger than the thickness of the boundary layer, is similar to that obtained with the original switching boundary. The boundary layer is implemented by replacing the original switched control law,

$$u = \tfrac{1}{2} \{u_{max}[1 + \operatorname{sgn}[S(\mathbf{x}, y_r, d)]] + u_{min}[1 - \operatorname{sgn}[S(\mathbf{x}, y_r, d)]]\}, \tag{9.46}$$

with the *saturated control law*,

$$u = \operatorname{sat}[K.S(\mathbf{x}, y_r, d), u_{min}, u_{max}], \tag{9.47}$$

9.5 Boundary Layer for Saturating Control Systems

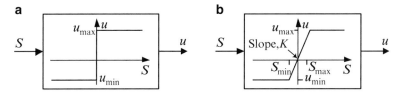

Fig. 9.12 S to u transfer characteristics. (**a**) Switching element (9.46). (**b**) Saturation element (9.47)

where the saturation function is defined as

$$\operatorname{sat}(x,\, b,\, a) \triangleq \begin{cases} x & \text{for } b \leq x \leq a \\ b & \text{for } x < b \\ a & \text{for } x > a \end{cases}. \tag{9.48}$$

For comparison, the transfer characteristics of the switching element and the saturation element that replaces it are shown in Fig. 9.12.

The two *saturation boundaries* in the state space are defined by $S = S_{\min}$ and $S = S_{\max}$ in Fig. 9.12b, defining, respectively, all the states for which the control saturates at $u_{\min}$ and $u_{\max}$. The equations of the saturation boundaries are

$$S\,(\mathbf{x},\, y_\text{r},\, d) - S_{\min} = 0 \quad \text{written as} \quad S_{\min}^{\text{sat}}\,(\mathbf{x},\, y_\text{r},\, d) = 0 \tag{9.49}$$

and

$$S\,(\mathbf{x},\, y_\text{r},\, d) - S_{\max} = 0 \quad \text{written as} \quad S_{\max}^{\text{sat}}\,(\mathbf{x},\, y_\text{r},\, d) = 0 \tag{9.50}$$

Graphic illustrations of these boundaries and the state trajectories within the boundary layer are given in Sect. 9.8.5.2. Within the boundary layer, the control law (9.47) becomes

$$u = K.S\,(\mathbf{x},\, y_\text{r},\, d)\,. \tag{9.51}$$

In order for the system to perform well in the region of the desired state, it must be ensured that (9.51) yields a stable non-oscillatory closed-loop system so that the state trajectory moves monotonically towards the desired state within the boundary layer. Modifications of $S(\mathbf{x}, y_\text{r}, d)$ are sometimes necessary to ensure this. In general, these modifications entail increasing the relative weightings of the output derivatives, which are state variables, to increase the damping within the boundary layer. The details of the required modifications vary from one case to the next, and therefore these are left to specific examples.

9.6 Supporting Theory

9.6.1 Background

Control theory supporting the design of feedback control systems with switched actuators or continuous actuators operated in saturation is less mature than linear control theory. The pulse modulation approach of Sect. 9.2 enables linear control system design methods to be employed with switched actuators but only within the saturation limits of the equivalent transfer characteristic of the modulator. In certain applications, better performance can be obtained with switched state feedback control laws of the form introduced in Sect. 9.3, which purposely operate with control saturation. This section presents some control theory enabling nonlinear control laws to be determined for certain plants that yield optimal performance in some sense, such as time-optimal control in which the control saturation limits are fully exploited to achieve shorter settling times than could be attained with linear control methods operating within these saturation constraints.

9.6.2 Optimal Control Through Pontryagin's Maximum Principle

9.6.2.1 Introduction

Consider the multivariable plant,

$$\dot{\mathbf{x}} = \mathbf{F}(\mathbf{x}, \mathbf{u}), \quad \mathbf{y} = \mathbf{H}(\mathbf{x}), \tag{9.52}$$

where $\mathbf{F}(\mathbf{x}, \mathbf{u})$ and $\mathbf{H}(\mathbf{x})$ are continuous functions of their arguments and the state and control vectors are $\mathbf{x} \in \Re^n$ and $\mathbf{u} \in \Re^r$. This is sufficiently general to cover most of the applications addressed in this book, the exceptions being the few containing transport delays. There are an infinite number of different control functions, $\mathbf{u}(t)$, that take the plant from an arbitrary initial state, $\mathbf{x}(t_0)$, to a specified final state, $\mathbf{x}(t_f)$. An *optimal control function*, denoted $\mathbf{u}^*(t)$, is one that achieves this change of state while minimising a *cost functional* that could be of the form

$$J = \int_{t_0}^{t_f} L[\mathbf{x}(t), \mathbf{u}(t)] \, dt. \tag{9.53}$$

where $L[\mathbf{x}(t), \mathbf{u}(t)]$ is the incremental cost function, the choice of which should be such that the optimal control is the best for a specific application. More general forms of cost functional may be found in the literature on optimal control theory [5], but (9.53) is sufficient to cover the applications considered here.

The measurement equation is included in (9.52) for completeness, but it should be emphasised that only the state differential equation (s.d.e.) and the cost functional

9.6 Supporting Theory

are used for the determination of $\mathbf{u}^*(t)$. Also no attempt is made to eliminate interaction between the control channels for multivariable plants ($r > 1$).

The application of Pontryagin's maximum principle [6] leads to a means of determining $\mathbf{u}^*(t)$ under the control saturation constraints

$$u_{i\ min} \leq u_i \leq u_{i\ max}, \quad i = 1, 2, \ldots, r, \tag{9.54}$$

but this is calculated offline and applied later in real time. Hence, the method leads directly to open-loop control with the attendant disadvantage of sensitivity to plant modelling errors and external disturbances. It is introduced here, however, since it leads to useful optimal or suboptimal switched or saturating state feedback control laws for a limited range of plants.

Pontryagin's method is analogous to finding the minimum values of a function defined by an algebraic expression subject to algebraic constraints using Lagrange multipliers. Hence, the minimum of the cost functional (9.53) is found subject to the constraints imposed by the plant s.d.e. of (9.52). To achieve this, the Hamiltonian,

$$H = \mathbf{p}^T \mathbf{F}(\mathbf{x}, \mathbf{u}) - L(\mathbf{x}, \mathbf{u}), \tag{9.55}$$

in which the components of the vector, $\mathbf{p} \in \Re^n$, are equivalent to the Lagrange multipliers, is *maximised* with respect to $\mathbf{u}$. The proof of the maximum principle [5, 6] reveals that $\mathbf{p}$ obeys

$$\dot{\mathbf{p}} = -\frac{\partial H}{\partial \mathbf{x}} \equiv \dot{p}_i = -\frac{\partial H}{\partial x_i}, \quad i = 1, 2, \ldots, n, \tag{9.56}$$

which is a state differential equation with $\mathbf{p}$ as the state vector. From (9.55),

$$\frac{\partial H}{\partial \mathbf{p}} \triangleq \left[\frac{\partial H}{\partial p_1} \frac{\partial H}{\partial p_2} \cdots \frac{\partial H}{\partial p_n} \right]^T = \left[\left[\frac{\partial}{\partial p_1} \frac{\partial}{\partial p_2} \cdots \frac{\partial}{\partial p_n} \right] \sum_{i=1}^{n} p_i f_i(\mathbf{x}, \mathbf{u}) \right]^T.$$

Thus,

$$\frac{\partial H}{\partial \mathbf{p}} = [f_1(\mathbf{x}, \mathbf{u}) \ f_2(\mathbf{x}, \mathbf{u}) \cdots f_n(\mathbf{x}, \mathbf{u})]^T = \mathbf{F}(\mathbf{x}, \mathbf{u}). \tag{9.57}$$

The plant s.d.e. of (9.52) may therefore be expressed similarly to (9.56), as follows:

$$\dot{\mathbf{x}} = \frac{\partial H}{\partial \mathbf{p}} \equiv \dot{x}_i = \frac{\partial H}{\partial p_i}, \quad i = 1, 2, \ldots, n. \tag{9.58}$$

Equation (9.56) is referred to as the costate differential equation (or alternatively the adjoint system) in view of its similarity with the plant state differential Eq. (9.58). Hence, $\mathbf{p}$ is referred to as the costate or adjoint system state. Equations (9.56) and (9.58) together are often referred to as a *Hamiltonian system*.

The maximum of H w.r.t. **u** may be searched for first by solving $\partial H/\partial u_i = 0$ for u_i, yielding u'_i, $i = 1, 2, \ldots, n$. Then

$$u_i^* = \begin{cases} u'_i & \text{if } u_{i\,\min} < u'_i < u_{i\,\max} \\ u_{i\,\min} & \text{if } u' \leq u_{i\,\min} \\ u_{i\,\max} & \text{if } u' \geq u_{i\,\max} \end{cases}. \tag{9.59}$$

It is most important to note, however, that it is necessary to determine the initial costate, $\mathbf{p}(0)$, that yields the specified final state, $\mathbf{x}(t_f)$, for a given initial state, $\mathbf{x}(t_0)$. This is referred to as the *two-point boundary value problem* which, in general, has to be solved by iterative numerical methods. Since the relationship between $\mathbf{p}(0)$ and $\mathbf{x}(t_f)$ is generally nonlinear, finding an iterative numerical algorithm that converges is not straightforward and the convergence rates are not predictable and will vary with $\mathbf{x}(t_0)$. Implementation of feedback control on this basis would therefore not be feasible. An alternative implementation of Pontryagin's method would be to compute the optimal control function, $u^*(t)$, off-line and apply it in real time later, but this *open-loop* system would be subject to cumulative errors due to plant modelling uncertainties and unknown external disturbances. If, however, optimal control is required and the plant has continuous actuators, a possible way forward is to apply $u^*(t)$ in real time to the model of the plant used for its computation and, at the same time, apply the controlled output of this model as the reference input to a feedback control loop applied to the real plant with dynamic lag pre-compensation (Chap. 12). In this chapter, however, the theory of Pontryagin's method will be used to assist in finding optimal or near-optimal *feedback* control laws.

9.6.2.2 Cost Functionals Tailored for Real Applications

Referring to (9.53), the simple incremental cost function, $L(\mathbf{x}, \mathbf{u}, t) = 1$, yields

$$J = \int_{t_0}^{t_f} 1 \cdot dt = t_f - t_0. \tag{9.60}$$

Thus, the plant state is taken from $\mathbf{x}(t_0)$ to $\mathbf{x}(t_f)$ in the minimum time of $T_{\text{opt}} = t_f - t_0$. This is *time-optimal control*. The largest subset of the set of closed-loop optimal control laws that have been derived to date are time-optimal ones [4].

One practical application of time-optimal control is the slewing (i.e. rotational movement through large angles) of scientific satellites, such as space telescopes, in which the maximum torques from the control actuators and the slewing rates are limited by relatively small available electrical power, often a few tens of Watts. The resulting slewing times for large angles (up to 180° about the Euler axis [7]) are typically many tens of seconds. Since such satellites are required to be frequently reorientated towards different objects in the celestial sphere, minimising the slewing times is highly advantageous to maximise the proportion of the total

9.6 Supporting Theory

mission time devoted to the scientific observations. This requires reaction wheels with regenerative electric drives so that the kinetic energy stored in the satellite body and the reaction wheels during a slew is nearly all returned to the onboard batteries by the end of the manoeuvre.

Time-optimal control could also be beneficial in the manufacturing industry on production lines entailing repeated movements of mechanisms in which maximisation of the throughput rate is important, but the friction between relatively moving parts would have to be small. Friction, however, is often very significant and unavoidable in some position control systems. In such cases, the peak velocities entailed in time-optimal control would waste an excessive amount of energy. A cost functional for minimisation of frictional energy and the derivation of the corresponding closed-loop optimal control law is presented in Chap. 12.

A general *quadratic form*-based cost functional that is frequently referred to is

$$J = \int_{t_0}^{t_F} \left[\mathbf{x}^\mathrm{T} \mathbf{Q} \mathbf{x} + \mathbf{u}^\mathrm{T} \mathbf{R} \mathbf{u} \right] \mathrm{d}t, \tag{9.61}$$

where $\mathbf{Q} \in \mathfrak{R}^{n \times n}$ and $\mathbf{R} \in \mathfrak{R}^{n \times n}$ are constant matrices chosen to suit the particular application. The first term in (9.61) leads to an optimal control that attempts to minimise the excursions of state variables with the most weightings according to the choice of $\mathbf{Q}$. As already highlighted, an important state variable to minimise is the relative velocity between two contacting surfaces of a mechanism to minimise frictional energy losses. The second term in (9.61) is often quoted as the *control energy* term to be considered in the optimisation, based on the assumption that more control activity requires more energy to be supplied. It is most important, however, to carefully examine a particular application to assess whether or not this really is the case and then choose the most appropriate cost functional. The most important question is *how optimal is the choice of the cost functional for the particular application*. This depends critically on the hardware used. The control variables may not *directly* contribute a significant proportion of the nett physical energy consumed by a control system and ultimately dissipated in the form of heat or kinetic energy of exhausted particles. Frictional energy is a case in point and is catered for by the 'Q' term in (9.61). As another example, the nett energy consumed by a modern electric locomotive with regenerative braking is mainly due to rolling friction and aerodynamic losses, which are state dependent. Increasing the acceleration to reduce the time to reach a cruising speed increases the peak in the control variable, $u(t)$, without substantially increasing the nett energy supplied. Suppose $u(t)$ produces a quadrature axis stator current, $i_\mathrm{q}(t) = K_\mathrm{i} u(t)$, for a vector-controlled induction motor drive (Chap. 2) where K_i is the transconductance constant, and then some power dissipation equal to $(K_\mathrm{i}^2/R_s) u^2(t)$, where R_s is the stator resistance, would occur, being catered for by the second term in (9.61), but in a well-engineered system, this would be a very small proportion of the nett energy supplied. A contrasting example is a spacecraft attitude control system actuated only by gas-jet thrusters that consume fuel, and therefore energy, at a rate proportional to $|u(t)|$.

There is very little other energy expenditure in such a system, and therefore if the cost functional (9.61) is used, then only the second term is needed. In this example, assuming a three-axis stabilised attitude control system, a more appropriate cost functional would be

$$J = \int_{t_0}^{t_f} (K_1 |u_1(t)| + K_2 |u_2(t)| + K_3 |u_3(t)|) \, dt \quad (9.62)$$

where K_1, K_2 and K_3 are the fuel rate constants for each axis. Finding controls that minimise such functionals is referred to as *fuel-optimal control*.

The quadratic form integrand in cost functional (9.61) gives the problem of determining $u^*(t)$ more mathematical tractability when applied to a linear time-invariant plant, but the control saturation constraints (9.54) have to be ignored. This is referred to as *linear quadratic (LQ) optimal control*. For the special case of $t_f \to \infty$, the optimal control simplifies to a linear state feedback control law [8].

9.6.2.3 Application for Time-Optimal Control of an LTI Plant

The equations needed to determine the time-optimal control, $\mathbf{u}^*(t)$, will now be derived for the general LTI multivariable plant with state differential equation

$$\dot{\mathbf{x}} = \mathbf{A}\mathbf{x} + \mathbf{B}\mathbf{u}. \quad (9.63)$$

With cost functional (9.60), the Hamiltonian of (9.55) becomes

$$H = \mathbf{p}^T [\mathbf{A}\mathbf{x} + \mathbf{B}\mathbf{u}] - 1. \quad (9.64)$$

Regarding the minimisation w.r.t. $\mathbf{u}$, (9.64) may be written as

$$H = \mathbf{p}^T \mathbf{A}\mathbf{x} + \sum_{i=1}^{r} [\mathbf{p}^T \mathbf{B}]_i u_i - 1 \quad (9.65)$$

Then, by inspection, the time-optimal control has to be a switched control sequence of the form of Fig. 9.1a, generated by the following control law:

$$u_i^* = \begin{cases} u_{i\,\min}, & [\mathbf{p}^T \mathbf{B}]_i \geq 0 \\ u_{i\,\max}, & [\mathbf{p}^T \mathbf{B}]_i < 0 \end{cases} = \tfrac{1}{2} \{u_{i\,\max} [1 - \text{sgn}[\mathbf{p}^T \mathbf{B}]_i] + u_{i\,\min} [1 + \text{sgn}[\mathbf{p}^T \mathbf{B}]_i]\}.$$
$$i = 1, 2, \ldots, r \quad (9.66)$$

The explanation of the asymmetry in the inequalities of this expression is the same as that given following (9.35). Often $u_{i\,\min} = -u_{i\,\max}$ and then (9.66) simplifies to

$$u_i^* = -u_{i\,\max} \, \text{sgn}[\mathbf{p}^T \mathbf{B}]_i, \quad i = 1, 2, \ldots, r. \quad (9.67)$$

9.6 Supporting Theory

It remains to find $\mathbf{p}(t)$, which is a solution of the adjoint system state differential Eq. (9.56) for this case, which is

$$\dot{\mathbf{p}} = -\frac{\partial}{\partial \mathbf{x}}\left[\mathbf{p}^T \mathbf{A} \mathbf{x}\right] \equiv \dot{p}_i = -\frac{\partial}{\partial x_i} \sum_{j=1}^{n}\left(p_j \sum_{k=1}^{n} a_{jk} x_k\right) = -\sum_{j=1}^{n}\left(p_j a_{ji}\right) = -\sum_{j=1}^{n}\left(a_{ji} p_j\right),$$

i.e.,
$$\dot{\mathbf{p}} = -\mathbf{A}^T \mathbf{p}. \tag{9.68}$$

Hence, in this case, the adjoint system is an unforced linear system whose eigenvalues are minus those of the plant, meaning that for every stable mode of the plant, there is an unstable mode of the adjoint system with poles that are mirrored in the imaginary axis, and vice versa. The general solution of (9.68) is

$$\mathbf{p}(t) = e^{-\mathbf{A}^T (t-t_0)} \mathbf{p}(t_0). \tag{9.69}$$

To calculate $\mathbf{u}^*(t)$ for a given $\mathbf{x}(t_0)$ and the demanded state, $\mathbf{x}(t_f)$, which will be called the reference state, $\mathbf{x}_r$, noting that t_f is not specified in advance, it would be necessary to find the $\mathbf{p}(t_0)$ for which control law (9.66) would cause the plant state trajectory, $\mathbf{x}(t)$, to reach $\mathbf{x}_r$, i.e. $\mathbf{x}(t) = \mathbf{x}_r$ for a certain value of t, which will be t_f. The mapping between $\mathbf{x}(t_0)$ and $\mathbf{p}(t_0)$ for a given $\mathbf{x}(t_f)$ is *nonlinear* despite the plant and the adjoint system being linear, and therefore, finding a general iterative algorithm that converges to the optimal solution is still a challenge.

It is important to note that if the application of $\mathbf{u}(t)$ as given by control law (9.66), after finding the correct $\mathbf{p}(t_0)$, was continued beyond the point where $\mathbf{x}(t) = \mathbf{x}_r$, control would be lost as the state trajectory would just pass through $\mathbf{x}_r$ and then move away from this desired point. This further illustrates the impracticality of open-loop control. In a real application, once the desired state is reached, $\mathbf{u}(t)$ has to be applied such that $\mathbf{x} = \mathbf{x}_r$ is maintained. The *feedback* control laws appearing in subsequent examples of this chapter are designed to achieve this.

Each component, $p_i(t)$, of (9.69) is a weighted sum of dynamic modes (Chap. 1) characterised by the eigenvalues of $\mathbf{A}$. In view of (9.66) or (9.67), the number of switches of $u_i^*(t)$ is equal to the number of zero crossings of $p_i(t)$ for $t > t_1$. This information actually enables time-optimal feedback control laws to be determined for some simple linear plants, using the back-tracing technique of the following section. For a plant with n real poles, the maximum number of switches in each control component is $n - 1$, but this can exceed $n - 1$ if the plant has any complex conjugate poles, this number generally increasing with the magnitude of the change of state demanded [4].

9.6.2.4 Back Tracing

Back tracing can be described as running the plant model and its inputs in reversed time commencing with the desired final state, to determine the initial states from which the desired state can be reached. If the control input is restricted to bang–bang control sequences, as illustrated in Fig. 9.1a, then if the number of switches is allowed to be infinite, the complete set of the initial states maps out the controllability region under the control saturation constraints referred to in Sect. 9.1.1.

Since time-varying external disturbances are usually unknown, they cannot be directly included in a plant model used for developing a controller on the basis of predicted state trajectories. Instead, they are set to zero in the model, and the action of the feedback loop to be formed is relied upon to provide a degree of robustness against the disturbance. This can be tested by simulation. An exception, however, is a constant but initially unknown external disturbance. In this case, an observer can be used to estimate the disturbance so a constant disturbance could be included in the back-tracing process, but for simplicity, it will be regarded zero in the following.

The reversed time, τ, is defined by

$$d\tau = -dt, \quad (9.70)$$

where t is the forward time, which implies that

$$\tau = t_f - t \quad (9.71)$$

where t_f is an arbitrary final time. If the forward time plant model is expressed as

$$d\mathbf{x}(t)/dt = \mathbf{f}[\mathbf{x}(t), u(t)], \quad y(t) = h[\mathbf{x}(t)], \quad (9.72)$$

then in view of (9.70), the reversed time plant model is obtained from (9.72) through replacing t by τ and changing the sign of the derivative, yielding

$$d\mathbf{x}(\tau)/d\tau = -\mathbf{f}[\mathbf{x}(\tau), u(\tau)], \quad y(\tau) = h[\mathbf{x}(\tau)]. \quad (9.73)$$

It should be noted that if the uncontrolled plant model (9.72) is stable, then its reversed time model is *unstable* and vice versa.

The back tracing is commenced with one or the other of the two control levels held, in theory, for an infinite time. Let this be $u = u_{max}$. The continuum of states in the n- dimensional state space mapped out by the resulting state trajectory is then a *one*-dimensional subspace of the state space. Then, for every state in this one-dimensional subspace, let the back tracing be continued by applying $u = u_{min}$, again for an infinite time. The resulting continuum of state trajectories then forms a *two*-dimensional subspace. Similarly if for every state within this two-dimensional subspace, the back tracing is continued with $u = u_{max}$ again, the resulting continuum of trajectories forms a *three*-dimensional subspace. This back tracing with alternating control levels is continued until the subspace from which the back tracing is continued is of dimension, $n - 1$, the resulting continuum of trajectories, obtained with $u = u_{max}$ for n odd or $u = u_{min}$ for n even, is an

n-dimensional subspace within the n-dimensional state space. Suppose at this stage the whole process is repeated but commencing with $u = u_{min}$ and ending with $u = u_{min}$ for n odd or $u = u_{max}$ for n even. The combined n-dimensional subspace comprising the two subspaces formed by the back-tracing process is the region of the state space from which the desired state can be reached with a control sequence having $n - 1$ switches. For certain plants, such as linear ones without any oscillatory modes, this is the controllability region under the control saturation constraints. On the other hand, if the back-tracing process is carried out with an infinite number of switches in the control sequences, the region of initial states mapped out is the controllability region under the control saturation constraints for any plant. This process can only be approximated using a computer to determine a finite number of state trajectories over a finite duration.

The back-tracing process leads to graphical visualisation for first- and second-order plants and can yield useful switched feedback control laws.

9.7 Feedback Control of First-Order Plants

9.7.1 Time-Optimal Feedback Control: Analytical Method

The equations of Pontryagin's method given in Sect. 9.6.2 applied to the general first-order plant,

$$\dot{y} = f(y, u - d), \tag{9.74}$$

are as follows. The cost function for time-optimal control is

$$J = \int_{t_0}^{t_f} L[y(t), u(t)]\, dt = \int_{t_0}^{t_f} 1 . dt = t_f - t_0. \tag{9.75}$$

The Hamiltonian is

$$H = pf(y, u - d) - L[y(t), u(t)] = pf(y, u - d) - 1 \tag{9.76}$$

The costate differential equation is then

$$\dot{p} = -\frac{\partial H}{\partial y} = -p \frac{\partial}{\partial y} f(y, u - d). \tag{9.77}$$

For arbitrary $d(t)$ and/or $y_r(t)$, or a nonlinear RHS of (9.74), analytical treatment is not generally possible, but is for the linear first-order plant model,

$$\dot{x}_1 = a_1 x_1 + b_1 (u - d), \quad y = c_1 x_1 \equiv \dot{y} = ay + b(u - d), \tag{9.78}$$

where $a = a_1$ and $b = c_1 b_1$, which covers many practical applications. Let (9.78) be expressed in terms of the error, $\varepsilon = y - y_r$, where y_r is constant, yielding

$$\dot{\varepsilon} = a(\varepsilon + y_r) + b(u - d). \tag{9.79}$$

Then (9.77) becomes

$$\dot{p} = -p \frac{\partial}{\partial \varepsilon}[a(\varepsilon + y_r) + b(u - d)] = -ap. \tag{9.80}$$

This has the general solution,

$$p(t) = e^{-at} p(0), \tag{9.81}$$

which is a particular case of (9.69). The Hamiltonian of (9.76) becomes

$$H = p[a(\varepsilon + y_r) + b(u - d)] - 1. \tag{9.82}$$

The constant, b, is assumed positive without loss of generality. Then by inspection of (9.82), H is minimised by

$$\begin{aligned} u &= \{u_{\min} \text{ if } p(t) \geq 0 \text{ or } u_{\max} \text{ if } p(t) < 0\} \\ &= \tfrac{1}{2}\{u_{\min}(1 + \mathrm{sgn}[p(t)]) + u_{\max}(1 - \mathrm{sgn}[p(t)])\}. \end{aligned} \tag{9.83}$$

According to (9.81), $p(t)$ does not change sign, so the time-optimal control is simply $u = \text{const}$. The back tracing method of Sect. 9.6.2.4 may then be used to find a switched feedback control law (the term switched being retained because switching of u must occur to maintain the plant state, in theory, equal to the demanded state once it has been reached). The plant model (9.79) may be written as

$$\dot{\varepsilon} = a\varepsilon + a[y_r + (b/a)(u - d)]. \tag{9.84}$$

With reference to Sect. 9.6.2.4, the reversed time plant model for back tracing is

$$\frac{d\varepsilon}{d\tau} = -a\varepsilon - a[y_r + (b/a)(u - d)]. \tag{9.85}$$

The solution to (9.85) for constant u and the required $\varepsilon(0) = 0$ is then

$$\varepsilon(\tau) = -[y_r + (b/a)(u - d)](1 - e^{-a\tau}). \tag{9.86}$$

If, as is usually the case, the plant is stable with $a < 0$, then provided

$$y_r + (b/a)(u_{\max} - d) < 0 \quad \text{and} \quad y_r + (b/a)(u_{\min} - d) < 0, \tag{9.87}$$

9.7 Feedback Control of First-Order Plants

all possible values of $\varepsilon(\tau)$ can be reached. So in forward time, the control objective of $\varepsilon = 0$ can be reached from any initial value of ε. The time-optimal feedback control law under conditions (9.87) can be deduced by inspection of (9.86). For $u = u_{\max}$, with $b > 0$ and $a < 0$, then since $1 - e^{-a\tau} < 0$, $\varepsilon(\tau) < 0$. Similarly for $u = u_{\min}$, $\varepsilon(\tau) > 0$. The time-optimal feedback control law is therefore

$$u = \{u_{\min} \text{ if } \varepsilon \geq 0 \text{ or } u_{\max} \text{ if } \varepsilon < 0\} = \tfrac{1}{2}\{u_{\min}[1 + \operatorname{sgn}(\varepsilon)] + u_{\max}[1 - \operatorname{sgn}(\varepsilon)]\}. \tag{9.88}$$

If the control levels are balanced, meaning $u_{\min} = -u_{\max}$, then (9.88) reduces to

$$u = -u_{\max} \operatorname{sgn}(\varepsilon). \tag{9.89}$$

If the plant is unstable with $a > 0$, then provided

$$y_r + (b/a)(u_{\max} - d) > 0 \quad \text{and} \quad y_r + (b/a)(u_{\min} - d) < 0, \tag{9.90}$$

the control objective of $\varepsilon = 0$ is reached from only a limited range of ε, the limits of which are the steady-state values of (9.86) with $u = u_{\max}$ and $u = u_{\min}$. Thus,

$$\varepsilon \in (\varepsilon_{\min}, \varepsilon_{\max}) = (-[y_r + (b/a)(u_{\max} - d)], -[y_r + (b/a)(u_{\min} - d)]). \tag{9.91}$$

The time-optimal feedback control law under conditions (9.90) and (9.91) can again be deduced by inspection of (9.86), in this case, for $u = u_{\max}$, with $b > 0$ and $a > 0$. Then since $1 - e^{-a\tau} > 0$, $\varepsilon(\tau) < 0$. Similarly for $u = u_{\min}$, $\varepsilon(\tau) > 0$. The time-optimal feedback control law is therefore as previously, i.e. (9.88) or (9.89), if $u_{\min} = -u_{\max}$.

Remarkably, in contrast with the continuous control laws met in the previous chapters, (9.88) and (9.89) are independent of the plant parameters, but their successful operation depends on conditions (9.87) or (9.90) and (9.91). In the form of (9.34), control law (9.89) has the following switching function:

$$S(x_1, y_r) = -\varepsilon = y_r - y = y_r - c_1 x_1. \tag{9.92}$$

The corresponding switching boundary equation is therefore

$$y_r - c_1 x_1 = 0, \tag{9.93}$$

which can be regarded as a surface of zero dimension in a one-dimensional state space, which is the point, $x_1 = y_r/c_1$. Figure 9.13 shows the system block diagram.

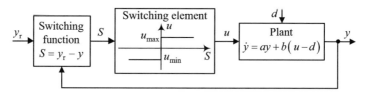

Fig. 9.13 Closed-loop time-optimal control of first-order linear plant

9.7.2 Time-Optimal Feedback Control: Graphical Approach

Time-optimal control laws for first-order plants can be derived using a graphical method that could be applied in cases where an analytical solution of the plant differential equation, such as (9.86), does not exist. This will be demonstrated for the plant (9.84) whose control laws have been derived in Sect. 9.7.1 and in Sect. 8.4 for second-order plants. First families of responses for $u = u_{max}$ and $u = u_{min}$ are produced by a simulation, the results of which are shown in Fig. 9.14. Figure 9.14a, b, d and e are for two plants taken from (9.78), one with $a = -1$ (open-loop stable) and the other with $a = +1$ (open-loop unstable), $b = 1$, $u_{max} = 10$, $u_{min} = -10$, $d = u_{max}/2$ and $y_r = 1$. These families of responses are referred to here as *response portraits* since they present *pictures* of the plant dynamic behaviour over a range of states for selected inputs. As already predicted, the responses of Fig. 9.14a, b, d and e are monotonic and therefore cross the desired value, zero, once or not at all. It may be observed that all the responses approaching the line, $\varepsilon = 0$, from below are for $u = u_{max}$. Similarly all the responses approaching the line, $\varepsilon = 0$, from above are for $u = u_{min}$. It has already been established theoretically in Sect. 9.7.1 that the time-optimal control is constant at one saturation limit.

A control law, satisfying this requirement, that automatically drives ε towards zero, at least for some initial states, is apparent by observation of the response portraits. This indicates that the line, $\varepsilon = 0$, should be the switching boundary, above which u is set to u_{min} and below which u is set to u_{max}, as shown in Fig. 9.14c, f. This is identical to control law (9.88). For $a = -1$, Fig. 9.14c shows the trajectories terminating, as required, on $\varepsilon = 0$, in agreement with the theoretical predictions. For Fig. 9.14f, inserting the numerical plant parameters in (9.91) yields

$$\varepsilon \in (-[1 + (1/1)(10 - 5)], -[1 + (1/1)(-10 - 5)]),$$

i.e., $\quad \varepsilon \in (u_{min}, u_{max}) = (-6, +14).$ \hfill (9.94)

It is evident in Fig. 9.14f that for $\varepsilon(0) < -6$, the trajectories do not cross $\varepsilon = 0$, in agreement with (9.94). The limit of $\varepsilon(0) = 14$ is outside the range of Fig. 9.14f.

In contrast with continuous linear control systems, if the reference input is $y_r(t) = Y_r = const.$, the control error, $y(t) - Y_r$, does not tend to zero asymptotically but, in theory, precisely reaches zero in the time-optimal settling time, T_{opt}.

9.7 Feedback Control of First-Order Plants

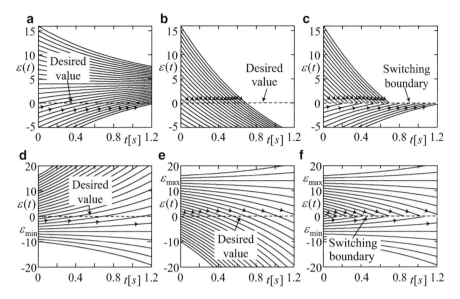

Fig. 9.14 Response portraits of first-order plants. (**a**) $u = u_{max}$, $a = -1$. (**b**) $u = u_{min}$, $a = -1$. (**c**) Closed loop, $a = -1$. (**d**) $u = u_{max}$, $a = 1$. (**e**) $u = u_{min}$, $a = 1$. (**f**) Closed-loop, $a = 1$

9.7.3 Limit Cycling and Its Control

In theory, a continuous control variable can maintain the desired plant state precisely, but switched controls can only do so approximately by switching repeatedly resulting in cyclic errors. In switched feedback control, it is therefore important to investigate the behaviour of the closed-loop system in the region of the desired state, where it will spend most of its time. The control variables can only switch at a finite frequency determined by the hardware, and it is sometimes necessary to add refinements to ensure that sufficient accuracy is obtained while the control actuators are not harmed by too high a switching frequency.

Continuing with the initial study of the switched control of the first-order plant (9.78) expressed in terms of the error, $\varepsilon = y - y_r$, in (9.79), a switch in u between u_{min} and u_{max} causes a step change in $\dot{\varepsilon}$, which is evident by inspection of (9.79). In the region of the switching boundary, this also makes $\dot{\varepsilon}$ change sign. The action of the basic feedback control laws (9.88) or (9.89) will now be considered assuming that the system is mathematically perfect, i.e. ignoring all implementation imperfections such as sampling of the digital processor. With reference to Fig. 9.14 (c) or (f) in which the response portraits are directed towards the switching boundary, $\varepsilon = 0$, from above and below, once ε changes sign for the first time, u will switch. Since this changes the sign of $\dot{\varepsilon}$, ε will immediately change sign again. This action will be repeated, u switching at an infinite frequency with a continuously varying mark–space ratio to hold $\varepsilon = 0$.

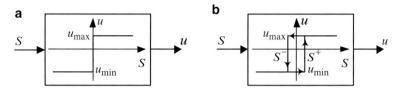

Fig. 9.15 Switching elements for bang–bang or on–off controllers. (**a**) Basic switching element. (**b**) Hysteresis switching element

This phenomenon is similar to the *sliding mode* introduced in Chap. 10. If control law (9.88) or (9.89) is implemented by a real digital processor, however, u would be piecewise constant at $-u_{max}$ or $+u_{max}$ for consecutive periods of h seconds, where h is the sampling period. Then following each change of sign of ε, the resulting change of the control sign would be delayed by a duration less than h. Hence, ε would oscillate continually about zero. Such behaviour is referred to as a *limit cycle*. For practical reasons, as discussed in the following examples, $f_{max} = 1/(2h)$ could be too high. Measures can be taken to reduce it, but at the expense of increasing the amplitude of the limit cycle. One way of reducing f_{max} that historically stemmed from hardware realisation using a Schmidt trigger circuit is to replace the signum function of the control law with a hysteresis element, as in Fig. 9.15.

Provided the digital processor sampling frequency is much higher than the limit cycle frequency, the hysteresis element enables control of the error, ε, to keep between two prescribed limits corresponding to S^+ and S^- (usually $S^- = -S^+$) thereby guaranteeing a specified control accuracy, the switching frequency varying with the set point. A bang–bang controller using this switching element is often referred to as a *hysteresis controller*. An alternative is a sample and hold element with sampling period, $H = qh$, where q is an integer, which has the advantage of guaranteeing a maximum switching frequency of $f_{max} = 1/(2H)$, but the peak values of ε during the limit cycle depend on the set point and may not be equal and opposite, allowing a steady-state error in the mean value. The following example demonstrates these two methods of limit cycle control.

Example 9.2 On–off temperature control of a horticultural green house

In this example, plant model (9.78) is applicable with $b = -a = 1/T_h$, where T_h is the heating time constant; y is the greenhouse temperature measurement scaled in the control computer to be numerically equal to the temperature; $d = -\Theta_a$, where Θ_a is the temperature of the ambient surroundings; $u_{min} = 0$ (fan heater off); and u_{max} is numerically equal to the steady-state temperature that would be reached with the fan heater turned on permanently with $d = 0$. Figure 9.16 shows simulations with $-d = 10$ [°C], $y_r = 30$ [°C], $u_{max} = 50$ [°C] and $T_h = 50$ [s].

With the basic switching element, very precise control is achieved once the set point is reached, but the controller attempts to switch the fan heater on and off at an infinite frequency as shown in Fig. 9.16a which is impracticable. Electrical relays are sometimes used in preference to power electronics to reduce cost, but these would be subject to premature wear with too high a limit cycling frequency.

9.7 Feedback Control of First-Order Plants

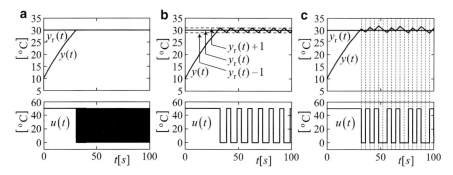

Fig. 9.16 Responses of a greenhouse temperature control system. (**a**) Basic switching element. (**b**) Hysteresis switching element. (**c**) Timed sample/hold

The hysteresis element used to produce the response of Fig. 9.16b has offsets of $S^+ = +1°C$ and $S^- = -1°C$, which give a controlled limit cycle in which $y(t)$ oscillates between the limits of $y_r + 1°C$ and $y_r - 1°C$, noting that the sampling period of a modern digital controller, typically 0.1 to 10 ms, will be orders of magnitude smaller than the limit cycle period. The mean value of $y(t)$ during the limit cycle, however, is slightly greater than y_r due to the piecewise exponential, rather than piecewise linear, variations of $y(t)$. For control system design purposes, an approximate formula for the limit cycle period, T_{lc}, may be derived using a piecewise linear approximation to $y(t)$. At the crossing points in Fig. 9.16b where $y(t) = y_r$, according to (9.78), the values of $\dot{y}$ are

$$\dot{y}^+ = ay_r + b(u_{max} - d) \quad \text{and} \quad \dot{y}^- = ay_r + b(0 - d) \tag{9.95}$$

The approximate limit cycle period is therefore

$$T_{lc} \cong \frac{2S^+}{\dot{y}^+} + \frac{2S^-}{\dot{y}^-} = \frac{2S^+}{\dot{y}^+} - \frac{2S^+}{\dot{y}^-}$$

$$= 2S^+ \left(\frac{1}{ay_r + b(u_{max} - d)} + \frac{1}{bd - ay_r} \right) = \frac{2\varepsilon^+ b u_{max}}{[ay_r + b(u_{max} - d)](bd - ay_r)}$$

$$= \frac{2\varepsilon^+ u_{max}/T_h}{(u_{max} - d - y_r)(d + y_r)/T_h^2} = \frac{2\varepsilon^+ u_{max} T_h}{(u_{max} - d - y_r)(d + y_r)} \tag{9.96}$$

For this example, this yields $T_{lc} \cong \dfrac{2 * 1 * 50 * 50}{[50 - (-10) - 30](-10 + 30)} = 8.\dot{3} \text{ [s]}$.

For the controller producing the response of Fig. 9.16c, the sampling period of the sample and hold unit is $H = 4$ s, the sampling instants being marked by the vertical dotted lines. Since the sampling instants are not state dependent and a control switch does not occur at every switching instant, pseudo-irregularity of the limit cycle is possible, as exhibited in this case.

9.7.4 Control with Time-Varying Reference Inputs

The general first-order plant model can be expressed in the form

$$\dot{y} = f(y, u - d). \tag{9.97}$$

Sometimes the reference input and/or the external disturbance is time varying. Then if the control magnitudes are sufficiently large and if the reference input derivatives are sufficiently small for the response portraits to be directed towards the graph of $y_r(t)$ from both sides, as illustrated in Fig. 9.17 along segment a–b, the 'ideal' system using the basic switching element will follow the reference input with zero error, i.e. with $\varepsilon(t) = y(t) - y_r(t) = 0$, which is the equation of the switching boundary.

In a practicable system, either with a hysteresis switching element or a timed sample and hold unit, $\varepsilon(t)$ will oscillate between positive and negative peaks that can be kept to relatively small proportions in a well-designed system. This is an example of control system operation with zero dynamic lag (Chap. 12). First-order switched control systems are able to operate in this way, but higher-order switched control systems and continuous control systems require the aid of a special precompensator in order to achieve this.

The condition that the response portraits are directed towards the graph of $y_r(t)$, which is the switching boundary, from both sides may be expressed mathematically for the general first-order plant by observing that along segment a–b, of Fig. 9.17, $\dot{\varepsilon} = \dot{y} - \dot{y}_r$ has to be negative for $u = u_{\min}$ and positive for $u = u_{\max}$, if $y = y_r$. Applying this in (9.97) yields

$$f(y_r, u_{\min} - d) - \dot{y}_r < 0, \quad f(y_r, u_{\max} - d) - \dot{y}_r > 0 \Rightarrow$$
$$[f(y_r, u_{\min} - d) - \dot{y}_r][f(y_r, u_{\max} - d) - \dot{y}_r] < 0. \tag{9.98}$$

This condition is a particular case of that for *sliding motion* derived in Chap. 10.

Example 9.3 Current control loop for power electronic drive applications

Switched mode power electronics is usually used to control the power flow to electrical loads requiring some form of feedback control, such as the armature

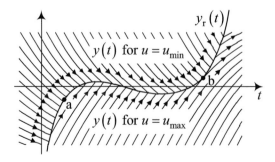

Fig. 9.17 Closed-loop response portrait for first-order bang–bang control system

9.7 Feedback Control of First-Order Plants

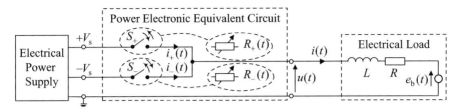

Fig. 9.18 Equivalent circuit of power electronics driving an inductive electrical load

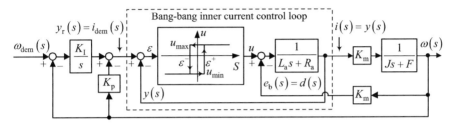

Fig. 9.19 Bang–bang current control loop in cascade speed control of a DC motor-based drive

winding of a DC motor, each phase winding of the stator of a synchronous or induction motor, or the set of control windings on a tokomak fusion reactor. These loads are usually fed from a balanced power supply with equal and opposite voltage levels via two switches, one turned on and the other turned off at any instant of time, at frequencies typically ranging from 1 kHz to 100 kHz, as shown in Fig. 9.18.

This arrangement is usually employed rather than a continuously operated device, such as a class A power amplifier, for the following reasons: a continuous power amplifier or the pair of electronic switches may be modelled by two variable resistances, $R_+(t)$ and $R_-(t)$, as shown. In the case of the continuous power amplifier $R_+(t)$ and $R_-(t)$ would be varied continuously to effect the required control, but this would entail a substantial energy loss, $W_L = \int \left[i_+^2(t) R_+(t) + i_-^2(t) R_-(t) \right] dt$.

Using the power electronic switches instead would reduce W_L several orders of magnitude since R_+ or R_- would be nearly zero with, respectively, switches S_+ or S_- on and with switches S_+ or S_- off, i_+ or i_- would, respectively, be zero.

The mark–space ratio of the anti-phase switching waveform of S_+ and S_- is varied to effect the required control. If the cascade control structure is employed (Chap. 4), then the first-order bang–bang control loop would control the load current, $i(t)$, to follow a demanded reference, $i_{\text{dem}}(t)$, with nominally zero dynamic lag, thereby forming the innermost loop and $i_{\text{dem}}(t)$ would be the control variable for the next loop in the cascaded structure, as shown in Fig. 9.19 for the speed control of a DC motor-based drive (Chap. 4).

Following common practice, Laplace transfer functions are used but bearing in mind that linear systems theory cannot be applied to the inner control loop due to the presence of the nonlinear bang–bang element. The outer IP control loop, however, can be designed by pole assignment (Chaps. 4 and 5) on the assumption that $i(t) \cong i_{\text{dem}}(t)$ during normal operation. Since the focus is on the inner bang–bang control loop, the design calculations for the outer loop will not be presented here. The IP outer loop is included to provide a realistic reference input, $y_r(t) = i_{\text{dem}}(t)$, and disturbance input, $d(t) = e_b(t)$, to the inner bang–bang control loop.

Figure 9.20 shows some simulations of the system of Fig. 9.19 with plant parameters $L_a = 0.002$ [H], $R_a = 0.5$ [Ω], $K_m = 0.06$ [Nm/A $\equiv$ V/(rad/s)], $J = 0.0005$ [Kg m^2] and $F = 0.0002$ [V/(rad/s]. Plant model (9.78) is applicable for the inner loop sub-plant with $a = -R_a/L_a$ and $b = 1/L_a$. The power supply voltages are $u_{\max} = -u_{\min} = 100$ [V]. The outer second-order loop is designed to be critically damped with a settling time of $T_s = 0.5$ [s]. $\omega_{\text{dem}}(t)$ is a 400 [rad/s] step input. The results are on different timescales to reveal (a) the overall system behaviour (0 – 1 [s]), (b) medium-term limit cycle amplitude changes (0 – 0.01 [s]) and (c) short-term limit cycling behaviour (0 – 50 [μs] and 0.8 [s] – {0.8 [s] + 20 [μs]}). The current reference from the outer speed control loop is sufficiently rate limited, and $u_{\max}$ is sufficiently large for the zero dynamic lag condition illustrated in Fig. 9.17 to be satisfied during the whole transient, which is indicated by the limit cycling. The evidence of the zero dynamic lag being achieved is the filtered tracking error plots, $\bar{\varepsilon}(t)$, where

$$\bar{\varepsilon}(s) = \varepsilon(s)/(1 + sT_f), \quad \varepsilon(s) = y(s) - y_r(s). \tag{9.99}$$

The filter is only for displaying the filtered tracking error and is designed to smooth out the relatively short-term limit cycling errors which have a maximum period of the order of 10^{-5} [s].

With the time constant set to $T_f = 0.01$ [s], the dynamic lag introduced by the filter is negligible. The errors in Fig. 9.20a–c are all small compared with the peak armature current of $\cong 12$ [A], and therefore no significant dynamic lag is introduced by the inner current control loop.

Figure 9.20a, b demonstrate the effect of increasing the hysteresis limits, ε^+ and ε^-, from ± 0.2 [A] to ± 1 [A]. Of course, the limit cycle amplitude and frequency, respectively, increase and decrease. In Fig. 9.20a, the limit cycling frequency, which is the switching frequency of the power electronics, is about 600 [kHz], which would normally restrict the implementation to power electronic devices suited to relatively low power applications, less than 1 [kW], say. For higher power applications, the switching frequency would have to be lowered and Fig. 9.20b indicates about 120 [kHz]. Despite the fivefold increase in the limit cycle amplitude, the filtered tracking error, $\bar{\varepsilon}(t)$, is not noticeably increased. This is due to the mean value of the error over one limit cycle period being nearly zero due to $\varepsilon(t)$ being very nearly piecewise linear and with peak values equal to ε^+ and $\varepsilon^- = -\varepsilon^+$. With a sample/hold period of $H = 20$ [μs], the controller producing the results of Fig. 9.20c yields about the same error envelope width as the system of Fig. 9.20b

9.7 Feedback Control of First-Order Plants

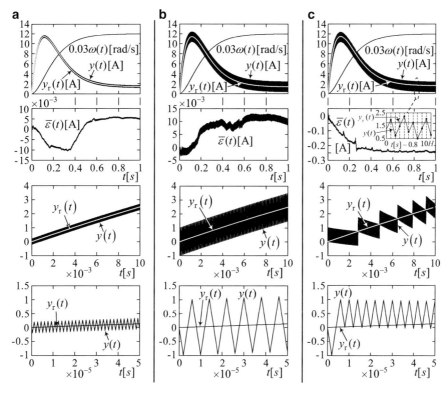

Fig. 9.20 Simulation of bang–bang current control loop in a DC motor speed control system. (**a**) Hysteresis controller $x_{2b} = \frac{1}{2}u\left(\tau_{s1}^2 + 2\tau_{s1}\tau_{s2} - \tau_{s2}^2\right)$ [A]. (**b**) Hysteresis controller $x_{3b} = u(\tau_{s2} - \tau_{s1})$ [A]. (**c**) Basic bang–bang element sample/hold $H = 2 \times 10^{-5}$ [s]

but a much greater filtered tracking error, this being largely a steady-state error due to the switch points, indicated in graph of $y(t)$ inset in the graph of $\bar{\varepsilon}(t)$ of Fig. 9.20c, being at the sampling times, indicated by the vertical dotted lines, rather than being determined by the error, $\varepsilon(t)$, as in Fig. 9.20a, b. The source of the steady-state error is evident from the study of the inset graph of $\bar{\varepsilon}(t)$ and is due to the negative slopes of $y(t)$, produced by $u = u_{\min}$, being larger in magnitude than the positive slopes of $y(t)$, produced by $u = u_{\max}$. This, however, does not impair the overall system performance as it is compensated by the integral term of the outer speed control loop, thereby avoiding any steady-state error in the step response, $\omega(t)$. The system of Fig. 9.20c also exhibits a shuffling effect that can be seen on the medium-term time scale of $0-0.01$ [s]. The reason for this is evident from the study of the detailed limit cycling behaviour on the short-term time scale of $0 - 50$ [μs], again due to the switch points being at the sampling times. Although this appears undesirable compared with the uniform limit cycling behaviour of the systems of Fig. 9.20a, b, the outer speed control loop (Fig. 9.19) filters out the relatively short-term errors due to the shuffling in $y(t)$, and consequently the control of $\omega(t)$ is not impaired.

It is evident from the last example that a first-order switched control loop using the basic bang–bang element with the sample/hold function would perform less well *alone* than the hysteresis controller but does not compromise the overall performance of a control system with the cascade structure in which it is the inner loop. It should also be noted that the sample/hold function with the sampling period, H, is not needed if the digital processor is operated with the same sampling period, since it acts as a sample/hold (Chap. 6). Under these circumstances, a hysteresis controller must operate with a much higher digital processor sampling frequency than the controller with the basic bang–bang switching element if it is to produce the same overall limit cycle peak-to-peak envelope width. Hence, cost savings may be made by using the controller with the basic bang–bang switching element.

9.7.5 Continuous Control with Saturation

9.7.5.1 Unintentional Saturation

The effects of control saturation on a first-order control loop designed to normally operate within the control saturation constraints will now be determined. Figure 9.21 shows a forced dynamic control law (Chap. 7) applied to plant (9.78), i.e.

$$\dot{y} = ay + b(u - d) \qquad (9.100)$$

indicating the control saturation constraints. For the purpose of this discussion, it will be assumed that $\hat{d} = d$, where $\hat{d}$ is the estimate of the external disturbance, d. Within the control saturation constraints, $u = u'$ and the closed-loop system obeys the linear differential equation

$$\dot{y} = (3/T_s)(y_r - y), \qquad (9.101)$$

where T_s is the settling time (5 % criterion).

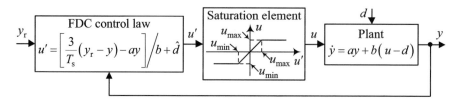

Fig. 9.21 First-order continuous control system with saturation

9.7 Feedback Control of First-Order Plants

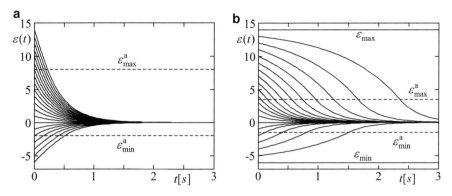

Fig. 9.22 Closed-loop response portrait for control system of Fig. 9.21. (**a**) a=−1: plant open-loop stable. (**b**) a=+1: plant open-loop unstable

In terms of the error, $\varepsilon = y - y_r$, the control law of Fig. 9.21, assuming $\widehat{d} = d$, is

$$u' = -[(a + 3/T_s)\varepsilon + ay_r]/b + d \quad (9.102)$$

There are two saturation boundaries associated with this control law. One boundary is $\varepsilon = \varepsilon_{\min}^{sat}$, which is the minimum error where the control just saturates at $u' = u_{\max}$. The other boundary is $\varepsilon = \varepsilon_{\max}^{sat}$, which is the maximum error where the control just saturates at $u' = u_{\min}$. On the graph of $\varepsilon(t)$, if y_r is constant, these boundaries are horizontal straight lines obtained from (9.102) as

$$\begin{cases} u_{\max} = -[(a+3/T_s)\varepsilon_{\min}^{sat} + ay_r]/b + d \Rightarrow \varepsilon_{\min}^{sat} = [b(d-u_{\max}) - ay_r]/(a+3/T_s) \\ u_{\min} = -[(a+3/T_s)\varepsilon_{\max}^{sat} + ay_r]/b + d \Rightarrow \varepsilon_{\max}^{sat} = [b(d-u_{\min}) - ay_r]/(a+3/T_s) \end{cases}.$$

$$(9.103)$$

Inside the region between these boundaries, the graphs of $\varepsilon(t)$ are governed by the FDC law (9.102) and obey (9.101), which may be written as $\dot{\varepsilon} = -(3/T_s)\varepsilon \Rightarrow$

$$\varepsilon(t) = e^{-t/T_s}\varepsilon(0) \quad (9.104)$$

if y_r is constant. Outside these boundaries, $u = -u_{\max}\,\text{sgn}(\varepsilon)$ and the graphs of $\varepsilon(t)$ follow the time-optimal closed-loop response portraits of Fig. 9.14 (c) or (f). Figure 9.22 shows the results of simulations with the same plant parameters as for Fig. 9.14, $T_s = 1$ s and a range of initial values of ε.

The exponential decay of $\varepsilon(t)$ towards zero, according to (9.104), between the saturation boundaries, i.e. for $\varepsilon(t) \in (\varepsilon_{\min}^{sat}, \varepsilon_{\max}^{sat})$, is evident. For $\varepsilon(t) \notin (\varepsilon_{\min}^{sat}, \varepsilon_{\max}^{sat})$, the dynamic character of the time-optimal response portrait is not easily distinguishable in Fig. 9.22a because the plant responses with $u = \pm u_{\max}$ are stable exponentials similar to the stable responses under the FDC law, and there is no discontinuity in $d\varepsilon/dt$ where the responses cross the boundaries, although the time constant changes from $-1/a = 1$ s to $T_s/3$ as the FDC law takes over. The continuity of $d\varepsilon/dt$ is also explained by the steady-state values of the exponential responses changing from $(b/a)(d \pm u_{\max}) - y_r$ (given by (9.91)) to zero as the FDC law takes over.

The different dynamic character of the time-optimal response portrait is more apparent in Fig. 9.22b where the plant responses with $u = \pm u_{\max}$ are unstable. They are directed towards the saturation boundaries for $\varepsilon(t) \notin (\varepsilon_{\min}^{sat}, \varepsilon_{\max}^{sat})$ and $\varepsilon(t) \in (\varepsilon_{\min}, \varepsilon_{\max})$, where $\varepsilon_{\min}$ and $\varepsilon_{\max}$ are the stability boundaries given by (9.91), so that the FDC law is able to take over rendering the overall system stable, but for $\varepsilon(t) \notin (\varepsilon_{\min}, \varepsilon_{\max})$ the trajectories under $u = \pm u_{\max}$ can never cross the saturation boundaries and the system is unstable. This situation does not usually pertain, however, since first most uncontrolled first-order plants are stable.

9.7.5.2 Saturating Control Systems with a Boundary Layer

Consider again the time-optimal control of the plant,

$$\dot{y} = ay + b(u - d), \qquad (9.105)$$

but with a continuous actuator and a boundary layer to avoid limit cycling as the desired state is approached. The time-optimal control law proven in Sect. 9.7.1 is

$$u = u_{\max} \operatorname{sgn}(y_r - y). \qquad (9.106)$$

In terms of Sect. 9.3, this first-order time-optimal control system has a zero-dimensional switching boundary in the one-dimensional state space of y, which is the point, $y = y_r$. With reference to Sect. 9.5, the signum function of control law (9.106) is replaced by a saturation function, yielding the saturated control law,

$$u = \operatorname{sat}[K.S(y, y_r), -u_{\max}, u_{\max}] = \operatorname{sat}[K(y_r - y), -u_{\max}, u_{\max}]. \qquad (9.107)$$

The boundaries forming the extremities of the boundary layer are the two points, y_+^a and y_-^a, given by $K(y_r - y_+^a) = u_{\max}$ and $K(y_r - y_-^a) = -u_{\max}$, i.e. $y_+^a = y_r - \frac{u_{\max}}{K}$ and $y_-^a = y_r + \frac{u_{\max}}{K}$. In terms of the error, $\varepsilon = y - y_r$, these boundary points are at $\varepsilon_-^a = u_{\max}/K$ and $\varepsilon_+^a = -u_{\max}/K$. As K increases, the two boundaries become closer and coincide with the time-optimal boundary as $K \to \infty$. Within the boundary layer, (9.107) becomes the simple proportional control law,

$$u = K(y_r - y). \qquad (9.108)$$

9.7 Feedback Control of First-Order Plants

This eases the analysis within the boundary layer. Thus, the closed-loop differential equation is obtained by substituting for u in (9.105) using (9.108) to yield

$$\dot{y} = ay + b\left[K\left(y_r - y\right) - d\right] \tag{9.109}$$

In terms of the error, $\varepsilon = y - y_r$, with y_r constant, (9.109) becomes

$$\dot{\varepsilon} = (a - bK)\left(\varepsilon + \frac{ay_r - bd}{a - bK}\right), \tag{9.110}$$

The solution for constant d is

$$\varepsilon(t) = e^{(a-bK)t}\varepsilon(0) + \left[1 - e^{(a-bK)t}\right]\left(\frac{ay_r - bd}{a - bK}\right). \tag{9.111}$$

If $a - bK < 0$, then $\varepsilon(t)$ decays exponentially to a steady-state value of

$$\varepsilon_{ss} = \frac{ay_r - bd}{a - bK} \tag{9.112}$$

with a time constant of $1/(bK - a)$. If K is made sufficiently large, ε_{ss} can be reduced to negligible proportions.

Figure 9.23 shows simulations for $a = \pm 1$, $b = 1$, $K = 100$, $u_{\max} = 10$ and $d = u_{\max}/2$.

The two closed-loop response portraits are as would be expected in the ideal time-optimal control, indicating the error being driven to approximately zero in a finite time dependent on the initial error and the plant parameter, a. The responses on the right are magnified samples from the response portraits, also indicating the initial control saturation. As would be expected, $\varepsilon(t)$ does not reach precisely zero in a finite time as would be the case with the ideal time-optimal control but, as predicted by (9.111), exponentially decays towards a steady-state value within the boundary layer, which is dependent upon the constant external disturbance. Likewise the control exponentially settles to the value needed to maintain a constant steady-state error and counteract the external disturbance.

If K is increased further, the short exponential transients visible in Fig. 9.23 within the boundary layer (shaded) diminish and the steady-state error almost vanishes. Then $\varepsilon(t)$ will *appear* to reach zero in a finite time, as in the ideal time-optimal control. Setting $K = 1,000$ achieves this, but a lower value of K is used in Fig. 9.23 to enable the transient behaviour to be seen.

It is evident from the above that only the introduction of the boundary layer via the saturation function is needed for first-order systems. Further refinements are needed for higher-order systems.

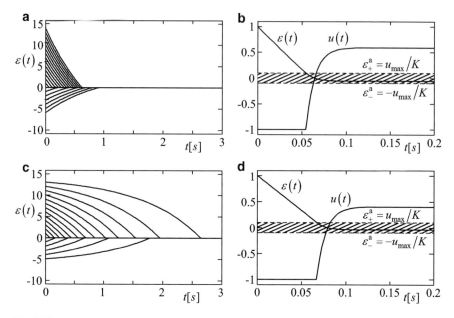

Fig. 9.23 Near-time-optimal control of first-order plant with continuous actuators and boundary layer. (**a**) Closed-loop response portrait for $a=-1$. (**b**) Behaviour in boundary layer for $a=-1$. (**c**) Closed-loop response portrait for $a=1$. (**d**) Behaviour in boundary layer for $a=1$

9.8 Feedback Control of Second-Order Plants

9.8.1 Introduction

Attention is now turned to the control of second-order plants modelled by a pair of state differential equations and a measurement equation of the general form,

$$\dot{x}_1 = f_1(x_1, x_2, u), \tag{9.113}$$

$$\dot{x}_2 = f_2(x_1, x_2, u) \tag{9.114}$$

$$y = g(x_1, x_2) \tag{9.115}$$

The range of plants considered is restricted to applications in which switched control actuators are used or control saturation is expected.

9.8.2 State Trajectories and State Portraits

The state of the plant may be displayed by plotting one state variable against the other as Cartesian coordinates of a point on a graph. The fact that time is an implicit

9.8 Feedback Control of Second-Order Plants

parameter, in contrast to the first-order systems considered in the previous section, is not a serious disadvantage, since the state variables contain all the information about the dynamic behaviour of the plant. The plane on which the graph is plotted is called the *state plane*, which is the two-dimensional state-space of a second-order plant. In many cases, one state variable may be the controlled output (or measurement variable) and the other state variable may be taken as the first output derivative. During the early developments of the state space theory of dynamical systems, state variables chosen in this way were referred to as *phase variables* and the graph on which they were plotted was referred to as the *phase plane*. This term is still in common use and may be used if $\dot{x}_1 = x_2$. The general term 'state plane' may be used.

Before developing phase (or state) plane methods, more nomenclature in common use will be defined. This will be done with reference to the double integrator,

$$\dot{x}_1 = x_2, \qquad (9.116)$$

$$\dot{x}_2 = b_0 u, \quad u \in [u_{\min}, u_{\max}] \qquad (9.117)$$

and

$$y = x_1 \qquad (9.118)$$

which models some real plants, such as a single-attitude control axis of a three-axis-stabilised spacecraft with negligible inter-axis coupling (Chap. 2).

First, a *state (or phase) trajectory* is the path taken by the point, $[x_1(t), x_2(t)]$, in the state (or phase) plane, sometimes referred to as the *state point*. It is the graph of $x_2(t)$ plotted against $x_1(t)$ with t as a parameter.

As for the first-order plants of the previous section, the plant behaviour with a constant control input of $u = \pm u_{\max}$ will be considered. The solution of the state Eqs. (9.116) and (9.117) is then

$$x_2(t) = b_0 u t + x_2(0) \qquad (9.119)$$

$$x_1(t) = \tfrac{1}{2} b_0 u t^2 + x_2(0)t + x_1(0). \qquad (9.120)$$

Since $x_2(t)$ and $x_1(t)$ are, respectively, linear and parabolic with respect to t, it is clear that the graph of $x_2(t)$ against $x_1(t)$ is a parabola. The equation of this trajectory could be derived by eliminating t between (9.119) and (9.120), but a more straightforward approach is to solve the *state trajectory differential equation* which is obtained directly by dividing one state differential equation by the other. In the general case, (9.114)/(9.113) yields

$$\frac{\dot{x}_2}{\dot{x}_1} = \frac{dx_2}{dx_1} = \frac{f_2(x_1, x_2, u)}{f_1(x_1, x_2, u)}. \qquad (9.121)$$

For the double integrator, (9.117)/(9.116) yields the following.

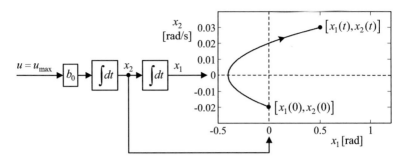

Fig. 9.24 State trajectory for a double integrator plant

$$\frac{dx_2}{dx_1} = \frac{b_0 u}{x_2}. \tag{9.122}$$

For constant u, this can be solved by the method of separation of variables. Thus,

$$b_0 u \int dx_1 = \int x_2 dx_2 \Rightarrow x_1 = \frac{1}{2b_0 u} x_2^2 + C, \tag{9.123}$$

which is a parabola symmetrical about the x_1 axis, where C is an arbitrary constant of integration. This constant can be expressed in terms of the initial state, $[x_1(0), x_2(0)]$, using (9.123), as follows:

$$C = x_1(0) - \frac{1}{2b_0 u} x_2^2(0). \tag{9.124}$$

Substituting for C in (9.123) using (9.124) then yields the *state trajectory equation*,

$$x_1 = x_1(0) + \frac{1}{2b_0 u} \left[x_2^2 - x_2^2(0) \right]. \tag{9.125}$$

A state trajectory is plotted in Fig. 9.24 for a single-attitude control axis of a space satellite regarded as a rigid body with moment of inertia, $J = 200 \; [\text{kg m}^2]$, controlled by a reaction wheel with torque constant, $K_w = 0.01 \; [\text{Nm/V}]$.

Plant model (9.116) through (9.118) applies with $b_0 = K_w/J$ and control saturation levels of $u_{max} = -u_{min} = 10$ V. The state trajectory shown is for $u = u_{max}$.

In order to make use of the state trajectory for control system analysis or design, it is important to determine the *direction* of motion of the state point along the trajectory as t increases. This can be found from the state differential Eqs. (9.113) and (9.114) by selecting any point, (x_1, x_2), on the trajectory, inserting the values of x_1 and x_2 in the right-hand sides and then examining the *signs* of $\dot{x}_1$ or $\dot{x}_2$, either of which determine the direction of motion. As an example, for the point, $(x_1, x_2) = (0, 0.2)$ in Fig. 9.24, $\dot{x}_1 = 0.2 > 0$, meaning that the projection of the

9.8 Feedback Control of Second-Order Plants

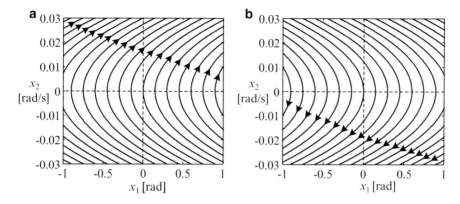

Fig. 9.25 Open-loop phase portraits for two control levels of a double integrator plant. (**a**) $u = +u_{max}$. (**b**) $u = -u_{max}$.

state point on the x_1 axis of Fig. 9.24 must be moving from left to right, as indicated by the arrow. For *phase variables*, x_2 is the time derivative of x_1. In this case, if $x_2 > 0$, x_1 must be increasing and conversely if $x_2 < 0$, x_1 must be decreasing, which is useful for determining the direction of motion on the phase portrait.

Also, if the state variables are phase variables, (9.113) reduces to $\dot{x}_1 = x_2$, (9.121) becomes $\frac{dx_2}{dx_1} = \frac{x_2}{f_1(x_1, x_2, u)}$ and therefore provided $f_1(x_1, x_2, u) \neq 0$, for $x_2 = 0$, the state trajectory crosses the x_2 axis at right angles, as exemplified in Fig. 9.24.

Useful information for the design of switched control systems may be gained from a family of state trajectories for each available control level. These families of trajectories give a complete *picture* of the dynamic behaviour of the plant under the selected control values and are referred to as *state portraits* or *phase portraits*. Since these state portraits are for the constant control levels without any feedback control, they are referred to as *open-loop state portraits*. Examples are shown in Fig. 9.25 for the space satellite example used to generate Fig. 9.24. Since any control system is required to respond to a reference input, which in this case will be assumed constant, the plant model of (9.116) and (9.117) will first be formulated in terms of the error state variables, $x_{1e} = x_1 - y_r$ and $x_{2e} = x_2 - \dot{y}_r = x_2$. Then the control objective is to bring x_{1e} and x_{2e} simultaneously to zero in the minimum time. The required plant model is obtained simply by replacing x_1 and x_2 in (9.116) and (9.117) by x_{1e} and x_{2e}.

9.8.3 Time-Optimal Feedback Control of the Double Integrator Plant

The equations of Pontryagin's method given in Sect. 9.6.2 will be applied, where the cost functional to be minimised is (9.75), so the model is required in the matrix–vector form,

$$\begin{bmatrix} \dot{x}_{1e} \\ \dot{x}_{2e} \end{bmatrix} = \begin{bmatrix} 0 & 1 \\ 0 & 0 \end{bmatrix} \begin{bmatrix} x_{1e} \\ x_{2e} \end{bmatrix} + \begin{bmatrix} 0 \\ b_0 \end{bmatrix} u. \quad (9.126)$$

Then the costate Eq. (9.68) is

$$\begin{bmatrix} \dot{p}_1 \\ \dot{p}_2 \end{bmatrix} = -\begin{bmatrix} 0 & 0 \\ 1 & 0 \end{bmatrix} \begin{bmatrix} p_1 \\ p_2 \end{bmatrix}. \quad (9.127)$$

The Hamiltonian is

$$H = \mathbf{p}^T \mathbf{F}(\mathbf{x}, \mathbf{u}) - L(\mathbf{x}, \mathbf{u}) = p_1 \dot{x}_{1e} + p_2 \dot{x}_{2e} - 1 = p_1 x_{2e} + p_2 b_0 u. \quad (9.128)$$

The optimal control that maximises H is therefore

$$u(t) = u_{\max} \operatorname{sgn}[p_2(t)], \quad (9.129)$$

noting that $b_0 > 0$. The nature of the optimal control switching can then be investigated by finding the general solution of (9.127), which is

$$\begin{cases} p_1(t) = p_1(0) = const. \\ p_2(t) = p_1(0)t + p_2(0) \end{cases}. \quad (9.130)$$

Since $p_2(t)$ can change sign only once, the time-optimal control for this plant has only one switch at time, $t_{s1} = -p_2(0)/p_1(0)$, calculation of which requires the determination of a pair of initial costates, that are not unique, corresponding to the initial plant state. This, however, would only be needed for application of the optimal control on an open-loop basis. The only information needed, in addition to the plant model, to derive the far more preferable time-optimal *feedback* control law is the fact that the optimal control has only one switch. The back tracing method of Sect. 9.6.2.4 may be used to achieve this readily. The reversed time plant model obtained by setting $t = -\tau$ in (9.116) and (9.117) is given by

$$dx_{1e}/d\tau = -x_{2e} \quad (9.131)$$

$$dx_{2e}/d\tau = -b_0 u \quad (9.132)$$

The initial error state with respect to the reversed time, τ, is $[x_{1e}(0), x_{1e}(0)] = [0, 0]$, and for u constant ($\pm u_{\max}$), the solution is

$$x_{2e}(\tau) = -b_0 u \tau \quad (9.133)$$

and

$$x_{1e}(\tau) = \tfrac{1}{2} b_0 u \tau^2 \quad (9.134)$$

These equations define two state trajectories with τ as a parameter, one for $u = +u_{\max}$ and the other for $u = -u_{\max}$. Since the time-optimal switching function in the state space is required and this is independent of time, it will be useful to eliminate τ between (9.133) and (9.134) to obtain the trajectory equations as follows:

9.8 Feedback Control of Second-Order Plants

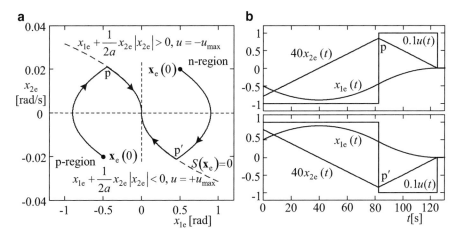

Fig. 9.26 Time-optimal switching boundary and trajectories for double integrator plant. (**a**) Switching boundary and state trajectories. (**b**) State and control variables

$$x_{1e} = \frac{1}{2b_0 u_{\max}} x_{2e}^2 \quad \text{for } u = u_{\max}. \tag{9.135}$$

$$x_{1e} = -\frac{1}{2b_0 u_{\max}} x_{2e}^2 \quad \text{for } u = -u_{\max}. \tag{9.136}$$

These two trajectories include the last segment of any time-optimal state trajectory.

It is clear that the single switch of any time-optimal control must occur on one of them. Hence, they are two segments of the time-optimal switching boundary. Let $[x_{1s}, x_{2s}]$ be any error state lying on this boundary. Then the equation of the switching boundary is obtained by combining (9.135) and (9.136) while taking note of the sign of x_{2e} using (9.133) and then replacing $[x_{1e}, x_{2e}]$ by $[x_{1s}, x_{2s}]$. Thus,

$$x_{1s} = -\frac{1}{2b_0 u_{\max}} x_{2s}^2 \operatorname{sgn}(x_{2s}) \Rightarrow x_{1s} + \frac{1}{2b_0 u_{\max}} x_{2s} |x_{2s}| = 0 \tag{9.137}$$

Noting that $x_{1e} = x_1 - x_{1r}$, the corresponding switching function and the control law in the form of (9.45) are therefore

$$S(\mathbf{x}_e) = S(x_{1e}, x_{2e}) = -x_{1e} - \frac{1}{2b_0 u_{\max}} x_{2e} |x_{2e}| \tag{9.138}$$

and

$$u = u_{\max} \operatorname{sgn}[S(\mathbf{x}_e)]. \tag{9.139}$$

Continuing with the satellite example used to generate Figs. 9.25 and 9.26 shows the graph of the time-optimal switching boundary together with time-optimal state trajectories for two different initial states.

Commencing with an arbitrary initial state, a constant control level of $u = \pm u_{max}$ is applied until the switching boundary is reached. The control switch occurs just as the state trajectory crosses the boundary, such as at point, p, or point, p′, in Fig. 9.26a, after which the state trajectory remains on the same side of the boundary but an infinitesimal distance away, following it to the origin of the phase plane, as shown.

9.8.4 Time-Optimal Control Law Synthesis Using State Portraits

A switched control law may be found by studying the state portraits for the available control levels and using this information to divide the state plane into regions where these control levels should be applied. The closed-loop behaviour may then be predicted by creating the *closed-loop state portrait* in which the boundaries separating the aforementioned regions are marked and the portions of the appropriate open-loop state portrait are shown in each region. The closed-loop state trajectories and the associated control switches may then be predicted for any initial state within the range of the closed-loop state portrait. In order to produce a control algorithm for implementation, however, equations for the switching boundaries must be found.

The method will now be demonstrated for the double integrator plant by study of the phase portraits of Fig. 9.25. Here, an observation may be made that there is a unique parabolic state trajectory segment terminating on the origin of the phase plain for $u = u_{max}$ and the trajectories for $u = -u_{max}$ cross this trajectory segment along its semi-infinite length. Similarly, there is a unique parabolic state trajectory segment terminating on the origin of the phase plain for $u = -u_{max}$ and the trajectories for $u = u_{max}$ cross this trajectory segment along its semi-infinite length. An interesting candidate-switching boundary would therefore consist of these two trajectory segments combined as shown in Fig. 9.27a. In fact, this is the time-optimal

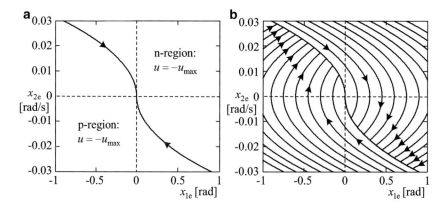

Fig. 9.27 Time-optimal switching boundary and closed-loop phase portrait for a double integrator plant. (**a**) State trajectories forming switching boundary. (**b**) Closed-loop phase portrait

9.8 Feedback Control of Second-Order Plants

switching boundary of Fig. 9.26a. The closed-loop-phase portrait consists of an infinite continuum of time-optimal state trajectories, only a few of which are shown in Fig. 9.26b. In this case, the parabolic state trajectory equations are well known and lead directly to (9.137), (9.138) and (9.139).

Example 9.4 Time-optimal control of a heating process with two dominant time constants

A kiln is modelled by the state differential equations and measurement equation,

$$\{\dot{x}_1 = a_1(x_2 - x_1), \quad \dot{x}_2 = a_2(K_{dc}u - x_2), \quad y = x_1\}, \quad (9.140)$$

where x_1 is the workpiece temperature to be controlled and x_2 is the temperature of the surrounding refractory bricks, $a_1 = 1/T_1$, $a_2 = 1/T_2$. T_1, T_2 and K_{dc} are, respectively, the two time constants characterising the process and the plant DC gain. Given the minimum control level, $u_{min} = 0$, the maximum control level, $u_{max} = 10$ [V], $T_1 = 100$ [s], $T_2 = 50$ [s] and $K_{dc} = 100$ °C/V, it is required to find the phase portraits for the two available control levels, use these to determine the time-optimal switching boundary and determine the time-optimal feedback control law for a constant reference input, y_r.

First, the plant model (9.140) will be reformulated in terms of the error state variables, $x_{1e} = x_1 - y_r$ and $x_{2e} = x_2 - y_r$, noting that the steady-state temperature for constant u is $x_{1ss} = x_{2ss} = K_{dc}u$. Thus, from (9.140),

$$\dot{x}_{1e} = a_1(x_{2e} - x_{1e}) \quad (9.141)$$

$$\dot{x}_{2e} = a_2[K_{dc}u - (x_{2e} + y_r)] \quad (9.142)$$

Figure 9.28 shows the state portraits within the operating envelope, which is the range of temperatures between the steady-state values with $u = u_{min}$ and $u = u_{max}$. It will be recalled from Sect. 9.6.2.3 that the number of switches in the time-optimal control is $n - 1$ if the plant has real poles. This is true here, as the plant poles are at $s_{1,2} = -a_1, -a_2$, so there is only one switch. In Fig. 9.28 there are two trajectory segments reaching the origin of the error state plane, P_1P_0 for $u = u_{max}$ and Q_1Q_0 for $u = 0$. Now suppose that Fig. 9.28a, b are superimposed.

Then the segment P_1P_0 for $u = u_{max}$ is crossed by trajectories for $u = 0$ along its entire length and the segment Q_1Q_0 for $u = 0$ is similarly crossed by trajectories for $u = u_{max}$. The time-optimal switching boundary is therefore formed by joining these two segments.

Next, the equations of these trajectory segments terminating at the origin will be derived. For back tracing, Eqs. (9.141) and (9.142) become

$$dx_{1e}/d\tau = a_1(x_{1e} - x_{2e}) \quad (9.143)$$

$$dx_{2e}/d\tau = a_2[x_{2e} + y_r - K_{dc}u] \quad (9.144)$$

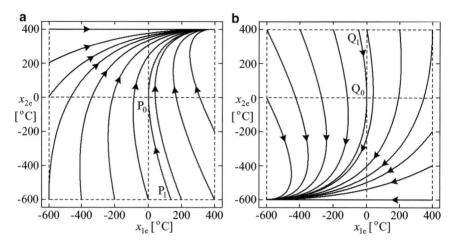

Fig. 9.28 Open-loop state portraits for heating process. (**a**) $u = u_{\max}$. (**b**) $u = 0$

The solution with u constant, is weighted sums of the exponentials, $e^{a_1\tau}$, $e^{a_2\tau}$, and constants. Noting that x_{2e} is the only state variable in (9.144), and $x_1(0) = x_2(0) = 0$ for the back tracing, the solution is of the form

$$x_{2e}(\tau) = A\left(e^{a_2\tau} - 1\right) \tag{9.145}$$

and

$$x_{1e}(\tau) = B\left(e^{a_1\tau} - 1\right) + C\left(e^{a_2\tau} - 1\right), \tag{9.146}$$

where A, B and C are constants to be determined. Differentiating (9.145) and (9.146) w.r.t. τ and substituting for $dx_{1e}(\tau)/d\tau$, $dx_{2e}(\tau)/d\tau$, $x_{1e}(\tau)$ and $x_{2e}(\tau)$ in (9.143) and (9.144) yields

$$A = y_r - K_{dc}u, \quad B = \frac{a_2}{a_2 - a_1}A \quad \text{and} \quad C = \frac{a_1}{a_1 - a_2}A. \tag{9.147}$$

The required trajectory equations are then obtained by eliminating τ between (9.145) and (9.146). From (9.145),

$$e^{a_2\tau} - 1 = x_{2e}/A \tag{9.148}$$

Also

$$e^{a_1\tau} = e^{(a_1/a_2)a_2\tau} = (e^{a_2\tau})^{a_1/a_2} = \left(\frac{x_{2e}}{A} + 1\right)^{a_1/a_2}. \tag{9.149}$$

9.8 Feedback Control of Second-Order Plants

Then substituting for $e^{a_2\tau} - 1$ and $e^{a_1\tau}$ in (9.146) using (9.148) and (9.149) yields

$$x_{1e} = B\left[\left(\frac{x_{2e}}{A} + 1\right)^{a_1/a_2} - 1\right] + (C/A) x_{2e}. \tag{9.150}$$

Substituting for A, B and C in (9.150) using (9.147) then yields

$$x_{1e} = \frac{1}{a_2 - a_1}\left\{a_2(y_r - K_{dc}u)\left[\left(\frac{x_{2e}}{y_r - K_{dc}u} + 1\right)^{a_1/a_2} - 1\right] - a_1 x_{2e}\right\} \tag{9.151}$$

With reference to Fig. 9.28, for $x_{2e} < 0$, the trajectory segment, P_1P_0, coincides with the switching boundary and therefore $u = u_{max}$. Similarly, for $x_{2e} > 0$, the trajectory segment, Q_1Q_0, coincides with the switching boundary and therefore $u = 0$. Then (9.151) becomes the time-optimal switching boundary with

$$u = u_s = \tfrac{1}{2}[1 - \text{sgn}(x_{2e})] u_{max}. \tag{9.152}$$

Given that $x_{1e} = x_1 - x_{1r}$, the corresponding switching function is

$$S(\mathbf{x}_e, y_r) = -x_{1e} + \frac{1}{a_2 - a_1}\left\{a_2(y_r - K_{dc}u_s)\left[\left(\frac{x_{2e}}{y_r - K_{dc}u_s} + 1\right)^{a_1/a_2} - 1\right] - a_1 x_{2e}\right\}. \tag{9.153}$$

The time-optimal control law corresponding to (9.45) is then

$$u = \frac{1}{2}\{1 + \text{sgn}[S(\mathbf{x}_e, y_r)]\} u_{max}. \tag{9.154}$$

Figure 9.29a shows the time-optimal switching boundary, P0Q, and the closed-loop state portrait formed from Fig. 9.28a, b. Although this is different in shape from that of the double integrator plant shown in Fig. 9.27b, it is topologically similar in that after the first switch the state trajectory follows the switching boundary (in theory an infinitesimal distance away) to the origin of the error state plane. This is true for any second-order plant with real poles. If, however, the uncontrolled plant has an oscillatory mode, then the time-optimal control has more than one switch and the movement along the switching boundary occurs only after the last switch. It should be noted that these statements apply to the state trajectory leading to the demanded state, but after this has been reached, only approximately in a real system with finite sampling frequency of the digital processor and plant modelling inaccuracies, further switching will occur just to maintain the system close to the demanded state. This can be seen in Fig. 9.29b, which shows the system response to a step reference temperature of $y_r = 600\,°C$ with initial state variables of $x_1(0) = x_2(0) = 0 \Rightarrow x_{1e}(0) = x_{2e}(0) = -y_r$.

The time-optimal control can be seen to automatically raise the temperature, $x_2(t)$, of the refractory bricks well above the demanded workpiece temperature.

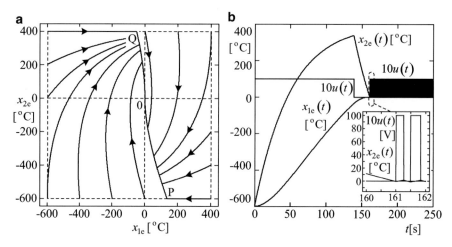

Fig. 9.29 Time-optimal control of heating process. (**a**) Closed-loop state portrait. (**b**) Step response

Then after the heating element is turned off, heat transfer to the workpiece continues until it is at the required temperature. After this, a limit cycle occurs indicated by the solid black band of $u(t)$ in Fig. 9.29b due to the switching being too frequent to be visible on the 250 s timescale. In fact, the basic switching element of Fig. 9.15a has been replaced by the hysteresis element of Fig. 9.15b with $S^+ = -S^- = 0.001°$. Despite the relatively narrow hysteresis band (defined as $S^+ - S^-$), the insert on the expanded timescale indicates the switching frequency to be about 2 Hz, which is well within the capability of power electronic switching devices. The mark–space ratio automatically adjusts so that $x_1(t)$ and $x_2(t)$ both oscillate about a constant steady-state value of 600 °C.

9.8.5 Continuous Control with Saturation

9.8.5.1 Unintentional Saturation

In general, if the uncontrolled plant is stable, unintentional control saturation of a control system designed to operate with a continuous control variable will not be problematic but may require more attention if the uncontrolled plant is unstable or oscillatory with only light damping. The following method of analysis based on the phase plane is generally applicable for continuous control of second-order plants with non-dynamic controllers. It will be introduced via its application to a controllable pendulum with unlimited angle of movement and negligible damping. This type of open-loop behaviour is encountered in the control and guidance of

9.8 Feedback Control of Second-Order Plants

spacecraft launch vehicles and robotic mechanisms. The state-space model of the plant to be considered is

$$\begin{cases} \dot{x}_1 = x_2 \\ \dot{x}_2 = bu - a\,\sin(x_1) \\ y = x_1 \end{cases} \quad (9.155)$$

where x_1 is the pendulum angular displacement from the vertical hanging position, x_2 is the angular velocity, $a = g/L$ and $b = K_T/(ML^2)$. Here, $g \cong 9.81$ m/s/s is the acceleration due to gravity, L is the pendulum length, M is the pendulum mass and K_T is the motor torque constant. It is assumed that the angle measurement, y, is scaled in the control computer so that it is numerically equal to the pendulum angle. Equal and opposite control saturation limits of $\pm u_{\max}$ will be assumed.

A feedback linearising (forced dynamic) control law will be designed to achieve a critically damped closed-loop system with a 5 % settling time of T_s seconds. The control law can be very quickly derived from the second-order plant differential equation, obtained from (9.155) as

$$\ddot{y} = bu - a\,\sin(y) \quad (9.156)$$

and the desired closed-loop differential equation,

$$\ddot{y} = p_0(y_r - y) - p_1\dot{y}, \quad (9.157)$$

where $p_0 = 81/(4T_s^2)$ and $p_1 = 9/T_s$. Thus, equating the RHS of (9.156) and (9.157) yields the control law

$$u = \frac{1}{b}[p_0(y_r - y) - p_1\dot{y} + a\,\sin(y)] = \frac{1}{b}[p_0(y_r - x_1) - p_1 x_2 + a\,\sin(x_1)]. \quad (9.158)$$

Without control saturation, the pendulum can be controlled to any angle with the closed-loop dynamics of (9.157). With control saturation, study of the two phase portraits, one for $u = u_{\max}$ and the other for $u = -u_{\max}$, will enable the system behaviour in saturation to be investigated. These are used in conjunction with the two *saturation boundaries* of the control law, which bound the region of the state space in which the control system operates without saturation, which will be called the *continuous region*. The saturation boundaries are the loci of all points in the phase plane for which the control law demands a control that just reaches one of the two saturation limits. For $u = +u_{\max}$, the saturation boundary is called the 'p' saturation boundary and its equation is just (9.158) with $u = +u_{\max}$. For plotting purposes, x_2 can be made the subject of this equation to yield

$$x_2 = \frac{1}{p_1}[p_0(y_r - x_1) + a\,\sin(x_1) - bu_{\max}]. \quad (9.159)$$

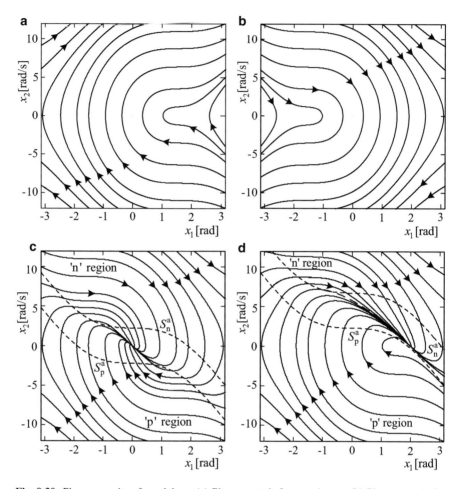

Fig. 9.30 Phase portraits of pendulum. (**a**) Phase portrait for $u = +u_{max}$. (**b**) Phase portrait for $u = -u_{max}$. (**c**) Closed-loop phase portrait for $y_r = 0$ rad. (**d**) Closed-loop phase portrait for $y_r = 2$ rad

Similarly, for $u = -u_{max}$, the 'n' saturation boundary is given by

$$x_2 = \frac{1}{p_1} \left[p_0 \left(y_r - x_1 \right) + a \, \sin(x_1) + b u_{max} \right]. \tag{9.160}$$

The phaseportraits of the uncontrolled pendulum have been computed for $M = 1$ [Kg], $L = 0.5$ [m], $K_T = 0.5$ [Nm/V] and $u_{max} = 10$ [V] and are plotted in Fig. 9.30a, b. The values of K_T and u_{max} have been chosen so that the maximum control torque is only just sufficient to move the pendulum to angles between $\pm \pi/2$ [rad] and $\pm \pi/2$ [rad] so that saturation will be likely during loop closure.

9.8 Feedback Control of Second-Order Plants

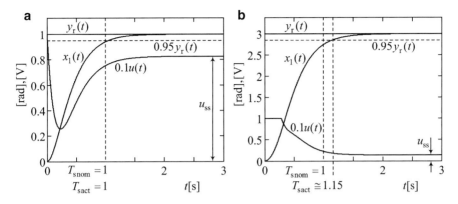

Fig. 9.31 Step responses of feedback linearising control of a pendulum. (**a**) Without control saturation. (**b**) With control saturation

Figure 9.30c, d are closed-loop-phase portraits produced from simulations with various initial states. The 'p' and 'n' saturation boundaries are superimposed on the closed-loop phase portraits, shown dotted and indicated, respectively, as S_p^a and S_n^a. As expected, the state trajectories converge to the desired state points $(x_1, x_2) = (0, 0)$ in Fig. 9.30c and $(x_1, x_2) = (2, 0)$ in Fig. 9.30d within the continuous region between the saturation boundaries where the forced dynamic control law operates as intended. The control saturates at $u = +u_{\max}$ in the 'p' region below the boundary, B_p, and saturates at $u = -u_{\max}$ in the 'n' region above the boundary, B_n. This is confirmed by observing that the state trajectories follow the phase portrait of Fig. 9.30a for $u = +u_{\max}$ in the 'p' region and they follow the phase portrait of Fig. 9.30b for $u = -u_{\max}$ in the 'n' region. For $y_r = 0$, as would be expected, every feature of Fig. 9.30c is reflected in the origin, which is the desired state point. For $y_r = 0$, however, the continuous region moves to the right, but the desired state point is very close to the saturation boundary, B_p, which is a consequence of the maximum torque being only just sufficient to lift the pendulum to an angle of 2 rad. The overall closed-loop behaviour, however, can be seen to be stable despite the saturation.

Figure 9.31 shows control system step responses with zero initial states, with and without saturation depending on the magnitude of the demanded angle.

These confirm that without the saturation, the settling time specification is met, but with the saturation, the actual settling time, $T_{s\,act}$, exceeds the nominal value, $T_{s\,nom}$. The steady-state value, u_{ss}, of the control is larger for the smaller reference angle due to the greater gravitational torque to be counteracted, considering that $y_r = 3$ rad is nearly an inverted attitude.

9.8.5.2 Saturating Control Systems with a Boundary Layer

Attention is now turned to control systems designed to operate with continuous actuators in saturation for relatively large changes of state, usually to minimise the settling time. The boundary layer introduced in Sect. 9.5 is a means to operate the actuators continuously as the desired state is approached in order to avoid limit cycling. In fact, the boundary layer is similar to the continuous region of Sect. 9.5 but with the difference that its shape in the phase plane is determined by the control system designer. Here, a continuous control law is replaced by a switching function designed to yield a certain performance with control saturation. The boundary layer is kept as small as possible to ensure that the performance for relatively large changes of state is almost the same as that of the 'ideal' system operating with hard switching according to the switching boundary. The focus here, however, is the control system behaviour within the boundary layer leading to the desired state. This will now be studied in detail by continuing the study of the rigid-body spacecraft attitude control system (single axis) whose basic time-optimal control law was derived in Sect. 9.8.3. The error state equations are

$$\begin{cases} \dot{x}_{1e} = x_{2e} \\ \dot{x}_{2e} = b_0 u \end{cases}, \qquad (9.161)$$

where $x_{1e} = x_1 - x_{1r}$, x_1 is the attitude angle to be controlled, x_{1r} is the constant reference input, $x_{2e} = x_2$, x_2 is the angular velocity, $b_0 = K_w/J$, K_w is the reaction wheel torque constant, J is the spacecraft moment of inertia and u is the reaction wheel drive input voltage. The time-optimal control law is then

$$u = u_{\max} \, \text{sgn}\left[S\left(x_{1e}, x_{2e}\right)\right] \qquad (9.162)$$

where $u_{\max}$ is the control magnitude saturation limit and the switching function is

$$S\left(x_{1e}, x_{2e}\right) = -x_{1e} - \frac{1}{2b_0 u_{\max}} x_{2e} \left|x_{2e}\right|. \qquad (9.163)$$

The double integrator example is taken because it contains no natural damping, in contrast with an earthbound positioning mechanism subject to friction. Due to this, refinements are needed further to the boundary layer. An understanding of these refinements is facilitated by following the steps of development leading to satisfactory performance. The first step is to introduce a boundary layer by replacing the signum function of (9.162) by a saturation function (Sect. 9.5). Thus,

$$u = \text{sat}\left[K.S\left(x_{1e}, x_{2e}\right), -u_{\max}, u_{\max}\right]. \qquad (9.164)$$

This introduces two saturation boundaries: $S_n^a(x_{1e}, x_{2e}) = 0$ for saturation at $u = -u_{\max}$ and $S_p^{\text{sat}}(x_{1e}, x_{2e}) = 0$ for saturation at $u = +u_{\max}$. These boundaries are obtained by translating the switching boundary, $S(x_{1e}, x_{2e}) = 0$, along the x_{1e} axis by S_-^{sat} to yield $S_n^{\text{sat}}(x_{1e}, x_{2e}) = 0$ and by S_+^{sat} to yield $S_p^a(x_{1e}, x_{2e}) = 0$. The equations of the two saturation boundaries are therefore

9.8 Feedback Control of Second-Order Plants

$$\begin{cases} S_n^{sat}(x_{1e}, x_{2e}) = S(x_{1e}, x_{2e}) - S_-^{sat} = 0 \\ S_p^{sat}(x_{1e}, x_{2e}) = S(x_{1e}, x_{2e}) - S_+^{sat} = 0 \end{cases}. \qquad (9.165)$$

where $S(x_{1e}, x_{2e})$ is given by (9.163). With reference to the saturation transfer characteristic of Fig. 9.12b, $S_-^{sat} = -u_{max}/K$ and $S_+^{sat} = +u_{max}/K$. The boundary layer width, i.e. the translational displacement between the saturation boundaries along the x_{1e} axis, is therefore

$$\Delta S^{sat} = S_+^{sat} - S_-^{sat} = \frac{2u_{max}}{K}. \qquad (9.166)$$

The two saturation boundaries are therefore placed on each side of the switching boundary they replace and as K is increased, they become closer to that switching boundary, the boundary layer diminishing. Hence, as $K \to \infty$, $S_n^{sat}(x_{1e}, x_{2e}) \to S(x_{1e}, x_{2e})$ and $S_p^{sat}(x_{1e}, x_{2e}) \to S(x_{1e}, x_{2e})$, and therefore the system behaviour approaches that of the basic system operating with hard switching. Figure 9.32 shows some simulation results with $J = 200 \,[\mathrm{kg\,m^2}]$, $K_w = 0.01$ [Nm/V] and $u_{max} = 10$ [V].

Figure 9.32a shows the variables during a slewing manoeuvre of π rad using the basic time-optimal control law of (9.162) with (9.163) for comparison purposes. The rapid switching of $u(t)$ during the limit cycle occurring after the system first approaches zero error state is indicated by the black area. This is referred to as *control chatter* in the context of sliding mode control, and the boundary layer is one of the methods used in Chap. 9 to eliminate it.

Figure 9.32b shows the result of introducing a boundary layer using (9.164) with (9.163). The boundary layer gain is set to $K = 200$, giving a boundary layer width of $2u_{max}/K = 0.1$ rad. While the initial transient behaviour is similar to that of the ideal time-optimal system, an issue is the oscillatory behaviour occurring in the region of the desired zero error state. It will now be shown that this is due to the quadratic term, $x_{2e}|x_{2e}|$, in the switching function. Within the boundary layer, the control law defined by (9.163) and (9.164) becomes

$$u = -K\left(x_{1e} + \frac{1}{2b_0 u_{max}} x_{2e}|x_{2e}|\right). \qquad (9.167)$$

Applying the method of linearisation about the operating point covered in Chap. 7, about an arbitrary point $(\overline{x}_{1e}, \overline{x}_{2e}, \overline{u})$ within the boundary layer, yields the equations for small changes, $\tilde{x}_{1e}, \tilde{x}_{2e}$ and $\tilde{u}$, with respect to the operating point

$$\begin{aligned}\tilde{u} &= -K\left\{\tilde{x}_{1e} + \tfrac{1}{2b_0 u_{max}}\left[x_{2e}\,\mathrm{sgn}(x_{2e}) + |x_{2e}|.1\right]\big|_{x_{2e}=\overline{x}_{2e}}.\tilde{x}_{2e}\right\} \\ &= -K\left(\tilde{x}_{1e} + \tfrac{1}{b_0 u_{max}}|\overline{x}_{2e}|\tilde{x}_{2e}\right).\end{aligned} \qquad (9.168)$$

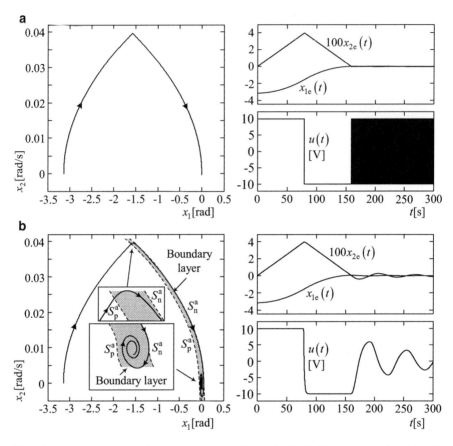

Fig. 9.32 Effect of boundary layer on time-optimal control of a double integrator plant (**a**) with basic time-optimal control law. (**b**) With time-optimal switching function and boundary layer

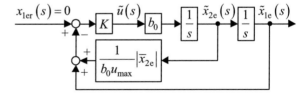

Fig. 9.33 Block diagram of linearised system within the boundary layer

The plant (9.161) is already linear, and therefore, linearisation will yield the same state equations with x_{1e} and x_{2e} replaced, respectively, by $\tilde{x}_{1e}$ and $\tilde{x}_{2e}$. Figure 9.33 shows the block diagram of the closed-loop system for these small changes.

The characteristic polynomial is

$$s^2\left\{1-\frac{Kb_0}{s}\left[-\frac{1}{b_0 u_{\max}}|\bar{x}_{2e}|-\frac{1}{s}\right]\right\}=s^2+\frac{K}{u_{\max}}|\bar{x}_{2e}|s+Kb_0=s^2+2\zeta\omega_n s+\omega_n^2$$

(9.169)

9.8 Feedback Control of Second-Order Plants

The closed-loop damping ratio and undamped natural frequency are therefore

$$\omega_n = \sqrt{Kb_0} \qquad (9.170)$$

and

$$\zeta = \frac{K}{2\omega_n u_{\max}} |\bar{x}_{2e}|. \qquad (9.171)$$

This explains the oscillatory behaviour about the desired error state, $(x_{1e}, x_{2e}) = (0, 0)$. For linearisation about this point, i.e. $(\bar{x}_{1e}, \bar{x}_{2e}) = (0, 0)$, (9.171) yields $\zeta = 0$, indicating oscillatory behaviour with zero damping. The spiral state trajectory shown in the lower left insert of Fig. 9.33b and the decaying oscillations of $u(t)$, however, indicate a small amount of damping. This is due to the small amount of velocity feedback brought about by $x_{2e}(t) \ne 0$ during the oscillations. This damping occurs since if relinearisation were to be performed at each point on the state trajectory encircling the origin of the phase plane, for $x_{2e} \ne 0$, then (9.171) would yield $\zeta > 0$. If (9.167) included a term linear with respect to x_{2e}, i.e. $K_d x_{2e}$, where K_d is an adjustable velocity feedback gain, then the linearisation would yield a constant component of ζ, thereby ensuring adequate damping with a suitable value of K_d. If the switching function (9.163) is modified to

$$S_m(x_{1e}, x_{2e}) = -x_{1e} - \frac{1}{2b_0 u_{\max}} x_{2e} |x_{2e}| - K_d x_{2e}, \qquad (9.172)$$

then (9.168) becomes

$$\tilde{u} = -K\left(\tilde{x}_{1e} + \frac{1}{b_0 u_{\max}} |\bar{x}_{2e}| \tilde{x}_{2e} + K_d \tilde{x}_{2e}\right) \qquad (9.173)$$

and consequently the closed-loop characteristic polynomial (9.169) becomes

$$s^2 + Kb_0\left(\frac{|\bar{x}_{2e}|}{u_{\max}} + K_d\right)s + Kb_0 = s^2 + 2\zeta\omega_n s + \omega_n^2. \qquad (9.174)$$

This enables the damping ratio in the region of the desired state to be chosen and K_d calculated to achieve this, once the boundary layer width is set using the gain, K, via (9.166). Setting $|\bar{x}_{2e}| = 0$ in (9.174) for linearisation about $(\bar{x}_{1e}, \bar{x}_{2e}) = (0, 0)$, equating the coefficients of s in (9.174) and using (9.170) then yield

$$K_d = \frac{2\zeta\omega_n}{Kb_0} = \frac{2\zeta}{\sqrt{Kb_0}}. \qquad (9.175)$$

This modification alone, however, has the effect of 'bending' the switching boundary towards the x_{1e} axis with the result that the state trajectory enters the boundary layer too soon, causing premature deceleration and consequently increasing the settling time beyond the time-optimal value. This will be demonstrated shortly. A further modification is possible, however, to overcome this problem. To arrive at this, attention is drawn to the first insert in Fig. 9.32b that shows the state trajectory crossing the S_p^{sat} saturation boundary into the boundary layer and moving monotonically towards the opposite S_n^{sat} saturation boundary, nearly reaching it well before approaching the origin of the phase plane. This causes an overshoot of almost $u_{max}/K = 0.05$ rad in magnitude. To show that the state trajectory moves towards the S_n^{sat} saturation boundary for $x_{2e} > 0$ within the boundary layer and similarly towards the S_p^{sat} for $x_{2e} < 0$, consider the slope of the state trajectory. From (9.161), this is $dx_{2e}/dx_{1e} = \dot{x}_{2e}/\dot{x}_{1e} = b_0 u/x_{2e}$. On the S_n^{sat} boundary, for $x_{2e} > 0$, $u = -u_{max}$, and therefore the trajectory slope is $dx_{2e}/dx_{1e} = -b_0 u_{max}/x_{2e}$. Similarly on the S_p^{sat} boundary, for $x_{2e} < 0$, $u = +u_{max}$, giving a trajectory slope of $dx_{2e}/dx_{1e} = +b_0 u_{max}/x_{2e}$. Within the boundary layer, $|u| < u_{max}$ and therefore $|dx_{2e}/dx_{1e}| < b_0 u_{max}/|x_{2e}|$. This implies that for $x_{2e} > 0$, the state trajectory must be moving towards the S_n^{sat} boundary and that for $x_{2e} < 0$, the state trajectory must be moving towards the S_p^{sat} boundary. Once within the boundary layer, the trajectory cannot 'escape' from it by crossing the saturation boundary it is approaching because the trajectory slope tends asymptotically towards the slope of the saturation boundary as it approaches it. This behaviour is evident in the phase plane of Fig. 9.32b.

Starting with the basic time-optimal switching boundary, a modification that eliminates the overshoot is to displace the upper segment for $x_{2e} \in (0, \infty)$ along the x_{1e} axis by $-u_{max}/K$ and similarly displace the lower segment for $x_{2e} \in [0, -\infty)$ by $+u_{max}/K$. This is implemented by replacing the switching function (9.163) by

$$S'(x_{1e}, x_{2e}) = -\left[x_{1e} + \frac{u_{max}}{K} \operatorname{sgn}(x_{2e})\right] - \frac{1}{2b_0 u_{max}} x_{2e} |x_{2e}|. \quad (9.176)$$

Figure 9.34a shows the saturation boundaries resulting when using (9.176) with the saturation function (9.164).

These are similar to the modified boundary, $S'(x_{1e}, x_{2e}) = 0$, displaced by $\pm u_{max}/K$ along the x_{1e} axis. The result is that for $x_{2e} > 0$, the $S_n'^{sat}$ saturation boundary passes through the origin of the phase plane, as does the $S_p'^{sat}$ saturation boundary for $x_{2e} < 0$. These are the boundary segments that the state trajectory will approach closely from within the boundary layer, and therefore, the overshoot

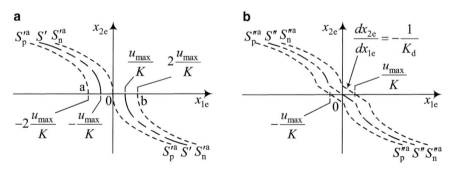

Fig. 9.34 Modified saturation boundaries to prevent overshooting and introduce damping. (**a**) Introduction of skew offset. (**b**) Introduction of linear segment

9.8 Feedback Control of Second-Order Plants

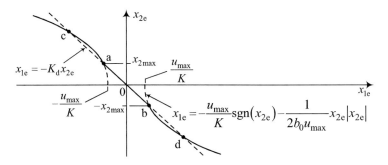

Fig. 9.35 Determination of the inner intersection points for segmented switching boundary

of Fig. 9.32b cannot occur. With reference to Fig. 9.34a, however, the discontinuity at $x_{2e} = 0$ introduced by the term $\frac{u_{max}}{K}$ sgn (x_{2e}) in the switching function (9.176) causes the saturation boundaries to have two complementary horizontal straight line segments, a–0 for $S_n'^{sat}$ and 0–b for $S_p'^{sat}$. Although the state acquisition would be nearly time optimal, the boundary layer would vanish in the first and third segments of the phase plane, causing control chatter similar to that of Fig. 9.32a. This is avoided by introducing the linear term $K_d x_{2e}$ in the switching function, as in (9.172), but using this *alone* as a linear segment passing through $(x_{1e}, x_{2e}) = (0, 0)$. Boundary, S', becomes active outside the two points where the linear segment intersects it. With reference to Fig. 9.35, for implementation, it is necessary to calculate the values, $\pm x_{2\,max}$, of x_{2e} at the two intersection points, a and b, closest to the origin of the phase plane.

For $|x_{2e}| > x_{2\,max}$, the switching boundary, S', is active regardless of points c and d.

For $x_{2e} > 0$, the equation of the switching boundary, S', is

$$x_{1e} = -\frac{u_{max}}{K} - \frac{1}{2b_0 u_{max}} x_{2e}^2. \quad (9.177)$$

The equation of the linear switching boundary segment is

$$x_{1e} = -K_d x_{2e}. \quad (9.178)$$

Equating the RHS of (9.177) and (9.178) then yields

$$x_{2e}^2 - 2b_0 u_{max} K_d x_{2e} + \frac{2b_0 u_{max}^2}{K} = 0 \quad (9.179)$$

The solution, $x_{2\,max}$, is then the smaller of the two roots, i.e.,

$$x_{2\,max} = b_0 u_{max} K_d - \sqrt{b_0^2 u_{max}^2 K_d^2 - \frac{2b_0 u_{max}^2}{K}} = b_0 u_{max} \left(K_d - \sqrt{K_d^2 - \frac{2}{b_0 K}} \right). \quad (9.180)$$

The required switching function for use with the saturation function (9.164) is then

$$S''(x_{1e}, x_{2e}) = \begin{cases} -(x_{1e} + K_d x_{2e}) & \text{for } |x_{2e}| \leq x_{2\max} \\ S'(x_{1e}, x_{2e}) & \text{for } |x_{2e}| > x_{2\max} \\ \text{where} \quad S'(x_{1e}, x_{2e}) = -\left[x_{1e} + \frac{u_{\max}}{K} \text{sgn}(x_{2e})\right] - \frac{1}{2b_0 u_{\max}} x_{2e} |x_{2e}| \end{cases}.$$
(9.181)

This resurrects the boundary layer around $(x_{1e}, x_{2e}) = (0, 0)$ as shown in Fig. 9.34b and gives the system damping that can be specified using (9.175). Figure 9.36a shows simulation results of the system based on (9.172). This demonstrates the premature deceleration, reduced control magnitude and increased settling time already mentioned. Figure 9.36b shows the equivalent results obtained

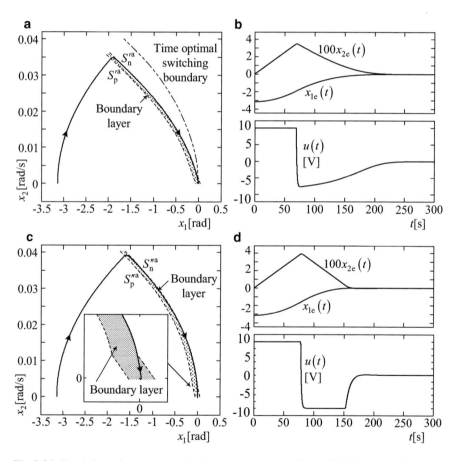

Fig. 9.36 Simulation of near-time-optimal slew manoeuvre with modified boundary layer. (**a**) Basic switching function with additional linear term. (**b**) Offset switching function with linear segment

with the recommended switching function (9.181). After the state trajectory enters the linear region of the boundary layer surrounding the origin of the phase plane, a small overshoot of $x_{1e}(t)$ occurs, but this is negligible. In fact the boundary layer width has been set to a larger value than necessary in order to display the behaviour of the state trajectory within the boundary layer.

It could be set to a much narrower value in practice by increasing K. For example, increasing K to 1,000 yields responses indistinguishable from the ideal time-optimal system when viewed on the scales of Fig. 9.36.

9.8.6 Limit Cycling Control

As for the first-order systems, limit cycling can be controlled by introducing a hysteresis element and adjusting it to achieve an acceptable limit cycle amplitude. A more sophisticated version for spacecraft attitude control is presented in Appendix A9, which ensures that the positive and negative excursions of the attitude error are equal and opposite with a specified magnitude and adapts automatically to the external disturbance torque, which can be estimated using an observer.

9.9 Feedback Control of Third and Higher-Order Plants

9.9.1 Overview

There is no general theory of switched feedback control for plants of more than second order, except the pulse modulation method of Sect. 9.2. Specific feedback control techniques, however, may be developed for individual applications, and this section contains two examples.

9.9.2 Time-Optimal Control of the Triple Integrator Plant

The plant state-space model is

$$\begin{bmatrix} \dot{x}_1 \\ \dot{x}_2 \\ \dot{x}_3 \end{bmatrix} = \begin{bmatrix} 0 & 1 & 0 \\ 0 & 0 & 1 \\ 0 & 0 & 0 \end{bmatrix} \begin{bmatrix} x_1 \\ x_2 \\ x_3 \end{bmatrix} + \begin{bmatrix} 0 \\ 0 \\ 1 \end{bmatrix} u, \quad |u| \leq u_{\max}. \quad (9.182)$$

If the reference input is $x_{1r} = const.$, the error state coordinates are $x_{1e} = x_1 - x_{1r}$, $x_{2e} = x_2$ and $x_{3e} = x_3$ and the error state-space model is

$$\begin{bmatrix} \dot{x}_{1e} \\ \dot{x}_{2e} \\ \dot{x}_{3e} \end{bmatrix} = \begin{bmatrix} 0 & 1 & 0 \\ 0 & 0 & 1 \\ 0 & 0 & 0 \end{bmatrix} \begin{bmatrix} x_{1e} \\ x_{2e} \\ x_{3e} \end{bmatrix} + \begin{bmatrix} 0 \\ 0 \\ 1 \end{bmatrix} u, \quad |u| \le u_{\max}. \tag{9.183}$$

The problem is then to determine the switched control law that brings the plant to the origin of the error state space in the minimum time. It has already been established in Sect. 9.6.2.3 that the number of control switches for an arbitrary initial state is $n - 1$, where n is the plant order, for any linear plant with poles lying on the real axis of the s plane. In this case, there are just two switches, and therefore, the back tracing method can be used to derive the equation of the switching boundary in the error state space, which leads to the required state feedback control law. The error state differential equation for the back tracing is

$$\frac{d}{d\tau}\begin{bmatrix} x_{1e} \\ x_{2e} \\ x_{3e} \end{bmatrix} = -\begin{bmatrix} 0 & 1 & 0 \\ 0 & 0 & 1 \\ 0 & 0 & 0 \end{bmatrix} \begin{bmatrix} x_{1e} \\ x_{2e} \\ x_{3e} \end{bmatrix} - \begin{bmatrix} 0 \\ 0 \\ 1 \end{bmatrix} u, \quad u = \pm u_{\max}. \tag{9.184}$$

The corresponding state transition equation for constant u is then

$$\begin{bmatrix} x_{1e}(\tau_{i+1}) \\ x_{2e}(\tau_{i+1}) \\ x_{3e}(\tau_{i+1}) \end{bmatrix} = \begin{bmatrix} 1 & -\tau & \frac{1}{2}\tau^2 \\ 0 & 1 & -\tau \\ 0 & 0 & 1 \end{bmatrix} \begin{bmatrix} x_{1e}(\tau_i) \\ x_{2e}(\tau_i) \\ x_{3e}(\tau_i) \end{bmatrix} + \begin{bmatrix} -\frac{1}{6}\tau^3 \\ \frac{1}{2}\tau^2 \\ -\tau \end{bmatrix} u. \tag{9.185}$$

where $\tau_{i+1} = \tau_i + \tau$. Let the back tracing start at the origin of the error state space at time, $\tau_0 = 0$, and let the time to the first switch along this trajectory (which is switch two in real time), be τ_{s1}. Then the error state at τ_{s1} is yielded by (9.185) as

$$\begin{bmatrix} x_{1s} \\ x_{2s} \\ x_{3s} \end{bmatrix} = \begin{bmatrix} 1 - \tau_{s1} & \frac{1}{2}\tau_{s1}^2 \\ 0 & 1 & -\tau_{s1} \\ 0 & 0 & 1 \end{bmatrix} \begin{bmatrix} 0 \\ 0 \\ 0 \end{bmatrix} + \begin{bmatrix} -\frac{1}{6}\tau_{s1}^3 \\ \frac{1}{2}\tau_{s1}^2 \\ -\tau_{s1} \end{bmatrix} u = \begin{bmatrix} -\frac{1}{6}u\tau_{s1}^3 \\ \frac{1}{2}u\tau_{s1}^2 \\ -u\tau_{s1} \end{bmatrix} \tag{9.186}$$

Let the time between the first and second switches along the back traced trajectory (which is the first switch in real time) be τ_{s2}. Then the error state at this switching time (which, for arbitrary τ_{s1} and τ_{s2}, is an arbitrary point on the time-optimal switching boundary) is yielded by (9.185) as

$$\begin{bmatrix} x_{1b} \\ x_{2b} \\ x_{3b} \end{bmatrix} = \begin{bmatrix} 1 - \tau_{s2} & \frac{1}{2}\tau_{s2}^2 \\ 0 & 1 & -\tau_{s2} \\ 0 & 0 & 1 \end{bmatrix} \begin{bmatrix} x_{1s} \\ x_{2s} \\ x_{3s} \end{bmatrix} + \begin{bmatrix} -\frac{1}{6}\tau_{s2}^3 \\ \frac{1}{2}\tau_{s2}^2 \\ -\tau_{s2} \end{bmatrix} (-u), \tag{9.187}$$

noting that u changes sign along this second segment of the back traced trajectory. It is now necessary to eliminate the switch state, $\mathbf{x}_s = [x_{1s}\ x_{2s}\ x_{3s}]^T$, the two switching times, τ_{s1} and τ_{s2}, and u between (9.186) and (9.187). First, substituting for $\mathbf{x}_s$ in (9.187) using (9.186) yields

9.9 Feedback Control of Third and Higher-Order Plants

$$\begin{bmatrix} x_{1b} \\ x_{2b} \\ x_{3b} \end{bmatrix} = \begin{bmatrix} 1 & -\tau_{s2} & \frac{1}{2}\tau_{s2}^2 \\ 0 & 1 & -\tau_{s2} \\ 0 & 0 & 1 \end{bmatrix} \begin{bmatrix} -\frac{1}{6}u\tau_{s1}^3 \\ \frac{1}{2}u\tau_{s1}^2 \\ -u\tau_{s1} \end{bmatrix} + \begin{bmatrix} \frac{1}{6}u\tau_{s2}^3 \\ -\frac{1}{2}u\tau_{s2}^2 \\ u\tau_{s2} \end{bmatrix} \quad (9.188)$$

which gives

$$x_{1b} = \tfrac{1}{6}u\left(\tau_{s2}^3 - 3\tau_{s2}^2\tau_{s1} - 3\tau_{s2}\tau_{s1}^2 - \tau_{s1}^3\right) \quad (9.189)$$

$$x_{2b} = \tfrac{1}{2}u\left(\tau_{s1}^2 + 2\tau_{s1}\tau_{s2} - \tau_{s2}^2\right) \quad (9.190)$$

$$x_{3b} = u\left(\tau_{s2} - \tau_{s1}\right) \quad (9.191)$$

Next τ_{s1} and τ_{s2} are eliminated between (9.189), (9.190) and (9.191). From (9.191),

$$\tfrac{1}{2}x_{3b}^2 = \tfrac{1}{2}u^2\left(\tau_{s2}^2 - 2\tau_{s2}\tau_{s1} + \tau_{s1}^2\right). \quad (9.192)$$

Then $u \times$ (9.190) + (9.192) gives

$$ux_{2b} + \tfrac{1}{2}x_{3b}^2 = u^2\tau_{s1}^2. \quad (9.193)$$

Again from (9.191),

$$\tfrac{1}{6}x_{3b}^3 = \tfrac{1}{6}u^3\left(\tau_{s2}^3 - 3\tau_{s2}^2\tau_{s1} + 3\tau_{s2}\tau_{s1}^2 - \tau_{s1}^3\right). \quad (9.194)$$

Then (9.194) $- u^2 \times$ (9.189) gives

$$\tfrac{1}{6}x_{3b}^3 - u^2 x_{1b} = u^3\tau_{s2}\tau_{s1}^2. \quad (9.195)$$

Then (9.195)/(9.193) gives

$$\frac{\tfrac{1}{6}x_{3b}^3 - u^2 x_{1b}}{ux_{2b} + \tfrac{1}{2}x_{3b}^2} = u\tau_{s2}. \quad (9.196)$$

Taking the square root of both sides of (9.193) yields

$$\left(ux_{2b} + \tfrac{1}{2}x_{3b}^2\right)^{\frac{1}{2}} = \pm|u\tau_{s1}|, \quad (9.197)$$

but since $|u\tau_{s1}| = u\tau_{s1}\,\text{sgn}(u\tau_{s1}) = u\tau_{s1}\,\text{sgn}(u)$, (9.197) may be written as

$$\left(ux_{2b} + \tfrac{1}{2}x_{3b}^2\right)^{\frac{1}{2}}\text{sgn}(u) = u\tau_{s1}, \quad (9.198)$$

provided the positive square root is taken. Then subtracting (9.198) from (9.196) and using (9.191) yield

$$\begin{aligned} x_{3b} &= \frac{\frac{1}{6}x_{3b}^3 - u^2 x_{1b}}{u x_{2b} + \frac{1}{2} x_{3b}^2} - \left(u x_{2b} + \frac{1}{2} x_{3b}^2\right)^{\frac{1}{2}} \operatorname{sgn}(u) \Rightarrow \\ u^2 x_{1b} &- \tfrac{1}{6} x_{3b}^3 + \left[\left(u x_{2b} + \tfrac{1}{2} x_{3b}^2\right)^{\frac{1}{2}} \operatorname{sgn}(u) + x_{3b}\right]\left(u x_{2b} + \tfrac{1}{2} x_{3b}^2\right) = 0 \end{aligned} \quad (9.199)$$

From (9.195),

$$\operatorname{sgn}(u) = \operatorname{sgn}\left(u^3\right) = \operatorname{sgn}\left(\tfrac{1}{6} x_{3b}^3 - u^2 x_{1b}\right). \quad (9.200)$$

Since $u = |u|\operatorname{sgn}(u)$ and for the time-optimal control $u = \pm u_{\max}$, in view of (9.200),

$$u = u_{\max} \operatorname{sgn}\left(\tfrac{1}{6} x_{3b}^3 - u^2 x_{1b}\right). \quad (9.201)$$

The switching boundary equation is then given by (9.199) with $u^2 = u_{\max}^2$, $\operatorname{sgn}(u)$ given by (9.200) and u given by (9.201), which may be written

$$\left.\begin{aligned} u_{\max}^2 x_{1b} - \tfrac{1}{6} x_{3b}^3 + \left(x^{\frac{1}{2}} S' + x_{3b}\right) x &= 0 \\ \text{where } S' &= \operatorname{sgn}\left(\tfrac{1}{6} x_{3b}^3 - u_{\max}^2 x_{1b}\right) \\ \text{and } x &= u_{\max} x_{2b} S' + \tfrac{1}{2} x_{3b}^2 \end{aligned}\right\}. \quad (9.202)$$

The time-optimal state feedback control law is based on (9.202), which may be written as $S(x_{1b}, x_{2b}, x_{3b}) = 0$. This boundary divides the error state space into the 'n' region in which the time-optimal control is $u^* = -u_{\max}$ and the 'p' region in which $u^* = u_{\max}$. In order to allocate these regions correctly, consider an initial error state, $[x_{1e}(0), 0, 0]$, where $x_{1e}(0) \neq 0$. By inspection of (9.202),

$$\operatorname{sgn}[S(x_{1e}(0), 0, 0)] = \operatorname{sgn}[x_{1e}(0)]. \quad (9.203)$$

The solution of (9.183) with this initial state and constant u is

$$\begin{bmatrix} x_{1e}(t) \\ x_{2e}(t) \\ x_{3e}(t) \end{bmatrix} = \begin{bmatrix} 1 & t & \tfrac{1}{2}t^2 \\ 0 & 1 & t \\ 0 & 0 & 1 \end{bmatrix} \begin{bmatrix} x_{1e}(0) \\ 0 \\ 0 \end{bmatrix} + \begin{bmatrix} \tfrac{1}{6}t^3 \\ \tfrac{1}{2}t^2 \\ t \end{bmatrix} u \Rightarrow \\ x_{1e} = x_{1e}(0) + \tfrac{1}{6} u t^3; \quad x_{2e} = \tfrac{1}{2} u t^2; \quad x_{3e} = u t. \quad (9.204)$$

With reference to (9.202), the switching function is

$$\left.\begin{aligned} S(x_{1e}, x_{2e}, x_{3e}) &= u_{\max}^2 x_{1e} - \tfrac{1}{6} x_{3e}^3 + \left(x^{\frac{1}{2}} S' + x_{3e}\right) x \\ \text{where } S' &= \operatorname{sgn}\left(\tfrac{1}{6} x_{3e}^3 - u_{\max}^2 x_{1e}\right) \\ \text{and } x &= u_{\max} x_{2e} S' + \tfrac{1}{2} x_{3e}^2 \end{aligned}\right\}. \quad (9.205)$$

9.9 Feedback Control of Third and Higher-Order Plants

Substituting for x_{1e}, x_{2e} and x_{3e} in (9.205) using (9.204) then yields the following:

$$S' = \text{sgn}\left\{\tfrac{1}{6}u^3 t^3 - u_{\max}^2\left[x_{1e}(0) + \tfrac{1}{6}ut^3\right]\right\} = \text{sgn}\left[-u_{\max}^2 x_{1e}(0)\right] = -\text{sgn}\left[x_{1e}(0)\right], \quad (9.206)$$

since $|u| = u_{\max}$ and $\text{sgn}(u) = \text{sgn}(u^3)$.

$$x = -u_{\max} \tfrac{1}{2} ut^2 \, \text{sgn}\left[x_{1e}(0)\right] + \tfrac{1}{2} u_{\max}^2 t^2. \quad (9.207)$$

Then the switching function evaluation is

$$S(x_{1e}, x_{2e}, x_{3e}) = u_{\max}^2\left[x_{1e}(0) + \tfrac{1}{6}ut^3\right] - \tfrac{1}{6}u^3 t^3 + \left(-x^{\tfrac{1}{2}} \, \text{sgn}\left[x_{1e}(0)\right] + ut\right)x$$

$$= u_{\max}^2 x_{1e}(0) + \left(ut - x^{\tfrac{1}{2}} \, \text{sgn}\left[x_{1e}(0)\right]\right)x, \quad (9.208)$$

since $u^3 = u_{\max}^2 u$. If $u = u_{\max} \, \text{sgn}\left[x_{1e}(0)\right]$, then according to (9.207), $x = 0 \Rightarrow$

$$S(x_{1c}, x_{2e}, x_{3e}) = u_{\max}^2 x_{1e}(0) = const. \quad (9.209)$$

Hence, in this case, the switching boundary cannot be reached. For the other option,

$$u = -u_{\max} \, \text{sgn}\left[x_{1e}(0)\right] \quad (9.210)$$

(9.207) yields $x = u_{\max}^2 t^2$ and (9.208) becomes

$$S(x_{1e}, x_{2e}, x_{3e}) = u_{\max}^2 x_{1e}(0) + \left(\left(-u_{\max} \, \text{sgn}\left[x_{1e}(0)\right]\right)t - u_{\max} t \, \text{sgn}\left[x_{1e}(0)\right]\right) u_{\max}^2 t^2$$

$$= u_{\max}^2 x_{1e}(0) - 2u_{\max}^3 t^3 \, \text{sgn}\left[x_{1e}(0)\right]. \quad (9.211)$$

In this case, $S(x_{1e}, x_{2e}, x_{3e}) = 0$ when $t = \{|x_{1e}(0)| / (2u_{\max})\}^{1/3}$ proving that the switching boundary is reached in a finite time.

In view of (9.203) and (9.210), the time-optimal control for the initial error state, $[x_{1e}(0), 0, 0]$, is

$$u = -u_{\max} \, \text{sgn}\{S\left[x_{1e}(0), \, 0, \, 0\right]\}. \quad (9.212)$$

It follows that the time-optimal control for an arbitrary error state is

$$u = -u_{\max} \, \text{sgn}\left[S(x_{1e}, x_{2e}, x_{3e})\right] \quad (9.213)$$

where the switching function is given by (9.205). An alternative switching function is obtained by multiplying $S(x_{1e}, x_{2e}, x_{3e})$ by -1, yielding

$$\left. \begin{array}{l} u = u_{\max} \operatorname{sgn}\left[S''(x_{1e}, x_{2e}, x_{3e})\right] \\ \text{where } S''(x_{1e}, x_{2e}, x_{3e}) = -u_{\max}^2 x_{1e} + \tfrac{1}{6} x_{3e}^3 - \left(x^{\frac{1}{2}} S' + x_{3e}\right) x, \\ S' = \operatorname{sgn}\left(\tfrac{1}{6} x_{3e}^3 - u_{\max}^2 x_{1e}\right) \text{ and } x = u_{\max} x_{2e} S' + \tfrac{1}{2} x_{3e}^2 \end{array} \right\}. \quad (9.214)$$

Example 9.5 Time-optimal attitude control of spacecraft by solar sailing

In conventional spacecraft attitude control systems, solar radiation pressure gives rise to a force vector that, due to the inevitable asymmetries of the spacecraft geometry and the nonuniform mass distribution, is not directed through the centre of mass. This causes a disturbance torque that has to be counteracted by the attitude control actuators, entailing direct fuel consumption in a system employing only mass expulsion actuators. If momentum exchange actuators such as reaction wheels or control moment gyros are employed, then in the process of maintaining the demanded attitude, these actuators will accrue angular momentum equal to the time integral of the disturbance torque from the instant of the initial loop closure. The finite angular momentum storage capacity of these actuators necessitates the inclusion of momentum dumping thrusters, and therefore, in the long term, fuel consumption is unavoidable. In solar sailing systems, however, the attitude control torque originates from the solar radiation pressure and is varied by changing the geometry of relatively large appendages such as solar panels, an example of which is shown in Fig. 9.37.

For simplicity of illustration, only one of the three attitude control axes is considered, and the interaction with the other two axes is assumed to be negligible.

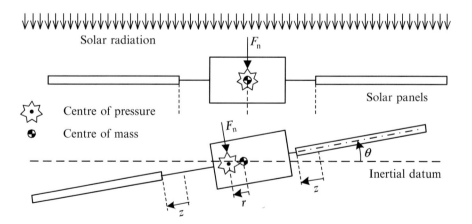

Fig. 9.37 Pitch attitude control of satellite by solar panel displacement along the yaw axis

9.9 Feedback Control of Third and Higher-Order Plants

It will be supposed that a motor with a maximum speed limit drives the solar panels along the longitudinal (yaw) axis so that

$$\dot{z} = K_v v \tag{9.215}$$

where z is the displacement of the panel with respect to the position where the centre of pressure and centre of mass are aligned, K_v is the panel drive speed constant and v is the control voltage of the panel motor drive electronics, subject to the saturation constraint, $|v| \leq v_{\max}$. It is assumed that a panel displacement, z, gives rise to a proportional displacement, r, between the centre of pressure and the centre of mass along the yaw axis. Thus,

$$r = K_d z, \tag{9.216}$$

where K_d is the displacement ratio. This gives rise to a control torque,

$$\Gamma_c = F_n r \tag{9.217}$$

where F_n is the solar radiation force component normal to the spacecraft yaw and pitch axes. The pitch attitude angle, θ, is then related to the control torque by

$$\ddot{\theta} = \frac{1}{J}\Gamma_c \tag{9.218}$$

where J is the moment of inertia of the spacecraft body about the pitch axis. Let the pitch attitude measurement be

$$y = K_s \theta \tag{9.219}$$

The spacecraft model is then reduced to a single differential equation by combining Eqs. (9.215) through (9.219), commencing by forming $\dddot{y}$, as the plant is of third order, as follows:

$$\dddot{y} = K_s \dddot{\theta} = K_s \frac{1}{J}\dot{\Gamma}_c = K_s \frac{1}{J} F_n \dot{r} = K_s \frac{1}{J} F_n K_d \dot{z} = K_s \frac{1}{J} F_n K_d K_v v. \tag{9.220}$$

Let y be scaled in the computer to be nominally equal to θ in radians. This will be the state variable, x_1. Then

$$x_1 = \frac{1}{K_s} y. \tag{9.221}$$

It follows from (9.220) that

$$\dddot{x}_1 = \frac{1}{J} F_n K_d K_v v. \tag{9.222}$$

Let the control variable, v, be scaled to form a redefined control variable, u, numerically equal to the angular jerk, i.e. the third derivative of the attitude angle. Then the spacecraft model reduces to

$$\dddot{x}_1 = u \tag{9.223}$$

The control law will calculate u. Then the physical control variable is given by

$$v = \frac{J}{F_n K_d K_v} u. \tag{9.224}$$

In view of (9.223), the plant state-space model is given by (9.182), and if the reference attitude angle, x_{1r}, is constant, then the error state-space model is the same as (9.183). It must be noted, however, that the maximum control torque levels attainable with solar sailing are at least an order of magnitude less than those attainable with other actuators. This severely increases the minimum settling times attainable with linear feedback controllers. For this reason, time-optimal control is recommended. Hence, the time-optimal state feedback control law defined by (9.213) and (9.205) will be applied.

The parameters taken for the simulation are as follows: $J = 20 \, [\text{Kg m}^2]$;, $K_{tp} = F_n K_d = 0.001$ [N] (panel torque generation constant); $K_v = 0.01$ [m/s/V]; $v_{max} = 10$ [V], giving $u_{max} = 10^{-4}$ [N] through (9.224). The simulations of Fig. 9.38 are for zero initial state variables and a step reference attitude angle of $x_{1r} = 0.1$ [rad].

The error state variables and control variable are shown in Fig. 9.38a for the basic time-optimal state feedback control system. The control torque is actually a state variable, being given by $\Gamma_c = J\ddot{\theta} = J x_{3e}$. The result is as predicted by the theory, $\Gamma_c(t)$ being piecewise linear, $x_{2e}(t)$ being piecewise parabolic and $x_{1e}(t)$ piecewise

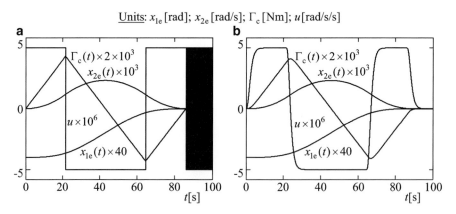

Fig. 9.38 Spacecraft attitude control by solar sailing. (**a**) Basic time-optimal control. (**b**) LSF control driven by time-optimal model

9.9 Feedback Control of Third and Higher-Order Plants

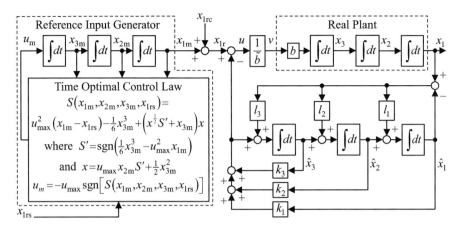

Fig. 9.39 Near-time-optimal control with reference input generator and LSF control law

cubic, all the three error state variables being brought to zero together with two control switches. After this, the state feedback induces a limit cycle in which $u(t)$ switches rapidly between $\pm u_{\max}$, keeping the state error to negligible proportions.

In such an application, the system may be required to respond to continuously varying reference inputs while the time-optimal control is intended for step changes in the reference input. This requirement, however, is easily catered for by first closing a linear control loop around the spacecraft and then driving this control loop with a time-varying reference input, $x_{1m}(t)$, identical to $x_{1e}(t)$ of Fig. 9.38a, obtained from a model of the time-optimal attitude control run in real time in the onboard control computer, as shown in Fig. 9.39.

Here, $b = F_n K_d K_v / J$ and $\tilde{b}$ is the estimate of b. In the simulations of Fig. 9.38, $\tilde{b} = b$ as the spacecraft parameters would be known with high accuracy, but in the development of such a control system, the control system designer would be advised to carry out a robustness test by mismatching b with respect to $\tilde{b}$ in further simulations. The linear control loop is closed via a linear state feedback control law aided by an observer in this example. If the linear control loop has a sufficiently small settling time, T_s, then the true spacecraft error state will follow that of the reference input generator with only a small dynamic lag of the order of T_s, as shown in Fig. 9.38b, where $T_s = 5$ s. The slewing time is not much greater than the time-optimal one, but, if required, this dynamic lag could be removed with the aid of the dynamic lag pre-compensation method presented in Chap. 12. With reference to Fig. 9.39, relatively large-angle slew manoeuvres are carried out with step reference attitude angle inputs, x_{1rs}, applied to the reference input generator with $x_{1rc} = 0$. Continuously variable attitude manoeuvres satisfying $|u| < u_{\max}$ are carried out with $x_{1rc}(t)$ as the input.

9.9.3 Posicast Control of Fourth-Order Plants with Oscillatory Modes

9.9.3.1 Origin

The posicast control technique was originated by O J M Smith [9]. The principle may be explained with reference to the operation of gantry cranes. Such a crane consists of a motorised truck running on an overhead gantry from which hangs a cradle containing the load, as illustrated in Fig. 9.40. The load is free to swing like a pendulum. Since the damping of the oscillatory pendulum motion is negligible, the crane operators developed a technique for moving the crane to the required position while minimising the swinging of the load at the end of the manoeuvre. With reference to Fig. 9.40, starting at position (i) with the load stationary, the truck is moved to the halfway point (ii) as fast as possible so that the load remains close to its starting position due to its inertia. Then, with the truck held at position (ii), the load is allowed to swing one-half cycle of oscillation to position (iii).

Then the truck is moved as fast as possible to the final position (iv). At the end of this manoeuvre, the load is hanging approximately vertically, and therefore, the residual swinging is minimal. This is the basis of posicast control.

9.9.3.2 Applicability

Posicast control systems for plants exhibiting similar dynamic characteristics to the gantry crane have been proposed on an open-loop basis in which the control input required to perform the manoeuvre is calculated offline and then applied in real time. This, however, suffers from the drawback of offering no counteraction of external disturbances and/or compensation for plant modelling errors. To overcome these problems, posicast feedback control is possible. If there is significant residual swinging at the end of the manoeuvre due to external disturbances or plant modelling errors, then posicast feedback control automatically repeats the manoeuvre on a smaller scale until the swinging is negligible. Furthermore, the plant can be

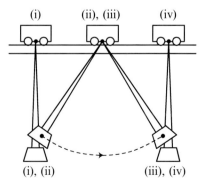

Fig. 9.40 The steps of posicast control of a gantry crane

9.9 Feedback Control of Third and Higher-Order Plants

brought to the desired state from an arbitrary initial state. Before proceeding further, however, it must be stated that posicast feedback control is only really advantageous with switched actuators. The author has not yet found a practical application with continuous actuators in which posicast control offers a significant advantage over other continuous control techniques. Examples of suitable applications are to be found in power electronics [10] and the attitude control of flexible spacecraft [11].

Plants that can be modelled by the transfer function,

$$\frac{y(s)}{u(s)} = \frac{c_3 s^3 + c_2 s^2 + c_1 s + c_0}{(s^2 + a_1 s)\left(s^2 + 2\zeta_1 \omega_1 s + \omega_1^2\right)}, \qquad (9.225)$$

will be considered in which the damping ratio, ζ_1, is relatively small (≤ 0.2).

9.9.3.3 Development

The stages of development of a posicast feedback control system intended originally for the manoeuvring of flexible spacecraft [12] will now be presented. First suppose that a partial fraction expansion is carried out to separate the plant model (9.225) into two second-order sub-plants as follows:

$$\frac{y(s)}{u(s)} = \underbrace{\frac{P_1 s + P_0}{s^2 + a_1 s}}_{\text{Sub-plant 1}} + \underbrace{\frac{Q_1 s + Q_0}{s^2 + 2\zeta_1 \omega_1 s + \omega_1^2}}_{\text{Sub-plant 2}} \qquad (9.226)$$

where P_1, P_0, Q_1 and Q_0 are the partial fraction coefficients. Expressions for these in terms of a_1, ζ_1 and ω_1 are obtainable using standard algebraic procedures and are given here as they will be needed subsequently. Thus,

$$P_0 = \frac{c_0}{\omega_1^2}, \quad P_1 = \frac{c_1 - a_1 c_2 + (a_1 - 2\zeta_1 \omega_1) P_0}{\omega_1^2 - 2a_1 \zeta_1 \omega_1},$$

$$Q_0 = c_2 - P_0 - 2\zeta_1 \omega_1 P_1 \quad \text{and} \quad Q_1 = c_3 - P_1. \qquad (9.227)$$

Sub-plant 1 is analogous to the truck of the gantry crane and sub-plant 2 is analogous to the pendular load. Next, a plant state-space model is formed with sub-plants 1 and 2 each in the control canonical form. Thus,

$$\begin{cases} \left.\begin{aligned} \dot{x}_1 &= x_2 \\ \dot{x}_2 &= u - p_1 x_2 \end{aligned}\right\} \text{Sub-plant 1} \\ \left.\begin{aligned} \dot{x}_3 &= x_4 \\ \dot{x}_4 &= u - 2\zeta_1 \omega_1 x_4 - \omega_1^2 x_3 \end{aligned}\right\} \text{Sub-plant 2} \\ y = P_0 x_1 + P_1 x_2 + Q_0 x_1 + Q_1 x_2 \end{cases} \qquad (9.228)$$

Fig. 9.41 Plant state-variable block diagram for posicast state feedback control

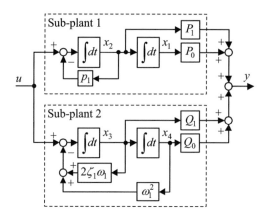

The corresponding state-variable block diagram is shown in Fig. 9.41.

Next, an approximate time-optimal control is formed for sub-plant 1. Assuming switched control with $u = \pm u_{\max}$, then if

$$|p_1 x_2| \ll u_{\max} \tag{9.229}$$

the state equations of sub-plant 1 in (9.228) approximate those of a double integrator, i.e. $\dot{x}_1 = x_2$ and $\dot{x}_2 = u$. This enables the substate, $(x_1\ x_2)$, of sub-plant 1 to be changed from $(x_1(0), 0)$ to $(x_{1r}, 0)$ in nearly the minimum time using a near-time-optimal double integrator control law, where x_{1r} is a constant reference input.

Since switched actuators are used, then a near-time-optimal control law with limit cycle control could be used.

In the region of $(x_3, x_4) = (0, 0)$ such that

$$\left|2\zeta_1\omega_1 x_4 + \omega_1^2 x_3\right| \ll u_{\max}, \tag{9.230}$$

the state equations of sub-plant 2 in (9.228) also approximate those of a double integrator, i.e. $\dot{x}_3 = x_4$ and $\dot{x}_4 = u$. The plant behaviour is examined now by simulation using the near-time-optimal control law with the switching function,

$$S(x_1, x_2, x_{1r}) = x_{1r} - x_1 - \frac{1}{2u_{\max}}|x_2|x_2, \tag{9.231}$$

and a hysteresis switching element for limit cycle control having hysteresis limits of $S^+ = -S^- = 0.001$. The plant parameters are set to $p_1 = 1\ \text{s}^{-1}$, $\omega_1 = 1\ \text{rad/s}$ and $\zeta_1 = 0.1$. These do not represent a particular plant and are chosen merely for illustrative purposes. Figure 9.42 shows the results for two different reference inputs. The state trajectories of both sub-plants are superimposed on the phase plane for comparison. The resemblance of the state trajectory, 0–a, of sub-plant 1 to the piecewise parabolic state trajectory of the double integrator time-optimal control is

9.9 Feedback Control of Third and Higher-Order Plants

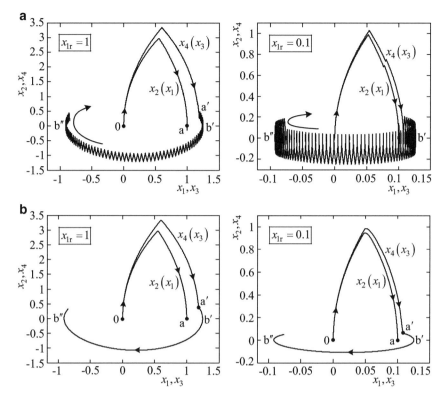

Fig. 9.42 Trajectories in simultaneous phase planes of fourth-order plant (**a**) with switched control and hysteresis element and (**b**) with saturating continuous near time optimal control

apparent. The corresponding state trajectory, $0 - a'$, of the uncontrolled sub-plant 2 is of a similar shape but differs from the state trajectory, $0-a$, due to the differences between the two sub-plants.

It is evident, however, that points a and a', which are reached at the same time, become closer together as x_{1r} is reduced. This is due to the two trajectories diverging less as $a_1 x_2$ and $2\zeta_1 \omega_1 x_4 + \omega_1^2 x_3$ become smaller and therefore, according to (9.229) and (9.230), more dominated by the control while it is $\pm u_{\max}$. After point 'a' is reached, u drops to nearly zero in Fig. 9.42b as sub-plant 1 reaches the neighbourhood of the desired state, $(x_1, x_2) = (0, 0)$. In Fig. 9.42a, sub-plant 2, however, representing the oscillatory mode, continues to oscillate with zero control input due to its non-zero state. After reaching a positive peak at point b', x_3 comes to a negative peak at point b// after one-half cycle of oscillation has occurred. So x_3 is analogous to the pendulum angle in the posicast control of the gantry crane. Point b// is closer to point '0' than point b', however, due to the modal damping. In Fig. 9.42a, the behaviour is similar except for the limit cycling of $u(t)$ about zero causing the oscillations of the state trajectory, $x_4(x_3)$, about a smooth nearly elliptical path as in Fig. 9.42b, which is clearly visible and larger in proportion for the smaller reference input.

For zero damping, the modal oscillations would continue at a constant amplitude. In this case, $\zeta = 0$ and the sub-plant 2 state equations become

$$\dot{x}_3 = x_4 \tag{9.232}$$

$$\dot{x}_4 = u - \omega_{n2}^2 x_3 \tag{9.233}$$

It will now be proven that for $u = 0$, the state trajectory for an arbitrary initial state is an ellipse centred on $(x_3, x_4) = (0, 0)$. In this case, the state trajectory differential equation obtained by dividing (9.233) by (9.232) is

$$\frac{dx_4}{dx_3} = -\omega_{n2}^2 \frac{x_3}{x_4}. \tag{9.234}$$

The solution by the method of separation of the variables is then

$$\int x_4 dx_4 = -\omega_{n2}^2 \int x_3 dx_3 \Rightarrow \tfrac{1}{2} x_4^2 = -\omega_{n2}^2 \tfrac{1}{2} x_3^2 + A \tag{9.235}$$

where A is an arbitrary constant of integration. Equation (9.235) is also valid for the initial state. Hence, $A = \tfrac{1}{2} x_4^2(0) + \omega_{n2}^2 \tfrac{1}{2} x_3^2(0)$. Hence, the general solution is

$$x_4^2 + \omega_{n2}^2 x_3^2 = x_4^2(0) + \omega_{n2}^2 x_3^2(0), \tag{9.236}$$

which is an ellipse centred on $(x_3, x_4) = (0, 0)$. The state trajectories of Fig. 9.42 beyond the points, a′, resemble ellipses but spiral towards $(0, 0)$ due the damping.

To formulate a closed-loop control strategy, continuous saturated time-optimal control will be considered initially and adaptation for switched control with hysteresis introduced subsequently. Consider first the hypothetical plant in which the points, 'a' and 'a''', of Fig. 9.42b are coincident and also $\zeta_1 = 0$. Then the following basic control strategy will bring the error state variables to nearly zero for a constant reference input, x_{1r}, in just three steps. The near-time-optimal control law for sub-plant 1 is presented with its own reference input, x'_{1r}, set as follows:

If $|x_4| < \varepsilon$ then $x'_{1r} = \tfrac{1}{2}(x_{1r} + x_1 - x_3)$, otherwise x'_{1r} is held constant. (9.237)

Here, $0 < \varepsilon \ll |x_{4pk}|$, where x_{4pk} is the peak value of x_4 during a state change. As will be seen, the behaviour emulates posicast control of the gantry crane. Figure 9.43 shows the resulting simultaneous sub-plant state trajectories for $x_{1r} > 0$.

Let the initial plant state be $(x_1, x_2, x_3, x_4) = (0, 0, 0, 0)$ at point 'a'. Then (9.237) immediately comes into play and sets $x'_{1r} = \tfrac{1}{2}(x_{1r} + 0 - 0) = \tfrac{1}{2} x_{1r}$. Then near-time-optimal control takes substate 1 from point 'a' to point 'b' in Fig. 9.43a, and at the same time, substate 2 follows an identically shaped trajectory to point 'b' in Fig. 9.43b, at the end of which $(x_1, x_2) = (\tfrac{1}{2} x_{1r}, 0)$ and $(x_3, x_4) = (\tfrac{1}{2} x_{1r}, 0)$. Since x_{4e} reaches zero again, (9.237) sets $x'_{1r} = \tfrac{1}{2}(x_{1r} + \tfrac{1}{2} x_{1r} - \tfrac{1}{2} x_{1r}) = \tfrac{1}{2} x_{1r}$. This does not change from its previous value because both x_1 and x_3 changed by the

9.9 Feedback Control of Third and Higher-Order Plants

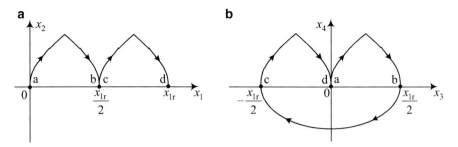

Fig. 9.43 State trajectories for posicast control of hypothetical undamped plant: (**a**) sub-plant 1 (**b**) sub-plant 2

same amount during the near-time-optimal change of substate 1. Then the near-time-optimal control law holds substate 1 at $(x_1, x_2) = (\frac{1}{2}x_{1r}, 0)$ with $u = 0$, but sub-state 2 follows the elliptical trajectory according to (9.236) until point 'c' is reached with $(x_3, x_4) = (-\frac{1}{2}x_{1r}, 0)$. Since x_4 again returns to zero, (9.237) sets $x'_{1r} = \frac{1}{2}(x_{1r} + \frac{1}{2}x_{1r} - (-\frac{1}{2}x_{1r})) = x_{1r}$. This causes the near-time-optimal control law to change substate 1 from $(x_1, x_2) = (\frac{1}{2}x_{1r}, 0)$ to $(x_1, x_2) = (x_{1r}, 0)$, following the trajectory from point 'c' to point 'd'. At the same time, substate 2 follows an identically shaped trajectory to $(x_3, x_4) = (0, 0)$, so the desired substate $(x_1, x_2, x_3, x_4) = (x_{1r}, 0, 0, 0)$ is reached.

With switched actuators, the limit cycling would cause zero crossings of x_4 to which (9.237) would respond prematurely. This can be avoided by employing the following modified posicast feedback control strategy,

$$\text{If } |x_4 - x_2| < \varepsilon \text{ then } x'_{1r} = \frac{1}{2}(x_{1r} + x_1 - x_3), \text{ otherwise } x'_{1r} \text{ is held.} \tag{9.238}$$

since the nearly identical limit cycle oscillations in x_2 cancel those in x_4.

To accommodate modal damping, the inward spiralling motion of sub-plant 2 state trajectory for $u = 0$ due to modal damping, already observed in Fig. 9.42, can be accommodated by a simple modification to control strategy (9.238) as follows.

$$\text{If } |x_4 - x_2| < \varepsilon \text{ then } x'_{1r} = ax_{1r} + (1-a)(x_1 - x_3), \text{ otherwise } x'_{1r} \text{ is held.} \tag{9.239}$$

where $0.5 \leq a < 1$. With the assumption that the phase-plane trajectories of subplants 1 and 2 are identical under control saturation, the ideal double phase-plane motion is as shown in Fig. 9.44.

At point 'a', (9.239) sets $x'_{1r} = ax_{1r} + (1-a)(0-0) = ax_{1r}$. Hence, the near-time-optimal controller takes the phase-plane trajectories to point 'b'. Here, (9.239) sets $x'_{1r} = ax_{1r} + (1-a)(ax_{1r} - ax_3) = ax_{1r}$, and therefore, the near-time-optimal controller keeps $x_1 = ax_{1r}$ with $u = 0$, while the trajectory, $x_4(x_3)$, moves on the spiral path to point 'c'. Here, (9.239) sets $x'_{1r} = $

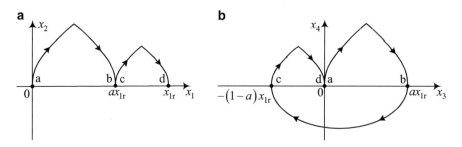

Fig. 9.44 State trajectories for posicast control of hypothetical damped plant: (**a**) sub-plant 1 (**b**) sub-plant 2

$ax_{1r} + (1-a)\{ax_{1r} - [-(1-a)x_{1r}]\} = x_{1r}$, and therefore, the near-time-optimal controller moves the plant to the required point 'd'.

Next, the constant parameter, a, will be expressed in terms of the damping ratio, ζ_1, of the oscillatory mode. Since the exponentially decaying envelope function of the oscillatory mode is $e^{-\zeta_1 \omega_1 t}$, and its half period is $T_1 = \pi/\omega_{1d}$, where $\omega_{1d} = \omega_1 \sqrt{1-\zeta_1^2}$, then the ratio between the magnitudes of x_3 at points a and b is

$$\frac{1-a}{a} = e^{-\zeta_1 \omega_1 T_1} = e^{-\frac{\zeta_1 \pi}{\sqrt{1-\zeta_1^2}}} \quad (9.240)$$

Let $e^{-\zeta_1 \pi/\sqrt{1-\zeta_1^2}} = \lambda_1$. Then (9.240) yields $1 - a = a\lambda_1 \Rightarrow$

$$a = \frac{1}{1+\lambda_1}. \quad (9.241)$$

It is important to realise that point 'd' in Fig. 9.44 will not be precisely reached in practice due to the differences in the substate trajectories under the saturated control, but this error will be reduced repeatedly by similar sequences of state changes induced automatically by control strategy (9.239) by its feedback action, so that the demanded state $(x_1, x_2, x_3, x_4) = (x_{1r}, 0, 0, 0)$ is approached until the errors are negligible.

A refinement that considerably reduces the residual errors at the end of the posicast control sequences is a variable threshold, ε, in (9.239) that starts at a relatively large value and diminishes with the control error, $x_{1e} = x_1 - x_{1r}$. This triggers the change of the variable reference input, x'_{1r}, before x_3 reaches point, b', in Fig. 9.42b. With reference to the saturation function defined by (9.48), the variable threshold is

$$\varepsilon = \varepsilon_{\min} + |\text{sat}(x_{1e}, 0, \varepsilon_{\max})|, \quad (9.242)$$

where the upper and lower limits, $\varepsilon_{\max}$ and $\varepsilon_{\min}$, are chosen to suit the application.

9.9 Feedback Control of Third and Higher-Order Plants

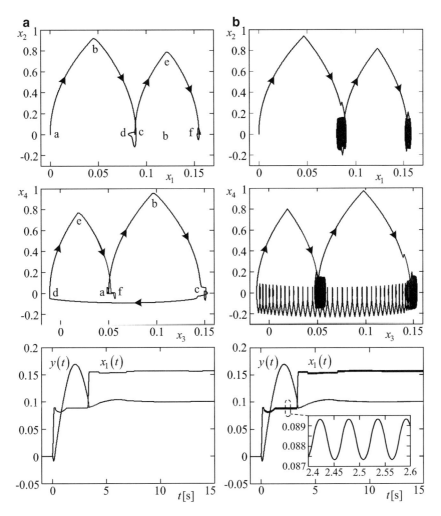

Fig. 9.45 Posicast state feedback control of a fourth order plant with damping allowance: (**a**) continuous saturated control, (**b**) switched control with hysteresis

Figure 9.45 shows a simulation with the plant parameters used for Fig. 9.42, with $x_{1r} = 0.1$, $\varepsilon_{max} = 0.05$ and $\varepsilon_{min} = 0.001$.

The basic posicast behaviour depicted in Fig. 9.44 is followed by this closed-loop system. As evident in Fig. 9.45b, replacing the continuous saturating near-time-optimal double integrator control law by the switched one with hysteresis does not impair the overall performance, the trajectories being similar to those of Fig. 9.45a but with a superposed limit cycle. This causes the oscillations of $x_1(t)$ shown in the inset of Fig. 9.45b, whose behaviour is not sinusoidal but appears so due to the filtering properties of the plant.

References

1. Perret R (2013) Power electronics semiconductor devices. Wiley, Hoboken, New Jersey
2. Sidi MJ (2002) Spacecraft dynamics & control. Cambridge University Press, Cambridge
3. Lai JS, Peng FZ (1996) Multilevel converters – a new breed of power converters. IEEE Trans Ind Appl 32(3):509–517
4. Ryan EP (1982) Optimal relay and saturating control system synthesis. P. Peregrinus, London
5. Athans M, Falb PL (2007) Optimal control: an introduction to the theory and its applications. Dover Publications Inc, Mineola, New York. ISBN 13: 9780486453286
6. Pontryagin LS et al (1987) Selected Works Vol. 4: The mathematical theory of optimal processes. ISBN 2-88124-077-1. Gordon and Breach Science Publishers, Montreux, Switzerland
7. Junkins JL, Turner D (1986) Optimal spacecraft rotational maneuvers. Elsevier, New York
8. Geering HP (2007) Optimal control with engineering applications. Springer, Berlin
9. Smith OJM (1957) Posicast control of damped oscillatory systems. Proc IRE 45:1249–1255
10. Hung JY (2003) Feedback control with posicast. IEEE Trans Ind Electron 50(1):795–8111
11. Singhose WE et al (1997) Slewing flexible spacecraft with deflection limiting input shaping. J Guid Control Dyn 20(2):291–298
12. Dodds SJ, Williamson SE (1984) A signed switching time bang-bang attitude control law for fine pointing of flexible spacecraft. Int J Control 40(4):795–8111

Chapter 10
Sliding Mode Control and Its Relatives

10.1 Introduction

10.1.1 Purpose and Origin

Sliding mode control (SMC) [1, 2] is a technique for achieving high robustness regarding plant parametric uncertainties and external disturbances. The technique originated from (a) work on switched control techniques (Chap. 9) and (b) the more general approach of variable structure systems [3] in which the control variable switches between the outputs of two differently structured controllers. Its discovery was linked to the observation of rapid switching, similar to that of a pulse modulator, in switched feedback control systems together with closed-loop system dynamics invariant with respect to changes in the plant parameters or the introduction of external disturbances. This led researchers to produce a general robust control technique deliberately inducing this behaviour. The overall aim is to achieve a prescribed closed-loop dynamics, while the only knowledge of the plant model is its relative degree.

10.1.2 Basic Principle

In its basic form, sliding mode control is switched state feedback control as introduced in Chap. 9, in which the control variable, u, switches between two limits, u_{min} and u_{max}, usually of opposite sign and often $u_{min} = -u_{max}$.

The term, *sliding mode*, is used to describe a mode of behaviour of a switched control system in which the state trajectory appears to slide along the switching boundary. A switched control law designed to operate in a sliding mode will be referred as a *sliding mode control law*. It will be recalled from Chap. 9 that a switched state feedback control law sometimes executes high-frequency switching between the two control levels, causing the controlled output to oscillate about the reference input. This is a simple example of a sliding mode.

In the remainder of this section, the basic concept of sliding mode control is presented. The remaining sections then develop sliding mode control laws in detail, commencing with SISO second-order plants for which the state trajectories can be easily visualised and displayed in two dimensions.

Sliding mode control is restricted to plants of finite order, i.e., those that do not contain transport delays or need distributed parameter models. Hence, the most general SISO plant to be considered has the state-space model,

$$\dot{\mathbf{x}} = \mathbf{f}(\mathbf{x}, u - d) \tag{10.1}$$

$$y = h(\mathbf{x}) \tag{10.2}$$

where, $\mathbf{x} \in \mathfrak{R}^n$ is the state vector, u is the control variable, y is the measured output to be controlled, u is the control variable and d is an external disturbance referred to u and $\mathbf{f}(\cdot)$ and $h(\cdot)$ are continuous functions of their arguments. If the control levels are $\pm u_{\max}$, then the general switched control law (Chap. 9) is

$$u = u_{\max} \, \text{sgn}\left[S\left(\mathbf{x}, \, y_r\right)\right], \tag{10.3}$$

where y_r is the reference input. For the closed-loop system formed by (10.1) and (10.3), the infinite continuum of points in the state space at which u switches between $\pm u_{\max}$ is defined by the switching boundary

$$S\left(\mathbf{x}, \, y_r\right) = 0. \tag{10.4}$$

It will be recalled from Chap. 9 that under control law (10.3), the switching boundary divides the whole of the state space into two regions, i.e., the 'p' region in which u is positive and the 'n' region in which u is negative. After the state trajectory crosses the switching boundary and penetrates the region on the opposite side by an infinitesimal amount, one of two events takes place. Either:

(a) The state trajectory changes direction and stays within the region it has penetrated.
(b) The state trajectory changes direction such that the switching boundary is immediately crossed again. Then the control changes back to the original sign, causing the event to be immediately repeated.

In event (b), the control switches, in theory, at infinite frequency, while the state trajectory is held on the switching boundary, so (10.4) remains satisfied. During this period, the state point appears to *slide* on the switching boundary. This is *sliding motion*. The switched control system is therefore said to be operating in a *sliding mode*. The necessary conditions for sliding motion will be derived in the following section. The two possible events are illustrated in Fig. 10.1.

10.1 Introduction

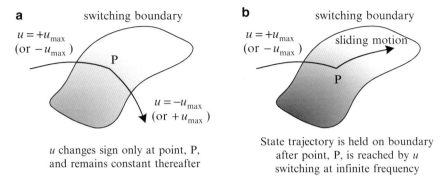

Fig. 10.1 Two possible events at a point on the switching boundary

In this figure, the boundaries appear as two-dimensional surfaces in a three-dimensional space, but they could be of higher dimensions and similar effects take place.

Sliding modes were first discovered in Russia as early as the 1930s and bang–bang controllers designed to exhibit these modes were tried utilising electrical relays for the implementation. The problem of contact wear, particularly due to sparking with inductive loads, however, slowed progress but much later advances in technology led to power electronics for the control switching, which promoted control engineers to think about the exploitation of the robustness offered by sliding mode control. They observed that for a second-order plant of full rank, if the state variables consisted of the controlled output, y, and its first derivative, $\dot{y}$, then, in contrast to orthodox controllers, the closed-loop dynamics, i.e. the differential equation relating y and $\dot{y}$ to the reference input, y_r, is almost entirely independent of any plant parameter changes and not affected by external disturbances. The question then arises of whether or not this robustness also occurs for the more general closed-loop system formed by the plant of (10.1) and (10.2) together with (10.3) when it is operating in the sliding mode. As will be seen in the following section, this depends on the state representation.

10.1.3 Implementation for Robustness

10.1.3.1 Output Derivative State Representation

Suppose that the closed-loop system formed by (10.1), (10.2) and (10.3) is operating in the sliding mode so that (10.4) is satisfied. First, assuming that the plant is of full relative degree (Chap. 3) it will be shown that the closed-loop dynamics is expressible as a differential equation relating y and its derivatives up to order, $n-1$, to y_r, where n is the plant order. This is done by changing to the output derivative state representation as follows.

$$y = h_0(\mathbf{x}), \dot{y} = h_1(\mathbf{x}), \ldots, y^{(n-1)} = h_{n-1}(\mathbf{x}) \tag{10.5}$$

In compact form, this can be written

$$\mathbf{y} = \mathbf{h}(\mathbf{x}) \tag{10.6}$$

This can be regarded as n simultaneous equations that can be solved for the n components of $\mathbf{x}$, so provided that they are not functionally dependent, i.e. no equation can be derived by manipulating a subset of the remaining equations, then an inverse set of equations exists giving

$$\mathbf{x} = \mathbf{h}^{-1}(\mathbf{y}), \tag{10.7}$$

i.e.

$$x_1 = h_0^{-1}(\mathbf{y}), x_2 = h_1^{-1}(\mathbf{y}), \ldots, x_n = h_{n-1}^{-1}(\mathbf{y}). \tag{10.8}$$

Substituting for $\mathbf{x}$ in (10.4) using (10.7) then yields

$$S\left(\mathbf{h}^{-1}(\mathbf{y}), y_r\right) = 0, \tag{10.9}$$

i.e.

$$S\left(\mathbf{h}^{-1}\left(y, \dot{y}, \ldots, y^{(n-1)}\right), y_r\right) = 0. \tag{10.10}$$

This is the required differential equation of the closed-loop system in the sliding mode. Importantly, the state transformation (10.6) and therefore the inverse transformation (10.7) *depend on the plant parameters*. The differential equation (10.10) therefore depends on the plant parameters. This means that the response of the closed-loop system to the reference input depends upon the plant parameters. In this case the system would *not* be robust. If, however, $\mathbf{y}$ is the chosen state vector to form the switching function, then the switching boundary equation (10.4) becomes

$$S(\mathbf{y}, y_r) = 0, \tag{10.11}$$

this being obeyed due to the action of the control law

$$u = u_{\max} \operatorname{sgn}\left[S(\mathbf{y}, y_r)\right]. \tag{10.12}$$

Then (10.11) is the differential equation of the closed-loop system when expressed as

$$S\left(y, \dot{y}, \ldots, y^{(n-1)}, y_r\right) = 0. \tag{10.13}$$

10.1 Introduction

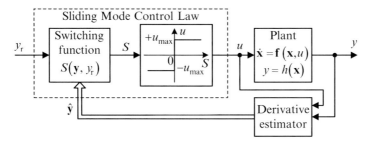

Fig. 10.2 SISO sliding mode control system using the output derivative state representation

A block diagram of the basic sliding mode control system is shown in Fig. 10.2.

Provided the output derivatives are either measured directly or estimated by an algorithm that does not depend on an accurate plant model, then the closed-loop dynamics is independent of the plant parameters as well as the external disturbance and therefore the desired robustness is attained. The control signal, u, is shown as an input of the derivative estimator as this could be a special plant model independent observer, as described in Sect. 10.1.3.2. The output derivative state representation will be assumed in the following sections.

An important observation is that the closed-loop system in the sliding mode is of order, $n - 1$, although the plant is of order, n. This is due to the sliding mode control law forcing the state trajectory to move in the $n - 1$ dimensional subspace of the switching boundary. Then one state variable may be expressed as a function of the remaining $n - 1$ state variables. The closed-loop system therefore has only $n - 1$ independent state variables and is therefore of order, $n - 1$.

Example 10.1 Impact of state representation on robustness for second-order heating process

This simple example demonstrates the importance of the choice of the state representation in achieving robustness when applying sliding mode control. Consider a second-order heating process having the following state-space model.

$$\dot{x}_1 = \frac{1}{T_1}(bu - x_1) \tag{10.14}$$

$$\dot{x}_2 = \frac{1}{T_2}(x_1 - x_2) \tag{10.15}$$

$$y = K_t x_2 \tag{10.16}$$

Suppose that the bang–bang control law

$$u = u_{\max} \operatorname{sgn}\left[S\left(x_1, x_2, y_r\right)\right] = \operatorname{sgn}\left(C_1 x_1 + C_2 x_1 + y_r\right) \tag{10.17}$$

is applied, where C_1 and C_2 are constants that may be chosen to yield the desired closed-loop dynamics if the system is operating in the sliding mode, requiring

$$C_1 x_1 + C_2 x_1 + y_r = 0. \tag{10.18}$$

To determine the resulting closed-loop dynamics in terms of a differential equation relating y and $\dot{y}$ to y_r, the state variables, x_1 and x_2, must first be expressed in terms of y and $\dot{y}$. Then substitutions are made in (10.18). Differentiating (10.16) and substituting for $\dot{x}_2$ using (10.15) yields

$$\dot{y} = \frac{K_t}{T_2}(x_1 - x_2) \tag{10.19}$$

Equations (10.16) and (10.19) then constitute the state transformation (10.6). The next step is to solve (10.16) and (10.19) for x_1 and x_2. From (10.16),

$$x_2 = \frac{1}{K_t} y. \tag{10.20}$$

Making x_1 the subject of (10.19) and substituting for x_2 using (10.20) yields

$$x_1 = x_2 + \frac{T_2}{K_t}\dot{y} = \frac{1}{K_t} y + \frac{T_2}{K_t}\dot{y}. \tag{10.21}$$

Equations (10.20) and (10.21) then constitute the inverse state transformation (10.8). Substituting for x_1 and x_2 in (10.18) using (10.20) and (10.21) then yields the required closed-loop differential equation as follows:

$$C_1\left(\frac{1}{K_t} y + \frac{T_2}{K_t}\dot{y}\right) + C_2 \frac{1}{K_t} y + y_r = 0 \Rightarrow \dot{y} = -\left(1 + \frac{C_2}{C_1}\right)\frac{1}{T_2} y - \frac{K_t}{C_1 T_2} y_r \tag{10.22}$$

Now the controller parameters, C_1 and C_2, may be chosen to give the specified closed-loop dynamics. For a settling time of the step response of T_s seconds (5 % criterion), the closed-loop differential equation is

$$\dot{y} = \frac{3}{T_s}(y_r - y) \tag{10.23}$$

Comparing (10.22) and (10.23) then yields

$$C_1 = \frac{K_t T_s}{3 T_2} \tag{10.24}$$

10.1 Introduction

and

$$\left(1 + \frac{C_2}{C_1}\right)\frac{1}{T_2} = \frac{3}{T_s} \Rightarrow C_2 = C_1\left(\frac{3T_2}{T_s} - 1\right) = \frac{K_tT_s}{3T_2}\left(1 - \frac{3T_2}{T_s}\right) \Rightarrow$$
$$C_2 = K_t\left(\frac{T_s}{3T_2} - 1\right) \tag{10.25}$$

It is now clear that the controller parameters, C_1 and C_2, depend on the plant parameters, T_2 and K_t. An accurate plant model is therefore needed for the closed-loop dynamics of (10.23) to be realised. Hence, the control system would not be robust with respect to errors in the assumed values of these plant parameters.

If instead the output derivative state, $(y, \dot{y})$, is used and control law (10.17) is replaced by

$$u = u_{max}\,\text{sgn}\left[S\left(y, \dot{y}, y_r\right)\right] = u_{max}\,\text{sgn}\left[\frac{3}{T_s}(y_r - y) - \dot{y}\right], \tag{10.26}$$

then in the sliding mode, $(3/T_s)(y_r - y) - \dot{y}$, and this is equivalent to the desired closed-loop differential equation (10.23). This is independent of the plant parameters and therefore yields the required robustness operating in the sliding mode.

10.1.3.2 Output Derivative Estimation

Several approaches are possible for estimating output derivatives, such as software differentiation and filtered differentiators. The method presented here is a special observer in which the accurate plant model of the observers introduced in Chap. 8 is replaced by a chain of integrators equal in number to the relative degree of the plant. The integrator outputs are then the required output derivatives and an accurate plant model is unnecessary, in keeping with the requirement for the control system to be robust with respect to plant modelling uncertainties. Also, the chain of integrators is driven by the control input via an adjustable gain, b, that can reduce the dynamic lag between the output derivatives and their estimates, recalling that in a conventional observer, the control input to the real-time model completely eliminates this dynamic lag with an ideal plant model. Minimisation of this lag is important as it can cause instability through the switching element in the forward path (Fig. 10.2) having a similar effect to a high gain. This can be understood by considering that infinitesimal changes of S about zero cause finite changes of u between u_{min} and u_{max}. Figure 10.3 shows a sliding mode control system using a triple integrator observer for estimation of $\dot{y}$ and $\ddot{y}$, the design of which will be carried out. This should be sufficient to enable the reader to design multiple integrator observers of different orders.

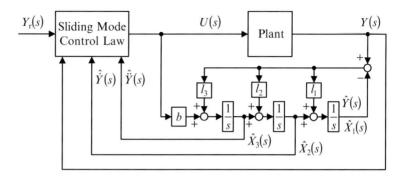

Fig. 10.3 Sliding mode control system incorporating a multiple integrator observer for output derivative estimation

The transfer function relationship of the observer is

$$\begin{bmatrix} \widehat{X}_1(s) \\ \widehat{X}_2(s) \\ \widehat{X}_3(s) \end{bmatrix} = \frac{\begin{bmatrix} l_1 s^2 + l_2 s + l_3 & b \\ l_2 s^2 + l_3 s & b(s + l_1) \\ l_3 s^2 & b(s^2 + l_1 s + l_2) \end{bmatrix} \begin{bmatrix} Y(s) \\ U(s) \end{bmatrix}}{s^3 + l_1 s^2 + l_2 s + l_3} \qquad (10.27)$$

where the observer gains are determined by pole placement to yield a triple pole with a filtering time constant of T_f, requiring

$$s^3 + l_1 s^2 + l_2 s + l_3 = \left(s + \tfrac{1}{T_f}\right)^3 = s^3 + \tfrac{3}{T_f} s^2 + \tfrac{3}{T_f^2} s + \tfrac{1}{T_f^3} \Rightarrow$$
$$l_1 = \tfrac{3}{T_f}, \quad l_2 = \tfrac{3}{T_f^2}, \quad \text{and} \quad l_3 = \tfrac{1}{T_f^3}. \qquad (10.28)$$

Then (10.27) becomes

$$\begin{bmatrix} \widehat{X}_1(s) \\ \widehat{X}_2(s) \\ \widehat{X}_3(s) \end{bmatrix} = \frac{\begin{bmatrix} 3T_f^2 s^2 + 3T_f s + 1 & bT_f^3 \\ (3T_f s + 1)s & b(sT_f^3 + 3T_f^2) \\ s^2 & b(s^2 T_f^3 + 3T_f^2 s + 3T_f) \end{bmatrix} \begin{bmatrix} Y(s) \\ U(s) \end{bmatrix}}{T_f^3 s^3 + 3T_f^2 s^2 + 3T_f s + 1}. \qquad (10.29)$$

It follows that

$$\lim_{T_f \to 0} \begin{bmatrix} \widehat{X}_1(s) \\ \widehat{X}_2(s) \\ \widehat{X}_3(s) \end{bmatrix} = \begin{bmatrix} Y(s) \\ sY(s) \\ s^2 Y(s) \end{bmatrix} \Rightarrow \lim_{T_{so} \to 0} \begin{bmatrix} \widehat{x}_1(t) \\ \widehat{x}_2(t) \\ \widehat{x}_3(t) \end{bmatrix} = \begin{bmatrix} y(t) \\ \dot{y}(t) \\ \ddot{y}(t) \end{bmatrix}. \qquad (10.30)$$

Hence, if T_f is made sufficiently small and the measurement noise levels are suitably low, the state estimate is a good approximation to the output and its first two derivatives, as required, and this result is *independent* of the plant model. In the output derivative estimator, y is fed back directly instead of $\hat{y}$ to avoid introducing unnecessary dynamic lag due to the non-zero T_f.

10.2 Control of SISO Second-Order Plants of Full Relative Degree

10.2.1 The Plant Model

The most general state-space model of a second-order SISO plant of full relative degree in the control canonical form is as follows:

$$\dot{x}_1 = x_2 \tag{10.31}$$

$$\dot{x}_2 = f(x_1, x_2, u - d_e) \tag{10.32}$$

$$y = x_1 \tag{10.33}$$

where $f(\cdot)$ is a continuous function of its arguments and d_e is an external disturbance referred to the control input. Most real second-order plants of full relative degree fall into a subclass with (10.32) in the form,

$$\dot{x}_2 = g(x_1, x_2) + h(x_1, x_2)(u - d_e), \tag{10.34}$$

where $g(\cdot)$ and $h(\cdot)$ are continuous functions of their arguments. In order to simplify the analysis of the sliding mode control system, however, (10.34) will be rearranged as

$$\dot{x}_2 = h(x_1, x_2)\left[\frac{g(x_1, x_2)}{h(x_1, x_2)} + u - d_e\right] = h(x_1, x_2)[u - d(x_1, x_2)], \tag{10.35}$$

where $d(x_1, x_2) = d_e - \frac{g(x_1, x_2)}{h(x_1, x_2)}$ is a newly defined state-dependent disturbance. An example is a DC motor-based electric drive (Chap. 2) which forms the basis of many closed-loop control systems. If the driven mechanical load is a balanced rigid body fixed to the motor output shaft, the state-space model is as follows.

$$\dot{x}_1 = x_2 \tag{10.36}$$

$$\dot{x}_2 = -ax_2 + b(u - d_e) = b(u - d) \tag{10.37}$$

$$y = x_1 \tag{10.38}$$

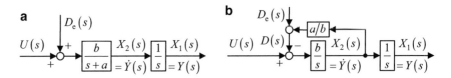

Fig. 10.4 State-variable block diagrams of DC motor and mechanical load. (**a**) From transfer function relationship. (**b**) In controller canonical form

Here, x_1 and x_2 are, respectively, the rotor angle and angular velocity and $d = d_e + (a/b)\,x_2$. The transfer function relationship of the model is

$$Y(s) = \frac{b\,[U(s) - D_e(s)]}{s^2 + as} \tag{10.39}$$

Figure 10.4 shows state-variable block diagrams corresponding to (10.39).

Figure 10.4a corresponds to (10.34) and Fig. 10.4b corresponds to (10.35), showing that the plant may be considered as a double integrator with a state-dependent disturbance, $d(t) \equiv D(s)$, referred to the control input, which is more convenient for discussing the sliding mode control. In this model, $a = F/J$ and $b = B/J$, where F is the viscous friction coefficient, B is the electric drive torque constant and J is the total moment of inertia presented to the motor output shaft. The limits of u taken will be the maximum and minimum control voltages, $\pm u_{\max}$, input to the drive electronics. Typically, $u_{\max} = 10$ [V]. The DC motor example will be used in the following sections.

10.2.2 Phase Portraits

The operation of the basic sliding mode control system will be studied by examining the state trajectories in the phase plane (Chap. 9). The state trajectory equation is a solution to the state trajectory differential equation obtained by dividing (10.37) by (10.36). Thus,

$$\frac{dx_2}{dx_1} = \frac{b\,(u - d)}{x_2} \tag{10.40}$$

Suppose first that d is constant, which would be valid for a drive with negligible friction and constant external disturbance torque, and $u = \text{const.} = \pm u_{\max}$. Then the solution of (10.40) by separation of the variables yields the state trajectory equation

$$x_1 = x_1(0) + \frac{1}{2b\,(u - d)}\left[x_2^2 - x_2^2(0)\right]. \tag{10.41}$$

This is a parabola in the phase plane symmetrical about the x_1 axis.

10.2 Control of SISO Second-Order Plants of Full Relative Degree

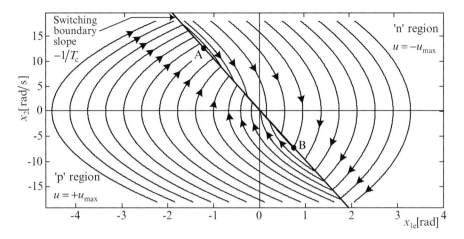

Fig. 10.5 Closed-loop phase portrait for sliding mode control of DC electric drive with constant load torque

Next, control law (10.3) will be chosen to yield a linear switching boundary. For any second-order plant of full relative degree, this is

$$u = u_{\max} \operatorname{sgn} [S(x_1, x_2, y_r)] \qquad (10.42)$$

where

$$S(x_1, x_2, y_r) = y_r - x_1 - T_c x_2. \qquad (10.43)$$

Here, y_r is the reference input, i.e. the demanded value of x_1. As will be seen, the closed-loop system is linear in the sliding mode and of first order with time constant, T_c. The switching boundary,

$$y_r - x_1 - T_c x_2 = 0, \qquad (10.44)$$

is a straight line with slope, $-1/T_c$, passing through the origin of the error state space, $(x_{1e}, x_{2e}) = (x_1 - y_r, x_2)$. The control objective is to bring the state trajectory to the origin. Figure 10.5 shows this switching boundary together with the closed-loop phase portrait (i.e. the infinite continuum of state trajectories occurring for all possible initial states) for $d(s) = \frac{s}{1+sT_f}$ constant. This is based on a simulation for which $a = 0$, $b = 10 \left[\mathrm{rad/s/s}\right]/\mathrm{V}$, $u_{\max} = 10$ [V], $d = 3$ [V] and $T_c = 0.1$ [s].

The difference between the scaling of the parabolic families of curves in the 'n' and 'p' regions of the phase portrait is due to the constant non-zero disturbance, d, producing a lower constant acceleration magnitude in the 'p' region than in the 'n' region.

10.2.3 Sliding Motion

Along segment AB of the switching boundary in Fig. 10.5, the state trajectories are directed towards the boundary from both sides. Once the state trajectory reaches this segment of the boundary, it may be reasoned that the control rapidly switches between $+u_{max}$ and $+u_{max}$, *in theory* at an infinite frequency with a continuously varying mark–space ratio so as to maintain the trajectory on the boundary. This is an example of the *sliding mode* already introduced and illustrated in Fig. 10.1 (event (b)). A full understanding of the behaviour of the control system in a practical situation will now be given by means of a step-by-step explanation of its operation assuming that the sliding mode control law is implemented using a control algorithm programmed on a digital processor.

The behaviour of the sliding mode control system is illustrated in Fig. 10.6 for the case where the rotor shaft angle commences from rest at zero angle, corresponding to the origin of the phase plane, and then the reference input, y_r, steps from its current value of zero to a new constant value of $Y_r > 0$. When this step reference input is applied, the state point in Fig. 10.6 jumps from the origin of the phase plane to point (i) in the '*p*' region. The sliding mode control law then sets the control to $u = +u_{max}$, which remains saturated at this value, while the angular velocity, $x_2(t)$, increases linearly with time and the rotor shaft angle increases as a parabolic function of time. Thus, $x_2(t) = b\,(u_{max} - d)\,t$ yielding $x_1(t) = \frac{1}{2}b\,(u_{max} - d)\,t^2$, the state point moving on a parabolic path towards the switching boundary in Fig. 10.6a until it is reached at point (ii).

Referring to the magnified inset of Fig. 10.6a, the control continues at $u = +u_{max}$ until the next iteration of the control algorithm, while the state trajectory moves a small distance into the '*n*' region. Then the control law sets $u = -u_{max}$, causing the state point to follow the phase portrait in the '*n*' region back towards the switching boundary. The control continues at $u = -u_{max}$ until the next iteration of the control algorithm, while the state trajectory crosses the switching boundary again but this time a little closer to the origin and moves a relatively small distance back into the '*p*' region. Upon the next iteration, the control law sets $u = +u_{max}$. The whole process then repeats itself as the control law drives the state point onto the switching boundary along segment A-B in Fig. 10.5. The crossing points move closer and closer to the origin, which is the control objective. This is the sliding mode.

It should be noted that in Fig. 10.6a, hysteresis of ± 0.005 [rad] has been introduced in the simulation so that the zigzag motion of the state trajectory in the sliding mode is visible in the magnified inset. In Fig. 10.6b the hysteresis has been increased to ± 0.1 [rad] so that the repeated switches of $u(t)$ and the piecewise linear segments of $x_2(t)$ in the sliding mode are visible.

The analysis of the motion towards the origin in the sliding mode will now be carried out, but for the hypothetical ideal system with infinite sampling frequency. In the sliding mode, the switching boundary equation directly yields the closed-loop differential equation because the state point is held precisely on the boundary. Substituting for x_2 in (10.44) using (10.31) yields

10.2 Control of SISO Second-Order Plants of Full Relative Degree

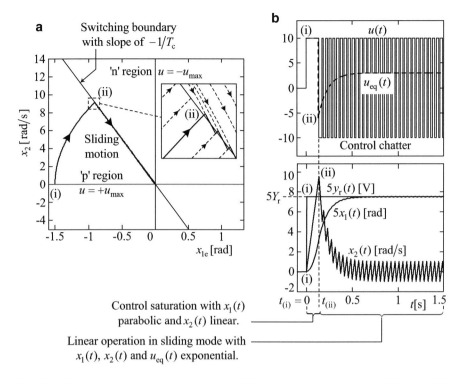

Fig. 10.6 State trajectory and time response for sliding mode position control of DC drive. (**a**) state trajectory in the phase plane (**b**) time responses

$$y_r - x_1 - T_c \dot{x}_1 = 0 \Rightarrow \dot{x}_1 = \frac{1}{T_c}(y_r - x_1). \tag{10.45}$$

Since $y_r = Y_r = \text{const.}$, the solution is

$$x_1(t) = x_1\left(t_{(ii)}\right) e^{-[t-t_{(ii)}]/T_c} + Y_r \left\{1 - e^{-[t-t_{(ii)}]/T_c}\right\} \tag{10.46}$$

and $x_2(t) = \dot{x}_1(t) \Rightarrow$

$$x_2(t) = \frac{Y_r - x_1\left(t_{(ii)}\right)}{T_c} e^{-[t-t_{(ii)}]/T_c} = x_2\left(t_{(ii)}\right) e^{-[t-t_{(ii)}]/T_c}, \tag{10.47}$$

where $t_{(ii)}$ is the time at which the state point first reaches the switching boundary in Fig. 10.6a. The exponential behaviour of the system indicated by (10.46) and (10.47) in the sliding mode entirely agrees with the graphs of Fig. 10.6b.

It is important to emphasise that the control system robustness is obtained only in the sliding mode and not during the acquisition phase in which the control is saturated, as exemplified by the state trajectory between points (i) and (ii) in

Fig. 10.6a, that depends upon the plant parameter, b, and the external disturbance, d_e. A sliding mode control system may not approach ideal robustness in the step response, since this will always cause initial control saturation, but as will be seen, it can do so with continuously varying reference inputs that do not cause control saturation.

10.2.4 The Equivalent Control

An important concept is the *equivalent control*, $u_{eq}(t)$, as this is not only an analytical tool in sliding mode control but can be used to form useful variants on the basic sliding mode control system that overcome some practical issues. This is the *continuously* varying control variable that is *equivalent* to the rapidly switching $u(t)$ yielded by the basic sliding mode control law in that it would maintain the state trajectory precisely on the switching boundary if applied instead. It is also the short-term mean value of $u(t)$ switching at infinite frequency in the ideal system. Also, the switching control, $u(t)$, can be regarded as $u_{eq}(t)$ plus an oscillatory signal with fundamental and harmonic components at infinite frequencies, the effects of which will be zero due to the low-pass filtering action of the plant. The equivalent control is shown dotted in Fig. 10.6b for the electric drive example. It exponentially converges with the time constant, T_c, towards a constant value equal to the external disturbance input, d_e, as needed to decelerate the drive and counteract d_e. Making u the subject of equation (10.37) and substituting for $\dot{x}_2$ using the derivative of $x_2(t)$ given by equation (10.47) yields an expression for the equivalent control,

$$\begin{aligned} u_{eq}(t) &= d + \tfrac{1}{b}\dot{x}_2 = d + \tfrac{1}{b}\cdot\tfrac{d}{dt}\left\{x_2\left(t_{(ii)}\right)e^{-[t-t_{(ii)}]/T_c}\right\} \\ &= \underline{d - \tfrac{1}{bT_c}x_2\left(t_{(ii)}\right)e^{-[t-t_{(ii)}]/T_c}} \end{aligned} \quad (10.48)$$

This expression agrees with the dotted graph in Fig. 10.6b. The equivalent control will be referred to frequently in the remainder of this chapter.

10.2.5 Control Chatter

The rapid switching of $u(t)$ exemplified in Fig. 10.6b is referred to as *control chatter*. In many applications, control chatter is considered undesirable due to induced vibrations and heating of the actuators. This has been viewed as a serious drawback regarding the practical application of sliding mode control. Techniques for preventing control chatter, however, are available that yield smooth control functions, and these are presented in some of the following subsections. In relatively high-powered electric drives, however, the power electronics has to be operated in a switched mode to minimise power loss in the devices used to regulate the electrical power input to the motor, and this provides the possibility of directly using the sliding mode control law without the need for the conventional pulse width modulator. This would be feasible if the frequency of switching in the sliding mode was sufficiently high to avoid exciting mechanical vibration modes while being sufficiently low to keep switching losses in the power electronics (due to the integral of the product of the voltage drop across the electronic switch and the load current during switching state transitions) to acceptable levels. This could be accomplished by the introduction of hysteresis (Chap. 9) with adaptive hysteresis levels. Other than this, a pulse width modulator would have to be employed together with the control chatter avoidance techniques presented later in this chapter.

10.2.6 Conditions for the Existence of Sliding Motion

In the process of designing a sliding mode control system, it is necessary to be sure that upon the initial loop closure, the system will reach the sliding mode, meaning that the situation depicted in Fig. 10.1 leading to event (b) pertains, in contrast to event (a). The basic *necessary and sufficient* condition for the existence of sliding motion is already evident from the examination of the closed-loop phase portrait of Fig. 10.5. Thus

> For sliding motion to exist along a segment of the switching boundary, $S(\mathbf{y}, y_r) = 0$, the infinite continuum of state trajectories of the closed-loop phase portrait has to be directed towards the switching boundary from both sides.

This is self-evident for a second-order system of full rank by inspection of the behaviour in the phase plane, but by analogy, it is true for systems of higher order. Another observation may be made in Fig. 10.5. Consider a close neighbourhood of the boundary along segment AB. Below this segment, $S(\mathbf{y}, y_r) > 0$, and $\dot{S}(\mathbf{y}, y_r)$ must be reducing towards zero as the boundary is approached, meaning that $\dot{S}(\mathbf{y}, y_r) < 0$. Similarly, in a close neighbourhood of the boundary above segment, AB, $S(\mathbf{y}, y_r) < 0$ and $S(\mathbf{y}, y_r)$ must be increasing towards the zero as the boundary is approached, meaning that $\dot{S}(\mathbf{y}, y_r) > 0$. So $S(\mathbf{y}, y_r)$ and $\dot{S}(\mathbf{y}, y_r)$ have to be opposite in sign in a close neighbourhood of the boundary. This leads to the following general statement of the necessary and sufficient condition for sliding motion to exist along a segment of the switching boundary.

$$S(\mathbf{y}, y_r)\dot{S}(\mathbf{y}, y_r) < 0. \tag{10.49}$$

A necessary but *not sufficient* condition for the existence of sliding motion can be obtained by calculating the equivalent control, u_{eq}, along the boundary. Then sliding motion *may exist* if u_{eq} lies between the two control saturation limits, which, for $u_{min} = -u_{max}$, can be expressed simply by

$$|u_{eq}| < u_{max} \qquad (10.50)$$

Applying this to the electric drive example, (10.37) yields

$$u = \frac{1}{b}\dot{x}_2 + d \qquad (10.51)$$

On the switching boundary (10.44), $u = u_{eq}$ and $x_2 = \frac{1}{T_c}(y_r - x_1)$, and therefore for constant y_r and in view of (10.36), $\dot{x}_2 = -\frac{1}{T_c}\dot{x}_1 = -\frac{1}{T_c}x_2$. Then (10.51) becomes

$$u_{eq} = \frac{1}{bT_c}x_2 + d \qquad (10.52)$$

and therefore condition (10.50) becomes

$$\left|\frac{1}{bT_c}x_2 - d\right| < u_{max} \qquad (10.53)$$

The upper and lower limits of x_2 defining points, A and B, in Fig. 10.5 are therefore obtained as follows.

$$\frac{1}{bT_c}x_{2\,min} - d = -u_{max} \Rightarrow \underline{x_{2\,min} = bT_c(d - u_{max})} \qquad (10.54)$$

$$\frac{1}{bT_c}x_{2\,max} - d = +u_{max} \Rightarrow \underline{x_{2\,max} = bT_c(d + u_{max})} \qquad (10.55)$$

As would be expected, the segment AB increases in size as u_{max} is increased.

10.2.7 Reaching the Sliding Condition

Suppose that the switching function, $S(\mathbf{y}, \mathbf{y}_r)$, is chosen in the first place to yield the required closed-loop dynamics in the sliding mode. Then if condition (10.49) is satisfied at every point in the state space, the state trajectories would always be attracted to the switching boundary and sliding motion would commence in a finite time. This is not, however, usually the case, condition (10.49) only being satisfied in finite regions of the state space adjacent to the sliding regions of the switching boundary. These attraction regions of the state space may or may not

10.2 Control of SISO Second-Order Plants of Full Relative Degree

be reached from other regions of the state space. Mathematical analysis to answer this question is difficult and some approaches to this problem may be found in [4]. Practical control system development would be greatly aided by simulation to ensure satisfactory operation over the operation envelope of the application. It is possible, however, that the switching boundary could be designed to force (10.49) to be satisfied while retaining acceptable closed-loop dynamics. These points will be illustrated using the DC drive example. The remainder of this section is devoted to the determination of the attraction regions analytically and the investigation of the behaviour outside these regions by examination of the closed-loop phase portraits.

In view of the plant state differential equations (10.36) and (10.37), control law (10.42) and switching function (10.43), inequality (10.49) becomes

$$(y_r - x_1 - T_c x_2)(-\dot{x}_1 - T_c \dot{x}_2) < 0 \Rightarrow (y_r - x_1 - T_c x_2)[-x_2 - bT_c(u - d)] < 0$$
$$\Rightarrow (y_r - x_1 - T_c x_2)\{-x_2 - bT_c[u_{\max} \operatorname{sgn}(y_r - x_1 - T_c x_2) - d]\} < 0 \Rightarrow$$
$$\Rightarrow -bT_c u_{\max} |y_r - x_1 - T_c x_2| - (y_r - x_1 - T_c x_2)(x_2 - bT_c d) < 0 \tag{10.56}$$

Clearly inequality (10.56) cannot be satisfied everywhere in the phase plane, and the boundary beyond which it is not satisfied is defined by the equation

$$-bT_c u_{\max} |y_r - x_1 - T_c x_2| - (y_r - x_1 - T_c x_2)(x_2 - bT_c d) = 0 \tag{10.57}$$

In the region below the switching boundary, $y_r - x_1 - T_c x_2 > 0$ and therefore, $y_r - x_1 - T_c x_2 = |y_r - x_1 - T_c x_2|$; whence, (10.57) simplifies to

$$-bT_c u_{\max} |y_r - x_1 - T_c x_2| - |y_r - x_1 - T_c x_2|(x_2 - bT_c d) = 0 \Rightarrow$$
$$x_2 = bT_c(d + u_{\max}) \tag{10.58}$$

Similarly, in the region above the switching boundary, $y_r - x_1 - T_c x_2 > 0$, and therefore, $y_r - x_1 - T_c x_2 = -|y_r - x_1 - T_c x_2|$; whence, (10.57) simplifies to

$$-bT_c u_{\max} |y_r - x_1 - T_c x_2| + |y_r - x_1 - T_c x_2|(x_2 - bT_c d) = 0 \Rightarrow$$
$$x_2 = bT_c(d - u_{\max}) \tag{10.59}$$

The values of x_2 given by (10.58) and (10.59) are precisely the values of $x_{2\max}$ and $x_{2\min}$ given by (10.54) and (10.55) valid on the switching boundary, but (10.58) and (10.59) are valid for any values of $x_1 - y_r$ below and above the switching boundary, respectively. The boundaries within which $S(\mathbf{y}, y_r) \dot{S}(\mathbf{y}, y_r) < 0$ are therefore horizontal straight line segments (shown dotted), and these, together with the switching boundary itself, define two skew symmetric semi-infinite triangular regions in which inequality (10.49) is satisfied as illustrated in Fig. 10.7.

The two regions are shown by populating them with the state trajectories of the closed-loop phase portrait, and these are seen to be directed towards the switching boundary. The slopes of the trajectories along the dotted lines are equal to the

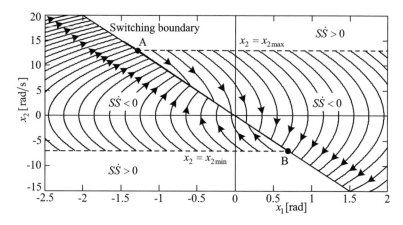

Fig. 10.7 Regions in the phase plane in which $S(\mathbf{y}, y_r) \dot{S}(\mathbf{y}, y_r) < 0$ for sliding mode position control of DC electric drive

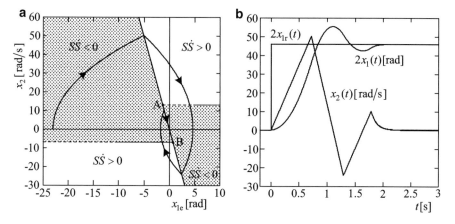

Fig. 10.8 Form of state trajectory for sliding mode position control of DC electric drive with relatively large step reference input. (**a**) State trajectory. (**b**) Time responses

slope of the switching boundary. The combined regions may be referred to as the *convergence region*. Examination of this figure, however, indicates that even if the initial state is within the convergence region, the state trajectory may or may not remain in this region after the first switch point on the boundary. Figure 10.8 shows the state trajectory for a relatively large step reference input and the corresponding time responses. Outside the convergence region, the magnitude of $S(\mathbf{y}, y_r)$ *increases* along every trajectory of the phase portrait.

For arbitrarily large initial error states, the state trajectories will move outside the convergence region, but it is evident from the examination of the geometry of the phase portrait that initially the control will saturate with a finite number of control switches for a finite initial state, but the successive switch points will become closer

10.2 Control of SISO Second-Order Plants of Full Relative Degree

to the origin of the phase plane, the state trajectory spiralling in towards it until ultimately the sliding segment, AB, is reached as shown in Fig. 10.8a. The piecewise linear behaviour of $x_2(t)$ during the saturated operation followed by the exponential decay to zero in the sliding mode is clearly visible in Fig. 10.8b. So even if inequality (8.3.15) is not satisfied, it does not mean that the sliding segment of the switching boundary cannot be reached.

For higher-order plants and/or time-varying reference inputs and disturbances, such analysis to determine if the sliding region of the switching boundary is reachable would be highly complex, and a more practicable approach would be simulation to examine the control system behaviour with realistic initial states.

10.2.8 Closed-Loop Dynamics in the Sliding Mode

During sliding motion, Eq. (10.44) governs the closed-loop behaviour and since $x_1 = y$ and $x_2 = \dot{y}$, the closed-loop differential equation is

$$\dot{y} = \frac{1}{T_c}(y_r - y) \qquad (10.60)$$

with the corresponding transfer function

$$\frac{Y(s)}{Y_r(s)} = \frac{1}{1 + sT_c} \qquad (10.61)$$

The closed-loop system is of first order in the sliding mode despite the plant being of second order due to the state trajectory being constrained to follow the switching boundary, which is only one dimensional. Again it is emphasised that this first-order closed-loop dynamics is only valid in the sliding mode, and, as will be seen, the reference input must be continuous and not changing too rapidly in order to avoid control saturation which would take the system out of the sliding mode.

10.2.9 Control with Time-Varying Disturbances and Reference Inputs

Now consider the operation of the sliding mode control law (10.42) with the linear switching function (10.43) and the more general second-order plant having the state-space model (10.31), (10.32), and (10.33). If the reference input, y_r, or the external disturbance, d_e, are both constant, then the shape of the trajectories of the closed-loop phase portrait will, in general, differ from the parabolic ones of the double integrator plant and will depend on the function, $f(x_1, x_2, u - d_e)$. The positions of the end points, A and B, of the sliding segment of the switching boundary

Fig. 10.9 Nonstationary closed-loop phase portrait for external disturbance and/or reference input time varying

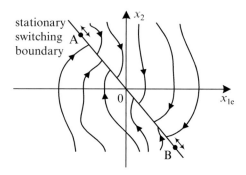

will also depend on this function. The closed-loop phase portrait will, however, still be stationary, i.e. an infinite continuum of trajectory curves that never change. On the other hand, if either y_r, d_e or both are time varying, a *nonstationary* phase portrait will result, a 'snapshot' of which is illustrated in Fig. 10.9. The shapes of the trajectories shown here are only illustrative and will be changing. If a movie could be shown, then the appearance of the closed-loop phase portrait could be likened to the motion of a forest of seaweed with long, thin stems anchored to the seabed (analogous to the stationary switching boundary) and subject to changing water currents and wave motion. In correspondence with the motion of the phase portrait, the points, A and B, will continually move along the switching boundary. The sliding motion, however, will be maintained provided the state point remains on the segment, AB.

The following salient points may now be made regarding the sliding mode control of second-order plants of full relative degree with linear switching boundaries.

(a) The system is extremely robust in the sliding mode since transfer function (10.61) is independent of the plant parameters and the external disturbance.
(b) The closed-loop system is only of first order despite the plant being of second order. This is due to one degree of freedom of movement in the two-dimensional state space being removed by the control law forcing the trajectory to move along the one-dimensional switching boundary.
(c) If the settling time, T_s, is chosen and the closed-loop time constant set to $T_c = T_s/3$ (5 % criterion) or $T_c = T_s/4$ (2 % criterion), then the step response will not have precisely this settling time due to the initial control saturation, but in the sliding mode, the response to a continuously varying reference input will be that of a first-order system with the specified settling time.

It is important to realise that ideal robustness cannot be attained in practice, even by a sliding mode control system, due to the finite sampling frequency of the digital implementation and any dynamic lags that are not taken into account such as the armature time constant in the DC drive example. The performance of a practicable sliding mode control system must be predicted by an accurate simulation before it is commissioned.

10.2.10 Rate-Limiting Switching Boundary for Zero Overshoot

Consider again the DC electric drive. A relatively simple modification of the linear switching boundary will eliminate the overshooting of a stepped reference input (Fig. 10.8) regardless of its magnitude and ensure sliding motion is maintained after the state trajectory first reaches the switching boundary. Two points, A' and B', are selected on the sliding segment of the previous switching boundary between points A and B, where the angular velocities are $\lambda x_{2\,min}$ and $\lambda x_{2\,max}$, with $0 < \lambda < 1$. Then the remainder of the switching boundary is replaced by two horizontal semi-infinite linear segments intersecting the points, A' and B', as shown in Fig. 10.10.

For forming the sliding mode control algorithm, the MATLAB®–SIMULINK®-compatible saturation function (Chap. 9) may be used to realise this switching boundary. Thus,

$$S(x_{1e}, x_2) = \text{sat}(-x_{1e}/T_c, \lambda x_{2\,min}, \lambda x_{2\,max}) - x_2 \qquad (10.62)$$

where

$$\text{sat}(-x_{1e}/T_c, \lambda x_{2\,min}, \lambda x_{2\,max}) \triangleq \begin{cases} -x_{1e}/T_c, & \lambda x_{2\,min} \leq -x_{1e}/T_c \leq \lambda x_{2\,max} \\ \lambda x_{2\,min}, & -x_{1e}/T_c < \lambda x_{2\,min} \\ \lambda x_{2\,max}, & -x_{1e}/T_c > \lambda x_{2\,max} \end{cases}$$

$$(10.63)$$

The resulting closed-loop phase portrait and corresponding family of step responses are shown in Fig. 10.11.

As can be seen in Fig. 10.11a, the condition for sliding motion is satisfied along the whole switching boundary, i.e. the infinite continuum of trajectories in the 'n' and 'p' regions is directed towards these segments of the boundary from both sides. Figure 10.11b shows a family of step responses for different reference input levels. The overshooting is seen to be eliminated due to the rate limiting. The exponential behaviour of $x_1(t)$ and $x_2(t)$ towards the desired values may be seen as the sliding motion moves onto segment A' B' of the switching boundary.

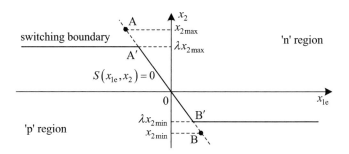

Fig. 10.10 Piecewise linear switching boundary for zero overshoot in the step response

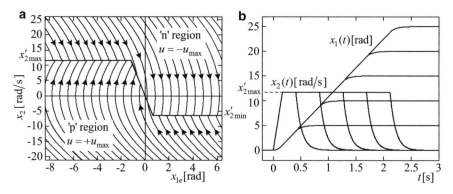

Fig. 10.11 Responses of sliding mode control system with rate-limiting switching boundary. (**a**) Closed-loop phase portrait. (**b**) Step responses

The factor of λ is chosen with the aim of maintaining the condition for sliding motion along the whole of the segment, $A'B'$, of the switching boundary. A value of $\lambda = 0.9$ is selected for this demonstration but in practice its choice is a matter of engineering judgement through knowledge of the application.

It must be noted that the control system is nonlinear except in the sliding mode along segment $A'B'$ of the switching boundary. One consequence of this is that the settling time is nearly proportional to the magnitude of the step reference input.

For arbitrary initial states, x_2 can be outside the range $(x_{2\min}, x_{2\max})$ as shown in Fig. 10.11a, but sliding motion is always attained that leads to linear operation on the segment, $A'B'$, of the switching boundary.

It must also be realised that according to (10.54) and (10.55), $x_{2\max}$ and $x_{2\min}$ are functions of the disturbance input, d, and the plant parameter, b. Implementation would therefore require an observer to estimate d.

10.2.11 Sub-Time-Optimal Control

For applications in which it is desirable to eliminate overshoot in the step response while achieving shorter settling times than the rate-limiting method of Sect. 10.2.10, a nonlinear switching boundary similar to the time-optimal ones of Chap. 9 may be employed but which are modified so that the state trajectories of the phase portrait are directed towards it from both sides over its whole length. The class of plants to which this may be applied has the state-space model,

$$\left.\begin{aligned}\dot{x}_1 &= x_2 \\ \dot{x}_2 &= f(x_1, x_2) + b(u - d_e)\end{aligned}\right\} \quad (10.64)$$

where b is a positive constant, $f(x_1, x_2)$ is continuous with $f(0, 0) = 0$ and d_e is an external disturbance. Since the purpose of sliding mode control is to achieve

10.2 Control of SISO Second-Order Plants of Full Relative Degree

robustness, the control law should be based on an approximate plant model not requiring knowledge of $f(x_1, x_2)$ and d_e. For $u = \pm u_{max}$, there exists a limited region of the state space enclosing the origin in which $|f(x_1, x_2)| \ll |bu|$. Then if $d_e \ll u_{max}$, the plant model may be approximated by the double integrator model,

$$\left. \begin{array}{l} \dot{x}_1 = x_2 \\ \dot{x}_2 = bu \end{array} \right\} \tag{10.65}$$

The time-optimal switching function (Chap. 9) for this approximate model is

$$S(x_{1e}, x_2) = -x_{1e} - \frac{1}{2bu_{max}} |x_2| x_2. \tag{10.66}$$

where $x_{1e} = x_1 - y_r$. The condition of sliding motion cannot be guaranteed over the operational segment for (10.66), i.e. the segment of the switching boundary within the operational envelope of the plant in the state space. If, however, a scaling factor, λ, where $0 < \lambda < 1$, is introduced so that (10.66) becomes

$$S(x_{1e}, x_2) = -x_{1e} - \frac{1}{2\lambda bu_{max}} |x_2| x_2, \tag{10.67}$$

so the graph of x_2 against x_{1e} becomes 'compressed' in the x_2 direction, then below a certain λ, the state trajectories of the phase portrait are directed towards the boundary from both sides over the operational envelope. This is reasonable since as λ approaches zero, the switching boundary approaches the x_{1e} axis, which the trajectories of the phase portrait approach from both sides at right angles. This follows from analysis of the state trajectory differential equation,

$$\frac{dx_2}{dx_1} = \frac{f(x_1, x_2) + b(u - d_e)}{x_2}. \tag{10.68}$$

If the switching boundary is the horizontal line, $x_2 = 0$, with the 'n' region above and the 'p' region below, then given

$$u_{max} > |f(x_1, x_2)/b - d_e| \tag{10.69}$$

then $dx_2/dx_1 \to \pm\infty$ as $x_2 \to 0$ so the state trajectories of the phase portrait are perpendicular to the x_2 axis. Also, the switching function is $S = -x_2$, giving

$$u = u_{max} \operatorname{sgn}(S) = -u_{max} \operatorname{sgn}(x_2) \tag{10.70}$$

so that the second of equations (10.64) becomes

$$\dot{x}_2 = f(x_1, x_2) - bd_e - bu_{max} \operatorname{sgn}(x_2). \tag{10.71}$$

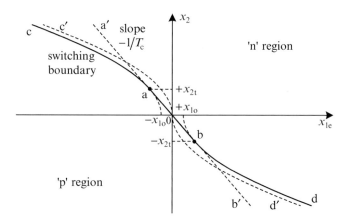

Fig. 10.12 Switching boundary for near-time-optimal sliding mode control

Then

$$S\dot{S} = (-x_2)(-\dot{x}_2) = (-x_2)(bu_{max}\,\text{sgn}(x_2) - f(x_1, x_2) + bd_e) < 0, \quad (10.72)$$

indicating the condition for sliding motion is satisfied at every point along the x_2 axis. Then as λ is increased, it would be expected that the segment or segments of switching boundary (10.67) along which $S\dot{S} < 0$ is satisfied become finite but will lie within the operation envelope of the state space for a given application up to a certain maximum value of λ.

The slope, dx_2/dx_{1e}, of switching boundary,

$$x_{1e} = -\frac{1}{2\lambda bu_{max}}|x_2|x_2, \quad (10.73)$$

is infinite at the origin, which does not yield a useful closed-loop dynamics in the sliding mode in the region of the origin, where the control system will sometimes be required to operate with time-varying reference inputs. A linear closed-loop performance is desirable in this region, and this can be achieved by having a linear segment with finite negative slope as in the previous section. Such a switching boundary is sketched in Fig. 10.12. The starting point is boundary (10.73) indicated by c′0d′.

Then the straight line, a′b′, is formed that includes the segment, 'ab', to be included in the switching boundary. Then the upper and lower parabolic segments of the boundary, c′0d′, are translated, respectively, by $-x_{1o}$ and $+x_{1o}$ units along the x_1 axis so they are tangential to the line, a′b′, at points 'a' and 'b'. This results in the required switching boundary, cabd, as shown. Next, the equation of this switching boundary will be derived as required for controller implementation. Let the values of x_{1e} at the tangent points, a and b, be $\pm x_{1et}$, $x_{1et} > 0$. Then since the tangent points are on the straight line segment with slope, $-1/T_c$, the velocity magnitude, x_{2t}, is

10.2 Control of SISO Second-Order Plants of Full Relative Degree

$$x_{2t} = \frac{1}{T_c} x_{1et} \Rightarrow x_{1et} = T_c x_{2t}. \tag{10.74}$$

Since the tangents are also on the offset parabolic curves, then

$$x_{1et} = \frac{1}{2A} x_{2t}^2 + x_{1o} \tag{10.75}$$

where

$$A = \lambda b u_{max} \tag{10.76}$$

Equating the RHS of (10.74) and (10.75) then yields

$$\frac{1}{2A} x_{2t}^2 + x_{1o} = T_c x_{2t} \Rightarrow x_{2t}^2 - 2AT_c x_{2t} + 2A x_{1o} \tag{10.77}$$

from which

$$x_{2t} = AT_c \pm \sqrt{A^2 T_c^2 - 2A x_{1o}} \tag{10.78}$$

Since the two roots are coincident at the tangent points,

$$A^2 T_c^2 - 2A x_{1o} \Rightarrow x_{1o} = \tfrac{1}{2} A T_c^2 \tag{10.79}$$

then

$$x_{2t} = AT_c. \tag{10.80}$$

The switching boundary equation may then be written as

$$x_{1e} = \begin{cases} -T_c x_2, & |x_2| \le x_{2t} \\ -\left[x_2^2/(2A) + x_{1o}\right] sgn(x_2), & |x_2| > x_{2t} \end{cases}. \tag{10.81}$$

The near-time-optimal sliding mode control law is then

$$u = u_{max} \, sgn\left[S(x_{1e}, x_2)\right], \tag{10.82}$$

where

$$S(x_{1e}, x_2) = \begin{cases} -x_{1e} - T_c x_2, & |x_2| \le x_{2t} \\ -x_{1e} - \left[x_2^2/(2A) + x_{1o}\right] sgn(x_2), & |x_2| > x_{2t} \end{cases}. \tag{10.83}$$

Returning to the electric drive example, Fig. 10.13a shows the switching boundary together with the closed-loop phase portrait for $\lambda = 0.6$ and Fig. 10.13b shows the variables for a step response.

It should be noted that the boundary layer method (Chap. 9) is employed in which the signum function of (10.83) is replaced by the saturation function.

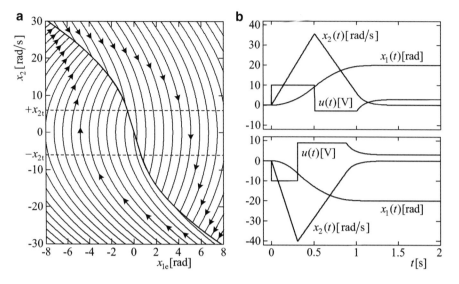

Fig. 10.13 Sub-time-optimal sliding mode position control of DC electric drive. (**a**) Closed-loop phase portrait. (**b**) Positive and negative step responses

The saturation function is sat $(Kx_2, -u_{max}, +u_{max})$, where K is a relatively high gain. $K = 1,000$ in this example. Then $u(t)$, instead of switching rapidly in the sliding mode, closely follows the equivalent control, $x_1(t)$ and $x_2(t)$ being indistinguishable from those that would occur with $u(t)$ switching.

Despite the significant external disturbance $(0.3\, u_{max})$, it is evident that the condition for sliding motion is satisfied over the whole portion of the switching boundary shown in Fig. 10.13a, as intended. The phase portrait certainly resembles that of the double integrator time-optimal control (Chap. 9). The straight line segment of the switching boundary is visible between the lines, $x_2 = \pm x_{2t}$.

When compared with the 20 [rad] step response of the rate-limited SMC in Fig. 10.11b, the step responses of Fig. 10.13b show the advantage of the sub-time-optimal SMC in shortening the settling time.

10.3 Control of SISO Plants of Arbitrary Order

10.3.1 Control of Plants Having Full Relative Degree

The general SISO plant with the state-space model defined by (10.1) and (10.2) is considered in which the state variables are the output derivatives, i.e.

$$\mathbf{x} = \begin{bmatrix} x_1 & x_2 & x_3 & \cdots & x_n \end{bmatrix}^T = \begin{bmatrix} y & \dot{y} & \ddot{y} & \cdots & y^{(n-1)} \end{bmatrix}^T \quad (10.84)$$

10.3 Control of SISO Plants of Arbitrary Order

The control law is then

$$u = u_{\max} \operatorname{sgn}\left[S\left(y - y_r, \dot{y}, \ddot{y}, \ldots, y^{(n-1)}\right)\right]. \tag{10.85}$$

If the switching boundary is

$$S(\mathbf{x}, y_r) = y_r - y - \sum_{i=1}^{n-1} w_i y^{(i)} = y_r - x_1 - \sum_{i=1}^{n-1} w_i x_{i+1} \tag{10.86}$$

then in the sliding mode,

$$S(\mathbf{x}, y_r) = 0 \Rightarrow y_r = y + \sum_{i=1}^{n-1} w_i y^{(i)} \Rightarrow \frac{Y(s)}{Y_r(s)} = \frac{1}{1 + \sum_{i=1}^{n-1} w_i s^i} \tag{10.87}$$

This enables the system to be designed by pole assignment. If the desired closed-loop poles are $s_i = -p_i$, $i = 1, 2, \ldots, n - 1$, then the corresponding coefficients, w_i, $i = 1, 2, \ldots, n - 1$, may be determined by solving the equation

$$1 + \sum_{i=1}^{n-1} w_i s^i = \prod_{i=1}^{n-1} (s + p_i). \tag{10.88}$$

If it is desired to achieve a specified settling time with non-oscillatory dynamics in the sliding mode, then one of the settling time formulae (Chap. 4) may be used, in which case (10.88) becomes

$$1 + \sum_{i=1}^{n-1} w_i s^i = (s + p)^{n-1}, \tag{10.89}$$

where $p = 1.5n/T_s$ (5 % criterion), or $p = 1.6(0.5 + n)/T_s$ (2 % criterion).

First, third-order plants are considered as the motion is in a three-dimensional state space, $(x_1, x_2, x_3) = (y, \dot{y}, \ddot{y})$, which can be visualised. The switching boundary becomes a two-dimensional surface,

$$S(\mathbf{y}, y_r) = S(y - y_r, \dot{y}, \ddot{y}) = 0, \tag{10.90}$$

in the three-dimensional state space, with the switched control law,

$$u = u_{\max} \operatorname{sgn}\left[S(y - y_r, \dot{y}, \ddot{y})\right]. \tag{10.91}$$

The two points, A and B, defining the segment of the switching boundary for a second-order plant along which sliding motion takes place, exemplified in Fig. 10.5, become a one-dimensional boundary defining the sliding region on the two-dimensional switching boundary. As a demonstration, consider the triple integrator plant,

$$\begin{aligned}\dot{x}_1 &= x_2 \\ \dot{x}_2 &= x_3 \quad y = x_1, \\ \dot{x}_3 &= bu\end{aligned} \tag{10.92}$$

under control law (10.91) with the linear switching boundary,

$$S(y - y_r, \dot{y}, \ddot{y}) = y_r - y - w_1\dot{y} - w_2\ddot{y} = y_r - x_1 - w_1x_2 - w_2x_3. \tag{10.93}$$

This yields a linear closed-loop system in the sliding mode with $S(y - y_r, \dot{y}, \ddot{y}) = 0$, the closed-loop transfer function being

$$\frac{Y(s)}{Y_r(s)} = \frac{1}{1 + w_1 s + w_2 s^2}. \tag{10.94}$$

Design for a specified settling time with zero overshoot would require $w_1 = 2T_c$ and $w_2 = T_c^2$, where $T_c = T_s/4.5$ (5 % criterion) or $T_c = T_s/5.6$ (2 % criterion). The switching boundary, whose equation is

$$S(x_{1e}, x_2, x_3) = -(x_{1e} + w_1 x_2 + w_2 x_3) = 0, \tag{10.95}$$

where $x_{1e} = x_1 - y_r$, is an infinite plane passing through the origin with the positive x_{1e}, x_2 and x_3 axes on one side since $w_1 > 0$ and $w_2 > 0$. A nominal settling time of $T_s = 1$ [s] (5 % criterion) is chosen for the simulation results presented in Fig. 10.14. The closed-loop phase portrait is a three-dimensional continuum of state trajectories and is therefore difficult to present, so instead Fig. 10.14 (a) shows a selection of twelve state trajectories on an isometric projection together with the state trajectory of a step response commencing with $(x_{1e}, x_2, x_3) = (0, 0, 0)$. The initial states have been calculated so that the trajectories meet the switching boundary equidistantly spaced along the edges of the sliding region.

The shaded portion, 'pqrs', of the switching boundary is part of the region in which the condition for sliding motion, $S\dot{S} < 0$, is satisfied in a close neighbourhood of the switching boundary, where S is short for $S(x_{1e}, x_2, x_3)$. In terms of (10.95), this is

$$\begin{aligned}&S(-\dot{x}_1 - w_1\dot{x}_2 - w_2\dot{x}_3) < 0 \Rightarrow S(-x_2 - w_1 x_3 - w_2 bu) < 0 \Rightarrow \\ &S[-x_2 - w_1 x_3 - w_2 bu_{\max} \operatorname{sgn}(S)] < 0 \Rightarrow \\ &|S|\operatorname{sgn}(S)[-x_2 - w_1 x_3 - w_2 bu_{\max} \operatorname{sgn}(S)] < 0 \Rightarrow \\ &-(x_2 + w_1 x_3)\operatorname{sgn}(S) - w_2 bu_{\max} < 0.\end{aligned} \tag{10.96}$$

10.3 Control of SISO Plants of Arbitrary Order

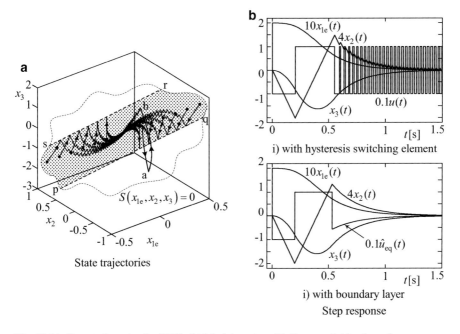

Fig. 10.14 State trajectories for SMC of triple integrator with linear switching boundary

Let ε be a positive infinitesimal. Then if $S = \varepsilon$, meaning all points in a close neighbourhood of the switching boundary on one side of it, condition (10.96) becomes

$$-x_2 - w_1 x_3 - w_2 b u_{\max} < 0 \qquad (10.97)$$

Similarly, if $x_{1e} + w_1 x_2 + w_2 x_3 = -\varepsilon$, meaning all points in a close neighbourhood of the switching boundary on the opposite side to that above, (10.96) becomes

$$x_2 + w_1 x_3 - w_2 b u_{\max} < 0. \qquad (10.98)$$

There are therefore two bounds of the sliding region of the switching boundary derived from (10.97) and (10.98). From (10.97),

$$-x_{2n} - w_1 x_{3n} - w_2 b u_{\max} = 0 \Rightarrow x_{2n} = -w_1 x_{3n} - w_2 b u_{\max} \qquad (10.99)$$

and from (10.98),

$$x_{2p} + w_1 x_{3p} - w_2 b u_{\max} = 0 \Rightarrow x_{2p} = -w_1 x_{3p} + w_2 b u_{\max}. \qquad (10.100)$$

Equations (10.99) and (10.100) are, respectively, the equations of the two parallel straight lines, 'rs' and 'qp', in Fig. 10.14 (a), bounding the sliding region, which is an infinite strip. To generate the twelve trajectories with starting points indicated by '•', the first twelve points on the boundaries of (10.99) and (10.100) are determined by choosing $x_{3p} = x_{3n} = \pm 0.5D, \pm 1.5D$ and $\pm 2.5D$, with $D = 0.3$. Then x_{2p} and x_{2n} are given by (10.99) and (10.100). The switching boundary is then used to compute

$$x_{1ep} = -w_1 x_{2p} - w_2 x_{3p} \tag{10.101}$$

$$x_{1en} = -w_1 x_{2n} - w_2 x_{3n} \tag{10.102}$$

This yields six equally spaced points on each boundary of the sliding region. The starting points are then found by back tracing (Chap. 9) by a fixed time, T, with the sign of $u_{\max}$ that brings the state point towards the boundary in forward time. Thus,

$$\begin{bmatrix} x_{1ep}(0) \\ x_{2p}(0) \\ x_{3p}(0) \end{bmatrix} = \begin{bmatrix} 1 & T & \frac{1}{2}T^2 \\ 0 & 1 & T \\ 0 & 0 & 1 \end{bmatrix} \begin{bmatrix} x_{1ep} \\ x_{2p} \\ x_{3p} \end{bmatrix} + \begin{bmatrix} \frac{1}{6}T^3 \\ \frac{1}{2}T^2 \\ T \end{bmatrix} bu_{\max} \tag{10.103}$$

and

$$\begin{bmatrix} x_{1en}(0) \\ x_{2n}(0) \\ x_{3n}(0) \end{bmatrix} = \begin{bmatrix} 1 & T & \frac{1}{2}T^2 \\ 0 & 1 & T \\ 0 & 0 & 1 \end{bmatrix} \begin{bmatrix} x_{1en} \\ x_{2n} \\ x_{3n} \end{bmatrix} - \begin{bmatrix} \frac{1}{6}T^3 \\ \frac{1}{2}T^2 \\ T \end{bmatrix} bu_{\max} \tag{10.104}$$

where $T < 0$. In this demonstration, $T = -0.1$ [s]. The result is the twelve trajectories of Fig. 10.14a that meet the switching boundary on the edges of the sliding regions, which are the two straight lines defined by (10.99), (10.100), (10.101), and (10.102), after which sliding motion occurs indicated by the zigzag motion, made visible by setting the hysteresis level (Chap. 9) of the switching element to $\delta S = 0.004$.

The trajectory of the step response in Fig. 10.14a commences with the control at $u = -u_{\max}$ until it crosses the switching boundary at point 'a' outside the sliding region, after which the control switches to $u = -u_{\max}$, remaining at this value until the trajectory crosses the switching boundary again, but this time within the sliding region at point 'b', after which sliding motion occurs and the system dynamics is governed by (10.94). Figure 10.14b shows the step response variables as functions of time, the periods of control saturation and sliding motion being clearly visible. Figure 10.14b shows the step response again but with the hysteresis switching element replaced by a high-gain (10^5) with saturation at $\pm u_{\max}$ to give a very narrow boundary layer (Chap. 9). In this case, the continuous control, $\hat{u}_{eq}(t)$, is a very close approximation to the equivalent control, $u_{eq}(t)$.

Returning to the general third-order system, if sliding motion can be guaranteed, then the control system design is very straightforward via (10.94). The theoretical difficulty in the control system design, however, is ensuring that the sliding region will be reached from any initial state within the operational envelope of a particular application. If some initial states are found for which control law (10.91) with the linear switching boundary (10.95) cannot bring the state to the sliding region, then it will be impossible to achieve the closed-loop dynamics characterised by (10.94) and the system is usually unstable. This is, for example, the case for the triple integrator plant with step reference inputs exceeding a certain threshold. While being difficult to deal with theoretically due to the nonlinear operation of the system, this problem can, however, be quickly investigated by simulation. If the operational envelope is found to include states leading to instability with control law (10.91) with the linear switching boundary (10.95), then it is possible to modify the control law to achieve acceptable performance. For example, a larger value of T_s could be used to calculate w_1 and w_2, which would tend to increase the regions of the state space from which the sliding region of the switching boundary can be reached.

Example 10.2 Sliding mode control of a vertical electromagnetic bearing axis

The controlled plant consists of an electromagnet exerting a controllable vertical force acting against the force of gravity as shown in Fig. 10.15.

The inductance of the electromagnet is

$$L(y) = \frac{\mu_0 A N^2}{2(G - y + y_0)}, \tag{10.105}$$

where μ_0 is the permeability of free space, A is the cross-sectional area of the pole piece, N is the number of coil turns, G is the nominal air gap, y is the upward vertical bearing displacement with respect to the nominal position and y_0 is a constant giving the correct inductance for zero gap. The displacement, y, is also the measured displacement to be controlled, scaled in the control computer to be numerically equal to the displacement. The energy stored in the magnetic field is then

$$W = \tfrac{1}{2} L(y) i^2 \tag{10.106}$$

where i is the coil current. The attractive force of the electromagnet is then

$$f = \frac{dW}{dy} = \frac{d}{dy}\left[\frac{\mu_0 A N^2 i^2}{4(G - y + y_0)}\right] = \frac{\mu_0 A N^2 i^2}{4(G - y + y_0)^2}. \tag{10.107}$$

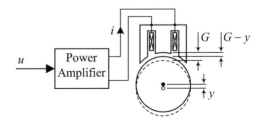

Fig. 10.15 Controlled electromagnet for vertical axis of electromagnetic bearing

The equation of motion of the bearing vertical axis is therefore

$$\ddot{y} = \frac{f}{M} = \frac{C}{M} \cdot \frac{i^2}{(G - y + y_0)^2} - g \quad (10.108)$$

where $C = \mu_0 A N^2 / 2$ and $g = 9.81$ [m/s/s] is the acceleration due to gravity. The plant model is then completed with the differential equation modelling the electric circuit of the electromagnet. If K_a is the voltage gain of the power amplifier, this is $\frac{d}{dt}[L(y)i] + Ri = K_a u$ in which $L(y) = \frac{C}{G-y+y_0} i$. This yields

$$\frac{di}{dt} = \frac{G - y + y_0}{C}(K_a u - Ri) - \frac{1}{G - y + y_0} i \dot{y}, \quad (10.109)$$

where R is the coil resistance. The plant model is therefore of third order as it is the sum of the orders of the two constituent Eqs. (10.108) and (10.109). Despite the plant being highly nonlinear, it will be demonstrated that a simple sliding mode control law with the linear switching boundary (10.95) will produce effective control without knowledge of the plant parameters. The relative degree is determined by differentiating (10.108) and substituting for di/dt on the RHS using (10.109), whereupon it is evident that $\dddot{y}$ depends directly on u. Thus, the relative degree is three, which is equal to the plant order, confirming that the plant is of full relative degree.

For the simulation, the plant parameters are taken as $A = 0.01$ [m²], $N = 1000$, $G = 0.002$ [m], $\mu_0 = 4\pi \times 10^{-7}$ [A²/m], $R = 1$ [Ω] and $K_a = 10$. The reference input is $y_r = 0$ as this commands an air gap of G. Rather than attempt to determine the boundary of the sliding region analytically, the simulation will be run with a range of initial conditions that are possible prior to the initial loop closure in the real application. This is as follows. The initial coil current, $i(0)$, and the initial mechanical load velocity, $\dot{y}(0)$, are zero. The settling time in the sliding mode is set to $T_s = 0.2$ [s] to give a non-overshooting step response with approximately this settling time. The control saturation level is $u_{max} = 10$ [V] and the hysteresis levels of the switching element are set to $\delta S = \pm 0.0001$ [m]. Five responses are simulated with $y(0)$ set to $-G, -G/2, 0, +G/2$ and $+G$, giving an initial gap ranging between $2G$ and zero, with $y_r = 0$. Then to test the step response, $y_r(t)$ is stepped to $-G$ at $t = 2T_s$, then back to 0 at $t = 4T_s$. Figure 10.16 shows the results.

In Fig. 10.16a, the output derivatives, $\dot{y}$ and $\ddot{y}$, are used directly in the control law, as if they were measured, to assess the operation of the ideal control system. For all five initial positions, it is evident that the demanded position is reached approximately in the specified settling time of $T_s = 0.2$ [s]. This occurs since the switching functions, $S_i(t)$, $i = 1, 2, 3, 4, 5$, all reach zero and stay approximately zero in about 2.5 [ms], which is the initial period of control saturation, prior to sliding motion, which is only a small proportion of T_s. The step responses are as specified. The last two graphs of $S_i(t)$, $i = 1, 2, 3, 4, 5$, in Fig. 10.16a clearly show the expected jumps to non-zero values at $t = 0.4$ [s] and $t = 0.8$ [s]

10.3 Control of SISO Plants of Arbitrary Order

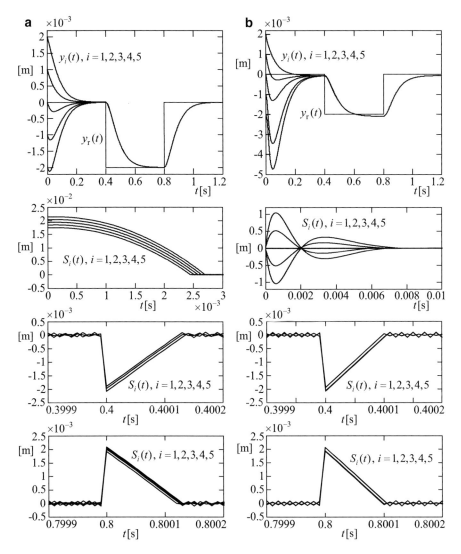

Fig. 10.16 Position error transients for sliding mode control of an electromagnetic bearing. (**a**) With exact output derivatives (**b**) With estimated output derivatives

as the steps of $y_r(t)$ are applied, followed by return to approximately zero and then oscillations about zero indicating reacquisition of the sliding mode. Since these reacquisition times are very small compared with T_s, the step responses are indistinguishable from those of a second-order critically damped system with a settling time of T_s.

In this application, however, it is not practicable to measure $\dot{y}$ and $\ddot{y}$, so the multiple integrator observer of Sect. 10.1.3.2 is applied to provide estimates, $\widehat{\dot{y}}$ and

$\widehat{y}$, using three integrators. The dynamic lag in $\widehat{\dot{y}}$ and $\widehat{\ddot{y}}$ is unavoidable but as the results of Fig. 10.16b show, with $T_f = 0.001$ [s], the performance of the sliding mode controller is not really compromised. The larger initial transients of $y_i(t)$, $i = 1, 2, 3, 4, 5$, are due to the settling transient of the observer but are still acceptable. The initial transients of $S_i(t)$, $i = 1, 2, 3, 4, 5$, are very much larger due to the observer settling transient. The peaks of just over 1 [m], however, are not physical displacements, being only in the controller software and due to $\widehat{\dot{y}}(t)$ and $\widehat{\ddot{y}}(t)$ during the initial observer transient. Once the observer has settled, the recovery of the sliding mode following application of the two steps in $y_r(t)$, indicated by the last two graphs of $S_i(t)$, $i = 1, 2, 3, 4, 5$, is similar to that of the ideal system shown in Fig. 10.16a, indicating that the introduction of the observer will not significantly affect the normal operation of the system.

10.3.2 Control of Plants Less Than Full Relative Degree

For a plant modelled by (10.1) and (10.2), it would be possible to determine a switching function, $S(\mathbf{x}, y_r)$, such that the switched control law, $u = u_{\max} \operatorname{sgn}[S(\mathbf{x}, y_r)]$, operates in a sliding mode but, as has been emphasised in Sect. 10.1.3, the robustness of sliding mode control can only be fully exploited if the output derivative state representation is used, meaning

$$\mathbf{x} = \left[y, \dot{y}, \ldots, y^{(n)}\right] \quad (10.110)$$

For a plant of relative degree, r, however, the complete set of output derivative state variables is limited to r in number and may be represented by the substate vector

$$\mathbf{y} = \left[y, \dot{y}, \ldots, y^{(r-1)}\right] \quad (10.111)$$

The derivatives, $y^{(i)}$, $i = r, r+1, \ldots, n-1$, which would increase the dimension of $\mathbf{y}$ from r to n, are not state variables due to their dependences on u and its derivatives. It would be possible, however, to obtain a sliding mode with the control law,

$$u = u_{\max} \operatorname{sgn}[S(\mathbf{y}, y_r)], \quad (10.112)$$

because the rate of change of the highest derivative, $y^{(r-1)}$, i.e. $y^{(r)}$, depends directly on u, allowing u to drive the substate trajectory towards the switching boundary from both sides. So this means that the system can control the substate, $\mathbf{y}$, of the plant but another substate, $\mathbf{z}$, of the plant is not controlled. This is precisely the same situation as described in Chap. 7, Sect. 7.3.3, in which there exists a sub-plant with state, $\mathbf{z}$, that is uncontrolled using feedback linearising or forced dynamic control, referred to as the *zero dynamics*. One of the remaining challenges in sliding mode control is the control of plants with unstable zero dynamics. A solution to a similar problem is

10.3 Control of SISO Plants of Arbitrary Order

presented in Chap. 7 that entails the creation of an artificial auxiliary output, z, with respect to which the plant has full relative degree. Implementation of this to generate z within the control computer, however, would require an observer with an accurate plant model, thereby defeating the object of sliding mode control of achieving robustness. It is now clear, however, that sliding mode control can be applied, with robustness, using the substate vector, $\mathbf{y}$, to plants not of full relative degree provided the zero dynamics is stable, as demonstrated by the following example. The robust pole placement of Sect. 10.5 will also be considered as a possible solution for robust control of plants with zero dynamics that is not stable.

Example 10.3 Sliding Mode Attitude Control of a Flexible Spacecraft

A spacecraft with a large solar panel is considered that has two significant vibration modes. The damping ratios of these modes would usually be negligible if the solar panel mounting were to be rigidly joined to the centre body. This would require a sophisticated control law with active model damping as well as control of the rigid-body mode. To enable a very simple sliding mode control law to be applied, aimed at accurate and robust attitude control of the centre body, passive damping is added to prevent the modal vibrations, which will not be actively controlled, to acceptable amplitudes. One attitude control axis is considered with the assumption that the inter-axis coupling is negligible. Figure 10.17 shows the vibration mode shapes and a model. In the model, J is the centre-body moment of inertia about the rotation axis. Each of the two vibration modes is represented by a rotational mass–spring–damper systems with moments of inertia, J_1 and J_2; spring constants, K_{s1} and K_{s2}; and damping coefficients, D_1 and D_2. The variables are the centre-body attitude angle, α, (which is to be controlled), the mode 1 displacement angle, β_1, the mode 2 displacement angle, β_2, the control torque from the reaction wheel, γ_c, and the external disturbance torque, γ_d. The measurement from the attitude sensor is $y = \alpha$.

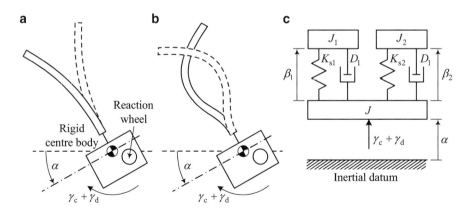

Fig. 10.17 Flexible spacecraft requiring centre-body attitude control. (**a**) First vibration mode (**b**) Second vibration mode (**c**) Lumped parameter model

The mode shapes of Fig. 10.17a, b are highly exaggerated for clarity of presentation. Also these figures are drawn with, as is intended for a constant attitude reference input, the centre-body stationary. Without the attitude control, the centre body would be 'rocking' in anti-phase with the modal vibrations of the solar panel. With reference to Fig. 10.17c, the differential equations of the model are as follows.

$$J\ddot{\alpha} = K_{s1}\beta_1 + D_1\dot{\beta}_1 + K_{s2}\beta_2 + D_2\dot{\beta}_2 + \gamma_c + \gamma_d, \quad (10.113)$$

$$J_1\left(\ddot{\beta}_1 + \ddot{\alpha}\right) = -K_{s1}\beta_1 - D_1\dot{\beta}_1, \quad (10.114)$$

and

$$J_2\left(\ddot{\beta}_2 + \ddot{\alpha}\right) = -K_{s2}\beta_2 - D_2\dot{\beta}_2. \quad (10.115)$$

In preparation for the simulation, these may be expressed in the standard form:

$$\ddot{\alpha} = v_1^2 r_1 \beta_1 + 2\eta_1 v_1 r_1 \dot{\beta}_1 + v_2^2 r_2 \beta_2 + 2\eta_2 v_2 r_2 \dot{\beta}_2 + b(u+d), \quad (10.116)$$

$$\ddot{\beta}_1 = -\gamma_1^2 \beta_1 - 2\eta_1 v_1 \gamma_1 - \ddot{\alpha} \quad (10.117)$$

and

$$\ddot{\beta}_2 = -\gamma_2^2 \beta_2 - 2\eta_2 v_2 \gamma_2 - \ddot{\alpha}, \quad (10.118)$$

where γ_1 and γ_2 are the undamped modal encastre frequencies, i.e. the modal vibration frequencies that would occur with the centre body held fixed with respect to inertia space with no damping, and η_1 and η_2 are the encastre modal damping ratios. Regarding the choice of symbols, note that ω_{n1}, ω_{n2}, ζ_1 and ζ_2 are the undamped natural frequencies and damping ratios that would occur with the centre body left free and no applied control. These are not relevant to this example. The parameters, $r_1 = J_1/J$ and $r_2 = J_2/J$, are the modal moment of inertia ratios. Finally, $\gamma_c = K_w u$ where K_w is the reaction wheel torque constant, u is the control variable and $b = K_w/J$. The external disturbance referred to u is $d = (K_w/J)\gamma_d$.

It is evident by inspection of (10.116) that the plant has relative degree, $r = 2$, and since the differential equations, (10.116), (10.117) and (10.118) are each of second order, the plant order is $n = 6$. In the sliding mode, the system will therefore have uncontrolled zero dynamics of order, $n - r = 4$. In this example, the uncontrolled zero dynamic subsystem comprises (10.117) and (10.118) with $\ddot{\alpha} = 0$ since the sliding mode control system will control only the state variables, α, and $\dot{\alpha}$, and if $y_r = \alpha_r = $ const., in the steady state, $\alpha = \alpha_r = $ const., $\Rightarrow \dot{\alpha} = 0$ and $\ddot{\alpha} = 0$.

Figure 10.18 shows a simulation of the control system with the control law

$$u = u_{\max} \text{sgn}\left[S(y, \dot{y}, y_r)\right], \quad S(y, \dot{y}, y_r) = y_r - y - w_1 \dot{y} \quad (10.119)$$

10.3 Control of SISO Plants of Arbitrary Order

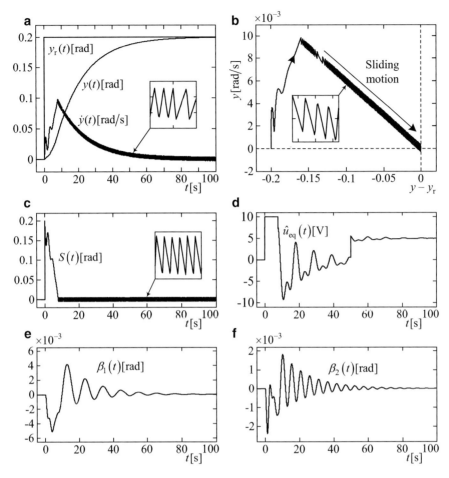

Fig. 10.18 Sliding mode attitude control of flexible spacecraft. (**a**) Centre body variables (**b**) centre body substate trajectory (**c**) switching function (**d**) control variable (**e**) vibration mode 1 displacement (**f**) vibration mode 2 displacement

where $w_1 = T_s/3$, $T_s = 50$ [s] (5 % crit.). The plant parameters are $J = 200$ $[Kg\, m^2]$, $v_1 = 0.6$ $[rad/s]$, $v_2 = 1.2$ $[rad/s]$, $\eta_1 = 0.1$, $\eta_1 = 0.05$, $K_w = 0.1$ [Nm/V], $r_1 = 10$ and $r_2 = 5$.

The hysteresis levels of the switching function are set to $\delta S = \pm 0.005$ [rad]. Also, a step disturbance torque equivalent to a step disturbance of $u_{\max}/2$ is applied at $t = T_s$, typical of that occurring due to the operation of orbit and change thruster whose force vector is not perfectly directed through the spacecraft centre of mass.

The oscillations in $\dot{y}(t)$ visible in Fig. 10.18a, are due to the oscillatory modes and occur during control saturation that can be seen in Fig. 10.18d. These oscillations may also be seen in the substate trajectory of Fig. 10.18b before it reaches the switching boundary, which is clearly indicated by the linear sliding part of the trajectory. The first-order dynamics in the sliding mode is evident due to the exponential behaviour of $y(t)$ and $\dot{y}(t)$ in Fig. 10.18a following the peak in $\dot{y}(t)$. The superimposed high-frequency oscillations in $\dot{y}(t)$ due to the non-zero hysteresis levels are shown in the inset magnified graphs of Fig. 10.18a, b. As expected, these oscillations also occur in $S(t)$ as shown in Fig. 10.18c. Importantly, the vibration mode displacements shown in Fig. 10.18e, f continue to oscillate during the sliding motion since these modes constitute the uncontrolled zero dynamic subsystem. The physical situation is that the vibrations of the solar panel exert torques on the centre body, which are treated as disturbance torques and counteracted just as effectively as the step disturbance torque referred to the control input. They are therefore left to oscillate uncontrolled by the sliding mode control system but, as evident in Fig. 10.18e, f, they decay due to the inclusion of active damping in the solar panel mount. If the plant transfer function, $\alpha(s)/U(s)$, were to be derived, then it would show two pairs of complex conjugate zeros in the left half of the s-plane due to the damping, which become the *poles* of the zero dynamic subsystem.

10.4 Methods for Elimination of Control Chatter

10.4.1 The Boundary Layer Method

A common method for eliminating the control chatter exhibited by a basic sliding mode control system is to replace the signum function of the control law,

$$u = u_{max} \, \text{sgn} \, [S(\mathbf{y}, y_r)], \tag{10.120}$$

that has an infinite slope at the origin, with a high-gain saturation function to yield

$$u = \text{sat}(KS, -u_{max}, u_{max}) \triangleq \begin{cases} u_{max} & \text{for } KS > u_{max} \\ KS & \text{for } -u_{max} \leq KS \leq u_{max} \\ -u_{max} & \text{for } KS < -u_{max} \end{cases} \tag{10.121}$$

as shown in Fig. 10.19.

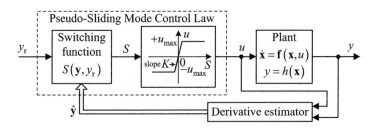

Fig. 10.19 SISO pseudo sliding mode control system using the boundary layer method

10.4 Methods for Elimination of Control Chatter

The saturation function can be regarded as an approximation to the signum function if its slope, K, at the origin is made sufficiently high. This method has already been introduced for switched control systems in Chap. 9, Sect. 9.5, and presented in detail in Sect. 9.8.5.3 for second-order plants. It follows from this that the switching boundary in the state space is replaced by a boundary layer where the control is unsaturated, sandwiched between two saturation boundaries defined by

$$S(\mathbf{y}, y_r) + u_{max}/K = 0 \quad \text{and} \quad S(\mathbf{y}, y_r) - u_{max}/K = 0 \quad (10.122)$$

If the infinite continuum of state trajectories of the closed-loop state portrait is directed towards the boundary layer from both sides, then the state trajectory will be 'trapped' by the boundary layer. It is clear that as $K \to 0$, the two saturation boundaries approach the original switching boundary from both sides and the state trajectory within the boundary layer approaches that of the basic sliding mode control system.

One of the properties of a control system operating in an ideal sliding mode [1], i.e. the one in which the control switches at an infinite frequency with continuously varying mark–space ratio to hold the state trajectory precisely on the switching boundary, is that it is impossible to back trace the state trajectory using the state transition method to determine the initial state. This is not the case once the boundary layer has been introduced since any set of state trajectories starting from arbitrary points outside the boundary layer are unique and never cross one another. Since the inability to determine the initial state once in an ideal sliding mode has come to be accepted as a mandatory property of a sliding mode control system, this term cannot be used *strictly* to describe the control system once the boundary layer has been introduced. For this reason the resulting system will be referred to as a *pseudo sliding mode control system* as its state trajectory is similar to that of the ideal sliding mode control system. The condition for pseudo sliding motion is as follows.

> For pseudo sliding motion to exist in the boundary layer, the infinite continuum of trajectories of the closed-loop phase portrait has to be directed towards both of the neighbouring saturation boundaries, $S(\mathbf{y}, y_r) \pm u_{max}/K = 0$.

This guarantees that the state trajectory will be trapped within the boundary layer after entering it.

Regarding robustness, even if the condition for pseudo sliding motion is satisfied over the ranges of uncertain plant parameters and external disturbances to be encountered in a particular application, the trajectories within the boundary layer will depend upon these plant parameters and disturbances and therefore the robustness is nonideal. Increasing K to narrow the boundary layer will improve the robustness, but, in practice, caution should be exercised to ensure that this gain is not made so large that the system breaks into oscillation, defeating the object.

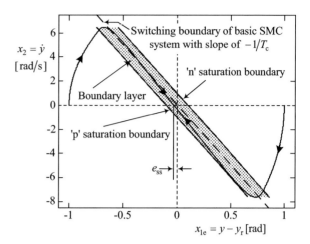

Fig. 10.20 State trajectory and time response for pseudo sliding mode position control of DC drive with an exaggerated boundary layer

This control chatter is due to the inevitable un-modelled small dynamic lags of sensors and actuators and/or the sampling process of the digital implementation allowing the state trajectory to move away from the boundary layer.

In theory, as $K \to \infty$, the state trajectory is trapped within the infinitesimal boundary layer and $u(t)$ remains continuous. This must be the equivalent control, $u_{eq}(t)$. The boundary layer method has already been used to compute close approximations to u_{eq}, denoted $\hat{u}_{eq}$, first for the DC electric drive example in Figs. 10.6 and 10.13, then for a third-order system in Fig. 10.14 and finally in Example 10.3 (Fig. 10.18). In these cases, the time responses and state trajectories with the boundary layers are indistinguishable from those of the basic sliding mode control system except for the oscillations of $S(t)$ and the highest derivative of $y(t)$ in the basic system due to the control chatter. To further illustrate the operation of the boundary layer method, however, the DC electric drive sliding mode control system considered in Sect. 10.2.3 is revisited with the same parameters used to produce the results of Fig. 10.6. A relatively large gain of $K = 10,000$ is set for the estimation of u_{eq} in Fig. 10.6b. To show the state trajectory behaviour clearly within the boundary layer, K is reduced to 100, producing the state trajectories of Fig. 10.20.

The two state trajectories, one for $y_r = 1$ [rad] and the other for $y_r = -1$ [rad], depart, as expected, from the switching boundary of the basic sliding mode control system but remain within the boundary layer. Due to the finite gain, K, a steady-state error, e_{ss}, is caused by the constant external disturbance, but the control smoothing method of the following section includes an integral term that eliminates this.

The general pseudo sliding mode control law with control saturation limits, $\pm u_{max}$, and a boundary layer based on a linear switching function is defined by the following equations when applied to an SISO plant of relative degree, r.

$$u = \text{sat}\left[KS\left(\mathbf{y}, \mathbf{y}_r\right), -u_{max}, u_{max}\right] = \begin{cases} KS\left(\mathbf{y}, \mathbf{y}_r\right), & |KS\left(\mathbf{y}, \mathbf{y}_r\right)| < u_{max} \\ u_{max} \, \text{sgn}\left[KS\left(\mathbf{y}, \mathbf{y}_r\right)\right], & |KS\left(\mathbf{y}, \mathbf{y}_r\right)| \geq u_{max} \end{cases}$$
(10.123)

10.4 Methods for Elimination of Control Chatter

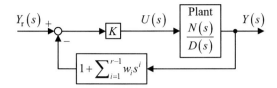

Fig. 10.21 Equivalent linear system within the boundary layer for an LTI plant

where

$$S(\mathbf{y}, y_r) = y_r - y - \sum_{i=1}^{r-1} w_i y^{(i)}. \tag{10.124}$$

When $S(\mathbf{y}, y_r) \cong 0$, the dynamics of the system closely approximates that of a linear system with closed-loop transfer function:

$$\frac{Y(s)}{Y_r(s)} = \frac{1}{1 + \sum_{i=1}^{r-1} w_i s^i}. \tag{10.125}$$

This may be designed by pole assignment to yield values of the derivative weighting coefficients, w_i, $i = 1, 2, \ldots, r-1$, that produce a specified closed-loop dynamics.

For an LTI plant with transfer function, $N(s)/D(s)$, within the boundary layer, the control system becomes equivalent to the linear one shown in Fig. 10.21.

Then the closed-loop transfer function may be written as

$$\frac{Y(s)}{Y_r(s)} = \frac{KN(s)/D(s)}{1 + K\left(1 + \sum_{i=1}^{r-1} w_i s^i\right) N(s)/D(s)} = \frac{1}{1 + \sum_{i=1}^{r-1} w_i s^i + \frac{D(s)}{KN(s)}}. \tag{10.126}$$

Then as $K \to \infty$, equivalent to diminishing the boundary layer to yield, in theory, the ideal sliding mode controller, (10.126) becomes the same as (10.125). This result may be substantiated by considering the root locus w.r.t. K. As $K \to \infty$, a subset of the loci terminate on the zeros of the open-loop transfer function,

$$G_{ol}(s) = K \frac{N(s)}{D(s)} \left(1 + \sum_{i=1}^{r-1} w_i s^i\right), \tag{10.127}$$

which are the roots of the polynomial equation

$$N(s) \left(1 + \sum_{i=1}^{r-1} w_i s^i\right) = 0. \tag{10.128}$$

Since $\deg[N(s)] = n - r$, then $\deg\left[N(s)\left(1 + \sum_{i=1}^{r-1} w_i s^i\right)\right] = n - r + r - 1 = n - 1$. There are therefore $n - 1$ branches of the root locus that terminate on the zeros that are the roots of (10.128) and a single branch following an asymptote that is the negative real axis of the s-plane. As K increases, the closed-loop pole following this asymptote becomes large and negative and therefore dominated by the $n-1$ finite closed-loop poles that are the roots of (10.128). Since $N(s)$ is also a factor of the closed-loop transfer function (10.126), evident when it is expressed in the standard polynomial form,

$$\frac{Y(s)}{Y_r(s)} = \frac{KN(s)}{D(s) + KN(s)\left(1 + \sum_{i=1}^{r-1} w_i s^i\right)} \tag{10.129}$$

Then $n-r$ of the closed-loop poles are cancelled by $N(s)$, leaving $n-1-(n-r) = r-1$ poles, i.e. the roots of $\left(1 + \sum_{i=1}^{r-1} w_i s^i\right) = 0$, to characterise the closed-loop system, agreeing with transfer function (10.125). So classical control theory provides an alternative explanation of the order reduction associated with sliding mode control.

10.4.2 The Control Smoothing Integrator Method

10.4.2.1 Basic Single Smoothing Integrator Version

A straightforward method is presented here that retains the robustness of the basic sliding mode control system while alleviating the effects of the control chatter for plants with continuous actuators. An integrator is inserted at the control input of the plant. The integrator and the plant are then regarded as an augmented plant, for which a basic sliding mode control law is designed. Then the input of the integrator becomes the *primary control variable*, u', that will be switched between two finite limits, $\pm u'_{max}$. Then the physical control variable, $u(t)$, is continuous. A block diagram of the system is shown in Fig. 10.22.

In contrast with the basic sliding mode control system, the control saturation limits, $\pm u'_{max}$, are not dependent upon the plant hardware, but the physical control limits, $\pm u_{max}$, that depend on the plant hardware must still be taken into account.

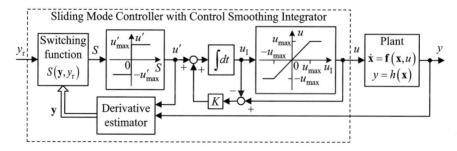

Fig. 10.22 Sliding mode control of general SISO plant with control smoothing integrator

This is done by including a saturation element in the controller that keeps the computed $u(t)$ within these limits together with an anti-windup loop with gain, K, (Chap. 1, Sect. 1.3.5). Since the control smoothing integrator is regarded as part of the plant for the controller design, this will raise the relative degree and order by one unit in normal operation for which $|u| < u_{\max}$. Hence, one more output derivative is needed than for the basic sliding mode control system, the maximum order being the relative degree, r. The determination of the switching boundary parameters may then be carried out as previously. To maintain the required robustness, it must be ensured that $|u(t)| < u_{\max}$ in normal operation, the anti-windup loop being included only as a 'safety net'. To design the system to avoid physical control saturation, it is recommended to carry out simulations of the ideal closed-loop system with the desired closed-loop dynamics and use the variables of these simulations to compute the continuous, $u(t)$ needed. To achieve this, equations for $u(t)$ in terms of the ideal closed loop system variables, the external disturbances and the plant parameters are required. Several computations should be carried out, each for the extreme values of the expected range for each plant parameter and for the largest expected external disturbances, covering all the combinations. If $|u(t)| < u_{\max}$ for every case, then the planned closed-loop dynamics is feasible. If not then the closed-loop dynamics must be altered by, for example, increasing the settling time, which will require smaller excursions of $u(t)$.

10.4.2.2 Determination of the Primary Control Saturation Limit

Since the control smoothing integrator is implemented in the software, there is no upper limit for $u'_{\max}$ but a lower limit is set by the maximum magnitude of the equivalent primary control, u'_{eq}. This could be determined by implementing one of the equivalent control estimation methods to be presented in Sects. 10.4.2.4 and 10.4.2.5 and then recording $\max |u'_{eq}(t)|$ during the simulations referred to in the previous section. Then $u'_{\max} = \lambda \max |u'_{eq}(t)|$, where $\lambda > 1$.

10.4.2.3 Adaptive Primary Saturation Limit

In view of the non-zero iteration period of the control algorithm, $u(t)$, though continuous, will contain an oscillatory component due to the switching of $u'(t)$ between $-u'_{\max}$ and $+u'_{\max}$ in the sliding mode, similar to the zigzag motion of the highest output derivative in the basic sliding mode control system. The amplitude of this oscillation will increase as $u'_{\max}$ is increased, and therefore, a useful refinement of the control system would be to employ an algorithm to calculate an on-line estimate, $\widehat{u}'_{eq}(t)$, of $u'_{eq}(t)$ and then continuously update the primary saturation limit using

$$u'_{\max} = \mathrm{sat}\left(\lambda \left|u'_{eq}(t)\right|, u'_{\max 1}, u'_{\max 2}\right), \qquad (10.130)$$

where $u'_{\max 1}$ and $u'_{\max 2}$ are, respectively, upper and lower values of $u'_{\max}$, and $\lambda > 1$ is a contingency factor, noting that $u'_{eq}(t) \to 0$ as the control system approaches steady state with a constant reference input and the minimum limit, and $u'_{\min}$ is needed to maintain loop closure. This holds with a constant external disturbance referred to the control input, because the control smoothing integrator would build up a constant component of u counteracting the disturbance, which would operate with $u'_{eq} = 0$ in the steady state. This action of the control smoothing integrator is similar to the action of the integral term in any of the continuous controllers presented in the previous chapters. The basic sliding mode controller needs a constant component of u_{eq} to counteract a constant disturbance, entailing considerable control chatter with its attendant disadvantages. In contrast, the sliding mode control system with the control smoothing integrator and the adaptive saturation limiting of (10.130) would minimise the oscillations in $u(t)$ for a given control algorithm iteration period, thereby maximising the control accuracy.

10.4.2.4 Model-Based Equivalent Control Estimation

In the following section, a method of estimating the equivalent control is presented that is practicable for real-time implementation as it does not require a plant model, bearing in mind that the purpose of a sliding mode control system is to obtain a prescribed closed-loop dynamics having extreme robustness with respect to plant model uncertainties as well as unknown external disturbances. In the development of a control system, however, it is advisable to compute the equivalent control precisely using a plant model so that the precision of the non-model-based estimation may be assessed comparatively by simulation. An algorithm for achieving this is derived here.

The augmented SISO plant model to be considered is that of (10.1) and (10.2) with the control smoothing integrator and the external disturbance referred to the primary control input, the 'physical' control input, u, becoming a state variable. Thus,

$$\dot{\mathbf{x}} = \mathbf{f}(\mathbf{x}, x_{n+1}) \qquad (10.131)$$

$$\dot{x}_{n+1} = u' - d' \qquad (10.132)$$

$$y = h(\mathbf{x}) \qquad (10.133)$$

The output derivatives used by the sliding mode control law are then

$$y = h_0(\mathbf{x}), \dot{y} = h_1(\mathbf{x}), \ldots, y^{r-1} = h_{r-1}(\mathbf{x}), y^r = h_r(\mathbf{x}, x_{n+1}), \qquad (10.134)$$

noting that while y^r is not a state variable of the physical plant, it is a state variable of the augmented plant. One further derivative then yields dependence on u'. Thus,

10.4 Methods for Elimination of Control Chatter

$$y^{r+1} = h_{r+1}\left(\mathbf{x},\ x_{n+1},\ u'\right). \tag{10.135}$$

This enables an equation for u' to be formed in terms of $\mathbf{x}$, x_{n+1} and y^{r+1}. Thus,

$$u' = g\left(\mathbf{x},\ x_{n+1},\ y^{r+1}\right) \tag{10.136}$$

If all the variables on the RHS are known from a simulation, then when the system is in the sliding mode, using (10.136) to calculate u' would yield the rapidly switching control, not the continuous equivalent control as required, the oscillations originating from the highest output derivative, y^{r+1}. This problem may be solved, however, by using the switching boundary equation,

$$S\left(\mathbf{y},\ \mathbf{y}_\mathrm{r}\right) = S\left(y_\mathrm{r},\ y,\ \dot{y},\ldots y^{(r)}\right) = 0 \tag{10.137}$$

as this is obeyed in the sliding mode. Making $y^{(r)}$ the subject of (10.137) gives

$$y^{(r)} = P\left(y_\mathrm{r},\ y,\ \dot{y},\ldots y^{(r-1)}\right). \tag{10.138}$$

Differentiating (10.138) then yields

$$y^{(r+1)} = Q\left(y_\mathrm{r},\ \dot{y}_\mathrm{r},\ y,\ \dot{y},\ldots y^{(r)}\right) \tag{10.139}$$

The highest derivative, $y^{(r)}$, on the RHS of (10.139) is continuous, exhibiting only the zigzag form of oscillation. Then substituting for $y^{(r+1)}$ on the RHS of (10.136) using (10.139) yields an expression whose arguments are all continuous and therefore the control calculated can be regarded as an estimate of the equivalent control. Thus,

$$\widehat{u}'_\mathrm{eq} = q\left(\mathbf{x},\ x_{n+1},\ y,\ \dot{y},\ldots,\ y^{(r)}\right). \tag{10.140}$$

This method, however, is not perfect as, even in the simulation, the switching frequency in the sliding mode cannot be infinite, allowing the state trajectory to oscillate about, rather than lie on, the switching boundary. Consequently $y^{(r)}$, and therefore $\widehat{u}'_\mathrm{eq}(t)$ computed using (10.140), exhibits a zigzag form of oscillatory error. Also, the method is only applicable in the sliding mode, not giving meaningful results during control saturation.

An alternative way of estimating $u'_\mathrm{eq}(t)$ without oscillatory errors that also gives correct results during control saturation would be to simulate the control system using the boundary layer method. This method is, however, subject to a continuous error due to the dynamic lag between the ideal sliding mode control system response and the boundary layer pseudo sliding mode system response. It is therefore necessary to reduce the boundary layer thickness as far as possible to minimise this dynamic lag. It will be recalled that this method has already been used in the previous sections.

10.4.2.5 Model-Free Equivalent Control Estimation

A straightforward way of approximating $u'_{eq}(t)$ for real-time implementation without a plant model is to pass $u'(t)$ through a low-pass filter, such as

$$\widehat{U}'_{eq}(s) = \left(\frac{1}{1+sT_e}\right)^n U'(s) \qquad (10.141)$$

as this filters out the high-frequency components of $u'(t)$, leaving a continuously varying signal that approximates $u'_{eq}(t)$. Setting $n = 2$ should be sufficient. Using only a first-order filter ($n = 1$) would leave a residual zigzag oscillation in $\widehat{u}'_{eq}(t)$. Passing this through a second identical first-order filter, equivalent to $n = 2$ in (10.141), would attenuate this oscillation and smooth the sharp peaks (due to the discontinuous first derivatives), but the filtering time constant, T_e, would have to be made large enough to attenuate the oscillations to a sufficiently small amplitude without introducing excessive dynamic lag.

If a step external disturbance is applied, the state trajectory will momentarily depart from the switching boundary requiring u'_{max} to be rapidly increased to return the system to the sliding mode as soon as possible to keep the resulting transient error in $y(t)$ acceptably small for the application in hand. Having to increase T_e will delay this process due to the dynamic lag in (10.141), but this can be compensated for, to an extent, by increasing λ in (10.130).

10.4.2.6 Application to the DC Electric Drive

The augmented plant model of the system to be demonstrated consists of (10.36), (10.37) and (10.38) and the control smoothing integrator, whose output is the physical control variable, u, which becomes an additional state variable, denoted x_3. Also the external disturbance is referred to the primary control input, u'. Thus,

$$\dot{x}_1 = x_2, \quad \dot{x}_2 = -ax_2 + bx_3, \quad \dot{x}_3 = u' - d'_e \qquad (10.142)$$

and

$$y = x_1 \qquad (10.143)$$

The output derivatives required for the control law are then

$$\dot{y} = \dot{x}_1 = x_2 \quad \text{and} \quad \ddot{y} = \dot{x}_2 = -ax_2 + bx_3 \qquad (10.144)$$

The third derivative depends on u'. Thus,

$$\dddot{y} = -a\dot{x}_2 + b\dot{x}_3 = -ax_3 + b\left(u' - d'_e\right) \qquad (10.145)$$

10.4 Methods for Elimination of Control Chatter

So the relative degree of the augmented plant is $r' = 3$. The SMC law is then

$$u' = u'_{\max} \, \text{sgn}\,[S\,(y_r,\, y,\, \dot{y},\, \ddot{y})] \tag{10.146}$$

where a linear switching boundary is chosen as

$$S\,(y_r,\, y,\, \dot{y},\, \ddot{y}) = y_r - y - w_1 \dot{y} - w_2 \ddot{y} = 0. \tag{10.147}$$

To give critical damping with a settling time of T_s (5 % criterion) in the sliding mode, the transfer function corresponding to (10.147) is

$$\frac{Y(s)}{Y_r(s)} = \frac{1}{1 + w_1 s + w_2 s^2} = \frac{1}{(1 + a T_s s)^2} \tag{10.148}$$

where $a = 2/9$, from which

$$w_1 = 2 a T_s \quad \text{and} \quad w_2 = a^2 T_s^2. \tag{10.149}$$

To derive the equation for the model-based equivalent control estimation, first u' is made the subject of (10.145) to yield

$$u' = \frac{1}{b}\left(\dddot{y} + a x_3\right) + d'_e \tag{10.150}$$

Then the switching boundary equation (10.147) is used to obtain an expression for $\dddot{y}$ in terms of continuous variables. For this, $\ddot{y}$ is made the subject of (10.147) and the resulting equation differentiated. Thus,

$$\ddot{y} = \frac{1}{w_2}(y_r - y - w_1 \dot{y}) \Rightarrow \dddot{y} = \frac{1}{w_2}(\dot{y}_r - \dot{y} - w_1 \ddot{y}) \tag{10.151}$$

Substituting for $\dddot{y}$ in (10.150) using (10.151) then gives the required equation for estimating the equivalent control. Thus,

$$\widehat{u}'_{eq} = \frac{1}{b}\left[\frac{1}{w_2}(\dot{y}_r - \dot{y} - w_1 \ddot{y}) + au\right] + d'_e \tag{10.152}$$

The plant parameters for the simulation of Fig. 10.23 are as in Sect. 10.2.

A step reference input, $y_r(t)$, of 1 [rad] is applied, followed by a step external disturbance, $d_e(t)$, of magnitude, $u_{\max}/3$, applied at $t = T_s = 0.3$ [s]. The filtering time constants are $T_f = 0.0005$ [s] and $T_e = 0.005$ [s]. The primary control contingency factor is set to $\lambda = 10$. The maximum and minimum primary control limit magnitudes are set to $u'_{\max 1} = 50$ [V/s] and $u'_{\max 2} = 1,000$ [V/s].

First, focusing on Fig. 10.23a, the step response, $y(t)$, is as specified. The transient error due to the step external disturbance is due to the momentary departure of the state trajectory from the switching boundary. This transient error is only just visible

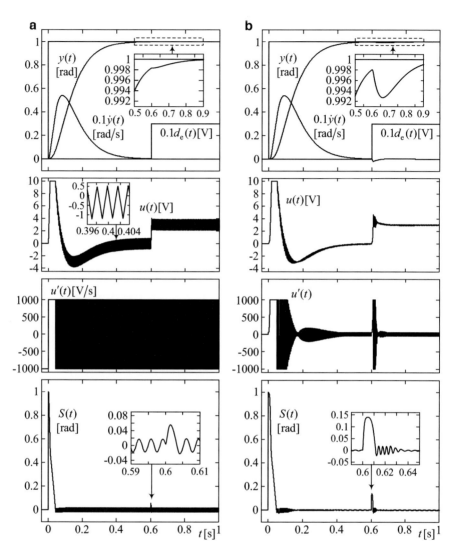

Fig. 10.23 Sliding mode position control of DC motor drive with control smoothing integrator. (**a**) Fixed primary control limits (**b**) Adaptive primary control limits

in the magnified inset graph and is therefore negligible. The physical control, $u(t)$, is not entirely free of the effects of the control chatter, exhibiting a zigzag oscillation as shown in the inset graph, but its peak-to-peak amplitude of about 1.5 [V] is only approximately 7.5 % of the 20 [V] peak-to-peak square wave of the standard SMC system, and it is continuous. The high-frequency square wave oscillations of $u'(t)$ in the sliding mode following the short initial period of control saturation are visible.

10.4 Methods for Elimination of Control Chatter

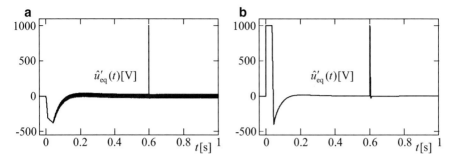

Fig. 10.24 Comparison of model-based equivalent control estimates. (**a**) Model based equation method (**b**) Pseudo SMC with boundary layer

The initial positive peak of the switching function, $S(t)$, corresponds to the control saturation. The small-amplitude oscillations of $S(t)$ with zero mean during the sliding mode, magnified in the inset graph, are nearly sinusoidal due to the smoothing effect of the filtered differentiators used to approximate the output derivatives used to form $S(t)$. This indicates an oscillation of the state trajectory about the switching boundary. The step disturbance applied at $t = 0.6$ [s] causes a positive-going impulse of $S(t)$ to be superimposed on the oscillations, indicating the momentary departure of the state trajectory from the switching boundary, the steady zero mean oscillations of the sliding mode being recovered after only half a cycle.

Moving to Fig. 10.23b, the adaptive primary control limits sacrifices a certain amount of robustness, as the transient error in $y(t)$ displayed on the magnified inset graph is larger than in Fig. 10.23a. The much reduced high-frequency oscillations of $u(t)$ and the corresponding time-varying oscillation amplitude of $u'(t)$ are visible. The increased positive peak of $S(t)$ due to the step disturbance is visible in the inset graph. On the scale of the step response, the transient error due to the external disturbance is still small enough to be considered negligible.

Finally, the model-based equivalent control estimation method based on (10.152) is applied to the system simulated in Fig. 10.23a and compared with the alternative estimation method obtained from a simulation using a saturation function instead of the signum function with a 'thin' boundary layer obtained by using a saturation element with a gain of 10,000. The results are shown in Fig. 10.24.

The pseudo SMC method with the boundary layer correctly shows the initial control saturation, while the model-based equation method is correct only in the sliding mode, where it exhibits high-frequency oscillations.

10.4.2.7 Multiple Smoothing Integrator Version

As has been demonstrated in Sect. 10.4.2.6, the single smoothing integrator achieves a continuous physical control with reduced oscillation amplitude. The zigzag oscillations, however, are rich in harmonics and potentially harmful.

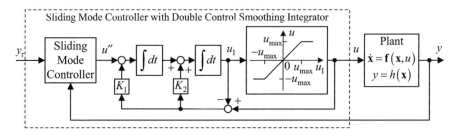

Fig. 10.25 Sliding mode control of SISO plant with double control smoothing integrator

Hence, the amplitude of the oscillations was minimised with only a small reduction in robustness by means of the adaptive primary saturation limit of Sect. 10.4.2.3. An alternative is presented in this section in which further smoothing is obtained by replacing the single integrator with a chain of integrators. In principle, any number of integrators can be used but just two should suffice for most applications, as shown in Fig. 10.25.

An essential feature is the anti-windup loop as this effect would be extreme in a chain of integrators in comparison with a single integrator. It is a simple matter to design this loop with a relatively short settling time to yield high gains giving effective control that only allows u_1 to exceed the saturation limits by negligible proportions. In this case, the highest order of output derivative needing estimation exceeds that of the basic sliding mode controller by two. For the DC motor example, the relative degree of the augmented plant is $r = 4$, and the SMC law becomes

$$u' = u'_{\max} \operatorname{sgn}\left[S\left(y_r, y, \dot{y}, \ddot{y}, \dddot{y} \right) \right] \qquad (10.153)$$

where

$$S\left(y_r, y, \dot{y}, \ddot{y}, \dddot{y} \right) = y_r - y - w_1 \dot{y} - w_2 \ddot{y} - w_3 \dddot{y} = 0, \qquad (10.154)$$

so that the system can be designed to have a specified settling time of T_s in the sliding mode, for which the ideal closed-loop system output, $y_{\text{ideal}}(t)$, obeys

$$\frac{Y_{\text{ideal}}(s)}{Y_r(s)} = \frac{1}{1 + w_1 s + w_2 s^2 + w_3 s^3} = \frac{1}{(1 + aT_s s)^3}, \qquad (10.155)$$

where, for the 5 % criterion, $a = 1/6$.

Since an external step disturbance referred to the physical control variable is equivalent to a double Dirac delta impulse referred to the primary control input, then it is impossible to completely eliminate transient errors due to the disturbance. They can, however, be reduced to acceptable proportions by utilising sufficiently large primary control saturation limits, $\pm u'''_{\max}$, and a sufficiently small time constant,

10.4 Methods for Elimination of Control Chatter

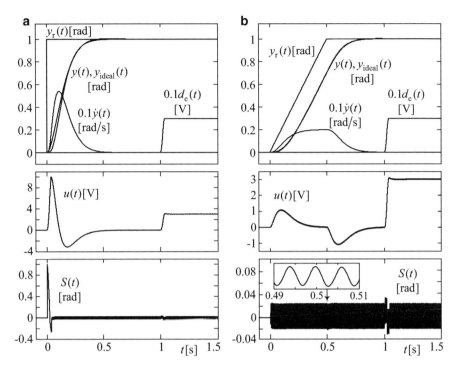

Fig. 10.26 Simulation of SMC of DC motor drive with double smoothing integrator. (**a**) Step response. (**b**) Ramp response

T_f, in the filtering of the approximate differentiation. For the DC motor control simulations presented in Fig. 10.26, $u''_{max} = 20,000$ [V/s] and $T_f = 0.0005$ [s].

In both simulations, an external disturbance input is applied that closely approximates a step function, rising rapidly with a constant finite second derivative followed by an equal and opposite second derivative, reaching a constant value of 3 [V] referred to the physical control input, in 0.04 [s]. This is generated using a switched input to a double integrator, giving a response similar to a time-optimal control. Note that in practice, external disturbances cannot step in zero time, so this disturbance is realistic. The theoretical primary control input, $u''(t)$, required to counteract this disturbance then falls within $\pm u''_{max}$ so that the actual primary control input is only limited by the time constant, T_f. The result is a negligible transient error. Simulations with a true step external disturbance (not shown) indicate a transient error of about 0.02 [rad], which is acceptable.

The success of the method in producing a smooth $u(t)$ is immediately apparent.

The switching function, $S(t)$, may be seen to go through a transient at the beginning of the step response of Fig. 10.26a, indicating the usual initial departure of the state trajectory from the switching boundary, giving a small initial difference between $y(t)$ and $y_{ideal}(t)$. The sliding motion following this is evident as $S(t)$

oscillates about zero. This oscillation is not of the zigzag form due to the filtering of the output derivative estimator. The ramp reference input is seen to cause negligible departure from the sliding condition.

10.4.3 Higher-Order Sliding Mode Control

10.4.3.1 Introduction

The standard sliding mode controller drives the state trajectory onto the switching boundary, $S(y_r, \mathbf{y}) = 0$, by means of the control law $u = \text{sgn}[S(y_r, \mathbf{y})]$, and it will be recalled that in a practicable controller, the switching frequency is finite in the sliding mode, during which the state trajectory executes a zigzag motion about the switching boundary. In this case, the derivative, $\dot{S}(y_r, \mathbf{y}) = S_1(y_r, \dot{y}_r \mathbf{y}, u)$, is oscillatory and discontinuous, as is the control variable, $u(t)$. This constitutes the control chatter that has to be eliminated and, like the control smoothing integrator method, the higher-order sliding mode method seeks to do this without loss of robustness. Various schemes have been devised to achieve this [5], some of which apply to specific plant models and embody exact differentiators (a special form of observer with a finite convergence time to the exact derivative in absence of measurement noise) for use with these plant models. In this section, the emphasis is on deriving a general and practicable controller that is capable of chatter-free operation with a similar level of robustness as achievable with a standard sliding mode controller and, in common with the other control techniques presented in this book, yields a prescribed closed-loop dynamics.

10.4.3.2 The Definition of Higher-Order Sliding Mode

The approach to eliminating control chatter here is to eliminate the zigzag motion of the state trajectory about the switching boundary in the sliding mode by creating a system that not only drives the switching function to zero but also drives at least its first derivative to zero, in a finite time, and thereafter maintains this condition. Thus, in the higher-order sliding mode,

$$S(y_r, \mathbf{y}) = 0, \quad \dot{S}(y_r, \mathbf{y}) = 0, \ldots, \quad S^{(p-1)}(y_r, \mathbf{y}) = 0. \quad (10.156)$$

The order of the sliding mode is defined as p. Thus, a standard sliding mode controller that only drives $S(y_r, \mathbf{y})$ to zero operates in a *first-order sliding mode*. In a *higher-order sliding mode*, $p > 1$. The ideal first-order (standard) sliding mode controller maintains $S(y_r, \mathbf{y}) = 0$. It may be argued that if $S(y_r, \mathbf{y}) = 0$, then (10.156) holds, but this is not the case. Strictly, $S(y_r, \mathbf{y})$ oscillates at an infinite frequency but with an infinitesimal amplitude under the action of the control variable, $u(t)$, that switches between its limits at the same infinite frequency with a

10.4 Methods for Elimination of Control Chatter

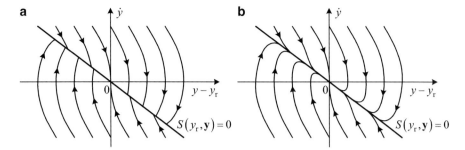

Fig. 10.27 Illustration of closed-loop phase portraits in the vicinity of the switching boundary. (**a**) Standard sliding mode control. (**b**) Higher-order sliding mode control

time-varying mark–space ratio. In this case $\dot{S}(y_r, \mathbf{y}) = S_1(y_r, \dot{y}_r \mathbf{y}, u)$ oscillates at an infinite frequency but with a non-zero amplitude and also is discontinuous due to its direct dependence on u. Hence, $\dot{S}(y_r, \mathbf{y}) \neq 0$. This may be seen easily by considering the practical system that, through the non-zero iteration period, h, of the digital processor, allows the zigzag oscillations of $S(y_r, \mathbf{y})$ about zero in the sliding mode. It is then clear that $\dot{S}(y_r, \mathbf{y})$ is a square wave oscillation that is discontinuous. The ideal first-order sliding mode control system is then obtained by letting $h \to 0$, in which case $\dot{S}(y_r, \mathbf{y})$ is a square wave at infinite frequency. The ideal pth order sliding mode control system drives the switching function and its derivatives to zero according to (10.156) but $S^{(p)}(y_r, \mathbf{y})$ oscillates at an infinite frequency. Figure 10.27 illustrates pictorially the difference between the closed-loop phase portraits produced by the standard first-order and higher-order sliding mode controllers for a plant with a relative degree of $r = 2$ so that the output derivative substate-space variables taking part in the sliding mode may be viewed on the phase plane.

In contrast with the standard sliding mode controller, a higher-order sliding mode (HOSM) produces substate trajectories that are tangential to the switching boundary, touching it after a finite time. Under these circumstances, the zigzag motion associated with the control chatter cannot occur in a practical realisation.

To summarise, the order of a HOSM controller is the order of the lowest derivative of $S(y_r, \mathbf{y})$ that is non-zero and discontinuous in the sliding mode.

10.4.3.3 A Higher-Order Sliding Mode Control System

First, the switching function, $S(y_r, \mathbf{y})$, is set up so that the differential equation,

$$S(y_r, \mathbf{y}) = S\left(y_r, y, \dot{y}, \ldots, y^{(r-1)}\right) = 0, \qquad (10.157)$$

is that of the specified closed-loop dynamics, where r is the relative degree of the plant, as for the standard SMC. Then for a pth order sliding mode controller, a system has to be devised that satisfies (10.156) in a finite time starting with an initial state from which the sliding boundary can be reached. Since the highest derivative of y in $S(y_r, \mathbf{y})$ is $y^{(r-1)}$, where r is the relative degree, differentiating this once yields

$$\dot{S}(y_r, \mathbf{y}) = S_1(y_r, \dot{y}_r \mathbf{y}, u). \qquad (10.158)$$

In the standard SMC, the control law, $u = \text{sgn}[S(y_r, \mathbf{y})]$, therefore makes $\dot{S}(y_r, \mathbf{y})$ non-zero, discontinuous and oscillating in the sliding mode. To make $\dot{S}(y_r, \mathbf{y})$ continuous and controllable to zero, it is necessary to introduce a dynamical subsystem in the controller software whose output is the physical plant control input, u, and whose input is a new control input, u_1. Then u becomes a state variable. The lowest order for this subsystem is one and the simplest is a pure integrator, requiring

$$\dot{u} = u_1. \qquad (10.159)$$

This is the same as the integrator introduced at the plant input in the single control smoothing integrator method of Sect. 10.4.2. Differentiating (10.158) yields

$$\dot{S}_1(y_r, \dot{y}_r, \mathbf{y}, u) = S_2(y_r, \dot{y}_r, \ddot{y}_r \mathbf{y}, u, \dot{u}) = S_2(y_r, \dot{y}_r, \ddot{y}_r \mathbf{y}, u, u_1). \qquad (10.160)$$

At this point, the notation will be simplified by omitting the state variables from the argument lists, so that (10.160) is written

$$\dot{S}_1 = S_2(u_1). \qquad (10.161)$$

The control law,

$$u_1 = u_{1\max} \text{sgn}(S_1 - S_{1r}), \qquad (10.162)$$

can then bring S_1 to any reference value, S_{1r}, in a finite time if the limit, $u_{1\max}$, is made sufficiently large. To determine a suitable value of $u_{1\max}$, the most practical approach is to run a simulation of the control system to be derived and increase $u_{1\max}$ until the system works correctly over the operational range of states for the application in hand. Once $S_1 = S_{1r}$, u_1 will switch, in theory, at infinite frequency between $\pm u_{1\max}$ with a time-varying mark–space ratio to maintain $S_1 = S_{1r}$, which can be time varying. Let $S(y_r, \mathbf{y})$ be written as S_0. Then $\dot{S}_0 = S_1 \Rightarrow$

$$\dot{S}_0 = S_{1r}. \qquad (10.163)$$

This may be regarded as a single integrator plant with state variable, S_0, and control input, S_{1r}. It remains to devise a control law that will bring S_0 to zero in a finite time, for then S_{1r} will also be zero through (10.163), S_1 being brought to zero at the same time due to the continuing action of control law (10.162). A very simple control law achieving this would be $S_{1r} = -S_{1r\max} \text{sgn}(S_0)$ but in a discrete implementation that would limit cycle about zero instead of holding a zero value. Instead, since

10.4 Methods for Elimination of Control Chatter

the single integrator plant is linear, a discrete linear control law giving a dead-beat response will be used. The z-transfer function corresponding to (10.163) is

$$\frac{\overline{S}_0(z)}{\overline{S}_{1r}(z)} = \frac{T}{z-1}. \tag{10.164}$$

where T is the chosen sampling time. So the control law,

$$\overline{S}_{1r}(z) = K_s \left[\overline{S}_{0r}(z) - \overline{S}_0(z) \right], \tag{10.165}$$

may be formed, where $\overline{S}_{0r}(z)$ is the reference input, which will be set to zero. The closed-loop transfer function is then

$$\frac{\overline{S}_0(z)}{\overline{S}_{0r}(z)} = \frac{K_s \frac{T}{z-1}}{1 + K_s \frac{T}{z-1}} = \frac{K_s T}{z - 1 + K_s T}. \tag{10.166}$$

The closed-loop pole is placed at the origin of the z-plane, requiring

$$K_s = \frac{1}{T}. \tag{10.167}$$

Then the control law to be applied in real time is

$$S_{1r}(k) = \frac{1}{T} \left[S_{0r}(k) - S_0(k) \right] = -\frac{1}{T} S_0(k) \tag{10.168}$$

since $S_{0r}(k) = 0$. The difference equation of the plant is

$$S_0(k+1) = S_0(k) + T S_{1r}(k). \tag{10.169}$$

Then substituting for $S_{1r}(k)$ in (10.169) using control law (10.168) yields

$$S_0(k+1) = S_0(k) + T \left[-\frac{1}{T} S_0(k) \right] = 0, \tag{10.170}$$

indicating that, in one iteration, S_0 is brought to zero starting with an arbitrary value. Then according to (10.168),

$$S_{1r}(k+1) = -\frac{1}{T} S_0(k+1) = 0. \tag{10.171}$$

As $S_1 = S_{1r} = 0$, then S_1 and S_0 are brought to zero in a finite time, T [s], as required.

Commencing from an arbitrary initial state, control law (10.162) brings S_1 to S_{1r} in a finite time, τ, which is dependent on the initial state.

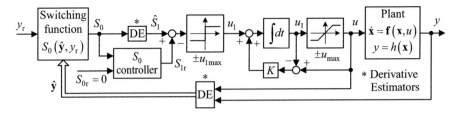

Fig. 10.28 General structure of second-order sliding mode control system

Simulation of the control system over the operation envelope of states for the application in hand will enable the maximum value, τ_{max}, of τ to be determined. The total time to reach the sliding mode is then

$$\tau_s = \tau + T. \tag{10.172}$$

The control system developed so far is a *second-order* sliding mode control system since both $S_1 = \dot{S}$ and $S_0 = S$ are brought to zero in a finite time and $S_2 = \ddot{S}$ oscillates at an infinite frequency with a non-zero amplitude since it depends directly on u_1 through (10.161). Figure 10.28 shows the control system structure.

The implementation embodies the essential features of anti-windup control of the integrator producing u and the derivative estimators producing estimates, $\hat{y}$ and $\hat{S}_1$, of y and S_1. The derivative estimator producing $\hat{y}$ is the multiple integrator type of Sect. 10.1.3.2 as dynamic lag in the derivative estimates can be compensated to an extent by using the available control signal, u. A straightforward filtered differentiator is found to be sufficient for $\hat{S}_1$. As in the control smoothing integrator method with one integrator, $u(t)$ is continuous but has a zigzag oscillation. This can be improved by increasing the order of the sliding mode beyond two since, as will be seen shortly, it entails inserting more integrators between the rapidly switching control that produces the robustness and the physical control, u, as in the control smoothing integrator method.

The third-order SMC is derived by first differentiating (10.160) to obtain

$$\dot{S}_2\left(y_r, \dot{y}_r, \ddot{y}_r\, \mathbf{y}, u, u_1\right) = S_3\left(y_r, \dot{y}_r, \ddot{y}_r, \dddot{y}_r\, \mathbf{y}, u, u_1, \dot{u}_1\right) \tag{10.173}$$

Then another integrator is inserted with output, u_1, and input, u_2, that is the new control variable, u_1 becoming a state variable. Thus,

$$\dot{u}_1 = u_2 \tag{10.174}$$

so (10.173) becomes

$$\dot{S}_2\left(y_r, \dot{y}_r, \ddot{y}_r\, \mathbf{y}, u, u_1\right) = S_3\left(y_r, \dot{y}_r, \ddot{y}_r, \dddot{y}_r\, \mathbf{y}, u, u_1, u_2\right) \tag{10.175}$$

10.4 Methods for Elimination of Control Chatter

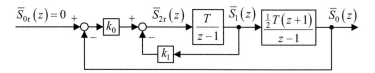

Fig. 10.29 Switching boundary control system for third-order sliding mode controller

which may again be simplified by omitting the state variables. Thus,

$$\dot{S}_2 = S_3(u_2) \tag{10.176}$$

Then with a sufficiently large value of $u_{2\max}$, the control law,

$$u_2 = u_{2\max} \operatorname{sgn}(S_2 - S_{2r}), \tag{10.177}$$

will bring S_2 to any reference value, S_{2r}, in a finite time and then maintain $S_2 = S_{2r}$ by infinite frequency switching with a varying mark–space ratio. The switching function derivatives then obey the following differential equations.

$$\left. \begin{array}{l} \dot{S}_0 = S_1 \\ \dot{S}_1 = S_{2r} \end{array} \right\} \tag{10.178}$$

This may be regarded as a double integrator plant with state variables, S_0 and S_1, with control input, S_{2r}. To bring S_0, S_1 and S_2 to zero in a finite time to achieve the third-order sliding mode, a discrete state feedback control law,

$$S_{2r}(k) = k_0 [S_{0r} - S_0] - k_1 S_1 \tag{10.179}$$

can be applied with both closed-loop poles placed at the origin of the z-plane to produce a dead-beat response reaching the desired zero values in $2T$ seconds from the instant control law. (10.177) brings S_2 to S_{2r}, where T is the sampling interval, if $S_{0r} = 0$. The z-transfer function relationships corresponding to (10.178) are

$$\frac{\overline{S}_0(z)}{\overline{S}_{2r}(z)} = \frac{\frac{1}{2}T^2(z+1)}{(z-1)^2} \quad \text{and} \quad \frac{\overline{S}_1(z)}{\overline{S}_{2r}(z)} = \frac{T}{z-1} \tag{10.180}$$

The complete linear state feedback control system, which will be called the switching boundary control system, or in short, the S control system, is shown in Fig. 10.29.

The closed-loop characteristic polynomial is

$$(z-1)^2 \left[1 + k_1 \tfrac{T}{z-1} + k_0 \tfrac{\frac{1}{2}T^2(z+1)}{(z-1)^2} \right] = (z-1)^2 + k_1 T(z-1) + \tfrac{1}{2} k_0 T^2 (z+1)$$
$$= z^2 + \left(\tfrac{1}{2} k_0 T^2 + k_1 T - 2 \right) z + \tfrac{1}{2} k_0 T^2 - k_1 T + 1 \tag{10.181}$$

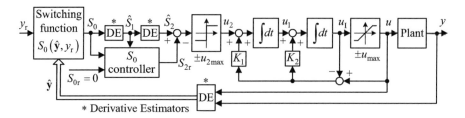

Fig. 10.30 General structure of third-order sliding mode control system

For the dead-beat response, the desired closed-loop characteristic polynomial is z^2, requiring

$$\left.\begin{array}{l}\frac{1}{2}k_0T^2 + k_1T = 2 \\ \frac{1}{2}k_0T^2 - k_1T = -1\end{array}\right\} \quad (10.182)$$

from which

$$k_0 = \frac{1}{T^2} \quad \text{and} \quad k_1 = \frac{3}{2T}. \quad (10.183)$$

The structure of the third-order sliding mode control system is shown in Fig. 10.30.

The reader will now be able to extend the above to create a pth order sliding mode controller for any p, but in most applications, sufficient smoothing of $u(t)$ should be attainable with $p = 2$ or 3.

The degree of attainable smoothing of $u(t)$ with $p = 2$ and $p = 3$ will be similar, respectively, to that attainable with the control smoothing integrator method with, respectively, one and two integrators. The fundamental difference between the two techniques, however, is as follows. The order of the closed-loop system in the sliding mode for the control smoothing integrator method increases beyond $r - 1$ by the number of integrators in the smoothing integrator chain. For the higher-order sliding mode method, the order of the closed-loop system in the sliding mode is only $r - 1$, which is the same as that of the standard SMC.

Some simulations of the HOSM of the DC electric drive are presented in Figs. 10.31 and 10.32. The plant parameters are as for Sect. 10.2.

A step reference input, $y_r(t)$, of 1 [rad] is applied, followed by a step external disturbance, $d_e(t)$, of magnitude, $u_{max}/3$, applied at $t = 1$ [s]. The filtered differentiation time constant is $T_f = 0.0002$ [s] for the second-order SMC and $T_f = 0.0001$ [s] for the third-order SMC, such small values being found necessary to obtain control of $S_0(t)$, $\hat{S}_1(t)$ and $\hat{S}_2(t)$ bringing them all to nearly zero in a finite time. The primary control limit magnitudes are set to $u_{1\,max} = 10.000$ [V/s] for the second-order SMC and $u_{2\,max} = 500,000$ [V/s^2] for the third-order SMC.

10.4 Methods for Elimination of Control Chatter

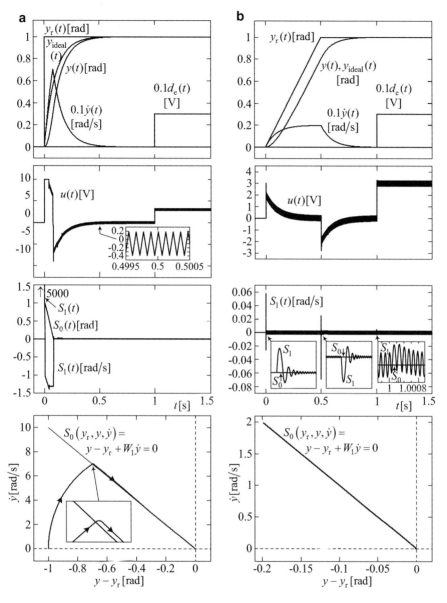

Fig. 10.31 Second-order sliding mode control of DC electric drive. (**a**) Step response. (**b**) Ramp response

Note that for the standard SMC of Sect. 10.2, $u_{max} = 10$ [V], the primary control and the physical control becoming identical. Such large increases in the primary control magnitude with the order of the sliding mode are found necessary to maintain the sliding mode over the operating envelope of the plant, due to the integrators inserted in the control channel.

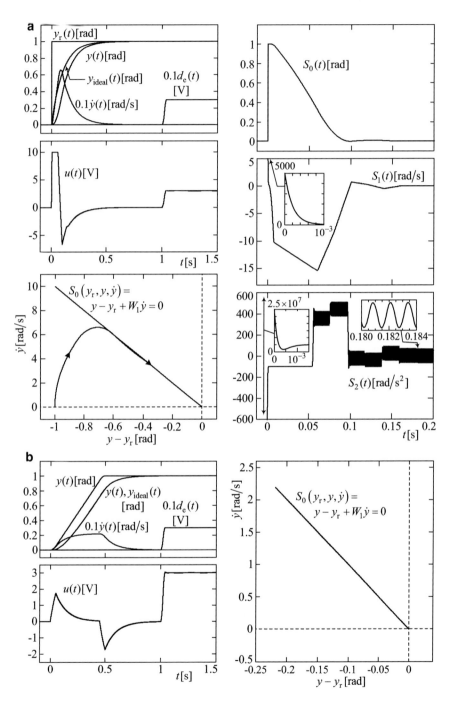

Fig. 10.32 Third-order sliding mode control of DC electric drive. (**a**) Step response. (**b**) Ramp response

10.4 Methods for Elimination of Control Chatter

First, the results for the second-order SMC of Fig. 10.31 will be discussed. The step response, $y(t)$ of Fig. 10.31a exhibits a lag with respect to the ideal response, $y_{ideal}(t)$, due to the initial saturation of the physical control, $u(t)$, that cannot be avoided. This accompanied by the initial transient of $S_0(t)$ and $\hat{S}_1(t)$, which are brought to nearly zero in a finite time. The plant state trajectory in the phase plane initially overshoots the switching boundary due to the initial transient of $S_0(t)$ but, as shown by the magnified inset graph, does not chatter, as required. As $S_0(t)$ and $\hat{S}_1(t)$ reach zero, the state trajectory meets the switching boundary at about $y - y_r = -0.3 \ [rad]$ and then follows it towards the origin without chatter. In the sliding mode, the physical control, $u(t)$, is continuous but executes a zigzag oscillation, as shown by the inset magnified graph, as was the case with the control smoothing integrator method using one integrator. For the ramp response of Fig. 10.31b, a positive jump of $\hat{S}_1(t)$ occurs when $\dot{y}_r(t)$ steps to a constant positive value, which causes $S_0(t)$ to begin ramping positively. The S controller then takes corrective action, returning both $S_0(t)$ and $\hat{S}_1(t)$ to zero in a finite time. This is followed by a similar sequence of events following a negative jump in $\hat{S}_1(t)$ of the same magnitude when $\dot{y}_r(t)$ steps back to zero. The two equal and opposite transients of $\hat{S}_1(t)$ are shown in the first two inset graphs of Fig. 10.31 (b). When the step external disturbance torque referred to u is applied, this causes a step in $\hat{S}_1(t)$ and a corresponding ramp in $S_0(t)$, followed by more corrective action by the S_0 controller. The transient in $\hat{S}_1(t)$, however, is so small that it is almost masked by the oscillations due to the switching of $u_1(t)$, as shown in the third inset magnified graph of Fig. 10.31b. The three transients described above, however, do not significantly affect the overall performance since the plant state trajectory is confined to the switching boundary as shown.

Turning attention now to the third-order SMC, the phase-plane trajectory of Fig. 10.32a clearly shows the desirable chatter-free approach to the switching boundary in the phase plane and the subsequent sliding motion towards the origin. As for the second-order SMC system, the lag in the response $y(t)$ with respect to the ideal response $y_{ideal}(t)$ is due to the initial saturation of the physical control, $u(t)$. The effectiveness of the S controller in bringing $S_0(t)$ and $\hat{S}_1(t)$ to zero in a finite time is evident. The control variable produced by this controller is $\hat{S}_2(t)$, and this oscillates about zero as shown by the second magnified inset graph, due to the high-frequency switching of $u_2(t)$ between $\pm u_{2\,max}$ in the sliding mode, once $S_0(t)$ and $\hat{S}_1(t)$ have been brought to zero. The very large initial values of $\hat{S}_1(t)$ and $\hat{S}_2(t)$, shown in the inset graphs, are due to the step in the reference input. In order to avoid interrupting the sliding mode at the beginning and end of the ramp response of Fig. 10.32b, the second derivatives of $y_r(t)$ are kept finite by generating the ramp as the output of a double integrator with an appropriately switched input. In this way, the large initial values of $\hat{S}_2(t)$ and $\hat{S}_1(t)$, such as caused by the step reference input in Fig. 10.32a, are avoided. Also, the disturbance input is continuous with finite second derivative, approximating a step function as the output of a double integrator driven by a switched input similar to the time-optimal control, to avoid a large transient spike in $\hat{S}_2(t)$ which would occur with a step disturbance. This is not unrealistic as real physical disturbances will have non-zero rise times. Consequently, the system remains in the sliding mode throughout, as is evident by the state trajectory in Fig. 10.32b, which adheres to the switching boundary.

The success of the controller in eliminating chatter in the physical control, $u(t)$, is evident in Fig. 10.32a, b.

10.5 Controllers with Robust Pole Placement

10.5.1 Introduction

A family of robust controllers stems from the pseudo sliding mode control using the boundary layer method of Sect. 10.4.1 for the control chatter avoidance, producing similar robustness. It will be recalled that for a linear SISO plant and a linear switching function, the operation of the control system in the pseudo sliding mode within the boundary layer is identical to that of the linear control system formed by replacing the saturation element by a gain, K. Root locus analysis then reveals that for an nth-order plant, as $K \to \infty$, $n - 1$ of the closed-loop poles terminate on the zeros, $n - r$ of which cancel any finite plant zeros, where r is the relative degree, the remaining $n - r$ closed-loop poles being determined by the switching boundary coefficients. There remains one closed-loop pole that follows the negative real axis of the s-plane towards $-\infty$ as $K \to \infty$. Since the first-order mode associated with this pole has a relatively fast decay due to its small time constant, it will be termed the *fast pole*. For finite but sufficiently large K, the $n - 1$ closed-loop poles can be brought close to the desired locations while dominating the fast pole which becomes large and negative, thereby realising the prescribed closed-loop dynamics. Consider a linear controller that can be designed by pole assignment for the same plant, using a perfect plant model, to realise the same set of closed-loop poles. Then this must yield identical closed-loop dynamics to the pseudo SMC. This controller must also exhibit the same robustness as the pseudo SMC. This implies that carrying out the pole placement design using a *nominal* but not necessarily accurate plant model will yield similar results. Hence, the complete set of linear controllers that can be so designed may be regarded as relatives of the sliding mode controller and will be referred to as *robust pole placement controllers*. As will be seen, this range of controllers can be extended to include those with a greater number of closed-loop poles.

Another robust control technique achieving robustness through high gains, but in an observer, is the observer-based robust control (OBRC) (Appendix A10).

Since a boundary layer-based pseudo sliding mode controller can also be applied to nonlinear plants, the related robust pole placement controllers may also be applied to nonlinear plants, giving prescribed linear closed-loop dynamics. So this can be regarded as a special technique for the control of nonlinear plants, in addition to the two model-based techniques of Chap. 7.

10.5.2 Output Derivative State Feedback Controller

10.5.2.1 Equivalence to Pseudo Sliding Mode Control

For an SISO plant with relative degree, r, the equation of the pseudo sliding mode control law of Sect. 10.4.1 operating within the boundary layer is

$$u = K\left(y_r - y - \sum_{i=1}^{r-1} W_i \hat{y}^{(i)}\right), \qquad (10.184)$$

where $\hat{y}^{(i)}$, $i = 1, 2, \ldots, r-1$ are the output derivative estimates from the filtered differentiators. A similar controller parameterised to render pole placement straightforward is given by

$$u = R y_r - k_0 y - \sum_{i=1}^{r-1} k_i \hat{y}^{(i)}. \qquad (10.185)$$

The equivalence of control law (10.185) to control law (10.184) may be demonstrated by setting $R = K$, $k_0 = K$ and $k_i = K W_i$, $l = 1, 2, \ldots, r-1$ but different, though similar, settings of these controller parameters will result from the pole placement procedure of the following section.

10.5.2.2 Pole Assignment Without an Accurate Plant Model

Since control law (10.184) is known to be robust with respect to relatively large plant parameter uncertainties as well as external disturbances, it is arguable that if control law (10.185) is designed by robust pole assignment using a chosen plant model with the same relative degree as the real plant and then applied to the real linear plant, the dominant group of $r-1$ poles will not change by large proportions from the nominal values and the remaining negative real pole will remain dominated by them.

In practice, if a plant model is available, then it should be used for the pole placement design. The resulting control system will then be very tolerant of changes of the plant parameters over its lifetime, as well as minimising the effect of external disturbances. If the available plant model is nonlinear, then its linearised model using the operating point method of Chap. 7 may be used for the pole placement design. On the other hand, in view of the extreme robustness of the method, a simple design can be produced using the simplest plant model, which is a chain of r pure integrators with a forward path gain of K_f. Thus

$$\frac{Y(s)}{U(s)} = \frac{K_f}{s^r}, \qquad (10.186)$$

where K_f should be set equal to an estimate of the forward path gain of the real plant if it is available. Note that the forward path gain is the scalar gain of the path between $U(s)$ and $Y(s)$ that contains the least number of pure integrators.

As with any controller implementation, it is recommended that the control variable is limited by a saturation function in the digital processor to $\pm u_{max}$ to protect the actuators. This will be included in the simulations of this section as this will also enable comparisons to be made with the performance of the pseudo SMC.

Robust pole placement design of a linear controller using a simplified plant model of order equal to the plant relative degree will work satisfactorily for plants with finite zeros but provided they all lie in the left half of the s-plane. Attempts at control of uncertain non-minimum phase plants using this method will fail due to the closed-loop poles that nearly cancel the right half plane zeros rendering the closed-loop system unstable. This might lead a control system designer to attempt robust pole placement design using a full-order model containing the zeros but unfortunately the closed-loop system is invariably sensitive to errors in the assumed zero locations and can be unstable in certain cases. This therefore remains a matter for future research.

Example 10.4 Robust pole placement controller for a motorised pendulum

This example demonstrates the ability of the control technique under study to handle plants that are both nonlinear and unstable. The model is given by

$$\ddot{y} = bu - a\sin(y) - c\dot{y}, \qquad (10.187)$$

where $b = K_T/(ML^2)$, $a = g/L$ and $c = D/(ML^2)$, M and L are the mass and length of the pendulum, $g = 9.81 \ [\text{m/s}^2]$ is the acceleration due to gravity, K_T is the motor drive torque constant, D is the damping coefficient and the measurement and y is numerically equal to the pendulum angle. The relative degree is $r = 2$. It will be supposed that this is the only available information and therefore the plant model for the pole placement is (10.186). Figure 10.33 shows the block diagrams of the linear system for the pole placement design and the control system to be implemented.

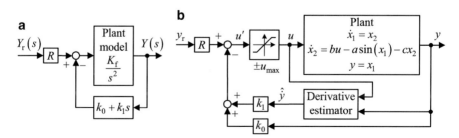

Fig. 10.33 Control system block diagrams. (**a**) System for pole assignment. (**b**) Control system for implementation

10.5 Controllers with Robust Pole Placement

The derivative estimator is of the multiple integrator observer type presented in Sect. 10.1.3.2. The closed-loop transfer function of Fig. 10.33a is

$$\frac{Y(s)}{Y_r(s)} = \frac{RK_f}{s^2 + K_f(sk_1 + k_0)}. \tag{10.188}$$

For a unity closed-loop DC gain, $R = k_0$. For a step response of T_s seconds and a 'fast' pole at $s = -1/T_{cf}$, the desired closed-loop characteristic polynomial is

$$s^2 + B(sk_1 + k_0) = \left(s + \frac{3}{T_s}\right)\left(s + \frac{1}{T_{cf}}\right) \tag{10.189}$$

from which

$$k_1 = \frac{1}{K_f}\left(\frac{3}{T_s} + \frac{1}{T_{cf}}\right) \quad \text{and} \quad k_0 = \frac{3}{K_f T_s T_{cf}}. \tag{10.190}$$

To aid in the comparison with SMC, a virtual switching function, $S(y_r, y, \dot{y})$, will be formed. For $r = 1$, control law (10.185) is

$$u = Ry_r - k_0 y - k_1 \widehat{\dot{y}}. \tag{10.191}$$

This may be written as

$$S\left(y_r, y, \widehat{\dot{y}}\right) = \frac{R}{k_0}y_r - y - \frac{k_1}{k_0}\widehat{\dot{y}} = y_r - y - \frac{k_1}{k_0}\widehat{\dot{y}}, \quad u = k_0 S\left(y_r, y, \widehat{\dot{y}}\right) \tag{10.192}$$

Figure 10.34 shows results for $L = 0.5$ [m], $D = 0.25$ [Nm/(rad/s)] and $K_T = 2$ [Nm/V]. The robustness of the control system is tested by repeating the simulations for $M = 1$ [Kg], 2 [Kg] and 3 [Kg]. In all cases, the pendulum commences at rest with $y = 0$ [rad] and is commanded to reach a constant angle of 2 [rad]. The fast pole time constant is chosen as $T_{cf} = 0.001$ [s]. A behaviour similar to a sliding mode control system is evident. In the step responses of Fig. 10.34a, the differences are due to the initial control saturation, but $y(t)$ is not very different from the ideal first-order step response, $y_{ideal}(t)$ in all cases. The sliding mode like behaviour is particularly noticeable in the graphs of the virtual switching functions, $S(t)$, which reach approximately zero in a finite time under control saturation and are then maintained approximately zero, and also in the phase-plane trajectories which closely follow the virtual switching boundary after reaching it.

The extreme robustness of the control system is evident in the ramp responses of Fig. 10.34b in which the control variables remain within the saturation limits of ± 10 [V]. The nearly coincident response graphs, $y(t)$, are almost identical to the

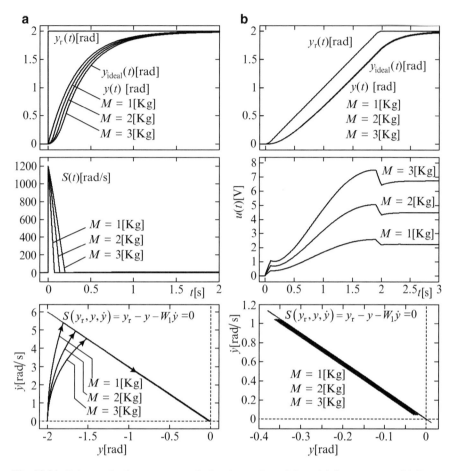

Fig. 10.34 Robust pole placement control of an inverted pendulum. (**a**) Step response. (**b**) Ramp response

ideal first-order response, $y_{ideal}(t)$. Also, the state trajectory remains fairly close to the virtual switching boundary. The small departures from this straight line would be brought to negligible proportions by reducing T_{cf} further.

10.5.3 Dynamic Controllers with Robust Pole Placement

For a plant with relative degree, r, and a controller containing dynamic elements that raises the order of the system from the plant order, n, to N, it is impossible to *exactly* replicate the behaviour of a pseudo sliding mode controller with a boundary layer, as in Sect. 10.5.2 but very similar behaviour can be obtained with robust pole

10.5 Controllers with Robust Pole Placement

placement. Even if only the plant relative degree is known and its parameters are uncertain, robust pole placement using a simplified plant model of order, r, such as a multiple integrator chain, is possible. In replicating the behaviour of an SISO sliding mode control system, only one fast pole has been introduced, but in view of the theory of pole-to-pole dominance introduced in Chap. 1, it is also possible to imitate the behaviour of a standard sliding mode control system by having $N - r + 1$ fast poles. The desired characteristic polynomial for the pole placement would then be

$$D(s) = (s + p_{cf})^{N-r+1}(s + p_c)^{r-1}, \qquad (10.193)$$

where $p_{cf} \gg p_c$. The settling time formulae can be used to yield specified non-overshooting step responses with a settling time of T_s by setting the dominant pole magnitude to $p_c = 1.5r/T_s$ (5 % criterion) or $p_c = 1.6(0.5 + r)/T_s$ (2 % criterion).

Regarding the choice of controller, a polynomial controller is attractive as it has the advantage of not needing output derivative estimation.

This is a known area of difficulty for sliding mode control. This will be investigated in a subsequent example.

With the pole placement of (10.193), the virtual switching function for comparison with SMC is the one which yields the first-order dynamics with time constant, $T_c = 1/p_c$ in the sliding mode, given by

$$S(y_r, y, \dot{y}) = y_r - y - T_c \dot{y}. \qquad (10.194)$$

Example 10.5 Polynomial controller for a motorised pendulum with robust pole placement

The same plant is used for as for Example 10.4, and the specified settling time and fast pole time constant are also the same, being set to $T_s = 1$ [s] and $T_{co} = 0.001$ [s], so that direct comparisons of performance may be made with the basic non-dynamic robust pole placement controller. The polynomial controller of minimum order is of first order and raises the system order to $N = 3$. Figure 10.35 shows the appropriate block diagrams.

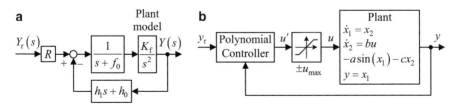

Fig. 10.35 Polynomial control system block diagrams. (**a**) System for pole assignment. (**b**) Control system for implementation

As for any controller, the implementation includes control saturation limits at $\pm u_{\max}$. The closed-loop transfer function for Fig. 10.35a is

$$\frac{Y(s)}{Y_r(s)} = \frac{RK_f}{s^3 + f_0 s^2 + K_f(h_1 s + h_0)} \qquad (10.195)$$

To attempt the same closed-loop dynamics as a sliding mode controller, the closed-loop characteristic polynomial is

$$s^3 + f_0 s^2 + K_f(h_1 s + h_0) = (s + p_c)(s + p_{cf})^2 = (s + p_c)\left(s^2 + 2p_{cf}s + p_{cf}^2\right)$$

$$= s^3 + (p_c + 2p_{cf})s^2 + \left(p_{cf}^2 + 2p_c p_{cf}\right)s + p_c p_{cf}^2, \qquad (10.196)$$

from which the polynomial controller parameters are

$$f_0 = p_c + 2p_{cf}, \quad h_1 = \frac{p_{cf}^2 + 2p_c p_{cf}}{K_f} \quad \text{and} \quad h_0 = \frac{p_c p_{cf}^2}{K_f}, \qquad (10.197)$$

where $p_c = 3/T_s$ and $p_{cf} = 1/T_{cf}$, where T_{cf} is the fast pole time constant. For a unity closed-loop DC gain, $R = h_0$.

Figure 10.36 shows a set of simulation results corresponding to those of Example 10.4 shown in Fig. 10.34. Comparison reveals them to be nearly identical.

This is due to both controllers being designed to yield closed-loop dynamics closely approximating that of the first-order system

$$\dot{y} = p_c(y_r - y) \qquad (10.198)$$

From the plant differential equation (10.187), the control variable is

$$u = \frac{1}{b}[\ddot{y} + a\sin(y) + c\dot{y}]. \qquad (10.199)$$

Since y and $\dot{y}$, and therefore $\ddot{y}$, are nearly governed by (10.198), according to (10.199), $u(t)$ will be nearly the same for both controllers, as is evident by comparing Fig. 10.34b and 10.36b although the controllers are different in structure. Comparing the phase-plane trajectories, however, reveals that the basic non-dynamic robust pole placement controller keeps the plant state closer to the virtual switching boundary than the polynomial controller. This is due to the fast mode of the polynomial controller being a second-order polynomial exponential mode with time constant, $T_{co} = 0.001$ [s], while the fast mode of the basic controller is only a first-order exponential mode with the same time constant. From a practical viewpoint, however, the polynomial controller can be considered to perform as well, but T_{co} could be reduced to bring the plant state closer to the switching boundary.

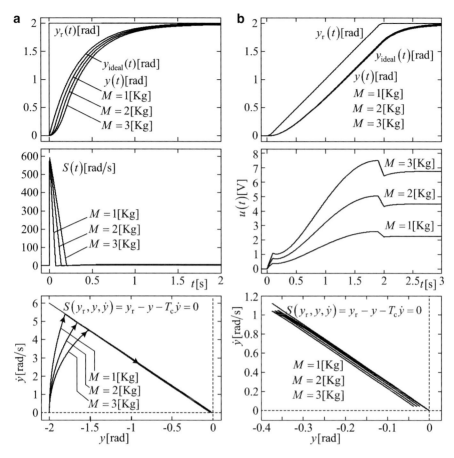

Fig. 10.36 Polynomial control of an inverted pendulum with robust pole placement. (**a**) Step response. (**b**) Ramp response

10.6 Multivariable Sliding Mode Control: An Introduction

10.6.1 Overview

Within limited space, an introduction is provided together with a simple method that should be feasible for many practical applications. This is followed by an approach to discrete sliding mode control of LTI plants that automatically includes multivariable plants. Although this is model based, it is able to accommodate relatively large plant parametric errors.

10.6.2 Simple Approach with Minimum Plant Information

For sliding mode control of SISO plants, the minimal information required is the relative degree. For multivariable plants, it is assumed that the number of controlled outputs is equal to the number of control variables. Then the minimal information is the relative degree with respect to each controlled output and also information that enables which control variable should be used to control which output. In many applications, this information will be known through knowledge of the plant hardware. Suppose that the following state-space model is given. Thus,

$$\dot{\mathbf{x}} = \mathbf{f}(\mathbf{x}, \mathbf{u} - \mathbf{d}) \qquad (10.200)$$

and

$$\mathbf{y} = \mathbf{h}(\mathbf{x}), \qquad (10.201)$$

where $\mathbf{x} \in \Re^n$ is the state vector, $\mathbf{u} \in \Re^m$ is the control vector $\mathbf{u} \in \Re^m$ and $\mathbf{d} \in \Re^m$ is an external disturbance vector referred to the control vector, and $\mathbf{f}(\cdot)$ and $\mathbf{h}(\cdot)$ are continuous functions of their arguments. The lowest-order output derivative components directly dependent on $\mathbf{u}$ may be written as

$$y_i^{(r_i)} = h_i(\mathbf{x}, \mathbf{u}_i), \quad i = 1, 2, \ldots, m, \qquad (10.202)$$

where r_i is the relative degree (Chap. 3) with respect to the output, y_i, and $\mathbf{u}_i$, $1, 2, \ldots, m$, represent sub-control vectors whose sets of elements may or may not comprise all m components of $\mathbf{u}$. For the multivariable SMC method presented here, each control input has to be paired with an output that it will control. It must therefore be possible to select a *different* component of $\mathbf{u}$ from each of $\mathbf{u}_i$, $1, 2, \ldots, m$. Let the selected control components be denoted u'_i, $1, 2, \ldots, m$, where $u'_i = u_k, k \in [1, 2, \ldots, m]$. Then it will be possible to form m sliding mode control loops of the same form as used for the SISO plants in the previous sections. For the standard sliding mode control method, the control law would comprise the following set of component control laws.

$$u'_i = u'_{i\,\max} \operatorname{sgn}\left[S_i\left(y_{ri}, y_i, \ldots, y_i^{(r_i-1)}\right)\right], \quad i = 1, 2, \ldots, m, \qquad (10.203)$$

Each control law component will treat the interaction from the remaining $m - 1$ control channels as disturbances, which it will reject in the sliding mode, since the set of switching functions contain no common variables. Then once every control channel is operating in the sliding mode, ideal multivariable control without interaction and with prescribed closed-loop dynamics will be attained. It is usual to refer to the complete set of switching boundaries,

$$S_i\left(y_{ri}, y_i, \ldots, y_i^{(r_i-1)}\right) = 0, \quad i = 1, 2, \ldots, m, \qquad (10.204)$$

10.6 Multivariable Sliding Mode Control: An Introduction

as a *switching manifold* or *sliding manifold*, as it comprises a set of nonintersecting surfaces in the state space.

It is highly probable that the set of control components, u'_i, $i = 1, 2, \ldots, m$, that can be selected from $\mathbf{u}_i$ is not unique, in which case the best set of control components should be determined such that u'_i has more influence than u'_k, $k \neq i$ on y_i.

As for SISO plants, if the plant is not of full relative degree, then the zero dynamics must be stable.

Example 10.6 Pseudo sliding mode large-angle attitude control of a rigid-body spacecraft

The attitude dynamics and kinematics model of the three-axis-stabilised rigid-body spacecraft of Chap. 7, Example 7.12, is taken, as it contains considerable state-dependent interaction due to the reaction wheel angular momentums and quaternion kinematics and control-dependent interaction due to the off-diagonal terms of the moment of inertia matrix. The plant inputs are the three reaction wheel drive voltages, u_1, u_2 and u_3, and the measured/controlled variables are the three quaternion components, $y_1 = q_1$, $y_2 = q_2$ and $y_3 = q_3$. The details of the plant model are given in Example 7.12. To determine the relative degrees, r_1, r_2 and r_3, the following output derivative equation was derived:

$$\begin{bmatrix} \ddot{y}_1 \\ \ddot{y}_2 \\ \ddot{y}_3 \end{bmatrix} = \frac{1}{2} \begin{bmatrix} \dot{q}_0 & -\dot{q}_3 & \dot{q}_2 \\ \dot{q}_3 & \dot{q}_0 & -\dot{q}_1 \\ -\dot{q}_2 & \dot{q}_1 & \dot{q}_0 \end{bmatrix} \begin{bmatrix} \omega_x \\ \omega_y \\ \omega_z \end{bmatrix} + \frac{1}{2} \underbrace{\begin{bmatrix} q_0 & -q_3 & q_2 \\ q_3 & q_0 & -q_1 \\ -q_2 & q_1 & q_0 \end{bmatrix}}_{\mathbf{Q}_3(t)} \underbrace{\begin{bmatrix} J_{xx} & J_{xy} & J_{xz} \\ J_{yx} & J_{yy} & J_{yz} \\ J_{zx} & J_{zy} & J_{zz} \end{bmatrix}^{-1}}_{\mathbf{J}}$$

$$\times \left\{ \begin{bmatrix} \Delta J_{yz}\omega_z\omega_y + \omega_z l_{wy} - \omega_y l_{wz} \\ \Delta J_{zx}\omega_x\omega_z + \omega_x l_{wz} - \omega_z l_{wx} \\ \Delta J_{xy}\omega_y\omega_x + \omega_y l_{wx} - \omega_x l_{wy} \end{bmatrix} + K_w \begin{bmatrix} u_1 \\ u_2 \\ u_3 \end{bmatrix} \right\} \quad (10.205)$$

By inspection, $r_1 = r_2 = r_3 = 2$. Due to the off-diagonal elements of the matrix, $\mathbf{Q}_3(t)\mathbf{J}$, in terms of (10.202), $\mathbf{u}_1 = \mathbf{u}_2 = \mathbf{u}_3 = \mathbf{u}$, and therefore, there are six possible selections of $\mathbf{u}'$, i.e.

$$\begin{bmatrix} u'_1 \\ u'_2 \\ u'_3 \end{bmatrix} = \begin{bmatrix} u_1 \\ u_2 \\ u_3 \end{bmatrix}, \begin{bmatrix} u_1 \\ u_3 \\ u_2 \end{bmatrix}, \begin{bmatrix} u_2 \\ u_3 \\ u_1 \end{bmatrix}, \begin{bmatrix} u_2 \\ u_1 \\ u_3 \end{bmatrix}, \begin{bmatrix} u_3 \\ u_1 \\ u_2 \end{bmatrix} \text{ or } \begin{bmatrix} u_3 \\ u_2 \\ u_1 \end{bmatrix}, \quad (10.206)$$

where u'_1, u'_2 and u'_3 are, respectively, paired with y_1, y_2 and y_3. To determine the best choice, the elements of the matrix, $\mathbf{Q}_3(t)\mathbf{J}$, have to be compared in magnitude. In this application, it is usual for $\mathbf{J}$ to be diagonally dominant, i.e. $|J_{ii}| > |J_{ij}|$, $j \neq i$, $i = 1, 2, 3$. At the initial zero attitude, the quaternion matrix, $\mathbf{Q}_3(0)$, is a unit matrix and for relatively small attitudes it also has diagonal dominance, as will $\mathbf{Q}_3(t)\mathbf{J}$. For large attitude angles, since $q_0^2 + q_1^2 + q_2^2 + q_3^2 = 1$ and $|q_i| \leq 1$, $i = 1, 2, 3$, $\mathbf{Q}_3(t)$, and therefore $\mathbf{Q}_3(t)\mathbf{J}$ will not be diagonally dominant. For small attitudes the first choice of $\mathbf{u}'$ is best. In this example, this choice will be maintained even for large attitudes and the system checked by simulation. In a real implementation, however, an interesting possibility is to switch between the choices of $\mathbf{u}'$ in (10.206) that will interchange the rows of $\mathbf{Q}_3(t)$ to maintain the diagonal dominance of $\mathbf{Q}_3(t)\mathbf{J}$. This could be called 'adaptive input-output pairing'.

A step response simulation with zero initial conditions will now be carried out using the boundary layer-based pseudo SMC law,

$$\left. \begin{array}{l} S_i\left(y_{ri},\, y_i,\, \dot{y}_i\right) = y_{ri} - y_i - \dfrac{T_s}{3}\dot{y}_i \\ u_i = K.\operatorname{sat}\left[S_i\left(y_{ri},\, y_i,\, \dot{y}_i\right),\, -u_{\max},\, +u_{\max}\right] \end{array} \right\},\quad i = 1,\, 2,\, 3 \qquad (10.207)$$

with $K = 1{,}000$ and $u_{\max} = 10$ [V] and a settling time in the sliding mode of $T_s = 50$ [s]. This control law is chosen since the control variables s closely approximate the equivalent controls of the standard SMC and their action in counteracting the interaction and the disturbances may therefore be seen. The stepped attitude reference inputs are given by

$$\mathbf{y}_r^T(t) = \left[y_{r1}(t)\ y_{r2}(t)\ y_{r3}(t) \right] = \left[0.4h\,(t-10)\ -0.4h\,(t-30)\ 0.3h\,(t-50) \right].$$

The different step delays are included to test the interaction rejection. Also a step disturbance torque of $\left[d_1(t)\ d_2(t)\ d_3(t) \right]^T = \left[2\,[\mathrm{V}]\ -1.5\,[\mathrm{V}]\ 1.8\,[\mathrm{V}] \right]^T h\,(t-100)$ is applied, which is typical of that caused by orbit change thruster misalignment with respect to the spacecraft centre of mass. The results are shown in Fig. 10.37. The attitude quaternions are dimensionless and therefore no units are shown with these variables.

First, the controlled outputs, $y_1(t)$, $y_2(t)$ and $y_3(t)$, shown in Fig. 10.37a lag behind the ideal first-order responses, $y_{1i}(t)$, $y_{2i}(t)$ and $y_{3i}(t)$, due to the control saturation that occurs upon the steps of the reference inputs, $y_{1r}(t)$, $y_{2r}(t)$ and $y_{3r}(t)$. This occurs in the step response of any sliding mode control system. The control saturation may be seen clearly in Fig. 10.37c. The other discontinuities in $u_1(t)$, $u_2(t)$ and $u_3(t)$ are due to counteraction of the control-dependent interaction. In other words, when one of the control variables jumps to $\pm u_{\max}$, this causes corresponding step interaction torques in the other two control channels that are immediately counteracted by the control variables of those channels. The desired close following of the switching boundaries by the substate trajectories upon reaching the boundaries is evident in Fig. 10.37b–d. The corresponding switching functions reach zero in a finite time after jumping to non-zero values upon application of the reference input steps, thereafter being maintained approximately zero, implying that the closed-loop dynamics is as desired.

It should be noted that in keeping with the previous sliding mode control systems presented in this chapter, control law (10.207) has been formulated without knowledge of the plant parameters, implying extreme robustness.

10.6 Multivariable Sliding Mode Control: An Introduction

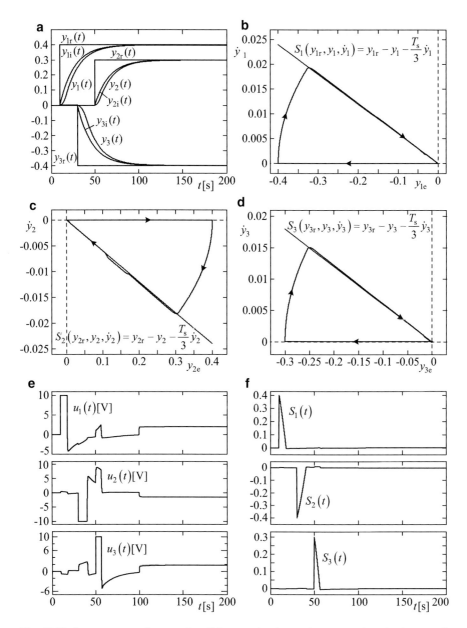

Fig. 10.37 Step response for pseudo sliding mode three-axis spacecraft attitude control. (**a**) Attitude quaternions. (**b**) Roll phase plane. (**c**) Pitch phase plane. (**d**) Yaw phase plane. (**e**) Control variables. (**f**) Switching functions

This robustness is the reason for negligible attitude error transients due the step disturbance torque, which are invisible in Fig. 10.37a. The steps in $u_1(t)$, $u_2(t)$ and $u_3(t)$ at $t = 100$ [s] counteracting the disturbance may be seen in Fig. 10.37c.

10.6.3 Discrete Sliding Mode Control

10.6.3.1 Introduction

Discrete sliding mode control is tailored especially for digital processor implementation and yields a piecewise constant control approximating the equivalent control of the ideal sliding mode. It therefore constitutes another method for eliminating control chatter, in addition to those of Sect. 10.4. In view of the desired robustness properties, the state variables required for the control law, like those of the previous sections, are the controlled outputs and their derivatives. If the plant is of full relative degree, this set of variables will be called the *output state*, otherwise the *output substate*. The switching boundary or manifold is formulated to yield the desired closed-loop dynamics when in the sliding mode, as previously. Since the control law is not of the switched type but operates in a *discrete sliding mode*, to be described shortly, the switching boundary, or manifold, will be referred to as a sliding boundary or manifold. Similarly the switching function will be referred to as the *sliding function*.

The discrete sliding mode control law will be developed for a general multivariable plant, this being applicable to SISO plants by setting the dimension of the control vector to $m = 1$.

10.6.3.2 Principle

A discrete model of the plant is used to calculate the constant value of the unconstrained control, $\mathbf{u}_b$, required to reach the sliding manifold in the sampling period, h. If every component of $\mathbf{u}_b$ lies within the control saturation limits, i.e. $-u_{i\,\max} < u_i < +u_{i\,\max}$, $i = 1, 2, \ldots, m$, then the real plant control, $\mathbf{u}$, is set equal to $\mathbf{u}_b$ and applied to the real plant for h seconds. If, on the other hand, any component, u_{bj}, of $\mathbf{u}_b$ exceeds these limits, then the corresponding component of $\mathbf{u}$ is set to $u_j = u_{j\,\max} \operatorname{sgn}(u_{bj})$ before it is applied. The process is repeated continually. If the plant model is exact and every component of $\mathbf{u}$ is within the saturation limits, then, the output state trajectory executes a motion roughly analogous to that of a bouncing ball on a hard horizontal surface with a horizontal velocity component. It departs by a small amount from the switching manifold at the beginning of each iteration, returning to the manifold at the end of each iteration, and therefore moves close to the path that would be taken in the ideal sliding mode, but touching the manifold at a discrete set of points. As $h \to 0$, the output state trajectory tends to the ideal sliding mode trajectory and $\mathbf{u} \to \mathbf{u}_{eq}$, where $\mathbf{u}_{eq}$ is the equivalent control.

10.6 Multivariable Sliding Mode Control: An Introduction

If, as will be the case in practice, there is a mismatch between the real plant and its model, the state trajectory will not coincide precisely with the switching manifold for $h > 0$ but will move in the vicinity of the manifold. This is a *discrete sliding mode*. Although the method is model based, the robustness against plant model parametric errors increases as h is reduced, analogous to the effect of reducing the boundary layer thickness of a pseudo sliding mode controller. It is therefore possible to accommodate large model mismatches to the extent of using a multiple integrator model for each control channel of order equal to the relative degree with respect to the output of that control channel, as was demonstrated in Sect. 10.5.3. If a more accurate model is available in a particular application, then it should be used, but for the discrete sliding mode control system, if the plant is not of full relative degree, then a lower-order model of full relative degree must be used with the relative degree with respect to each output matching that of the real plant.

10.6.3.3 Derivation of the Discrete SMC Algorithm

The control law will be formulated using the general LTI plant model. First the continuous state-space model will be used to obtain the equation for the output derivative state variables that are needed to implement the sliding manifold. Then the control law will be derived using the corresponding discrete state-space model with the selected iteration period, h.

The continuous-time LTI state-space model is

$$\dot{\mathbf{x}} = \mathbf{A}\mathbf{x} + \mathbf{B}\mathbf{u} \qquad (10.208)$$

$$\mathbf{y} = \mathbf{C}\mathbf{x} \qquad (10.209)$$

where $\mathbf{A} \in \Re^{n \times n}$, $\mathbf{B} \in \Re^{n \times m}$ and $\mathbf{C} \in \Re^{m \times n}$. The state representation is arbitrary but the relative degree, r_i, $i = 1, 2, \ldots, m$, with respect to each output, should match that of the physical plant. Thus, an available plant model can be used if the plant is known to be of full relative degree.

The procedure to obtain the output derivative state is the one based on Lie derivatives to find the relative degrees (Chap. 3). It will also be recalled that this was used to derive the forced dynamic control laws for LTI plants in Chap. 7, Sect. 7.3.4. Essentially each output, y_i, $i = 1, 2, \ldots, m$, is repeatedly differentiated and the state differential equations used to eliminate state-variable derivatives appearing on the RHS, until the derivative of order, r_i, directly depends on any control input component. First, the measurement Eq. (10.209) is written in the component form.

Then the repeated differentiation procedure yields the complete set of output derivative state variables as

$$y_i = \mathbf{c}_i^T \mathbf{x}, \quad \dot{y}_i = \mathbf{c}_i^T \mathbf{A}\mathbf{x}, \quad \ldots, \quad y_i^{(r_i-1)} = \mathbf{A}^{r_i-1}\mathbf{x}, \quad i = 1, 2, \ldots, m. \qquad (10.210)$$

To facilitate the discrete sliding mode control law derivation, these derivatives are arranged as follows.

$$\mathbf{y}_i \triangleq \begin{bmatrix} y_i \\ \dot{y}_i \\ \vdots \\ y_i^{(r_i-1)} \end{bmatrix} = \begin{bmatrix} \mathbf{c}_i^T \mathbf{x} \\ \mathbf{c}_i^T \mathbf{A} \mathbf{x} \\ \vdots \\ \mathbf{c}_i^T \mathbf{A}^{r_i-1} \mathbf{x} \end{bmatrix} = \begin{bmatrix} \mathbf{c}_i^T \\ \mathbf{c}_i^T \mathbf{A} \\ \vdots \\ \mathbf{c}_i^T \mathbf{A}^{r_i-1} \end{bmatrix} \mathbf{x} \triangleq [\mathbf{C}]_i \mathbf{x}, \quad i = 1, 2, \ldots, m. \tag{10.211}$$

The sliding manifold comprises m sliding boundaries, each of which operates on one of the output derivative substate vectors of (10.211) together with the reference input $\dot{\mathbf{y}}$. The sliding boundary equations are of the same form as previously. Thus,

$$y_{ri} - y_i - w_{1i}\dot{y}_i + \cdots + w_{r_i-1, i} y_i^{(r_i-1)} = 0, \quad i = 1, 2, \ldots, m. \tag{10.212}$$

These may be written in terms of the substate vectors, $\mathbf{y}_i$, of (10.211) as

$$y_{ri} - \mathbf{w}_i^T \mathbf{y}_i = 0, \quad i = 1, 2, \ldots, m. \tag{10.213}$$

where $\mathbf{w}_i^T = \begin{bmatrix} 1 & w_{1i} & w_{2i} & \cdots & w_{r_i-1, i} \end{bmatrix}$. As previously, $\mathbf{w}_i$, $i = 1, 2, \ldots, m$, are chosen to yield the desired closed-loop dynamics by pole assignment. As in Sect. 10.6.2, the control law will eliminate interaction between the ith reference input and the jth output, $i \neq j$, because there are no common output state variables between the switching boundary equations, meaning that the boundaries are nonintersecting.

The following description of the control system operation pertains to the hypothetical case of the plant model of (10.208) and (10.209) being perfectly matched to the real plant. Commencing with an arbitrary initial state, the discrete sliding mode control law is designed to make the output state trajectory reach the sliding manifold *in just one iteration period of h seconds* by applying a constant control vector, provided all of the required components lie within the control saturation limits. After the output state trajectory reaches the sliding manifold, the control law recalculates the control to reach the manifold again in a further h seconds. If the control vector is non-saturated, the output state trajectory will move away from the sliding manifold during the iteration period by an amount that increases with h and return to it at the end of the period.

The discrete state-space plant model corresponding to the continuous model of (10.208) and (10.209) is

$$\mathbf{x}(k+1) = \mathbf{\Phi}(h)\mathbf{x}(k) + \mathbf{\Psi}(h)\mathbf{u}(k) \tag{10.214}$$

$$\mathbf{y}(k) = \mathbf{C}\mathbf{x}(k) \tag{10.215}$$

10.6 Multivariable Sliding Mode Control: An Introduction

where $\mathbf{\Phi}(h) = e^{\mathbf{A}h}$, $\mathbf{\Psi}(h) = \int_0^h \mathbf{\Phi}(\tau) \mathbf{B} \, d\tau$, k is the iteration index. From (10.211),

$$\mathbf{y}_i(k+1) = [\mathbf{C}]_i \mathbf{x}(k+1). \tag{10.216}$$

If the switching manifold is reached at the end of each iteration of the control law, then, from (10.213)

$$y_{ri}(k+1) - \mathbf{w}_i^T \mathbf{y}_i(k+1) = 0, \quad i = 1, 2, \ldots, m. \tag{10.217}$$

Substituting for $\mathbf{y}_i(k+1)$ in (10.217) using (10.216) gives

$$y_{ri}(k+1) - \mathbf{w}_i^T [\mathbf{C}]_i \mathbf{x}(k+1) = 0 \tag{10.218}$$

Then substituting for $\mathbf{x}(k+1)$ in (10.218) using (10.214) yields

$$y_{ri}(k+1) - \mathbf{w}_i^T [\mathbf{C}]_i [\mathbf{\Phi}(h)\mathbf{x}(k) + \mathbf{\Psi}(h)\mathbf{u}(k)] = 0 \Rightarrow$$
$$y_{ri}(k+1) - \mathbf{w}_i^T [\mathbf{C}]_i \mathbf{\Phi}(h)\mathbf{x}(k) = \mathbf{w}_i^T [\mathbf{C}]_i \mathbf{\Psi}(h)\mathbf{u}(k) \Rightarrow$$

$$\begin{bmatrix} y_{r1}(k+1) \\ y_{r2}(k+1) \\ \vdots \\ y_{rm}(k+1) \end{bmatrix} - \begin{bmatrix} \mathbf{w}_1^T [\mathbf{C}]_1 \\ \mathbf{w}_2^T [\mathbf{C}]_2 \\ \vdots \\ \mathbf{w}_m^T [\mathbf{C}]_m \end{bmatrix} \mathbf{\Phi}(h) \mathbf{x}(k) = \begin{bmatrix} \mathbf{w}_1^T [\mathbf{C}]_1 \\ \mathbf{w}_2^T [\mathbf{C}]_2 \\ \vdots \\ \mathbf{w}_m^T [\mathbf{C}]_m \end{bmatrix} \mathbf{\Psi}(h) \mathbf{u}(k) \tag{10.219}$$

Since the reference input vector is an external input, $y_{ri}(k+1)$ may be replaced by $y_{ri}(k)$ without affecting the closed-loop dynamics. Solving (10.219) for $\mathbf{u}(k)$ then yields the unconstrained boundary reaching control, $\mathbf{u}_b(k)$, as follows.

$$\mathbf{u}_b(k) = \left[\begin{bmatrix} \mathbf{w}_1^T [\mathbf{C}]_1 \\ \mathbf{w}_2^T [\mathbf{C}]_2 \\ \vdots \\ \mathbf{w}_m^T [\mathbf{C}]_m \end{bmatrix} \mathbf{\Psi}(h) \right]^{-1} \left[\begin{bmatrix} y_{r1}(k) \\ y_{r2}(k) \\ \vdots \\ y_{rm}(k) \end{bmatrix} - \begin{bmatrix} \mathbf{w}_1^T [\mathbf{C}]_1 \\ \mathbf{w}_2^T [\mathbf{C}]_2 \\ \vdots \\ \mathbf{w}_m^T [\mathbf{C}]_m \end{bmatrix} \mathbf{\Phi}(h) \mathbf{x}(k) \right]. \tag{10.220}$$

This, however, is a state feedback control law using the state, $\mathbf{x}$, with an arbitrary state representation, which would require an observer with an accurate plant model for state estimation. Variants of control law (10.220) are possible, however, which use only the output derivative vector, $\mathbf{Y}$, defined as

$$\mathbf{Y} = \begin{bmatrix} \mathbf{y}_1^T & \mathbf{y}_2^T & \cdots & \mathbf{y}_m^T \end{bmatrix}^T. \tag{10.221}$$

Then from (10.211),

$$\mathbf{Y} = \left[\mathbf{x}^T [\mathbf{C}]_1^T \; \mathbf{x}^T [\mathbf{C}]_2^T \; \cdots \; \mathbf{x}^T [\mathbf{C}]_m^T \right]^T = \left[\mathbf{x}^T \left[[\mathbf{C}]_1^T \; [\mathbf{C}]_2^T \; \cdots \; [\mathbf{C}]_m^T \right] \right]^T$$
$$= \left[[\mathbf{C}]_1^T \; [\mathbf{C}]_2^T \; \cdots \; [\mathbf{C}]_m^T \right]^T \mathbf{x}$$

which may be written as

$$\mathbf{Y} = [\mathbf{C}] \mathbf{x}. \tag{10.222}$$

Since the plant model of (10.208) and (10.209) is of full relative degree, the output derivative matrix, [**C**], is square and non-singular. Hence

$$\mathbf{x} = [\mathbf{C}]^{-1}\mathbf{Y} \Rightarrow \mathbf{x}(k) = [\mathbf{C}]^{-1}\mathbf{Y}(k). \tag{10.223}$$

This enables the state, **x**, to be eliminated from (10.220), yielding

$$\mathbf{u}_b(k) = \begin{bmatrix} \mathbf{w}_1^T[\mathbf{C}]_1 \\ \hdashline \mathbf{w}_2^T[\mathbf{C}]_2 \\ \hdashline \vdots \\ \hdashline \mathbf{w}_m^T[\mathbf{C}]_m \end{bmatrix}^{-1} \left(\begin{bmatrix} y_{r1}(k) \\ y_{r2}(k) \\ \vdots \\ y_{rm}(k) \end{bmatrix} - \begin{bmatrix} \mathbf{w}_1^T[\mathbf{C}]_1 \\ \hdashline \mathbf{w}_2^T[\mathbf{C}]_2 \\ \hdashline \vdots \\ \hdashline \mathbf{w}_m^T[\mathbf{C}]_m \end{bmatrix} \mathbf{\Phi}(h)[\mathbf{C}]^{-1}\mathbf{Y}(k) \right). \tag{10.224}$$

If the iteration period, h, is sufficiently small for $\mathbf{\Phi}(h) \cong \mathbf{I}_n$, however, a simpler control law may be formed with the aid of the following manipulation:

$$\begin{bmatrix} \mathbf{w}_1^T[\mathbf{C}]_1 \\ \hdashline \mathbf{w}_2^T[\mathbf{C}]_2 \\ \hdashline \vdots \\ \hdashline \mathbf{w}_m^T[\mathbf{C}]_m \end{bmatrix} = \begin{bmatrix} \mathbf{w}_1^T & 0 & \cdots & 0 \\ \hdashline 0 & \mathbf{w}_2^T & \ddots & \vdots \\ \hdashline \vdots & \ddots & \ddots & 0 \\ \hdashline 0 & \cdots & 0 & \mathbf{w}_m^T \end{bmatrix} \begin{bmatrix} [\mathbf{C}]_1 \\ [\mathbf{C}]_2 \\ \vdots \\ [\mathbf{C}]_m \end{bmatrix} = \begin{bmatrix} \mathbf{w}_1^T & 0 & \cdots & 0 \\ \hdashline 0 & \mathbf{w}_2^T & \ddots & \vdots \\ \hdashline \vdots & \ddots & \ddots & 0 \\ \hdashline 0 & \cdots & 0 & \mathbf{w}_m^T \end{bmatrix} [\mathbf{C}]. \tag{10.225}$$

Then in view of (10.221), (10.224) simplifies to

$$\mathbf{u}_b(k) = \underbrace{\begin{bmatrix} \mathbf{w}_1^T[\mathbf{C}]_1 \\ \hdashline \mathbf{w}_2^T[\mathbf{C}]_2 \\ \hdashline \vdots \\ \hdashline \mathbf{w}_m^T[\mathbf{C}]_m \end{bmatrix}^{-1} \mathbf{\Psi}(h)}_{\text{Gain matrix, } \mathbf{K}} \underbrace{\begin{bmatrix} y_{r1}(k) - \mathbf{w}_1^T \mathbf{y}_1(k) \\ y_{r2}(k) - \mathbf{w}_2^T \mathbf{y}_2(k) \\ \vdots \\ y_{rm}(k) - \mathbf{w}_m^T \mathbf{y}_m(k) \end{bmatrix}}_{\text{Sliding functions}}. \tag{10.226}$$

10.6 Multivariable Sliding Mode Control: An Introduction

Since $\boldsymbol{\Psi}(h) \to \boldsymbol{0}$ as $h \to 0$, the gain matrix, $\mathbf{K}$, becomes large and enables the control law to drive the sliding functions to negligible proportions, thereby keeping the output state trajectory close to the sliding boundaries, as required.

Finally, to implement the control saturation limits for actuator protection, the required discrete sliding mode control comprises (10.224) or (10.226) and

$$u_i = \begin{cases} u_{pi} \text{ if } |u_{pi}| < u_{max} \\ u_{max} \, \text{sgn}\left(u_{pi}\right) \text{ if } |u_{pi}| \geq u_{max} \end{cases}. \tag{10.227}$$

10.6.3.4 Discrete Derivative Estimation

If output derivative estimates are required, a discrete observer can be used based on a linear plant model if it is available. If only the relative degree of the plant is known with respect to each output, then a discrete version of the multiple integrator observer of Sect. 10.1.3.2 may be used, a separate observer being used for each output of a multivariable plant with the number of integrators in the chain being equal to the relative degree, r_i. It will be recalled from Sect. 10.1.3.2 that the correction loop settling time was made as small as possible to minimise the errors due to the mismatching between the real plant and the model. Essentially, the relatively large correction loop gains drive the error between the measurement, y_i, and the model output, $\hat{y}_i$, to negligible proportions so that the outputs of the integrators in the chain approximate the required derivatives. In the discrete case, the minimum settling time that can be attained is the dead-beat settling time of $r_i h$. As an example, the gains of a discrete observer with a triple integrator model, for $r_i = 3$, will now be derived. The observer equation is

$$\begin{bmatrix} \hat{y}_i(k+1) \\ \hat{\dot{y}}_i(k+1) \\ \hat{\ddot{y}}_i(k+1) \end{bmatrix} = \begin{bmatrix} 1 & h & \tfrac{1}{2}h^2 \\ 0 & 1 & h \\ 0 & 0 & 1 \end{bmatrix} \begin{bmatrix} \hat{y}_i(k) \\ \hat{\dot{y}}_i(k) \\ \hat{\ddot{y}}_i(k) \end{bmatrix} + b_i \begin{bmatrix} \tfrac{1}{6}h^3 \\ \tfrac{1}{2}h^2 \\ h \end{bmatrix} u_q(k) + \begin{bmatrix} l_1 \\ l_2 \\ l_3 \end{bmatrix} [y_i(k) - \hat{y}_i(k)]$$
(10.228)

where is the control input upon which $\dddot{y}$ is known to directly depend and b_i is the triple integrator forward path gain provided as an adjustable parameter to minimise the error, $y_i(k) - \hat{y}_i(k)$. The observer characteristic equation is

$$\begin{vmatrix} z-1+l_1 & -h & -\tfrac{1}{2}h^2 \\ l_2 & z-1 & -h \\ l_3 & 0 & z-1 \end{vmatrix} = (z-1)\left[(z-1+l_1)(z-1)+l_2 h\right] + l_3\left[h^2 + \tfrac{1}{2}h^2(z-1)\right]$$
$$= u_q(k)(z-1)^3 + l_1(z-1)^2 + \left(l_2 h + \tfrac{1}{2}l_3 h^2\right)(z-1) + l_3 h^2 = 0$$
(10.229)

For dead-beat settling, this should be simply $z^3 = 0$, which, in terms of $z - 1$, is

$$[(z-1)+1]^3 = (z-1)^3 + 3(z-1)^2 + 3(z-1) + 1 = 0 \qquad (10.230)$$

Equating the RHS of (10.229) then yields

$$l_1 = 3, \quad l_3 = 1/h^2 \quad \text{and} \quad l_2 = 5/(2h) \qquad (10.231)$$

10.6.3.5 Simulations

First, to illustrate the form of the state trajectory in the discrete sliding mode, a simulation will be carried out for the double integrator plant,

$$\begin{bmatrix} \dot{x}_1 \\ \dot{x}_2 \end{bmatrix} = \begin{bmatrix} 0 & 1 \\ 0 & 0 \end{bmatrix} \begin{bmatrix} x_1 \\ x_2 \end{bmatrix} + \begin{bmatrix} 0 \\ b \end{bmatrix} u, \quad y = \begin{bmatrix} 1 & 0 \end{bmatrix} \begin{bmatrix} x_1 \\ x_2 \end{bmatrix}, \qquad (10.232)$$

first with a perfectly matched plant model with a forward path gain of $b_m = b$ and then mismatching the model, first with $b_m = 1.4b$ and then with $b_m = 0.6b$.

The relative degree is $r = 2$. The output derivative equation is therefore

$$\mathbf{Y} = \begin{bmatrix} y \\ \dot{y} \end{bmatrix} = \underbrace{\begin{bmatrix} 1 & 0 \\ 0 & 1 \end{bmatrix}}_{[\mathbf{C}]} \begin{bmatrix} x_1 \\ x_2 \end{bmatrix} \qquad (10.233)$$

The discrete plant model corresponding to (10.232) is

$$\begin{bmatrix} x_1(k+1) \\ x_2(k+1) \end{bmatrix} = \underbrace{\begin{bmatrix} 1 & h \\ 0 & 1 \end{bmatrix}}_{\boldsymbol{\Phi}(h)} \begin{bmatrix} x_1(k) \\ x_2(k) \end{bmatrix} + \underbrace{\begin{bmatrix} \tfrac{1}{2}bh^2 \\ bh \end{bmatrix}}_{\boldsymbol{\psi}(h)} u(k), \quad y(k) = \begin{bmatrix} 1 & 0 \end{bmatrix} \begin{bmatrix} x_1(k) \\ x_2(k) \end{bmatrix}.$$

$$(10.234)$$

The sliding boundary equation is

$$y_r - \underbrace{\begin{bmatrix} 1 & w_1 \end{bmatrix}}_{\mathbf{w}^T} \underbrace{\begin{bmatrix} y \\ \dot{y} \end{bmatrix}}_{\mathbf{Y}} - 0. \qquad (10.235)$$

Then the exact boundary reaching control law (10.224) is

$$\begin{aligned} u_b(k) &= \left[\mathbf{w}^T \boldsymbol{\psi}(h)\right]^{-1} \left[y_r(k) - \mathbf{w}^T [\mathbf{C}]\boldsymbol{\Phi}(h)\mathbf{Y}(k)\right] \\ &= \left[\begin{bmatrix} 1 & w_1 \end{bmatrix} \begin{bmatrix} \tfrac{1}{2}bh^2 \\ bh \end{bmatrix}\right]^{-1} \left[y_r(k) - \begin{bmatrix} 1 & w_1 \end{bmatrix} \begin{bmatrix} 1 & 0 \\ 0 & 1 \end{bmatrix} \begin{bmatrix} 1 & h \\ 0 & 1 \end{bmatrix} \begin{bmatrix} y(k) \\ \dot{y}(k) \end{bmatrix}\right] \\ &= \tfrac{2}{bh(h+2w_1)} \left[y_r(k) - y(k) - (w_1 + h)\dot{y}(k)\right] \end{aligned}$$

$$(10.236)$$

10.6 Multivariable Sliding Mode Control: An Introduction

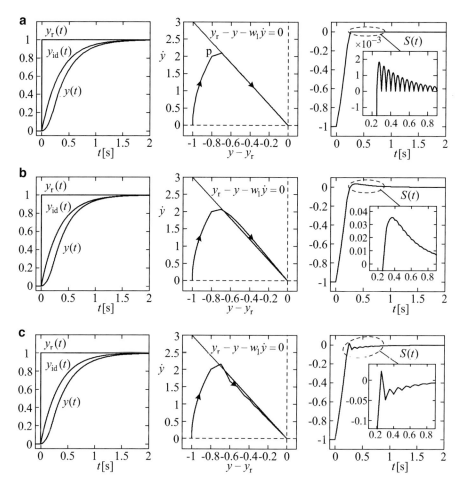

Fig. 10.38 Discrete sliding mode control of a double integrator plant. (**a**) Perfectly matched plant model: $b_m = b$. (**b**) Mismatched plant model: $b_m = 1.4b$. (**c**) Mismatched plant model: $b_m = 0.6b$

In the simulation, $\dot{y}$ is assumed available as a direct measurement. The results are shown in Fig. 10.38. The period of initial control saturation can, as usual, be detected in the phase-plane trajectories by the continuously curved segment approaching the sliding boundary, in this case, a parabolic curve. In Fig. 10.38a, the control saturation at $u = +u_{\max}$ ceases at point p on the first iteration of the control algorithm where $|u_b| < u_{\max}$ and therefore $u = u_b$. The trajectory changes direction due to this change of u and reaches the sliding boundary at the end of the iteration. This segment may be seen quite clearly in Fig. 10.38a, but this is due to the relatively long iteration period of $h = 0.05$ and would be almost invisible in many cases. After reaching the sliding boundary, the state trajectory appears to follow it perfectly, but the 'bouncing ball' behaviour described previously may be seen clearly in the magnified inset graph of $S(t)$. The fact that this only occurs with a well-matched plant model is evident in $S(t)$ of Fig. 10.38b, c. The oscillatory $S(t)$ for plant parameter mismatching in one direction and non-oscillatory $S(t)$ for mismatching in the other direction is often observed.

In both cases, however, $S(t)$ is driven to relatively small proportions in about the same time as for the perfectly matched plant model. The result is that despite the large $\pm 40\%$ plant model mismatches, $y(t)$ is not noticeably affected, as shown, indicating considerable robustness even with the relatively long iteration period of $h = 0.05$ [s].

Next, a multivariable control example is presented that demonstrates the ability of the control technique to eliminate interaction between the control channels as well as yielding the prescribed dynamic responses to the reference inputs.

Example 10.7 Discrete sliding mode control of gantry crane and cradle

In this application, the gantry crane consists of a truck running on an overhead rail carrying a payload cradle, as illustrated in Fig. 10.39.

The truck position and the cradle swing angle are to be actively controlled. The relevant quantities, including those indicated in the figure, are defined in Table 10.1.

Applying the Lagrangian method of Chap. 2, Sect. 2.2.2.8, yields the following plant differential equations:

$$\left. \begin{array}{l} \ddot{x} = a\ddot{\theta} + bf \\ \ddot{\theta} = c\ddot{x} - d\theta + e\gamma \end{array} \right\}, \qquad (10.237)$$

where $a = \frac{M_2 R}{M_1 + M_2}$, $b = \frac{1}{M_1 + M_2}$, $c = \frac{1}{R}$, $d = \frac{g}{R}$ and $e = \frac{1}{M_2 R^2}$. To form a state-space model, Eq. (10.237) is solved algebraically for $\ddot{x}$ and $\ddot{\theta}$ to obtain

$$\left. \begin{array}{l} \ddot{x} = -a\omega_n^2 \theta + b_{11} u_1 + b_{12} u_2 \\ \ddot{\theta} = -\omega_n^2 \theta + b_{21} u_1 + b_{22} u_2 \end{array} \right\}, \qquad (10.238)$$

where $\omega_n^2 = \frac{d}{1-ac}$, $b_{11} = \frac{K_f b}{1-ac}$, $b_{12} = \frac{K_t a e}{1-ac}$, $b_{21} = \frac{K_f c b}{1-ac}$ and $b_{22} = \frac{K_t e}{1-ac}$.

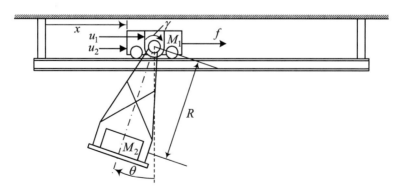

Fig. 10.39 Controlled overhead gantry crane and cradle

10.6 Multivariable Sliding Mode Control: An Introduction

Table 10.1 Quantities of gantry crane and cradle system

Quantity	Description	Units
M_1	Truck mass	Kg
M_2	Combined mass of payload and cradle	Kg
R	Radius of gyration of payload and cradle	m
u_1	Truck drive control input	V
$f = K_f u_1$	Truck control force	N
K_f	Truck drive force constant	N/V
x	Truck displacement	m
u_2	Cradle swing drive control input	V
$\gamma = K_T u_2$	Cradle control torque	Nm
K_T	Cradle swing drive torque constant	Nm/V
θ	Cradle swing angle	rad

If the state variables are chosen as $x_1 = x$, $x_2 = \dot{x}$, $x_3 = \theta$ and $x_4 = \dot{\theta}$, the state-space model is as follows:

$$\begin{bmatrix} \dot{x}_1 \\ \dot{x}_2 \\ \dot{x}_3 \\ \dot{x}_4 \end{bmatrix} = \underbrace{\begin{bmatrix} 0 & 1 & 0 & 0 \\ 0 & 0 & -a\omega_n^2 & 0 \\ 0 & 0 & 0 & 1 \\ 0 & 0 & -\omega_n^2 & 0 \end{bmatrix}}_{A} \begin{bmatrix} x_1 \\ x_2 \\ x_3 \\ x_4 \end{bmatrix} + \underbrace{\begin{bmatrix} 0 & 0 \\ b_{11} & b_{12} \\ 0 & 0 \\ b_{21} & b_{22} \end{bmatrix}}_{B} \begin{bmatrix} u_1 \\ u_2 \end{bmatrix} \qquad (10.239)$$

and

$$\begin{bmatrix} y_1 \\ y_2 \end{bmatrix} = \underbrace{\begin{bmatrix} 1 & 0 & 0 & 0 \\ 0 & 0 & 1 & 0 \end{bmatrix}}_{C} \begin{bmatrix} x_1 \\ x_2 \\ x_3 \\ x_4 \end{bmatrix}, \qquad (10.240)$$

The reason for considering sliding mode control for this application is that M_2 changes whenever the payload is changed and a robust control technique is needed to guarantee a specified closed-loop dynamics.

It may be seen from (10.238) that the relative degrees with respect to y_1 and y_2 are, respectively, $r_1 = 2$ and $r_2 = 2$. The output derivative equations are therefore

$$\mathbf{y}_1 = [C]_1 \mathbf{x}, \text{ i.e., } \begin{bmatrix} y_1 \\ \dot{y}_1 \end{bmatrix} = \begin{bmatrix} 1 & 0 & 0 & 0 \\ 0 & 1 & 0 & 0 \end{bmatrix} \begin{bmatrix} x_1 \\ x_2 \\ x_3 \\ x_4 \end{bmatrix} \qquad (10.241)$$

and

$$\mathbf{y}_2 = [\mathbf{C}]_2\mathbf{x}, \text{ i.e., } \begin{bmatrix} y_2 \\ \dot{y}_2 \end{bmatrix} = \begin{bmatrix} 0 & 0 & 1 & 0 \\ 0 & 0 & 0 & 1 \end{bmatrix} \begin{bmatrix} x_1 \\ x_2 \\ x_3 \\ x_4 \end{bmatrix}. \quad (10.242)$$

If the specified settling times in the sliding mode for $y_1(t)$ and $y_2(t)$ are, respectively, T_{s1} and T_{s2} (5 % criterion), then the sliding boundary coefficient matrices are

$$\mathbf{w}_1^T = \begin{bmatrix} 1 & w_1 \end{bmatrix} \text{ and } \mathbf{w}_2^T = \begin{bmatrix} 1 & w_2 \end{bmatrix}, \quad (10.243)$$

where $w_1 = T_{s1}/3$ and $w_2 = T_{s2}/3$.

Next, the discrete state difference equation,

$$\mathbf{x}(k+1) = \mathbf{\Phi}(h)\mathbf{x}(k) + \mathbf{\Psi}(h)\mathbf{u}(k), \quad (10.244)$$

corresponding to the state differential equation of (10.239), has to be derived. This will be done using the method of Chap. 3, Sect. 3.4.2.1. First the characteristic polynomial of the plant is needed. This is

$$|s\mathbf{I}_4 - \mathbf{A}| = \begin{vmatrix} s & -1 & 0 & 0 \\ 0 & s & a\omega_n^2 & 0 \\ 0 & 0 & s & -1 \\ 0 & 0 & \omega_n^2 & s \end{vmatrix} = s \begin{vmatrix} s & a\omega_n^2 & 0 \\ 0 & s & -1 \\ 0 & \omega_n^2 & s \end{vmatrix} = s^2 \left(s^2 + \omega_n^2\right). \quad (10.245)$$

So the plant has a polynomial exponential mode having basis functions, $e^{-t/T}$ and $te^{-t/T}$ with $T \to \infty$, i.e. 1 and t, and an undamped oscillatory mode with basis functions, $\sin(\omega_n t)$ and $\cos(\omega_n t)$. The discrete system matrix is therefore

$$\mathbf{\Phi}(h) = \mathbf{M}_0 \cdot 1 + \mathbf{M}_1 h + \mathbf{M}_2 \sin(\omega h) + \mathbf{M}_3 \cos(\omega h) \quad (10.246)$$

where $\mathbf{M}_i$, $i = 1, 2, 3, 4$, are constant matrices to be determined. In preparation, differentiating three times w.r.t. h yields

$$\dot{\mathbf{\Phi}}(h) = \mathbf{M}_1 + \omega_n \mathbf{M}_2 \cos(\omega_n h) - \omega_n \mathbf{M}_3 \sin(\omega_n h) \quad (10.247)$$

$$\ddot{\mathbf{\Phi}}(h) = -\omega_n^2 \mathbf{M}_2 \sin(\omega_n h) - \omega_n^2 \mathbf{M}_3 \cos(\omega_n h) \quad (10.248)$$

and

$$\dddot{\mathbf{\Phi}}(h) = -\omega_n^3 \mathbf{M}_2 \cos(\omega_n h) + \omega_n^3 \mathbf{M}_3 \sin(\omega_n h). \quad (10.249)$$

10.6 Multivariable Sliding Mode Control: An Introduction

Then the simultaneous equations for the matrices are determined as follows.
From (10.246),

$$\mathbf{\Phi}(0) = \mathbf{I}_4 \Rightarrow \mathbf{M}_0 + \mathbf{M}_3 = \mathbf{I}_4. \tag{10.250}$$

From (10.247),

$$\dot{\mathbf{\Phi}}(0) = \mathbf{A} \Rightarrow \mathbf{M}_1 + \omega_n \mathbf{M}_2 = \mathbf{A}. \tag{10.251}$$

From (10.248),

$$\ddot{\mathbf{\Phi}}(0) = \mathbf{A}^2 \Rightarrow -\omega_n^2 \mathbf{M}_3 = \mathbf{A}^2. \tag{10.252}$$

From (10.249),

$$\dddot{\mathbf{\Phi}}(0) = \mathbf{A}^3 \Rightarrow -\omega_n^3 \mathbf{M}_2 = \mathbf{A}^3. \tag{10.253}$$

Then from (10.252) and (10.253),

$$\mathbf{M}_3 = -\frac{1}{\omega_n^2}\mathbf{A}^2, \quad \mathbf{M}_2 = -\frac{1}{\omega_n^3}\mathbf{A}^3, \tag{10.254}$$

and then from (10.250) and (10.251),

$$\mathbf{M}_0 = \mathbf{I}_4 - \mathbf{M}_3 \quad \text{and} \quad \mathbf{M}_1 = \mathbf{A} - \omega_n \mathbf{M}_2. \tag{10.255}$$

Using (10.246), the discrete drive matrix is

$$\begin{aligned}\mathbf{\Psi}(h) &= \int_0^h \mathbf{\Phi}(\tau) \mathbf{B} \, d\tau = \int_0^h [\mathbf{M}_0 + \mathbf{M}_1 h + \mathbf{M}_2 \sin(\omega_n h) + \mathbf{M}_3 \cos(\omega_n h)] \mathbf{B} \, d\tau \\ &= \left[\mathbf{M}_0 h + \tfrac{1}{2}\mathbf{M}_1 h^2 + \tfrac{1}{\omega_n}[1 - \cos(\omega_n h)] \mathbf{M}_2 + \tfrac{1}{\omega_n}[\sin(\omega_n h)]\mathbf{M}_3\right] \mathbf{B}\end{aligned} \tag{10.256}$$

The above equations will be used to evaluate $\mathbf{\Phi}(h)$ and $\mathbf{\Psi}(h)$ numerically.
The discrete sliding mode control law (10.224) for this example is

$$\begin{bmatrix}u_{b1}(k)\\u_{b2}(k)\end{bmatrix} = \underbrace{\left[\begin{bmatrix}\mathbf{w}_1^T[\mathbf{C}]_1\\\mathbf{w}_2^T[\mathbf{C}]_2\end{bmatrix}\mathbf{\Psi}(h)\right]^{-1}}_{\text{Gain matrix, }\mathbf{K}}\left\{\begin{bmatrix}y_{r1}(k)\\y_{r2}(k)\end{bmatrix} - \begin{bmatrix}\mathbf{w}_1^T[\mathbf{C}]_1\\\mathbf{w}_2^T[\mathbf{C}]_2\end{bmatrix}\mathbf{\Phi}(h)\begin{bmatrix}[\mathbf{C}]_1\\[\mathbf{C}]_2\end{bmatrix}^{-1}\begin{bmatrix}\mathbf{y}_1(k)\\\mathbf{y}_2(k)\end{bmatrix}\right\}.$$

$$\tag{10.257}$$

The parameter values are given in Table 10.2.

Table 10.2 Plant parameters for gantry crane and cradle control simulation

Parameter	Value	Units	Parameter	Value	Units
M_1		Kg	M_2	Empty cradle: 100 Maximum: 300	Kg
K_f	10	N/V	K_t	10	Nm/V
R	3	m			

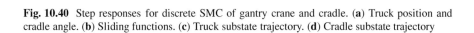

Fig. 10.40 Step responses for discrete SMC of gantry crane and cradle. (**a**) Truck position and cradle angle. (**b**) Sliding functions. (**c**) Truck substate trajectory. (**d**) Cradle substate trajectory

The control saturation limits are $u_{1\,\text{max}} = u_{2\,\text{max}} = 20$ [V]. The sliding mode settling times are set to $T_{s1} = 2$ [s] and $T_{s2} = 1$ [s]. The iteration interval is $h = 0.01$ [s].

Figure 10.40 shows the results of a simulation with the plant model perfectly matched to the real plant when M_2 is at the minimum value.

It should be noted that although the robustness is maintained when mismatching the plant relative to its model by reducing M_2, this would result in control chatter due to the real substate trajectories overshooting the sliding boundaries, thereby defeating the object of using discrete sliding mode control. This is the reason for calculating the controller parameters using a plant model with the cradle unladen.

In this application, smooth reference inputs would usually be applied to protect the load against excessive accelerations and keep the system in the sliding mode and therefore realise the prescribed closed-loop dynamics. Here, step reference inputs are applied to demonstrate the ability of the system to acquire the sliding mode after the substates are forced away from the sliding boundaries. The system commences in the sliding mode with zero reference inputs. When a step truck position input of 0.02 [m] is applied at $t = 0$ [s], control saturation occurs, as usual, and the system leaves the sliding mode. Although the cradle reference input is zero, the plant interaction causes a small disturbance of $y_2(t)$, as can be seen in Fig. 10.40a, as the system has not yet returned to the sliding mode. Figure 10.40b shows the resulting spike in $S_1(t)$ and also a smaller disturbance of $S_2(t)$ due to the plant interaction. Correspondingly, the substate trajectories pass through points 1 and 2 and return to their respective sliding boundaries at point 3. The system stays in the sliding mode until a cradle angle step reference of 0.01 [rad] is applied at $t = 1$ [s], causing control saturation and the system once more leaves the sliding mode, the spike this time occurring in $S_2(t)$ and a smaller disturbance of $S_1(t)$, again due to the plant interaction. The substate trajectories pass through points 5 and 6, both reaching the sliding boundaries again at point 7. Then the system remains in the sliding mode. The magnified inset graphs of Fig. 10.40 show clearly the 'bouncing ball' phenomena occurring simultaneously in the substates in the discrete sliding mode.

The reader may wonder why such small reference inputs were chosen for the simulations of Fig. 10.40. The reason is to be able to show clearly the sliding modes in the substate trajectories. With much larger reference inputs, if the phase-plane scales were large enough to view the complete substate trajectories, the sliding portions would appear very diminished in proportion.

Figure 10.41 shows simulations with a more realistic ramp reference input of the truck, reaching 1 [m] with a constant acceleration to a constant velocity and a constant deceleration to the final value. For this, the cradle reference angle input is zero so that the ability of the controller to eliminate the effects of plant interaction in the sliding mode may be assessed by observing any deviations of this angle from zero. In Fig. 10.41a, the cradle is unladen so that the mass, M_2, is at the minimum value of 100 [Kg]. In Fig. 10.41, the cradle is full so that $M_2 = 300$ [Kg]. The extreme robustness of the system is evident through the graphs of $y_1(t)$ following the ideal first-order responses, $y_{1i}(t)$, with no visible error in both cases. As the deviations of $y_2(t)$ are invisible on this scale, they are shown separately below. The extremely small deviations of $y_2(t)$ in Fig. 10.41a are attributed to imperfections in the numerical integration of the simulation since the plant model is perfectly matched to the real plant. Even in Fig. 10.41b, the deviation of $y_2(t)$ is negligible.

The changes in $u_1(t)$ and $u_2(t)$ needed to maintain the same response despite the added payload mass may be seen by comparing Fig. 10.41a, b. Also the fact that the control variables both keep within the control saturation limits is evidence that the sliding mode is maintained throughout.

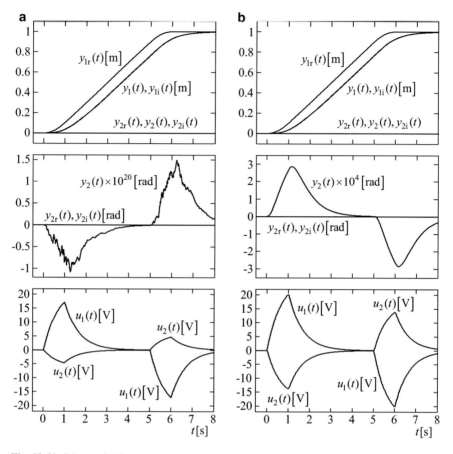

Fig. 10.41 Discrete SMC of gantry crane and cradle with realistic truck reference input. (**a**) Cradle unladen. (**b**) Cradle full

References

1. Utkin VI (1992) Sliding modes in control and optimization. Springer, Berlin/New York
2. Edwards C, Spurgeon SK (1998) Sliding mode control: theory and applications. Taylor and Francis, London
3. Itkis U (1976) Control systems of variable structure. Wiley, New York
4. Zinober ASI (ed) (1990) Deterministic control of uncertain systems. Peregrinus, London
5. Shtessel Y et al (2014) Sliding mode control and observation. Springer Science+Business Media, New York

Chapter 11
Motion Control

11.1 Introduction

Motion control is defined here as the feedback control of any plant whose controlled variable or variables are positions or position derivatives of mechanical components. Many of the examples of the previous chapters, such as spacecraft attitude control and ship stabilisation, fall into this category, but the main applications of concern in this chapter have the special requirement of zero dynamic lag explained in Sect. 11.3. These include positioning devices on production lines, being either tailor-made mechanisms for special operations or general-purpose jointed-arm robots as described in the following section.

11.2 Controlled Mechanisms

11.2.1 The General-Purpose Jointed-Arm Robot

There are many different controlled mechanisms throughout the manufacturing industry, many of which are specialised and typically repeat the same motion. Industrial robots also perform repetitive tasks but are structured so that they can be adapted to new tasks when necessary. There are several industrial robot configurations, one of which, the jointed-arm robot, is illustrated in Fig. 11.1.

The robot is structured to be able to control all six degrees of freedom of motion of the end effector, considered as a rigid body. As in vehicle control (automobiles, ships, aircraft and spacecraft) there are three translational degrees of freedom defining the position of a defined point on the end effector in a fixed frame of reference. The coordinates, x, y and z, of this point are controlled by varying the waist, shoulder and elbow joint angles, α, β and γ, by means of actuators, usually DC brushless motors or synchronous motors. Regarding the three rotational degrees

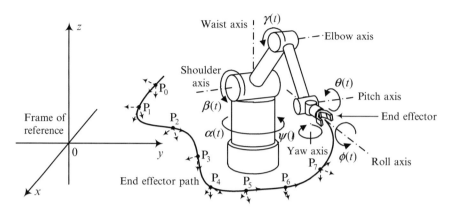

Fig. 11.1 Jointed-arm robot and the degrees of freedom of movement

of freedom, the orientation of the end effector relative to the frame of reference is controlled by three more actuators varying the roll, pitch and yaw attitude angles, ϕ, θ and ψ, as shown. The required translational motion of the end effector and its orientation, i.e. $[x_r(t), y_r(t), z_r(t), \phi_r(t), \theta_r(t), \psi_r(t)]$, is defined here as the *reference path*. The reference path is required to pass through a sequence of time tagged values $[x_i(t_i), y_i(t_i), z_i(t_i), \phi_i(t_i), \theta_i(t_i), \psi_i(t_i)]$, $i = 0, 1, \ldots, N_p - 1$, at the corresponding points, P_i, $i = 0, 1, \ldots, N_p - 1$, as illustrated in Fig. 11.1. In such robots, feedback control is implemented with the six joint angles as the controlled measurement variables. The required reference input vector is therefore $[\alpha_r(t), \beta_r(t), \gamma_r(t), \phi_r(t), \theta_r(t), \psi_r(t)]$. The nonlinear simultaneous equations,

$$\begin{cases} x_i = f_x(\alpha_i, \beta_i, \gamma_i, \phi_i, \theta_i, \psi_i) \\ y_i = f_y(\alpha_i, \beta_i, \gamma_i, \phi_i, \theta_i, \psi_i) \\ z_i = f_z(\alpha_i, \beta_i, \gamma_i, \phi_i, \theta_i, \psi_i) \end{cases} \quad (11.1)$$

which are known from the geometry of a robot, are solved for α_i, β_i and γ_i, enabling the points, $[\alpha_i(t_i), \beta_i(t_i), \gamma_i(t_i), \phi_i(t_i), \theta_i(t_i), \psi_i(t_i)]$, $i = 0, 1, \ldots, N_p - 1$, that the smooth reference input vector must pass through to be determined. Means of generating such reference inputs are presented in Appendix A11. If a feedback linearising control law (Chap. 7) is used, then the closed-loop dynamics is known accurately, enabling the dynamic lag pre-compensation of Sect. 11.3.2 to be applied so that the planned reference path is followed accurately.

11.2.2 General Model

If the elasticity of each link of a mechanism such as the robotic manipulator illustrated in Fig. 11.1 is negligible, then it may be regarded as a system of

11.2 Controlled Mechanisms

interconnected rigid bodies. In this case each degree of freedom of motion may be modelled by a second-order differential equation. The following example illustrates this.

Example 11.1 Dynamics and kinematics model of a two degree of freedom mechanism

Consider a mechanism comprising a prismatic joint and a revolute joint, each with an actuator, as illustrated in Fig. 11.2.

This two degree-of-freedom mechanism is simpler than a jointed-arm robot (typically six degrees of freedom) but contains similar features and will be useful in explaining in a straightforward manner the dynamics of controlled mechanisms.

Consider first pure translational motion of the slider produced by the force, f, with the angle, α, maintained constant. The equation of motion for this degree of freedom may be obtained by the force balance method and is

$$M_s \ddot{x} = f - M_s g \sin(\alpha) \Rightarrow M_s \ddot{x} + M_s g \sin(\alpha) = f \quad (11.2)$$

If only the rotational motion produced by the torque, γ, is considered, with x maintained constant, then the equation of motion for this degree of freedom may be obtained by the torque balance method and is

$$(J_c + J_s)\ddot{\alpha} = \gamma - (M_c g R + M_s g x)\cos(\alpha) \Rightarrow (J_c + J_s)\ddot{\alpha}$$
$$+ (M_c R + M_s x) g \sin(\alpha) = \gamma \quad (11.3)$$

Equation (11.2) is linear since α is constant in this instance but (11.3) is nonlinear due to the term $\sin(\alpha)$ in the gravity-dependent torque varying with α. Such nonlinear terms are common in the models of mechanisms. It is usual, of course, for all the degrees of freedom of a mechanism to be exercised simultaneously. In the mechanism of Fig. 11.2, the force, f, and the torque, γ, would be applied simultaneously and, as will be seen, the equations of motion are considerably more complicated than Eqs. (11.2) and (11.3) taken together. This is due to dynamic interaction between the degrees of freedom of motion. In other words, motion in a given degree of freedom will cause forces and torques that effect the motion in other degrees of freedom. The simple torque balance or force balance approaches are inadequate to deal with such mechanical systems as already demonstrated in Chap. 2, Sect. 2.2.2.8, on Lagrangian mechanics and this method will now be used to derive the complete equations of motion. First, the Lagrangian is

$$L = T - V \quad (11.4)$$

where T and V are the total kinetic and potential energies. Referring to Fig. 11.2,

$$T = \frac{1}{2} M_s \dot{x}^2 + \frac{1}{2}[J_c + J_s]\dot{\alpha}^2. \quad (11.5)$$

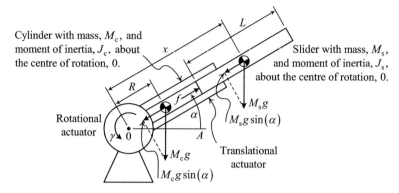

Fig. 11.2 Two degree of freedom mechanism with actuators

Here, $J_c = M_c R^2$ and $J_s = 1/12 M_s L^2 + M_s x^2$. Note that this expression for J_s is obtained by means of the parallel axis theorem. Thus, it is equal to the moment of inertia, $1/12 M_s L^2$, about the axis parallel to the axis of rotation and passing through the centre of mass, plus the moment of inertia, $M_s x^2$, of the slider as if it was a point mass, M_s, at a distance, x from the centre of rotation. The total potential energy may be formulated with respect to an arbitrary horizontal, inertial, datum, since its expression will ultimately be differentiated. Thus taking the line, $0A$, as the datum, $V = M_c g R \sin(\alpha) + M_s g x \sin(\alpha)$. Then the Lagrangian is

$$L = \frac{1}{2} M_s \dot{x}^2 + \frac{1}{2}\left[M_c R^2 + M_s \left(\frac{1}{12} L^2 + x^2\right)\right] \dot{\alpha}^2 - (M_c R + M_s x) g \sin(\alpha). \tag{11.6}$$

The equations of motion are then obtained from

$$\frac{d}{dt}\left(\frac{\partial L}{\partial \dot{x}}\right) - \frac{\partial L}{\partial x} = f \tag{11.7}$$

and

$$\frac{d}{dt}\left(\frac{\partial L}{\partial \dot{\alpha}}\right) - \frac{\partial L}{\partial \alpha} = \gamma \tag{11.8}$$

Thus

$$\frac{d}{dt}(M_s \dot{x}) - M_s x \dot{\alpha}^2 - [-M_s g \sin(\alpha)]$$
$$= f \Rightarrow \underbrace{M_s \ddot{x}}_{\text{inertial force}} - \underbrace{M_s x \dot{\alpha}^2}_{\text{centrifugal force}} + \underbrace{M_s g \sin(\alpha)}_{\text{gravitational force}} = \underbrace{f}_{\text{control force}} \tag{11.9}$$

11.2 Controlled Mechanisms

and

$$\frac{d}{dt}\left\{[M_cR^2 + M_s\left(\tfrac{1}{12}L^2 + x^2\right)]\dot{\alpha}\right\} - [-(M_cR + M_sx)g\cos(\alpha)] = \gamma \Rightarrow$$

$$\underbrace{[M_cR^2 + M_s\left(\tfrac{1}{12}L^2 + x^2\right)]\ddot{\alpha}}_{\text{inertial torque}} + \underbrace{2M_sx\dot{\alpha}\dot{x}}_{\text{Coriolis torque}} + \underbrace{(M_cR + M_sx)g\cos(\alpha)}_{\text{gravitational torque}} = \underbrace{\gamma}_{\text{control torque}}.$$

(11.10)

It may be seen that for $\dot{\alpha} = 0$, Eq. (11.9) reduces to Eq. (11.2). Similarly, for $\dot{x} = 0$, Eq. (11.10) reduces to Eq. (11.3).

The Coriolis torque introduced in Example 11.1 requires some explanation. First the Coriolis force is a force due to translational motion of a mass in a rotating frame of reference. A practical example of this may be found in a particular type of children's' roundabout, comprising a raised rotating platform often-found in public parks. Occasionally, against others' advice, a child stands at the centre of the spinning roundabout and then starts to walk towards the outside. Although the child may lean backwards to counteract the centrifugal force experienced in a fixed position away from the centre of the roundabout, a sideways force is experienced which may be unexpected. This is the Coriolis force, which may be understood by viewing the situation from an inertial frame of reference. Although the path taken is a straight line with respect to the rotating roundabout, the motion of the roundabout forces the path to be curved with respect to inertial space. This causes a sideways force similar to the centrifugal force experienced when travelling around a bend in a car. In the mechanism of Fig. 11.2, simultaneous rotation and translation gives rise to a Coriolis force of magnitude, $2M_s\dot{\alpha}\dot{x}$, acting at right angles to the slider axis and therefore causing a torque, $2M_s\dot{\alpha}\dot{x}x$, which is referred to as the Coriolis torque.

Returning now to equations (11.9) and (11.10) of Example 11.1, with $x = q_1$, $\alpha = q_2$, $f = \tau_1$ and $\gamma = \tau_2$, expressing these equations in the matrix form yields

$$\underbrace{\begin{bmatrix} M_s & 0 \\ 0 & M_cR^2 + M_s\left(\tfrac{1}{12}L^2 + q_1^2\right) \end{bmatrix}}_{\text{mass-inertia matrix}} \begin{bmatrix} \ddot{q}_1 \\ \ddot{q}_2 \end{bmatrix} + \underbrace{\begin{bmatrix} 0 & M_sq_1\dot{q}_2 \\ 2M_sq_1\dot{q}_2 & 0 \end{bmatrix}}_{\text{Coriolis-centrifugal forces \& torques}} \begin{bmatrix} \dot{q}_1 \\ \dot{q}_2 \end{bmatrix}$$

$$+ \underbrace{\begin{bmatrix} M_sg\sin(q_2) \\ (M_cR + M_sq_1)g\cos(q_2) \end{bmatrix}}_{\text{gravitational forces \& torques}} = \underbrace{\begin{bmatrix} \tau_1 \\ \tau_2 \end{bmatrix}}_{\text{control forces \& torques}}$$

(11.11)

This is now in the standard form for controlled mechanisms that can be modelled as interconnected actuated rigid bodies. The general equation is

$$\mathbf{M}(\mathbf{q})\,\ddot{\mathbf{q}} + \mathbf{C}(\mathbf{q},\dot{\mathbf{q}})\,\dot{\mathbf{q}} + \mathbf{g}(\mathbf{q}) = \boldsymbol{\tau},$$

(11.12)

where $\mathbf{q} \in \Re^d$ is the position coordinate vector, $\boldsymbol{\tau} \in \Re^d$ is the control force/torque vector, $\mathbf{M} \in \Re^{d \times d}$, $\mathbf{C} \in \Re^{d \times d}$, $\mathbf{g} \in \Re^d$ and d is the number of degrees of freedom. Figure 11.3 shows the model defined by (11.12) in the form of a block diagram.

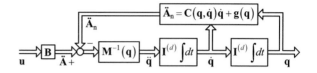

Fig. 11.3 General block diagram model of a controlled mechanism

The nonlinear torque vector, τ_n, comprises the Coriolis and gravitational torque components. The diagonal matrix, **B**, represents the actuators with negligible dynamic lag, meaning that the control forces and torques are assumed to be directly proportional to the input voltages of the actuator drive electronics that constitute the control vector, **u**. This is possible with current loops with bandwidths far greater than those of the position control loops, assuming electromagnetic actuators.

11.2.3 Feedback Linearising Control Law

The model of (11.12), which will be referred to as the plant, lends itself readily to the feedback linearising control technique of Chap. 7, Sect. 7.3. It is evident by inspection that the relative degree of the plant with respect to each controlled position, q_i, is $r_i = 2$, $i = 1, 2, \ldots, d$, and that the plant is of full relative degree, since $\sum_{i=1}^{d} q_i = 2d$ is the plant order. Assuming that the d control loops are to be decoupled from one another, the most general desired closed-loop differential equation is

$$\ddot{\mathbf{q}} = \mathbf{W}[\mathbf{q}_r - \mathbf{q}] - \mathbf{Z}\dot{\mathbf{q}} \qquad (11.13)$$

where

$$\mathbf{W} = \mathrm{diag}\left(\omega_{n1}^2, \omega_{n2}^2, \ldots, \omega_{nd}^2\right) \qquad (11.14)$$

and

$$\mathbf{Z} = \mathrm{diag}\left(2\zeta_1\omega_{n1}, 2\zeta_2\omega_{n2}, \ldots, 2\zeta_d\omega_{nd}\right). \qquad (11.15)$$

If zero overshooting is required, the settling time formulae may be used. Then

$$\mathbf{W} = \mathrm{diag}\left(p_1^2, p_2^2, \ldots, p_d^2\right) \qquad (11.16)$$

and

$$\mathbf{Z} = \mathrm{diag}\left(2p_1, 2p_2, \ldots, 2p_d\right). \qquad (11.17)$$

where $p_i = 4.5/T_{si}$ (5 % criterion) or $p_i = 5.6/T_{si}$ (2 % criterion), $i = 1, 2, \ldots, d$, where T_{si} are the desired settling times. The next step in the control

11.2 Controlled Mechanisms

law derivation is to pre-multiply both sides of (11.13) by the matrix, $\mathbf{M}(\mathbf{q})$. Thus

$$\mathbf{M}(\mathbf{q})\ddot{\mathbf{q}} = \mathbf{M}(\mathbf{q})\{\mathbf{W}[\mathbf{q}_r - \mathbf{q}] - \mathbf{Z}\dot{\mathbf{q}}\}. \tag{11.18}$$

Then (11.12) may be re-arranged as

$$\mathbf{M}(\mathbf{q})\ddot{\mathbf{q}} = \tau - \mathbf{C}(\mathbf{q}, \dot{\mathbf{q}})\dot{\mathbf{q}} - \mathbf{g}(\mathbf{q}). \tag{11.19}$$

Now the LHS of (11.19) is the same as that of (11.18), enabling the feedback linearising control law to be obtained by equating the RHS of these equations. Hence

$$\tau - \mathbf{C}(\mathbf{q}, \dot{\mathbf{q}})\dot{\mathbf{q}} - \mathbf{g}(\mathbf{q}) = \mathbf{M}(\mathbf{q})\{\mathbf{W}[\mathbf{q}_r - \mathbf{q}] - \mathbf{Z}\dot{\mathbf{q}}\} \Rightarrow \tau$$
$$= \mathbf{M}(\mathbf{q})\{\mathbf{W}[\mathbf{q}_r - \mathbf{q}] - \mathbf{Z}\dot{\mathbf{q}}\} + \mathbf{C}(\mathbf{q}, \dot{\mathbf{q}})\dot{\mathbf{q}} + \mathbf{g}(\mathbf{q}) \tag{11.20}$$

This is similar to the so-called computed torque control law established for jointed-arm robots, but the material on dynamic lag pre-compensation of Sect. 11.3 is required in order to derive this, at the end of which it will be presented.

11.2.4 Simplified Model for Mechanisms with Geared Actuators

11.2.4.1 Introduction

Many mechanisms employ reduction gearboxes between electric motors and interface with the mechanism (Chap. 2, Sect. 2.2.3.1). In this case, many motor shaft revolutions are required to turn the output shaft interfacing with the mechanism through one revolution, the ratio typically varying between 20 to 1 and 200 to 1. The purpose of this is to obtain control torques considerably larger than the maximum motor torques. This enables standard, off-the-shelf servomotors to be used. The principal disadvantages of this arrangement, however, are (a) limited joint speeds and (b) impaired accuracy due to gearbox backlash and stick-slip friction. Directly controlled mechanisms eliminate these problems but require specially designed high torque motors, which are usually brushless DC motors or synchronous motors (Chap. 2, Sect. 2.2.5). These motors have become available through the introduction of rare earth magnets producing high flux densities, but they are more expensive than conventional servomotors and gearboxes. As will be shown, the reduction gearbox simplifies the model needed for control system design to a simple linear one, enabling linear controllers to be employed. The following subsection introduces the inverse dynamic representation of the mechanical load in preparation for forming this simplified model.

11.2.4.2 Inverse Dynamic Representation of Mechanical Load

Consider an electric motor having drive electronics that provides an electromagnetic torque, γ_e, that is directly proportional to the control variable, u, so that $\gamma_e = bu$, where b is a constant. Let the motor rotor moment of inertia be J_r and the bearing friction be negligible. Then if ω_r is the rotor angular velocity, the equation of motion and the corresponding transfer function are

$$J_r \dot{\omega}_r = bu. \qquad (11.21)$$

and

$$\frac{\Omega_r(s)}{U(s)} = \frac{1}{J_r s}. \qquad (11.22)$$

Let a rigid, balanced mechanical load with moment of inertia, J_L, be directly driven by the motor. Then the equation of motion and the transfer function become

$$(J_r + J_L) \dot{\omega}_r = bu \qquad (11.23)$$

and

$$\frac{\Omega_r(s)}{U(s)} = \frac{b}{(J_r + J_L) s}. \qquad (11.24)$$

Equation (11.23) may be written as

$$J_r \dot{\omega}_r = bu - J_L \dot{\omega}_r. \qquad (11.25)$$

The transfer function block diagram representation of (11.25) is shown in Fig. 11.4.

The motor must supply two torque components, i.e. the dynamic load torque, γ_d, and the inertial load torque, γ_I, required to accelerate or decelerate the motor rotor. Hence, $\gamma_e = \gamma_d + \gamma_I$. Recall that the term *dynamics* is used to describe a model of a mechanical system or subsystem that gives the velocity resulting from an applied force or torque. The block in the feedback path of Fig. 11.4 performs the inverse of this function for the mechanical load, i.e. it gives the torque that

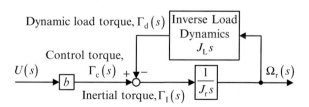

Fig. 11.4 An example of load representation by inverse dynamics in the feedback path

11.2 Controlled Mechanisms

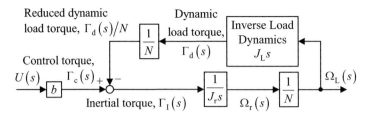

Fig. 11.5 Rigid-body inertial load driven by a motor via a gearbox with ratio, N

must be applied to produce a given velocity. Such an element is therefore described by the term *inverse dynamics*. To confirm the correctness of Fig. 11.4, it is easily shown that the transfer function of the block diagram is given by (11.24). Another observation is that the inverse dynamics transfer function is the reciprocal of the dynamics transfer function. In general, if the motor drives any linear mechanical load, then if its inverse dynamics transfer function is inserted in the feedback path of Fig. 11.4, a valid model of the combined motor and load is obtained.

Now suppose that a reduction gearbox of ratio, N, with $N > 1$, is inserted between the motor output shaft and the mechanical load (Chap. 2, Sect. 2.2.3.1). The moments of inertia of the gear-train components would then be referred to the motor shaft and added to the rotor moment of inertia, giving a total moment of inertia of J_m. Then if the load speed is ω_L and the dynamic load torque acting on the gearbox output shaft is γ_d, the equations of motion are

$$J_m \dot{\omega}_r = bu - \gamma_d/N, \; \gamma_d = J_L \dot{\omega}_L \text{ and } \omega_r = N\omega_L. \tag{11.26}$$

Figure 11.5 shows the corresponding transfer function block diagram.

The corresponding transfer function is

$$\frac{\Omega_r(s)}{U(s)} = \frac{b}{(J_m + J_L/N^2)s}. \tag{11.27}$$

Comparing this with (11.24), the effect of the reduction gearbox is to reduce the load moment of inertia reflected to the gearbox. If N is sufficiently high, then the motor rotor moment of inertia dominates. As will be seen, reduction gearboxes can greatly simplify the model needed for the control system design.

Now, returning to the general model of Fig. 11.3, without reduction gears, the motor rotor inertias are included in the inertia matrix, $\mathbf{M}(\mathbf{q})$. If this diagram is reformulated with the motor rotor inertias separated, then the inertia matrix, $\mathbf{M}(\mathbf{q})$, is that of the mechanism without the motors, and (11.12) inserted in the feedback path with $\boldsymbol{\tau}$ becoming the dynamic load torque, $\boldsymbol{\tau}_d$, as shown in Fig. 11.6.

For a d degree of freedom mechanism, $\mathbf{J}_m = \text{diag}(J_{m1}, J_{m2}, \ldots, J_{md})$, where $J_{mi}, i = 1, 2, \ldots, d$, are the actuator motor moments of inertia including the gear trains. The equations of the driven robot corresponding to Fig. 11.6 are as follows.

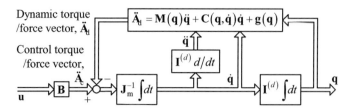

Fig. 11.6 Model of mechanism in the inverse dynamic form with motor inertias separated

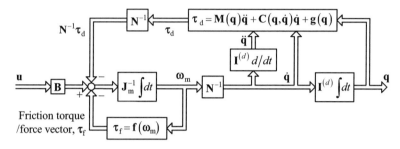

Fig. 11.7 General model of mechanism driven via reduction gearboxes

$$\mathbf{J}_m \ddot{\mathbf{q}} = \mathbf{Bu} - \boldsymbol{\tau}_d, \quad \boldsymbol{\tau}_d = \mathbf{M}(\mathbf{q}) \, \ddot{\mathbf{q}} + \mathbf{C}(\mathbf{q}, \dot{\mathbf{q}}) \, \dot{\mathbf{q}} + \mathbf{g}(\mathbf{q}) \qquad (11.28)$$

Suppose that gearboxes with reduction ratios, $N_1, N_2, \ldots, N_d$, are introduced. Then if $\mathbf{N} = \mathbf{diag}(N_1, N_2, \ldots, N_d)$, the block diagram of the mechanism can be formed from Fig. 11.6 by introducing the inverse reduction ratios, by analogy with the formation of Fig. 11.5 from Fig. 11.4. Also, since gearboxes also introduce considerable friction, the motor torques must overcome this. Hence, a friction force/torque vector, $\boldsymbol{\tau}_f = \mathbf{f}(\boldsymbol{\omega}_m)$, is added as shown in Fig. 11.7.

The corresponding equations of motion of the mechanism are as follows:

$$\mathbf{J}_m \dot{\boldsymbol{\omega}}_m = \mathbf{Bu} - \mathbf{N}^{-1} \boldsymbol{\tau}_d - \boldsymbol{\tau}_f \qquad (11.29)$$

$$\mathbf{M}(\mathbf{q}) \, \ddot{\mathbf{q}} + \mathbf{C}(\mathbf{q}, \dot{\mathbf{q}}) \, \dot{\mathbf{q}} + \mathbf{g}(\mathbf{q}) = \boldsymbol{\tau}_d \qquad (11.30)$$

$$\boldsymbol{\omega}_m = \mathbf{N} \dot{\mathbf{q}} \qquad (11.31)$$

$$\boldsymbol{\tau}_f = \mathbf{f}(\boldsymbol{\omega}_m) \qquad (11.32)$$

Substituting for $\boldsymbol{\omega}_m$ in (11.29) using (11.31) and noting that because the matrix $\mathbf{N}$ is diagonal, $\mathbf{J}_m \mathbf{N} = \mathbf{N} \mathbf{J}_m$ yields $\mathbf{N} \mathbf{J}_m \ddot{\mathbf{q}} = \mathbf{Bu} - \mathbf{N}^{-1} \boldsymbol{\tau}_d - \boldsymbol{\tau}_f$, and pre-multiplying both sides of this equation by $\mathbf{N}$ then yields

$$\mathbf{N}^2 \mathbf{J}_m \ddot{\mathbf{q}} = \mathbf{NBu} - \boldsymbol{\tau}_d - \mathbf{N} \boldsymbol{\tau}_f \qquad (11.33)$$

11.2 Controlled Mechanisms

Adding (11.30) and (11.33) to eliminate τ_d then yields

$$\left[\mathbf{M}(\mathbf{q}) + \mathbf{N}^2 \mathbf{J}_m\right] \ddot{\mathbf{q}} + \mathbf{C}(\mathbf{q}, \dot{\mathbf{q}}) \dot{\mathbf{q}} + \mathbf{g}(\mathbf{q}) = \mathbf{NBu} - \mathbf{N}\tau_f \qquad (11.34)$$

Then pre-multiplying both sides by $\mathbf{N}^{-2} = \left[\mathbf{N}^{-1}\right]^2$ yields

$$\left[\mathbf{N}^{-2}\mathbf{M}(\mathbf{q}) + \mathbf{J}_m\right] \ddot{\mathbf{q}} + \mathbf{N}^{-2}\left[\mathbf{C}(\mathbf{q}, \dot{\mathbf{q}}) \dot{\mathbf{q}} + \mathbf{g}(\mathbf{q})\right] = \mathbf{N}^{-1}\mathbf{Bu} - \mathbf{N}^{-1}\tau_f. \qquad (11.35)$$

Finally, the gearbox friction term, $\tau_f = \mathbf{f}(\omega_m) = \mathbf{f}(\mathbf{N}\dot{\mathbf{q}})$, will be simplified to $\mathbf{F}(\dot{\mathbf{q}})$. Then the required equation of the driven mechanism is

$$\left[\mathbf{N}^{-2}\mathbf{M}(\mathbf{q}) + \mathbf{J}_m\right] \ddot{\mathbf{q}} + \mathbf{N}^{-2}\left[\mathbf{C}(\mathbf{q}, \dot{\mathbf{q}}) \dot{\mathbf{q}} + \mathbf{g}(\mathbf{q})\right] = \mathbf{N}^{-1}\mathbf{Bu} - \mathbf{N}^{-1}\mathbf{F}(\dot{\mathbf{q}}). \qquad (11.36)$$

Recall the single degree of freedom mechanism of Fig. 11.5 with transfer function (11.27). The equivalent moment of inertia, $J_m + N^{-2}J_L$, referred to the motor output shaft, is analogous to the term $\mathbf{J}_m + \mathbf{N}^{-2}\mathbf{M}(\mathbf{q})$ in the forward path of Fig. 11.7.

The gear reduction ratios, N_i, $i = 1, 2, \ldots, d$, are usually much greater than 1, typically in the range 20–200. Under these circumstances, the motor dynamics dominates. This may be seen by inspection of Fig. 11.7. The nonlinear dynamic force/torques, $\mathbf{N}^{-1}\tau_d$, which the motors must overcome, are reduced by the gear reduction ratios and are generally much smaller than the component forces and torques of τ_c and τ_f. As an approximation, therefore, the upper feedback path may be removed from Fig. 11.7. In addition, generally $[\mathbf{J}_m]_{ii} \gg \left[\mathbf{N}^{-2}\mathbf{M}(\mathbf{q})\right]_{ii}$ and therefore *only the motor rotor moments of inertia are significant*. The gearbox ratios are considered *large* if the following conditions are simultaneously satisfied:

$$\left. \begin{array}{l} \left|\left[\mathbf{N}^{-1}\mathbf{C}(\mathbf{q}, \dot{\mathbf{q}}) \dot{\mathbf{q}} + \mathbf{g}(\mathbf{q})\right]\right|_i \ll \left|\left[\mathbf{Bu} - \mathbf{F}(\dot{\mathbf{q}})\right]\right|_i \\ \text{and } \left[\mathbf{N}^{-2}\mathbf{M}(\mathbf{q})\right]_{ii} \gg [\mathbf{J}_m]_{ii}, \ i = 1, 2, \ldots, d \end{array} \right\}. \qquad (11.37)$$

The off-diagonal terms of $\mathbf{J}_m + \mathbf{N}^{-2}\mathbf{M}(\mathbf{q})$ are contributed only by $\mathbf{N}^{-2}\mathbf{M}(\mathbf{q})$. Then, if the second of conditions (11.37) is satisfied, the driven mechanism of Fig. 11.7 simplifies to that of Fig. 11.8, which comprises d decoupled subsystems. To summarise, the inclusion of reduction gearboxes has eliminated the nonlinear Coriolis, centrifugal and gravitational terms in the model needed for control system design but at the expense of limiting the speed of movement of the mechanism and introducing significant friction including the troublesome stick-slip friction.

The speed limitation is principally due to the upper limits of the electric motor speeds. These are set by the finite power supply voltage that limits the maximum back e.m.f. that is proportional to the motor speed, which can be overcome by the power electronics in order to generate the required torque-producing currents.

The attainable accuracy is limited by backlash in the gear train and stick-slip friction. In some applications, special gear trains with two gear wheels per stage, sprung apart by torsion springs, can be employed to eliminate the backlash, but at additional cost and with reduced limits on the maximum motor torques.

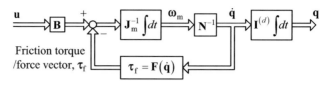

Fig. 11.8 Simplified model of mechanism driven via reduction gearboxes

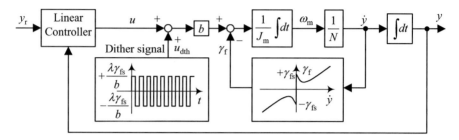

Fig. 11.9 Control dither to reduce the effects of stick-slip friction

Regarding control techniques, if controllers without integral terms are employed, then the stick-slip friction causes steady-state errors. These can be reduced by increasing the controller gains or employing instead robust control techniques (Chap. 10). With integral terms, whenever the controlled position becomes 'stuck', the control error causes the output of the integral term to ramp until the actuator torque exceeds the stick-slip limit. Then the mechanism moves but inevitably overshoots, reversing the sign of the error, requiring further integral action to overcome the stick-slip friction again, thereby giving rise to a limit cycle. These effects of the stick-slip friction can be greatly reduced by means of *control dither*, which is an oscillating signal superimposed on the control signal, $\mathbf{u}(t)$, with sufficient amplitude to *just* exceed the stick-slip torque limits, $\pm\gamma_{fs}$ (Chap. 2, Sect. 2.2.2.5), as shown in Fig. 11.9 applied to a single degree of freedom mechanism.

The contingency factor, λ, has to be greater than unity by an amount commensurate with the uncertainty of the value of γ_{fs}. The variation of γ_{fs} with temperature is often significant. There will always be a certain residual oscillation of the controlled output about the reference input, therefore causing an error, but less than would occur without the control dither.

If higher accuracy of control is required than attainable with reduction gears, then direct drives are needed and more sophisticated control laws such as feedback linearising control, which can be accommodated easily by modern digital processors.

11.3 Dynamic Lag Pre-compensation

11.3.1 Definition of Dynamic Lag

It is desirable to design controllers for motion control applications, such as jointed-arm robots and machine tool positioning, to follow time-varying reference inputs with minimal error. All the control systems dealt with so far have controlled outputs, $y(t)$, that are unable to follow any changes in the reference input, $y_r(t)$, without a transient error. This must occur if the closed-loop system is a dynamic one in which the relationship between $y(t)$ and $y_r(t)$ is a differential equation such as

$$\ddot{y} = \omega_n^2 (y_r - y) - 2\zeta\omega_n \dot{y} \quad (11.38)$$

for a second order linear system. The only input-output relationship for which these transient errors do not occur is the non-dynamic one of

$$y = y_r. \quad (11.39)$$

Two terms are used in connection with the transient error, and these are defined in Fig. 11.10 which show the response of the system governed by (11.38) with $\omega_n = 2$ rad/s and $\zeta = 0.7$ to a reference input of

$$y_r(t) = 1 - \cos(t) \text{ for } 0 \le t \le \pi, \ 2 \text{ for } t > \pi. \quad (11.40)$$

If $y(t)$ is the position of a cutting tool in a machining process, the dynamic lag causes insufficient material to be removed. Moreover, the small overshoot following $t = t_2$ would cause too much material to be removed. Such issues are avoided by reducing the dynamic lag in a control system to negligible proportions.

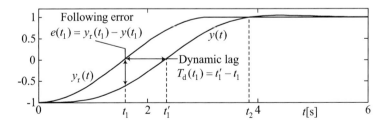

Fig. 11.10 Definitions of dynamic lag and following error

11.3.2 Derivative Feedforward Pre-compensation

11.3.2.1 Basic Pre-compensator

Let an LTI SISO control loop have a reference input $y'_r(t)$. It is possible for $y(t)$ to follow *another* reference input, $y_r(t)$, with negligible dynamic lag by applying $y'_r(t)$ as provided by the *dynamic lag pre-compensator* of Fig. 11.11.

To attempt realisation of (11.39),

$$P(s) = 1/\tilde{G}_{cl}(s), \tag{11.41}$$

where $\tilde{G}_{cl}(s)$ is an estimate of $G_{cl}(s)$. For the closed-loop system (11.38), the transfer function and the corresponding pre-compensator are, respectively,

$$G_{cl}(s) = \frac{\omega_n^2}{s^2 + 2\zeta\omega_n s + \omega_n^2} \tag{11.42}$$

and

$$P(s) = 1 + \frac{2\tilde{\zeta}}{\tilde{\omega}_n}s + \frac{1}{\tilde{\omega}_n^2}s^2. \tag{11.43}$$

In the time domain, the equation of the pre-compensator is

$$y'_r(t) = y_r(t) + \frac{2\tilde{\zeta}}{\tilde{\omega}_n}\dot{y}_r(t)s + \frac{1}{\tilde{\omega}_n^2}\ddot{y}_r(t). \tag{11.44}$$

This is weighted derivative feedforward. If the system has to respond to an arbitrary $y_r(t)$, then approximate numerical differentiation is needed. If, however, the reference input derivatives have been determined analytically in advance, the method of the following sub-subsection can be applied.

11.3.2.2 Implementation Using Reference Input Derivatives

In the interests of accuracy, the reference input derivatives should be calculated analytically, if possible, together with the reference input. For example, the spline method [1] presented in Appendix A11 can provide reference input derivatives up to an order equal to the degree of the polynomials constituting the spline function.

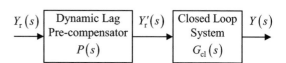

Fig. 11.11 Application of dynamic lag pre-compensator

11.3 Dynamic Lag Pre-compensation

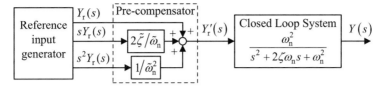

Fig. 11.12 Realisation of dynamic lag pre-compensator by calculated derivatives

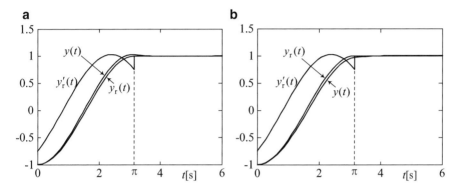

Fig. 11.13 Simulation of dynamic lag pre-compensation. (**a**) $\omega_n = 1.05\tilde{\omega}_n$, $\zeta = 0.95\tilde{\zeta}$. (**b**) $\omega_n = 0.95\tilde{\omega}_n$, $\zeta = 1.05\tilde{\zeta}$

Continuing with the second-order example, the block diagram of Fig. 11.11 is replaced with that of Fig. 11.12.

Figure 11.13 shows a simulation of the system of Fig. 11.12 with realistically mismatched parameters. The nominal (estimated) parameters are $\tilde{\omega}_n = 2$ rad/s and $\tilde{\zeta} = 0.7$. The mismatched (real) parameters, ω_n and ζ, are as indicated. The mismatches of ω_n and ζ are oppositely signed, i.e. $\pm 5\,\%$ and $\mp 5\,\%$, respectively, to yield worst-case errors in the first derivative weighting coefficient, $2\tilde{\zeta}/\tilde{\omega}_n$. The discontinuity in $y'_r(t)$ at $t = \pi$ is due to the discontinuity in $\ddot{y}_r(t)$ which, according to (11.40), jumps from $\cos(\pi) = -1$ to zero at $t = \pi$. In Fig. 11.13a, the mismatching causes overcompensation, i.e. a slightly negative dynamic lag, and an overshoot, while in Fig. 11.13b, a little positive dynamic lag remains. It is sometimes advisable to purposely mismatch the pre-compensator parameters in such directions that, for example, zero overshoot is guaranteed if this is critical.

11.3.2.3 Polynomial Controller with Dynamic Lag Pre-compensation

Since the polynomial controller introduced in Chap. 5 has a built-in pre-compensator, its use for dynamic lag pre-compensation will be addressed here. This, however, is limited to nominally linear plants that can be accurately modelled by transfer functions. To simplify the presentation, mismatches between the real

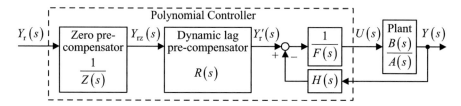

Fig. 11.14 Generic polynomial controller used for dynamic lag pre-compensation

and estimated plant parameters will be ignored, but the reader should be aware of the errors caused by such mismatches and is advised to simulate them as in the previous sub-subsection to assess their significance when developing a real control system.

For convenience, the general block diagram of Fig. 5.27 is reproduced in Fig. 11.14 but with the pre-compensator separated into two factors, one for plant zero pre-compensation, if necessary, and the other for dynamic lag pre-compensation.

The overall transfer function is

$$\frac{Y(s)}{Y_r(s)} = \frac{R(s)}{Z(s)} \cdot \frac{B(s)}{A(s)F(s) + B(s)H(s)}. \tag{11.45}$$

If the plant has no finite zeros, $B(s) = b_0$ and $Z(s) = 1$. Then

$$Y_{rz}(s) = Y_r(s). \tag{11.46}$$

Once the pole assignment has been carried out, then

$$\frac{Y(s)}{Y_r(s)} = \frac{b_0 R(s)}{D(s)}, \tag{11.47}$$

where $D(s)$ is the desired closed-loop characteristic polynomial which is realised by choosing the controller transfer functions, $F(s)$ and $H(s)$, such that

$$A(s)F(s) + B(s)H(s) = D(s). \tag{11.48}$$

Zero dynamic lag therefore requires

$$R(s) = D(s)/b_0. \tag{11.49}$$

Then the closed-loop pole placement is made to achieve a degree of robustness. For this purpose, the robust pole placement of Chap. 10 might be considered.

11.3 Dynamic Lag Pre-compensation

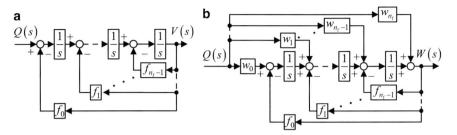

Fig. 11.15 Forming the state-variable block diagram of the controller in two steps. (**a**) Basic dynamic element without zeros. (**b**) Feedforward giving zero relative degree

With $Z(s) = 1$ in Fig. 11.14,

$$Y'_r(s) = R(s)Y_r(s) = \left(\sum_{k=0}^{n_r} r_k s^k\right) Y_r(s) \Rightarrow y'_r(t) = \sum_{k=0}^{n_r} r_k y_r^{(k)}(t). \quad (11.50)$$

In view of (11.49), $n_r = n_d$, which is the maximum order of derivative of $y_r(t)$ needed. Not every derivative, however, requires calculation. This is made possible by weighted feedforward of the reference input around the integrators implementing the dynamic part of the controller, i.e. $1/F(s)$. For this, first, a state-variable block diagram of the dynamic part of the controller is formed with transfer function,

$$\frac{V(s)}{Q(s)} = \frac{1}{F(s)} = \frac{1}{s^{n_f} + f_{n_f-1}s^{n_f-1} \cdots + f_1 s + f_0}, \quad (11.51)$$

in the *observer* canonical form, as shown in Fig. 11.15a. The reader may readily confirm that the transfer function is (11.51).

Then the input, $Q(s)$, is fed forward via a set of weighting coefficients, including the output of the last integrator in the chain, as shown in Fig. 11.15b. This yields the following transfer function with as many zeros as poles, i.e. zero relative degree:

$$\frac{W(s)}{Q(s)} = \frac{w_{n_f}s^{n_f} + w_{n_f-1}s^{n_f-1} \cdots + w_1 s + w_0}{s^{n_f} + f_{n_f-1}s^{n_f-1} \cdots + f_1 s + f_0}. \quad (11.52)$$

Comparison with (11.51) then reveals

$$W(s) = \left(w_{n_f}s^{n_f} + w_{n_f-1}s^{n_f-1} \cdots + w_1 s + w_0\right) V(s) \Rightarrow w(t) = \sum_{i=0}^{n_f} w_i v^{(i)}. \quad (11.53)$$

Hence, by combining the required derivative action with the implementation of the dynamic part of the controller, all the input derivatives up to order, n_f, can be realised.

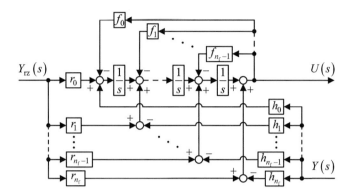

Fig. 11.16 Controller implementation with restricted reference input derivative feedforward

This avoids software differentiation. The next step is to implement the controller transfer function relationship,

$$U(s) = \frac{R(s)Y_{rz}(s) - H(s)Y(s)}{F(s)}, \quad (11.54)$$

as for (11.52) twice, once for the input, $Y_{rz}(s)$, and once for the input, $Y(s)$, using the principle of superposition. First, the maximum orders of the derivatives of $y(t)$ and $y_{rz}(t)$ to be realised will be restricted to n_f or below. The inequality constraint,

$$n_f \geq n_h = n_a - 1, \quad (11.55)$$

of the polynomial degrees will be recalled from Chap. 4. The maximum order of derivative of $y(t)$ that is needed is n_h, which is valid since from (11.55) $n_h \leq n_f$. For zero nominal dynamic lag, however, the maximum order of derivative of $y_{rz}(t)$ is $n_d = n_f + n_a$, but this problem will be resolved in the final step. With the degree restriction, the implementation block diagram of (11.54) is shown in Fig. 11.16.

The transfer function relationship of this controller is then

$$U(s) = \frac{\left(r_{n_f}s^{n_f} + r_{n_f-1}s^{n_f-1} + \cdots + r_1s + r_0\right)Y_{rz}(s) - \left(h_{n_f}s^{n_f} + h_{n_f-1}s^{n_f-1} + \cdots + h_1s + h_0\right)Y(s)}{s^{n_f} + f_{n_f-1}s^{n_f-1} + \cdots + f_1s + f_0}, \quad (11.56)$$

which is (11.54). Finally, to obtain the further reference input derivatives between the orders of $n_f + 1$ and $n_d = n_f + n_a$, it is necessary to provide a calculated or estimated derivative of $y_r(t)$, and therefore $y_{rz}(t)$ through (11.46), of order n_a.

This may be shown by expressing the complete weighted sum of reference input derivatives as the following expansion of $R(s)Y_{rz}(s)$.

11.3 Dynamic Lag Pre-compensation

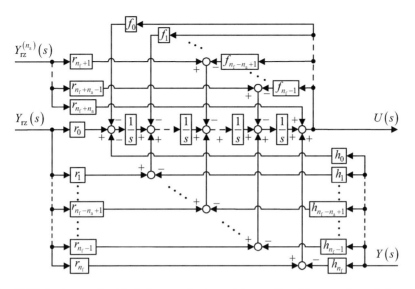

Fig. 11.17 Implementation block diagram of polynomial controller giving zero dynamic lag

$$\begin{aligned}
&\left(r_{n_f+n_a}s^{n_a+n_f} + r_{n_f+n_a-1}s^{n_a+n_f-1} + \cdots + r_{n_f+1}s^{n_f+1} + r_{n_f}s^{n_f}\right. \\
&\left. + r_{n_f-1}s^{n_f-1} + \cdots + r_1 s + r_0\right) Y_{rz}(s) \\
&= \left(r_{n_f+n_a}s^{n_f} + r_{n_f+n_a-1}s^{n_f-1} + \cdots + r_{n_f+1}s^{n_f-(n_a-1)}\right) s^{n_a} Y_{rz}(s) \\
&\quad + (r_{n_f}s^{n_f} + \cdots + r_1 s + r_0) Y_{rz}(s) \\
&= \left(r_{n_f+n_a}s^{n_f} + r_{n_f+n_a-1}s^{n_f-1} + \cdots + r_{n_f+1}s^{n_f-(n_a-1)}\right) Y_{rz}^{(n_a)}(s) \\
&\quad + (r_{n_f}s^{n_f} + \cdots + r_1 s + r_0) Y_{rz}(s).
\end{aligned} \qquad (11.57)$$

Both parts of (11.57) can be implemented in a controller of similar structure to that of Fig. 11.16 but with an additional input, $y_{rz}^{(n_a)}$, fed forward via the coefficients in the first part, as shown in Fig. 11.17.

If the zero pre-compensator is needed, then in view of the controller structure of Fig. 11.17, it would have to be implemented separately for the reference input and its derivative of order, n_a, to yield

$$Y_{rz}(s) = [1/Z(s)] Y_r(s) \text{ and } Y_{rz}^{(n_a)}(s) = [1/Z(s)] Y_r^{(n_a)}(s). \qquad (11.58)$$

The block diagram corresponding to Fig. 11.14 is shown in Fig. 11.18.

Here $\begin{cases} R'(s) = r_{n_f}s^{n_f} + r_{n_f-1}s^{n_f-1} + \cdots + r_1 s + r_0 \\ R''(s) = r_{n_f+n_a}s^{n_f} + r_{n_f+n_a-1}s^{n_f-1} + \cdots + r_{n_f+2}s^{n_f-n_a+2} + r_{n_f+1}s^{n_f-n_a+1}. \end{cases}$

(11.59)

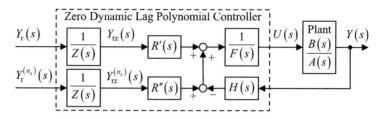

Fig. 11.18 Generic polynomial control with derivative feedforward and zero compensation

After pole placement according to (11.48), (11.45) becomes

$$\frac{Y(s)}{Y_r(s)} = \frac{R(s)}{Z(s)} \cdot \frac{B(s)}{D(s)} \tag{11.60}$$

and therefore zero dynamic lag requires $R(s) = D(s)$ and $Z(s) = B(s)$. Then in view of (11.58), dynamic lag pre-compensation is only practicable for plants with finite zeros in the left half of the s-plane. It is possible, however, that the dynamic lag could be reduced, but not eliminated, for plants with right half-plane zeros by setting $R(s) = D(s)$ but forming $Z(s)$ according to the method of Sect. 5.2.7.5.

Example 11.2 Position control of a flexible drive with zero dynamic lag

The plant here is the one of Example 5.10 but with an additional kinematic integrator (Chap. 2) for position control. The plant transfer function is therefore

$$\frac{y(s)}{u(s)} = \frac{b_0}{s^4 + a_2 s^2} \tag{11.61}$$

with the plant parameters b_0 and a_2, the same as, respectively, b_0 and a_1 in Example 4.10. The specified settling time is the same, i.e. $T_s = 0.2$ s. For the minimal order controller, the degree inequality constraint (11.55) becomes the equation $n_f = n_h = n_a - 1 = 4 - 1 = 3$ and therefore the closed-loop system order is $N = n_a + n_f = 4 + 3 = 7$. With multiple pole assignment using the 5 % settling time formula, the desired closed-loop characteristic polynomial is

$$s^7 + d_6 s^6 + d_5 s^5 + d_4 s^4 + d_3 s^3 + d_2 s^2 + d_1 s + d_0 = (s+d)^7 \tag{11.62}$$
$$= s^7 + 7ds^6 + 21d^2 s^5 + 35d^3 s^4 + 35d^4 s^3 + 21d^5 s^2 + 7d^6 s + d^7.$$

where $d = 1.5(1+N)/T_s|_{N=7} = 12/T_s$. The controller polynomials are

$$H(s) = h_3 s^3 + h_2 s^2 + h_1 s + h_0, \quad F(s) = f_3 s^3 + f_2 s^2 + f_1 s + f_0,$$
$$Z(s) = 1, \quad R(s) = D(s)/b_0. \tag{11.63}$$

11.3 Dynamic Lag Pre-compensation

where $f_3 = 1$. For dynamic lag compensation, feedforward implementation reduces the reference input derivative calculation requirement to $\ddddot{y}_r \equiv s^4 Y_r(s)$ based on

$$R(s)Y_r(s) = \left(r_7 s^3 + r_6 s^2 + r_5 s + r_4\right) s^4 Y_r(s) + \left(r_3 s^3 + r_2 s^2 + r_1 s + r_0\right) Y_r(s). \tag{11.64}$$

Equating the characteristic polynomial of the loop in Fig. 11.14 to that desired yields

$$A(s)F(s) + B(s)H(s) = D(s), \tag{11.65}$$

from which a linear matrix equation of the form, $\mathbf{Pg} = \mathbf{d}$, results, where $\mathbf{P}$ is the square matrix of plant parameters, $\mathbf{g}$ is the column vector of controller parameters to be determined and $\mathbf{d}$ is the column vector of desired characteristic polynomial coefficients. As in Example 4.10, the plant has no finite zeros and therefore $\mathbf{P}$ is lower triangular. For this example, the equation is

$$\begin{bmatrix} 1 & 0 & 0 & 0 & 0 & 0 & 0 & 0 \\ 0 & 1 & 0 & 0 & 0 & 0 & 0 & 0 \\ a_2 & 0 & 1 & 0 & 0 & 0 & 0 & 0 \\ 0 & a_2 & 0 & 1 & 0 & 0 & 0 & 0 \\ 0 & 0 & a_2 & 0 & b_0 & 0 & 0 & 0 \\ 0 & 0 & 0 & a_2 & 0 & b_0 & 0 & 0 \\ 0 & 0 & 0 & 0 & 0 & 0 & b_0 & 0 \\ 0 & 0 & 0 & 0 & 0 & 0 & 0 & b_0 \end{bmatrix} \begin{bmatrix} f_3 \\ f_2 \\ f_1 \\ f_0 \\ h_3 \\ h_2 \\ h_1 \\ h_0 \end{bmatrix} = \begin{bmatrix} 1 \\ d_6 \\ d_5 \\ d_4 \\ d_3 \\ d_2 \\ d_1 \\ d_0 \end{bmatrix} \tag{11.66}$$

Solution by back substitution is as follows:

$$\left. \begin{array}{l} f_3 = 1, \quad f_2 = d_6, \quad a_2 f_3 + f_1 = d_5 \Rightarrow \underline{f_1 = d_5 - a_2 f_3}, \\ a_2 f_2 + f_0 = d_4 \Rightarrow \underline{f_0 = d_4 - a_2 f_2}, \quad a_2 f_1 + b_0 h_3 = d_3 \Rightarrow \underline{h_3 = (d_3 - a_2 f_1)/b_0}, \\ a_2 f_0 + b_0 h_2 = d_2 \Rightarrow \underline{h_2 = (d_2 - a_2 f_0)/b_0}, \quad b_0 h_1 = d_1 \Rightarrow \underline{h_1 = d_1/b_0}, \quad \underline{h_0 = d_0/b_0} \end{array} \right\} \tag{11.67}$$

Finally, the reference input polynomial coefficients are

$$\underline{r_i = d_i/b_0}, \quad i = 1, 2, \ldots, 7. \tag{11.68}$$

The complete control system is shown in Fig. 11.19, with the controller,

$$U(s) = \frac{\left[\begin{array}{c}\left(r_7 s^3 + r_6 s^2 + r_5 s + r_4\right) s^4 Y_r(s) + \left(r_3 s^3 + r_2 s^2 + r_1 s + r_0\right) Y_r(s) \\ - \left(h_3 s^3 + h_2 s^2 + h_1 s + h_0\right) Y(s)\end{array}\right]}{s^3 + f_2 s^2 + f_1 s + f_0}. \tag{11.69}$$

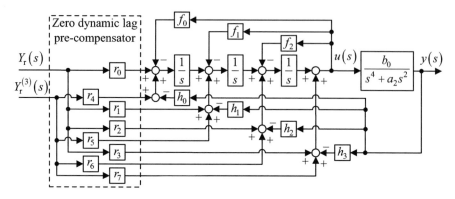

Fig. 11.19 Polynomial control of flexible drive with dynamic lag pre-compensation

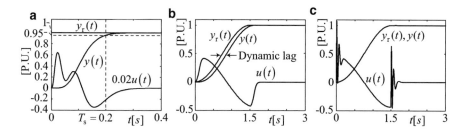

Fig. 11.20 Simulation of polynomial control of flexible drive. (**a**) $y_r(t) = h(t)$; no pre-compensation. (**b**) $y_r(t)$ continuous; no pre-compensation. (**c**) $y_r(t)$ continuous; with pre-compensation

The controller block diagram is in the observer canonical form for direct implementation.

Figure 11.20a shows the step response without the dynamic lag pre-compensation, which confirms that the settling time and step response shape are correct.

Figure 11.20b shows the response of the same system to a continuous and monotonically increasing reference input,

$$y_r(t) = \begin{cases} \frac{1}{2}\{1 - \cos[(\pi/T_r)t]\}, & 0 \le t \le T_r \\ 1, & t > T_r \end{cases}. \tag{11.70}$$

This demonstrates the dynamic lag. Finally, Fig. 11.20c shows the response of the pre-compensated system in which the calculated fourth reference input derivative is

$$\overset{....}{y_r}(t) = \begin{cases} \frac{1}{2}(\pi/T_r)^4 \cos[(\pi/T_r)t], & 0 \le t \le T_r \\ 0, & t > T_r \end{cases}. \tag{11.71}$$

Note that the step response of Fig. 11.20a requires a normalised control variable with $|u(t)| \le 1$ that could not be realised in practice unless the step magnitude was

11.3 Dynamic Lag Pre-compensation

reduced by two orders of magnitude. The high acceleration and deceleration for the monotonically increasing continuous reference input (11.70) are much smaller.

Hence, in Fig. 11.20b, c, $u(t)$ is realisable.

11.3.2.4 Inclusion of Robustness

In applications requiring dynamic lag pre-compensation in which robustness against plant parametric uncertainties and/or external disturbances is required, then any robust control technique can be employed yielding known linear closed-loop dynamics such as those in Chap. 10. Then dynamic lag pre-compensation can be applied as described in Sects. 11.3.2.1 and 11.3.2.2. Since sliding mode control or robust pole assignment reduces the order of the transfer function describing the closed-loop system, the dynamic lag pre-compensator is simpler than it would be with purely model-based control for the same plant.

11.3.2.5 Constant Disturbance Rejection

If any control technique is employed embodying an integral term for ensuring zero steady-state error with a constant external disturbance, then provided the closed-loop dynamics is linear, dynamic lag pre-compensation as described in Sects. 11.3.2.1 and 11.3.2.2 can be applied, but bearing in mind the need for generation of a full set of reference input derivatives up to an order equal to that of the closed-loop transfer function. If the plant is linear, however, then the polynomial control technique of Sect. 11.3.2.3 could be applied to ease the requirement for reference input derivatives, but introducing the integral term in a different way to that described in Sect. 4.3.5. The outer integral control loop described in Chap. 4 would render the derivative feedforward technique of Sect. 11.3.2.3 inapplicable, but instead, a pure integrator can be inserted in the control channel, as for the sliding mode control with a control smoothing integrator (Chap. 9). The general polynomial control system of Fig. 11.18 is then replaced by that of Fig. 11.21, where $D_s(s)$ is the external disturbance referred to the control input.

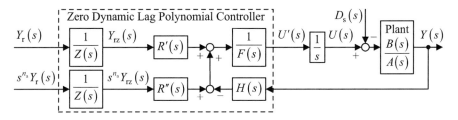

Fig. 11.21 Generic polynomial control with derivative feedforward and an integral term

To demonstrate that a constant disturbance causes no control error, it is sufficient to show that the steady-state value of $y(t)$, with all inputs zero except $d_s(t) = Dh(t)$, $D = \text{const.} \Rightarrow D_s(s) = D/s$, is zero. Hence,

$$y(s) = \frac{-sB(s)F(s)}{sA(s)F(s) + B(s)H(s)} \cdot \frac{D}{s} \Rightarrow y_{ss} = \lim_{s \to 0} sy(s) = 0. \quad (11.72)$$

The method of determination of the controller parameters is the same as for the generic polynomial controller of Sect. 11.3.2.3, but with the plant transfer function, $B(s)/A(s)$, replaced by $B(s)/[sA(s)]$.

Example 11.3 Acceleration control of a goods vehicle Diesel engine with zero dynamic lag

The accelerator pedal of a conventional motor vehicle does not directly control the acceleration. Since keeping the pedal in a fixed position results in a constant steady-state speed on a level terrain, it is really a speed control. This led a commercial vehicle manufacturer to research into the possibility of designing the engine control system so that the accelerator pedal properly controls the acceleration and carry out experiments to investigate how this would feel to the driver. It is apparent that once the vehicle is in motion, the system would have to be arranged such that zero acceleration would be commanded by keeping the pedal in an intermediate position. Then accelerating the vehicle to a higher speed would require a momentary further depression of the foot pedal and slowing the vehicle down would require momentarily raising the pedal, afterwards returning to the zero acceleration position. To implement this scheme would require a control system in which acceleration is the reference input, provided by the pedal position measurement. To avoid the control loop dynamics affecting the driver's feel, dynamic lag pre-compensation will also be required.

A common model of a Diesel engine, relating the torque equivalent control input, $u_t(t)$, to the crankshaft speed, $\omega_c(t)$, and the mechanical load represented by a load torque, $d_L(t)$, referred to the control input, is the transfer function relationship:

$$\Omega_c(s) = \left(\frac{b}{s+a}\right)[U_f(s) - D_L(s)]. \quad (11.73)$$

The mechanical load may be assumed purely inertial and equivalent to a mass bolted to the crank shaft with a moment of inertia of up to 20 times the equivalent moment of inertia of the moving engine parts referred to the crank shaft. This load moment of inertia is assumed to be unknown as it will depend upon the variable mass of the goods being transported and the gear selected.

The plant parameters are given as $b = 180$ [r/m/Nm] and $a = -0.05$ $[s^{-1}]$, noting that Diesel engines are open-loop unstable so that feedback control is essential.

11.3 Dynamic Lag Pre-compensation

Two different control systems are developed below. The first is direct acceleration control using an approximate crankshaft angular acceleration measurement obtained by applying software differentiation of the angular velocity measurement, yielding

$$Y(s) = \left(\frac{b}{s+a}\right)\left(\frac{s}{1+sT_f}\right)[U_f(s) - D_L(s)]. \tag{11.74}$$

where low-pass filtering with time constant, T_f, is included to attenuate high-frequency components of measurement noise. Then $y(t) \cong \dot{\omega}_c(t)$ for sufficiently small T_f. The DC blocking effect of the differentiator, however, would prevent the control of the acceleration to reach a constant steady-state value using $u_f(t)$ directly. This problem could be solved by introducing an integrator with input, $u(t)$, and output, $u_f(t)$, giving an augmented plant transfer function relationship

$$Y(s) = \left(\frac{b}{s+a}\right)\left(\frac{s}{1+sT_{f1}}\right)\left[\frac{1}{s}U(s) - D_L(s)\right]$$

$$= \left(\frac{b}{s+a}\right)\left(\frac{1}{1+sT_{f1}}\right)[U(s) - sD_L(s)]. \tag{11.75}$$

The pole–zero cancellation between the differentiator and the integrator renders the integrator state unobservable (Chap. 3) with respect to the controlled output, $\alpha_c(t)$. The consequence is that any acceleration controller designed using model (11.75) could not control the vehicle speed. For example, a zero acceleration reference could be reached precisely but at an arbitrary constant speed dependent upon a constant value of the unobservable integrator output. This, however, would not be a problem since the driver would close an additional loop, viewing the speedometer and controlling the speed (and therefore the integrator output) through the acceleration demand, $y_r(t)$, as shown in Fig. 11.22.

The second approach yields a slightly simpler system but is not based on direct feedback control of the crankshaft acceleration. First the simplest possible speed control loop is formed that will guarantee zero steady-state error with a constant speed demand and have closed-loop dynamics almost independent of plant modelling errors and the unknown mechanical load.

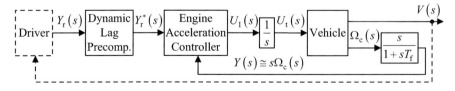

Fig. 11.22 Direct acceleration control of a road vehicle Diesel engine

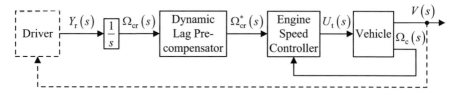

Fig. 11.23 Acceleration control of a road vehicle Diesel engine via an integrator inserted in the reference input channel of the speed control loop

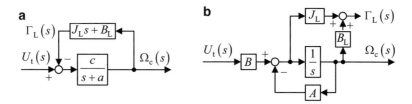

Fig. 11.24 Vehicle model comprising engine and inertial mechanical load. (**a**) Inverse dynamic form, (**b**) equivalent form without algebraic loop

Once this has been achieved, the loop is easily adapted to respond to a time-varying acceleration demand, by simply providing the speed reference input from an integrator whose input is the acceleration reference input, denoted by $Y_r(s)$, as in Fig. 11.22. This is shown in Fig. 11.23.

If the hardware implemented part of the control system were to be shown alone, then the pure integrator outside the loop would seem counter-intuitive but it is seen to be the integral term of the outer loop completed by the driver.

In preparation for the control system design, the plant model parameters will be converted to SI units, so b is replaced by

$$c = \pi b/30. \qquad (11.76)$$

For simulation, a vehicle model will be formed including the mechanical load, represented by a mass with moment of inertia, J_L, and a viscous drag coefficient, B_L, using the inverse dynamic form shown in Fig. 11.24a.

MATLAB®–SIMULINK® would be intolerant of the algebraic loop present in Fig. 11.24a, so this can be removed by deriving the transfer function relationship from the block diagram of Fig. 11.24b, which is free of algebraic loops, while having the same transfer function relationship.

The load torque is made available in Fig. 11.24b for monitoring purposes, not feedback control:

11.3 Dynamic Lag Pre-compensation

$$\begin{bmatrix} \Omega_c(s) \\ \Gamma_L(s) \end{bmatrix} = \frac{\begin{bmatrix} \frac{c}{s+a}(J_L s + B_L) \\ \frac{c}{s+a}(J_L s + B_L) \end{bmatrix}}{1 + \frac{c}{s+a}(J_L s + B_L)} = \frac{\begin{bmatrix} c(J_L s + B_L) \end{bmatrix}}{s + a + cJ_L s + cB_L}$$

$$= \frac{\begin{bmatrix} \frac{c}{1+cJ_L}(J_L s + B_L) \\ \frac{c}{1+cJ_L}(J_L s + B_L) \end{bmatrix}}{s + \frac{a+cB_L}{1+cJ_L}} = \frac{\begin{bmatrix} B \\ B(J_L s + B_L) \end{bmatrix}}{s + A}.$$
(11.77)

Since the load moment of inertia, J_L, is unknown, the load torque, $\Gamma_L(s)$, will be treated as external and the two plant transfer function relationships, (11.73) and (11.75), will be used for the design of the controllers. If T_f is small enough to be ignored in Fig. 11.23, then these transfer function relationships are the same except for the disturbances, which are $\gamma_L(t)$ for the velocity control loop and $\dot{\gamma}_L(t)$ for the acceleration control loop. In this example, the control systems will be made insensitive to the disturbances by robust pole placement rather than disturbance estimation and cancellation. Then the following transfer function, obtained by setting $\Gamma_L(s) = 0$ in (11.73) and (11.75), can be used for the design of both controllers:

$$\frac{Y(s)}{U_1(s)} = \frac{\Omega_c(s)}{U_t(s)} = \frac{c}{s+a}.$$
(11.78)

The simplest controller guaranteeing zero-steady state error with a constant reference input is the IP controller, shown applied to plant model (11.78) in Fig. 11.25.

For the acceleration control loop of Fig. 11.22, $Z(s) = Y(s)$, $Z_r^*(s) = Y_r^*(s)$ and $U(s) = U_1(s)$. For the speed control loop of Fig. 11.23, $Z(s) = \Omega_c(s)$, $Z_r^*(s) = \Omega_{cr}^*(s)$ and $U(s) = U_t(s)$. The closed-loop transfer function is

$$\frac{Z(s)}{Z_r^*(s)} = \frac{\frac{c}{s+a} \cdot \frac{K_I}{s}}{1 + \left(\frac{c}{s+a}\right)\left(K_p + \frac{K_I}{s}\right)} = \frac{cK_I}{s^2 + (a + cK_p)s + cK_I}.$$
(11.79)

For robust pole placement (Chap. 10), the desired closed-loop transfer function is

$$\frac{Y(s)}{U(s)} = G_{cl}(s) = \frac{pq}{(s+p)(s+q)} = \frac{pq}{s^2 + (p+q)s + pq}.$$
(11.80)

Fig. 11.25 IP control loop common to both control systems

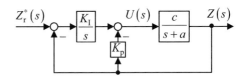

Here, $p = 3/T_s$ is the dominant pole magnitude, T_s is the settling time (5 % criterion) and $q = 1/T_r$ is the robust pole magnitude, where $T_r \ll T_s$. For the pole placement, equating the denominators of (11.79) and (11.80) yields the gains

$$K_p = (p + q - a)/c \text{ and } K_1 = pq/c. \tag{11.81}$$

For both control systems, the dynamic lag pre-compensator transfer function is

$$\frac{Z_r^*(s)}{Z_r(s)} = P(s) = \frac{1}{G_{cl}(s)} = 1 + P_1 s + P_2 s^2, \tag{11.82}$$

where $P_1 = (p + q)/(pq)$ and $P_2 = 1/(pq)$.

To test the control systems by simulation, an acceleration demand profile will be generated that emulates the action of a driver in accelerating a vehicle from rest to a constant speed. The pre-compensator requires finite first and second derivatives of the acceleration demand. This can be achieved by first generating a second derivative signal that switches between zero and equal and opposite constant values of $\pm D$ and then integrating this signal twice, as shown in Fig. 11.26.

This produces a piecewise linear jerk profile comprising two equal and opposite triangular pulses, as shown. The demanded acceleration profile then consists of start and finish phases that are two contiguous parabolic segments, with a constant intermediate phase. Suppose that the second derivative pulse duration, T_d, and the maximum demanded acceleration, $y_{r\max}$, are chosen initially. Then the required second derivative pulse magnitude, D, is calculated as follows. By inspection of Fig. 11.26, $\ddot{y}_{r\max}$ is the area under the first square pulse of $\dddot{y}_r(t)$, which is

$$\ddot{y}_{r\max} = DT_d. \tag{11.83}$$

Then $\dot{y}_{r\max}$ is the area under the first triangular pulse of $\ddot{y}_r(t)$, which is

$$\dot{y}_{r\max} = \ddot{y}_{r\max} T_d. \tag{11.84}$$

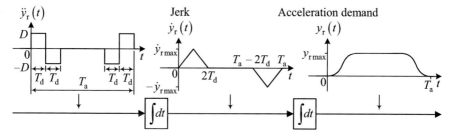

Fig. 11.26 Acceleration demand profile for Diesel engine acceleration control system

11.3 Dynamic Lag Pre-compensation

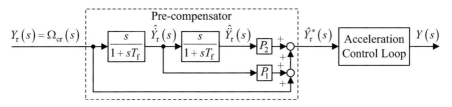

Fig. 11.27 Pre-compensator estimating first and second derivative using position measurement

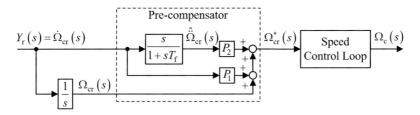

Fig. 11.28 Pre-compensator estimating only second derivative, using velocity measurement

Substituting for $y_{r\,max}$ in (11.84) using (11.83) then yields

$$y_{r\ max} = DT_d^2 \Rightarrow D = y_{r\ max}/T_d^2. \tag{11.85}$$

The dynamic lag pre-compensators for the acceleration control loop of Fig. 11.22 and the velocity control loop of Fig. 11.23 are identical and given by (11.82), but their implementations are slightly different through utilising the integrator of Fig. 11.23, whose input is already the required first derivative of the reference input. The second derivative is provided by software differentiation for which a filtered differentiator identical to that of Fig. 11.22 will suffice, as shown in Fig. 11.27a.

The hats above some of the quantities indicate estimates that are approximate due to the small lag introduced by the low-pass filters built into the differentiators. For the acceleration control loop, one such differentiator is required, as shown in Fig. 11.28.

Figure 11.29a, b show simulations of the control systems of Figs. 11.22 and 11.23 with the same demanded acceleration profile and controller parameters. In both systems, the mechanical load referred to the crankshaft is represented by a mass with moment of inertia, $J_L = 1\,[\text{Kg m}^2]$, which is about 20 times the equivalent moment inertia of the unloaded engine moving parts and a viscous rolling friction coefficient of $B_L = 0.6\,[\text{Nms/rad}]$. The settling time of the control loops are set to $T_s = 0.5\,[s]$. The filtering time constants of the approximate differentiators are set to $T_{fl} = 0.005\,[s]$. The time constant of the robust pole is also set to $T_r = 0.005\,[s]$. The performance of both control systems is satisfactory and almost identical. The peak acceleration following error of the speed control loop-based control system with only one filtered differentiator is slightly greater than that of the acceleration-based control system with three filtered differentiators.

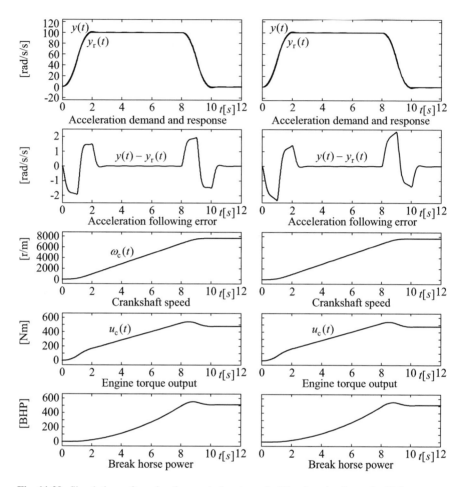

Fig. 11.29 Simulations of acceleration control systems for Diesel engine in road vehicle

This result is attributed to the fact that the filtered output differentiator producing the crankshaft acceleration estimate of the acceleration loop-based control system is in the feedback loop and therefore introduces a factor, $1 + sT_f$, in the *numerator* of the closed-loop transfer function, which tends to reduce the following error.

11.3.3 Implementation with Feedback Linearising Control

11.3.3.1 Closed-Loop Transfer Function Relationship

The general feedback linearising (FL) control law presented in Sect. 11.2.3 may be used with a dynamic lag pre-compensator for mechanisms containing continuous nonlinearities. To summarise the control law, given the mechanism model,

11.3 Dynamic Lag Pre-compensation

$$M(q)\ddot{q} + C(q,\dot{q})\dot{q} + g(q) = \tau, \tag{11.86}$$

the control law is

$$\tau = M(q)\{W[q_r - q] - Z\dot{q}\} + C(q,\dot{q})\dot{q} + g(q). \tag{11.87}$$

where $q \in \Re^d$ is the position vector and d is the number of degrees of freedom of the controlled mechanism. As it stands, the closed-loop system formed by (11.87) and (11.86) obeys the differential equation,

$$\ddot{q} = W[q_r - q] - Z\dot{q}, \tag{11.88}$$

and therefore has dynamic lag between the demanded position vector, $q_r(t)$, and the controlled position vector, $q(t)$. This dynamic lag can be nearly eliminated with a precision depending upon the accuracy of the model of (11.86). For SISO plants in the previous subsections, the pre-compensator transfer function is the reciprocal of the closed-loop transfer function, which is converted to derivative feedforward equations in the time domain. Taking a similar approach here, the closed-loop transfer function relationship corresponding to (11.88) is given by

$$[I^{(d)}s^2 + Zs + W]Q(s) = WQ_r(s) \Rightarrow Q(s) = [I^{(d)}s^2 + Zs + W]^{-1}WQ_r(s), \tag{11.89}$$

i.e.,

$$Q(s) = G_{cl}(s)Q_r(s) \tag{11.90}$$

where

$$G_{cl}(s) = [I^{(d)}s^2 + Zs + W]^{-1}W. \tag{11.91}$$

11.3.3.2 Dynamic Lag Pre-compensator

With an exact model, the dynamic lag can by eliminated by replacing $Q_r(s)$ with $Q'_r(s)$ in (11.90), so that

$$Q(s) = G_{cl}(s)Q'_r(s) \tag{11.92}$$

retaining $Q_r(s)$ as the reference input vector to be followed and finding a pre-compensator with a transfer function matrix, $P(s)$, such that

$$Q'_r(s) = P(s)Q_r(s) \tag{11.93}$$

and $Q(s) = Q_r(s)$ in the hypothetical case of a perfect model.

Substituting for $\mathbf{Q}'_r(s)$ in (11.92) using (11.93) yields

$$\mathbf{Q}(s) = \mathbf{G}_{cl}(s)\mathbf{P}(s)\mathbf{Q}_r(s). \tag{11.94}$$

Hence, if $\mathbf{Q}(s) = \mathbf{Q}_r(s)$, then

$$\mathbf{G}_{cl}(s)\mathbf{P}(s) = \mathbf{I}^{(d)} \Rightarrow \mathbf{P}(s) = \mathbf{G}_{cl}^{-1}(s) \tag{11.95}$$

and in view of (11.91), the required pre-compensator transfer function matrix is

$$\mathbf{P}(s) = \mathbf{W}^{-1}\left[\mathbf{I}^{(d)}s^2 + \mathbf{Z}s + \mathbf{W}\right]. \tag{11.96}$$

Substituting for $\mathbf{P}(s)$ in (11.93) using (11.96) then yields

$$\mathbf{Q}'_r(s) = \mathbf{W}^{-1}\left[\mathbf{I}^{(d)}s^2 + \mathbf{Z}s + \mathbf{W}\right]\mathbf{Q}_r(s) = \left[\mathbf{I}^{(d)} + \mathbf{W}^{-1}\mathbf{Z}s + \mathbf{W}^{-1}s^2\right]\mathbf{Q}_r(s). \tag{11.97}$$

In the time domain this is

$$\mathbf{q}'_r = \mathbf{q}_r + \mathbf{W}^{-1}\mathbf{Z}\dot{\mathbf{q}}_r + \mathbf{W}^{-1}\ddot{\mathbf{q}}_r \tag{11.98}$$

So the diagonal matrices, $\mathbf{Z}$ and $\mathbf{W}$, formed in Sect. 11.2.3 to yield the desired closed-loop responses without the pre-compensator, are used to form the derivative feedforward weighting matrices of the pre-compensator.

11.3.3.3 Combination of FL Control Law and Pre-compensator

An algorithm that combines the feedback linearising control law and the derivative feedforward dynamic lag pre-compensator may be formed by first replacing $\mathbf{q}_r$ by $\mathbf{q}'_r$ in (11.87) and then substituting for $\mathbf{q}'_r$ using (11.98). Thus,

$$\begin{aligned}\boldsymbol{\tau} &= \mathbf{M}(\mathbf{q})\left\{\mathbf{W}\left[\mathbf{q}'_r - \mathbf{q}\right] - \mathbf{Z}\dot{\mathbf{q}}\right\} + \mathbf{C}(\mathbf{q},\dot{\mathbf{q}})\,\dot{\mathbf{q}} + \mathbf{g}(\mathbf{q}) \\ &= \mathbf{M}(\mathbf{q})\left[\ddot{\mathbf{q}}_r + \mathbf{Z}(\dot{\mathbf{q}}_r - \dot{\mathbf{q}}) + \mathbf{W}(\mathbf{q}_r - \mathbf{q})\right] + \mathbf{C}(\mathbf{q},\dot{\mathbf{q}})\,\dot{\mathbf{q}} + \mathbf{g}(\mathbf{q}).\end{aligned} \tag{11.99}$$

This is identical to the *computed torque control law* [2] that originated in the field of robotics for direct control of jointed-arm manipulators without reduction gears. In the computed torque control law, the matrices, $\mathbf{Z}$ and $\mathbf{W}$, are presented, respectively, as the derivative gain matrix, $\mathbf{K}_d$, and the proportional gain matrix, $\mathbf{K}_p$.

11.3.3.4 Modelling Errors, External Disturbances and Noise

When applying control law (11.99), then in the ideal system with a perfect model and without external disturbances, plant noise, measurement noise and control saturation, the performance of the control system will be unaffected by $\mathbf{Z}$ and $\mathbf{W}$ provided, of course, that the 2d poles of the closed-loop system defined by (11.88) have negative real parts. In practice, however, adequate robustness against external disturbances, parametric uncertainties and the effects of plant and measurement noise has to be provided. This can be achieved by increasing ω_{ni} in (11.14) and (11.15) or p_i in (11.16) and (11.17), $i = 1, 2, \ldots, d$, as this increases the feedback gains applied to the errors, $\mathbf{q}_r - \mathbf{q}$ and $\dot{\mathbf{q}}_r - \dot{\mathbf{q}}$, in (11.99). To analyse the effects of these imperfections, let the plant Eq. (11.86) be rewritten as

$$\tilde{\mathbf{M}}(\mathbf{q})\ \ddot{\mathbf{q}} + \tilde{\mathbf{C}}(\mathbf{q},\dot{\mathbf{q}})\ \dot{\mathbf{q}} + \tilde{\mathbf{g}}(\mathbf{q}) = \boldsymbol{\tau} + \mathbf{n}_p \qquad (11.100)$$

where

$$\tilde{\mathbf{M}} = \mathbf{M} + \Delta\mathbf{M},\ \tilde{\mathbf{C}} = \mathbf{C} + \Delta\mathbf{C}\ \text{and}\ \tilde{\mathbf{g}} = \mathbf{g} + \Delta\mathbf{g}, \qquad (11.101)$$

$\Delta\mathbf{M}$, $\Delta\mathbf{C}$ and $\Delta\mathbf{g}$ are the plant modelling errors and $\mathbf{n}_p$ is the plant noise. Any external disturbance referred to the plant input, $\boldsymbol{\tau}$, is additive to $\mathbf{n}_p$, and therefore, for the purpose of this analysis, it is sufficient to consider it part of $\mathbf{n}_p$.

The control law (11.99) will be rewritten as

$$\boldsymbol{\tau} = \mathbf{M}(\mathbf{q}_m)\left[\ddot{\mathbf{q}}_r + \mathbf{Z}(\dot{\mathbf{q}}_r - \dot{\mathbf{q}}_m) + \mathbf{W}(\mathbf{q}_r - \mathbf{q}_m)\right] + \mathbf{C}(\mathbf{q}_m,\dot{\mathbf{q}}_m)\ \dot{\mathbf{q}}_m + \mathbf{g}(\mathbf{q}_m), \qquad (11.102)$$

where

$$\mathbf{q}_m = \mathbf{q} + \mathbf{n}_m, \qquad (11.103)$$

$\mathbf{n}_m$ being the measurement noise. Applying (11.102) to plant (11.100) then yields

$$\tilde{\mathbf{M}}(\mathbf{q})\ \ddot{\mathbf{q}} + \tilde{\mathbf{C}}(\mathbf{q},\dot{\mathbf{q}})\ \dot{\mathbf{q}} + \tilde{\mathbf{g}}(\mathbf{q}) - \mathbf{M}(\mathbf{q}_m)\left[\ddot{\mathbf{q}}_r + \mathbf{Z}(\dot{\mathbf{q}}_r - \dot{\mathbf{q}}_m) + \mathbf{W}(\mathbf{q}_r - \mathbf{q}_m)\right]$$
$$+ \mathbf{C}(\mathbf{q}_m,\dot{\mathbf{q}}_m)\ \dot{\mathbf{q}}_m + \mathbf{g}(\mathbf{q}_m) + \mathbf{n}_p. \qquad (11.104)$$

Let $\mathbf{Z}$ and $\mathbf{W}$ be given, respectively, by (11.17) and (11.16). Then $\mathbf{W} = \frac{1}{4}\mathbf{Z}^2$. Rearranging (11.104) to make the terms involving $\mathbf{W}$ and $\mathbf{Z}$ the subject then yields

$$\mathbf{Z}(\dot{\mathbf{q}}_r - \dot{\mathbf{q}}_m) + \frac{1}{4}\mathbf{Z}^2(\mathbf{q}_r - \mathbf{q}_m) = \mathbf{M}^{-1}(\mathbf{q}_m)\left[\tilde{\mathbf{M}}(\mathbf{q})\ \ddot{\mathbf{q}} + \tilde{\mathbf{C}}(\mathbf{q},\dot{\mathbf{q}})\ \dot{\mathbf{q}} + \tilde{\mathbf{g}}(\mathbf{q})\right.$$
$$\left. - \mathbf{C}(\mathbf{q}_m,\dot{\mathbf{q}}_m)\ \dot{\mathbf{q}}_m - \mathbf{g}(\mathbf{q}_m) - \mathbf{n}_p\right] - \ddot{\mathbf{q}}_r. \qquad (11.105)$$

Pre-multiplying throughout by $4\mathbf{Z}^{-2}$ then yields

$$4\mathbf{Z}^{-1}(\dot{\mathbf{q}}_r - \dot{\mathbf{q}}_m) + \mathbf{q}_r - \mathbf{q}_m = 4\mathbf{Z}^{-2}\left\{\mathbf{M}^{-1}(\mathbf{q}_m)\left[\tilde{\mathbf{M}}(\mathbf{q})\ddot{\mathbf{q}} + \tilde{\mathbf{C}}(\mathbf{q},\dot{\mathbf{q}})\dot{\mathbf{q}} + \tilde{\mathbf{g}}(\mathbf{q})\right.\right.$$
$$\left.\left. - \mathbf{C}(\mathbf{q}_m,\dot{\mathbf{q}}_m)\dot{\mathbf{q}}_m - \mathbf{g}(\mathbf{q}_m) - \mathbf{n}_p\right] - \ddot{\mathbf{q}}_r\right\}. \tag{11.106}$$

Recalling from Sect. 11.2.3 that $\mathbf{Z} = \mathbf{diag}\,(2p_1,\,2p_2,\ldots,\,2p_d)$, if the control loops are 'tightened' by increasing p_i, $i = 1, 2, \ldots, d$, which are the closed-loop pole magnitudes, the term $\mathbf{q}_r - \mathbf{q}_m$ becomes more significant and in the extreme, as $p_i \to \infty$, $\mathbf{Z}^{-1} \to 0$ and therefore $\mathbf{Z}^{-2} \to 0$, so that (11.106) becomes

$$\mathbf{q}_r - \mathbf{q}_m = 0 \tag{11.107}$$

In view of (11.103), the corresponding position control error becomes

$$\mathbf{e} = \mathbf{q}_r - \mathbf{q} = \mathbf{q}_m - \mathbf{q} = \mathbf{n}_m. \tag{11.108}$$

It may be concluded that increasing the control loop gains improves the robustness against modelling errors, plant noise and external disturbances but causes the control error to approach the measurement noise. This should not pose a problem in practice, however, since modern encoders used for position measurement generate negligible random noise and have negligible quantisation noise due to limited resolution.

11.4 Optimal Control for Minimising Frictional Energy Loss

11.4.1 Motivation

A large proportion of the energy needed to supply motion control systems is wasted in the form of nonrecoverable heat due to friction between moving surfaces. Consider the simple illustration of two surfaces in contact, such as in a mechanism on a production line executing repeated motions for extended periods. Let the relative velocity be $v(t)$ and a friction force $f(t)$. Then the frictional power loss is $p(t) = f(t)v(t)$. With linear viscous friction, $f(t) = K_v v(t)$ where K_v is the viscous friction coefficient. Then $p(t) = K_v v^2(t)$. In production line situations, it is common for a rest-to-rest position change manoeuvre to be repeatedly carried out in a fixed, predetermined time, T_m, an example being positioning of a workpiece for drilling operations. If a traditional controller of the PID pedigree is used with step reference position inputs, then the velocity, $v(t)$, typically rises rapidly at the beginning of the manoeuvre to a relatively large peak value after which it falls in an exponential fashion towards zero at the end of the manoeuvre. Due to the dependence of $p(t)$ on $v^2(t)$, this velocity peak imposes a severe penalty in energy

11.4 Optimal Control for Minimising Frictional Energy Loss

loss, which is $w(T_m) = K_f \int_0^{T_m} v^2(t) dt$. A closed-loop position control system that avoids such a velocity peak is therefore desirable. This will be derived in the following subsections.

11.4.2 Formulation of Optimal Control

In practice, the friction in mechanisms to be positioned often has significant nonlinearities in the force-velocity characteristic (Chap. 2). To cater for this, the state-space model of the simple one degree of freedom mechanism to be considered here is

$$\begin{aligned} \dot{x}_1 &= x_2 \\ \dot{x}_2 &= bu - f(x_2), \end{aligned} \quad (11.109)$$

where x_1 and x_2 are, respectively, the position and velocity, and the function, $f(x_2)$, which will not yet be specified, represents the effect of the friction. The problem is then to find an optimal state feedback control law, $u^*(x_{1r}, x_1, x_2)$, subject to the saturation constraints,

$$-u_{max} \le u \le +u_{max}, \quad (11.110)$$

that minimises the cost functional,

$$J = \int_{t_0}^{t_f} x_2^2(t) dt, \quad (11.111)$$

which is proportional to the frictional energy, for initial and final states of $[x_1(t_0), x_2(t_0)] = [0, 0]$ and $[x_1(t_f), x_2(t_f)] = [x_{1r}, 0]$, for a fixed manoeuvre time,

$$T_m = t_f - t_0. \quad (11.112)$$

First, the method of Pontryagin [3–5] will be applied. This only gives an open-loop solution but will be useful to investigate the nature of the optimal control as this information enables a feedback control system to be created, with its advantages of robustness against plant parameter uncertainties and external disturbances.

The summary of Pontryagin's method for a general plant will be given here and then applied to the plant of (11.109). The optimisation problem can be stated as follows. It is required to minimise the cost functional,

$$J = \int_{t_0}^{t_f} L[\mathbf{x}(t), \mathbf{u}(t)] dt, \quad (11.113)$$

where $\mathbf{x} \in \Re^n$ and $\mathbf{u} \in \Re^r$, with respect to $\mathbf{u}(t)$, subject to the constraints imposed by the plant state equation,

$$\dot{\mathbf{x}}(t) = \mathbf{f}[\mathbf{x}(t), \mathbf{u}(t)], \qquad (11.114)$$

and the control saturation constraints,

$$u_{i\,\text{min}} \le u_i \le u_{i\,\text{max}}, \quad i = 1, 2, \ldots, r, \qquad (11.115)$$

with initial and final states of, respectively, $\mathbf{x}(t_0)$ and $\mathbf{x}(t_f)$, $T_m = t_f - t_0$ being fixed. The optimal control will be denoted, $\mathbf{u}^*(t)$. The method is analogous to the minimisation of a function subject to constraint equations using Lagrange multipliers. Thus, the Hamiltonian,

$$H[\mathbf{p}(t), \mathbf{x}(t), \mathbf{u}(t)] = \mathbf{p}^T(t)\mathbf{f}[\mathbf{x}(t), \mathbf{u}(t)] + L[\mathbf{x}(t), \mathbf{u}(t)], \qquad (11.116)$$

is formed, where $\mathbf{p}(t)$ is the Lagrange multiplier vector, which is also the state of a second dynamical system called the adjoint system. The state, $\mathbf{p}(t)$, is referred to as the *costate* or the *adjoint system state*. Pontryagin's minimum principle states that the required optimal control function, $\mathbf{u}^*(t)$, and the corresponding optimal state trajectory, $\mathbf{x}^*(t)$, minimise (11.116). In terms of the Hamiltonian, the plant state equations are given by

$$\dot{\mathbf{x}} = \frac{\partial H}{\partial \mathbf{p}} \triangleq \left[\frac{\partial H}{\partial p_1} \frac{\partial H}{\partial p_2} \cdots \frac{\partial H}{\partial p_n}\right]^T \qquad (11.117)$$

and the aforementioned adjoint system state equation is given similarly by

$$\dot{\mathbf{p}} = -\frac{\partial H}{\partial \mathbf{x}} \triangleq \left[\frac{\partial H}{\partial x_1} \frac{\partial H}{\partial x_2} \cdots \frac{\partial H}{\partial x_n}\right]^T. \qquad (11.118)$$

For the plant (11.109) and cost functional (11.111), the Hamiltonian is

$$H = \begin{bmatrix} p_1 & p_2 \end{bmatrix} \begin{bmatrix} x_2 \\ bu - f(x_2) \end{bmatrix} + x_2^2 = p_1 x_2 + p_2 [bu - f(x_2)] + x_2^2 \qquad (11.119)$$

Then (11.117) yields

$$\begin{bmatrix} \dot{x}_1 \\ \dot{x}_2 \end{bmatrix} = \begin{bmatrix} \partial H/\partial p_1 \\ \partial H/\partial p_2 \end{bmatrix} = \begin{bmatrix} x_2 \\ bu - f(x_2) \end{bmatrix} \qquad (11.120)$$

which agrees with (11.109). Next, (11.118) yields

$$\begin{bmatrix} \dot{p}_1 \\ \dot{p}_2 \end{bmatrix} = -\begin{bmatrix} \partial H/\partial x_1 \\ \partial H/\partial x_2 \end{bmatrix} = \begin{bmatrix} 0 \\ -p_1 + p_2 f'(x_2) - 2x_2 \end{bmatrix} \qquad (11.121)$$

11.4 Optimal Control for Minimising Frictional Energy Loss

where $f'(x_2) = \partial f(x_2)/\partial x_2$. Consider the hypothetical case without the control saturation constraints (11.115) and denote all variables for the optimal control with the superscript †. Then, in view of (11.119), the optimal control, $u^\dagger(t)$, satisfies

$$\frac{\partial H}{\partial u} = 0 \Rightarrow p_2^\dagger = 0 \quad (11.122)$$

so that (11.121) yields

$$2x_2^\dagger = -p_1^\dagger = \text{const.} \Rightarrow x_2^\dagger = \text{const.} \quad (11.123)$$

The optimal control therefore yields a constant velocity of the mechanism. With reference to (11.109), the control needed to sustain this constant velocity is

$$u^\dagger = f\left(x_2^\dagger\right)/b. \quad (11.124)$$

This must be just sufficient to reach the demanded position, x_{1r}, in the specified manoeuvre time, T_m. Since $x_{1r} = \int_0^{T_m} x_2(t)dt$ is the area under the graph of $x_2(t)$ against t, and T_m is fixed, then any deviation of $x_2(t)$ from the constant optimal value, $x_2^\dagger$, would result in a peak value of $x_2(t)$ larger than $x_2^\dagger$ and therefore a frictional energy penalty. Hence, $x_2^\dagger$ is the extreme value of the velocity for the optimal control. This will be denoted by $x_{2\,\text{ext}}^\dagger$. The term 'extreme' is preferred to 'maximum' because it is applicable to both signs of $x_{2\,\text{ext}}^\dagger$. Although (11.124) would have to satisfy (11.115), however, the jump from $x_2 = 0$ to $x_2 = x_{2\,\text{ext}}^\dagger$ at the beginning of the manoeuvre and from $x_2 = x_{2\,\text{ext}}^\dagger$ to $x_2 = 0$ at the end of the manoeuvre would require $u^\dagger(t)$ to have oppositely signed Dirac delta impulses, and since these are of infinite magnitude albeit for an infinitesimal time, constraint (11.115) would be violated. Respecting this constraint, however, let the constant velocity during the manoeuvre of the practical optimal control be denoted, $x_{2\,\text{ext}}^*$. The optimal control must first saturate at $u^* = u_{\max} \,\text{sgn}(x_{1r})$ to give the system the maximum possible acceleration to reach $x_2 = x_{2\,\text{ext}}^*$ and at the end of the manoeuvre saturate at $u^* = -u_{\max} \,\text{sgn}(x_{1r})$ to give the system the maximum possible deceleration to reach $x_2 = 0$. The magnitude of $x_{2\,\text{ext}}^*$ would have to be slightly greater than $x_{2\,\text{ext}}^\dagger$ but would be minimised in magnitude by allowing control saturation during the acceleration and the deceleration. The form of the optimal control, velocity and position profiles is sketched in Fig. 11.30.

Remarkably, the hypothetical optimal state trajectory of Fig. 11.30a without the control constraints is independent of the plant parameters, including the friction. The state trajectory of Fig. 11.30b, however, *does* depend on the plant parameters during the acceleration and deceleration phases when the control is saturated, but if the control magnitude is such that the acceleration and deceleration periods are much less than the demanded manoeuvre time, then $x_{2\,\text{ext}}^* \cong x_{2\,\text{ext}}^\dagger$ and the optimal state trajectory is almost the same as for Fig. 11.30a.

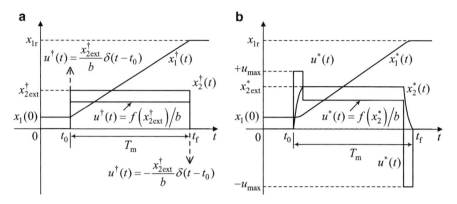

Fig. 11.30 Forms of optimal trajectories for minimising frictional energy loss. (**a**) Hypothetical case without control saturation, (**b**) practical case with control saturation

11.4.3 Minimum Frictional Energy State Feedback Control Law

A state feedback control law realising the behaviour of Fig. 11.30b will now be synthesised, first as a bang–bang control law by forming an appropriate switching boundary in the phase plane. This draws on the material of Chaps. 9 and 10. As for the time-optimal control, the two segments of the switching boundary including the origin of the phase plane are the two state trajectories, one for $u = +u_{\max}$ and the other for $u = -u_{\max}$, that terminate on the origin and start at $x_2 = x_{2\,\text{ext}}^* = \pm x_{2\,\max}^*$. These are solutions of the state trajectory differential equation,

$$\frac{dx_2}{dx_1} = [bu - f(x_2)]/x_2, \qquad (11.125)$$

formed from (11.109) for $u = \pm u_{\max}$. The switching boundary is then continued from the end points of the central segments by two horizontal straight line segments at $x_2 = \pm x_{2\,\max}^*$. The switching boundary, the resulting closed-loop phase portrait and the state trajectories for two rest-to-rest manoeuvres are shown in Fig. 11.31. Since the trajectories of the phase portrait are directed towards the horizontal segments of the switching boundary from both sides, sliding motion will take place and hold the velocity at $x_{2\,\text{ext}}^* = \pm x_{2\,\max}^*$, as required. Examination of Fig. 11.31 indicates that the closed-loop system exhibits stable behaviour for any initial state.

In Fig. 11.31b, the system starts from at rest at point 'a'. The control saturates at $u = +u_{\max}$ immediately at time, t_a, to accelerate the system. Once the state trajectory reaches point 'b' at time t_b, sliding motion commences, and the system is forced to stop accelerating and move at the constant velocity, $x_2 = x_{2\,\max}^*$, until point 'c' is reached at time, t_c, when the control again saturates at $u = -u_{\max}$ to decelerate the system, which comes to rest at point, 'h' at time, t_h. The second manoeuvre takes the system back to the original position via the

11.4 Optimal Control for Minimising Frictional Energy Loss

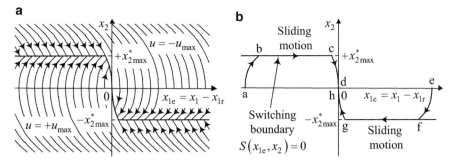

Fig. 11.31 Bang–bang sliding mode form of minimal frictional energy control. (**a**) Closed-loop phase portrait, (**b**) state trajectories for two rest-to-rest manoeuvres

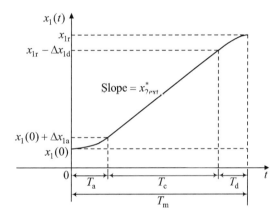

Fig. 11.32 Rest-to-rest position response of minimum friction energy control system

trajectory, 'efgh', and is trajectory 'abcd' reflected in the origin of the phase plane, with $x_{2\,\text{ext}}^* = -x_{2\,\text{max}}^*$.

The user of the control system has to specify the required manoeuvre time, T_m as well as x_{1r}. Then the software of the controller calculates $x_{2\,\text{ext}}^*$. The way in which this is done can be explained with the aid of Fig. 11.32.

Note that Figs. 11.32 and 11.30 pertain to $x_{1r} > x_1(0)$. For $x_{1r} < x_1(0)$, the graphs would be inverted forms of those shown. In principle, the solutions to (11.109) with the appropriate initial conditions yield relationships between the acceleration distance, Δx_{1a}, the acceleration time, T_a, the deceleration distance, Δx_{1d}, the deceleration time, T_d; and the extreme velocity, $x_{2\,\text{ext}}^*$. Thus,

$$\Delta x_{1a} = f_a(T_a), \quad \Delta x_{1d} = f_d(T_d), \quad x_{2\,\text{ext}}^* = g_a(T_a) \text{ and } x_{2\,\text{ext}}^* = g_d(T_d) \tag{11.126}$$

and

$$x_{2\,\text{ext}}^* = \frac{x_{1r} - x_1(0) - (\Delta x_{1a} + \Delta x_{1d})}{T_m - (T_a + T_d)}. \tag{11.127}$$

The solution of (11.126) and (11.127) for $x^*_{2\,\text{ext}}$, however, depends upon the plant model (11.109) and would be subject to errors due to uncertainties in the friction function, $f(x_2)$, exacerbated by its variation with temperature in most applications. Even if a reliable friction model is available, a further inconvenience would be the lack of existence of an analytical solution to (11.127), necessitating a suitable numerical method to be built into the controller, since sometimes $\dot{x}_{1r}$ and/or T_m has to be changed. In many cases, however, the optimal acceleration and deceleration times are much smaller than the manoeuvre time. As already remarked, the optimal state trajectory then closely approximates that of the hypothetical system without control constraints, which is independent of the plant parameters. Under these circumstances, the much simpler and more versatile method presented in Sect. 11.4.5 can be employed to obtain a control that is almost optimal.

11.4.4 Higher-Order Mechanisms

The model (11.109) is the simplest one for a single degree of freedom mechanism, and this assumes rigid components. All materials from which physical systems are constructed, however, have elasticity with the consequence that sometimes vibration modes are significant. An example is the electric drive with a flexible coupling already used to demonstrate various control techniques. Another feature that is ignored in model (11.109) is the actuator dynamics, such as that introduced by the inductance and resistance of the coil of an electromagnetic actuator. These features, if modelled, would increase the number of state variables and considerably complicate the formulation of the optimal control using the approach of Sect. 11.4.2 for model (11.109). As will be seen in the following subsection, however, a system can be created starting with a controller designed to keep all the state variables under control. Then any vibration modes will be actively damped, and any significant actuator dynamics will be automatically controlled to deliver the required force or torque to the mechanism. Under these circumstances, a reference input of the same form as in Fig. 11.32 will yield a near-minimal frictional energy loss, as in the following section.

11.4.5 Near-Optimal Control Using a Reference Input Generator

11.4.5.1 Basic Scheme

First, let a position control system be designed yielding nominally zero dynamic lag between the position measurement, $y(t)$ and the reference input, $y_r(t)$. It will be assumed that $y(t)$ is scaled in the digital processor so that it is numerically equal to the actual position, i.e. $y(t) = x_1(t)$. Then if a time-varying reference input, $y_r(t)$, is

11.4 Optimal Control for Minimising Frictional Energy Loss

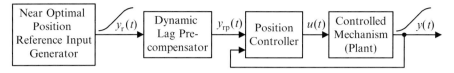

Fig. 11.33 Basic scheme for near-minimum friction energy position control

applied that is the same as $x_1(t)$ of Fig. 11.32, provided the system operates within the control saturation constraints, the plant output, $y(t)$, will precisely follow $y_r(t)$. If the plant is given by (11.109), the control variable, $u(t)$, will precisely follow the optimal control, $u^*(t)$. If, on the other hand $y_r(t)$ is not the same as $x_1(t)$ of Fig. 11.32 during the acceleration and deceleration phases but reaches the final value in the same specified manoeuvre time, T_m, and has a constant velocity segment with $\dot{y}_r \cong x_{2\,\text{ext}}^*$, the frictional energy loss would be approximately the same as produced by the optimal control law of the previous section. This can be achieved by means of a *near-optimal* reference input generator as presented in the following sub-subsection. Figure 11.33 shows the basic scheme.

11.4.5.2 Near-Optimal Reference Input Generator

Of the many possible means of generating near-optimal reference inputs, the one presented here is a chain of integrators controlled by a piecewise constant input, $r(t) = \pm R$ or 0, where R is a positive constant depending on T_m, T_a and x_{1r}. $y_r(t)$ is the output of the last integrator in the chain. This method is used since:

(a) It enables $y_r(t)$ to have a precisely defined acceleration/deceleration time, T_a,
(b) It allows $y_r(t)$ to reach the demanded position, x_{1r}, precisely in the planned manoeuvre time, T_m,
(c) The outputs of the integrators constitute a set of reference input derivatives for feedforward in the dynamic lag pre-compensator,
(d) No impulses that could cause control saturation can occur due to reference input differentiation, since the finite $r(t)$ is the highest derivative, the remaining derivatives and the reference input being obtained by repeated *integrations*.

The highest-order derivative available is the number of integrators in the chain. The piecewise constant input comprises two groups, the first for the acceleration phase and the second for the deceleration phase, as in Fig. 11.34.

The intermediate constant velocity phase requires the output, $\dot{y}_r(t)$, of the penultimate integrator to be constant and therefore the outputs of all the integrators in the portion of the chain driving it and the input function to be zero. There are an infinite number of suitable choices of $r(t)$, the differences in $y_r(t)$ only occurring during the acceleration and deceleration phases, which will not have a great impact on the frictional energy loss if $T_a \ll T_m$. A specific approach is taken here in which $r(t)$ comprises an acceleration pulse group of duration, T_a, followed by a null

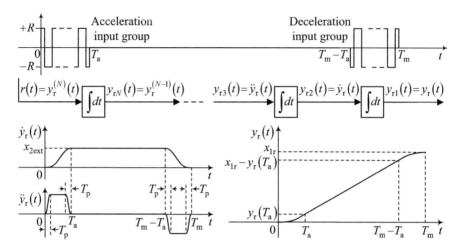

Fig. 11.34 Multiple integrator reference input generator and its variables

period and ending with a deceleration pulse group of duration, T_a, and ending at the manoeuvre time, T_m. The acceleration pulse group commences with an acceleration start sequence of duration, T_p [s], $T_p \ll T_a$, switching between $\pm R$ to bring $\dddot{y}_r(t)$ to a constant value, hold this value for $T_a - 2T_p$ seconds, and then bring $\dddot{y}_r(t)$ back to zero at $t = T_a$ with an acceleration finish sequence, which is an inverted version of the start sequence. The deceleration pulse group is an inverted version of the acceleration pulse group so that substate trajectory $[y_{r2}(t), y_{r3}(t), \ldots, y_{rN}(t)]$ takes a return journey from $[\dot{y}_{rext}, 0, \ldots, 0]$ to $[0, 0, \ldots, 0]$ during the deceleration phase along the same path taken in the acceleration phase. This strategy could be described as open-loop bang–off–bang control of a multiple integrator plant, no feedback being needed as there are no external disturbances and no modelling uncertainties.

Figure 11.35 shows the forms of the input variables and output derivatives for $N = 2$, 3 and 4, all of which yield $y_r(t)$ of the required form shown in Fig. 11.34.

It should be noted that for many applications, the actuators are sufficient to impart such high accelerations that T_a is a much smaller proportion of T_m than shown in Fig. 11.35. This proportion has been increased in the figure for clarity of illustration.

In order to implement the reference input generator, formulae for the parameters, R and T_a, are needed in terms of the demanded position change, x_{1r}; the demanded manoeuvre time, T_m; and the peak acceleration magnitude, a.

With reference to Fig. 11.32, no generality is lost in taking $x_1(0) = 0$, since the reference input for the next manoeuvre can be set relative to any zero position datum.

The parameter, a, should be set so that the control variable, u, approaches but does not exceed the control saturation limits during the acceleration and deceleration phases in order to minimise T_a and consequently minimise $|x_{2ext}|$ and the frictional energy loss. This could easily be accomplished by running a simulation with an initial guess at a suitable value of a and then making adjustments if necessary.

11.4 Optimal Control for Minimising Frictional Energy Loss

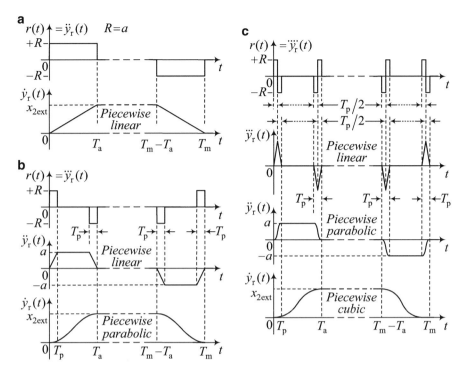

Fig. 11.35 Multiple integrator reference input generator variables for different orders. (**a**) $N = 2$, (**b**) $N = 3$, (**c**) $N = 4$

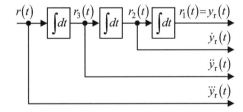

Fig. 11.36 Triple integrator reference input generator

In the following subsection, comparisons of the near-optimal controller and conventional controllers will be made to demonstrate the significant energy savings that can be made. This requires a reference input generator with $N = 3$ and the formulae will be derived for this case. Figure 11.36 shows the state variables, r_1, r_2 and r_3, which are needed as intermediate variables in the derivation.

With reference to Fig. 11.35b, values of T_a, T_p, a and R have to be determined that cause $y_r(t)$ to reach the required demanded position, x_{1r}, in the required manoeuvre time, T_m. A set of equations for T_a, T_p and R will now be derived. These are based on the state transition equations for calculation of the triple integrator state variables, r_1, r_2 and r_3, at the end of each of the periods of constant input, $r = \pm R$ or 0, during the acceleration phase. Since $r(t)$ is piecewise constant, the following state transition equation may be applied to obtain expressions for the state, $\mathbf{r} = \begin{bmatrix} r_1 & r_2 & r_3 \end{bmatrix}^T$, of the reference input generator at the times, T_p, $T_a - T_p$ and T_a, from which equations may be derived in terms of the demanded values of T_m and x_{1r}. If the input, r, is constant, then the general state transition equation yielding the state at time, t_2, given the state at time, t_1, with $t_2 > t_1$ is

$$\mathbf{r}(t_2) = \mathbf{\Phi}(t_2 - t_1)\mathbf{r}(t_1) + \mathbf{\Psi}(t_2 - t_1)r \tag{11.128}$$

where

$$\mathbf{\Phi}(t) = \begin{bmatrix} 1 & t & \frac{1}{2}t^2 \\ 0 & 1 & t \\ 0 & 0 & 1 \end{bmatrix} \text{ and } \mathbf{\Psi}(t) = \begin{bmatrix} \frac{1}{6}t^3 \\ \frac{1}{2}t^2 \\ t \end{bmatrix}. \tag{11.129}$$

Let

$$r = R \operatorname{sgn}(x_{1r}) \tag{11.130}$$

Then, with reference to Fig. 11.35b starting with $\mathbf{r}(0) = \mathbf{0}$, the state transitions for the acceleration phase are given by the following:

$$\begin{bmatrix} r_1(T_p) \\ r_2(T_p) \\ r_3(T_p) \end{bmatrix} = \begin{bmatrix} \frac{1}{6}T_p^3 \\ \frac{1}{2}T_p^2 \\ T_p \end{bmatrix} r, \tag{11.131}$$

$$\begin{bmatrix} r_1(T_a - T_p) \\ r_2(T_a - T_p) \\ r_3(T_a - T_p) \end{bmatrix} = \begin{bmatrix} 1 & T_a - 2T_p & \frac{1}{2}(T_a - 2T_p)^2 \\ 0 & 1 & T_a - 2T_p \\ 0 & 0 & 1 \end{bmatrix} \begin{bmatrix} \frac{1}{6}T_p^3 \\ \frac{1}{2}T_p^2 \\ T_p \end{bmatrix} r$$

$$= \begin{bmatrix} \frac{7}{6}T_p^3 - \frac{3}{2}T_aT_p^2 + \frac{1}{2}T_a^2T_p \\ -\frac{3}{2}T_p^2 + T_aT_p \\ T_p \end{bmatrix} r, \tag{11.132}$$

and

$$\begin{bmatrix} r_1(T_a) \\ r_2(T_a) \\ r_3(T_a) \end{bmatrix} = \begin{bmatrix} 1 & T_p & \frac{1}{2}T_p^2 \\ 0 & 1 & T_p \\ 0 & 0 & 1 \end{bmatrix} \begin{bmatrix} \frac{7}{6}T_p^3 - \frac{3}{2}T_aT_p^2 + \frac{1}{2}T_a^2T_p \\ -\frac{3}{2}T_p^2 + T_aT_p \\ T_p \end{bmatrix} r - \begin{bmatrix} \frac{1}{6}T_p^3 \\ \frac{1}{2}T_p^2 \\ T_p \end{bmatrix} r$$

$$= \begin{bmatrix} \frac{1}{2}T_aT_p(T_a - T_p) \\ T_p(T_a - T_p) \\ 0 \end{bmatrix} r. \tag{11.133}$$

11.4 Optimal Control for Minimising Frictional Energy Loss

Since $r_2(T_a) = x_{2\text{ ext}}$, it follows from Eq. (11.133) that

$$x_{2\text{ ext}} = T_p (T_a - T_p) r. \tag{11.134}$$

With reference to Fig. 11.34, let the position change, $r_1(T_a)$, in $r_1(t)$ over the acceleration phase be denoted by x_{1a}. Then (11.133) yields

$$x_{1a} = \frac{1}{2} T_a T_p (T_a - T_p) r \tag{11.135}$$

The change in the position over the constant velocity phase is

$$x_{1r} - 2x_{1a} = x_{2\text{ ext}} (T_m - 2T_a). \tag{11.136}$$

By inspection of (11.134) and (11.135),

$$x_{1a} = \frac{1}{2} T_a x_{2\text{ ext}} \Rightarrow 2x_{1a} = T_a x_{2\text{ ext}}. \tag{11.137}$$

Substituting for $2x_{1a}$ in (11.136) using (11.137) then yields

$$x_{1r} - T_a x_{2\text{ ext}} = x_{2\text{ ext}} (T_m - 2T_a) \Rightarrow x_{1r} = x_{2\text{ ext}} (T_m - T_a) \Rightarrow x_{2\text{ ext}} = \frac{x_{1r}}{T_m - T_a} \tag{11.138}$$

Equating the RHS of equations (11.138) and (11.134) then gives

$$\frac{x_{1r}}{T_m - T_a} = (T_a - T_p) T_p r \Rightarrow (T_m - T_a)(T_a - T_p) = \frac{x_{1r}}{T_p r}. \tag{11.139}$$

With reference to (11.131) and (11.130), the peak acceleration is given by

$$a = |r_3(T_p)| = T_p R = T_p r \operatorname{sgn}(x_{1r}). \tag{11.140}$$

Also, T_p may be made a small proportion of T_a by setting

$$T_p = \beta T_a \tag{11.141}$$

where β is a chosen constant such as 0.1. Then substituting for $T_p r$ and T_p in (11.139), using, respectively, (11.140) and (11.141), yields

$$(T_m - T_a)(1 - \beta) T_a = \frac{|x_{1r}|}{a} \Rightarrow T_a^2 - T_m T_a + \frac{|x_{1r}|}{(1 - \beta) a}, \tag{11.142}$$

which can be solved for T_a, giving

$$T_a = \tfrac{1}{2}\left(T_m - \sqrt{T_m^2 - \frac{4|x_{1r}|}{(1-\beta)a}}\right), \quad (11.143)$$

the negative square root being correct as $x_{1r} = 0$ for $T_a = 0$. Then from (11.140),

$$R = a/T_p. \quad (11.144)$$

To summarise, β and a are first chosen. Then T_a is calculated using (11.143), T_p using (11.141) and R using (11.144). Then a simulation of the control system is run and a adjusted so that $u(t)$ approaches but does not exceed $\pm u_{\max}$.

11.4.5.3 Scaling Laws for Frictional Energy Cost Function

In this subsection, the influence of x_{1r} and T_m on the frictional energy cost,

$$J = \int_0^{T_m} x_2^2(t)\,dt, \quad (11.145)$$

will be determined. For this, the controller does not have to be optimal. Here,

$$x_{1r} = \int_0^{T_m} x_2(t)\,dt. \quad (11.146)$$

First let T_m be fixed and x_{1r} scaled by a factor, λ, such that

$$x'_{1r} = \lambda x_{1r}. \quad (11.147)$$

If the controller is linear, then changing x_{1r} in this way will not change the form of $x_2(t)$ but scale it to

$$x'_2(t) = \lambda x_2(t). \quad (11.148)$$

Then (11.145) becomes

$$J' = \int_0^{T_m} [\lambda x_2(t)]^2\,dt = \lambda^2 \int_0^{T_m} x_2^2(t)\,dt = \lambda^2 J. \quad (11.149)$$

From (11.147) and (11.149),

$$\lambda^2 = \frac{J'}{J} = \left(\frac{x'_{1r}}{x_{1r}}\right)^2 \Rightarrow J \propto x_{1r}^2. \quad (11.150)$$

11.4 Optimal Control for Minimising Frictional Energy Loss

The constant of proportionality depends upon the controller, and the minimum one would be for the hypothetical optimal controller referred to in Sect. 11.4.2 that, with an unconstrained control variable, is linear and is equivalent to a control loop able to follow a reference input of the form of Fig. 11.34, but with $T_a \to 0 \Rightarrow R \to \infty$. Relationship (11.150) also applies to the near-optimal control system since T_m and T_a are both fixed and therefore (11.147) through (11.149) hold.

Suppose now that x_{1r} is fixed and T_m is scaled by a factor, μ, such that

$$T'_m = \mu T_m. \tag{11.151}$$

Then from (11.146),

$$x_{1r} = \int_0^{T_m} x_2(t) \, dt = \int_0^{\mu T_m} x'_2(t) \, dt, \tag{11.152}$$

and from (11.145),

$$J' = \int_0^{\mu T_m} x'^2_2(t) \, dt. \tag{11.153}$$

If the controller is linear, then the form of $x'_2(t)$ is unaltered. Only its time and magnitude scales are changed. Thus,

$$x'_2(t) = \frac{1}{\mu} x_2(t/\mu). \tag{11.154}$$

It is evident that (11.152) holds with $x'_2(t)$ given by (11.154). Then from (11.153),

$$J' = \int_0^{\mu T_m} \frac{1}{\mu^2} x_2^2(t/\mu) \, dt = \int_0^{T_m} \frac{1}{\mu} x_2^2(t/\mu) \, d(t/\mu) = \frac{1}{\mu} J. \tag{11.155}$$

From (11.151) and (11.155),

$$\mu = \frac{J}{J'} = \frac{T'_m}{T_m} \Rightarrow J \propto \frac{1}{T_m}. \tag{11.156}$$

In contrast to (11.150), (11.156) does not strictly apply to the near-optimal control system because T_a is fixed, and therefore the form of $x_2(t)$ varies with T_m. If $T_a \ll T_m$, however, this variation is not very great and therefore (11.156) *approximately* holds. Combining (11.150) and (11.156) yields

$$J \propto x_{1r}^2 / T_m. \tag{11.157}$$

11.4.5.4 Performance Improvement Over Conventional Control

Here, the frictional energy loss reduction achieved by the near-optimal reference input with dynamic lag pre-compensation relative to traditional feedback control with step reference inputs will be assessed by simulation using plant model (11.109) with linear viscous friction so that $f(x_2) = cx_2$, where c is the viscous friction coefficient.

Let J_{min} be the frictional loss cost of the hypothetical optimal control system without control saturation for given values of x_{1r} and T_m. This is the absolute minimum and will be used as a standard of comparison. Similarly, let J_1 be the frictional loss cost of linear system 1. Then the percentage increase of J_1 relative to J_{min} is

$$\Delta J_{r1\%} = \left(\frac{J_1 - J_{min}}{J_{min}}\right) \cdot 100 \ \%. \tag{11.158}$$

This will be referred to as the *relative frictional loss*. From (11.157),

$$J_{min} = K_{min}\left(x_{1r}^2/T_m\right) \quad \text{and} \quad J_1 = K_1\left(x_{1r}^2/T_m\right). \tag{11.159}$$

where K_{min} and K_1 are the constants of proportionality. Then substituting for J_{min} and J_1 in (11.158) using (11.159) yields

$$\Delta J_{r1\%} = \left(\frac{K_1 - K_{min}}{K_{min}}\right) \cdot 100 \ \%. \tag{11.160}$$

Since this is independent of x_{1r} and T_m, it is only necessary to run a single simulation to determine (11.158) for selected values of x_{1r} and T_m.

For given x_{1r} and T_m, the increase of J, which will be denoted $J_{n\,min}$, for the near-optimal control system, becomes significant as the ratio, T_a/T_m, increases. The term 'near optimal' only really applies if this ratio is relatively small. For this reason, the standard of comparison is J_{min} for the hypothetical optimal control system, although this is unattainable in practice. Also, the value of J obtained with traditional controllers is variable for common x_{1r}, T_m and settling time. Thus, the performance comparison has to be made with more than one traditional control system.

Proceeding now towards the simulations, for the near-optimal control, any linear controller can be used with a suitable dynamic lag pre-compensator. The generic polynomial controller of Sect. 11.3.2.5 is selected for this in view of its versatility, and the fact that constant disturbance rejection is good practice in motion control. The control system block diagram is shown in Fig. 11.37.

As stated previously, the dynamics of the closed-loop system does not influence the overall system response with dynamic lag pre-compensation, so in this, case the polynomial controller parameters will be simply determined using the 5 % settling

11.4 Optimal Control for Minimising Frictional Energy Loss

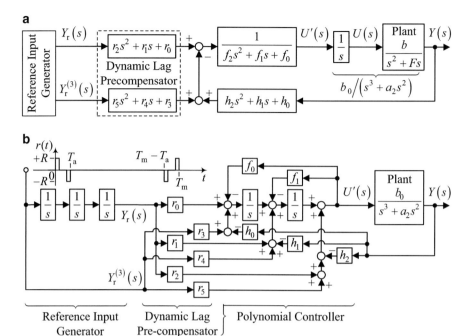

Fig. 11.37 Position control system for near-minimal frictional energy. (**a**) Block diagram in polynomial form, (**b**) implementation block diagram

time formula. The order of the closed-loop system is $N = 5$, so the desired closed-loop characteristic polynomial is

$$(s + p)^5 = s^5 + d_4 s^4 + d_3 s^3 + d_2 s^2 + d_1 s + d_0 \\
= s^5 + 5ps^4 + 10p^2 s^3 + 10p^3 s^2 + 5p^4 s + p^5 \quad (11.161)$$

where $p = 1.5(1 + N)/T_s = 9/T_s$. The matrix equation for the controller parameters, formulated according to Chap. 5, Sect. 5.3.4, is then

$$\begin{bmatrix} 1 & 0 & 0 & 0 & 0 & 0 \\ a_2 & 1 & 0 & 0 & 0 & 0 \\ a_1 & a_2 & 1 & 0 & 0 & 0 \\ a_0 & a_1 & a_2 & b_0 & 0 & 0 \\ 0 & a_0 & a_1 & 0 & b_0 & 0 \\ 0 & 0 & a_0 & 0 & 0 & b_0 \end{bmatrix} \begin{bmatrix} f_2 \\ f_1 \\ f_0 \\ h_2 \\ h_1 \\ h_0 \end{bmatrix} = \begin{bmatrix} 1 \\ d_4 \\ d_3 \\ d_2 \\ d_1 \\ d_0 \end{bmatrix} \quad (11.162)$$

where $b_0 = b$, $a_0 = a_1 = 0$ and $a_2 = F$. The solution by back substitution yields

$$f_2 = 1, \quad f_1 = d_4 - a_2 f_2, \quad f_0 = d_3 - a_2 f_1,$$
$$h_2 = (d_2 - a_2 f_0)/b_0, \quad h_1 = d_1/b_0, \quad h_0 = d_0/b_0 \quad (11.163)$$

For unity closed-loop DC gain, the dynamic lag pre-compensator coefficients are given by (11.49), yielding

$$r_i = d_i/b_0, \quad i = 0, 1, 2, 3, 4, \quad r_5 = 1/b_0. \quad (11.164)$$

The performance of the control system of Fig. 11.37 will be compared with that of four control systems that should include any traditional control system likely to be applied. These are shown in Fig. 11.38.

The comparison is by means of (11.158) with $J_1 = J_{PD}, J_{DP}, J_{PID}$ and J_{IPD}. Next, design formulae will be derived for the four control systems yielding settling times equal to the selected value of T_m. Since the near-optimal control system settles *precisely* in the manoeuvre time, T_m, a fairer comparison with the systems of Fig. 11.38 will be obtained by designing them to have a settling time of $T_s = T_m$ according to the 2 % criterion than the 5 % criterion. For each control system, there exists an infinite number of combinations of closed-loop poles yielding the same 2 % settling time. A fair comparison would entail finding the pole placement that minimises J, and it could be argued that this is likely to be multiple pole placement as follows. Starting with all the closed-loop poles at a point on the negative real axis of the s-plane yielding a settling time of T_s, let one of the poles be moved nearer the origin. This alone would increase T_s and necessitate moving at least one of the other poles further from the origin to restore T_s to its original value. The faster mode or modes associated with these further poles would have the general

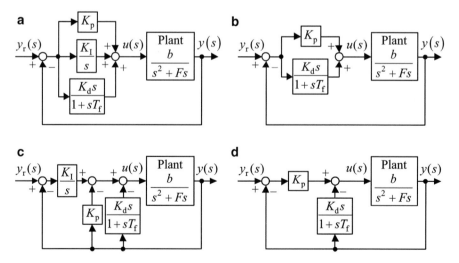

Fig. 11.38 Traditional position control systems for frictional energy loss comparison. (**a**) PID control system, (**b**) PD control system, (**c**) IPD control system, (**d**) DP control system

11.4 Optimal Control for Minimising Frictional Energy Loss

effect of increasing the peak of $x_2(t)$ and therefore increasing J. Moving two of the poles into complex conjugate locations and, in the case of the PID and IPD control systems, the real pole, to maintain the same settling time would cause overshooting and possibly oscillations, again causing lager excursions of $x_2(t)$. So multiple pole placement will be applied for the performance comparison.

It is straightforward to show that the characteristic polynomials of control systems (a) and (c) of Fig. 11.38 are identical for the same settings of the controller parameters and similarly for control systems (b) and (d). The pole placement design equations for systems (a) and (c) are therefore the same, similarly for systems (b) and (d). The noise filtering time constant, T_f, will be set sufficiently small for its effect on the closed-loop dynamics to be negligible, so it will be set to zero in the pole placement process. For systems (a) and (c), the three poles are placed at $s_{1,2,3} = -q = -1.6(1.5+n)/T_m = -7.2/T_m$ to yield a 2 % settling time of $T_s = T_m$ seconds. Thus,

$$K_d = (3q - F)/b, \quad K_p = 3q^2/b, \quad K_I = q^3/b. \tag{11.165}$$

For systems (b) and (d), the two poles are placed at $s_{1,2} = -q' = -1.6(1.5+n)/T_m = -5.6/T_m$ requiring

$$K_d = (2q' - F)/b, \quad K_p = q'^2/b. \tag{11.166}$$

The friction coefficient, F, will be chosen to yield a time constant of $1/F = 2\,\text{s} \Rightarrow F = 0.5\,\text{s}^{-1}$. The choice of this parameter is considered to be realistic for many applications but is not very critical as it will not affect the percentage differences between the frictional energy wastage of the two control systems for the same reference input. Translational position control will be assumed and an analogue control voltage with saturation limits of $\pm u_{max} = \pm 10$ V, giving a linear acceleration magnitude of 50 m/s/s. Hence, $b = 50/10 = 5$ [m/s/s/V]. The maximum position limits of the mechanism are taken to be $\pm y_{max} = \pm 1$ m.

First, the near-optimal control system performance will be determined for increasing T_a/T_m, which would be chosen according to the control saturation limits of the particular application, for the same value of x_{1r}. The parameter, β, is chosen as 0.1. Then the pulse time is given by (11.141) as $T_p = \beta T_a$. Then (11.139) gives

$$r = x_{1r}/\left[(T_m - T_a)(T_a - T_p)T_p\right] \tag{11.167}$$

and since $\text{sgn}^2(x_{1r}) = 1$ and $x_{1r}\,\text{sgn}(x_{1r}) = |x_{1r}|$, multiplying both sides of (11.167) by $\text{sgn}(x_{1r})$ and observing (11.130) gives

$$R = |x_{1r}|/\left[(T_m - T_a)(T_a - T_p)T_p\right] \tag{11.168}$$

The demanded manoeuvre parameters are $T_m = 1$ [s] and $x_{1r} = 1$ [m]. Simulations of the near-optimal control system for six values of T_a/T_m are shown in Fig. 11.39 with a simulation of the hypothetical optimal control system where $T_a = 0$.

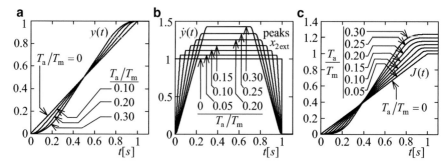

Fig. 11.39 Relative frictional energy losses of near-optimal control system. (**a**) Position responses, (**b**) velocities, (**c**) frictional energy costs

Table 11.1 Variation of relative frictional energy loss with acceleration time ratio

T_a/T_m	0.05	0.01	0.15	0.20	0.25	0.30
$\Delta J_{r\%}$	3.59 %	7.39 %	11.38 %	15.57 %	19.92	24.07

The position responses of Fig. 11.39a cross at a common point, $(T_m/2, x_{1r}/2)$, because of the symmetry of the velocity–time profiles, $\dot{y}(t)$, about the vertical line (not drawn) at $t = T_m/2$. The velocity plots of Fig. 11.39b clearly show the increase of $x_{2\,ext}$, i.e. the extreme values of $\dot{y}(t)$, with increase of T_a/T_m necessary for the system to 'catch up' to complete the manoeuvre in the specified T_m seconds.

The area under each curve is $\int_0^{T_m} \dot{y}(t)\mathrm{d}t = x_{1r}$, which is constant. The consequential increase in the total amount of frictional energy wasted is evident in Fig. 11.39c. The corresponding relative frictional losses are shown in Table 11.1. Recall that $\Delta J_{r\%} = 0\%$ cannot be achieved in practice due to the control saturation constraints.

The value of $\Delta J_{r\%}$ for the attainable optimal control is variable from one application to another, being dependent upon the minimum attainable value of T_a, and would be expected to be less than 10 % in most cases. The control variables corresponding to Fig. 11.39 are shown in Fig. 11.40.

In a real application, the peak of $u(t)$ would be examined for each value of T_a/T_m and the selected one would be that approaching but not exceeding the control saturation limits. In this example, if the control saturation limit is 13 [V], then a ratio of $T_a/T_m = 0.1$ would be suitable.

Next, the performance of the four traditional control systems of Fig. 11.38 will be assessed with step position reference inputs of $x_{1r} = 1.h(t)$ m. Figure 11.41 shows the results for the PID and PD controller-based systems.

The overshooting due to the controller zeros (Chap. 1) causes a large peak velocity resulting in excessive frictional energy loss relative to those of the near-optimal systems shown in Fig. 11.39. The absence of the controller zeros in the IPD and DP controller-based systems enables greatly improved performance, however,

11.4 Optimal Control for Minimising Frictional Energy Loss

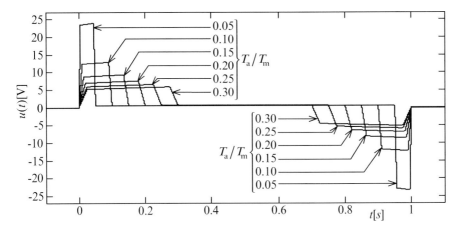

Fig. 11.40 Near-optimal control variables for different acceleration time ratios

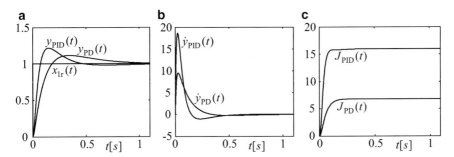

Fig. 11.41 Relative frictional energy losses of traditional PID and PD control systems. (**a**) Position responses, (**b**) velocities, (**c**) frictional energy costs

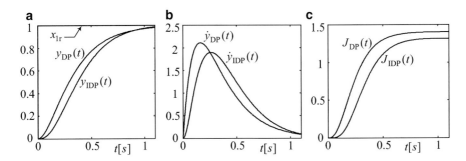

Fig. 11.42 Relative frictional energy losses of traditional IPD and DP control systems. (**a**) Position responses, (**b**) velocities, (**c**) frictional energy costs

for the same gain settings as can be seen by comparing Fig. 11.41 with Fig. 11.42. The corresponding relative frictional energy losses are shown in Table 11.2.

Table 11.2 Relative frictional energy loss with traditional position control

Controller	PID	PD	DP	IDP
$\Delta J_{r\%}$	1498.6 %	578.7	40.42	31.78

In conclusion, the traditional PID and PD controllers are entirely unsuitable when used with step changes in the reference position. It has to be pointed out, however, that if these controllers were to be used with a suitably designed dynamic lag pre-compensator *including poles to cancel the zeros*, then similar performance to that presented in Table 11.1 could be obtained.

Comparing Table 11.1 with Table 11.2, it can be concluded that to respect the needs for environmental protection by minimising the demands on energy resources, it is highly advantageous to employ near-optimal position control for frictional energy wastage minimisation.

It is also important to point out that stepper motor drives [6] typically move mechanical loads from one demanded position to another at constant velocities through being operated at a constant pulse repetition frequency. It follows that if the manoeuvre time is the maximum that can be tolerated, the control is near optimal with respect to frictional energy loss. This is, however, conditional on the fluctuations in velocity due to the torque oscillating at the pulse repetition frequency being minimal due to the load inertia. Details of stepper motor control systems are not included here as, in their basic form, they are not feedback control systems, only sometimes employing a rudimentary form of feedback control consisting of the injection of extra pulses in case previous pulses fail to produce angle increments of the output shaft.

References

1. Schumaker L (2007) Spline functions: basic theory. Cambridge University Press, Cambridge, UK
2. Kelly R et al (2006) Control of robotic manipulators in joint space. Springer, London
3. Pontryagin LS (1987) Selected Works Vol. 4: The mathematical theory of optimal processes. ISBN 2-88124-077-1. Gordon and Breach Science Publishers, Montreux, Switzerland
4. Fuller AT (1963) Bibliography of Pontryagin's maximum principle. J Electron Control 15(5):513–517
5. Athans M, Falb F (2007) Optimal control: an introduction to the theory and its applications. Dover, Mineola, New York
6. Acarnley P (2007) Stepping motors: a guide to theory and practice, 4th edn. IET, London

Tables

Laplace Transforms and z-Transfer Functions

Table 1 Laplace Transforms

$f(t)$ for $t \geq 0$	$\mathcal{L}\{f(t)\} = \int_0^\infty f(t)\,\mathrm{e}^{-st}\,\mathrm{d}t = F(s)$
Unit impulse function, $\delta(t) \triangleq \lim_{\Delta t \to 0} \dfrac{1}{\Delta t}[h(t) - h(t - \Delta t)]$	1
Unit step function, $h(t) = \begin{cases} 0 \text{ for } t < 0 \\ 1 \text{ for } t \geq 0 \end{cases}$	$\dfrac{1}{s}$
Ramp function, $t.\,h(t)$	$\dfrac{1}{s^2}$
$\dfrac{1}{(n-1)!}t^{n-1}h(t)$	$\dfrac{1}{s^n}$
e^{-at} or $\dfrac{1}{T}\mathrm{e}^{-\frac{t}{T}}$	$\dfrac{1}{s+a}$ or $\dfrac{1}{1+sT}$
$\dfrac{1}{a}(1 - \mathrm{e}^{-at})$ or $1 - \mathrm{e}^{-\frac{t}{T}}$	$\dfrac{1}{s(s+a)}$ or $\dfrac{1}{s(1+sT)}$
$\dfrac{1}{b-a}\left(\mathrm{e}^{-at} - \mathrm{e}^{-bt}\right)$	$\dfrac{1}{(s+a)(s+b)}$
$\dfrac{1}{(n-1)!}t^{n-1}\mathrm{e}^{-at}$	$\left(\dfrac{1}{s+a}\right)^n$
$\dfrac{1}{T_1 - T_2}\left(\mathrm{e}^{-\frac{t}{T_1}} - \mathrm{e}^{-\frac{t}{T_2}}\right)$	$\dfrac{1}{(1+sT_1)(1+sT_2)}$

(continued)

Table 1 (continued)

$f(t)$	$F(s)$
$1 - \dfrac{T_1}{T_1 - T_2}e^{-\frac{t}{T_1}} - \dfrac{T_2}{T_2 - T_1}e^{-\frac{t}{T_2}}$	$\dfrac{1}{s(1 + sT_1)(1 + sT_2)}$
$\dfrac{1}{(n-1)!}\dfrac{t^{n-1}}{T^n}e^{-\frac{t}{T}}$	$\dfrac{1}{(1 + sT)^n}$
$1 - \left(1 + \dfrac{t}{T}\right)e^{-\frac{t}{T}}$	$\dfrac{1}{s(1 + sT)^2}$
$1 - \sum_{k=0}^{n-1} \dfrac{1}{k!}\left(\dfrac{1}{T}\right)^k e^{-\frac{t}{T}}$	$\dfrac{1}{s(1 + sT)^n}$
$\dfrac{\omega_n}{\sqrt{1-\zeta^2}}e^{-\zeta\omega_n t}\sin\left(\omega_n\sqrt{1-\zeta^2}\,t\right)$	$\dfrac{\omega_n^2}{s^2 + 2\zeta\omega_n s + \omega_n^2}$
$1 - \dfrac{1}{\sqrt{1-\zeta^2}}e^{-\zeta\omega_n t}\sin\left(\omega_n\sqrt{1-\zeta^2}\,t + \phi\right)$ where $\phi = \cos^{-1}(\zeta)$	$\dfrac{1}{s}\left(\dfrac{\omega_n^2}{s^2 + 2\zeta\omega_n s + \omega_n^2}\right)$
$\sin(\omega t)$	$\dfrac{\omega}{s^2 + \omega^2}$
$\cos(\omega t)$	$\dfrac{s}{s^2 + \omega^2}$
$1 - \cos(\omega t)$	$\dfrac{1}{s}\left(\dfrac{\omega^2}{s^2 + \omega^2}\right)$
$e^{-at}\cos(\omega t)$	$\dfrac{s + a}{(s + a)^2 + \omega^2}$
$e^{-at}\sin(\omega t)$	$\dfrac{\omega}{(s + a)^2 + \omega^2}$

Table 2 Laplace Transform Relationships

$f(t)$ for $t \geq 0$	$F(s)$
$f(t) + g(t)$	$F(s) + G(s)$
$\lambda f(t),\ \lambda \in \Re$	$\lambda F(s)$
$\dfrac{d}{dt}f(t)$	$sF(s) - f(0)$
$\dfrac{d^q}{dt^q}f(t)$	$s^q F(s) - s^{q-1}f(0) - s^{q-2}\dfrac{df}{dt}(0) - \cdots - \dfrac{d^{q-1}f}{dt^{q-1}}(0)$
$g(t) = \displaystyle\int_0^t f(\tau)\,d\tau$	$G(s) = \dfrac{1}{s}F(s)$
$f(\lambda t),\ \lambda > 0$	$(1/\lambda)\cdot F(s/\lambda)$

(continued)

Table 2 (continued)

$e^{-at} f(t)$	$F(s+a)$
$t^q f(t)$	$(-1)^q \dfrac{d^q}{ds^q} F(s)$
$g(t) = \begin{cases} 0 & 0 \le t < \tau \\ f(t-\tau) & t \ge \tau \end{cases}, \quad \tau > 0$	$G(s) = e^{-s\tau} F(s)$

The first column of Table 3 shows the plant Laplace transfer function. The second column shows the z-transfer function of the same plant viewed through a sampling process with a sampling period of h seconds, but without a first order hold and therefore referred to as the pulse transfer function. This is the z-transfer function commonly presented in tables but cannot be used directly for control system design. The third column, however, shows the corresponding z-transfer function including the first order hold that is implicitly included in the sampling process of a digital processor. This can therefore be used directly in a control system design.

Table 3 Plant Models: Laplace and z-Transfer Functions

Laplace Transfer Function, $G(s)$	Pulse Transfer Function, $P(z)$	z-Transfer Function, $Q(z)$
$\dfrac{1}{s}$	$\dfrac{z}{z-1}$	$\dfrac{h}{z-1}$
$\dfrac{1}{s^2}$	$\dfrac{hz}{(z-1)^2}$	$\dfrac{h^2(z+1)}{2(z-1)^2}$
$\dfrac{1}{s^3}$	$\dfrac{h^2 z(z+1)}{2(z-1)^3}$	$\dfrac{h^3(z^2+4z+1)}{6(z-1)^3}$
$\dfrac{a}{s+a}$	$\dfrac{az}{z-e^{-ah}}$	$\dfrac{1-e^{-ah}}{z-e^{-ah}}$
$\dfrac{a}{s(s+a)}$	$\dfrac{z\left(1-e^{-ah}\right)}{(z-1)(z-e^{-ah})}$	$\dfrac{h}{z-1} - \dfrac{1-e^{-ah}}{a(z-e^{-ah})}$
$\dfrac{1}{(s+a)(s+b)}$	$\dfrac{z\left(e^{-ah}-e^{-bh}\right)}{(b-a)(z-e^{-ah})(z-e^{-bh})}$	$\dfrac{1}{ab}\left[1 + \dfrac{z-1}{a-b}\left(\dfrac{b}{z-e^{-ah}} - \dfrac{a}{z-e^{-bh}}\right)\right]$
$\dfrac{s}{(s+a)(s+b)}$	$\dfrac{z}{b-a}\left(\dfrac{b}{z-e^{-bh}} - \dfrac{a}{z-e^{-ah}}\right)$	$\dfrac{1}{b-a}\left(\dfrac{1-e^{-bh}}{z-e^{-bh}} - \dfrac{1-e^{-ah}}{z-e^{-ah}}\right)$
$\dfrac{1}{s(s+a)(s+b)}$	$\dfrac{z}{ab}\left[\dfrac{1}{z-1} + \dfrac{1}{a-b}\left(\dfrac{b}{z-e^{-ah}} - \dfrac{a}{z-e^{-bh}}\right)\right]$	$\dfrac{1}{a^2 b^2}\left[\dfrac{abh}{z-1} + \dfrac{b^2}{a-b}\cdot\dfrac{1-e^{-ah}}{z-e^{-ah}} - \dfrac{a^2}{a-b}\cdot\dfrac{1-e^{-bh}}{z-e^{-bh}}\right]$

Characteristic Polynomial Coefficients of the Settling Time Formulae

Consider any control system (or subsystem such as an observer) whose dynamic behaviour is determined by a characteristic polynomial,

$$s^n + d_{n-1}s^{n-1} + \cdots + d_1 s + d_0, \tag{1}$$

where the constant coefficients, d_i, $i = 0, 1, \ldots, n-1$, may be chosen freely. This section presents tables for these coefficients that can be used in software for setting up control systems prior to loop closure, thereby relieving the control system designer of time-consuming derivations on paper. This is based on the settling time formulae of Chap. 4. If the settling times according to the 5 % and 2 % criteria are denoted, respectively, by $T_{s5\%}$ and $T_{s2\%}$, then (1) may be written as

$$(s + \alpha)^n \tag{2}$$

where

$$\alpha = 1.5\,(1+n)\,/\,T_{s5\%} \quad \text{or} \quad \alpha = 1.6\,(1.5+n)\,/\,T_{s2\%}. \tag{3}$$

Table 4 gives the coefficients corresponding to (2) and (3) for orders between and including $n = 1$ and $n = 8$. The reader may easily generate further rows of the table for higher orders with the aid of the binomial expansion.

If more robustness is needed, the control loop stiffness can be increased [Chap. 10], by making one of the closed loop poles sufficiently large for it to be dominated by the remaining poles so that (2) and (3) are replaced, respectively, by

$$(s + \alpha)^{n-1}\,(s + r_{pp}\alpha) \tag{4}$$

Table 4 Coefficients of closed loop differential equation based on settling time formulae

r	1	2	3	4	5	6	7	8
d_0	α	α^2	α^3	α^4	α^5	α^6	α^7	α^8
d_1	–	2α	$3\alpha^2$	$4\alpha^3$	$5\alpha^4$	$6\alpha^5$	$7\alpha^6$	$8\alpha^7$
d_2	–	–	3α	$6\alpha^2$	$10\alpha^3$	$15\alpha^4$	$21\alpha^5$	$28\alpha^6$
d_3	–	–	–	4α	$10\alpha^2$	$20\alpha^3$	$35\alpha^4$	$56\alpha^5$
d_4	–	–	–	–	5α	$15\alpha^2$	$35\alpha^3$	$70\alpha^4$
d_5	–	–	–	–	–	6α	$21\alpha^2$	$56\alpha^3$
d_6	–	–	–	–	–	–	7α	$28\alpha^2$
d_7	–	–	–	–	–	–	–	8α

where

$$\alpha = \frac{1.5n}{T_{s5\%}} \quad \text{or} \quad \alpha = \frac{1.6(0.5+n)}{T_{s2\%}}, \qquad (5)$$

and where $r_{pp} \geq r_{pp\ min}$. Here, r_{pp} is the pole-to-pole dominance ratio [Chap. 1] and $r_{pp\ min}$ is the minimum value such that the pole at $s = -r_{pp}\alpha$ has a negligible effect on the closed loop dynamics, which is virtually the same as that of a system (or subsystem) with characteristic polynomial,

$$(s+\alpha)^{n-1}. \qquad (6)$$

Hence (5) is (3) with n replaced by $n-1$. Table 5 gives the coefficients of (4) for orders between and including $n = 2$ and $n = 9$, so that the corresponding range of dynamic responses is almost the same as that for Table 4, provided $r_{pp} \geq r_{pp\ min}$.

Table 5 Coefficients of closed loop differential equation based on robust pole placement

r	2	3	4	5
d_0	$r_{pp}\alpha^2$	$r_{pp}\alpha^3$	$r_{pp}\alpha^4$	$r_{pp}\alpha^5$
d_1	$(1+r_{pp})\alpha$	$(1+2r_{pp})\alpha^2$	$(1+3r_{pp})\alpha^3$	$(1+4r_{pp})\alpha^4$
d_2	–	$(2+r_{pp})\alpha$	$3(1+r_{pp})\alpha^2$	$2(2+3r_{pp})\alpha^3$
d_3	–	–	$(3+r_{pp})\alpha$	$2(3+2r_{pp})\alpha^2$
d_4	–	–	–	$(4+r_{pp})\alpha$
r	6	7	8	9
d_0	$r_{pp}\alpha^6$	$r_{pp}\alpha^7$	$r_{pp}\alpha^8$	$r_{pp}\alpha^9$
d_1	$(1+5r_{pp})\alpha^5$	$(1+6r_{pp})\alpha^6$	$(1+7r_{pp})\alpha^7$	$(1+8r_{pp})\alpha^8$
d_2	$5(1+2r_{pp})\alpha^4$	$3(2+5r_{pp})\alpha^5$	$7(1+3r_{pp})\alpha^6$	$4(2+7r_{pp})\alpha^7$
d_3	$10(1+r_{pp})\alpha^3$	$5(3+4r_{pp})\alpha^4$	$7(3+5r_{pp})\alpha^5$	$28(1+2r_{pp})\alpha^6$
d_4	$5(2+r_{pp})\alpha^2$	$5(4+3r_{pp})\alpha^3$	$35(1+r_{pp})\alpha^4$	$14(4+5r_{pp})\alpha^5$
d_5	$(5+r_{pp})\alpha$	$3(5+2r_{pp})\alpha^2$	$7(5+3r_{pp})\alpha^3$	$14(5+4r_{pp})\alpha^4$
d_6	–	$(6+r_{pp})\alpha$	$7(3+r_{pp})\alpha^2$	$28(2+r_{pp})\alpha^3$
d_7	–	–	$(7+r_{pp})\alpha$	$4(7+2r_{pp})\alpha^2$
d_8	–	–	–	$(8+r_{pp})\alpha$

Appendices

A2 Appendix to Chap. 2

A2.1 Kinematics of Vehicle Attitude Control

A2.1.1 The Direction Cosine Matrix

The direction cosine matrix is fundamental to three degree of freedom rotational kinematics since it entirely defines the attitude of a rigid body with respect to a reference frame and is common to the two attitude representations to be derived.

Figure A2.1 shows two sets of unit vectors, $\left(\hat{\mathbf{i}}_r, \hat{\mathbf{j}}_r, \hat{\mathbf{k}}_r\right)$ and $\left(\hat{\mathbf{i}}_b, \hat{\mathbf{j}}_b, \hat{\mathbf{k}}_b\right)$ directed, respectively, along the mutually orthogonal axes of the reference frame, (x_r, y_r, z_r), and the body fixed frame, (x_b, y_b, z_b). Then the orientation of the frame, (x_b, y_b, z_b), with respect to the frame (x_r, y_r, z_r) is equivalent to the orientation of $\left(\hat{\mathbf{i}}_b, \hat{\mathbf{j}}_b, \hat{\mathbf{k}}_b\right)$ with respect to $\left(\hat{\mathbf{i}}_r, \hat{\mathbf{j}}_r, \hat{\mathbf{k}}_r\right)$. Then $\hat{\mathbf{i}}_b$ may be expressed as a weighted sum of the unit vectors of the reference frame, the weightings being the direction cosines of $\hat{\mathbf{i}}_r$ with respect to the body-fixed frame, which will be denoted by c_{xx}, c_{xy} and c_{xz}. Then

$$\hat{\mathbf{i}}_b = c_{xx}\hat{\mathbf{i}}_r + c_{xy}\hat{\mathbf{j}}_r + c_{xz}\hat{\mathbf{k}}_r. \tag{A2.1}$$

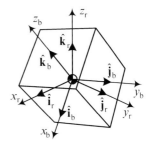

Fig. A2.1 Unit vectors in reference frame and body fixed frame

Similarly,

$$\hat{\mathbf{j}}_b = c_{yx}\hat{\mathbf{i}}_r + c_{yy}\hat{\mathbf{j}}_r + c_{yz}\hat{\mathbf{k}}_r \qquad (A2.2)$$

and

$$\hat{\mathbf{k}}_b = c_{zx}\hat{\mathbf{i}}_r + c_{zy}\hat{\mathbf{j}}_r + c_{zz}\hat{\mathbf{k}}_r. \qquad (A2.3)$$

where

$$\begin{bmatrix} c_{xx} & c_{xy} & c_{xz} \\ c_{yx} & c_{yy} & c_{yz} \\ c_{zx} & c_{zy} & c_{zz} \end{bmatrix} = \left\{ \begin{array}{c} \hat{\mathbf{i}}_b \\ \hat{\mathbf{j}}_b \\ \hat{\mathbf{k}}_b \end{array} \right\} \bullet \left\{ \hat{\mathbf{i}}_r \ \hat{\mathbf{j}}_r \ \hat{\mathbf{k}}_r \right\} \triangleq \begin{bmatrix} \hat{\mathbf{i}}_b.\hat{\mathbf{i}}_r & \hat{\mathbf{i}}_b.\hat{\mathbf{j}}_r & \hat{\mathbf{i}}_b.\hat{\mathbf{k}}_r \\ \hat{\mathbf{j}}_b.\hat{\mathbf{i}}_r & \hat{\mathbf{j}}_b.\hat{\mathbf{j}}_r & \hat{\mathbf{j}}_b.\hat{\mathbf{k}}_r \\ \hat{\mathbf{k}}_b.\hat{\mathbf{i}}_r & \hat{\mathbf{k}}_b.\hat{\mathbf{j}}_r & \hat{\mathbf{k}}_b.\hat{\mathbf{k}}_r \end{bmatrix} \qquad (A2.4)$$

Here the sets of unit vectors are assembled to form 3×1 matrices. In this notation [1], matrices whose elements are vectors are identified by means of the special brackets, the standard brackets being reserved for matrices whose elements are scalars.

The usual rules of matrix addition, subtraction and multiplication are valid, but with the dot or cross product operator placed between the matrices, such as in (A2.4). If no operator is shown, scalar multiplication is implied. Thus (A2.1), (A2.2) and (A2.3) may be written as the matrix–vector equation,

$$\left\{ \begin{array}{c} \hat{\mathbf{i}}_b \\ \hat{\mathbf{j}}_b \\ \hat{\mathbf{k}}_b \end{array} \right\} = \begin{bmatrix} c_{xx} & c_{xy} & c_{xz} \\ c_{yx} & c_{yy} & c_{yz} \\ c_{zx} & c_{zy} & c_{zz} \end{bmatrix} \left\{ \begin{array}{c} \hat{\mathbf{i}}_r \\ \hat{\mathbf{j}}_r \\ \hat{\mathbf{k}}_r \end{array} \right\}, \qquad (A2.5)$$

which may be written compactly as

$$\{\mathbf{I}\}_b = \mathbf{C}\{\mathbf{I}\}_r. \qquad (A2.6)$$

Similarly,

$$\left\{ \begin{array}{c} \hat{\mathbf{i}}_r \\ \hat{\mathbf{j}}_r \\ \hat{\mathbf{k}}_r \end{array} \right\} = \begin{bmatrix} c_{xx} & c_{yx} & c_{zx} \\ c_{xy} & c_{yy} & c_{zy} \\ c_{xz} & c_{yz} & c_{zz} \end{bmatrix} \left\{ \begin{array}{c} \hat{\mathbf{i}}_b \\ \hat{\mathbf{j}}_b \\ \hat{\mathbf{k}}_b \end{array} \right\}, \qquad (A2.7)$$

which may be written as

$$\{\mathbf{I}\}_r = \mathbf{C}^T\{\mathbf{I}\}_b. \qquad (A2.8)$$

This, incidentally, proves that the matrix, $\mathbf{C}$, is orthogonal, meaning $\mathbf{C}^{-1} = \mathbf{C}^T$. $\mathbf{C}$ is the *direction cosine matrix* and since it completely defines the orientation of the body with respect to the reference frame it is sometimes called the attitude matrix or the rotation matrix. It directly provides an attitude representation for which *nine* KDEs may be written whose solutions obey *six* constraint equations because the body orientation has only three degrees of freedom. With limited space, however, only the fairly widely used quaternion based KDEs are derived, preceded by the direction cosine based KDEs, that are needed as part of this derivation.

An important use of the direction cosine matrix is the determination of the components of any vector, **v**, in the reference frame, given its components in the body-fixed frame, and vice versa. This will also be needed for the translational three degree of freedom dynamic and kinematic models. So

$$\mathbf{v} = v_{xr}\hat{\mathbf{i}}_r + v_{yr}\hat{\mathbf{j}}_r + v_{zr}\hat{\mathbf{k}}_r = v_{xb}\hat{\mathbf{i}}_b + v_{yb}\hat{\mathbf{j}}_b + v_{zb}\hat{\mathbf{k}}_b, \quad (A2.9)$$

which may be written as

$$\begin{bmatrix} v_{xb} & v_{yb} & v_{zb} \end{bmatrix} \{\mathbf{I}\}_b = \begin{bmatrix} v_{xr} & v_{yr} & v_{zr} \end{bmatrix} \{\mathbf{I}\}_r \quad \text{or} \quad \mathbf{v}_b^T \{\mathbf{I}\}_b = \mathbf{v}_r^T \{\mathbf{I}\}_r. \quad (A2.10)$$

Then substituting for $\{\mathbf{I}\}_r$ in (A2.10) using (A2.8) yields

$$\mathbf{v}_b^T \{\mathbf{I}\}_b = \mathbf{v}_r^T \{\mathbf{I}\}_r \Rightarrow \mathbf{v}_b^T \{\mathbf{I}\}_b = \mathbf{v}_r^T \mathbf{C}^T \{\mathbf{I}\}_b \Rightarrow$$
$$\mathbf{v}_b = \mathbf{C} \mathbf{v}_r. \quad (A2.11)$$

It follows that

$$\mathbf{v}_r = \mathbf{C}^T \mathbf{v}_b. \quad (A2.12)$$

A2.1.2 The Direction Cosine Matrix Based Kinematic Differential Equations

Let the rigid body be rotating with an instantaneous angular velocity vector of $\hat{\boldsymbol{\omega}}$, relative to the reference frame (x_0, y_0, z_0), as shown in Fig. A2.2. In (A2.8), $\{\mathbf{I}\}_r$ represents the set of mutually orthogonal unit vectors that are fixed with respect to the (x_r, y_r, z_r) frame. Although the mutually orthogonal unit vectors represented by

Fig. A2.2 Representation of rigid body in rotational motion

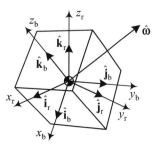

$\{\mathbf{I}\}_b$ are fixed with respect to the (x_b, y_b, z_b) frame, they rotate at angular velocity, $\hat{\boldsymbol{\omega}}$, w.r.t. the (x_r, y_r, z_r) frame. The right hand side of (A2.8), however, expresses these unit vectors as linear weighted sums of the unit vectors in the (x_r, y_r, z_r) frame. With respect to this frame $\{\mathbf{I}\}_b$ is time varying due to the rotation of the body. The direction cosine matrix, of course, is also time-varying. Differentiating both sides of (A2.6) then yields

$$\frac{d}{dt}\{\mathbf{I}\}_b = \dot{\mathbf{C}}\{\mathbf{I}\}_r. \tag{A2.13}$$

Since the changes of the body fixed unit vectors in the (x_r, y_r, z_r) frame are only due to their rotation at the angular velocity, $\hat{\boldsymbol{\omega}}$, their individual rates of change are $\frac{d\hat{\mathbf{i}}_b}{dt} = \hat{\boldsymbol{\omega}} \wedge \hat{\mathbf{i}}_b$, $\frac{d\hat{\mathbf{j}}_b}{dt} = \hat{\boldsymbol{\omega}} \wedge \hat{\mathbf{j}}_b$ and $\frac{d\hat{\mathbf{k}}_b}{dt} = \hat{\boldsymbol{\omega}} \wedge \hat{\mathbf{k}}_b$, which can be written collectively as

$$\frac{d}{dt}\{\mathbf{I}\}_b = \underset{\sim}{\boldsymbol{\omega}} \wedge \{\mathbf{I}\}_b. \tag{A2.14}$$

Equating the right hand sides of (A2.13) and (A2.14) then yields

$$\dot{\mathbf{C}}\{\mathbf{I}\}_r = \hat{\boldsymbol{\omega}} \wedge \{\mathbf{I}\}_b. \tag{A2.15}$$

Expanding the RHS then gives

$$\left[\omega_x\hat{\mathbf{i}}_b + \omega_y\hat{\mathbf{j}}_b + \omega_z\hat{\mathbf{k}}_b\right] \wedge \begin{Bmatrix} \hat{\mathbf{i}}_b \\ \hat{\mathbf{j}}_b \\ \hat{\mathbf{k}}_b \end{Bmatrix} = \begin{Bmatrix} \omega_z\hat{\mathbf{j}}_b - \omega_y\hat{\mathbf{k}}_b \\ \omega_x\hat{\mathbf{k}}_b - \omega_z\hat{\mathbf{i}}_b \\ \omega_y\hat{\mathbf{i}}_b - \omega_x\hat{\mathbf{j}}_b \end{Bmatrix} = \begin{bmatrix} 0 & \omega_z & -\omega_y \\ -\omega_z & 0 & \omega_x \\ \omega_y & -\omega_x & 0 \end{bmatrix} \begin{Bmatrix} \hat{\mathbf{i}}_b \\ \hat{\mathbf{j}}_b \\ \hat{\mathbf{k}}_b \end{Bmatrix}. \tag{A2.16}$$

Let $\boldsymbol{\Omega}_3 = \begin{bmatrix} 0 & \omega_z & -\omega_y \\ -\omega_z & 0 & \omega_x \\ \omega_y & -\omega_x & 0 \end{bmatrix}$. Then (A2.15) can be written as

$$\dot{\mathbf{C}}\{\mathbf{I}\}_r = \boldsymbol{\Omega}_3\{\mathbf{I}\}_b \tag{A2.17}$$

and in view of (A2.6), this yields the direction cosine based KDE, as follows.

$$\dot{\mathbf{C}}\{\mathbf{I}\}_r = \boldsymbol{\Omega}_3\mathbf{C}\{\mathbf{I}\}_r \Rightarrow \dot{\mathbf{C}} = \boldsymbol{\Omega}_3\mathbf{C} \tag{A2.18}$$

which, in expanded form, is

$$\begin{bmatrix} \dot{c}_{xx} & \dot{c}_{xy} & \dot{c}_{xz} \\ \dot{c}_{yx} & \dot{c}_{yy} & \dot{c}_{yz} \\ \dot{c}_{zx} & \dot{c}_{zy} & \dot{c}_{zz} \end{bmatrix} = \begin{bmatrix} 0 & \omega_z & -\omega_y \\ -\omega_z & 0 & \omega_x \\ \omega_y & -\omega_x & 0 \end{bmatrix} \begin{bmatrix} c_{xx} & c_{xy} & c_{xz} \\ c_{yx} & c_{yy} & c_{yz} \\ c_{zx} & c_{zy} & c_{zz} \end{bmatrix}. \tag{A2.19}$$

Appendices

A2.1.3 Introduction to the Quaternion

First, the quaternion due to the mathematician, Hamilton, may be regarded as a generalisation of the complex number that has three, rather than just one imaginary part. Thus

$$\mathbf{q} = q_0 + q_1 i + q_2 j + q_3 k, \qquad (A2.20)$$

where the three basis elements, i, j and k, satisfy an equation that was considered so important by Hamilton that he carved it on a canal bridge in Ireland for safe keeping! It is as follows.

$$i^2 = j^2 = k^2 = ijk = -1. \qquad (A2.21)$$

It is important to note that the multiplication is non-commutative. This can be demonstrated as follows. From (A2.21),

$$ijk^2 = -k \Rightarrow -ij = -k \Rightarrow ij = k \qquad (A2.22)$$

$$i^2 jk = -i \Rightarrow -jk = -i \Rightarrow jk = i \qquad (A2.23)$$

From (A2.22)

$$ij^2 = kj \Rightarrow kj = -i \qquad (A2.24)$$

Comparing (A2.23) and (A2.24) shows that $jk = -kj$. Similar manipulations lead to the following complete set of relationships.

$$ij = -ji = k; \quad jk = -kj = i; \quad ki = -ki = j. \qquad (A2.25)$$

Remarkably, these products are similar to the cross products of mutually orthogonal unit vectors but the three self products on the LHS of (A2.21) are different from the zero self cross products of the unit vectors.

A2.1.4 A Complex Number Based Single Degree of Freedom Analogy

To understand the nature of the singularity free quaternion based attitude representation, an equivalent attitude representation for single degree of freedom rotations will be described, using a complex number based representation, obtained by setting $q_2 = q_3 = 0$. This yields the complex number

$$\mathbf{q} = q_0 + q_1^i. \qquad (A2.26)$$

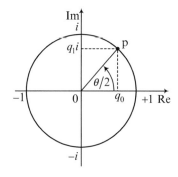

Fig. A2.3 Complex number representation for a single degree of freedom rotation

This complex number will be used to define the attitude of a rigid body rotating about a single axis with angular velocity, ω. The relationship between the attitude angle, θ, and the real and imaginary parts, q_0 and q_1, is defined in the Argand diagram of Fig. A2.3. Constraining the point, p, to lie on a circle of fixed radius reduces the number of degrees of freedom of the pair, (q_0, q_1), from 2 to 1. Choosing the unit circle simplifies the attitude representation. The argument of the complex number is $\theta/2$ rather than θ to comply with the quaternion representation for three rotational degrees of freedom presented below.

The KDE in terms of θ is simply

$$\dot{\theta} = \omega \tag{A2.27}$$

The corresponding KDEs in terms of the new attitude coordinates, q_0 and q_1, are obtained as follows. From Fig. A2.3,

$$q_0 = \cos\left(\frac{1}{2}\theta\right) \tag{A2.28}$$

$$q_1 = \sin\left(\frac{1}{2}\theta\right) \tag{A2.29}$$

Differentiating (A2.28) and (A2.29) then yields

$$\dot{q}_0 = \left[-\sin\left(\frac{1}{2}\theta\right)\right] \cdot \frac{1}{2}\dot{\theta} \tag{A2.30}$$

$$\dot{q}_1 = \left[\cos\left(\frac{1}{2}\theta\right)\right] \cdot \frac{1}{2}\dot{\theta} \tag{A2.31}$$

Substituting for $\dot{\theta}$, $\cos\left(\frac{1}{2}\theta\right)$ and $\sin\left(\frac{1}{2}\theta\right)$ in (A2.30) and (A2.31) using, respectively, (A2.27), (A2.28) and (A2.29) yields

$$\dot{q}_0 = -\frac{1}{2}\omega q_1 \tag{A2.32}$$

$$\dot{q}_1 = \frac{1}{2}\omega q_0 \qquad (A2.33)$$

Although there are two KDEs, they only represent one degree of freedom of motion since the solution is automatically constrained to the unit circle, with appropriate initial conditions. This may be shown by dividing (A2.33) by (A2.32) and solving the resulting differential equation by the method of separation of variables. Thus

$$\frac{\dot{q}_1}{\dot{q}_0} = \frac{dq_1}{dq_0} = -\frac{q_0}{q_1} \Rightarrow \int q_1 dq_1 = -\int q_0 dq_0 \Rightarrow \frac{1}{2}q_1^2 = -\frac{1}{2}q_0^2 + A, \qquad (A2.34)$$

where A is an arbitrary constant. This can be determined by taking the initial conditions at $\theta = 0$ for which $(q_0, q_1) = (1, 0)$. Substituting in (A2.34) then yields $A = 1/2$. The solution is therefore

$$q_1^2 + q_0^2 = 1, \qquad (A2.35)$$

which is the equation of the unit circle.

In the matrix–vector form, (A2.32) and (A2.33) may be written as

$$\begin{bmatrix} \dot{q}_0 \\ \dot{q}_1 \end{bmatrix} = \frac{1}{2} \begin{bmatrix} 0 & -\omega \\ \omega & 0 \end{bmatrix} \begin{bmatrix} q_0 \\ q_1 \end{bmatrix}. \qquad (A2.36)$$

There is no advantage of (A2.36) over (A2.27) for a single rotational degree of freedom. This is only intended to provide insight into the quaternion-based attitude representation for three rotational degrees of freedom, which yields a KDE of similar form to (A2.36) with no singularities, but with an angular velocity matrix of dimension, 4×4.

Mathematically, a circle may be regarded as a sphere of two dimensions whose surface has just one dimension. So the complex number based attitude representation for a single rotational degree of freedom represents the attitude by a point on the one dimensional surface of a two dimensional sphere of unit radius defined by (A2.35). By analogy, the quaternion-based attitude representation for three rotational degrees of freedom represents the attitude by a point on the three dimensional surface of a four dimensional sphere of unit radius defined by

$$q_0^2 + q_1^2 + q_2^2 + q_3^2 = 1. \qquad (A2.37)$$

The derivation of the KDEs requires the preliminaries presented below.

A2.1.5 Euler's Rotation Theorem

Euler's rotation theorem states that a rigid body may be moved to an arbitrary orientation by a simple rotation about an axis, referred to as the *Euler axis*, fixed in the body and in the reference frame. The angle of rotation about the Euler axis

Fig. A2.4 Principal axis rotation illustrating Euler's rotation theorem

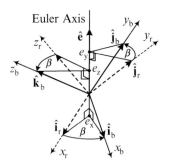

is referred to as the *principal angle* (or the *Euler angle*), β, and the direction of the Euler axis is that of the *principal unit vector*, $\hat{\mathbf{e}}$, as shown in Fig. A2.4.

A2.1.6 Rotation of a Vector About a Fixed Axis

Consider an arbitrary vector fixed in the body, which is about to be rotated about the Euler axis, as described in the previous section. Let this vector be $\widehat{\mathbf{r}}_r$ before the rotation and $\widehat{\mathbf{r}}_b$ after the rotation (viewed in the frame of reference of the initial orientation) as illustrated in Fig. A2.5.

An equation will now be derived that expresses $\widehat{\mathbf{r}}_b$ in terms $\widehat{\mathbf{r}}_r$, the principal unit vector, $\hat{\mathbf{e}}$, and the rotation angle, β. $\widehat{\mathbf{r}}_r$ will be expressed as a weighted sum of the three mutually orthogonal unit vectors, $\hat{\mathbf{e}}$, $\hat{\mathbf{i}}_r$ and $\hat{\mathbf{j}}_r$, fixed in the body in its initial orientation. These vectors rotate with the body through the angle, β, about $\hat{\mathbf{e}}$ and when viewed in the frame of reference of the initial orientation become $\hat{\mathbf{e}}$ (unaltered), $\hat{\mathbf{i}}_b$ and $\hat{\mathbf{j}}_b$. From Fig. A2.5,

$$\widehat{\mathbf{r}}_r = r\cos(\gamma)\hat{\mathbf{e}} + r\sin(\gamma)\hat{\mathbf{i}}_r \tag{A2.38}$$

$$\widehat{\mathbf{r}}_b = r\cos(\gamma)\hat{\mathbf{e}} + r\sin(\gamma)\hat{\mathbf{i}}_b \tag{A2.39}$$

$$\hat{\mathbf{i}}_b = \cos(\beta)\hat{\mathbf{i}}_r + \sin(\beta)\hat{\mathbf{j}}_r \tag{A2.40}$$

Fig. A2.5 Rotation of an arbitrary body-fixed vector during an Euler axis rotation

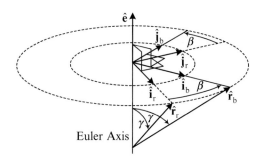

Appendices

The vector, $\hat{\mathbf{e}} \wedge \hat{\mathbf{r}}_r$, has the same direction as the unit vector, $\hat{\mathbf{j}}_r$, and therefore

$$\hat{\mathbf{j}}_r = \frac{\hat{\mathbf{e}} \wedge \hat{\mathbf{r}}_r}{|\hat{\mathbf{e}} \wedge \underset{\sim}{\mathbf{r}}_r|} = \frac{1}{r \sin(\gamma)} \left(\hat{\mathbf{e}} \wedge \hat{\mathbf{r}}_r\right). \tag{A2.41}$$

Also, $\hat{\mathbf{i}}_r = \hat{\mathbf{j}}_r \wedge \hat{\mathbf{e}}$. Then substituting for $\hat{\mathbf{j}}_r$ using (A2.41) yields

$$\hat{\mathbf{i}}_r = \frac{\hat{\mathbf{e}} \wedge \hat{\mathbf{r}}_r}{|\hat{\mathbf{e}} \wedge \underset{\sim}{\mathbf{r}}_r|} = \frac{1}{r \sin(\gamma)} \left(\hat{\mathbf{e}} \wedge \hat{\mathbf{r}}_r\right) \wedge \hat{\mathbf{e}}. \tag{A2.42}$$

Applying the vector triple product identity, $(\hat{\mathbf{v}}_1 \wedge \hat{\mathbf{v}}_2) \wedge \hat{\mathbf{v}}_3 = (\hat{\mathbf{v}}_3.\hat{\mathbf{v}}_1)\hat{\mathbf{v}}_2 - (\hat{\mathbf{v}}_3.\hat{\mathbf{v}}_2)\mathbf{v}_1$, then yields

$$\hat{\mathbf{i}}_r = \frac{1}{r \sin(\gamma)} \left[(\hat{\mathbf{e}}.\hat{\mathbf{e}})\hat{\mathbf{r}}_r - (\hat{\mathbf{e}}.\hat{\mathbf{r}}_r)\hat{\mathbf{e}}\right] = \frac{1}{r \sin(\gamma)} \left[\hat{\mathbf{r}}_r - (\hat{\mathbf{e}}.\hat{\mathbf{r}}_r)\hat{\mathbf{e}}\right] \tag{A2.43}$$

The required relationship is then obtained by first substituting for $\hat{\mathbf{i}}_b$ in (A2.39) using (A2.40), then substituting for $\hat{\mathbf{j}}_r$ using (A2.41) and finally substituting for $\hat{\mathbf{i}}_r$ using (A2.43). Hence

$$\begin{aligned}
\hat{\mathbf{r}}_b &= r \cos(\gamma)\hat{\mathbf{e}} + r \sin(\gamma) \left\{\cos(\beta)\hat{\mathbf{i}}_r + \sin(\beta)\hat{\mathbf{j}}_r\right\} \\
&= r \cos(\gamma)\hat{\mathbf{e}} + r \sin(\gamma) \left\{\cos(\beta) \frac{1}{r \sin(\gamma)} \left[\hat{\mathbf{r}}_r - (\hat{\mathbf{e}}.\hat{\mathbf{r}}_r)\hat{\mathbf{e}}\right]\right. \\
&\quad \left. + \sin(\beta) \frac{1}{r \sin(\gamma)} \left(\hat{\mathbf{e}} \wedge \hat{\mathbf{r}}_r\right)\right\} \\
&= r \cos(\gamma)\hat{\mathbf{e}} + \left\{\cos(\beta) \left[\hat{\mathbf{r}}_r - (\hat{\mathbf{e}}.\hat{\mathbf{r}}_r)\hat{\mathbf{e}}\right] + \sin(\beta) \left(\hat{\mathbf{e}} \wedge \hat{\mathbf{r}}_r\right)\right\} \\
&= (\hat{\mathbf{e}}.\hat{\mathbf{r}}_r)\hat{\mathbf{e}} + \left\{\cos(\beta) \left[\hat{\mathbf{r}}_r - (\hat{\mathbf{e}}.\hat{\mathbf{r}}_r)\hat{\mathbf{e}}\right] + \sin(\beta) \left(\hat{\mathbf{e}} \wedge \hat{\mathbf{r}}_r\right)\right\},
\end{aligned} \tag{A2.44}$$

Noting that since $|\hat{\mathbf{e}}| = 1$, $\hat{\mathbf{e}}.\hat{\mathbf{r}}_r = r \cos(\gamma)$. Then (A2.44) simplifies to

$$\hat{\mathbf{r}}_b = [1 - \cos(\beta)](\hat{\mathbf{e}}.\hat{\mathbf{r}}_r)\hat{\mathbf{e}} + \cos(\beta)\hat{\mathbf{r}}_r + \sin(\beta)\left(\hat{\mathbf{e}} \wedge \hat{\mathbf{r}}_r\right). \tag{A2.45}$$

A2.1.7 The Quaternion-Based Kinematic Differential Equations

The first step in the derivation is to observe that the unit vector pairs, $(\hat{\mathbf{i}}_r, \hat{\mathbf{i}}_b)$, $(\hat{\mathbf{j}}_r, \hat{\mathbf{j}}_b)$ and $(\hat{\mathbf{k}}_r, \hat{\mathbf{k}}_b)$ in Fig. A2.4 are particular cases of the vector pair, $(\hat{\mathbf{r}}_r, \hat{\mathbf{r}}_b)$, of Fig. A2.5 and therefore (A2.45) applies to each of these as follows.

$$\hat{\mathbf{i}}_b = (1-c_\beta)\left(\hat{\mathbf{e}}.\hat{\mathbf{i}}_r\right)\hat{\mathbf{e}} + c_\beta\hat{\mathbf{i}}_r + s_\beta\left(\hat{\mathbf{e}} \wedge \hat{\mathbf{i}}_r\right)$$
$$\hat{\mathbf{j}}_b = (1-c_\beta)\left(\hat{\mathbf{e}}.\hat{\mathbf{j}}_r\right)\hat{\mathbf{e}} + c_\beta\hat{\mathbf{j}}_r + s_\beta\left(\hat{\mathbf{e}} \wedge \hat{\mathbf{j}}_r\right) \qquad (A2.46)$$
$$\hat{\mathbf{k}}_b = (1-c_\beta)\left(\hat{\mathbf{e}}.\hat{\mathbf{k}}_r\right)\hat{\mathbf{e}} + c_\beta\hat{\mathbf{k}}_r + s_\beta\left(\hat{\mathbf{e}} \wedge \hat{\mathbf{k}}_r\right)$$

where $c_\beta = \cos(\beta)$ and $s_\beta = \sin(\beta)$. The direction cosines, e_x, e_y and e_z, introduced in Sect. A2.1.5 are the components of $\hat{\mathbf{e}}$ along, respectively, the axes, x_r, y_r and z_r (and also along the axes, x_b, y_b and z_b). Then

$$\hat{\mathbf{e}} = e_x\hat{\mathbf{i}}_b + e_y\hat{\mathbf{j}}_b + e_z\hat{\mathbf{k}}_b \qquad (A2.47)$$

Substituting for $\hat{\mathbf{e}}$ in (A2.46) using (A2.47) and noting that $\hat{\mathbf{e}}.\hat{\mathbf{i}}_r = e_x$, $\hat{\mathbf{e}}.\hat{\mathbf{j}}_r = e_y$, $\hat{\mathbf{e}}.\hat{\mathbf{k}}_r = e_z$, $\hat{\mathbf{e}} \wedge \hat{\mathbf{i}}_r = e_z\hat{\mathbf{j}}_r - e_y\hat{\mathbf{k}}_r$, $\hat{\mathbf{e}} \wedge \hat{\mathbf{j}}_r = e_x\hat{\mathbf{k}}_r - e_z\hat{\mathbf{i}}_r$ and $\hat{\mathbf{e}} \wedge \hat{\mathbf{k}}_r = e_y\hat{\mathbf{i}}_r - e_x\hat{\mathbf{j}}_r$, yields

$$\hat{\mathbf{i}}_b = (1-c_\beta) e_x \left(e_x\hat{\mathbf{i}}_r + e_y\hat{\mathbf{j}}_r + e_z\hat{\mathbf{k}}_r\right) + c_\beta\hat{\mathbf{i}}_r + s_\beta\left(e_z\hat{\mathbf{j}}_r - e_y\hat{\mathbf{k}}_r\right)$$
$$\hat{\mathbf{j}}_b = (1-c_\beta) e_y \left(e_x\hat{\mathbf{i}}_r + e_y\hat{\mathbf{j}}_r + e_z\hat{\mathbf{k}}_r\right) + c_\beta\hat{\mathbf{j}}_r + s_\beta\left(e_x\hat{\mathbf{k}}_r - e_z\hat{\mathbf{i}}_r\right) \qquad (A2.48)$$
$$\hat{\mathbf{k}}_b = (1-c_\beta) e_z \left(e_x\hat{\mathbf{i}}_r + e_y\hat{\mathbf{j}}_r + e_z\hat{\mathbf{k}}_r\right) + c_\beta\hat{\mathbf{k}}_r + s_\beta\left(e_y\hat{\mathbf{i}}_r - e_x\hat{\mathbf{j}}_r\right)$$

Grouping the coefficients of $\hat{\mathbf{i}}_r, \hat{\mathbf{j}}_r$ and $\hat{\mathbf{k}}_r$

$$\hat{\mathbf{i}}_b = \left[(1-c_\beta) e_x^2 + c_\beta\right]\hat{\mathbf{i}}_r + \left[(1-c_\beta) e_x e_y + s_\beta e_z\right]\hat{\mathbf{j}}_r + \left[(1-c_\beta) e_x e_z - s_\beta e_y\right]\hat{\mathbf{k}}_r$$
$$\hat{\mathbf{j}}_b = \left[(1-c_\beta) e_y e_x - s_\beta e_z\right]\hat{\mathbf{i}}_r + \left[(1-c_\beta) e_y^2 + c_\beta\right]\hat{\mathbf{j}}_r + \left[(1-c_\beta) e_y e_z + s_\beta e_x\right]\hat{\mathbf{k}}_r$$
$$\hat{\mathbf{k}}_b = \left[(1-c_\beta) e_z e_x - s_\beta e_z\right]\hat{\mathbf{i}}_r + \left[(1-c_\beta) e_z e_y - s_\beta e_x\right]\hat{\mathbf{j}}_r + \left[(1-c_\beta) e_z^2 + c_\beta\right]\hat{\mathbf{k}}_r$$
$$(A2.49)$$

In matrix form, this may be written

$$\begin{bmatrix}\hat{\mathbf{i}}_b \\ \hat{\mathbf{j}}_b \\ \hat{\mathbf{k}}_b\end{bmatrix} = \begin{bmatrix} (1-c_\beta) e_x^2 + c_\beta & (1-c_\beta) e_x e_y + s_\beta e_z & (1-c_\beta) e_x e_z - s_\beta e_y \\ (1-c_\beta) e_y e_x - s_\beta e_z & (1-c_\beta) e_y^2 + c_\beta & (1-c_\beta) e_y e_z + s_\beta e_x \\ (1-c_\beta) e_z e_x - s_\beta e_z & (1-c_\beta) e_z e_y - s_\beta e_x & (1-c_\beta) e_z^2 + c_\beta \end{bmatrix} \begin{bmatrix}\hat{\mathbf{i}}_r \\ \hat{\mathbf{j}}_r \\ \hat{\mathbf{k}}_r\end{bmatrix}.$$
$$(A2.50)$$

It then follows by comparison with (A2.5) that the matrix on the RHS of (A2.50) is the direction cosine matrix, **C**. It is defined in terms of the four attitude parameters of Fig. A2.4. As stated at the end of Sect. A2.1.5, the quaternion parameters are obtained from a nonlinear transformation of these attitude parameters and this is defined as follows.

Appendices

$$q_0 = \cos(\beta/2)$$
$$q_1 = e_x \sin(\beta/2)$$
$$q_2 = e_y \sin(\beta/2)$$
$$q_3 = e_z \sin(\beta/2)$$
(A2.51)

Next, **C** will be expressed in terms of these parameters. First note that

$$\sin(\beta) = 2\sin(\beta/2)\cos(\beta/2) \qquad (A2.52)$$

and

$$\cos(\beta) = \cos^2(\beta/2) - \sin^2(\beta/2) = 2\cos^2(\beta/2) - 1. \qquad (A2.53)$$

Then from (A2.51)

$$\cos(\beta) = 2q_0^2 - 1 \quad \text{and} \quad \sin(\beta) = \frac{2}{e_x}q_1q_0 = \frac{2}{e_y}q_2q_0 = \frac{2}{e_z}q_3q_0 \qquad (A2.54)$$

Then squaring and adding equations (A2.51) yields another important relationship.

$$q_0^2 + q_1^2 + q_2^2 + q_3^2 = \cos^2(\beta/2) + \left(e_x^2 + e_y^2 + e_x^2\right)\sin^2(\beta/2) \qquad (A2.55)$$

Since $e_x^2 + e_y^2 + e_z^2 = 1$ and $\cos^2(\beta/2) + \sin^2(\beta/2) = 1$, then

$$q_0^2 + q_1^2 + q_2^2 + q_3^2 = 1. \qquad (A2.56)$$

This the constraint equation (A2.37) already introduced in Sect. A2.1.4 that leaves three degrees of freedom for the attitude representation. From (A2.51),

$$q_1^2 + q_2^2 + q_3^2 = \left(e_x^2 + e_y^2 + e_x^2\right)\sin^2(\beta/2) = \sin^2(\beta/2) \qquad (A2.57)$$

and

$$e_x = \frac{q_1}{\sin(\beta/2)}, \quad e_y = \frac{q_2}{\sin(\beta/2)} \quad \text{and} \quad e_z = \frac{q_3}{\sin(\beta/2)}. \qquad (A2.58)$$

Hence

$$e_x = \frac{q_1}{\sqrt{q_1^2 + q_2^2 + q_3^2}}, \quad e_y = \frac{q_2}{\sqrt{q_1^2 + q_2^2 + q_3^2}} \quad \text{and} \quad e_z = \frac{q_3}{\sqrt{q_1^2 + q_2^2 + q_3^2}}.$$
(A2.59)

The direction cosine matrix, **C**, will now be obtained in terms of the quaternion parameters by substituting for $\cos(\beta)$ in the elements of (A2.50) using (A2.51), for $\sin(\beta)$ using choices of (A2.54) that cancel e_x, e_y, e_z and using (A2.56) and (A2.59) where appropriate. After substitution using (A2.59), all the elements of **C** in (A2.50) will contain the factor, $\frac{1-c_\beta}{q_1^2+q_2^2+q_3^2}$ which, through (A2.54), becomes $\frac{2(1-q_0^2)}{q_1^2+q_2^2+q_3^2}$ but a considerable simplification comes from (A2.56), since this gives $1-q_0^2 = q_1^2 + q_2^2 + q_3^2$ and hence $\frac{2(1-q_0^2)}{q_1^2+q_2^2+q_3^2} = 2$. Hence

$$\mathbf{C} = \begin{bmatrix} 2q_1^2 + 2q_0^2 - 1 & 2q_1q_2 + 2q_3q_0 & 2q_1q_3 - 2q_2q_0 \\ 2q_2q_1 - 2q_3q_0 & 2q_2^2 + 2q_0^2 - 1 & 2q_2q_3 + 2q_1q_0 \\ 2q_3q_1 + 2q_2q_0 & 2q_3q_2 - 2q_1q_0 & 2q_3^2 + 2q_0^2 - 1 \end{bmatrix}. \quad (A2.60)$$

The three diagonal component equations will now be used together with (A2.56) to obtain equations for q_0, q_1, q_2 and q_3, in terms of c_{ij}, $i = x, y, z$, $j = x, y, z$, in preparation for the differentiations that will lead to the required kinematic differential equations. Adding the leading diagonal terms and then using (A2.56) yields

$$c_{xx} + c_{yy} + c_{zz} = 2\left(q_1^2 + q_2^2 + q_3^2\right) + 6q_0^2 - 3 = 2\left(1 - q_0^2\right) + 6q_0^2 - 3 = 4q_0^2 - 1 \Rightarrow$$

$$q_0^2 = \frac{1}{4}\left(c_{xx} + c_{yy} + c_{zz} + 1\right). \quad (A2.61)$$

The first diagonal component equation yields

$$q_1^2 = \frac{1}{2}\left(-2q_0^2 + c_{xx} + 1\right). \quad (A2.62)$$

Substituting for q_0^2 using (A2.61) then yields:

$$q_1^2 = \frac{1}{2}\left[-\frac{1}{2}\left(c_{xx} + c_{yy} + c_{zz} + 1\right) + c_{xx} + 1\right] = \frac{1}{2}\left[\frac{1}{2}\left(c_{xx} - c_{yy} - c_{zz}\right) + \frac{1}{2}\right] \Rightarrow$$

$$q_1^2 = \frac{1}{4}\left(c_{xx} - c_{yy} - c_{zz} + 1\right). \quad (A2.63)$$

Similarly, the second diagonal component equation yields

$$q_2^2 = \frac{1}{2}\left[-\frac{1}{2}\left(c_{xx} + c_{yy} + c_{zz} + 1\right) + c_{yy} + 1\right] = \frac{1}{2}\left[\frac{1}{2}\left(-c_{xx} + c_{yy} - c_{zz}\right) + \frac{1}{2}\right] \Rightarrow$$

$$q_2^2 = \frac{1}{4}\left(-c_{xx} + c_{yy} - c_{zz} + 1\right), \quad (A2.64)$$

Appendices

and the third diagonal component equation yields

$$q_3^2 = \frac{1}{2}\left[-\frac{1}{2}(c_{xx} + c_{yy} + c_{zz} + 1) + c_{zz} + 1\right] = \frac{1}{2}\left[\frac{1}{2}(-c_{xx} - c_{yy} + c_{zz}) + \frac{1}{2}\right] \Rightarrow$$

$$q_3^2 = \frac{1}{4}(-c_{xx} - c_{yy} + c_{zz} + 1). \tag{A2.65}$$

Differentiating (A2.61), (A2.63), (A2.64) through (A2.65) will yield terms in $\dot{c}_{ii}$, $i = x, y, z$, but these may be expressed in terms of c_{ii}, ω_x, ω_y and ω_z using (A2.19). Thus

$$2q_0\dot{q}_0 = \frac{1}{4}(\dot{c}_{xx} + \dot{c}_{yy} + \dot{c}_{zz}) = \frac{1}{4}[(\omega_z c_{yx} - \omega_y c_{zx}) + (-\omega_z c_{xy} + \omega_x c_{zy}) + (\omega_y c_{xz} - \omega_x c_{yz})] \Rightarrow$$

$$2q_0\dot{q}_0 = \frac{1}{4}[\omega_x(c_{zy} - c_{yz}) + \omega_y(c_{xz} - c_{zx}) + \omega_z(c_{yx} - c_{xy})], \tag{A2.66}$$

$$2q_1\dot{q}_1 = \frac{1}{4}(\dot{c}_{xx} - \dot{c}_{yy} - \dot{c}_{zz}) = \frac{1}{4}[(\omega_z c_{yx} - \omega_y c_{zx}) - (-\omega_z c_{xy} + \omega_x c_{zy}) - (\omega_y c_{xz} - \omega_x c_{yz})] \Rightarrow$$

$$2q_1\dot{q}_1 = \frac{1}{4}[\omega_x(c_{yz} - c_{zy}) - \omega_y(c_{xz} + c_{zx}) + \omega_z(c_{yx} + c_{xy})], \tag{A2.67}$$

$$2q_2\dot{q}_2 = \frac{1}{4}(-\dot{c}_{xx} + \dot{c}_{yy} - \dot{c}_{zz}) = \frac{1}{4}[-(\omega_z c_{yx} - \omega_y c_{zx}) + (-\omega_z c_{xy} + \omega_x c_{zy}) - (\omega_y c_{xz} - \omega_x c_{yz})] \Rightarrow$$

$$2q_2\dot{q}_2 = \frac{1}{4}[\omega_x(c_{yz} + c_{zy}) + \omega_y(c_{zx} - c_{xz}) - \omega_z(c_{yx} + c_{xy})], \tag{A2.68}$$

$$2q_3\dot{q}_3 = \frac{1}{4}(-\dot{c}_{xx} - \dot{c}_{yy} + \dot{c}_{zz}) = \frac{1}{4}[-(\omega_z c_{yx} - \omega_y c_{zx}) - (-\omega_z c_{xy} + \omega_x c_{zy}) + (\omega_y c_{xz} - \omega_x c_{yz})] \Rightarrow$$

$$2q_3\dot{q}_3 = \frac{1}{4}[-\omega_x(c_{yz} + c_{zy}) + \omega_y(c_{zx} + c_{xz}) + \omega_z(c_{xy} - c_{yx})]. \tag{A2.69}$$

Finally substituting for c_{ij}, $i = x, y, z$, $j = x, y, z$, using (A2.19) yields

$$2q_0\dot{q}_0 = \frac{1}{4}\begin{Bmatrix} \omega_x\left[(2q_3q_2 - 2q_1q_0) - (2q_2q_3 + 2q_1q_0)\right] \\ + \omega_y\left[(2q_1q_3 - 2q_2q_0) - (2q_3q_1 + 2q_2q_0)\right] \\ + \omega_z\left[(2q_2q_1 - 2q_3q_0) - (2q_1q_2 + 2q_3q_0)\right] \end{Bmatrix}$$

$$= \frac{1}{4}[\omega_x(-4q_1q_0) + \omega_y(-4q_2q_0) + \omega_z(-4q_3q_0)] \Rightarrow$$

$$\dot{q}_0 = -\frac{1}{2}(\omega_x q_1 + \omega_y q_2 + \omega_z q_3). \tag{A2.70}$$

$$2q_1\dot{q}_1 = \frac{1}{4}\left\{\begin{array}{l}\omega_x\left[(2q_3q_2+2q_1q_0)-(2q_2q_3-2q_1q_0)\right]\\-\omega_y\left[(2q_1q_3-2q_2q_0)+(2q_3q_1+2q_2q_0)\right]\\+\omega_z\left[(2q_2q_1-2q_3q_0)+(2q_1q_2+2q_3q_0)\right]\end{array}\right\}$$
$$= \frac{1}{4}\left[\omega_x(4q_1q_0)-\omega_y(4q_3q_1)+\omega_z(4q_2q_1)\right] \Rightarrow$$
$$\dot{q}_1 = \frac{1}{2}\left(\omega_x q_0 - \omega_y q_3 + \omega_z q_2\right) \tag{A2.71}$$

$$2q_2\dot{q}_2 = \frac{1}{4}\left\{\begin{array}{l}\omega_x\left[(2q_3q_2+2q_1q_0)+(2q_2q_3-2q_1q_0)\right]\\+\omega_y\left[(2q_1q_3+2q_2q_0)-(2q_3q_1-2q_2q_0)\right]\\-\omega_z\left[(2q_2q_1-2q_3q_0)+(2q_1q_2+2q_3q_0)\right]\end{array}\right\}$$
$$= \frac{1}{4}\left[\omega_x(4q_2q_3)+\omega_y(4q_2q_0)-\omega_z(4q_2q_1)\right] \Rightarrow$$
$$\dot{q}_2 = \frac{1}{2}\left(\omega_x q_3 + \omega_y q_0 - \omega_z q_1\right) \tag{A2.72}$$

$$2q_3\dot{q}_3 = \frac{1}{4}\left\{\begin{array}{l}-\omega_x\left[(2q_3q_2+2q_1q_0)+(2q_2q_3-2q_1q_0)\right]\\+\omega_y\left[(2q_1q_3+2q_2q_0)+(2q_3q_1-2q_2q_0)\right]\\+\omega_z\left[(2q_1q_2+2q_3q_0)-(2q_2q_1-2q_3q_0)\right]\end{array}\right\}$$
$$= \frac{1}{4}\left[-\omega_x(4q_3q_2)+\omega_y(4q_3q_1)+\omega_z(4q_3q_0)\right] \Rightarrow$$
$$\dot{q}_3 = \frac{1}{2}\left(-\omega_x q_2 + \omega_y q_1 + \omega_z q_0\right). \tag{A2.73}$$

Equations (A2.70) through (A2.73) are the required kinematic differential equations, which may be written as a single equation in the matrix form, as follows.

$$\begin{bmatrix}\dot{q}_0\\\dot{q}_1\\\dot{q}_2\\\dot{q}_3\end{bmatrix} = \frac{1}{2}\begin{bmatrix}0 & -\omega_x & -\omega_y & -\omega_z\\\omega_x & 0 & \omega_z & -\omega_y\\\omega_y & -\omega_z & 0 & \omega_x\\\omega_z & \omega_y & -\omega_x & 0\end{bmatrix}\begin{bmatrix}q_0\\q_1\\q_2\\q_3\end{bmatrix}. \tag{A2.74}$$

Such kinematic differential equations would be applied by numerical integration on the control computer, as the variables, $q_i(t)$, $i = 0, 1, 2, 4$, are not measureable. It would, however, be considered part of the controlled plant in a control system. An example using this is given in Chap. 7.

A2.2 Plant Model Determination from Frequency Response

A2.2.1 Relatively Close Real Poles and Zeros

If, as is often the case, two adjacent corner frequencies are separated by less than an order of magnitude, then for reasons given at the end of Sect. 2.3.3.5 in Chap. 2, the graph of the piecewise linear function, $\widehat{L}_{dB1}(\omega)$, has to be found by an alternative method to the straightforward tangent fitting method of Sect. 2.2.3.6. in Chap. 2. Two different methods are developed below, one applying if both corner frequencies are associated with poles or zeros and the other if one corner frequency is associated with a pole and the other with a zero. Features of the graph of $M_{dB}(\omega)$ may be detected that enable an appropriate choice of method to be made.

First consider the combination of two terms isolated from (2.201) in Chap. 2, one contributed by a pole and the other by a zero, as follows.

$$M_{dB1}(\omega) = 10\log_{10}\left(1 + \frac{\omega^2}{v_1^2}\right) - 10\log_{10}\left(1 + \frac{\omega^2}{\omega_1^2}\right) \tag{A2.75}$$

Consider the graph of $M_{dB1}(\omega)$ and the graph of the piecewise linear approximating function, $L_{dB1}(\omega)$. In the direction of increasing ω, the first segment of $L_{dB1}(\omega)$ has zero slope. If $\omega_1 < v_1$, then the first change of slope of $L_{dB1}(\omega)$ is due to the pole and will be -20 [dB/decade], occurring at $\omega = \omega_1$ and the second change of slope is due to the zero and will be $+20$ [dB/decade], occurring at $\omega = v_1$. The result is that the net change of slope of $L_{dB1}(\omega)$ is zero. The third linear segment has the same slope as the first but is translated downwards by $20[\log_{10}(v_1) - \log_{10}(\omega_1)]$ [dB]. Conversely, if $v_1 < \omega_1$, the net change of slope of $L_{dB1}(\omega)$ is again zero but the third linear segment is translated upwards by $20[\log_{10}(\omega_1) - \log_{10}(v_1)]$ [dB] relative to the third segment. This feature may be readily recognised in a graph of $M_{dB}(\omega)$, an example being visible in Fig. 2.26 of Chap. 2 for $1 < \omega < 10^5$ [rad/s]. Figure A2.6a shows families of graphs obtained for different values of v_1/ω_1, equally separated by imposing the constraint, $v_1\omega_1 = 1$. For each value of v_1/ω_1, both $M_{dB1}(\omega)$ and $L_{dB1}(\omega)$ are reflected in their point of intersection, which lies on the vertical dotted line in Fig. A2.6. For any value of v_1/ω_1 satisfying $0 < v_1/\omega_1 < \infty$, $M_{dB1}(\omega)$ is not tangential to $L_{dB1}(\omega)$ but intersects it at a non-zero angle. As the separation between v_1 and ω_1 approaches two orders of magnitude, this angle is so small that $M_{dB1}(\omega)$ *appears* to be tangential to $L_{dB1}(\omega)$ which can be seen in Fig. A2.6a.

This enables the simple tangent fitting method of Sect. 2.3.3.6 of Chap. 2 to be used to determine every segment of $L_{dB1}(\omega)$ but for closer v_1 and ω_1, the middle segment has to be determined differently. In Fig. A2.6b, c the non-zero intersection angle between $M_{dB1}(\omega)$ and $L_{dB1}(\omega)$ is clearly visible.

To determine the corner frequencies of a pole and a zero that can be arbitrarily close to one another, using the graph of $M_{dB1}(\omega)$, first, an attempt is made to fit $\widehat{L}_{dB1}(\omega)$ to $M_{dB}(\omega)$ as in Sect. 2.3.3.6 of Chap. 2. Recalling that every segment of $\widehat{L}_{dB1}(\omega)$ has a slope that is an integer multiple of -20 [dB/decade], if the 'close' pole-zero pair is well separated from the remaining poles and zeros, the segment of $M_{\mathrm{dB}}(\omega)$ associated with them will be sandwiched between two parallel segments of $L_{\mathrm{dB}}(\omega)$. It will be impossible to find a segment to bridge between the parallel ones that is nearly tangential to $M_{\mathrm{dB}}(\omega)$. The correct bridging tangent, however, is that with the closest admissible slope to that of $M_{\mathrm{dB}}(\omega)$ at the midpoint between the two parallel segments. This bridging segment is drawn to pass through the midpoint. Then the required corner frequencies are at the intersections between the bridging segment and the parallel segments, as illustrated in Fig. A2.6b, c. Starting with the graph of $M_{\mathrm{dB1}}(\omega)$, the piecewise linear function, $\hat{L}_{\mathrm{dB1}}(\omega)$, is first drawn. Then the required values of v_1 and ω_1 are the corner frequencies of $\hat{L}_{\mathrm{dB1}}(\omega)$. It must be stressed, however, that the first segment of $L_{\mathrm{dB1}}(\omega)$ may not have zero slope due to the presence of the other poles and zeros. In general, the segment of $M_{\mathrm{dB}}(\omega)$ could be as $M_{\mathrm{dB1}}(\omega)$ in Fig. A2.6 (b) or (c) or with both graphs decreased in slope at every point by an integer multiple of -20 [dB/decade], as illustrated in Fig. A2.7 for a plant with the transfer function,

$$\frac{Y(s)}{U(s)} = \frac{10(1+s/2)}{s(1+s)} \quad (A2.76)$$

Figure A2.7a shows the basic plot, $M_{\mathrm{dB}}(\omega)$, sandwiched between the asymptotes, A_1 and A_3, which have a slope of -20 [dB/decade]. Then the slope of the intermediate asymptote, A_2, is chosen as the smallest that exceeds that of $M_{\mathrm{dB}}(\omega)$ in magnitude

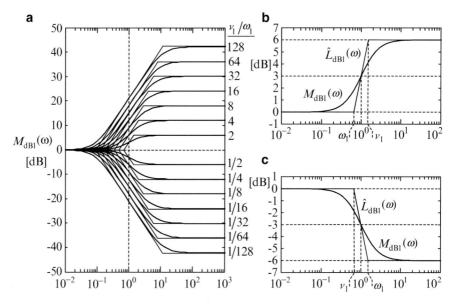

Fig. A2.6 Contributions of pole-zero combinations to Bode magnitude plot. (**a**) Variation with corner frequency ratio. (**b**) $v_1 = 2\omega_1$. (**c**) $\omega_1 = 2v_1$

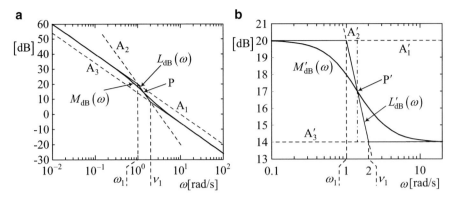

Fig. A2.7 Determination of relatively close pole and zero corner frequencies: (**a**) Basic plots with skewing. (**b**) Unskewed plots

at the midway point, P, between the asymptotes, A_1 and A_3. Then the asymptote is placed to pass through this point. It will be apparent, however, that this will be difficult to accomplish accurately. To overcome this practical problem, it is recommended to process the frequency response data to yield the Bode magnitude plot of $s^q G(s)$, where q is an integer and the slope of the asymptotes, A_1 and A_3, is $-20q$ [dB/decade]. With an appropriate choice of q, this unskews the plots and asymptotes together to yield, $M'_{dB}(\omega)$, $L'_{dB}(\omega)$ and asymptotes, A'_1 and A'_3, with zero slope without altering the corner frequencies, as shown in Fig. A2.7b.

Next the determination of the corner frequencies of two real poles (or two real zeros) that can be arbitrarily close will be considered, assuming that the remaining poles and zeros are sufficiently far removed to have, at most, a skewing effect without altering the error,

$$E_{dB}(\omega) = M_{dB}(\omega) - \widehat{L}_{dB}(\omega), \quad (A2.77)$$

at the corner frequencies.

A segment of $M_{dB}(\omega)$ with two adjacent corner frequencies that are too close to be determined by the simple tangent fitting method may be recognised by first attempting this method. Then the error of (A2.77) should be observed. If all the poles and zeros are real and distinct, then according to Sect. 2.3.3.5 of Chap. 2, the magnitude of the error should not exceed the limits of ± 3 [dB] if the simple tangent fitting method is successful. If not, then error peaks will be found of at least ± 6 [dB], the locations of which will be at the intersections of segments of $\widehat{L}_{dB}(\omega)$ differing in slope by ± 12 [dB/octave] rather than ± 6 [dB/octave]. Each of these error peaks is straddled by a pair of corner frequencies associated with two poles (or two zeros) that are not sufficiently separated. These larger errors occur because no segment of $\widehat{L}_{dB}(\omega)$ is formed between the relatively close corner frequencies. Instead a vertex of $\widehat{L}_{dB}(\omega)$ occurs at a frequency equal to their geometric mean, i.e., the midway point between the corner frequencies on the logarithmic scale. Let the frequency at which the ith 'large' error peak occurs be denoted q_{pi} and the straddling corner frequencies be q_i and q_{i+1} with $q_{i+1} > q_i$ ($q = \omega$ for zeros and $q = \nu$ for poles). Then

$$q_{pi} = \sqrt{q_{i+1} q_i} \Rightarrow \frac{q_{pi}}{q_i} = \frac{q_{i+1}}{q_{pi}} = r_i = \sqrt{\frac{q_{i+1}}{q_i}} \Rightarrow q_i = \frac{q_{pi}}{r_i} \text{ and } q_{i+1} = q_{pi} r_i \Rightarrow$$

$$\log_{10}(q_{i+1}) = \log_{10}(q_{pi}) + \log_{10}(r_i) \tag{A2.78}$$

$$\log_{10}(q_i) = \log_{10}(q_{pi}) - \log_{10}(r_i) \tag{A2.79}$$

Once q_{pi} is read from the plot, all that is needed is to determine r_i. Then the corner frequencies will be known from (A2.78) and (A2.79). The quantity, $\log_{10}(r_i)$, is a fixed difference on the logarithmic frequency scale of the Bode magnitude plot that can be used to directly read off the corner frequencies, q_i and q_{i+1}, now q_{pi} is known. The frequency ratio, r_i, can be found from the value of the error peak, as the following analysis will show.

Consider first the isolated contribution of a single pole or zero defined by (2.211) of Chap. 2 with the concatenated asymptote function (2.214) of Chap. 2 and depicted in Fig. 2.24 of Chap. 2. The corresponding error function is then

$$E_{dBi}(\omega) = M_{dBi}(\omega) - A_i(\omega). \tag{A2.80}$$

Before proceeding further, it will be proven that $E_{dBi}(\omega)$ is a symmetrical function reflected in the vertical line, $\omega = q_i$, when ω is on a logarithmic scale. Using, in Chap. 2, (2.211), (2.214), (2.212) and (2.213) yields

$$E_{dBi}(\omega) = \begin{cases} 10Q\log_{10}(1 + \omega^2/q_i^2) - 0, & 0 < \omega < q_i \\ 10Q\log_{10}(1 + \omega^2/q_i^2) - 10Q\log_{10}(\omega^2/q_i^2), & \omega \geq q_i \end{cases}$$

$$= \begin{cases} 10Q\log_{10}(1 + \omega^2/q_i^2), & 0 < \omega < q_i \\ 10Q\log_{10}(1 + q_i^2/\omega^2), & \omega \geq q_i \end{cases}. \tag{A2.81}$$

Let $\omega = \lambda q_i$ for $0 < \omega < q_i$ and $\omega = q_i/\lambda$ for $\omega \geq q_i$, where $0 < \lambda < 1$. Then (A2.81) yields

$$E_{dBi}(\lambda q_i) = 10Q\log_{10}(1 + \lambda^2), \quad 0 < \omega < q_i \tag{A2.82}$$

and

$$E_{dBi}(q_i/\lambda) = 10Q\log_{10}(1 + \lambda^2), \quad \omega \geq q_i \tag{A2.83}$$

Since (A2.82) and (A2.83) yield the same value, the symmetry of $E_{dBi}(\omega)$ referred to above is true.

Appendices

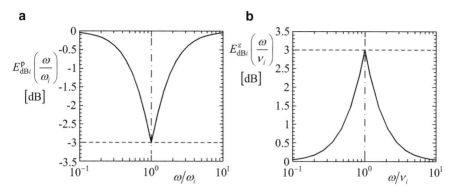

Fig. A2.8 Error functions of asymptotic approximations (**a**) for contribution from real pole (**b**) for contribution from real zero

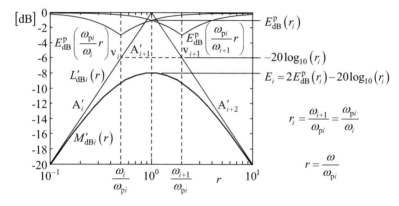

Fig. A2.9 Anti-skewed Bode magnitude plot for two relatively close poles

Figure A2.8 shows the graphs of these error functions expressed as functions of the frequency ratio, ω/ω_i, for poles and ω/v_i for zeros. They are identical for every corner frequency, with $E_{dB}^p(r) = -E_{dB}^z(r)$. Expressed as functions of ω, rather than $r = \omega/\omega_i$ or ω/v_i, they are translated along the ω axis without changing shape to be centred on ω_i or v_i.

Now consider the common situation of two 'close' poles. At the outset, the Bode magnitude plot, $E_{dB}(\omega)$, will be replaced by

$$E'_{dBi}(\omega) = K_i s^q E_{dBi}(\omega), \qquad (A2.84)$$

shown graphically in Fig. A2.9, where the corner frequencies are ω_i and ω_{i+1}.

Here the integer, q, is chosen to remove any skewing due to the other poles and zeros by making $E'_{dBi}(\omega)$ reflected in the vertical straight line, $r = 1 \Rightarrow \omega = \omega_{pi}$, where ω_{pi} is the frequency of the peak error, E_i, of the initial attempt at asymptote fitting that uses only the asymptotes, A'_i and A'_{i+2}. The constant, K_i, is included so

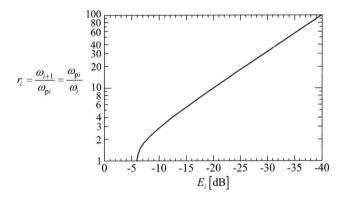

Fig. A2.10 Peak error curve for corner frequency determination

that the intersection between the asymptotes, A'_i and A'_{i+2}, at $r = 1$, lies on the 0[dB] line, as shown. Let $L''_{dBi}(r)$ be the piecewise linear approximation function that would be obtained if the segment contained in the asymptote, A'_{i+1}, joining the vertices, v_i and v_{i+1}, was included. Then the error would be

$$E''_{dBi}(r) = M'_{dBi}(r) - L''_{dBi}(r) = E^p_{dBi}\left(\frac{\omega_{pi}}{\omega_i}r\right) + E^p_{dBi}\left(\frac{\omega_{pi}}{\omega_{i+1}}r\right) \Rightarrow$$

$$E''_{dBi}(1) = E^p_{dBi}\left(\frac{\omega_{pi}}{\omega_i}\right) + E^p_{dBi}\left(\frac{\omega_{pi}}{\omega_{i+1}}\right) = 2E^p_{dBi}(r_i). \quad \text{(A2.85)}$$

Observing Fig. A2.9, without including the asymptote, A'_{i+1}, the peak error is

$$E_i = E''_{dBi}(1) - 20\log_{10}(r_i). \quad \text{(A2.86)}$$

Then in view of (A2.85),

$$E_i = 2E^p_{dBi}(r_i) - 20\log_{10}(r_i). \quad \text{(A2.87)}$$

Figure A2.10 shows this relationship graphically.

To summarise, a piecewise linear approximation, $L_{dB}(\omega)$, to the Bode magnitude plot, $M_{dB}(\omega)$ is made. Then any adjacent pair of asymptotes with a change of slope of 40[dB/decade] is identified. The peak error, E_i[dB], is then evaluated at the frequency, ω_{pi}, of the intersection between the asymptotes.

The corresponding frequency ratio, r_i, is then determined from the graph of Fig. A2.10. The required corner frequencies are then

$$\omega_i = \omega_{pi}/r_i \quad \text{and} \quad \omega_{i+1} = \omega_{pi}r_i. \quad \text{(A2.88)}$$

It is advisable to de-skew the graphs of $M_{dB}(\omega)$ and $L_{dB}(\omega)$ at each of the points identified to enable E_i [dB] to be determined more accurately.

It should be noted that the smallest peak error of -6 [dB] gives $r_i = 1$, meaning that the two poles are coincident.

For a pair of zeros, the method is the same as for the poles except for Figs. A2.9 and A2.10 that will both be inverted.

A2.2.2 The Resonance Peak Function

The resonance peak function has been introduced by the author to enable the parameters of a plant having complex conjugate pairs of poles and with arbitrarily close corner frequencies to be determined using the asymptotic approximation to the Bode magnitude plot.

The value of a resonance or anti-resonance peak in $M_{dB}(\omega)$ of (2.201) of Chap. 2 due to an individual contribution,

$$M_{dBi}(\omega) = -10\log_{10}\left[\left(1 - \frac{\omega^2}{\omega_{ni}^2}\right)^2 + 4\frac{\eta_i^2 \omega^2}{\omega_{ni}^2}\right] \text{ or } 10\log_{10}\left[\left(1 - \frac{\omega^2}{\nu_{ni}^2}\right)^2 + 4\frac{\eta_i^2 \omega^2}{\nu_{ni}^2}\right], \quad (A2.89)$$

will differ significantly from that of $M_{dBi}(\omega)$ taken in isolation, if the corner frequency, ω_{ni} or ν_{ni}, is sufficiently close to the other corner frequencies. This phenomenon will be called *resonance peak interaction*, with the understanding that it includes interaction with anti-resonance dips due to complex conjugate zeros. Then if the method of Sect. 2.3.3.8 of Chap. 2 were to be applied directly, the estimates of the parameters, ω_{ni} and ζ_i, or ν_{ni} and η_i, would be incorrect. To overcome this problem, a method is developed in Sect. A2.2.3 that enables the resonance peak magnitudes of $M_{dBi}(\omega)$ taken in isolation to be estimated. This makes use of a family of *resonance peak functions*.

Each complex conjugate pole factor,

$$G_i(j\omega) = \frac{K_{dci}}{1 + \frac{2\zeta_i j\omega}{\omega_{ni}} + \frac{(j\omega)^2}{\omega_{ni}^2}}, \quad (A2.90)$$

of the plant frequency transfer function, $G(j\omega)$, has an associated resonance peak function defined as

$$P_{dBi}(\omega, \zeta_i) \triangleq M_{dBi}(\omega, \zeta_i) - A_i(\omega), \quad (A2.91)$$

where

$$M_{dBi}(\omega, \zeta_i) = 20\log_{10}|G_i(j\omega)| = 20\log_{10}K_{dci} - 20\log_{10}\sqrt{\left(1 - \frac{\omega^2}{\omega_{ni}^2}\right)^2 + 4\frac{\zeta_i^2\omega^2}{\omega_{ni}^2}}$$
(A2.92)

is the Bode magnitude plot of (A2.90) and $A_i(\omega)$ is the corresponding concatenated asymptote function already introduced in Sect. 2.3.3.5 of Chap. 2, which can be expressed mathematically as

$$A_i(\omega) \triangleq \begin{cases} 20\log_{10}(K_{dci}) & \text{for } \omega \leq \omega_n \\ 20\log_{10}(K_{dci}) - 20\log_{10}(\omega^2/\omega_{ni}^2) & \text{for } \omega > \omega_{ni} \end{cases}.$$
(A2.93)

It will now be proven that $P_{dBi}(\omega, \zeta_i)$ is symmetrical about the vertical line, $\omega = \omega_{ni}$, when ω is plotted on a logarithmic scale. Substituting for $M_{dBi}(\omega, \zeta_i)$ and $A_i(\omega)$ on the RHS of (A2.91) using (A2.92) and (A2.93) yields the following.

For $\omega \leq \omega_n$,

$$P_{dBi}(\omega, \zeta_i) = -20\log_{10}\sqrt{\left(1 - \frac{\omega^2}{\omega_{ni}^2}\right)^2 + \frac{4\zeta_i^2\omega^2}{\omega_{ni}^2}} \Rightarrow$$

$$P_i(\omega, \zeta_i) = \frac{1}{\sqrt{\left(1 - \frac{\omega^2}{\omega_{ni}^2}\right)^2 + \frac{4\zeta_i^2\omega^2}{\omega_{ni}^2}}}.$$
(A2.94)

and for $\omega > \omega_n$,

$$P_{dBi}(\omega, \zeta_i) = -20\log_{10}\sqrt{\left(1 - \frac{\omega^2}{\omega_{ni}^2}\right)^2 + \frac{4\zeta_i^2\omega^2}{\omega_{ni}^2}} + 20\log_{10}\left(\frac{\omega^2}{\omega_{ni}^2}\right) \Rightarrow$$

$$P(\omega, \zeta_i) = \frac{\frac{\omega^2}{\omega_{ni}^2}}{\sqrt{\left(1 - \frac{\omega^2}{\omega_{ni}^2}\right)^2 + \frac{4\zeta_i^2\omega^2}{\omega_{ni}^2}}} = \frac{1}{\sqrt{\left(1 - \frac{\omega_{ni}^2}{\omega^2}\right)^2 + \frac{4\zeta_i^2\omega_{ni}^2}{\omega^2}}}.$$
(A2.95)

Then for $0 < \lambda < 1$, setting $\omega = \lambda\omega_{ni}$ in (A2.94) and $\omega = \omega_{ni}/\lambda$ in (A2.95) produces the same value of $P(\omega, \zeta_i)$. Hence if ω is plotted on a logarithmic scale, $P_{dBi}(\omega, \zeta_i)$ is reflected in the vertical line, $\omega = \omega_{ni}$. Figure A2.11 shows the resonance peak function for $\zeta_i = 0.4$ formed graphically. There appears to be a single maximum at $\omega = \omega_{ni}$ but there are actually two close together. From Sect. 2.3.3.8 of Chap. 2, there is a maximum at the resonance frequency,

$$\omega_{ri} = \omega_{ni}\sqrt{1 - 2\zeta_i^2}.$$
(A2.96)

Since $P_{dBi}(\omega, \zeta_i)$ is reflected in the vertical line, $\omega = \omega_{ni}$, there must be another maximum with the same magnitude at an angular frequency of

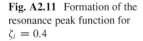

Fig. A2.11 Formation of the resonance peak function for $\zeta_i = 0.4$

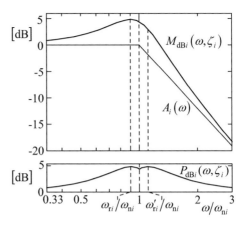

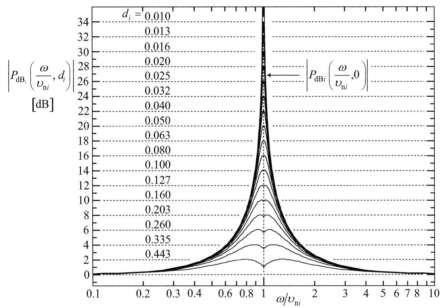

Fig. A2.12 Family of resonance peak function magnitudes

$$\omega'_{ri} = \omega_{ni}/\sqrt{1 - 2\zeta_i^2}. \tag{A2.97}$$

This will be called the *mirrored resonance frequency*. This is relevant to the parameter estimation method presented in Sect. 2.3.3.9 of Chap. 2 and Sect. A2.2.3.

By inspection of (A2.96) and (A2.97), as $\zeta_i \to 0$, $\omega'_{ri} \to \omega_{ri}$ but there are two equal peaks that become more separated as ζ_i increases, as shown in Fig. A2.12. The contribution to the Bode magnitude plot from all (A2.89) is as follows.

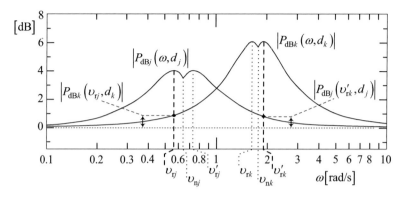

Fig. A2.13 Resonance peak interaction illustrated using resonance peak functions

$$\sum_{i=1}^{m} M_{\mathrm{dB}i}(\omega, d_i) = \sum_{i=1}^{m} A_i(\omega) + \sum_{i=1}^{m} P_{\mathrm{dB}i}(\omega, d_i). \quad (A2.98)$$

The term, $\sum_{i=1}^{m} A_i(\omega)$, is a piecewise linear function whose vertices are at the corner frequencies, which are the undamped natural frequencies, but without any resonance or anti-resonance peaks. These are all included in the term, $\sum_{i=1}^{m} P_{\mathrm{dB}i}(\omega, d_i)$, which may therefore be referred to as the total resonance function. For any pair, $P_{\mathrm{dB}j}(\omega, d_j)$ and $P_{\mathrm{dB}k}(\omega, d_k)$, the resonance peak interaction may be quantified by the value of each function at a peak frequency of the other function. If $\upsilon_{nk} > \upsilon_{nj}$ then these values are $P_{\mathrm{dB}j}(\upsilon_{rk}, d_j)$ and $P_{\mathrm{dB}k}(\upsilon'_{rj}, d_k)$, as illustrated in Fig. A2.13.

Note that the peak frequency selected for the resonance peak function with the lower undamped natural frequency is its resonance frequency while that selected for the resonance peak function with the higher undamped natural frequency is its mirrored resonance frequency. This 'symmetrical' selection is relevant to the method presented in Sects. A2.2.3 and A2.2.4.

It is evident that the resonance peak interaction is negligible if υ_{nj} and υ_{nk} are separated by more than an order of magnitude since, from Fig. A2.12, every resonance peak function falls to negligible proportions if $\omega < \upsilon_{ni}/10$ or $\omega > 10\upsilon_{ni}$.

A2.2.3 Relatively Close Complex Conjugate Pole Pairs

This section addresses the determination of the coefficients of the second degree factors in (2.188) in Chap. 2 for a plant having two pairs of complex conjugate poles that may not be well separated in that the corner frequencies differ from one another by less than an order of magnitude. Study of this case should enable the reader to extend the method to cases involving more than two pairs of complex conjugate poles.

Consider the factor contributed to the measured Bode magnitude plot by two neighbouring pairs of complex conjugate poles, i.e.,

$$G_{m1}(j\omega) = G_1(j\omega) G_2(j\omega), \quad (A2.99)$$

where

$$G_1(j\omega) = \frac{1}{1 + \frac{2\zeta_1 j\omega}{\omega_{n1}} + \frac{(j\omega)^2}{\omega_{n1}^2}}, \quad G_2(j\omega) = \frac{1}{1 + \frac{2\zeta_2 j\omega}{\omega_{n2}} + \frac{(j\omega)^2}{\omega_{n2}^2}} \quad (A2.100)$$

and $\omega_{n2} > \omega_{n1}$. The plot will be presented with the skewing removed, as in Sect. A2.2.1, by basing it on $G'_{m1}(s) = s^q G_m(s)$ with $q = 2$ giving

$$G'_{m1}(j\omega) = G'_1(j\omega) G'_2(j\omega) \quad (A2.101)$$

where

$$G'_1(j\omega) = \frac{\frac{(j\omega)^2}{\omega_{n1}^2}}{1 + \frac{2\zeta_1 j\omega}{\omega_{n1}} + \frac{(j\omega)^2}{\omega_{n1}^2}} \quad \text{and} \quad G'_2(j\omega) = \frac{1}{1 + \frac{2\zeta_2 j\omega}{\omega_{n2}} + \frac{(j\omega)^2}{\omega_{n2}^2}} \quad (A2.102)$$

Figure A2.14a shows the Bode magnitude plot, $M'_{dBm1}(\omega) = 20\log_{10}|G'_{m1}(j\omega)|$, together with the Bode magnitude plots of the isolated factors, i.e., $M'_{dB1}(\omega) = 20\log_{10}|G'_1(j\omega)|$ and $M'_{dB2}(\omega) = 20\log_{10}|G'_2(j\omega)|$. Figure A2.14b shows the corresponding resonance peak functions, $P'_{dB1}(\omega, \zeta_1)$ and $P'_{dB2}(\omega, \zeta_2)$. For this demonstration, $\omega_{n1} = 1$ [rad/s], $\omega_{n2} = 2$ [rad/s], $\zeta_1 = 0.3$ and $\zeta_2 = 0.2$. In view of definition (A2.91),

$$M'_{dBm1}(\omega) = M'_{dB1}(\omega) + M'_{dB2}(\omega) = P'_{dB1}(\omega, \zeta_1) + P'_{dB2}(\omega, \zeta_2) + A'_1(\omega) + A'_2(\omega). \quad (A2.103)$$

Thus $M'_{dBm1}(\omega)$ is obtained by adding the net resonance function, $P'_{dB1}(\omega, \zeta_1) + P'_{dB2}(\omega, \zeta_2)$, to the piecewise linear approximation, $A'_1(\omega) + A'_2(\omega)$.

It is clear from Fig. A2.14 that the measured peak magnitudes, $P^{\wedge}_{m1}$ and $P^{\wedge}_{m2}$, of $M'_{dBm1}(\omega)$ are greater than the corresponding peaks, $P^{\wedge}_1$ and $P^{\wedge}_2$, of $M'_{dB1}(\omega)$ and $M'_{dB2}(\omega)$ for the factors of $G'_{p1}(j\omega)$ given by (A2.102). This is due to the overlapping tails of the resonance peak functions shown in Fig. A2.14b and discussed at the end of Sect. A2.2.2. Hence the following relationships come from (A2.103) and Fig. A2.14.

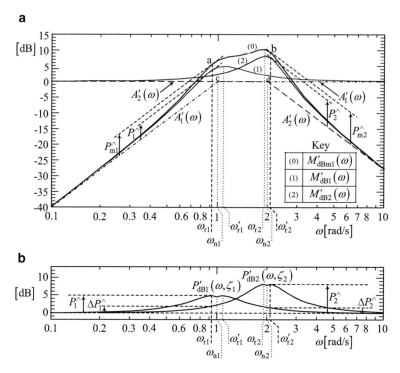

Fig. A2.14 Resonance peak interaction between two complex conjugate pole pairs. (**a**) Bode magnitude plots. (**b**) Resonance peak functions

$$P_{m1}^{\wedge} = P_1^{\wedge} + \Delta P_1^{\wedge} \quad \text{(from tail of } P'_{dB2}(\omega, \zeta_2)\text{)} \tag{A2.104}$$

and

$$P_{m2}^{\wedge} = P_2^{\wedge} + \Delta P_2^{\wedge} \quad \text{(from tail of } P'_{dB1}(\omega, \zeta_1)\text{)}. \tag{A2.105}$$

The process of extracting the plant parameters from the Bode magnitude plot, $M'_{dBm1}(\omega)$, is as follows.

(i) The $+40$ [dB/decade] and -40 [dB/decade] asymptotes are drawn, which coincide with the sloping segments of $A'_1(\omega)$ and $A'_2(\omega)$.
(ii) The peaks, $P_{m1}^{\wedge}$ and $P_{m2}^{\wedge}$, are measured.
(iii) The resonance and reflected resonance frequencies, ω_{r1} and ω'_{r2}, are measured.
(iv) To find estimates of $P_1^{\wedge}$ and $P_2^{\wedge}$, two resonance peak functions are selected from Fig. A2.12 that satisfy (A2.104) and (A2.105), as shown in Fig. A2.14b.
(v) The peak values of $P_1^{\wedge}$ and $P_2^{\wedge}$ determined in step (iv) are used to determine the damping ratios, ζ_1 and ζ_2, using the graph of Fig. 2.28 of Chap. 2.

(vi) The values of ω_{n1} and ω_{n2} are determined using, respectively, (A2.96) and (A2.97), i.e., $\omega_{n1} = \omega_{r1}/\sqrt{1 - 2\zeta_1^2}$ and $\omega_{n2} = \omega'_{r2}\sqrt{1 - 2\zeta_2^2}$.

In the example, the peak values, $\left|P_1^\wedge\right| = 5\,[\text{dB}]$ and $\left|P_2^\wedge\right| = 8\,[\text{dB}]$, yield the correct damping ratio values of $\zeta_1 = 0.3$ and $\zeta_2 = 0.2$ for step (v). For the measured values, $\omega_{r1} = 0.93$ [rad/s] and $\omega'_{r2} = 2.1$ [rad/s], step (vi) yields $\omega_{n1} = 1.03$ [rad/s] and $\omega_{n2} = 2.01$ [rad/s], which is of acceptable accuracy for a graphical method.

A2.2.4 Relatively Close Complex Conjugate Pole and Zero Pairs

The next consideration is the contribution of a pair of complex conjugate zeros together with a pair of complex conjugate poles, given by the factor,

$$G_{m1}(j\omega) = G_1(j\omega)\, G_2(j\omega) \tag{A2.106}$$

where

$$G_1(j\omega) = \frac{1}{1 + \dfrac{2\zeta_1 j\omega}{\omega_{n1}} + \dfrac{(j\omega)^2}{\omega_{n1}^2}} \quad \text{and} \quad G_2(j\omega) = 1 + \frac{2\eta_1 j\omega}{\nu_{n1}} + \frac{(j\omega)^2}{\nu_{n1}^2}$$

$$\tag{A2.107}$$

This situation is often found in mechanical structures requiring active vibration control. The second order subsystems modelling the vibration modes of such structures often have complex conjugate zeros as well as complex conjugate poles. This depends upon the placement of the actuators relative to the sensors measuring the displacement on the structure. If the sensors and actuators are collocated, then these zeros will be present. It is also common for the separation between the resonance frequency, ω_{ri}, and the anti-resonance frequency, ν_{ri}, to be less than an order of magnitude, with $\nu_{ri} < \omega_{ri}$. In this case, there is significant *subtractive* interaction between the resonance and anti-resonance components of the Bode magnitude plot resulting in the measured peak and dip magnitudes being *less* than they would be without the interaction. Direct use of these measurements with the method of Sect. 2.3.3.9 of Chap. 2 would therefore yield inaccurate plant parameter estimates.

As for the 'all poles' case of Sect. A2.2.3, a plot, $G'_{m1}(s) = s^q G_m(s)$, will be taken with the skewing removed by appropriate choice of q so that the asymptote between the two corner frequencies is horizontal, giving

$$G'_{m1}(j\omega) = G'_1(j\omega)\, G'_2(j\omega). \tag{A2.108}$$

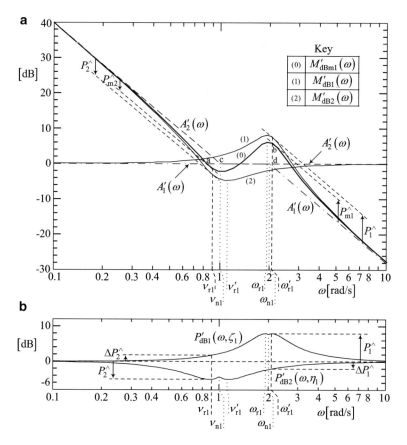

Fig. A2.15 Resonance peak interaction between a complex conjugate pole pair and a complex conjugate zero pair: (**a**) Bode magnitude plots. (**b**) Resonance peak functions

Here,

$$G'_1(j\omega) = \frac{1}{1 + \frac{2\zeta_1 j\omega}{\omega_{n1}} + \frac{(j\omega)^2}{\omega_{n1}^2}} \quad \text{and} \quad G'_2(j\omega) = \frac{1 + \frac{2\eta_1 j\omega}{\nu_{n1}} + \frac{(j\omega)^2}{\nu_{n1}^2}}{\frac{(j\omega)^2}{\nu_{n1}^2}}$$

(A2.109)

Figure A2.15 shows the Bode magnitude plot and resonance peak functions for $\omega_{n1} = 2$ [rad/s], $\nu_{n1} = 1$ [rad/s], $\zeta_1 = 0.2$ and $\eta_1 = 0.3$. As for the 'all poles' case, in view of definition (A2.91),

$$M'_{dBm1}(\omega) = M'_{dB1}(\omega) + M'_{dB2}(\omega) = P'_{dB1}(\omega, \zeta_1) + P'_{dB2}(\omega, \zeta_2) + A'_1(\omega) + A'_2(\omega).$$

(A2.110)

Appendices

It is clear from Fig. A2.15 that the measured peak magnitudes, $P_{m1}^{\wedge}$ and $P_{m2}^{\wedge}$, of $M'_{dBm1}(\omega)$ are, respectively, less than the corresponding peaks, $P_1^{\wedge}$ and $P_2^{\wedge}$, of $M'_{dB1}(\omega)$ and $M'_{dB2}(\omega)$ for the isolated factors of $G'_{m1}(j\omega)$ given by (A2.109).

This is due to the overlapping tails of the resonance peak functions shown in Fig. A2.15b. Hence, with reference to (A2.110) and Fig. A2.15,

$$P_{m1}^{\wedge} = P_1^{\wedge} + \Delta P_1^{\wedge} \text{ (from tail of } P'_{dB2}(\omega, \eta_1)) \tag{A2.111}$$

and

$$P_{m2}^{\wedge} = P_2^{\wedge} + \Delta P_2^{\wedge} \text{ (from tail of } P'_{dB1}(\omega, \zeta_1)). \tag{A2.112}$$

noting that $\Delta P_i^{\wedge}$ is opposite in sign to $P_i^{\wedge}$, $i = 1, 2$.

The process of extracting the plant parameters from the Bode magnitude plot, $M'_{dBm1}(\omega)$, is similar to that of the 'poles only' case and is as follows.

(i) The two -40 [dB/decade] asymptotes are drawn, which coincide with the sloping segments of $A'_1(\omega)$ and $A'_2(\omega)$.
(ii) The peaks, $P_{m1}^{\wedge}$ and $P_{m2}^{\wedge}$, are measured.
(iii) The resonance and reflected resonance frequencies, v_{r1} and ω'_{r1}, are measured.
(iv) To find estimates of $P_1^{\wedge}$ and $P_2^{\wedge}$, two resonance peak functions are selected from Fig. A2.12 that satisfy (A2.111) and (A2.112), as shown in Fig. A2.15b, noting that $P'_{dB2}(\omega, \eta_1)$ is negative and Fig. A2.12 shows only the magnitudes of the resonance peak functions.
(v) The peak values of $P_1^{\wedge}$ and $P_2^{\wedge}$ determined in step (iv) are used to determine the damping ratios, ζ_1 and ζ_2, using the graph of Fig. 2.28 of Chap. 2.
(vi) The values of v_{n1} and ω_{n1} are determined using, respectively, (A2.96) and (A2.97), i.e., $v_{n1} = v_{r1}/\sqrt{1 - 2\eta_1^2}$ and $\omega_{n1} = \omega'_{r1}\sqrt{1 - 2\zeta_1^2}$.

In the example, the peak values of $|P_1^{\wedge}| = 8$ [dB] and $|P_2^{\wedge}| = 5$ [dB] then yield the correct damping ratio values of $\zeta_1 = 0.2$ and $\eta_1 = 0.3$ after step (v). The measured values of $v_{r1} = 0.9$ [rad/s] and $\omega'_{r1} = 2.1$ [rad/s] yield $v_{n1} = 0.99$ [rad/s] and $\omega_{n1} = 2.01$ [rad/s], which is of acceptable accuracy for a graphical method.

A2.3 A Case Study of Plant Modelling Undertaken in Industry: Modelling for a Throttle Valve Servomechanism

A2.3.1 Introduction

This section presents modelling undertaken by Delphi Diesel Systems, UK, on a servomechanism using some of the theory and techniques of Chap. 2 and describes useful hardware features. A complete account of the work is contained in the thesis by J L Pedersen [3], which also presents simulation and experimental studies of the application of many of the control techniques and design procedures in this book. While space limitations prevent inclusion of every detail, sufficient information is given to be a practical aid for the development of servomechanisms.

The approach taken is to develop a linear model for linear control system design and a more accurate nonlinear model for simulation supporting the control system development prior to implementation.

A2.3.2 Overview of the Throttle System

The throttle system consists of a spring loaded throttle plate mechanically connected to a brushed DC motor through a gear system. A pre-stressed coil spring is included as a safety measure preventing the engine stalling in case of an electrical fault in which the motor is not energised by making the plate go to its open position. The plate position is measured by a potentiometer with an output range of 0.5–4.5 [V] corresponding to the mechanical limits of 0° (fully open) to 90° (fully closed). The voltage range does not include 0 [V] to facilitate detection of failure of the electrical supply.

A pictorial view of the throttle valve components is shown in Fig. A2.16. The throttle valve is of the butterfly type so that air flow will not create a load torque. The only load torque is from the retention spring proportional to the closing angle.

The throttle valve system model is divided into electrical and mechanical subsystems consisting, respectively, of the equations of the armature circuit of the DC motor and the equations modelling the mechanical load, including the moment of inertia, the gear system, the spring and friction. Figure A2.17 shows a schematic diagram of the throttle valve system whose parameters are defined in Table A2.1.

It should be noted that the gear train has four stages comprising the DC-motor gearwheel, two intermediate gearwheels and the throttle plate gearwheel. These are replaced by a simple equivalent train of two gearwheels in Fig. A2.17.

The modelling principles of all the components shown are covered in Chap. 2.

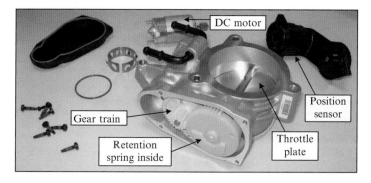

Fig. A2.16 A disassembled throttle valve

Appendices

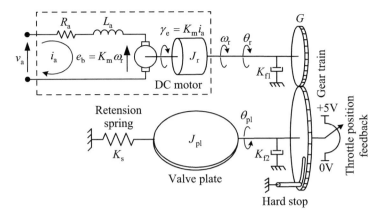

Fig. A2.17 Throttle valve schematic diagram

Table A2.1 Parameters and variables of throttle valve model

Quantity	Description	Units
v_a	DC motor armature voltage (control variable)	V
i_a	DC motor armature current	A
R_a	DC motor armature resistance	Ohm
L_a	DC motor armature inductance	H
K_m	DC motor torque and back e.m.f. constant	Nm/A & V/(rad/sec)
e_b	DC motor back e.m.f.	V
γ_e	Electromagnetic torque from the DC motor	Nm
θ_a	DC motor shaft angle	rad
ω_r	DC motor speed	rad/sec
θ_{pl}	Throttle valve plate angle	rad
J_{pl}	Valve plate moment of inertia	Kg m^2
J_m	DC motor rotor (armature) moment of inertia	Kg m^2
K_s	Retention spring constant	Nm/rad
K_{f1}	Lumped viscous friction coefficient for the DC motor bearings and half of the gear train	Nm sec/rad
K_{f2}	Lumped viscous friction coefficient for the throttle plate bearings and half of the gear train	Nm sec/rad
G	Gear ratio (motor shaft angle/output shaft angle)	–

A2.3.3 The Linear Plant Model

A2.3.3.1 The Electrical Subsystem

The model for the electrical part of the DC motor consists of the following differential equation for the circuit of Fig. A2.17.

$$\frac{di_a}{dt} = \frac{1}{L_a}(v_a - R_a i_a - K_m \omega_r). \qquad (A2.113)$$

Since, however, the mechanical subsystem of the following section is referred to the throttle plate side of the gear train, for the electrical subsystem to be compatible, its mechanical interface must also be referred to the throttle side using

$$\omega_r = G\dot{\theta}_{pl} \tag{A2.114}$$

Then (A2.113) becomes

$$\frac{di_a}{dt} = \frac{1}{L_a}\left(v_a - R_a i_a - K_m G \dot{\theta}_{pl}\right). \tag{A2.115}$$

A2.3.3.2 The Linear Mechanical Subsystem

The model presented here complements the electrical model of Sect. 2.3.3.1 of Chap. 2 to form a linear plant model for control system design. It corresponds to the part of the system comprising the motor armature, the gear train, the throttle plate and the retention spring shown in Fig. A2.17. It excludes the hard stop as this is a nonlinear element. The mechanical system and its model developed in Sect. 2.2.3.1 of Chap. 2 models the throttle valve system with $J_1 = J_r$, $\gamma_1 = \gamma_e$, $\omega_1 = \omega_r$, $\theta_1 = \theta_r$, $J_2 = J_{pl}$ and $\theta_2 = \theta_{pl}$. Then referring the parameters to the throttle plate side of the gear train, the differential equation modelling the subsystem is

$$J_L \ddot{\theta}_{pl} + K_f \dot{\theta}_{pl} + K_s \theta_{pl} = G\gamma_e \tag{A2.116}$$

where

$$J_L = J_{pl} + G^2 J_r \quad \text{and} \quad K_f = K_{f2} + G^2 K_{f1}. \tag{A2.117}$$

Finally, by introducing the plate angular velocity, $\omega_{pl} = \dot{\theta}_{pl}$, as an intermediate state variable, (A2.116) can be replaced by the following state differential equations.

$$\dot{\theta}_{pl} = \omega_{pl}, \quad \dot{\omega}_{pl} = \frac{1}{J_L}\left(-K_f \omega_{pl} - K_s \theta_{pl} + G\gamma_e\right). \tag{A2.118}$$

A2.3.3.3 Combining the Electrical and Mechanical Subsystems

Equation (A2.115) becomes another state differential equation when $\dot{\theta}_{pl}$ is replaced by ω_{pl}. This variable forms a connection between the electrical and mechanical subsystems. In addition, however, γ_e has to be expressed in terms of i_a using $\gamma_e = K_m i_a$. Then the complete set of plant state differential equations becomes

Appendices

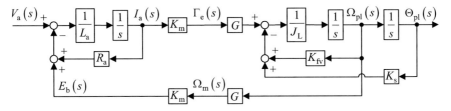

Fig. A2.18 Transfer function block diagram of linear throttle valve model

$$\frac{di_a}{dt} = \frac{1}{L_a}(v_a - R_a i_a - K_m G \omega_{pl})$$

$$\dot{\theta}_{pl} = \omega_{pl} \qquad (A2.119)$$

$$\dot{\omega}_{pl} = \frac{1}{J_L}(-K_{fv}\omega_{pl} - K_s \theta_{pl} + GK_m i_a).$$

The corresponding Laplace transform block diagram is shown in Fig. A2.18 from which the transfer function model is derived as follows.

$$\frac{\Theta_{pl}(s)}{V_a(s)} = \frac{b_0}{s^3 + a_2 s^2 + a_1 s + a_0}, \qquad (A2.120)$$

where $a_0 = \frac{R_a K_s}{L_a J_L}$, $a_1 = \frac{K_s}{J_L} + \frac{1}{L_a J_L}\left(\frac{K_m^2}{G^2} + R_a K_f\right)$, $a_2 = \frac{R_a}{L_a} + \frac{K_{fv}}{J_L}$ and $b_0 = \frac{K_m}{L_a J_L G^2}$. This model may be reduced to second order by noting that the transfer function relationship of the loop on the left of Fig. A2.18 is

$$I_a(s) = \frac{1/R_a}{1 + s\, L_a/R_a}[V_a(s) - E_b(s)]. \qquad (A2.121)$$

The armature time constant, L_a/R_a, is typically of the order of milliseconds or less and if this is very much smaller than the mechanical time constant, J_L/K_{fv}, which is typically hundreds of milliseconds, and also much smaller than the reciprocals of the magnitudes of the closed loop poles of the linear control system to be formed, then (A2.121) can be replaced by

$$I_a(s) = \frac{1}{R_a}[V_a(s) - E_b(s)], \qquad (A2.122)$$

resulting in the transfer function model,

$$\frac{\Theta_{pl}(s)}{V_a(s)} = \frac{b_0}{s^2 + a_1 s + a_0}, \qquad (A2.123)$$

where $a_0 = \frac{R_a K_s}{J_L}$, $a_1 = \frac{1}{J_L}\left(\frac{K_m^2}{G^2} + R_a K_{fv}\right)$ and $b_0 = \frac{K_m}{J_L G^2}$.

A2.3.4 The Nonlinear Plant Model

The throttle valve has significant static and Coulomb friction, which are nonlinear and defined in Sect. 2.2.2.5 of Chap. 2. These will therefore be combined with the linear viscous friction already present in the model of Sect. 2.3.3. This will create a more realistic model for predicting the performances of linear controllers based on the linear plant model. The friction velocity to torque transfer characteristic is similar to that of Sect. 2.2.2.5 of Chap. 2 but has a continuous, high slope, transition region instead of the discontinuity shown in Fig. 2.4b of Chap. 2. Figure A2.19 shows a block diagram of this model.

Another system nonlinearity consists of the two hard stops of the throttle valve plate that limit its movement, modelled according to Sect. 2.2.3.2 of Chap. 2 but with $K_{vh} = 0$ since the bearing friction is sufficient to prevent bouncing. The retention spring bias torque, which keeps the throttle open at a hard stop during an electrical failure, is modelled by a spring torque, $K_s \left(\theta_{pl} + \theta_b \right)$, where θ_b is the bias angle.

The nonlinear model of Fig. A2.20 is formed by first replacing the viscous friction block (coefficient, K_{fv}) in Fig. A2.18 by the nonlinear friction transfer characteristic between ω_{pl} and γ_f, of Fig. A2.19.

Then the hard stop simulation is added together with the spring bias. Finally a time delay, τ_d, is included in the model to represent the power electronics pulse width modulation sampling delay.

The parameters, A and B, of the static friction model in Fig. A2.19 are given by

$$B = \frac{\gamma_1 \omega_1 - \gamma_2 \omega_2}{\gamma_1 - \gamma_2} \quad \text{and} \quad A = \gamma_1 (\omega_1 + B) \quad \text{or} \quad \gamma_2 (\omega_2 + B) \quad (A2.124)$$

where (ω_1, γ_1) and (ω_2, γ_2) are the coordinates of two points on the static friction graph, $\gamma_{fs}(\omega_{pl})$, which is also reflected in the origin, as shown in Fig. A2.20.

Figure A2.20 is in the time domain as the Laplace domain, as used for Fig. A2.18, is strictly only appropriate for linear systems. The reader is reminded, however,

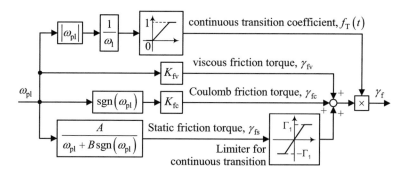

Fig. A2.19 Nonlinear friction model block diagram

Appendices

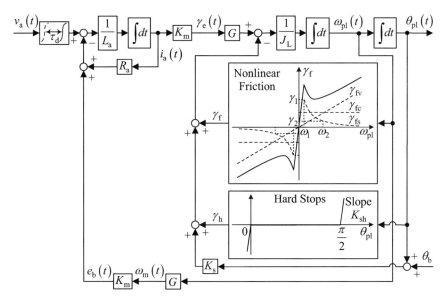

Fig. A2.20 State variable block diagram of nonlinear throttle valve model

that the notation of the Laplace domain is in common use in block diagram based simulation software including nonlinear elements such as Simulink®.

A2.3.5 Model Parameterisation: Overview

The approach taken is referred to as grey box parameter estimation. This is a cross between the white box and black box modelling referred to in Chap. 2, Sect. 2.1.4. Grey box estimation is usually applied where white box estimation is preferred but is only applied to a subset of the physical components for which it is practicable. A set of subsystems is then identified that contains all the remaining components and the black box approach is applied to these subsystems.

In pursuance of the white box approach, the throttle valve was disassembled to measure as many parameters as possible. These included the gear train ratio, G, and the DC motor torque/back e.m.f. constant, K_m. Other parameters, including the moments of inertia and the static friction, could not be measured directly due to lack of appropriate equipment or insufficient precision with the best available equipment. These parameters were determined indirectly using measurements of the plant inputs and outputs with the aid of the parameter estimation tool of the Simulink® design optimization toolbox from Mathworks®. This 'played back' a data logged set of measurements taken previously from the assembled throttle valve while driving the current plant model with the control variable, i.e., $v_a(t)$, taken from

Table A2.2 Gearwheel diameter measurements

Gearwheel	Measurement [mm]
On DC motor shaft	$D_m = 11.5$
Intermediate 1	$D_1 = 41.5$
Intermediate 2	$D_2 = 16$
Driving throttle plate	$D_{pl} = 51$

these measurements. The parameter estimation tool adjusted the model parameters until the model outputs followed those of the real plant within certain tolerances. This was repeated for different input voltage waveforms to enrich the information presented to the parameter estimator and the estimates averaged to improve the accuracy. It is important to note that initial parameter estimates are required for the parameter estimation tool. High accuracy is unnecessary but they should have the correct orders of magnitude for fast convergence towards more accurate values.

A2.3.6 Parameter Measurements

A2.3.6.1 Gear Train Ratio

The gearwheel diameters were measured as shown in Table A2.2.
The gear train ratio is then

$$G = \frac{D_{pl}}{D_2} \cdot \frac{D_1}{D_m} = 11.5. \tag{A2.125}$$

A2.3.6.2 DC Motor Back e.m.f./Torque Constant

With the available equipment, the following approach was adopted. The DC motor was disconnected from the gear train and a constant armature voltage, $v_a = v_{aconst}$, applied while the steady state value, ω_{mass} of the speed, ω_m was measured with the aid of a light reflection speed meter and a reflection pad attached to the motor shaft. The relevant equation of the DC motor model is

$$L_a \frac{di_a}{dt} + R_a i_a + K_m \omega_a = v_a \Rightarrow R_a i_{ass} + K_m \omega_{ass} = v_{aconst}, \tag{A2.126}$$

where i_{ass} is the steady state armature current. With no mechanical load, it is reasonable to assume that the voltage drop, $R_a i_{ass}$, is negligible compared with the back e.m.f., $K_m \omega_{ass}$, in which case (A2.126) may be approximated by

$$K_m \omega_{ass} = v_{aconst} \Rightarrow K_m = v_{aconst}/\omega_{ass}. \tag{A2.127}$$

Table A2.3 Measured armature inductance

Frequency [kHz]	Inductance [mH]
1	0.76
5	0.60
10	0.52
50	0.36
100	0.31

The test was carried out with $v_{\text{aconst}} = 2$ and 3 [V] and the corresponding speeds were $\omega_{\text{ass}} = 66.46$ and 114.61 [rad/sec]. Averaging the two estimates of K_m using (A2.127) yields $K_m = 0.028$ [V/(rad/sec)].

A2.3.6.3 Armature Resistance and Inductance

Measurements of R_a and L_a were made using a digital LCR meter based on a Wheatstone bridge, giving $R_a = 2.795$ [Ω]. The inductance was less straightforward due to its frequency dependence caused by the armature ferromagnetic circuit material as shown by the measurements of Table A2.3, which were taken with the sinusoidal meter voltage set to 1 [V] peak-to-peak. The PWM switch frequency used in the experiments for the parameterisation was 2 kHz and therefore an interpolation was carried out yielding $L_a = 0.72$ [mH]. This, however, could not be taken as very accurate in view of the PWM being a variable mark-space ratio square wave while the results of Table A2.3 were taken with sinusoidal excitation.

A2.3.6.4 Armature Moment of Inertia and Viscous Friction Coefficient

The available speed measurement equipment was unable to provide graphs of $\omega_a(t)$ and therefore a method using data logged graphs of $i_a(t)$ was employed. A simplified block diagram model of the DC motor is shown in Fig. A2.21 in which the armature inductance has been ignored on the basis that the electrical time constant, L_a/R_a, is much smaller than the mechanical time constant, J_m/K_{fm}. Here, K_{fm} is the viscous friction of the DC motor alone without any mechanical load. The parameters, K_m and R_a, are assumed to be already found using the methods of Sects. A2.3.6.2 and A2.3.6.3. From Fig. A2.21,

$$\frac{I_a(s)}{V_a(s)} = \frac{\frac{1}{R_a}}{1 + \left(\frac{K_m^2}{R_a} \cdot \frac{1}{J_m s + K_{fm}}\right)} = \frac{(J_m s + K_{fm})\left(\frac{1}{K_{fm} R_a + K_m^2}\right)}{1 + s\frac{J_m}{K_{fm} + K_m^2/R_a}}. \quad (A2.128)$$

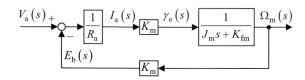

Fig. A2.21 Simplified DC motor model

Hence, in the steady state, with a constant armature voltage, v_{ass},

$$i_{ass} = \frac{K_{fm}}{R_a}\left(\frac{1}{K_{fm}+K_m^2/R_a}\right)v_{ass} \Rightarrow K_{fm} = \frac{K_m^2 i_{ass}}{v_{ass}-R_a i_{ass}}. \tag{A2.129}$$

The step response is needed to obtain an estimate of J_m as follows. The step response of the subsystem with transfer function,

$$\frac{X(s)}{V_a(s)} = \frac{1}{1+s\dfrac{J_m}{K_{fm}+K_m^2/R_a}}, \tag{A2.130}$$

given $v_a(t) = v_{ass}h(t) \Rightarrow V_a(s) = v_{ass}/s$, is

$$x(t) = v_{ass}\left(1-e^{-t/T}\right), \tag{A2.131}$$

where

$$T = J_m/\left(K_{fm}+K_m^2/R_a\right). \tag{A2.132}$$

In view of (A2.128), the corresponding armature current step response is

$$i_a(t) = \frac{1}{K_{fm}R_a+K_m^2}\left\{K_{fm}x(t)+J_m\frac{d}{dt}[x(t)]\right\}. \tag{A2.133}$$

Substituting for $x(t)$ in (A2.133) using (A2.131) then yields

$$i_a(t) = \frac{v_{ass}}{K_{fm}R_a+K_m^2}\left[K_{fm}\left(1-e^{-t/T}\right)+\frac{J_m}{T}e^{-t/T}\right] \tag{A2.134}$$

Hence

$$i_a(T) = \frac{v_{ass}}{K_{fm}R_a+K_m^2}\left[K_{fm}\left(1-e^{-1}\right)+\frac{J_m}{T}e^{-1}\right]. \tag{A2.135}$$

Finally, substituting for T in (A2.135) using (A2.132) yields

$$i_a(T) = \frac{K_{fm}R_a+K_m^2 e^{-1}}{K_{fm}R_a+K_m^2} \cdot \frac{v_{ass}}{R_a}. \tag{A2.136}$$

The data logged $i_a(t)$ is then used to determine T. Finally, from (A2.132),

$$J_m = \left(K_{fm}+K_m^2/R_a\right)T. \tag{A2.137}$$

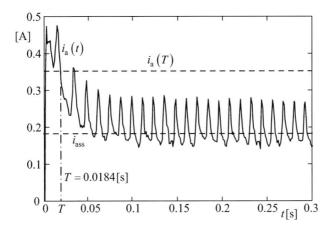

Fig. A2.22 Data logged armature current step response

The DC motor was disconnected from the gear train and an armature voltage step of $v_{ass} = 2$ [V] applied and $i_a(t)$ measured and data logged as shown in Fig. A2.22. The ripple is due to the pulse width modulated armature voltage with mean value, $v_{ass} = 2$ [V]. For the following calculations, the steady state armature current, i_{ass}, is the mean value of the waveform of $i_a(t)$ taken over the interval, 0.1 [s] $< t <$ 0.3 [s].

First, i_{ass}, and v_{ass}, are used to find the viscous friction coefficient from (A2.129), yielding $K_{fm} = 5.61 \times 10^{-5}$ [Nm/(rad/s)]. Then $i_a(T)$ was found from (A2.136) and marked as a horizontal straight line in Fig. A2.22. The intersection between this line and the graph of $i_a(t)$ then indicated the value of $T = 0.0184$ [s]. Then the armature moment of inertia was calculated using (A2.137), giving $J_m = 3.65 \times 10^{-6}$ [Kg m^2].

In view of the relatively large ripple amplitude of $i_a(t)$, an accuracy check was made, the result of which is shown in Fig. A2.23. This entailed using the calculated values of F_m and J_m to generate a simulated armature current, $i_{as}(t)$, using the system of Fig. A2.21, which also obeys (A2.134). The closeness of the intersections between the horizontal straight line, $i_a(T)$, the graph of $i_{as}(t)$ and the data logged graph of $i_a(t)$ would then be an accuracy indication. Figure A2.23 indicates high accuracy so that the values of K_{fm} and J_m arrived at were good initial estimates for the parameter estimation tool.

A2.3.6.5 Throttle Valve System Moment of Inertia

Tests on the completely assembled throttle valve for determination of the total moment of inertia would be impracticable due to the hard stops which could not be temporarily removed. Instead, approximate calculations of the moment of inertia contributions of the plate, shaft and gear wheels were carried out using their estimated masses and radii of gyration.

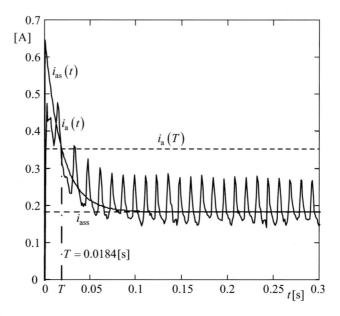

Fig. A2.23 DC motor friction and moment of inertia estimation accuracy assessment

A2.3.6.6 The Retention Spring Constant and Bias Angle

An estimate of the retention spring constant, K_s, was obtained using a hanging scale calibrated in [kg] and the throttle position sensor, as shown in Fig. A2.24. The DC motor was removed from the gear train to avoid errors due to static friction and cogging torque. A metal clamp was fixed to the throttle plate and a cord tied to its end for attachment to the hanging scale. This was used to measure the force, f_o, acting at a radius, $r = 0.144$ [m], from the plate rotation axis needed to *just* move the throttle plate away from the end stop, at the plate angle defined as $\theta_o = 0$ [rad]. The hanging scale indicated directly the weight of the equivalent mass, $m_o = 0.21$ [Kg].

Hence, given the acceleration, $g = 9.85$ [m/s/s], due to gravity, the offset torque is

$$\gamma_0 = f_o r = m_o g r = 0.21 \times 9.85 \times 0.144 = 0.297 \text{ [Nm]}. \quad (A2.138)$$

The retention spring constant was obtained using the above measurement and a second measurement at a plate position of $\theta_1 = 27° = 0.471$ [rad], assuming Hookes Law. This yielded a torque of

$$\gamma_1 = f_1 r = m_1 g r = 0.25 \times 9.85 \times 0.144 = 0.3546 \text{ [Nm]} \quad (A2.139)$$

where f_1 is the required force m_1 is the corresponding hanging scale reading.

Fig. A2.24 Retension spring torque measurement

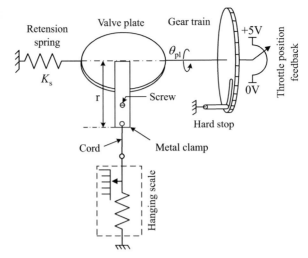

Hence

$$K_s = \frac{\gamma_1 - \gamma_0}{\theta_1 - \theta_0} = 0.122 \ [\text{Nm/rad}]. \tag{A2.140}$$

The spring torque is then given by

$$\gamma_s = K_s \left(\theta_{pl} + \theta_b\right). \tag{A2.141}$$

where θ_b is the bias angle. At $\theta_{pl} = \theta_0 = 0$, $\gamma_s = \gamma_0$. Hence

$$\theta_b = \gamma_0/K_s = 2.9 \ [\text{rad}]. \tag{A2.142}$$

A2.3.6.7 Hard Stops

With reference to Fig. A2.20, K_{sh} was set to 50. The maximum plate angle was measured to be $\approx 90°$. The minimum was set to $-0.1°$ for linear operation around $0°$.

A2.3.6.8 Throttle Valve Friction Parameters

Estimation of the viscous, static and Coulomb friction parameters experimentally is impracticable due to the hard stops. The parameter estimation tool was therefore relied upon with initial parameter values chosen within realistic bounds.

A2.3.7 Parameter Estimation Tool

A2.3.7.1 Introduction

The *Simulink Design Optimization* toolbox from Mathworks® was used to estimate the model parameters offline using data logged input and output variables from previously conducted tests, commencing with the parameters determined in Sect. 2.3.6 as initial estimates. The tool operates on an iterative basis in which the data logged input and output variables are repeatedly played back, the input variables being applied to a plant model running in Simulink. After each simulation run, the tool adjusts the model parameters to minimise the errors between the data logged real plant outputs and the model outputs. The iterations are stopped once these errors are reduced below specified thresholds.

To maximise the accuracy, the estimation was carried out in two stages. In stage 1, only the DC motor parameters were estimated. In stage 2, the remaining parameters were estimated with the DC motor parameters fixed at those estimated in stage 1.

A2.3.7.2 Stage 1: DC Motor Model Parameters

The parameters, K_m, L_a and R_a, were estimated using three logs of $i_a(t)$ and $\theta_{pl}(t)$ for different armature voltage functions, $v_a(t)$, as shown in Fig. A2.25. The estimation tool required $\omega_m(t)$, found by filtering and differentiating $\theta_{pl}(t)$ and multiplying by the gear ratio, G. This, together with $v_a(t)$, were the inputs to the DC motor model while $i_a(t)$ was the output reference. Figure A2.26 shows a schematic diagram of the parameter estimation process in which the hat placed over a quantity indicates it is an estimate. Note that $\widehat{q}$ denotes an estimate of a quantity, q.

A2.3.7.3 Stage 2: Throttle Valve Model Parameters

Estimating the throttle valve parameters entailed 17 different data logged sets following similar lines to those of Fig. A2.25, but with $v_a(t)$ as the input together with $i_a(t)$ and $\theta_{pl}(t)$ as the two outputs, forming two errors as shown in Fig. A2.27.

With reference to the friction transfer characteristic shown in Fig. A2.20, the point, (ω_1, γ_1) is the break-away point for which ω_1 is chosen and γ_1 is estimated. The point, (ω_2, γ_2), is an arbitrary point on the static friction transfer characteristic for which γ_2 is chosen and ω_2 is estimated.

A2.3.8 Model Validation in the Frequency Domain

The parameter estimation tool of A2.3.7 can be regarded as validating the model it produces since it carries out direct comparisons between the responses of the model and the real plant with common inputs. It is common practice, however, to carry out validation by comparing the frequency responses of the real plant and the model. This has been done for the throttle valve, not only as a additional model

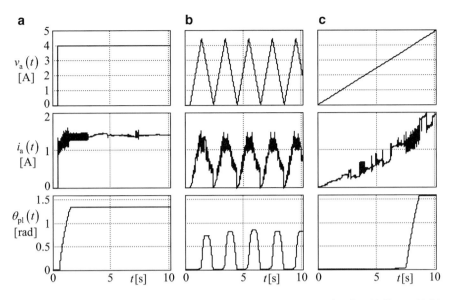

Fig. A2.25 Data logged variables for the DC motor model parameter estimation (**a**) Step $v_a(t)$ (**b**) triangular $v_a(t)$ (**c**) ramp $v_a(t)$

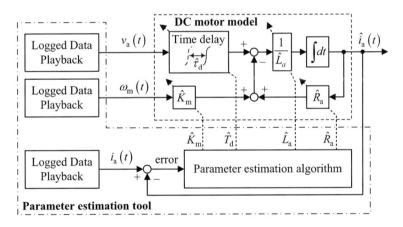

Fig. A2.26 Stage 1: Estimation of the DC motor parameters

validation but also to determine the bandwidth, which is needed to determine a suitable sampling period for control using digital processors (Chap. 6). To achieve this, the input and output signals are sampled and processed using fast Fourier transforms [3]. This information is then used in the System Identification Toolbox from Mathworks® to create frequency response plots.

The quality of the result obtained from the System Identification Toolbox depends on factors including the form of the input signal, the sampling period of the data acquisition and the signal-to-noise ratio. The experiment on the real plant to capture the input and output signals can be intrusive or nonintrusive.

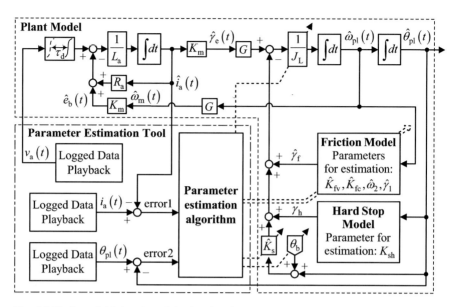

Fig. A2.27 Stage 2: Estimation of the throttle valve parameters

In the intrusive method the input signal, from the controller, is replaced by a step, impulse or sinusoidal signal. The nonintrusive method is principally intended for a controlled plant in normal operation. A signal is added to the controller output before entering the input of the real plant. The signal level should be low compared to the controller output to minimise the impact on the operating control loop. Common types of signals are sinusoids at different frequencies spanning the intended bandwidth of the control system to be ultimately designed, and the pseudo random binary sequence (PRBS) as described in Sect. 2.3.3.1 of Chap. 2, with a Fourier spectrum spanning this bandwidth. Changing the types of signals used for the system identification can result in slightly different frequency response plots. It is therefore advisable to carry out the procedure on a known plant model of similar form to the one expected from the identification process using the different input signal types and select the one that produces the closest approach to the known plant model. The PRBS is usually a good choice in view of its broad Fourier spectrum.

The non-intrusive method is also useful for laboratory based tests as it enables a controller to hold the plant at a desirable setpoint while the test is being conducted. This is the case for the throttle valve, for which a PRBS was used.

The factors that are important when using the PRBS are the signal amplitude, the PRBS sequence length, the plant sampling frequency and, in the case of the throttle valve, the power electronics switching frequency. Table A2.4 shows the parameters.

For the throttle valve test, it was found unnecessary to maintain a given plate angle through loop closure. Instead a DC armature voltage was applied and the level adjusted to maintain $\theta_{pl} \cong 45°$, which is about midway between the end stops to allow the maximum amplitude of movement for nominally linear operation. Then the PRPS signal was superimposed.

Table A2.4 Parameters for PRBS based frequency response for throttle valve

Chosen parameters	Dependent parameters
Data sampling frequency: $f_s = 500$ [Hz]	PRBS update period: $T = 1/f_s = 2$ [ms]
PRBS sequence length: $N_s = 2^{10} - 1 = 1023$	Minimum injected signal frequency: $f_{min} = \frac{1}{N_s T} = 0.49$ [Hz]
PRBS amplitude: ± 1.5 [V]	Minimum injected signal frequency: $f_{max} = \frac{1}{2T} = 250$ [Hz]
Experiment time: 40 [s] (20 PRBS cycles)	PRBS cycle length: $N_s T = 2.045$ [s]

From the experimental input and output data a transfer function,

$$\frac{\Theta_{exp}(s)}{V_a(s)} = G_{exp}(s), \qquad (A2.143)$$

was generated using the System Identification Toolbox from Mathworks. The same procedure was used to generate a transfer function,

$$\frac{\Theta_{mod}(s)}{V_a(s)} = G_{mod}(s) \qquad (A2.144)$$

from the nonlinear throttle valve model. The Bode plots of these transfer functions are shown in Fig. A2.28. The expected bandwidth of a throttle valve control system

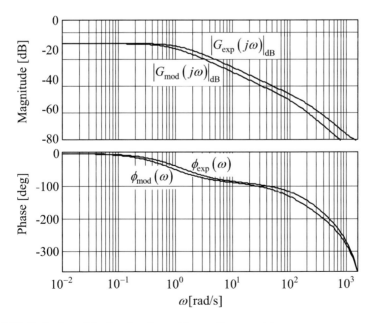

Fig. A2.28 Bode plots of the throttle valve and its non-linear model

is about 60 [rad/s] and the maximum error between the Bode magnitude plots of about 4 dB is too not large for the model to be useful for simulation based control system development.

Although, in theory, the transfer function applies only to linear systems, frequency response based identification is common in industry even for plants known to be nonlinear. Arguably, the transfer function obtained is similar to the one that would be derived analytically by the method of linearisation about the operating point (Chap. 7) if the plant contained continuous nonlinearities and a nonlinear model was available. The dominant stick slip friction in the throttle valve, however, is discontinuous. Unfortunately, there is no other known way to obtain a better transfer function model for control system design. The rapid increase of the phase angle for $\omega > 500$ [rad/s] in contrast to the asymptotic increase to a limit of $-270°$ predicted by the linear model is typical. The transfer function model cannot be heavily relied upon for plants containing discontinuous nonlinearities but can be useful for the initial controller design with the expectation of having to make adjustments following the first experimental trials.

A2.3.9 Summary of Throttle Valve Model Parameter Values

Table A2.5 shows the values of the throttle valve model parameters obtained, most of which were refined using the parameter estimation tool. The few direct measurements deemed sufficiently accurate are indicated with an asterisk. In addition, parameters indicated with a double asterisk are chosen rather than measured.

Table A2.5 Estimated and measured throttle valve model parameters

Quantity	Description	Value
R_a *	DC motor armature resistance	2.795 [Ω]
L_a	DC motor armature inductance	8.372×10^{-4} [H]
K_m	DC motor torque and back e.m.f. constant	0.0257 [Nm/A or Vs/rad]
J_L	Total moment of inertia referred to plate	8.521×10^{-4} [kgm^2]
K_s	Retention spring constant	0.0851 [Nm/rad]
K_{fv}	Viscous friction coefficient referred to plate	0.0022 [Nms/rad]
K_{fc}	Coulomb friction constant	0.0386 [Nm]
ω_1 **	Static friction velocity coordinate 1	0.01 [rad/s]
γ_1	Static friction torque coordinate 1	0.0539 [Nm]
ω_2	Static friction velocity coordinate 2	0.047 [rad/s]
γ_2 **	Static friction torque coordinate 2	0.001 [Nm]
G *	Gear ratio	11.5
θ_b	Retention spring bias angle	4.427 [rad]
$\theta_{pl\ max}$ *	Maximum plate angle	$\pi/2$ [rad]
$\theta_{pl\ min}$ *	Minimum plate angle	$-0.1° = -0.01745$ [rad]
K_{hs} **	Proportional gain of hard stop controller	50
τ_d	Pure time delay	0.0016 [s]

References

1. Junkins JL, Turner JD (1986) Optimal spacecraft attitude maneuvers. Elsevier, Amsterdam
2. James JF (2011) A students guide to Fourier transforms with applications in physics and engineering, 3rd edn. ISBN 978 0 521 17683 5, Cambridge University Press, New York
3. Pedersen JL (2013) Model based and robust control techniques for internal combustion engine throttle valves. PhD thesis, University of East London, London, UK

A4 Appendix to Chap. 4

A4.1 Application of Mason's Formula Using Block Diagrams

A4.1.1 Background

Mason's Formula, sometimes referred to as Mason's Rule, is a useful means of determining transfer functions and the characteristic equations of linear, time invariant (LTI) plant models and closed loop control systems algebraically using their block diagrams. In many cases, particularly related to SISO plants, the method is more straightforward than others such as block diagram reduction or using Laplace or z transforms, respectively, on differential or difference equations. Consider, for example, the continuous SISO state space model,

$$\left.\begin{aligned}\dot{\mathbf{x}} &= \mathbf{A}\mathbf{x} + \mathbf{b}u \\ y &= \mathbf{c}^T\mathbf{x}\end{aligned}\right\}, \quad (A4.1)$$

and its corresponding discrete time state space model,

$$\left.\begin{aligned}\mathbf{x}(k+1) &= \mathbf{\Phi}(h)\mathbf{x}(k) + \mathbf{\psi}(h)\mathbf{x}(k) \\ y(k) &= \mathbf{c}^T\mathbf{x}(k)\end{aligned}\right\} \quad (A4.2)$$

It is easy to write down the corresponding general transfer functions,

$$\frac{y(s)}{u(s)} = \mathbf{c}^T[sI - \mathbf{A}]^{-1}\mathbf{b} = \mathbf{c}^T\frac{\text{adj}\,[sI - \mathbf{A}]}{\det\,[sI - \mathbf{A}]}\mathbf{b} \quad (A4.3)$$

or

$$\frac{y(z)}{u(z)} = \mathbf{c}^T[zI - \mathbf{\Phi}]^{-1}\mathbf{\Psi} = \mathbf{c}^T\frac{\text{adj}\,[zI - \mathbf{\Phi}]}{\det\,[zI - \mathbf{\Phi}]}\mathbf{\Psi} \quad (A4.4)$$

but in individual cases, the determinant expansions and adjoint matrix evaluations can be quite tedious. In many cases, Mason's formula offers a simpler route to the required solution, provided a transfer function block diagram or a state variable block diagram is available.

Mason's formula may be regarded as a generalisation of the well-known formula for the transfer function,

$$\frac{Y(s)}{Z(s)} = \frac{G(s)}{1 + G(s)H(s)}, \quad (A4.5)$$

of the simple closed-loop system shown in Fig. A4.1.

Fig. A4.1 Simple single loop system

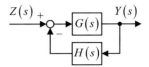

Fig. A4.2 General block diagram representation of time invariant, linear dynamical system

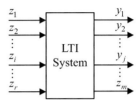

In many other applications, the control loop structure is more complex than this. Figure A4.2 represents a general LTI system that may be a plant to be controlled or a complete closed-loop system.

Masons formula enables the transfer function between any two points, such as $Y_j(s)/Z_i(s)$ for a continuous model and $Y_j(z)/Z_i(z)$ for a discrete model, or the characteristic polynomial, to be readily determined.

A4.1.2 The Rules

Since the method is applicable to continuous time and discrete time models, then the functional notation, (s) or (z) will be omitted in the presentation of Mason's formula and the rules for determining the transfer functions in particular cases. It should be noted that for historic reasons, the method is traditionally applied to the *signal flow graph*, which is equivalent to the block diagram. The method is applied here to the much more commonly used block diagram models, an example of which is shown in Fig. A4.3.

Here, G_i, $i = 1, \ldots, 6$, are the transfer functions of dynamic elements or the constant gains of non-dynamic elements. This figure will be useful in understanding the following terms that are relevant to application of the formula.

(a) *Forward path.* A forward path between an input and output is a path through the elements and connections of the block diagram commencing at the input and ending at the output. It must follow the direction indicated by the arrows on the diagram and may *pass through any point only once*. The block diagram of Fig. A4.3 has two forward paths marked p_1 and p_2.

(b) *Forward path gain.* This is the overall gain between the input and output of a forward path and is equal to the product of the individual gains of the elements in the path. The gain of the *i*th forward path is denoted p_i. The gains of the two forward paths between the input, Z, and the output, Y, in Fig. A4.3 are

$p_1 = G_1G_4G_5$ and $p_2 = G_3G_4G_5$. Note that the gain from the '+' input of a summing junction to its output is equal to +1.

(c) *Loop.* A loop is a *closed path* within the block diagram. It must follow the direction of the arrows and *pass through any point only once*. Six loops may be identified in Fig. A4.3 and these are marked, l_i, $i = 1, 2, \ldots, 6$.

(d) *Loop Gain.* This is the product of the individual gains of the elements in the loop, commencing at any point in the loop and ending at the same point. The gain of the ith loop is denoted l_i. The gains of the six loops in Fig. A4.3 are $l_1 = G_1G_2$, $l_2 = G_4G_5$, $l_3 = -G_1G_4G_6$, $l_4 = -G_1G_4G_5$, $l_5 = -G_3G_4G_6$ and $l_6 = -G_3G_4G_5$. Note that the gain from the '−' input of a summing junction to its output is equal to −1. This applies to the loops, l_i, $i = 3, 4, 5, 6$.

(e) *Touching Loops.* Two or more loops *touch* one another if they share at least one signal and therefore all pass through at least one point. In Fig. A4.3, loops l_1, l_3, and l_4 are touching loops. Similarly, loops l_2, l_3, l_4, l_5 and l_6 are touching loops.

(f) *Non-touching Loops.* Two or more loops are *non-touching* if they share *no* signals and therefore have *no* points in common. In Fig. A4.3, loops l_1 and l_2 are non-touching loops. Also loops l_1 and l_5 are non-touching loops. Similarly, l_1 and l_6 are non-touching loops.

(g) *Loops Touching Forward Paths.* A loop touches a forward path if it shares at least one signal with it and therefore passes through at least one point on it. In Fig. A4.3, all the loops, l_i, $i = 1, 2, \ldots, 6$, touch the forward path, p_1. Similarly, loops l_2, l_3, l_4, l_5 and l_6 touch the forward path, p_2.

(h) *Loops Non-touching Forward Paths.* A loop is non-touching with respect to a forward path if it shares *no* signals with it and therefore passes through no points on it. In Fig. A4.3, loop, l_1, is non-touching with respect to the forward path, p_2.

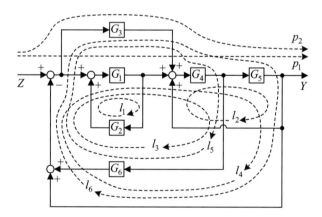

Fig. A4.3 Example of the block diagram of an LTI system with a multiple loop structure

Appendices

Given the above information, the transfer function between any two points in the block diagram of an LTI system is given by Mason's Formula, as follows.

$$G = \frac{\sum_k p_k \Delta_k}{\Delta}, \qquad (A4.6)$$

where Δ is the *determinant* of the system and Δ_k is the *cofactor* corresponding to the kth forward path, p_k. Here,

$$\Delta = 1 - \underbrace{\sum_i l_i}_{\substack{\text{sum of}\\\text{all the}\\\text{loop}\\\text{gains}}} + \underbrace{\sum_{i \neq j} l_i l_j}_{\substack{\text{sum of}\\\text{products}\\\text{of gains}\\\text{of pairs of}\\\text{non-}\\\text{touching}\\\text{loops}}} - \underbrace{\sum_{i \neq j \neq m} l_i l_j l_m}_{\substack{\text{sum of}\\\text{products}\\\text{of gains}\\\text{of non-}\\\text{touching}\\\text{loops taken}\\\text{three at a}\\\text{time}}} + \cdots \qquad (A4.7)$$

and Δ_k is obtained by commencing with Δ and then *setting to zero* the gains of all the loops that *touch* the forward path, p_k.

The proof of Mason's formula may be found in the original paper [1]. Some remarks follow, however, which should aid in gaining an intuitive understanding. The determinant, Δ, of Mason's formula (A4.6) is essentially the same as the determinants, $\det[sI - A]$ and $\det[zI - F]$, of the general transfer functions (A4.3) and (A4.4) obtained from the state space models, but is more easily obtained from (A4.6) in particular cases. Similarly, the cofactor term, $\sum_k p_k \Delta_k$, is essentially the same as the numerator terms, $\mathbf{c}^T \text{adj}[sI - A] \mathbf{b}$ and $\mathbf{c}^T \text{adj}[zI - \Phi] \Psi$, of the general transfer functions, (A4.3) and (A4.4) but again is more easily obtained.

Importantly, Mason's formula also provides a straightforward means of determining the characteristic equation of a LTI system. This is simply given by

$$\Delta = 0 \qquad (A4.8)$$

i.e., $\Delta(s) = 0$ for a continuous time block diagram model incorporating elements whose Laplace transfer functions are given and $\Delta(z) = 0$ for a discrete time block diagram model incorporating elements whose z-transfer functions are given.

To demonstrate the application of the formula, the transfer function, Y/Z, of the system shown in Fig. A4.3 is

$$\frac{Y}{Z} = \frac{\sum_k p_k \Delta_k}{\Delta} = \frac{p_1 \Delta_1 + p_2 \Delta_2}{1 - (l_1 + l_2 + l_3 + l_4 + l_5 + l_6) + (l_1 l_2 + l_1 l_5 + l_1 l_6)}$$

$$= \frac{p_1 \cdot 1 + p_2 (1 - l_1)}{1 - (l_1 + l_2 + l_3 + l_4 + l_5 + l_6) + (l_1 l_2 + l_1 l_5 + l_1 l_6)}$$

$$= \frac{G_4 G_5 [G_1 + G_3 (1 - G_1 G_2)]}{1 - G_1 G_2 - G_4 \{G_5 [1 - G_3 - G_1 (1 - G_2 (1 - G_3))] - G_6 [G_3 + G_1 (1 + G_2 G_3)]\}}.$$
(A4.9)

No distinction is made here between transfer functions of continuous or discrete time models and therefore the complex variables, s or z, of, respectively, the Laplace or z transforms do not appear.

Initially, the reader may wish to write down the individual expressions for the forward path gains, loop gains, cofactors and the determinant, but with practice, should be able to write down expressions for the transfer function or characteristic polynomial directly, as in the following examples.

Example A4.1 DC motor position control servomechanism with linear state feedback control

It is required to derive the transfer functions, $\frac{\Theta(s)}{\Theta_r(s)}$, $\frac{U(s)}{\Theta_r(s)}$, $\frac{E(s)}{\Gamma_L(s)}$ and the closed loop characteristic equation of the control system shown in Fig. A4.4.

Since the determinant, $\Delta(s)$, is common to any transfer function of a given LTI system and is used to determine the characteristic polynomial, it will be found first.

$$\Delta(s) = 1 - \left(\frac{1}{sL_a + R_a}\right) \left[-K_d K_i k_3 - \left(\frac{c\Phi}{Js + B}\right)\left(c\Phi + K_d K_v k_2 + K_d K_p k_1 \frac{1}{s}\right)\right]$$

$$= 1 + \left(\frac{1}{sL_a + R_a}\right) \left[K_d K_i k_3 + \left(\frac{c\Phi}{Js + B}\right)\left(c\Phi + K_d K_v k_2 + K_d K_p k_1 \frac{1}{s}\right)\right]$$
(A4.10)

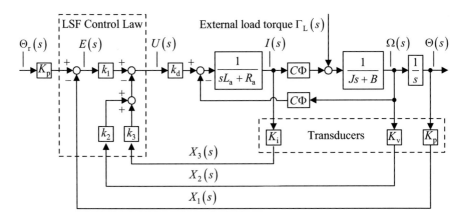

Fig. A4.4 A d.c. position control servo with linear state feedback control

The required transfer functions are then

$$\frac{\Theta(s)}{\Theta_r(s)} = \frac{K_d K_p k_1 \left(\dfrac{1}{sL_a + R_a}\right)\left(\dfrac{C\Phi}{Js+B}\right)\dfrac{1}{s}}{1 + \left(\dfrac{1}{sL_a + R_a}\right)\left[K_d K_i k_3 + \left(\dfrac{C\Phi}{Js+B}\right)\left(C\Phi + K_d K_v k_2 + K_d K_p k_1 \dfrac{1}{s}\right)\right]},$$

i.e.,

$$\frac{\Theta(s)}{\Theta_r(s)} = \frac{\dfrac{K_d K_p k_1 C\Phi}{L_a J}}{\dfrac{1}{L_a J}(sL_a + R_a)(Js+B)s} \quad \text{(A4.11)}$$
$$+ \frac{1}{L_a J}\left[K_d K_i k_3 (Js^2 + Bs) + C\Phi\left((C\Phi + K_d K_v k_2)s + K_d K_p k_1\right)\right]$$

$$\frac{U(s)}{\Theta_r(s)} = \frac{K_p k_1 \left[1 + \left(\dfrac{1}{sL_a + R_a}\right)\left(\dfrac{C^2\Phi^2}{Js+B}\right)\right]}{1 + \left(\dfrac{1}{sL_a + R_a}\right)\left[K_d K_i k_3 + \left(\dfrac{C\Phi}{Js+B}\right)\left(C\Phi + K_d K_v K_2 + K_d K_p K_1 \dfrac{1}{s}\right)\right]}$$

$$= \frac{\dfrac{K_p k_1}{L_a J}\left[(sL_a + R_a)(Js+B) + C^2\Phi^2\right]s}{\dfrac{1}{L_a J}(sL_a + R_a)(Js+B)s}$$
$$+ \frac{1}{L_a J}\left[K_d K_i k_3 (Js^2 + Bs) + C\Phi\left((C\Phi + K_d K_v k_2)s + K_d K_p k_1\right)\right] \quad \text{(A4.12)}$$

and

$$\frac{E(s)}{\Gamma_L(s)} = \frac{\left(\dfrac{1}{Js+B}\right)\dfrac{1}{s} K_p(-1)}{1 + \left(\dfrac{1}{sL_a + R_a}\right)\left[K_d K_i k_3 + \left(\dfrac{C\Phi}{Js+B}\right)\left(C\Phi + K_d K_v k_2 + K_d K_p k_1 \dfrac{1}{s}\right)\right]}$$

$$= \frac{-\dfrac{K_p}{L_a J}(sL_a + R_a)}{\dfrac{1}{L_a J}(sL_a + R_a)(Js+B)s}.$$
$$+ \frac{1}{L_a J}\left[K_d K_i k_3 (Js^2 + Bs) + C\Phi\left((C\Phi + K_d K_v k_2)s + K_d K_p k_1\right)\right] \quad \text{(A4.13)}$$

The characteristic equation is given by $\Delta(s) = 0$, and it is evident from the derivation of the above transfer functions that this equation is converted to a polynomial equation in the standard form in which the characteristic polynomial is a monic polynomial, meaning that the coefficient of the highest power of

s equal to unity, by multiplying both sides of the equation, $\Delta(s) = 0$, by $\frac{1}{L_a J} s \left(s L_a + R_a\right)(Js + B)$. Thus, $\frac{1}{L_a J} s \left(s L_a + R_a\right)(Js + B) \Delta(s) = 0$, yielding

$$\frac{1}{L_a J}(s L_a + R_a)(Js + B)s$$
$$+ \frac{1}{L_a J}\left[K_d K_i k_3 \left(Js^2 + Bs\right) + C\Phi\left((C\Phi + K_d K_v k_2)s + K_d K_p k_1\right)\right] = 0$$

and separating the terms of like degree yields

$$s^3 + \left(\frac{B}{J} + \frac{R_a + K_d K_i k_3}{L_a}\right)s^2$$
$$+ \left[\frac{R_a B + K_d K_i k_3 B + C\Phi(C\Phi + K_d K_v k_2)}{L_a J}\right]s + \frac{C\Phi K_d K_p k_1}{L_a J} = 0$$
(A4.14)

The forgoing example serves to illustrate the fact that, in general, the denominator polynomial of every transfer function of a given linear system is the same. This is a consequence of the basic modes of a linear system being reflected in all the transfer functions, the eigenvalues of these modes (i.e., the poles of the transfer functions) being the roots of the common characteristic equation, $\Delta(s) = 0$. The differences between the transfer functions of a given linear system are only in the numerator polynomials, and hence the zeros.

Example A4.2 Discrete three phase signal generator

Figure A4.5a shows the block diagram of a software-implemented discrete time three phase signal generator for AC electric drive applications.

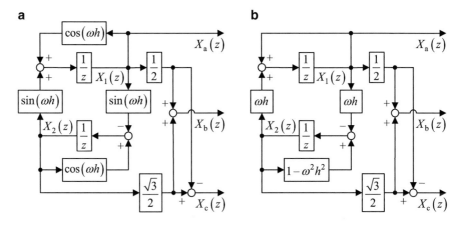

Fig. A4.5 Discrete time three phase signal generators. (**a**) Exact version (**b**) Approximate version

Appendices

This is an unforced linear dynamical system which, with non-zero initial conditions (i.e., initial states) generates two state variables, x_1 and x_2, which coincide precisely with two sinusoids of equal amplitude and mutually in quadrature and of frequency, ω [rad/s], at sampling times separated by a constant period of h [s]. These state variables form 'basis phasors' from which the three-phase sinusoids, $x_a(t)$, $x_b(t)$ and $x_c(t)$, are generated, these precisely coinciding with the ideal continuous three phase sinusoids at the sampling times. This scheme, however, requires on-line evaluation of sine and cosine functions if ω is to be variable. The scheme of Fig. A4.5b avoids this but at the expense of the *actual* frequency of oscillation, ω_a, being slightly different from the nominal frequency, ω and the phase angle between $x_1(t)$ and $x_2(t)$ differing slightly from 90°.

First, it is required to confirm, by means of the characteristic equation, that the signal generator of Fig. A4.5a oscillates at constant amplitude with a frequency of ω [rad/s]. Then it is similarly required to show that the signal generator of Fig. A4.5b oscillates at constant amplitude with a different frequency of ω_a and finally to derive an equation for ω_a in terms of ω and h.

Both generators have three loops, two of which are non-touching. The characteristic equation for Fig. A4.5a is therefore given by

$$\Delta(z) = 1 - \left[\frac{2}{z}\cos(\omega h) - \frac{1}{z^2}\sin^2(\omega h)\right] + \frac{1}{z^2}\cos^2(\omega h) = 0 \Rightarrow$$
$$1 - \frac{2}{z}\cos(\omega h) - \frac{1}{z^2} = 0 \Rightarrow z^2 - 2z\cos(\omega h) + 1 = 0 \quad \text{(A4.15)}$$

from which the roots are

$$z_{1,2} = \cos(\omega h) \pm j \sin(\omega h) \quad \text{(A4.16)}$$

This satisfies the necessary condition for unforced oscillations at a constant amplitude, i.e., complex conjugate poles lying precisely on the unit circle in the z-plane (corresponding to the $j\omega$ axis in the s-plane). Oscillations at a frequency of ω [*rad/s*] with constant amplitude are produced by a *continuous* unforced dynamical system having poles of the Laplace transfer function located at $s_{1,2} = \pm j\omega$ in the s-plane. Applying the transformation to yield the poles of the equivalent *discrete* unforced linear system in the z-plane yields:

$$z_{1,2} = e^{s_{1,2}h} = e^{\pm j\omega h} = \cos(\omega h) \pm j \sin(\omega h) \quad \text{(A4.17)}$$

This agrees with the roots of the characteristic equation above and therefore the system oscillates as required.

Applying Mason's formula to obtain the characteristic equation of the system of Fig. A4.5b yields

$$\Delta(z) = 1 - \left(\frac{1}{z} + \frac{1-\omega^2 h^2}{z} - \frac{\omega^2 h^2}{z^2}\right) + \frac{1-\omega^2 h^2}{z^2} = 0 \Rightarrow$$

$$1 - \frac{2-\omega^2 h^2}{z} + \frac{1}{z^2} = 0 \Rightarrow z^2 - \left(2-\omega^2 h^2\right)z + 1 = 0 \quad \text{(A4.18)}$$

from which the roots are

$$z_{1,2} = 1 - \frac{\omega^2 h^2}{2} \pm j\omega h \sqrt{1 - \frac{\omega^2 h^2}{4}} \quad \text{(A4.19)}$$

It may easily be confirmed that these roots have unity magnitude, as required. If the frequency of oscillation is ω_a [rad/s], then the roots of the characteristic equation must also be given by

$$z_{1,2} = e^{\pm j\omega_a h} = \cos(\omega_a h) \pm j \sin(\omega_a h). \quad \text{(A4.20)}$$

Comparing (A4.19) and (A4.20) then gives

$$\omega_a = \frac{1}{h}\cos^{-1}\left(1 - \omega^2 h^2\right) \quad \text{(A4.21)}$$

or alternatively

$$\omega_a = \frac{1}{h}\sin^{-1}\left(\omega h \sqrt{1 - \frac{\omega^2 h^2}{4}}\right). \quad \text{(A4.22)}$$

The RHS of (A4.21) and (A4.22) may be shown to be equal. For precise frequency control, the required frequency can be set to ω_a and ω calculated by making it the subject of (A4.21). Thus

$$\omega = \frac{1}{h}\sqrt{1 - \cos(\omega_a h)}. \quad \text{(A4.23)}$$

A4.2 Traditional Controller Zero Cancellation by Pole Assignment

A4.2.1 Background

In industry, a controller is often purchased rather than developed in-house. Invariably this will be a PID controller whose structure is fixed and only the three gains, K_P, K_I and K_D, can be adjusted, one of the two variants, i.e., the PI or the PD controller, being obtained simply by setting, respectively, $K_D = 0$ or $K_I = 0$. As already highlighted in Chap. 1, these traditional controllers introduce one or two zeros in the closed loop transfer function that can cause a single overshoot, possibly followed by a single undershoot in the step response, even if the closed loop poles are real and therefore the system contains no oscillatory modes. In applications requiring zero overshoot, the zero-less versions of these controllers, i.e., the IPD, IP or DP controllers would be preferable but if a PID controller has to be used, it may be feasible to eliminate the overshooting by placing a subset of the closed loop poles to cancel the zeros or, if sensitivity is an issue, similarly place the poles of an external pre-compensator (Chap. 5).

Pole-zero cancellation is often regarded inadvisable since plant parameteric uncertainty could cause a considerable departure from the specified step response, or even instability if the zeros are positioned relatively close to the imaginary axis of the s-plane. Pole-zero cancellation is certainly invalid if the zeros are in the right half of the s-plane since the poles cancelling them would create unstable modes despite the dynamic response of the controlled output, $y(t)$, to the reference input, $y_r(t)$, being as specified. This would be an example of a closed loop system that is unobservable in the sense that it would fail the observability test applied to plant models in Chap. 3, with the reference input replacing the control input. The result is that the unstable internal mode could not be detected *upon the initial loop closure* by observing the reference input and the output, but would manifest itself due to the control saturation limits in practical applications after a finite time. It must be realised that plant modelling errors will cause the actual closed loop pole locations to be different from the nominal ones. Hence if the zeros are in the left half of the s-plane but so close to the imaginary axis that the plant parameteric errors might move the actual compensating poles into the right half plane, then the method is invalid. On the other hand, if the zeros are situated sufficiently far into the left half plane for the actual poles intended to cancel the zeros to remain in the left half plane for worst case combinations of plant parametric errors, then the method may be applied safely. It must be understood, however, that the effects of the zeros may only be reduced rather than eliminated because the closed loop poles can never precisely coincide with the zeros.

The problem of possible instability could be circumvented by using an external pre-compensator with fixed and known poles instead of employing a subset of the closed loop poles for the zero cancellation. In any case, the control system designer is advised to analyse the system to assess the effect of plant modelling errors before proposing pole-zero cancellation, preferably including a simulation in which the plant parameters are varied within given tolerances while the controller gains are based on the nominal plant parameters.

A4.2.2 PI Control Loop

First consider the classical PI controller applied to the general SISO LTI plant of first order as shown in Fig. A4.6. The external disturbance will not be considered initially as it is unnecessary for the development of methods to cancel zeros.

Fig. A4.6 PI controller applied to a first order plant

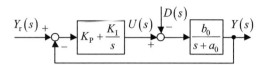

With $D(s) = 0$, the system input–output behaviour is determined by the closed loop transfer function,

$$\frac{Y(s)}{Y_r(s)} = \frac{\dfrac{K_P s + K_I}{s} \cdot \dfrac{b_0}{s + a_0}}{1 + \dfrac{K_P s + K_I}{s} \cdot \dfrac{b_0}{s + a_0}} = \frac{b_0 K_P \left(s + \dfrac{K_I}{K_P}\right)}{s^2 + (a_0 + b_0 K_P) s + b_0 K_I}. \quad (A4.24)$$

To avoid a single overshoot due to the zero at $s = -K_I/K_P$, the possibility of placing one of the closed loop poles to cancel the zero will be considered. To achieve this, the gains, K_P and K_I, have to be chosen such that $(s + K_I/K_P)$ is a factor of the closed loop characteristic polynomial. Then

$$\frac{Y(s)}{Y_r(s)} = \frac{b_0 K_P \left(s + \dfrac{K_I}{K_P}\right)}{\left(s + \dfrac{K_I}{K_P}\right)\left(s + \dfrac{1}{T_c}\right)} = \frac{b_0 K_P}{s + \dfrac{1}{T_c}}. \quad (A4.25)$$

where T_c is the time constant of the resulting first order closed loop transfer function that may be chosen to achieve the specified settling time, T_s (5 % criterion), by setting $T_c = T_s/3$. Expanding the characteristic polynomial of (A4.25) and equating it to that of (A4.24) yields

$$s^2 + \left(\frac{K_I}{K_P} + \frac{1}{T_c}\right)s + \frac{K_I}{K_P T_c} = s^2 + (a_0 + b_0 K_P)s + b_0 K_I \Rightarrow K_P = \frac{1}{b_0 T_c} \quad (A4.26)$$

and

$$K_I = \frac{a_0}{b_0 T_c} \quad (A4.27)$$

Substituting for K_P in (A4.25) yields

$$\frac{Y(s)}{Y_r(s)} = \frac{1/T_c}{s + 1/T_c}, \quad (A4.28)$$

confirming that the closed loop system has a unity DC gain due to the integral term. Importantly, however, although the closed loop transfer function is of first order, the closed loop *system* is still of second order. The zero in the closed loop transfer

Appendices

function is cancelled by the second *pole* at the same location of $s_2 = -K_1/K_p$ and substituting for the controller gains using (A4.26) and (A4.27) yields

$$s_2 = -a_0. \qquad (A4.29)$$

This closed loop pole is equal to the *plant pole* and is therefore fixed. The fact that this pole still plays a part in the closed loop dynamics of the system may be illustrated by re-introducing the disturbance input, $D(s)$. Applying the Principle of Superposition to Fig. A4.6 yields the following transfer function relationship.

$$Y(s) = \frac{b_0 K_p \left(s + \frac{K_1}{K_p}\right) Y_r(s) - b_0 s D(s)}{s^2 + (a_0 + b_0 K_p) s + b_0 K_1}. \qquad (A4.30)$$

Using (A4.25), (A4.26) and (A4.27), (A4.30) becomes

$$Y(s) = \frac{b_0 K_p \left(s + \frac{K_1}{K_p}\right) y_r(s) - b_0 s D(s)}{\left(s + \frac{K_1}{K_p}\right)\left(s + \frac{1}{T_c}\right)} = \frac{1}{s + \frac{1}{T_c}} Y_r(s) - \frac{b_0 s}{(s + a_0)\left(s + \frac{1}{T_c}\right)} D(s). \qquad (A4.31)$$

The disturbance therefore excites the mode corresponding to the pole at $s_2 = -a_0$. The zero cancellation method is therefore restricted to use with a stable plant, i.e., one with its pole located in the left half of the s-plane. It would also be necessary for the pole to be sufficiently far from the imaginary axis of the s-plane for the associated exponential mode, $e^{-a_0 t}$, to decay to negligible proportions on a time scale of the same order as T_c, or less. This implies that the pole-zero cancellation is conditional upon stability of the uncontrolled plant.

The factor, s, in the numerator of the transfer function between $D(s)$ and $Y(s)$ in (A4.31) appears whenever the controller contains an integral term and indicates the zero DC gain that ensures constant disturbances are rejected in the steady state.

A4.2.3 PID Control Loop

Consider a PID controller applied to a second order LTI plant as shown in Fig. A4.7.

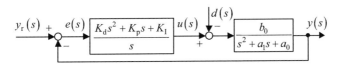

Fig. A4.7 PID control of a second order plant without zeros

The closed loop transfer function relationship is as follows.

$$Y(s) = \frac{\left(K_D s^2 + K_P s + K_I\right) b_0 Y_r(s) - b_0 s D(s)}{s^3 + (a_1 + K_D b_0) s^2 + (a_0 + K_P b_0) s + K_I b_0}$$

$$= \frac{\left(s^2 + \frac{K_P}{K_D} s + \frac{K_I}{K_D}\right) K_d b_0 Y_r(s) - b_0 s D(s)}{s^3 + (a_1 + K_D b_0) s^2 + (a_0 + K_P b_0) s + K_I b_0}. \quad (A4.32)$$

Placing two of the closed loop poles to cancel the zeros in the transfer function between $y_r(s)$ and $y(s)$ is achieved by making $s^2 + (K_p/K_d) s + K_I/K_d$ a factor of the closed loop characteristic polynomial, yielding

$$s^3 + (a_1 + K_d b_0) s^2 + (a_0 + K_P b_0) s + K_I b_0 = \left(s^2 + \frac{K_P}{K_D} s + \frac{K_I}{K_D}\right) \left(s + \frac{1}{T_c}\right)$$

$$= s^3 + \left(\frac{K_P}{K_D} + \frac{1}{T_c}\right) s^2 + \left(\frac{K_P}{K_D T_c} + \frac{K_I}{K_d}\right) s + \frac{K_I}{K_D T_c}, \quad (A4.33)$$

where, as in the previous case, the closed loop transfer function is (A4.28) and T_c is the time constant of the 'free' closed loop pole that may be chosen to yield the required settling time. Solving (A4.33) for the three controller gains yields

$$K_I b_0 = \frac{K_I}{K_D T_c} \Rightarrow K_D = \frac{1}{b_0 T_c}, \quad (A4.34)$$

$$a_0 + K_P b_0 = \frac{K_p}{K_D T_c} + \frac{K_I}{K_D} = \frac{b_0 T_c K_P}{T_c} + b_0 T_c K_I \Rightarrow K_I = \frac{a_0}{b_0 T_c} \quad (A4.35)$$

and

$$a_1 + K_D b_0 = \frac{K_p}{K_d} + \frac{1}{T_c} \Rightarrow a_1 + \frac{1}{b_0 T_c} b_0 = b_0 T_c K_P + \frac{1}{T_c} \Rightarrow K_P = \frac{a_1}{b_0 T_c} \quad (A4.36)$$

Then (A4.32) becomes $Y(s) = \dfrac{\left(s^2 + \frac{K_p}{K_D} s + \frac{K_I}{K_D}\right) K_D b_0 Y_r(s) - b_0 s D(s)}{\left(s^2 + \frac{K_p}{K_D} + \frac{K_I}{K_D}\right)\left(s + \frac{1}{T_c}\right)}$

and substituting for the controller gains using (A4.34), (A4.35) and (A4.36) yields

$$Y(s) = \frac{\frac{1}{T_c}}{s + \frac{1}{T_c}} Y_r(s) - \frac{b_0 s}{(s^2 + a_1 s + a_0)\left(s + \frac{1}{T_c}\right)} D(s). \quad (A4.37)$$

As previously, the transfer function between $Y_r(s)$ and $Y(s)$ is of first order with the time constant, T_c, and the transfer function between $D(s)$ and $Y(s)$ includes the plant poles. Again, since the denominator of the second transfer function in (A4.37) is that

Appendices

of the plant transfer function, the technique of cancelling the controller zeros with closed loop poles to circumvent their effects is conditional upon the uncontrolled plant being stable.

If one or more of the plant poles lie in the right half of the s-plane, indicating instability of the uncontrolled plant, then it is still possible to place some of the closed loop poles in stable locations such that the differentiating effect of the controller zeros is reduced. The pole-to-zero dominance ratio (Chap. 1) may then be used to arrive at suitable closed loop pole locations. This is addressed in Chap. 4 together with the option of using an external pre-compensator which might be necessary to keep the sensitivity within acceptable limits.

A4.2.4 Stability Analysis for the PID Control Loop with Plant Modelling Errors

In a practical situation, the controller gains of (A4.34), (A4.35) and (A4.36) are based upon the estimated plant parameters, $\tilde{b}_0$, $\tilde{a}_0$ and $\tilde{a}_1$, rather than the actual ones, b_0, a_0 and a_1. Routh's stability criterion will now be used to determine limits of the modelling errors, $b_0 - \tilde{b}_0$, $a_0 - \tilde{a}_0$ and $b_0 - \tilde{b}_0$, beyond which instability will occur when attempting pole zero cancellation using closed loop poles as in Sect. A4.2.3. The characteristic polynomial of (A4.33) then becomes

$$s^3 + c_2 s^2 + c_1 s + c_0 = s^3 + (a_1 + K_d b_0) s^2 + (a_0 + K_P b_0) s + K_I b_0$$

$$= s^3 + \left(a_1 + \frac{b_0}{\tilde{b}_0 T_c}\right) s^2 + \left(a_0 + \frac{b_0 \tilde{a}_1}{\tilde{b}_0 T_c}\right) s + \frac{b_0 \tilde{a}_0}{\tilde{b}_0 T_c} \quad \text{(A4.38)}$$

The Routh array is therefore as follows.

1	c_1	0	0
c_2	c_0	0	0
d_1	0	0	0
e_1	0	0	0

Here,

$$d_1 = c_2 c_1 - c_0 = \left(a_1 + \frac{b_0}{\tilde{b}_0 T_c}\right)\left(a_0 + \frac{b_0 \tilde{a}_1}{\tilde{b}_0 T_c}\right) - \frac{b_0 \tilde{a}_0}{\tilde{b}_0 T_c} \quad \text{(A4.39)}$$

and

$$e_1 = d_1 c_0 = \left[\left(a_1 + \frac{b_0}{\tilde{b}_0 T_c}\right)\left(a_0 + \frac{b_0 \tilde{a}_1}{\tilde{b}_0 T_c}\right) - \frac{b_0 \tilde{a}_0}{\tilde{b}_0 T_c}\right]\frac{b_0 \tilde{a}_0}{\tilde{b}_0 T_c} \quad \text{(A4.40)}$$

It will be assumed that $a_i > 0$, $\tilde{a}_i > 0$, $i = 0, 1$, $b_0 > 0$ and $\tilde{b}_0 > 0$. Then since

$$c_2 = a_1 + \frac{b_0}{\tilde{b}_0 T_c}, \qquad (A4.41)$$

it follows that $c_2 > 0$. With $T_c > 0$, inspection of (A4.39) and (A4.40) reveals that d_1 and e_1 are of the same sign. It follows that there is only one sign change in the sequence of elements of the first column of the array. Hence one closed loop pole lies in the right half plane if $d_1 < 0$. The condition for stability is therefore $d_1 > 0$, i.e.,

$$\left(a_1 + \frac{b_0}{\tilde{b}_0 T_c}\right)\left(a_0 + \frac{b_0 \tilde{a}_1}{\tilde{b}_0 T_c}\right) - \frac{b_0 \tilde{a}_0}{\tilde{b}_0 T_c} > 0 \Rightarrow$$

$$a_1 \left(a_0 + \frac{b_0 \tilde{a}_1}{\tilde{b}_0 T_c}\right) + \left(\frac{b_0}{\tilde{b}_0 T_c}\right)^2 \tilde{a}_1 + \frac{b_0}{\tilde{b}_0 T_c}(a_0 - \tilde{a}_0) > 0 \Rightarrow \qquad (A4.42)$$

$$\tilde{a}_0 < a_0 + a_1 \left(\frac{a_0 \tilde{b}_0 T_c}{b_0} + \tilde{a}_1\right) + \frac{b_0}{\tilde{b}_0 T_c} \tilde{a}_1.$$

Let the known upper and lower limits of a_0 be $a_{0\max}$ and $a_{0\min}$. Then

$$a_{0\min} < a_0 < a_{0\max}. \qquad (A4.43)$$

Then setting $\tilde{a}_0 = a_{0\min}$ will ensure (A4.42) is satisfied, since the second and third terms on the RHS of (A4.42) are positive, thereby ensuring closed loop stability. In many cases, these terms are sufficiently positive to ensure closed loop stability with any value of $\tilde{a}_0 \in [a_{0\min}, a_{0\max}]$, so that setting

$$\tilde{a}_0 = \frac{(a_{0\min} + a_{0\max})}{2} \qquad (A4.44)$$

is valid. Even with the plant poles approaching the imaginary axis of the s-plane from the left hand side, given by $b_0 \to 0^-$ and $\tilde{b}_0 \to 0^-$, (A4.42) would become

$$\tilde{a}_0 < a_0 + a_1 (a_0 r T_c + \tilde{a}_1) \qquad (A4.45)$$

where r is indeterminate but satisfies $r \in (0, \infty)$, so even for the worst case of $r = 0$, the condition for stability would be

$$\tilde{a}_0 < a_0 + a_1 \tilde{a}_1 \qquad (A4.46)$$

which is satisfied if $\tilde{a}_0 = a_{0\min}$.

In conclusion, zero cancellation with closed loop poles should not cause instability due to plant parametric uncertainty in many cases. The fears of instability stem from the era in which linear control systems invariably had the structure of Fig. 4.13 in Chap. 4. Then the approach to eliminate the effects of plant zeros was to design a series pre-compensator with poles to cancel them. The subsequent loop closure often resulted in instability, as often predicted by the root locus, with respect to the proportional gain, K, entering the right half of the s-plane due to plant modelling errors.

A4.3 Partial Pole Assignment for Traditional Controllers

If a traditional controller has to be used in a given application but the closed loop system order exceeds three, then complete pole assignment cannot be done. A satisfactory control system design might result, however, by placing as many of the closed loop poles as there are independently adjustable controller parameters. This process is referred to as *partial pole assignment*. The condition for this to be successful is that the remaining poles have no significant influence on the closed loop dynamics through being dominated by the placed poles, requiring the pole to pole dominance ratios (Chap. 1) to be sufficiently large. Achieving this could entail a compromise when attempting to satisfy a stringent performance specification in terms of settling time and minimal overshoot. If only closed loop stability is the main requirement, however, the classical compensator design methods may be sufficient.

Let the closed loop characteristic equation be

$$s^n + C_{n-1}(\mathbf{k}) s^{n-1} + \cdots + C_1(\mathbf{k}) s + C_0(\mathbf{k}) = 0, \quad (A4.47)$$

where the constant coefficients, $C_i(\mathbf{k})$, $i = 1, 2, ..., n$, can be adjusted only by means of the three traditional controller gains, since $\mathbf{k} = (K_P, K_I, K_D)$. Then three of the closed loop poles, $s_{1,2,3} = p_1, p_2, p_3$, can be placed as desired. The characteristic polynomial is then expressed as a product of two polynomials, one third degree and determined by the poles, p_1, p_2 and p_3, i.e., $(s - p_1)(s - p_2)(s - p_3) = s^3 + D_2 s^2 + D_1 s + D_0$, and the other that cannot be chosen, yielding

$$\left(s^3 + D_2 s^2 + D_1 s + D_0\right)\left(s^{n-3} + q_{n-4} s^{n-4} + \cdots + q_1 s + q_0\right) = 0. \quad (A4.48)$$

where n is the order of the closed loop system. The $n - 3$ roots of

$$s^{n-3} + q_{n-4} s^{n-4} + \cdots + q_1 s + q_0 = 0 \quad (A4.49)$$

are the closed loop poles, p_i, $i = 4, 5, \ldots, n$, that are dependent on the controller structure and the choice of p_1, p_2 and p_3. These will be termed the *dependent poles* and must, of course, lie in the left half of the s-plane and be dominated by the placed poles, p_1, p_2 and p_3, for the desired closed loop dynamics to be attained. This constraint, however, may compromise the choice of the placed poles and hence the attainable closed loop dynamics. In general, increasing the speed of response of the closed loop system to changes in the reference input increases the magnitude of the placed poles, resulting in reduction in the magnitudes of the dependent poles which, beyond a certain point, will cease to be dominated. Hence the attainable speed of response is limited.

Multiplying the two factor polynomials of (A4.48) to obtain the characteristic equation in the standard form yields

$$s^n + (D_2 + q_{n-4}) s^{n-1} + (D_1 + D_2 q_{n-4} + q_{n-5}) s^{n-2}$$
$$+ (D_0 + D_1 q_{n-4} + D_2 q_{n-5} + q_{n-6}) s^{n-3}$$
$$+ (D_0 q_{n-4} + D_1 q_{n-5} + D_2 q_{n-6} + q_{n-7}) s^{n-4} + \cdots \quad (A4.50)$$
$$+ (D_0 q_4 + D_1 q_3 + D_2 q_2 + q_1) s^4 + (D_0 q_3 + D_1 q_2 + D_2 q_1 + q_0) s^3$$
$$+ (D_0 q_2 + D_1 q_1 + D_2 q_0) s^2 + (D_0 q_1 + D_1 q_0) s + D_0 q_0 = 0.$$

Equating the characteristic polynomial of (A4.50) to that of (A4.47) yields the following set of simultaneous equations for determination of the controller gains, K_P, K_I and K_D, together with the factor polynomial coefficients, q_i, $i = 0, 1, \ldots, n-4$.

$$\left.\begin{aligned}
& C_0(K_P, K_I, K_D) = D_0 q_0, \quad C_1(K_P, K_I, K_D) = D_0 q_1 + D_1 q_0, \\
& C_2(K_P, K_I, K_D) = D_0 q_2 + D_1 q_1 + D_2 q_0, \\
& C_3(K_P, K_I, K_D) = D_0 q_3 + D_1 q_2 + D_2 q_1 + q_0, \\
& C_4(K_P, K_I, K_D) = D_0 q_4 + D_1 q_3 + D_2 q_2 + q_1, \\
& \vdots \\
& C_{n-4}(K_P, K_I, K_D) = D_0 q_{n-4} + D_1 q_{n-5} + D_2 q_{n-6} + q_{n-7} \\
& C_{n-3}(K_P, K_I, K_D) = D_0 + D_1 q_{n-4} + D_2 q_{n-5} + q_{n-6} \\
& C_{n-2}(K_P, K_I, K_D) = D_1 + D_2 q_{n-4} + q_{n-5}, \quad C_{n-1}(K_P, K_I, K_D) = D_2 + q_{n-4}
\end{aligned}\right\}$$
(A4.51)

This is a completely determined set of simultaneous equations since the number of unknowns is n, consisting of three controller gains and $n-3$ factor polynomial coefficients, and the number of equations is also n. Furthermore, the equations are linear with respect to the unknowns since the coefficients, C_i, $i = 0, 1, \ldots, n-1$, are each linear with respect to K_P, K_I and K_D. To be soluble, of course, these equations must be linearly independent, which is expected. Lack of solubility would imply lack of controllability of the plant, which is rarely encountered in practice.

Example A4.3 IPD controller applied to a third order plant without finite zeros

The system of Fig. A4.8 is of fourth order but only three of its closed loop poles may be placed freely.

The closed loop characteristic equation is

$$1 - \left[-\frac{b_0}{s^3 + a_2 s^2 + a_1 s + a_0} \left(K_D s + K_P + \frac{K_I}{s} \right) \right] = 0 \Rightarrow$$
$$s^4 + a_2 s^3 + (a_1 + b_0 K_D) s^2 + (a_0 + b_0 K_P) s + b_0 K_I = 0$$
(A4.52)

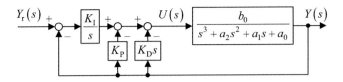

Fig. A4.8 IPD controller applied to third order plant without finite zeros

Since the chosen factor polynomial is $s^3 + D_2 s^2 + D_1 s + D_0$, then the remaining factor polynomial must be $s + q_0$, so the closed loop characteristic equation is also

$$\begin{aligned}(s^3 + D_2 s^2 + D_1 s + D_0)(s + q_0) &= 0 \Rightarrow \\ s^4 + (D_2 + q_0) s^3 + (D_2 q_0 + D_1) s^2 + (D_1 q_0 + D_0) s + D_0 q_0 &= 0\end{aligned} \quad \text{(A4.53)}$$

Equating the characteristic polynomials of (A4.52) and (A4.53) then yields the following set of four simultaneous equations to solve for K_p, K_I K_d and q_0.

$$a_2 = D_2 + q_0 \quad \text{(A4.54)}$$

$$a_1 + b_0 K_D = D_2 q_0 + D_1 \quad \text{(A4.55)}$$

$$a_0 + b_0 K_P = D_1 q_0 + D_0 \quad \text{(A4.56)}$$

$$b_0 K_I = D_0 q_0 \quad \text{(A4.57)}$$

As expected, these equations are linear with respect to the unknowns. They may also be solved by back substitution. Thus

$$\begin{aligned}q_0 = a_2 - D_2, \ K_d &= (D_2 q_0 + D_1 - a_1)/b_0, \\ K_p = (D_1 q_0 + D_0 - a_0)/b_0, \ K_I &= D_0 q_0 / b_0\end{aligned} \quad \text{(A4.58)}$$

Suppose that the desired closed loop poles are to be coincident at $s_{1,2,3} = s_c = -p_c$ to avoid any oscillations in the step response. Then

$$s^3 + D_2 s^2 + D_1 s + D_0 = (s + p_c)^3 = s^3 + 3p_c s^2 + 3p_c^2 s + p_c^3 \Rightarrow \quad \text{(A4.59)}$$

$$\{D_2 = 3p_c, \quad D_1 = 3p_c^2, \quad D_0 = p_c^3\} \quad \text{(A4.60)}$$

Since the dependent pole is at $s_d = -q_0$, then the first of equations (A4.58) becomes $s_d = D_2 - a_2$. Then substituting for D_2 using (A4.60) yields

$$s_d = -(3s_c + a_2). \tag{A4.61}$$

Since $s_c < 0$ and to maintain closed loop stability $s_d < 0$, attempting to speed up the system response by increasing $|s_c|$ reduces $|s_d|$. If this falls below a certain limit, the triple pole at $s_{1,2,3} = s_c$ cannot dominate the dependent pole. According to Sect. 1.4.3, this requires $|s_d| \geq r_{ppmin}|s_c|$, where r_{ppmin} is the minimum pole-to-pole dominance ratio. Referring to Fig. 1.28 of Chap. 1, for $n = 4$ and $n_d = 3$, $r_{ppmin} = 5.4$. Noting that $s_d < 0$ and $s_c < 0$, this inequality may be written as $s_d \leq r_{ppmin} s_c$. Substituting for s_d using (A4.61) then yields $-(3s_c + a_2) \leq r_{ppmin} s_c \Rightarrow$

$$-(3 + r_{ppmin}) s_c \leq a_2. \tag{A4.62}$$

Since $s_c < 0$, it follows that a_2 must be positive. It also follows that

$$|s_c| \leq \frac{a_2}{3 + r_{ppmin}}. \tag{A4.63}$$

This means that the speed of response of the closed loop system is severely limited for relatively small values of a_2.

A simulation will now be carried out of the application to a heating process with transfer function,

$$\frac{Y(s)}{U(s)} = \frac{b_0}{s^3 + a_2 s^2 + a_1 s + a_0} = \frac{K_t K_h}{(1 + sT_1)(1 + sT_2)(1 + sT_3)}$$
$$= \frac{K_t K_h / (T_1 T_2 T_3)}{s^3 + \left(\frac{1}{T_1} + \frac{1}{T_2} + \frac{1}{T_3}\right) s^2 + \left(\frac{1}{T_1 T_2} + \frac{1}{T_2 T_3} + \frac{1}{T_3 T_1}\right) s + \frac{1}{T_1 T_2 T_3}}, \tag{A4.64}$$

where the three dominant time constants and the DC gain are, respectively, $T_1 = 20$ s, $T_2 = 10$ s, $T_3 = 5$ s and $K_h = 100$ °C/V is the heating element temperature constant. Also $y = K_t \Theta$, where Θ is the controlled temperature and $K_t = 0.01$ V/°C is the temperature transducer coefficient. In this case,

$$a_2 = \frac{1}{T_1} + \frac{1}{T_2} + \frac{1}{T_3} = 0.35, \tag{A4.65}$$

and with reference to Fig. 1.28 of Chap. 1, for $n = 4$ and $n_d = 3$, $r_{pp\,min} = 5.4$, (A4.63) becomes

$$|s_c| \leq 0.0417. \tag{A4.66}$$

Fig. A4.9 Step responses of IPD control system designed by partial pole assignment

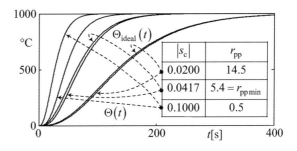

Figure A4.9 shows the ideal and attained step responses for three locations of the triple closed loop pole.

It is evident that the achieved response, $\Theta(t)$, is acceptably close to the ideal one for $|s_c| \leq 0.0417$ which corresponds to $r_{pp\,min}$ and approximately halving $|s_c|$ achieves even closer following of $\Theta_{ideal}(t)$ but at the expense of considerably slowing the system response. Attempting to speed up the response by approximately doubling $|s_c|$ does achieve a faster response but one that does not follow $\Theta_{ideal}(t)$ due to the influence of the dependent pole that has moved closer to the imaginary axis of the s-plane than the chosen triple pole location, indicated by the fractional value of r_{pp}.

Reference

1. Mason SJ (1956) Feedback theory: further properties of signal flow graphs. In: Proceedings of IRE, Vol. 44, No. 7, pp 920–926

A5 Appendix to Chap. 5

A5.1 Computer Aided Pole Assignment

A5.1.1 Introduction

The numerical methods described here facilitate the design of certain classes of linear control system or subsystem which contain a sufficient number of independently adjustable parameters for any desired set of poles to be realised within the limitations imposed by the hardware. The algebraic approaches of Chaps. 4 and 5 leading to design formulae provide a useful insight, especially for those studying state space methods for the first time, and enable effective controllers to be designed with little effort for many commonly occurring relatively simple plants or if convenient state representations, such as the controller or observer canonical forms, are used. The work can, however, be onerous for systems having a more complex structure or an arbitrary state representation. For those already having a good understanding of linear control systems, especially those working in industry, this chapter provides alternative numerical methods for controller parameter determination that minimises the time taken to complete a control system design. The classes of system to which these numerical methods may be applied are defined in the subsections to follow.

A5.1.2 Ackermann's Gain Formulae

A5.1.2.1 Applicability

Ackermann's formulae are for calculating the gains of a linear state feedback control law and the gains of a state observer if its plant model is sufficiently good for the separation theorem (Chap. 4) to apply.

A5.1.2.2 Gain Formula for the Linear State Feedback Control Law

Ackermann's control law gain formula enables the gain, $\mathbf{k}^T$, of the SISO linear state feedback control system defined by

$$\dot{\mathbf{x}} = \mathbf{A}\mathbf{x} + \mathbf{b}u, \ u = ry_r - \mathbf{k}^T\mathbf{x}, \ y = \mathbf{c}^T\mathbf{x} \tag{A5.1}$$

(Chapter 5) to be calculated that places the closed loop poles in desired locations, regardless of the state representation (Chap. 3). The formula is

$$\boxed{\mathbf{k}^T = [\,0 \cdots 0\ 1\,]\mathbf{M}_c^{-1}\mathbf{D}_c(\mathbf{A})} \tag{A5.2}$$

Appendices

where

$$\mathbf{M}_c = \begin{bmatrix} \mathbf{b} & \mathbf{Ab} & \cdots & \mathbf{A}^{n-1}\mathbf{b} \end{bmatrix} \quad (A5.3)$$

is the controllability matrix (Chap. 3) and

$$\mathbf{D}_c(\mathbf{A}) = \mathbf{A}^n + d_{c\,n-1}\mathbf{A}^{n-1} + \cdots + d_{c\,1}\mathbf{A} + d_{c\,0}\mathbf{I}, \quad (A5.4)$$

where the coefficients, $d_{c\,i}$, $i = 0, 1, ..., n-1$, are the same as those of the desired characteristic equation,

$$s^n + d_{c\,n-1}s^{n-1} + \ldots d_{c\,1}s + d_{c\,0} = 0, \quad (A5.5)$$

needed for the pole placement. The formula will now be proven for $n = 2$ and $n = 3$. Then the general formula (A5.2) will be inferred by inductive reasoning. First, applying the Cayley-Hamilton theorem to (A5.5) yields

$$\mathbf{D}_c(\mathbf{A}_{cl}) = \mathbf{A}_{cl}^n + d_{c\,n-1}\mathbf{A}_{cl}^{n-1} + \cdots + d_{c\,1}\mathbf{A}_{cl} + d_{c\,0}\mathbf{I} = \mathbf{0}. \quad (A5.6)$$

The proof is commenced by expanding (A5.6) using the equation for the closed loop system matrix

$$\mathbf{A}_{cl} = \mathbf{A} - \mathbf{bk}^T \quad (A5.7)$$

(Chapter 5), separating out $\mathbf{D}_c(\mathbf{A})$ of (A5.4) and then simplifying the remaining terms by recognising $\mathbf{A}_{cl}$ as a post-multiplying factor. Writing (A5.6) for $n = 2$ and substituting for $\mathbf{A}_{cl}$ yields

$$\mathbf{D}_c(\mathbf{A}_{cl}) = \mathbf{A}_{cl}^2 + d_{c\,1}\mathbf{A}_{cl} + d_{c\,0}\mathbf{I} = \left[\mathbf{A} - \mathbf{bk}^T\right]^2 + d_{c\,1}\left[\mathbf{A} - \mathbf{bk}^T\right] + d_{c0}\mathbf{I} = \mathbf{0}. \quad (A5.8)$$

Then expanding the quadratic term and simplifying using (A5.7) gives

$$\left[\mathbf{A} - \mathbf{bk}^T\right]^2 = \mathbf{A}^2 - \mathbf{Abk}^T - \mathbf{bk}^T\mathbf{A} + \left(\mathbf{bk}^T\right)^2 = \underline{\mathbf{A}^2 - \mathbf{Abk}^T - \mathbf{bk}^T\mathbf{A}_{cl}}. \quad (A5.9)$$

The term, $\mathbf{A}^2$, is retained as it is the leading term of (A5.4). Using this in (A5.8) and then grouping terms to separate $\mathbf{D}_c(\mathbf{A}) = \mathbf{A}^2 + d_{c\,1}\mathbf{A} + d_{c\,0}\mathbf{I}$ and arranging the remaining terms with the controllability matrix as a pre-multiplying factor, yields

$$\mathbf{D}_c(\mathbf{A}_{cl}) = \left[\mathbf{A}^2 - \mathbf{A}\mathbf{b}\mathbf{k}^T - \mathbf{b}\mathbf{k}^T\mathbf{A}_{cl}\right] + d_{c\,1}\left[\mathbf{A} - \mathbf{b}\mathbf{k}^T\right] + d_{c\,0}\mathbf{I}$$
$$= \left[\mathbf{A}^2 + d_{c\,1}\mathbf{A} + d_{c\,0}\mathbf{I}\right] - \mathbf{A}\mathbf{b}\mathbf{k}^T - \mathbf{b}\mathbf{k}^T\mathbf{A}_{cl} - d_{c\,1}\mathbf{b}\mathbf{k}^T$$
$$= \mathbf{D}_c(\mathbf{A}) - \mathbf{b}\left[d_{c\,1}\mathbf{k}^T + \mathbf{g}^T\mathbf{A}_{cl}\right] - \mathbf{A}\mathbf{b}\left[\mathbf{k}^T\right] = \mathbf{0} \Rightarrow$$
$$\mathbf{D}_c(\mathbf{A}) = \mathbf{b}\left[d_{c\,1}\mathbf{k}^T + \mathbf{k}^T\mathbf{A}_{cl}\right] + \mathbf{A}\mathbf{b}\left[\mathbf{k}^T\right] = \left[\mathbf{b}\ \mathbf{A}\mathbf{b}\right]\begin{bmatrix} d_{c\,1}\mathbf{k}^T + \mathbf{k}^T\mathbf{A}_{cl} \\ \mathbf{g}^T \end{bmatrix} \Rightarrow$$
$$\begin{bmatrix} d_{c\,1}\mathbf{k}^T + \mathbf{k}^T\mathbf{A}_{cl} \\ \mathbf{k}^T \end{bmatrix} = \left[\mathbf{b}\ \mathbf{A}\mathbf{b}\right]^{-1}\mathbf{D}_c(\mathbf{A}) \tag{A5.10}$$

Pre-multiplying by $\begin{bmatrix} 0 & 1 \end{bmatrix}$ then extracts the required gain vector. Thus

$$\begin{bmatrix} 0 & 1 \end{bmatrix}\begin{bmatrix} d_{c\,1}\mathbf{k}^T + \mathbf{k}^T\mathbf{A}_{cl} \\ \mathbf{k}^T \end{bmatrix} = \mathbf{k}^T \Rightarrow \underline{\mathbf{k}^T = \begin{bmatrix} 1 & 0 \end{bmatrix}\left[\mathbf{b}\ \mathbf{A}\mathbf{b}\right]^{-1}\mathbf{D}_c(\mathbf{A})} \tag{A5.11}$$

which is (A5.2) for $n = 2$.
For $n = 3$, (A5.6) is

$$\mathbf{D}_c(\mathbf{A}_{cl}) = \mathbf{A}_{cl}^3 + d_{c\,2}\mathbf{A}_{cl}^2 + d_{c\,1}\mathbf{A}_{cl} + d_{c\,0}\mathbf{I}$$
$$= \left[\mathbf{A} - \mathbf{b}\mathbf{k}^T\right]^3 + d_{c\,2}\left[\mathbf{A} - \mathbf{b}\mathbf{k}^T\right]^2 + d_{c\,1}\left[\mathbf{A} - \mathbf{b}\mathbf{k}^T\right] + d_{c\,0}\mathbf{I} \tag{A5.12}$$

Expanding the cubic term using (A5.9) and simplifying using (A5.7), yields

$$\left[\mathbf{A} - \mathbf{b}\mathbf{k}^T\right]^3 = \left[\mathbf{A} - \mathbf{b}\mathbf{k}^T\right]^2\left[\mathbf{A} - \mathbf{b}\mathbf{k}^T\right] = \left[\mathbf{A}^2 - \mathbf{A}\mathbf{b}\mathbf{k}^T - \mathbf{b}\mathbf{k}^T\mathbf{A}_{cl}\right]\left[\mathbf{A} - \mathbf{b}\mathbf{k}^T\right]$$
$$= \mathbf{A}^3 - \mathbf{A}\mathbf{b}\mathbf{k}^T\mathbf{A} - \mathbf{b}\mathbf{k}^T\mathbf{A}_{cl}\mathbf{A} - \mathbf{A}^2\mathbf{b}\mathbf{k}^T + \mathbf{A}\left[\mathbf{b}\mathbf{k}^T\right]^2 + \mathbf{b}\mathbf{k}^T\mathbf{A}_{cl}\mathbf{b}\mathbf{k}^T$$

i.e.,

$$\left[\mathbf{A} - \mathbf{b}\mathbf{k}^T\right]^3 = \mathbf{A}^3 - \mathbf{A}^2\mathbf{b}\mathbf{k}^T - \mathbf{A}\mathbf{b}\mathbf{k}^T\left[\mathbf{A} - \mathbf{b}\mathbf{k}^T\right] - \mathbf{b}\mathbf{k}^T\mathbf{A}_{cl}\left[\mathbf{A} - \mathbf{b}\mathbf{k}^T\right]$$
$$= \mathbf{A}^3 - \mathbf{A}^2\mathbf{b}\mathbf{k}^T - \mathbf{A}\mathbf{b}\mathbf{k}^T\mathbf{A}_{cl} - \mathbf{b}\mathbf{k}^T\mathbf{A}_{cl}^2. \tag{A5.13}$$

The term, $\mathbf{A}^3$, is retained as it is the leading term of $\mathbf{D}_c(\mathbf{A})$. Substituting for the cubic and quadratic terms in (A5.12) using, respectively, (A5.13) and (A5.9), then separating $\mathbf{D}_c(\mathbf{A}) = \mathbf{A}^3 + d_{c\,2}\mathbf{A}^2 + d_{c\,1}\mathbf{A} + d_{c\,0}\mathbf{I}$ and simplifying the remaining terms using (A5.7), with the controllability matrix as a pre-multiplying factor yields

Appendices

$$\begin{aligned}
\mathbf{D}_c\left(\mathbf{A}_{cl}\right) &= \mathbf{A}^3 - \mathbf{A}^2\mathbf{b}\mathbf{k}^T - \mathbf{A}\mathbf{b}\mathbf{k}^T\mathbf{A}_{cl} - \mathbf{b}\mathbf{k}^T\mathbf{A}_{cl}^2 \\
&\quad + d_{c\,2}\left[\mathbf{A}^2 - \mathbf{A}\mathbf{b}\mathbf{k}^T - \mathbf{b}\mathbf{k}^T\mathbf{A}_{cl}\right] + d_{c\,1}\left[\mathbf{A} - \mathbf{b}\mathbf{k}^T\right] + d_{c\,0}\mathbf{I} \\
&= \left[\mathbf{A}^3 + d_{c\,2}\mathbf{A}^2 + d_{c\,1}\mathbf{A} + d_{c\,0}\mathbf{I}\right] \\
&\quad - \mathbf{b}\left[\mathbf{k}^T\mathbf{A}_{cl}^2 + d_{c\,2}\mathbf{k}^T\mathbf{A}_{cl} + d_{c\,1}\mathbf{k}^T\right] - \mathbf{A}\mathbf{b}\left[\mathbf{k}^T\mathbf{A}_{cl} + d_{c\,2}\mathbf{k}^T\right] - \mathbf{A}^2\mathbf{b}\mathbf{k}^T = \mathbf{0} \Rightarrow
\end{aligned}$$

$$\mathbf{D}_c(\mathbf{A}) = \begin{bmatrix}\mathbf{b} & \mathbf{A}\mathbf{b} & \mathbf{A}^2\mathbf{b}\end{bmatrix}\begin{bmatrix}\mathbf{k}^T\mathbf{A}_{cl}^2 + d_{c\,2}\mathbf{k}^T\mathbf{A}_{cl} + d_{c\,1}\mathbf{k}^T \\ \mathbf{k}^T\mathbf{A}_{cl} + d_{c\,2}\mathbf{k}^T \\ \mathbf{k}^T\end{bmatrix} \Rightarrow$$

$$\begin{bmatrix}\mathbf{k}^T\mathbf{A}_{cl}^2 + d_{c\,2}\mathbf{k}^T\mathbf{A}_{cl} + d_{c\,1}\mathbf{k}^T \\ \mathbf{k}^T\mathbf{A}_{cl} + d_{c\,2}\mathbf{k}^T \\ \mathbf{k}^T\end{bmatrix} = \begin{bmatrix}\mathbf{b} & \mathbf{A}\mathbf{b} & \mathbf{A}^2\mathbf{b}\end{bmatrix}^{-1}\mathbf{D}_c(\mathbf{A}). \tag{A5.14}$$

Pre-multiplying by $\begin{bmatrix}0 & 0 & 1\end{bmatrix}$ then extracts the required gain vector. Thus

$$\begin{bmatrix}0 & 0 & 1\end{bmatrix}\begin{bmatrix}\mathbf{k}^T\mathbf{A}_{cl}^2 + d_{c\,2}\mathbf{k}^T\mathbf{A}_{cl} + d_{c\,1}\mathbf{k}^T \\ \mathbf{k}^T\mathbf{A}_{cl} + d_{c\,2}\mathbf{k}^T \\ \mathbf{k}^T\end{bmatrix} = \mathbf{k}^T = \begin{bmatrix}0 & 0 & 1\end{bmatrix}\begin{bmatrix}\mathbf{b} & \mathbf{A}\mathbf{b} & \mathbf{A}^2\mathbf{b}\end{bmatrix}^{-1}\mathbf{D}_c(\mathbf{A}), \tag{A5.15}$$

which is (A5.2) for $n = 3$.

It is evident that attempting the proof for an arbitrary system order would yield very unwieldy expressions, but it will be considered reasonable to assume that (A5.2) holds for $n \geq 3$ by comparison with (A5.15) and (A5.11).

In principle, Ackermann's formula could be used to derive algebraic expressions for the gains of a linear state feedback control law, as in Chap. 5, but the term, $\mathbf{M}_c^{-1}$ would render this task impracticable for systems of greater than second order, unless software is available capable of algebraic matrix inversion. In absence of this, the recommended approach is to implement (A5.2) as a computer-based algorithm with the plant parameters entered numerically.

A5.1.2.3 Gain Formula for the Linear State Observer

Ackermann's observer gain formula enables the gain, $\mathbf{l}$, of the SISO linear state observer defined by

$$\dot{\widehat{\mathbf{x}}} = \mathbf{A}_o\widehat{\mathbf{x}} + \mathbf{b}_o u + \mathbf{l}\left(y - \mathbf{c}_o^T\widehat{\mathbf{x}}\right) \tag{A5.16}$$

(Chapter 4) to be calculated that places the observer poles in desired locations, regardless of the state representation (Chap. 3).

In order for the observer block diagram to be equivalent to (A5.16) the model correction loop inputs must be applied directly to the inputs of the pure integrators of the model. In contrast, the algebraic method of Chap. 8 permitted the correction inputs to be applied to any selected set of first order subsystems of the plant model but an observer cannot be structured in this more general way if Ackermann's observer gain formula is to be applied, which is

$$\boxed{\mathbf{l} = \mathbf{D_o}(\mathbf{A_o})\mathbf{M_o}^{-1}\begin{bmatrix} 0 \cdots 0\ 1 \end{bmatrix}^T} \quad \text{(A5.17)}$$

where

$$\mathbf{M_o} = \begin{bmatrix} \mathbf{c_o} \\ \mathbf{c_o}\mathbf{A_o} \\ \vdots \\ \mathbf{c_o}\mathbf{A_o}^{n-1} \end{bmatrix} \quad \text{(A5.18)}$$

is the observability matrix (Chap. 3) and

$$\mathbf{D_o}(\mathbf{A_o}) = \mathbf{A_o}^n + d_{o\,n-1}\mathbf{A_o}^{n-1} + \cdots + d_{o\,1}\mathbf{A_o} + d_{o\,0}\mathbf{I}, \quad \text{(A5.19)}$$

where the coefficients, $d_{o\,i}$, $i = 0, 1, \ldots, n-1$ are the same as those of the desired observer correction loop characteristic equation (Chap. 8),

$$s^n + d_{o\,n-1}s^{n-1} + \ldots d_{o\,1}s + d_{o\,0} = 0. \quad \text{(A5.20)}$$

The proof follows similar lines to that presented in Sect. A5.1.2.2 for the linear state feedback gain formula but is not given here.

Example A5.1 Linear state feedback control system for flexible drive using an observer

This example has the same plant and LSF control law as Example 5.3 but includes an observer for state estimation. Figure A5.1 shows the state variable block diagram. Here the plant parameters are related to the time constants of the per unit model by

$$q_1 = 1/T_1, \quad q_s = 1/T_{sp}, \quad q_2 = 1/T_2 \quad \text{(A5.21)}$$

First, the plant state equation is

$$\begin{bmatrix} \dot{x}_1 \\ \dot{x}_2 \\ \dot{x}_3 \\ \dot{x}_4 \end{bmatrix} = \underbrace{\begin{bmatrix} 0 & 1 & 0 & 0 \\ 0 & 0 & q_2 & 0 \\ 0 & -q_s & 0 & q_s \\ 0 & 0 & -q_1 & 0 \end{bmatrix}}_{\mathbf{A}} \begin{bmatrix} x_1 \\ x_2 \\ x_3 \\ x_4 \end{bmatrix} + \underbrace{\begin{bmatrix} 0 \\ 0 \\ 0 \\ q_1 \end{bmatrix}}_{\mathbf{b}} u. \quad \text{(A5.22)}$$

Appendices

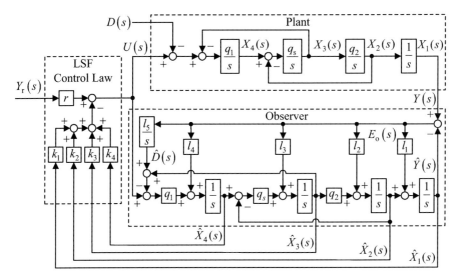

Fig. A5.1 Linear state feedback control system for flexible drive using an observer

The plant matrices, **A** and **b**, are then used to form the controllability matrix (A5.3). The desired characteristic polynomial to yield a non-overshooting step response with a settling time of T_s (5 % criterion) yields

$$s^4 + d_{c3}s^3 + d_{c2}s^2 + d_{c1}s + d_{c0} = (s+p)^4 = s^4 + 4ps^3 + 6p^2s^2 + 4p^3s + p^4 \quad (A5.23)$$

where $p = 1.5(1+n)/T_s|_{n=4} = 7.5/T_s$. Hence

$$d_{c0} = p^4, \ d_{c1} = 4p^3, \ d_{c2} = 6p^2 \text{ and } d_{c3} = 4p. \quad (A5.24)$$

Now the linear state feedback gain matrix, $\mathbf{k}^T$, can be calculated using (A5.2), (A5.3) and (A5.4) with $n = 4$ and numerical inputs from, (A5.21), (A5.22) and (A5.24).

Ackermann's linear state feedback formula does not, of course, cater for the reference input scaling coefficient and this has to be determined separately. The steady state conditions with a constant reference input, Y_r, are $\widehat{x}_{1ss} = x_{1ss} = Y_r$, $\widehat{x}_{2ss} = x_{2ss} = \widehat{x}_{3ss} = x_{3ss} = \widehat{x}_{4ss} = x_{4ss} = 0$ and $u_{ss} = 0$ since the input to the first integrator of the plant is $\dot{x}_{4ss}/q_1 = 0$. Applying these conditions at the reference input summing junction of Fig. A5.1 with $d(s) = 0$ then yields

$$rY_r - k_1 Y_r = 0 \Rightarrow r = k_1 \quad (A5.25)$$

The observer includes estimation of the external disturbance, d, referred to the control input, which is equivalent to external load torque estimation. Although the estimate, $\widehat{d}$, is not used in the standard linear state feedback control law (in contrast to the forced dynamic control of Chap. 7), it is included to avoid steady state errors in the state estimate with constant disturbances.

To reinforce the foregoing point made about the observer structure, the coefficients, q_1, q_s and q_2, have been separated from the integrators of the plant model in the observer of Fig. A5.1, to allow the model correction inputs to be applied directly to the integrator inputs. If the correction inputs were applied to the combined integrators and coefficients, as in the plant above, then Ackermann's formula would give incorrect gain values.

The observer state equation is

$$\begin{bmatrix} \dot{\hat{x}}_1 \\ \dot{\hat{x}}_2 \\ \dot{\hat{x}}_3 \\ \dot{\hat{x}}_4 \\ \dot{\hat{d}} \end{bmatrix} = \underbrace{\begin{bmatrix} 0 & 1 & 0 & 0 & 0 \\ 0 & 0 & q_2 & 0 & 0 \\ 0 & -1 & 0 & q_s & 0 \\ 0 & 0 & -q_1 & 0 & -q_1 \\ 0 & 0 & 0 & 0 & 0 \end{bmatrix}}_{\mathbf{A}_o} \begin{bmatrix} \hat{x}_1 \\ \hat{x}_2 \\ \hat{x}_3 \\ \hat{x}_4 \\ \hat{d} \end{bmatrix}$$

(A5.26)

$$+ \begin{bmatrix} 0 \\ 0 \\ 0 \\ q_1 \\ 0 \end{bmatrix} u + \begin{bmatrix} l_1 \\ l_2 \\ l_3 \\ l_4 \\ l_5 \end{bmatrix} \left(y - \underbrace{\begin{bmatrix} 1 & 0 & 0 & 0 & 0 \end{bmatrix}}_{\mathbf{c}_o^T} \begin{bmatrix} \hat{x}_1 \\ \hat{x}_2 \\ \hat{x}_3 \\ \hat{x}_4 \\ \hat{d} \end{bmatrix} \right).$$

The plant/disturbance model matrices, $\mathbf{A}_o$ and $\mathbf{c}_o^T$, are then used to form the observability matrix (A5.18). The desired observer characteristic polynomial to yield a correction loop settling time of T_{so} (5 % criterion) yields

$$s^5 + d_{o4}s^4 + d_{o3}s^3 + d_{o2}s^2 + d_{o1}s + d_{o0} = s^5 + 5qs^4 + 10q^2s^3 + 10q^3s^2 + 5q^4s + q^5.$$
(A5.27)

(the RHS being $(s+q)^5$) where $q = 1.5(1+n)/T_{so}|_{n=5} = 9/T_s$. Hence

$$d_{o0} = q^5, \ d_{o1} = 5q^4, \ d_{o2} = 10q^3, \ d_{o3} = 10q^2 \text{ and } d_{o4} = 5q. \quad (A5.28)$$

Now the observer gain matrix, $\mathbf{l}$, can be calculated using (A5.17), (A5.18) and (A5.19) with $n = 5$ and numerical inputs from (A5.21), (A5.26) and (A5.28). Figure A5.2 shows simulations with $T_1 = T_2 = 0.1$ s, $T_{sp} = 0.008$ s, $T_s = 0.2$ s and $T_{so} = 0.02$ s.

Figure A5.2a shows the response of the state variables to a step reference angle input of $y_r(t) = h(t)$ [rad] and Fig. A5.2b shows the convergence of the observer state estimation errors, $\hat{x}_{i\,e}(t) = \hat{x}_i(t) - x_i(t)$, $i = 1, 2, 3, 4, 5$, to zero following an initial state estimate mismatch. These demonstrate correct operation of the system.

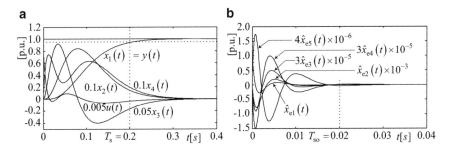

Fig. A5.2 Simulation of flexible drive control system designed with Ackermann's formulae. (**a**) step response with observer settled (**b**) observer error convergence, $x_{e1}(0) = 1$

A5.2 Linear Characteristic Polynomial Interpolation

A5.2.1 Introduction

In many linear systems that can be designed by pole assignment (Chap. 4), the characteristic polynomial

$$s^n + a_{n-1}(\mathbf{k}) s^{n-1} + \cdots + a_1(\mathbf{k}) s + a_0(\mathbf{k}) \qquad (A5.29)$$

has coefficients that are linear functions of the adjustable parameters, $k_1, k_2, \ldots$ and k_n, such as controller gains or observer gains, where $\mathbf{k} = \begin{bmatrix} k_1 & k_2 & \ldots & k_n \end{bmatrix}^T$ is a vector formed from these parameters. It is this class of linear system to which the author's linear characteristic polynomial interpolation (LCPI) method is applicable.

A5.2.2 Development of the LCPI Algorithm

The general linear relationship between the gains and the coefficients of the characteristic polynomial can be written

$$\mathbf{a} = \mathbf{M}\mathbf{k} + \mathbf{a}_0 \qquad (A5.30)$$

where $\mathbf{a} = \begin{bmatrix} a_0 & a_1 & \ldots & a_{n-1} \end{bmatrix}^T$ is the vector of coefficients of the characteristic polynomial, $s^n + \sum_{i=0}^{n-1} a_i s^i$, $\mathbf{M} \in \Re^{n \times n}$, is a constant matrix and $\mathbf{a}_0$ is a constant vector.

The practical aid needed for the application of the method based on (A5.30) is the MATLAB®-SIMULINK® linearisation routine that is normally used to produce a linear state space model of a nonlinear dynamical system about a specified operating point, given its SIMULINK® block diagram. Instead, this is used with the block diagram of the linear system or subsystem under development. The resulting linear state space model is then converted to a transfer function. The coefficients of the denominator polynomial are then assembled to form the vector, $\underline{a}$. The use of this routine is illustrated in Fig. A5.3.

As seen in the examples of Chaps. 4 and 5, it is relatively straightforward to determine the desired vector of polynomial coefficients, $\mathbf{a}_d$, to achieve a specified settling time with no overshooting, using the author's settling time formulae. For the 5 % criterion, the characteristic polynomial is

$$[s + 1.5(1+n)/T_s]^n = s^n + \sum_{i=0}^{n-1} a_{di} s^i \qquad (A5.31)$$

but even these could be determined by a method similar to that illustrated in Fig. A5.3 using a SIMULINK® block diagram consisting of n cascaded identical first order elements as shown in Fig. A5.4.

Any other method, however, may be applied to determine $\mathbf{a}_d$.

With reference to Fig. A5.3, the method consists of applying a number of different parameter vectors, $\mathbf{k}$, called test parameter vectors and noting the corresponding coefficient vectors, $\mathbf{a}$. Then this information is used to determine the desired value of $\mathbf{k}$, denoted $\mathbf{k}_d$, that yields $\mathbf{a} = \mathbf{a}_d$ by multivariable linear interpolation. As will be seen, the minimum number of parameter vectors is $n + 1$. Linear regression using more data than this is unnecessary because no random errors are involved.

Referring to (A5.30), once $\mathbf{M}$ and $\mathbf{a}_0$ have been determined, then the required parameter vector is obtained by setting $\mathbf{a} = \mathbf{a}_d$ and then making $\mathbf{k}$, which is then $\mathbf{k}_d$, the subject of the equation. Thus,

$$\mathbf{k}_d = \mathbf{M}^{-1}[\mathbf{a}_d - \mathbf{a}_0]. \qquad (A5.32)$$

First, $\mathbf{a}_0$ is the value of $\mathbf{a}$ with $\mathbf{k} = 0$. Applying this to Fig. A5.3 yields $\mathbf{a} = \mathbf{a}_0$.

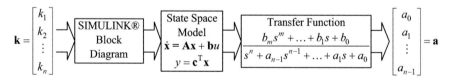

Fig. A5.3 Implementation tool for LCPI method

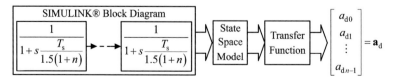

Fig. A5.4 A computer-aided method of desired characteristic polynomial determination

To determine $\mathbf{M}$, let $n+1$ test parameter vectors, $\mathbf{k}_{t1}, \mathbf{k}_{t2}, \ldots, \mathbf{k}_{t\,n+1}$, be chosen (how being determined shortly), then applied, one at a time, to Fig. A5.3 and the corresponding coefficient vectors, $\mathbf{a}_1, \mathbf{a}_2, \ldots, \mathbf{a}_{n+1}$, noted. To utilise this data, (A5.30) may be written down for each pair, $(\mathbf{k}_{t\,i}, \mathbf{a}_i)$. Thus,

$$\mathbf{a}_i = \mathbf{M}\mathbf{k}_{t\,i} + \mathbf{a}_0, \quad i = 1, 2, \ldots, n+1. \tag{A5.33}$$

Now $\mathbf{a}_0$ may be eliminated between n consecutive equation pairs taken from (A5.33) by subtracting one from the other, as follows.

$$\mathbf{a}_{i+1} - \mathbf{a}_i = \mathbf{M}[\mathbf{k}_{t\,i+1} - \mathbf{k}_{t\,i}], \quad i = 1, 2, \ldots, n. \tag{A5.34}$$

This may be written as

$$\Delta\mathbf{a}_i = \mathbf{M}\Delta\mathbf{k}_{t\,i}, \quad i = 1, 2, \ldots, n. \tag{A5.35}$$

Now (A5.35) may be written as a single matrix equation by assembling the coefficient difference vectors, $\Delta\mathbf{a}_i$, and the corresponding parameter difference vectors, $\Delta\mathbf{k}_{t\,i}$, as the columns of coefficient difference and parameter difference matrices, $\Delta\mathbf{A}$ and $\Delta\mathbf{K}_t$, as follows.

$$\underbrace{\left[\Delta\mathbf{a}_1 \mid \Delta\mathbf{a}_2 \mid \ldots \mid \Delta\mathbf{a}_n\right]}_{\Delta\mathbf{A}} = \mathbf{M}\underbrace{\left[\Delta\mathbf{k}_{t1} \mid \Delta\mathbf{k}_{t2} \mid \ldots \mid \Delta\mathbf{k}_{tn}\right]}_{\Delta\mathbf{K}_t} \tag{A5.36}$$

Since the test vectors, $\Delta\mathbf{k}_{t\,i}$, have n elements, the matrix, $\Delta\mathbf{K}_t$, is square and therefore the matrix, $\mathbf{M}$, may be determined as

$$\mathbf{M} = \Delta\mathbf{A}[\Delta\mathbf{K}_t]^{-1} \tag{A5.37}$$

provided $\Delta\mathbf{K}_t$ is non-singular. An arbitrary set of test parameter vectors could be chosen within the restriction of non-singularity of $\Delta\mathbf{K}_t$ but to guarantee numerical accuracy it is essential for $\Delta\mathbf{K}_t$ to be well conditioned, i.e., to have a relatively small condition number, close to unity in the range $[1, \infty]$, defined as $|\lambda_{max}|/|\lambda_{min}|$, where λ_{max} and λ_{min} are, respectively, the eigenvalues of $\Delta\mathbf{K}_t$ having the maximum and minimum magnitudes. An ideal choice is

$$\Delta\mathbf{K}_t = \begin{bmatrix} \lambda & 0 & \cdots & 0 \\ 0 & \lambda & & \vdots \\ \vdots & & \ddots & 0 \\ 0 & \cdots & 0 & \lambda \end{bmatrix} \tag{A5.38}$$

where λ is a real, non-zero, constant. This gives the smallest possible condition number (i.e., unity) as all the n eigenvalues are equal to λ and also avoids the matrix inverse in (A5.37) by replacing it with a scalar division by λ. Thus

$$\mathbf{M} = \frac{1}{\lambda}\Delta\mathbf{A} \tag{A5.39}$$

Let the columns of (A5.38) be written

$$\boldsymbol{\lambda}_1 = \begin{bmatrix} \lambda \\ 0 \\ \vdots \\ 0 \end{bmatrix}, \boldsymbol{\lambda}_2 = \begin{bmatrix} 0 \\ \lambda \\ \vdots \\ 0 \end{bmatrix}, \ldots, \boldsymbol{\lambda}_n = \begin{bmatrix} 0 \\ 0 \\ \vdots \\ \lambda \end{bmatrix} \tag{A5.40}$$

Then in view of (A5.34), (A5.35), (A5.36) and (A5.38),

$$\Delta\mathbf{k}_{t\,i} = \mathbf{k}_{t\,i+1} - \mathbf{k}_{t\,i} = \boldsymbol{\lambda}_i, \quad i = 1, 2, \ldots, n \tag{A5.41}$$

At the beginning, only one test parameter vector has to be chosen. Let this be $\mathbf{k}_{t\,1}$. Then the remaining n test parameter vectors can be found from (A5.41). Thus

$$\mathbf{k}_{t\,i+1} = \mathbf{k}_{t\,i} + \boldsymbol{\lambda}_i, \quad i = 1, 2, \ldots, n \tag{A5.42}$$

It remains to consider the choice of λ. It is not actually critical and the calculations are even simpler by choosing $\lambda = 1$ because (A5.39) becomes just

$$\mathbf{M} = \Delta\mathbf{A}. \tag{A5.43}$$

To summarise, the steps of the numerical pole placement procedure are as follows.

Step 1. Create a SIMULINK® block diagram of any system having the desired closed loop characteristic equation such as in Fig. A5.4.
Step 2. Run the appropriate MATLAB® routines to determine the vector of desired characteristic polynomial coefficients, $\mathbf{a}_d$, of the system created in Step 1.
Step 3. Choose the initial test parameter vector. This can be simply $\mathbf{k}_{t1} = \mathbf{0}$.
Step 4. Form the columns, $\boldsymbol{\lambda}_i$, $i = 1, 2, \ldots, n$, of $\Delta\mathbf{K}_t = \lambda\mathbf{I}$ with $\lambda = 1$ and use these to calculate the remaining n test parameter vectors as $\mathbf{k}_{t\,i+1} = \mathbf{k}_{t\,i} + \boldsymbol{\lambda}_i$, $i = 1, 2, \ldots, n$.
Step 5. Apply the $n + 1$ test gain vectors generated by Step 2 and Step 3 to the SIMULINK® block diagram of the closed loop system being designed, and use the appropriate MATLAB® routine to determine the corresponding characteristic polynomial coefficient vectors, $\mathbf{a}_i$, $i = 1, 2, \ldots, n + 1$.
Note that by setting $\mathbf{k}_{t1} = \mathbf{0}$ [Step 3] in (A5.33), the constant vector is $\mathbf{a}_0 = \mathbf{a}_1$.

Step 6. Calculate the coefficient difference vectors, $\Delta \mathbf{a}_i = \mathbf{a}_{i+1} - \mathbf{a}_i$, $i = 1, 2, \ldots, n$, from the polynomial coefficient vectors of Step 5 and then form the matrix, $\Delta \mathbf{A} = [\Delta \mathbf{a}_1 \vdots \Delta \mathbf{a}_2 \vdots \ldots \vdots \Delta \mathbf{a}_n]$.

Step 7. Calculate the vector of desired gains using $\mathbf{k_d} = [\Delta \mathbf{A}]^{-1} [\mathbf{a_d} - \mathbf{a_0}]$.

So the pole assignment algorithm accepts the vector of coefficients,

$$\mathbf{a_d} = \begin{bmatrix} a_{d0} & a_{d1} & \ldots & a_{d,\,n-1} \end{bmatrix}^T \quad (A5.44)$$

of the desired closed loop characteristic polynomial (A5.31) and returns the set of system or subsystem parameters,

$$\mathbf{k} = \begin{bmatrix} k_1 & k_2 & \ldots & k_n \end{bmatrix}^T \quad (A5.45)$$

that yield (A5.44). Each of the elements of $\mathbf{k}$ is a particular system/subsystem parameter allocated by the algorithm user. When forming a control system block diagram for application of the numerical pole assignment procedure, the notation of the parameters affecting the closed loop poles will be different from above. Hence correspondences between these parameters and the parameters of the numerical pole assignment procedure have to be established. This will be seen in the subsequent examples.

A5.2.3 Examples Illustrating Applicability and Non-applicability

A5.2.3.1 PID Control Loops

First, Fig. A5.5 shows two versions of the same control system. In Fig. A5.5a, the PID controller is parameterised in terms of the proportional, integral and derivative gains, K_p, K_I and K_d, while in Fig. A5.5b it is parameterised in terms of the proportional gain, K, the integral action time, T_I, and the derivative action time, T_d.

If $G(s) = b_0 / (s^2 + a_1 s + a_0)$, the characteristic polynomial of version 1 is

$$s^3 + (a_1 + b_0 K_d) s^2 + (a_0 + b_0 K_p) s + b_0 K_I \quad (A5.46)$$

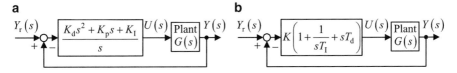

Fig. A5.5 Two versions of a PID control system. (**a**) Version 1 with separate gains (**b**) Version 2 with gain and action times

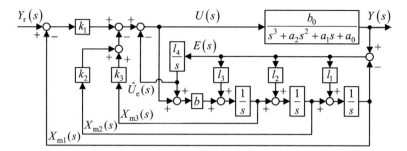

Fig. A5.6 Observer-based robust control of a third order plant without finite zeros

The characteristic polynomial of version 2 is

$$s^3 + (a_1 + b_0 K T_d) s^2 + (a_0 + b_0 K) s + b_0 K / T_1 \quad (A5.47)$$

The three coefficients of (A5.46) are all linear with respect to K_p, K_I and K_d, and therefore the numerical method of subsection 5.2.2 can be applied to version 1. In (A5.47), however, the coefficient of s^2 is nonlinear with respect to K and T_d due to the term, KT_d, and the constant term is nonlinear with respect to K and T_1 due to the term, K/T_1. Version 2 therefore does not qualify but in this case, version 1 can be used for the gain determination and version 2 used for the implementation with its parameters set to $K = K_p$, $T_1 = K_p/K_I$ and $T_d = K_d/K_p$.

A5.2.3.2 Observer-Based Robust Control and an Equivalent Polynomial Control

Figure A5.6 shows the block diagram of an observer-based robust controller (Appendix A10) applied to a third order linear plant without finite zeros.

This control system can be easily designed on the assumption that if the observer gains are sufficiently large through a suitable choice of the observer poles (with the observer disconnected), the error, $e(s)$, is driven to small proportions. If the triple integrator forward path gain, b, is very different from that of the plant, i.e., b_0, this pseudo separation theorem may not hold resulting in closed loop instability. Under these circumstances, the only suitable approach is to design the whole system by pole placement using an available plant model. Equating the determinant of Mason's formula (Appendix A4) to zero yields the closed loop characteristic equation of the system as

$$1 - \left\{ \left[-\frac{bk_3}{s} - \frac{bk_2}{s^2} - \frac{bk_1}{s^3} - \frac{l_1}{s} - \frac{l_2}{s^2} - \frac{l_3}{s^3} - \frac{bl_4}{s^4} \right] \right.$$

$$- \frac{b_0}{s^3 + a_2 s^2 + a_1 s + a_0} \left(\frac{l_4}{s} + \frac{bl_4 k_3}{s^2} + \frac{bl_4 k_2}{s^3} + \frac{bl_4 k_1}{s^4} \right.$$
$$\left. + \frac{l_3 k_3}{s} + \frac{l_3 k_2}{s^2} + \frac{l_3 k_1}{s^3} + \frac{l_2 k_2}{s} + \frac{l_2 k_1}{s^2} + \frac{l_1 k_1}{s} \right) \right\}$$

$$+ \left(-\frac{l_1}{s} \right) \left(-\frac{bk_3}{s} - \frac{bk_2}{s^2} \right) + \left(-\frac{l_2}{s^2} \right) \left(-\frac{bk_3}{s} \right) = 0 \Rightarrow$$

$$(s^3 + a_2 s^2 + a_1 s + a_0) \left[s^4 + (bk_3 + l_1) s^3 + (bk_2 + l_2) s^2 + (bk_1 + l_3) s + bl_4 \right]$$
$$+ b_0 \left[(l_4 + l_3 k_3 + l_2 k_2 + l_1 k_1) s^3 + (bl_4 k_3 + l_3 k_2 + l_2 k_1) s^2 + (bl_4 k_2 + l_3 k_1) s + bl_4 k_1 \right]$$
$$+ bl_1 k_3 s^2 + (bl_1 k_2 + bl_2 k_3) s = 0 \Rightarrow$$

$$s^7 + (a_2 + bk_3 + l_1) s^6 + (a_1 + a_2 (bk_3 + l_1) + (bk_2 + l_2)) s^5$$
$$+ (a_0 + a_1 (bk_3 + l_1) + a_2 (bk_2 + l_2) + (bk_1 + l_3)) s^4$$
$$+ (a_0 (bk_3 + l_1) + a_1 (bk_2 + l_2) + a_2 (bk_1 + l_3) + bl_4 + b_0 (l_4 + l_3 k_3 + l_2 k_2 + l_1 k_1)) s^3$$
$$+ (a_0 (bk_2 + l_2) + a_1 (bk_1 + l_3) + a_2 bl_4 + b_0 (bl_4 k_3 + l_3 k_2 + l_2 k_1) + bl_1 k_3) s^2$$
$$+ (a_0 (bk_1 + l_3) + a_1 bl_4 + b_0 (bl_4 k_2 + l_3 k_1) + b (l_1 k_2 + l_2 k_3)) s$$
$$+ (a_0 bl_4 + b_0 bl_4 k_1) = 0. \tag{A5.48}$$

In this case, the coefficients of s^3, s^2, s all contain product nonlinearities with respect to the observer and model state controller gains. The numerical method to be described therefore cannot be applied to the system of Fig. A5.6 as it stands. If (A5.48) were to be equated to the desired characteristic equation then the solution of the seven nonlinear simultaneous equations for the seven gains would be impossible analytically and onerous numerically. Instead, an alternative but equivalent controller to the observer-based robust controller with the required linear dependence of the closed loop characteristic polynomial coefficients on the controller parameters can be created. This is a polynomial controller (Chap. 5) designed by robust pole assignment (Chap. 10) and including derivative feedforward pre-compensation (Chap. 11). First, the transfer function relationship between $Y(s)$, $Y_r(s)$ and $U(s)$ is derived for the combined observer and model state controller, after disconnecting the plant. Thus, from Fig. A5.6,

$$U(s) = k_1 \frac{\left\{ 1 - \left[-\frac{l_1}{s} - \frac{l_2}{s^2} - \frac{l_3}{s^3} - \frac{bl_4}{s^4} \right] \right\} Y_r(s) - \left(\frac{l_1 k_1}{s} + \frac{l_2 k_2}{s} + \frac{l_2 k_1}{s^2} + \frac{l_3 k_3}{s} + \frac{l_3 k_2}{s^2} + \frac{l_3 k_1}{s^3} + \frac{bl_4 k_3}{s^2} + \frac{bl_4 k_2}{s^3} + \frac{bl_4 k_1}{s^4} + \frac{l_4}{s} \right) Y(s)}{1 - \left[-\frac{l_1}{s} - \frac{l_2}{s^2} - \frac{l_3}{s^3} - \frac{bl_4}{s^4} - \frac{bk_3}{s} - \frac{bk_2}{s^2} - \frac{bk_1}{s^3} \right] + \left(-\frac{l_1}{s} \right) \left(-\frac{bk_3}{s} - \frac{bk_2}{s^2} \right) + \left(-\frac{l_2}{s^2} \right) \left(-\frac{bk_3}{s} \right)}$$

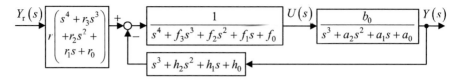

Fig. A5.7 Block diagram of equivalent robust control system

Hence

$$U(s) = \frac{k_1 \left(s^4 + l_1 s^3 + l_2 s^2 + l_3 s + bl_4\right) Y_r(s) - \begin{bmatrix} (l_1 k_1 + l_2 k_2 + l_3 k_3 + l_4) s^3 \\ + (l_2 k_1 + l_3 k_2 + bl_4 k_3) s^2 \\ + (l_2 k_1 + l_3 k_2 + bl_4 k_3) s^2 \\ + (l_3 k_1 + bl_4 k_2) s + bl_4 k_1 \end{bmatrix} Y(s)}{s^4 + (l_1 + bk_3) s^3 + (l_2 + bk_2 + l_1 bk_3) s^2 + (l_3 + bk_1 + bl_1 k_2 + bl_2 k_3) s + bl_4}.$$
(A5.49)

It is evident that (5.49) could be replaced by an equivalent controller with minimal parameterisation, as follows.

$$U(s) = \frac{r\left(s^4 + r_3 s^3 + r_2 s^2 + r_1 s + r_0\right) Y_r(s) - \left(s^3 + h_2 s^2 + h_1 s + h_0\right) Y(s)}{s^4 + f_3 s^3 + f_2 s^2 + f_1 s + f_0}.$$
(A5.50)

Figure A5.7 shows the corresponding block diagram.
Hence,

$$\frac{Y(s)}{Y_r(s)} = \frac{\dfrac{r\left(s^4 + r_3 s^3 + r_2 s^2 + r_1 s + r_0\right)}{s^4 + f_3 s^3 + f_2 s^2 + f_1 s + f_0} \cdot \dfrac{b_0}{s^3 + a_2 s^2 + a_1 s + a_0}}{1 + \dfrac{s^3 + h_2 s^2 + h_1 s + h_0}{s^4 + f_3 s^3 + f_2 s^2 + f_1 s + f_0} \cdot \dfrac{b_0}{s^3 + a_2 s^2 + a_1 s + a_0}} \Rightarrow$$

$$\frac{Y(s)}{Y_r(s)} = \frac{b_0 r \left(s^4 + r_3 s^3 + r_2 s^2 + r_1 s + r_0\right)}{(s^4 + f_3 s^3 + f_2 s^2 + f_1 s + f_0)(s^3 + a_2 s^2 + a_1 s + a_0) + b_0 (s^3 + h_2 s^2 + h_1 s + h_0)}$$

$$= \frac{b_0 r \left(s^4 + r_3 s^3 + r_2 s^2 + r_1 s + r_0\right)}{\begin{array}{l} s^7 + (a_2 + f_3) s^6 + (a_1 + a_2 f_3 + f_2) s^5 \\ + (a_0 + a_1 f_3 + a_2 f_2 + f_1) s^4 + (a_0 f_3 + a_1 f_2 + a_2 f_1 + f_0 + b_0) s^3 \\ + (a_0 f_2 + a_1 f_1 + a_2 f_0 + b_0 h_2) s^2 + (a_1 f_0 + a_0 f_1 + b_0 h_1) s + (a_0 f_0 + b_0 h_0) \end{array}}.$$
(A5.51)

where the parameters, f_i, $i = 0, 1, 2, 3$ and h_i, $i = 0, 1, 2$, can be adjusted to place the seven closed loop poles (four relatively large ones for the observer, that can be coincident, and three dominant ones for the desired closed loop dynamics). Thus, if T_s and T_{so} are, respectively, the specified control system settling time and the observer settling time (5 % criterion), the desired characteristic polynomial, whose coefficients are required for the numerical procedure, is

$$D(s) = (s+p)^3(s+q)^4 = \left(s^3 + 3ps^2 + 3p^2s + p^3\right)\left(s^4 + 4qs^3 + 6q^2s^2 + 4q^3s + q^4\right)$$

Hence

$$D(s) = s^7 + (3p + 4q)s^6 + \left(3p^2 + 12pq + 6q^2\right)s^5$$
$$+ \left(p^3 + 12p^2q + 18pq^2 + 4q^3\right)s^4 + \left(4p^3q + 18p^2q^2 + 12pq^3 + q^4\right)s^3$$
$$+ \left(6p^3q^2 + 12p^2q^3 + 3pq^4\right)s^2 + \left(4p^3q^3 + 3p^2q^4\right)s + p^3q^4.$$
(A5.52)

where $p = 1.5(1+n)/T_s|_{n=3} = 6/T_s$ and $q = 1.5(1+n)/T_{so}|_{n=4} = 7.5/T_{so}$. The reference input polynomial coefficients, r_i, $i = 0, 1, 2$, can be chosen so that the four zeros of the numerator polynomial cancel the aforementioned poles associated with the observer, to ensure a closed loop dynamics of third order character, as would occur with the original observer based robust controller. This simply means setting

$$s^4 + r_3s^3 + r_2s^2 + r_1s + r_0 = s^4 + 4qs^3 + 6q^2s^2 + 4q^3s + q^4 \Rightarrow$$
$$\underline{r_0 = q^4}, \ \underline{r_1 = 4q^3}, \ \underline{r_2 = 6q^2} \text{ and } \underline{r_3 = 4q}$$
(A5.53)

Most importantly, the coefficients of the closed loop characteristic polynomial of (A5.51) are linear with respect to the adjustable controller parameters and therefore the equivalent robust controller parameters, d_i, $i = 0, 1, 2, 3$ and f_i, $i = 0, 1, 2$, can be determined using the numerical method to be described. Once this has been done, the reference input scaling coefficient, r, is set to achieve a unity closed loop DC gain. Thus, by setting $s = 0$ in (A5.51),

$$\frac{b_0 r r_0}{a_0 f_0 + b_0 h_0} = 1 \Rightarrow r = \underline{\frac{a_0 f_0 + b_0 h_0}{b_0 r_0}}$$
(A5.54)

Figure A5.8 shows a block diagram for implementation that avoids the differentiations of the output and reference input indicated by (5.50) and Fig. A5.7, with the aid of the observer canonical form (Chaps. 1 and 4).

Application of Mason's formula to Fig. A5.8 with the plant removed confirms that the correct controller transfer function relationship of (A5.50) is realised.

The numerical method requires the user to start with a MATLAB®-SIMULINK® diagram of the control system or subsystem concerned and it is important to check that the relationship between the set of adjustable parameters and the set of coefficients of the characteristic polynomial is linear. This can be done by inspecting the block diagram as if the determinant of Mason's formula was to be formed. Every loop in the diagram must contain just one adjustable parameter as a multiplying gain, as in Fig. A5.8. Every set of non-touching loops must contain just one adjustable parameter, otherwise products of these parameters will occur in the characteristic polynomial coefficients, which is inadmissible.

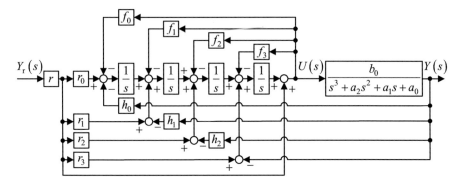

Fig. A5.8 Block diagram of equivalent robust polynomial control system for implementation

For the system of Fig. A5.8, the desired characteristic polynomial vector is obtained from (A5.52) as

$$\mathbf{a}_d = \begin{bmatrix} a_{d0} \\ a_{d1} \\ a_{d2} \\ a_{d3} \\ a_{d4} \\ a_{d5} \\ a_{d6} \end{bmatrix} = \begin{bmatrix} p^3 q^4 \\ 4p^3 q^3 + 3p^2 q^4 \\ 6p^3 q^2 + 12p^2 q^3 + 3pq^4 \\ 4p^3 q + 18p^2 q^2 + 12pq^3 + q^4 \\ p^3 + 12p^2 q + 18pq^2 + 4q^3 \\ 3p^2 + 12pq + 6q^2 \\ 3p + 4q \end{bmatrix} \quad (A5.55)$$

and the controller parameter vector, whose element order is not critical but must be maintained, is

$$\mathbf{k} = \begin{bmatrix} k_1 & k_2 & k_3 & k_4 & k_5 & k_6 & k_7 \end{bmatrix}^T = \begin{bmatrix} h_0 & h_1 & h_2 & f_0 & f_1 & f_2 & f_3 \end{bmatrix}^T. \quad (A5.56)$$

The dimensions of the vectors and matrices of the numerical pole assignment procedure is always equal to the system order.

A5.3 Routh Stability Criterion

A5.3.1 Applicability

Given the characteristic polynomial of a continuous LTI system, E J Routh's stability criterion is a set of mathematical operations on the polynomial coefficients for determination of the number of unstable poles, i.e., poles in the right half of the s-plane.

In continuous model based LTI control system design, the closed loop characteristic polynomial is set to yield not only a stable system but one having a specified transient performance. In practice, however, the plant model cannot be perfect. It is therefore necessary to investigate the effects of the worst case extremes of plant mismatching with respect to the plant model. For this, the closed loop system may be modelled with the controller parameters calculated using the nominal plant model while the closed loop transfer function is determined using the real plant modelled with the worst case errors for the particular application. Then simulations would enable the effects of the worst case plant modelling errors to be predicted.

From an analytical viewpoint, the coefficients of the closed loop characteristic polynomial will be known enabling the closed loop dynamic characteristics to be deduced by determining the closed loop poles, which are the roots of the closed loop characteristic equation. Nowadays these can be determined numerically by root finding algorithms. Before the advent of digital computers, however, this task was tedious and time consuming for systems of third and higher order but Routh's stability criterion, which enables the question to be answered of whether or not an LTI system is stable, was available from the late nineteenth century. It is presented in this appendix as, as it can be useful today for relatively simple LTI systems of greater than second order, if only stability analysis is required.

A5.3.2 The Criterion

Consider any continuous LTI system with a characteristic polynomial,

$$P(s) = a_n s^n + a_{n-1} s^{n-1} + \cdots + a_1 s + a_0, \; a_n \neq 0, \; a_0 \neq 0. \qquad (A5.57)$$

Then the Routh array is formed as follows.

Row	Array				
s^n	a_n	a_{n-2}	a_{n-4}	$\cdots$	0
s^{n-1}	a_{n-1}	a_{n-3}	a_{n-5}	$\cdots$	0
s^{n-2}	b_1	b_2	b_3	$\cdots$	0
s^{n-3}	c_1	c_2	c_3	$\cdots$	0
$\vdots$	$\vdots$	$\vdots$	$\vdots$		$\vdots$
s^1					0
s^0					0

The terms, s^n, s^{n-1}, ..., shown in the column labelled 'row' are a traditional way of labelling the $n + 1$ rows. They are labelled in this way as the elements of these rows are actually the coefficients of a set of polynomials of successively reducing degree, i.e.,

$$\left.\begin{array}{l} a_n s^n + a_{n-2} s^{n-2} + a_{n-3} s^{n-3} + \cdots \\ a_{n-1} s^{n-1} + a_{n-2} s^{n-3} + a_{n-3} s^{n-5} + \cdots \\ b_1 s^{n-2} + b_2 s^{n-4} + \cdots \\ c_1 s^{n-2} + c_2 s^{n-4} + \cdots \\ \vdots \\ p_1 s^1 \\ q_1 s^0 \end{array}\right\}. \qquad (A5.58)$$

These polynomials originate from Sturm's theorem upon which Routh's criterion is based.

The first two rows are formed from the characteristic polynomial coefficients, as shown, and these 'seed' the process of forming the remaining coefficients of the Routh array. Also, a final column of zeros is formed alongside the columns formed from the polynomial coefficients. The third row is formed from the first and second rows by means of the following set of determinants.

$$b_1 = -\frac{1}{a_{n-1}} \begin{vmatrix} a_n & a_{n-2} \\ a_{n-1} & a_{n-3} \end{vmatrix} = \frac{a_{n-2} a_{n-1} - a_n a_{n-3}}{a_{n-1}}, \qquad (A5.59)$$

$$b_2 = -\frac{1}{a_{n-1}} \begin{vmatrix} a_n & a_{n-4} \\ a_{n-1} & a_{n-5} \end{vmatrix} = \frac{a_{n-4} a_{n-1} - a_n a_{n-5}}{a_{n-1}}, \qquad (A5.60)$$

$$b_3 = -\frac{1}{a_{n-1}} \begin{vmatrix} a_n & a_{n-6} \\ a_{n-1} & a_{n-8} \end{vmatrix} = \frac{a_{n-6} a_{n-1} - a_n a_{n-8}}{a_{n-1}}. \qquad (A5.61)$$

and so forth. The fourth row is similarly formed from the second and third rows. Thus

$$c_1 = -\frac{1}{b_1} \begin{vmatrix} a_{n-1} & a_{n-3} \\ b_1 & b_2 \end{vmatrix} = \frac{a_{n-3} b_1 - a_{n-1} b_2}{b_1}, \qquad (A5.62)$$

$$c_2 = -\frac{1}{b_1} \begin{vmatrix} a_{n-1} & a_{n-5} \\ b_1 & b_3 \end{vmatrix} = \frac{a_{n-5} b_1 - a_{n-1} b_3}{b_1}, \qquad (A5.63)$$

$$c_3 = -\frac{1}{b_1} \begin{vmatrix} a_{n-1} & a_{n-7} \\ b_1 & b_4 \end{vmatrix} = \frac{a_{n-7} b_1 - a_{n-1} b_4}{b_1}. \qquad (A5.64)$$

and so forth. This is repeated until the last two rows are reached, which have single elements. Once the array is completed, then the signs of the sequence of first column coefficients, $a_n, a_{n-1}, b_1, c_1, \ldots, p_1, q_1$, are examined. If all the signs are the same, then all n roots of $P(s) = 0$ lie in the left half of the s-plane and therefore the system is stable. Any sign changes in this sequence of coefficients indicates at least one unstable root, i.e., a root in the right half of the s-plane, meaning instability. In fact, the total number of sign changes in the sequence is equal to the number of unstable roots.

There are, however, two occurrences for which the procedure has to be modified.

1. The first element of a row is zero and at least one other element in that row is non-zero. In this case, the modification is to first replace the first zero element by an infinitesimal element, ε, whose sign is the same as that of the element above. Then the procedure is continued as described above. When completed, the sign changes are counted in the limit as $\varepsilon \to 0$.
2. All elements of a row are zero. This will always be a row whose elements are the coefficients of a polynomial of odd degree in (A5.58). Then the row above the row of zeros are those of a polynomial, $Q(s)$, of even degree, which is referred to as the *auxiliary polynomial*. In this case, the row of zeros is replaced by the coefficients of the polynomial, $\frac{dQ(s)}{ds}$, and the procedure described above continued.

It should be noted that the roots of $Q(s) = 0$ always occur in equal and opposite real or imaginary pairs. Also $Q(s)$ is a factor of $P(s)$, so it is possible to apply the Routh criterion to the remainder polynomial, $R(s) = P(s)/Q(s)$, to complete the analysis, if $R(s)$ of third or higher degree.

A8 Appendix to Chap. 8

A8.1 An Approach for State Estimation for Nonlinear Plants

A8.1.1 Introduction

The success of the Luenberger observer (Chap. 8) is based on the guaranteed convergence of the estimated state to the plant state for an arbitrary initial state mismatch. It will be recalled that through the separation principle it is possible to form an error state differential equation with adjustable parameters that can be set to guarantee the convergence. This, however, is only valid for linear plants. Nonlinear plants present a greater challenge when attempting the same basic approach.

The most general plant state space equations to be considered are as follows.

$$\dot{\mathbf{x}} = \mathbf{f}(\mathbf{x}, \mathbf{u} + \mathbf{n}_p) \tag{A8.1}$$

and

$$\mathbf{y} = \mathbf{h}(\mathbf{x}) + \mathbf{n}_m, \tag{A8.2}$$

where $\mathbf{x} \in \Re^n$ is the state vector to be estimated, $\mathbf{u} \in \Re^m$ is the state vector, $\mathbf{y} \in \Re^m$ is the measurement vector, $\mathbf{f}(\cdot)$ and $\mathbf{h}(\cdot)$ are continuous and differentiable functions of their arguments, $\mathbf{n}_p \in \Re^m$ is the plant noise vector and $\mathbf{n}_m \in \Re^m$ is the measurement noise vector. If a conventional observer structure is attempted, then, assuming a perfectly known plant model, the observer equation corresponding to (A8.1) and (A8.2) would be

$$\dot{\hat{\mathbf{x}}} = \mathbf{f}(\hat{\mathbf{x}}, \mathbf{u}) + \mathbf{L}\left[\mathbf{h}(\mathbf{x}) - \mathbf{h}(\hat{\mathbf{x}}) + \mathbf{n}_m\right]. \tag{A8.3}$$

where $\hat{\mathbf{x}}$ is the state estimate and $\mathbf{L}$ is the model correction loop gain matrix. The differential equation for the state estimation error, $\boldsymbol{\varepsilon} = \hat{\mathbf{x}} - \mathbf{x}$, would then be obtained by subtracting (A8.1) from (A8.3) to yield

$$\dot{\boldsymbol{\varepsilon}} = \mathbf{f}(\hat{\mathbf{x}}, \mathbf{u}) - \mathbf{f}(\mathbf{x}, \mathbf{u} + \mathbf{n}_p) + \mathbf{L}\left[\mathbf{h}(\mathbf{x}) - \mathbf{h}(\hat{\mathbf{x}}) + \mathbf{n}_m\right]. \tag{A8.4}$$

It is not possible to express the RHS in terms of $\boldsymbol{\varepsilon}$ to yield a differential equation of the form, $\dot{\boldsymbol{\varepsilon}} = \mathbf{q}(\mathbf{e})$, due to the nonlinear functions. Hence the separation principle does not apply. The observer may well work for some $\mathbf{L}$ and over certain regions of the plant and model state spaces, but it is difficult to mathematically guarantee $\boldsymbol{\varepsilon} \to \mathbf{0}$ as $t \to \infty$ in absence of the noise signals.

The following three subsections present different solutions to the nonlinear state estimation problem with different limitations that may be taken into account when considering a specific application. The observer of Sect. A8.1.2 is applicable to any plant modelled by (A8.1) and (A8.2), but the plant and model states are restricted to a region in the neighbourhood of a selected operating point to guarantee convergence of the estimated state towards the plant state. The state estimator of Sect. A8.1.3 is free of this restriction but is subject to measurement noise amplification that increases with frequency. The observer of Sect. A8.1.4 combines the techniques of Sects. A8.1.2 and A8.1.3 to achieve dynamic lag free filtering without the operating point restriction.

A8.1.2 Observer Based on Linearised Plant Model

The traditional approach to estimating the state of a plant with continuous nonlinearities is to first linearise the plant model about a selected operating point, $(\overline{\mathbf{x}}, \overline{\mathbf{y}}, \overline{\mathbf{u}})$, using the method of Chap. 7, Sect. 7.2, yielding

$$\dot{\tilde{\mathbf{x}}} = \overline{\mathbf{A}}\tilde{\mathbf{x}} + \overline{\mathbf{B}}\tilde{\mathbf{u}}$$
$$\tilde{\mathbf{y}} = \overline{\mathbf{C}}\tilde{\mathbf{x}}. \tag{A8.5}$$

where $\tilde{\mathbf{x}} = \mathbf{x} - \overline{\mathbf{x}}$, $\tilde{\mathbf{y}} = \mathbf{y} - \overline{\mathbf{y}}$, $\tilde{\mathbf{u}} = \mathbf{u} - \overline{\mathbf{u}}$. The matrices, $\overline{\mathbf{A}}$, $\overline{\mathbf{B}}$ and $\overline{\mathbf{C}}$ are, respectively, the Jacobians, $\frac{\partial \dot{\mathbf{x}}}{\partial \mathbf{x}}$, $\frac{\partial \dot{\mathbf{x}}}{\partial \mathbf{u}}$ and $\frac{\partial \mathbf{y}}{\partial \mathbf{x}}$, evaluated at the operating point. The observer is then linear, being defined by the equation,

$$\dot{\hat{\mathbf{x}}} = \overline{\mathbf{A}}\hat{\mathbf{x}} + \overline{\mathbf{B}}\tilde{\mathbf{u}} + \mathbf{L}\left(\tilde{\mathbf{y}} - \overline{\mathbf{C}}\hat{\mathbf{x}}\right), \tag{A8.6}$$

and can be designed using the methods of the previous sections. It is important to remember, however, that the state estimate, $\hat{\mathbf{x}}$, is only an estimate of $\tilde{\mathbf{x}} = \mathbf{x} - \overline{\mathbf{x}}$. So this method is only suitable for control systems in which the reference input is constant or piecewise constant with gain scheduling and corresponding model updating for the different operating points. This method could also be implemented using the discrete version of the linearised plant model, as in Chap. 8, Sect. 8.3.

For applications requiring larger state excursions, it might be possible to employ observer (A8.3) instead of (A8.6), but with the same gain matrix, **L**, as in (A8.6). The extent of departure of the plant and model states from the operating point that would be tolerable, however, would have to be explored by simulation.

A8.1.3 Output Derivative Based State Estimator

A8.1.3.1 Introduction

The state estimator developed here circumvents the restrictions of an operating point by generating a state estimate that is independent of the plant model. Estimates of all the plant output derivatives are provided up to the maximum orders for which they are state variables. The method is therefore restricted to plants of full relative degree (Chap. 3, Sect. 3.2.10). As previously stated, however, extreme care must be taken to ensure that the measurement noise levels are not so high that their amplification at high frequencies due to the differentiation is unacceptable. The method presented here embodies low pass filtering to alleviate this problem but caution has to be taken to avoid destabilising the control loops by setting the filter cut-off frequencies too low. If the noise levels are still too high after the filtering, then dynamic lag free filtering with lower cut-off frequencies may be introduced by means of the observer of Sect. A8.1.4 that uses the outputs of the state estimator of this subsection as if they were raw measurements.

To clarify the notation used in this subsection, the plant model of (A8.1) and (A8.2) will be rewritten as follows.

$$\dot{\mathbf{x}} = \mathbf{f}(\mathbf{x}, \mathbf{u} + \mathbf{n}_p) \tag{A8.7}$$

$$\mathbf{z} = \mathbf{h}(\mathbf{x}), \quad \mathbf{y} = \mathbf{z} + \mathbf{n}_m. \tag{A8.8}$$

A8.1.3.2 The Filtered Multiple Derivative State Estimator

With reference to (A8.8), z_i, $i = 1, 2, \ldots, m$, are the hypothetical plant outputs that would occur without the measurement noise. The state estimator of this subsection produces estimates of the state variables, $d^q z_i / dt^q$, $i = 1, 2, \ldots, m$, $q = 1, 2, \ldots, r_i - 1$, where r_i is the relative degree with respect to z_i. These, together with z_i, constitute the plant state, $\mathbf{x}_d$, the state representation of which will be called the output derivative state representation. The estimates are formed using the available outputs, $y_i = z_i + n_{mi}$, $i = 1, 2, \ldots, m$, by combined differentiators and low pass filters to prevent high frequency components of the measurement noise being amplified to unacceptable levels. The resulting filtered state estimate will be denoted, $\mathbf{x}_{df}$. The transfer functions between the outputs and the derivatives are

$$\frac{\widehat{Y}_i^{(q)}(s)}{Y_i(s)} = \left(\frac{s}{1 + sT_f}\right)^q, \quad q = 1, 2, \ldots, r_i - 1, \tag{A8.9}$$

where T_f is the filtering time constant. Then, as $T_f \to 0$, $\widehat{y}_i^{(q)}(t) \to y_i^{(q)}(t)$, indicating ideal differentiation but in practice, $T_f > 2h$, where h is the sampling/iteration period of the digital processor (Chap. 6). The block diagram of the approximate differentiator for each output, which may easily be supported by an implementation medium such as dSPACE®, is shown in Fig. A8.1.

It is important to realise that the low pass filters introduce a dynamic lag between $\mathbf{x}_d(t)$ and $\mathbf{x}_{df}(t)$. Also, when the control system is turned on, any non-zero quantity, x, at a differentiator input will be treated as a step input and therefore cause a large transient spike of magnitude, x/T_f, that could be harmful. In the implementation, therefore, the recommendation is to block out the differentiator outputs (i.e., multiply them by zero) for a period of $5T_f$ at system start-up to ensure that these transient spikes are not allowed to propagate through the system.

Appendices

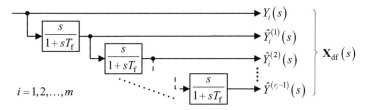

Fig. A8.1 Multiple filtered approximate differentiator

In the frequency domain, for the hypothetical case of ideal continuous differentiators, the noise amplification factor would be

$$G_{nr}(\omega) = \left| \frac{Y^{(r)}(j\omega)}{Y(j\omega)} \right| = \omega^r, \qquad (A8.10)$$

indicating noise amplification that would reach extreme values as the angular frequency, ω, increases. With the filtering, the noise amplification factor is

$$G_{nfr}(\omega) = \left| \frac{\widehat{Y}^{(r)}(j\omega)}{Y(j\omega)} \right| = \frac{\omega^r}{\left(1 + \omega^2 T_f^2\right)^{r/2}}, \qquad (A8.11)$$

showing that making the order of the filtering equal to the order of the derivative limits the noise amplification factor to a finite value as $\omega \to \infty$. There is a trade-off, however, in that as T_f is reduced, without the measurement noise the derivatives become more accurate, but the maximum noise level amplification factor is

$$\lim_{\omega \to \infty} G_{nfr}(\omega) = \frac{1}{T_f^r} \qquad (A8.12)$$

and therefore becomes large as T_f is reduced, increasing exponentially with r for a given T_f. The method therefore requires low measurement noise levels.

A8.1.3.3 Transformation to the State Representation of the Plant Model

A plant model formulated with the required state representation may be used to determine a transformation from the output derivative state representation of Sect. A8.1.3.2. This will be formulated ignoring the dynamic lag due to the filtering in Sect. A8.1.3.2 and is the inverse of the transformation to the output derivative state representation formed as a by-product of the process of determining the relative degree of a plant with respect to each output (Sect. 3.2.10 OF Chap. 3). Taking this plant model to be that of (A8.7) and (A8.8), with $\mathbf{n}_p = \mathbf{0}$ and $\mathbf{n}_m = \mathbf{0}$, it will be recalled that the qth derivative of the ith output is the qth order Lie derivative of $h_i(\mathbf{x})$ along $\mathbf{f}(\mathbf{x}, \mathbf{u})$, i.e., $y_i^{(q)} = L_\mathbf{f}^q h_i(\mathbf{x})$, and that the relative degree is the order, r_i, of the lowest derivative that directly depends upon **u**. For notational simplification, $L_f^q h_i(\mathbf{x})$ will be written as $h_{iq}(\mathbf{x})$. Then the set of derivative state variables is

$$x_{diq} = y_i^{(q)} = h_{iq}(\mathbf{x}), \, i = 1, 2, \ldots, m, \, q = 0, 1, \ldots, r_i - 1. \tag{A8.13}$$

Since this set of state variables is complete, $\sum_{i=1}^{m} r_i = n$. Let these variables be assembled to form a derivative state vector, $\mathbf{x}_d$. Then (A8.13) can be written

$$\mathbf{x}_d = \mathbf{h}_d(\mathbf{x}) \tag{A8.14}$$

This is a set of n equations that can be solved for $\mathbf{x}$, yielding

$$\mathbf{x} = \mathbf{h}_d^{-1}(\mathbf{x}_d). \tag{A8.15}$$

In the real system, the plant and measurement noise signals are present and therefore $\mathbf{x}_d$ in (A8.15) has to be replaced by $\mathbf{x}_{df}$ from the filtered multiple derivative state estimator of Sect. 8.1.3.2, the resulting estimate of $\mathbf{x}$ being denoted, $\mathbf{x}_f$. Thus

$$\mathbf{x}_f = \mathbf{h}_d^{-1}(\mathbf{x}_{df}). \tag{A8.16}$$

It should be noted that since the state estimator of Sect. A8.1.3.2 introduces a dynamic lag between $\mathbf{x}_d(t)$ and $\mathbf{x}_{df}(t)$, there must be a corresponding dynamic lag between $\mathbf{x}(t)$ and $\mathbf{x}_f(t)$. Provided the noise levels are sufficiently low, however, the filtering time constant, T_f, can be reduced to make the dynamic lag negligible.

A8.1.4 The Output Derivative Based Filtering Observer

A8.1.4.1 Introduction

This observer embodies the output derivative based state estimator of Sect. 8.1.3 and is therefore restricted to plants of full relative degree. The output, $\mathbf{x}_f$, of this estimator is treated as a measurement vector, which potentially contains high noise levels with significant measurement noise levels. These state measurements are filtered without dynamic lag by means of an observer with an unconventional structure that yields convergence of the estimated state to the real state with prescribed first order dynamics, commencing with an arbitrarily large state estimation error, despite the plant being nonlinear.

A8.1.4.2 Formulation of Observer

The general plant state space Eqs. (A8.1) and (A8.2) are repeated for convenience. Thus

$$\dot{\mathbf{x}} = \mathbf{f}(\mathbf{x}, \, \mathbf{u} + \mathbf{n}_p) \tag{A8.17}$$

and
$$\mathbf{y} = \mathbf{h}(\mathbf{x}) + \mathbf{n}_m. \quad (A8.18)$$

The real time plant model on which the observer is based is
$$\dot{\widehat{\mathbf{x}}} = \mathbf{f}(\widehat{\mathbf{x}}, \mathbf{u}) \quad (A8.19)$$

and
$$\widehat{\mathbf{y}} = \mathbf{h}(\widehat{\mathbf{x}}) \quad (A8.20)$$

The filtering observer is first formulated as
$$\dot{\widehat{\mathbf{x}}} = \mathbf{f}(\widehat{\mathbf{x}}, \mathbf{u}) + \mathbf{v} \quad (A8.21)$$

where v_i is the correction input of the real time model integrator whose output is $\widehat{x}_i$. The state estimation error is then
$$\boldsymbol{\varepsilon} = \widehat{\mathbf{x}} - \mathbf{x} \quad (A8.22)$$

Differentiating (A8.22) and substituting for $\dot{\widehat{\mathbf{x}}}$ and $\dot{\mathbf{x}}$ using (A8.21) and (A8.17) gives
$$\dot{\boldsymbol{\varepsilon}} = \dot{\widehat{\mathbf{x}}} - \dot{\mathbf{x}} = \mathbf{f}(\widehat{\mathbf{x}}, \mathbf{u}) - \mathbf{f}(\mathbf{x}, \mathbf{u} + \mathbf{n}_p) + \mathbf{v}. \quad (A8.23)$$

It will now be supposed that a control law for $\mathbf{v}$ exists yielding first order error dynamics for each error component. Thus
$$\dot{\boldsymbol{\varepsilon}} = -\boldsymbol{\Lambda}\boldsymbol{\varepsilon}, \quad (A8.24)$$

where
$$\boldsymbol{\Lambda} = \mathrm{diag}\left(\frac{3}{T_{so1}}, \frac{3}{T_{so2}}, \ldots, \frac{3}{T_{son}}\right). \quad (A8.25)$$

The individual settling times, T_{soi}, $i = 1, 2, \ldots, n$, may be chosen as desired. Equating the RHS of (A8.23) and (A8.24), then solving for $\mathbf{v}$ yields
$$\mathbf{v} = \mathbf{f}(\mathbf{x}, \mathbf{u} + \mathbf{n}_p) - \mathbf{f}(\widehat{\mathbf{x}}, \mathbf{u}) - \boldsymbol{\Lambda}\boldsymbol{\varepsilon} \quad (A8.26)$$

Substituting for $\mathbf{v}$ in (A8.23) using control law (A8.26) then yields
$$\dot{\widehat{\mathbf{x}}} = \mathbf{f}(\mathbf{x}, \mathbf{u} + \mathbf{n}_p) - \boldsymbol{\Lambda}\boldsymbol{\varepsilon} \quad (A8.27)$$

The final step is to replace $\mathbf{x}$ in (A8.27) by the estimated state, $\mathbf{x}_f$, from the output derivative based state estimator of Sect. A8.1.3, replace the state estimation error, $\boldsymbol{\varepsilon}$, by $\widehat{\mathbf{x}} - \mathbf{x}_f$ and replace the unknown plant noise signal, $\mathbf{n}_p$, by its mean value, which is assumed to be zero. The filtering observer equation is then

$$\dot{\widehat{\mathbf{x}}} = \mathbf{f}(\mathbf{x}_f, \mathbf{u}) - \Lambda\left(\widehat{\mathbf{x}} - \mathbf{x}_f\right). \tag{A8.28}$$

The second term in (A8.28) is similar to that of a conventional observer and the noise content, $\mathbf{n}_f$, of $\mathbf{x}_f$ will be low pass filtered in the usual way. Although the nonlinear term also contains a noise component, $\mathbf{n}_n$, originating from $\mathbf{x}_f$ and $\mathbf{u}$, this will be low pass filtered due to the RHS being integrated to obtain $\widehat{\mathbf{x}}$. This may be seen readily by expressing (A8.28) as

$$\dot{\widehat{\mathbf{x}}} = \mathbf{f}(\mathbf{x}, \mathbf{u}) + \mathbf{n}_n - \Lambda\left(\widehat{\mathbf{x}} - \mathbf{x} - \mathbf{n}_f\right) \Rightarrow \widehat{\mathbf{x}} = \int \left[\mathbf{f}(\mathbf{x}, \mathbf{u}) + \mathbf{n}_n - \Lambda\left(\widehat{\mathbf{x}} - \mathbf{x} - \mathbf{n}_f\right)\right] \mathrm{d}t. \tag{A8.29}$$

In this observer, all the nonlinear terms have been excluded from the model correction loop, leaving only the set of n plant model integrators, each with separate correction loops. The settling times of these correction loops may not be very critical in cases where the state estimation noise due to $\mathbf{n}_n$ exceeds that due to $\mathbf{n}_f$ as Λ will not affect the filtering of $\mathbf{n}_n$, otherwise it would be advantageous to reduce Λ to values for which the noise component in $\widehat{\mathbf{x}}$ due to $\mathbf{n}_n$ dominates over that due to $\mathbf{n}_f$.

The general structure of the observer is shown in Fig. A8.2. Since the correction loop is only closed around the set of n integrators of the plant model, which constitute a linear subsystem of the plant model, stable correction loop behaviour is guaranteed for arbitrarily large initial state estimation errors.

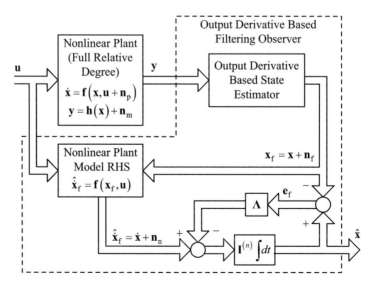

Fig. A8.2 Structure of output derivative based filtering observer

Appendices

Example A8.1 State estimation for large angle three axis spacecraft attitude control

A three axis stabilised unmanned spacecraft has an attitude measurement system consisting of a three axis star sensor with an on-board star map providing star coordinates from which the spacecraft attitude coordinates are calculated using the quaternion based attitude representation. Rigid body dynamics is assumed. The three mutually orthogonal control axes are aligned with the principal axes of inertia. Also the gyroscopic inter-axis coupling is assumed to be due entirely to the angular momentum vector of the reaction wheel set providing the control torque components (Sect. 2.2.4 of Chap. 2). The spacecraft dynamics differential equations are as follows.

$$\begin{bmatrix} J_{xx} & 0 & 0 \\ 0 & J_{yy} & 0 \\ 0 & 0 & J_{zz} \end{bmatrix} \begin{bmatrix} \dot{\omega}_x \\ \dot{\omega}_y \\ \dot{\omega}_z \end{bmatrix} + \begin{bmatrix} 0 & -\omega_z & \omega_y \\ \omega_z & 0 & -\omega_x \\ -\omega_y & \omega_x & 0 \end{bmatrix} \begin{bmatrix} l_x \\ l_y \\ l_z \end{bmatrix} = K_w \begin{bmatrix} u_x \\ u_y \\ u_z \end{bmatrix} \quad \text{(A8.30)}$$

$$\begin{bmatrix} \dot{l}_x \\ \dot{l}_y \\ \dot{l}_z \end{bmatrix} = -K_w \begin{bmatrix} u_x \\ u_y \\ u_z \end{bmatrix} \quad \text{(A8.31)}$$

where J_{xx}, J_{yy} and J_{zz} are the principal axis moments of inertia, l_x, l_y and l_z are the components of the angular momentum vector of the reaction wheel set along the principal axes, x, y and z, K_w is the reaction wheel torque constant, u_x, u_y and u_z, are reaction wheel input voltages and ω_x, ω_y and ω_z are the body angular velocities. The kinematic differential equation is

$$\begin{bmatrix} \dot{q}_0 \\ \dot{q}_1 \\ \dot{q}_2 \\ \dot{q}_3 \end{bmatrix} = \frac{1}{2} \begin{bmatrix} 0 & -\omega_x & -\omega_y & -\omega_z \\ \omega_x & 0 & \omega_z & -\omega_y \\ \omega_y & -\omega_z & 0 & \omega_x \\ \omega_z & \omega_y & -\omega_x & 0 \end{bmatrix} \begin{bmatrix} q_0 \\ q_1 \\ q_2 \\ q_3 \end{bmatrix} \quad \text{(A8.32)}$$

where q_i, $i = 0, 1, 2, 3$, are the attitude quaternions. It should be noted that the three attitude coordinates comprise q_1, q_2 and q_3, the solution to (A8.32) obey

$$q_0^2 + q_1^2 + q_2^2 + q_3^2 = Q^2 \quad \text{(A8.33)}$$

where Q is a constant, which is made equal to unity by setting

$$[q_0(0), q_1(0), q_2(0), q_3(0)] = [1, 0, 0, 0] \Rightarrow q_0 = \left[1 - \left(q_1^2 + q_2^2 + q_3^2\right)\right]^{\frac{1}{2}}. \quad \text{(A8.34)}$$

This constraint is due to the spacecraft body only having three rotational degrees of freedom. Despite q_0 being redundant as an attitude coordinate, it is retained as a measurement from the plant as this will avoid its calculation using (A8.34).

It will be assumed that the reaction wheel speeds are measured and since their rotor moments of inertia will be known, the angular momenta, l_x, l_y and l_z, can be treated as known measurements. Equation (A8.31) will therefore not be included in the real time model of the observer. The attitude quaternions, q_i, $i = 0, 1, 2, 3$, are the measurements derived from the star sensor data. For the purpose of deriving the observer equations, the measurement noise and plant noise signals are set to zero and then, using standard notation for control systems, the plant state space model is as follows.

$$\dot{x}_0 = -0.5\,(x_4 x_1 + x_5 x_2 + x_6 x_3), \tag{A8.35}$$

$$\dot{x}_1 = 0.5\,(x_4 x_0 + x_6 x_2 - x_5 x_3), \tag{A8.36}$$

$$\dot{x}_2 = 0.5\,(x_5 x_0 - x_6 x_1 + x_4 x_3), \tag{A8.37}$$

$$\dot{x}_3 = 0.5\,(x_6 x_0 + x_5 x_1 - x_4 x_2), \tag{A8.38}$$

$$\dot{x}_4 = a_1\,(l_2 x_6 - l_3 x_5) + b_1 u_1, \tag{A8.39}$$

$$\dot{x}_5 = a_2\,(l_3 x_4 - l_1 x_6) + b_2 u_2, \tag{A8.40}$$

$$\dot{x}_6 = a_3\,(l_1 x_5 - l_2 x_4) + b_3 u_3, \tag{A8.41}$$

$$\dot{l}_1 = -K_w u_1, \tag{A8.42}$$

$$\dot{l}_2 = -K_w u_2, \tag{A8.43}$$

$$\dot{l}_3 = -K_w u_3, \tag{A8.44}$$

$$y_0 = x_0 \tag{A8.45}$$

$$y_1 = x_1, \tag{A8.46}$$

$$y_2 = x_2 \tag{A8.47}$$

and

$$y_3 = x_3, \tag{A8.48}$$

where $x_i = q_i$, $i = 0, 1, 2, 3$, $x_4 = \omega_x$, $x_5 = \omega_y$, $x_6 = \omega_z$, $l_1 = l_x$, $l_2 = l_y$, $l_3 = l_z$, $a_1 = 1/J_{xx}$, $a_2 = 1/J_{yy}$, $a_3 = 1/J_{zz}$, $b_1 = K_w/J_{xx}$, $b_2 = K_w/J_{yy}$, $b_3 = K_w/J_{zz}$.

Next, to derive the output derivative state transformation, it is evident by inspection of (A8.35) through (A8.48) that the relative degrees of the plant with respect to y_1, y_2 and y_3 are $r_1 = r_2 = r_3 = 2$, since differentiating (A8.46), (A8.47) and (A8.48) twice and substituting for the state derivatives using (A8.36), (A8.37) and (A8.38) after the first differentiation and using (A8.35) through (A8.41) after the second differentiation reveals direct dependence of the second output derivatives on the control inputs, u_1, u_2 and u_3. In this case, the output derivative state transformation equations consist of (A8.46), (A8.47) and (A8.48) together with

$$\dot{y}_1 = 0.5\,(x_4 x_0 + x_6 x_2 - x_5 x_3), \tag{A8.49}$$

$$\dot{y}_2 = 0.5\,(x_5 x_0 - x_6 x_1 + x_4 x_3), \tag{A8.50}$$

and

$$\dot{y}_3 = 0.5\,(x_6 x_0 + x_5 x_1 - x_4 x_2). \tag{A8.51}$$

Adhering to the notation introduced in Sect. A8.1.3.3, the state variables in the original state representation yielded by the inverse state transformation are denoted by $x_{\mathrm{f}i}$, $i = 1, 2, \ldots, 6$. On the basis that the dynamic lags due to the low pass filters of Sect. A8.1.3.2 are negligible for sufficiently small T_{f}, x_i, $i = 1, 2, \ldots, 6$, in (A8.46) through (A8.51) are replaced by $x_{\mathrm{f}i}$, $i = 1, 2, \ldots, 6$ and $\dot{y}_i$, $i = 1, 2, 3$, are replaced by the filtered derivative estimates, $\hat{\dot{y}}_i$, $i = 1, 2, 3$. The inverse state transformation required for the observer is then

$$x_{\mathrm{f}1} = y_1, \tag{A8.52}$$

$$x_{\mathrm{f}2} = y_2, \tag{A8.53}$$

and

$$x_{\mathrm{f}3} = y_3 \tag{A8.54}$$

with the following solution of (A8.49), (A8.50) and (A8.51) for $x_{\mathrm{d}4}$, $x_{\mathrm{d}5}$ and $x_{\mathrm{d}6}$, also using (A8.52), (A8.53) and (A8.54).

$$2 \begin{bmatrix} \hat{\dot{y}}_1 \\ \hat{\dot{y}}_2 \\ \hat{\dot{y}}_3 \end{bmatrix} = \begin{bmatrix} y_0 & -y_3 & y_2 \\ y_3 & y_0 & -y_1 \\ -y_2 & y_1 & y_0 \end{bmatrix} \begin{bmatrix} x_{\mathrm{f}4} \\ x_{\mathrm{f}5} \\ x_{\mathrm{f}6} \end{bmatrix} \Rightarrow$$

$$\begin{bmatrix} x_{\mathrm{f}4} \\ x_{\mathrm{f}5} \\ x_{\mathrm{f}6} \end{bmatrix} = 2\,\frac{\begin{bmatrix} (y_0^2 + y_1^2) & (y_1 y_2 + y_3 y_0) & (y_3 y_1 - y_2 y_0) \\ (y_1 y_2 - y_3 y_0) & (y_0^2 + y_2^2) & (y_3 y_2 + y_1 y_0) \\ (y_3 y_1 + y_2 y_0) & (y_3 y_2 - y_1 y_0) & (y_0^2 + y_3^2) \end{bmatrix} \begin{bmatrix} \hat{\dot{y}}_1 \\ \hat{\dot{y}}_2 \\ \hat{\dot{y}}_3 \end{bmatrix}}{y_0\,(y_0^2 + y_1^2) + y_3\,(y_3 y_0 - y_2 y_1) + y_2\,(y_3 y_1 + y_2 y_0)} \Rightarrow$$

$$x_{\mathrm{f}4} = 2\left[(y_0^2 + y_1^2)\,\hat{\dot{y}}_1 + (y_1 y_2 + y_3 y_0)\,\hat{\dot{y}}_2 + (y_3 y_1 - y_2 y_0)\,\hat{\dot{y}}_3\right]/y_0 \tag{A8.55}$$

$$x_{f5} = 2\left[(y_1y_2 - y_3y_0)\hat{\dot{y}}_1 + (y_0^2 + y_2^2)\hat{\dot{y}}_2 + (y_3y_2 + y_1y_0)\hat{\dot{y}}_3\right]/y_0 \quad \text{(A8.56)}$$

$$x_{f6} = 2\left[(y_3y_1 + y_2y_0)\hat{\dot{y}}_1 + (y_3y_2 - y_1y_0)\hat{\dot{y}}_2 + (y_0^2 + y_3^2)\hat{\dot{y}}_3\right]/y_0 \quad \text{(A8.57)}$$

The equations of the filtering observer providing filtered versions of these state estimates are as follows, replacing y_0 by x_{f0} for notational uniformity.

$$\begin{aligned}
\hat{\dot{x}}_0 &= -0.5\,(x_{f4}x_{f1} + x_{f5}x_{f2} + x_{f6}x_{f3}) + \lambda_0\,(x_{f0} - \hat{x}_0) \\
\hat{\dot{x}}_1 &= 0.5\,(x_{f4}x_{f0} + x_{f6}x_{f2} - x_{f5}x_{f3}) + \lambda_1\,(x_{f1} - \hat{x}_1) \\
\hat{\dot{x}}_2 &= 0.5\,(x_{f5}x_{f0} - x_{f6}x_{f1} + x_{f4}x_{f3}) + \lambda_2\,(x_{f2} - \hat{x}_2) \\
\hat{\dot{x}}_3 &= 0.5\,(x_{f6}x_0 + x_{f5}x_{f1} - x_{f4}x_{f2}) + \lambda_3\,(x_{f3} - \hat{x}_3) \\
\hat{\dot{x}}_4 &= a_1\,(l_2 x_{f6} - l_3 x_{f5}) + b_1 u_1 + \lambda_4\,(x_{f4} - \hat{x}_4) \\
\hat{\dot{x}}_5 &= a_2\,(l_3 x_{f4} - l_1 x_{f6}) + b_2 u_2 + \lambda_5\,(x_{f5} - \hat{x}_5) \\
\hat{\dot{x}}_6 &= a_3\,(l_1 x_{f5} - l_2 x_{f4}) + b_3 u_3 + \lambda_6\,(x_{f6} - \hat{x}_6)
\end{aligned} \quad \text{(A8.58)}$$

where $\lambda_i = 3/T_{soi}$, $i = 0, 2, \ldots, 6$, T_{soi} being the individually chosen integrator correction loop settling times.

The output derivative state estimator block diagram is shown in Fig. A8.3.

To provide the plant with realistic inputs, three rudimentary DP attitude controllers will be applied to each axis, noting that q_i, $i = 1, 2, 3$, are roughly half the attitude angles, $\int \omega_i(t)dt$, $\omega_1 = \omega_x$, $\omega_2 = \omega_y$, $\omega_3 = \omega_z$, about each axis for moderately small angles (less than 20°), using the output derivative state variables, as shown in Fig. A8.4.

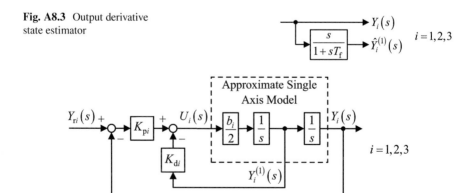

Fig. A8.3 Output derivative state estimator

Fig. A8.4 Rudimentary attitude controllers for the observer demonstration

Table A8.1 Variances, $\sigma^2_{n_{fi} n_{fi}}$ of $\mathbf{x}_f$ and $\sigma^2_{n_{ei} n_{ei}}$ of $\hat{\mathbf{x}}$ and selected T_{soi} for correction loops

i	0	1	2	3	4	5	6
$\sigma^2_{n_{fi} n_{fi}}$	6×10^{-12} [rad²]	5.9×10^{-12}	5.9×10^{-12}	5.8×10^{-12}	1.55×10^{-7} [(rad/s)²]	1.53×10^{-7}	1.57×10^{-7}
$\sigma^2_{n_{ei} n_{ei}}$	1.8×10^{-14} [rad²]	2.4×10^{-12}	2.4×10^{-12}	2.3×10^{-12}	2.2×10^{-13} [(rad/s)²]	2.2×10^{-13}	2.3×10^{-13}
T_{soi} [s]	5	5	5	5	20	20	20

For a critically damped step response with a settling time of T_s seconds (5 % criterion), the controller gains are

$$K_{di} = 18/(b_i T_s) \text{ and } K_{pi} = 81/(2 b_i T_s^2), \ i = 1, 2, 3. \tag{A8.59}$$

For a proper control system design, a more sophisticated control law based on the feedback linearising control technique (Chap. 7) would be used to eliminate inter-axis coupling and achieve a precisely specified dynamic response, but this is unnecessary in this example, the purpose of which is to design and demonstrate the observer, which operates independently of the control input.

For the simulation, the plant parameters are $J_{xx} = 300$ [Kg m²], $J_{yy} = 200$ [Kg m²], $J_{zz} = 150$ [Kg m²] and $K_w = 0.1$ [Nm/V]. The star sensor noise level is taken as 1 [arcsec²] $\equiv ((1/3, 600) \pi/180)^2 = 2.35 \times 10^{-11}$ [rad²], i.e., $\sigma^2_{n_s n_s} = 2.35 \times 10^{-11}/4 = 5.88 \times 10^{-12}$ [rad²], in quaternion units. The reaction wheel (plant) noise is $\sigma^2_{n_{pi} n_{pi}} = 10^{-4}$ [V²]. The nominal settling time of the three attitude control loops is set to $T_s = 80$ [s]. First a simulation of the system is carried out to determine the noise levels, $\sigma^2_{n_{ni} n_{ni}}$ and $\sigma^2_{n_{fi} n_{fi}}$, $i = 0, 2, \ldots, 6$ to determine suitable values of the correction loop settling times, T_{soi}. For this, the reference inputs and initial attitude coordinates are set to zero and the noise variances calculated as $\sigma^2_{xx} - \lim_{t \to \infty} \frac{1}{t} \int_0^t x^2(\tau) \, d\tau$.

The results, which are shown in Table A8.1, enable the state estimation noise variances of the output derivative based state estimator to be compared with those of the output derivative based filtering observer.

For the output derivative based state estimator, $\sigma^2_{n_{fi} n_{fi}}$, $i = 0, 1, 2, 3$, are simply the variances of the stochastic errors in the direct quaternion based attitude measurements, y_0, y_1, y_2 and y_3 while $\sigma^2_{n_{fi} n_{fi}}$, $i = 4, 5, 6$, are the variances of the stochastic errors in the body rate estimates, x_{f4}, x_{f5} and x_{f6}.

Regarding the output derivative based filtering observer, it is found that the variance of the attitude estimation, i.e., $\sigma^2_{n_{ei} n_{ei}}$, $i = 0, 1, 2, 3$, is almost independent of the correction loop settling times, T_{soi}, $i = 0, 1, 2, 3$, indicating, with reference to Fig. A8.2, that the state estimation error is dominated by the noise signals, n_{ni}, $i = 0, 1, 2, 3$, that are filtered directly by the integrators of the observer. So the correction loop settling times are set to values considerably less than the attitude control loop settling times, a value of 5[s] being practicable for this application.

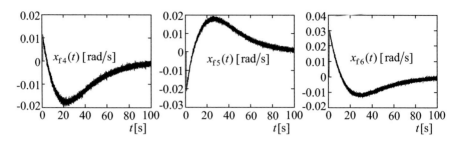

Fig. A8.5 Outputs from output derivative based state estimator

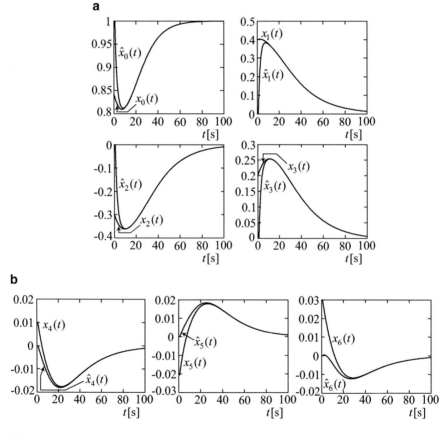

Fig. A8.6 Initial state estimate convergence. (a) Quaternion attitude coordinates, (b) Body angular velocities

On the other hand, $\sigma^2_{n_{ei} n_{ei}}$, $i = 4, 5, 6$, are found to be highly dependent on T_{soi}, $i = 4, 5, 6$, indicating a strong influence of the noise signals, n_{fi}, $i = 4, 5, 6$, for the body rate estimates, compared with the influence of n_{ni}. Under these circumstances, the correction loop settling times are set to the highest practicable values to minimise $\sigma^2_{n_{ei} n_{ei}}$. This is a matter of engineering judgement and a value of $T_{soi} = T_s/4 = 20$ [s] is chosen.

The following simulation emulates the initial attitude acquisition of the spacecraft by setting $[x_0(0), x_1(0), x_2(0), x_3(0)] = [0.71, 0.4, -0.3, 0.2]$ and $[x_4(0), x_5(0), x_6(0)] = [0.01, -0.02, 0.03]$ [rad/s]. Figure A8.5 shows the body rate estimates, $x_{fi}(t)$, $i = 4, 5, 6$.

The noise contaminating these estimates due principally to the measurement noise amplification of the filtered differentiators in the output derivative based state estimator is clearly visible in Fig. A8.5.

To assess the performance of the filtering observer with large initial state estimation errors, Fig. A8.6 shows $\widehat{\mathbf{x}}(t)$ and $\mathbf{x}(t)$ for $[\widehat{x}_0(0), \widehat{x}_1(0), \widehat{x}_2(0), \widehat{x}_3(0)] = [1, 0, 0, 0]$.

Convergence of the state estimates to towards the correct values in times commensurate with the set correction loop settling times is evident. Also, the effective filtering of the observer is apparent through the noise contaminating all the state estimates not being visible in Fig. A8.6, which is in keeping with the very low values of the variances, $\sigma^2_{n_{ei} n_{ei}}$, $i = 0, 1, \ldots, 6$, recorded in Table A8.1.

A9 Appendix to Chap. 9

A9.1 Limit Cycling Control for Second Order Plants

A9.1.1 Overview

As for the first order systems, features have to be introduced into a practicable controller that ensure satisfactory behaviour in the region of the demanded state, where any control system will spend the greatest portion of its lifetime. The hysteresis switching element is included in a switched control system to avoid a switching frequency too high for the actuators. In some applications, however, it would be desirable to design the controller so that the amplitude of the control error, $y(t) - y_r(t)$, during the limit cycle oscillates between precisely prescribed extreme limits to minimise the switching frequency while satisfying a control accuracy specification. A relevant application, to be studied shortly, is spacecraft attitude control using on-off thrusters in which the wear rate of the valves admitting fuel into the plenum chamber [1] must be minimised. In Chap. 9, Example 9.2, this is achieved by simply setting the hysteresis limits, S^+ and S^-, of the switching element to the extreme permitted limits of the control error. This, however, requires the switching function to be expressed with the control error, $y - y_r$, as an additive term. Let the first component of the state error, $\mathbf{x}_e$, be $x_{1e} = y - y_r$. Then the switching function is manipulated to be in the form

$$S(\mathbf{x}_e) = -x_{1e} + f(x_{2e}, x_{3e}, \ldots, x_{ne}). \tag{A9.1}$$

The following section demonstrates the need for this.

A9.1.2 The Influence of the Switching Function on the Limit Cycle

The basic switching element of Chap. 9, Fig. 9.15a together with the switching function, $S(\mathbf{x}_e)$, realises the switching boundary, $S(\mathbf{x}_e) = 0$, but the introduction of the hysteresis switching element of Fig. 9.15b of Chap. 9 replaces this single switching boundary with *two* switching boundaries defined by

$$S_p(\mathbf{x}_e) = S(\mathbf{x}_e) - S^+ = 0 \tag{A9.2}$$

and

$$S_n(\mathbf{x}_e) = S(\mathbf{x}_e) - S^- = 0. \tag{A9.3}$$

The corresponding switching function evaluations are

$$S = S(\mathbf{x}_e) - S^+ \tag{A9.4}$$

and

$$S = S(\mathbf{x}_e) - S^-. \tag{A9.5}$$

The hysteresis switching element then ensures the following:

$$\text{If } S > S^+, \text{ then } A(9.2) \text{ is active.} \tag{A9.6}$$

$$\text{If } S < S^-, \text{ then } A(9.3) \text{ is active.} \tag{A9.7}$$

$$\text{If } S^- \leq S \leq S^+, \text{ then the current boundary is kept active.} \tag{A9.8}$$

Consider a different switching function, $S'(\mathbf{x}_e)$, yielding the same switching boundary, meaning $S(\mathbf{x}_e) = 0 \equiv S'(\mathbf{x}_e) = 0$ without the hysteresis element. After introduction of the hysteresis element, however, for given values of S^+ and S^-, replacing $S(\mathbf{x}_e)$ by $S'(\mathbf{x}_e)$ in (A9.4) and (A9.5) yields a *different* pair of switching boundaries. It follows that for given values of S^+ and S^-, the limit cycling behaviour caused by the hysteresis element depends upon the switching function. Now suppose that the switching boundary equation is manipulated into the form

$$f_i(x_1, \ldots, x_{i-1}, x_{i+1}, x_n) - x_i = 0. \tag{A9.9}$$

Then the switching boundaries of (A9.4) and (A9.5) become

$$f_i(x_1, \ldots, x_{i-1}, x_{i+1}, x_n) - x_i - S_b = 0, \quad S_b = S^+, S^-. \tag{A9.10}$$

Regarding this in a geometrical sense, it is evident that the two boundaries defined by (A9.10) are obtained by translational displacement of the boundary defined by (A9.9) 'distances' of S^+ and S^- along the x_i axis. Clearly, there are n different pairs of boundaries for $i = 1, 2, \ldots, n$.

A9.1.3 Application to the Attitude Control of a Rigid Body Spacecraft

To illustrate the approach introduced above with the different resulting limit cycles, the single axis attitude control of the space satellite will be studied using on-off thrusters instead of the reaction wheel. Also a constant or slowly varying external disturbance torque will be taken into account. The plant model is

$$\dot{x}_1 = x_2 \tag{A9.11}$$

$$\dot{x}_2 = u - d, \quad u = \pm u_{\max} \text{ or } u = 0 \tag{A9.12}$$

$$y = x_1 \tag{A9.13}$$

where x_1 is the attitude angle, x_2 is the angular velocity and u is the *control acceleration*, defined as the control torque from the thrusters divided by the moment of inertia. The corresponding plant model expressed in terms of the error state variables for a constant reference input, y_r, i.e., $x_{1e} = x_1 - y_r$ and $x_{2e} = x_2$, is

$$\dot{x}_{1e} = x_{2e} \tag{A9.14}$$

$$\dot{x}_{2e} = u - d, \quad u = \pm u_{\max} \text{ or } u = 0. \tag{A9.15}$$

The state variables are defined as before but u is the *control angular acceleration* in which $u_{\max} = \Gamma_t/J$, where Γ_t is the magnitude of the control torque produced by either thruster and, as before, J is the moment of inertia of the spacecraft about the control axis. Also, $d = \Gamma_d/J$ is the *disturbance angular acceleration* produced by the disturbance torque, Γ_d.

First the time optimal feedback control law for the plant of (A9.14) and (A9.15) will be derived and hence only the two active control levels, $\pm u_{\max}$, are used.

The state trajectory differential equation formed by dividing (A9.14) by (A9.15) is

$$\frac{dx_{1e}}{dx_{2e}} = \frac{x_{2e}}{u - d} \tag{A9.16}$$

and its general solution for $u = $ const. and $d = $ const. is the state trajectory equation,

$$x_{1e} = x_{1e}(0) + \frac{1}{2(u-d)} \left[x_{2e}^2 - x_{2e}^2(0) \right] \tag{A9.17}$$

Following similar reasoning to that of the time optimal control using the reaction wheel, (A9.16) and (A9.17) are also valid for the back tracing from the origin of the state plane and therefore the trajectory approaching the origin of the error state plane under $u = -u_{\max}$ is defined by (A9.17) with $x_{1e}(0) = x_{2e}(0) = 0$, for $x_{2e} > 0$ and the trajectory approaching the origin of the error state plane under $u = +u_{\max}$ is defined by (A9.17) with $x_{1e}(0) = x_{2e}(0) = 0$ for $x_{2e} < 0$. The time optimal switching boundary is the following concatenation of the two equations of these trajectories.

$$\left\{ \begin{array}{l} x_{1e} = \dfrac{1}{2(-u_{\max} - d)} x_{2e}^2 \text{ for } x_{2e} > 0 \\ x_{1e} = \dfrac{1}{2(u_{\max} - d)} x_{2e}^2 \text{ for } x_{2e} \leq 0 \end{array} \right\} \Rightarrow x_{1e} = -\dfrac{1}{2[u_{\max} \operatorname{sgn}(x_{2e}) + d]} x_{2e}^2 \tag{A9.18}$$

The corresponding switching function with $x_{1e} = x_1 - x_{1r}$ is then

$$S_{to}(\mathbf{x}_e) = -x_{1e} - \frac{1}{2[u_{max} + d\,\text{sgn}(x_{2e})]} |x_{2e}| x_{2e} \quad (A9.19)$$

and the control law is

$$u = u_{max} \text{sgn}[S_{to}(\mathbf{x}_e)]. \quad (A9.20)$$

It should be noted that for $d \neq 0$, the two parabolic segments of the time optimal switching boundary, $S_{to}(\mathbf{x}_e) = 0$, have different acceleration parameters, i.e., $u_{max} \pm d$, in contrast with the time optimal switching boundary of Fig. 9.27a of Chap. 9. Hence this boundary is not reflected in the origin of the error state plane unless $d = 0$.

Switching function (A9.19) is an example of that used in (A9.9) for $n = 2$ and $i = 1$. The alternative equation of the same boundary for $i = 2$ is obtained by making x_{2e} the subject of the trajectory equations of the LHS of (A9.18), as follows.

$$\begin{cases} x_{2e} = \sqrt{2(u_{max} + d)|x_{1e}|} \text{ for } x_{1e} \leq 0 \\ x_{2e} = -\sqrt{2(u_{max} - d)|x_{1e}|} \text{ for } x_{1e} > 0 \end{cases} \Rightarrow$$

$$x_{2e} = -\sqrt{2[u_{max}|x_{1e}| + x_{1e}d]} \text{sgn}(x_{1e}) \quad (A9.21)$$

This yields the following switching boundary equation and switching function.

$$S'_{to}(\mathbf{x}_e) = -x_{2e} - \sqrt{2[u_{max}|x_{1e}| + x_{1e}d]} \text{sgn}(x_{1e}) \quad (A9.22)$$

$$u = u_{max} \text{sgn}[S'_{to}(\mathbf{x}_e)] \quad (A9.23)$$

The two 'p' and 'n' boundaries produced by the hysteresis switching element with symmetrical hysteresis limits of S^+ and $S^- = -S^+$, in conjunction with (A9.19), (to be referred to as formulation 1) are defined by

$$\begin{cases} S_p(\mathbf{x}_e) = S_{to}(\mathbf{x}_e) - S^+ = 0 \\ S_n(\mathbf{x}_e) = S_{to}(\mathbf{x}_e) + S^+ = 0 \end{cases}$$

$$\Rightarrow \begin{cases} S_p(\mathbf{x}_e) = -x_{1e} - S^+ + \frac{1}{2[u_{max} + d\,\text{sgn}(x_{2e})]} |x_{2e}| x_{2e} = 0 \\ S_n(\mathbf{x}_e) = -x_{1e} + S^+ + \frac{1}{2[u_{max} + d\,\text{sgn}(x_{2e})]} |x_{2e}| x_{2e} = 0 \end{cases} \quad (A9.24)$$

while the equivalent boundaries produced in conjunction with (A9.22) (to be referred to as formulation 2) are

$$\left\{\begin{array}{l} S'_p(\mathbf{x}_e) = S'_{to}(\mathbf{x}_e) - S^+ = 0 \\ S'_n(\mathbf{x}_e) = S'_{to}(\mathbf{x}_e) + S^+ = 0 \end{array}\right\}$$

$$\Rightarrow \left\{\begin{array}{l} S'_p(\mathbf{x}_e) = -x_{2e} - S^+ - \sqrt{2\left[u_{\max}|x_{1e}| + x_{1e}d\right]}\mathrm{sgn}(x_{1e}) = 0 \\ S'_n(\mathbf{x}_e) = -x_{2e} + S^+ - \sqrt{2\left[u_{\max}|x_{1e}| + x_{1e}d\right]}\mathrm{sgn}(x_{1e}) = 0 \end{array}\right\} \quad (A9.25)$$

The two boundaries of (A9.24) can be obtained graphically by commencing with the curve of $S_{to}(\mathbf{x}_e) = 0$ (or $S'_{to}(\mathbf{x}_e) = 0$, which is the same) and then translating it parallel to the x_{1e} axis by $\pm S^+$. Similarly, the two boundaries of (A9.25) are obtained by translating the same curve parallel to the x_{2e} axis by $\pm S^+$. The results are different, as may be seen by comparing Fig. A9.1a, b. The corresponding closed loop phase portraits according to (A9.6), (A9.7) and (A9.8) are shown in Fig. A9.1c, d. With switching function formulation 1, it is evident that commencing with an arbitrary initial state error, following the first switch, the state trajectory follows the active switching boundary until reaching point 'b' or point 'd', where it leaves this boundary and crosses the other boundary at point 'c' or point 'a', which becomes active and a control switch occurs causing the state trajectory to follow this boundary, after which the events described above are repeated. Thus the system enters the limit cycle, 'a-b-c-d', of Fig. A9.1e after the first switch. This occurs due to the parabolic segment of the *active* 'p' or 'n' switching boundary met by the state trajectory being members of the set of trajectories of the phase portrait. Most importantly, the control error, $y(t) - y_r(t) = x_{1e}(t)$, oscillates precisely between peak values of $\pm S_{\max}$ during the limit cycle. With the switching function formulation 2, the angular velocity, $x_{2e}(t)$, oscillates precisely between peak values of $\pm S_{\max}$ during the limit cycle but the control error, $x_{1e}(t)$, oscillates between values dependent upon the external disturbance, d, that are unequal in magnitude, as illustrated in Fig. A9.1f, unless $d = 0$. Furthermore, study of Fig. A9.1d reveals that the state trajectory commencing with an arbitrary initial state has oscillatory convergence towards the limit cycle with asymmetrical positive and negative peak magnitudes of $x_{1e}(t)$. Hence switching function formulation 1 is preferable. This illustrates that for satisfactory limit cycle control using a hysteresis switching element, the switching function formulation has to be in the form of (A9.1) with x_{1e} equal to the control error, $y - y_r$.

While the foregoing study of the switched control of a double integrator plant has been useful to introduce limit cycle control, for the spacecraft application, unnecessary fuel would be used during limit cycling and further refinements are needed to arrive at a practicable attitude control system. To achieve this, the phase portraits for the plant model of (A9.14) and (A9.15), representing a rigid body spacecraft controlled by on-off thrusters, are used to aid the derivation of a practicable feedback control law that adapts automatically to different disturbance torque levels. Simulations are presented to confirm the correctness of the control law.

The attitude control system presented here is similar to that of ESA's x-ray astronomy satellite, Exosat, but to overcome digital processor computation speed

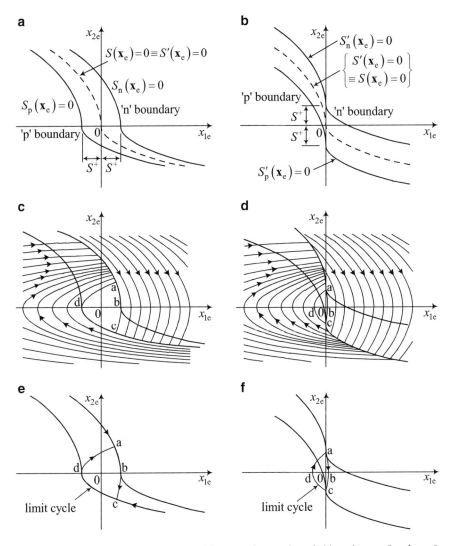

Fig. A9.1 Switching boundaries produced by same hysteresis switching element for $d > 0$. (**a**) Switching boundaries switching function formulation 1. (**b**) Switching boundaries switching function formulation 2. (**c**) Closed loop phase portraits with switching function formulation 1. (**d**) Closed loop phase portraits with switching function formulation 2. (**e**) Limit cycle with switching function formulation 1. (**f**) Limit cycle with switching function formulation 2

limitations, this calculated signed switching times for dedicated thruster firing electronics, at an iteration period of 0.25 [s], enabling firing times of the order of milliseconds [2].

For the system developed here, the performance specification is as follows.

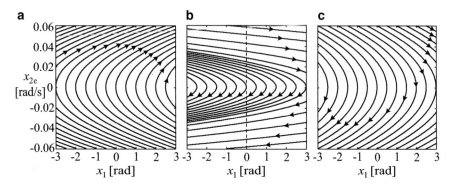

Fig. A9.2 Phase portraits for synthesis of thruster attitude control of rigid body spacecraft. (**a**) 'p' trajectories for $u = u_{max}$. (**b**) 'z' trajectories for $u = 0$. (**c**) 'n' trajectories for $u = -u_{max}$

1. With an arbitrary initial state, the system enters the limit cycle by turning one thruster on, then momentarily both thrusters off followed by the other thruster on. This limit cycle acquisition is nearly time optimal.
2. The nominal limit cycle excursions are $x_{1e} = \pm A$, where the limit cycle excursion, A, is given in a control system performance specification.
3. Limit cycle control is arranged to cater for a minimum thruster on-time, $T_{on\,min}$ below which the fuel usage would be inefficient.
4. The system is able to tolerate plant model mismatches without excessive repeated thruster switchings.

The moment of inertia about the control axis is taken as $J = 800$ kg m^2 and the thruster torque magnitude is $\Gamma_t = 0.1$ Nm, giving $u_{max} = \Gamma_t/J = 5 \times 10^{-3}$ rad/s/s. The minimum thruster on time is taken as $T_{on\text{-}min} = 3$ ms. Three constant external disturbance torque levels are taken, i.e., $\Gamma_d = 0.02$ Nm due to an imperfectly aligned trajectory control thruster, then $\Gamma_d = 10^{-4}$Nm and $\Gamma_d = 10^{-6}$Nm, both due to misalignment of the solar radiation force vector with respect to the spacecraft centre of mass. The specified limit cycle excursion is taken as $A = 5 \times 10^{-6}$ rad.

The parabolic phase portraits for the three control levels are shown in Fig. A9.2 with the higher of the disturbance torque levels.

Higher constant angular acceleration magnitudes yield steeper parabolic trajectories. This relationship is visible in Fig. A9.2a, where the disturbance torque opposes the thruster torque, in Fig. A9.2c where it aids the thruster torque and in Fig. A9.2b where it acts alone. For the lower level of disturbance torque, the difference in the acceleration magnitudes for $u = \pm u_{max}$ would be hardly visible and the parabolas of Fig. A9.2b would be so shallow that they would appear as horizontal straight lines. In the limit, as $d \rightarrow 0$, they are actually horizontal straight lines, as confirmed by setting $u = d = 0$ in (A9.15) giving $x_{2e} = $ const. If the sign of d is reversed but its magnitude kept the same, then the phase portraits would be the mirror images of those in Fig. A9.2, obtained by rotating all three graphs, maintained in a vertical plane, about the x_{2e} axis of Fig. A9.2b (shown dotted).

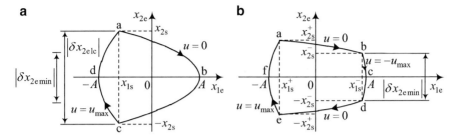

Fig. A9.3 Possible limit cycles for a positive constant disturbance torque. (**a**) Single sided limit cycle. (**b**) Double-sided limit cycle

The minimum thruster on time constraint sets a lower limit on the magnitude of the change in x_{2e}, i.e., $|\delta x_{2e}|$, that can be produced by a thruster operation. The solution to (A9.12) with $u = \pm u_{max}$, $d = $ const., and a thruster on-time of T_{on} is

$$\delta x_{2e} = (u - d) T_{on} \Rightarrow |\delta x_{2e}| = |u - d| T_{on}. \qquad (A9.26)$$

For a given value of $|\delta x_{2e}|$, the smallest value of T_{on} is required when the disturbance aids the thrusters. This will occur during the double-sided limit cycle to be introduced. Let $|\delta x_{2e\,min}|$ be the minimum value of $|\delta x_{2e}|$ for which $T_{on} = T_{on\text{-}min}$. Then (A9.26) becomes

$$|\delta x_{2e\,min}| = (u_{max} + |d|) T_{on\text{-}min}. \qquad (A9.27)$$

The starting point in deriving the three-state switched control law from the phase portraits is to identify the limit cycle trajectory that satisfies the performance specification. Let the maximum and minimum specified values of x_{1e} be $\pm A$. Examination of the phase portraits of Fig. A9.2a, b reveals the possibility of a limit cycle comprising two state trajectory segments, one for $u = u_{max}$ and the other for $u = 0$, as shown in Fig. A9.3a. Since only one thruster is being operated on one side of the limit cycle, it is referred to as a *single sided limit cycle*.

The positive torque thruster is turned on to produce trajectory segment 'cda', giving a positive and constant angular acceleration. It is turned off over trajectory segment 'abc' during which the disturbance torque alone produces a constant angular deceleration of magnitude, $|d|$. The smaller d, the smaller the value of $|\delta x_{2e\,lc}|$ in Fig. A9.3a. Below a critical magnitude, $|d|_{min}$, $|\delta x_{2e\,lc}|$ would fall below $|\delta x_{2e\,min}|$ of (A9.27). To avoid this while keeping the peak values of x_{1e} equal to $\pm A$, the other thruster can be brought into play. Study of the phase portraits of Fig. A9.2 reveals the possibility of creating the limit cycle of Fig. A9.3b. Since both thrusters operate alternately on the left and right sides of the limit cycle, it is referred to as a *double-sided limit cycle*. To minimise fuel consumption within the '$T_{on\text{-}min}$' constraint, the thruster aiding the disturbance torque, which always has a smaller on time than the thruster opposing the disturbance torque, is made to operate for precisely $T_{\text{on-min}}$ seconds, giving an angular velocity magnitude increment of $|\delta x_{2e\,\text{min}}|$ rad/s, as shown in Fig. A9.3b.

For negative constant disturbance torques, the same considerations apply as discussed above but the limit cycles are mirror images of those shown in Fig. A9.3 reflected in the x_{2e} axis, except for the direction of motion, indicated by the arrows, that remains clockwise.

As part of the control algorithm it is necessary to evaluate $|\delta x_{2e\,\text{lc}}|$ in terms of $u_{\max}$, d and A and compare it with $|\delta x_{2e\,\text{min}}|$ in order to set the control law for single or double-sided limit cycling. In fact, comparing $(\delta x_{2e\,\text{lc}})^2$ with $(\delta x_{2e\,\text{min}})^2$ is more convenient. An expression for $(\delta x_{2e\,\text{lc}})^2$ may be derived from the general state trajectory equation, which is the general solution of the state trajectory differential equation, obtained through dividing (A9.14) by (A9.15). Hence

$$\frac{dx_{1e}}{dx_{2e}} = \frac{x_{2e}}{u-d}, \quad u = \pm u_{\max} \text{ or } u = 0. \tag{A9.28}$$

For $u =$ const. and $d =$ const., the general solution is

$$x_{1e} = x_{1e}(0) + \frac{1}{2(u-d)}\left[x_{2e}^2 - x_{2e}^2(0)\right]. \tag{A9.29}$$

This is then applied to the two segments of the single-sided limit cycle of Fig. A9.3a. An observer would provide an estimate, $\widehat{d}$, of d for use in the control algorithm. Hence $\widehat{d}$ replaces d in the following working. For segment, bc,

$$x_{1e} = x_{1e}(0) - \frac{1}{2\widehat{d}}\left[x_{2e}^2 - x_{2e}^2(0)\right] \Rightarrow x_{1s} = A - \frac{1}{2\widehat{d}}x_{2s}^2 \tag{A9.30}$$

and for segment, da,

$$x_{1e} = x_{1e}(0) + \frac{1}{2\left(u_{\max} - \widehat{d}\right)}\left[x_{2e}^2 - x_{2e}^2(0)\right] \Rightarrow x_{1s} = -A + \frac{1}{2\left(u_{\max} - \widehat{d}\right)}x_{2s}^2. \tag{A9.31}$$

Eliminating x_{1s} between (A9.30) and (A9.31) yields an equation for x_{2s}. Since

$$|\delta x_{2e\,\text{lc}}| = 2x_{2s} \Rightarrow (\delta x_{2e\,\text{lc}})^2 = 4x_{2s}^2, \tag{A9.32}$$

this also yields the required equation for $(\delta x_{2e\,\text{lc}})^2$. Thus

$$x_{1s} = A - \frac{1}{2\widehat{d}}x_{2s}^2 = -A + \frac{1}{2\left(u_{\max} - \widehat{d}\right)}x_{2s}^2 \Rightarrow x_{2s}^2 = 4\widehat{d}\left(\frac{u_{\max} - \widehat{d}}{u_{\max}}\right)A. \tag{A9.33}$$

Then using (A9.32),

$$(\delta x_{2e\ lc})^2 = 16\widehat{d}\left(\frac{u_{max} - \widehat{d}}{u_{max}}\right) A. \qquad (A9.34)$$

If $\widehat{d}$ is changed in sign without changing $|\widehat{d}|$ then the result would be the same. It follows that the version of (A9.34) catering for both signs of $\widehat{d}$ is

$$(\delta x_{2e\ lc})^2 = 16|\widehat{d}|\left(\frac{u_{max} - |\widehat{d}|}{u_{max}}\right) A \qquad (A9.35)$$

The condition for the controller to be set for single-sided limit cycling is

$$(\delta x_{2e\ lc})^2 > (\delta x_{2e\ min})^2 \qquad (A9.36)$$

This can be expressed in terms of u_{max}, $\widehat{d}$, A and $T_{on\text{-}min}$ using (A9.35) and (A9.27) with $d = \widehat{d}$, as follows.

$$16|\widehat{d}|\left(\frac{u_{max} - |\widehat{d}|}{u_{max}}\right) A > \left(u_{max} + |\widehat{d}|\right)^2 T^2_{on\text{-}min} \Rightarrow$$

$$\frac{16A|\widehat{d}|\left(u_{max} - |\widehat{d}|\right)}{u_{max}\left(u_{max} + |\widehat{d}|\right)^2} > T^2_{on\text{-}min} \qquad (A9.37)$$

Next the phase portraits of Fig. A9.2 may be studied again to create a set of switching boundaries that satisfy the control system specification, shown in Fig. A9.4.

These are easily implemented as each is similar, one being obtained from the other by translation parallel to the x_{1e} axis. Since the double-sided limit cycle has four control switches per cycle, four switching boundaries will be formed. As will be seen, only two of these boundaries are used during the single-sided limit cycle, this having only two switches per cycle. These boundaries are similar to those of Fig. A9.1, the upper and lower parabolic segments having acceleration parameters of respectively, $-(u_{max} + d)$ and $u_{max} - d$. As previously, the illustrations are for $d > 0$. For $d < 0$, the diagrams would be similar but mirror images of those shown, except for the direction of motion indicated by the arrows, which remain clockwise on the limit cycles.

The switching functions realising the switching boundaries in Fig. A9.4 are generated from (A9.19) by means of an additional positioning term, S_b, as follows

$$S = S_{to} - S_b. \qquad (A9.38)$$

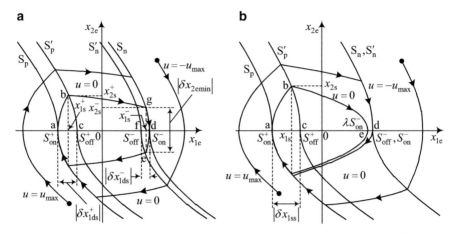

Fig. A9.4 Switching boundaries and limit cycle acquisition trajectories illustrated for $\hat{d} > 0$. (**a**) Double-sided limit cycle. (**b**) Single-sided limit cycle

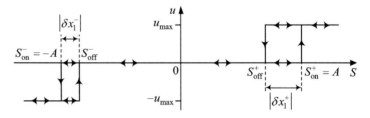

Fig. A9.5 Hysteresis switching element transfer characteristic for controlled limit cycle

Here,

$$S_{to} = -x_{1e} - \frac{1}{2\left[u_{max} + \hat{d}\,\text{sgn}\,(x_{2e})\right]} |x_{2e}| x_{2e} \tag{A9.39}$$

where $S_b = S_{on}^+$ for boundary, S_p, and $S_b = -S_{on}^-$ for boundary S_n. Boundaries, S_p' and S_n', are realised by applying hysteresis switching elements to the switching boundaries, S_p and S_n, yielding the combined transfer characteristic of Fig. A9.5. For a single sided limit cycle with $\hat{d} > 0$, as shown in Fig. A9.4b boundary S_n' is made to coincide with boundary S_n by setting $S_{off}^- = S_{on}^-$, this boundary only being used during the limit cycle acquisition from an arbitrary initial state. Similarly, for a single sided limit cycle with $\hat{d} < 0$, boundary S_p' is made to coincide with boundary S_p by setting $S_{off}^+ = S_{on}^+$.

The corresponding hysteresis band settings are given by Table A9.1.

Appendices

Table A9.1 Hysteresis band settings

	Single-sided operation		Double-sided operation
	$\widehat{d} > \left|\widehat{d}\right|_{min}$	$\widehat{d} < -\left|\widehat{d}\right|_{min}$	$\left|\widehat{d}\right| \leq \left|\widehat{d}\right|_{min}$
Right hand band, $\left|\delta x_1^+\right| =$	$\left|\delta x_{1\,ss}\right|$	0	$\delta x_{1\,ds}^+$
Left hand band, $\left|\delta x_1^-\right| =$	0	$\left|\delta x_{1\,ss}\right|$	$\delta x_{1\,ds}^-$

The peak of the *ideal* single sided limit cycle for $u = 0$ is precisely at $x_{1e} = S_{\text{off}}^- = S_{\text{on}}^- = A$ for $\widehat{d} > 0$ (point 'd' in Fig. A9.4b) and at $x_{1e} = S_{\text{off}}^+ = S_{\text{on}}^+ = -A$ for $\widehat{d} < 0$.

Plant parametric errors, such as the difference between the true spacecraft moment of inertia and its estimate used in the control algorithm, and stochastic variations of the thruster valve opening times about the nominal one, however, could cause the state trajectory to just cross, rather than precisely reach the boundary that should be inactive during the limit cycle, causing the thruster aiding the disturbance torque to turn on and waste fuel. To overcome this issue, the hysteresis band, $|\delta x_{1\,ss}|$, is calculated so that the peak of $x_{1e}(t)$ during the ideal limit cycle occurs at $\lambda S_{\text{on}}^- = \lambda A$ for $\widehat{d} > 0$ and at $\lambda S_{\text{on}}^+ = -\lambda A$ for $\widehat{d} < 0$, where $\lambda = $ const., $0 < \lambda < 1$, is a safety factor. Usually $\lambda = 0.9$ is sufficient to prevent sporadic double-sided thruster operation.

It remains to determine expressions for the hysteresis bands in terms of $u_{\max}$, $\widehat{d}$ and A. For single-sided operation, $|\delta x_{1\,ss}|$ is determined as follows. Applying (A9.17) to segment 'ab' of the state trajectory in Fig. A9.4b with $u = u_{\max}$ yields

$$x_{1s} = S_{\text{on}}^+ + \frac{1}{2\left(u_{\max} - \widehat{d}\right)} x_{2s}^2 = -A + \frac{1}{2\left(u_{\max} - \widehat{d}\right)} x_{2s}^2. \qquad (A9.40)$$

Applying (A9.17) to segment 'bc' of boundary S_p' with $u = -u_{\max}$ (*as if* the negative torque thruster operates) yields

$$S_{\text{off}}^+ = -A + |\delta x_{1\,ss}| = x_{1s} + \frac{1}{2\left(-u_{\max} - \widehat{d}\right)} \left(-x_{2s}^2\right). \qquad (A9.41)$$

Eliminating x_{2s}^2 between (A9.40) and (A9.41) gives

$$x_{2s}^2 = 2\left(u_{\max} - \widehat{d}\right)(x_{1s} + A) = 2\left(u_{\max} + \widehat{d}\right)[|\delta x_{1\,ss}| - (x_{1s} + A)] \Rightarrow \quad (A9.42)$$

$$|\delta x_{1\,ss}| = \frac{2u_{\max}(x_{1s} + A)}{u_{\max} + \widehat{d}}. \qquad (A9.43)$$

To obtain another equation for elimination of x_{1s}, (A9.17) is applied to segment 'be' of the limit cycle with $u = 0$ and $d = \widehat{d}$, yielding

$$\lambda S_{on}^{-} = \lambda A = x_{1s} - \frac{1}{2\widehat{d}}\left(-x_{2s}^{2}\right). \tag{A9.44}$$

Eliminating x_{2s}^2 between (A9.40) and (A9.44) gives

$$x_{2s}^{2} = 2\widehat{d}\,(\lambda A - x_{1s}) = 2\left(u_{\max} - \widehat{d}\right)(x_{1s} + A) \tag{A9.45}$$

from which

$$x_{1s} = \frac{\left[(\lambda + 1)\widehat{d} - u_{\max}\right]A}{u_{\max}}. \tag{A9.46}$$

Substituting for x_{1s} in (A9.43) using (A9.46) then yields

$$|\delta x_{1\,ss}| = \frac{2(\lambda + 1)\widehat{d}\,A}{u_{\max} + \widehat{d}}. \tag{A9.47}$$

For $\widehat{d} < 0$, the result would be identical for the same value of $\left|\widehat{d}\right|$ and therefore to cater for both signs of disturbance, $\widehat{d}$ is replaced by $\left|\widehat{d}\right|$ in (A9.47). Thus

$$|\delta x_{1\,ss}| = \frac{2(\lambda + 1)\left|\widehat{d}\right|A}{u_{\max} + \left|\widehat{d}\right|}. \tag{A9.48}$$

Also, boundaries S_p and S_p' become coincident and boundaries S_n and S_n' are separated by $|\delta x_{1\,ss}|$ along the x_{1e} axis.

A similar approach will now be taken to derive expressions for the hysteresis bands, $\left|\delta x_{1ds}^{+}\right|$ and $\left|\delta x_{1ds}^{-}\right|$, of the double sided limit cycle. Repeated working is avoided by comparing segments 'ab' and 'bc' in Fig. A9.4a with segments 'ab' and 'bc' in Fig. A9.4b that led to (A9.43). First, by analogy with (A9.43),

$$\left|\delta x_{1\,ds}^{+}\right| = \frac{2u_{\max}\left(x_{1s}^{+} + A\right)}{u_{\max} + \widehat{d}}. \tag{A9.49}$$

Second, an expression for $\left(x_{2s}^{+}\right)^2$ will be needed and this may be written down by analogy with (A9.42). Thus

$$\left(x_{2s}^{+}\right)^{2} = 2\left(u_{\max} - \widehat{d}\right)\left(x_{1s}^{+} + A\right). \tag{A9.50}$$

Next, x_{2s}^- is fixed by the minimum thruster on time, $T_{\text{on-min}}$, via (A9.27) with $d = \widehat{d}$, since $|\delta x_{2e\text{ min}}| = 2x_{2s}^-$. Thus

$$x_{2s}^- = \frac{1}{2}\left(u_{\max} + \left|\widehat{d}\right|\right) T_{\text{on-min}}. \tag{A9.51}$$

Then applying (A9.17) to trajectory segment 'de' with $u = -u_{\max}$ and segment 'ef' of the S_n' boundary with $u = u_{\max}$ yields

$$x_{1s}^- = S_{\text{on}}^- + \frac{1}{2\left(-u_{\max} - \widehat{d}\right)}\left(x_{2s}^-\right)^2 = A - \frac{1}{2\left(u_{\max} + \widehat{d}\right)}\left(x_{2s}^-\right)^2 \tag{A9.52}$$

and

$$S_{\text{off}}^- = x_{1s}^- + \frac{1}{2\left(u_{\max} - \widehat{d}\right)}\left[-\left(x_{2s}^-\right)^2\right] \Rightarrow A - |\delta x_{1ds}^-| = x_{1s}^- - \frac{1}{2\left(u_{\max} - \widehat{d}\right)}\left(x_{2s}^-\right)^2. \tag{A9.53}$$

Substituting for x_{1s}^- in (A9.53) using (A9.52) then yields

$$A - |\delta x_{1ds}^-| = A - \frac{1}{2(u_{\max}+\widehat{d})}\left(x_{2s}^-\right)^2 - \frac{1}{2(u_{\max}-\widehat{d})}\left(x_{2s}^-\right)^2 \Rightarrow$$
$$|\delta x_{1ds}^-| = \frac{u_{\max}}{u_{\max}^2 - \widehat{d}^2}\left(x_{2s}^-\right)^2. \tag{A9.54}$$

Applying (A9.17) to trajectory 'bg' in Fig. A9.4a with $u = 0$ gives

$$x_{1s}^- = x_{1s}^+ - \frac{1}{2\widehat{d}}\left[\left(x_{2s}^-\right)^2 - \left(x_{2s}^+\right)^2\right]. \tag{A9.55}$$

Substituting for x_{1s}^- and $\left(x_{2s}^+\right)^2$ in (A9.55) using (A9.52) and (A9.50), yields

$$A - \frac{1}{2\left(u_{\max} + \widehat{d}\right)}\left(x_{2s}^-\right)^2 = x_{1s}^+ - \frac{1}{2\widehat{d}}\left[\left(x_{2s}^-\right)^2 - 2\left(u_{\max} - \widehat{d}\right)\left(x_{1s}^+ + A\right)\right] \Rightarrow$$

$$A + \left[\frac{1}{2\widehat{d}} - \frac{1}{2\left(u_{\max} + \widehat{d}\right)}\right]\left(x_{2s}^-\right)^2 = \left(1 + \frac{u_{\max}}{\widehat{d}} - 1\right)x_{1s}^+ + \left(\frac{u_{\max}}{\widehat{d}} - 1\right)A \Rightarrow$$

$$\left(2 - \frac{u_{\max}}{\widehat{d}}\right)A + \frac{u_{\max}}{2\widehat{d}\left(u_{\max} + \widehat{d}\right)}\left(x_{2s}^-\right)^2 = \frac{u_{\max}}{\widehat{d}}x_{1s}^+ \Rightarrow$$

$$x_{1s}^+ = \left(\frac{2\widehat{d}}{u_{\max}} - 1\right)A + \frac{1}{2\left(u_{\max} + \widehat{d}\right)}\left(x_{2s}^-\right)^2. \tag{A9.56}$$

Substituting for x_{1s}^+ in (A9.49) using (A9.56) then yields

$$|\delta x_{1\,ds}^+| = \frac{2u_{max}}{u_{max} + \widehat{d}} \left[\frac{2\widehat{d}A}{u_{max}} + \frac{1}{2\left(u_{max} + \widehat{d}\right)} (x_{2s}^-)^2 \right]. \tag{A9.57}$$

If the sign of $\widehat{d}$ is changed without altering $|\widehat{d}|$, then x_{2s}^- calculated by (A9.51) and used in (A9.54) and (A9.57) becomes x_{2s}^+. Since this is the smaller of the two switch point velocities, it will be denoted by $x_{2s\,min}$ in the general algorithm catering for both signs of $\widehat{d}$. The hysteresis bands calculated by (A9.54) and (A9.57) are also exchanged between the boundary pair, S_p, S_p', and the boundary pair, S_n, S_n' if the sign of $\widehat{d}$ is changed without altering $|\widehat{d}|$. Hence to cater for both signs of disturbance, $\widehat{d}$ is replaced by $|\widehat{d}|$ in (A9.57) and, with reference to (A9.51), (A9.54) and (A9.57), the hysteresis bands are set as follows.

$$\left.\begin{array}{l} x_{2s\,min} = \frac{1}{2}\left(u_{max} + |\widehat{d}|\right) T_{on-\,min}, \\[6pt] h_{ds1} = \frac{u_{max}}{u_{max}^2 - \widehat{d}^2} x_{2s\,min}^2, \quad h_{ds2} = \frac{2u_{max}}{u_{max} + |\widehat{d}|} \left[\frac{2|\widehat{d}|A}{u_{max}} + \frac{1}{2(u_{max}+|\widehat{d}|)} x_{2s\,min}^2 \right] \\[6pt] \left(|\delta x_{1\,ds}^+|, |\delta x_{1\,ds}^-|\right) = \begin{cases} (h_{ds2}, h_{ds1}) & \text{for } |\widehat{d}_{min}| > \widehat{d} > 0 \\ (h_{ds1}, h_{ds2}) & \text{for } -|\widehat{d}_{min}| < \widehat{d} < 0 \end{cases} \end{array}\right\} \tag{A9.58}$$

Figure A9.6 shows a block diagram of the control system with some implementation details. Separate hysteresis elements generate the thruster drive logic signals, L^+ and L^-. With the offset, A, they realise the transfer characteristic of Fig. A9.5. A star sensor alone provides a continuous attitude measurement, y, on the basis that enough stars of sufficient brightness appear in the star sensor field of view to obtain attitude measurements for all the required spacecraft attitudes. Otherwise a rate integrating gyro pack is needed for continuous attitude measurement, together with drift correction using the star sensor attitude measurement whenever it is available. An observer, however, will still be needed for disturbance estimation.

Figure A9.7 shows simulations of the step responses of the control system and the limit cycling behaviour. The observer is not included in this simulation, as it is intended just to demonstrate the limit cycle and the initial state trajectory leading to it.

The magnitude of the step reference input of Fig. A9.7a has been chosen sufficiently large to demonstrate the near time optimal limit cycle acquisition using both thrusters but sufficiently small for the single-sided limit cycle under the higher level disturbance torque to be visible. Both can be seen in the state trajectory of Fig. A9.7a (i), which indicates similar behaviour to that predicted in Fig. A9.4a. The limit cycle acquisition is near time optimal rather than time optimal due to the relatively short interval in which both thrusters are turned off ($u = 0$) between

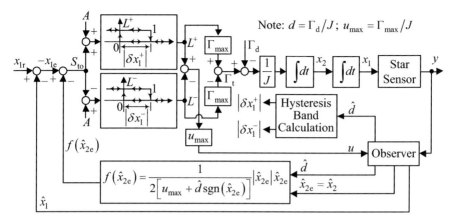

Fig. A9.6 Block diagram of thruster based spacecraft attitude control system

$u = u_{\max}$ and $u = -u_{\max}$, indicated by the short segment of smaller slope at the peak of the state trajectory. In Fig. A9.7a (ii), the acquisition transient followed by the required oscillations of $x_{1e}(t)$ between the prescribed limits during the limit cycling are also visible.

As expected, $x_{2e}(t)$ is piecewise linear in Fig. A9.7a (iii). Figure A9.7a (iv) indicates the positive torque thruster accelerating the spacecraft shortly followed by the negative torque thruster decelerating it, and only the positive torque thruster pulsing during the limit cycle to counteract the disturbance torque with an equal and opposite average value.

Figure A9.7b, c focuses on the limit cycling action at the two lower level disturbance torque levels, with a relatively small attitude reference input of 1.04A to check the limit cycle acquisition is correct. In Fig. A9.7b, the disturbance torque is sufficient for single-sided limit cycling. The nearly vertical segments of the state trajectory in Fig. A9.7b (i) are, in fact, relatively steep parabolas during the thruster on periods. The much shallower parabola of the limit cycle during the thruster off period is clearly visible. The reader's attention is drawn to the long limit cycle period of about 10 s, evident in Fig. A9.7b (ii), (iii) and (iv), that is typical of this application. In both Fig. A9.7a (iii), b (iii), the angular velocity is reversed during the relatively short thruster on period and reversed again due to the action of the disturbance torque over the relatively long thruster off period. Hence the saw-tooth appearance of $x_{2e}(t)$ during the limit cycle. The thruster on period is only about 4 ms during the limit cycle and therefore $u(t)$ appears as a train of impulses on the time scale of Fig. A9.7b (iv).

In Fig. A9.7c, the disturbance torque level is so small that the system behaves almost as if the disturbance torque is zero, i.e., piecewise linear attitude angle error in Fig. A9.7c (ii) and piecewise constant angular velocity in Fig. A9.7c (iii). The alternate positive and negative thruster operations during the double-sided limit cycle are clearly visible in Fig. A9.7c (iv) and these appear as a train of impulses of alternating sign as the thruster on periods are only the minimum value of 3 ms.

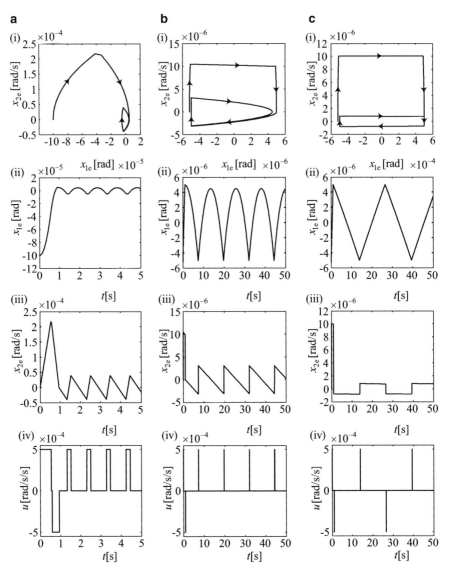

Fig. A9.7 Step responses and limit cycles of thruster-based spacecraft attitude control. (a) $x_{1r} = 20A$, $\Gamma_d = 0.02$ Nm (b) $x_{1r} = 1.04A$, $\Gamma_d = 10^{-4}$ Nm (c) $x_{1r} = 1.04A$, $\Gamma_d = 10^{-6}$ Nm

References

1. Turner, MJL (2009) Rocket and spacecraft propulsion: principles, practice and new developments. Springer-Praxis, Praxis Publishing, Chichester, ISBN 978-3-540-69203-4
2. Dodds SJ (1981) Adaptive, high precision, satellite attitude control for microprocessor implementation. Automatica 17(1)

A10 Appendix to Chap. 10

A10.1 Observer Based Robust Control

A10.1.1 Background

Observer-based robust control (OBRC) can be regarded as a distant relative of sliding mode control through its reliance upon relatively high gains, in common with the pseudo sliding mode controller with the boundary layer described in Chap. 10, Sect. 10.4.1. It resulted from the author's work on spacecraft attitude control and the control of electric drives. In these applications, suddenly applied external disturbances often occur that can cause unacceptable transient control errors. The author's approach to the problem was to design an observer for estimation of the external disturbance referred to the control input, as well as for estimation of the plant state. The disturbance estimate was then used to form a control component to counteract the actual disturbance to minimise its effect.

A very welcome 'side effect' was a substantial increase in robustness with respect to plant parametric uncertainties. This will now be demonstrated by means of a simulation of a simple single axis, rigid body, spacecraft attitude control system. The plant state space model is as follows.

$$\left.\begin{aligned} \dot{x}_1 &= x_2 \\ \dot{x}_2 &= b(u-d) \\ \dot{d} &= 0 \\ y &= x_1 \end{aligned}\right\} \quad (A10.1)$$

where x_1 is the attitude angle, x_2 is the angular velocity, d is the external disturbance typically caused by the torque from an orbit change thruster misaligned with respect to the spacecraft centre of mass and $b = K_w/J$, where K_w is the reaction wheel torque constant and J is the spacecraft moment of inertia. In this model, d is regarded as a state variable as in Chap. 8, Sect. 8.2.4, and although it is assumed constant in the model, the observer,

$$\left.\begin{aligned} \dot{\widehat{x}}_1 &= \widehat{x}_2 + l_1 e \\ \dot{\widehat{x}}_2 &= \tilde{b}(u-d) + l_2 e \\ \dot{\widehat{d}} &= l_3 e \\ \widehat{y} &= \widehat{x}_1 \\ e &= y - \widehat{y} \end{aligned}\right\}, \quad (A10.2)$$

will cause $\widehat{d}(t)$ to follow $d(t)$ with a dynamic lag that can be reduced to acceptable proportions by increasing the eigenvalues of the observer, which can be achieved conveniently by reducing the correction loop settling time, T_{so}, if one of the settling time formulae is used. Here l_1, l_2 and l_3 are the correction loop gains and $\tilde{b} = K_w/\tilde{J}$

is an estimate of b, where $\tilde{J}$ is estimate of J. The error in the assumed value of K_w is considered to be negligible. The basic linear state feedback control law is

$$u = ry_r - k_1\hat{x}_1 - k_2\hat{x}_2, \quad (A10.3)$$

where r is the reference input scaling coefficient and k_i, $i = 1, 2$, are the linear state feedback gains. The modified control law that uses the disturbance estimate to counteract the real disturbance is

$$u = ry_r - k_1\hat{x}_1 - k_2\hat{x}_2 + \hat{d}. \quad (A10.4)$$

Figure A10.1 shows the block diagram of the control system with a switch, S, that enables either control law (A10.3) or control law (A10.4) to be selected.

The following simulations show step responses with both control laws. For each control law, two step responses are shown, one being the ideal one with a perfectly matched model, i.e., $J = \tilde{J}$, and the other with an extreme underestimation of the moment of inertia, i.e., $J = 5\tilde{J}$.

The observer correction loop and the linear state feedback control law will both be designed by pole assignment using the 5 % settling time formula, having respective settling times of T_{so} and T_s. For the observer correction loop, equating the actual and desired characteristic polynomials yields

$$s^3\left[1 - \left(-\frac{l_1}{s} - \frac{l_2}{\tilde{b}s^2} + \frac{l_3}{\tilde{b}s^3}\right)\right] = s^3 + l_1 s^2 + l_2 \tilde{b}s - l_3 \tilde{b} = (s+q)^3$$

$$= s^3 + 3qs^2 + 3q^2 s + q^3 \Rightarrow$$

$$l_1 = 3q, \, l_2 = \frac{3}{\tilde{b}}q^2 \text{ and } l_3 = -\frac{1}{\tilde{b}}q^3, \quad (A10.5)$$

where $q = 6/T_{so}$.

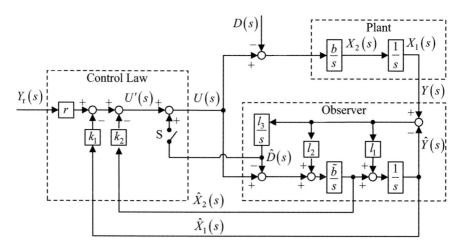

Fig. A10.1 Single axis, rigid body spacecraft attitude control system with an observer

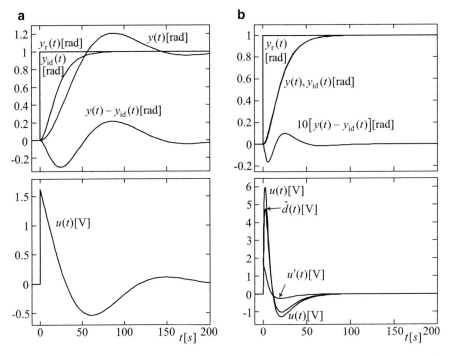

Fig. A10.2 Step responses of rigid body spacecraft attitude control system with $J = 5\tilde{J}$. (a) Basic LSF control law. (b) LSF control law augmented by $\hat{d}(t)$

For the linear state feedback control loop, equating the actual and desired closed loop transfer functions yields

$$\frac{Y(s)}{Y_r(s)} = \frac{\frac{r\tilde{b}}{s^2}}{1 - \left[-\frac{k_2\tilde{b}}{s} - \frac{k_1\tilde{b}}{s^2}\right]} = \frac{r\tilde{b}}{s^2 + k_2\tilde{b}s + k_1\tilde{b}} = \frac{p^2}{(s+p)^2} = \frac{p^2}{s^2 + 2p + p^2} \Rightarrow$$

$$r = \frac{1}{\tilde{b}}p^2, k_1 = \frac{1}{\tilde{b}}p^2 \quad \text{and} \quad k_2 = \frac{2}{\tilde{b}}p,$$

(A10.6)

where $p = 4.5/T_s$.

The nominal plant parameters are $\tilde{J} = 100 \; [\text{Kg m}^2]$, $K_w = 0.5 \; [\text{Nm/V}]$. The specified settling time of the LSF control loop is $T_s = 50 \; [\text{s}]$. Although the recommended ratio between the eigenvalue magnitude of the observer correction loop and that of the LSF control loop is often quoted as 5, which is also T_s/T_{so}, a much larger ratio is found necessary to quickly counteract step external disturbances to minimise the resulting error transients. In this example, $T_s/T_{so} = 100$. Figure A10.2 shows the results.

As would be expected, the basic linear state feedback control system suffers from a sluggish, overshooting response, as evident in Fig. A10.2a. In contrast, once the switch, S, is closed, the control system becomes remarkably robust, the response, $y(t)$, being hardly distinguishable from the ideal response, $y_{id}(t)$. In Fig. A10.2b, $\widehat{d}(t)$ is seen to be about four times greater than the control component, $u'(t)$, provided by the LSF control law and gives rise to the extra torque needed to compensate for the increased moment of inertia to produce the specified step response. This result provided the incentive to carry out the development of the generic observer based robust controller [1, 2], which is explained in the following sections.

A10.1.2 The Model Mismatch Equivalent Input Premise

The most general plant considered obeys the state differential and output equations,

$$\left.\begin{array}{l}\dot{\mathbf{x}} = \mathbf{F}(\mathbf{x},\ \mathbf{u}) \\ \mathbf{y} = \mathbf{H}(\mathbf{x})\end{array}\right\}, \qquad (A10.7)$$

where $\mathbf{x} \in \Re^n$ is the state vector, $\mathbf{u} \in \Re^r$ is the control vector and $\mathbf{y} \in \Re^m$ is the measured and controlled vector. $\mathbf{F}(\cdot)$ and $\mathbf{H}(\cdot)$ are continuous functions of their arguments and $\mathbf{H}(\mathbf{0}) = \mathbf{0}$. Let a plant model,

$$\left.\begin{array}{l}\dot{\mathbf{x}}_m = \mathbf{F}_m(\mathbf{x}_m,\ \mathbf{u}_m) \\ \mathbf{y}_m = \mathbf{H}_m(\mathbf{x}_m)\end{array}\right\}, \qquad (A10.8)$$

be formed, where $\mathbf{x}_m \in \Re^N$, $\mathbf{u}_m \in \Re^r$ and $\mathbf{y}_m \in \Re^r$, $\mathbf{F}_m(\cdot)$ and $\mathbf{H}_m(\cdot)$ are continuous functions of their arguments and $\mathbf{H}_m(\mathbf{0}) = \mathbf{0}$. External physical disturbances are not introduced at this stage as robustness with respect to mismatches between (A10.7) and (A10.8) are the main concern. Since any external disturbance set is equivalent to an external disturbance referred to the control input, direct counteraction by its estimate is possible, as already exemplified in Sect. A10.1.1. In fact, the system developed below achieves this automatically.

Of considerable interest is the possibility of accommodating model order uncertainty, i.e., $N \neq n$, as this is not achieved by sliding mode control to date. This will be found possible but with certain restrictions.

Suppose that the real plant (A10.7) and its model (A10.8) are fed by the same arbitrary control vector, $\mathbf{u}_m(t) = \mathbf{u}(t)$, with zero initial states, $\mathbf{x}_m(0) = \mathbf{0}$ and $\mathbf{x}(0) = \mathbf{0}$. Then $\mathbf{y}_m(0) = \mathbf{y}(0) = \mathbf{0}$ but it is clear that $\mathbf{y}_m(t) \neq \mathbf{y}(t)$ for $t > 0$. Now suppose the existence of a *plant model mismatch equivalent input*, $\mathbf{u}_e(t)$, such that if $\mathbf{u}_m(t) = \mathbf{u}(t) - \mathbf{u}_e(t)$, then $\mathbf{y}_m(t) = \mathbf{y}(t)\ \forall t > 0$. Then $\mathbf{u}_e(t)$ converts the mismatched plant model into an *exact representation of the real plant* when both are viewed as black boxes, i.e., viewed only via their inputs and outputs, without regard to their internal states, as shown in Fig. A10.3.

Appendices

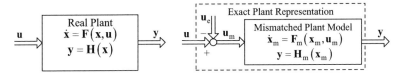

Fig. A10.3 Real plant and its exact representation using a mismatched plant model and $\mathbf{u}_e$

The existence of $\mathbf{u}_e$ may be determined in individual cases. For example, let the real plant be a tumbling but controllable rigid body spacecraft obeying Euler's equations of rigid body dynamics, which are of the form,

$$\left. \begin{array}{l} \dot{x}_1 = a_1 x_2 x_3 + b_1 u_1 \quad y_1 = x_1 \\ \dot{x}_2 = a_2 x_3 x_1 + b_2 u_2 \quad y_2 = x_2 \\ \dot{x}_3 = a_3 x_1 x_2 + b_2 u_2 \quad y_3 = x_3 \end{array} \right\} \quad (A10.9)$$

where x_1, x_2 and x_3 are the body angular velocity components about the principal axes of inertia and a_i, b_i, $i = 1, 2, 3$, are constants. Let the mismatched (and simplified) representation be three separate integrators together with $\mathbf{u}_e$, as follows.

$$\left. \begin{array}{l} \dot{x}_{m1} = u_{m1} - u_{e1} \quad y_{m1} = x_{m1} \\ \dot{x}_{m2} = u_{m2} - u_{e2} \quad y_{m2} = x_{m2} \\ \dot{x}_{m3} = u_{m3} - u_{e3} \quad y_{m3} = x_{m3} \end{array} \right\} \quad (A10.10)$$

If $u_{mi} = u_i$ gives $y_{mi} = y_i$, $i = 1, 2, 3$, then $x_{mi} = x_i$, $i = 1, 2, 3$ and

$$\left. \begin{array}{l} a_1 x_2 x_3 + b_1 u_1 = u_1 - u_{e1} \Rightarrow u_{e1} = (1 - b_1) u_1 - a_1 x_2 x_3 \\ a_2 x_3 x_1 + b_2 u_2 = u_2 - u_{e2} \Rightarrow u_{e2} = (1 - b_2) u_2 - a_2 x_3 x_1 \\ a_3 x_1 x_2 + b_2 u_2 = u_3 - u_{e3} \Rightarrow u_{e3} = (1 - b_3) u_3 - a_3 x_1 x_2 \end{array} \right\}. \quad (A10.11)$$

A10.1.3 Converting the Problem of Controlling an Uncertain Plant to That of Controlling a Known Plant Model

Starting with Fig. A10.3, suppose that the plant input is formed as $\mathbf{u} = \mathbf{u}' + \mathbf{u}_e$, as shown in Fig. A10.4.

Then since the externally added $\mathbf{u}_e$ cancels the internally subtracted $\mathbf{u}_e$ in the plant model, the overall control input, $\mathbf{u}'$, must equal the model control input, $\mathbf{u}_m$, as shown. Hence, if $\mathbf{u}_e$ is known, the problem of controlling the uncertain real plant has been converted to the relatively simple problem of controlling the known plant model, since the model state is accessible.

A10.1.4 Development of the General Controller Structure

Since $\mathbf{u}_e$ acts similarly to an external disturbance referred to the control input, it may be estimated by means of an observer and counteracted, as $\widehat{d}$ is estimated and counteracted in Fig. A10.1. Although general plant model state differential equations are indicated in Fig. A10.4 only linear time invariant plant models will be considered henceforth in the interests of simplicity, no disadvantage being envisaged. A generic control system with this structure is shown in Fig. A10.5.

Comparison with Fig. A10.4 shows that $\mathbf{u}_e$ has been replaced by its estimate, $\widehat{\mathbf{u}}_e$, from the observer. Increasing the observer eigenvalues sufficiently yields $\mathbf{e} \to \mathbf{0} \Rightarrow \widehat{\mathbf{u}}_e \to \mathbf{u}_e \Rightarrow \mathbf{y} \to \mathbf{y}_m$. So once the model state control law is designed to give the specified closed loop dynamic response of $\mathbf{y}_m(t)$ to $\mathbf{y}_r(t)$, if the observer drives $\mathbf{e}(t)$ to negligible proportions, then $\mathbf{y}(t)$ will also exhibit this dynamics.

The system of Fig. A10.5 could be modified by adopting an unconventional observer structure in which all the model correction takes place using one loop including $\widehat{\mathbf{u}}_e$, as shown in Fig. A10.6.

The conventional correction loop of any observer, which comprises individual loops acting on the correction inputs of the integrators of the model, can be replaced by a single loop as shown in Fig. A10.6. For example, the three loops of the observer in Fig. A10.1 can be replaced by a single loop as shown in Fig. A10.7.

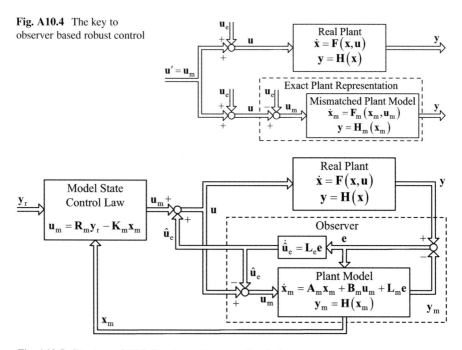

Fig. A10.4 The key to observer based robust control

Fig. A10.5 Structure of OBRC system with conventional observer

Appendices

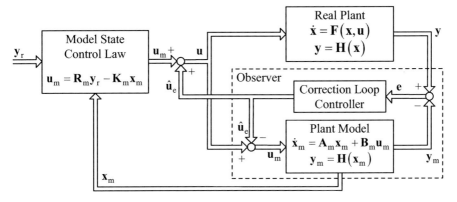

Fig. A10.6 Structure of OBRC system with modified observer

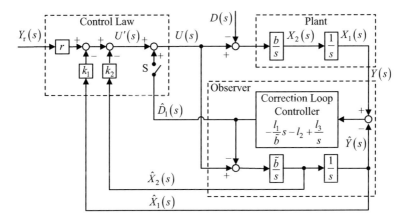

Fig. A10.7 Single axis, rigid body spacecraft attitude control system with modified observer

The performance of this observer in estimating x_1 and x_2 will be identical to that of Fig. A10.1 but the disturbance estimate, $\widehat{d}_1(t)$ will have slightly different dynamics than $\widehat{d}(t)$. Comparing the two observers yields

$$\widehat{D}_1(s) = \frac{\frac{l_3}{s} - l_2 - \frac{l_1}{b}s}{\frac{l_3}{s}} \widehat{D}(s) = \left(1 - \frac{l_2}{l_3}s - \frac{l_1}{bl_3}s^2\right)\widehat{D}(s)$$

$$= \left(1 + \frac{1}{2}T_{so}s + \frac{1}{12}T_{so}^2 s^2\right)\widehat{D}(s) \qquad (A10.12)$$

Analysis of Fig. A10.1 shows that

$$\widehat{D}(s) = \frac{1}{1 + \frac{1}{2}T_{so}s + \frac{1}{12}T_{so}^2 s^2 + \frac{1}{216}T_{so}^3 s^3} D(s). \qquad (A10.13)$$

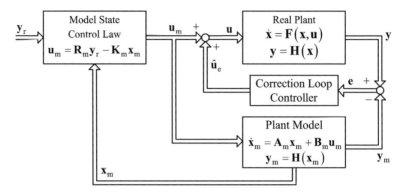

Fig. A10.8 Basic OBRC system

Hence

$$\widehat{D}_1(s) = \frac{1 + \frac{1}{2}T_{so}s + \frac{1}{12}T_{so}^2 s^2}{1 + \frac{1}{2}T_{so}s + \frac{1}{12}T_{so}^2 s^2 + \frac{1}{216}T_{so}^3 s^3} D(s) \qquad (A10.14)$$

Indicating that less dynamic lag of the disturbance estimate will occur with the modified observer than with the conventional one.

Returning to the system of Fig. A10.6, the connections and summing junctions between the blocks may be simplified to yield the system of Fig. A10.8, which is equivalent, although the observer correction loop has apparently vanished.

A10.1.5 Limitations

The question is now addressed of the restrictions on the plant model for a given plant. In every case, it should be possible to design the correction loop controller based only on the known plant model so that the error, $e(t)$, is kept to negligible proportions. It is clear that the model can be of the same order as the real plant and of the same form. Then the model parameters may differ considerably from those of the real plant, the differences being accommodated by $\hat{u}_e$. It has also been demonstrated by example that the real plant can be nonlinear and the model linear. Each case, however, would have to be examined individually for feasibility.

Suppose that the model order is less than that of the plant. Then $y(t)$ closely follows $y_m(t)$ so that the plant *appears* to be of the same order as the model. In reality, of course, it is of higher order, meaning that there must be uncontrolled states of a zero dynamic subsystem. This would have to be analysed for stability or investigated by simulation.

An interesting case is the model order being greater than that of the plant. There appears to be no restriction here, implying that the same controller might be used with a range of plants with different orders less or equal to the model order. A simulation will be presented that demonstrates this.

It is important to realise that the error, e(t), cannot be driven to precisely zero and this could cause instability in some cases. To combat this problem, it has been found necessary to include an adjustable forward path gain parameter in the plant model as described in the following section. Theoretical analysis may be intractable for nonlinear or higher order systems but in any case investigation by simulation can be carried out.

A10.1.6 The Multiple Integrator Plant Model and Correction Loop Controller

Given that OBRC is intended to accommodate large mismatches between the plant and its model, the simplest plant model that can be used is a chain of integrators at least equal in number to the plant order for a SISO plant. For a multivariable plant, with m control inputs and m measured/controlled outputs, the simplest plant model consists of m separate chains of integrators. This will automatically eliminate control channel interaction in the closed loop system. If the relative degree with respect to the ith output is r_i, then there should be at least r_i integrators in the associated integrator chain. If the plant is not of full relative degree, then there will be an uncontrolled zero dynamic subsystem that will have to be analysed or investigated by simulation to check for stability.

The observer design for a multiple integrator model is particularly straightforward in theory but a certain issue will be highlighted by considering the observer based on a triple integrator chain shown in Fig. A10.9.

Note that it is unnecessary to include an integrator for producing the estimate, $\hat{u}_e$, of the model mismatch equivalent input when forming the observer structure of Fig. A10.9a as the output of the equivalent correction loop controller in Fig. A10.9b automatically produces this estimate, as shown. Also the forward path gain, b, will have no effect on the hypothetical ideal controller, which would have $T_{so} = 0$, but for a realisable controller $T_{so} > 0$, and it has been found necessary to include b for adjustment to a value that minimises observer transient errors or even instability.

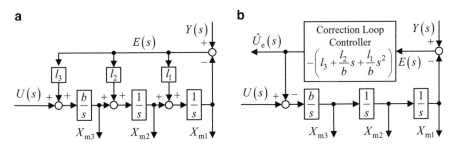

Fig. A10.9 Triple integrator based observer. (**a**) Conventional observer structure. (**b**) Basic structure for OBRC

Fig. A10.10 Observer for OBRC system using a polynomial correction loop controller

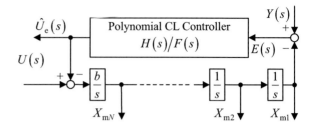

The issue referred to above is the implementation of the derivative and double derivative terms of the correction loop controller. It would be necessary to employ discrete approximations and preferably include low pass measurement noise filtering by replacing s and s^2 by, respectively, $s/(1+sT_f)$ and $s^2/(1+sT_f)^2$, where T_f is the filtering time constant. Since relatively high observer gains would be needed to achieve an observer settling time, T_{so}, sufficiently short to drive $e(t)$ to negligible proportions, then it would be necessary to set $T_f \ll T_{so}$ with the risk of rendering the filtering ineffective and placing high computational demands on the digital processor through requiring a sampling period satisfying $h \ll T_f$. It is here that the polynomial controller of Chap. 5 could be brought to the rescue, as it achieves complete pole placement without differentiators. A generic observer using this correction loop controller is shown in Fig. A10.10. Referring to Chap. 5, Sect. 5.3.3, in this case, the degrees, n_f and n_h, of the controller polynomials satisfy

$$n_f \geq n_h = N - 1. \tag{A10.15}$$

The minimum order controller will be chosen for which $n_f = n_h = N - 1$. So for a triple integrator plant model, $F(s) = f_2 s^2 + f_1 s + f_0$ and $H(s) = h_2 s^2 + h_1 s + h_0$, the coefficients of which may be calculated using

$$\begin{bmatrix} 1 & 0 & 0 & 0 & 0 & 0 \\ 0 & 1 & 0 & 0 & 0 & 0 \\ 0 & 0 & 1 & 0 & 0 & 0 \\ 0 & 0 & 0 & b & 0 & 0 \\ 0 & 0 & 0 & 0 & b & 0 \\ 0 & 0 & 0 & 0 & 0 & b \end{bmatrix} \begin{bmatrix} f_2 \\ f_1 \\ f_0 \\ h_2 \\ h_1 \\ h_0 \end{bmatrix} = \begin{bmatrix} 1 \\ d_4 \\ d_3 \\ d_2 \\ d_1 \\ d_0 \end{bmatrix} \Rightarrow \begin{cases} f_2 = 1 \\ f_1 = d_4 \\ f_0 = d_3 \\ h_2 = d_2/b \\ h_1 = d_1/b \\ h_0 = d_0/b \end{cases} \tag{A10.16}$$

where the desired characteristic polynomial is

$$s^5 + d_4 s^4 + d_3 s^3 + d_2 s^2 + d_1 s + d_0 = (s+q)^5 = s^5 + 5qs^4 + 10q^2 s^3 \\ + 10q^3 s^2 + 5q^4 s + q^5 \Rightarrow$$

$$d_0 = q^5, d_1 = 5q^4, d_2 = 10q^3, d_3 = 10q^2, \text{ and } d_4 = 5q. \tag{A10.17}$$

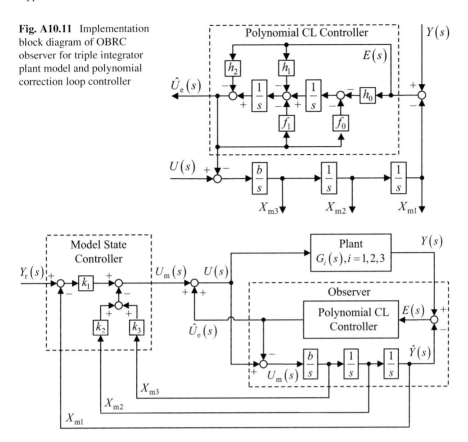

Fig. A10.11 Implementation block diagram of OBRC observer for triple integrator plant model and polynomial correction loop controller

Fig. A10.12 OBRC system for control of plants of up to third order

If the correction loop settling time is T_{so} for the 5 % criterion, then $q = 9/T_{so}$. Extension to other values of N may be undertaken by analogy with the example above.

For implementation, as shown in Fig. A10.11 (Sect. 5.3 of Chap. 5), the correction loop controller only contains integrators, summing junctions and gains.

A10.1.7 A Demonstration of Model Order Uncertainty Accommodation

Control of a first, second and third order plant using the same controller will now be simulated using the observer of Fig. A10.11 to obtain nominally the same third order response to a given reference input with a specified settling time of T_s seconds (5 % criterion). The complete control system is shown in Fig. A10.12.

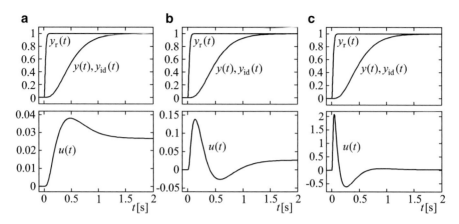

Fig. A10.13 Control of 1st, 2nd and 3rd order plants with the same robust controller. (**a**) Plant $G_1(s)$. (**b**) Plant $G_2(s)$. (**c**) Plant $G_3(s)$.

The model state controller gains are given by

$$s^3 + k_3bs^2 + k_2bs + k_1b = (s+p)^3 = s^3 + 3ps^2 + 3p^2s + p^3 \Rightarrow$$
$$k_1 = p^3/b, k_2 = 3p^2/b \quad \text{and} \quad k_3 = 3p/b, \tag{A10.18}$$

where $p = 6/T_s$. The three plant transfer functions are

$$G_1(s) = \frac{b_0}{s+a_0}, G_2(s) = \frac{b_0}{s^2 + a_1s + a_0}, \text{ and } G_3(s) = \frac{b_0}{s^3 + a_2s^2 + a_1s + a_0}, \tag{A10.19}$$

where, arbitrarily, $b_0 = 0.6$, $a_0 = 1.6$, $a_1 = 0.2$ and $a_2 = 1.5$. The controller parameters are $T_s = 1$ [s], $T_{so} = 0.01$ [s] and $b = 10$. Figure A10.13 shows the results.

It was found that step reference inputs excited short term initial observer transients which, while being harmless regarding the output responses, were visible in $u(t)$ and an order of magnitude larger than the remainder of $u(t)$, rendering observations difficult. These transients were avoided by smoothing the reference input using a filter with transfer function, $[100/(s+100)]^3$. Referring to Fig. A10.13, it is remarkable that there is no visible difference between the response, $y(t)$, for each plant and the ideal system response, $y_{id}(t)$. The controller produces the three very different control inputs, $u(t)$, automatically with no readjustments.

A10.1.8 A Multivariable Control Example

Finally, the capability of OBRC in controlling a nonlinear multivariable plant will be demonstrated. This is the large angle, three axis attitude control manoeuvre of the rigid body spacecraft of Chap. 10, Example 10.6, which demonstrated pseudo sliding mode control. The plant parameters and the specified settling time are the same as in this example. The OBRC system comprises three identical controllers, one for each plant input/output pair, which assures elimination of inter-axis cross coupling, i.e., control channel interaction.

In this case, the relative degree with respect to each output is $r_1 = r_2 = r_3 = 2$, so the observer model is a double integrator for each spacecraft control axis. The block diagram of one loop of the three axis control system is shown in Fig. A10.14.

The other two loops are identical.

For an observer settling time of T_{so} seconds (5 % criterion), the polynomial correction loop parameters are

$$f_0 = 3q, \quad h_0 = q^3/b, \quad \text{and} \quad h_1 = 3q^2/b \qquad (A10.20)$$

where $q = 3/T_{so}$. Similarly, for a specified attitude control settling time of T_s seconds (5 % criterion), the model state control loop gains are

$$k_1 = p^2/b \quad \text{and} \quad k_2 = 2p/b. \qquad (A10.21)$$

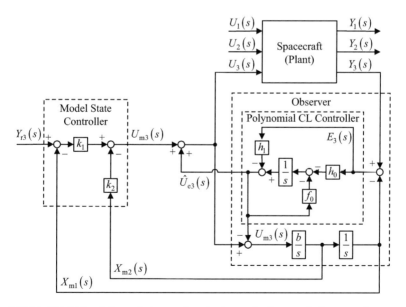

Fig. A10.14 Third (z) axis loop of observer-based robust attitude control of spacecraft

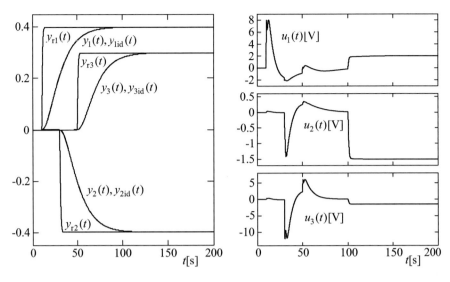

Fig. A10.15 Large angle slew manoeuvre of spacecraft under observer based robust control

Here, $p = 4.5/T_s$. Figure A10.15 shows a spacecraft attitude control simulation corresponding to that of Chap. 10, Example 10.6, in which staggered roll, pitch and yaw attitude demands are applied to demonstrate the elimination of inter-axis cross coupling and a disturbance torque is applied at $t = 100$ [s], a typical side effect of trajectory change thruster firings. The controller parameters are set to $T_s = 50$ [s], $T_{so} = 1$ [s] and $b = 10^{-5}$. Also, as in the demonstration of Fig. A10.13, the step reference inputs are smoothed by a low pass filter, in this case with transfer function, $[s/(s+2)]^2$, to avoid observer transients that, while being harmless regarding the attitude response, would mask the variations in the control inputs. This also applied to the disturbance torque inputs, which is not unrealistic considering the force build-up time of real trajectory change thrusters.

The outputs are shown dimensionless because they are measurements of attitude quaternions that themselves are dimensionless.

As for the pseudo sliding mode control (Example 10.6 of Chap. 10), the responses, $y_i(t)$, $i = 1, 2, 3$, are indistinguishable from the ideal responses, $y_{i\,id}(t)$, $i = 1, 2, 3$ and the control system is very effective at counteracting the interaction due to the inherent cross-coupling of the spacecraft dynamics and kinematics.

References

1. Dodds SJ (2008) Observer based robust control. In: Proceedings of the AC&T 2008. University of East London, London, pp 151–159. ISBN 0-9550008-3-1
2. Dodds SJ et al (2010) Observer based robust control of an electric drive with a flexible coupling, Speedam 2010, Pisa, Italy

A11 Appendix to Chap. 11

A11.1 Path Planning and Reference Input Trajectory Generation

A11.1.1 Basic Requirements

Software differentiation is not recommended for two reasons. First, an arbitrary continuous reference input designed to pass through the time tagged points may not be differentiable up to the required order. A suitable one is inevitably found in the form of a differentiable mathematical function and this enables the required derivatives to be determined analytically. Second, determining the derivatives analytically and computing them from the derivative functions is more accurate than software differentiation. The spline method, which is ideal for producing the required reference inputs and derivatives, is introduced in the following section.

A11.1.2 Introduction to Splines

The *spline* is a smooth curve passing through a number of specified points in which the segments between neighbouring points are defined by individual equations. The values yielded by the equations of two adjacent segments must, of course, be identical at the point they share. To ensure smoothness, the first derivatives of two adjacent segments must also be identical at the point they share. The simplest segment equation satisfying these requirements is a cubic equation whose four coefficients are calculated to yield the values and first derivatives specified at the end points. The spline formed in this way is referred to as a cubic spline. Considering the history, the original splines were not sets of segment equations but physical strips of wood or metal used in the boat and aircraft building industries that could be elastically stressed to pass through a number of fixed points on a structure to mark a smooth curve. Similar devices, called flexi-curve rulers, are in use today by draftsmen for marking smooth curves passing through given points.

Various functions can be used to form the segment equations of a spline [1]. The most convenient for the reference input generator application, however, is the polynomial function. The notation adopted is defined in Fig. A11.1.

Each spline segment has its own local time, τ, starting at $\tau = 0$ and ending at $\tau = \tau_i$ for the ith segment, to simplify the determination of the polynomial coefficients. The dotted curves are the continuation of the graph of the ith segment outside the range, $\tau \in [0, \tau_i)$, that contributes to $y_r(t)$. The number of polynomial coefficients is equal to the number of segment parameters specified. Since these are equally divided between the end points of the segment, then the number of polynomial coefficients has to be even, implying that the degree of the polynomial has to be odd. So the cubic spline, with four coefficients per segment, enables the two end point positions and two end point first derivatives to be specified.

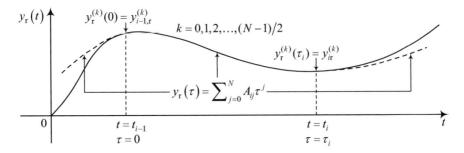

Fig. A11.1 Notation for polynomial spline function in reference input generator

The quintic spline enables, in addition, the specified second derivatives, one at each end point, to be realised. This ensures no discontinuities in $y_r^{(2)}$ when passing from one segment to the next, which is advantageous in many applications as this would ensure bump-less transitions from one segment to the next through continuous forces and torques. In general, a polynomial spline of Nth order, N odd, enables $y(0)$ and $y(\tau_i)$ to be realised together with the derivatives, $y^{(k)}(0)$ and $y^{(k)}(\tau_i)$, $k = 1, 2, \ldots, (N-1)/2$.

The coefficients of each segment are obtained as the solution of $N + 1$ linear simultaneous equations obtained by repeated differentiation of the segment equation and substitution of the end point times, $\tau = 0$ and $\tau = \tau_i$. Algorithms are developed in the following two sections for calculation of the cubic and quintic spline segment polynomial coefficients. These should be sufficient for most applications. If needed, however, the coefficient algorithms for polynomial splines of degree 6 or higher may be derived using the same method.

A11.1.3 The Cubic Spline Coefficients

The cubic equation of the ith segment is

$$y_{ir}^{(0)}(\tau) = A_{i0} + A_{i1}\tau + A_{i2}\tau^2 + A_{i3}\tau^3 \qquad (A11.1)$$

The first three derivatives are then

$$y_{ir}^{(1)}(\tau) = A_{i1} + 2A_{i2}\tau + 3A_{i3}\tau^2, \qquad (A11.2)$$

$$y_{ir}^{(2)}(\tau) = 2A_{i2} + 6A_{i3}\tau \qquad (A11.3)$$

$$y_{ir}^{(3)}(\tau) = 6A_{i3} \qquad (A11.4)$$

The choice of the local time scale for the segment enables two of the coefficients to be obtained simply. Setting $\tau = 0$ in (A11.1) and (A11.2) yields

$$A_{i0} = y_{ir}^{(0)}(0) \qquad (A11.5)$$

and

$$A_{i1} = y_{ir}^{(1)}(0). \qquad (A11.6)$$

Two simultaneous equations for the remaining coefficients are then obtained by substituting for A_{i0} and A_{i1} in (A11.1) and (A11.2) using (A11.5) and (A11.6) and setting $\tau = \tau_i$, yielding
$$\begin{cases} y_{ir}^{(0)}(\tau_i) = y_{ir}^{(0)}(0) + y_r^{(1)}(0)\tau_i + A_{i2}\tau_i^2 + A_{i3}\tau_i^3 \\ y_{ir}^{(1)}(\tau_i) = y_{ir}^{(1)}(0) + 2A_{i2}\tau_i + 3A_{i3}\tau_i^2 \end{cases} \Rightarrow$$

$$\begin{bmatrix} \tau_i^2 & \tau_i^3 \\ 2\tau_i & 3\tau_i^2 \end{bmatrix} \begin{bmatrix} A_{i2} \\ A_{i3} \end{bmatrix} = \begin{bmatrix} y_{ir}^{(0)}(\tau_i) - y_{ir}^{(0)}(0) - y_{ir}^{(1)}(0)\tau_i \\ y_{ir}^{(1)}(\tau_i) - y_{ir}^{(1)}(0) \end{bmatrix}.$$

Since $\begin{bmatrix} \tau_i^2 & \tau_i^3 \\ 2\tau_i & 3\tau_i^2 \end{bmatrix}^{-1} = \frac{1}{\tau_i} \begin{bmatrix} \tau_i & \tau_i^2 \\ 2 & 3\tau_i \end{bmatrix}^{-1} = \frac{1}{\tau_i^3} \begin{bmatrix} 3\tau_i & -\tau_i^2 \\ -2 & \tau_i \end{bmatrix}$, then

$$\begin{bmatrix} A_{i2} \\ A_{i3} \end{bmatrix} = \frac{1}{\tau_i^3} \begin{bmatrix} 3\tau_i & -\tau_i^2 \\ -2 & \tau_i \end{bmatrix} \begin{bmatrix} y_{ir}(\tau_i) - y_{ir}(0) - y_{ir}^{(1)}(0)\tau_i \\ y_{ir}^{(1)}(\tau_i) - y_{ir}^{(1)}(0) \end{bmatrix}. \qquad (A11.7)$$

This ensures $y_r(t)$ and $\dot{y}_r(t)$ are continuous through the whole spline function but there are discontinuities in $\ddot{y}_r(t)$ and $\dddot{y}_r(t)$ between one segment and the next.

A11.1.4 The Quintic Spline Coefficients

The equation of the ith quintic spline segment is

$$y_{ir}^{(0)}(\tau) = A_{i0} + A_{i1}\tau + A_{i2}\tau^2 + A_{i3}\tau^3 + A_{i4}\tau^4 + A_{i5}\tau^5 \qquad (A11.8)$$

The first five derivatives, which are available for dynamic lag pre-compensation, are

$$y_{ir}^{(1)}(\tau) = A_{i1} + 2A_{i2}\tau + 3A_{i3}\tau^2 + 4A_{i4}\tau^3 + 5A_{i5}\tau^4 \qquad (A11.9)$$

$$y_{ir}^{(2)}(\tau) = 2A_{i2} + 6A_{i3}\tau + 12A_{i4}\tau^2 + 20A_{i5}\tau^3 \qquad (A11.10)$$

$$y_{ir}^{(3)}(\tau) = 6A_{i3} + 24A_{i4}\tau + 60A_{i5}\tau^2 \qquad (A11.11)$$

$$y_{ir}^{(4)}(\tau) = 24A_{i4} + 120A_{i5}\tau \qquad (A11.12)$$

$$y_{ir}^{(5)}(\tau) = 120A_{i5}. \qquad (A11.13)$$

In this case, the specified segment parameters are $y_{ir}^{(0)}(0)$, $y_{ir}^{(1)}(0)$, $y_{ir}^{(2)}(0)$, $y_{ir}^{(0)}(\tau_i)$, $y_{ir}^{(1)}(\tau_i)$ and $y_{ir}^{(2)}(\tau_i)$. Then three of the spline coefficients are obtained easily from (A11.8), (A11.9) and (A11.10) by setting $\tau = 0$. Thus

$$A_{i0} = y_{ir}^{(0)}(0), \qquad (A11.14)$$

$$A_{i1} = y_{ir}^{(1)}(0), \qquad (A11.15)$$

and

$$A_{i2} = y_{ir}^{(2)}(0)/2. \qquad (A11.16)$$

Three simultaneous equations for the remaining coefficients are then obtained by substituting for A_{i0}, A_{i1} and A_{i2} in (A11.8), (A11.9) and (A11.10) using (A11.14), (A11.15) and (A11.16) and setting $\tau = \tau_i$, yielding

$$\begin{cases} y_{ir}^{(0)}(\tau_i) = y_{ir}^{(0)}(0) + y_{ir}^{(1)}(0)\tau_i + \left[y_{ir}^{(2)}(0)/2\right]\tau_i^2 + A_{i3}\tau_i^3 + A_{i4}\tau_i^4 + A_{i5}\tau_i^5 \\ y_{ir}^{(1)}(\tau_i) = y_{ir}^{(1)}(0) + 2\left[y_{ir}^{(2)}(0)/2\right]\tau_i + 3A_{i3}\tau_i^2 + 4A_{i4}\tau_i^3 + 5A_{i5}\tau_i^4 \\ y_{ir}^{(2)}(\tau_i) = 2\left[y_{ir}^{(2)}(0)/2\right] + 6A_{i3}\tau_i + 12A_{i4}\tau_i^2 + 20A_{i5}\tau_i^3 \end{cases} \Rightarrow$$

$$\begin{bmatrix} \tau_i^3 & \tau_i^4 & \tau_i^5 \\ 3\tau_i^2 & 4\tau_i^3 & 5\tau_i^4 \\ 6\tau_i & 12\tau_i^2 & 20\tau_i^3 \end{bmatrix} \begin{bmatrix} A_{i3} \\ A_{i4} \\ A_{i5} \end{bmatrix} = \begin{bmatrix} y_{ir}^{(0)}(\tau_i) - y_{ir}^{(0)}(0) - y_{ir}^{(1)}(0)\tau_i - \left[y_{ir}^{(2)}(0)/2\right]\tau_i^2 \\ y_{ir}^{(1)}(\tau_i) - y_{ir}^{(1)}(0) - y_{ir}^{(2)}(0)\tau_i \\ y_{ir}^{(2)}(\tau_i) - y_{ir}^{(2)}(0) \end{bmatrix} \Rightarrow$$

$$\begin{bmatrix} A_{i3} \\ A_{i4} \\ A_{i5} \end{bmatrix} = \frac{1}{\tau_i} \begin{bmatrix} \tau_i^2 & \tau_i^3 & \tau_i^4 \\ 3\tau_i & 4\tau_i^2 & 5\tau_i^3 \\ 6 & 12\tau_i & 20\tau_i^2 \end{bmatrix}^{-1} \begin{bmatrix} y_{ir}^{(0)}(\tau_i) - y_{ir}^{(0)}(0) - y_{ir}^{(1)}(0)\tau_i - \left[y_{ir}^{(2)}(0)/2\right]\tau_i^2 \\ y_{ir}^{(1)}(\tau_i) - y_{ir}^{(1)}(0) - y_{ir}^{(2)}(0)\tau_i \\ y_{ir}^{(2)}(\tau_i) - y_{ir}^{(2)}(0) \end{bmatrix}.$$

$$(A11.17)$$

Forming the matrix inverse algebraically produces rather unwieldy expressions and therefore numerical computation for each spline segment is recommended.

A11.1.5 Setting of the Derivatives at the Segment End Points

If one is free to choose any smooth curve passing through a finite set of points, then there are an infinite number of such curves to choose from. Similarly, there exists an

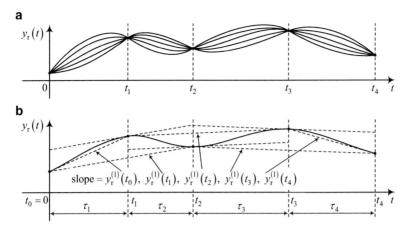

Fig. A11.2 Illustration of different splines with continuous first derivative. (**a**) Five different splines passing through the same points. (**b**) Splines with point derivatives calculated using neighbouring points

infinite set of polynomial splines that pass through a finite set of points, examples of which are shown in Fig. A11.2a.

Each of these has different derivatives at the end points. This is true even with the restriction that the derivatives that can be specified are common for shared points of adjacent spline segments. The possibility of this infinite set is illustrated in Fig. A11.2a for splines with four segments fitted to five points. The question then arises of which is the best choice. The shortest path connecting the adjacent points comprises contiguous straight line segments but would be unsuitable due to the first derivative being discontinuous. So suitable spline fits with, at least, continuous first derivatives, would be those that do not stray unnecessarily far from the shortest path. To eliminate unnecessary oscillations of the spline fit between the points, a method should be chosen such that in the special case of all the points lying on a straight line, then the spline becomes that straight line. Such a spline can be found simply by calculating the desired first derivatives at the interior points, P_i, $i = 1, 2, N_p - 2$, i.e., all the points excluding the end points, P_0 and P_{N_p-1}, as the slope of the straight line connecting the two neighbouring points, as follows.

$$y_r^{(1)}(t_i) = [y_r(t_{i+1}) - y_r(t_{i-1})] / (\tau_i + \tau_{i+1}), \quad i = 1, 2, N_p - 2. \quad (A11.18)$$

Neither the neighbouring point to the left of the first point nor the neighbouring point to the right of the last point exist, and therefore the derivatives at these points may be calculated as the slopes of the lines joining these points to their existing neighbouring points, as follows.

and

$$y_r^{(1)}(t_0) = [y_r(t_1) - y_r(t_0)]/\tau_1$$
$$y_r^{(1)}(t_{N_p-1}) = [y_r(t_{N_p-1}) - y_r(t_{N_p-2})]/\tau_{N_p-1}. \quad \text{(A11.19)}$$

Figure A11.2b shows an example. Once the specified first derivatives have been calculated using (A11.18) and (A11.19), then the same method can be applied to calculate the specified second derivatives and higher derivatives up to order, $(N-1)/2$. The complete set of specified derivatives for a polynomial spline of order N is then

$$\begin{cases} y_r^{(k)}(t_i) = \left[y_r^{(k-1)}(t_{i+1}) - y_r^{(k-1)}(t_{i-1})\right]/(\tau_i + \tau_{i+1}), \\ y_r^{(k)}(t_0) = \left[y_r^{(k-1)}(t_1) - y_r^{(k-1)}(t_0)\right]/\tau_1, \\ y_r^{(1)}(t_{N_p-1}) = \left[y_r^{(k-1)}(t_{N_p-1}) - y_r^{(k-1)}(t_{N_p-2})\right]/\tau_{N_p-1} \end{cases}, \quad \begin{array}{l} i = 1, 2, N_p - 2 \\ k = 1, 2, \ldots, \frac{N-1}{2} \end{array}.$$

(A11.20)

It should be noted that the segment periods, τ_i, $i = 1, 2, \ldots, N_p$, do not have to be equal but if they are then the segment coefficient computations would be greatly reduced since the matrix inversion, such as that in (A11.17), would only have to carried out once, rather than for every segment.

It is also important to note that the method of (A11.20) is not mandatory. It may be preferable to specify different point derivatives to better suit specific applications.

A11.1.6 Maximum Relative Degree of Controlled Plant

It has been established in the previous section that for a spline-based reference input, $y_r(t)$, comprising contiguous polynomials of degree, N, all the derivatives of $y_r(t)$ with orders up to a maximum of $(N-1)/2$ can be specified at both end points of each segment. Hence for all the interior points, i.e., all the points marking the beginnings and ends of segments, not including the end points of the spline, the derivatives at the end of each segment can be made equal to the derivative at the beginning of the following segment, thereby ensuring continuous derivatives of $y_r(t)$ up to order, $(N-1)/2$ over the complete spline. The next highest derivative of order, $(N-1)/2 + 1$, will be discontinuous with finite jumps at the interior points. Higher output derivatives will contain infinite impulses. If the plant has relative degree, R, with respect to the output in question (Chap. 3), then the output derivative of order, R, depends directly on the control input, $u(t)$. If the output, $y(t)$, of the plant is forced by the control input to precisely follow the spline by means of a dynamic lag pre-compensator and $R = (N-1)/2 + 1$, then $u(t)$ will exhibit finite jumps, which is acceptable provided control saturation doesn't occur. Thus, the maximum relative degree of plant controlled with zero dynamic lag using a spline of polynomial order, N, is

$$R_{\max} = (N-1)/2 + 1 = (N+1)2. \quad \text{(A11.21)}$$

Appendices

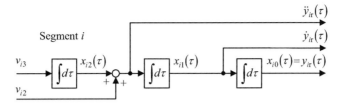

Fig. A11.3 Controlled integrator chain producing cubic spline reference input and derivatives

A11.1.7 Reference Input Generator Implementation

A11.1.7.1 The Cubic Spline Generator

The complete cubic spline function may be generated using a chain of three pure integrators, as shown in Fig. A11.3, driven by two piecewise constant control inputs,

$$v_j(t) = v_{ij} = const., \; j = 2, 3, t \in [t_{i-1}, t_i), i = 1, 2, \ldots, N_p - 1. \quad (A11.22)$$

With reference to Sect. A11.1.6, the coefficients of the spline segments are chosen to make $\dot{y}_r(t)$ continuous through the whole spline function. Hence no control input, v_{i1}, is provided at the input of the third integrator as changing this input from one segment to the next would cause discontinuities in $\dot{y}_r(t)$. Also, since the control system is expected to follow $y_r(t)$ without dynamic lag, the integrator chain is driven with zero initial conditions at the beginning of segment 1, requiring $y_r(0) = 0$. It is therefore assumed that the plant is started with zero initial state. For subsequent spline segments, in the local time, $\tau = t - t_{i-1}$, the initial conditions with respect to the local time, τ, i.e., $x_{ij}(0)$, $j = 0, 1, 2$, are those occurring at the end of the previous segment. In the following, note that $x_{i0}(\tau) = y_{ir}(\tau)$.

Equations enabling the piecewise constant integrator control levels, v_{i2} and v_{i3}, $i = 1, 2, \ldots, N_p - 1$, to be calculated will now be derived in terms of the cubic spline coefficients and the spline segment duration, τ_s, which is assumed to be the same for every segment. The state transition equation (Chap. 3) connecting the integrator chain states at the beginning and end of the spline segments, excluding segment 1, is

$$\begin{bmatrix} x_{i0}(\tau_s) \\ x_{i1}(\tau_s) \\ x_{i2}(\tau_s) \end{bmatrix} = \begin{bmatrix} 1 & \tau_s & \frac{1}{2}\tau_s^2 \\ 0 & 1 & \tau_s \\ 0 & 0 & 1 \end{bmatrix} \begin{bmatrix} x_{i-1,0}(\tau_s) \\ x_{i-1,1}(\tau_s) \\ x_{i-1,2}(\tau_s) \end{bmatrix} + \begin{bmatrix} \frac{1}{2}\tau_s^2 & \frac{1}{6}\tau_s^3 \\ \tau_s & \frac{1}{2}\tau_s^2 \\ 0 & \tau_s \end{bmatrix} \begin{bmatrix} v_{i2} \\ v_{i3} \end{bmatrix} \quad (A11.23)$$

for $i = 2, 3, \ldots, N_p - 1$. For segment 1, the state transition equation is

$$\begin{bmatrix} x_{10}(\tau_s) \\ x_{11}(\tau_s) \\ x_{12}(\tau_s) \end{bmatrix} = \begin{bmatrix} 1 & \tau_s & \frac{1}{2}\tau_s^2 \\ 0 & 1 & \tau_s \\ 0 & 0 & 1 \end{bmatrix} \begin{bmatrix} x_{10}(0) \\ x_{11}(0) \\ x_{12}(0) \end{bmatrix} + \begin{bmatrix} \frac{1}{2}\tau_s^2 & \frac{1}{6}\tau_s^3 \\ \tau_s & \frac{1}{2}\tau_s^2 \\ 0 & \tau_s \end{bmatrix} \begin{bmatrix} v_{12} \\ v_{13} \end{bmatrix}, \quad (A11.24)$$

of which the first component equation is

$$x_{10}(\tau_s) = x_{10}(0) + x_{11}(0)\tau_s + \frac{1}{2}[x_{12}(0) + v_{12}]\tau_s^2 + \frac{1}{6}v_{13}\tau_s^3 \quad (A11.25)$$

This must also be the equation of the cubic spline segment 1, i.e.,

$$x_{10}(\tau_s) = A_{10} + A_{11}\tau_s + A_{12}\tau_s^2 + A_{13}\tau_s^3. \quad (A11.26)$$

Equating the RHS of (A11.25) and (A11.26) then yields

$$x_{10}(0) = A_{10} \text{ and } x_{11}(0) = A_{11}, \quad (A11.27)$$

together with

$$v_{12} = 2A_{12} \text{ with } x_{12}(0) = 0 \text{ and } v_{13} = 6A_{13}. \quad (A11.28)$$

As stated above, the initial conditions of (A11.27) would normally be zero, the reference input starting from zero with zero initial plant state.

For the segment 2 calculations, the equations for the integrator state, $x_{12}(\tau_s)$, is needed and this is given by the third component equation of (A11.24), which is

$$x_{12}(\tau_s) = x_{12}(0) + v_{13}\tau_s. \quad (A11.29)$$

The first component equation of (A11.23) for segment 2 is

$$x_{20}(\tau_s) = x_{10}(\tau_s) + x_{11}(\tau_s)\tau_s + \frac{1}{2}[x_{12}(\tau_s) + v_{22}]\tau_s^2 + \frac{1}{6}v_{23}\tau_s^3, \quad (A11.30)$$

This must also be the equation of the segment 2 cubic spline polynomial,

$$x_{20}(\tau_s) = A_{20} + A_{21}\tau_s + A_{22}\tau_s^2 + A_{23}\tau_s^3. \quad (A11.31)$$

Equating the RHS of (A11.30) and (A11.31) then yields

$$x_{10}(\tau_s) = A_{20} \quad \text{and} \quad x_{11}(\tau_s) = A_{21} \quad (A11.32)$$

together with

$$v_{22} = 2A_{22} - x_{12}(\tau_s) \quad \text{and} \quad v_{23} = 6A_{23}. \quad (A11.33)$$

The coefficients, A_{20} and A_{21}, will have been chosen so that (A11.32) is satisfied. So only (A11.33) is needed.

For segment 3, the equation for the integrator state, $x_{22}(\tau_s)$, is needed and is given by the third component equation of (A11.23) with $i = 2$, which is

$$x_{22}(\tau_s) = x_{12}(\tau_s) + v_{23}\tau_s. \tag{A11.34}$$

The equations for calculating the integrator controls for the remaining spline segments are similar to those of segment 2 above. At the beginning of segment i, the integrator state, $x_{i-1,2}(\tau_s)$, will have been calculated. The complete algorithm for calculating the sequence of integrator controls is as follows, starting with the set of cubic spline coefficients and zero initial conditions. For segment 1,

$$v_{12} = 2A_{12} \quad \text{and} \quad v_{13} = 6A_{13} \tag{A11.35}$$

In preparation for segment 2,

$$x_{12}(\tau_s) = v_{13}\tau_s. \tag{A11.36}$$

For segment i,

$$v_{i2} = 2A_{i2} - x_{i-1,2}(\tau_s) \quad v_{i3} = 6A_{i3} \quad \text{and,} \quad i = 2, 3, \ldots, N_p - 1 \tag{A11.37}$$

In preparation for segment $i + 1$, from (A11.23),

$$x_{i2}(\tau_s) = x_{i-1,2}(\tau_s) + v_{i3}\tau_s, i = 2, 3, \ldots, N_p - 2 \tag{A11.38}$$

A11.1.7.2 The Quintic Spline Generator

The complete cubic spline function may be generated using a chain of five pure integrators driven by three piecewise constant control inputs,

$$v_j(t) = v_{ij} = const., j = 3, 4, 5, t \in [t_{i-1}, t_i), i = 1, 2, \ldots, N_p - 1, \tag{A11.39}$$

as shown in Fig. A11.4.

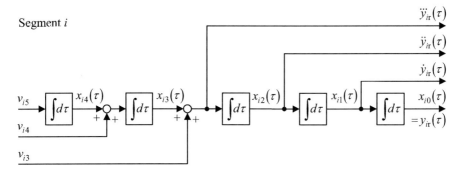

Fig. A11.4 Controlled integrator chain producing quintic spline reference input and derivatives

Since the coefficients of the spline segments are chosen to make $\dot{y}_r(t)$ and $\ddot{y}_r(t)$ continuous through the whole spline function, no control inputs, v_{i1} and v_{i2}, are provided at the inputs of the fourth and fifth integrators as changing these inputs from one segment to the next would cause discontinuities in $\dot{y}_r(t)$ and $\ddot{y}_r(t)$. As for the cubic spline generator, since the control system is expected to follow $y_r(t)$ without dynamic lag from $t = 0$, the integrator chain is driven with zero initial conditions at the beginning of segment 1, requiring $y_r(0) = 0$, assuming that the plant is started with zero initial state. For subsequent spline segments, the initial conditions with respect to the local time, $\tau = t - t_{i-1}$, are the integrator outputs occurring at the end of the previous segment. Again, $y_{i0}(\tau) = y_{ir}(\tau)$.

Equations enabling the piecewise constant integrator control levels, v_{i3}, v_{i4} and v_{i5}, $i = 1, 2, \ldots, N_p - 1$, to be calculated will now be derived in terms of the quintic spline coefficients and the spline segment duration, τ_s, which, is assumed to be the same for every segment. The state transition equation connecting the integrator chain states at the beginning and end of the spline segments, excluding segment 1, is

$$\begin{bmatrix} x_{i0}(\tau_s) \\ x_{i1}(\tau_s) \\ x_{i2}(\tau_s) \\ x_{i3}(\tau_s) \\ x_{i4}(\tau_s) \end{bmatrix} = \begin{bmatrix} 1 & \tau_s & \frac{1}{2}\tau_s^2 & \frac{1}{6}\tau_s^3 & \frac{1}{24}\tau_s^4 \\ 0 & 1 & \tau_s & \frac{1}{2}\tau_s^2 & \frac{1}{6}\tau_s^3 \\ 0 & 0 & 1 & \tau_s & \frac{1}{2}\tau_s^2 \\ 0 & 0 & 0 & 1 & \tau_s \\ 0 & 0 & 0 & 0 & 1 \end{bmatrix} \begin{bmatrix} x_{i-1,0}(\tau_s) \\ x_{i-1,1}(\tau_s) \\ x_{i-1,2}(\tau_s) \\ x_{i-1,3}(\tau_s) \\ x_{i-1,4}(\tau_s) \end{bmatrix} + \begin{bmatrix} \frac{1}{6}\tau_s^3 & \frac{1}{24}\tau_s^4 & \frac{1}{120}\tau_s^5 \\ \frac{1}{2}\tau_s^2 & \frac{1}{6}\tau_s^3 & \frac{1}{24}\tau_s^4 \\ \tau_s & \frac{1}{2}\tau_s^2 & \frac{1}{6}\tau_s^3 \\ 0 & \tau_s & \frac{1}{2}\tau_s^2 \\ 0 & 0 & \tau_s \end{bmatrix} \begin{bmatrix} v_{i3} \\ v_{i4} \\ v_{i5} \end{bmatrix}.$$
(A11.40)

for $i = 2, 3, \ldots, N_p - 1$. For segment 1, the state transition equation is

$$\begin{bmatrix} x_{10}(\tau_s) \\ x_{11}(\tau_s) \\ x_{12}(\tau_s) \\ x_{13}(\tau_s) \\ x_{14}(\tau_s) \end{bmatrix} = \begin{bmatrix} 1 & \tau_s & \frac{1}{2}\tau_s^2 & \frac{1}{6}\tau_s^3 & \frac{1}{24}\tau_s^4 \\ 0 & 1 & \tau_s & \frac{1}{2}\tau_s^2 & \frac{1}{6}\tau_s^3 \\ 0 & 0 & 1 & \tau_s & \frac{1}{2}\tau_s^2 \\ 0 & 0 & 0 & 1 & \tau_s \\ 0 & 0 & 0 & 0 & 1 \end{bmatrix} \begin{bmatrix} x_{10}(0) \\ x_{11}(0) \\ x_{12}(0) \\ x_{13}(0) \\ x_{14}(0) \end{bmatrix} + \begin{bmatrix} \frac{1}{6}\tau_s^3 & \frac{1}{24}\tau_s^4 & \frac{1}{120}\tau_s^5 \\ \frac{1}{2}\tau_s^2 & \frac{1}{6}\tau_s^3 & \frac{1}{24}\tau_s^4 \\ \tau_s & \frac{1}{2}\tau_s^2 & \frac{1}{6}\tau_s^3 \\ 0 & \tau_s & \frac{1}{2}\tau_s^2 \\ 0 & 0 & \tau_s \end{bmatrix} \begin{bmatrix} v_{13} \\ v_{14} \\ v_{15} \end{bmatrix},$$
(A11.41)

of which the first component equation is

$$x_{10}(\tau_s) = x_{10}(0) + x_{11}(0)\tau_s + \frac{1}{2}x_{12}(0)\tau_s^2$$
$$+ \frac{1}{6}[x_{13}(0) + v_{13}]\tau_s^3 + \frac{1}{24}[x_{14}(0) + v_{14}]\tau_s^4 + \frac{1}{120}v_{15}\tau_s^5. \quad \text{(A11.42)}$$

This must also be the equation of the quintic spline segment 1, i.e.,

$$x_{10}(\tau_s) = A_{10} + A_{11}\tau_s + A_{12}\tau_s^2 + A_{13}\tau_s^3 + A_{14}\tau_s^4 + A_{15}\tau_s^5. \quad \text{(A11.43)}$$

Equating the RHS of (A11.42) and (A11.43) then yields

$$x_{10}(0) = A_{10}, \ x_{11}(0) = A_{11} \text{ and } x_{12}(0) = 2A_{12} \quad \text{(A11.44)}$$

and

$$v_{13} = 6A_{13} \text{ with } x_{13}(0) = 0, \ v_{14} = 24A_{14} \text{ with } x_{14}(0) = 0 \text{ and } v_{15} = 120A_{15}. \quad \text{(A11.45)}$$

As stated towards the beginning of this section, the initial conditions of (A11.44) would usually be zero, $y_r(t)$, starting from zero with zero initial plant state.

In preparation for the segment 2 calculations, the equations for the integrator states, $x_{13}(\tau_s)$ and $x_{14}(\tau_s)$, are the fourth and fifth component equations of (A11.41) with $x_{13}(0) = 0$ and $x_{14}(0) = 0$. Thus

$$x_{13}(\tau_s) = v_{14}\tau_s + \frac{1}{2}v_{15}\tau_s^2 \quad \text{(A11.46)}$$

and

$$x_{14}(\tau_s) = v_{15}\tau_s. \quad \text{(A11.47)}$$

The first component equation of (A11.40) for segment 2 is

$$x_{20}(\tau_s) = x_{10}(\tau_s) + x_{11}(\tau_s)\tau_s + \frac{1}{2}x_{12}(\tau_s)\tau_s^2$$
$$+ \frac{1}{6}[x_{13}(\tau_s) + v_{23}]\tau_s^3 + \frac{1}{24}[x_{14}(\tau_s) + v_{24}]\tau_s^4 + \frac{1}{120}v_{25}\tau_s^5. \quad \text{(A11.48)}$$

This must also be the equation of the segment 2 quintic spline polynomial,

$$x_{20}(\tau_s) = A_{20} + A_{21}\tau_s + A_{22}\tau_s^2 + A_{23}\tau_s^3 + A_{24}\tau_s^4 + A_{25}\tau_s^5 \quad \text{(A11.49)}$$

Equating the RHS of (A11.48) and (A11.49) then yields

$$x_{10}(\tau_s) = A_{20}, \ x_{11}(\tau_s) = A_{21} \text{ and } x_{12}(\tau_s) = 2A_{22} \quad \text{(A11.50)}$$

with

$$v_{23} = 6A_{23} - x_{13}(\tau_s), \ v_{24} = 24A_{24} - x_{14}(\tau_s) \text{ and } v_{25} = 120A_{25}. \quad \text{(A11.51)}$$

The coefficients A_{20}, A_{21} and A_{22} will have been chosen so that (A11.50) is satisfied. So only (A11.51) is needed.

For segment 3, the equations for the states, $x_{23}(\tau_s)$ and $x_{24}(\tau_s)$ are needed. These are the fourth and fifth component equations of (A11.40) with $i = 2$, which are

$$x_{23}(\tau_s) = x_{13}(\tau_s) + x_{14}(\tau_s)\tau_s + v_{24}\tau_s + \frac{1}{2}v_{25}\tau_s^2$$

$$x_{24}(\tau_s) = x_{14}(\tau_s) + v_{25}\tau_s. \tag{A11.52}$$

The equations for calculating the integrator controls for the remaining spline segments are similar to those of segment 2 above. At the beginning of segment i, the integrator states, $x_{i-1,3}(\tau_s)$ and $x_{i-1,4}(\tau_s)$, will have been calculated. The complete algorithm for calculating the sequence of integrator controls is as follows, starting with the set of cubic spline coefficients and zero initial conditions.

For segment 1,

$$v_{13} = 6A_{13}, \quad v_{14} = 24A_{14} \quad \text{and} \quad v_{15} = 120A_{15}. \tag{A11.53}$$

In preparation for segment 2,

$$x_{13}(\tau_s) = v_{14}\tau_s + \frac{1}{2}v_{15}\tau_s^2 \quad \text{and} \quad x_{14}(\tau_s) = v_{15}\tau_s \tag{A11.54}$$

For segment i,

$$v_{i3} = 6A_{i3} - x_{i-1,3}(\tau_s), \quad v_{i4} = 24A_{i4} - x_{i-1,4}(\tau_s),$$

$$v_{i5} = 120A_{i5}, \quad i = 2, 3, \ldots, N_p - 1. \tag{A11.55}$$

In preparation for segment $i + 1$, from (A11.40),

$$x_{i3}(\tau_s) = x_{i-1,3}(\tau_s) + x_{i-1,4}(\tau_s)\tau_s + v_{i4}\tau_s + \frac{1}{2}v_{i5}\tau_s^2$$

$$\text{and} \quad x_{i4}(\tau_s) = x_{i-1,4}(\tau_s) + v_{i5}\tau_s, \quad i = 2, 3, \ldots, N_p - 2. \tag{A11.56}$$

A11.1.8 Checking the Actuator Capability for Reference Trajectory Tracking

Having planned a trajectory, it is important to check that a system designed with a dynamic lag pre-compensator will operate without control saturation, as this would cause the real trajectory to depart from the planned one. This could be done by means of a simulation of the complete system but if the plant is of full relative degree

(Chap. 3), there is an analytical method that can be applied prior to the controller design. Consider the plant state space model of the general form,

$$\dot{\mathbf{x}} = \mathbf{f}(\mathbf{x}, \mathbf{u}) \tag{A11.57}$$

$$\mathbf{y} = \mathbf{h}(\mathbf{x}) \tag{A11.58}$$

where the state, control and measurement vectors are, respectively, $\mathbf{x} \in \mathfrak{R}^n$, $\mathbf{u} \in \mathfrak{R}^r$ and $\mathbf{y} \in \mathfrak{R}^m$. The method is based on determining the control, $\mathbf{u}(t)$, that would be required to cause $\mathbf{y}(t)$ to follow a planned trajectory depending solely on the plant model of (A11.57) and (A11.58). Let the component equations of (A11.58) be repeatedly differentiated and substitutions for the state variable derivatives made using (A11.57) until algebraic dependences on the control vector, $\mathbf{u}$, are detected. It will be recalled that this is the procedure for determining the relative degree with respect to each controlled measurement variable (Chap. 3) or the formulation of forced dynamic or feedback linearising control laws (Chap. 7). Then if R_i is the relative degree with respect to y_i,

$$y_i^{(k)} = h_{ik}(\mathbf{x}), \quad k = 0, 1, \ldots, R_i - 1, \quad i = 1, 2, \ldots, m \tag{A11.59}$$

and
$$y_i^{(R_i)} = h_{iR_i}(\mathbf{x}, \mathbf{u}), \quad i = 1, 2, \ldots, m. \tag{A11.60}$$

Let $\mathbf{u}$ be made the subject of (A11.60), yielding

$$\mathbf{u} = \mathbf{f}_x\left(\mathbf{x}, y_1^{(R_1)}, y_2^{(R_2)}, \ldots, y_m^{(R_m)}\right). \tag{A11.61}$$

The number of output derivative equations in (A11.59) (including the measurement equations given by $k = 0$) is $\sum_{i=1}^{m} R_i$.

If the plant is of full rank, then $\sum_{i=1}^{m} R_i = n$, and therefore $\mathbf{x}$ can be made the subject of (A11.59) and substituted in (A11.61) to yield

$$\mathbf{u} = \mathbf{f}_y(\mathbf{y}_1, \mathbf{y}_2, \ldots, \mathbf{y}_m) \tag{A11.62}$$

where

$$\mathbf{y}_i = \left[y_i^{(0)}, y_i^{(1)}, \ldots, y_i^{(R_i)}\right]^T, \quad i = 1, 2, \ldots, m \tag{A11.63}$$

Then the reference trajectory generator can be simulated in real time to yield $\mathbf{y}_i(t)$, $i = 1, 2, \ldots, m$ and (A11.62) used to calculate the corresponding $\mathbf{u}(t)$ which can be monitored to check that the control saturation constraints are not exceeded.

Example 11.1 Comparison of elevator controls with cubic and quintic spline reference inputs

An elevator has the following state space model.

$$\dot{x}_1 = x_2 \quad (A11.64)$$

$$\dot{x}_2 = bu - ax_2 - g \quad (A11.65)$$

$$y = x_1 \quad (A11.66)$$

where x_1 is the position, x_2 is the velocity, y is the position measurement scaled in the control computer to be numerically in meters, b is the traction system control acceleration constant, u is the control input to the traction system, a is the viscous friction coefficient and g is the acceleration due to gravity. The elevator is to be controlled with the aid of a dynamic lag pre-compensator. The position trajectory when moving from one floor to the next is required to pass through the four encircled points in Fig. A11.5. In this application, rather than set the first derivatives at the points according to (A11.20), they are set to zero at points P_0 and P_3 to avoid infinite acceleration and deceleration demands at $t = 0$ s and $t = 6$ s.

At points P_1 and P_2 they are set equal to the slope of the line joining these two points so that the reference position trajectory follows this line. There is no control over the acceleration using the cubic spline, but with the quintic spline the acceleration can be specified at the points and is set to zero at points P_0 and P_3 to avoid jerks at $t = 0$ s and $t = 6$ s and has also to be set to zero at points P_1 and P_2, again to avoid jerks as the acceleration along the linear motion of segment 2 is zero.

There is a critical value of y_1 below which $y_r(t)$ first goes negative in segment 1 reaching a minimum for $t \in (0, \tau_s)$ and similarly, $y_r(t)$ overshoots Y_r in segment 3 reaching a maximum for $t \in (2\tau_s, 3\tau_s)$. This critical value of y_1 is calculated below.

The cubic spline coefficients will now be determined. The first two coefficients are given by (A11.5) and (A11.6), which are

$$A_{i0} = y_{ir}^{(0)}(0) \quad A_{i1} = y_{ir}^{(1)}(0), \quad i = 1, 2, 3. \quad (A11.67)$$

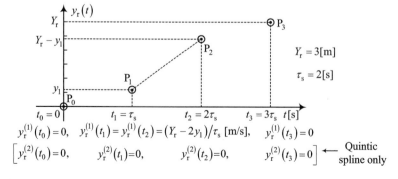

Fig. A11.5 Trajectory points for elevator motion

Since $\tau_1 = \tau_2 = \tau_3 = \tau_s$, the remaining two coefficients are given by (A11.7) as

$$\begin{bmatrix} A_{i2} \\ A_{i3} \end{bmatrix} = \frac{1}{\tau_s^3} \begin{bmatrix} 3\tau_s & -\tau_s^2 \\ -2 & \tau_s \end{bmatrix} \begin{bmatrix} y_{ir}^{(0)}(\tau_s) - y_{ir}^{(0)}(0) - y_{ir}^{(1)}(0)\tau_s \\ y_{ir}^{(1)}(\tau_s) - y_{ir}^{(1)}(0) \end{bmatrix}$$

$$= \frac{1}{\tau_s^3} \begin{bmatrix} 3\left[y_{ir}^{(0)}(\tau_s) - y_{ir}^{(0)}(0)\right]\tau_s - \left[2y_{ir}^{(1)}(0) + y_{ir}^{(1)}(\tau_s)\right]\tau_s^2 \\ 2\left[y_{ir}^{(0)}(0) - y_{ir}^{(0)}(\tau_s)\right] + \left[y_{ir}^{(1)}(0) + y_{ir}^{(1)}(\tau_s)\right]\tau_s \end{bmatrix}, \quad i = 1, 2, 3.$$

(A11.68)

The critical value of y_1 that ensures monotonic behaviour of $y_r(t)$ in segments 1 and 3 will now be determined. In segment 1,

$$y_{1r}^{(0)}(\tau) = A_{10} + A_{11}\tau + A_{12}\tau^2 + A_{13}\tau^3. \quad \text{(A11.69)}$$

The stationary points of $y_{1r}^{(0)}(\tau)$ satisfy

$$A_{11} + 2A_{12}\tau + 3A_{13}\tau^2 = 0. \quad \text{(A11.70)}$$

The information given in Fig. A11.5 together with (A11.67) and (A11.68) yields

$$A_{11} = y_{1r}^{(1)}(0) = 0, \quad \text{(A11.71)}$$

$$A_{12} = \left\{3\left[y_{1r}^{(0)}(\tau_s) - y_{1r}^{(0)}(0)\right]\tau_s - \left[2y_{1r}^{(1)}(0) + y_{1r}^{(1)}(\tau_s)\right]\tau_s^2\right\}/\tau_s^3$$
$$= \left\{3[y1 - 0]\tau_s - [0 + (Y_r - 2y_1)/\tau_s]\tau_s^2\right\}/\tau_s^3 = (5y_1 - Y_r)/\tau_s^2 \quad \text{(A11.72)}$$

and

$$A_{13} = \left\{2[0 - y_1] + [0 + (Y_r - y_1)/\tau_s]\tau_s\right\}/\tau_s^3 = (Y_r - 3y_1)/\tau_s^3. \quad \text{(A11.73)}$$

In view of (A11.71), the two roots of (A11.70) are $\tau_1 = 0$ and $\tau_2 = -\frac{2}{3}A_{12}/A_{13}$. So (A11.72) and (A11.73) yield

$$\tau_2 = \frac{2}{3}\frac{Y_r - 5y_1}{Y_r - 3y_1}\tau_s. \quad \text{(A11.74)}$$

Analysis shows that if $y_1 \in \left(0, \frac{1}{5}Y_r\right)$, then $\tau_2 \in \left(0, \frac{2}{3}\tau_s\right)$, indicating non-monotonic $y_r(t)$ in segment 1 and also segment 3 since, with reference to Fig. A11.5, $y_r(t)$ in segment 3 with the origin shifted to point, P_3, is a reflection in the origin of $y_r(t)$ in segment 1. If $y_1 \in \left[\frac{1}{5}Y_r, \frac{1}{3}Y_r\right)$, then $\tau_2 \in [0, -\infty)$, indicating monotonic behaviour of $y_r(t)$ in segments 1 and 3, as desired. The critical value of y_1 is therefore

$$y_{1\text{crit}} = Y_r/5. \tag{A11.75}$$

This is also the value yielding the longest distance with zero acceleration and will be selected for the simulation.

The remaining range not covered by the analysis above is $y_1 > \frac{1}{3}Y_r$, but this is considered too large to be useful.

Next the quantic spline coefficients will be determined. With $\tau_1 = \tau_2 = \tau_3 = \tau_s$, the matrix inverse in (A11.17) becomes

$$\mathbf{M} = \frac{1}{\tau_s} \begin{bmatrix} \tau_s^2 & \tau_s^3 & \tau_s^4 \\ 3\tau_s & 4\tau_s^2 & 5\tau_s^3 \\ 6 & \tau_s & \tau_s^2 \end{bmatrix}^{-1}, \tag{A11.76}$$

Then for the three segments, (A11.14), (A11.15), (A11.16) and (A11.17) yield the following for segment i.

$$A_{i0} = y_{ir}^{(0)}(0), \quad A_{i1} = y_{ir}^{(1)}(0), \quad A_{i2} = y_{ir}^{(2)}(0)/2$$

$$\begin{bmatrix} A_{i3} \\ A_{i4} \\ A_{i5} \end{bmatrix} = \mathbf{M} \begin{bmatrix} y_{ir}^{(0)}(\tau_s) - y_{ir}^{(0)}(0) - y_{ir}^{(1)}(0)\tau_s - \frac{1}{2}y_{ir}^{(2)}(0)\tau_s^2 \\ y_{ir}^{(1)}(\tau_s) - y_{ir}^{(1)}(0) - y_{ir}^{(2)}(0)\tau_s \\ y_{ir}^{(2)}(\tau_s) - y_{ir}^{(2)}(0) \end{bmatrix}, \quad i = 1, 2, 3.$$

$$\tag{A11.77}$$

These coefficients will be calculated in the computer used to produce the simulation.

The analytical determination of the expression of the critical value of y_1 below which non-monotonic behaviour of $y_r(t)$ in segments 1 and 3 occurs would be inordinately lengthy and therefore this will be found with the aid of the simulation.

For the determination of control required to force the plant output to follow the spline-generated reference input without dynamic lag, differentiating (A11.66) twice and replacing $\dot{x}_1$ and $\dot{x}_2$ by the RHS of (A11.64) and (A11.65) yields

$$\dot{y} = \dot{x}_1 = x_2 \tag{A11.78}$$

and

$$\ddot{y} = \dot{x}_2 = bu - ax_2 - g \tag{A11.79}$$

Making u the subject of (A11.79) then yields

$$u = \frac{1}{b}(\ddot{y} + ax_2 + g) \tag{A11.80}$$

The required equation for u in terms of the output derivatives is then obtained by substituting for x_2 on the RHS of (A11.80) using (A11.78).

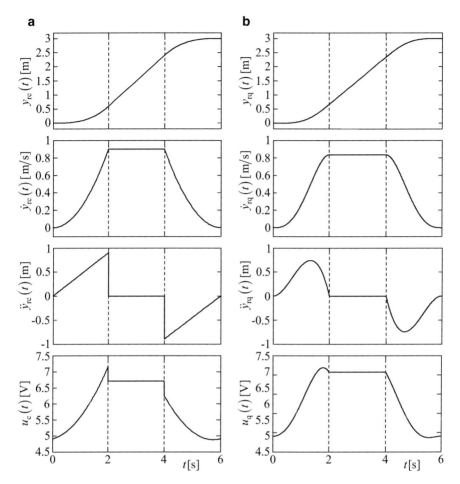

Fig. A11.6 Spline-generated elevator reference inputs and control inputs for zero dynamic lag. (**a**) Using cubic spline. (**b**) Using quintic spline

Thus

$$u = \frac{1}{b}(\ddot{y} + a\dot{y} + g). \qquad (A11.81)$$

Figure A11.6 shows the results of some simulations.

The cubic spline reference input, $y_{rc}(t)$, and the quintic spline reference input, $y_{rq}(t)$, are generated together with their first and second derivatives, thereby enabling the required control to be calculated using (A11.81). The plant parameters are $b = 2\ \left[\text{m/s}^2/V\right]$ and $a = 4\ \left[\text{s}^{-1}\right]$. The control saturation limit is 10[V]. First, the control variables, $u_c(t)$, and $u_q(t)$, both stay below the control saturation limit indicating that a control system with zero dynamic lag would be feasible using either the cubic or quintic spline generated reference inputs. As predicted in the theory, the demanded acceleration, $\ddot{y}_{rc}$, has discontinuities at the end of the acceleration phase of segment 1 and at the beginning of the deceleration phase of segment 2. With reference to the graph of $\ddot{y}_{rc}(t)$ the elevator passengers would experience a steady increase in the acceleration followed by an abrupt change to zero acceleration in segment 2. At the beginning of segment 3, they would experience a sudden jump to maximum deceleration. These jerks are alleviated in the quintic spline based reference input as is evident in Fig. A11.6b.

Regarding the subsequent control system design, it would be advantageous to base it on a robust control technique as the payload mass will vary by significant proportions depending on the number of passengers and their luggage.

Reference

1. Schumaker L (2007) Spline functions: basic theory. Cambridge University Press, Cambridge. ISBN 978-0-521-70512-7

Index

A

Ackermann's Gain Formulae
　applicability, 929
　linear state feedback control law, 929
　linear state observer, 923
AC motor models, 115
Aircraft
　control of, 380, 526
　model (unstable), 380
　state space model, 526
Anti-windup, 32–33
Asymptotes
　Bode plot, 131
Attitude control
　flexible spacecraft, 67
　rigid body spacecraft, 11, 25, 32
Auxiliary output to circumvent zero dynamics, 515, 521, 529, 533, 739

B

Back tracing, 649, 670, 688, 734, 956
Bandwidth
　of closed loop system, 127, 337
　of disturbance, 579, 583
　of noise signal, 604
　of plant model, 895
Bang-bang control. *See* Switched control
Biasing in identification, 156
Bilinear transformation, 285, 421
Bode magnitude plot, 131–150, 313, 338, 354, 867–881, 897
Bode phase angle plot, 143–150, 313, 350, 354, 897
Boundary layer, 642, 664, 679–687, 729, 742–746, 749, 766

C

Canonical forms, 201, 237
　controller canonical form, 57, 208, 237, 247
　observer canonical form, 210, 238, 249
　modal form, 203–208, 220–237
Cart and pole mechanism, 91
Cascade control, 301, 319, 388, 659
Cayley–Hamilton theorem, 177, 921
Closed loop
　dynamics, 298
　feedback control system, 360–364
　poles, 12
　transfer function, 300, 360, 457, 732, 745
Computer aided pole assignment
　Ackermann's Gain Formulae
　　linear state feedback control law, 920–923
　　linear state observer, 923–927
　linear characteristic polynomial interpolation
　　applicability and non-applicability, 931–936
　　LCPI Algorithm, 927–931
Condition number, 152

Control
    actions/terms, 6, 27
    dither, 804
    saturation, 30, 33, 37, 40, 201, 304, 318
Control moment gyro (CMG) (for spacecraft
        attitude control), 98
Control system, 3, 32
    design steps, 70
Controllability, 169
    continuous LTI plants, 179
    discrete LTI plants, 287
Controller zero cancellation
    method, 908–909
    PI Control Loop, 909–911
    PID Control Loop, 911–913
    PID Control Loop with Plant Modelling
        Errors, 913–915
Corner frequency, 132
Crane
    boom crane modelling and control, 486
    gantry crane control, 696, 786
    gantry crane model, 786
Cubic spline, 411, 413, 985, 998, 1001

D

Damping coefficient, 9
Damping ratio, 12, 307
DC gain, 7
    closed loop, 301, 308
    discrete LTI system, 419
    plant with integrators, 132–133
    plant without integrators, 131–132, 301
Dead beat response, 453
Dead space, 633
Degrees of freedom, 81
    of jointed arm robot, 793, 796
    translational and rotational, 82
Derivative
    control action, 6, 8, 18
Deterministic system, 596
Diagonalisation transformation, 221
Diesel driveline
    control, 389
    model, 389
Diesel engine
    control, 816
    model, 816
Discontinuous control. *See* Switched control
Discrete (digital) control, 415
    basic elements, 439
    bilinear transformation, 421
    computational delay allowance, 464
    control algorithms, 447
    equivalent continuous system, 455
    implementation block diagram, 475
    integral polynomial control, 472
    polynomial controller, 465
    RST controller, 465
    simulation block diagrams, 443
    small iteration intervals, 439
    switched actuators with pulse
        modulation, 635
    time delays in plants, 476
    unlimited iteration intervals, 452
Discrete control pole placement
    criterion for applicability of continuous
        system theory, 435
    dead beat response, 453, 470, 472
    discretisation zeros, 433
    first order plants, 457
    flow charts, 447
    integral polynomial control, 472
    IPD controller block diagram, algorithm
        and flow chart, 450
    linear state feedback control, 461
    LTI systems (not plant models), 431
    modes (exponential), 427
    modes (oscillatory), 428
    modes (polynomial exponential), 430
    negligible processing time, 456
    plant zeros, 433
    poles: influence on dynamics, 420, 425
    real time operation, 416
    stability analysis, 417
    third and higher order plants, 464
    using settling time formulae, 452
        polynomial control, 465
    unit circle stability boundary, 418
    zeros, effects of, 432
Discrete LTI plant models, 255
    change of period for z-transfer
        function, 282
    modal basis functions, 259
    state space model, 255
    Tustin's transformation, variation of, 283
    z-transfer function, 274
Discrete control, 415
Disturbance
    estimation, 575
    referred to control input, 5, 306
    steady state error, 574
Dither, 804
Dominance
    pole-to-pole, 49
    pole-to-zero, 54

Index 1005

D
DP control, 24
Drag forces and torques, 85
Dynamic lag, 805
Dynamic lag pre-compensation, 805
  incorporation in polynomial controller, 807
  reference input generator with derivatives, 806
Dynamical
  subsystem, 81
  system, 73
Dynamics
  of mechanical system/vehicle, 73, 81, 84, 97, 103
  of dynamical/control system, 73, 325, 417

E
Eigenvalues, 218
Eigenvectors, 218
  generalised, 232
  geometric multiplicity, 231
Electric drive control
  bang-bang control, 659
  flexible coupling (linear), 366, 404, 410, 516
  flexible coupling (nonlinear), 516
  flexible drive (polynomial control with dynamic lag pre-compensation), 812
  induction motor, 510
  series wound DC motor, 486
  sliding mode control, 717, 750, 763
  two/three phase oscillator for AC drives, 267
Electric drive model
  flexible coupling, 366, 404, 410
  standard DC motor based, 713
Electric motor
  DC motor model, 108
  description of basic types, 104
  induction motor d-q model, 115
  induction motor α-β model, 116
  series wound DC motor model, 198, 486
  synchronous motor d-q model, 116, 188
Electromagnetic levitation (Maglev)
  control of, 498, 732
  model, 498
Error
  steady state, 6–8, 19, 33, 88, 303, 309, 387, 395, 574, 661, 804
  stochastic, 611
Euler's rigid body equations, 506
Euler's rotation theorem, 859–860

F
Feedback linearisation, 492
Feedback linearising and forced dynamic control, 491, 495
  auxiliary output, 515
  discrete, 543
  multivariable plants, 502
  SISO plants, 486, 522
  zero dynamics, 509
Filtering
  Kalman, 603, 611, 613
  Kalman–Bucy, 622
  measurement noise, 25, 942, 944
  plant identification, 159
  time constant, 27
First order systems (linear), 300
  settling time formula, 302
  step responses, 304
Fluid systems, 117
Forced dynamic control, 491, 522
  auxiliary output, 533
  multivariable LTI plants, 523
Fourier transform, 128, 603, 629
Frequency domain, 26, 127, 191, 295, 297, 313, 326, 336–356
Frequency domain performance specifications, 337
  bandwidth, 337
  robustness, 343
  sensitivity, 339
Frequency domain stability analysis, 347
  delay margin, 352
  gain margin, 350
  LTI control system (general), 353
  Nyquist criterion, 348
  phase margin, 351
  relative stability, 350
Friction
  forces and torques, 86
  stick slip, Coulomb, 86, 803

G
Gain
  anti-integral wind-up, 35–36
  matrix
    for Kalman filter, 615–621
    for linear state feedback, 359, 525
    for observer, 564, 594
    observer/Kalman filter, 614
    traditional controller, 5, 16
Gear trains, 92, 164, 389, 799–804, 882–898
Generalised eigenvectors, 232

Gimbal mechanism, 82, 100
Greenhouse
  model, 306, 656
  temperature control of, 306, 656

H
Hamiltonian, 645
  system, 645
Hard stops, 96, 886
Harmonics, 603, 629, 753
Heat exchanger
  control, 478
  model, 478

I
Identification of plants, 121
  from frequency response, 127
  by recursive parameter estimation, 151
  from step response, 122
Inertial force and torque, 90
Instability, 35, 42, 43, 308, 349, 354, 383, 420–425, 446, 477
Integral
  anti-windup, 32–43
  term/control action, 6, 19, 33, 40, 321–322, 388, 392, 407, 472, 479, 486, 815
  wind up, 38
IP control, 22
IPD control, 19, 25, 450
IPD/IDP control, 343–344

J
Jacobean matrix, 485
Jointed arm robot, 705. *See also* Motion control
Jordan
  block, 230
  canonical form, 230

K
Kalman Bucy filter, 622
  for double integrator plant, 622
Kalman filter, 613
  derivation of gain algorithm, 617
  state difference and error equations, 615
  steady state, 621
Kiln/heating process
  control, 36–37, 160, 301, 362, 365, 625
  control with time delay, 589

  model, 36, 120, 256, 261, 301, 362, 365, 673
  model with time delay, 589
Kinematic
  subsystem, 81
  differential equations, 102

L
Lagrangian mechanics, 90, 795
Lagrange multipliers, 645, 828
Laplace transform, 4
  final value theorem, 309
  tables of Laplace transforms, 847
LCPI Algorithm, 927–931
Lie derivative, 193, 492, 493
Limit cycling
  analysis, 954
  influence of switching function, 954–955
  in spacecraft attitude control, 955–970
Linearisation
  about an operating point, 482
  feedback linearisation, 182, 492
  linear state space model, 484
  pulse modulator, 633
Linearity, 76
Linear state feedback control, 356
  aided by observer, 564
  control law, 357
  integral terms, 387, 392
  matrix vector formulation, 358
  output derivative form, 767
Loudspeaker state space model, 195–196

M
Mason's Formula
  rules, 901–908
  simple closed-loop system, 900–901
Matrix
  function of square matrix, 178
MIMO/multivariable
  control, 494, 502, 523, 533, 644, 773, 979, 983
  plant model, 75, 174, 183, 193, 217, 245, 288
  system, 3, 74
Modal decomposition, 46
Modal forms
  multivariable plants, 217–237
  SISO plants, 181, 203–208, 242, 374–375, 577

Index 1007

Modelling of plants, 73
  black box modelling, 81
  physical modelling, 81
  white box modelling, 80
Modes
  exponential modes, 44, 427
  of linear systems, 43
  oscillatory modes, 45, 263, 373, 426, 696
  polynomial exponential modes, 45, 430
  sliding modes, 707
  vibration modes, 535, 879
Momentum ellipsoid, 508
Motion control, 525–527, 793
  backlash and its elimination, 803
  block diagram, 798
  dynamic lag pre-compensator, 823
  feedback linearising control law, 798
  general mechanism model, 794–798
  inverse dynamic load representation, 535, 800–802, 818
  jointed arm robot, 793
  optimal control for frictional loss minimization, 826
  path planning, 794
  simplified model with geared actuator, 799
Multiple pole, 49, 56, 63, 204
  closed loop systems, 49, 56, 63, 299, 329–339, 598, 812
  plant models, 204, 268
Multiplicity
  algebraic multiplicity of eigenvalue, 231
  geometric multiplicity of eigenvector, 231
Multivariable control
  dynamic interaction, elimination, 540
  forced dynamic control, 523
  introduction, 494

N

Natural frequency
  damped, 11, 374, 428
  encastre, 67, 372, 375, 634
  free, 162, 241, 375, 434, 634
  undamped, 12, 21, 64, 124, 307, 428, 683, 876
Noise
  band limited white noise, 604, 617
  coloured noise, 604–605
  contamination in identification, 155
  covariance, 604, 617–618, 621–622
  differentiation of, 26
  filtering, 25, 613
  lumped noise sources, 597
  measurement noise, 25–32, 152, 155, 159, 164, 399, 401, 467, 562, 583, 596–623, 712, 817, 825, 940–948, 980
    definition of measurement noise, 597
  plant noise, 597
    definition of plant noise, 597
  spectrum, 26–28
  variance, 583, 602, 603, 605–617
Non-dynamical system, 73
Non-linearity, 76, 497, 886
Nutation of rigid body, 505

O

Observability, 169
  of continuous LTI plants, 183
  of discrete LTI plants, 290
Observer
  disturbance estimation, 575
  filtering property, 596
  LTI continuous plants, 562
  LTI discrete plants, 587, 614
    plant including pure time delay, 589
  LTI multivariable plants, 592
  model correction loop, 563–564, 567, 570–577, 588, 597–599, 602, 614, 940
  multiple integrator observer, 712
  separation principle, 565–566, 570, 590, 940
  settling time definition, 569
  transparency property, 565
Observer based on linearised plant model, 941
Observer-based robust control (OBRC), 971
  background, 971–974
  controlling uncertain plant *vs.* controlling known plant model, 975
  general controller structure, 976–978
  limitations, 978–979
  model mismatch equivalent input premise, 974–975
  model order uncertainty accommodation, 981–982
  multiple integrator plant model and correction loop controller, 979–981
  multivariable control example, 983–984
Observer design
  considering noise levels, 602
  constrained variance minimisation, 605
  deterministic, 566
  unconstrained variance minimisation, 609–612
  stochastic state estimation error variation with gains, 598

On-off control (switched control), 625
Open loop system, 312, 342, 348, 351, 354, 646
Operating point, 4, 482–494, 511, 681, 941
Optimal control
    co-state/adjoint system differential equations, 828
    cost functional, 644, 827
    of first order plants, 651–664
    fuel optimal control, 648
    of general linear plant, 648
    linear quadratic, 648
    near time optimal control, 550
    Pontryagin maximum principle, 644
    quadratic form cost functional, 647
    of second order plants, 552–554, 669–676, 680
        near/sub time optimal control, 680–686, 726–730
    time optimal control, 38, 627–628, 644–646
    of triple integrator, 688–696
    two point boundary value problem, 646
Optimal control: frictional loss minimisation, 826–846
    comparisons with traditional control, 840
    implementation block diagram, 841
    optimal trajectories, 829–830
    polynomial controller use, 841
    position response, 831
    reference input generator, 833
    scaling laws, 838
    sliding mode control law, 831
    variation with acceleration time, 844
Output derivative based filtering observer, 944–953
Output derivative based state estimator
    filtered multiple derivative state estimator, 942–943
    state representation, 943–944

P

Padé time delay approximation, 378, 477
Partial pole assignment for traditional controllers, 915–919
Path planning and reference input trajectory generation, 985
    actuator capability checking, 996–997
    cubic spline coefficients, 986–987
    maximum relative degree, controlled plant, 990–991
    quintic spline coefficients, 987–988
    reference input generator implementation
        cubic spline generator, 991–993
        quintic spline generator, 993–996
    requirements, 985
    setting the derivatives, 988–990
    splines (introduction), 985–986
PD control, 11–13
Pendulum, motorised
    control with saturation, 676–679
    model, 496, 677
    nonlinear control, 497
    robust control, 768
Per unit parameters
    of flexible electric drive, 366, 404, 924
Phase angle, 60, 127, 144–151
PI control, 13–14
PI Control Loop, 909–911
PID control, 4, 14–19
PID Control Loop, 911–913
PID Control Loop with Plant Modelling Errors, 913–915
Plant, 3
Plant modelling, 73
    direction cosine matrix, 853–856
    dynamics (rigid body, rotational/attitude), 84, 97
    dynamics (rigid body, translational), 84, 103
    Euler's rotation theorem, 859–860
    from frequency response, 127–150
        relatively close complex conjugate pole and zero pairs, 879–881
        relatively close complex conjugate pole pairs, 876–879
        relatively close real poles and zeros, 867–873
        resonance peak function, 873–876
    kinematics (rigid body, rotational/attitude), 100, 102, 853–866
    kinematics (rigid body, translational), 103
    quaternion, 101, 103, 558, 776
    quaternion-based kinematic differential equations, 861–866
    quaternion (introduction), 857–859
    single degree of freedom rotations, 857–859
    throttle valve servomechanism, 881–898
        linear plant model, 883–885
        model parameterisation, 887–888
        model validation in frequency domain, 894–898
        nonlinear plant model, 886–887
        parameter estimation tool, 894
        parameter measurements, 888–893

Index
1009

Pole
　closed loop pole, 368
　external pre-compensator pole, 368, 373
　open loop pole, 368
　plant pole, 368
　symbols for categories, 368
Pole assignment/placement, 297
　dynamic controller, 770
　LTI plants with significant zeros, 368
　Mason's formula (use of), 364
　multiple integrator plant model, 767
　robust, 766
　state feedback, 361–367
Poles and zeros, dominance, 43–69
Pole-zero cancellation, 369–377, 383–387
Polynomial control, 398–406
　continuous pole assignment design, 403
　determination of polynomial degrees, 401
　discrete pole assignment design, 465–476
　generic polynomial controller, 808–822
　implementation block diagram, 406, 412, 469, 475, 809–811, 814, 841, 981
　polynomial integral controller, 407–413, 472–476, 815
　pre-compensators, 407, 806
Posicast control, 696
　plant damping allowance, 701, 702
Power electronics, 384, 658–659
Power spectral density, 603
　of sum of two random signals, 604
Pre-compensator, 342
　dynamic lag, 806, 833
　external, 369–383, 389–392, 400, 412, 635, 806
　decoupling for multivariable control, 541
　implementation with feedback linearising control, 822
　implementation with polynomial controllers, 412, 810–811, 814
Principal axes of inertia, 80, 233, 504, 947
Proportional control action, 6
Proportional/P control, 22, 300, 305, 308
Pseudo-random binary sequence (PRBS), 129

Q
Quadratic form, 647
Quaternion kinematics, 103, 555, 775, 857–866
Quintic spline generator, 993–996

R
Rank
　of controllability matrix, 289
　of observability matrix, 293
　term for relative degree, 9
Reaction wheel (for spacecraft attitude control), 98
Recursive parameter estimation, 151
Reference input generator, 695, 807, 833–838, 841, 991
Relative degree
　in continuous domain, 75, 192, 401, 524
　in discrete domain, 544
Relay control. See Switched control
Resonance and anti-resonance, 135
Rigid body
　dynamics, 84–85, 97–100
　nutation, 505–506
Robustness, 296, 339, 343–347, 705, 707, 850, 971
Roll, pitch and yaw attitude angles, 83
Root locus, 302–303, 353–354, 423, 745
Routh's stability criterion, 413, 421, 575, 913–915, 936–939

S
Sampling process, 274
Saturating control, 627
　boundary layer, 664, 679
　continuous control variable, 662
　near time optimal control
　　of double integrator sub-plant using boundary layer, 698
　　of double integrator using forced dynamic control, 552–554
　　of triple integrator using reference input generator, 695
　posicast control of fourth order plant, 696–704
　response portrait of first order plant
　　closed loop, 655, 658, 663, 668
　　for constant control, 655
　saturation boundary, 642, 663, 677–679
　saturation function, 552, 642–643, 725, 742
　soft switching, 642
Saturation
　control, 30, 33, 37, 40, 305, 318, 627, 643, 662
　boundary, 642, 663, 677–679
　function, 552, 642–643, 725, 742

Scaling law
  for frictional energy cost function, 838
  Laplace to time domain, 327
Scaling property of linear systems, 77
Second order systems (linear), 307
  rate limiting switching boundary for zero overshoot, 725
  settling time formula for under-damped systems, 310–311
  step responses, 307
Sensitivity, 296, 339–343, 369, 380–382, 386, 478–480
Separation of variables, method of, 507, 668, 859
Settling time
  definition for step response, 299
  first order LTI system, 302
  general definition, 568
  second order, under-damped LTI system, 311
Settling time formulae for n^{th} order LTI system with multiple pole
  derivation, 331–332
  error correction, 333
  2% formula, 333
  5% formula, 332
  modified pole placement for specified overshoot and settling time, 335–336
Ship (surface)
  control, 369, 580–583
  disturbance estimation and correction, 581
  model, 369, 581
Signum function, 638–639
Similarity transformation, 216
Sliding mode, 706–707
Sliding mode control, 705
  arbitrary order plant, 730
  closed loop dynamics in sliding mode, 723
  closed loop phase portrait, 715, 724
  control chatter, 718
  equivalent control, 718
  equivalent control estimation, 748
  full relative degree, 730
  less than full relative degree, 738
  multivariable (introduction), 773
  rate limiting switching boundary, 725
  reaching sliding condition, 720
  robustness, 707
  second order plants, 713
  sliding manifold, 775, 780
  state trajectory, 716–717
  sub-time optimal control, 726, 730
  time varying disturbances and reference inputs, 723
  triple integrator plant, 732
  zero dynamics, 738
Sliding mode control chatter elimination
  boundary layer method, 742
  control smoothing integrator method, 746
  higher order sliding mode control (HOSM), 756
  discrete SMC, 778, 785
  output derivatives and estimation, 707, 711
  phase portrait, 757
  pseudo sliding mode control, 742
  second order HOSM, 760
  third order HOSM, 762
Sliding motion, 658, 706–707, 716
  conditions for, 719
Smith predictor, 477–480
Software differentiation/discrete derivative estimation, 440–441
Spacecraft (flexible)
  attitude/slewing control, 67, 374, 535, 542, 633–638, 739, 775
  continuous models, 241, 242, 256, 375, 434, 541, 634, 740
  discrete models, 434
  pulse modulated thrusters, 633
Spacecraft (rigid body)
  attitude/slewing control, 11, 17, 21, 27–30, 32, 162, 270, 315–319, 374, 395, 467–472, 504, 555, 571, 583–586, 611, 680–686, 692–696, 775, 955–974, 983
  continuous models, 11, 80, 97, 102–104, 186, 233, 235, 253–256, 315, 395, 504, 508, 542, 555, 571, 583–585, 600, 611, 680, 693, 775, 947, 955–956, 971, 983
  discrete models, 422, 467
  disturbance estimation and correction, 583
  solar sailing, time optimal control, 692
  spin control, 504
  state/disturbance estimation (observer), 570, 584, 600
  state estimation with noise variance minimisation, 606, 609
Spline, 985
  cubic spline, 986–987
  quintic spline, 987–988
Spring force and torque, 89
Stability/instability, 7, 9, 35, 39–43, 122, 296, 308, 313, 322, 340, 347–354

Index   1011

State differential equation, 172–174, 188,
    194–207, 221, 226–228, 233–235,
    266, 388, 489–496, 502, 510–511,
    556, 565, 570, 576–577, 645–651,
    666, 673, 688, 779, 884–885, 940
  solution for LTI model, 175
State estimation, 561
  for nonlinear plants, 940–953
  error transfer function relationship, 599
  error variation with observer gains, 598,
    609, 610
State representation, 194
  block diagrams, 201–203
  effect on robustness, 707
  modal forms, 203
  multivariable controller canonical form,
    245, 247–249
  multivariable observer canonical form, 245,
    249–251
  SISO controller canonical form, 208, 237
  SISO observer canonical form, 210, 237
  transformations, 199, 211, 239, 251
State space, 171
State space models, 170
  from transfer functions, 200
State trajectory, 171
State variable, 171
  block diagram, 187
Steady state
  analysis of linear state feedback control
    systems, 387
  error, 6–8, 19, 33, 88, 303, 309
Step response
  first order LTI system, 303
  second order LTI system, 12, 307, 310
  systems with multiple real poles, 330
Stochastic system, 596
Superposition property, 77
Switched control, 625
  first order plants, 651
  hard switching, 642
  hysteresis controller, 656, 659
  PM implementation, 629
  pulse modulation (PM), 628
  response portrait, 654–655
  soft switching, 642
  time optimal, 38, 469–472, 551–554,
    626–628, 644–648, 651–652, 654,
    664–682, 688, 690, 702, 726, 730,
    755, 765
  time varying reference input, 658
  transfer characteristics of linearising pulse
    modulators, 633

Switched control of 2nd order plants
  phase plane, 667
  phase portraits (*see below* state portraits)
  phase variables, 667
  state plane, 667
  state portraits, 666, 669–674
  state trajectories, 507–508, 627, 668
  state trajectory differential equation, 667,
    714, 727, 830, 956
  time optimal control of double integrator
    plant, 669
  time optimal control of LTI plant with two
    real poles, 673
Switched control of third and higher order
    plants, 687
  posicast control of fourth order plant, 696
  time optimal control of triple integrator
    plant, 688
Switched state feedback control of plants of
    arbitrary order, 638
  back tracing, 649
  limit cycling, 655–657, 686, 954–970
  switching boundary, 638, 640
  switching element, 656
  switching element (hysteresis), 656
  switching function, 639
  switching function sign convention, 641
  three-level and two-level, 626

T
Thermal systems, 119
Third order LTI system, 328
Throttle valve (internal combustion engines)
  cascade control, 320
  continuous model, 321, 881
  discrete model, 286
  discrete model (identified), 472
    control of, 444, 472
    recursive parameter estimation, 164
  discrete position control, 445, 473
Time constant
  filtering, 27
  plant, 13
Time optimal control, 551, 646
  approximate using forced dynamic
    control, 555
  of double integrator plant, 669–673
  of first order plants, 651–655
  of LTI plant, 648
  near time optimal control
    of double integrator sub-plant using
      boundary layer, 698

Time optimal control (*cont.*)
  of double integrator using forced dynamic control, 552–554
  of triple integrator using reference input generator, 695
  of second order plant with real poles, 673–676
  of triple integrator plant, 688–696
Tokomak fusion reactor, 383
  plasma current control, 383–387
Transfer function
  closed loop
    for discrete control, 453, 455, 461, 465
    with linear state feedback controller, 361
    with polynomial controller, 400
    with traditional controllers, 22–25
  from continuous state space model, 190
  matrix, 190, 223, 245, 246, 278, 540, 823
  SISO LTI plant, 20
  z-transfer function model, 274
    multivariable plant (matrix), 151
    SISO plant, 284
Type (plant and system), 7

U
Underwater vehicle modelling and control, 85, 97, 103, 547

V
Vacuum air bearing control, 39, 576
  disturbance estimation and correction, 577
  effect of modelling errors, 579
  model, 576

Variance, 603, 951
Vector control
  Clarke transformation, 111
  of electric drives, 112
  Park transformation, 112
Vehicle modelling, 97

Z
Zero dynamics
  auxiliary output to circumvent zero dynamics, 515–539, 738
  in feedback linearising control, 509
  in sliding mode control, 738
Zero order hold, 279–282
Zeros
  complex conjugate, 64, 127, 372, 742, 873
  derivative effect on step response, 21, 54, 528
  left half plane, 63
  mirroring, 378
  at origin of s-plane, 383
  plant zero invariance, 53
  pre-compensation, 369, 383, 390, 400, 634, 913
  right half plane, 63, 66, 143
Z-transfer function
  definition, 274
  from discrete state space model, 277
  general plant model, 151
  relationship with pulse transfer function, 279
  time shifting property, 276

Printed by Printforce, the Netherlands